OCT'91$X41 55

INTERMEDIATE ALGEBRA

THE JOHNSTON/WILLIS DEVELOPMENTAL MATHEMATICS SERIES

***Essential Arithmetic,* Sixth Edition** (paperbound, 1991)
Johnston/Willis/Lazaris

***Essential Algebra,* Sixth Edition** (paperbound, 1991)
Johnston/Willis/Lazaris

***Developmental Mathematics,* Third Edition** (paperbound, 1991)
Johnston/Willis/Hughes

***Elementary Algebra,* Third Edition** (hardbound, 1991)
Johnston/Willis/Buhr

***Intermediate Algebra,* Third Edition** (hardbound, 1991)
Johnston/Willis/Buhr

***Intermediate Algebra,* Fifth Edition** (paperbound, 1991)
Johnston/Willis/Lazaris

INTERMEDIATE ALGEBRA
FIFTH EDITION

C. L. JOHNSTON
ALDEN T. WILLIS
Formerly of East Los Angeles College

JEANNE LAZARIS
East Los Angeles College

Wadsworth Publishing Company
Belmont, California
A Division of Wadsworth, Inc.

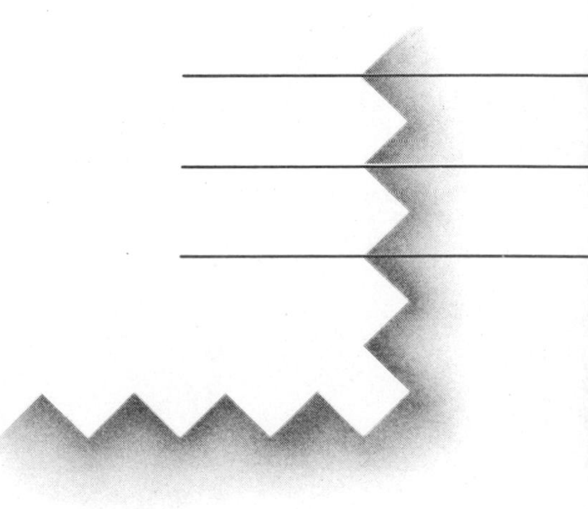

Mathematics Editor: Anne Scanlan-Rohrer
Assistant Editor: Tamiko Verkler
Special Projects Editor: Alan Venable
Editorial Assistant: Leslie With
Production: Ruth Cottrell
Print Buyer: Randy Hurst
Designer: Julia Scannell
Copy Editor: Mary Roybal
Technical Illustrators: Carl Brown, Reese Thornton, and Pat Rogondino
Compositor: Polyglot Pte. Ltd.
Cover: Frank Miller
Signing Representative: Kenneth King

© 1991, 1988, 1983 by Wadsworth, Inc.

© 1979, 1976, 1975 by Wadsworth Publishing Company, Inc. All rights reserved. No part of this book may be reproduced, stored in a retrieval system, or transcribed, in any form or by any means, electronic, mechanical, photocopying, recording, or otherwise, without the prior written permission of the publisher, Wadsworth Publishing Company, Belmont, California 94002, a division of Wadsworth, Inc.

Printed in the United States of America

1 2 3 4 5 6 7 8 9 10—95 94 93 92 91

Library of Congress Cataloging-in-Publication Data

Johnston, C. L. (Carol Lee), 1911–
Intermediate algebra / C. L. Johnston, Alden T. Willis, Jeanne Lazaris.—5th ed.
p. cm.
Includes index.
ISBN 0-534-14328-8
1. Algebra. I. Willis, Alden T. II. Lazaris, Jeanne, 1932–
III. Title
QA154.2.J63 1991
512.9—dc20 90-41143
CIP

This book is dedicated to our students,
who inspired us to do our best
to produce a book worthy of their time.

Contents

4 POLYNOMIALS

5 FACTORING

6 RATIONAL EXPRESSIONS AND EQUATIONS

7 EXPONENTS AND RADICALS

8 GRAPHING AND FUNCTIONS

9 EXPONENTIAL AND LOGARITHMIC FUNCTIONS

10 NONLINEAR EQUATIONS AND INEQUALITIES

Preface

Intermediate Algebra, Fifth Edition, can be used in an intermediate algebra course in any community college or four-year college, in either a lecture format or a learning laboratory setting, or it can be used for self-study. This book will prepare the student for college algebra, statistics, business calculus, and science courses.

Features of This Book

The major features of this book include the following:

1. The book uses a one-step, one-concept-at-a-time approach. The topics are divided into small sections, each with its own examples and exercises. This approach allows students to master each section before proceeding confidently to the next section.
2. Many concrete, annotated examples illustrate the general algebraic principles covered in each section. To prevent confusion, each example ends with these symbols: ——— and ■.
3. Important concepts and algorithms are enclosed in boxes for easy identification and reference.
4. The approach to solving word problems utilizes problem solving and includes a detailed method for translating a word statement into an algebraic equation or inequality.
5. Visual aids such as shading, color, and annotations guide students through worked-out problems.
6. In special "Words of Caution," students are warned against common algebraic errors.
7. The importance of checking solutions is stressed throughout the book.
8. A review section with Set I and Set II exercise sets appears at the end of each chapter, and some chapters also have a midchapter review that includes Set I and Set II exercise sets. The Set II review exercises allow space for working problems and for answers; they can be removed from the text for grading. (Removal of these pages will not interrupt the continuity of the text.)
9. This book contains more than 6,000 exercises.

   Set I Exercises. The complete solutions for odd-numbered Set I exercises are included in the back of the book, together with the answers for all of the Set I exercises. In most cases (except in the review exercises), the even-numbered exercises are matched with the odd-numbered exercises. Thus, students can use the solutions for odd-numbered exercises as study aids for doing the even-numbered Set I exercises.

   Set II Exercises. No answers for Set II exercises are given in the text; answers to all these exercises are included in the Instructor's Manual. The odd-numbered exercises of Set II are, for the most part (except in the review exercises), matched to the odd-numbered exercises of Set I, while the even-numbered exercises of Set II are *not* so matched. Thus, while students can use the odd-numbered exercises of Set I as study aids for doing the odd-numbered exercises of Set II, they are on their own in doing the even-numbered exercises of Set II.

10. A diagnostic test at the end of each chapter can be used for study and review or as a pretest. Complete solutions to all problems in these diagnostic tests, together with section references, appear in the answer section of the book.
11. A set of Cumulative Review Exercises is included at the end of each chapter except Chapter 1; the answers are in the answer section of this book.

Using This Book

Intermediate Algebra, Fifth Edition, can be used in three types of instructional programs: lecture, laboratory, and self-study.

The conventional lecture course. This book has been class-tested and used successfully in conventional lecture courses by the authors and by many other instructors. It is not a workbook, and therefore it contains enough material to stimulate classroom discussion. Examinations for each chapter are provided in the Instructor's Manual, and two different kinds of computer software enable instructors to create their own tests. One software program utilizes a test bank, and the other is a random-access test generator. Tutorial software is available to help students who require extra assistance.

The learning laboratory class. This text has also been used successfully in many learning labs. The format of explanation, example, and exercises in each section of the book and the tutorial software make the book easy to use in laboratories. Students can use the diagnostic test at the end of each chapter as a pretest or for review and diagnostic purposes. Because several forms of each chapter test are available in the Instructor's Manual, and because test generators are available, a student who does not pass a test can review the material covered on that test and can then take a different form of the test.

Self-study. This book lends itself to self-study because each new topic is short enough to be mastered before continuing, and because more than 900 examples and over 1,500 completely solved exercises show students exactly how to proceed. Students can use the diagnostic test at the end of each chapter to determine which parts of that chapter they need to study and can thus concentrate on those areas in which they have weaknesses. The random test generator, which provides answers and cross-references to the text and permits the creation of individualized work sheets, and the tutorial software extend the usefulness of this text in laboratory and self-study settings.

Changes in the Fifth Edition

The fifth edition includes changes that resulted from many helpful comments by users of the first four editions as well as from the authors' own classroom experience in teaching from the book. The major changes in the fifth edition include the following:

1. More word problems have been added, and, at the request of several reviewers, examples are *not* given for *some* of these problems, so the student must use general problem-solving techniques to devise his or her own solutions.
2. The section on scientific notation has been rewritten, with more motivation being included and with more problems (including word problems) being added.
3. Interval notation is now discussed, and students are asked to write the algebraic statement that describes a graph in interval notation as well as in set-builder notation; students are also asked to write the solutions of some of the linear inequalities in interval notation.
4. The section on the binomial theorem has been rewritten, with factorial notation and the standard notation $\binom{n}{r}$ for the binomial coefficient being included.
5. While the topics of factoring by completing the square of a polynomial (Section 5.7) and factoring by using synthetic division (Section 5.8) are still

included, in *subsequent* sections of the text, those examples and exercises that required the use of one of these methods have been deleted. Thus, instructors who wish to omit Sections 5.7 and 5.8 can do so.

6. The terminology has been changed in Chapter 6 (Rational Expressions and Equations), with the expression "algebraic fraction" being replaced by "rational expression."
7. Because the logarithmic function is the inverse of the exponential function, placing the chapter on logarithms immediately after the chapter that includes finding inverse functions gives immediate reinforcement of the discussion of inverse functions. The chapter on logarithms (now Chapter 9) is a completely independent chapter and can be omitted without loss of continuity by instructors who don't have time to cover logarithms. As in the fourth edition, the main focus in the chapter is on the use of logarithms in solving exponential and logarithmic equations, with the calculator being used for finding the logs. The sections covering the finding of logs by using tables and the solving of arithmetic problems by using logs remain in Appendix B.
   The rearrangement of chapters also gives those students who *do* cover the chapter on logarithms a better chance to absorb this completely new topic before the end of the course.
8. The technique of completing the square in order to sketch the graphs of circles and parabolas is now included in Section 10.8 (Conic Sections).
9. A new, independent subsection on solving systems of equations by a *matrix* method (the Gaussian elimination method) is now included in the chapter on solving systems of equations.
10. In Chapter 12, summation notation has been added to the discussion of series.
11. At the request of several reviewers, the numbering of the problems in the diagnostic tests (and in the tests in the Instructor's Manual) has been changed so that, for easier grading, each test now contains 10, 20, or 25 problems.

Ancillaries

The following ancillaries are available with this text:

1. The Instructor's Manual contains five different tests for each chapter, two forms of three midterm examinations, and two final examinations that can be easily removed and duplicated for class use. These tests are prepared with adequate space for students to work the problems. Answer keys for these tests are provided in the manual, as are the answers to the Set II exercises. The manual also contains essays to help the instructor teach development mathematics students. Essays cover such topics as: writing in the mathematics classroom, running a lab, cooperative learning, and more.
2. The test bank is also available in a computerized format entitled EXP-Test. EXP-Test is a fast, highly flexible computerized testing system for the IBM PC and compatibles. Instructors can edit and scramble test items or create their own tests.
3. In addition, Wadsworth offers the *Johnston/Willis/Lazaris Computerized Test Generator* (JeWeL TEST) software for Apple II and IBM PC or compatible machines. This software, written by Ron Staszkow of Ohlone College, allows instructors to produce many different forms of the same test for quizzes, work sheets, practice tests, and so on. Answers and cross-references to the text provide additional instructional support.

4. An "intelligent" tutoring software system is available for the IBM PC and compatibles. *Expert Tutor*, written by Sergei Ovchinnikov of San Francisco State University, uses a highly interactive format and sophisticated techniques to tailor lessons to the specific algebra and prealgebra learning problems of students. The result is individualized tutoring strategies with specific page references to problems, examples, and explanations in the textbook. The software has been fully revised for the new edition.
5. Fifteen videotapes, created by John Jobe of Oklahoma State University, review the most essential and difficult topics from the textbook.

To obtain additional information about these supplements, contact your Wadsworth–Brooks/Cole representative.

Acknowledgments

We wish to thank the members of the editorial staff at Wadsworth Publishing Company for their help with this edition. Special thanks go to Anne Scanlan-Rohrer, Tamiko Verkler, Alan Venable, Leslie With, Ruth Cottrell, and Mary Roybal.

We are deeply grateful to Gale Hughes for preparing the Instructor's Manual, and we also thank the following reviewers for their helpful comments: Beverly Abila, Rio Hondo College; Sharon Bird, Richland College; Cynthia Fleck, Wright State University; Jeannine Hugill, Highland Community College; Katherine J. Huppler, St. Cloud State University; Barbara McLachlan, Mesa College; Jack D. Murphy, Pennsylvania College of Technology; Ann B. Oaks, Hobart and William Smith College; and Judith Willoughby, Minneapolis Community College. We'd also like to thank the following who checked the text for mathematical accuracy: Sharon Bird, Richland College; Bradford Bynum, Westminster College; Barbara McLachlan, Mesa College; Richard Spangler, Tacoma Community College; and Todd Zimmermann.

1 Review of Elementary Topics

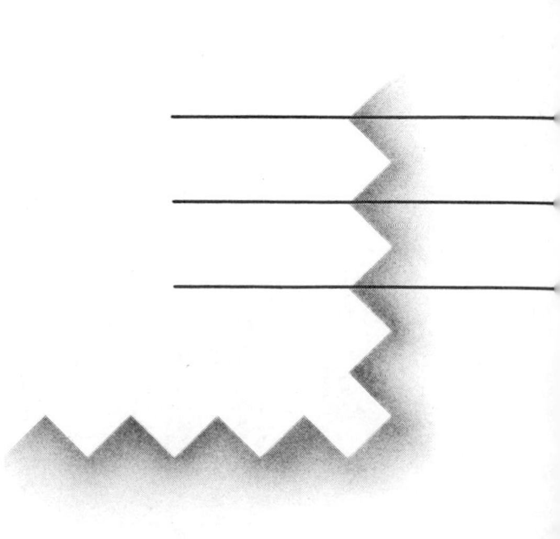

Intermediate algebra is made up of two types of topics: (1) those that were introduced in beginning algebra and are expanded in this course, and (2) new topics not covered in beginning algebra. With most topics, we begin with the ideas learned in beginning algebra and then develop these ideas further.

In Chapter 1 we review sets, the properties of real numbers, integral exponents, and simplifying and evaluating algebraic expressions.

1.1 Sets

Since ideas in all branches of mathematics—arithmetic, algebra, geometry, calculus, statistics, and so on—can be explained in terms of sets, a basic understanding of sets is helpful.

1.1A Basic Definitions

Set A **set** is a collection of objects or things.

Example 1 Examples of sets:

a. The set of students registered at your college on May 7, 1990, at 8 A.M.

b. The set of letters in our alphabet; that is, a, b, c, d, and so forth. ■

Element of a Set The objects or things that make up a set are called its **elements** or **members**. A set may contain just a few elements, many elements, or no elements at all.

Roster Method of Representing a Set A **roster** is a list of the members of a group. The method of representing a set by listing its elements is called the **roster method**. Whenever the roster method is used, it is customary to put the elements inside a pair of braces, { }. We *never* use parentheses for sets. Thus, (3, 4, 5) does *not* denote a set.

Example 2 Examples to show the elements of sets and the roster method:

a. Set $\{5, 7, 9\}$ has elements 5, 7, and 9.

b. Set $\{a, f, h, k\}$ has elements a, f, h, and k. ■

NOTE It is customary, though not necessary, to arrange numbers and letters in sets in numerical and alphabetical order (to make reading easier) and to represent elements of a set with lowercase letters. ☑

Naming a Set A set is usually named by a capital letter. The expression "$A = \{1, 5, 7\}$" can be read "A is the set whose elements are 1, 5, and 7."

Set of Digits One important set of numbers is the set of **digits**. This set contains the numerals 0, 1, 2, 3, 4, 5, 6, 7, 8, and 9. These symbols make up our entire number system; *any* number can be written by using some combination of these numerals.

Modified Roster Method If the number of elements in a set is so large that it is either inconvenient or impossible to list all its members, we modify the roster notation. For example, the set of digits could be represented as follows:

$$D = \{0, 1, 2, \ldots, 9\}$$

This is read "D is the set whose elements are 0, 1, 2, and so on up to 9." The three dots to the right of the number 2 indicate that the remaining numbers are to be found in the same way we have begun, namely by adding 1 to each number to find the next number, until we reach 9, the last number in the set.

Natural Numbers The numbers 1, 2, 3, 4, 5, 6, and so on are called **natural numbers** or **counting numbers**. Since this set has no largest number, we represent it as follows:

$$\{1, 2, 3, 4, \ldots\}$$

This is read "the set whose elements are 1, 2, 3, 4, and so on." This set will be called N throughout this text; that is,

$$N = \{1, 2, 3, 4, \ldots\}$$

Even Numbers Natural numbers that end in 0, 2, 4, 6, or 8 are called **even natural numbers**.

Odd Numbers Natural numbers that end in 1, 3, 5, 7, or 9 are called **odd natural numbers**.

Consecutive Numbers Numbers that follow one another without interruption in sequence are called **consecutive numbers**. For example, 15, 16, 17, and 18 are four consecutive numbers beginning with 15.

The Meaning of the Equal Sign The equal sign ($=$) in a statement means that the expression on the left side of the equal sign *has the same value or values as* the expression on the right side of the equal sign.

Equal sets Two sets are said to be equal if they both contain exactly the same elements. We will use the equal sign between such sets. If two sets are *not* equal to each other, we will use the sign $\neq$, which is read "is not equal to."

Example 3 Examples of equal sets and unequal sets:

a. $\{1, 5, 7\} = \{5, 1, 7\}$. Notice that both sets have exactly the same elements, even though the elements are not listed in the same order.

b. $\{1, 5, 5, 5\} = \{5, 1\}$. Notice that both sets have exactly the same elements, even though 5 is repeated several times in one of the rosters.

c. $\{7, 8, 11\} \neq \{7, 11\}$. These sets are not equal because they do not both have exactly the same elements. ■

The Set-builder (or Rule) Method of Representing a Set A set can also be represented by a *rule* describing its members in such a way that we definitely know whether a particular element is in that set or is not in that set. In this notation, the vertical bar is read "such that."

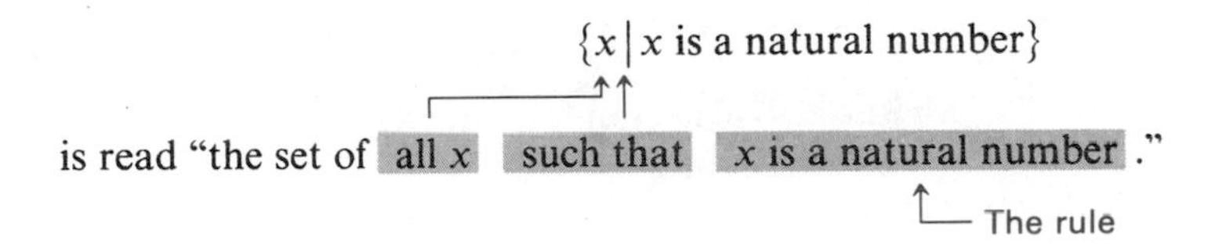

Example 4 Example of changing from set-builder notation to roster notation: Write $\{x \mid x \text{ is an even natural number}\}$ in roster notation.

Solution The natural numbers are 1, 2, 3, 4, 5, 6, . . . ; selecting the *even* ones gives us the set $\{2, 4, 6, \ldots\}$. Therefore,

$$\{x \mid x \text{ is an even natural number}\} = \{2, 4, 6, \ldots\}$$ ■

Example 5 Examples of changing from roster notation to set-builder notation:

a. Write $\{1, 3, 5, \ldots\}$ in set-builder notation.

Solution We have to find a new way of describing this set, a way that starts "The set of all x such that . . . ." Notice that all of the numbers listed are natural numbers and that they are also all odd numbers. Therefore, we can say

$$\{1, 3, 5, \ldots\} = \{x \mid x \text{ is a natural number and } x \text{ is an odd number}\}$$

or

$$\{1, 3, 5, \ldots\} = \{x \mid x \text{ is an odd natural number}\}$$

NOTE It is important to realize that other answers could also be correct. ☑

b. Write $\{0, 3, 6, 9\}$ in set-builder notation.

Solution We might notice that all of the numbers are digits and that they are all divisible by 3.* We could then say

$$\{0, 3, 6, 9\} = \{x \mid x \text{ is a digit and } x \text{ is divisible by } 3\}$$

or

$$\{0, 3, 6, 9\} = \{x \mid x \text{ is a digit and } x \text{ is 0 or is a multiple of } 3^{\dagger}\}$$ ■

The Symbol ∈ If we wish to show that a number or object is a member of a given set, we use the symbol ∈, which is read "is an element of." Thus, the expression "$2 \in A$" is read "2 is an element of set A." If we wish to show that a number or object is *not* a member of a given set, we use the symbol ∉, which is read "is not an element of."

Example 6 Examples to show the use of ∈ and ∉:

a. If $A = \{2, 3, 4\}$, we can say $2 \in A$, $5 \notin A$, $3 \in A$, and $4 \in A$.

b. If $F = \{x \mid x \text{ is an even natural number}\}$, then $2 \in F$, $5 \notin F$, $12 \in F$, and $\frac{2}{3} \notin F$. ■

The Empty Set or Null Set A set with *no* elements in it is said to be the **empty set** (or **null set**). We use the symbols { } or ∅ to represent the empty set.

A WORD OF CAUTION $\{\emptyset\}$ is *not* the correct symbol for the empty set. ☑

Example 7 Examples of empty sets:

a. The set of all people in your math class who are 10 ft tall.

b. The set of all the digits greater than 10. ■

Finite Set If, in counting the elements of a set, the counting comes to an end, the set is called a **finite set**.

* When we say that a number is "divisible by n," we mean that when we divide the number by n, the remainder is 0.

† When we say that a number is a "multiple of n," we mean that it is one of the numbers n, $2n$, $3n$, $4n$, $5n$, and so on, where n is any natural number.

Example 8 Examples of finite sets:

a. $A = \{5, 9, 10, 13\}$

b. $D =$ The set of digits

c. $\varnothing = \{\ \ \}$ ■

Infinite Set If, in counting the elements of a set, the counting never comes to an end, the set is called an **infinite set**. A set is infinite if it is not finite.

Example 9 Examples of infinite sets:

a. $N = \{1, 2, 3, \ldots\}$ The natural numbers

b. $\{x \mid x \text{ is an odd natural number}\}$ ■

Subsets A set A is called a **subset** of set B if every member of A is also a member of B. "A is a subset of B" is written "$A \subseteq B$."

NOTE The symbol $\subseteq$ is used to indicate that one *set* is a *subset* of another set. The symbol $\in$ is used to indicate that a particular *element* is a *member* of a particular set. ☑

Example 10 Examples of subsets:

a. If $A = \{3, 5\}$ and $B = \{3, 5, 7\}$, then $A \subseteq B$ because every member of A is also a member of B.

A WORD OF CAUTION The statement "$3, 5 \subseteq B$" is a *false* statement; a subset must be a *set*, and, because there are no braces around "3, 5," "3, 5" does not name a set. ☑

b. If $C = \{10, 7, 5\}$ and $H = \{5, 7, 10\}$, then $C \subseteq H$ because every member of C is also a member of H.

NOTE *Every set is a subset of itself*. Also, mathematicians agree that *the empty set is a subset of every set*. ☑

c. If $E = \{4, 7\}$ and $F = \{7, 8, 5\}$, then E is not a subset of F because $4 \in E$, but $4 \notin F$. The symbol for "is not a subset of" is $\nsubseteq$. Therefore, $E \nsubseteq F$. ■

EXERCISES 1.1A

In all exercises, N refers to the set of natural numbers and D to the set of digits.

Set I

For Exercises 1 and 2, write "True" if the statement is always true; otherwise, write "False."

1. a. The collection consisting of *, &, $, and 5 is a set.

b. $\{x \mid x \text{ is a natural number}\} = \{1, 2, 3\}$

c. $\{2, 6, 1, 6\} = \{1, 2, 6\}$

d. $23 \subseteq N$

e. $0 \in N$

f. $\{2\} \in D$

g. $\{2, 6\} \subseteq N$

h. $\{\ \ \} \subseteq N$

i. $10 \in D$ j. $0 \in \varnothing$

k. If $A = \{5, 11, 19\}$, then $11 \in A$.

l. The collection consisting of all the natural numbers between 0 and 1 is a set.

2. a. The collection consisting of a, 3, ^, and # is a set.

b. $\{0, 3, 6, \ldots\} = \{x \mid x \text{ is a natural number divisible by } 3\}$

c. $\{2, 2, 7, 7\} = \{7, 2\}$ d. $2 \in N$

e. $\{5\} \in N$ f. $5 \subseteq N$

g. $\{1, 3, 5\} \subseteq N$ h. $\{\ \} \subseteq D$

i. $0 \in D$ j. $\{3\} \in D$

k. If $B = \{a, b, c, d, e\}$, then $\{a, b\} \subseteq B$.

l. The collection consisting of all the digits greater than 12 is a set.

In Exercises 3 and 4, state which of the sets are finite and which are infinite.

3. a. the set of even natural numbers

b. the set of days in the week

c. the set of books in the Huntington Library

4. a. the set of natural numbers divisible by 7

b. the set of months in the year

c. the set of odd digits

5. List all the subsets of the set $\{a, b, c\}$.

6. List all the subsets of the set $\{1, 2\}$.

7. If $A = \{3, 5, 10, 11\}$, $B = \{3, 5, 12\}$, and $C = \{5, 3\}$, state which of the following statements are true and which are false.

a. $B \subseteq A$ b. $C \subseteq B$ c. $C \nsubseteq A$

8. If $E = \{a, b, h, k\}$, $F = \{a, b, c, k\}$, and $G = \{b, k\}$, state which of the following statements are true and which are false.

a. $F \nsubseteq E$ b. $G \nsubseteq F$ c. $G \subseteq E$

9. Write each of the given sets in roster notation.

a. $\{x \mid x \text{ is an even digit}\}$

b. $\{x \mid x \text{ is a natural number divisible by } 5\}$

10. Write each of the given sets in roster notation.

a. $\{x \mid x \text{ is an odd natural number}\}$

b. $\{x \mid x \text{ is a natural number that is a multiple of } 11\}$

11. Write each of the given sets in set-builder notation.

a. $\{a, b, c, d, e\}$ b. $\{4, 8, 12\}$ c. $\{10, 20, 30, \ldots\}$

12. Write each of the given sets in set-builder notation.

a. $\{0, 2, 4, 6, 8\}$ b. $\{x, y, z\}$ c. $\{5, 10, 15, \ldots\}$

Set II

1. Write "True" if the statement is always true; otherwise, write "False."

a. The collection consisting of 3, w, %, and @ is a set.

b. $\{x \mid x \text{ is a digit}\} = \{0, 1, 2, 9\}$

c. $\{5, x, 5\} = \{5, x\}$ d. $\{\ \} \subseteq \{1, 3, 5\}$

e. $\{0\} \in D$ f. $12 \in D$

g. $0 \notin N$ h. $\{0, 5\} \nsubseteq N$

i. $0 \notin \{\ \}$ j. $3 \subseteq D$

k. If $C = \{a, 3, d, 5\}$, then $2 \notin C$.

l. The collection consisting of all the digits less than 0 is a set.

2. Write all the elements of the set $\{2, a, 3\}$.

In Exercises 3 and 4, state which of the sets are finite and which are infinite.

3. a. the set of even digits

b. the set of letters in our alphabet

c. the set of books in the Library of Congress

4. a. the set of natural numbers divisible by 6

b. the set of digits that are multiples of 6

c. $\{1, 2, 3, \ldots, 100\}$

5. List all the subsets of the set {*, $, &}.

6. List all the subsets of the set $\{0, 1, 2, 3\}$.

7. If $A = \{2, 4, 7, 11\}$, $B = \{4, 7, 12\}$, and $C = \{4, 7, 11\}$, state which of the following statements are true and which are false.

a. $B \subseteq A$ b. $C \subseteq A$ c. $C \nsubseteq B$

8. If $E = \{1, 2, 3, \ldots, 8\}$, $F = \{2, 3, 5\}$, and $G = \{5\}$, state which of the following statements are true and which are false.

a. $F \subseteq E$ b. $G \nsubseteq E$ c. $E \subseteq F$

9. Write each of the given sets in roster notation.

a. $\{x \mid x \text{ is a digit divisible by } 4\}$

b. $\{x \mid x \text{ is a natural number that is a multiple of } 8\}$

10. Write each of the given sets in roster notation.

a. $\{x \mid x \text{ is a natural number that is a multiple of } 6\}$

b. $\{x \mid x \text{ is a digit that is divisible by } 5\}$

11. Write each of the given sets in set-builder notation.

a. $\{0, 3, 6, 9\}$ b. $\{7, 14, 21, \ldots\}$ c. $\{a, b, c\}$

12. Write each of the given sets in set-builder notation.

a. {Sunday, Monday, Tuesday, Wednesday, Thursday, Friday, Saturday}

b. $\{0, 5\}$ c. $\{4, 8, 12, \ldots\}$

1.1B Union and Intersection of Sets

Union of Sets The **union** of sets A and B, written $A \cup B$, is the set that contains all the elements of set A and all the elements of set B. For an element to be in $A \cup B$, it must be in set A *or* set B *or* both.

Example 11 Examples of set union:

a. If $A = \{2, 3, 4\}$ and $B = \{1, 3, 5\}$, then $A \cup B = \{1, 2, 3, 4, 5\}$. (These are the elements that are in set A or set B or both.)

b. If $C = \{3, 4, 8\}$ and $D = \{4, 3, 8\}$, then $C \cup D = \{3, 4, 8\}$. ■

Intersection of Sets The **intersection** of sets A and B, written $A \cap B$, is the set that contains only those elements that are in *both* A and B. For an element to be in $A \cap B$, it must be in set A *and* in set B. (The symbol $\cap$ suggests the letter A for *and*.)

Example 12 Examples of intersection of sets:

a. If $A = \{2, 3, 4\}$ and $B = \{1, 3, 5\}$, then $A \cap B = \{3\}$. (3 is the only element in A *and* in B.)

b. If $E = \{b, c, g\}$ and $F = \{1, 2, 5, 7\}$, then $E \cap F = \{\ \}$. (None of the elements of E are in F.) ■

Example 13 If $A = \{1, 4, 7\}$, $B = \{1, 6, 9\}$, and $C = \{4, 7\}$, find $(A \cap B) \cup C$.

Solution Because there are parentheses around $A \cap B$, we first find $A \cap B$: $A \cap B = \{1\}$. *Then* we find the union of $(A \cap B)$ with C:

$$\{1\} \cup \{4, 7\} = \{1, 4, 7\}$$

Therefore,

$$(A \cap B) \cup C = \{1, 4, 7\}$$ ■

EXERCISES 1.1B

Set I

1. Given $A = \{1, 2, 3, 4\}$ and $B = \{2, 4, 5\}$, find the following.

a. $A \cup B$ b. $A \cap B$ c. $B \cup A$ d. $B \cap A$

2. Given $C = \{2, 5, 6, 12\}$ and $D = \{7, 8, 9\}$, find the following.

a. $C \cap D$ b. $C \cup D$ c. $D \cap C$ d. $D \cup C$

3. Given $X = \{2, 5, 6, 11\}$, $Y = \{5, 7, 11, 13\}$, and $Z = \{0, 3, 4, 6\}$, find the following.

a. $X \cap Y$ b. $Y \cup Z$ c. $X \cap Z$

d. $Y \cap Z$ e. $X \cup Y$ f. $Z \cup Y$

g. $(X \cap Y) \cap Z$ h. $X \cap (Y \cap Z)$

i. $(X \cup Y) \cup Z$ j. $X \cup (Y \cup Z)$

4. Given $K = \{a, 4, 7, b\}$, $L = \{m, 4, 6, b\}$, and $M = \{n, 3, 5, t\}$, find the following.

a. $K \cap L$ b. $K \cup L$ c. $K \cup M$

d. $L \cap M$ e. $K \cap M$ f. $L \cup M$

g. $(K \cap L) \cap M$ h. $K \cap (L \cap M)$

i. $(K \cup L) \cup M$ j. $K \cup (L \cup M)$

Set II

1. Given $H = \{1, 3, 5, 7\}$ and $K = \{2, 4, 6\}$, find the following.

a. $H \cap K$ b. $H \cup K$ c. $K \cap H$ d. $K \cup H$

2. Given $S = \{1, 5, 10\}$ and $T = \{5\}$, find the following.

a. $S \cap T$ b. $S \cup T$ c. $T \cap S$ d. $T \cup S$

3. Given $A = \{a, 5, 7, b\}$, $B = \{c, 6, 7, b\}$, and $C = \{3, 4, 6\}$, find the following.

a. $A \cap B$ b. $A \cup B$ c. $A \cap C$

d. $A \cup C$ e. $B \cap C$ f. $B \cup C$

g. $(A \cap B) \cap C$ h. $A \cap (B \cap C)$

i. $(A \cup B) \cup C$ j. $A \cup (B \cup C)$

4. Given $E = \{a, b, c\}$, $F = \{c, d, e, f, g\}$, and $G = \{g\}$, find the following.

a. $E \cap F$ b. $F \cap E$ c. $E \cup F$

d. $E \cup G$ e. $E \cap G$ f. $G \cap E$

g. $E \cap (F \cup G)$ h. $(E \cap F) \cup (E \cap G)$

i. $E \cup (F \cap G)$ j. $(E \cup F) \cap (E \cup G)$

1.2 Real Numbers

We have already identified two important sets of numbers, namely, the set of *digits*, $D = \{0, 1, 2, \ldots, 9\}$, and the set of *natural numbers*, $N = \{1, 2, 3, \ldots\}$.

Whole Numbers When zero is included with the natural numbers, we have the set of **whole numbers**, which we call W. Therefore, $W = \{0, 1, 2, \ldots\}$. The set of natural numbers is a subset of the set of whole numbers; that is, $N \subseteq W$.

The Number Line The whole numbers can be represented by numbered points equally spaced along a straight line, as in Figure 1.2.1; such a line is called a **number line**. We put an arrowhead at the right of the number line, indicating the direction in which the numbers get larger; some authors put arrowheads at both ends of the number line.

The Graph of a Number We **graph** a number by placing a dot on the number line above that number. In Figure 1.2.1, we show the graphs of the first ten whole numbers (the digits).

Negative Numbers We can extend the number line to the left of zero and continue with the set of equally spaced points (see Figure 1.2.1). Numbers used to name the points to the left of zero on the number line are called **negative numbers**.

The Negative of a Number The *negative* of a nonzero number is found by changing its sign. Thus, the negative of b is $-b$, and the negative of $-b$ is b. Zero is neither a positive nor a negative number. However, we *can* say that the negative of zero is zero; that is, $-0 = 0$.

Integers The union of the set of whole numbers (W) with the set of the negatives of the natural numbers is the **set of integers**, which we will call J. In roster notation,

$$J = \{\ldots, -3, -2, -1, 0, 1, 2, 3, \ldots\}$$

The set of whole numbers is a subset of the set of integers; that is, $W \subseteq J$.

FIGURE 1.2.1

Rational Numbers The set of **rational numbers** is the set

$$Q = \left\{ \frac{a}{b} \,\middle|\, a, b \in J, b \neq 0 \right\}$$

The set of integers is a subset of the set of rational numbers (that is, $J \subseteq Q$), because any integer can be written as a rational number; for example,

$$5 = \frac{5}{1}, \qquad 17 = \frac{17}{1}, \qquad 0 = \frac{0}{1}, \qquad -7 = \frac{-7}{1}$$

The set of *terminating decimals* (decimals that have only zeros to the right of some specific decimal place) is a subset of the set of rational numbers, because any terminating decimal can be written as a rational number; for example,

$$0.1 = \frac{1}{10}, \qquad -1.03 = -\frac{103}{100}$$

The set of *nonterminating, repeating decimals** is a subset of the set of rational numbers, because any nonterminating, repeating decimal can be written as a rational number. A *bar* written above a block of digits indicates that that block repeats. For example,

$$0.666\ldots = 0.\overline{6} = \frac{2}{3}, \qquad -0.272727\ldots = -0.\overline{27} = -\frac{3}{11}$$

The set of *mixed numbers* is a subset of the set of rational numbers, because any mixed number can be written as a rational number; for example,

$$2\frac{1}{3} = \frac{7}{3}, \qquad -5\frac{1}{2} = -\frac{11}{2}$$

It is also true that $N \subseteq Q$ and $W \subseteq Q$; that is, the set of natural numbers and the set of whole numbers are both subsets of the set of rational numbers.

Irrational Numbers One very important set of numbers exists that is not a subset of any of the sets of numbers we've discussed so far. When we change any one of these numbers into its decimal equivalent, we always get a *nonterminating, nonrepeating decimal.* This set is the set of **irrational numbers**, which we will call H. A few examples of irrational numbers are $\sqrt{2}$, $\sqrt{8}$, $\sqrt[3]{55}$, and π.

The digits go on forever, never terminating and never repeating (the "…")

$\sqrt{2} = 1.414213562\ldots \approx 1.414$ Rounded off to three decimal places

$\approx$ means "is approximately equal to"

$\sqrt{8} = 2.828427125\ldots \approx 2.828$ Rounded off to three decimal places

$\sqrt[3]{55} = 3.802952461\ldots \approx 3.803$ Rounded off to three decimal places

$\pi = 3.141592654\ldots \approx 3.1416$ Rounded off to four decimal places

---

* These repeating decimals repeat a *number* or a *block of numbers.* (In 0.666 . . . , the 6 repeats, and in −0.272727 . . . , the block 27 repeats.) In Chapter 12, we discuss how to change 0.272727 . . . to $\frac{3}{11}$.

There are infinitely many more irrational numbers, each of which becomes a nonterminating, nonrepeating decimal when it is changed to its decimal equivalent.

Irrational numbers cannot be put into the form $\frac{a}{b}$ where a and b are integers and $b \neq 0$.

Real Numbers The union of the set of rational numbers and the set of irrational numbers comprises the set of **real numbers**, which we call R. In set notation, $Q \cup H = R$. (We will be concerned only with real numbers until Chapter 7.)

The set of rational numbers is a subset of the set of real numbers, and the set of irrational numbers is a subset of the set of real numbers. *No number can be a rational number and also an irrational number*. That is, $H \cap Q = \{\ \}$.

We can graph all of the real numbers on the *real number line*. There is a point on the number line that corresponds to each real number; we show the graphs of a few real numbers in Figure 1.2.2.

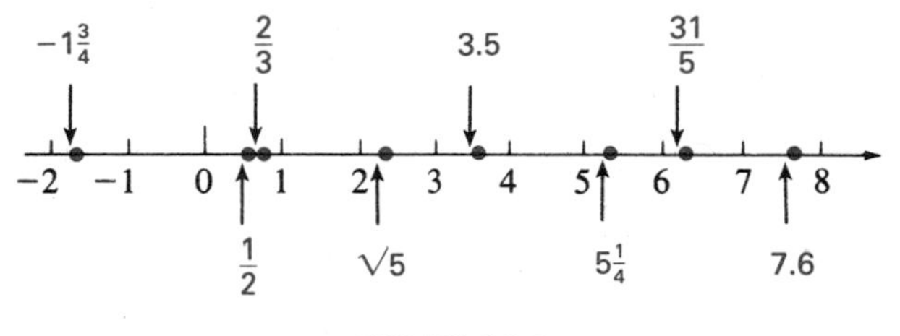

FIGURE 1.2.2

Example 1 Graph the numbers -4, 2.23606797 . . . , 3, and $-1\frac{3}{4}$ on the real number line and determine which are real numbers, which are integers, which are natural numbers, which are irrational numbers, and which are rational numbers.

Solution

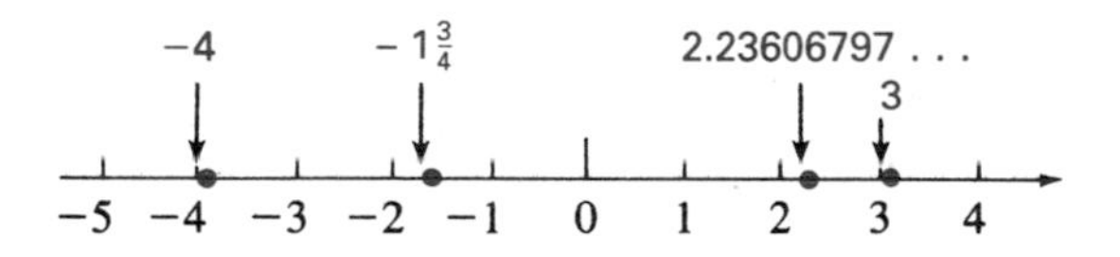

-4 is a rational number $\left(\text{it could be expressed as } -\frac{4}{1}\right)$, a real number, and an integer. 2.23606797 . . . is a nonterminating, nonrepeating decimal; therefore, it is an irrational number and a real number. (Remember, every irrational number is a real number.) 3 is a real number, an integer, a natural number, and a rational number. $-1\frac{3}{4}$ is a real number and a rational number. Therefore, the following relationships hold:

The real numbers are -4, 2.23606797 . . . , 3, and $-1\frac{3}{4}$.

The integers are -4 and 3.

The natural number is 3.

The irrational number is 2.23606797 . . . .

The rational numbers are -4, 3, and $-1\frac{3}{4}$. ■

The relationships among the sets of numbers discussed above are shown in Figure 1.2.3.

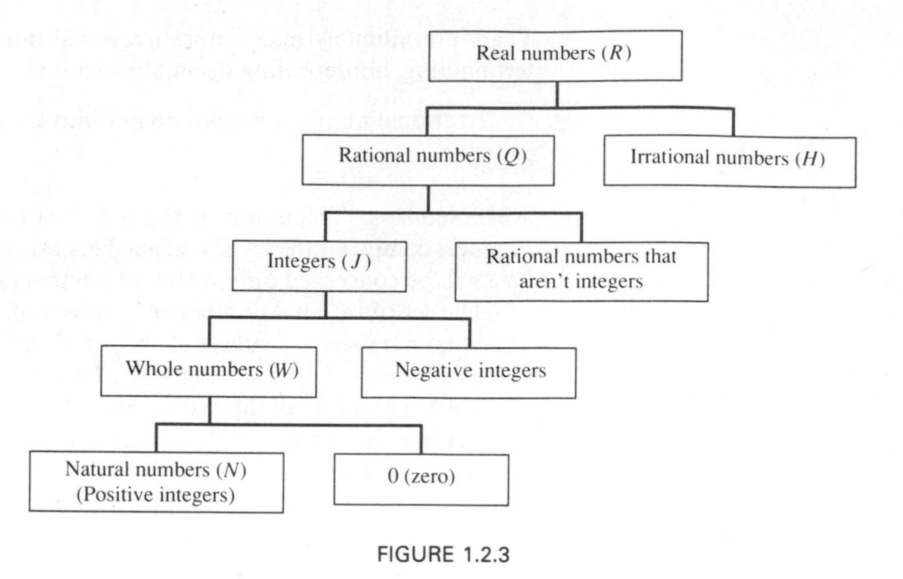

FIGURE 1.2.3

NOTE $N \subseteq W \subseteq J \subseteq Q \subseteq R$, $H \subseteq R$, and $Q \cap H = \{\ \}$. ☑

EXERCISES 1.2

Set I

1. Graph the numbers 7, 2.449489734 . . . , -5, 10, $-\frac{3}{4}$, and $0.\overline{2}$ on a number line and determine the following.

a. Which are real numbers?

b. Which are integers?

c. Which are natural numbers?

d. Which are irrational numbers?

e. Which are rational numbers?

2. Graph the numbers $-2.64575131\ldots$, -11, $\frac{5}{6}$, $3\frac{1}{3}$, $1.\overline{6}$, and 4 on a number line and determine the following.

a. Which are real numbers?

b. Which are integers?

c. Which are natural numbers?

d. Which are irrational numbers?

e. Which are rational numbers?

In Exercises 3 and 4, replace the question mark with either $\subseteq$ or $\in$ to form a true statement. The letters refer to the sets defined in this section.

3. a. $\{1, 3\}\ ?\ J$ b. $2\ ?\ N$ c. $N\ ?\ R$

d. $\frac{3}{4}\ ?\ Q$ e. $\frac{3}{4}\ ?\ R$ f. $\{3\}\ ?\ R$

4. a. $J\ ?\ Q$ b. $\{2, 3, 4\}\ ?\ R$ c. $\frac{2}{3}\ ?\ R$

d. $-4\ ?\ J$ e. $J\ ?\ R$ f. $\{-5\}\ ?\ J$

In Exercises 5 and 6, write "True" if the statement is always true; otherwise write "False." The letters refer to the sets defined in this section.

5. a. Every real number is a rational number.

b. Every integer is a real number.

c. $R \subseteq W$ d. $3 \in J$

e. $3 \in H$ f. $H \subseteq R$

g. $3.31662479\ldots \in H$ h. $-3.872983346\ldots \in R$

6. a. Every irrational number is a real number.

b. Every rational number is an integer.

c. $5 \in H$ d. $5 \in Q$

e. $Q \subseteq R$ f. $J \subseteq Q$

g. $0.\overline{13} \in H$ h. $-4.123105626\ldots \in H$

Set II **1.** Graph the numbers $-5.196152432\ldots$, $\frac{5}{7}$, $-2\frac{4}{9}$, $4.\overline{4}$, 0, and 2 on a number line and determine the following.

a. Which are real numbers?

b. Which are integers?

c. Which are natural numbers?

d. Which are irrational numbers?

e. Which are rational numbers?

2. Graph the numbers -3, $0.\overline{12}$, 6, $2\frac{2}{3}$, $-1\frac{1}{2}$, and $-4.2163461\ldots$ on a number line and determine the following.

a. Which are real numbers?

b. Which are integers?

c. Which are natural numbers?

d. Which are irrational numbers?

e. Which are rational numbers?

In Exercises 3 and 4, replace the question mark with either $\subseteq$ or $\in$ to form a true statement. The letters refer to the sets defined in this section.

3. a. $\{0, 5\}$? W b. 12 ? W c. $\{12\}$? N

d. $\frac{8}{9}$? R c. W ? J f. $\{1, 3\}$? D

4. a. 0 ? W b. 12 ? N c. $\{0\}$? W

d. 5 ? D e. $\{12\}$? R f. $\frac{1}{4}$? R

In Exercises 5 and 6, write "True" if the statement is always true; otherwise write "False." The letters refer to the sets defined in this section.

5. a. Every whole number is an integer.

b. Every integer is a rational number.

c. $W \subseteq H$ d. $7 \in Q$

e. $7 \in R$ f. $Q \subseteq N$

g. $-0.\overline{23} \in H$ h. $3.741657387\ldots \in H$

a. Every rational number is a real number.

b. The set of digits is a subset of the set of natural numbers.

c. $R \subseteq Q$ d. $0 \in H$

e. $\frac{1}{5} \in H$ f. $H \subseteq Q$

g. $0.16513278\ldots \in H$ h. $0.16513278\ldots \in R$

1.3 Definitions and Symbols

1.3A Definitions

Constant A **constant** is an object or symbol that does not change its value in a particular problem or discussion. It is usually represented by a number symbol, but it is also often represented by one of the first few letters of the alphabet. Thus, in the expression $4x + 3y - 5$, the constants are 4, 3, and -5. In the expression $ax + by + c$, the constants are understood to be a, b, and c. The Greek letter π (≈ 3.14159) is also a constant.

Variable A **variable** is an object or symbol that acts as a placeholder for a number that is unknown. A variable is usually represented by a letter, and it may assume different values in a particular problem or discussion. An equation or inequality (see Chapter 2) is usually, but not always, *true* for some values of the variables and *false* for other values. In the expression $ax + by + c$, it is understood that the variables are x and y.

Algebraic Expression An **algebraic expression** consists of numbers, variables, signs of operation (such as $+$, $-$, $\times$, $\div$), and signs of grouping. For example, $4x + 3y - 5$ is an algebraic expression.

Factors The numbers that are multiplied together to give a product are called the **factors** of that product. The number 1 may be considered to be a factor of any number.

Example 1 Examples of identifying the factors in an expression:

a. $3 \cdot 5 = 15$ — 3 and 5 are factors of 15

b. $7abc$ — 7, a, b, and c are the factors of $7abc$

c. x — 1 (understood) and x are the factors of x ■

Terms A **term** of an algebraic expression consists of *factors only*. This means that the only operations indicated within a *term* are multiplication and division $\left(\text{we include division here because the division } a \div b \text{ can be interpreted as the multiplication } a \cdot \frac{1}{b}\right)$. *Exception*: An expression *within grouping symbols* is considered as a single term, even though it may contain one or more terms *within* the grouping symbols. A *bar* (for example, a fraction bar) is one example of a grouping symbol.

If a term is preceded by a minus sign, that sign is *part of* the term.

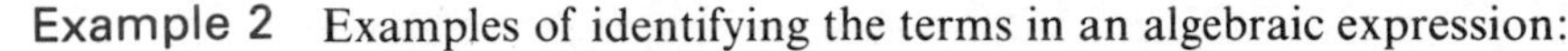

Example 2 Examples of identifying the terms in an algebraic expression:

a. $3x - 9x(2y + 5z)$

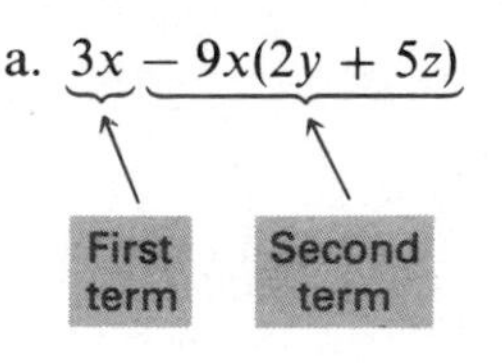

$3x$ consists only of *factors*; therefore it is a term

One factor of the second term is $-9x$, and the other factor is the expression within the parentheses

The part inside the parentheses is considered as a single unit or term, and since the part inside the parentheses is being multiplied by $-9x$, $-9x(2y + 5z)$ is considered to be one term. Therefore, this expression has only two terms, as shown.

b. $\dfrac{2-x}{xy} + (2x - 1)$ $\qquad \dfrac{2-x}{xy}$ can be written as $(2-x)\cdot\dfrac{1}{x}\cdot\dfrac{1}{y}$

This expression also has only two terms, since the fraction bar is a grouping symbol. The first term is $\dfrac{2-x}{xy}$, and the second term is $(2x - 1)$. ■

Coefficients In a term with two factors, the **coefficient** of one factor is the other factor. In a term with more than two factors, the coefficient of each factor is the product of all the other factors in the term.

Numerical Coefficients A **numerical coefficient** is a coefficient that is a *number* rather than a *letter*. When we refer to "the coefficient" of a term, it is understood that we mean the numerical coefficient of that term. If a term has no numerical coefficient showing, then the numerical coefficient is understood to be 1.

Example 3 Examples of identifying the coefficients of a term:

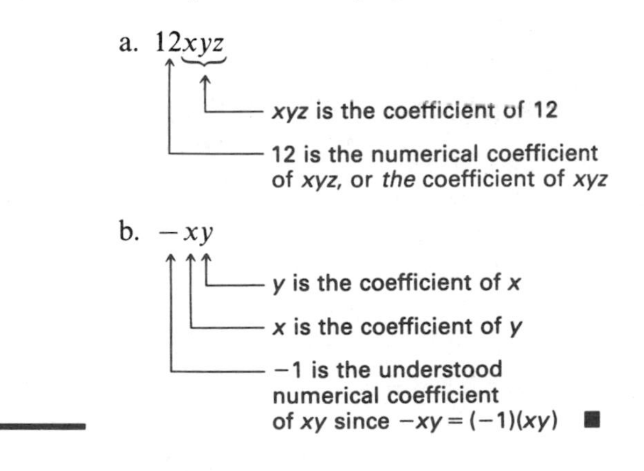

■

EXERCISES 1.3A

Set I In Exercises 1 and 2, (a) list the constants and (b) list the variables.

1. $xy + 3$ **2.** $4z - 2y$

In Exercises 3–10, (a) determine the number of terms, (b) write the *second* term if there is one, and (c) list all of the factors of the *first* term.

3. $E + 5F - 3$ **4.** $R + 2T - 6$

5. $(R + S) - 2(x + y)$ **6.** $(A + 2B) - 5(W + V)$

7. $3XYZ + 4$ **8.** $4ab + 3x$

9. $2A + \frac{3B - C}{DE}$ **10.** $3st + \frac{w + z}{xyz}$

In Exercises 11–14, (a) write the numerical coefficient of the first term and (b) write the variable part of the second term.

11. $2R - 5RT + 3T$ **12.** $4x - 3xy + z$

13. $-x - y + z$ **14.** $-a + 2b - c$

Set II In Exercises 1 and 2, (a) list the constants and (b) list the variables.

1. $4X - Z$ **2.** $3x - 2st$

In Exercises 3–10, (a) determine the number of terms, (b) write the *second* term if there is one, and (c) list all of the factors of the *first* term.

3. $X + 7Y - 8$ **4.** $3(x - 5y + 7z)$

5. $(2X - Y) - 4(a + b)$ **6.** $-4x + \frac{1}{6st} - 3z$

7. $7st + 3uv$ **8.** $x + 2(y + z + 4) - w + s$

9. $3A + \frac{2B - 3C}{2AB}$ **10.** $6 - 3xy + \frac{z}{2}$

In Exercises 11–14, (a) write the numerical coefficient of the first term and (b) write the variable part of the second term.

11. $2X + Y - 3$ **12.** $9x - 5y + z$

13. $4(X + 2) - 5Y$ **14.** $z - 3w$

1.3B Inequalities and Absolute Values

"Greater Than" and "Less Than" Symbols The symbols $>$ and $<$ are called **inequality symbols.** Let X be any number on the number line. Then numbers to the right of X on the number line are said to be *greater than* X, written "$> X$." Numbers to the left of X on the number line are said to be *less than* X, written "$< X$."

Example 4 Examples of "greater than" and "less than" symbols:

a. $-1 > -4$ is read "-1 is greater than -4"

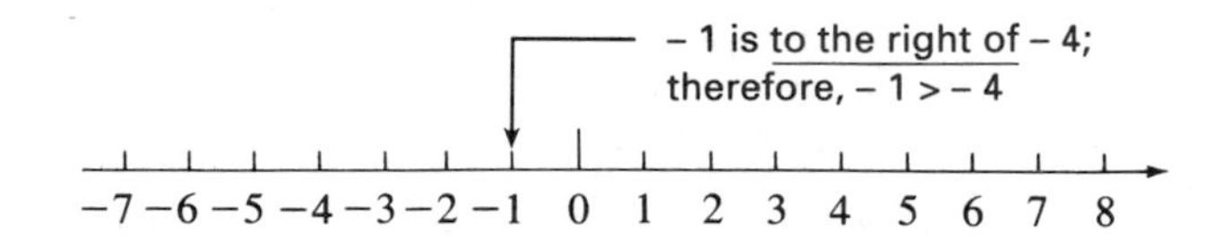

b. $-2 < 1$ is read "-2 is less than 1"

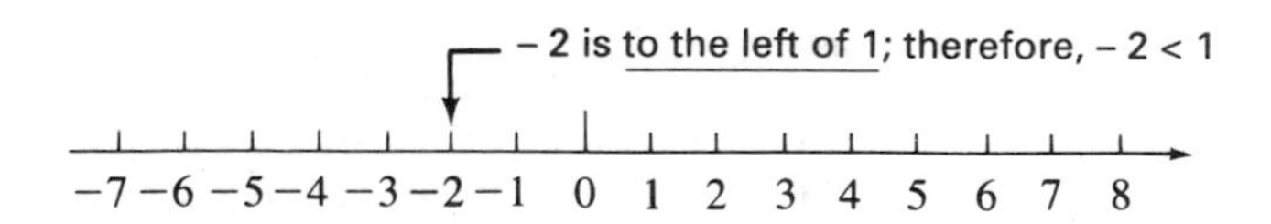

Note that $1 > -2$ and $-2 < 1$ give the same information even though they are read differently. ■

In general,

$$a < b \text{ may always be replaced by } b > a$$

and

$$a > b \text{ may always be replaced by } b < a.$$

"Less Than or Equal to" Symbol The inequality $a \leq b$ is read "a is less than *or* equal to b. This means that if $\left\{\begin{matrix}\text{either } a < b \\ \text{or} \quad a = b\end{matrix}\right\}$ is true, then $a \leq b$ is true. For example, $2 \leq 3$ is true because $2 < 3$ is true (even though $2 = 3$ is *not* true). Remember, only *one* of the two statements $\left\{\begin{matrix}2 < 3 \\ 2 = 3\end{matrix}\right\}$ need be true in order that $2 \leq 3$ be true.

"Greater Than or Equal to" Symbol The inequality $a \geq b$ is read "a is greater than *or* equal to b." This means that if $\left\{\begin{matrix}\text{either } a > b \\ \text{or} \quad a = b\end{matrix}\right\}$ is true, then $a \geq b$ is true. For example, $5 \geq 1$ is true because $5 > 1$ is true (even though $5 = 1$ is *not* true).

Continued Inequality Symbols An inequality is *valid* if it is true *and* written in the correct form. A **continued inequality** has *two* valid inequality symbols in it; *they must both be "less than" symbols or both be "greater than" symbols.* If the continued inequality $a < b < c$ is to be valid, the following two inequalities must *both* be true: $a < b$ *and* $b < c$.*

Example 5 Examples of continued inequalities:

a. $4 < 7 < 9$ is read "4 is less than 7, which is less than 9." It can also be read "7 is greater than 4 *and* less than 9." (Remember than $4 < 7$ can be replaced by $7 > 4$.) $4 < 7 < 9$ means that 7 is *between* 4 and 9.

b. $10 > 0 > -3$ is read "10 is greater than 0, which is greater than -3," or "0 is less than 10 and greater than -3." (Remember that $10 > 0$ can be replaced by $0 < 10$.) $10 > 0 > -3$ means that 0 is between 10 and -3.

NOTE Even though the statement $10 > 0 > -3$ is correct as written, it is customary to write such continued inequalities so that the numbers are in the same order as on the number line, that is, as $-3 < 0 < 10$. ■

A WORD OF CAUTION Notice that $-3 < 0 > -2$ is *not* correct because the inequality symbols are not both less than or both greater than. ☑

Example 6 Examples of determining whether a continued inequality is valid or invalid:

a. $10 < 12 > 5$

The inequality is *invalid* because one symbol is $<$ and the other is $>$.

b. $10 < 15 < 32$

The inequality is *valid* because both symbols are $<$ *and* it is true that $10 < 15$, $15 < 32$, *and* $10 < 32$.

* It must also be true that a must be less than c because of the *transitive property of inequalities*. The transitive property states that if $x < y$ and $y < z$, then $x < z$.

c. $2 > 5 > 8$

The inequality is *invalid* because $2 > 5$, $5 > 8$, and $2 > 8$ are *false* statements.

d. $9 > 3 < 5$

The inequality is *invalid* because one symbol is $>$ and the other is $<$. ■

The Slash Line Symbol A slash line drawn through a symbol puts a "not" in the meaning of the symbol. We've seen this already in such symbols as $\neq$, $\notin$, and $\not\subseteq$. The symbol $\not<$ is read "is not less than" and is equivalent to the symbol $\geq$. The symbol $\not>$ is read "is not greater than" and is equivalent to the symbol $\leq$.

Example 7 Examples showing the use of the slash line:

a. $3 \not< -2$ is read "3 is *not* less than -2." It is equivalent to $3 \geq -2$.

b. $-6 \not> -5$ is read "-6 is *not* greater than -5." It is equivalent to $-6 \leq -5$. ■

Absolute Value The symbol for the **absolute value** of a real number x is $|x|$. The formal definition of the absolute value of a real number x is

$$|x| = \begin{cases} x \text{ if } x \geq 0 \\ -x \text{ if } x < 0 \end{cases}$$

$-x$ is *not* a negative number in this case because x is a negative number, and the negative of a negative number is positive; for example, if $x = -2$, then $-x = -(-2) = 2$

We can think of the absolute value of a number as being the distance between that number and zero on the number line *with no regard for direction.* See Figure 1.3.1.

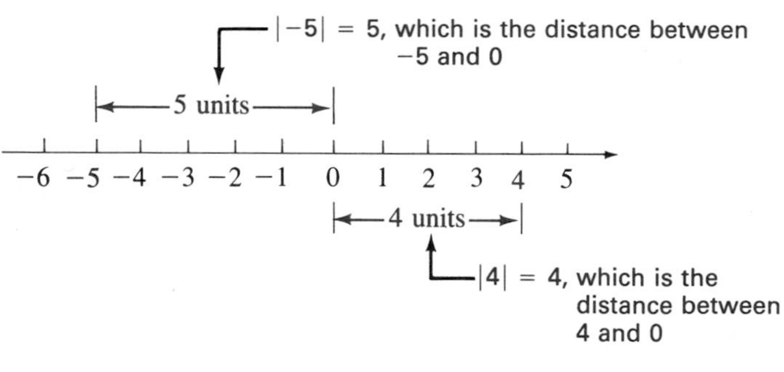

FIGURE 1.3.1

NOTE The absolute value of a number is always *nonnegative*; that is, it is always positive or zero. ✓

Example 8 Examples of absolute values of real numbers:

a. $|9| = 9$ The distance between 0 and 9 is 9

b. $|-4| = -(-4) = 4$ The distance between -4 and 0 is 4

c. $-|-5| = -(+5) = -5$ Substituting $+5$ for $|-5|$

d. $-|3| = -(+3) = -3$ Substituting $+3$ for $|3|$

e. $|0| = 0$

NOTE In (c) and (d), negative signs were located *outside* the absolute value symbols. ■

EXERCISES 1.3B

Set I In Exercises 1 and 2, evaluate each expression.

1. a. $|7|$ b. $|-5|$ c. $-|-12|$

2. a. $|-34|$ b. $-|5|$ c. $-|-3|$

In Exercises 3 and 4, determine which of the two symbols, $<$ or $>$, should be used to make each statement true.

3. a. $-15 ? -13$ b. $-15 ? |-13|$ c. $|-15| ? -13$ d. $|2| ? |-3|$

4. a. $-19 ? -23$ b. $-19 ? |-23|$ c. $|-19| ? -23$ d. $|-19| ? |-23|$

In Exercises 5 and 6, determine whether each inequality is a valid or an invalid continued inequality. If it is invalid, state *why* the inequality is invalid.

5. a. $3 < 5 < 10$ b. $-2 > 1 > 7$ c. $0 < 4 > 2$

d. $8 > 3 > -1$ e. $8 > 5 < 9$ f. $0 > 3 > 8$

6. a. $4 > 0 > -2$ b. $-2 < 6 > 5$ c. $3 > 1 < 7$

d. $-1 < 3 < 6$ e. $4 > 7 > 11$ f. $5 < 0 < -2$

In Exercises 7 and 8, determine whether the expression is true or false.

7. $3 \nless 8$ **8.** $-4 \ngtr 5$

Set II In Exercises 1 and 2, evaluate each expression.

1. a. $|-15|$ b. $|35|$ c. $-|-22|$

2. a. $-|-5|$ b. $-|13|$ c. $|-2|$

In Exercises 3 and 4, determine which of the two symbols, $<$ or $>$, should be used to make each statement true.

3. a. $-18 ? -12$ b. $-63 ? |-104|$ c. $|7| ? |-8|$ d. $|-3| ? -5$

4. a. $0 ? |-5|$ b. $|15| ? |-3|$ c. $-8 ? |-6|$ d. $|10| ? |-14|$

In Exercises 5 and 6, determine whether the inequality is a valid or an invalid continued inequality. If it is invalid, state *why* the inequality is invalid.

5. a. $-1 < 7 > 5$ b. $0 < 4 < 6$ c. $1 > 3 > 7$

d. $3 > 2 < 5$ e. $8 < 4 < 2$ f. $8 > 2 > -4$

6. a. $-3 < 5 > 4$ b. $18 > 5 > -6$ c. $2 > 0 < 8$

d. $-8 < -6 < -2$ e. $-3 > 1 > 5$ f. $8 > 5 > 1$

In Exercises 7 and 8, determine whether the statement is true or false.

7. $-2 \ngtr 0$ **8.** $3 \nless -5$

1.4 Operations on Signed Numbers

Addition of Signed Numbers

The answer to an addition problem is called the **sum**. The rules for addition of signed numbers are summarized in the box at the top of page 20.

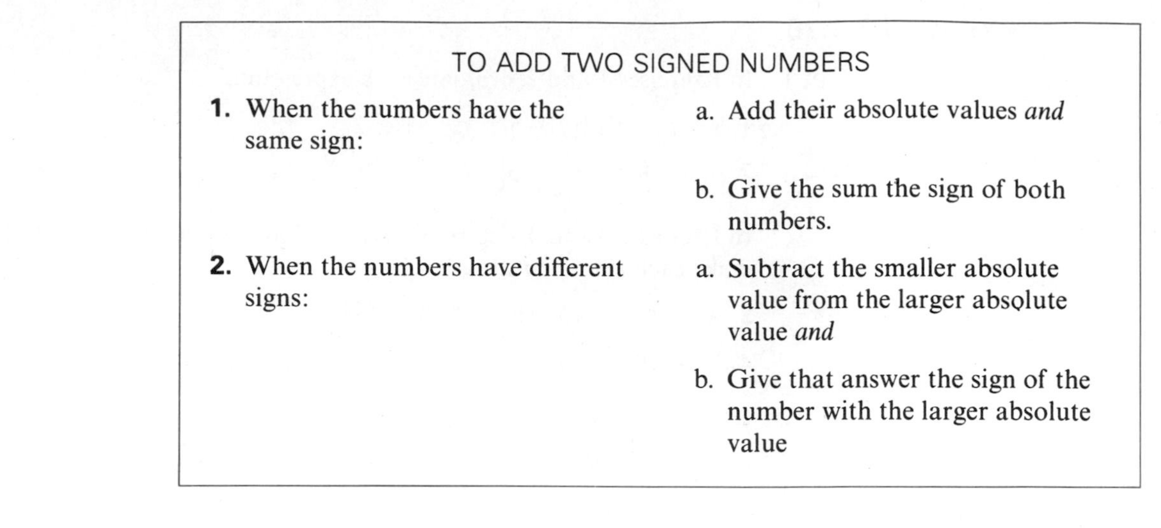

TO ADD TWO SIGNED NUMBERS

1. When the numbers have the same sign:
   a. Add their absolute values *and*
   b. Give the sum the sign of both numbers.
2. When the numbers have different signs:
   a. Subtract the smaller absolute value from the larger absolute value *and*
   b. Give that answer the sign of the number with the larger absolute value

Example 1 Examples of adding signed numbers:

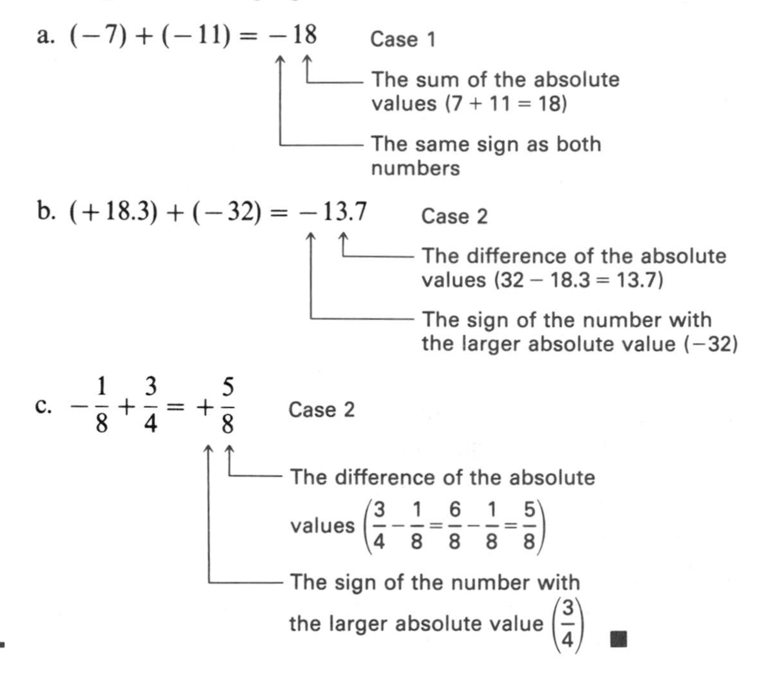

a. $(-7)+(-11)=-18$ Case 1

b. $(+18.3)+(-32)=-13.7$ Case 2

c. $-\frac{1}{8}+\frac{3}{4}=+\frac{5}{8}$ Case 2

■

Additive Identity Since adding 0 to a number gives the *identical* number we started with (for example, $7 + 0 = 7$), we call 0 the **additive identity**.

THE ADDITIVE IDENTITY IS 0

If a represents any real number, then

$$a + 0 = a \quad \text{and} \quad 0 + a = a$$

Additive Inverse When the sum of two numbers is 0, we say they are the **additive inverses** of each other. Therefore, the *negative* of a number can also be called the *additive inverse* of that number.

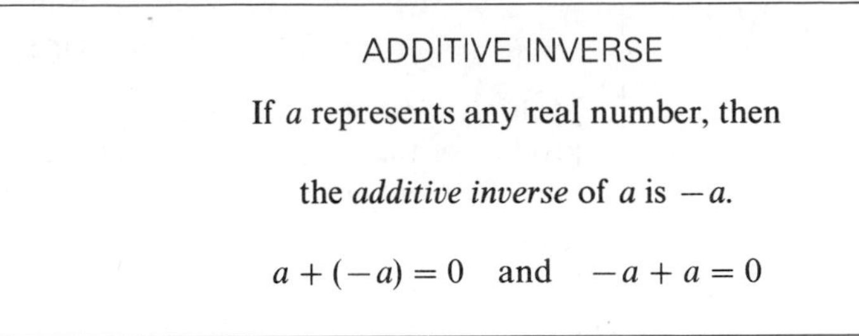

ADDITIVE INVERSE

If a represents any real number, then

the *additive inverse* of a is $-a$.

$$a + (-a) = 0 \quad \text{and} \quad -a + a = 0$$

Subtraction of Signed Numbers

Subtraction is the *inverse* operation of addition. That is, it "undoes" addition. The answer to a subtraction problem is called the **difference**.

DEFINITION OF SUBTRACTION

$$a - b = a + (-b)$$

This definition means that we perform a subtraction in two steps. In step 1 we change the subtraction symbol to an addition symbol *and* we also change the sign of the number being subtracted. Then, in the second step, we do the resulting *addition* problem.

The definition of subtraction also leads to the following facts regarding subtraction involving zero:

$a - 0 = a$ — This is true because $a - 0 = a + (-0) = a + 0 = a$.

$0 - a = -a$ — This is true because $0 - a = 0 + (-a) = -a$.

A WORD OF CAUTION "Subtract a from b" means $b - a$. ✓

Example 2 Examples of subtracting signed numbers:

a. Subtract -3 from 0. [This means $0 - (-3)$.]

$$0 - (-3) = -(-3) = 3 \quad or \quad 0 - (-3) = 0 + (+3) = 3$$

b. Subtract 4 from -2. [This means $(-2) - (+4)$.]

$$(-2) - (+4) = -2 + (-4) = -6$$

c. $2.345 - 11.6 = 2.345 + (-11.6) = -9.255$

d. $\frac{2}{3} - \left(-\frac{2}{5}\right) = \frac{2}{3} + \left(+\frac{2}{5}\right) = \frac{10}{15} + \frac{6}{15} = \frac{16}{15}$ or $1\frac{1}{15}$ ■

Multiplication of Signed Numbers

The answer to a multiplication problem is called the **product**. Recall from Section 1.3A that the numbers that are multiplied together are called the *factors* of the product. The rules for multiplying two signed numbers are summarized as follows:

TO MULTIPLY TWO SIGNED NONZERO NUMBERS

Multiply their absolute values *and* attach the correct sign to the left of the product of the absolute values. The sign is *positive* when the numbers have the same sign and *negative* when the numbers have different signs.

Multiplicative Identity Since multiplying any real number by 1 gives the *identical* number we started with (for example, $0.15 \times 1 = 0.15$), we call 1 the **multiplicative identity**.

THE MULTIPLICATIVE IDENTITY IS 1

If a represents any real number, then

$$a \cdot 1 = a \quad \text{and} \quad 1 \cdot a = a$$

Multiplicative Inverse (Reciprocal) When the product of two numbers is 1, we say that the numbers are the **multiplicative inverses**, or **reciprocals**, of each other.

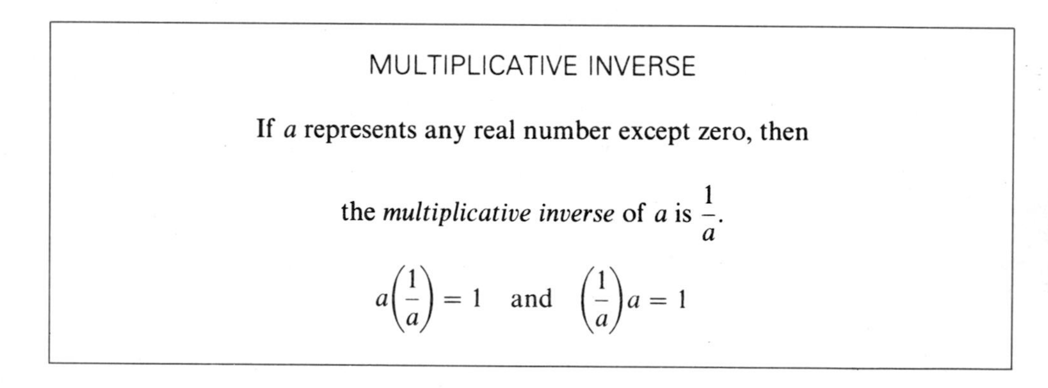

MULTIPLICATIVE INVERSE

If a represents any real number except zero, then

the *multiplicative inverse* of a is $\frac{1}{a}$.

$$a\left(\frac{1}{a}\right) = 1 \quad \text{and} \quad \left(\frac{1}{a}\right)a = 1$$

Multiplication Involving Zero Multiplying any real number by zero gives a product of zero.

THE MULTIPLICATION PROPERTY OF ZERO

If a represents any real number, then

$$a \cdot 0 = 0 \quad \text{and} \quad 0 \cdot a = 0$$

Example 3 Examples of multiplying signed numbers:

a. $(-0.14)(-10) = +1.40$

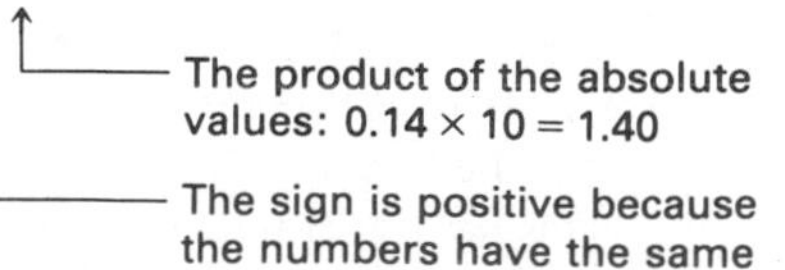

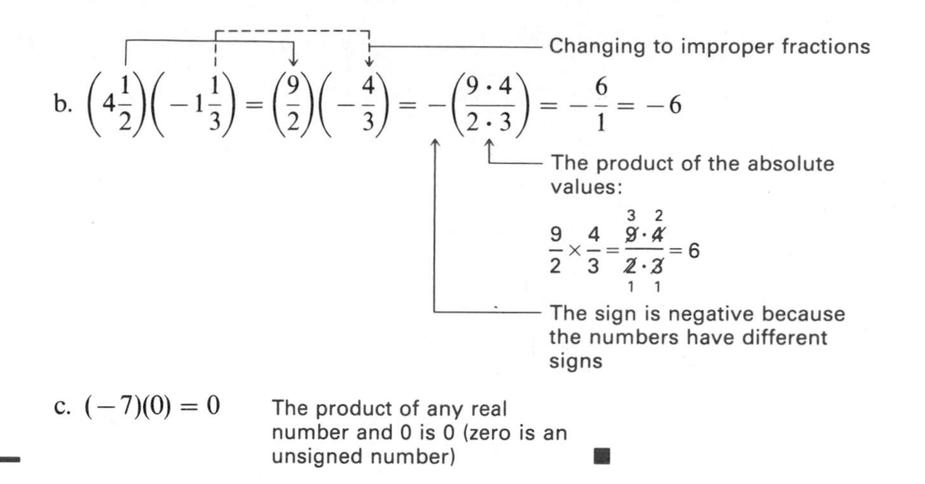

c. $(-7)(0) = 0$ The product of any real number and 0 is 0 (zero is an unsigned number) ■

Division of Signed Numbers

Division is the *inverse* of multiplication because it "undoes" multiplication. The answer to a division problem is called the **quotient**. The number we are dividing *by* is called the **divisor**, and the number we are dividing *into* is called the **dividend**. If the divisor does not divide exactly into the dividend, the part that is left over is called the **remainder**.

Factors When the remainder is 0, we can say that the divisor and the quotient are both **factors** or **divisors** of the dividend.

Because of the inverse relation between division and multiplication, the rules for finding the sign of a quotient are the same as those used for finding the sign of a product.

> TO DIVIDE ONE SIGNED NUMBER BY ANOTHER
>
> Divide the absolute value of the dividend by the absolute value of the divisor *and* attach the correct sign to the left of the quotient of the abolute values. The sign is *positive* when the numbers have the same sign and *negative* when the numbers have different signs.

To check the answer in a division problem,

> Divisor × Quotient + Remainder = Dividend

or, if the divisor is a *factor* of the dividend,

> Divisor × Quotient = Dividend

Division Involving Zero

Division of zero by a number other than zero is possible, and the quotient is always 0. That is, $0 \div 2 = \frac{0}{2} = 0$, or $2\overline{)0}$ with quotient 0, because $0 \times 2 = 0$.

Division of a nonzero number by zero is impossible. Let's try to divide some nonzero number by zero. For example, let's try $4 \div 0$, or $\frac{4}{0}$. Suppose the quotient is some unknown number we call q. Then $4 \div 0 = q$ means that we must find a number q such that $q \times 0 = 4$. But $q \times 0 = 4$ is impossible, since any number times zero is zero. Therefore, no answer q exists, and $4 \div 0$ has no answer (or is *undefined*). (It is for this reason that zero has no reciprocal.)

Division of zero by zero cannot be determined. Consider $0 \div 0$ or $\frac{0}{0}$. Suppose that the quotient is 1. Then $0 \div 0 = 1$. This means that 0×1 has to equal zero, and, in fact, it does. This might lead us to assume that the quotient is indeed 1. But now let us suppose that the quotient is zero. (That is, suppose $0 \div 0 = 0$.) This could be true only if 0×0 equals zero, which it does. Furthermore, we could say $0 \div 0 = 5$, since $0 \times 5 = 0$. The quotient could also be -3 or π or 156, or *any* number! Therefore, we say that $0 \div 0$ cannot be determined.

For these reasons, we say that division by zero is *undefined*. The important thing to remember about division involving zero is that you cannot divide *by* zero. Division involving zero can be summarized as follows:

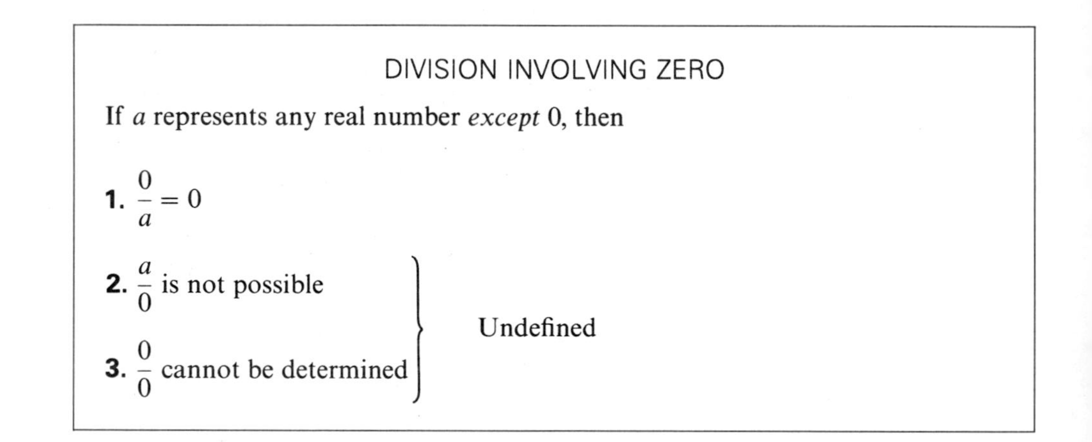

DIVISION INVOLVING ZERO

If a represents any real number *except* 0, then

1. $\frac{0}{a} = 0$

2. $\frac{a}{0}$ is not possible

3. $\frac{0}{0}$ cannot be determined

(2 and 3: Undefined)

Example 4 Examples of dividing signed numbers:

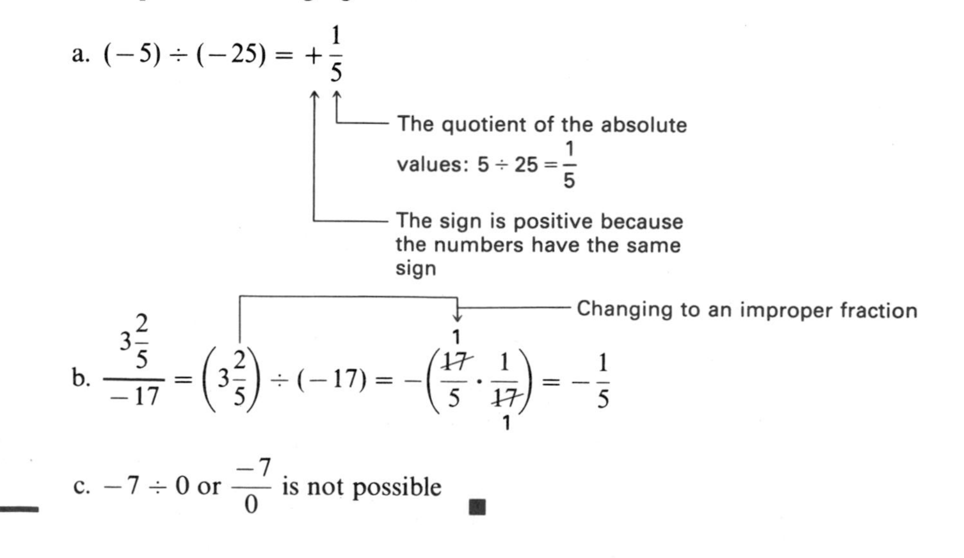

a. $(-5) \div (-25) = +\frac{1}{5}$

b. $\frac{3\frac{2}{5}}{-17} = \left(3\frac{2}{5}\right) \div (-17) = -\left(\frac{17}{5} \cdot \frac{1}{17}\right) = -\frac{1}{5}$

c. $-7 \div 0$ or $\frac{-7}{0}$ is not possible ■

EXERCISES 1.4

Set I Perform the indicated operations (or write "undefined").

1. $5 + (-9)$ **2.** $-10 + 7$ **3.** $-12 + (-7)$

4. $-\frac{7}{8} + \left(-\frac{3}{8}\right)$ **5.** $-\frac{7}{12} + \left(-\frac{1}{6}\right)$ **6.** $-\frac{3}{4} + \left(-\frac{3}{8}\right)$

7. $5\frac{1}{4} + \left(-2\frac{1}{2}\right)$ **8.** $-1\frac{7}{10} + 3\frac{2}{5}$ **9.** $-13.5 - (-8.06)$

10. $-1.5 - (-1.13)$ **11.** $2.4 - (-13)$ **12.** $7.2 - (-89)$

13. $\frac{1}{3} - \left(-\frac{1}{2}\right)$ **14.** $17\frac{5}{6} - \left(-8\frac{1}{3}\right)$ **15.** $-5\frac{3}{4} - 2\frac{1}{2}$

16. $-1.56 - 9.7$ **17.** $16.71 - (-18.9)$ **18.** $8.9 - (-3.71)$

19. $0 - 3$ **20.** $0(-3)$ **21.** $(-26)(-10)$

22. $(-1.1)(-7)$ **23.** $\frac{3}{8}\left(-\frac{4}{9}\right)$ **24.** $\frac{-150}{10}$

25. $-12 \div 36$ **26.** $\frac{-15}{-6}$ **27.** $7\frac{1}{2} \div \left(-\frac{1}{2}\right)$

28. $-\frac{5}{4} \div 0$ **29.** $0 \div \left(-\frac{3}{8}\right)$ **30.** $0 \div 0$

31. $-\frac{5}{4} \div \left(-\frac{3}{8}\right)$ **32.** $-25{,}000 \div (-100)$

33. Subtract -3 from -8. **34.** Subtract 0.3 from -1.26.

35. Subtract $-\frac{2}{3}$ from 5. **36.** $|-8| + (-15)$

37. $|-10| - (-25)$ **38.** $|-2| - |-9|$

39. $-3|-4|$ **40.** $|-15| - 4$

Set II Perform the indicated operations (or write "undefined").

1. $-59 + 74$ **2.** $6 + (-15)$ **3.** $\frac{5}{8} + \left(-\frac{3}{4}\right)$

4. $\frac{1}{2} + \frac{1}{3}$ **5.** $-4\frac{1}{2} + 1\frac{3}{4}$ **6.** $10\frac{3}{8} + \left(-2\frac{3}{4}\right)$

7. $-395.7 + (84.91)$ **8.** $-1\frac{2}{3} + \left(-3\frac{2}{5}\right)$ **9.** $281 - 960$

10. $3.62 + (-1.634)$ **11.** $28.61 - (-37.9)$ **12.** $-8 - (-3.2)$

13. $-18\frac{3}{10} - 7\frac{2}{5}$ **14.** $16\frac{2}{3} - \left(-3\frac{7}{12}\right)$ **15.** $-6.483 - (-2.7)$

16. $3.28 - 7.6$ **17.** $(-26)(10)$ **18.** $-5.2 - (7.3)$

19. $0(-13)$ **20.** $0 - 13$ **21.** $(-34)(12)$

22. $-6(0)$ **23.** $\frac{3}{7}\left(\frac{-14}{15}\right)$ **24.** $\frac{-180}{10}$

25. $-15 \div 60$ **26.** $-4 \div 12$ **27.** $\frac{0}{0}$

28. $\frac{2}{3} \div 0$ **29.** $0 \div 7$ **30.** $1.6 \div 0$

31. Subtract -1 from -2.

32. Subtract 1.26 from 0.3.

33. $\frac{160}{-40}$

34. Subtract -5.2 from -1.63.

35. $|-3| + (-7)$

36. $|-6| - |-10|$

37. $|-5| - (-6)$

38. $4 - |-8|$

39. $|8| + |-18|$

40. $-2 - |-10|$

1.5 Commutative, Associative, and Distributive Properties of Real Numbers

Commutative Properties

Addition Is Commutative If we change the order of the two numbers in an addition problem, we get the same sum. This property is called the **commutative property of addition**.

> COMMUTATIVE PROPERTY OF ADDITION
>
> If a and b represent any real numbers, then
>
> $$a + b = b + a$$

Example 1 Verify that the commutative property holds for the sum of 7 and 5.

Solution $7 + 5 = 12$, and $5 + 7 = 12$. Since $7 + 5$ and $5 + 7$ both equal 12, they must equal each other. Therefore, $7 + 5 = 5 + 7$. ■

Subtraction is not commutative, as a single example will prove.

Example 2 Show that subtraction is not commutative by showing that $3 - 2 \neq 2 - 3$.

Solution $3 - 2 = 1$, and $2 - 3 = -1$. Since $3 - 2$ and $2 - 3$ do not both equal the same number, $3 - 2 \neq 2 - 3$. By finding one subtraction problem for which the commutative property does *not* hold, we've shown that subtraction is *not* commutative. ■

Multiplication Is Commutative If we change the order of the two numbers in a multiplication problem, we get the same product. This property is called the **commutative property of multiplication**.

> COMMUTATIVE PROPERTY OF MULTIPLICATION
>
> If a and b represent any real numbers, then
>
> $$a \cdot b = b \cdot a$$

Example 3 Verify that the commutative property holds for the product of 4 and 5.

Solution $4 \times 5 = 20$, and $5 \times 4 = 20$. Since 4×5 and 5×4 both equal 20, they must equal each other. Therefore, $4 \times 5 = 5 \times 4$. ■

Division is not commutative, as a single example will prove.

Example 4 Show that division is not commutative by showing that $10 \div 5 \neq 5 \div 10$.

Solution $10 \div 5 = 2$, and $5 \div 10 = \frac{1}{2}$. Since $10 \div 5$ and $5 \div 10$ do not both equal the same number, $10 \div 5 \neq 5 \div 10$. By finding one division problem for which the commutative property does *not* hold, we've shown that division is *not* commutative. ■

Associative Properties

Addition Is Associative In adding three numbers, we obtain the same sum when we add the first two numbers together first as when we add the last two numbers together first. This property is called the **associative property of addition**.

> ASSOCIATIVE PROPERTY OF ADDITION
>
> If a, b, and c represent any real numbers, then
>
> $$(a + b) + c = a + (b + c)$$

Example 5 Verify that the associative property holds for the sum of 2, 3, and 4.

Solution $(2 + 3) + 4 = 5 + 4 = 9$, and $2 + (3 + 4) = 2 + 7 = 9$. Since $(2 + 3) + 4$ and $2 + (3 + 4)$ both equal 9, they must equal each other. Therefore, $(2 + 3) + 4 = 2 + (3 + 4)$. ■

Subtraction is not associative, as a single example will prove.

Example 6 Show that subtraction is not associative by showing that $(7 - 4) - 8 \neq 7 - (4 - 8)$.

Solution $(7 - 4) - 8 = 3 - 8 = -5$, and $7 - (4 - 8) = 7 - (-4) = 7 + 4 = 11$. Since $(7 - 4) - 8$ and $7 - (4 - 8)$ do not both equal the same number, $(7 - 4) - 8 \neq 7 - (4 - 8)$. By finding one subtraction problem for which the associative property does *not* hold, we've shown that subtraction is *not* associative. ■

Multiplication Is Associative In multiplying three numbers, we obtain the same product when we multiply the first two numbers together first as when we multiply the last two numbers together first. This property is called the **associative property of multiplication**.

> ASSOCIATIVE PROPERTY OF MULTIPLICATION
>
> If a, b, and c represent any real numbers, then
>
> $$(a \cdot b) \cdot c = a \cdot (b \cdot c)$$

Example 7 Verify that the associative property holds for the product of 3, 4, and 2.

Solution $(3 \cdot 4) \cdot 2 = 12 \cdot 2 = 24$, and $3 \cdot (4 \cdot 2) = 3 \cdot 8 = 24$. Since $(3 \cdot 4) \cdot 2$ and $3 \cdot (4 \cdot 2)$ both equal 24, they must equal each other. Therefore, $(3 \cdot 4) \cdot 2 = 3 \cdot (4 \cdot 2)$. ■

Division is not associative, as a single example will prove.

Example 8 Show that division is not associative by showing that $(16 \div 4) \div 2 \neq 16 \div (4 \div 2)$.

Solution $(16 \div 4) \div 2 = 4 \div 2 = 2$, and $16 \div (4 \div 2) = 16 \div 2 = 8$. Since $(16 \div 4) \div 2$ and $16 \div (4 \div 2)$ do not both equal the same number, $(16 \div 4) \div 2 \neq 16 \div (4 \div 2)$. By finding one division problem for which the associative property does *not* hold, we've shown that division is *not* associative. ■

How to Determine Whether Commutativity or Associativity Has Been Used In commutativity, the numbers or variables actually exchange places (commute).

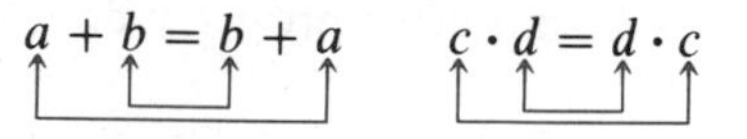

The first element occupies the second place and vice versa.

In associativity, the numbers or variables stay in their original places, but the grouping is changed.

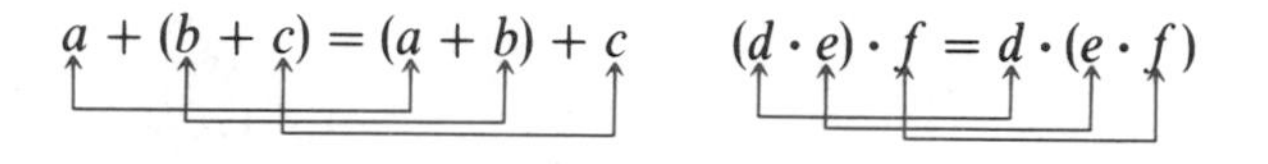

Distributive Property

The distributive property is one of the most important and most often used properties in mathematics.

Multiplication Is Distributive over Addition The **distributive property** can be verified by substituting *any* real numbers for a, b, and c. We may distribute either from the right or from the left.

> MULTIPLICATION IS DISTRIBUTIVE OVER ADDITION
>
> If a, b, and c represent any real numbers, then
>
> $$a(b + c) = ab + ac$$
>
> $$(b + c)a = ba + ca$$

NOTE The distributive properties may be extended to include any number of terms inside the parentheses, and the terms may be *subtracted* as well as added. That is,

$$a(b + c - d + e) = ab + ac - ad + ae$$

and so forth. ☑

Example 9 Verify that the distributive property holds for $a = 3$, $b = 5$, and $c = 7$.

Solution

$$3(5 + 7) \stackrel{?}{=} 3 \cdot 5 + 3 \cdot 7$$

$$3(\ 12\ ) \stackrel{?}{=} 15 + 21$$

$$36 = 36$$ ■

A WORD OF CAUTION A mistake students often make is to think that the distributive property applies to expressions such as $2(3 \cdot 4)$.

The distributive property applies only when this symbol [the · in $2(3 \cdot 4)$] is an addition or subtraction symbol

$$2(3 \cdot 4) \neq (2 \cdot 3)(2 \cdot 4)$$
$$2(12) \neq 6 \cdot 8$$
$$24 \neq 48$$

☑

NOTE We accept the commutative, associative, and distributive properties as true without proof. ☑

Example 10 State whether each of the following is true or false. If the statement is *true*, give the reason.

	Solutions
a. $(1.2 + 5) + 2.4 = 1.2 + (5 + 2.4)$	True (Addition is associative.)
b. $\frac{2}{3}\left(-\frac{5}{8}\right) = -\frac{5}{8}\left(\frac{2}{3}\right)$	True (Multiplication is commutative.)
c. $-8(1) = -8$	True (Multiplicative identity.)
d. $0.1 - 5 = 5 - 0.1$	False
e. $17 + (-17) = 0$	True (Additive inverse.)
f. $3(7 + 9) = 3(7) + 3(9)$	True (Distributive property.)
g. $2 \div \left(\frac{3}{5}\right) = \left(\frac{3}{5}\right) \div 2$	False
h. $(y \div z) \div x = y \div (z \div x)$	False
i. $4(2 \cdot 3) = 4(2) \cdot 4(3)$	False
j. $(1 + 2) + 3 = (2 + 1) + 3$	True (Addition is commutative. Notice that the *order* was changed, not the grouping.)
k. $(3 + 7) + 8 = 8 + (7 + 3)$	True (Commutative property was used *twice*. That is, $(3 + 7) + 8 = 8 + (3 + 7) = 8 + (7 + 3)$.)
l. $0 = 8(0)$	True (Multiplication property of zero.)
m. $\left(-\frac{1}{8}\right)(-8) = 1$	True (Multiplicative inverse.) ■

EXERCISES 1.5

Set I In Exercises 1–32, determine whether each statement is true or false. If the statement is *true*, give the reason.

1. $5 + 10 = 10 + 5$

2. $x + y = y + x$

3. $(3)(8 + 2) = (2 + 8)(3)$

4. $(2)(5 + 6) = (6 + 5)(2)$

5. $3(8 + 2) = 3 \cdot 8 + 3 \cdot 2$

6. $2(5 + 6) = 2 \cdot 5 + 2 \cdot 6$

7. $6 - (4 - 2) = (6 - 4) - 2$

8. $(12 \div 6) \div 3 = 12 \div (6 \div 3)$

9. $14(7 \cdot 2) = 14(7) \cdot 14(2)$

10. $25(2 \cdot 6) = 25(2) \cdot 25(6)$

11. $16 - 4 = 4 - 16$

12. $\left(\frac{2}{3}\right) \div \left(\frac{1}{2}\right) = \left(\frac{1}{2}\right) \div \left(\frac{2}{3}\right)$

13. $6 + (2 \cdot 4) = 6 + (4 \cdot 2)$

14. $(6)(3) + 4 = 4 + (3)(6)$

15. $19 + 0 = 19$

16. $0 + (-5) = -5$

17. $8 + (2 + 4) = (8 + 2) + 4$

18. $(8)(2) - 10 = 10 - (2)(8)$

19. $3(0) = 3$

20. $0(-1) = -1$

21. $(7 + 5)(3) = 7(3) + 5(3)$

22. $(6 + 2)(5) = 6(5) + 2(5)$

23. $1(-6) = -6$

24. $5(1) = 5$

25. $4 \div 12 = 12 \div 4$

26. $31 + 2 = 2 + 31$

27. $-8 + 8 = 0$

28. $15 + (-15) = 0$

29. $(6 + 3) + 7 = 7 + (3 + 6)$

30. $8 + (3 + 5) = (5 + 3) + 8$

31. $\left(\frac{1}{7}\right)(7) = 1$

32. $-11 + 11 = 0$

In Exercises 33–44, complete each statement by using the property indicated.

33. Commutative property: $-4(-5) =$ __________

34. Commutative property: $8(-3) =$ __________

35. Associative property: $7 + [(-2) + 8] =$ __________

36. Associative property: $-9 + [12 + (-3)] =$ __________

37. Distributive property: $-3(2 + 7) =$ __________

38. Distributive property: $5(-3 + 6) =$ __________

39. Commutative property: $-7 + 8 =$ __________

40. Commutative property: $-5 + (-4) =$ __________

41. Associative property: $9(-4 \cdot 7) =$ __________

42. Associative property: $9(-3 \cdot 8) =$ __________

43. Distributive property: $(-8 \cdot 6) + (-8 \cdot 3) =$ __________

44. Distributive property: $(9 \cdot 6) + (9 \cdot 8) =$ __________

Set II In Exercises 1–32, determine whether each statement is true or false. If the statement is *true*, give the reason.

1. $8 + 3 = 3 + 8$

2. $6 \cdot 3 + 6 \cdot 8 = 6(3 + 8)$

3. $(7)(2 + 3) = (3 + 2)(7)$

4. $\frac{7}{8} \div 8 = 8 \div \frac{7}{8}$

5. $4(9 + 3) = 4(9) + 4(3)$

6. $4(2 + 9) = 4(9 + 2)$

7. $4(9 \cdot 3) = 4(9) \cdot 4(3)$

8. $3(7) = 7(3)$

9. $(7)(8) = (8)(7)$

10. $(6 \cdot 7)5 = (6 \cdot 5) \cdot (7 \cdot 5)$

11. $16 \div 5 = 5 \div 16$

12. $2 - 8 = 8 - 2$

13. $16 - 23 = 23 - 16$

14. $(1 + 4)(5) = 1(5) + 4(5)$

15. $18 + 0 = 18$

16. $-6 + 6 = 0$

17. $5 + (4 + 2) = (2 + 4) + 5$

18. $\frac{1}{2}(6 + 7) = \frac{1}{2}(6) + \frac{1}{2}(7)$

19. $0(-6) = -6$

20. $1(-6) = -6$

21. $3(5 \cdot 7) = (3 \cdot 5)(7)$

22. $17 \div 1 = 1 \div 17$

23. $(-4)(1) = -4$

24. $-4 + 4 = 0$

25. $(4 + 9)(2) = 4(2) + 9(2)$

26. $0 \div 5 = 5 \div 0$

27. $13 + (-13) = 0$

28. $13(0) = 0$

29. $(5 + 2) + 6 = 6 + (2 + 5)$

30. $4 + (9 + 2) = (2 + 9) + 4$

31. $(-4)\left(-\frac{1}{4}\right) = 1$

32. $8 + 0 = 8$

In Exercises 33–44, complete each statement by using the property indicated.

33. Commutative property: $7 + (-2) =$ ______________

34. Commutative property: $-7(-3) =$ ______________

35. Associative property: $4 + [(-6) + 8] =$ ______________

36. Associative property: $6(-5 \cdot 8) =$ ______________

37. Distributive property: $-4(11 + 6) =$ ______________

38. Distributive property: $2(-9 + 7) =$ ______________

39. Commutative property: $-8 + 4 =$ ______________

40. Commutative property: $-7(9) =$ ______________

41. Associative property: $2(-4 \cdot 6) =$ ______________

42. Associative property: $-4 + [5 + (-3)] =$ ______________

43. Distributive property: $(-2 \cdot 5) + (-2 \cdot 7) =$ ______________

44. Distributive property: $(7 \cdot 4) + (7 \cdot 13) =$ ______________

1.6 Powers of Rational Numbers

The shortened notation for a product such as $3 \cdot 3 \cdot 3 \cdot 3$ is 3^4. That is, by definition,

$$3^4 = 3 \cdot 3 \cdot 3 \cdot 3 = 81$$

Base and Exponent

In the expression 3^4, 3 is called the **base** and 4 is called the **exponent**. The number 4 (the exponent) indicates that 3 (the basc) is to be used as a *factor* 4 times. An exponent is written as a small number above and to the right of the base. The entire symbol 3^4 is called an *exponential expression* and is commonly read as "three to the fourth power" or "three to the fourth." See Figure 1.6.1.

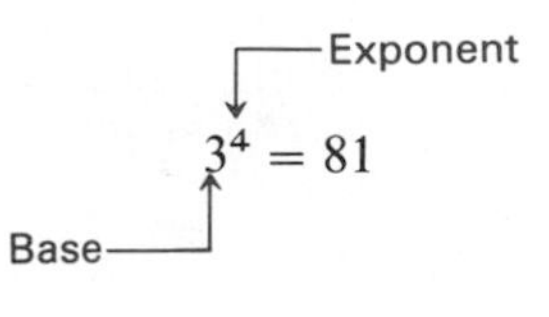

FIGURE 1.6.1

A WORD OF CAUTION When you write an exponential number, *be sure that your exponents look like exponents.* For example, be sure that 3^4 does not look like 34. ☑

When the exponent is 2, as in b^2, we usually say "b squared" rather than "b to the second power"; likewise, when the exponent is 3, as in b^3, we usually say "b cubed" rather than "b to the third power."

A WORD OF CAUTION *The exponent always applies only to the immediately preceding symbol.* Students often think that expressions such as $(-6)^2$ and -6^2 are the same. They are *not* the same. For example, in the expression

$$(-6)^2 = (-6)(-6) = 36$$

the exponent is immediately to the right of a set of parentheses; therefore, the exponent applies to everything inside the parentheses. However, in the expression

$$-6^2 = -(6 \cdot 6) = -36$$

the exponent applies *only* to the 6. Therefore, $-6^2 \neq (-6)^2$. ☑

Even Power If a base has an exponent that is an even number, we say that it is an **even power** of the base. For example, 3^2, 5^4, and $(-2)^6$ are even powers.

Odd Power If a base has an exponent that is an odd number, we say that it is an **odd power** of the base. For example, 3^1, 10^3, and $(-4)^5$ are odd powers.

Example 1 Examples of finding powers of numbers:

a. $2^3 = 2 \cdot 2 \cdot 2 = 8$ — $2^3 = 8$ is read "two cubed equals 8"; it is an odd power

b. $(-4)^2 = (-4)(-4) = 16$ — $(-4)^2 = 16$ is read "the square of negative four equals 16"; it is an even power

c. $(-0.2)^3 = (-0.2)(-0.2)(-0.2) = -0.008$

NOTICE: An *odd* power of a negative real number is always negative

NOTICE: An *even* power of a negative real number is always positive

d. $\left(-\frac{1}{2}\right)^2 = \left(-\frac{1}{2}\right)\left(-\frac{1}{2}\right) = \frac{1}{4}$

e. $1^5 = 1 \cdot 1 \cdot 1 \cdot 1 \cdot 1 = 1$

f. $(-1)^9 = -1$

g. $0^4 = 0 \cdot 0 \cdot 0 \cdot 0 = 0$ — Any nonzero power of zero equals zero ■

In the examples above, the exponents were all natural numbers. Cases in which exponents are numbers other than natural numbers will be discussed in Section 1.11B and in Chapter 7.

NOTE It is suggested that you memorize the squares of the first sixteen whole numbers and the cubes of the first five whole numbers. ☑

Finding Powers with a Calculator Many calculators have an x^2 key, [x^2], and some have a y^x or x^y key, [y^x] or [x^y]. (You may need to press the [INV] or [2nd] key before pressing one of these keys.)

Example 2 Find 38^2 with a calculator.

Solution On most calculators, to find 38^2 press these keys (in order): [3] [8] [x^2]. The display shows 1444. Therefore, $38^2 = 1,444$. (If your calculator uses Reverse Polish Notation (RPN), consult your instruction manual.) ■

Example 3 Find 4^5 with a calculator (if your calculator has a [y^x] key).

Solution Press these keys (in order): [4] [y^x] [5] [=]. The display shows 1024. Therefore, $4^5 = 1,024$. (Again, if your calculator uses RPN, consult your manual.) ■

EXERCISES 1.6

Set I Find the value of each of the following expressions. In Exercises 1–15, do *not* use a calculator.

1. 4^3	**2.** 7^2	**3.** $(-3)^4$	**4.** $(-2)^4$
5. -2^4	**6.** -3^4	**7.** 0^5	**8.** 0^6
9. $(-1)^{49}$	**10.** $(-1)^{50}$	**11.** $\left(\frac{1}{2}\right)^4$	**12.** $\left(\frac{7}{8}\right)^2$
13. $(-0.1)^3$	**14.** $(0.1)^4$	**15.** $\left(\frac{1}{10}\right)^3$	**16.** 17.3^2
17. 9.2^3	**18.** $(-1.5)^4$	**19.** $(-2.5)^4$	**20.** $(-5.3)^3$

Set II Find the value of each of the following expressions. In Exercises 1–15, do *not* use a calculator.

1. 5^3	**2.** $(-2)^6$	**3.** -2^6	**4.** 0^4
5. $(-15)^2$	**6.** $(-3)^3$	**7.** $(-1)^{79}$	**8.** $(-1)^{64}$
9. $\left(\frac{1}{10}\right)^4$	**10.** $(-2)^3$	**11.** -2^7	**12.** $(-1)^{83}$
13. $(-13)^2$	**14.** -2^3	**15.** $\left(\frac{1}{10}\right)^5$	**16.** $(-1.4)^3$
17. $(-1.26)^4$	**18.** $(-3.8)^4$	**19.** $(3)^9$	**20.** 18^3

1.7 Roots

1.7A Square Roots

Just as subtraction is the inverse operation of addition and division is the inverse operation of multiplication, finding roots is an inverse operation of raising to powers. Thus, finding the **square root** of a number is the inverse operation of squaring a number.

Every positive number has both a positive and a negative square root. The positive square root is called the **principal square root**.

The Square Root Symbol

The notation for the principal square root of p is $\sqrt{p}$, which is read "the square root of p." When we are asked to find $\sqrt{p}$, we must find some *positive* number whose square is p. For example, "find $\sqrt{9}$" means we must find a *positive* number whose square is 9. The answer is, of course, that $\sqrt{9} = 3$.

The entire expression $\sqrt{p}$ is called a **radical expression** or, more simply, a **radical**. The parts of a radical are shown in Figure 1.7.1.

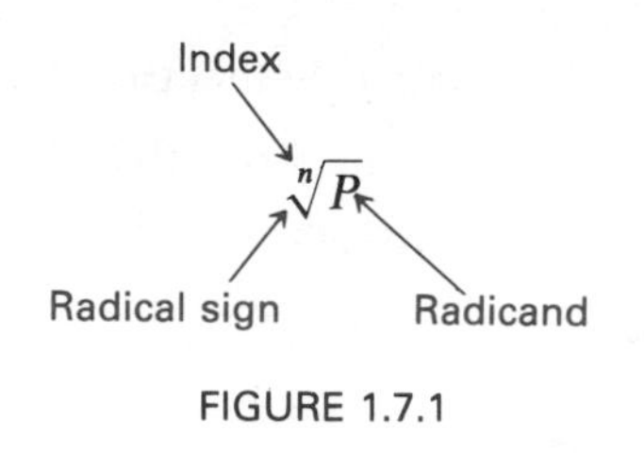

FIGURE 1.7.1

When no index is given, the index is understood to be 2, and the radical is a square root.

A WORD OF CAUTION When the symbol $\sqrt{p}$ is used, it *always* represents the *principal square root* of p. Since the principal square root is positive, $\sqrt{p}$ is always positive (or zero). ☑

In this section (1.7A), the radicand will always be the square of some integer.* You will find such problems easy to do if you have memorized the squares of the first sixteen whole numbers.

Example 1 Examples of finding the principal square roots:

a. $\sqrt{36} = 6$ This is true because $6^2 = 36$

b. $\sqrt{0} = 0$ This is true because $0^2 = 0$

c. $\sqrt{1} = 1$ This is true because $1^2 = 1$

d. $-\sqrt{9} = -(\sqrt{9}) = -3$ Since $\sqrt{9} = 3$, $-\sqrt{9} = -3$ ■

A WORD OF CAUTION Note that in Example 1d the negative sign was *outside* the radical sign. Had the problem been $\sqrt{-9}$, we could not have solved it at this time. No *real* number exists whose square is -9, since the square of every real number is always positive or zero. We will discuss square roots of negative numbers in Chapter 7. ☑

EXERCISES 1.7A

Set I Simplify each expression.

1. $-\sqrt{36}$ **2.** $\sqrt{49}$ **3.** $-\sqrt{25}$ **4.** $\sqrt{64}$ **5.** $-\sqrt{100}$

6. $-\sqrt{144}$ **7.** $\sqrt{81}$ **8.** $\sqrt{121}$ **9.** $-\sqrt{256}$ **10.** $-\sqrt{16}$

Set II Simplify each expression.

1. $-\sqrt{169}$ **2.** $-\sqrt{49}$ **3.** $\sqrt{225}$ **4.** $\sqrt{289}$ **5.** $-\sqrt{400}$

6. $\sqrt{144}$ **7.** $-\sqrt{900}$ **8.** $-\sqrt{64}$ **9.** $\sqrt{100}$ **10.** $-\sqrt{121}$

* In Section 1.7B we will consider square roots in which the radicand is *not* the square of an integer.

1.7B Finding Square Roots with a Calculator

We can easily find the square root of a number such as 762,129 by using a calculator that has a square root key, [√]. We simply press the following keys (in order): [7] [6] [2] [1] [2] [9] [√]. (You may have to press the [INV] or [2nd] key before the [√] key.) The display shows 873. Thus, $\sqrt{762{,}129} = 873$. We can also use the calculator to *approximate* square roots of numbers when the radicands are not the squares of rational numbers.

Example 2 Example of approximating a square root with the calculator:

Find $\sqrt{2}$. Since this is an *irrational* number—that is, since no rational number exists whose square is 2—its decimal approximation will not terminate and will not repeat. We press these keys: [2] [√]. The display shows 1.414213562. (Your calculator may not show the same number of digits.) If we round off this answer to three decimal places, we see that $\sqrt{2} \approx 1.414$. ■

EXERCISES 1.7B

Set I Find the exact or approximate square roots, using a calculator. (Round off approximate roots to three decimal places.)

1. $\sqrt{209{,}764}$	**2.** $\sqrt{389{,}376}$	**3.** $\sqrt{12}$	**4.** $\sqrt{17}$
5. $\sqrt{184}$	**6.** $\sqrt{191}$	**7.** $\sqrt{2.8}$	**8.** $\sqrt{9.38}$

Set II Find the exact or approximate square roots, using a calculator. (Round off approximate roots to three decimal places.)

1. $\sqrt{393{,}129}$	**2.** $\sqrt{395{,}641}$	**3.** $\sqrt{18}$	**4.** $\sqrt{23}$
5. $\sqrt{172}$	**6.** $\sqrt{1.63}$	**7.** $\sqrt{85.2}$	**8.** $\sqrt{73.2}$

1.7C Higher Roots

Roots other than square roots are called **higher roots**.

Radicals A **radical** is any indicated root of a number. Some examples of radicals are $\sqrt{9}$, $\sqrt{2}$, $\sqrt{8}$, $\sqrt[3]{8}$, $\sqrt[3]{55}$, and $\sqrt[5]{-32}$.

Some Symbols Used to Indicate Roots The symbol $\sqrt[3]{p}$ indicates the *cubic root* of *p*. When the *index* of a radical is 3, we must find a number whose *cube* is *p*; for example, $\sqrt[3]{8} = 2$ because $2^3 = 8$. You will find these problems easier to do if you have memorized the cubes of the first few whole numbers.

The notation $\sqrt[4]{p}$ indicates the *fourth root* of *p*. When the *index* is 4, we must find a number whose *fourth power* is *p*. The symbol $\sqrt[5]{p}$ indicates a *fifth root*, and so forth.

When the index is an even number, we call the index an *even index*. When the index is an odd number, we call the index an *odd index*.

Principal Roots Every even-index radical with a positive radicand has *both* a positive and a negative real root; the positive root is called the *principal root*. (An even-index radical with a negative radicand is not real.) Every odd-index radical has a single real root called the *principal root*; that principal root is *positive* when the radicand is positive and *negative* when the radicand is negative.

NOTE Mathematicians agree that the radical symbol is to stand for the principal root. ☑

Principal roots can be summarized as follows:

PRINCIPAL ROOTS

The symbol $\sqrt[n]{p}$ always represents the principal nth root of p. $\sqrt[n]{p} = r$ means $r^n = p$.

If the radicand is positive, the principal root is positive.

If the radicand is negative,

1. when the index is *odd*, the principal root is negative;
2. when the index is *even*, the principal root is *not a real number*.

Example 3 Examples of finding higher roots:

a. Find $\sqrt[4]{16}$.

We must find a positive number whose *fourth* power is 16. That is, we must solve $(?)^4 = 16$. If we have memorized that $2^4 = 16$, then this problem is easy to do. If we haven't memorized that fact, then we must use the "trial and error" method. That is, we start with 1 and check to see if $1^4 = 16$. It doesn't. Next, we check to see if $2^4 = 16$. It does. Therefore, $\sqrt[4]{16} = 2$.

b. Find $-\sqrt[3]{-343}$.

Since the index is odd and the radicand is negative, the principal root of $\sqrt[3]{-343}$ will be negative. We must find a negative number whose cube is -343. We will use the "trial and error" method. We probably know that $(-3)^3 = -27$ and that $(-4)^3 = -64$. Does $(-5)^3 = -343$? No; $(-5)^3 = -125$. Does $(-6)^3 = -343$? No; $(-6)^3 = -216$. Does $(-7)^3 = -343$? Yes! Therefore, $\sqrt[3]{-343} = -7$. But the problem was to find the *negative* of $\sqrt[3]{-343}$, that is, $-\sqrt[3]{-343}$. $-\sqrt[3]{-343} = -(-7) = 7$. Therefore, $-\sqrt[3]{-343} = 7$. ■

All roots of positive numbers (and zero) and all odd roots of negative numbers are *real numbers* and therefore can be represented by points on the number line; we graph a few such points in Figure 1.7.2. We find decimal approximations for irrational numbers (by trial and error or by using a calculator) in order to graph them.

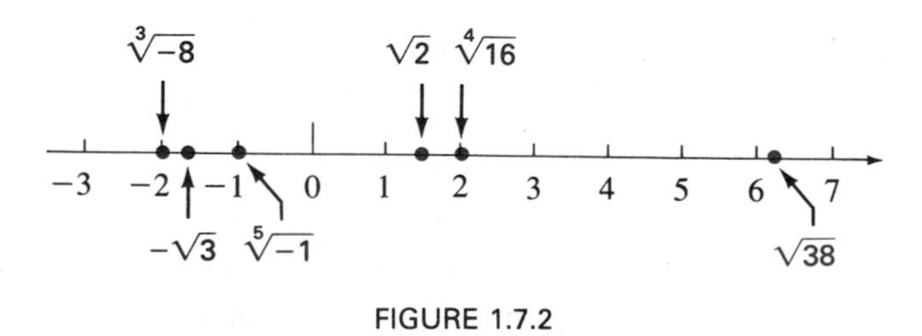

FIGURE 1.7.2

EXERCISES 1.7C

Set I Find each indicated root either by inspection or by trial and error.

1. $\sqrt[3]{8}$	**2.** $\sqrt[5]{1}$	**3.** $-\sqrt[3]{27}$	**4.** $-\sqrt[4]{16}$
5. $\sqrt[3]{-64}$	**6.** $\sqrt[3]{-27}$	**7.** $-\sqrt[5]{32}$	**8.** $-\sqrt[3]{64}$
9. $\sqrt[3]{-1{,}000}$	**10.** $\sqrt[3]{-216}$	**11.** $-\sqrt[5]{-32}$	**12.** $-\sqrt[3]{-125}$

Set II Find each indicated root either by inspection or by trial and error.

1. $\sqrt[4]{16}$	**2.** $-\sqrt[7]{-1}$	**3.** $\sqrt[5]{-243}$	**4.** $\sqrt[3]{-125}$
5. $-\sqrt[4]{81}$	**6.** $-\sqrt[6]{64}$	**7.** $-\sqrt[3]{216}$	**8.** $-\sqrt[3]{-8}$
9. $\sqrt[3]{27}$	**10.** $\sqrt[9]{1}$	**11.** $\sqrt[3]{-1}$	**12.** $\sqrt[4]{10{,}000}$

1.8 Order of Operations

In evaluating expressions with more than one operation, the following order of operations is used:

> ORDER OF OPERATIONS
>
> **1.** If operations are indicated inside grouping symbols, those operations within the grouping symbols should be performed first. A fraction bar and the bar of a radical sign are grouping symbols.
>
> **2.** The evaluation *then* proceeds *in this order*:
>
> *First*: Powers and roots are done.
>
> *Next*: Multiplication and division are done *in order from left to right.*
>
> *Last*: Addition and subtraction are done *in order from left to right.*

Example 1 Evaluate $8 - 6 - 4 + 7$.

Solution It is important to realize that an expression such as $8 - 6 - 4 + 7$ is evaluated by doing the addition and subtraction *in order from left to right*, because subtraction is neither associative nor commutative. If the expression is considered as a *sum*—that is, as $8 + (-6) + (-4) + 7$—then the terms can be added in any order, because addition is commutative and associative.

Both methods are correct

Evaluated left to right	*Added in any order*
$8 - 6 - 4 + 7$	$(8) + (-6) + (-4) + (7)$
$= 2 - 4 + 7$	$= (8) + (7) + (-6) + (-4)$
$= -2 + 7$	$= 15 + (-10)$
$= 5$	$= 5$ ■

Example 2 Evaluate $7 + 3 \cdot 5$.

Solution

$$7 + 3 \cdot 5$$
$$= 7 + 15 = 22 \quad ■$$

Multiplication must be done *before* addition

Example 3 Evaluate $4^2 + \sqrt{25} - 6$.

Solution

$$4^2 + \sqrt{25} - 6$$
$$= 16 + 5 - 6$$
$$= 21 - 6 = 15 \quad ■$$

Powers and roots must be done first

Example 4 Evaluate $16 \div 2 \cdot 4$.

Solution

$$\underbrace{16 \div 2}\cdot 4$$

Division must be done first because it is on the left

$$= 8 \cdot 4$$

$$= 32 \quad ■$$

Example 5 Evaluate $(-3)^2\sqrt[3]{-8} - 2(-6)$.

Solution

$$(-3)^2\sqrt[3]{-8} - 2(-6)$$

Powers and roots must be done first

$$= (9)(-2) - 2(-6)$$

Multiplication must be done before subtraction

$$= -18 - (-12)$$

$$= -18 + 12 = -6 \quad ■$$

Example 6 Evaluate $\dfrac{(-4)+(-2)}{8-5}$.

Solution

$$\frac{(-4)+(-2)}{8-5}$$

This bar is a grouping symbol for both $\underline{(-4)+(-2)}$ and $\overline{8-5}$; notice that the bar can be used either above or below the numbers being grouped ■

$$= \frac{-6}{3} = -2$$

Example 7 Evaluate $6 - 4(2 \cdot 3^2 - 12 \div 2)$.

Solution

$$6 - 4(2 \cdot 3^2 - 12 \div 2)$$

We first evaluate the expression inside the parentheses

$$= 6 - 4(2 \cdot 9 - 12 \div 2)$$

We raise to powers before multiplying

$$= 6 - 4(18 - 6)$$

Multiplication and division inside parentheses are done next

$$= 6 - 4(12)$$

Subtraction inside parentheses is done next

$$= 6 - 48$$

Multiplication must be done before the final subtraction

$$= -42 \quad ■$$

Example 8 Evaluate $20 - [5 - (3 - 7)]$.

Solution

$$20 - [5 - (3 - 7)]$$

$$= 20 - [5 - (-4)]$$

$$- 20 - [5 + 4]$$

$$= 20 - 9 = 11 \quad ■$$

When grouping symbols appear within other grouping symbols, evaluate the *inner* grouping first

---

EXERCISES 1.8

Set I Evaluate each expression; be sure to perform the operations in the correct order.

1. $16 - 9 - 4$ **2.** $18 - 6 - 3$ **3.** $12 \div 6 \div 2$

4. $16 \div 4 \div 2$ **5.** $10 \div 2(-5)$ **6.** $15 \div 5(-3)$

7. $3 \cdot 2^4$ **8.** $5 \cdot 3^2$ **9.** $8 + 6 \cdot 5$

10. $3 + 2 \cdot 4$ **11.** $7 + 5 \div 3$ **12.** $5 + 3 \div 2$

13. $10 - 3 \cdot 2$ **14.** $12 - 2 \cdot 4$ **15.** $10(-15)^2 - 4^3$

16. $3(-4)^2 - 2^4$ **17.** $\frac{1}{2} - 0.02 \times 10^3$ **18.** $\frac{1}{3} - 0.04 \times 10^2$

19. $10^2(5)\sqrt{16}$ **20.** $10^2(3)\sqrt{25}$ **21.** $2 + 3 \cdot 100 \div 25$

22. $3 + 2 \cdot 100 \div 4$ **23.** $28 + 14/7$ **24.** $36 + 18/3$

25. $2(2^3 - 5)\sqrt{9}$ **26.** $5(4^2 - 8)\sqrt{36}$

27. $(-18) \div (-3)(-6)$ **28.** $(-16) \div (-4)(-2)$

29. $(-10)^3 - 5(10^2)\sqrt[3]{-27}$ **30.** $(-10)^2 - 5(10^3)\sqrt[3]{-8}$

31. $20 - [5 - (7 - 10)]$ **32.** $16 - [8 - (2 - 7)]$

33. $\frac{7 + (-12)}{8 - 3}$ **34.** $\frac{-14 + (-2)}{9 - 5}$

35. $8 - [5(-2)^3 - \sqrt{16}]$ **36.** $10 - [3(-3)^2 - \sqrt{25}]$

37. $(3 \cdot 5^2 - 15 \div 3) \div (-7)$ **38.** $(3 \cdot 4^3 - 72 \div 6) \div (-9)$

39. $15 - \{4 - [2 - 3(6 - 4)]\}$ **40.** $17 - \{6 - [9 - 2(7 - 2)]\}$

Set II Evaluate each expression; be sure to perform the operations in the correct order.

1. $15 - 7 - 2$ **2.** $3 + 2 \cdot 6$ **3.** $81 \div 9 \div 3$

4. $12 + 6 \div 6$ **5.** $18 \div 3(-6)$ **6.** $8 - 24 \div 8$

7. $2 \cdot 5^2$ **8.** $8 \cdot 3^2$ **9.** $9 + 6 \cdot 3$

10. $2(-5) + 6 \div 24$ **11.** $9 + 7 \div 4$ **12.** $64 \div 2 \div 2$

13. $17 - 3 \cdot 5$ **14.** $3(2^3 - 1)\sqrt{64}$ **15.** $5(-3)^2 - 3^3$

16. $6(4) - 8 \div 24$ **17.** $\frac{1}{2} - 0.2 \times 10^2$ **18.** $10^2\sqrt[3]{8} \div 10 \cdot 2$

19. $10^2(4)\sqrt{9}$ **20.** $75 \div 5^2 \cdot 4 + 12 \cdot 5$ **21.** $4 + 2 \cdot 100 \div 5$

22. $100 \div 5^2 \cdot 6 + 8 \cdot 75$ **23.** $24 + 12/6$ **24.** $10^4 \cdot 4^2 + 100(15)$

25. $4(3^2 - 4)\sqrt{25}$ **26.** $8^2\sqrt{25} \div 10 \cdot 4$

27. $(-15) \div (-3)(-5)$ **28.** $10^4 \cdot 2^3 + 10(140)$

29. $(-10)^3 - 4(10^2)\sqrt[3]{-8}$ **30.** $6 + 3 \cdot 30 \div 5$

31. $18 - [9 - (3 - 8)]$ **32.** $100 \div 2^2 \cdot 5 + 9 \cdot 25$

33. $\frac{8 + (-16)}{12 - 4}$ **34.** $\frac{3}{5} + 72 \div 9 \cdot 2 - \frac{1}{3}$

35. $(-8)/2 \times (-4)/(-1)$

36. $18 - 11 - 3$

37. $(2 \cdot 3^3 - 63 \div 7) \div (-9)$

38. $10^2 \sqrt[3]{27} \div 10 \cdot 3$

39. $20 - \{5 - [9 - 3(6 - 2)]\}$

40. $\frac{2}{3} + 77 \div 11 \cdot 2 - \frac{1}{2}$

1.9 Prime Numbers and Factorization of Natural Numbers; LCM

Prime Numbers A **prime number** is a natural number greater than 1 that cannot be written as a product of two natural numbers except as the product of itself and 1. (That is, a prime number has no natural number factors other than itself and 1.)

A partial list of prime numbers is 2, 3, 5, 7, 11, 13, 17, 19, 23, 29, ... There is no largest prime number.

Composite Numbers A **composite number** is a natural number greater than 1 that is not prime; it is a natural number that can be exactly divided by some natural number besides itself and 1.

NOTE One (1) is neither prime nor composite. ☑

Factoring a Natural Number Recall from Section 1.3A that numbers multiplied together to give a product are called *factors*. To **factor** a natural number means to rewrite the number, if possible, as a *product* of smaller numbers (its factors). Recall also that when the remainder in a division problem is zero, the divisor and quotient are both *factors* of the dividend. We make use of these facts in *factoring a number*.

Prime Factorization of Natural Numbers The **prime factorization** of a natural number greater than 1 is the indicated product of all of its factors that are themselves prime numbers.

Example 1 Find the prime factorization of 18.

Solution

$$\left.\begin{aligned} 18 &= 2 \cdot 9 \\ 18 &= 3 \cdot 6 \end{aligned}\right\}$$ These are *not prime* factorizations because 9 and 6 are not prime numbers

$$\left.\begin{aligned} 18 &= 2 \cdot 9 = 2 \cdot 3 \cdot 3 = \boxed{2 \cdot 3^2} \\ 18 &= 3 \cdot 6 = 3 \cdot 2 \cdot 3 = \boxed{2 \cdot 3^2} \end{aligned}\right\}$$ These *are prime* factorizations because all the factors are prime numbers

Note that the two ways we factored 18 led to the *same* prime factorization ($2 \cdot 3^2$). The prime factorization of any positive integer (greater than 1) is unique. ■

A systematic method for finding the prime factorization will be demonstrated in Examples 2, 3, and 4.

Example 2 Find the prime factorization of 315.

Solution We first try to divide 315 by the smallest prime, 2. Two does not divide exactly into 315, and so we try to divide 315 by the next prime number, which is 3. Three *does* divide exactly into 315 and gives a quotient of 105. We again try 3 as a divisor of the quotient, 105. Three *does* divide exactly into 105 and gives a new quotient of 35. We then try to divide *that* quotient (35) by 3. Three does not divide exactly into 35, and so we try to divide 35 by the next prime number, which is 5. Five *does* divide exactly into 35 and gives a quotient of 7. The process then ends because the quotient, 7, is itself a prime.

The work of finding the prime factorization of a number can be conveniently arranged by putting the quotients under the number we're dividing into, as follows:

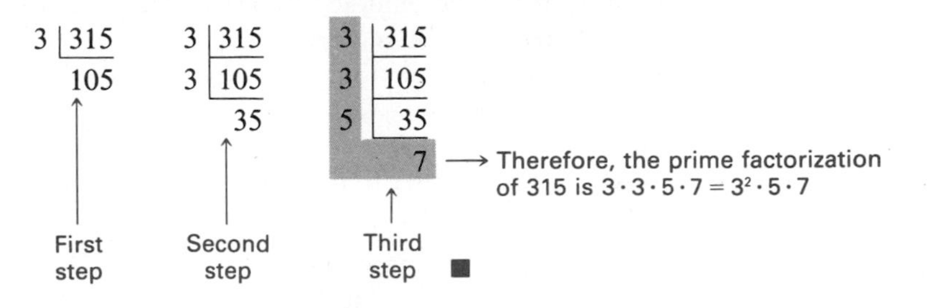

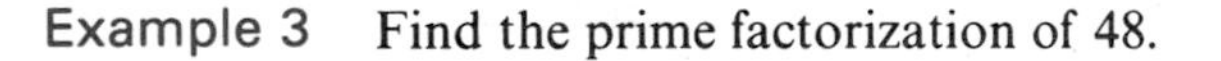

Example 3 Find the prime factorization of 48.

Solution

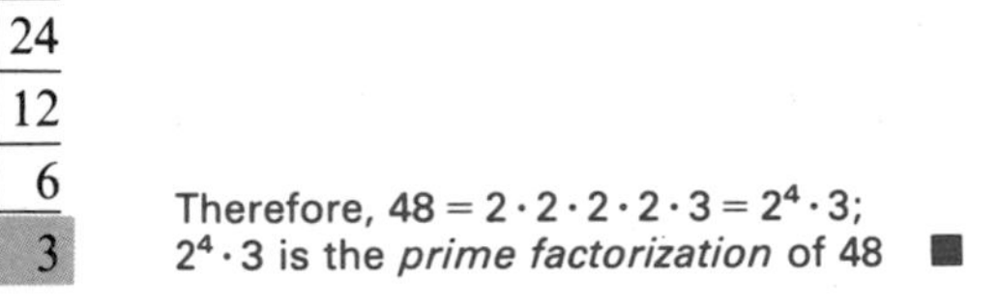

Therefore, $48 = 2 \cdot 2 \cdot 2 \cdot 2 \cdot 3 = 2^4 \cdot 3$;
$2^4 \cdot 3$ is the *prime factorization* of 48 ■

When trying to find the prime factors of a number, we do not need to try any prime that has a square greater than that number. (See Example 4.)

Example 4 Find the prime factorization of 97.

Solution

Primes in order of size

2	does not divide 97
3	does not divide 97
5	does not divide 97
7	does not divide 97
11	and larger primes need not be tried because $11^2 = 121$, which is greater than 97

Therefore, 97 is a prime number, and the prime factorization of 97 is simply 97. ■

Finding All the Integral Factors of a Number Sometimes we need to find *all* the integers that divide exactly into a given number. A systematic method for doing this is demonstrated in Example 5.

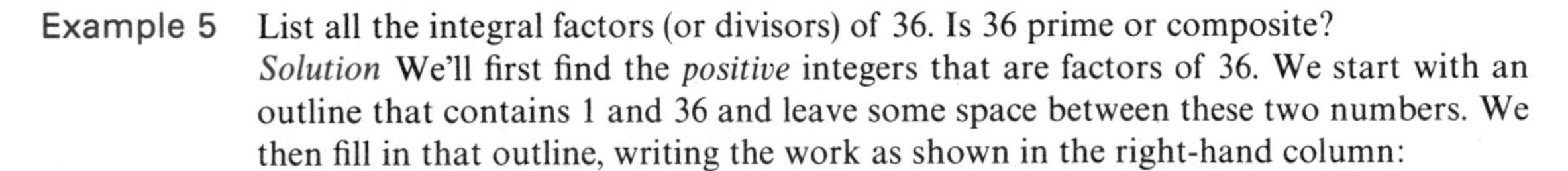

Example 5 List all the integral factors (or divisors) of 36. Is 36 prime or composite?
Solution We'll first find the *positive* integers that are factors of 36. We start with an outline that contains 1 and 36 and leave some space between these two numbers. We then fill in that outline, writing the work as shown in the right-hand column:

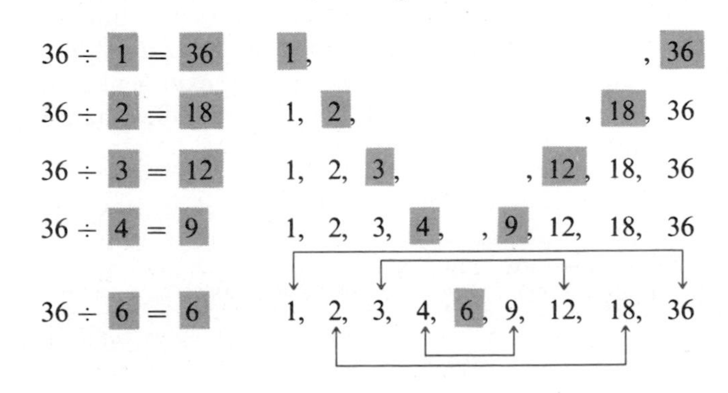

By checking each positive integer, in turn, to see if it is a factor of 36 (5 is, of course, *not* a factor of 36) and by "pairing up" the factors in this manner, continuing until we reach the "middle" of the list of factors, we can be sure we have not omitted any factors. (Because 36 is the square of an integer, when we "pair up" the factors, there is a single, unpaired number left in the center.) We then find *all the integral factors* by placing the symbol $\pm$* in front of each positive factor; thus, the *integral* factors of 36 are ± 1, ± 2, ± 3, ± 4, ± 6, ± 9, ± 12, ± 18, and ± 36. Since 36 has natural number factors other than 1 and 36, it is a composite number. ■

Example 6 List all the integral factors of 31. Is 31 prime or composite?

Solution The integral factors of 31 are ± 1 and ± 31. Since the only natural number factors are 1 and 31, 31 is a prime number. ■

Least Common Multiple The **least common multiple (LCM)** of two or more numbers is the *smallest* (least) number that is a multiple of each of the numbers. The LCM is used in solving fractional equations and in adding and subtracting fractions. The least common multiple can be found as follows:

TO FIND THE LCM OF TWO OR MORE NUMBERS

1. Express each number in prime factored form; repeated factors must be expressed in exponential form.
2. Write down each *different* base that appears in any of the numbers.
3. The exponent on each base must be the largest exponent that appears on that base in *any* of the numbers.
4. The LCM is the *product* of the exponential numbers found in step 3.

Example 7 Find the LCM of 18 and 48.

Solution $18 = 2 \cdot 3^2$ and $48 = 2^4 \cdot 3$. The bases that appear in one or both numbers are 2 and 3. Therefore, the bases in the LCM are 2 and 3. The *largest* exponent that appears on the 2 (in either of the numbers) is 4. Therefore, the exponent on the 2 in the LCM must be 4. The *largest* exponent that appears on the 3 in either of the numbers is 2. Therefore, the exponent on the 3 in the LCM must be 2. The LCM, then is $2^4 \cdot 3^2 = 144$.

NOTE 144 is the *smallest* number that is a multiple of both 18 and 48. We could also say that 144 is the smallest number that 18 and 48 both divide into. ■

Example 8 Find the LCM of 360, 378, and 108.

Solution $360 = 2^3 \cdot 3^2 \cdot 5$, $378 = 2 \cdot 3^3 \cdot 7$, and $108 = 2^2 \cdot 3^3$. Bases in the LCM must be 2, 3, 5, and 7. The exponent on the 2 must be 3; the exponent on the 3 must be 3; the exponent on the 5 must be 1; the exponent on the 7 must be 1. The LCM, therefore, is $2^3 \cdot 3^3 \cdot 5 \cdot 7 = 7{,}560$. This is the smallest number that 360, 378, and 108 all divide into. ■

* The symbol $\pm$ is read "plus or minus" (positive or negative). For example, ± 1 is read "plus or minus 1" and means $+1$ or -1.

EXERCISES 1.9

Set I In Exercises 1–12, find the prime factorization of each number.

1. 28	**2.** 30	**3.** 32	**4.** 33	**5.** 43
6. 35	**7.** 84	**8.** 75	**9.** 144	**10.** 180
11. 156	**12.** 221			

In Exercises 13–22, state whether each of the numbers is prime or composite, and give the set of *all* integral factors for each number.

13. 5	**14.** 8	**15.** 13	**16.** 15	**17.** 12
18. 11	**19.** 51	**20.** 42	**21.** 111	**22.** 101

In Exercises 23–26, find the LCM of the numbers.

23. 144, 360	**24.** 84, 35	**25.** 270, 900, 75	**26.** 140, 105, 98

27. List any prime numbers greater than 17 and less than 37 that yield a remainder of 5 when divided by 7.

28. List any prime numbers less than 43 and greater than 13 that yield a remainder of 1 when divided by 6.

Set II In Exercises 1–12, find the prime factorization of each number.

1. 21	**2.** 31	**3.** 45	**4.** 87	**5.** 186
6. 238	**7.** 19	**8.** 40	**9.** 36	**10.** 27
11. 72	**12.** 228			

In Exercises 13–22, state whether each of the numbers is prime or composite, and give the set of *all* integral factors for each number.

13. 21	**14.** 23	**15.** 55	**16.** 41	**17.** 49
18. 9	**19.** 44	**20.** 30	**21.** 17	**22.** 87

In Exercises 23–26, find the LCM of the numbers.

23. 87, 58	**24.** 93, 155	**25.** 280, 84, 350	**26.** 85, 102, 170

27. List any prime numbers greater than 7 and less than 29 that yield a remainder of 3 when divided by 5.

28. List any prime numbers less than 37 and greater than 11 that yield a remainder of 2 when divided by 9.

1.10 Review: 1.1–1.9

Sets 1.1A A **set** is a collection of objects or things.

The *elements of a set* are the objects that make up the set.

$5 \in A$ is read "5 is an element of set A"

$9 \notin B$ is read "9 is not an element of set B"

The method of representing a set by listing its members is called the *roster method.*

The *modified roster method* is used when the number of elements in the set is so large that it is not convenient or even possible to list all its members.

The *rule* (or *set-builder*) *method* can also be used to represent a set.

Equal sets are sets that have exactly the same members.

The *empty set* is a set that has no elements.

$\varnothing$ is read "the empty set"

$\{\ \ \}$ is read "the empty set"

If, in counting the elements of a set, the counting comes to an end, the set is a *finite set.*

If, in counting the elements of a set, the counting never comes to an end, the set is an *infinite set.*

Set A is a *subset* of set B if every member of A is also a member of B.

$P \subseteq Q$ is read "P is a subset of Q"

$P \nsubseteq Q$ is read "P is not a subset of Q"

Union and Intersection of Sets 1.1B

The union of sets A and B, written $A \cup B$, is the set that contains all the elements that are in A or B or both.

The intersection of sets A and B, written $A \cap B$, is the set that contains all the elements that are in *both* A and B.

Sets of Numbers 1.2

All the numbers that can be represented by points on the number line are **real numbers.** Some subsets of the set of real numbers are:

The set of digits, $D = \{0, 1, 2, 3, 4, 5, 6, 7, 8, 9\}$

The set of natural numbers, $N = \{1, 2, 3, \ldots\}$

The set of whole numbers, $W = \{0, 1, 2, \ldots\}$

The set of integers, $J = \{\ldots, -3, -2, -1, 0, 1, 2, 3, \ldots\}$

The set of rational numbers, $Q = \left\{\frac{a}{b} \middle| a, b \in J; b \neq 0\right\}$ The decimal form of a rational number is always a terminating or repeating decimal.

The set of irrational numbers, $H = \{x \mid x \in R, x \notin Q\}$ The decimal form of an irrational number is always a nonterminating, nonrepeating decimal.

Definitions 1.3A

Constant An object or symbol that does not change its value in a particular problem or discussion.

Variable An object or symbol that is a placeholder for a number that is unknown.

Algebraic expression An expression that consists of numbers, letters, signs of operation ($+$, $-$, $\times$, $\div$, powers, roots), and signs of grouping.

Factors The numbers that are multiplied together to give a product are called the **factors** of that product.

Terms A **term** of an algebraic expression consists of factors only. If a term is preceded by a minus sign, that sign is *part of* the term. An expression within grouping symbols is considered as a single term.

Inequality Symbols 1.3B

Greater than: The statement $a > b$ is read "a is greater than b."

Greater than or equal to: The statement $c \geq d$ is read "c is greater than or equal to d."

Not greater than: The statement $e \ngtr f$ is read "e is not greater than f."

Less than: The statement $g < h$ is read "g is less than h."

Less than or equal to: The statement $i \leq j$ is read "i is less than or equal to j."

Not less than: The statement $k \nless \ell$ is read "k is not less than ℓ."

Not equal to: The statement $m \neq n$ is read "m is not equal to n."

Continued inequalities: The statement $a < b < c$ is a valid continued inequality only if the inequalities $a < b$, $b < c$, and $a < c$ are all true.

Absolute Value 1.3B

The absolute value of a real number x is written $|x|$, where

$$|x| = \begin{cases} x & \text{if } x \geq 0 \\ -x & \text{if } x < 0 \end{cases}$$

Operations on Signed Numbers 1.4

To add signed numbers:

1. When the numbers have the same sign
   a. Add their absolute values *and*
   b. Give the sum the sign of both numbers.
2. When the numbers have different signs
   a. Subtract the smaller absolute value from the larger absolute value *and*
   b. Give that answer the sign of the number with the larger absolute value.

To subtract one signed number from another:

Change the subtraction symbol to an addition symbol *and* change the sign of the number being subtracted; add the resulting signed numbers.

To multiply two signed numbers:

Multiply their absolute values *and* attach the correct sign to the product. The sign is *positive* when the numbers have the same sign and *negative* when the numbers have different signs.

To divide one signed number by another:

Divide the absolute value of the dividend by the absolute value of the divisor *and* attach the correct sign to the quotient. The sign is *positive* when the numbers have the same sign and *negative* when the numbers have different signs.

Operations with Zero 1.4

If a represents any real number,

1. $a + 0 = 0 + a = a$
2. $a - 0 = a$
3. $0 - a = 0 + (-a) = -a$
4. $a \cdot 0 = 0 \cdot a = 0$

If a represents any real number *except* 0,

5. $\dfrac{0}{a} = 0$

6. $\dfrac{a}{0}$ is not possible — Undefined

7. $\dfrac{0}{0}$ cannot be determined — Undefined

Identity and Inverse Properties 1.4

If a represents any real number,
The **additive identity** element is 0; that is, $a + 0 = 0 + a = a$.

The **multiplicative identity** element is 1; that is, $a \cdot 1 = 1 \cdot a = a$.

The **additive inverse** of a is $-a$; that is, $a + (-a) = -a + a = 0$.

If $a \neq 0$, the **multiplicative inverse** of a is $\dfrac{1}{a}$; that is, $a\left(\dfrac{1}{a}\right) = \left(\dfrac{1}{a}\right)a = 1$.

Commutative Properties (Order Changed) 1.5

Addition: $a + b = b + a$
Multiplication: $a \cdot b = b \cdot a$
$a, b \in R$

Subtraction and *division* are not commutative.

Associative Properties (Grouping Changed) 1.5

Addition: $(a + b) + c = a + (b + c)$
Multiplication: $(a \cdot b) \cdot c = a \cdot (b \cdot c)$
$a, b, c \in R$

Subtraction and *division* are not associative.

Distributive Properties 1.5

If $a, b, c \in R$,

$a(b + c) = ab + ac$
$(b + c)a = ba + ca$

These properties can be extended to have any number of terms within the parentheses and to include subtraction as well as addition

Powers of Signed Numbers 1.6

$(-2)^3 = (-2)(-2)(-2) = -8$

Exponent: 3; Base: -2

Square Roots 1.7

The *square root* of a number N is a number that, when squared, gives N. Every positive number has both a positive and a negative square root. The positive square root is called the *principal square root*. The principal square root of N is written $\sqrt{N}$. When the symbol $\sqrt{N}$ is used, it always represents the principal square root of N.

Higher Roots 1.7C

Roots other than square roots are called *higher roots*.

Order of Operations 1.8

1. If operations are indicated inside grouping symbols, those operations within the grouping symbols should be performed first. A fraction bar and the bar of a radical sign are grouping symbols.
2. The evaluation then proceeds *in this order*:

First: Powers and roots are done.

Next: Multiplication and division are done *in order from left to right*.

Last: Addition and subtraction are done *in order from left to right*.

Prime and Composite Numbers 1.9

A **prime number** is a natural number greater than 1 that cannot be written as a product of two natural numbers except as the product of itself and 1. A prime number has no factors other than itself and 1.

A **composite number** is a natural number that can be exactly divided by some natural number other than itself and 1.

Prime Factorization of Natural Numbers 1.9

The **prime factorization** of a natural number greater than 1 is the indicated product of all its factors that are themselves prime numbers.

Least Common Multiple (LCM) 1.9

To find the LCM of two or more numbers, proceed as follows:

1. Factor each number completely; repeated factors must be expressed in exponential form.
2. Write down each *different* base that appears in any of the numbers.
3. Raise each base to the highest power it has in *any* of the numbers.
4. The LCM is the product of all the powers found in step 3.

Review Exercises 1.10 Set I

1. Is the set of natural numbers a finite set or an infinite set?

2. Is the set of digits a finite set or an infinite set?

3. Write the following set in roster notation:

$$\{x \mid x \text{ is a digit less than } 5\}$$

4. $P = \{5, x, z\}$. Is $y \in P$?

5. Given $A = \{5, 7, 8\}$, $B = \{2, 5, 7\}$, and $C = \{1, 3, 4, 6\}$, find the following.

a. $A \cup B$ b. $A \cap B$ c. $C \cap B$ d. $(A \cup B) \cap C$

6. Given the numbers -2, 4.53, $0.\overline{16}$, $\frac{2}{3}$, $2.6457513\ldots$, and 0, answer the following questions.

a. Which are real numbers?

b. Which are integers?

c. Which are natural numbers?

d. Which are irrational numbers?

e. Which are rational numbers?

In Exercises 7–17, find the value of each expression or write "undefined."

7. $|4|$ **8.** $|0|$ **9.** $|-10|$ **10.** $(-6)^2$

11. -5^2 **12.** $\frac{0}{5}$ **13.** $\frac{7}{0}$ **14.** 0^3

15. $6 + 2\sqrt{16} - 2^3$

16. $20 - \{5 - 2[-3(4 - 6) + 2] - 3\}$ **17.** $6 + 18 \div 6 \div 3$

18. a. Write all of the integral factors of 28.

b. Write the prime factorization of 28.

c. Write the prime factorization of 168.

d. Find the least common multiple of 28 and 168.

In Exercises 19–21, if your answer is no, give a reason.

19. Is $-2 < 0 < 8$ a valid continued inequality?

20. Is $1 > 5 > 7$ a valid continued inequality?

21. Is $5 < 8 > 6$ a valid continued inequality?

In Exercises 22–30, state whether each of the following is true or false. If the statement is *true*, give the reason.

22. $8 - (-15) = -15 - 8$

23. $7 \cdot 1 = 7$

24. $\left(\frac{5}{6} + \frac{1}{8}\right) + \frac{2}{3} = \frac{5}{6} + \left(\frac{1}{8} + \frac{2}{3}\right)$

25. $6 + (-6) = 0$

26. $\left(\frac{1}{2} \div \frac{3}{8}\right) \div \frac{2}{7} = \frac{1}{2} \div \left(\frac{3}{8} \div \frac{2}{7}\right)$

27. $\frac{1}{2} + 0 = \frac{1}{2}$

28. $6 + (2 + 5) = 6 + (5 + 2)$

29. $(-5) \cdot 0 = -5$

30. $\frac{1}{2}\left(\frac{5}{8} + \frac{1}{3}\right) = \frac{1}{2} \cdot \frac{5}{8} + \frac{1}{2} \cdot \frac{1}{3}$

Review Exercises 1.10 Set II

NAME ______

1. Is the empty set a subset of the set of integers?
2. Write the following set in roster notation:

$$\{x \mid x \text{ is a whole number less than } 6\}$$

3. Is every irrational number a real number?
4. Given $G = \{1, 4, 8\}$, is $12 \in G$?
5. Is 10 a digit?
6. Given $A = \{2, 4, 6\}$, $B = \{1, 3, 5\}$, and $C = \{3, 5, 7\}$, find the following.
   a. $A \cup B$
   b. $A \cap B$
   c. $C \cap B$
   d. $(A \cap B) \cup C$
7. Is 0 a real number?
8. Given the numbers $6.1644140\ldots$, -3, 4.1, $\frac{3}{7}$, 5, and $0.\overline{65}$, determine the following.
   a. Which are real numbers?
   b. Which are integers?
   c. Which are natural numbers?
   d. Which are irrational numbers?
   e. Which are rational numbers?

In Exercises 9–20, find the value of each expression or write "undefined."

9. $|-14|$ **10.** $-|-23|$ **11.** $(-11)^2$

12. $(-5)^3$ **13.** -4^2 **14.** $\frac{0}{7}$

ANSWERS

1. ______
2. ______
3. ______
4. ______
5. ______
6a. ______
6b. ______
6c. ______
6d. ______
7. ______
8a. ______
8b. ______
8c. ______
8d. ______
8e. ______
9. ______
10. ______
11. ______
12. ______
13. ______
14. ______

15. $\frac{0}{0}$

16. $\frac{8}{0}$

17. $10 - 2\sqrt[3]{64} - 3^2$

18. $20 \div 10 \cdot 2$

19. $4 + 16 \div 4 \div 2$

20. $17 + 12 \cdot 2$

21. a. Write all of the integral factors of 132.

b. Write the prime factorization of 132.

c. Write the prime factorization of 110.

d. Find the least common multiple of 132 and 110.

22. Is $7 < 10 > 8$ a valid continued inequality? If your answer is no, give a reason.

In Exercises 23–30, state whether each of the following is true or false. If the statement is *true*, give the reason.

23. $\frac{4}{5} \cdot \left(\frac{2}{3} \cdot \frac{6}{11}\right) = \left(\frac{4}{5} \cdot \frac{2}{3}\right) \cdot \frac{6}{11}$

24. $\frac{6}{7} \div \frac{5}{2} = \frac{5}{2} \div \frac{6}{7}$

25. $\left(\frac{1}{2} - \frac{1}{8}\right) - \frac{1}{3} = \frac{1}{2} - \left(\frac{1}{8} - \frac{1}{3}\right)$

26. $1 \cdot 17 = 17$

27. $3 \cdot [(-4) \cdot 6] = 3 \cdot [6 \cdot (-4)]$

28. $6(-2 + 6) = 6 \cdot (-2) + 6 \cdot 6$

29. $7 + (4 \cdot 5) = (5 \cdot 4) + 7$

30. $\frac{2}{3} + \frac{1}{6} = \frac{1}{6} + \frac{2}{3}$

ANSWERS

15. ____________

16. ____________

17. ____________

18. ____________

19. ____________

20. ____________

21a. ____________

21b. ____________

21c. ____________

21d. ____________

22. ____________

23. ____________

24. ____________

25. ____________

26. ____________

27. ____________

28. ____________

29. ____________

30. ____________

1.11 Properties of Exponents

1.11A Positive Integer Exponents

In Section 1.6, we discussed powers of rational numbers. We now discuss powers of variables. Just as 3^4 means $3 \cdot 3 \cdot 3 \cdot 3$, x^4 means $x \cdot x \cdot x \cdot x = xxxx$. In the exponential expression x^n, the *base* is x, the *exponent* is n, and x^n means that x is multiplied by itself n times. If a variable has no exponent, then the exponent is understood to be 1.

An exponent always applies *only* to the immediately preceding symbol, as shown in the following example:

Example 1 Rewrite each of the following without exponents.

a. $x^2y^3 = xxyyy$

b. $xy^3 = xyyy$ The exponent applies only to the y

c. $(xy)^3 = (xy)(xy)(xy) = xxxyyy$

This exponent applies to whatever is inside the parentheses

d. $y^1 = y$ ■

The Properties of Exponents

The first five basic properties of exponents are the natural result of the definition of an exponent.

PROPERTY 1.1

$$x^m \cdot x^n = x^{m+n}$$

PROPERTY 1.2

$$(x^m)^n = x^{mn}$$

PROPERTY 1.3

$$(xy)^n = x^ny^n$$

PROPERTY 1.4

$$\frac{x^m}{x^n} = x^{m-n} \qquad (x \neq 0\text{, and in } \textit{this} \text{ section, } m > n.)$$

PROPERTY 1.5

$$\left(\frac{x}{y}\right)^n = \frac{x^n}{y^n} \qquad (y \neq 0)$$

NOTE We must add the restrictions $x \neq 0$ and $y \neq 0$ to avoid dividing by zero, which is not defined. ☑

Example 2 Rewrite each of the following with the base appearing only once, if possible.

a. $w \cdot w^7 = w^1 \cdot w^7 = w^{1+7} = w^8$. (When a base is written without an exponent, its exponent is understood to be 1.)

b. $x^3 \cdot y^2$ cannot be rewritten. (Property 1.1 does not apply because the bases are different.)

c. $x^3 + x^2$ cannot be simplified. (Property 1.1 does not apply to a *sum* of two exponential numbers.)

d. $2^a \cdot 2^b = 2^{a+b}$ ■

A WORD OF CAUTION A mistake students often make when the bases are constants is to add the exponents and *also* multiply the bases.

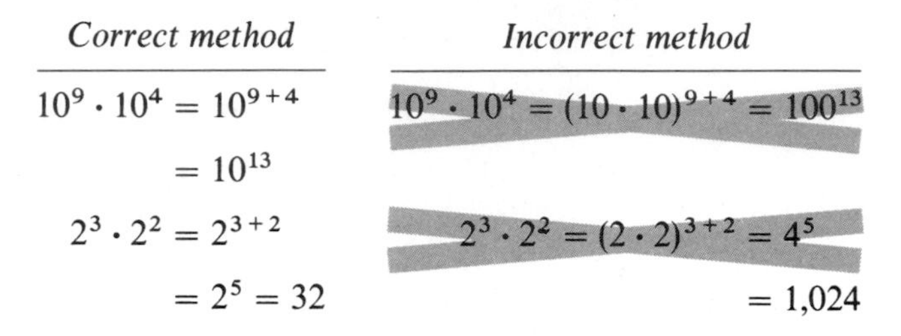

When multiplying powers of the same base, just add the exponents; do *not* multiply the bases as well. ☑

Example 3 Rewrite each of the following without parentheses.

a. $(x)^4 = (x^1)^4 = x^{1 \cdot 4} = x^4$

b. $(10^6)^2 = 10^{6 \cdot 2} = 10^{12}$

c. $(2^a)^b = 2^{a \cdot b} = 2^{ab}$

d. $(2x)^3 = 2^3x^3$ or $8x^3$

e. $(xy)^7 = x^7y^7$ ■

A WORD OF CAUTION Remember that an exponent applies only to the symbol immediately preceding it. For example,

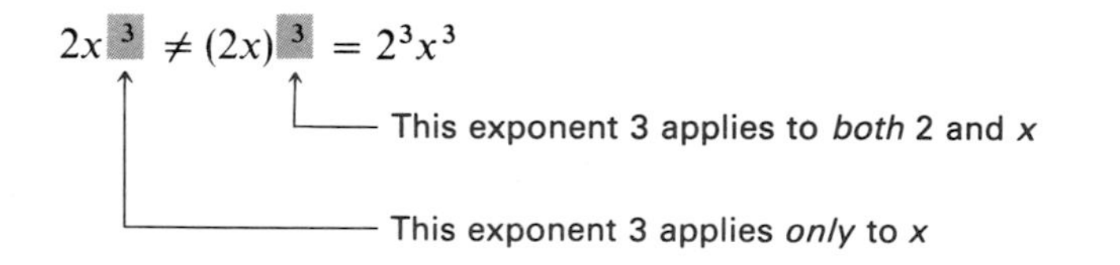

☑

Throughout this book, unless otherwise noted, we do not allow any of the variables to have a value that makes a denominator zero.

Example 4 Rewrite each of the following with the base appearing only once, if possible.

a. $\dfrac{x^5}{x^3} = x^{5-3} = x^2$

b. $\dfrac{10^7}{10^3} = 10^{7-3} = 10^4$

c. $\dfrac{y^3}{y} = \dfrac{y^3}{y^1} = y^{3-1} = y^2$

d. $\dfrac{2^a}{2^b} = 2^{a-b}$

e. $\dfrac{x^5}{y^2}$ This cannot be rewritten, since Property 1.4 does not apply when the bases are different.

f. $x^5 - x^2$ This cannot be rewritten, since Property 1.4 does not apply to subtraction problems. (Try this with numbers; does $3^5 - 3^2 = 3^3$?)

g. $\dfrac{x^5 - y^3}{x^2}$ The expression can either be left as it is or be changed as follows:

$$\frac{x^5 - y^3}{x^2} = \frac{x^5}{x^2} - \frac{y^3}{x^2} = x^3 - \frac{y^3}{x^2} \quad ■$$

A WORD OF CAUTION A common mistake is to divide *only* x^5 by x^2 instead of dividing *both* x^5 and y^3 by x^2 in Example 4g. However, $\dfrac{x^5 - y^3}{x^2} \neq x^3 - y^3$. Expressions such as $\dfrac{x^5 - y^3}{x^2}$ are discussed further in Section 4.5. ☑

Example 5 Rewrite each of the following without parentheses.

a. $\left(\dfrac{a}{b}\right)^4 = \dfrac{a^4}{b^4}$

b. $\left(\dfrac{3}{z}\right)^2 = \dfrac{3^2}{z^2}$ or $\dfrac{9}{z^2}$ ■

NOTE The proofs of Properties 1.1, 1.2, 1.3, 1.4, and 1.5 are beyond the scope of this book. ☑

EXERCISES 1.11A

Set I Use the properties of exponents to simplify each of the following expressions, if possible.

1. $10^2 \cdot 10^4$
2. $2^3 \cdot 2^2$
3. $x^2 \cdot x^5$
4. $y^6 \cdot y^3$
5. $x^4 + x^6$
6. $y^3 + y^7$
7. $2^x \cdot 2^y$
8. $3^m \cdot 3^n$
9. $\dfrac{a^8}{a^3}$
10. $\dfrac{x^5}{x^2}$
11. $x^6 - x$
12. $y^4 - y$
13. $(10^3)^2$
14. $(5^2)^4$
15. $(3^a)^b$
16. $(2^m)^n$
17. $(xy)^5$
18. $(uv)^4$
19. $(2x)^6$
20. $(3x)^4$
21. $x^8 \cdot y^5$
22. $s^2 \cdot t^4$
23. $3^{4x} \cdot 3^{7x}$
24. $5^x \cdot 5^{3x}$
25. $\dfrac{a^x}{a^y}$
26. $\dfrac{x^a}{x^b}$
27. $\dfrac{x^4}{y^2}$
28. $\dfrac{a^5}{b^3}$
29. $\left(\dfrac{x}{y}\right)^3$
30. $\left(\dfrac{u}{v}\right)^7$
31. $\left(\dfrac{3}{x}\right)^4$
32. $\left(\dfrac{x}{5}\right)^2$

Set II Use the properties of exponents to simplify each of the following expressions, if possible.

1. $5^2 \cdot 5^3$	**2.** $y^3 + y$	**3.** $X^4 \cdot X^2$	**4.** $x^4 + x^3$
5. $u^4 + u^7$	**6.** $4 \cdot 4^3$	**7.** $5^a \cdot 5^b$	**8.** $(2v)^3$
9. $\frac{P^5}{P^3}$	**10.** $\frac{x^4}{y^3}$	**11.** $z^7 + z^4$	**12.** $y^5 - y^2$
13. $(4^2)^3$	**14.** xx^5	**15.** $(2^x)^y$	**16.** xy^5
17. $(ab)^7$	**18.** $(2^3)^2$	**19.** $(4x)^2$	**20.** $(xy)^4$
21. $a^4 \cdot b^2$	**22.** $3^2 \cdot 3^3$	**23.** $4^{3x} \cdot 4^{5x}$	**24.** $x^2 \cdot y^5$
25. $\frac{U^x}{U^y}$	**26.** $\frac{x^6}{y^2}$	**27.** $\frac{r^6}{t^2}$	**28.** $\frac{z^6}{z^2}$
29. $\left(\frac{a}{b}\right)^6$	**30.** $\left(\frac{4}{x}\right)^3$	**31.** $\left(\frac{x}{2}\right)^5$	**32.** $\frac{a^{14}}{a^9}$

1.11B Zero and Negative Exponents

The Zero Exponent

When we used Property 1.4 in Section 1.11A, the exponent of the numerator was always larger than the exponent of the denominator. We now consier the case where the exponents of the numerator and the denominator are the same.

We know that $\frac{x^n}{x^n} = 1$ (if $x \neq 0$) because any nonzero number divided by itself is 1. However, if we use Property 1.4 in the same problem, we have

$$\frac{x^n}{x^n} = x^{n-n} = x^0$$

We want x^0 to equal 1; therefore, we *define* x^0 to be 1 (if $x \neq 0$).

PROPERTY 1.6

$$x^0 = 1 \qquad (x \neq 0)$$

NOTE $x \neq 0$ in Property 1.6 because $0^0 = 0^{1-1} = \frac{0}{0}$, which cannot be determined. ☑

Example 6 Rewrite each of the following with no zero exponents.

a. $a^0 = 1$ provided $a \neq 0$.

b. $10^0 = 1$

c. $(-7)^0 = 1$ The zero power of *any* expression is *positive* 1

d. $(3y)^0 = 1$ provided $y \neq 0$.

e. $6x^0 = 6 \cdot 1 = 6$ provided $x \neq 0$.

↑ Remember: The exponent applies *only* to the x, not also to the 6 ■

Negative Exponents

We now consider Property 1.4 when the exponent of the numerator is *less than* the exponent of the denominator.

Consider the expression $\frac{x^3}{x^5}$.

$$\frac{x^3}{x^5} = \frac{xxx}{xxxxx} = \frac{xxx \cdot 1}{xxx \cdot xx} = \frac{xxx}{xxx} \cdot \frac{1}{xx} = 1 \cdot \frac{1}{xx} = \frac{1}{x^2}$$

The value of this fraction is 1

If we use Property 1.4,

$$\frac{x^3}{x^5} = x^{3-5} = x^{-2}$$

Therefore, we want x^{-2} to equal $\frac{1}{x^2}$. Also, according to Properties 1.1 and 1.6, $x^n \cdot x^{-n} = x^{n+(-n)} = x^0 = 1$. Therefore, since the product of x^n and x^{-n} is 1, x^{-n} must be the multiplicative inverse of x^n. Therefore, we *define* x^{-n} to equal $\frac{1}{x^n}$.

PROPERTY 1.7a

$$x^{-n} = \frac{1}{x^n} \qquad (x \neq 0)$$

NOTE To see why $x \neq 0$ in Property 1.7a, consider 0^{-3}. $0^{-3} = \frac{1}{0^3}$, which is undefined. ✓

Property 1.7a is valid whether n is a positive or a negative number, and x^{-n} may be either a positive or a negative number. If x^n is positive, x^{-n} is positive also; if x^n is negative, x^{-n} is negative also.

Example 7 Rewrite each of the following with no negative exponents.

a. $6^{-2} = \frac{1}{6^2} = \frac{1}{36}$

b. $(-5)^{-3} = \frac{1}{(-5)^3} = -\frac{1}{125}$ ■

Proofs of the following properties are provided in Appendix A. Your instructor may recommend that you memorize these properties.

PROPERTY 1.7b

$$\frac{1}{x^{-n}} = x^n \qquad (x \neq 0)$$

PROPERTY 1.7c

$$\left(\frac{x}{y}\right)^{-n} = \left(\frac{y}{x}\right)^n \qquad (x \neq 0, y \neq 0)$$

Example 8 Rewrite each of the following with positive exponents only.

a. $x^{-5} = \dfrac{1}{x^5}$ Property 1.7a

b. $10^{-4} = \dfrac{1}{10^4}$ Property 1.7a

c. $(3x)^{-1} = \dfrac{1}{(3x)^1}$ Property 1.7a

d. $\dfrac{1}{y^{-3}} = y^3$ Property 1.7b

e. $\left(\dfrac{a}{b}\right)^{-4} = \left(\dfrac{b}{a}\right)^4$ Property 1.7c

f. $x^{-6a} = \dfrac{1}{x^{6a}}$ Property 1.7a ■

A *factor* can be moved either from the numerator to the denominator or from the denominator to the numerator of a fraction simply by moving it and changing the sign of its exponent. If a factor has no exponent, then the exponent is understood to be 1.

NOTE Moving a factor from the numerator to the denominator (or from the denominator to the numerator) by changing the sign of the exponent does *not* change the sign of the *expression*. For example,

$$(-6)^{-1} = \frac{1}{(-6)^1} = \frac{1}{-6}$$ The (-6) *stays* negative ☑

Example 9 Rewrite $\dfrac{a^{-2}b^4}{c^5d^{-3}}$ with positive exponents only.

Solution

$$\frac{a^{-2}b^4}{c^5d^{-3}} = \frac{b^4d^3}{a^2c^5}$$

Notice that a^{-2} (which was a factor of the numerator) was moved to the denominator and its exponent changed to $+2$, and that d^{-3} (which was a factor of the denominator) was moved to the numerator and its exponent changed to $+3$. ■

NOTE If a single number or variable appears in the numerator or denominator of a fraction, that number can still be considered a factor of the numerator or denominator. For example,

$$\frac{5}{x} = \frac{1 \cdot 5}{1 \cdot x}$$

Therefore, we can say that 5 is a factor of the numerator and x is a factor of the denominator of $\dfrac{5}{x}$. ☑

Example 10 Rewrite each of the following with no negative exponents.

a. $\dfrac{h^5}{k^{-4}} = \dfrac{h^5}{1k^{-4}} = \dfrac{h^5 \cdot k^4}{1} = h^5k^4$

NOTE If an expression has no denominator, then the denominator is understood to be 1. See Examples 10b and 10c. ☑

b. $6x^{-8} = \dfrac{6x^{-8}}{1} = \dfrac{6}{1 \cdot x^8} = \dfrac{6}{x^8}$

In the expression $6x^{-8}$, the denominator is understood to be 1. The factor x^{-8} moves to the denominator, and its exponent becomes $+8$. (Remember: The exponent applies only to the x, not to the 6.)

c. $(3x)^{-5} = \dfrac{1}{(3x)^5} = \dfrac{1}{3^5 x^5}$

In this case, the -5 applies to everything inside the parentheses. ■

A WORD OF CAUTION An expression that is a *term* of (rather than a factor of) a numerator *cannot* be moved from the numerator to the denominator of a fraction simply by moving it and changing the sign of the exponent. For example,

$$\frac{a^{-2} + b^5}{c^4} \neq \frac{b^5}{a^2 c^4}$$

Notice that since a^{-2} is a term of rather than a factor of the numerator, we cannot move it to the denominator and change the exponent to $+2$. Instead, we proceed as follows:

$$\frac{a^{-2} + b^5}{c^4} = \frac{\frac{1}{a^2} + b^5}{c^4}$$

← The expression on the right is still not simplified; such expressions will be simplified in Section 6.6 ☑

The properties listed in Section 1.11A for multiplying exponential numbers, raising exponential numbers to powers, and dividing exponential numbers also apply when the exponents are negative integers or zero.

Example 11 Use the properties of exponents to rewrite each of the following with positive exponents only or no exponents and with the base appearing only once.

a. $y^3 \cdot y^{-2} = y^{3-2} = y^1 = y$

b. $10^7 \cdot 10^{-3} = 10^{7-3} = 10^4$ *or* 10,000

c. $\dfrac{x^4}{x^7} = x^{4-7} = x^{-3} = \dfrac{1}{x^3}$

d. $(x^3)^{-2} = x^{-6} = \dfrac{1}{x^6}$

e. $5^4 \cdot 5^{-4} = 5^{4-4} = 5^0 = 1$

f. $(2^3)^{-1} = 2^{-3} = \dfrac{1}{2^3}$ *or* $\dfrac{1}{8}$

g. $\dfrac{x^{-3}}{x^2} = x^{-3-2} = x^{-5} = \dfrac{1}{x^5}$ *or* $\dfrac{x^{-3}}{x^2} = \dfrac{1}{x^3 x^2} = \dfrac{1}{x^5}$ ■

Example 12 Examples of evaluating expressions with numerical bases:

a. $10^8 \cdot 10^{-3} = 10^{8-3} = 10^5 = 100{,}000$

b. $(2^{-3})^2 = 2^{-6} = \dfrac{1}{2^6} = \dfrac{1}{64}$ ■

Example 13 Write each of the following without denominators, using negative exponents if necessary:

a. $\dfrac{m^5}{n^2} = m^5n^{-2}$

Notice that n^2 (a factor of the denominator) was moved to the numerator and its exponent changed to -2.

b. $\dfrac{y^3}{z^{-2}} = y^3z^2$

c. $\dfrac{a}{2c^2} = 2^{-1}ac^{-2}$

Notice that 2 and c^2 were moved to the numerator and now have negative exponents. ■

EXERCISES 1.11B

Set I In Exercises 1–16, write each expression using only positive exponents or no exponents.

1. a^{-3} **2.** x^{-2} **3.** 10^{-3} **4.** 10^{-5}

5. $5b^{-7}$ **6.** $3y^{-2}$ **7.** $(5b)^{-2}$ **8.** $(3y)^{-2}$

9. $x^{-3}y^2z^0$ **10.** $r^3s^{-4}t^0$ **11.** $xy^{-2}z^{-3}w^0$ **12.** $z^{-4}bc^{-5}a^0$

13. $\dfrac{a^3}{b^{-4}}$ **14.** $\dfrac{c^4}{d^{-5}}$ **15.** $\dfrac{x^{-3}}{y^{-2}}$ **16.** $\dfrac{P^{-2}}{Q^{-4}}$

In Exercises 17–40, use the properties of exponents to rewrite each expression with no parentheses, with the base appearing only once if possible, and with positive exponents or no exponents.

17. $x^{-3} \cdot x^{-4}$ **18.** $y^{-1} \cdot y^{-4}$ **19.** $(a^3)^{-2}$ **20.** $(b^2)^{-4}$

21. $x^8 \cdot x^{-2}$ **22.** $a^{-3} \cdot a^5$ **23.** $(x^{2a})^{-3}$ **24.** $(y^{3c})^{-2}$

25. $\dfrac{y^3}{y^{-2}}$ **26.** $\dfrac{x^8}{x^{-4}}$ **27.** $\dfrac{x^{3a}}{x^{-a}}$ **28.** $\dfrac{a^{4x}}{a^{-2x}}$

29. $(3x)^0$ **30.** $(2^a)^0$ **31.** $5x^0$ **32.** $2y^0$

33. $x^{-3} + x^{-5}$ **34.** $y^{-2} + y^{-6}$ **35.** $x^7 - x^{-5}$ **36.** $y^{10} - y^{-3}$

37. $x^0 + y^0$ **38.** $a^0 + b^0 + c^0$ **39.** $(x + y)^0$ **40.** $(a + b + c)^0$

In Exercises 41–48, evaluate each expression.

41. $10^5 \cdot 10^{-2}$ **42.** $2^4 \cdot 2^{-2}$ **43.** $(3^{-2})^{-2}$ **44.** $(10^{-1})^{-3}$

45. $(10^0)^5$ **46.** $(3^0)^4$ **47.** $\dfrac{10^2 \cdot 10^{-1}}{10^{-3}}$ **48.** $\dfrac{2^{-3} \cdot 2^2}{2^{-4}}$

In Exercises 49–54, write each expression without denominators, using negative exponents if necessary.

49. $\dfrac{y}{x^3}$ **50.** $\dfrac{x}{y^2}$ **51.** $\dfrac{x}{a^{-4}}$ **52.** $\dfrac{m^2}{n^{-3}}$

53. $\dfrac{x^4y^{-3}}{z^{-2}}$ **54.** $\dfrac{a^{-1}b^3}{c^{-4}}$

Set II In Exercises 1–16, write each expression using only positive exponents or no exponents.

1. x^{-4} **2.** 16^{0} **3.** 10^{-2} **4.** $(-2)^{-3}$

5. $2a^{-4}$ **6.** $4x^{0}$ **7.** $(2a)^{-2}$ **8.** $(-4x)^{-2}$

9. $x^{0}y^{-1}z^{2}$ **10.** $3x^{0}y^{-1}$ **11.** $x^{-3}yz^{-2}$ **12.** $y^{3}z^{-1}w^{0}$

13. $\dfrac{a^{4}}{b^{-3}}$ **14.** $\dfrac{x^{-1}y}{z^{-2}}$ **15.** $\dfrac{x^{-5}}{y^{-4}}$ **16.** $\dfrac{a^{0}b^{-3}c}{d^{-4}}$

In Exercises 17–40, use the properties of exponents to rewrite each expression with no parentheses, with the base appearing only once if possible, and with positive exponents or no exponents.

17. $a^{-3}\cdot a^{-5}$ **18.** $x^{-3}\cdot x^{5}$ **19.** $(a^{-2})^{3}$ **20.** $(-x^{2})^{-3}$

21. $x^{8}x^{-2}$ **22.** $(a^{-3})^{2}$ **23.** $(z^{4b})^{-4}$ **24.** $s^{-4}\cdot s^{2}$

25. $\dfrac{x}{y^{-3}}$ **26.** $\dfrac{a^{5}}{b^{2}}$ **27.** $\dfrac{y^{2b}}{y^{-b}}$ **28.** $\dfrac{x^{4}}{x^{-2}}$

29. $(3R)^{0}$ **30.** $6a^{0}$ **31.** $4X^{0}$ **32.** $(3xy)^{0}$

33. $a^{-7}+a^{-2}$ **34.** $x^{-2}-x^{3}$ **35.** $u^{4}-u^{-5}$ **36.** $3x^{-5}$

37. $a^{0}+b^{0}+c^{0}+d^{0}$ **38.** $x+5^{0}$

39. $(a+b+c+d)^{0}$ **40.** $a^{3}b^{-2}$

In Exercises 41–48, evaluate each expression.

41. $10^{-3}\cdot 10^{5}$ **42.** -2^{-3} **43.** $(2^{-3})^{-2}$ **44.** $(10^{2})^{-4}$

45. $(7^{0})^{8}$ **46.** $(8^{0})^{2}$ **47.** $\dfrac{10^{-3}\cdot 10}{10^{-4}}$ **48.** $\dfrac{2^{-1}}{2^{6}\cdot 2^{-4}}$

In Exercises 49–54, write each expression without denominators, using negative exponents if necessary.

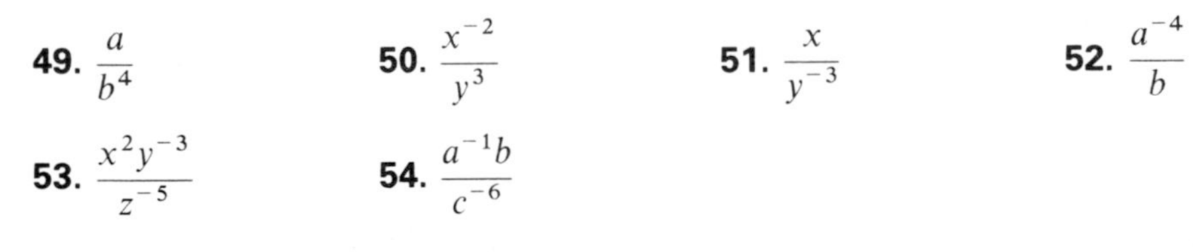

49. $\dfrac{a}{b^{4}}$ **50.** $\dfrac{x^{-2}}{y^{3}}$ **51.** $\dfrac{x}{y^{-3}}$ **52.** $\dfrac{a^{-4}}{b}$

53. $\dfrac{x^{2}y^{-3}}{z^{-5}}$ **54.** $\dfrac{a^{-1}b}{c^{-6}}$

1.11C General Property of Exponents

The combination of all the properties of exponents gives us the following general property:

PROPERTY 1.8

$$\left(\frac{x^{a}y^{b}}{z^{c}}\right)^{n}=\frac{x^{an}y^{bn}}{z^{cn}}$$

None of the variables can have a value that makes the denominator zero.

In applying Property 1.8, notice the following:

1. x^a, y^b, and z^c are *factors* of, not separate *terms* of, the expression within the parentheses.
2. The exponent of *each* factor within the parentheses is multiplied by the exponent outside the parentheses.

Example 14 Use the General Property of Exponents (Property 1.8) to rewrite each of the following with positive exponents only and with no parentheses, if possible.

The understood exponent of 2 is 1, and that 1 must be multiplied by 3

a. $\left(\frac{2a^{-3}b^2}{c^5}\right)^3 = \frac{2^{1\cdot 3}a^{(-3)3}b^{2\cdot 3}}{c^{5\cdot 3}} = \frac{2^3a^{-9}b^6}{c^{15}} = \frac{8b^6}{a^9c^{15}}$

b. $\left(\frac{3^2c^{-4}}{d^3}\right)^{-1} = \frac{3^{2(-1)}c^{(-4)(-1)}}{d^{3(-1)}} = \frac{3^{-2}c^4}{d^{-3}} = \frac{c^4d^3}{3^2} = \frac{c^4d^3}{9}$

If we use Property 1.7c first, we have

$$\left(\frac{3^2c^{-4}}{d^3}\right)^{-1} = \left(\frac{d^3}{3^2c^{-4}}\right)^1 = \frac{d^{3\cdot 1}}{3^{2\cdot 1}c^{-4\cdot 1}} = \frac{c^4d^3}{9}$$

c. $(x^2 + y^3)^4$

Property 1.8 cannot be used here because the plus sign means that x^2 and y^3 are terms, not factors, of the expression inside the parentheses. Expressions of this kind will be simplified in Section 4.4C. ■

Simplified Form of Expressions with Exponents When we wish to remove parentheses in an expression in which a *single term* is inside each set of grouping symbols, that is, as in the expression $(3x^2)(7x^3)$, we first multiply the numerical coefficients together and then use the rules of exponents to simplify the literal part of the expression (see Example 15). This is possible because multiplication is both associative and commutative.

Example 15 Examples of simplifying expressions with exponents:

a. $(3x^2)(7x^3) = 3 \cdot 7 \cdot x^2 \cdot x^3 = 21x^{2+3} = 21x^5$

b. $(2x^3y^2)(4xy^5z)(-3x^2) = 2(4)(-3)x^3y^2xy^5zx^2 = -24x^6y^7z$

c. $(-8a^3b^2c^6)(-5ab^3) = 40a^4b^5c^6$ ■

An expression with exponents is considered *simplified* when (1) there are no parentheses, (2) each different base appears only once in each separate term, and (3) the exponent on each base is a single natural number.

Example 16 Examples of simplifying expressions:

a. $x^{-2} \cdot x^7 = x^{-2+7} = x^5$

b. $(83x^5)^0 = 1$

c. $\frac{x^5y^2}{x^3y^{-1}} = x^{5-3}y^{2-(-1)} = x^2y^3$

d. $(2x)^{-1} = \frac{1}{2x}$

e. $2x^{-1} = \frac{2}{x}$

f. $(x^3y^{-1})^5 = x^{15}y^{-5} = \frac{x^{15}}{y^5}$

g. $\left(\frac{-3^{-7}x^{10}}{y^{-4}}\right)^0 = 1$ The answer is positive 1 even though the expression inside the parentheses is negative

h. $\left(\frac{x^5y^4}{x^3y^7}\right)^2 = (x^2y^{-3})^2 = x^4y^{-6} = \frac{x^4}{y^6}$

Simplify the expression within the parentheses first whenever possible

i. $(5^0h^{-2})^{-3} = (1h^{-2})^{-3} = (h^{-2})^{-3} = h^6$ ■

EXERCISES 1.11C

Set I Write each expression in simplest form.

1. $(5x)^{-3}$ **2.** $(4y)^{-2}$ **3.** $7x^{-2}$ **4.** $3y^{-4}$

5. $\left(\frac{3}{x}\right)^3$ **6.** $\left(\frac{7}{y}\right)^2$ **7.** $\frac{8^2}{z}$ **8.** $\frac{2^3}{x}$

9. $(a^2b^3)^2$ **10.** $(x^4y^5)^3$ **11.** $(m^{-2}n)^4$ **12.** $(p^{-3}r)^5$

13. $(x^{-2}y^3)^{-4}$ **14.** $(w^{-3}z^4)^{-2}$ **15.** $(10^0k^{-4})^{-2}$ **16.** $(6^0z^{-5})^{-2}$

17. $(2x^2y^{-4})^3$ **18.** $(3a^{-1}b^5)^2$ **19.** $(5m^{-3}n^5)^{-2}$ **20.** $(8x^8y^{-2})^{-1}$

21. $\left(\frac{xy^4}{z^2}\right)^2$ **22.** $\left(\frac{a^3b}{c^2}\right)^3$ **23.** $\left(\frac{M^{-2}}{N^3}\right)^4$ **24.** $\left(\frac{R^5}{S^{-4}}\right)^3$

25. $\left(\frac{x^{-5}}{y^4z^{-3}}\right)^{-2}$ **26.** $\left(\frac{a^{-4}}{b^2c^{-5}}\right)^{-3}$ **27.** $\left(\frac{r^7s^8}{r^9s^6}\right)^0$ **28.** $\left(\frac{t^5u^6}{t^8u^7}\right)^0$

29. $\left(\frac{3x^2}{y^3}\right)^2$ **30.** $\left(\frac{2a^4}{b^2}\right)^4$ **31.** $\left(\frac{4a^{-2}}{b^3}\right)^{-1}$ **32.** $\left(\frac{m^{-4}}{5n^3}\right)^{-2}$

33. $\left(\frac{x^{-1}y^2}{x^4}\right)^{-2}$ **34.** $\left(\frac{u^{-4}v^3}{v^4}\right)^{-3}$ **35.** $\left(\frac{x^{3n}x^{2n}}{x^{4n}}\right)^2$ **36.** $\left(\frac{x^{2n-1}y^{3n}}{x^ny^{2n+2}}\right)^2$

37. $(2x^5y^2)(-3x^4)$ **38.** $(-4a^2b^3)(7b^2)$ **39.** $(-4x^{-2}y^3)(-2xy^{-1})$

40. $(-5a^4b^{-3})(-2a^{-5}b^4)$ **41.** $(2s^3t^0u^{-4})(3su^3)(-s)$ **42.** $(4xy^0z^{-3})(-z)(2x^3z)$

Set II Write each expression in simplest form.

1. $(2x)^{-4}$ **2.** $2x^{-4}$ **3.** $3y^{-3}$ **4.** $(3y)^{-3}$

5. $\left(\frac{5}{z}\right)^2$ **6.** $\frac{5^2}{z}$ **7.** $\frac{2^4}{y}$ **8.** $\left(\frac{6z}{x}\right)^2$

9. $(x^4y^6)^2$ **10.** $(2xy^5)^5$ **11.** $(x^{-3}y^4)^3$ **12.** $(xy^4)^3$

13. $(a^{-3}b^2)^{-3}$ **14.** $(2x^{-4})^{-4}$ **15.** $(x^0y^{-3})^{-5}$ **16.** $(3x^{-6}z)^0$

17. $(4x^4y^{-2})^4$ **18.** $(3x^4yz^2)^0$ **19.** $(2x^{-1}y^5)^{-3}$ **20.** $(3xy^2)^{-1}$

21. $\left(\frac{a^3b}{c^4}\right)^2$ **22.** $\left(\frac{5^0x^4}{3^{-1}y^2}\right)^2$ **23.** $\left(\frac{u^{-2}}{v^4}\right)^3$ **24.** $\left(\frac{x^3y^5}{z^4}\right)^{-1}$

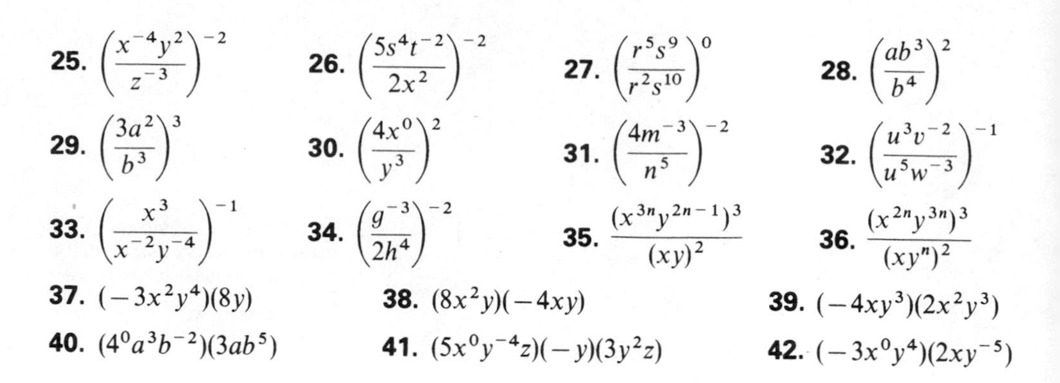

25. $\left(\frac{x^{-4}y^2}{z^{-3}}\right)^{-2}$ **26.** $\left(\frac{5s^4t^{-2}}{2x^2}\right)^{-2}$ **27.** $\left(\frac{r^5s^9}{r^2s^{10}}\right)^0$ **28.** $\left(\frac{ab^3}{b^4}\right)^2$

29. $\left(\frac{3a^2}{b^3}\right)^3$ **30.** $\left(\frac{4x^0}{y^3}\right)^2$ **31.** $\left(\frac{4m^{-3}}{n^5}\right)^{-2}$ **32.** $\left(\frac{u^3v^{-2}}{u^5w^{-3}}\right)^{-1}$

33. $\left(\frac{x^3}{x^{-2}y^{-4}}\right)^{-1}$ **34.** $\left(\frac{g^{-3}}{2h^4}\right)^{-2}$ **35.** $\frac{(x^{3n}y^{2n-1})^3}{(xy)^2}$ **36.** $\frac{(x^{2n}y^{3n})^3}{(xy^n)^2}$

37. $(-3x^2y^4)(8y)$ **38.** $(8x^2y)(-4xy)$ **39.** $(-4xy^3)(2x^2y^3)$

40. $(4^0a^3b^{-2})(3ab^5)$ **41.** $(5x^0y^{-4}z)(-y)(3y^2z)$ **42.** $(-3x^0y^4)(2xy^{-5})$

1.12 Scientific Notation and the Calculator

Now that we have discussed zero and negative exponents, we can introduce *scientific notation*, a notation used in many sciences and often seen in calculator displays. Scientific notation gives us a shorter way of writing those very large and very small numbers often seen in chemistry and physics, such as Avogadro's number ($\approx$ 602,000,000,000,000,000,000,000) and Boltzmann's constant (0.000 000 000 000 000 138) (see Example 4). We can also use scientific notation to simplify the arithmetic in certain problems that arise in the sciences (see Example 6).

In this text (and in most texts), a positive number written in **scientific notation** is in the form

$$a \times 10^n, \text{ where } 1 \leq a < 10 \text{ and } n \text{ is an integer}$$

For example, 4.32×10^3 is written in scientific notation, because 4.32 is a number between 1 and 10, and 10^3 is an integral power of 10. Any positive decimal number can be written in scientific notation.

Since a is to be greater than or equal to 1 but less than 10, it must have *exactly one digit to the left of its decimal point.*

When we change a number to scientific notation, we do not change its value, because we are using the facts that $10^n \times 10^{-n} = 1$ and that multiplying a number by 1 doesn't change its value. Consider the following examples.

Example 1 Change 0.0372 to scientific notation.

Solution When 0.0372 is written in scientific notation, the decimal point must be between the 3 and the 7. Since the decimal point must be moved two places to the right, we'll multiply 0.0372 by $10^2 \times 10^{-2}$.

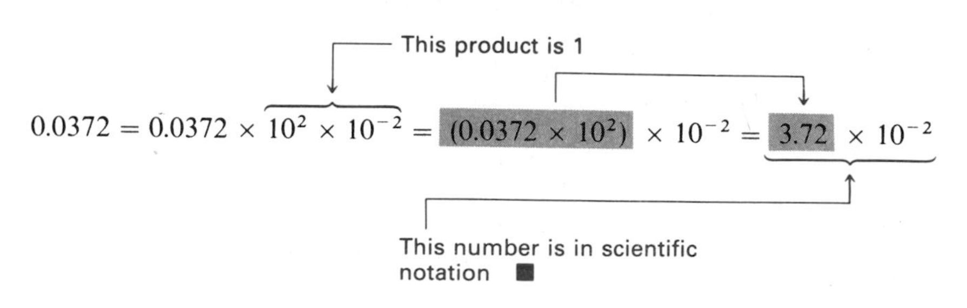

Example 2 Change 38,500 to scientific notation.

Solution When 38,500 is written in scientific notation, the decimal point must be between the 3 and the 8. Since the decimal point must be moved four places to the left, we'll multiply 38,500 by $10^{-4} \times 10^4$.

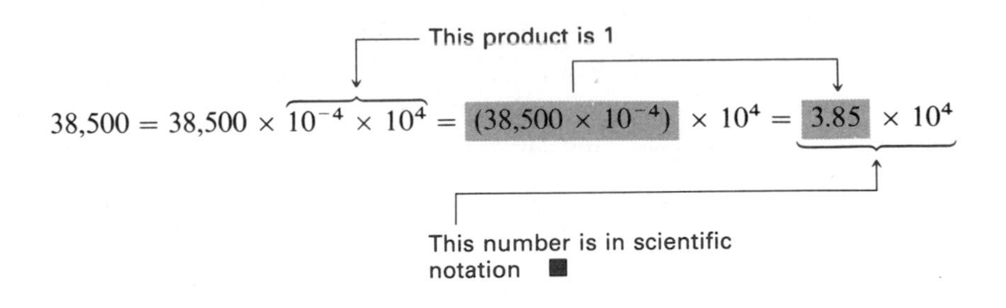

In practice, we usually do not show any of the intermediate steps shown in Examples 1 and 2; instead, we use the following method:

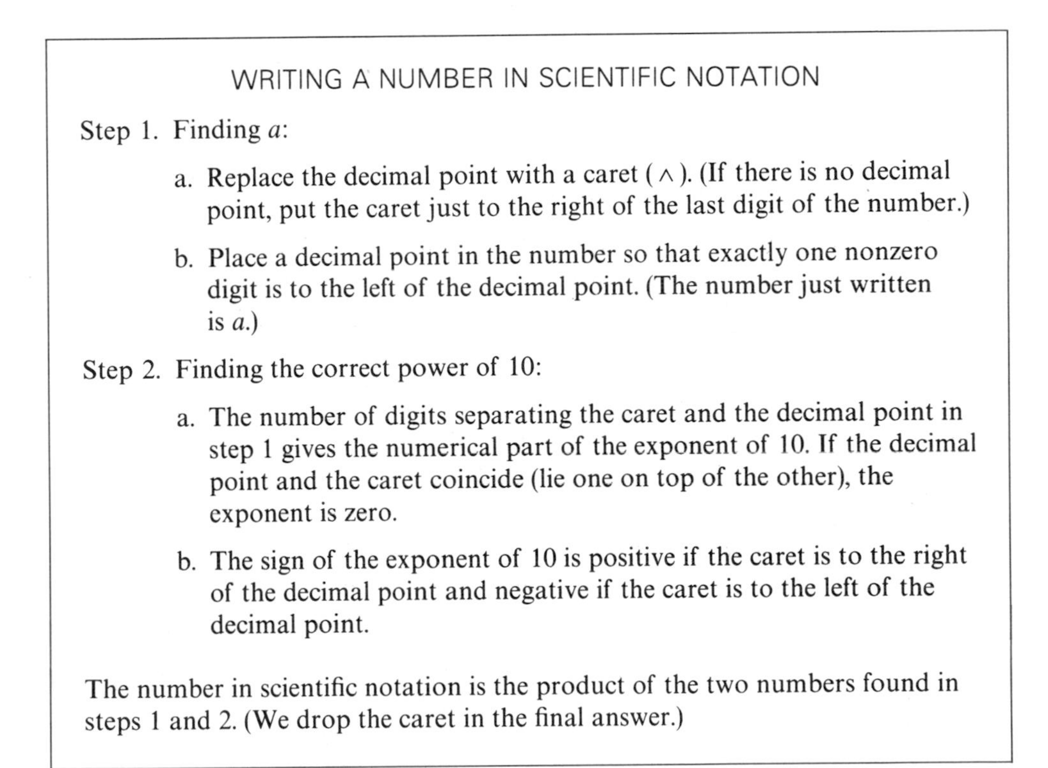

WRITING A NUMBER IN SCIENTIFIC NOTATION

Step 1. Finding a:

a. Replace the decimal point with a caret (‸). (If there is no decimal point, put the caret just to the right of the last digit of the number.)

b. Place a decimal point in the number so that exactly one nonzero digit is to the left of the decimal point. (The number just written is a.)

Step 2. Finding the correct power of 10:

a. The number of digits separating the caret and the decimal point in step 1 gives the numerical part of the exponent of 10. If the decimal point and the caret coincide (lie one on top of the other), the exponent is zero.

b. The sign of the exponent of 10 is positive if the caret is to the right of the decimal point and negative if the caret is to the left of the decimal point.

The number in scientific notation is the product of the two numbers found in steps 1 and 2. (We drop the caret in the final answer.)

These rules imply that if the number to be converted to scientific notation is greater than or equal to 10, the exponent of the 10 will be positive; if the number is less than 1, the exponent of the 10 will be negative. If the number is between 1 and 10, the exponent of the 10 will be zero.

Example 3 Examples of writing decimal numbers in scientific notation:

	Decimal notation	*Finding a*	*Scientific notation*
a.	2,450	2.450‸ $\times 10^?$	2.45×10^3
b.	2.45	2.45 (caret under decimal point) $\times 10^?$	2.45×10^0
c.	0.0245	0‸02.45 $\times 10^?$	2.45×10^{-2}
d.	92,900,000	9.290 000 0‸ $\times 10^?$	9.29×10^7
e.	0.000 561 8	0‸0005.618 $\times 10^?$	5.618×10^{-4}

a, the number between 1 and 10 — Power of 10

Notice that in parts (c) and (e) the caret is to the *left* of the decimal point and the exponent on the 10 is *negative*, and in parts (a) and (d) the caret is to the *right* of the decimal point and the exponent on the 10 is *positive*. In part (b), the decimal point and the caret coincide, and the exponent on the 10 is zero. ■

Example 4 Express (a) Avogadro's number and (b) Boltzmann's constant in scientific notation.
Solution

a. Avogadro's number is approximately 602,000,000,000,000,000,000,000.

The caret is to the right of the decimal point; there are 23 digits between the caret and the decimal point

$$602{,}000{,}000{,}000{,}000{,}000{,}000{,}000 = 6.\overbrace{020\,000\,000\,000\,000\,000\,000\,00}_{\wedge} \times 10^{23}$$
$$= 6.02 \times 10^{23}$$

b. Boltzmann's constant is 0.000 000 000 000 000 138.

The caret is to the left of the decimal point; there are 16 digits between the caret and the decimal point

$$0.000\,000\,000\,000\,000\,138 = 0_{\wedge}\overbrace{000\,000\,000\,000\,000\,1}.38 \times 10^{-16}$$
$$= 1.38 \times 10^{-16}$$ ■

To change from scientific notation to decimal notation, simply multiply by the power of 10. For example,

$$4.32 \times 10^3 = 4{,}320$$
$$2.3 \times 10^{-3} = 0.0023$$

It is sometimes necessary to convert a number such as 732.4×10^3 to scientific notation or to decimal notation. Example 5 demonstrates how to do this.

Example 5 Convert 732.4×10^3 to scientific notation and then to decimal notation.
Solution

$$732.4 \times 10^3 = (7.324 \times 10^2) \times 10^3 = 7.324 \times 10^5 \text{ in scientific notation}$$
$$732.4 \times 10^3 = 732{,}400 \text{ in decimal notation}$$ ■

Example 6 Use scientific notation in solving this problem: $\dfrac{30{,}000{,}000 \times 0.0005}{0.0000006 \times 80{,}000}$.

Solution

$$\frac{30{,}000{,}000 \times 0.000\,5}{0.000\,000\,6 \times 80{,}000} = \frac{(3 \times 10^7) \times (5 \times 10^{-4})}{(6 \times 10^{-7}) \times (8 \times 10^4)}$$ Writing each factor in scientific notation

$$= \frac{(3 \times 5) \times (10^7 \times 10^{-4})}{(6 \times 8) \times (10^{-7} \times 10^4)}$$ Collecting the powers of 10

$$= \frac{5}{16} \times \frac{10^3}{10^{-3}}$$ Simplifying what's in the parentheses *and* rewriting the expression as a product of two fractions

$= 0.3125 \times (10^3 \times 10^3)$ — Writing 5/16 in decimal form and writing $1/10^{-3}$ as 10^3

Changing 0.3125 to scientific notation

$= (3.125 \times 10^{-1}) \times 10^6$

$= 3.125 \times 10^5$ — Combining powers of 10

$= 312{,}500$ — Decimal notation ■

Scientific calculators usually express very large or very small answers in scientific notation (see Examples 7 and 8). On the calculator, however, numbers in scientific notation are displayed in a different (and possibly misleading) way. The calculator display [2.45 ^{04}] does *not* mean 2.45 to the fourth power. It means 2.45×10^4. (The calculator displays [2.45 04] and [2.45 E 4] also mean 2.45×10^4.)

Example 7 Use a calculator to find $600{,}000 \times 300{,}000$.

Solution The display probably shows [1.8 11], [1.8 ^{11}], or [1.8 E 11]. These displays all mean 1.8×10^{11}, *not* 1.8^{11}. Observe that while $1.8 \times 10^{11} = 180{,}000{,}000{,}000$, 1.8^{11} is approximately 642.6841007. ■

Example 8 Use a calculator to find $0.00006 \div 500$.

Solution The display probably shows [1.2 −07], [1.2 $^{-07}$], or [1.2 E −07]. These displays all mean $1.2 \times 10^{-7} = 0.00000012$. ■

Most calculators will give the *wrong answer* if very large (or very small) numbers are entered *unless* the numbers are entered in scientific notation (see Example 9). The keystrokes used to indicate operations vary, of course, with the brand of calculator. We will assume, in Example 9, that the key for indicating that you're *entering* a number in scientific notation is marked [EE], but it might be marked [EXP]. You must consult your calculator manual for details about *your* calculator.

Example 9 Given that the formula for simple interest is $I = prt$, use a calculator to find the (simple) interest for one year on \$2,600,000,000,000 if the interest rate is $7\frac{3}{4}\%$.

Solution P is 2,600,000,000,000; $t = 1$; and r is 0.0775 (0.0775 is $7\frac{3}{4}\%$ in decimal form). If we change P to scientific notation, we have

$$2{,}600{,}000{,}000{,}000. = 2.600\,000\,000\,000_{\wedge} \times 10^{12}$$
$$= 2.6 \times 10^{12}$$
$$I = prt = (2.6 \times 10^{12})(0.0775)(1)$$

This calculator display means 2.015×10^{11}, or 201,500,000,000. Therefore, the interest is \$201,500,000,000. ■

A WORD OF CAUTION In Example 9, watch your calculator display carefully if you *start* entering the problem as follows:

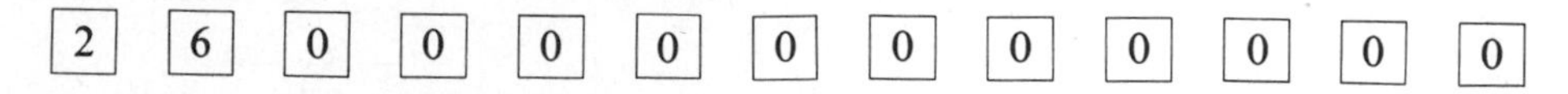

The calculator may *ignore* the last few zeros; if it does, you will get an incorrect answer. ☑

The even-numbered word problems at the end of Exercises 1.12 are not "matched" to the odd-numbered problems.

EXERCISES 1.12

Set I

In Exercises 1–12, write each number in scientific notation.

1. 28.56 **2.** 375.4 **3.** 0.06184

4. 0.003056 **5.** 78,000 **6.** 1,400

7. 0.2006 **8.** 0.000095 **9.** 0.362×10^{-2}

10. 0.6314×10^{-3} **11.** 245.2×10^{-5} **12.** 31.7×10^{-4}

In Exercises 13 and 14, perform the calculations by first changing all the factors to scientific notation.

13. $\dfrac{0.00006 \times 800{,}000{,}000}{50{,}000{,}000 \times 0.0003}$ **14.** $\dfrac{65{,}000{,}000{,}000 \times 0.0007}{0.0013 \times 1{,}400{,}000}$

In Exercises 15–18, perform the indicated operations with a calculator and express each answer in scientific notation.

15. $560{,}000 \times 23{,}000$ **16.** $0.00006 \div 20{,}000$

17. $\sqrt{0.00000256}$ **18.** $\sqrt{0.00000081}$

In Exercises 19–22, use a scientific calculator.

19. By definition (in chemistry and physics), 1 mole of any substance contains approximately 6.02×10^{23} molecules. How many molecules will 600 moles of oxygen contain? (Express your answer in scientific notation.)

20. The indebtedness of one of the developing nations is $120,000,000,000. If the annual (simple) interest rate is 7.8%, what is the interest on this debt for one year?

21. If the spacecraft Voyager traveled, 4,400,000,000 mi in 12 years, what was its average speed in miles per hour? (Use "rate = distance/time." Assume that each year has 365 days. Round off the answer to the nearest mile per hour.)

22. The speed of light is about 186,000 miles per second. How many miles does light travel in one day? (Use "rate × time = distance." Express your answer in scientific notation and in decimal notation.)

Set II

In Exercises 1–12, write each number in scientific notation.

1. 50.48 **2.** 0.0878 **3.** 4500.9

4. 0.000505 **5.** 289.3 **6.** 2,478,000

7. 0.00612 **8.** 0.00001 **9.** 63.7×10^{4}

10. 0.0357×10^{-5} **11.** 0.492×10^{-3} **12.** 0.251×10^{4}

In Exercises 13 and 14, perform the calculations by first changing all the factors to scientific notation.

13. $\dfrac{0.00006 \times 800{,}000{,}000}{50{,}000{,}000 \times 0.0003}$

14. $\dfrac{65{,}000{,}000{,}000 \times 0.0007}{0.0013 \times 1{,}400{,}000}$

In Exercises 15–18, perform the indicated operations with a calculator and express each answer in scientific notation.

15. $340{,}000 \times 680{,}000$

16. $0.00025 \div 500$

17. $\sqrt{0.00000289}$

18. $\sqrt{0.00000001}$

In Exercises 19–22, use a scientific calculator.

19. The indebtedness of one of the developing nations is \$52,000,000,000. If the annual (simple) interest rate is 7.4%, what is the interest on this debt for two years?

20. The planet Neptune is about 2,700,000,000 mi from the earth. What is this distance in kilometers? (One mile $\approx$ 1.61 km.)

21. A unit used in measuring the length of light waves is the Ångström. One micron is a millionth of a meter, and one Ångström is one ten-thousandth of a micron. One Ångström is what part of a meter? (Express your answer in scientific notation.)

22. The speed of light is about 186,000 miles per second. How long will it take light to reach us from the planet Neptune? (Use "time = distance/rate." Assume that Neptune is about 2,700,000,000 miles from the earth. Round off your answer to the nearest hour.)

1.13 Evaluating and Substituting in Algebraic Expressions and Formulas

1.13A Evaluating Algebraic Expressions

The process of substituting a numerical value for a variable in an algebraic expression is called *evaluating*. If the number being substituted is a *negative* number, it is almost always necessary to enclose the number in parentheses (see Example 1).

Example 1 Examples of evaluating expressions:

a. Find the value of $3x$ when $x = -2$.
Solution When $x = -2$, $3x = 3(-2) = -6$.

NOTE Writing $3 - 2$ is *incorrect*; $3x$ means 3 *times* x, and $3 - 2$ means *subtract* 2 from 3. The correct notation is $3(-2)$. ☑

b. Find the value of x^2 when $x = -4$.
Solution When $x = -4$, $x^2 = (-4)^2 = 16$.

NOTE Writing -4^2 is *incorrect*, since -4^2 means $-(4)^2 = -(16)$. We want -4 to be squared; therefore, the correct notation is $(-4)^2$. ☑

c. Find the value of $3x^2 - 5y^3$ if $x = -4$ and $y = -2$.
Solution When $x = -4$ and $y = -2$,

$$\begin{aligned} &3x^2 - 5y^3 \\ &= 3(-4)^2 - 5(-2)^3 \\ &= 3(16) - 5(-8) \\ &= 48 + 40 \\ &= 88 \quad \blacksquare \end{aligned}$$

Example 2 Find the value of $2a - [b - (3x - 4y)]$ if $a = -3$, $b = 4$, $x = -5$, and $y = -2$.
Solution

$$\begin{aligned} &2a - [b - (3x - 4y)] \\ &= 2(-3) - [(4) - \{3(-5) - 4(-2)\}] \longleftarrow \text{Braces are used in place of parentheses to clarify the grouping} \\ &= 2(-3) - [4 - \{-15 + 8\}] \\ &= 2(-3) - [4 - \{-7\}] \\ &= 2(-3) - [4 + 7] \\ &= 2(-3) - [11] \\ &= -6 - 11 \\ &= -17 \quad \blacksquare \end{aligned}$$

Example 3 Evaluate $b - \sqrt{b^2 - 4ac}$ when $a = 3$, $b = -7$, and $c = 2$.

Solution

$$\begin{aligned} &b - \sqrt{b^2 - 4ac} \quad \text{(This bar is a grouping symbol for } b^2 - 4ac\text{)} \\ &= (-7) - \sqrt{(-7)^2 - 4(3)(2)} \\ &= (-7) - \sqrt{49 - 24} \\ &= (-7) - \sqrt{25} \\ &= (-7) - 5 = -12 \quad \blacksquare \end{aligned}$$

A WORD OF CAUTION In Example 3, *two* errors are made if you write

$$b - \sqrt{b^2 - 4ac} \neq -7 - \sqrt{-7^2 - 4(3)(2)} \neq -7 - \sqrt{49 - 24}.$$

It is incorrect to omit the parentheses around -7 here, and $-7^2 = -49$, not 49. ☑

Substitutions in Algebraic Expressions

It is sometimes desirable to make substitutions of one variable for another variable in an algebraic expression. If the expression being substituted has more than one term, *be sure to enclose the expression in parentheses* (see Example 5).

Example 4 Substitute x for $2a^2 - 3a$ in the expression $(2a^2 - 3a)^2 + 5(2a^2 - 3a) - 15$.
Solution When we let $x = 2a^2 - 3a$ in the expression $(2a^2 - 3a)^2 + 5(2a^2 - 3a) - 15$, we get $x^2 + 5x - 15$. $\blacksquare$

Example 5 Substitute $x^3 - 7x$ for b in the expression $2b^2 - 5$.

Solution When we let $b = x^3 - 7x$ in the expression $2b^2 - 5$, we get $2(x^3 - 7x)^2 - 5$.* Notice that it is necessary to put $x^3 - 7x$ *inside parentheses* when we make the substitution. ■

EXERCISES 1.13A

Set I In Exercises 1–18, evaluate each expression, given that $a = \frac{1}{3}$, $b = -5$, $c = -1$, $x = 5$, $y = -6$, $D = 0$, $E = -1$, $F = 5$, $G = -15$, $H = -4$, and $J = 2$.

1. $2y^2 + 3x$ **2.** $c^3 - y$ **3.** $b - 12a^2$

4. $y - 24a^2$ **5.** $b^2 - 4xy$ **6.** $y^2 - 5cx$

7. $(x + y)^2$ **8.** $(b + c)^2$ **9.** $x^2 + 2xy + y^2$

10. $b^2 + 2bc + c^2$ **11.** $x^2 + y^2$ **12.** $b^2 + c^2$

13. $\dfrac{3D}{F + G}$ **14.** $\dfrac{5D}{G + H}$

15. $2E - [F - (D - 5G)]$ **16.** $3G - [D - (F - 2H)]$

17. $-E - \sqrt{E^2 - 4HF}$ **18.** $-E + \sqrt{E^2 - 4GJ}$

In Exercises 19–22, make the indicated substitutions.

19. Substitute b for $x^3 - 7x$ in the expression $5(x^3 - 7x)^2 - 4(x^3 - 7x) + 8$.

20. Substitute c for $a^5 + a$ in the expression $4(a^5 + a) + 3$.

21. Substitute $x^2 - 4x$ for a in the expression $2a^2 - 3a + 7$. (Do not attempt to simpify the result.)

22. Substitute $y^4 + 2$ for b in the expression $b^2 - 2b$. (Do not attempt to simplify the result.)

Set II In Exercises 1–18, evaluate each expression, given that $a = \frac{3}{4}$, $b = 5$, $c = -4$, $d = 3$, $e = -2$, and $f = -8$.

1. $4c^2 + 5e$ **2.** $c^2 + e$ **3.** $b - 14a^2$

4. $(b + c)^2$ **5.** $c^2 - 4bc$ **6.** $b^2 + c^2$

7. $(b + e)^2$ **8.** $b^2 + 2bc + c^2$ **9.** $b^2 + 2be + e^2$

10. $(f + e)^2$ **11.** $b^2 + e^2$ **12.** $f^2 + 2ef + e^2$

13. $\dfrac{5c}{e + f}$ **14.** $f^2 + e^2$

15. $2b - [f - (c - b)]$ **16.** $c - \{b - (d - f) - e\}$

17. $-e + \sqrt{e^2 - 4df}$ **18.** $(b + c + d)^2$

* We cannot yet simplify this expression.

In Exercises 19–22, make the indicated substitutions.

19. Substitute d for $x^2 + 2x$ in the expression $3(x^2 + 2x)^2 + 4(x^2 + 2x) - 5$.

20. Substitute a for $x^3 - x$ in the expression $7(x^3 - x)^2 - (x^3 - x) - 3$.

21. Substitute $z^2 + 3z - 5$ for c in the expression $3c^4 - 2c + 4$. (Do not attempt to simplify the result.)

22. Substitute $a^2 + 5a$ for x in the expression $-2x^3 - 4x^2 + 3x - 2$. (Do not attempt to simplify the result.)

1.13B Using Formulas

One reason for studying algebra is to prepare to use *formulas*. Students encounter formulas in many college courses, as well as in real-life situations. In the examples and exercises in this section, we have listed the subject areas in which the given formulas are used.

A formula is used or evaluated in the same way in which we evaluate any expression having numbers and variables.

Example 6 Given the formula $A = \frac{1}{2}h(a + b)$, find A when $h = 5$, $a = 3$, and $b = 7$. (Geometry)

Solution

$$A = \frac{1}{2}h(a + b)$$

$$A = \frac{1}{2}(5)(3 + 7)$$

$$A = \frac{1}{2}(5)(10) = 25 \quad ■$$

Example 7 Given the formula $T = \pi\sqrt{\frac{L}{g}}$, find T when $\pi \approx 3.14$, $L = 96$, and $g = 32$. (Physics)

Solution

$$T = \pi\sqrt{\frac{L}{g}}$$

$$T \approx (3.14)\sqrt{\frac{96}{32}} = (3.14)\sqrt{3} \approx (3.14)(1.732)$$

$$T \approx 5.44 \quad \text{Rounded off to 2 decimal places} \quad ■$$

Formulas will be discussed further in Chapter 6.

EXERCISES 1.13B

Set I Use each formula, substituting the given values for the variables. Answers that are not exact may be rounded off to two decimal places.

For Exercises 1 and 2 find q, using this formula from nursing: $q = \frac{DQ}{H}$.

1. $D = 5, H = 30, Q = 420$

2. $D = 25, H = 90, Q = 450$

For Exercises 3 and 4, find A, using this formula from business: $A = P(1 + rt)$.

3. $P = 500, r = 0.09, t = 2.5$

4. $P = 400, r = 0.07, t = 3.5$

For Exercises 5 and 6, find A, using this formula from business: $A = P(1 + i)^n$.

5. $P = 600, i = 0.085, n = 2$

6. $P = 700, i = 0.075, n = 2$

For Exercises 7 and 8, find C, using this formula from chemistry: $C = \frac{5}{9}(F - 32)$.

7. $F = -10$

8. $F = -7$

For Exercises 9 and 10, find s, using this formula from physics: $s = \frac{1}{2}gt^2$.

9. $g = 32, t = 8\frac{1}{2}$

10. $g = 32, t = 4\frac{3}{4}$

For Exercises 11 and 12, find Z, using this formula from physics: $Z = \frac{Rr}{R + r}$.

11. $R = 22, r = 8$

12. $R = 55, r = 25$

For Exercises 13 and 14, find S, using this formula from geometry: $S = 2\pi r^2 + 2\pi rh$. (Use $\pi \approx 3.14$.)

13. $r = 3, h = 20$

14. $r = 6, h = 10$

Set II Use each formula, substituting the given values for the variables. Answers that are not exact may be rounded off to two decimal places.

For Exercises 1 and 2, find q, using this formula from nursing: $q = \frac{DQ}{H}$.

1. $D = 15, H = 80, Q = 320$

2. $D = 20, H = 24, Q = 300$

For Exercises 3 and 4, find A, using this formula from business: $A = P(1 + rt)$.

3. $P = 450, r = 0.08, t = 2.5$

4. $P = 1{,}000, r = 0.06, t = 5.5$

For Exercises 5 and 6, find A, using this formula from business: $A = P(1 + i)^n$.

5. $P = 900, i = 0.095, n = 2$

6. $P = 400, i = 0.055, n = 3$

For Exercises 7 and 8, find C, using this formula from chemistry: $C = \frac{5}{9}(F - 32)$.

7. $F = 95$

8. $F = 14$

For Exercises 9 and 10, find s, using this formula from physics: $s = \frac{1}{2}gt^2$.

9. $g = 32, t = 5\frac{1}{2}$

10. $g = 32, t = 6\frac{1}{4}$

For Exercises 11 and 12, find Z, using this formula from physics: $Z = \frac{Rr}{R + r}$.

11. $R = 150, r = 25$

12. $R = 80, r = 35$

For Exercises 13 and 14, find S, using this formula from geometry: $S = 2\pi r^2 + 2\pi rh$. (Use $\pi \approx 3.14$.)

13. $r = 3.6, h = 5.1$

14. $r = 6.5, h = 1.2$

1.14 Simplifying Algebraic Expressions

1.14A Square Roots (of Perfect Squares) with Variables

We discussed the method for finding the square roots of positive numbers or zero in Section 1.7. In this section, we find square roots of algebraic expressions that are products of factors.

The algebraic expression x^2 has *two* square roots, x and $-x$. Since x itself can be positive, negative, or zero and since $\sqrt{x^2}$ (the *principal* square root of x^2) must always be positive or zero, we don't know whether $\sqrt{x^2} = x$ or $\sqrt{x^2} = -x$. However, the statement $\sqrt{x^2} = |x|$ is true for all x, because $|x|$ is always positive or zero. Therefore, the principal square root of x^2 is $|x|$.

NOTE In this chapter, we will assume that all variables represent positive numbers. For this reason, the absolute value symbols need not be used, and we will write $\sqrt{x^2} = x$. ☑

All of the radicands in this section are perfect squares. This means that any numerical coefficients are squares of integers and that the exponents on any variables are always *even* numbers. In Chapter 7, we consider simplifying radicals in which the radicands are *not* perfect squares.

TO FIND THE PRINCIPAL SQUARE ROOT OF A PRODUCT OF FACTORS

1. The principal square root of the numerical coefficient is found by inspection or by trial and error.
2. The square root of each literal factor is found by dividing its exponent by 2.

Example 1 Examples of finding principal square roots:

a. $\sqrt{9x^2} = 3x$ because $(3x)^2 = 9x^2$.

b. $\sqrt{25x^2} = 5x$ because $(5x)^2 = 25x^2$.

c. $\sqrt{100a^6b^{10}} = 10a^{6/2}b^{10/2} = 10a^3b^5$ because $(10a^3b^5)^2 = 100a^6b^{10}$. ■

EXERCISES 1.14A

Set I Find the principal square root of each expression.

1. $\sqrt{4x^2}$	**2.** $\sqrt{9y^2}$	**3.** $\sqrt{m^4n^2}$	**4.** $\sqrt{u^{10}v^6}$
5. $\sqrt{25a^4b^2}$	**6.** $\sqrt{100b^4c^2}$	**7.** $\sqrt{x^{10}y^4}$	**8.** $\sqrt{x^{12}y^8}$
9. $\sqrt{100a^{10}y^2}$	**10.** $\sqrt{121a^{24}b^4}$	**11.** $\sqrt{81m^8n^{16}}$	**12.** $\sqrt{49c^{18}d^{10}}$

Set II Find the principal square root of each expression.

1. $\sqrt{100a^8}$	**2.** $\sqrt{49b^6}$	**3.** $\sqrt{36e^8f^2}$	**4.** $\sqrt{81h^{12}k^{14}}$
5. $\sqrt{9a^4b^2c^6}$	**6.** $\sqrt{144x^8y^2z^6}$	**7.** $\sqrt{121a^4b^8}$	**8.** $\sqrt{144x^6y^2}$
9. $\sqrt{169a^8b^6c^4}$	**10.** $\sqrt{4s^6t^6}$	**11.** $\sqrt{16u^2v^8}$	**12.** $\sqrt{x^{12}y^{16}}$

1.14B Using the Distributive Properties to Simplify an Expression

In Section 1.11C, we simplified multiplication problems in which each factor contained only one term. In this section, we use the distributive properties

$$a(b + c) = ab + ac, (b + c)a = ba + ca, \quad \text{and} \quad a(b - c) = ab - ac$$

(first discussed in Section 1.5) to simplify a multiplication problem when one of the factors contains *two or more* terms. To use one of the distributive properties, multiply *each term* inside the grouping symbols by the factor that is outside them.

Example 2 Examples of using the distributive property to remove parentheses:

a. $4x(x^2 - 2xy + y^2) = (4x)(x^2) - (4x)(2xy) + (4x)(y^2)$
$\quad = 4x^3 - 8x^2y + 4xy^2$

b. $(-2x^2 + xy^2)(-3xy) = (-2x^2)(-3xy) + (xy^2)(-3xy)$
$\quad = 6x^3y - 3x^2y^3$ ■

A WORD OF CAUTION $(-2x^2 + xy^2)(-3xy) \neq (-2x^2 + xy^2) - 3xy$

$-3xy$ is being *multiplied* by $(-2x^2 + xy^2)$

$(-3xy)$ *is* a factor following the parentheses

$3xy$ is being *subtracted* from $(-2x^2 + xy^2)$

$-3xy$ is *not* a factor following the parentheses

☑

EXERCISES 1.14B

Set I Simplify.

1. $3a(6 + x)$
2. $5b(7 + y)$
3. $(x - 5)(-4)$
4. $(y - 2)(-5)$
5. $-3(x - 2y + 2)$
6. $-2(x - 3y + 4)$
7. $x(xy - 3)$
8. $a(ab - 4)$
9. $3a(ab - 2a^2)$
10. $4x(3x - 2y^2)$
11. $(3x^3 - 2x^2y + y^3)(-2xy)$
12. $(4z^3 - z^2y - y^3)(-2yz)$
13. $(-2ab)(3a^2b)(6abc^3)$
14. $(5x^2y)(-2xy^3)(3xyz^2)$
15. $4xy^2(3x^3y^2 - 2x^2y^3 + 5xy^4)$
16. $-2x^2(5x^4y - 2x^3y^2 - 3xy^3)$
17. $(3mn^2)(-2m^2n)(5m^2 - n^2)$
18. $(6a^2b)(-3ab^2)(2a^2 - b^2)$
19. $(2x^2)(-4xy^2z)(3xy - 2xz + 5yz)$
20. $(7a^2b)(3a - 2b - c)(ab^2)$
21. $(7x + 3y) - 2x^2$
22. $(3a - b) - 4c$

Set II Simplify.

1. $2R(R + S)$
2. $7x(x - 5)$
3. $(x - 7)(-3)$
4. $(x - 7) - y$
5. $-4(x - 5y + 3)$
6. $(a + 7b - c) - 3$
7. $u(uv - 5)$
8. $(a + 7b - c)(-3)$
9. $4xy^2z(3x^2 - xy)$
10. $2a(a^2 + 3b - c^3)$

11. $(2x^3 - 5xy + y^2)(-3xy)$

12. $(3x^3y - 4x + 2z) - 3y$

13. $(5xy^2z^2)(-3yz)(2xz^2)$

14. $(-2xy^3)(4xz)(-2yz^4)$

15. $2x^2y(-4xy^2z + 3xz - y^2z)$

16. $3a^2b(-2ab^2 + 8ab - 1)$

17. $(4xy^2)(-2xy)(5x^2 - xy)$

18. $9x^2y(2y^2)(3x^3 - y^2)$

19. $4u^2v(-3uv^2)(u - 3v - w)$

20. $2x^3z(-3xz)(8x^2 + 2z^2)$

21. $(8c^3 - 3bc) - b^2$

22. $(8c - 3bc)(-b^2)$

1.14C Removing Grouping Symbols

The common grouping symbols are

( ) Parentheses

[ ] Brackets

{ } Braces

— Bar (generally used with fractions and radicals)

An algebraic expression is not simplified unless all grouping symbols are removed.

If there are grouping symbols in an *addition* problem, we can often simply drop the grouping symbols (see Examples 3a and 3b); however, if the first term inside the grouping symbols has a *written* sign, we must also drop any addition symbols in front of the grouping symbols (see Example 3c).

Example 3 Remove the grouping symbols.

a. $(3x - 5) + 6y$

Solution This is an *addition* problem; we simply drop the parentheses

$$(3x - 5) + 6y = 3x - 5 + 6y$$

b. $5z + (4y + 7)$

Solution This is an *addition* problem; we simply drop the parentheses.

$$5z + (4y + 7) = 5z + 4y + 7$$

c. $5z + (-3y + 7)$

Solution This is an *addition* problem; because the first term inside the parentheses has a written sign, we drop the parentheses *and* the addition symbol in front of them.

$$5z + (-3y + 7) = 5z - 3y + 7 \quad \blacksquare$$

The Negative of an Algebraic Expression We can find the negative of an algebraic expression by multiplying that expression by -1. For example, the negative of $(8y + 2z - 5w)$ is $-1(8y + 2z - 5w)$, or $-8y - 2z + 5w$.

If an expression inside grouping symbols is being *subtracted* from another expression, we use the definition of subtraction:

$$a - b = a + (-b)$$

That is, we *add* the negative of the expression we're subtracting to the expression we're subtracting from (see Examples 4a and 4b).

Example 4 Remove the grouping symbols.

a. $3x - (8y + 2z - 5w)$

Solution

In this subtraction problem, we add the negative of the expression being subtracted

$$3x - (8y + 2z - 5w) = 3x + (-1)(8y + 2z - 5w)$$
$$= 3x + (-8y - 2z + 5w)$$
$$= 3x - 8y - 2z + 5w$$

b. $2x - (-8 - 6z + w)$

Solution

We add the negative of the expression being subtracted

$$2x - (-8 - 6z + w) = 2x + (-1)(-8 - 6z + w)$$
$$= 2x + (8 + 6z - w)$$
$$= 2x + 8 + 6z - w \quad ■$$

We use the distributive property to remove grouping symbols in a *multiplication* problem when one of the factors contains two or more terms; if grouping symbols occur *within* other grouping symbols, we remove the *innermost* grouping symbols first (see Example 5).

Example 5 Remove the grouping symbols in $3 + 2[a - 5(x - 4y)]$.

Solution

$$3 + 2[a - 5(x - 4y)] = 3 + 2[a + (-5)(x) + (-5)(-4y)]$$

Removing the *inner* grouping symbols first by multiplying each term inside them by -5

$$= 3 + 2[a - 5x + 20y]$$

Simplifying the expression inside the brackets

$$= 3 + 2(a) - 2(5x) + 2(20y)$$

Removing the brackets by multiplying each term inside them by 2

$$= 3 + 2a - 10x + 40y \quad ■$$

A WORD OF CAUTION The following are some common errors made in removing grouping symbols:

	Correct	*Common error*
a.	$-(x - 2y) = -x + 2y$	~~$-(x - 2y) = -x - 2y$~~ $-1(-2y) = +2y$, not $-2y$
b.	$6(y - 3) = 6y - 18$	~~$6(y - 3) = 6y - 3$~~ -3 was not multiplied by the 6
c.	$3 + 2(x + y) = 3 + 2x + 2y$	~~$3 + 2(x + y) = 5(x + y)$~~ Order of operations was not followed

EXERCISES 1.14C

Set I Remove the grouping symbols.

1. $10 + (4x - y)$
2. $8 + (3a - b)$
3. $7 - (-4R - S)$
4. $9 - (-3m - n)$
5. $6 - 2(a - 3b)$
6. $12 - 3(2R - S)$
7. $3 - 2x(x - 4y)$
8. $2 - 5x(2x - 3y)$
9. $-(x - y) + (2 - a)$
10. $-(a - b) + (x - 3)$
11. $(a - b)(2) - 6$
12. $(x - y)(3) - 5$
13. $x - [a + (y - b)]$
14. $y - [m + (x - n)]$
15. $5 - 3[a - 4(2x - y)]$
16. $7 - 5[x - 3(2a - b)]$
17. $2 - [a - (b - c)]$
18. $5 - [x - (y + z)]$
19. $9 - 2[-3a - 4(2x - y)]$
20. $P - \{x - [y - (4 - z)]\}$

Set II Remove the grouping symbols.

1. $5 + (2x - y)$
2. $7 - (a - 3b)$
3. $4 - (-3a - b)$
4. $8 - 5(z - w)$
5. $7 - 2(6R - S)$
6. $2 - 2(3x + y)$
7. $4 - 6x(3x - 2y)$
8. $(9 - z) - x$
9. $-(c - d) + (a - b)$
10. $(a + b) - (c - d)$
11. $(c - d)(5) - 4$
12. $(a + b)(3) - 2$
13. $z - [a + (b - c)]$
14. $s - \{t - (u - v) - w\}$
15. $6 - 4[x - 3(a - 2b)]$
16. $5 - 2(3x - [y - a])$
17. $3 - [x - (y - z)]$
18. $(x + y)(2) - 3z$
19. $a - \{b - [c - (2 - d)]\}$
20. $x - \{y - [z + w]\}$

1.14D Combining Like Terms

Like Terms Terms that have *equal* variable parts are called **like terms.**

Unlike Terms Terms that do not have *equal* variable parts are called **unlike terms.**

Example 6 Examples of like terms:

a. $23x$, $5.4x$, x, $0.7x$, and $\frac{1}{2}x$ are like terms. (They can be called "x-terms.")

$yx^2 = x^2y$ and $xyx = x^2y$; therefore the terms are like terms

b. $34x^2y$, $8yx^2$, xyx, $2.8x^2y$, and $\frac{1}{5}x^2y$ are like terms. ■

Example 7 Examples of unlike terms:

a. $5x^2$ and $5x$ are unlike terms. (The variable parts, x^2 and x, are different.)

b. $4x^2$ and $10x^2y$ are unlike terms. ■

Because of the distributive property, combining like terms is possible. For example,

This step need not be shown

$$3x + 5x = (3 + 5)x = 8x$$

> **TO COMBINE (ADD) LIKE TERMS**
>
> Add the numerical coefficients *of the like terms*; the number so obtained is the numerical coefficient of the sum. The variable part of the sum is the same as the variable part of any *one* of the like terms.

When we combine like terms, we are usually changing the grouping and the order in which the terms appear. The commutative and associative properties of addition guarantee that when we do this the sum remains unchanged.

Example 8 Examples of combining like terms:

$x^2y = 1x^2y$

a. $5x^2y - 8x^2y + x^2y = (5 - 8 + 1)x^2y = -2x^2y$

$ba = ab$; therefore, all the terms are like terms

b. $4ab + ba - 6ab = (4 + 1 - 6)ab = -1ab = -ab$

c. $12a - 7b - 9a + 4b$

$= (12a - 9a) + (-7b + 4b)$ ← Only like terms may be combined

$= 3a - 3b$

d. $7x - 2y + 9 - 11x + 3 - 4y = -4x - 6y + 12$ ■

NOTE While it is possible to *combine* (add or subtract) only *like* terms, it is possible to *multiply* unlike terms together; for example, $3x + 3x^2$ cannot be simplified, but $(3x)(3x^2) = 9x^3$. ☑

EXERCISES 1.14D

Set I In each exercise, simplify each term and combine the like terms.

1. $5x - 8x + x$
2. $3a - 5a + a$
3. $8x^2y - 2x^2y$
4. $10ab^2 - 3ab^2$
5. $6xy^2 + 8x^2y$
6. $5a^2b - 4ab^2$
7. $2xy - 5yx + xy$
8. $8mn - 7nm + 3mn$
9. $5xyz^2 - 2(xyz)^2$
10. $3(abc)^3 - 4abc^3$
11. $5xyz^2 - xyz^2 - 4xyz^2$
12. $7a^2bc - a^2bc - 5a^2bc$
13. $7x^2y - 2xy^2 - 4x^2y$
14. $4xy^2 - 5x^2y - 2y^2x$
15. $3ab - a + b - ab$
16. $5xy - x - y + xy$
17. $2x^3 - 2x^2 + 3x - 5x$
18. $5y^2 - 3y^3 + 2y - 4y$

19. $4x - 3y + 7 - 2x + 4 - 6y$

20. $3b - 5a - 9 - 2a + 4 - 5b$

21. $a^2b - 5ab + 7ab^2 - 3a^2b + 4ab$

22. $xy^2 + y - 5x^2y + 3xy^2 + x^2y$

Set II In each exercise, simplify each term and combine the like terms.

1. $4y + y - 10y$

2. $10a - 16a + a$

3. $a^2b - 3a^2b$

4. $2x^3y - 9yx^3$

5. $4a^2b + 6ab^2$

6. $12x^3y + 8xy^3$

7. $5uv - 2vu + uv$

8. $ab + 3ab - 18ba$

9. $3(stu)^2 - 5stu^2$

10. $2 + 4xy$

11. $8ab^2c - ab^2c - 4ab^2c$

12. $6fg^2 - 2(fg)^2$

13. $5ab^2c - 7a^2bc - 2ab^2c$

14. $17x(yz)^2 - 12xy^2z^2$

15. $6st + s - t - st$

16. $8 - 2xy - 6yx$

17. $5R^3 - 2R + 3R^2 - R$

18. $9x + 2y - 3z$

19. $6x - 6 + 4y - 3 + 2x - 7y$

20. $4 + 2a - 3b - 7 - 6b - 2a$

21. $xy - 2xy^2 - 5x^2y - 3xy + 2x^2y - 4xy^2$

22. $st^2 - 3s + 2t - s^2t - 5s + 3t$

1.14E Simplifying Algebraic Expressions

TO SIMPLIFY AN ALGEBRAIC EXPRESSION

1. Remove all grouping symbols.
2. Combine powers of variables in each term.
3. Remove all zero and negative exponents.
4. Simplify all radicals.*
5. Combine all like terms.

NOTE We will add to this list in later chapters. ☑

Example 9 Examples of simplifying algebraic expressions:

a. $x(x^2 + xy + y^2) - y(x^2 + xy + y^2)$

$= x^3 + x^2y + xy^2 - x^2y - xy^2 - y^3$ Using the distributive property

$= x^3 + x^2y - x^2y + xy^2 - xy^2 - y^3$ Collecting like terms

$= x^3 + 0 + 0 - y^3$ Combining like terms

$= x^3 - y^3$

*In this section, all radical signs will be completely removed, since all radicands will be perfect squares.

b. $\sqrt{16} - \{8[-5(3x-2)+13] - 11x\}$

$= 4 - \{8[-15x + 10 + 13] - 11x\}$ — Using the distributive property

$= 4 - \{8[-15x + 23] - 11x\}$ — Combining like terms

$= 4 - \{-120x + 184 - 11x\}$ — Using the distributive property

$= 4 - \{-131x + 184\}$ — Combining like terms

$= 4 + 131x - 184$ — Removing grouping symbols

$= 131x - 180$ — Combining like terms

c. $\sqrt{9x^2} + x^{-1}x^2 - 4x^0(12x) - (3x)^0$

$= 3x + x^1 - 4(1)(12x) - 1$ — Simplifying each term

$= 3x + x - 48x - 1$

$= -44x - 1$ — Combining like terms ■

Example 10 Rewrite $-3(a-2b) + 2(-a-3b)$ when $a = x + y$ and $b = x - y$ and then simplify.
Solution If $a = x + y$ and $b = x - y$, then we have

$-3(a-2b) + 2(-a-3b)$

$= -3([x+y] - 2[x-y]) + 2(-[x+y] - 3[x-y])$ — Substituting

$= -3(x + y - 2x + 2y) + 2(-x - y - 3x + 3y)$

$= -3(-x + 3y) + 2(-4x + 2y)$ — Combining like terms

$= 3x - 9y - 8x + 4y$ — Using the distributive property

$= -5x - 5y$ — Combining like terms ■

EXERCISES 1.14E

Set I In Exercises 1–28, simplify the algebraic expressions. Assume that the variables represent positive numbers.

1. $2h(3h^2 - k) - k(h - 3k^3)$

2. $4x(2y^2 - 3x) - x(2x - 3y^2)$

3. $(3x - 4) - 5x$

4. $(5x - 7) - 8x$

5. $(3x - 4)(-5x)$

6. $(5x - 7)(-8x)$

7. $2 + 3x$

8. $5 + 8y$

9. $3x - [5y - (2x - 4y)]$

10. $2x - [7y - (3x - 2y)]$

11. $-10[-2(3x - 5) + 17] - 4x$

12. $-20[-3(2x - 4) + 20] - 5x$

13. $8 - 2(x - [y - 3x])$

14. $9 - 4(u - [t - 2u])$

15. $2x(4 + 5x) - \sqrt{16x^2}$

16. $5y(2 + 5y) - \sqrt{36y^2}$

17. $(3u - v) - \{2u - (10 - v) - 20\} - \sqrt{64v^2}$

18. $(5x - y) - \{3x - (8 - y) - 15\} - \sqrt{49x^2}$

19. $50 - \{-2t - [5t - (6 - 2t)]\} + 7^0$

20. $24 - \{-4x - [2x - (3 - 5x)]\} + 3^0$

21. $100v - 3\{-4[-2(-4 - v) - 5v]\}$

22. $60z - 4\{-3[-4(-2 - z) - 3z]\}$

23. $w^2(w^2 - 4) + 4(w^2 - 4)$

24. $x^2(x^2 - 9) + 9(x^2 - 9)$

25. $3x(5 \cdot 4x^2)(2x^3)$

26. $2y(2 \cdot 3y^3)(5y^4)$

27. $5X^{-4}X^6 + 3X^0$

28. $3Y^3Y^{-1} + 2Y^0$

In Exercises 29–32, substitute $x + 2y$ for a and $3x - y$ for b and simplify.

29. $3a - 5b$ **30.** $4a - 2b$ **31.** $2(3a - b)$ **32.** $3(5b - a)$

Set II In Exercises 1–28, simplify the algebraic expressions. Assume that the variables represent positive numbers.

1. $3x(2x^2 - y) - y(x - 4y^2)$

2. $5a^2(4a - ab + b^2)$

3. $(4s - 7) - 2s$

4. $(2x + 3y) - 9x$

5. $(4y - 7)(-2y)$

6. $(2x + 3y)(-9x)$

7. $4 + 5z$

8. $(8 + 3x)y$

9. $2s - [4t - (3s - 5t)]$

10. $(8 + 3x) + y$

11. $-5[-2(2x - 4) + 15] - 2x$

12. $\sqrt{121x^6z^4} + 3x^3z^2$

13. $6 - 4(y - [x - 3y])$

14. $8 + 2(x - [y - 4])$

15. $2z(3 + 7z) - \sqrt{16z^2}$

16. $3x + 7y - 2z$

17. $(2x - y) - \{5x - (6 - y) - 12\} - \sqrt{121y^2}$

18. $16 - 4\{6 - 2(x - 3y) + x\}$

19. $36 - \{-2x - [6x - (5 - 2x)]\} + 3x^0$

20. $(3x + 14y - 3z^2[3z + x])^0$

21. $50x - 5\{-2[-3(-3 - x) - 2x]\}$

22. $x^0 + y^0 + z^0 + 8^0$

23. $y^2(y^2 - 16) + 16(y^2 - 16)$

24. $(x + y + z + 8)^0$

25. $4x(3 \cdot 5x^2)(6x^4)$

26. $3x(2x^2)(-16xy^3)$

27. $4z^{-7}z^{10} + 7z^0$

28. $-3x^{-3}x^5 - 18x^0$

In Exercises 29–32, substitute $x + 3y$ for a and $2x - y$ for b and simplify.

29. $7a - 2b$ **30.** $12a - b$ **31.** $5(2a - b)$ **32.** $b - 3a$

1.15 Review: 1.11–1.14

The Properties of Exponents 1.11

1.1 $x^m \cdot x^n = x^{m+n}$

1.2 $(x^m)^n = x^{mn}$

1.3 $(xy)^n = x^n y^n$

1.4 $\dfrac{x^m}{x^n} = x^{m-n} \quad (x \neq 0)$

1.5 $\left(\dfrac{x}{y}\right)^n = \dfrac{x^n}{y^n} \quad (y \neq 0)$

1.6 $x^0 = 1 \quad (x \neq 0)$

1.7a $x^{-n} = \dfrac{1}{x^n} \quad (x \neq 0)$

1.7b $\dfrac{1}{x^{-n}} = x^n \quad (x \neq 0)$

1.7c $\left(\dfrac{x}{y}\right)^{-n} = \left(\dfrac{y}{x}\right)^n \quad (x \neq 0, y \neq 0)$

1.8 $\left(\dfrac{x^a y^b}{z^c}\right)^n = \dfrac{x^{an}y^{bn}}{z^{cn}} \quad (z \neq 0)$

None of the variables can have a value that makes the denominator zero.

Scientific Notation 1.12 A number written in scientific notation is written as the product of some number between 1 and 10 and a power of 10. That is, it must be of the form

$$a \times 10^n, \text{ where } 1 \le a < 10 \text{ and } n \text{ is an integer}$$

To Evaluate Algebraic Expressions 1.13

1. First replace each variable by its numerical value.
2. Then carry out all operations in the correct order.

To Simplify Algebraic Expressions 1.14

1. Remove all grouping symbols.
2. Combine powers of variables in each term.
3. Remove all zero and negative exponents.
4. Simplify all radicals.
5. Combine all like terms.

Review Exercises 1.15 Set I

In Exercises 1–18, simplify each expression if possible, using only positive exponents (or no exponents) in your answers.

1. $x^3 \cdot x^5$ **2.** $x^4 + x^2$ **3.** $(N^2)^3$ **4.** $s^6 - s^2$

5. $\dfrac{a^5}{a^2}$ **6.** $\dfrac{x^6}{y^4}$ **7.** $\left(\dfrac{2a}{b^2}\right)^3$ **8.** x^4y^{-2}

9. $\left(\dfrac{x^{-4}y}{x^{-2}}\right)^{-1}$ **10.** $s^0 + t^0$ **11.** $(s + t)^0$ **12.** xy^4

13. $3c^3d^2(c - 4d)$ **14.** $8 - 2(3x - y)$

15. $(-10x^2y^3)(-8x^3)(-xy^2z^4)$ **16.** $(4x^3 + 2x - y)(-2x)$

17. $5 - 2[3 - 5(x - y) + 4x - 6]$ **18.** $(4x^3 + 2x - y) - 2x$

19. Write 148.6 in scientific notation.

20. Write 3.17×10^{-3} in decimal notation.

In Exercises 21–24, use a calculator and use the values of the variables given with the formula.

21. $A = P(1 + rt)$ Find A when $P = 550$, $r = 0.09$, $t = 2.5$.

22. $C = \dfrac{5}{9}(F - 32)$ Find C when $F = 104$.

23. $S = R\left[\dfrac{(1 + i)^n - 1}{i}\right]$ Find S when $R = 750$, $i = 0.09$, $n = 3$.

24. $S = \dfrac{a(1 - r^n)}{1 - r}$ Find S when $a = 7.5$, $r = 2$, $n = 6$.

Review Exercises 1.15 Set II

NAME ______________

ANSWERS

1. ______________
2. ______________
3. ______________
4. ______________
5. ______________
6. ______________
7. ______________
8. ______________
9. ______________
10. ______________
11. ______________
12. ______________
13. ______________
14. ______________
15. ______________
16. ______________

In Exercises 1–18, simplify each expression if possible, using only positive exponents (or no exponents) in your answers.

1. $y^6 + y^5$

2. $x^4 \cdot x^8$

3. $(z^4)^3$

4. $\dfrac{a^5}{b^2}$

5. $\dfrac{y^6}{y^4}$

6. $s^6 - s^2$

7. $\left(\dfrac{x}{2b^2}\right)^3$

8. $\left(\dfrac{x^{-4}y}{x^{-2}}\right)^{-2}$

9. x^4x^{-2}

10. x^0y^0

11. $(x + y)^0$

12. $x^0 + y^0$

13. $5 - 2(x - 6y)$

14. $(3x^2 - 2) - 3x$

15. $(5ab^2)(-2a^2)(-3a^3b)$

16. $3xy^2(4x^2 - 5y^3)$

17. $2x(4x^2 + 2xy + y^2) - y(4x^2 + 2xy + y^2)$

18. $20 - 3\{x - 2[5x - 3(x - y) - 3y] + 4x\}$

In Exercises 19–21, use a calculator and use the values of the variables given with the formula.

19. $I = Prt$ Find I when $P = 1{,}250$, $r = 0.08$, $t = 3.5$.

20. $F = \frac{9}{5}C + 32$ Find F when $C = 25.5$.

21. $A = P(1 + i)^n$ Find A when $P = 950$, $i = 0.09$, $n = 4$.
(Round off the answer to two decimal places.)

22. Write 0.000538 in scientific notation.

23. Write 1.452×10^5 in decimal notation.

24. Use scientific notation to solve this problem:

$$\frac{700{,}000{,}000 \times 0.00009}{0.0000003 \times 40{,}000}$$

ANSWERS

17. ______________

18. ______________

19. ______________

20. ______________

21. ______________

22. ______________

23. ______________

24. ______________

Chapter 1 Diagnostic Test

The purpose of this test is to see how well you understand the basic ideas of sets, operations with real numbers, and simplification and evaluation of algebraic expressions. We recommend that you work this diagnostic test *before* your instructor tests you on this chapter. Allow yourself about 50 minutes.

Complete solutions for all the problems on this test, together with section references, are given in the answer section at the end of the book. For the problems you do incorrectly, study the sections cited.

In Problems 1–3, write "true" if the statement is always true; otherwise, write "false."

1. a. $\{5, 3, 3, 5\}$ and $\{3, 5\}$ are equal sets.

b. 0 is a real number.

c. Every irrational number is a real number.

d. $\frac{9}{0} = 0$

2. a. $(7 \cdot 5) \cdot 2 = 7(5 \cdot 2)$ illustrates the commutative property of multiplication.

b. Division is commutative.

c. The set of digits is a subset of the set of natural numbers.

d. Every integer is a real number.

3. a. $-2 < 7 < 10$ is a valid continued inequality.

b. Every real number is a rational number.

c. $4 < 8 > 12$ is a valid continued inequality.

d. $3 \cdot (7 \cdot 2) = (3 \cdot 7) \cdot (3 \cdot 2)$

4. Given $A = \{x, w, k, z\}$, $B = \{x, y, z, k\}$, and $C = \{k, x\}$, state whether the following statements are true or false.

a. $B \subseteq A$ b. $C \subseteq B$

5. Given the numbers $-3, 2.4, 0, 5, 2.8652916\ldots, \frac{1}{2}$, and $0.\overline{18}$, determine the following.

a. Which are real numbers?

b. Which are integers?

c. Which are natural numbers?

d. Which are irrational numbers?

e. Which are rational numbers?

6. Given $A = \{x, z, w\}$, $B = \{x, y, w\}$, and $C = \{y, r, s\}$, find the following.

a. $A \cup C$ b. $A \cap C$ c. $A \cap B$ d. $A \cup (B \cap C)$

In Problems 7–11, find the value of each expression.

7. a. $|-17|$ b. $(-5)^2$ c. $30 \div (-5)$ d. $-35 - 2$

8. a. $-27 - (-17)$ b. $-9(-8)$ c. $-19(0)$ d. $\frac{-40}{-8}$

9. a. $-9 + (-13)$ b. -6^2 c. $(-2)^0$ d. $-|-3|$

10. a. $\sqrt[3]{-27}$ b. $\sqrt[4]{16}$ c. $(3^{-2})^{-1}$ d. $10^{-3} \cdot 10^5$

11. a. $\dfrac{2^{-4}}{2^{-7}}$ b. $16 \div 4 \cdot 2$ c. $2\sqrt{9} - 5$ d. $\sqrt{81}$

12. Perform the following calculation by first changing all the factors to scientific notation: $\dfrac{81{,}000{,}000 \times 0.00000003}{0.00004 \times 600{,}000{,}000}$. Express your answer in scientific notation and in decimal notation.

13. List all the prime numbers greater than 7 and less than 29 that yield a remainder of 2 when divided by 5.

14. Write each number in prime factored form.

a. 78 b. 65

15. Find the LCM of 78 and 65.

In Problems 16–23, simplify each expression.

16. $x^2 \cdot x^{-5}$

17. $(N^2)^4$

18. $\left(\dfrac{2X^3}{Y}\right)^2$

19. $\left(\dfrac{xy^{-2}}{y^{-3}}\right)^{-1}$

20. $\dfrac{1}{a^{-3}}$

21. $7x - 2(5 - x) + \sqrt{81x^2}$

22. $6x(2xy^2 - 3x^3) - 3x^2(2y^2 - 6x^2)$

23. $7x - 2\{6 - 3[8 - 2(x - 3) - 2(6 - x)]\}$

24. Given the formula $S = \dfrac{a(1 - r^n)}{1 - r}$, find S when $a = -8$, $r = 3$, and $n = 2$.

25. Use a scientific calculator and the formula $I = prt$ to find the (simple) interest for 3 years on a debt of \$3,400,000,000 if the annual interest rate is $7\frac{1}{2}\%$.

2 First-Degree Equations and Inequalities in One Unknown

Most problems in algebra are solved by the use of equations or inequalities. In this chapter, we show how to solve equations and inequalities that have only one unknown. We discuss other types of equations and inequalities in later chapters.

2.1 Types of First-Degree Equations, Their Solutions, and Their Graphs

An **equation** is a *statement* that two quantities are equal. It may be a true statement, a false statement, or a statement that is sometimes true and sometimes false.

First-Degree Equation with Only One Unknown

A **first-degree equation with only one unknown** is an equation with only one variable in which the highest power of that variable is the first power; for example, $3x + 8 = 17$ is a first-degree equation with one unknown.

Domain The set of all the numbers that can be used in place of a variable is called the **domain** of the variable. For example, in the expression $1/x$, x cannot be 0 because we cannot allow division by 0. We would say, therefore, that 0 is not in the domain of the variable; we could also say that the domain is the set of all real numbers except 0.

In some applications, the solution set must be restricted to some subset of the set of real numbers. For example, if x were to represent the number of dimes in a collection of coins, it wouldn't make sense to say that we had 1.72 dimes. In this case, we would need to restrict the domain of the variable to the set of whole numbers.

In this chapter, if the domain of the variable is not mentioned, the domain is understood to be the set of real numbers, R.

Solutions of Equations

To **solve** an equation means to find values of the variable that make the equation (the *statement*) true. Therefore, a **solution** of an equation is a number from the domain of the variable that, when substituted for the variable, makes the two sides of the equation equal. A solution of an equation is also called a **root** or a **zero** of the equation.

Kinds of Equations

Conditional Equations A **conditional equation** is a statement that is true for certain values of the variable (the solutions) but is not true for others. For example, $3x + 8 = 17$ is a conditional equation; the statement is true if 3 is substituted for x but is *not* true if any other number is substituted for x. When we solve a first-degree conditional equation in one variable, our final equation should be in the form $x = a$, where a is some number. When the equation is in this form, we say that we have *isolated* the variable; that is, we have the variable by itself on one side of the equal sign, and that variable does not appear on the other side of the equal sign.

Identities An **identity** is a statement that is true when *any* value in the domain is substituted for the variable. For example, $2(x + 3) = 2x + 6$ is an identity; the statement is true if we substitute 7 for x, 0 for x, -23 for x, or any other real number for x.

Equations with No Solution An **equation with no solution** is a statement that is false for all values of the variable. For example, $x + 1 = x + 2$ is such an equation; no values of x will make the statement true.

Solution Sets of Equations

The **solution set** of an equation is the set of all numbers that are solutions of that equation. For example, the solution set of the equation $3x + 8 = 17$ is $\{3\}$. The solution set of an identity is the set of all the numbers in the domain of the variable. The solution set of an equation with no solution is the empty set.

Equivalent Equations

Equations that have the same solution set are called **equivalent equations**.

Example 1 Examples of solutions, solution sets, and equivalent equations:

The *solution* of the equation $3x + 8 = 17$ is 3, because $3(3) + 8 = 17$. The *solution set* of this equation is $\{3\}$. (Notice that 3 is written within braces, because we are discussing a set.) The equations $3x + 8 = 17$, $3x = 9$, and $x = 3$ are *equivalent equations* because they all have the same solution set. ■

Solving First-Degree Equations

When we solve a first-degree equation, we want to *isolate the variable*. That is, we want to find an equation in the form $x = a$ that is equivalent to the original equation. We use the following principles of equality in writing equations equivalent to the given equation:

THE ADDITION PROPERTY OF EQUALITY

If the same number is added to both sides of an equation, the new equation is equivalent to the original equation.

THE SUBTRACTION PROPERTY OF EQUALITY

If the same number is subtracted from both sides of an equation, the new equation is equivalent to the original equation.

THE MULTIPLICATION PROPERTY OF EQUALITY

If both sides of the equation are multiplied by the same nonzero number, the new equation is equivalent to the original equation.

THE DIVISION PROPERTY OF EQUALITY

If both sides of the equation are divided by the same nonzero number, the new equation is equivalent to the original equation.

THE SYMMETRIC PROPERTY OF EQUALITY

The equation $a = b$ is equivalent to the equation $b = a$.

The symmetric property allows us to rewrite the equation $a = x$ as $x = a$.

When we solve a first-degree equation with one variable using the principles of equality listed above, three outcomes are possible:

1. If the equation can be reduced to the form $x = a$, where a is some number, the equation is a *conditional equation* (see Example 2). If there are restrictions on the domain, the equation may have *no solution* even though it reduces to the form $x = a$ (see Example 3).
2. If the two sides of the equation reduce to the same constant so that we obtain a true statement (for example, $0 = 0$), the equation is an *identity* (see Example 4).
3. If the two sides of the equation reduce to unequal constants so that we obtain a false statement (for example, $2 = 5$), the equation is an *equation with no solution* (see Example 5).

Although there is no single correct way to solve an equation, you may find the following suggestions helpful. (The first five steps can be done in any order.) Remember: Your goal is to *isolate the variable*, if possible.

TO SOLVE A FIRST-DEGREE EQUATION WITH ONLY ONE UNKNOWN

Always write each new equation *under* the previous one.

1. Remove fractions by multiplying both sides of the equation by the least common multiple (LCM) of all the denominators.
2. Remove all grouping symbols.
3. Combine like terms on each side of the equal sign.
4. Move all the terms that contain the variable to one side of the equal sign (usually the *left* side) and all the constants to the other side by adding the appropriate terms to both sides of the equation.
5. Divide both sides of the equation by the coefficient of the variable.
6. Determine whether the equation is a conditional equation, an identity, or an equation with no solution.
7. If the equation is a conditional equation, check the solution.

To check the solution of an equation, perform the following steps:

1. Determine whether the *apparent* solution (the number found by following steps 1–5 in the box above) is in the domain of the variable. If it is not, the apparent solution is *not* a solution. If the apparent solution *is* in the domain, continue with the following steps.
2. Replace the variable in the given equation by the apparent solution.
3. Perform the indicated operations on both sides of the equal sign.
4. If the resulting numbers on both sides of the equal sign are the same, the solution checks.

Example 2 Find the solution set for $8x - 3[2 - (x + 4)] = 4(x - 2)$ and graph the solution on the number line.

Solution

$$8x - 3[2 - (x + 4)] = 4(x - 2)$$

$$8x - 3[2 - x - 4] = 4x - 8$$ Removing grouping symbols and simplifying

$$8x - 3[-x - 2] = 4x - 8$$ Combining like terms

$$8x + 3x + 6 = 4x - 8$$

$$\begin{array}{rcl} 11x + 6 & = & 4x - 8^* \\ -4x - 6 & & -4x - 6 \\ \hline 7x & = & -14 \end{array}$$ Adding $-4x - 6$ to both sides to get the x-term on one side and the constant on the other

$$x = -2$$ Dividing both sides by 7

* It is incorrect to write $11x + 6 = 4x - 8 = 7x = -14 = x = -2$.

An equal sign here implies that $4x - 8 = 7x$ (arrow to the first highlighted equal sign)

An equal sign here implies that $-14 = -2$ (arrow to the second highlighted equal sign)

The apparent solution of the equation is -2. Since the equation reduced to the form $x = a$, the equation is a conditional equation.
Check The domain is the set of all real numbers, so the apparent solution, -2, is in the domain.

$$\begin{aligned} 8x - 3[2 - (x + 4)] &= 4(x - 2) \\ 8(-2) - 3[2 - (\{-2\} + 4)] &\stackrel{?}{=} 4(\{-2\} - 2) \\ -16 - 3[2 - (2)] &\stackrel{?}{=} 4(-4) \\ -16 - 3[0] &\stackrel{?}{=} -16 \\ -16 &= -16 \end{aligned}$$

The solution checks. Therefore, $x = -2$ is the condition necessary to make the two sides of the equation equal to each other, and $\{-2\}$ is the solution set.

The *graph* of the solution (or of the solution set) is

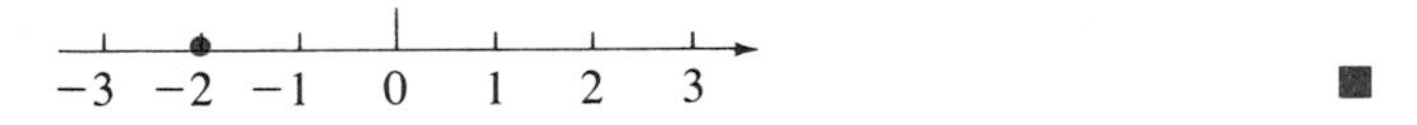

■

Example 3 Find $\{x \mid 2(3x + 5) = 14, x \in J\}$.
Solution In this example, the domain is restricted by the statement $x \in J$; that is, we are interested only in integer solutions to the equation.

$$\begin{aligned} 2(3x + 5) &= 14 \\ 6x + 10 &= 14 \\ 6x &= 4 \\ x &= \frac{4}{6} = \frac{2}{3} \end{aligned}$$

The only apparent solution, $\frac{2}{3}$, is not in the domain; therefore, the equation has no solution, and $\{x \mid 2(3x + 5) = 14, x \in J\} = \{\ \}$. If we had been asked to graph the solution set, there would be no points to graph. ■

Example 4 Solve $\frac{x+3}{6} - \frac{2x-3}{9} = \frac{5}{6} - \frac{x}{18}$ and graph its solution set on the number line.
Solution

$$\frac{18}{1}\left(\frac{x+3}{6} - \frac{2x-3}{9}\right) = \frac{18}{1}\left(\frac{5}{6} - \frac{x}{18}\right) \qquad \text{Multiplying both sides by the LCM of 6, 9, and 18, which is 18}$$

$$\frac{18}{1}\left(\frac{x+3}{6}\right) - \frac{18}{1}\left(\frac{2x-3}{9}\right) = \frac{18}{1}\left(\frac{5}{6}\right) - \frac{18}{1}\left(\frac{x}{18}\right) \qquad \text{Using the distributive property}$$

$$3(x + 3) - 2(2x - 3) = 15 - x \qquad \text{Simplifying}$$

$$3x + 9 - 4x + 6 = 15 - x \qquad \text{Using the distributive property}$$

$$\begin{aligned} -x + 15 &= 15 - x && \text{Combining like terms} \\ +x - 15 & \quad\ -15 + x && \text{Adding } x - 15 \text{ to both sides} \\ \hline 0 &= 0 && \text{A true statement} \end{aligned}$$

If we had added just $+x$ to both sides, our final statement would have been $15 = 15$, which is also a true statement. *Note that no variable appears in the last step.* Since the variable disappeared completely and we were left with a *true* statement, this equation is an identity. *Any* real number will make the two sides equal. We will show the check for two real numbers selected at random.

Check The domain is the set of all real numbers; for checking, we'll try -1 and 7. If we substitute -1 for x, we have

$$\frac{-1+3}{6}-\frac{2(-1)-3}{9} \stackrel{?}{=} \frac{5}{6}-\frac{(-1)}{18}$$

$$\frac{8}{9}=\frac{8}{9}$$ Verify that both sides simplify to $\frac{8}{9}$

If we substitute 7 for x, we have

$$\frac{7+3}{6}-\frac{2(7)-3}{9} \stackrel{?}{=} \frac{5}{6}-\frac{(7)}{18}$$

$$\frac{4}{9}=\frac{4}{9}$$ Verify that both sides simplify to $\frac{4}{9}$

The solution set is $\{x \mid x \in R\}$—the set of all real numbers. The graph of the solution set of this equation is

−5 −4 −3 −2 −1 0 1 2 3 4 5 ■

Example 5 Solve $4(x-1)-3(4-x)=7x-13$.

Solution

$$\begin{aligned} 4(x-1)-3(4-x) &= 7x-13 \\ 4x-4-12+3x &= 7x-13 \\ 7x-16 &= 7x-13 \\ -7x \quad & \quad -7x \\ -16 &= -13 \end{aligned}$$

Adding $-7x$ to both sides

A false statement

Note that no variable appears in the last step. Since the variable disappeared completely and we were left with a *false* statement, this equation is an equation with *no solution*. No number can be found that will make the two sides of the equation equal.

The solution set is the empty set, { }. If we had been asked to graph the solution set, there would be no points to graph. ■

EXERCISES 2.1

Set I In each exercise, find the solution set. Identify any equation that is not a conditional equation as either an identity or an equation with no solution. Graph the solution set of each conditional equation on the real number line.

1. $4x+3(4+3x)=-1$

2. $5x+2(4+x)=-6$

3. $7y-2(5+4y)=8$

4. $5y-3(4+2y)=3$

5. $4x+12=2(6+2x)$

6. $6x+15=3(5+2x)$

7. $3[5-2(5-z)]=2(3z+7)$

8. $5[4-3(2-z)]=3(5z+6)$

9. $\frac{x}{3}-\frac{x}{6}=18$

10. $\frac{x}{4}-\frac{x}{8}=16$

11. $\frac{y+3}{8} - \frac{3}{4} = \frac{y+6}{10}$

12. $\frac{y+7}{12} - \frac{5}{6} = \frac{y+4}{9}$

13. $\{z \mid 5z - 3(2 + 3z) = 6, z \in N\}$

14. $\{z \mid 8z + 6 = 2(7z + 9), z \in N\}$

15. $\{x \mid 7x - 2(5 + 4x) = 8, x \in J\}$

16. $\{x \mid 7x + 15 = 3(3x + 5), x \in J\}$

17. $\{x \mid 2(3x - 6) - 3(5x + 4) = 5(7x - 8), x \in J\}$

18. $\{z \mid 4(7z - 9) - 7(4z + 3) = 6(9z - 10), z \in J\}$

19. $\frac{2(y-3)}{5} - \frac{3(y+2)}{2} = \frac{7}{10}$

20. $\frac{5(x-4)}{6} - \frac{2(x+4)}{9} = \frac{5}{18}$

In Exercises 21 and 22, use a calculator and round off each solution to two decimal places.

21. $6.23x + 2.5(3.08 - 8.2x) = -14.7$

22. $9.84 - 4.6x = 5.17(9.01 - 8.23x)$

Set II In each exercise, find the solution set. Identify any equation that is not a conditional equation as either an identity or an equation with no solution. Graph the solution set of each conditional equation on the real number line.

1. $6x + 3(2 + 2x) = -6$

2. $7x + 5(2 + x) = -2$

3. $8y - 3(2 + 3y) = 5$

4. $8y + 4(6 - 2y) = -2$

5. $8x + 12 = 4(2x + 3)$

6. $14x - 3(4 - x) = 2(x + 9)$

7. $3[15 - (5 - 3x)] = 9(x + 18)$

8. $3 + 5(x - 4) = 9x + 1$

9. $\frac{z}{9} - \frac{z}{2} = \frac{7}{3}$

10. $\frac{x}{5} - \frac{x}{10} = 8$

11. $\frac{x+1}{4} - \frac{5}{8} = \frac{x-1}{8}$

12. $\frac{x+3}{5} - \frac{x}{2} = \frac{5-3x}{10}$

13. $\{x \mid 4x + 5(-4 - 5x) = 22, x \in N\}$

14. $\{x \mid 3x - 4(2 - x) = 6, x \in N\}$

15. $\{y \mid 10 - 7y = 4(11 - 6y), y \in J\}$

16. $\{y \mid -37 - 9(y + 1) = 4(2y - 3), y \in J\}$

17. $\{x \mid 6(3x - 5) = 3[4(1 - x) - 7], x \in J\}$

18. $\{x \mid 16 - 8(2x - 2) = 5(5x - 10), x \in N\}$

19. $\frac{2(x-3)}{3} - \frac{3(x+2)}{4} = \frac{5}{6}$

20. $\frac{3(x-4)}{4} - \frac{5(x+2)}{6} = \frac{7}{12}$

In Exercises 21 and 22, use a calculator and round off each solution to two decimal places.

21. $7.02(5.3x - 4.28) = 11.6 - 2.94x$

22. $7.23 - 6.1x = 3.2(1.08 - 5.3x)$

2.2 Simple First-Degree Inequalities, Their Solutions, and Their Graphs

In this section, we discuss simple conditional first-degree inequalities that have the symbols $<$, $>$, $\leq$, or $\geq$ in them. **Simple** (or **singular**) **inequalities** are inequalities that contain only *one* statement with only *one* inequality symbol.

The Sense of an Inequality

The **sense** of an inequality symbol refers to whether the inequality symbol is a *greater than* symbol or a *less than* symbol.

$\left.\begin{matrix} a > b \\ c > d \end{matrix}\right\}$	Same sense (both are >)	$\left.\begin{matrix} a < b \\ c > d \end{matrix}\right\}$	Opposite sense (one is >, one is <)
$\left.\begin{matrix} a \geq b \\ c \leq d \end{matrix}\right\}$	Opposite sense	$\left.\begin{matrix} a \leq b \\ c \leq d \end{matrix}\right\}$	Same sense

Properties of Inequalities

In Section 2.1, we solved equations by adding the same number to both sides of the equation, multiplying both sides of the equation by the same number, and so on. We solve simple *inequalities* by using any of the following properties of inequalities:

THE ADDITION AND SUBTRACTION PROPERTIES OF INEQUALITIES

If an inequality contains one of the symbols >, <, ≥, or ≤, the sense of the inequality is unchanged if the same number is added to or subtracted from both sides of the inequality. For example,

$$\text{if } a < b, \text{ then } a + c < b + c \text{ and } a - c < b - c$$

Senses are the same

where a, b, and c are real numbers.

THE MULTIPLICATION AND DIVISION PROPERTIES OF INEQUALITIES

If an inequality contains one of the symbols >, <, ≥, or ≤, the sense of the inequality is unchanged if both sides of the inequality are multiplied or divided by the same *positive* number. For example,

$$\text{if } a < b, \text{ and } c > 0, \text{ then } ac < bc \text{ and } \frac{a}{c} < \frac{b}{c}$$

where a, b, and c are real numbers.

However, the sense of the inequality is *changed* if both sides of the inequality are multiplied or divided by the same *negative* number. For example,

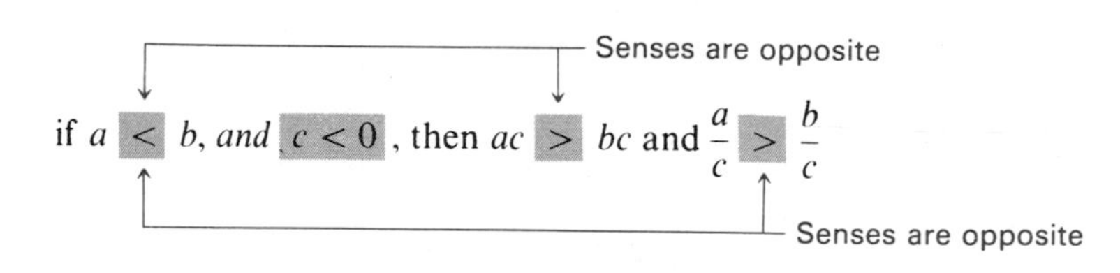

where a, b, and c are real numbers.

Example 1 Examples verifying that the properties of inequalities are valid for the inequality $8 < 12$:

a. Add 5 to both sides.

$$\begin{array}{rcr} 8 & < & 12 \\ +5 & & +5 \\ \hline 13 & ? & 17 \\ 13 & < & 17 \end{array}$$

The sense of the inequality is *unchanged* when we add the same number to both sides of the inequality.

b. Subtract 9 from both sides.

$$\begin{array}{rcr} 8 & < & 12 \\ -9 & & -9 \\ \hline -1 & ? & 3 \\ -1 & < & 3 \end{array}$$

The sense of the inequality is *unchanged* when we subtract the same number from both sides of the inequality.

c. Multiply both sides by 3 (a *positive* number).

$$\begin{array}{rcr} 8 & < & 12 \\ \times 3 & & \times 3 \\ \hline 24 & ? & 36 \\ 24 & < & 36 \end{array}$$

The sense of the inequality is *unchanged* when we multiply. both sides of the inequality by the same *positive* number.

d. Divide both sides by 2 (a *positive* number).

$$8 < 12$$

$$\frac{8}{2} ? \frac{12}{2}$$

$$4 ? 6$$

$$4 < 6$$

The sense of the inequality is *unchanged* when we divide both sides of the inequality by the same *positive* number.

e. Multiply both sides by -3 (a *negative* number).

$$\begin{array}{rcr} 8 & < & 12 \\ \times(-3) & & \times(-3) \\ \hline -24 & ? & -36 \\ -24 & > & -36 \end{array}$$

Senses are opposite

The sense of the inequality is *changed* when we multiply both sides of the inequality by the same *negative* number.

f. Divide both sides by -2 (a *negative* number).

$$8 < 12$$

$$\frac{8}{-2} ? \frac{12}{-2}$$

$$-4 ? -6$$

$$-4 > -6$$

Senses are opposite ■

The sense of the inequality is *changed* when we divide both sides of the inequality by the same *negative* number.

Solutions and Solution Sets of an Inequality

A *solution* of a conditional inequality is any number that, when substituted for the variable, makes the inequality a true statement.

The *solution set* of an inequality is the set of all numbers that are solutions of the inequality. While the solution set of a first-degree conditional *equation* normally has just one element in it, the solution set of an *inequality* usually contains infinitely many numbers.

Solving an Inequality

When we solve a simple first-degree inequality, we must find *all* of the values of the variable that satisfy the inequality. Therefore, our final inequality will be in the form $x < a$, $x > a$, $x \leq a$, or $x \geq a$.

The method used to solve inequalities may be summarized as follows:

> TO SOLVE AN INEQUALITY
>
> Proceed in the same way used to solve equations, with the exception that the sense must be changed when multiplying or dividing both sides by a negative number.

A WORD OF CAUTION We solve an inequality by a method very much like the method used for solving an equation. For this reason, some students confuse the solution of an inequality with that of an equation.

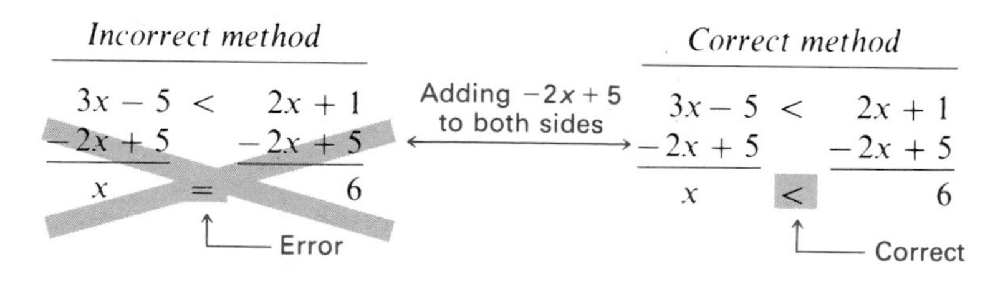

Remember that when we add the same number to both sides of an inequality we get an *inequality* with the sense unchanged. We do *not* get an equation. An infinite number of numbers satisfy the above inequality, since any real number less than 6 is a solution. For example, 3, −6, −201, $5\frac{3}{4}$, and 5. 999 are all solutions. The graph of the solution set for $x < 6$ is shown below.

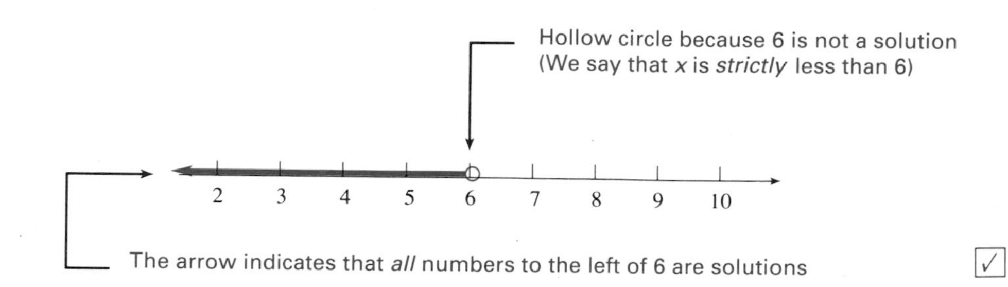

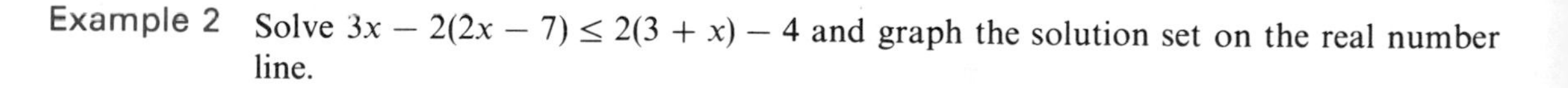

Example 2 Solve $3x - 2(2x - 7) \leq 2(3 + x) - 4$ and graph the solution set on the real number line.

Solution $3x - 2(2x - 7) \leq 2(3 + x) - 4$ — Note that no restrictions were put on the domain of the variable

$$3x - 4x + 14 \leq 6 + 2x - 4$$

$$\begin{array}{rcr} -x + 14 & \leq & 2 + 2x \\ x - 2 & & -2 + x \\ \hline 12 & \leq & 3x \end{array}$$

← Adding $x - 2$ to both sides to get the x-term on one side and the constant on the other*

$$\frac{12}{3} \leq \frac{3x}{3}$$

The sense is not changed if we divide both sides by 3

$$4 \leq x$$

or

$$x \geq 4$$

Remember: $4 \leq x$ can be replaced by $x \geq 4$

The solution set is $\{x \mid x \geq 4\}$.

Because the solution set of the inequality is the set of all real numbers *greater than or equal to* 4, when we graph the solution set we must have a solid circle at the 4 and must shade in that part of the number line that lies to the right of 4.

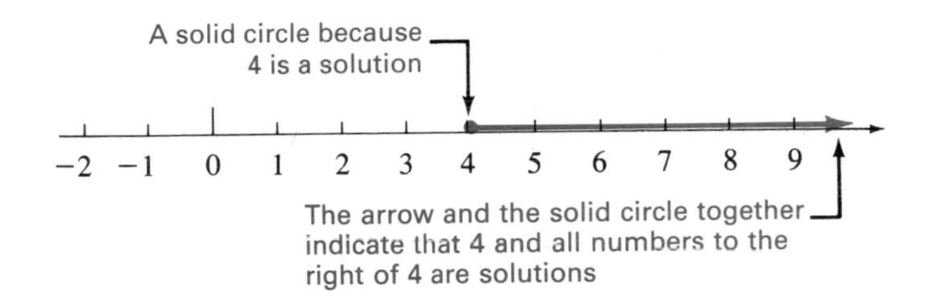

■

Example 3 Solve $\frac{x + 2}{15} > \frac{x + 3}{5} - \frac{1}{3}$ and graph the solution set.

Solution

$$\frac{x + 2}{15} > \frac{x + 3}{5} - \frac{1}{3}$$

$$\frac{15}{1} \cdot \frac{x + 2}{15} > \frac{15}{1} \cdot \left(\frac{x + 3}{5} - \frac{1}{3}\right)$$

The LCM of 15, 5, and 3 is 15

$$x + 2 > 3(x + 3) - 5$$

$$x + 2 > 3x + 9 - 5$$

$$\begin{array}{rcr} x + 2 & > & 3x + 4 \\ -x - 4 & & -x - 4 \\ \hline -2 & > & 2x \end{array}$$

$$\frac{-2}{2} > \frac{2x}{2}$$

$$-1 > x \leftarrow \text{Equivalent statements}$$

$$x < -1$$

Alternate solution

or →

$$\begin{array}{rcr} x + 2 & > & 3x + 4 \\ -3x - 2 & & -3x - 2 \\ \hline -2x & > & 2 \end{array}$$

$$\frac{-2x}{-2} < \frac{2}{-2}$$

← The sense of the inequality changes when we divide both sides by −2

$$x < -1$$

The solution set is $\{x \mid x < -1\}$.

---

* We could have added $-2x - 14$ to both sides rather than $x - 2$.

To graph the solution set on the real number line, we put a *hollow* circle at -1 to indicate that -1 itself is *not* to be included, but that numbers such as -1.1, -1.01, -1.00001, and so on *are* to be included. We heavily shade in all of the number line to the *left* of -1.

−6 −5 −4 −3 −2 −1 0 1

■

Example 4 Rewrite $\{x \mid 4(3x - 5) < 10, x \in N\}$ in roster notation and graph it on the real number line.

Solution We must first solve $4(3x - 5) < 10$.

$$12x - 20 < 10$$

$$12x < 30 \quad \text{Adding 20 to both sides}$$

$$x < \frac{30}{12} \quad \text{Dividing both sides by 12}$$

$$x < \frac{5}{2}$$

The *natural numbers* that are less than 5/2 are 1 and 2. Therefore,

$$\{x \mid 4(3x - 5) < 10, x \in N\} = \{1, 2\}$$

The graph is

−1 0 1 2 3 4 5

■

EXERCISES 2.2

Set I

In Exercises 1–18, solve each inequality and graph the solution set on the real number line.

1. $3x - 1 < 11$

2. $7x - 12 < 30$

3. $17 \geq 2x - 9$

4. $33 \geq 5 - 4x$

5. $2y - 16 > 17 + 5y$

6. $6y + 7 > 4y - 3$

7. $4z - 22 < 6(z - 7)$

8. $8(a - 3) > 15a - 10$

9. $9(2 - 5m) - 4 \geq 13m + 8(3 - 7m)$

10. $18k - 3(8 - 4k) \leq 7(2 - 5k) + 27$

11. $10 - 5x > 2[3 - 5(x - 4)]$

12. $3[2 + 4(y + 5)] < 30 + 6y$

13. $\frac{z}{3} > 7 - \frac{z}{4}$

14. $\frac{t}{5} - 8 > -\frac{t}{3}$

15. $\frac{1}{3} + \frac{w + 2}{5} \geq \frac{w - 5}{3}$

16. $\frac{u - 2}{3} - \frac{u + 2}{4} \geq -\frac{2}{3}$

In Exercises 17 and 18, use a calculator and round off each answer to three decimal places.

17. $14.73(2.65x - 11.08) - 22.51x \geq 13.94x(40.27)$

18. $1.065 - 9.801x \leq 5.216x - 2.740(9.102 - 7.641x)$

In Exercises 19–22, rewrite the set in roster notation and graph it on the real number line.

19. $\{x \mid x + 3 < 10, x \in N\}$ **20.** $\{x \mid x + 5 < 8, x \in N\}$

21. $\{x \mid 2(x + 3) \leq 11, x \in N\}$ **22.** $\{x \mid 3(x + 1) \leq 17, x \in N\}$

In Exercises 23–26, write, in set-builder notation, the algebraic statement that describes the set of numbers graphed.

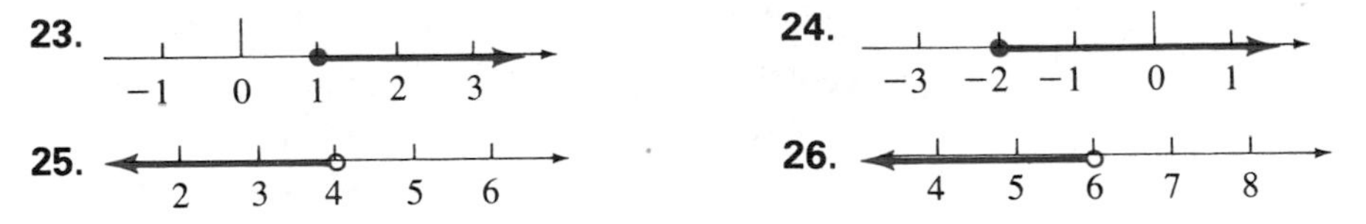

Set II

In Exercises 1–18, solve the inequality and graph the solution set on the real number line.

1. $-3 \leq x + 4$ **2.** $3x + 2 \geq 8$

3. $18 - 7y > -3$ **4.** $5 - 3y \leq 8$

5. $11z - 7 < 5z - 13$ **6.** $3x - 8 \leq 7 - 2x$

7. $3(2 + 3x) \geq 5x - 6$ **8.** $4(x - 1) \leq 7x + 2$

9. $6(10 - 3t) + 25 \geq 4t - 5(3 - 2t)$ **10.** $6x - 4(3 - 2x) > 5(3 - 4x) + 7$

11. $6z < 2 - 4[2 - 3(z - 5)]$ **12.** $28 - 7x \geq [6 - 2(x - 1)]$

13. $\dfrac{w}{3} > 12 - \dfrac{w}{6}$ **14.** $\dfrac{x}{2} \leq 5 - \dfrac{x}{3}$

15. $\dfrac{1}{2} + \dfrac{u + 9}{5} \geq \dfrac{u + 1}{2}$ **16.** $\dfrac{x - 4}{2} - \dfrac{x + 4}{4} < -\dfrac{3}{2}$

In Exercises 17 and 18, use a calculator and round off each answer to three decimal places.

17. $54.7x - 48.2(20.5 - 37.6x) \leq 81.9(60.3x - 19.1) + 97.4$

18. $3.7 - 1.06x < 8.62 - 1.4(6.2 - 3.2x)$

In Exercises 19–22, rewrite the set in roster notation and graph it on the real number line.

19. $\{x \mid x + 2 < 8, x \in N\}$ **20.** $\{x \mid x - 1 < 4, x \in N\}$

21. $\{x \mid 4(x + 4) \leq 42, x \in N\}$ **22.** $\{x \mid 3(x + 5) < 27, x \in N\}$

In Exercises 23–26, write, in set-builder notation, the algebraic statement that describes the set of numbers graphed.

2.3 Combined Inequalities and Their Graphs

Combined Inequalities

Combined inequalities result when we connect two or more simple inequalities with the words *or* or *and.*

Example 1 Examples of combined inequalities:

a. $x > 3$ or $x \leq -1$ b. $x \leq -1$ and $x \geq -4$ c. $-3 < x < 4$

A *continued inequality*, first discussed in Section 1.3B, is a combined inequality, because the statement $a < x < b$ is equivalent to the compound statement $a < x$ *and* $x < b$. If $a < b$, then $a < x < b$ is a valid continued inequality, and so is $b > x > a$. ■

Intervals and Interval Notation The set $\{x \mid a < x < b\}$ (that is, the set of *all real numbers* between a and b) is called an **open interval** because neither endpoint is included; the set $\{x \mid c \leq x \leq d\}$ is called a **closed interval** because both endpoints *are* included. It is possible to use a notation called **interval notation** to describe an interval; we do this by writing a *pair* of numbers (with the smaller number *always* on the left) separated by a comma and enclosed within brackets and/or parentheses; a *bracket* indicates that an endpoint is included, while a *parenthesis* indicates that an endpoint is not included. Therefore, interval notation for the open interval $\{x \mid a < x < b\}$ is (a, b) and interval notation for the closed interval $\{x \mid c \leq x \leq d\}$ is $[c, d]$. The intervals $\{x \mid f < x \leq g\}$ and $\{x \mid j \leq x < k\}$ are called **half-open** (or **half-closed**) **intervals**; interval notation for $\{x \mid f < x \leq g\}$ is $(f, g]$, and for $\{x \mid j \leq x < k\}$ it is $[j, k)$.

We use a similar notation for sets such as $\{x \mid x > a\}$ and $\{x \mid x \leq b\}$, making use of the symbols for positive infinity, $+\infty$, and negative infinity, $-\infty$. (The infinity symbols are *not* symbols for real numbers and cannot be treated as such.) We consider $-\infty$ to be smaller than any real number and $+\infty$ to be larger than any real number, and we *always* use a parenthesis—never a bracket—with an infinity symbol. Therefore, interval notation for $\{x \mid x > a\}$ is $(a, +\infty)$, or (a, ∞); for $\{x \mid x \leq b\}$, it is $(-\infty, b]$. Interval notation for the set of *all* real numbers is $(-\infty, +\infty)$.

Writing Algebraic Statements to Describe a Graph When we are asked to write an algebraic statement to describe a graph, we can use set notation or (sometimes) interval notation. (We can*not* use interval notation to describe a graph if the graph consists of separate *points.*)

When we're using set notation and when the graph to be described is an *unbroken line segment*, we can describe the graph with a *continued inequality* or with *interval notation* (see Example 2).

Example 2 Write the algebraic statement that describes the set of numbers that is graphed (a) using set notation and (b) using interval notation.

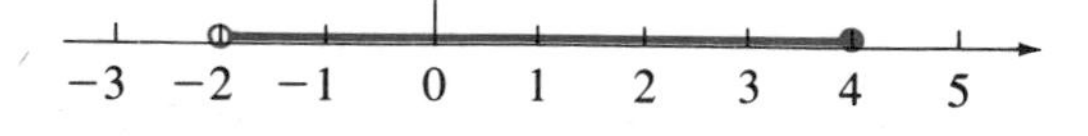

Solution

a. The graph is an unbroken line segment; the set is $\{x \mid -2 < x \leq 4\}$ which can be written as $\{x \mid x > -2\} \cap \{x \mid x \leq 4\}$.

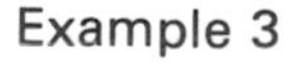

b. In interval notation: $(-2, 4]$. ■

Example 3 Write the algebraic statement that describes the set of numbers that is graphed (a) using set notation and (b) using interval notation.

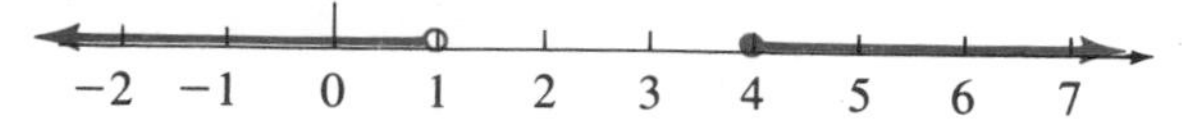

Solution

a. The graph occupies two separate portions of the number line, so we must write the algebraic description as *two separate statements*. The set graphed here is $\{x \mid x < 1\} \cup \{x \mid x \geq 4\}$. This set could also be written as $\{x \mid x < 1$ *or* $x \geq 4\}$.

b. In interval notation: $(-\infty, 1) \cup [4, +\infty)$. ■

A WORD OF CAUTION Students often write the algebraic statement for the graph of Example 3 as $\{x \mid 1 > x \geq 4\}$, which is incorrect. The statement $1 > x \geq 4$ is an *invalid* inequality, because $1 \geq 4$ is a false statement.

Students also often write the algebraic statement for the graph of Example 3 as $\{x \mid 1 < x \geq 4\}$ or as $\{x \mid 1 > x \leq 4\}$. These are incorrect, because in a continued inequality both inequalities must have the *same sense*. ☑

Solution Sets and Graphs of Combined Inequalities The solution set of a combined inequality containing the word *or* is the *union* of the solution sets of the two simple inequalities.

Example 4 Find and graph the solution set for $x - 1 > 2$ *or* $x - 1 \leq -2$.

Solution We must solve each inequality:

$$x - 1 > 2 \quad \text{or} \quad x - 1 \leq -2$$
$$x > 3 \quad \text{or} \quad x \leq -1$$

The solution set is $\{x \mid x > 3$ or $x \leq -1\}$, which can also be written $\{x \mid x > 3\} \cup \{x \mid x \leq -1\}$, or, in interval notation, $(-\infty, -1] \cup (3, +\infty)$. We must graph all the numbers greater than 3 or less than or equal to -1.

−3 −2 −1 0 1 2 3 4 5 6 ■

The solution set of a combined inequality containing the word *and* is the *intersection* of the solution sets of the two simple inequalities; likewise, the solution set of a continued inequality is the intersection of the solution sets of the two simple inequalities.

Example 5 Graph the solution set for $x > -2$ *and* $x < 4$.

Solution We must find the set of all the numbers greater than -2 *and at the same time* less than 4. We find this set by graphing each inequality separately and then finding where the two graphs overlap.

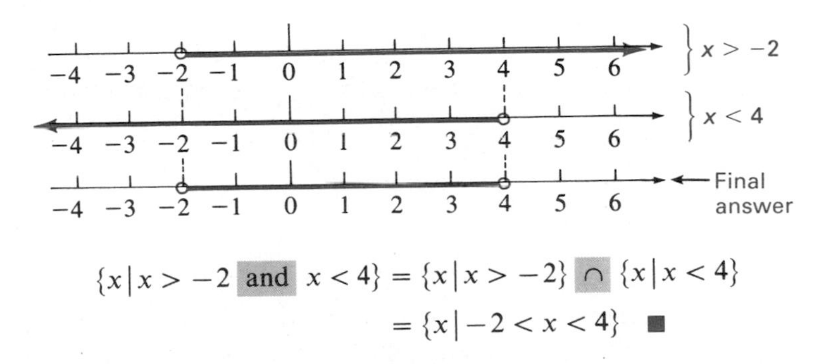

$$\{x \mid x > -2 \text{ and } x < 4\} = \{x \mid x > -2\} \cap \{x \mid x < 4\}$$
$$= \{x \mid -2 < x < 4\} \quad ■$$

If $a > b$, then $a < x < b$ is an *invalid* continued inequality, and the solution set for the inequality is the empty set (see Example 6).

Example 6 Graph the set $\{x \mid 7 < x < 2\}$.

Solution The statement $7 < x < 2$ is an *invalid* statement because $7 < 2$ is *false*. The solution set is $\{\ \}$, and the graph has no points on it. ■

Example 7 Solve $x > 1$ *or* $x < 4$ graphically, and describe the solution set in set-builder notation.

Solution We graph each inequality separately and then include in our final answer any number that is in one graph *or* the other *or* both.

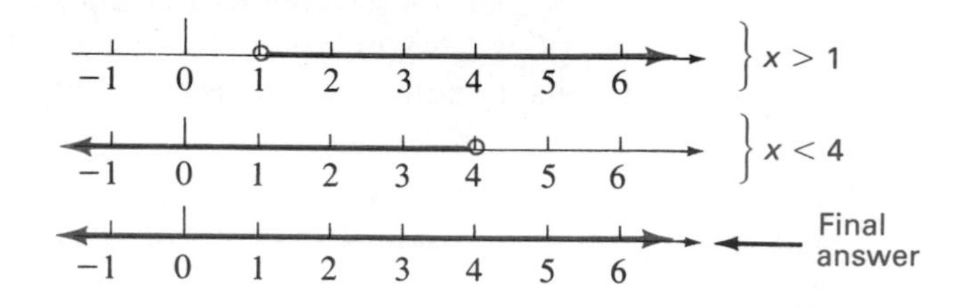

The solution set for $x > 1$ or $x < 4$ is $\{x \mid x \in R\}$. ■

Solving Continued Inequalities

When we solve a continued inequality, we want to rewrite the inequality so that x is all by itself between the two inequality symbols. That is, we want our answer to be in the form $a < x < b$ (if $a < b$) or $a > x > b$ (if $a > b$).

Example 8 Find and graph the solution set for $2 < x + 5 \leq 9$.

Solution Notice first of all that $2 \leq 9$ is a true statement, so the continued inequality is valid. Also notice that no restrictions were put on the domain of the variable.

Since $\{x \mid 2 < x + 5 \leq 9\} = \{x \mid x + 5 > 2\} \cap \{x \mid x + 5 \leq 9\}$, we could first find the solution set of each of the inequalities separately and then find the intersection of the two solution sets. However, the continued inequality may be more conveniently solved as follows:

$$\begin{array}{rcl} 2 < x + 5 \leq & 9 & \\ -5 \quad\ \ -5 & -5 & \text{Adding } -5 \text{ to all three parts of the inequality} \\ \hline -3 < x \quad\ \leq & 4 & \end{array}$$

The solution set is $\{x \mid -3 < x \leq 4\}$, or, in interval notation, $(-3, 4]$. The graph will include the 4 (we will have a solid circle at 4) and the set of all points on the number line that lie between -3 and 4. We will have a hollow circle at -3.

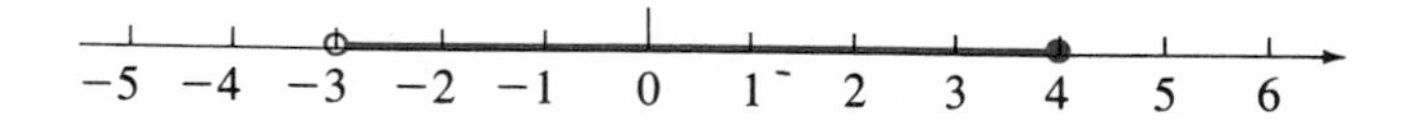

■

Example 9 Solve $-7 \leq 2x + 1 \leq 5$ and graph the solution set on the number line.

Solution

$$\begin{array}{rcl} -7 \leq 2x + 1 \leq & 5 & \\ -1 \qquad\ -1 & -1 & \text{Adding } -1 \text{ to all three parts of the inequality} \\ \hline -8 \leq 2x \qquad \leq & 4 & \\ -4 \leq x \qquad\ \leq & 2 & \leftarrow \text{Dividing all parts of a continued inequality by the } \textit{positive} \text{ number 2 does } \textit{not} \text{ change the sense of the inequality} \end{array}$$

The solution set is $\{x \mid -4 \leq x \leq 2\}$, or, in interval notation, $[-4, 2]$, and its graph is as follows:

■

Example 10 Write $\{x \mid -7 \leq 2x + 1 \leq 5,\ x \in N\}$ in roster notation and graph the set.

Solution The inequality is the same inequality as in Example 9, but now a restriction has been put on the domain. The only *natural numbers* that satisfy the inequality $-4 \leq x \leq 2$ are 1 and 2; therefore,

$$\{x \mid -7 \leq 2x + 1 \leq 5, x \in N\} = \{1, 2\}$$

■

NOTE We could *not* use interval notation for the result in Example 10, since the solution set does not include *all real numbers* between 1 and 2. ✓

EXERCISES 2.3

Set I In Exercises 1–8, write, in set-builder notation and in interval notation, the algebraic statement that describes the set that is graphed.

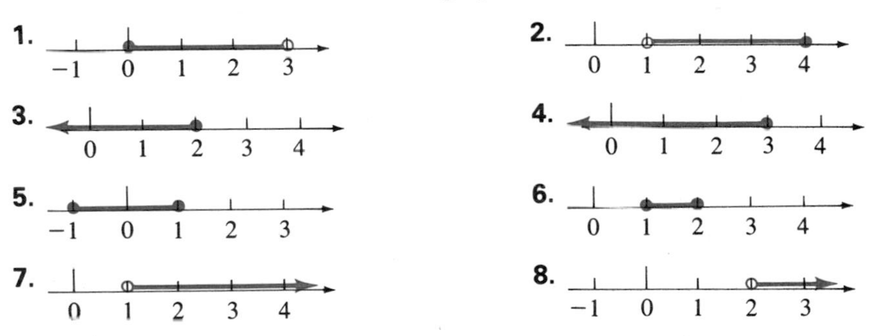

In Exercises 9–22, solve the inequalities, writing the solutions in interval notation, and graph their solution sets.

9. $5 > x - 2 \geq 3$

10. $7 > x - 3 \geq 4$

11. $-5 \geq x - 3 \geq 2$

12. $-3 \geq x - 2 \geq 4$

13. $-4 < 3x - 1 \leq 7$

14. $-6 < 4x - 2 \leq 5$

15. $x - 1 > 3$ or $x - 1 < -3$

16. $x - 2 > 5$ or $x - 2 < -5$

17. $2x + 1 \geq 3$ or $2x + 1 \leq -3$

18. $3x - 2 \geq 5$ or $3x - 2 \leq -5$

19. $x > 4$ and $x \geq 2$

20. $x < 3$ and $x < -1$

21. $x > 4$ or $x \geq 2$

22. $x < 3$ or $x < -1$

In Exercises 23–28, rewrite the inequalities in roster notation and graph.

23. $\{x \mid -5 \leq x - 3 \leq 2, x \in N\}$

24. $\{x \mid -3 \leq x - 2 \leq 4, x \in N\}$

25. $\{x \mid 4 \geq x - 3 > -5, x \in J\}$

26. $\{x \mid 6 \geq x - 2 > -4, x \in J\}$

27. $\{x \mid -3 \leq 2x + 1 \leq 7, x \in N\}$

28. $\{x \mid -5 \leq 2x + 3 \leq 5, x \in N\}$

Set II In Exercises 1–8, write, in set-builder notation and in interval notation, the algebraic statement that describes the set that is graphed.

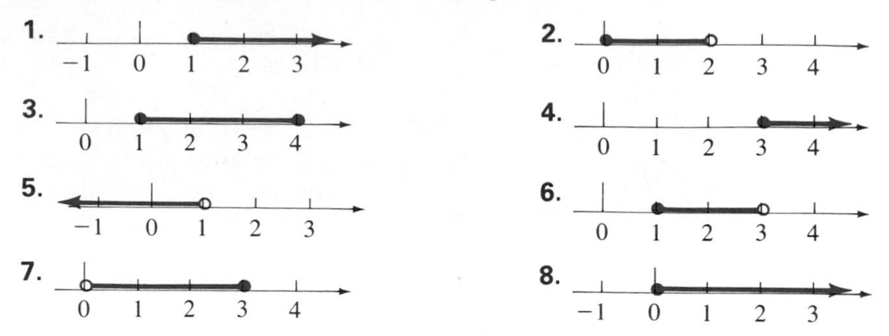

In Exercises 9–22, solve the inequalities, writing the solutions in interval notation, and graph their solution sets.

9. $8 > x - 1 \geq 2$

10. $-3 < x + 2 < 5$

11. $-6 \geq x - 4 \geq 3$

12. $7 > x + 3 \geq -1$

13. $-5 < 3x - 2 \leq 4$

14. $0 \leq 2x + 4 < 8$

15. $x - 3 > 4$ or $x - 3 < -4$

16. $x + 5 > 1$ or $x + 5 < -1$

17. $4x - 1 \geq 2$ or $4x - 1 \leq -2$

18. $3x + 2 > 5$ or $3x + 2 < -5$

19. $x < 6$ and $x < -2$

20. $x > 3$ and $x < 6$

21. $x < 6$ or $x < -2$

22. $x > 3$ or $x < 6$

In Exercises 23–28, rewrite the inequalities in roster notation and graph.

23. $\{x \mid -4 \geq x - 2 > 1, x \in N\}$

24. $\{x \mid 2 \leq x + 4 < 8, x \in N\}$

25. $\{x \mid 5 \geq x - 2 > -3, x \in J\}$

26. $\{x \mid -2 \leq x - 1 < 4, x \in J\}$

27. $\{x \mid -2 \leq 2x + 4 \leq 10, x \in N\}$

28. $\{x \mid -6 < 2x - 3 \leq 5, x \in J\}$

2.4 Conditional Equations and Inequalities with Absolute Value Symbols

In this section, we consider conditional equations and inequalities that contain absolute value symbols. Recall that the definition of $|N|$ is

$$|N| = \begin{cases} N & \text{if } N \geq 0 \\ -N & \text{if } N < 0 \end{cases}$$

Equations: $|N| = a$

We first consider equations of the form $|N| = a$, where $a \geq 0$.* Rule 2.1 (which is proved in Appendix A) permits us to rewrite such equations with no absolute value symbols in them.

* If $a < 0$, then $|N| = a$ will have no solution, because no positive number can equal a negative number.

RULE 2.1

If $a \geq 0$, and

$$\text{if} \quad |N| = a, \quad \text{then} \quad N = a \quad \text{or} \quad N = -a$$

where N is any algebraic expression.

Example 1 Solve $\left|\dfrac{3 - 2x}{5}\right| = 2$ and graph its solution.

Solution Because the equation is in the form $|N| = a$, we can use Rule 2.1. According to Rule 2.1, if

$$\left|\frac{3 - 2x}{5}\right| = 2,$$

then

$$\frac{3 - 2x}{5} = 2 \quad \text{or} \quad \frac{3 - 2x}{5} = -2$$

$$3 - 2x = 10 \quad \text{or} \quad 3 - 2x = -10$$

$$-2x = 7 \quad \text{or} \quad -2x = -13$$

$$x = -\frac{7}{2} \quad \text{or} \quad x = \frac{13}{2}$$

Therefore, the solution set is $\{-3\frac{1}{2}, 6\frac{1}{2}\}$. The graph of the solution set is

−4 −3 −2 −1 0 1 2 3 4 5 6 7 8

(The check is left to the student.) ■

Example 2 Solve $|x| = -3$.

Solution We know that $|x| \geq 0$, and no number that is greater than or equal to 0 can equal -3. Therefore, there is *no* solution. ■

Inequalities: $|N| < a$

We now consider conditional inequalities of the form $|N| < a$ or $|N| \leq a$. Rule 2.2 (which is proved in Appendix A) permits us to rewrite such inequalities as continued inequalities with no absolute value symbols in them.

RULE 2.2

If a is a positive real number,* and

$$\text{if} \quad |N| < a, \quad \text{then} \quad -a < N < a$$

$$\text{if} \quad |N| \leq a, \quad \text{then} \quad -a \leq N \leq a$$

where N is any algebraic expression. The graph of the solution set will *always* be an uninterrupted portion of the number line.

* If $a < 0$ in Rule 2.2, then the solution set is the empty set. For example, $|x| < -2$ has no solution, since $|x|$ is always greater than or equal to 0 and cannot be < -2.

Example 3 Find the solution set for $|x| < 3$.

Solution Using Rule 2.2, we rewrite $|x| < 3$ as $-3 < x < 3$, and this continued inequality has been solved.

If we consider the absolute value to mean the distance between a number and zero, the solution set of the inequality $|x| < 3$ will be the set of all real numbers whose distance from zero is less than 3 units (see Figure 2.4.1).

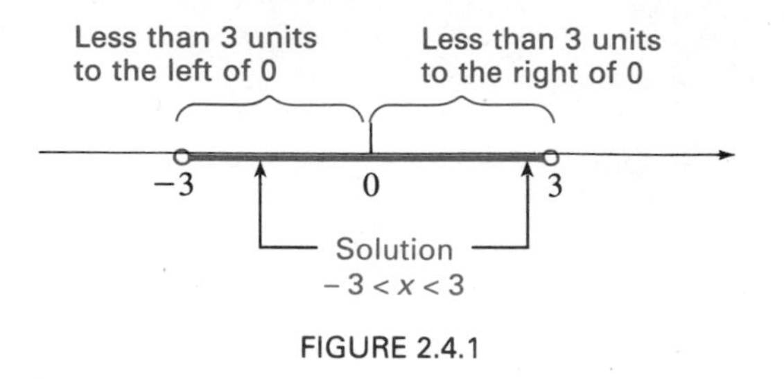

FIGURE 2.4.1

The solution set is $\{x \mid -3 < x < 3\}$ or $(-3, 3)$. ■

Example 4 Solve $|2x - 3| \leq 5$ and graph its solution set.

Solution We use Rule 2.2 because our inequality is in the form $|N| \leq a$. Rule 2.2 instructs us to rewrite our inequality as

$$-5 \leq 2x - 3 \leq 5$$

We then solve that inequality as follows:

$$-5 \leq 2x - 3 \leq 5$$

$$-2 \leq 2x \leq 8 \qquad \text{Adding 3 to all three parts}$$

$$-1 \leq x \leq 4 \qquad \text{Dividing all three parts by 2}$$

The solution set is $\{x \mid -1 \leq x \leq 4\}$ or $[-1, 4]$. The graph is

−2 −1 0 1 2 3 4 5 ■

Example 5 Find the solution set for $|x + 3| < -5$.

Solution Since $|x + 3|$ is always positive or zero, it can't be less than -5. Therefore, the solution set is the empty set. ■

Example 6 Solve $\left|\frac{4 - 3x}{2}\right| \leq 6$ and graph its solution set.

Solution We use Rule 2.2, because the inequality is in the form $|N| \leq a$.

$$-6 \leq \frac{4 - 3x}{2} \leq 6$$

$$-12 \leq 4 - 3x \leq 12 \qquad \text{Multiplying all three parts by 2}$$

$$-16 \leq -3x \leq 8 \qquad \text{Adding } -4 \text{ to all three parts}$$

$$\frac{-16}{-3} \geq \frac{-3x}{-3} \geq \frac{8}{-3} \qquad \text{Dividing all three parts by } -3 \text{ changes the } \textit{sense}$$

$$\frac{16}{3} \geq x \geq -\frac{8}{3}$$

Therefore, the solution set is $\left\{x \,\middle|\, -\frac{8}{3} \le x \le \frac{16}{3}\right\}$ or $\left[-\frac{8}{3}, \frac{16}{3}\right]$. The graph of the solution set is

−4 −3 −2 −1 0 1 2 3 4 5 6 ■

Inequalities: $|N| > a$

Now let us examine conditional inequalities of the form $|N| > a$ or $|N| \ge a$. Rule 2.3, which follows, is proved in Appendix A.

> RULE 2.3
>
> If a is a positive real number,* and
>
> if $|N| > a$, then $N > a$ or $N < -a$
>
> if $|N| \ge a$, then $N \ge a$ or $N \le -a$
>
> where N is any algebraic expression. The graph of the solution set will *always* be along two separate portions of the number line.

We use Rule 2.3 in rewriting any inequality of the form $|N| > a$ or $|N| \ge a$ as a combined inequality with no absolute value symbols in it. We then solve the combined inequality according to the methods learned in Section 2.2 or 2.3.

Example 7 Express the solution set for $|x| > 3$ in set-builder notation.

Solution Using Rule 2.3 (because the inequality is in the form $|N| > a$), we rewrite $|x| > 3$ as $x > 3$ *or* $x < -3$.

If we consider the absolute value to mean the distance between a number and zero, then $|x| > 3$ can be interpreted as the set of points whose distance from zero is *greater than* 3 units (see Figure 2.4.2).

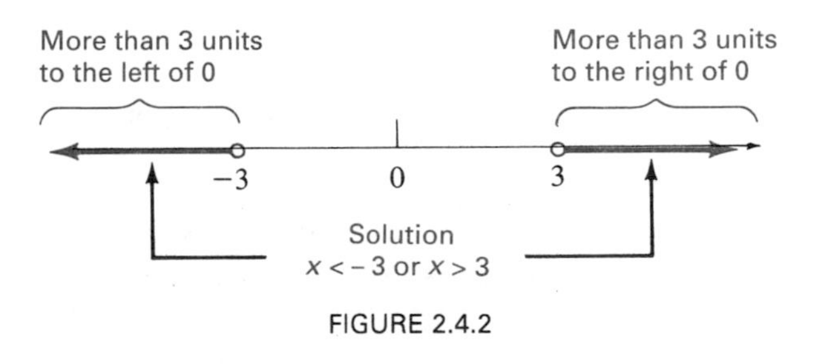

FIGURE 2.4.2

The solution set is the set of all real numbers less than -3 *or* greater than $+3$. In set-builder notation, the solution set is $\{x \mid x > 3 \text{ or } x < -3\}$, which can also be written as

$$\{x \mid x > 3\} \cup \{x \mid x < -3\} \quad ■$$

* If $a < 0$ in Rule 2.3, then the solution set is R. For example, if we solve $|x| > -2$, we can see that *any* real number will be a solution, since the absolute value of any real number is greater than or equal to zero, and zero is, of course, greater than -2. Thus, the solution set is the set of all real numbers.

Example 8 Express the solution set for $|5 - 2x| \geq 3$ in set-builder notation and graph the solution set.

Solution We use Rule 2.3, because the inequality is in the form $|N| \geq a$. Rule 2.3 instructs us to rewrite the original inequality as

$$5 - 2x \geq 3 \quad \text{or} \quad 5 - 2x \leq -3$$

It is *incorrect* to use the word *and* here

We then solve each of these inequalities as follows:

$$\begin{array}{rcl} 5 - 2x \geq 3 & \text{or} & 5 - 2x \leq -3 \\ -5 \quad\; -5 & & -5 \quad\; -5 \\ \hline -2x \geq -2 & \text{or} & -2x \leq -8 \\ \dfrac{-2x}{-2} \leq \dfrac{-2}{-2} & \text{or} & \dfrac{-2x}{-2} \geq \dfrac{-8}{-2} \\ x \leq 1 & \text{or} & x \geq 4 \end{array}$$

Dividing both sides by -2 changes the sense

The solution set is $\{x \mid x \leq 1\} \cup \{x \mid x \geq 4\}$. The graph of the solution set is

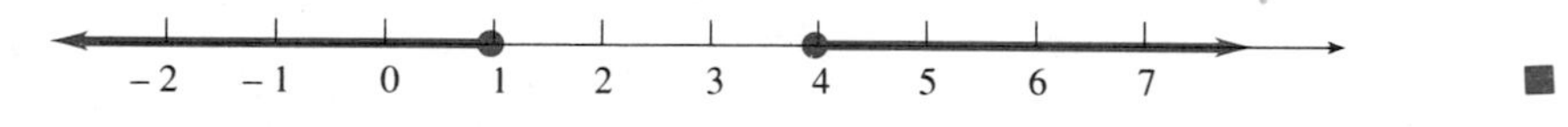

■

Example 9 Find the solution set for $|3x - 1| \geq -4$.

Solution Since $|3x - 1|$ is always positive or zero, it will *always* be greater than -4. Therefore, the solution set is the set of all real numbers; in interval notation, this is $(-\infty, +\infty)$. ■

Example 10 Solve $\left|3 - \dfrac{x}{2}\right| > 5$ and graph its solution set.

Solution We use Rule 2.3, since the inequality is in the form $|N| > a$. Rule 2.3 instructs us to rewrite the inequality as

$$\begin{array}{rcl} 3 - \dfrac{x}{2} > 5 & \text{or} & 3 - \dfrac{x}{2} < -5 \\ 2\left(3 - \dfrac{x}{2}\right) > 2(5) & \text{or} & 2\left(3 - \dfrac{x}{2}\right) < 2(-5) \\ 6 - x > 10 & \text{or} & 6 - x < -10 \\ -x > 4 & \text{or} & -x < -16 \\ x < -4 & \text{or} & x > 16 \end{array}$$

Dividing both sides by -1 changes the sense

The solution set is $\{x \mid x < -4\} \cup \{x \mid x > 16\}$. In interval notation, this is written $(-\infty, -4) \cup (16, +\infty)$. The graph of the solution set is

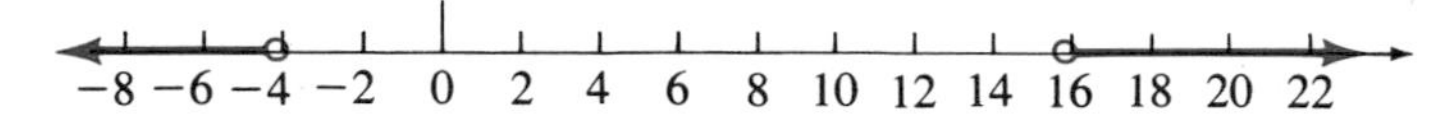

■

EXERCISES 2.4

Set I Solve the following equations and inequalities and graph their solution sets.

1. $\lvert x\rvert = 3$	**2.** $\lvert x\rvert = 5$	**3.** $\lvert 3x\rvert = 12$
4. $\lvert 2x\rvert = 10$	**5.** $\lvert x\rvert < 2$	**6.** $\lvert x\rvert < 7$
7. $\lvert 4x\rvert < 12$	**8.** $\lvert 3x\rvert < 9$	**9.** $\lvert 5x\rvert \leq 25$
10. $\lvert 2x\rvert \leq 2$	**11.** $\lvert x\rvert > 2$	**12.** $\lvert x\rvert > 3$

13. $|3x| \geq -3$ **14.** $|4x| \geq -5$ **15.** $|x + 2| = 5$

16. $|x + 3| = 7$ **17.** $|x - 3| < 2$ **18.** $|x - 4| < 1$

19. $|x + 4| \leq -3$ **20.** $|x + 2| \leq -4$ **21.** $|x + 1| > 3$

22. $|x - 2| > 4$ **23.** $|x + 5| \geq 2$ **24.** $|x + 4| \geq 1$

25. $|3x + 4| = 3$ **26.** $|4x + 3| = 5$ **27.** $|2x - 3| < 4$

28. $|3x - 1| < 5$ **29.** $|3x - 5| \geq 6$ **30.** $|4x - 1| \geq 3$

31. $|1 - 2x| \leq 5$ **32.** $|2 - 3x| < 4$ **33.** $|2 - 3x| > 4$

34. $|5 - 2x| \geq 6$ **35.** $\left|\frac{5x + 2}{3}\right| \geq 2$ **36.** $\left|\frac{4x + 3}{5}\right| \geq 3$

37. $\left|\frac{3x - 4}{5}\right| < 1$ **38.** $\left|\frac{5x - 1}{2}\right| < 2$ **39.** $\left|\frac{1 - x}{2}\right| = 6$

40. $\left|\frac{3 - x}{2}\right| = 4$ **41.** $\left|\frac{5 - x}{2}\right| \leq 7$ **42.** $\left|\frac{4 - x}{3}\right| \leq 3$

43. $\left|3 - \frac{x}{2}\right| > 4$ **44.** $\left|4 - \frac{x}{3}\right| > 1$ **45.** $\left|4 - \frac{x}{2}\right| < 3$

46. $\left|5 - \frac{x}{3}\right| < 2$

Set II Solve the following equations and inequalities and graph their solution sets.

1. $|x| = 4$ **2.** $|x| = -3$ **3.** $|4x| = 16$

4. $|x| = 0$ **5.** $|x| < 4$ **6.** $|x| < -1$

7. $|3x| < 12$ **8.** $|2x| \leq 4$ **9.** $|3x| \leq 15$

10. $|8x| < 12$ **11.** $|x| > 1$ **12.** $|x| > 0$

13. $|2x| \geq -6$ **14.** $|5x| < 15$ **15.** $|x + 7| = 5$

16. $|x - 3| = 2$ **17.** $|x - 5| < 3$ **18.** $|x + 1| \leq 1$

19. $|x + 3| \leq -2$ **20.** $|x - 2| < 0$ **21.** $|x + 3| > 5$

22. $|x + 2| \geq -3$ **23.** $|x + 5| \geq 2$ **24.** $|x - 1| > 1$

25. $|2x + 3| = 5$ **26.** $|3x - 1| = 2$ **27.** $|3x - 4| < 7$

28. $|5x + 3| \leq -8$ **29.** $|5x - 3| \geq 4$ **30.** $|4x + 2| > 2$

31. $|3 - 4x| \leq 7$ **32.** $|4 - 2x| < 1$ **33.** $|2 - 4x| \geq 5$

34. $|7 - 3x| > 0$ **35.** $\left|\frac{2x + 5}{5}\right| \geq 3$ **36.** $\left|\frac{5x - 2}{3}\right| \leq 2$

37. $\left|\frac{3x - 4}{2}\right| < 1$ **38.** $\left|\frac{4x + 7}{3}\right| \geq 1$ **39.** $\left|\frac{3 - x}{4}\right| = 5$

40. $\left|\frac{4 - x}{3}\right| = 2$ **41.** $\left|\frac{3 - x}{2}\right| \leq 5$ **42.** $\left|\frac{6 - 2x}{3}\right| \leq 4$

43. $\left|5 - \frac{x}{2}\right| > 1$ **44.** $\left|7 - \frac{x}{3}\right| > 5$ **45.** $\left|4 - \frac{x}{3}\right| < 1$

46. $\left|8 - \frac{x}{2}\right| \leq 2$

2.5 Review: 2.1–2.4

Types of Equations 2.1

A **conditional equation** is an equation whose two sides are equal only when certain numbers (called *solutions*) are substituted for the variable.

An **identity** is an equation whose two sides are equal no matter what permissible number is substituted for the variable.

No solution exists for an equation whose two sides are unequal no matter what permissible number is substituted for the variable.

To Solve a First-Degree Equation with Only One Unknown 2.1

1. Remove fractions by multiplying both sides by the least common multiple (LCM) of all the denominators.
2. Remove grouping symbols.
3. Combine like terms on each side of the equation.
4. Move all the terms that contain the variable to one side of the equal sign and all constants to the other side.
5. Divide both sides of the equation by the coefficient of the variable.
6. Determine whether the equation is a conditional equation, an identity, or an equation with no solution.
7. If the equation is a conditional equation, check the solution.

To Solve a First-Degree Inequality with Only One Unknown 2.2

Proceed in the same way used to solve equations, with the exception that the *sense* must be changed when multiplying or dividing both sides by a negative number.

To Graph Solutions on the Number Line 2.2 and 2.3

To graph the set $\{x \mid a < x \le b\}$:

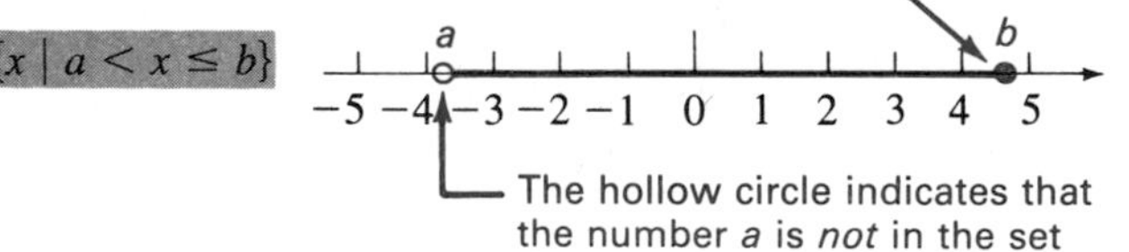

To graph the set $\{x \mid x > c\}$:

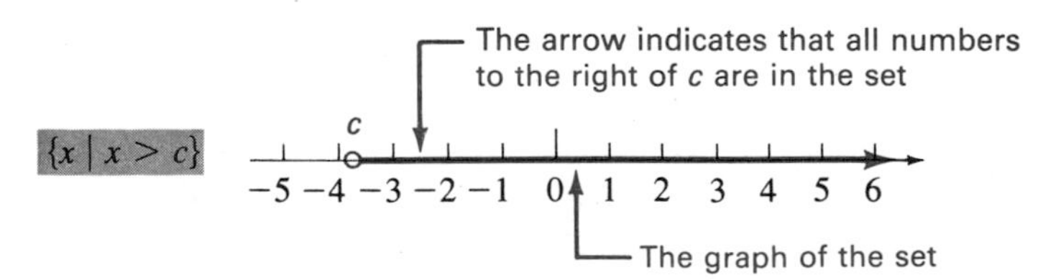

The solution sets of inequalities such as $x < c, x \le c, a \le x < b$, and so on, are graphed using similar procedures.

Intervals and Interval Notation 2.3

$\{x \mid a < x < b\}$ is an **open interval**; **interval notation** is (a, b). $\{x \mid c \le x \le d\}$ is a **closed interval**; interval notation is $[c, d]$. Interval notation for $\{x \mid f < x \le g\}$ is $(f, g]$, and for $\{x \mid j \le x < k\}$, it is $[j, k)$; these are **half-open**—or **half-closed**—**intervals.** Interval notation for $\{x \mid x > a\}$ is $(a, +\infty)$; for $\{x \mid x \le b\}$, it is $(-\infty, b]$.

To Solve an Equation with Absolute Value Symbols 2.4

An equation that contains absolute value symbols is equivalent to *two* equations without absolute value symbols.

If $|N| = a$, then $N = a$ or $N = -a$. (Rule 2.1)

To Solve an Inequality with Absolute Value Symbols 2.4

If $|N| < a$, then $-a < N < a$. (Rule 2.2)

If $|N| > a$, then $N > a$ or $N < -a$. (Rule 2.3)

Review Exercises 2.5 Set I

In Exercises 1–5, find the solution set for each equation. Identify any equation that is not a conditional equation as either an identity or an equation with no solution.

1. $7 - 2(M - 4) = 5$

2. $5[-13z - 8(4 - 2z) + 20] = 15z - 17$

3. $\frac{3(x + 3)}{4} - \frac{2(x - 3)}{3} = 1$

4. $\frac{x - 1}{4} + \frac{x - 8}{2} = x - 4 - \frac{x + 1}{4}$

5. $2[-7y - 3(5 - 4y) + 10] = 10y - 12$

In Exercises 6–10, solve each inequality.

6. $10 - 3(x + 2) \geq 9 - 2(4 - 3x)$

7. $\frac{3z}{5} - \frac{2z}{3} < \frac{1}{2}$

8. $\frac{2w}{3} - \frac{5w}{6} < \frac{7}{12}$

9. $\frac{2(x + 6)}{10} + \frac{3x}{20} < 3$

10. $\frac{5(x - 3)}{7} - \frac{3x}{2} < 5$

In Exercises 11–25, find each set or solution set and graph it. *Whenever possible*, express the answer in interval notation.

11. $\{x \mid 3x + 2 = 11, x \in J\}$

12. $|3x| = 9$

13. $|x| \leq -3$

14. $|x| \geq 2$

15. $|6 - 2x| = 10$

16. $3|x - 2| \leq 6$

17. $2|x - 3| \leq 4$

18. $\{x \mid -3 < 2x - 1 < 8\}$

19. $\{x \mid -2 < x + 1 < 0, x \in J\}$

20. $\{x \mid -5 < 3x + 1 < 2\}$

21. $\{x \mid -3 < x - 1 < 0, x \in J\}$

22. $\left\{x \,\middle|\, \frac{2(x - 1)}{6} + \frac{x}{9} = 1, x \in J\right\}$

23. $\left\{x \,\middle|\, \frac{3(x - 1)}{4} + \frac{x}{8} = 1, x \in J\right\}$

24. $\left|2 - \frac{x}{3}\right| \leq 1$

25. $\left|\frac{2x - 4}{3}\right| \geq 2$

In Exercises 26–28, write, in set-builder notation, the algebraic statement that describes each graph.

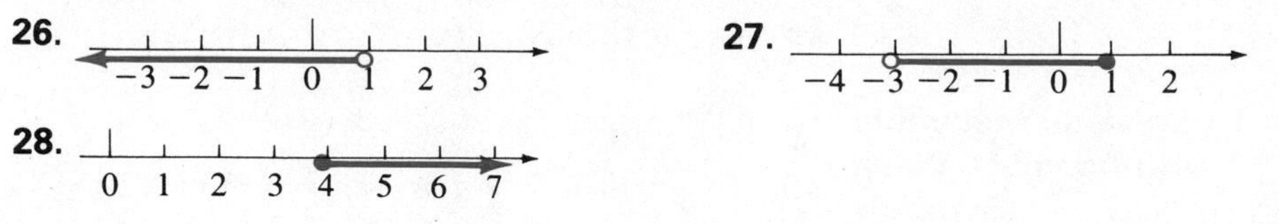

Review Exercises 2.5 Set II

NAME ______________________

ANSWERS

1. ______________________
2. ______________________
3. ______________________
4. ______________________
5. ______________________
6. ______________________
7. ______________________
8. ______________________
9. ______________________
10. ______________________
11. ______________________
12. ______________________

In Exercises 1–5, find the solution set for each equation. Identify any equation that is not a conditional equation as either an identity or an equation with no solution.

1. $13 = 8 - 5(2a + 13)$

2. $5(t - 3) + 1 = 3t - 2(7 - t)$

3. $3[-8x - 5(3 - 4x) + 10] = 36x + 2$

4. $\dfrac{x + 3}{4} = \dfrac{x - 2}{3} + \dfrac{1}{4}$

5. $\dfrac{3(x - 3)}{4} + 1 = \dfrac{8(x + 2)}{7}$

In Exercises 6–10, solve each inequality.

6. $2(x - 4) - 5 \geq 7 + 3(2x - 1)$

7. $15 - 4(m - 6) \leq 2(2 - 5m) + 8$

8. $\dfrac{2w}{4} - \dfrac{7}{9} < \dfrac{5w}{6}$

9. $\dfrac{5x}{3} < 11 - \dfrac{7(x - 9)}{12}$

$\dfrac{x}{3} - \dfrac{x + 2}{5} < 1$

In Exercises 11–25, find each set or solution set. *Whenever possible,* express the answer in interval notation.

11. $|8x| = 12$

12. $\{x \mid 13 + 2x = 1, x \in N\}$

13. $|11 - 7x| = 17$

14. $|x| > -5$

15. $\{x \mid -2 < x + 3 < 1, x \in J\}$

16. $\left\{x \,\middle|\, \frac{13}{15} - \frac{x}{5} = \frac{2(8 - x)}{15}, x \in J\right\}$

17. $\left|4 - \frac{x}{2}\right| \leq 2$

18. $8|2 - x| < 40$

19. $\left|\frac{3x + 2}{4}\right| > 2$

20. $\{x \mid -1 < 2x + 3 < 4\}$

21. $\{x \mid 5 < 3x - 1 < 2\}$

22. $\left|5 - \frac{x}{3}\right| \geq 1$

23. $\{x \mid -3 \leq 2x - 1 \leq 7, x \in N\}$

24. $|x - 4| = -3$

25. $\{x \mid -4 < 3 - x < 4\}$

In Exercises 26–28, write, in set-builder notation, the algebraic statement that describes each graph.

26.

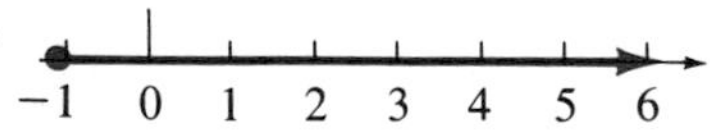

27.

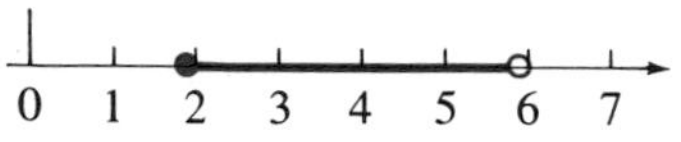

28.

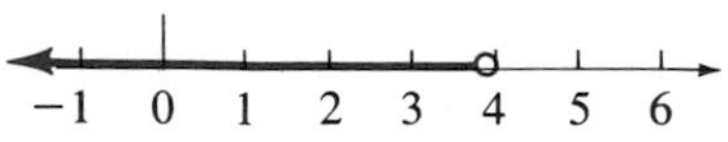

ANSWERS

13. ______________________

14. ______________________

15. ______________________

16. ______________________

17. ______________________

18. ______________________

19. ______________________

20. ______________________

21. ______________________

22. ______________________

23. ______________________

24. ______________________

25. ______________________

26. ______________________

27. ______________________

28. ______________________

Chapter 2 Diagnostic Test

The purpose of this test is to see how well you understand the solution of first-degree equations and inequalities in one unknown. We recommend that you work this diagnostic test *before* your instructor tests you on this chapter. Allow yourself about 50 minutes.

Complete solutions for all the problems on this test, together with section references, are given in the answer section at the end of the book. For the problems you do incorrectly, study the sections cited.

In Problems 1 and 2, find the solution set for each equation. Identify any equation that is not a conditional equation as an identity or as an equation with no solution.

1. $\frac{x}{6} - \frac{x+2}{4} = \frac{1}{3}$

2. a. $3(x - 6) = 5(1 + 2x) - 7(x - 4)$

b. $2[7x - 4(1 + 3x)] = 5(3 - 2x) - 23$

3. Solve each inequality.

a. $5w + 2 \leq 10 - w$

b. $13h - 4(2 + 3h) \geq 0$

In Problems 4–10, find each set or solution set and graph it. *Whenever possible,* express the answer in interval notation.

4. a. $\{x \mid -3 < x + 1 < 5, x \in J\}$

b. $\{x \mid 4 \geq 3x + 7 > -2, x \in R\}$

5. $\left\{x \,\middle|\, \frac{5(x-2)}{3} + \frac{x}{4} \leq 12, x \in R\right\}$

6. $\left\{x \,\middle|\, \left|\frac{2x+3}{5}\right| = 1\right\}$

7. $|3x - 1| > 2$

8. $|7 - 3x| \geq 6$

9. $\left|\frac{5x+1}{2}\right| \leq 7$

10. $\{x \mid |2x - 5| < 11\}$

Cumulative Review Exercises: Chapters 1–2

In Exercises 1–30, find the value of each expression or write "Undefined."

1. $(-14) - (-22)$

2. $(-12)(-4)$

3. $(-1)^7$

4. $(-6) + (-8)$

5. $(-2)^4$

6. $\sqrt{64}$

7. $\frac{9}{0}$

8. $\frac{0}{-7}$

9. $\frac{|-20|}{-5}$

10. $(-18) - (7)$

11. $\sqrt[3]{-64}$

12. -6^2

13. $(35) \div (-7)$

14. $(-10)(0)(8)$

15. $\sqrt[9]{-1}$

16. $10^{-2} \cdot 10^5$

17. $(4^0)^3$

18. $(2^{-3})^{-1}$

19. $-3 - 2^2 \cdot 6$

20. $7 - [4 - (13 - 5)]$

21. $(-11) + 15$

22. $(14)(-2)$

23. $|0|$

24. $\sqrt{81}$

25. $\sqrt[4]{81}$

26. $3 + 2 \cdot 5$

27. $\frac{0}{0}$

28. $16 \div (-2)^2 - \frac{7-1}{2}$

29. $0 \div 15$

30. $36 \div 18 \times 2$

31. Write the prime factorization of 78.

32. Find all the prime numbers between 11 and 31 that yield a remainder of 3 when divided by 4.

33. What is the additive identity element?

34. Complete this statement by using the associative property:

$$3 + (2 + 7) = \underline{\qquad\qquad\qquad}$$

35. Complete this statement by using the commutative property:

$$(8 \times 17) \times 6 = (\underline{\qquad\qquad}) \times 6$$

36. Is $\sqrt{17}$ a real number?

In Exercises 37–40, simplify the expression.

37. $y - 2(x - y) - 3(1 - y) - \sqrt{4y^2}$

38. $2x(x^2 + 1) - x(x^2 + 3x - 2)$

39. $(5x)^2(3x^2)^3$

40. $(2x^3 - 4y^2) - 5x^3$

In Exercises 41–50, find the solution set for each equation or inequality. Identify any *equation* that is not a conditional equation as either an identity or an equation with no solution.

41. $8x - 4(2 + 3x) = 12$

42. $2[-5y - 6(y - 7) < 6 + 4y$

43. $6(3x - 5) + 7 = 9(3 + 2x) - 1$

44. $5x - 7 \leq 8x + 4$

45. $2x + 3 = 2(2x + 5)$

46. $\frac{x}{4} + \frac{x-3}{10} = \frac{29}{20}$

47. $-5 < x + 4 \leq 3$

48. $8\{4 + 3(x - 2)\} = 3(8x + 3) - 25$

49. $|2x + 3| > 1$

50. $|x - 4| \leq 3$

3 Word Problems

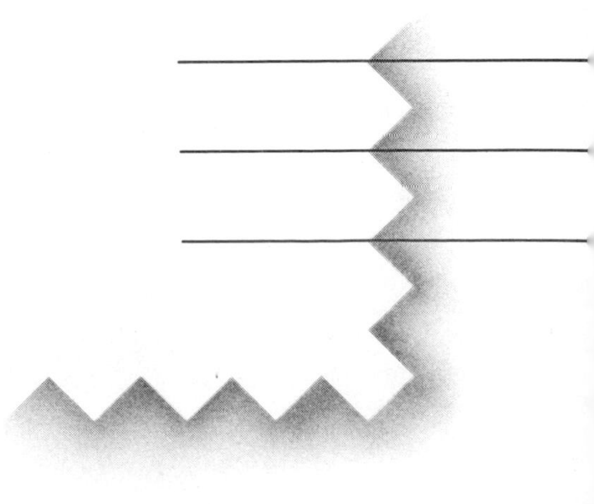

The main reason for studying algebra is to equip oneself with the tools necessary to solve problems. Most problems are expressed in words. In this chapter, we show methods for solving some traditional word problems. The skills learned in this chapter can be applied to solving mathematical problems encountered in many fields of learning as well as in real-life situations.

3.1 Method of Solving Word Problems

While we can't give you a definite set of rules that will enable you to solve all word problems, we do suggest a procedure that should help you get started. In this chapter, we discuss several different "types" of word problems (money problems, mixture problems, distance-rate-time problems, and so forth) in separate sections, because we feel that this technique is the most helpful to beginning students. However, the general *method* of attacking word problems is the same for *all* types of word problems, and it is this *method* on which you should concentrate.

In the following suggestions for solving mathematical word problems, we use the notation "Step 1," "Step 2," and so forth for the steps you will be *writing*.

SUGGESTIONS FOR SOLVING WORD PROBLEMS

Read — To solve a word problem, first read it very carefully. *Be sure you understand the problem.* Read it several times, if necessary.

Think — Determine what *type* of problem it is (money problem, distance-rate-time problem, and so forth), if possible. Determine what is unknown. What is being asked for is often found in the last sentence of the problem, which may begin with "What is the . . ." or "Find the . . . ." Is enough information given so that you *can* solve the problem? Do you need a special formula? What operation(s) must be used? What is the domain of the variable? What kind of answer is reasonable?

Sketch — Draw a sketch *with labels*, if possible.

Step 1. — Represent one unknown number by a variable, and *declare* the meaning in a sentence of the form "Let $x =$ . . . ." Then reread the problem to see how you can represent any other unknown numbers in terms of the same variable.

Reread — Reread the entire word problem, breaking it up into small pieces.

Step 2. — Translate each English phrase into an algebraic expression and fit these expressions together into an equation or inequality.

Step 3. — Using the methods described in Chapter 2, solve the equation or inequality.

Step 4. — Solve for *all* the unknowns asked for in the problem.

Step 5. — Check the solution(s) *in the word statement.*

Step 6. — State your results clearly.

The following list of key word expressions and their corresponding algebraic operations can help you translate English sentences into an algebraic equation or inequality:

$+$	$-$	$\times$	$\div$	$=$
the sum of	minus	times	divided by	is
added to	decreased by	the product of	the quotient of	equals
increased by	less than	multiplied by	per	is equal to
plus	subtracted from	by		
more than	the difference of			
all				
total				

NOTE Because subtraction is not commutative, care must be taken to get the numbers in a subtraction word problem in the correct order. For example, while the statements "m minus n," "m decreased by n," and "the difference of m and n" are translated as $m - n$, the expressions "m subtracted from n" and "m less than n" are translated as $n - m$. ☑

Example 1 Three times an unknown number, decreased by 5, is 13. What is the unknown number?

Step 1. Let $x =$ the unknown number.

Reread	Three	times	an unknown number	decreased by	5	is	13
Step 2.	3	$\cdot$	x	$-$	5	$=$	13

Step 3. $3x - 5 = 13$

$3x = 18$ Adding 5 to both sides

Step 4. $x = 6$ Dividing both sides by 3

Step 5. *Check*	Three	times	an unknown number	decreased by	5	is	13
	3	$\cdot$	(6)	$-$	5	$=$	13

$3(6) - 5 \stackrel{?}{=} 13$ The unknown number is replaced by 6

$18 - 5 \stackrel{?}{=} 13$

$13 = 13$

Step 6. Therefore, the unknown number is 6. ■

NOTE To check a word problem, you must check the solution in the *word statement*. Any error that may have been made in writing the equation will not be discovered if you simply substitute the solution in the equation. ☑

Example 2 Find three consecutive integers such that the sum of the first two is 23 less than 3 times the third.

Think The domain is the set of integers.

Step 1. Let $x =$ the first integer
$x + 1 =$ the second integer
$x + 2 =$ the third integer

Reread

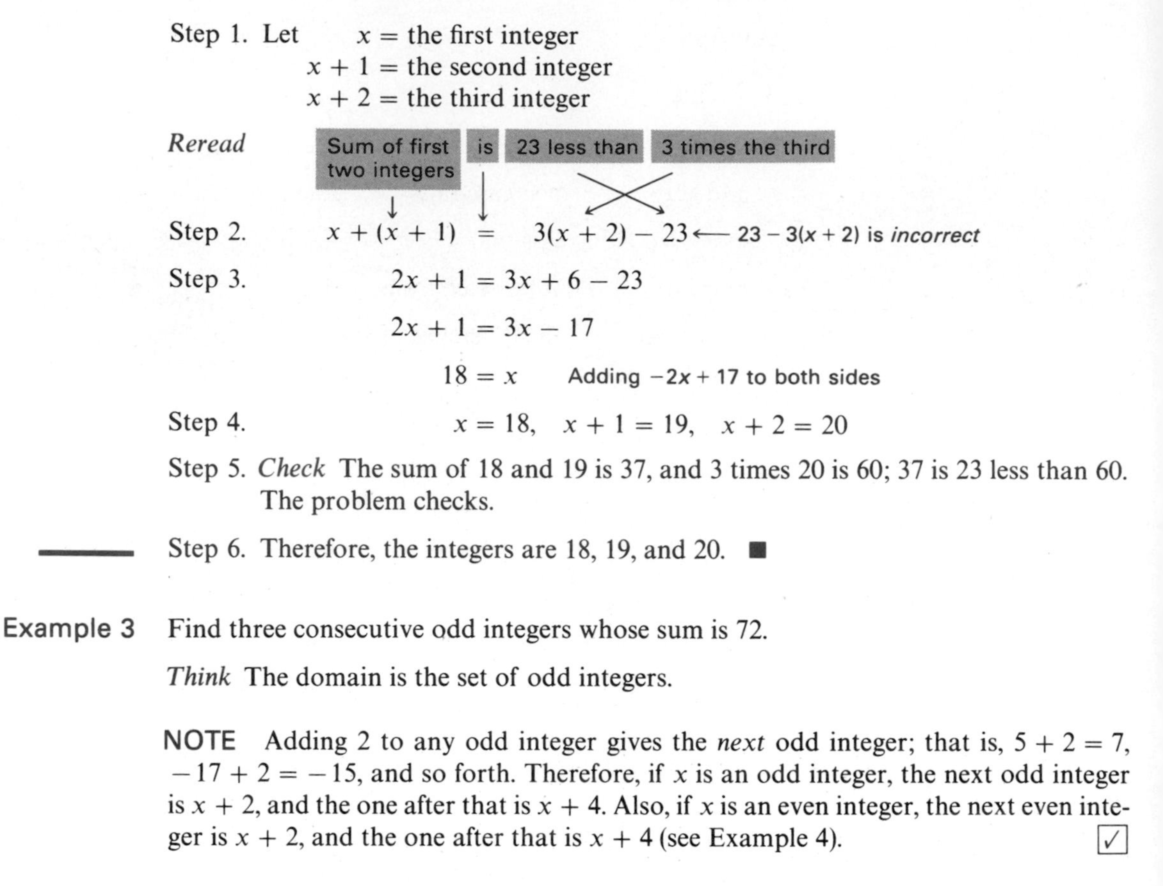

Step 2. $x + (x + 1) = 3(x + 2) - 23$ ← $23 - 3(x + 2)$ is *incorrect*

Step 3. $2x + 1 = 3x + 6 - 23$

$2x + 1 = 3x - 17$

$18 = x$ Adding $-2x + 17$ to both sides

Step 4. $x = 18,\quad x + 1 = 19,\quad x + 2 = 20$

Step 5. *Check* The sum of 18 and 19 is 37, and 3 times 20 is 60; 37 is 23 less than 60. The problem checks.

Step 6. Therefore, the integers are 18, 19, and 20. ■

Example 3 Find three consecutive odd integers whose sum is 72.

Think The domain is the set of odd integers.

NOTE Adding 2 to any odd integer gives the *next* odd integer; that is, $5 + 2 = 7$, $-17 + 2 = -15$, and so forth. Therefore, if x is an odd integer, the next odd integer is $x + 2$, and the one after that is $x + 4$. Also, if x is an even integer, the next even integer is $x + 2$, and the one after that is $x + 4$ (see Example 4). ☑

Step 1. Let $x =$ the first odd integer
$x + 2 =$ the second odd integer
$x + 4 =$ the third odd integer

Reread Sum of the integers | is | 72

Step 2. $x + (x + 2) + (x + 4) = 72$

Step 3. $3x + 6 = 72$ Combining like terms

$3x = 66$ Adding -6 to both sides

Step 4. $x = 22$ Dividing both sides by 3

Steps 5 and 6. Since 22 is not an odd integer, it is not in the domain. Therefore, there is no solution. ■

When a problem deals with *inequalities* rather than with quantities that are *equal* to each other, you will almost always find more than one correct answer (see Example 4).

Example 4 Find four consecutive even integers whose sum is between 21 and 45.

Think The domain is the set of even integers.

Step 1. Let $x =$ the first even integer
$x + 2 =$ the second even integer
$x + 4 =$ the third even integer
$x + 6 =$ the fourth even integer

Because the sum of the integers is to be *between* 21 and 45, the algebraic statement can be written as a continued inequality.

Step 2. $21 < x + (x + 2) + (x + 4) + (x + 6) < 45$

Step 3. $21 < 4x + 12 < 45$ Combining like terms

$9 < 4x < 33$ Adding -12 to all parts

$\frac{9}{4} < x < \frac{33}{4}$ Dividing all parts by 4

The even integers between 9/4 and 33/4 are 4, 6, and 8; therefore, we will have *three* sets of answers:

Step 4. If $x = 4$	If $x = 6$	If $x = 8$
$x + 2 = 6$	$x + 2 = 8$	$x + 2 = 10$
$x + 4 = 8$	$x + 4 = 10$	$x + 4 = 12$
$x + 6 = 10$	$x + 6 = 12$	$x + 6 = 14$

Step 5. *Check* $4 + 6 + 8 + 10 = 28$, $6 + 8 + 10 + 12 = 36$, and $8 + 10 + 12 + 14 = 44$. All these sums of four consecutive even integers are between 21 and 45.

Step 6. Therefore, four consecutive even integers whose sum is between 21 and 45 are 4, 6, 8, and 10; 6, 8, 10, and 12; *and* 8, 10, 12, and 14. ■

EXERCISES 3.1

Set I In *all* exercises, set up the problem algebraically. Be sure to state what your variables represent.

In Exercises 1–6, solve for the unknown number and check your solution.

1. Seven more than twice an unknown number is 23.
2. Eleven more than 3 times an unknown number is 38.
3. Four times an unknown number, decreased by 7, is 25.
4. Five times an unknown number, decreased by 6, is 49.
5. When an unknown number is decreased by 7, the difference is half the unknown number.
6. When an unknown number is decreased by 12, the difference is half the unknown number.

In Exercises 7–18, solve for the unknowns and check your solutions.

7. A 12-cm length of string is cut into two pieces. The first piece is 3 times as long as the second piece. How long is each piece?
8. A 12-m length of rope is cut into two pieces. The first piece is twice as long as the second piece. How long is each piece?
9. The sum of three consecutive integers is 19. Find the integers.
10. The sum of three consecutive integers is 40. Find the integers.
11. A 42-cm piece of wire is cut so that the first piece is 8 cm longer than the second piece. How long is each piece?
12. A 50-m piece of hose is cut so that the first piece is 6 m longer than the second piece. How long is each piece?

13. The sum of the first two of three consecutive integers less the third integer is 10. What are the integers?

14. The sum of the first two of three consecutive integers less the third is 17. What are the integers?

15. Find three consecutive odd integers such that 3 times the sum of the last two is 40 more than 5 times the first.

16. Find three consecutive odd integers such that 5 times the sum of the last two is 60 more than 5 times the first.

17. David buys 4 more cans of corn than cans of peas and 3 times as many cans of green beans as cans of peas. If he buys 24 cans of these three vegetables altogether, how many cans of each kind does he buy?

18. John buys 6 more cans of peaches than cans of pears and 3 times as many cans of cherries as cans of pears. If he buys 21 cans of these three fruits altogether, how many cans of each kind does he buy?

In Exercises 19–22, find the unknown number and check your solution.

19. Three times the sum of 8 and an unknown number is equal to twice the sum of the unknown number and 7.

20. Four times the sum of 5 and an unknown number is equal to 3 times the sum of the unknown number and 9.

21. When twice the sum of 5 and an unknown number is subtracted from 8 times the unknown number, the result is equal to 4 times the sum of 8 and twice the unknown number.

22. When 7 times the sum of 2 and an unknown number is subtracted from 4 times the sum of 3 and twice the unknown number, the result is equal to 0.

In Exercises 23–26, find *all possible solutions* for each problem.

23. Find three consecutive even integers whose sum is between 21 and 45.

24. Find three consecutive even integers whose sum is between 81 and 93.

25. In a certain mathematics class, a student needs between 560 and 640 points in order to receive a C. The final exam is worth 200 points. If Clark has 396 points just before the final, what range of scores on the final exam will give him a C for the course?

26. In a certain English class, a student needs between 720 and 810 points in order to receive a B. The final exam is worth 200 points. If Cathy has 584 points just before the final, what range of scores on the final exam will give her a B grade for the course?

Set II In *all* exercises, set up the problem algebraically. Be sure to state what your variables represent.

In Exercises 1–6, solve for the unknown number and check your solution.

1. Nine more than 4 times an unknown number is 33.

2. Seven plus an unknown number is equal to 17 decreased by the unknown number.

3. Five times an unknown number, decreased by 8, is 12.

4. Four plus an unknown number is equal to 20 decreased by the unknown number.

5. When an unknown number is decreased by 12, the difference is one-third the unknown number.

6. Two plus an unknown number is equal to 8 decreased by the unknown number.

In Exercises 7–18, solve for the unknowns and check your solutions.

7. A 36-cm length of cord is cut into two pieces. The first piece is 5 times as long as the second piece. How long is each piece?

8. A 32-cm piece of string is cut so that the first piece is one-third the length of the second piece. How long is each piece?

9. The sum of three consecutive integers is 73. Find the integers.

10. The sum of three consecutive integers is 84. What are the integers?

11. A 52-m piece of rope is cut so that the first piece is 14 m longer than the second piece. How long is each piece?

12. A piece of pipe 60 cm long is cut so that one piece is 34 cm shorter than the other piece. How long is each piece?

13. The sum of the first two of three consecutive integers less the third is 20. What are the integers?

14. The sum of three consecutive even integers is -54. What are the integers?

15. Find three consecutive odd integers such that 4 times the sum of the last two is 48 more than 7 times the first.

16. The sum of the first two of three consecutive integers less the third is 85. What are the integers?

17. Tom buys 4 more cans of tomato soup than cans of split pea soup and 3 times as many cans of vegetable soup as cans of split pea soup. If he buys 29 cans of these three soups altogether, how many cans of each kind does he buy?

18. The total receipts for a concert were \$7,120. Some tickets were \$8.50 each, and the rest were \$9.50 each. If 800 tickets were sold altogether, how many of each kind of ticket were sold?

In Exercises 19–22, solve for the unknown number and check your solution.

19. Four times the sum of 6 and an unknown number is equal to 3 times the sum of the unknown number and 10.

20. When an unknown number is subtracted from 42, the difference is one-sixth the unknown number.

21. When 3 times the sum of 4 and an unknown number is subtracted from 8 times the unknown number, the result is equal to 4 times the sum of 3 and twice the unknown number.

22. When an unknown number is subtracted from 16, the difference is one-third the unknown number.

In Exercises 23–26, find *all possible solutions* for each problem.

23. Find three consecutive even integers whose sum is between 39 and 51.

24. Find four consecutive odd integers whose sum is between -5 and 20.

25. In a certain history class, a student needs between 600 and 675 points in order to receive a B. The final exam is worth 150 points. If Manuel has 467 points just before the final, what range of scores on the final exam will give him a B for the course?

26. In a certain physics class, a student needs at least 450 points in order to receive an A. The final exam is worth 100 points. If Cindy has 367 points just before the final, what range of scores on the final exam will give her an A for the course?

3.2 Ratio Problems

Ratio is another word for *fraction*. When the numerators and denominators are integers, fractions are often called *ratio*nal numbers.

The ratio of a to b is written $\frac{a}{b}$ or $a:b$.

The ratio of b to a is written $\frac{b}{a}$ or $b:a$.

We call a and b the *terms* of the ratio. The terms of a ratio may be any kind of number; the only restriction is that the denominator cannot be zero.

The key to solving ratio problems is to use the given ratio to help represent the unknown numbers, as shown in the examples.

> TO REPRESENT THE UNKNOWNS IN A RATIO PROBLEM
>
> **1.** Multiply each term of the ratio by x.
>
> **2.** Let the resulting products represent the unknowns.

Example 1 Two numbers are in the ratio of 3 to 5. Their sum is -80. Find the numbers.*

Think $3:5$ The ratio

$3x:5x$ Multiply each term of the ratio by x

Let the resulting products, $3x$ and $5x$, represent the unknown numbers.

Step 1. Let $3x$ = one number

$5x$ = the other number

Reread Their sum is -80

Step 2. $\overbrace{3x + 5x} = -80$

Step 3. $8x = -80$

$x = -10$

Although we have solved for x, we *are not finished yet*. Now we must replace x by -10 in the expressions $3x$ and $5x$, since the unknown numbers were $3x$ and $5x$.

Step 4. Therefore, $3x = 3(-10) = -30$ One number

$5x = 5(-10) = -50$ The other number

Step 5. *Check* $-30 = 3(-10)$ and $-50 = 5(-10)$; therefore, -30 and -50 are in the ratio of 3 to 5. The sum of -30 and -50 is -80.

Step 6. Therefore, one number is -30 and the other is -50. ■

* The same problem could have been worded as follows: Divide -80 into two parts whose ratio is $3:5$, or separate -80 into two parts whose ratio is $3:5$.

Certain problems relating to rectangles, squares, and triangles lead to ratio word problems in algebra. Figure 3.2.1 shows a collection of facts from geometry that are helpful in solving such problems. Recall that the *perimeter* of a geometric figure is the distance around it.

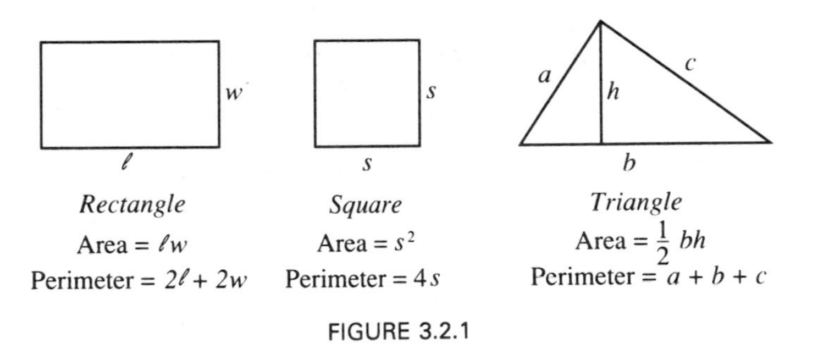

FIGURE 3.2.1

Example 2 The three sides of a triangle are in the ratio 2:3:4. The perimeter is 63. Find the lengths of the three sides.

Think $2\ :3\ :4$ The ratio
$2x:3x:4x$ Multiply each term of the ratio by x

Let the resulting products represent the unknowns.

Step 1. Let $2x =$ first side
$3x =$ second side
$4x =$ third side

Reread The perimeter is 63

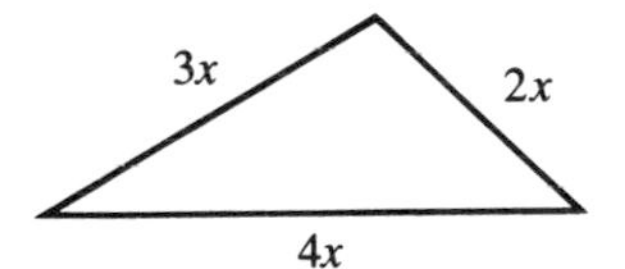

Step 2. $\overbrace{2x + 3x + 4x} = 63$

Step 3. $9x = 63$

$x = 7$

Step 4. Therefore, $2x = 2(7) = 14$ First side

$3x = 3(7) = 21$ Second side

$4x = 4(7) = 28$ Third side

Step 5. *Check* $14 = 2(7)$, $21 = 3(7)$, and $28 = 4(7)$; therefore, 14, 21, and 28 are in the ratio 2:3:4, and $14 + 21 + 28 = 63$, so the perimeter is 63.

Step 6. Therefore, the lengths of the sides are 14, 21, and 28. ■

Example 3 The length and width of a rectangle are in the ratio 5:3. The perimeter is to be less than 48. What values can the length of the rectangle have?

Think $5\ :3$ The ratio
$5x:3x$ Multiply each term of the ratio by x

Step 1. Let $3x =$ the width
$5x =$ the length

Think The perimeter is $2(3x) + 2(5x)$; it must be greater than *0 and* less than 48.

Step 2. $0 < 2(3x) + 2(5x) < 48$

Step 3. $0 < 6x + 10x < 48$

$0 < 16x < 48$

$0 < x < 3$ Dividing all three parts by 16

Step 4. $0 < 5x < 15$ The *length* is $5x$

Step 5. *Check* We show checks for lengths of 10 and 8 (two values for the length chosen arbitrarily between 0 and 15). If the length were 10, the width would be 6 and the perimeter would be 32. If the length were 8, the width would be 24/5 and the perimeter would be 128/5, or 25.6. (You check the problem using several other values between 0 and 15 for the length of the rectangle.)

Step 6. Therefore, the length must be greater than 0 *and* less than 15. ■

EXERCISES 3.2

Set I In each exercise, set up the problem algebraically and solve. Be sure to state what your variables represent. In Exercises 1–10, check your solutions.

1. Two numbers are in the ratio of 4 to 5. Their sum is 81. Find the numbers.
2. Two numbers are in the ratio of 8 to 3. Their sum is 77. Find the numbers.
3. The three sides of a triangle are in the ratio 3:4:5. The perimeter is 108. Find the lengths of the three sides.
4. The three sides of a triangle are in the ratio 4:5:6. The perimeter is 120. Find the lengths of the three sides.
5. Fifty-four hours of a student's week are spent in study, class, and work. The times spent in these activities are in the ratio 4:2:3. How many hours are spent in each activity?
6. Forty-eight hours of a student's week are spent in study, class, and work. The times spent in these activities are in the ratio 4:2:6. How many hours are spent in each activity?
7. The length and width of a rectangle are in the ratio of 7 to 6. The perimeter is 78. Find the length and width.
8. The length and width of a rectangle are in the ratio of 9 to 5. The perimeter is 196. Find the length and width.
9. Divide (separate) 143 into two parts whose ratio is 7:6.
10. Divide (separate) 221 into two parts whose ratio is 8:5.
11. The three sides of a triangle are in the ratio 4:5:6. The perimeter is to be less than 60. What is the range of values that the *shortest* side can have?
12. The three sides of a triangle are in the ratio 8:9:10. The perimeter is to be less than 81. What is the range of values that the *shortest* side can have?

Set II In each exercise, set up the problem algebraically and solve. Be sure to state what your variables represent. In Exercises 1–10, check your solutions.

1. Two numbers are in the ratio of 2 to 7. Their sum is 99. Find the numbers.
2. Three numbers are in the ratio 4:7:3. Their sum is 56. Find the numbers.
3. The sides of a triangle are in the ratio 5:6:7. The perimeter is 72. Find the lengths of the sides.

4. The ratio of pineapple juice to orange juice to lemon juice in a punch recipe is 11 to 6 to 1. If 36 ℓ of punch are to be made (with only these three juices), how much pineapple juice should be used?

5. The formula for a particular shade of green paint calls for mixing 3 parts of blue paint with 1 part of yellow paint. Find the number of liters of blue and the number of liters of yellow needed to make 14 ℓ of the desired shade of green paint.

6. In Mrs. Aguilar's bread recipe, flour and water are to be mixed in the ratio of 3 parts water to 5 parts flour. How many parts of flour are needed to make a mixture of 64 parts of water and flour?

7. The length and width of a rectangle are in the ratio of 8 to 3. The perimeter is 88. Find the length and the width.

8. Nick needs to mix gasoline and oil in the ratio of 50 to 1 for his motorcycle. If he needs to mix 382.5 oz altogether, how much gasoline should he use? How much oil?

9. Divide (separate) 207 into two parts whose ratio is 5:4.

10. Todd has nickels, dimes, and quarters in the ratio 5:8:7. If he has 140 of these coins altogether, how many of each kind does he have?

11. The three sides of a triangle are in the ratio 4:5:8. The perimeter is to be less than 85. What is the range of values that the *shortest* side can have?

12. The three sides of a triangle are in the ratio 3:5:7. The perimeter is to be greater than 60. What is the range of values that the *longest* side can have?

3.3 Distance-Rate-Time Problems

Distance-rate-time problems are used in any field involving motion. A physical law relating distance traveled d, rate of travel r, and time of travel t is

$$d = r \cdot t \qquad \text{or} \qquad r \cdot t = d$$

For example, if you're driving your car at an average speed of 50 mph, then

you travel a distance of 100 mi in 2 hr: $50\frac{\text{mi}}{\cancel{\text{hr}}}(2\,\cancel{\text{hr}}) = 100\text{ mi}$

you travel a distance of 150 mi in 3 hr: $50\frac{\text{mi}}{\cancel{\text{hr}}}(3\,\cancel{\text{hr}}) = 150\text{ mi}$

and so on. Students often find a chart helpful in solving distance-rate-time problems.

"CHART" METHOD FOR SOLVING DISTANCE-RATE-TIME PROBLEMS

Read — Read the problem carefully. *Be sure you understand the problem.*

Think — If this is a distance-rate-time problem, a chart can be used.

Step 1. — Represent one unknown number by a variable and declare its meaning in a sentence of the form "Let $x = \ldots$." Express other unknowns in terms of x, if possible.

Chart Draw a blank chart, as follows:

	r	$\cdot$ t	$=$ d
One object			
Other object			

Fill in Fill in as many of the blanks as possible, using the information given.

Reread Fill in the remaining blanks, using x (or some other variable) and the formula $r \cdot t = d$.

Step 2. Are the distances equal? If so, you obtain the equation by setting the two d-values from the chart equal to each other. Do the distances differ by some constant? If so, write the equation using this information. Are the *times* equal, or do you know something about the *sum* of the times? If so, write the equation using this information.

Step 3. Using the methods described in Chapter 2, solve the equation.

Step 4. Solve for *all* the unknowns asked for in the problem.

Step 5. Check the solution(s) *in the word statement.*

Step 6. State your results clearly.

Example 1 A carload of campers leaves Los Angeles for Lake Havasu at 8:00 A.M. A second carload of campers leaves Los Angeles at 8:30 A.M. and drives 10 mph faster over the same road. If the second car overtakes the first at 10:30 A.M., what is the average speed of each car?

Think This is a distance-rate-time problem (the formula is $r \cdot t = d$).

Step 1. Let x = rate of first car (the slower car) in miles per hour
$x + 10$ = rate of second car (the faster car) in miles per hour

Chart (partially filled in)

	r	$\cdot$ t	$=$ d
First car	$x \frac{\text{mi}}{\text{hr}}$	$\frac{5}{2}$ hr	
Second car	$(x + 10)\frac{\text{mi}}{\text{hr}}$	2 hr	

First car leaves at 8:00 A.M. and is overtaken at 10:30 A.M. therefore, time is 5/2 hr

Second car leaves at 8:30 A.M. and overtakes the first car at 10:30 A.M.; therefore, time is 2 hr

Use the formula $r \cdot t = d$ to fill in these blanks as follows:

Chart (completed)

	r ·	t =	d	$r \cdot t = d$
First car	$x \frac{\text{mi}}{\text{hr}}$	$\frac{5}{2}$ hr	$x\left(\frac{5}{2}\right)$ mi	$x\left(\frac{5}{2}\right) = d$
Second car	$(x + 10)\frac{\text{mi}}{\text{hr}}$	2 hr	$(x + 10)(2)$ mi	$(x + 10)(2) = d$

Since both cars start at the same place and end at the same place, they have traveled the same distance. Therefore,

Step 2. $\frac{5}{2}x = 2(x + 10)$ — "=" because the distances are equal; We *drop* the units

Step 3. $5x = 4(x + 10)$ — Multiplying both sides by 2

$5x = 4x + 40$

Step 4. $x = 40$ — Rate of first car

$x + 10 = 40 + 10 = 50$ — Rate of second car

Step 5. *Check* The first car travels $\left(2\frac{1}{2}\,\cancel{\text{hr}}\right)\left(40\frac{\text{mi}}{\cancel{\text{hr}}}\right) = 100$ miles.

The second car travels $(2\,\cancel{\text{hr}})\left(50\frac{\text{mi}}{\cancel{\text{hr}}}\right) = 100$ miles.

Step 6. Therefore, the average speed of the first car is 40 mph, and the average speed of the second car is 50 mph. ■

We often see problems in which a boat is running in moving water. In this case, if the water is moving at w miles per hour and the speed of the boat in still water is b miles per hour, then when the boat is going *upstream* (*against* the current, or against the movement of the water), the actual speed of the boat will be $b - w$ miles per hour, and when it's going downstream (with the current), the actual speed will be $b + w$ miles per hour (see Example 2).

Similarly, if the wind is blowing at w miles per hour and an airplane is flying with a speed that would be a miles per hour in still air, then when the plane flies *with* the wind, its actual speed will be $a + w$ miles per hour, and when it's flying *against* the wind, its actual speed will be $a - w$ miles per hour.

Example 2 A boat cruises downstream for 4 hr before heading back. After traveling upstream for 5 hr, it is still 16 mi short of the starting point. If the speed of the stream is 4 mph, find the speed of the boat in still water.

Think This is a distance-rate-time problem; the formula is $r \cdot t = d$.

Step 1. Let x = speed of boat in still water (in miles per hour)
$x + 4$ = speed of boat downstream (in miles per hour)
$x - 4$ = speed of boat upstream (in miles per hour)

Chart

	r ·	t =	d
Downstream	$(x + 4)\frac{\text{mi}}{\text{hr}}$	4 hr	$(x + 4)(4)$ mi
Upstream	$(x - 4)\frac{\text{mi}}{\text{hr}}$	5 hr	$(x - 4)(5)$ mi

$r \cdot t = d$
$(x + 4)(4) = d$

$r \cdot t = d$
$(x - 4)(5) = d$

We now use the "unused fact" that the distance downstream is 16 mi plus the distance upstream, as shown in the diagram below:

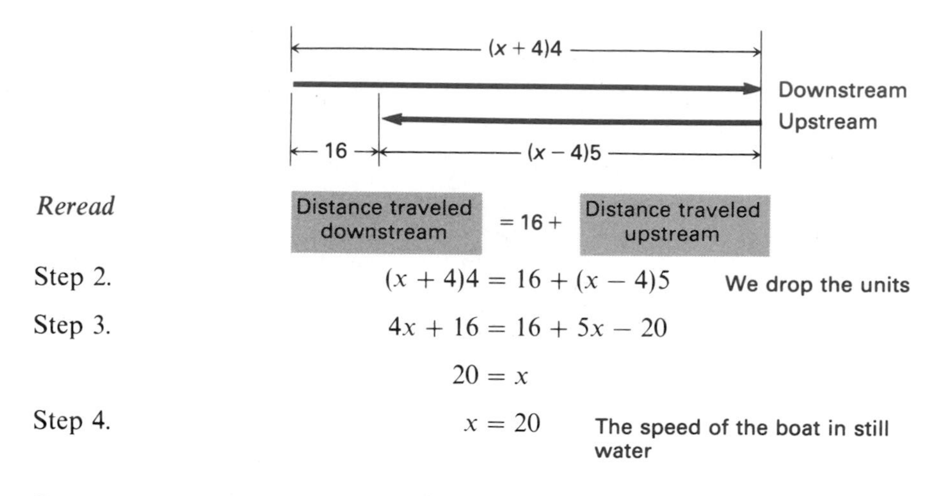

Reread Distance traveled downstream = 16 + Distance traveled upstream

Step 2. $(x + 4)4 = 16 + (x - 4)5$ We drop the units

Step 3. $4x + 16 = 16 + 5x - 20$

$20 = x$

Step 4. $x = 20$ The speed of the boat in still water

Step 5. *Check* Speed downstream is 24 mph. Speed upstream is 16 mph.

$$(4\,\cancel{\text{hr}})\left(24\frac{\text{mi}}{\cancel{\text{hr}}}\right) \stackrel{?}{=} 16\text{ mi} + (5\,\cancel{\text{hr}})\left(16\frac{\text{mi}}{\cancel{\text{hr}}}\right)$$

$$96\text{ mi} \stackrel{?}{=} 16\text{ mi} + 80\text{ mi}$$

$$96\text{ mi} = 96\text{ mi}$$

Step 6. Therefore, the speed of the boat in still water is 20 mph. ■

Example 3 Joe drove from his home to Los Angeles to pick up his wife, Linda. He drove 45 mph. Linda, driving at 54 mph, drove on the return trip. If the total driving time for the round trip was 11 hr, how far from their home is Los Angeles?

Step 1. Let x = the number of hours Joe drove
$11 - x$ = the number of hours Linda drove

Chart

	r ·	t =	d
Joe drove	$45\frac{\text{mi}}{\text{hr}}$	x hr	$45x$ mi
Linda drove	$54\frac{\text{mi}}{\text{hr}}$	$(11 - x)$ hr	$54(11 - x)$ mi

We now use the "unused fact" that the distances are equal.

Step 2. $45x = 54(11 - x)$ We drop the units

Step 3. $45x = 594 - 54x$

$99x = 594$

Step 4. $x = 6$ Number of hours Joe drove

$45x = 270$ Number of miles Joe lives from Los Angeles

Step 5. *Check* Hours Linda drove: $11 - 6 = 5$

Miles Linda drove: $\left(54\frac{\text{mi}}{\cancel{\text{hr}}}\right)(5\cancel{\text{hr}}) = 270 \text{ mi}$

Step 6. Therefore, Joe and Linda live 270 mi from Los Angeles. ■

EXERCISES 3.3

Set I

In each exercise, set up the problem algebraically, solve, and check. Be sure to state what your variables represent.

1. The Malone family left San Diego by car at 7 A.M., bound for San Francisco. Their neighbors, the King family, left in their car at 8 A.M., also bound for San Francisco. By traveling 9 mph faster, the Kings overtook the Malones at 1 P.M.

   a. Find the average speed of each car.

   b. Find the total distance traveled by each car before they met.

2. The Duran family left Ames, Iowa, by car at 6 A.M., bound for Yellowstone National Park. Their neighbors, the Silva family, left in their car at 8 A.M., also bound for Yellowstone. By traveling 10 mph faster, the Silvas overtook the Durans at 4 P.M.

   a. Find the average speed of each car.

   b. Find the total distance traveled before they met.

3. Eric hiked from his camp to a lake in the mountains and returned to camp later in the day. He walked at a rate of 2 mph going to the lake and 5 mph coming back. The trip to the lake took 3 hr longer than the trip back.

   a. How long did it take him to hike to the lake?

   b. How far is it from his camp to the lake?

4. Lee hiked from her camp up to an observation tower in the mountains and returned to camp later in the day. She walked up at the rate of 2 mph and jogged back at the rate of 6 mph. The trip to the tower took 2 hr longer than the return trip.

   a. How long did it take her to hike to the tower?

   b. How far is it from her camp to the tower?

5. Fran and Ron live 54 mi apart. Both leave their homes at 7 A.M. by bicycle, riding toward one another. They meet at 10 A.M. If Ron's average speed is four-fifths of Fran's, how fast does each cycle?

6. Tran and Atour live 60 mi apart. Both leave their homes at 10 A.M. by bicycle, riding toward one another. They meet at 2 P.M. If Atour's average speed is two-thirds of Tran's, how fast does each cycle?

7. Colin paddles a kayak downstream for 3 hr. After having lunch, he paddles upstream for 5 hr. At that time, he is still 6 mi short of his starting point. If the speed of the stream is 2 mph, how fast did Colin's kayak move in still water? How far downstream did he travel?

8. Jessica paddles a kayak downstream for 4 hr. After having lunch, she paddles upstream for 6 hrs. At that time, she is still 4 mi short of her starting point. If the speed of the stream is 1 mph, how fast did Jessica's kayak move in still water? How far downstream did she travel?

9. Mr. Zaleva flew his private plane from his office to his company's storage facility bucking a 20 mph head wind all the way. He flew home the same day with the same wind at his back. The round trip took 10 hr of flying time. If the speed of the plane would have been 100 mph in still air, how far is the storage facility from his office?

10. Mr. Summers drove his motorboat upstream a certain distance while pulling his son Brian on a water ski. He returned to the starting point pulling his other son Derek. The round trip took 25 min of skiing time. On both legs of the trip, the speedometer read 30 mph. If the speed of the current is 6 mph, how far upstream did he travel?

Set II In each exercise, set up the problem algebraically, solve, and check. Be sure to state what your variables represent.

1. The Dent family left Chino by car at 6 A.M., bound for Lake Mojave. Their neighbor, Mr. Scott, left in his car at 7 A.M., also bound for Lake Mojave. By traveling 10 mph faster, Mr. Scott overtook the Dents at noon.

   a. Find the average speed of each car.

   b. Find the total distance traveled before they met.

2. Bill and Andrew live 34 mi apart. Both leave their homes at 9 A.M., walking toward each other. They meet at 1 P.M. If Andrew's average speed is $\frac{1}{2}$ mph faster than Bill's, how fast does each walk? How far from Bill's house did they meet?

3. Lori hiked from her camp to a waterfall and returned to camp later in the day. She walked at a rate of $1\frac{1}{2}$ mph going to the waterfall and 3 mph coming back. The trip to the waterfall took 1 hr longer than the trip back.

   a. How long did it take her to hike to the waterfall?

   b. How far is it from her camp to the waterfall?

4. David jogged from his home to a park at the rate of 7 mph. He later walked home at the rate of 5 mph. If the trip home took 0.6 hours longer than the trip to the park, how long did it take David to get from his home to the park? How far is the park from his home?

5. Matthew and Lucas live 63 mi apart. Both leave their homes at 8 A.M. by bicycle, riding toward one another. They meet at 11 A.M. If Lucas's average speed is three-fourths of Matthew's, how fast does each cycle?

6. Anthony and Mark leave a marina at the same time, traveling in the same direction. The speed of Mark's boat is $\frac{7}{8}$ the speed of Anthony's boat. In 3 hr, Anthony is 12 mi ahead of Mark. How fast is each going?

7. In their houseboat, the Powitzsky family motors upstream for 4 hr. After lunch, they motor downstream for 2 hr. At that time, they are still 12 mi away from the marina where they began. If the speed of the houseboat in still water is 15

mph, what is the speed of the stream? How far upstream did the Powitzskys travel?

8. Plane A can travel a distance of 630 mi in $2\frac{1}{2}$ hr. The speed of plane B is $\frac{6}{7}$ that of plane A. If both planes leave an airport at the same time and fly in opposite directions, how long will it take them to be 1,872 mi apart?

9. Mr. Lee flew his private plane from his home to a nearby recreation area, bucking a 25 mph head wind all the way. He flew back home the same day with the same wind at his back. The round trip took 8 hr. If the speed of the plane would have been 100 mph in still air, how far is his home from the recreation area?

10. Two trains are traveling in the same direction, with the first train going 6 mph slower than the second train. The first train leaves a depot at 4 A.M. The second leaves the same depot at 5 A.M. and passes the first train at 10 A.M. How fast is each train going?

3.4 Mixture Problems

Mixture problems, used in business, usually involve mixing two or more dry ingredients. This type of problem can be regarded as a *money problem*, since the cost of each item is generally important.

Students often find a chart helpful in solving mixture problems. The general method is the same as that shown in Section 3.3, although the chart itself is different and we use the following relationships rather than a specific formula.

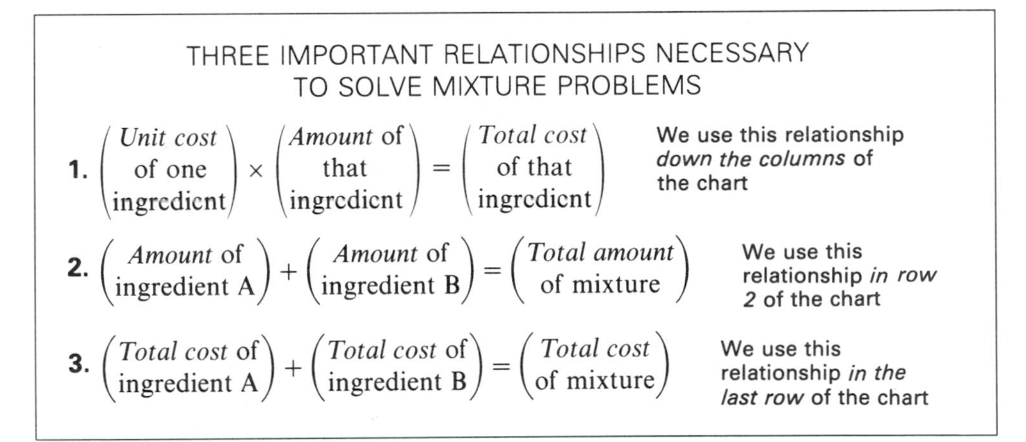

THREE IMPORTANT RELATIONSHIPS NECESSARY TO SOLVE MIXTURE PROBLEMS

1. $\begin{pmatrix}\textit{Unit cost}\\ \text{of one}\\ \text{ingredient}\end{pmatrix} \times \begin{pmatrix}\textit{Amount}\text{ of}\\ \text{that}\\ \text{ingredient}\end{pmatrix} = \begin{pmatrix}\textit{Total cost}\\ \text{of that}\\ \text{ingredient}\end{pmatrix}$ We use this relationship *down the columns* of the chart

2. $\begin{pmatrix}\textit{Amount}\text{ of}\\ \text{ingredient A}\end{pmatrix} + \begin{pmatrix}\textit{Amount}\text{ of}\\ \text{ingredient B}\end{pmatrix} = \begin{pmatrix}\textit{Total amount}\\ \text{of mixture}\end{pmatrix}$ We use this relationship *in row 2* of the chart

3. $\begin{pmatrix}\textit{Total cost}\text{ of}\\ \text{ingredient A}\end{pmatrix} + \begin{pmatrix}\textit{Total cost}\text{ of}\\ \text{ingredient B}\end{pmatrix} = \begin{pmatrix}\textit{Total cost}\\ \text{of mixture}\end{pmatrix}$ We use this relationship *in the last row* of the chart

The *last row* of the chart gives us the equation to use. The chart is as follows:

	Ingredient A	Ingredient B	Mixture	
Unit cost				← This square is sometimes empty
Amount		+	=	
Total cost		+	=	← This row gives us the equation

Example 1 A wholesaler makes up a 50-lb mixture of two kinds of coffee, one costing \$3.40 per pound and the other costing \$3.60 per pound. How many pounds of each kind of coffee must be used if the mixture is to cost \$3.54 per pound?

Step 1. Let x = number of pounds of less expensive coffee

$50 - x$ = number of pounds of more expensive coffee

A WORD OF CAUTION

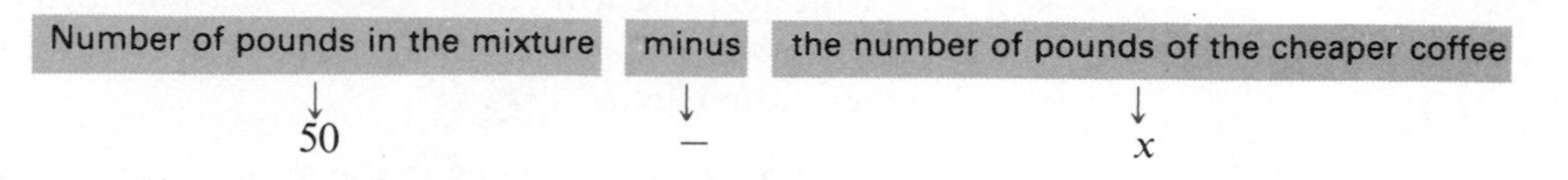

$$50 \quad - \quad x$$

equals the number of pounds left over (the number of pounds of the more expensive coffee). Students sometimes write $x - 50$ or $x + 50$ as the amount of the more expensive coffee; however, in neither case would there be 50 lb in the mixture:

	Amount of ingredient A	+	Amount of ingredient B	=	Total amount of mixture	
Trying $x - 50$:	x	+	~~$(x - 50)$~~	=	~~$2x - 50$,~~	not 50
Trying $x + 50$:	x	+	~~$(x + 50)$~~	=	~~$2x + 50$,~~	not 50
Using $50 - x$:	x	+	$(50 - x)$	=	50	✓

Chart

	Less expensive coffee	More expensive coffee	Mixture
Unit cost	\$3.40	\$3.60	\$3.54
Amount	x +	$50 - x$ =	50
Total cost	\$3.40x +	\$3.60$(50 - x)$ =	\$3.54(50)

Reread

Notice that Amount of less expensive coffee + Amount of more expensive coffee = Amount of mixture

$$x + (50 - x) = 50$$ Second row of chart

and that Total cost of less expensive coffee + Total cost of more expensive coffee = Total cost of mixture

Step 2. $3.40x + 3.60(50 - x) = 3.54(50)$ Last row of chart

Step 3. $340x + 360(50 - x) = 354(50)$ Multiplying both sides by 100 to clear decimals

$$340x + 18{,}000 - 360x = 17{,}700$$

$$-20x = -300$$

Step 4. $x = 15$ — Less expensive coffee

$50 - x = 35$ — More expensive coffee

Step 5. *Check* Total cost of less expensive coffee is 15($3.40) = $51.00
Total cost of more expensive coffee is 35($3.60) = +126.00
$177.00

Total cost of mixture is 50($3.54) = $177.00

Step 6. Therefore, the wholesaler should use 15 lb of the $3.40 coffee and 35 lb of the $3.60 coffee. ■

Example 2 Bill has $3.85 in nickels, dimes, and quarters. If he has twice as many nickels as quarters and 34 coins altogether, how many of each kind of coin does he have?

Step 1. Let

x = the number of quarters

$2x$ = the number of nickels — There are twice as many nickels as quarters

$34 - [x + 2x]$ = the number of dimes — There are 34 coins in all, so when we subtract $[x + 2x]$ (the sum of the number of quarters and the number of nickels), we have the number of dimes left over

Chart

	Quarters		Nickels		Dimes		Mixture
Value of each coin	$0.25		$0.05		$0.10		XXXXXXX XXXXXXX
Number of coins	x	+	$2x$	+	$34 - 3x$	=	34
Total value	$\$0.25x$	+	$\$0.05(2x)$	+	$\$0.10(34 - 3x)$	=	$3.85

Step 2. $0.25x + 0.05(2x) + 0.10(34 - 3x) = 3.85$ — Bottom row of chart

Step 3. $25x + 5(2x) + 10(34 - 3x) = 385$ — Multiplying both sides by 100

$25x + 10x + 340 - 30x = 385$ — Simplifying

$5x = 45$ — Combining like terms

Step 4. $x = 9$ — The number of quarters

$2x = 18$ — The number of nickels

$34 - 3x = 7$ — The number of dimes

Step 5. *Check*
9 quarters: $2.25
18 nickels: 0.90
7 dimes: 0.70
34 coins $3.85

There are twice as many nickels as quarters and 34 coins altogether

The total value is $3.85

Step 6. Therefore, Bill has 9 quarters, 18 nickels, and 7 dimes. ■

EXERCISES 3.4

Set I In each exercise, set up the problem algebraically, solve, and check. Be sure to state what your variables represent.

1. A dealer makes up a 100-lb mixture of Colombian coffee costing $3.90 per pound and Brazilian coffee costing $3.60 per pound. How many pounds of each kind must be used in order for the mixture to cost $3.72 per pound?
2. A dealer makes up a 50-lb mixture of cashews and peanuts. If the cashews cost $7.60 a pound and the peanuts $2.60 a pound, how many pounds of each kind of nut must be used in order for the mixture to cost $4.20 a pound?
3. A 10-lb mixture of almonds and walnuts costs $39.72. If walnuts cost $4.50 a pound and almonds $3.00 a pound, how many pounds of each kind are there?
4. Mrs. Diederich paid $15.81 for a 6-lb mixture of granola and dried apple chunks. If the granola cost $2.10 a pound and the dried apple chunks cost $4.24 a pound, how many pounds of each kind did she buy?
5. Doris has 17 coins with a total value of $1.15. If all the coins are nickels and dimes, how many of each kind of coin does she have?
6. Heather has 17 coins with a total value of $3.35. If the coins are all dimes and quarters, how many of each kind of coin does she have?
7. Dianne has $3.20 in nickels, dimes, and quarters. If she has 7 more dimes than quarters and 3 times as many nickels as quarters, how many of each kind of coin does she have?
8. Trisha has $4.15 in nickels, dimes, and quarters. If she has 5 more nickels than dimes and twice as many quarters as dimes, how many of each kind of coin does she have?
9. Mrs. Robinson mixes 15 lb of English toffee candy costing $6.60 a pound with peanut brittle costing $4.20 a pound. How many pounds of peanut brittle must she use to make a mixture costing $5.10 a pound?
10. Mrs. Reid mixes 24 lb of macadamia nuts costing $8.30 a pound with peanuts costing $3.30 a pound. How many pounds of peanuts must she use to make a mixture costing $6.30 a pound?
11. Joyce spent $108.50 for 23 tickets to a movie. Some were for children and some for adults. If each child's ticket cost $3.50 and each adult's ticket cost $5.25, how many of each kind of ticket did she buy?
12. Jerry spent $9.18 for 45 stamps. Some were 18¢ stamps and some were 22¢ stamps. How many of each kind did he buy?

Set II In each exercise, set up the problem algebraically, solve, and check. Be sure to state what your variables represent.

1. A 100-lb mixture of Delicious and Spartan apples costs 96¢ a pound. If the Delicious apples cost 99¢ a pound and the Spartan apples cost 89¢ a pound, how many pounds of each kind should be used?
2. A delicatessen makes up a 40-pt mixture of peaches and pears for a fruit salad. Peaches cost 71¢ a pint and pears cost 79¢ a pint. If the 40-pt mixture costs $29.76, how many pints of each kind of fruit must be used?
3. Mrs. Koontz paid $36.85 for a 10-lb mixture of Brand A and Brand B coffee. If

Brand A cost \$3.85 a pound and Brand B cost \$3.60 a pound, how many pounds of each brand did she buy?

4. A 5-lb mixture of caramels and nougats costs \$11.27. If nougats cost \$2.30 a pound and caramels \$2.20 a pound, how many pounds of each kind are there?

5. Michelle has 14 coins with a total value of \$1.85. If all the coins are dimes and quarters, how many of each kind of coin does she have?

6. Wing has 18 coins consisting of nickels, dimes, and quarters. If he has 5 more dimes than quarters and if the total value of the coins is \$1.90, how many of each kind of coin does he have?

7. Michael has \$2.25 in nickels, dimes, and quarters. If he has 3 fewer dimes than quarters and as many nickels as the sum of the dimes and quarters, how many of each kind of coin does he have?

8. A 50-lb mixture of Delicious and Granny Smith apples costs \$47.70. If the Delicious apples cost 99¢ a pound and the Granny Smith apples cost 89¢ a pound, how many pounds of each kind are there?

9. Mrs. Curtis mixes 16 lb of Brand A coffee costing \$3.50 a pound with Brand B coffee costing \$3.60 a pound. If the mixture is to cost \$3.56 a pound, how many pounds of Brand B should she use?

10. Jennifer's piggy bank contains nickels, dimes, and quarters. She has 4 more dimes than quarters and 3 times as many nickels as dimes. The total value of the coins is \$5.00. How many of each kind of coin does she have?

11. A 50-lb mixture of Delicious and Golden Delicious apples costs \$53.30. If the Delicious apples cost \$1.29 a pound and the Golden Delicious apples cost 89¢ a pound, how many pounds of each kind are there?

12. A dealer makes up a 10-lb mixture of dried tropical fruit and raisins. The tropical fruit costs \$1.58 a pound, and the raisins cost \$1.38 a pound. How many pounds of each should be used if the mixture is to cost \$1.49 a pound?

3.5 Solution Problems

Another type of mixture problem, found in chemistry and nursing, involves the mixing of liquids. Such problems are often called *solution problems* because a mixture of two or more liquids is, under certain conditions, a solution.*

A "60% solution of alcohol" is a mixture that is 60% pure alcohol and 40% water. To find the number of milliliters of pure alcohol in, for example, 80 ml of a 60% solution of alcohol, change 60% to a decimal (0.6) and multiply that number by 80 ml (answer: 48 ml of the solution is pure alcohol). If we add pure alcohol to a solution to get a *stronger* solution, we are adding a 100% solution of alcohol. If we add water to obtain a *weaker* solution, we are adding a 0% solution of alcohol.

An alloy of metals is often a solution, too. To find the amount of pure tin in 10 oz of a lead-tin alloy that is 18% tin, change 18% to 0.18 and multiply 0.18 by 10 oz (answer: 1.8 oz of the solution is pure tin).

Solution problems can be solved by using a chart method similar to that used for solving other mixture problems. We use the following relationships rather than a specific formula.

* A **solution** is a homogenous mixture of two or more substances.

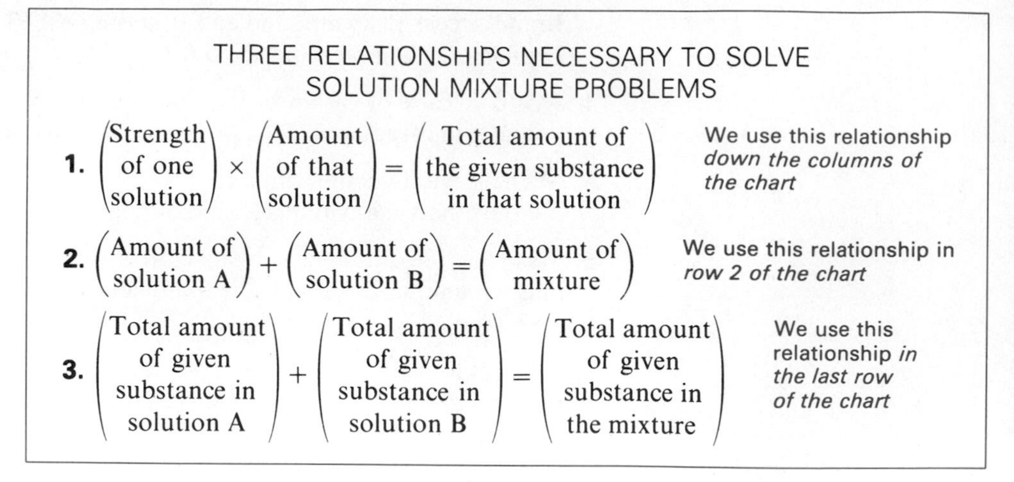

THREE RELATIONSHIPS NECESSARY TO SOLVE SOLUTION MIXTURE PROBLEMS

1. $\begin{pmatrix}\text{Strength}\\ \text{of one}\\ \text{solution}\end{pmatrix} \times \begin{pmatrix}\text{Amount}\\ \text{of that}\\ \text{solution}\end{pmatrix} = \begin{pmatrix}\text{Total amount of}\\ \text{the given substance}\\ \text{in that solution}\end{pmatrix}$ We use this relationship *down the columns of the chart*

2. $\begin{pmatrix}\text{Amount of}\\ \text{solution A}\end{pmatrix} + \begin{pmatrix}\text{Amount of}\\ \text{solution B}\end{pmatrix} = \begin{pmatrix}\text{Amount of}\\ \text{mixture}\end{pmatrix}$ We use this relationship in *row 2 of the chart*

3. $\begin{pmatrix}\text{Total amount}\\ \text{of given}\\ \text{substance in}\\ \text{solution A}\end{pmatrix} + \begin{pmatrix}\text{Total amount}\\ \text{of given}\\ \text{substance in}\\ \text{solution B}\end{pmatrix} = \begin{pmatrix}\text{Total amount}\\ \text{of given}\\ \text{substance in}\\ \text{the mixture}\end{pmatrix}$ We use this relationship *in the last row of the chart*

The chart is as follows:

	Solution A	Solution B	Mixture
Strength (percent)			
Amount		+	=
Total amount of pure substance		+	=

← This row gives us the equation (last row)

Example 1 How many milliliters (ml) of water must be added to 60 ml of a 5% glycerin solution to reduce it to a 2% solution?

Step 1. Let x = number of ml of water to be added.

Chart

(Water is 0% glycerin)

	5% solution	Water	Mixture
Strength (percent)	0.05	0.0	0.02
Amount	60 +	x =	$60 + x$
Total amount of pure glycerin	0.05(60) +	$0.0x$ =	$0.02(60 + x)$

Reread Amount of glycerin in 5% solution + Amount of glycerin in water added = Amount of glycerin in 2% solution

Step 2. $0.05(60) + 0 = 0.02(60 + x)$

Step 3. $5(60) + 0 = 2(60 + x)$

$$300 = 120 + 2x$$

$$180 = 2x$$

Step 4. $x = 90$

Step 5. *Check* In 60 ml of a 5% glycerin solution, there are (60 ml)(0.05) or 3 ml of glycerin. The total volume of the mixture is 90 ml + 60 ml or 150 ml. If there are 3 ml of glycerin in 150 ml of solution (no more glycerin was added), the solution is a $\left(\frac{3 \text{ ml}}{150 \text{ ml}} \times 100\right)$% or 2% solution.

Step 6. Therefore, 90 ml of water must be added to reduce the 5% solution to a 2% solution. ■

Example 2 How many liters of a 20% alcohol solution must be added to 3 ℓ of a 90% alcohol solution to make an 80% solution?

Step 1. Let x = number of liters of 20% solution.

Chart

	20% solution	90% solution	Mixture
Strength	0.20	0.90	0.80
Amount	x +	3 =	$x + 3$
Total	$0.20x$ +	$0.90(3)$ =	$0.80(x + 3)$

Reread Amount of alcohol in 20% solution + Amount of alcohol in 90% solution = Amount of alcohol in 80% solution

Step 2. $0.20x + 0.90(3) = 0.80(x + 3)$

Step 3. $2x + 9(3) = 8(x + 3)$

$2x + 27 = 8x + 24$

$3 = 6x$

Step 4. $x = \frac{1}{2}$

Step 5. *Check* Three liters of a 90% alcohol solution contains (3 ℓ)(0.90) or 2.7 ℓ of pure alcohol. One-half liter of a 20% alcohol solution contains ($\frac{1}{2}$ ℓ)(0.20) or 0.1 ℓ of pure alcohol. Therefore, there are 2.8 ℓ of pure alcohol in the mixture. The total volume of the mixture is 3.5 ℓ, and the strength of the mixture is $\left(\frac{2.8 \ \ell}{3.5 \ \ell} \times 100\right)$% or 80%.

Step 6. Therefore, 1/2 ℓ of the 20% alcohol solution must be added. ■

EXERCISES 3.5

Set I In each exercise, set up the problem algebraically, solve, and check. Be sure to state what your variables represent.

1. How many milliliters of water must be added to 500 ml of a 40% solution of hydrochloric acid to reduce it to a 25% solution?

2. How many liters of water must be added to 10 ℓ of a 30% solution of antifreeze to reduce it to a 20% solution?

3. How many liters of pure alcohol must be added to 10 ℓ of a 20% solution of alcohol to make a 50% solution?

4. How many milliliters of pure alcohol must be added to 1,000 ml of a 10% solution of alcohol to make a 40% solution?

5. How many cubic centimeters (cc) of a 20% solution of sulfuric acid must be mixed with 100 cc of a 50% solution to make a 25% solution of sulfuric acid?

6. How many pints of a 2% solution of disinfectant must be mixed with 5 pt of a 12% solution to make a 4% solution of disinfectant?

7. If 100 gal of 75% glycerin solution are made up by combining a 30% glycerin solution with a 90% glycerin solution, how much of each solution must be used?

8. Two copper alloys, one containing 65% copper and the other 20% copper, are to be mixed to produce an alloy that is 35% copper. How much of the 65% alloy should be used to obtain 1,200 g of an alloy that is 35% copper?

9. A chemist has two solutions of hydrochloric acid. One is a 40% solution, and the other is a 90% solution. How many liters of each should be mixed to get 10 ℓ of a 50% solution?

10. If whole milk contains 3.25% butterfat and nonfat milk contains 0.5% butterfat, how much whole milk should be added to 1,000 ml of nonfat milk to obtain milk with 2% butterfat?

Set II In each exercise, set up the problem algebraically, solve, and check. Be sure to state what your variables represent.

1. How many cubic centimeters of water must be added to 600 cc of a 20% solution of potassium chloride to reduce it to a 5% solution?

2. How many milliliters of a 36% solution of hydrochloric acid should be added to 1,000 ml of a 60% solution to make a 56% solution?

3. How many liters of pure alcohol must be added to 200 ℓ of a 5% solution to make a 62% solution?

4. A 30% solution of antifreeze is to be mixed with an 80% solution of antifreeze to make 1,000 ml (1 ℓ) of a 40% solution. How many milliliters of each should be used?

5. How many cubic centimeters of a 25% solution of nitric acid must be mixed with 100 cc of a 5% solution to make a 21% solution?

6. A 30% acetic acid solution must be mixed with a 5% acetic acid solution to obtain 100 ℓ of a 25% acetic acid solution. How many liters each of the 30% solution and the 5% solution must be used?

7. A chemist has two solutions of hydrochloric acid, one a 40% solution and the other a 90% solution. How many liters of each should be mixed to get 20 ℓ of a 75% solution?

8. If 1,500 cc of a 10% dextrose solution is made up by combining a 20% solution with a 5% solution, how much of each solution must be used?

9. A certain fertilizer has a 20% nitrogen content. Another fertilizer is 40% nitrogen. How much of each should be mixed together to get a 100-kg mixture with a 32% nitrogen content?

10. How many liters of a 24% acetic acid solution must be mixed with 3 ℓ of a 32% solution to make a 28.8% solution?

3.6 Miscellaneous Word Problems

Some word problems involve the individual digits of a number. Consider the number 59. In that number, we say that 5 is the *tens digit* and 9 is the *units digit*. The number itself equals $5(10) + 9$. In general, a two-digit number equals the tens digit times ten, *plus* the units digit. A three-digit number equals the hundreds digit times 100, *plus* the tens digit times 10, *plus* the units digit.

If we reverse the digits in a two-digit number, the new number will be the old units digit times 10, *plus* the old tens digit. For example, if we reverse the digits of the number 75, which is $7(10) + 5$, we get the number 57, which is $5(10) + 7$.

Example 1 The sum of the digits of a two-digit number is 8. The number formed by reversing the digits is 36 more than the original number. Find the original number.

Step 1. Let $x =$ units digit
$8 - x =$ tens digit

Think The sum of the digits is 8. (Notice that $x + (8 - x) = 8$.)
The value of the original number: $10(8 - x) + x$
The value of the number with the digits reversed: $10x + (8 - x)$

Reread [Value of the original number] + 36 = [Value of the number with digits reversed]

Step 2. $[10(8 - x) + x] + 36 = 10x + (8 - x)$

Step 3. $80 - 10x + x + 36 = 10x + 8 - x$

$-9x + 116 = 9x + 8$

$108 = 18x$ Adding $9x - 8$ to both sides

Step 4. $x = 6$ Units digit

$8 - x = 8 - 6 = 2$ Tens digit

The number appears to be 26.

Step 5. *Check* The sum of the digits of "26" is $2 + 6$, or 8. The number formed by reversing the digits is "62," and 62 is 36 more than 26.

Step 6. Therefore, the original number is 26. ■

The only way to learn to solve word problems is to attempt to solve lots of them! In this section, we want you to try your skills at general problem solving, so we provide you with a number of different kinds of word problems to solve. Some may be money problems, some percent problems, and so forth, but some problems are different from any you have seen so far; you must devise your own solutions. You can (and should) solve all the problems by using the *algebraic methods* that have been covered in this book. The odd and even problems are not necessarily "matched."

EXERCISES 3.6

Set I In each exercise, set up the problem algebraically, solve, and check. Be sure to state what your variables represent.

1. The sum of the digits of a two-digit number is 13. The number formed by reversing the digits is 27 more than the original number. Find the original number.

2. Karla can reach Rachelle's house from her house in 3 hr if she averages 45 mph. If Susan lives 35 mi closer to Karla than Rachelle does, what is the distance between Karla's house and Susan's house?

3. Heather has an octagonally shaped aquarium that contains some water. Right now it is one-third full. If she adds 3,000 cubic inches of water, it will be three-fourths full. What is the volume of the tank (in cubic inches)?

4. The sum of the digits of a two-digit number is 7. The number formed by reversing the digits is 27 less than the original number. Find the original number.

5. The units digit of a three-digit number is twice the hundreds digit. The sum of the digits is 6. If the digits are reversed, the new number is 198 more than the original number. Find the original number.

6. Rebecca is supposed to practice the piano for a certain number of minutes each day. So far today, she has practiced two-fifths of that time. If she spends 12 min more at the piano, she will have practiced two-thirds of her allotted time. How many minutes is she supposed to practice each day?

For Exercises 7 and 8, use the following: Nickel coinage is composed of copper and nickel in the ratio of 3 to 1.

7. How much pure copper should be melted with 20 lb of a copper-nickel alloy containing 60% copper to obtain the alloy for nickel coinage?

8. How much pure copper should be melted with 30 lb of a copper-nickel alloy containing 40% copper to obtain the alloy for nickel coinage?

9. Farah invested $13,000. Part of the money was placed in an account that paid 5.25% interest per year, and the rest was put into an account that paid 5.75% interest per year. If the interest Farah earned the first year was $702.50, how much did she place in each account?

10. The units digit of a three-digit number is twice the hundreds digit. The sum of the digits is 11. If the digits are reversed, the new number is 297 more than the original number. Find the original number.

11. A druggist buys 18 cameras. If she had bought 10 cameras of a higher quality, she would have paid $48 more per camera for the same total expenditure. Find the price of each type of camera.

12. Fifty-six percent of the children enrolled in kindergarten at Jill's school are boys. Last Monday, 12.5% of the boys and 25% of the girls went on a field trip. If 246 children did *not* go on the field trip, how many children are enrolled in kindergarten at Jill's school?

Set II In each exercise, set up the problem algebraically, solve, and check. Be sure to state what your variables represent.

1. The sum of the digits of a two-digit number is 12. The number formed by reversing the digits is 36 more than the original number. Find the original number.

2. The ratio of the tens digit to the units digit of a two-digit number is 3 to 2. The sum of the digits is 10. What is the number?

3. The units digit of a three-digit number is 3 times the hundreds digit. The sum of the digits is 12. If the digits are reversed, the new number is 594 more than the original number. Find the original number.

4. The tens digit of a two-digit number is 4 times the units digit. If the digits are reversed, the new number is 54 less than the original number. Find the original number.

5. How much pure copper should be melted with 50 lb of a copper-nickel alloy containing 40% copper to obtain an alloy that is 50% copper?

6. George has received exam scores of 83, 75, 62, and 87. What score must he receive on the fifth exam in order to have an average score of 80?

7. Jim invested $20,000. Some of the money was invested at 7.4% interest per year, and the rest was invested at 6.8%. The amount of interest he earned for the year was $1,469.20. How much did he invest at each rate?

8. Tickets for some seats at a concert sold for $9 each, and tickets for better seats sold for $12 each. If 1,000 tickets were sold in all and if the amount of money brought in was $10,320, how many of the $9 tickets were sold?

9. A dealer buys 22 cameras. If he had bought 15 cameras of a higher quality, he would have paid $35 more per camera for the same total expenditure. Find the price of each type of camera.

10. Scott is 6 times as old as Sharon. In 20 years, Scott will be only twice as old as Sharon. How old is Scott now?

11. Jason wants to enclose a rectangular area with 336 ft of fencing. If the rectangle is to be three times as long as it is wide, what must the length and width of the rectangle be?

12. Kevin had a basket that contained only red and green balls, 40% of which were green. Then 25% of the red balls and 75% of the green balls were removed. If 33 balls remained in the basket, how many balls did Kevin have in the basket originally?

3.7 Review: 3.1–3.6

Method for Solving Word Problems 3.1–3.6

Read	To solve a word problem, first read it very carefully. *Be sure* you understand the problem. Read it several times, if necessary.
Think	Determine what *type* of problem it is, if possible. Determine what is unknown. Is enough information given so that you *can* solve the problem? Do you need a special formula? What operations must be used? What is the domain of the variable? What kind of answer is reasonable?
Sketch	Draw a sketch *with labels*, if possible.
Step 1.	Represent one unknown number by a variable, and declare the meaning in a sentence of the form "Let $x = \ldots$." Then reread the problem to see how you can represent any other unknown numbers in terms of the same variable.
Reread	Reread the entire word problem, breaking it up into small pieces.
Step 2.	Translate each English phrase into an algebraic expression and fit these expressions together into an equation or inequality.
Step 3.	Using the methods described in Chapter 2, solve the equation or inequality.

Step 4. Solve for *all* the unknowns asked for in the problem.

Step 5. Check the solution(s) *in the word statement.*

Step 6. State your results clearly.

(For more details on the "chart" method for solving distance-rate-time problems, see Section 3.3; for mixture problems, see Section 3.4; and for solution mixture problems, see Section 3.5.)

Review Exercises 3.7 Set I

In each exercise, set up the problem algebraically, solve, and check. Be sure to state what your variables represent.

1. When 5 times an unknown number is subtracted from 28 and the result is divided by 3, the quotient is 3 times the unknown number. Find the unknown number.
2. The sum of three consecutive integers is 23. What are the integers?
3. The sides of a triangle are in the ratio 3:7:8. The perimeter is to be less than 36. What is the range of values that the *shortest* side can have?
4. Gwen has $2.20 in nickels, dimes, and quarters. If she has 3 more dimes than quarters and 2 fewer nickels than quarters, how many of each kind of coin does she have?
5. A dealer makes up a 15-lb mixture of different candies costing $2.20 and $2.60 a pound. How many pounds of each kind of candy must be used for the mixture to cost $2.36 a pound?
6. How many cubic centimeters of water must be added to 10 cc of a 17% solution of a disinfectant to reduce it to a 0.2% solution?
7. Mr. Sontag takes 30 min to drive to work in the morning, but he takes 45 min to return home over the same route during the evening rush hour. If his average morning speed is 10 mph faster than his average evening speed, how far is it from his home to his work?
8. The sum of the digits of a two-digit number is 11. If the digits are reversed, the new number is 45 less than the original number. What is the original number?
9. Mr. Curtis invested part of $27,000 at 12% interest and the remainder at 8%. His total yearly income from these two investments is $2,780. How much is invested at each rate?
10. Three court reporters are planning to buy some equipment (a computer, the necessary software, and a laser printer) to use in their work. They plan to share the cost equally. If they allow one more person to join their group and to share the cost equally with them, the cost for each of the original three reporters will be reduced by $1,125. What is the total cost of the equipment?
11. Two airplanes leave an airport at the same time, flying in opposite directions. One is flying 25 mph faster than the other. If the planes are 570 mi apart after 2 hr, what is the rate of speed for each airplane?

Review Exercises 3.7 Set II

NAME

In each exercise, set up the problem algebraically, solve, and check. Be sure to state what your variables represent.

ANSWERS

1. ______________

____

2. ______________

3. ______________

4. ______________

5. ______________

____

1. Dimitri has 19 coins in her pocket with a total value of \$3.10. If all the coins are dimes and quarters, how many of each does she have?

2. How many cubic centimeters of a 50% phenol solution must be added to 400 cc of a 5% solution to make it a 10% solution?

3. After sailing downstream for 2 hr, it takes a boat 7 hr to return to its starting point. If the speed of the boat in still water is 9 mph, what is the speed of the stream?

4. The sum of the digits of a two-digit number is 10. If the digits are reversed, the new number is 72 less than the original number. Find the original number.

5. Mrs. McMahon paid \$5.63 for a total of 12 cans of beef soup and tomato soup. If the beef soup costs 65¢ a can and the tomato soup 34¢ a can, how many cans of each kind of soup did she buy?

6. The sum of three consecutive integers is -222. Find the integers.

7. The ratio of the length to the width of a rectangle is 7 to 3. The perimeter is to be less than or equal to 40. What is the range of values that the width can have?

8. When 7 times an unknown number is subtracted from 120 and the result is divided by 2, the quotient is 46. Find the unknown number.

9. Mr. Reid had $15,000 to invest. Part of the money was invested at 7.8% interest per year, and the rest was invested at 7.2% interest per year. If the amount of money earned the first year was $1,138.80, how much was invested at each rate?

10. Lupe is twice as old as Juana. In 9 yr, Juana will be five-sevenths as old as Lupe is then. How old is Juana now?

11. One car leaves Boston at 8 A.M. traveling at 40 mph. One and a half hours later, a second car starts from the same point, traveling at 45 mph along the same route. How long will it take the second car to overtake the first car?

ANSWERS

6. ______________

____

____

7. ______________

8. ______________

9. ______________

____

10. ______________

11. ______________

Chapter 3 Diagnostic Test

The purpose of this test is to see how well you understand solving word problems. We recommend that you work this diagnostic test *before* your instructor tests you on this chapter. Allow yourself about 50 minutes.

Complete solutions for all the problems on this test, together with section references, are given in the answer section at the end of the book. For the problems you do incorrectly, study the sections cited.

Set up each problem algebraically and solve. Be sure to state what your variables represent.

1. When 23 is added to 4 times an unknown number, the sum is 31. Find the unknown number.
2. The sum of two consecutive integers is 55. Find the integers.
3. The length and width of a rectangle are in the ratio of 5 to 4. The perimeter is 90. Find the length and the width.
4. Linda has 14 coins with a total value of \$2.15. If all the coins are dimes and quarters, how many of each kind of coin does she have?
5. A grocer makes up a 60-lb mixture of cashews and peanuts. If the cashews cost \$7.40 a pound and the peanuts cost \$2.80 a pound, how many pounds of each kind of nut must be used for the mixture to cost \$4.18 a pound?
6. How many cubic centimeters of water must be added to 600 cc of a 20% solution of potassium chloride to reduce it to a 15% solution?
7. The units digit of a two-digit number is twice the tens digit. If the digits are reversed, the value is increased by 27. Find the original number.
8. Mrs. Rice invested part of \$23,000 at 8% interest and the remainder at 10%. Her total yearly income from these two investments is \$2,170. How much is invested at each rate?
9. Roy hikes from his camp to a mountain lake one day and returns the next day. He walks up to the lake at a rate of 3 mph and returns to his camp at a rate of 5 mph. The hike to the lake takes Roy 2 hr longer than the return trip.

   a. How long does it take him to hike to the lake?

   b. How far is the lake from his camp?
10. A discount store buys 25 stereos. If it had bought 14 stereos of a higher quality, it would have paid \$55 more per stereo for the same total expenditure. Find the price of each type of stereo.

Cumulative Review Exercises: Chapters 1–3

1. Given the numbers $-2, \frac{1}{2}, 4.5, 1.4142136\ldots, 10, 0.$ and $0.\overline{234}$:

a. Which are natural numbers?

b. Which are real numbers?

c. Which are rational numbers?

d. Which are integers?

e. Which are irrational numbers?

2. Simplify. Write the answer using only positive exponents or no exponents.

$$\left(\frac{a^2b^2}{a^5b}\right)^{-2}$$

3. Given the formula $s = \frac{1}{2}gt^2$, find s if $g = 32$ and $t = 2$.

In Exercises 4 and 5, remove grouping symbols and combine like terms.

4. $2xy(4xy - 5x + 2) - 3y(2x^2y - x)$

5. $6 - \{4 - [3x - 2(5 - 3x)]\}$

In Exercises 6–10, find the solution set for each equation or inequality.

6. $\frac{x}{2} + \frac{x+3}{3} = \frac{1}{6}$

7. $12 - 3(4x - 5) \geq 9(2 - x) - 6$

8. $-3 < 2x + 1 < 5$

9. $\left|\frac{2x-6}{4}\right| = 1$

10. $|4x - 2| < 10$

In Exercises 11–14, set up the problem algebraically, solve, and check. Be sure to state what your variables represent.

11. The ratio of the length of a rectangle to its width is 7:4. The perimeter is 66 in. Find the length and the width.

12. Bud invested \$12,000. Part of the money was invested at 7.4% interest per year, and the rest was invested at 6.8% interest. If the amount of money earned the first year was \$861.00, how much was invested at each rate?

13. A metal tank is in the shape of a right circular cylinder, and it contains some water. Right now it is one-fourth full. If we add 80 cu. in. of water, it will be one-third full. What is the volume of the tank?

14. Patricia is filling an 18-cu.-ft planter with a mixture of potting soil and perlite. She wants to use twice as many cubic feet of potting soil as perlite. The potting soil comes in $1\frac{1}{2}$-cu.-ft bags, and the perlite comes in 2-cu.-ft bags. How many bags of potting soil should she buy? How many bags of perlite?

4 Polynomials

In this chapter, we look in detail at a particular type of algebraic expression called a *polynomial.* Polynomials have the same importance in algebra that whole numbers have in arithmetic. Most of the work in arithmetic involves operations with whole numbers. In the same way, most of the work in algebra involves operations with polynomials.

4.1 Basic Definitions

Polynomials

A **polynomial in one variable** is an algebraic expression that has only terms of the form ax^n, where a stands for any real number, n stands for any positive integer (or zero), and x stands for any variable. For example, $3x^3 - 4x^2 + 2x - 5$ is a polynomial in x, because each of its terms is of the form ax^n. Reminder: $-5 = -5x^0$.

A polynomial with only one term is called a **monomial**, a polynomial with two terms is called a **binomial**, and a polynomial with three terms is called a **trinomial**. We will not use special names for polynomials with more than three terms.

Example 1 Examples of algebraic expressions that are polynomials in one variable:

a. $4z^4 - 2z^2$ This polynomial is a binomial in z.

b. $7x^2 - 5x + 2$ This polynomial is a trinomial.

c. 5 This is a monomial; it is of the form $5x^0$ $(5x^0 = 5 \cdot 1 = 5)$. ■

If an algebraic expression in simplified form contains terms with negative (or rational*) exponents on the variables or if it contains terms with variables in a denominator or under a radical sign, then the algebraic expression is *not* a polynomial.

Example 2 Examples of algebraic expressions that are *not* polynomials:

a. $4x^{-2}$ This expression is *not* a polynomial because it has a negative exponent on a variable.

b. $\dfrac{2}{x-5}$ This is *not* a polynomial because the variable is in the denominator.

c. $\sqrt{x-5}$ This is *not* a polynomial because the variable is under a radical sign. ■

An algebraic expression with two variables is a **polynomial in two variables** if (1) it contains no negative or rational exponents on the variables, (2) no variables are in denominators, and (3) no variables are under radical signs.

Example 3 Examples of algebraic expressions that are polynomials in two variables:

a. $x^2y\sqrt{5} + \frac{1}{2}xy^2$ This polynomial is a binomial; note that *constants* can be under radical signs and in denominators.

b. $4u^2v^2 - 7uv + 6$ This polynomial is a trinomial in u and v.

c. $(x - y)^2 - 2(x - y) - 8$ ■

* Rational exponents are discussed in Section 7.1.

Degree of a Term of a Polynomial

To find the **degree of any term** of a polynomial, we first write the polynomial in simplest form. Then, if the polynomial has only one variable, the degree of any of its terms is the exponent *on the variable* in that term; if the polynomial has more than one variable, the degree of any of its terms is the *sum* of the exponents *on the variables* in that term.

Example 4 Examples of finding the degree of a term:

a. 5^2x^3 3rd degree Only exponents of *variables* determine the degree of the term

b. $6x^2y$ 3rd degree because $6x^2y = 6x^2y^1$ $2 + 1 = 3$

c. 14 0 degree because $14 = 14x^0$ ■

Degree of a Polynomial

The **degree of a polynomial** is defined to be the degree of its highest-degree term. Therefore, to find the degree of a polynomial, first find the degree of each of its terms. The *largest* of these numbers will be the degree of the polynomial.*

Example 5 Examples of finding the degree of a polynomial:

a. $9x^3 - 7x + 5$

0 degree term
1st degree term
3rd degree term ← Highest-degree term

Therefore, $9x^3 - 7x + 5$ is a 3rd degree polynomial.

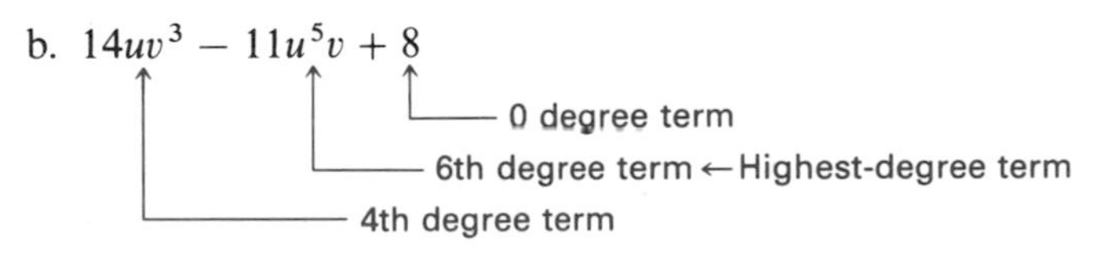

Therefore, $14uv^3 - 11u^5v + 8$ is a 6th degree polynomial. ■

Leading Coefficient The **leading coefficient** of a polynomial is the numerical coefficient of its highest-degree term.

Descending Powers When polynomials must be written in **descending powers** of one of the variables, the exponents of that variable must get smaller as we read from left to right.

Example 6 Arrange $5 - 2x^2 + 4x$ in descending powers of x and name the leading coefficient.
Solution $-2x^2 + 4x + 5$. The leading coefficient is -2. ■

A polynomial with more than one variable can be arranged in descending powers of *any one* of its variables.

* Mathematicians define the zero polynomial, 0, as having *no* degree.

Example 7 Arrange $3x^3y - 5xy + 2x^2y^2 - 10$ (a) in descending powers of x, (b) in descending powers of y.

Solution

a. $3x^3y + 2x^2y^2 - 5xy - 10$ Descending powers of x

b. $2x^2y^2 + 3x^3y - 5xy - 10$ Descending powers of y

Since y is to the same power in both terms, the higher-degree term is written first ■

Polynomial Equations

A **polynomial equation** is an equation that has a polynomial on both sides of the equal sign. The polynomial on one side of the equal sign can be the zero polynomial, 0. The **degree of the equation** equals the degree of the term in the equation with the *highest degree.*

Example 8 Examples of polynomial equations:

a. $5x - 3 = 3x + 4$ This equation is a 1st degree polynomial equation in one variable.

b. $2x^2 - 4x + 7 = 0$ This equation is a 2nd degree equation in one variable. It is also called a *quadratic equation.* ■

EXERCISES 4.1

Set I In Exercises 1–16, if the expression is a polynomial, find its degree. If it is *not* a polynomial, write "Not a polynomial."

1. $2x^2 + \frac{1}{3}x$ **2.** $\frac{1}{9}y^2 - 5$ **3.** $x^{-2} + 5x^{-1} + 4$

4. $y^{-3} + y^{-2} + 6$ **5.** 10 **6.** 20

7. $\frac{4}{x} + 7x - 3$ **8.** $\frac{3}{x} - 2x + 5$ **9.** $x\sqrt{5} + 6$

10. $y\sqrt{2} - 5$ **11.** $\sqrt{x+4}$ **12.** $\sqrt{z-3}$

13. $\frac{1}{2y^2 - 5y} - 3y$ **14.** $\frac{1}{2x^2 + 4x} + 5x$

15. $x^3y^3 - 3^7x^2y + 3^4xy^2 - y^3$ **16.** $2^7y^2z^3 - 3yz^5 + 6z^4$

In Exercises 17–20, write each polynomial in descending powers of the indicated variable and find the leading coefficient.

17. $7x^3 - 4x - 5 + 8x^5$ Powers of x

18. $10 - 3y^5 + 4y^2 - 2y^3$ Powers of y

19. $3x^2y + 8x^3 + y^3 - y^5$ Powers of y

20. $6y^3 + 7x^2 - 4y^2 + y$ Powers of y

Set II In Exercises 1–16, if the expression is a polynomial, find its degree. If it is *not* a polynomial, write "Not a polynomial."

1. $4x^3 - \frac{1}{2}x$ **2.** $x\sqrt{3} - 5$ **3.** $x^{-3} - 3x^2 + 3$

4. $4x^3 + 2x^2z^2 - 6$ **5.** 7 **6.** $\sqrt{8 - 3x}$

7. $\frac{3}{x} - 4x + 2$ **8.** $\frac{3}{4}x^5 - 2x^3y^3 + 3$ **9.** $y\sqrt{3} + 2$

10. $\frac{3}{x^3 + 2x^2 - 5x + 1}$ **11.** $\sqrt{x - 2}$ **12.** $3^5x + \sqrt{5}$

13. $5z + \frac{1}{3x^3 + 7x}$ **14.** $y^{-3} - 2y^{-2} + 3y$

15. $x^4y^3 + 2^8x^3y^2 - 3xy^5$ **16.** x

In Exercises 17–20, write each polynomial in descending powers of the indicated variable and find the leading coefficient.

17. $12b^2 - 14b^4 + 8 - 7b$ Powers of b

18. $8 - 2x + 13x^2y + 2x^4$ Powers of x

19. $3xy + 4y^5 - 3y^2 - 3$ Powers of y

20. $4st^2 - 9s^2t + 3s^2t^3 - 5t^6$ Powers of t

4.2 Addition and Subtraction of Polynomials

Addition

Polynomials can be added horizontally by removing the grouping symbols and combining like terms, as was done in Section 1.14. It is often helpful to underline all like terms with the same kind of line before adding.

Example 1 Example of adding polynomials.

$$(5x^3y^2 - 3x^2y^2 + 4xy^3) + (4x^2y^2 - 2xy^2) + (-7x^3y^2 + 6xy^2 - 3xy^3)$$
$$= \underline{5x^3y^2} - \underline{\underline{3x^2y^2}} + 4xy^3 + \underline{\underline{4x^2y^2}} - 2xy^2 - \underline{7x^3y^2} + 6xy^2 - 3xy^3$$
$$= -2x^3y^2 + x^2y^2 + xy^3 + 4xy^2 \quad \blacksquare$$

Vertical addition is sometimes desirable. In this case, it is important to have all like terms lined up vertically.

> TO ADD POLYNOMIALS VERTICALLY
>
> 1. Arrange the polynomials under one another so that like terms are in the same vertical line.
> 2. Find the sum of the terms in each vertical line by adding the numerical coefficients.

Example 2 Add $(3x^2 + 2x - 1) + (2x + 5) + (4x^3 + 7x^2 - 6)$.

Solution

$$\begin{array}{r} 3x^2 + 2x - 1 \\ 2x + 5 \\ \underline{4x^3 + \ \ 7x^2 \qquad\quad - 6} \\ 4x^3 + 10x^2 + 4x - 2 \end{array} \quad \blacksquare$$

Subtraction

Polynomials can be subtracted horizontally by removing grouping symbols and combining like terms (see Section 1.14). Remember that "subtract 3 from 5" means $5 - 3$, *not* $3 - 5$.

Example 3 Subtract $(-4x^2y + 10xy^2 + 9xy - 7)$ from $(11x^2y - 8xy^2 + 7xy + 2)$.

Solution

$$(11x^2y - 8xy^2 + 7xy + 2) - (-4x^2y + 10xy^2 + 9xy - 7)$$
$$= 11x^2y - 8xy^2 + 7xy + 2 + 4x^2y - 10xy^2 - 9xy + 7$$
$$= 15x^2y - 18xy^2 - 2xy + 9 \quad \blacksquare$$

It is necessary to know how to subtract polynomials vertically in order to be able to use long division in dividing one polynomial by another.

TO SUBTRACT POLYNOMIALS VERTICALLY

1. Write the polynomial being subtracted *under* the polynomial it is being subtracted from. Write like terms in the same vertical line.
2. *Mentally* change the sign of each term in the polynomial being subtracted.*
3. Find the sum of the *resulting* terms in each vertical line by adding the numerical coefficients.

Example 4 Subtract $(4x^2y^2 + 7x^2y - 2xy + 9)$ from $(6x^2y^2 - 2x^2y + 5xy + 8)$ vertically.

Solution

Signs changed mentally

$$\begin{array}{r} 6x^2y^2 - 2x^2y + 5xy + 8 \\ \underline{4x^2y^2 + 7x^2y - 2xy + 9} \\ 2x^2y^2 - 9x^2y + 7xy - 1 \end{array}$$

Alternate method
Sign changes shown

$$\begin{array}{r} 6x^2y^2 - 2x^2y + 5xy + 8 \\ \ominus \quad\quad \ominus \quad\quad \oplus \quad\quad \ominus \\ \underline{4x^2y^2 + 7x^2y - 2xy + 9} \\ 2x^2y^2 - 9x^2y + 7xy - 1 \end{array} \quad \blacksquare$$

EXERCISES 4.2

Set I In Exercises 1–4, perform the indicated operations.

1. $(-3x^4 - 2x^3 + 5) + (2x^4 + x^3 - 7x - 12)$
2. $(-5y^3 + 3y^2 - 3y) + (3y^3 - y + 4)$
3. $(7 - 8v^3 + 9v^2 + 4v) - (9v^3 + 6 - 8v^2 + 4v)$
4. $(3x + 7x^3 - x^2 - 5) - (3x^2 - 6x^3 + 2 - 5x)$

In Exercises 5 and 6, add the two polynomials.

5. $$\begin{array}{r} 4x^3 + 7x^2 - 5x + 4 \\ \underline{2x^3 - 5x^2 + 5x - 6} \end{array}$$

6. $$\begin{array}{r} 3y^4 - 2y^3 + 4y + 10 \\ \underline{-5y^4 + 2y^3 + 4y - 6} \end{array}$$

* Your instructor may allow or require you to *show* the sign changes. See Example 4—Alternate method.

7. Subtract $(6 + 3x^5 - 4x^2)$ from $(4x^3 + 6 + x)$.

8. Subtract $(7 - 4x^4 + 3x^3)$ from $(x^3 + 7 - 3x)$.

9. Subtract $(3y^4 + 2y^3 - 5y)$ from $(2y^4 + 4y^3 + 8)$.

10. Subtract $(2y^3 - 3y^2 + 4)$ from $(7y^3 + 5y^2 - 5y)$.

In Exercises 11–14, subtract the bottom polynomial from the top polynomial.

11. $$\begin{array}{r} -3x^4 - 2x^3 \qquad + 4x - 3 \\ \underline{2x^4 + 5x^3 - x^2 \qquad - 1} \end{array}$$

12. $$\begin{array}{r} -2x^4 \qquad + 3x^2 - x + 1 \\ \underline{- x^3 - 2x^2 + x - 5} \end{array}$$

13. $$\begin{array}{r} 5x^3 - 3x^2 + 7x - 17 \\ \underline{-2x^3 - 7x^2 \qquad - 6} \end{array}$$

14. $$\begin{array}{r} 2x^3 + 7x^2 - 3x - 12 \\ \underline{-5x^3 + 9x^2 - 2x \qquad} \end{array}$$

15. Subtract $(-3m^2n^2 + 2mn - 7)$ from the sum of $(6m^2n^2 - 8mn + 9)$ and $(-10m^2n^2 + 18mn - 11)$.

16. Subtract $(-9u^2v + 8uv^2 - 16)$ from the sum of $(7u^2v - 5uv^2 + 14)$ and $(11u^2v + 17uv^2 - 13)$.

17. Subtract the sum of $(x^3y + 3xy^2 - 4)$ and $(2x^3y - xy^2 + 5)$ from the sum of $(5 + xy^2 + x^3y)$ and $(-6 - 3xy^2 + 4x^3y)$.

18. Subtract the sum of $(2m^2n - 4mn^2 + 6)$ and $(-3m^2n + 5mn^2 - 4)$ from the sum of $(5 + m^2n - mn^2)$ and $(3 + 4m^2n + 2mn^2)$.

In Exercises 19 and 20, perform the indicated operations.

19. $(8.586x^2 - 9.030x + 6.976) - [1.946x^2 - 41.45x - (7.468 - 3.914x^2)]$

20. $(24.21 - 35.28x - 73.92x^2) - [82.04x - 53.29x^2 - (64.34 - 19.43x^2)]$

Set II In Exercises 1–4, perform the indicated operations.

1. $(-3x^5 + 2x^3 - 4x + 1) + (-4x^5 + 2x^2 - 4x - 6)$

2. $(-5x^4 - 2x^3 + 6x^2 - 3) - (-2x^4 + 4x^3 + 3x + 4)$

3. $(4x^5 + 3x^4 - 2x^2 + 5) - (-x^4 + x^3 - x^2 + x)$

4. $(-8y^3 - y^2 + 7y - 6) - (-8y^3 - y^2 + 7y - 6)$

In Exercises 5 and 6, add the two polynomials.

5. $$\begin{array}{r} 13h^3 - 8h^2 + 16h - 14 \\ \underline{6h^3 + 5h^2 - 18h + 9} \end{array}$$

6. $$\begin{array}{r} 3x^3 - 2x^2 + 5x - 3 \\ \underline{-7x^3 \qquad - 9x - 4} \end{array}$$

7. Subtract $(22 + 8y^2 - 14y)$ from $(11y^2 - 5y - 12y^3)$.

8. Subtract $(5x^3 - 3x - x^4)$ from $(-3x^2 - x^4 + 7 - x)$.

9. Subtract $(w - 12w^3 - 15 - 18w^2)$ from $(18w^3 + 5 - 9w)$.

10. Subtract $(z^3 - 3z^2 + 2 - z)$ from $(5z - z^3 + z^2)$.

In Exercises 11–14, subtract the bottom polynomial from the top polynomial.

11. $$\begin{array}{r} 2x^3 - 5x^2 \qquad - 3 \\ \underline{5x^3 + 2x^2 - 3x + 7} \end{array}$$

12. $$\begin{array}{r} -4y^4 + 6y^3 + 2y^2 \qquad - 6 \\ \underline{3y^4 \qquad + 6y^2 - y + 5} \end{array}$$

13. $$\begin{array}{r} 7z^3 - 6z^2 + 4z \qquad \\ \underline{-4z^3 - 8z^2 + 8z - 2} \end{array}$$

14. $$\begin{array}{r} -2x^3 + x^2 \qquad \qquad \\ \underline{-6x^3 - 3x^2 - 5x + 4} \end{array}$$

15. Subtract $(5xy - 12 + 3xy^2)$ from the sum of $(7x^2y + 4xy^2 - 5)$ and $(8xy^2 - 7x^2y + xy)$.

16. Subtract the sum of $(2x^3 - 5x + 4)$ and $(-5x^3 + 2x^2 - 2)$ from $(x^3 - 2x^2 - 3x)$.

17. Given the polynomials $(10a^3 - 8a + 12)$, $(11a^2 + 9a - 14)$, and $(-6a^3 + 17a)$, subtract the sum of the first two from the sum of the last two.

18. Subtract the sum of $(4m^3n^3 - 10mn)$ and $(-10m^2n^2 - 15mn)$ from the sum of $(5m^3n^3 - 8m^2n^2 + 20mn)$ and $(-14m^2n^2 + 5mn)$.

In Exercises 19 and 20, perform the indicated operations.

19. $(55.26x - 41.37 - 72.84x^2) - [28.10 - 19.05x - (89.91x^2 - 13.33)]$

20. $(23.1x^2 - 3.4x + 2) - [3.05x - 4.6x^2 - (5.13x^2 + 8.1)]$

4.3 Multiplication of Polynomials

When we multiply a monomial by a monomial, we use the techniques described in Section 1.11C (see Example 15 in that section). For example, $(3xy^2)(2x^3y^3) = 6x^4y^5$.

When we multiply a monomial by a polynomial with *more* than one term, we use the distributive property (see Section 1.14B). For example, $7xy^3(3x - 1) = 21x^2y^3 - 7xy^3$.

4.3A Products of Two Binomials

Since we often need to find the product of two binomials, it is helpful to be able to find their product by inspection. First, however, the step-by-step procedure for multiplying two binomials is shown.

Example 1 Multiply $(2x + 3)(x + 5)$.

Solution In step 1, $(2x + 3)$ is distributed over $x + 5$.

$$(2x + 3)(x + 5) = (2x + 3)(x) + (2x + 3)(5) \quad \text{Step 1}$$
$$= 2x^2 + 3x + 10x + 15 \quad \text{Step 2}$$
$$= 2x^2 + 13x + 15 \quad \text{Step 3}$$

The distributive rule is used again in step 2; in that step, the two *middle terms* are $3x$ and $10x$. We call $3x$ (the product of the two "inside" terms) the *inner product*, and we call $10x$ (the product of the two "outside" terms) the *outer product*.

Outer product $= 10x$

$(2x + 3)(x + 5)$

Inner product $= 3x$

Product of two first terms is $(2x)(x) = 2x^2$.

Product of two last terms is $3(5) = 15$. ■

When a multiplication problem is in the form $(ax + by)(cx + dy)$, as in Example 1, the multiplication can be done very quickly using the rules at the top of p. 157.

TO MULTIPLY $(ax + by)(cx + dy)$

1. The *first term* of the product is the product of the first terms of the binomials.
2. The *middle term* of the product is the sum of the inner and outer products.
3. The *last term* of the product is the product of the last terms of the binomials.

When we use this method of multiplying binomials, we find the product of the two **F**irst terms, the **O**uter product, the **I**nner product, and the product of the two **L**ast terms. For this reason, this procedure is often call the **FOIL** method.

Example 2 Multiply $(3x + 2y)(4x - 5y)$.

Solution

$$(3x + 2y)(4x - 5y) = 12x^2 - 7xy - 10y^2$$

$12x^2$: The product of the first terms

$-10y^2$: The product of the last terms

$-7xy$: The sum of the inner and outer products ■

Practice this procedure until you can write the answer without having to write down any intermediate steps. We call this *finding the product by inspection.*

We can use the FOIL method to find the product of two binomials even when the problem is *not* in the form $(ax + by)(cx + dy)$. See Examples 3 and 4.

Example 3 Multiply $(2x + 3)(5y - 7)$.

Solution $(2x + 3)(5y - 7) = 10xy - 14x + 15y - 21$ ■

Example 4 Multiply $(3x^2 + 2)(x + 5)$.

Solution $(3x^2 + 2)(x + 5) = 3x^3 + 15x^2 + 2x + 10$ ■

In Examples 3 and 4, the inner and outer products are not like terms, so they cannot be combined.

EXERCISES 4.3A

Set I Find the products.

1. $(x + 5)(x + 4)$
2. $(y + 3)(y + 7)$
3. $(y + 8)(y - 9)$
4. $(z + 10)(z - 3)$
5. $(3x + 4)(2x - 5)$
6. $(2y + 5)(4y - 3)$
7. $(4x + 5y)(4x - 5y)$
8. $(s - 2t)(s + 2t)$
9. $(2x - 3y)(5x - y)$
10. $(6u - v)(3u - 4v)$
11. $(7z - 2)(8w + 3)$
12. $(9z - 1)(4x + 5)$
13. $(8x - 9y)(8x + 9y)$
14. $(7w + 2x)(7w - 2x)$
15. $(2s^2 + 5)(s + 1)$
16. $(3u^2 + 2)(u + 1)$
17. $(2x - y)(2x - y)$
18. $(5z - w)(5z - w)$

19. $(2x + 3)(3y - 4)$ 20. $(4u - 2)(3v - 5)$ 21. $(7x^2 - 3)(7x^2 - 3)$
22. $(4x^2 - 5)(4x^2 - 5)$ 23. $(3x - 4)(2x^2 + 5)$ 24. $(5y - 2)(3y^2 + 1)$
25. $3x(2x - 4)(x - 1)$ 26. $7x(x + 3)(2x - 5)$

Set II Find the products.

1. $(z + 2)(z + 5)$ 2. $(x - 9)(x + 2)$ 3. $(z + 4)(z - 5)$
4. $(y - 6)(y - 4)$ 5. $(6x + 2)(2x - 6)$ 6. $(3x^2 + 3)(x - 1)$
7. $(3x + 4y)(3x - 4y)$ 8. $(x + 8y)(x + 8y)$ 9. $(3x - y)(5x - 7y)$
10. $(2x - 3)(5 - y)$ 11. $(8x + 3)(3y - 2)$ 12. $(4x + 5)(2x^2 + 7)$
13. $(6y - 5z)(6y + 5z)$ 14. $(7x - y)(7x - y)$ 15. $(8x^2 + 7)(x + 1)$
16. $(3x^3 - 4)(2x^3 + 3)$ 17. $(3x - y)(3x - y)$ 18. $(4x^2 - 1)(4x^2 + 1)$
19. $(2x - 1)(y - 4)$ 20. $(2x - 1)(x - 4)$ 21. $(2x^2 - 6)(2x^2 - 6)$
22. $(x - 6)(2x - 3)$ 23. $(7x - 1)(2x^2 + 3)$ 24. $(x + 1)(x + 1)$
25. $5x(3x + 2)(2x - 4)$ 26. $3x^2(4x - 1)(x + 2)$

4.3B Multiplying a Polynomial by a Polynomial

When each polynomial has several terms, we must apply the distributive property at least twice. The problems can be done horizontally or vertically. If the work is done vertically, we must be sure to write only like terms in the same vertical line when we write the partial products.

Example 5 Multiply $(x^2 - 3x + 2)(x - 5)$.

Solution

$$(x^2 - 3x + 2)(x - 5) = (x^2 - 3x + 2)x + (x^2 - 3x + 2)(-5)$$
$$= x^3 - 3x^2 + 2x - 5x^2 + 15x - 10$$
$$= x^3 - 8x^2 + 17x - 10$$

We call $x^3 - 3x^2 + 2x$ and $-5x^2 + 15x - 10$ the *partial products*

When the multiplication is done vertically, we can multiply either from right to left or from left to right. When one of the polynomials contains more terms than the other, it is usually easier to begin by writing the polynomial with more terms *above* the other one.

Multiplying from right to left

$$\begin{array}{r} x^2 - 3x + 2 \\ x - 5 \\ \hline -5x^2 + 15x - 10 \\ x^3 - 3x^2 + 2x \phantom{{}- 10} \\ \hline x^3 - 8x^2 + 17x - 10 \end{array}$$

Partial product $(x^2 - 3x + 2)(-5)$

Partial product $(x^2 - 3x + 2)(x)$

Notice that like terms *in the partial products* are in the same vertical line

Multiplying from left to right

$$\begin{array}{l} x^2 - 3x + 2 \\ x - 5 \\ \hline x^3 - 3x^2 + 2x \\ - 5x^2 + 15x - 10 \\ \hline x^3 - 8x^2 + 17x - 10 \end{array}$$

■

Missing Terms Consider the polynomial $ax^3 + bx + c$. Because there is no x^2 term written, we say that we have a **missing term**; that is, the coefficient of x^2 is understood to be 0. When we multiply and divide polynomials, it is usually desirable to write in any missing terms *with a coefficient of 0*. For example, we would write $2m^4 - 3m^2 + 2m - 5$ as $2m^4 + 0m^3 - 3m^2 + 2m - 5$ (see Example 6).

Example 6 Multiply $(2m + 2m^4 - 5 - 3m^2)(2 + m^2 - 3m)$.

Solution

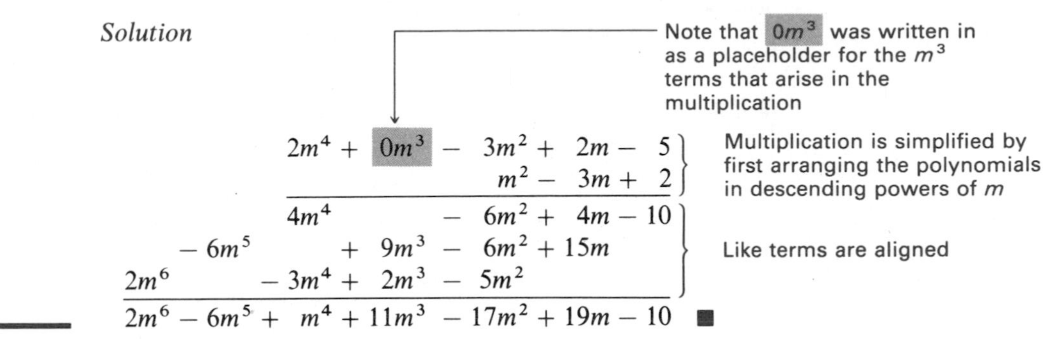

Example 7 Multiply $(3a^2b - 6ab^2)(2ab - 5)$.

Solution

$$
\begin{array}{r}
3a^2b - 6ab^2 \\
2ab \quad - 5 \\
\hline
-15a^2b + 30ab^2 \\
6a^3b^2 - 12a^2b^3 \qquad\qquad\qquad \\
\hline
6a^3b^2 - 12a^2b^3 - 15a^2b + 30ab^2
\end{array}
$$

Notice that the second line is moved over far enough so that we won't have unlike terms in the same vertical line ■

EXERCISES 4.3B

Set I Find the products.

1. $(2h - 3)(4h^2 - 5h + 7)$
2. $(5k - 6)(2k^2 + 7k - 3)$
3. $(4 + a^4 + 3a^2 - 2a)(a + 3)$
4. $(3b - 5 + b^4 - 2b^3)(b - 5)$
5. $(4 - 3z^3 + z^2 - 5z)(4 - z)$
6. $(3 + 2v^2 - v^3 + 4v)(2 - v)$
7. $(3u^2 - u + 5)(2u^2 + 4u - 1)$
8. $(2w^2 + w - 7)(5w^2 - 3w - 1)$
9. $-2xy(-3x^2y + xy^2 - 4y^3)$
10. $-3xy(-4xy^2 - x^2y + 3x^3)$
11. $(a^3 - 3a^2b + 3ab^2 - b^3)(-5a^2b)$
12. $(m^3 - 3m^2p + 3mp^2 - p^3)(-4mp^2)$
13. $(x^2 + 2x + 3)^2$
14. $(z^2 - 3z - 4)^2$
15. $[(x + y)(x^2 - xy + y^2)][(x - y)(x^2 + xy + y^2)]$
16. $[(a - 1)(a^2 + a + 1)][(a + 1)(a^2 - a + 1)]$

Set II Find the products.

1. $(4a - 3)(6a^2 - 8a + 15)$
2. $(3x - 5)(4x^2 - 5x + 2)$
3. $(8 + 2z^4 - z^2 - 9z)(z - 4)$
4. $(3x + 1)(x - 5 - 2x^3 + x^2)$
5. $(6 - k)(4k - 7 + k^4 - 5k^3)$
6. $(3x + 2y - 1)(3x + 2y + 1)$
7. $(5h^2 - h + 8)(h^2 + 3h - 2)$
8. $(2x^2 + 3x - 4)(2x^2 + 3x - 4)$
9. $-4yz(6y^3 - y^2z + 3yz^2)$
10. $(5xy)(-2y^2)(3x^2y)$
11. $(2e^3 - 4e^2f + 7ef^2 - f^3)(-5ef^2)$
12. $(2x - y - 1)(2x + y + 1)$
13. $(3p^2 - 5p + 4)^2$
14. $(x + 1)(x^4 - x^3 + x^2 - x + 1)$
15. $[(2a - b)(4a^2 + 2ab + b^2)][(2a + b)(4a^2 - 2ab + b^2)]$
16. $[(x + 3y)(x - 3y)][(x^2 - 3xy + 9y^2)(x^2 + 3xy + 9y^2)]$

4.4 Special Products and the Binomial Theorem

4.4A The Product of the Sum and Difference of Two Terms

RULE 4.1

$$(a + b)(a - b) = a^2 - b^2$$

Proof of Rule 4.1

$$(a + b)(a - b) = a^2 - ab + ab - b^2 = a^2 - b^2$$

Because $a + b$ is the *sum* of two terms and $a - b$ is the *difference* of two terms, Rule 4.1 can be stated in words as follows:

The product of the sum and difference of two terms is equal to the square of the first term minus the square of the second term.

Example 1 Examples of finding products of the sum and difference of two terms:

a. $(2x + 3y)(2x - 3y) = (2x)^2 - (3y)^2 = 4x^2 - 9y^2$

b. $(x + y + 2)(x + y - 2)$

At first glance, this does not appear to be the product of two binomials at all. However, if we group the variables as $[x + y]$, we then have a product of the sum and difference of two terms:

$$\begin{aligned}(x + y + 2)(x + y - 2) &= ([x + y] + 2)([x + y] - 2)\\ &= (x + y)^2 - 2^2\\ &= (x + y)(x + y) - 4\\ &= x^2 + 2xy + y^2 - 4 \quad \blacksquare\end{aligned}$$

EXERCISES 4.4A

Set I Find the products.

1. $(2u + 5v)(2u - 5v)$
2. $(3m - 7n)(3m + 7n)$
3. $(2x^2 - 9)(2x^2 + 9)$
4. $(10y^2 - 3)(10y^2 + 3)$
5. $(x^5 - y^6)(x^5 + y^6)$
6. $(a^7 + b^4)(a^7 - b^4)$
7. $(7mn + 2rs)(7mn - 2rs)$
8. $(8hk + 5ef)(8hk - 5ef)$

9. $(12x^4y^3 + u^7v)(12x^4y^3 - u^7v)$

10. $(11a^5b^2 + 9c^3d^6)(11a^5b^2 - 9c^3d^6)$

11. $(a + b + 2)(a + b - 2)$

12. $(x + y + 5)(x + y - 5)$

13. $(x^2 + y + 5)(x^2 + y - 5)$

14. $(u^2 - v + 7)(u^2 - v - 7)$

15. $7x(x - 1)(x + 1)$

16. $3y(2y - 5)(2y + 5)$

Set II Find the products.

1. $(6x - 5)(6x + 5)$

2. $(4 + 3z)(4 - 3z)$

3. $(7u - 8v)(7u + 8v)$

4. $(x - 8y)(x + 8y)$

5. $(x^6 - y^5)(x^6 + y^5)$

6. $(x^2 + y^3)(x^2 - y^3)$

7. $(2xy + 3uv)(2xy - 3uv)$

8. $(3abc - ef)(3abc + ef)$

9. $(13z^2w^3 + 5v^5u^7)(13z^2w^3 - 5v^5u^7)$

10. $(16x^3y^2 - 11z^4)(16x^3y^2 + 11z^4)$

11. $(s + t - 1)(s + t + 1)$

12. $(2x - y - 3)(2x + y + 3)$

13. $(x^2 + y + 5)(x^2 + y - 5)$

14. $(x^2 - y + 5)(x^2 + y - 5)$

15. $3z(5z - 1)(5z + 1)$

16. $4ab^2(a - 2)(a + 2)$

4.4B The Square of a Binomial

Rules 4.2 and 4.3 help us quickly square a binomial.

THE SQUARE OF A BINOMIAL

RULE 4.2 $(a + b)^2 = a^2 + 2ab + b^2$

RULE 4.3 $(a - b)^2 = a^2 - 2ab + b^2$

Proof of Rule 4.2

$$(a + b)^2 = (a + b)\,(a + b) = a^2 + 2ab + b^2$$

ab

ab

$$a^2 + \overline{2ab} + b^2$$

Proof of Rule 4.3

$$(a - b)^2 = (a - b)\,(a - b) = a^2 - 2ab + b^2$$

$-ab$

$-ab$

$$a^2 - \overline{2ab} + b^2$$

These two rules can be stated in words as follows:

TO SQUARE A BINOMIAL

1. The *first term* of the product is the square of the first term of the binomial.
2. The *middle term* of the product is twice the product of the two terms of the binomial.
3. The *last term* of the product is the square of the last term of the binomial.

Example 2 Examples of squaring binomials:

a. $(5m + 2n^2)^2 = (5m)^2 + 2(5m)(2n^2) + (2n^2)^2$
$= 25m^2 + 20mn^2 + 4n^4$

b. $(3 - [x + y])^2 = 3^2 - 2(3)[x + y] + [x + y]^2$
$= 9 - 6x - 6y + x^2 + 2xy + y^2$ ■

A WORD OF CAUTION Using the rules for exponents, students remember that

$$(ab)^2 = a^2b^2$$

Here, a and b are *factors*

They try to apply this rule of exponents to the expression $(a + b)^2$, but

$(a + b)^2$ cannot be found simply by squaring a and b.

Here, a and b are *terms*

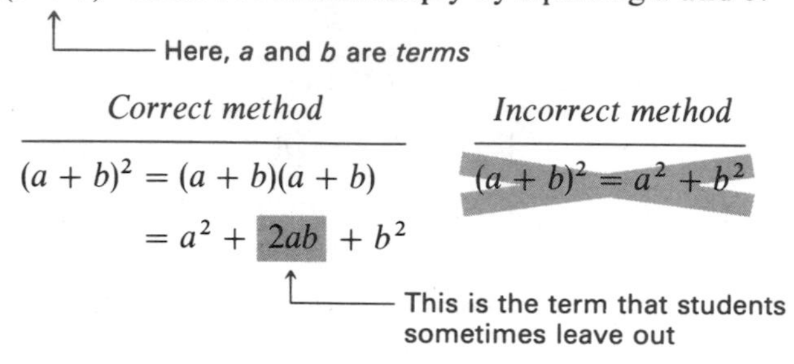

✓

EXERCISES 4.4B

Set I Simplify.

1. $(2x + 3)^2$
2. $(6x + 5)^2$
3. $(5x - 3)^2$
4. $(9x - 6)^2$
5. $(7x - 10y)^2$
6. $(4u - 9v)^2$
7. $([u + v] + 7)^2$
8. $([s + t] + 4)^2$
9. $([2x - y] - 3)^2$
10. $([3z - w] - 7)^2$
11. $(x + [u + v])^2$
12. $(s + [y + 2])^2$
13. $(y - [x - 2])^2$
14. $(x - [z - 5])^2$
15. $5(x - 3)^2$
16. $7(x + 1)^2$

Set II Simplify.

1. $(3x + 4)^2$
2. $(6x - 5y)^2$
3. $(5x - 2)^2$
4. $(8yz + 2)^2$
5. $(6x - 7y)^2$
6. $(3 - [x - y])^2$
7. $([x + y] + 5)^2$
8. $([x + y] + [a + b])^2$
9. $([7x - y] - 3)^2$
10. $(4x + [a - b])^2$
11. $(s + [x + 3])^2$
12. $3x(x - 1)^2$
13. $(z - [y - 3])^2$
14. $(3x + 4)^2x^2$
15. $8(a + 5)^2$
16. $3x(-4x)^2(2x^3)$

4.4C The Binomial Theorem

Before we discuss raising a binomial to a power, we introduce two new symbols.

Factorial Notation Mathematicians define ***n*!** (read "***n factorial***") as follows:

$0! = 1$

$1! = 1$

$n! = n(n - 1)(n - 2) \cdot \cdots \cdot 3 \cdot 2 \cdot 1$, where n is an integer greater than 1

Example 3 Examples of evaluating factorial notation:

a. $3! = 3 \cdot 2 \cdot 1 = 6$

b. $5! = 5 \cdot 4 \cdot 3 \cdot 2 \cdot 1 = 120$ ■

The Binomial Coefficient Mathematicians define the **binomial coefficient**, $\binom{n}{r}$, as follows:

$$\binom{n}{0} = 1$$

$$\binom{n}{r} = \frac{n(n-1)(n-2)\cdot\cdots\cdot(n-r+1)}{r!}$$, where n and r are *positive* integers and $r \le n$

Example 4 Examples of evaluating the binomial coefficient:

a. $\binom{8}{5}$ n is 8, r is 5, and $n - r + 1 = 8 - 5 + 1 = 4$. This means that the *last* factor in the numerator will be 4; therefore,

$$\binom{8}{5} = \frac{8 \cdot 7 \cdot 6 \cdot 5 \cdot 4}{5!} = \frac{8 \cdot 7 \cdot \overset{1}{\cancel{6}} \cdot \overset{1}{\cancel{5 \cdot 4}}}{\underset{1}{\cancel{5 \cdot 4}} \cdot \underset{1}{\cancel{3 \cdot 2}} \cdot 1} = 56$$

The first factor in the numerator is 8

The last factor in the numerator is 4

The denominator is 5!, since r is 5

b. $\binom{4}{0}$ By definition, $\binom{4}{0} = 1$. ■

A number of theorems about the binomial coefficient can be proved. Some of them are

$$\binom{n}{r} = \frac{n!}{r!(n-r)!}$$ ← You may find this easier to remember than the definition

$$\binom{n}{n} = 1; \quad \binom{n}{1} = n; \quad \binom{n}{n-1} = n; \quad \binom{n}{r} = \binom{n}{n-r}$$

Example 5 Examples of evaluating $\binom{n}{r}$ using the formula $\binom{n}{r} = \frac{n!}{r!(n-r)!}$:

a. $\binom{5}{2}$ n is 5 and r is 2; $\binom{5}{2} = \frac{5!}{2!(5-2)!} = \frac{5!}{2!(3!)} = \frac{5 \cdot \overset{2}{\cancel{4}} \cdot \overset{1}{\cancel{3 \cdot 2 \cdot 1}}}{(\underset{1}{\cancel{2}} \cdot 1)(\underset{1}{\cancel{3 \cdot 2 \cdot 1}})} = 10$

b. $\binom{4}{3}$ n is 4 and r is 3; $\binom{4}{3} = \frac{4!}{3!(4-3)!} = \frac{4!}{3!(1!)} = \frac{4 \cdot \overset{1}{\cancel{3 \cdot 2 \cdot 1}}}{(\underset{1}{\cancel{3 \cdot 2 \cdot 1}})(1)} = 4$

We found that $\binom{4}{3} = 4$; therefore, we have verified that $\binom{n}{n-1} = n$ for $n = 4$.

c. $\binom{8}{3}$ n is 8 and r is 3; $\binom{8}{3} = \frac{8!}{3!(8-3)!} = \frac{8!}{3!(5!)}$

$$= \frac{8 \cdot 7 \cdot \not{6} \cdot \not{5} \cdot \not{4} \cdot \not{3} \cdot \not{2} \cdot \not{1}}{(\not{3} \cdot \not{2} \cdot \not{1})(\not{5} \cdot \not{4} \cdot \not{3} \cdot \not{2} \cdot \not{1})} = 56$$

■

NOTE Observe that $\binom{8}{8-3} = \binom{8}{5}$. We found in Example 5c that $\binom{8}{3} = 56$ and in Example 4a that $\binom{8}{5} = 56$. Therefore, $\binom{8}{3} = \binom{8}{8-3}$. This verifies that $\binom{n}{r} = \binom{n}{n-r}$ for $n = 8$ and $r = 3$. ☑

Powers of Binomials

Powers of binomials occur so frequently that it is convenient to have a method for expanding *any* power of *any* binomial. We know that $(a+b)^0 = 1$, $(a+b)^1 = a+b$, and $(a+b)^2 = a^2 + 2ab + b^2$. Let's find $(a+b)^3$ and $(a+b)^4$.

$$\begin{aligned}
(a+b)^3 &= (a+b)^2(a+b) \\
&= (a^2 + 2ab + b^2)(a+b) \\
&= (a^2 + 2ab + b^2)(a) + (a^2 + 2ab + b^2)(b) \\
&= a^3 + 2a^2b + ab^2 + a^2b + 2ab^2 + b^3 \\
&= a^3 + 3a^2b + 3ab^2 + b^3 \\
(a+b)^4 &= \{(a+b)^2\}^2 \\
&= (a^2 + 2ab + b^2)^2
\end{aligned}$$

Let's finish the multiplication vertically:

$$\begin{array}{r}
a^2 + 2ab + b^2 \\
a^2 + 2ab + b^2 \\
\hline
a^2b^2 + 2ab^3 + b^4 \\
2a^3b + 4a^2b^2 + 2ab^3 \qquad \\
a^4 + 2a^3b + a^2b^2 \qquad\qquad \\
\hline
a^4 + 4a^3b + 6a^2b^2 + 4ab^3 + b^4
\end{array}$$

Now let's consider the following binomial expansions. (We suggest that you verify our results for $(a+b)^5$ and $(a+b)^6$ by further multiplication.)

$$\begin{aligned}
(a+b)^0 &= 1 \\
(a+b)^1 &= 1a + 1b \\
(a+b)^2 &= 1a^2 + 2ab + 1b^2 \\
(a+b)^3 &= 1a^3 + 3a^2b + 3ab^2 + 1b^3 \\
(a+b)^4 &= 1a^4 + 4a^3b + 6a^2b^2 + 4ab^3 + 1b^4 \\
(a+b)^5 &= a^5 + 5a^4b + 10a^3b^2 + 10a^2b^3 + 5ab^4 + b^5 \\
(a+b)^6 &= a^6 + 6a^5b + 15a^4b^2 + 20a^3b^3 + 15a^2b^4 + 6ab^5 + b^6
\end{aligned}$$

A careful examination of the above expansions shows that $(a + b)^n$ is always a polynomial in a and b; that there are always $n + 1$ terms; that each *term* has the form $C_r a^s b^r$, where C_r is the numerical coefficient; and that the degree of each term is n (that is, $r + s = n$).

We can also see that when we raise a binomial to the nth power, the first and last coefficients are always 1 and the second and the next-to-last coefficients are always n. Likewise, the third coefficient always equals the coefficient of the term that is third from last, and so forth. Therefore, we often say that the coefficients are *symmetrical.*

Let's look carefully at the way in which $(a + b)^6$ *could* be written:

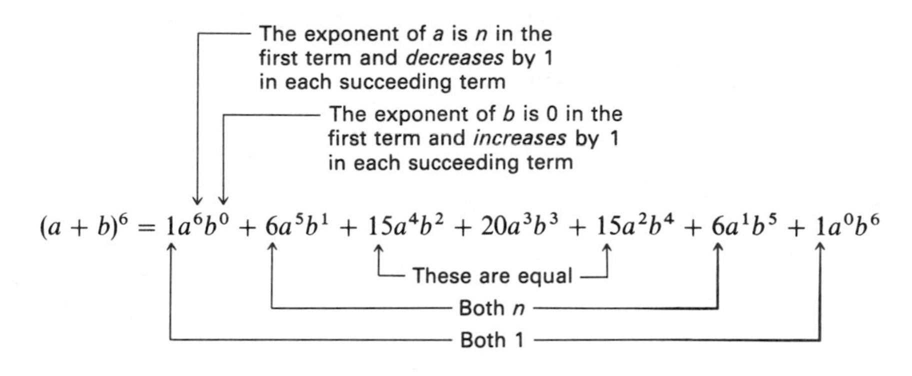

There are seven terms, and every term is a sixth-degree term.

The Binomial Theorem

The proof of the **binomial theorem** requires the use of mathematical induction and is, therefore, beyond the scope of this text.

RULE 4.4: THE BINOMIAL THEOREM

If n is any positive integer and if $r = 0, 1, 2, 3, \ldots, n$, then

$$(a + b)^n = \binom{n}{0}a^n b^0 + \binom{n}{1}a^{n-1}b^1 + \binom{n}{2}a^{n-2}b^2 + \cdots + \binom{n}{r}a^{n-r}b^r + \cdots + \binom{n}{n-2}a^2 b^{n-2} + \binom{n}{n-1}a^1 b^{n-1} + \binom{n}{n}a^0 b^n$$

The coefficient of the $(r + 1)$st term is $\binom{n}{r}$.

NOTE Observe that the first term in the binomial expansion simplifes to a^n and the second term to $na^{n-1}b$; also observe that the next-to-last term simplifies to nab^{n-1} and the last term to b^n. ☑

NOTE If you do not care to memorize one of the formulas given for finding the binomial coefficients, you can instead memorize the following pattern of coefficients:

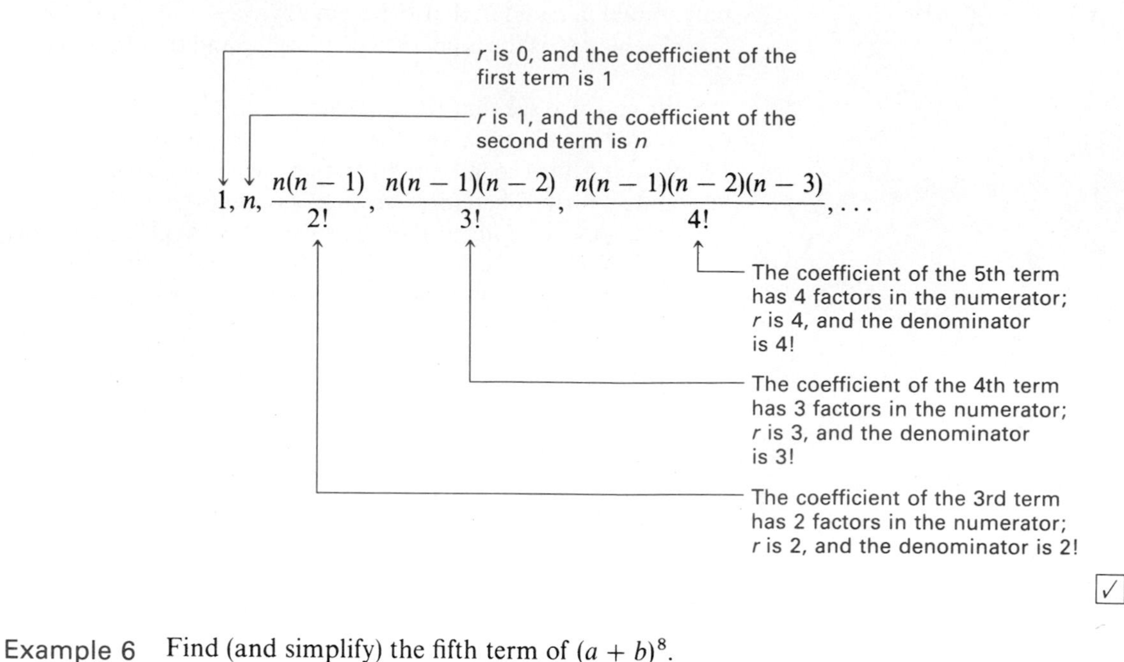

Example 6 Find (and simplify) the fifth term of $(a+b)^8$.

Solution For the *fifth* term, r is 4; n is 8. Therefore, the fifth term is $\binom{8}{4}a^{8-4}b^4$, and

$\binom{8}{4} = \frac{8!}{4!(8-4)!} = \frac{\overset{2}{\cancel{8}} \cdot 7 \cdot \overset{1}{\cancel{6}} \cdot 5 \cdot \overset{1}{\cancel{4 \cdot 3 \cdot 2 \cdot 1}}}{\underset{1}{(\cancel{4} \cdot \cancel{3 \cdot 2 \cdot 1})} \underset{1}{\cancel{(4 \cdot 3 \cdot 2 \cdot 1)}}} = 70$. Therefore, the fifth term of $(a+b)^8$ is $70a^4b^4$.

NOTE We could have shown the work as follows:

$$\binom{8}{4} = \frac{8!}{4!(8-4)!} = \frac{\overset{2}{\cancel{8}} \cdot 7 \cdot \overset{1}{\cancel{6}} \cdot 5 \cdot \overset{1}{\cancel{(4!)}}}{\underset{1}{\cancel{4}} \cdot \underset{1}{\cancel{3 \cdot 2}} \cdot 1 \cdot \underset{1}{\cancel{(4!)}}} = 70$$

■

One possible procedure for raising a binomial to a power is summarized in Rule 4.5.

RULE 4.5: USING THE BINOMIAL THEOREM

To expand $(a+b)^n$, where n is a positive integer, first make a blank outline with $(n+1)$ terms:

$$\underbrace{(\quad)a^{\square}b^{\square} + (\quad)a^{\square}b^{\square} + (\quad)a^{\square}b^{\square} + (\quad)a^{\square}b^{\square} + \cdots + (\quad)a^{\square}b^{\square}}_{(n+1)\text{ terms}}$$

1. a. Fill in the powers of a and b, as shown below.

b. Fill in the coefficients (only the first half must be calculated; the last half are symmetrical to these).

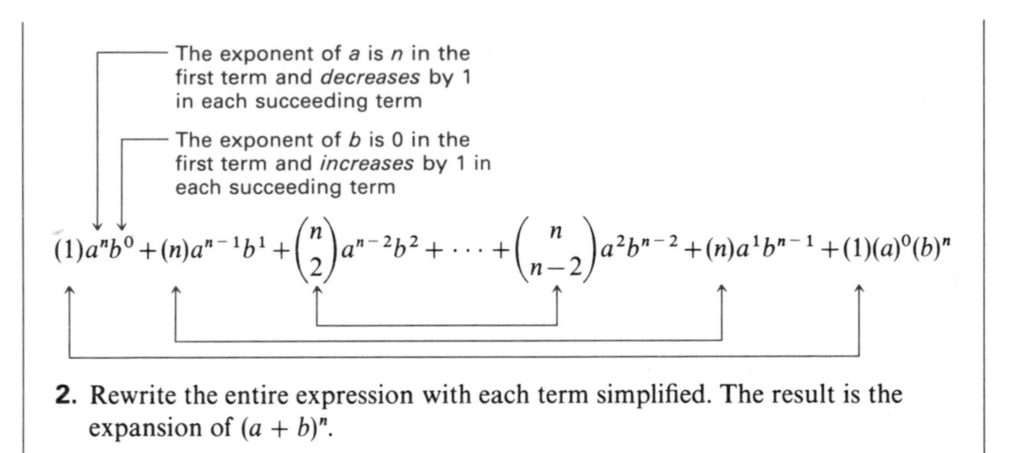

2. Rewrite the entire expression with each term simplified. The result is the expansion of $(a + b)^n$.

Example 7 Expand $(a + b)^8$.

Solution There will be nine terms. The blank outline is

$$\left.\begin{aligned}(a+b)^8 &= (\ \ )a^{\square}b^{\square} + (\ \ )a^{\square}b^{\square} + (\ \ )a^{\square}b^{\square} + (\ \ )a^{\square}b^{\square} + (\ \ )a^{\square}b^{\square}\\ &\quad + (\ \ )a^{\square}b^{\square} + (\ \ )a^{\square}b^{\square} + (\ \ )a^{\square}b^{\square} + (\ \ )a^{\square}b^{\square}\end{aligned}\right\}\ \text{Nine terms}$$

Step 1. $(a + b)^8 = 1a^8b^0 + 8a^7b^1 + \binom{8}{2}a^6b^2 + \binom{8}{3}a^5b^3 + \binom{8}{4}a^4b^4 + \binom{8}{5}a^3b^5$

$$+ \binom{8}{6}a^2b^6 + 8a^1b^7 + 1a^0b^8$$

Step 2. We found in Example 6 that $\binom{8}{4} = 70$. We must calculate $\binom{8}{2}$ and $\binom{8}{3}$:

For $\binom{8}{2}$, n is 8 and r is 2; therefore, $\binom{8}{2} = \dfrac{8!}{2!(8-2)!} = \dfrac{8 \cdot 7 \cdot (6!)}{2(6!)} = 28$; also,

$\binom{8}{6} = \binom{8}{2}$.

For $\binom{8}{3}$, n is 8 and r is 3; therefore, $\binom{8}{3} = \dfrac{8!}{3!(8-3)!} = \dfrac{8 \cdot 7 \cdot 6 \cdot (5!)}{3 \cdot 2 \cdot (5!)} = 56$;

also, $\binom{8}{5} = \binom{8}{3}$.

Therefore, $(a + b)^8 = a^8 + 8a^7b + 28a^6b^2 + 56a^5b^3 + 70a^4b^4 + 56a^3b^5 + 28a^2b^6 + 8ab^7 + b^8$ ■

An alternate method of determining the numerical coefficients is to use Pascal's triangle, as shown on page 168.

Pascal's Triangle

If the binomial expansions for the first five values of n are written, we have

$$(a+b)^0 = 1$$
$$(a+b)^1 = 1a + 1b$$
$$(a+b)^2 = 1a^2 + 2ab + 1b^2$$
$$(a+b)^3 = 1a^3 + 3a^2b + 3ab^2 + 1b^3$$
$$(a+b)^4 = 1a^4 + 4a^3b + 6a^2b^2 + 4ab^3 + 1b^4$$
$$\vdots$$

If we delete the variables and the plus signs from the right sides of the equations, we get a triangular array of numbers known as **Pascal's triangle**:

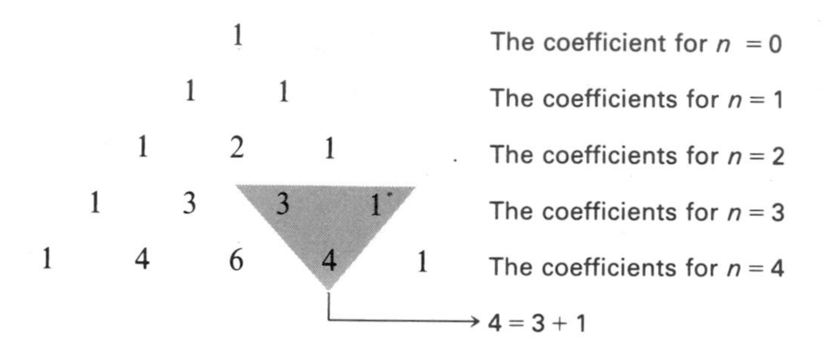

A close examination of the numbers in Pascal's triangle reveals that

1. The first and last numbers in any row are 1.
2. Any other number in Pascal's triangle is the sum of the two closest numbers in the row above it (see the shaded triangle).

Pascal's triangle can be used to find the coefficients in a binomial expansion. The triangle we have shown can be extended to any size by using the two rules given above. Try extending the triangle by one more row (that is, until the second number is a 5); if you do this correctly, you will get the coefficients for $(a+b)^5$.

The terms of the binomial being expanded often consist of something other than the single variables a and b (see Example 8).

Example 8 Examples of binomials with terms that consist of more than a single variable:

a. $(3x^2 + 5y)^3$ In this binomial $\begin{cases} a = 3x^2 \\ b = 5y \end{cases}$

b. $(2e^3 - f^2)^5$ In this binomial $\begin{cases} a = 2e^3 \\ b = -f^2 \end{cases}$

c. $\left(\frac{x}{2} - 4\right)^4$ In this binomial $\begin{cases} a = \frac{x}{2} \\ b = -4 \end{cases}$ ■

Example 9 Expand $(3x^2 + 5y)^3$, using any of the methods discussed for finding the coefficients.

Step 1. The coefficients are 1, 3, 3, 1:

$$1(3x^2)^3(5y)^0 + 3(3x^2)^2(5y)^1 + 3(3x^2)^1(5y)^2 + 1(3x^2)^0(5y)^3$$

Step 2. $(3x^2 + 5y)^3 = 3^3x^6 + 3(3^2x^4)(5y) + 3(3x^2)(5^2y^2) + 5^3y^3$

$= 27x^6 + 27(5)x^4y + 9(25)x^2y^2 + 125y^3$

$= 27x^6 + 135x^4y + 225x^2y^2 + 125y^3$ ■

NOTE When the binomial being expanded contains a *difference* of terms, the terms of the expansion alternate in sign (see Examples 10 and 11). ☑

Example 10 Expand $(2e^3 - f^2)^5$.

Step 1. If we extend Pascal's triangle by one more row (or use any other method), we see that the coefficients are 1, 5, 10, 10, 5, and 1.

$$1(2e^3)^5(-f^2)^0 + 5(2e^3)^4(-f^2)^1 + 10(2e^3)^3(-f^2)^2 + 10(2e^3)^2(-f^2)^3 + 5(2e^3)^1(-f^2)^4 + 1(2e^3)^0(-f^2)^5$$

Step 2. $(2e^3 - f^2)^5 = 2^5e^{15} + 5(2^4e^{12})(-f^2) + 10(2^3e^9)f^4 + 10(2^2e^6)(-f^6) + 5(2e^3)f^8 + (-f^{10})$

$= 32e^{15} - 5(16)e^{12}f^2 + 10(8)e^9f^4 - 10(4)e^6f^6 + 10e^3f^8 - f^{10}$

$= 32e^{15} - 80e^{12}f^2 + 80e^9f^4 - 40e^6f^6 + 10e^3f^8 - f^{10}$ ■

Example 11 Expand $\left(\frac{x}{2} - 4\right)^4$.

Step 1.

$$1\left(\frac{x}{2}\right)^4(-4)^0 + 4\left(\frac{x}{2}\right)^3(-4)^1 + \frac{4!}{(2!)(2!)}\left(\frac{x}{2}\right)^2(-4)^2 + 4\left(\frac{x}{2}\right)^1(-4)^3 + 1\left(\frac{x}{2}\right)^0(-4)^4$$

Step 2. $\left(\frac{x}{2} - 4\right)^4 = \frac{x^4}{2^4} - \frac{16x^3}{2^3} + \frac{6(16)x^2}{2^2} - \frac{4(64)x}{2} + 256$

$= \frac{x^4}{16} - 2x^3 + 24x^2 - 128x + 256$ ■

Example 12 Write the first five terms of the expansion of $(x + y^2)^{12}$.

Step 1. $1(x)^{12}(y^2)^0 + 12(x)^{11}(y^2)^1 + \frac{12!}{(2!)(10!)}(x)^{10}(y^2)^2 + \frac{12!}{(3!)(9!)}(x)^9(y^2)^3 + \frac{12!}{(4!)(8!)}(x)^8(y^2)^4 + \cdots$

Step 2.

$$(x + y^2)^{12} = x^{12} + 12x^{11}y^2 + 66x^{10}y^4 + 220x^9y^6 + 495x^8y^8 + \cdots$$ ■

NOTE In Example 11, notice that $\frac{4!}{(2!)(2!)} = \frac{4 \cdot 3 \cdot \cancel{2!}}{2 \cdot 1 \cdot \cancel{2!}} = \frac{4 \cdot 3}{1 \cdot 2}$, and in Example 12, notice that $\dfrac{12!}{(2!)(10!)} = \dfrac{12 \cdot 11 \cdot \cancel{10!}}{2 \cdot 1 \cdot \cancel{10!}} = \dfrac{12 \cdot 11}{1 \cdot 2}$, that $\dfrac{12!}{(3!)(9!)} = \dfrac{12 \cdot 11 \cdot 10 \cdot \cancel{9!}}{3 \cdot 2 \cdot 1 \cdot \cancel{9!}} = \dfrac{12 \cdot 11 \cdot 10}{1 \cdot 2 \cdot 3}$,

and that $\dfrac{12!}{(4!)(8!)} = \dfrac{12 \cdot 11 \cdot 10 \cdot 9 \cdot \cancel{8!}}{4 \cdot 3 \cdot 2 \cdot 1 \cdot \cancel{8!}} = \dfrac{12 \cdot 11 \cdot 10 \cdot 9}{1 \cdot 2 \cdot 3 \cdot 4}$. ☑

In this section, we discussed the binomial expansion for *natural number powers* only. Other numbers can be used for powers of binomials, but binomial expansions using such numbers as exponents are not discussed in this book.

EXERCISES 4.4C

Set I In Exercises 1–6, evaluate each expression.

1. $4!$ **2.** $6!$ **3.** $\binom{3}{2}$

4. $\binom{7}{3}$ **5.** $\binom{6}{4}$ **6.** $\binom{4}{0}$

7. Find the sixth term of $(a+b)^9$.

8. Find the fourth term of $(a+b)^7$.

In Exercises 9–24, raise each binomial to the indicated power.

9. $(x+y)^5$ **10.** $(r+s)^4$ **11.** $(x-2)^5$

12. $(y-3)^4$ **13.** $(3r+s)^6$ **14.** $(2x+y)^6$

15. $(x+y^2)^4$ **16.** $(u+v^2)^5$ **17.** $(2x-\frac{1}{2})^5$

18. $(3x-\frac{1}{3})^4$ **19.** $(\frac{1}{3}x+\frac{3}{2})^4$ **20.** $(\frac{1}{5}x+\frac{5}{2})^4$

21. $(4x^2-3y^2)^5$ **22.** $(3x^2-2y^3)^5$ **23.** $(x+x^{-1})^4$

24. $(x^{-1}-x)^4$

25. Write the first four terms of the expansion of $(x+2y^2)^{10}$.

26. Write the first four terms of the expansion of $(x+3y^2)^8$.

27. Write the first four terms of the expansion of $(x-3y^2)^{10}$.

28. Write the first four terms of the expansion of $(x-2y^3)^{11}$.

Set II In Exercises 1–6, evaluate each expression.

1. $7!$ **2.** $2!$ **3.** $\binom{5}{4}$

4. $\binom{6}{6}$ **5.** $\binom{9}{5}$ **6.** $\binom{7}{4}$

7. Find the fourth term of $(a+b)^9$.

8. Find the third term of $(a+b)^{10}$.

In Exercises 9–24, raise each binomial to the indicated power.

9. $(s+t)^6$ **10.** $(x-y)^6$ **11.** $(x+4)^4$

12. $(2x-3y)^3$ **13.** $(4x-y)^5$ **14.** $(s-t)^5$

15. $(x + y^2)^3$ **16.** $(x - y)^5$ **17.** $(2x - \frac{1}{2})^6$

18. $(y + y^{-1})^5$ **19.** $(\frac{1}{4}x + \frac{3}{2})^4$ **20.** $(x + x^{-1})^6$

21. $(3s^2 - 3t^3)^5$ **22.** $(x^{-1} - x)^3$ **23.** $(3x + 2y^{-1})^4$

24. $(y^{-1} - y)^7$

25. Write the first four terms of the expansion of $(2x - y^2)^{14}$.

26. Write the first four terms of the expansion of $(3x + y^2)^{12}$.

27. Write the first four terms of the expansion of $(3 - x^2)^{10}$.

28. Write the first four terms of the expansion of $(2h - k^2)^{11}$.

4.5 Division of Polynomials

4.5A Division of a Polynomial by a Monomial

TO DIVIDE A POLYNOMIAL BY A MONOMIAL

Divide *each* term in the polynomial by the monomial; then add the resulting quotients.

Example 1 Perform the indicated divisions.

a. $\dfrac{4x^3 - 6x^2}{2x}$

Solution

$$\frac{4x^3 - 6x^2}{2x} = \frac{1}{2x}(4x^3 - 6x^2)$$

$$= \left(\frac{1}{2x}\right)(4x^3) + \left(\frac{1}{2x}\right)(-6x^2)$$ Using the distributive property

$$= \frac{4x^3}{2x} + \frac{-6x^2}{2x}$$ Dividing each term of the polynomial by the monomial

$$= 2x^2 - 3x$$

b. $\dfrac{4x^2y - 8xy^2 + 12xy}{-4xy}$

Solution $$\frac{4x^2y - 8xy^2 + 12xy}{-4xy} = \frac{4x^2y}{-4xy} + \frac{-8xy^2}{-4xy} + \frac{12xy}{-4xy}$$

$$= -x + 2y - 3 \quad ■$$

EXERCISES 4.5A

Set I Perform the indicated divisions.

1. $\dfrac{18x^5 - 24x^4 - 12x^3}{6x^2}$ **2.** $\dfrac{16y^4 - 36y^3 + 20y^2}{-4y^2}$

3. $\dfrac{55a^4b^3 - 33ab^2}{-11ab}$

4. $\dfrac{26m^2n^4 - 52m^3n}{-13mn}$

5. $\dfrac{-15x^2y^2z^2 - 30xyz}{-5xyz}$

6. $\dfrac{-24a^2b^2c^2 - 16abc}{-8abc}$

7. $(13x^3y^2 - 26x^5y^3 + 39x^4y^6) \div (13x^2y^2)$

8. $(21m^2n^3 - 35m^3n^2 - 14m^3n^5) \div (7m^2n^2)$

Set II Perform the indicated divisions.

1. $\dfrac{24h^6 - 56h^4 - 40h^3}{8h^2}$

2. $\dfrac{5x^4 - 40x^3 + 10x^2}{-5x^2}$

3. $\dfrac{60e^5f^3 - 84e^2f^4}{12ef^2}$

4. $\dfrac{7y^5 - 35y^4 + 14y^3}{7y^2}$

5. $\dfrac{-45r^3s^2t^4 + 63r^2s^3t^3}{-9rs^2t}$

6. $\dfrac{42u^4 - 56u^2 + 28u^6}{-14u^2}$

7. $(-30m^4n^2 + 60m^2n^3 - 45m^3n^4) \div (15m^2n^2)$

8. $(32x^3y^2z - 64xy^2z + 48xy^3z^2) \div (-16xy^2z)$

4.5B Division of a Polynomial by a Polynomial

The method used to divide one polynomial (the dividend) by a polynomial (the divisor) with two or more terms is similar to the method used to divide one whole number by another (using long division) in arithmetic. The long-division procedure can be summarized as follows [Reading Hint: Read the steps in the summary, and *while you are doing so*, look also at the steps in Example 2 to see that the authors are following the step-by-step method of the summary.]:

TO DIVIDE ONE POLYNOMIAL BY ANOTHER

1. Arrange the divisor and the dividend in descending powers of one variable. In the *dividend*, use placeholders (or leave spaces) for any missing terms.
2. Find the first term of the quotient by dividing the first term of the dividend by the first term of the divisor.
3. Multiply the *entire* divisor by the first term of the quotient. Place the product under the dividend, lining up like terms.
4. Subtract the product found in step 3 from the dividend, bringing down at least one term. This difference is the *remainder*. If the degree of the remainder is not less than the degree of the divisor, continue with steps 5–8.
5. Find the next term of the quotient by dividing the first term of the remainder by the first term of the divisor.
6. Multiply the entire divisor by the term found in step 5.
7. Subtract the product found in step 6 from the polynomial above it, bringing down at least one more term.

8. Repeat steps 5 through 7 until the remainder is 0 *or* until the degree of the remainder is less than the degree of the divisor. If there is a nonzero remainder, it can be written as the numerator of a fraction whose denominator is the *divisor* of the division problem. This fraction (which *must* be preceded by either a plus sign or a minus sign) is written as the last term of the final answer.

9. Check your answer. (Divisor × Quotient + Remainder = Dividend)

A WORD OF CAUTION Most errors in division problems are *subtraction* errors. ☑

Example 2 Divide $(27x + 19x^2 + 6x^3 + 10)$ by $(5 + 3x)$.

Step 1. $3x + 5\overline{)6x^3 + 19x^2 + 27x + 10}$ ← In descending powers

↑ In descending powers (the divisor $3x + 5$)

The first term in the quotient is $\frac{6x^3}{3x} = 2x^2$

Step 2.
$$\begin{array}{r} 2x^2 \\ 3x + 5\overline{)6x^3 + 19x^2 + 27x + 10} \end{array}$$

Step 3.
$$\begin{array}{r} 2x^2 \qquad\qquad\qquad \\ 3x + 5\overline{)6x^3 + 19x^2 + 27x + 10} \\ 6x^3 + 10x^2 \qquad\qquad\qquad \end{array}$$
← This is $2x^2(3x + 5)$

Step 4.
$$\begin{array}{r} 2x^2 \qquad\qquad\qquad \\ 3x + 5\overline{)6x^3 + 19x^2 + 27x + 10} \\ \underline{6x^3 + 10x^2} \qquad\qquad\qquad \\ 9x^2 + 27x \qquad\quad \end{array}$$
← Subtract and bring down a term

Remainder (the $9x^2 + 27x$)

The division must be continued because the degree of the remainder is not less than the degree of the divisor.

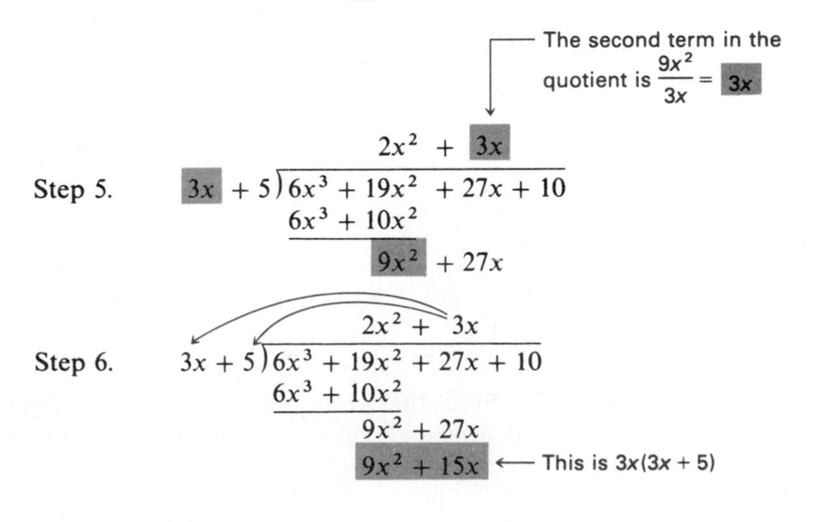

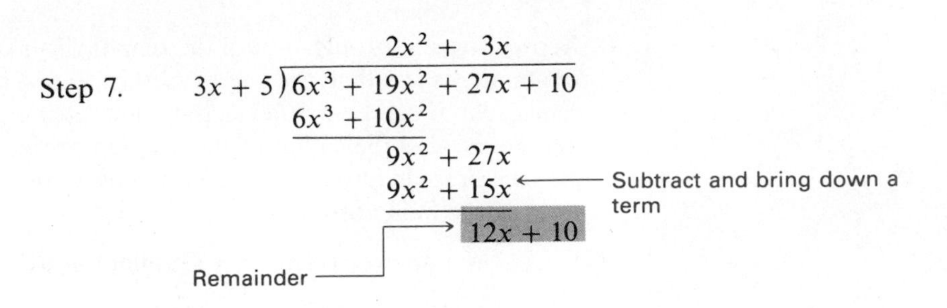

The degree of the remainder is *still* not less than the degree of the divisor. Therefore, we must repeat steps 5, 6, and 7.

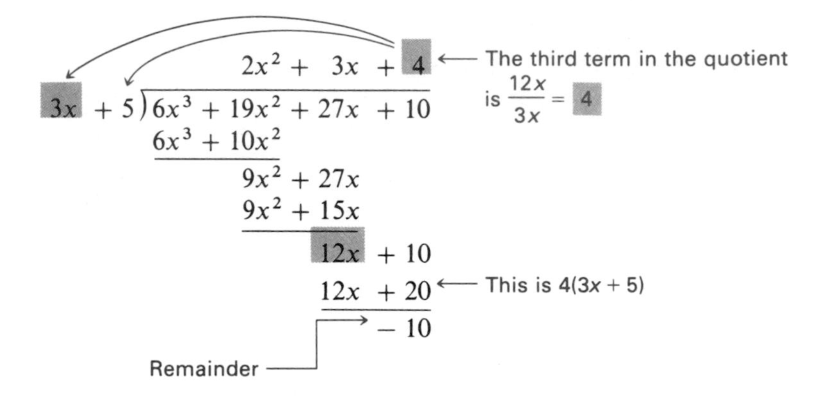

Step 8. The division is finished because the degree of the remainder is less than the degree of the divisor.

Step 9. *Check* $(3x + 5)(2x^2 + 3x + 4) + (-10) = 6x^3 + 19x^2 + 27x + 10$

Answer $2x^2 + 3x + 4 - \dfrac{10}{3x + 5}$ or $2x^2 + 3x + 4 + \dfrac{-10}{3x + 5}$ ■

Example 3 Divide $(x^3 - 3 + x^4 - 5x)$ by $(x^2 - 1 - x)$.

Solution Arrange the terms of the dividend and the divisor in descending powers of the variable before beginning the division. (Notice that an x^2 term is missing from the dividend, so we write $0x^2$ as a placeholder.)

$$
\begin{array}{r}
x^2 + 2x + 3 \\
x^2 - x - 1 \overline{)\, x^4 + x^3 + 0x^2 - 5x - 3} \\
x^4 - x^3 - x^2 \qquad\qquad \\
\hline
2x^3 + x^2 - 5x \quad\;\; \\
2x^3 - 2x^2 - 2x \quad\;\; \\
\hline
3x^2 - 3x - 3 \\
3x^2 - 3x - 3 \\
\hline
0
\end{array}
$$

In descending powers (divisor) — In descending powers (dividend)

Check $(x^2 + 2x + 3)(x^2 - x - 1) = x^4 + x^3 - 5x - 3$

Answer $x^2 + 2x + 3$ ■

Factors Recall that when the final remainder is 0, we can say that the *divisor* is a *factor* of the dividend and that the *quotient* is a *factor* of the dividend. Therefore, in Example 3, we can say that $x^2 + 2x + 3$ and $x^2 - x - 1$ are factors of $x^4 + x^3 - 5x - 3$.

It is sometimes necessary to bring down more than one term from the dividend (see Example 4).

Example 4 Solve $(6x^3 + 4x^4 - 1 + 2x) \div (3x - 1 + 2x^2)$.

Solution

$$
\begin{array}{r l}
 & 2x^2 \qquad\qquad + 1 \\
2x^2 + 3x - 1 \,\big) & 4x^4 + 6x^3 + 0x^2 + 2x - 1 \\
 & \underline{4x^4 + 6x^3 - 2x^2} \\
 & \qquad\qquad 2x^2 + 2x - 1 \leftarrow \text{Subtracting and bringing down two terms} \\
 & \qquad\qquad \underline{2x^2 + 3x - 1} \\
 & \qquad\qquad\qquad -x
\end{array}
$$

Check $(2x^2 + 1)(2x^2 + 3x - 1) - x = 4x^4 + 6x^3 + 2x - 1$

Answer $2x^2 + 1 + \dfrac{-x}{2x^2 + 3x - 1}$ or $2x^2 + 1 - \dfrac{x}{2x^2 + 3x - 1}$ ■

Example 5 Solve $(17ab^2 + 12a^3 - 10b^3 - 11a^2b) \div (3a - 2b)$.

Solution

$$
\begin{array}{r l}
 & 4a^2 \quad - \quad ab \quad + \quad 5b^2 \\
3a - 2b \,\big) & 12a^3 - 11a^2b + 17ab^2 - 10b^3 \\
 & \underline{12a^3 - 8a^2b} \\
 & \qquad - 3a^2b + 17ab^2 \\
 & \qquad \underline{- 3a^2b + 2ab^2} \\
 & \qquad\qquad\qquad 15ab^2 - 10b^3 \\
 & \qquad\qquad\qquad \underline{15ab^2 - 10b^3} \\
 & \qquad\qquad\qquad\qquad\qquad 0
\end{array}
$$

Writing the divisor and the dividend in descending powers of *a*

You should verify that if we had written the divisor and the dividend in descending powers of b, we would have obtained the same answer.

Check $(4a^2 - ab + 5b^2)(3a - 2b) = 12a^3 - 11a^2b + 17ab^2 - 10b^3$

Answer $4a^2 - ab + 5b^2$

Since the remainder is 0, we can say that the divisor and quotient are both factors of the dividend. ■

Example 6 Solve $(1 - x^5) \div (1 - x)$.

Solution

$$
\begin{array}{r l}
 & x^4 + x^3 + x^2 + x + 1 \\
-x + 1 \,\big) & -x^5 + 0x^4 + 0x^3 + 0x^2 + 0x + 1 \\
 & \underline{-x^5 + x^4} \\
 & \qquad -x^4 + 0x^3 \\
 & \qquad \underline{-x^4 + x^3} \\
 & \qquad\qquad -x^3 + 0x^2 \\
 & \qquad\qquad \underline{-x^3 + x^2} \\
 & \qquad\qquad\qquad -x^2 + 0x \\
 & \qquad\qquad\qquad \underline{-x^2 + x} \\
 & \qquad\qquad\qquad\qquad -x + 1 \\
 & \qquad\qquad\qquad\qquad \underline{-x + 1} \\
 & \qquad\qquad\qquad\qquad\qquad 0
\end{array}
$$

Check $(1 - x)(x^4 + x^3 + x^2 + x + 1) = -x^5 + 1 = 1 - x^5$

Answer $x^4 + x^3 + x^2 + x + 1$

Note that $x^4 + x^3 + x^2 + x + 1$ and $1 - x$ are factors of $1 - x^5$. ■

EXERCISES 4.5B

Set I Perform the indicated divisions.

1. $(x^2 + 10x - 5) \div (x + 7)$
2. $(x^2 + 9x - 5) \div (x + 4)$
3. $(6z^3 - 13z^2 - 4z + 15) \div (3z - 5)$
4. $(6y^3 + 7y^2 - 11y - 12) \div (2y + 3)$
5. $(v^4 - 3v^3 - 8v^2 - 9v - 5) \div (v + 1)$
6. $(w^4 - 2w^3 - 12w^2 - 13w - 10) \div (w + 2)$
7. $(8x - 4x^3 + 10) \div (2 - x)$
8. $(12x - 15 - x^3) \div (3 - x)$
9. $(17xy^2 + 12x^3 - 10y^3 - 11x^2y) \div (3x - 2y)$
10. $(13x^2y - 11xy^2 + 10x^3 - 12y^3) \div (2x + 3y)$
11. $(x^4 - 7x + 6) \div (x - 2)$
12. $(x^4 + 6x + 3) \div (x - 1)$
13. $(x + 8x^2 + 4x^4 - 3x^3 + 2) \div (x + 1 + 4x^2)$
14. $(11x^2 + 10 + 2x^4 - x^3 + 3x) \div (x + 2x^2 + 2)$
15. $(2x^4 + 7x^2 + 5) \div (x^2 + 2)$
16. $(3x^4 + 11x^2 + 1) \div (x^2 + 3)$
17. $(2x^5 - 10x^3 + 3x^2 - 15) \div (3 + 2x^3)$
18. $(3x^5 - 6x^3 + x^2 - 2) \div (1 + 3x^3)$
19. $(3x^4 + 14x^3 + 2x^2 + 3x + 2) \div (3x^2 - x + 1)$
20. $(2x^4 + 13x^3 - 10x^2 + 9x - 2) \div (2x^2 - x + 1)$

Set II Perform the indicated divisions.

1. $(x^2 - 3x - 11) \div (x + 3)$
2. $(6x^3 + 23x^2y + 16xy^2 + 3y^3) \div (3x + y)$
3. $(20w^3 - 23w^2 - 29w + 14) \div (5w - 2)$
4. $(2x^4 + 3x^3 + 2x^2 - x + 3) \div (2x + 3)$
5. $(z^4 - 2z^3 - 4z^2 - 10z - 9) \div (z + 1)$
6. $(x^4 - 1) \div (x - 1)$
7. $(8x - x^3 + 25) \div (4 - x)$
8. $(2x^2 + 6 + 3x + x^3) \div (x + 2)$
9. $(15x^3 - 5xy^2 + 12y^3 - 23x^2y) \div (3x - 4y)$
10. $(4x + x^3 + 4 + x^2) \div (x + 1)$
11. $(x^4 - 29x + 4) \div (x - 3)$
12. $(x^5 - 1) \div (x^2 + x - 1)$
13. $(2x + 3x^4 - 2x^3 + 3 + 9x^2) \div (x + 3x^2 + 1)$

14. $(3x^4 + 2x^3 - 6x^2 - x + 5) \div (3x + 2)$

15. $(2x^4 + 7x^2 + 3) \div (x^2 + 1)$

16. $(x^3 + x^2 + x + 3) \div (x + 1)$

17. $(3x^2 + 2x + 6 + x^3) \div (x + 3)$

18. $(2x^4 + x^3 + 4x^2 - 4x - 5) \div (2x + 1)$

19. $(3x^5 - 12x^3 + 2x^2 - 8) \div (2 + 3x^3)$

20. $(2x^4 + 7x^3 + 10x^2 + x - 4) \div (2x^2 + x - 1)$

4.6 Synthetic Division

Synthetic division may be used when we divide a polynomial by a *first-degree* binomial.

Synthetic Division and Long Division

Before we show how synthetic division is done, we will show how synthetic division is derived from long division.

Example 1 Divide $(19x + 2x^3 - 5 - 11x^2)$ by $(x - 3)$, using *long division.*

Solution

$$\begin{array}{r l}
 & 2x^2 - 5x + 4 \quad \text{R } 7 \\
x - 3 \,) & \overline{2x^3 - 11x^2 + 19x - 5} \\
 & \underline{2x^3 - 6x^2} \\
 & -5x^2 + 19x \\
 & \underline{-5x^2 + 15x} \\
 & 4x - 5 \\
 & \underline{4x - 12} \\
 & 7
\end{array}$$

■

We note (1) that when we divide any polynomial by a *first-degree* polynomial, the degree of the quotient is *always one less than* the degree of the dividend and (2) that the quotient is in descending powers of the variable. Therefore, in order to shorten the division process, we can *temporarily* omit the variables. (We will put them back in later.) When we omit the variables, we must put in zeros for any missing terms. If we leave out the variables and the plus signs in Example 1, we have

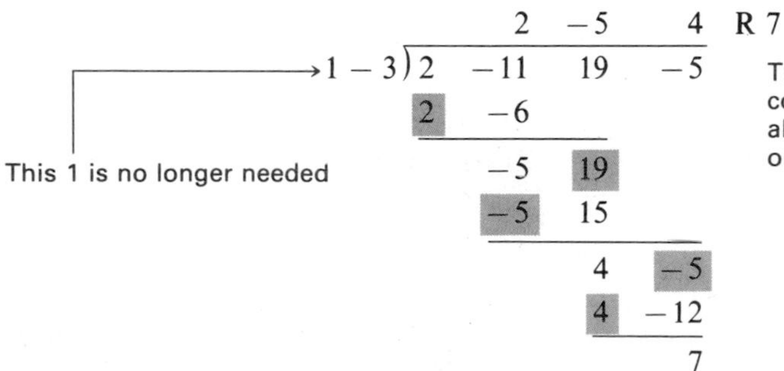

The shaded numbers are copies of the numbers directly above them, and can be omitted

We see that most of the numerical coefficients are repeated several times. We can abbreviate the long-division process considerably by leaving out the *repetitions* of the

numerical coefficients. This gives us

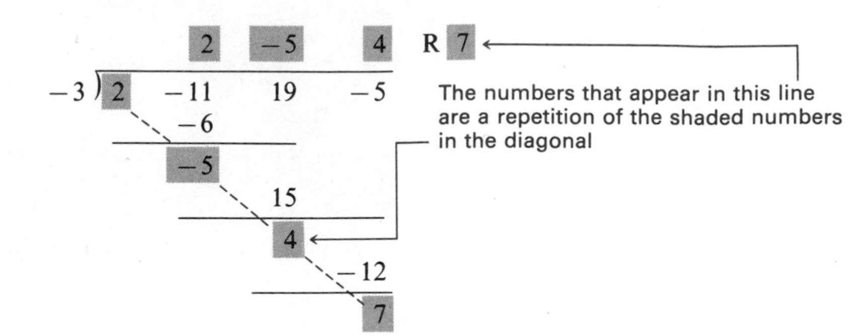

If we omit the last three numbers at the top, remove the empty spaces, and "bring down" the *first* number from the top, the work is brought closer together, as follows:

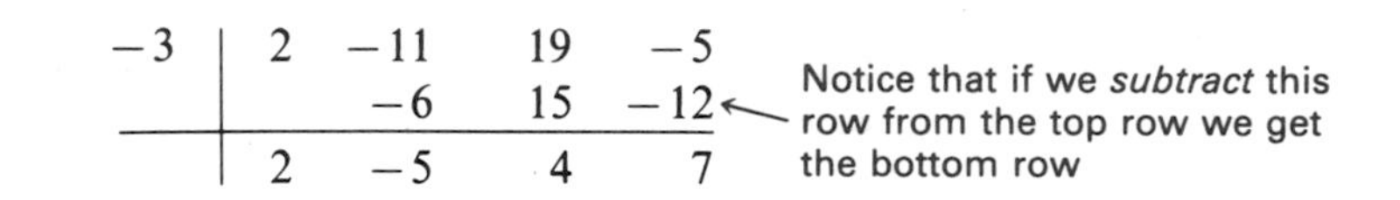

Also notice that if we multiply each of the first three numbers in the bottom row by the -3, we get the numbers in the second row. Changing the -3 to a $+3$ would permit us to *add* instead of *subtract*.

The Synthetic Division Process

We now demonstrate the synthetic division procedure for the division problem in Example 1. The numbers for the problem $(19x + 2x^3 - 5 - 11x^2) \div (x - 3)$ are arranged as follows when we are getting ready to use synthetic division:

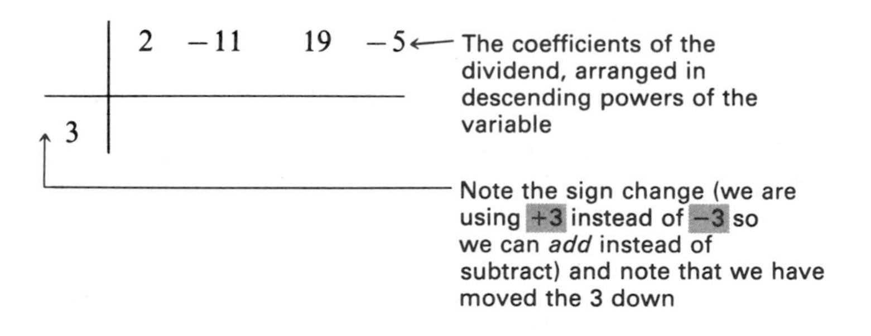

We next bring down the *first* number, 2, from the top row.

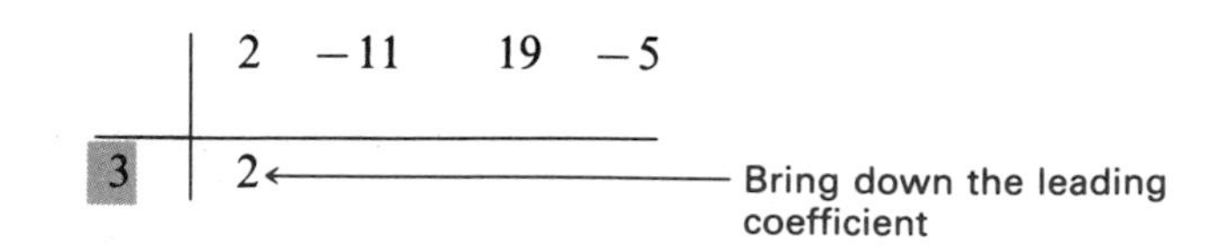

We now multiply and add as follows:

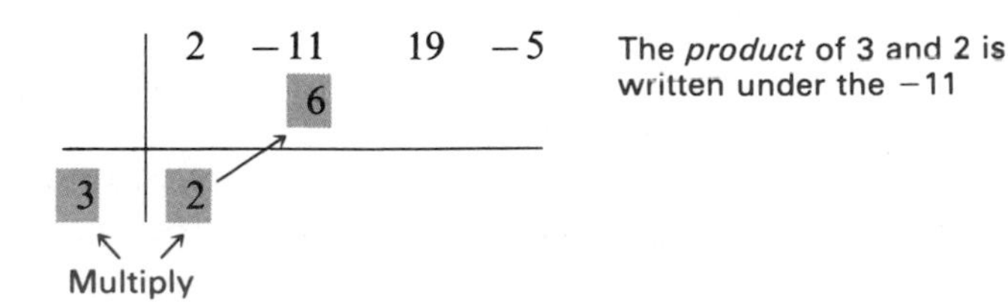

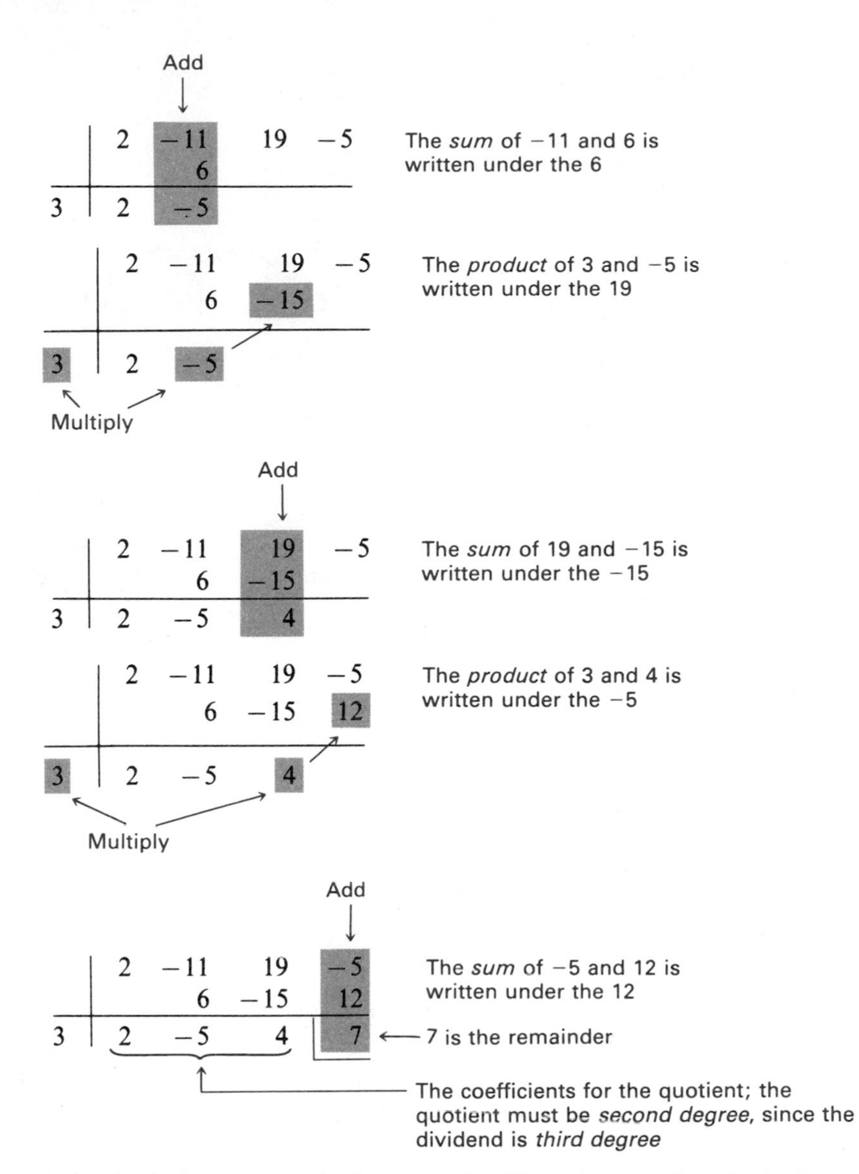

The last number in the bottom row is the *remainder*. The other numbers in the last row are the coefficients for the quotient. We now supply the missing variables, remembering that the *degree* of the quotient is *one less than* the degree of the dividend. The quotient is (apparently) $2x^2 - 5x + 4$, with a remainder of 7.

Check $(x - 3)(2x^2 - 5x + 4) + 7 = 2x^3 - 11x^2 + 19x - 5$

Therefore, $(19x + 2x^3 - 5 - 11x^2) \div (x - 3) = 2x^2 - 5x + 4 + \dfrac{7}{x - 3}$.

Example 2 Divide $2x^4 - 3x^2 + 5$ by $x - 1$.

Solution

	2	0	−3	0	5
		2	2	−1	−1
1	2	2	−1	−1	4
	$2x^3$	$+2x^2$	$-1x$	-1	R 4

The dividend is fourth degree; zeros are used for all missing powers

The quotient must be third degree

Sign changed (We're dividing by $x - 1$)

Check $(x - 1)(2x^3 + 2x^2 - x - 1) + 4 = 2x^4 - 3x^2 + 5$.

Therefore, $\dfrac{2x^4 - 3x^2 + 5}{x - 1} = 2x^3 + 2x^2 - x - 1 + \dfrac{4}{x - 1}$. ■

Example 3 Divide $\dfrac{x^5 + 32}{x + 2}$.

Solution

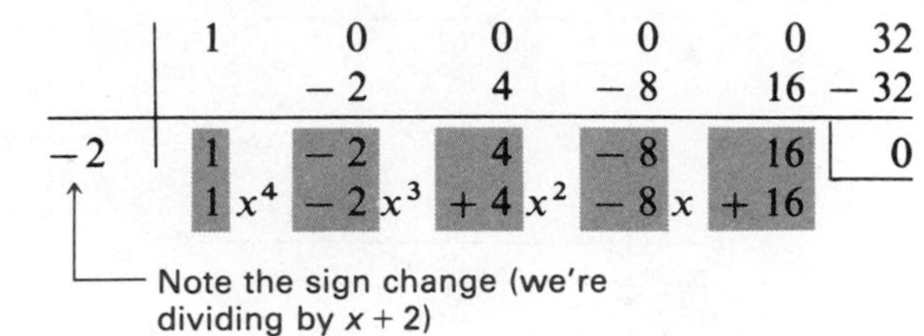

Check $(x + 2)(x^4 - 2x^3 + 4x^2 - 8x + 16) = x^5 + 32$

Therefore, $\dfrac{x^5 + 32}{x + 2} = x^4 - 2x^3 + 4x^2 - 8x + 16$. Since the remainder is 0, we can say that $x + 2$ and $x^4 - 2x^3 + 4x^2 - 8x + 16$ are *factors* of $x^5 + 32$. ■

Example 4 Divide $8x^4 - 6x^3 - 4x^2 + 7x - 3$ by $x - \dfrac{3}{4}$.

Solution

	8	−6	−4	7	−3
		6	0	−3	3
$\frac{3}{4}$	8	0	−4	4	0
	$8x^3$ +	$0x^2$	$-4x$ +	4	

The dividend is fourth degree

The quotient must be third degree

Check $(x - \frac{3}{4})(8x^3 - 4x + 4) = 8x^4 - 6x^3 - 4x^2 + 7x - 3$

Therefore, $(8x^4 - 6x^3 - 4x^2 + 7x - 3) \div (x - \frac{3}{4}) = 8x^3 - 4x + 4$. ■

Synthetic division *can* be used when the divisor is of the form $ax + b$ or $ax - b$; however, in this book we use it only when the divisor is of the form $x + a$ or $x - a$, where a is an integer, a common fraction, or a decimal. (Synthetic division *cannot* be used if the divisor is not a first-degree binomial.)

Synthetic division is a very useful tool; we will use it in both *factoring* (Section 5.8) and *graphing* (Section 8.8).

EXERCISES 4.6

Set I Use synthetic division to perform the divisions.

1. $(x^2 + 2x - 18) \div (x - 3)$

2. $(x^2 + 4x - 10) \div (x - 2)$

3. $(x^3 + 3x^2 - 5x + 6) \div (x + 4)$

4. $(x^3 + 6x^2 + 4x - 7) \div (x + 5)$

5. $(x^4 + 6x^3 - x - 4) \div (x + 6)$

6. $(2x^4 + 5x^3 + 10x - 2) \div (x + 3)$

7. $\dfrac{x^4 - 16}{x - 2}$

8. $\dfrac{x^7 - 1}{x - 1}$

9. $\dfrac{x^6 - 3x^5 - 2x^2 + 3x + 5}{x - 3}$

10. $\dfrac{x^6 - 3x^4 - 7x - 2}{x - 2}$

11. $(3x^4 - x^3 + 9x^2 - 1) \div (x - \frac{1}{3})$

12. $(3x^4 - 4x^3 - x^2 - x - 2) \div (x + \frac{2}{3})$

13. $\dfrac{x^5 + x^4 - 45x^3 - 45x^2 + 324x + 324}{x + 3}$

14. $\dfrac{x^5 + x^4 - 45x^3 - 45x^2 + 324x + 324}{x + 6}$

15. $\dfrac{4x^4 - 45x^2 + 3x + 100}{x - 2}$

16. $\dfrac{9x^4 - 13x^2 + 2x + 6}{x - 1}$

In Exercises 17–20, use a calculator.

17. $(2.6x^3 + 1.8x - 6.4) \div (x - 1.5)$

18. $(3.8x^3 - 1.4x^2 - 23.9) \div (x - 2.5)$

19. $(2.7x^3 - 1.6x + 3.289) \div (x - 1.2)$

20. $(3x^3 + 1.2x^2 - 1.5) \div (x - 1.6)$

Set II Use synthetic division to perform the divisions.

1. $(x^2 + 7x + 10) \div (x + 3)$

2. $(x^2 + 6x + 12) \div (x - 2)$

3. $(x^3 - 8x^2 + 11x - 14) \div (x - 6)$

4. $(x^4 + 5x^3 - x^2 - 4x + 7) \div (x + 5)$

5. $(2x^4 - 10x^3 - 6x + 15) \div (x - 5)$

6. $(2x^3 + 3x^2 + x - 15) \div (x - \frac{3}{2})$

7. $\dfrac{x^4 - 81}{x - 3}$

8. $(x^5 - 4x^4 + 4x^3 + 4x^2 - 11x + 8) \div (x - 2)$

9. $(x^6 - 4x^5 - 12x^3 + 48x^2 - 15x + 43) \div (x - 4)$

10. $(x^5 - 32) \div (x - 2)$

11. $(8x^4 - 2x^3 - 6x^2 + 4x + 3) \div (x + \frac{3}{4})$

12. $(2x^4 + 15x^3 + 6x^2 - 8x - 7) \div (x + 7)$

13. $\dfrac{x^5 + x^4 - 45x^3 - 45x^2 + 324x + 324}{x - 6}$

14. $(x^5 + 1) \div (x + 1)$

15. $\dfrac{4x^4 - 13x^2 + 2x + 6}{x - 2}$

16. $(2x^3 + 5x^2 + 5x - 4) \div \left(x - \dfrac{1}{2}\right)$

In Exercises 17 and 19, use a calculator.

17. $(1.4x^3 - 2.6x^2 + 56.3) \div (x + 3.5)$

18. $(3x^3 + 8x^2 + 6x - 3) \div \left(x - \dfrac{1}{3}\right)$

19. $(2.4x^3 + 1.32x^2 - 3.6) \div (x - 1.1)$

20. $(x^5 - 5x^4 + 10x^3 - 10x^2 + 5x - 1) \div (x - 1)$

4.7 Review: 4.1–4.6

Polynomials 4.1 A **polynomial in x** is an algebraic expression that has only terms of the form ax^n, where a is any real number and n is a positive integer (or zero). A **polynomial in x and y** is an algebraic expression that can be written so that it has only terms of the form $ax^n y^m$, where a is any real number and n and m are positive integers (or zero).

A *monomial* is a polynomial with only one term.

A *binomial* is a polynomial with two terms.

A *trinomial* is a polynomial with three terms.

Degree of a Polynomial 4.1

The **degree of a term** in a polynomial is the sum of the exponents of its variables. The **degree of a polynomial** is the same as that of its highest-degree term.

To Add Polynomials 4.2

Add *like* terms.

To Subtract Polynomials 4.2

Change the sign of each term in the polynomial being subtracted; then add the resulting *like* terms.

To Multiply Two Binomials 4.3A

To multiply two binomials, the FOIL method can be used.

To Multiply a Polynomial by a Polynomial 4.3B

Multiply the first polynomial by *each term* of the second polynomial; then add the resulting products.

Special Products 4.4

$$(a+b)(a-b) = a^2 - b^2 \quad \text{(Section 4.4A)}$$

$$\left.\begin{array}{l}(a+b)^2 = a^2 + 2ab + b^2 \\ (a-b)^2 = a^2 - 2ab + b^2\end{array}\right\} \quad \text{The square of a binomial (Section 4.4B)}$$

$$(a+b)^n \quad \text{Use the binomial theorem (Section 4.4C)}$$

Factorial Notation 4.4C

$0! = 1$

$1! = 1$

$n! = n(n-1)(n-2) \cdot \cdots \cdot 3 \cdot 2 \cdot 1$, where n is an integer greater than 1

The Binomial Coefficient 4.4C

$$\binom{n}{0} = 1$$

$\binom{n}{r} = \dfrac{n(n-1)(n-2) \cdot \cdots \cdot (n-r+1)}{r!} = \dfrac{n!}{r!(n-r)!}$, where n and r are *positive* integers and $r \leq n$

To Divide a Polynomial by a Monomial 4.5A

Divide *each term* in the polynomial by the monomial; then add the resulting quotients.

To Divide a Polynomial by a Polynomial 4.5B

We use a method like that of long division of whole numbers in arithmetic.

Synthetic Division 4.6

Synthetic division can be used to divide a polynomial by a first-degree binomial.

Review Exercises 4.7 Set I

In Exercises 1–4, if the expression is a polynomial, find its degree. Otherwise, write "Not a polynomial."

1. $7x^2 - \frac{3}{5}xy^2 + 2$

2. $\frac{16}{3z - w} - 2w^2$

3. $\sqrt{7 + x}$

4. $8x^{-2} + x^{-1} - 4$

In Exercises 5 and 6, perform the indicated operations and simplify the results.

5. $(13x - 6x^3 + 14 - 15x^2) + (-17 - 23x^2 + 4x^3 + 11x)$

6. $(5x^2y + 3xy^2 - 4 + y^2) - (8 - 4x^2y + 2xy^2 - y^2)$

In Exercises 7 and 8, subtract the bottom polynomial from the top one.

7.
$$\begin{array}{l} x^3 \qquad\qquad - 4x + 2 \\ \underline{3x^3 + 2x^2 + \ x - 5} \end{array}$$

8.
$$\begin{array}{l} x^4 \qquad\quad + x^2 \qquad + 1 \\ \underline{\qquad x^3 \qquad\quad + x\qquad} \end{array}$$

9. Subtract $(3k^2 - 5k - 6)$ from the sum of $(2k^3 - 7k + 11)$ and $(4k^3 + k^2 - 9k)$.

In Exercises 10–22, perform the indicated operations.

10. $(5a^2b + 3a - 2c)(-3ab^2)$

11. $(5x - 4y)(7x - 2y)$

12. $(x^2 + 3)^4$

13. $(4x^2 - 5x + 1)(2x^2 + x - 3)$

14. $(7x + 5)(7x - 5)$

15. $(2x^2 + \frac{1}{2})^5$

16. $(3 + x + y)^2$

17. $(a - 5)^2$

18. $(z + 3)(z^2 - 3z + 9)$

19. $(a + b + 4)^2$

20. $(z^{-2} + z^2)^4$

21. $4x(x^2 - y^2)(x^2 + y^2)$

22. $(\frac{1}{3} + 3x^2)^4$

In Exercises 23–27, perform the divisions.

23. $\frac{-15a^2b^3 + 20a^4b^2 - 10ab}{-5ab}$

24. $(6x^2 - 9x + 10) \div (2x - 3)$

25. $(10a^2 + 23ab - 5b^2) \div (5a - b)$

26. $(3x^4 - 2x^3 + 2x^2 + 2x - 15) \div (3x^2 - 2x + 5)$

27. $\frac{x^4 - 81}{x + 3}$

In Exercises 28–30, use synthetic division.

28. $(3x^5 + 7x^4 - 4x^2 + 4) \div (x + 2)$

29. $(5x^5 - 2x^4 + 10x^3 - 4x^2 + 2) \div (x - \frac{2}{5})$

30. $(2x^5 + 6x^4 + x + 5) \div (x + 3)$

Review Exercises 4.7 Set II

NAME ______________________

In Exercises 1–4, if the expression is a polynomial, find its degree. Otherwise, write "Not a polynomial."

1. $\frac{1}{2}a^2b^3 - 2^3ab^2 + 5$

2. $9m^3 - \dfrac{8}{m - 5n}$

3. $\sqrt{z + 4}$

4. $x^{-3} + 2x^{-1} - 4$

In Exercises 5 and 6, perform the indicated operations and simplify the results.

5. $(7z - 20 + 11z^3 - 9z^2) + (-16z^3 + 15 - 3z^2 + 12z)$

6. $(18ef^2 - 14 - 6e^2f + 4f) - (17 + 13f - 14e^2f + 10ef^2)$

In Exercises 7 and 8, subtract the bottom polynomial from the top one.

7.
$$\begin{array}{r} x^3 - 2x^2 + 1 \\ \underline{5x^3 + 2x^2 - x + 3} \end{array}$$

8.
$$\begin{array}{r} x^4 + x^3 + 1 \\ \underline{ x^2 + x } \end{array}$$

9. Subtract $(13h^3 - 2h^2 + 7)$ from the sum of $(12h^2 - 5h - 18)$ and $(9h^3 + 10h - 6)$.

In Exercises 10–22, perform the indicated operations.

10. $-4ab^2(7ab^2 - 5b + 4c)$

11. $(y^2 - 3y + 5)(2y - 3)$

12. $(3n^2 - 6n + 2)(n^2 + 4n - 5)$

13. $(w - 2x - 5)^2$

14. $(a^2 + 5b)(a^2 - 5b)$

15. $(2x - 3y)^2$

16. $(4m + 3n)(2m - 5n)$

17. $2x(x^4 + y^4)(x^4 - y^4)$

ANSWERS

1. ______________________
2. ______________________
3. ______________________
4. ______________________
5. ______________________
6. ______________________
7. ______________________
8. ______________________
9. ______________________
10. ______________________
11. ______________________
12. ______________________
13. ______________________
14. ______________________
15. ______________________
16. ______________________
17. ______________________

18. $(x^2 + 1)^6$

19. $(x^2 + x^{-1})^4$

20. $\left(4c^2 - \frac{1}{2}\right)^5$

21. $(m^{-1} - m^2)^6$

22. $(2x - 3)(4x^2 + 6x + 9)$

In Exercises 23–27, perform the divisions.

23. $\dfrac{24m^4n^3 - 30m^2n^2 + 18m^4n}{-6m^2n}$

24. $(21c^2 - 29c - 18) \div (3c - 5)$

25. $(12a^2 - 7ab - 10b^2) \div (4a - 5b)$

26. $(20k^4 - 8k^3 - 39k^2 + 6k - 9) \div (5k^2 - 2k - 6)$

27. $\dfrac{x^6 - 64}{x - 2}$

In Exercises 28–30, use synthetic division.

28. $(4x^5 - 16x^3 - 6x + 3) \div (x - 2)$

29. $(12x^5 + x^4 - 6x^3 + 8x + 11) \div \left(x + \frac{3}{4}\right)$

30. $(2x^5 + 10x^4 - 3x^2 - 14x + 1) \div (x + 5)$

ANSWERS

18. ____________

19. ____________

20. ____________

21. ____________

22. ____________

23. ____________

24. ____________

25. ____________

26. ____________

27. ____________

28. ____________

29. ____________

30. ____________

Chapter 4 Diagnostic Test

The purpose of this test is to see how well you understand operations with polynomials. We recommend that you work this diagnostic test *before* your instructor tests you on this chapter. Allow yourself about 50 minutes.

Complete solutions for all the problems on this test, together with section references, are given in the answer section at the end of the book. For the problems you do incorrectly, study the sections cited.

1. In the polynomial $2x^2y^3 - 5^2x^2y^2 - \frac{1}{3}xy$, find:

a. The degree of the third term

b. The degree of the polynomial

2. a. Add:
$$\begin{array}{r} -4x^3 - 3x^2 \qquad\quad + 5 \\ 2x^2 + 6x - 10 \\ \underline{3x^3 \qquad\quad - 2x + \ \ 8} \end{array}$$

b. Subtract the bottom polynomial from the top one.

$$\begin{array}{l} 3x^4 - x^3 \qquad\quad + x - 2 \\ \underline{5x^4 + x^3 + x^2 - 3x - 5} \end{array}$$

3. a. Subtract $(8 - z + 4z^2)$ from $(-3z^2 - 6z + 8)$.

b. Simplify: $(3ab^2 - 5ab) - (a^3 + 2ab) + (4ab - 7ab^2)$

In Problems 4–10, perform the indicated operations.

4. a. $-3ab(6a^2 - 2ab^2 + 5b)$

b. $\dfrac{9z^3w + 6z^2w^2 - 12zw^3}{3zw}$

5. a. $(x - 2)(x^2 + 2x + 4)$

b. $(2x^4 + 3)(2x^4 - 3)$

6. a. $(5m - 2)(3m + 4)$

b. $(3R^2 - 5)^2$

7. a. $(12y^2 - 4y + 1) \div (3y + 2)$

b. Use synthetic division: $(2x^4 + 3x^3 - 7x^2 - 5) \div (x + 3)$

8. $(m^2 - 2m + 5)^2$

9. $(6x^4 - 3x^3 + 2x^2 + 5x - 7) \div (2x^2 - x + 4)$

10. $(2x + 1)^5$

Cumulative Review Exercises: Chapters 1–4

1. Evaluate $10 - (\sqrt[3]{-27} - 5^2)$.

2. Simplify $\left(\dfrac{x^2y^{-1}}{y^{-4}}\right)^{-1}$. Write your answer using only positive exponents.

3. Find the LCM of 108 and 360.

4. Simplify: $5a - \{9 - 2(3 - a) - 3a\}$

In Exercises 5–9, solve each equation or inequality.

5. $4(x - 3) = 4 - (x + 6)$

6. $1 - x \leq 2(x - 4)$

7. $|5 - x| = 6$

8. $|3x - 5| > 7$

9. $|2x - 1| \leq 4$

In Exercises 10–12, perform the indicated operations and simplify the results.

10. $(3x^4 + 7x^2 - 2x)(x + 2)$

11. $(x - 1)^6$

12. $(x^3 + 2x^2 - 13x - 16) \div (x + 4)$

In Exercises 13–18, set up the problem algebraically and solve it. Be sure to state what your variables represent.

13. The three sides of a triangle are in the ratio 2:4:5. The perimeter is 88. Find the lengths of the three sides.

14. If raisins costs \$1.38 per pound and granola costs \$2.10 per pound, how many pounds of raisins must be mixed with 15 lb of granola for the mixture to cost \$37.71?

15. A boat has a speed of 30 mph in still water. It travels downstream for 40 min and then starts back upstream. In 40 more minutes, it is still 4 mi short of its starting point. What is the speed of the stream?

16. Debbie invested \$1,400. Part of the money earned 6.25% per year, and the rest earned 5.75%. At the end of one year, the interest paid was \$83.50. How much did Debbie place in each account?

17. A boater is running her boat's engine so that in still water she would be going 28 mph; the speed of the current is 4 mph. She wants to go some distance up the river (against the current). How far up the river can she go if she must be back at her starting point in $3\frac{1}{2}$ hr?

18. Mr. Liao must mix a 2% and a 10% solution of disinfectant in order to obtain 12 ℓ of a 5% solution. How much of the 2% solution should he use? How much of the 10% solution?

5 Factoring

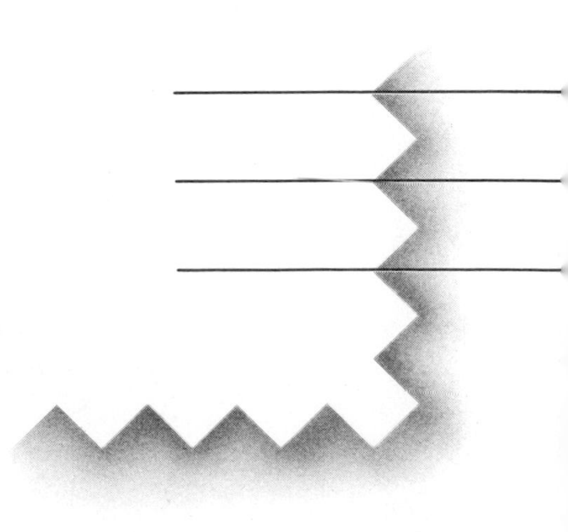

This chapter deals with factoring algebraic expressions. We will review the methods of factoring learned in elementary algebra and discuss additional types of factoring. We will also solve equations whose solution involves factoring and solve word problems that lead to such equations.

5.1 Greatest Common Factor (GCF)

It is essential that the techniques of factoring be mastered, because factoring is used extensively in work with algebraic fractions, in graphing, and in solving equations.

Factoring an algebraic expression means rewriting it (if possible) as a single term that is a *product of prime factors*. That is, when we *factor* an algebraic expression, we are "*undoing*" multiplication. In Chapter 1, we used the distributive property to rewrite a product of factors as a sum of terms, that is, $a(b + c) = ab + ac$. We now use the distributive property to rewrite a sum of terms as a product of factors (that is, as a single term) whenever possible. When we use the distributive property to get

$$ab + ac = a(b + c)$$

we say that we are factoring out the greatest common factor. Notice that the right side of this equation has only *one term.*

Greatest Common Factor (GCF)

The **greatest common factor (GCF)** of two or more terms in an algebraic expression is the *largest* term that is a factor of all the terms; that is, it is the *largest* term that *divides exactly* into all the terms. We find the GCF as follows:

TO FIND THE GREATEST COMMON FACTOR (GCF)

1. Write each numerical coefficient in prime factored form. Repeated factors must be expressed in exponential form.
2. Write down each different base, numerical or literal, *that is common to all terms.*
3. Raise each of the bases written down in step 2 to the *lowest* power to which it occurs in any of the terms.
4. The *greatest common factor* (GCF) is the *product* of all the expressions found in step 3. It may be positive or negative.

Example 1 Find the GCF for the terms of $-35xy^3z^3 + 21x^3y^3z^3 - 28x^2y^2z^4 + 7xy^2z^3$.

Solution

$$-35xy^3z^3 + 21x^3y^3z^3 - 28x^2y^2z^4 + 7xy^2z^3$$
$$= -5 \cdot 7xy^3z^3 + 3 \cdot 7x^3y^3z^3 - 2^2 \cdot 7x^2y^2z^4 + 7xy^2z^3$$

The bases common to all four terms are 7, x, y, and z. The lowest power on the 7 and on the x is an understood 1, the lowest power on the y is 2, and the lowest power on the z is 3. Therefore, the GCF (the *largest* expression that divides exactly into all four terms) is $7xy^2z^3$ *or* $-7xy^2z^3$. ■

To rewrite the *sum of terms* $ab + ac$ as the *product of factors* $a(b + c)$—that is, to "undo" the multiplication—we can follow the suggestions in the following box. They are based on the fact that we're asking ourselves a question such as "If we had had $-35xy^3z^3 + 21x^3y^3z^3 - 28x^2y^2z^4 + 7xy^2z^3$ *as the answer to a multiplication problem*, what would the original factors have looked like?"

TO FACTOR AN EXPRESSION WITH A COMMON FACTOR

1. Combine like terms, if there are any.
2. Find the GCF for all of the terms. It will often, but not always, be a monomial.
3. Find the *polynomial factor** by dividing each term of the polynomial being factored by the GCF. The polynomial factor will always have as many terms as the expression in step 1. It should have only *integer* coefficients.
4. Rewrite the expression as the product of the factors found in steps 2 and 3.
5. Check the result by using the distributive property to remove the parentheses; you should get back the polynomial you started with.

NOTE If an expression has been factored, *it has only one term.* However, even if an expression has only one term, it still may not be factored completely. Any expression inside parentheses should always be examined carefully to see whether it could be factored further. ☑

When we use the rules in the box above, we are factoring *over the integers*; that is, all the constants in the new polynomials are *integers.* In this chapter, when we say to factor a polynomial, we mean to factor *over the integers.*

Factorization of polynomials over the integers is unique; that is, a polynomial can be completely factored over the integers in one and only one way (except for the order in which the factors are written).

Example 2 Factor $-35xy^3z^3 + 21x^3y^3z^3 - 28x^2y^2z^4 + 7xy^2z^3$.

Solution We found in Example 1 that the GCF is $7xy^2z^3$ or $-7xy^2z^3$. To find the polynomial factor, we divide each term of the given polynomial by the GCF. If we use $7xy^2z^3$ as the GCF, the polynomial factor is

$$\frac{-35xy^3z^3}{7xy^2z^3} + \frac{21x^3y^3z^3}{7xy^2z^3} - \frac{28x^2y^2z^4}{7xy^2z^3} + \frac{7xy^2z^3}{7xy^2z^3} = -5y + 3x^2y - 4xz + 1$$

Therefore,
$$-35xy^3z^3 + 21x^3y^3z^3 - 28x^2y^2z^4 + 7xy^2z^3$$
$$= 7xy^2z^3(-5y + 3x^2y - 4xz + 1)$$

Check
$$7xy^2z^3(-5y + 3x^2y - 4xz + 1)$$
$$= -35xy^3z^3 + 21x^3y^3z^3 - 28x^2y^2z^4 + 7xy^2z^3$$

* We call this factor the polynomial factor because it will be a polynomial and will always have more than one term.

If we use $-7xy^2z^3$ as the GCF, the polynomial factor is

$$\frac{-35xy^3z^3}{-7xy^2z^3} + \frac{21x^3y^3z^3}{-7xy^2z^3} - \frac{28x^2y^2z^4}{-7xy^2z^3} + \frac{7xy^2z^3}{-7xy^2z^3} = 5y - 3x^2y + 4xz - 1$$

Therefore,
$$-35xy^3z^3 + 21x^3y^3z^3 - 28x^2y^2z^4 + 7xy^2z^3$$
$$= -7xy^2z^3(5y - 3x^2y + 4xz - 1)$$

Check
$$-7xy^2z^3(5y - 3x^2y + 4xz - 1)$$
$$= -35xy^3z^3 + 21x^3y^3z^3 - 28x^2y^2z^4 + 7xy^2z^3$$

Note that the answers $7xy^2z^3(-5y + 3x^2y - 4xz + 1)$ and $-7xy^2z^3(5y - 3x^2y + 4xz - 1)$ each have just one term and that in both answers the expression within the parentheses has no common factor left. Both answers are correct and acceptable. ■

Prime (or Irreducible) Polynomials

A polynomial is said to be **prime**, or **irreducible**, over the integers if it cannot be expressed as a product of polynomials of lower degree such that all the constants in the new polynomials are integers.

Example 3 Factor $3x - 7y$.

Solution Although 3 and x are factors of the first term of $3x - 7y$, they are not factors of the second term. In fact, $3x$ and $-7y$ have no common integral factor. While it is true that

$$3x - 7y = 3(x - \tfrac{7}{3}y)$$

and that

$$3x - 7y = 7(\tfrac{3}{7}x - y)$$

the parentheses now contain constants that are *not* integers. We must conclude that $3x - 7y$ is *not factorable* over the integers. It is a prime polynomial. ■

Sometimes an expression has a GCF that is not a monomial. Such an expression can still be factored using the same rules as above. The type of factoring seen in Examples 4 and 5 is used in *factoring by grouping* (covered in Section 5.2) and is also used extensively in calculus.

Example 4 Factor $a(x + y) + b(x + y)$.

Solution This expression has two terms and thus is *not* in factored form. The common factor, $x + y$, is not a monomial; it is a binomial. Nevertheless, $x + y$ is the GCF. We factor as follows:

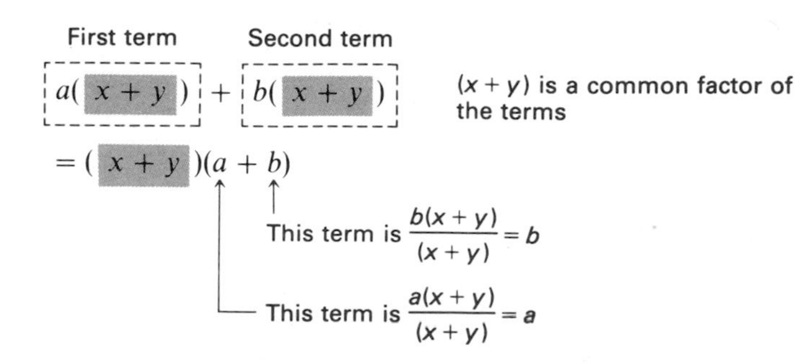

Therefore, $a(x + y) + b(x + y) = (x + y)(a + b)$. Our answer has one term, so it is in factored form.

Check $(x + y)(a + b) = (a + b)(x + y) = a(x + y) + b(x + y)$ ■

Example 5 Factor $72x^2(m^2+n)^2 + 36(m^2+n)$.

Solution The GCF is $2^2 \cdot 3^2(m^2+n)^1$. Therefore,

This exponent means that there are *two* factors of (m^2+n)

$$72x^2(m^2+n)^2 + 36(m^2+n) = 36(m^2+n)(2x^2[m^2+n]+1)$$

When we remove the inner grouping symbols, we have

$$72x^2(m^2+n)^2 + 36(m^2+n) = 36(m^2+n)(2x^2m^2+2x^2n+1)$$

Check $36(m^2+n)(2x^2m^2+2x^2n+1) = 36(m^2+n)(2x^2[m^2+n]+1)$

$$= 72x^2(m^2+n)^2 + 36(m^2+n) \quad ■$$

EXERCISES 5.1

Set I Factor each of the following expressions or write "Not factorable."

1. $54x^3yz^4 - 72xy^3$

2. $225ab^5c - 105a^3b^2c^6$

3. $16x^3 - 8x^2 + 4x$

4. $27a^4 - 9a^2 + 3a$

5. $3x^3 - 4y^2 + 5z$

6. $7z^3 + 3y^2 - 4x$

7. $6my + 15mz - 5n - 4n$

8. $4nx + 8ny + 16z - 4z$

9. $-35r^7s^5t^4 - 55r^8s^9u^4 + 40p^8r^9s^8$

10. $-120a^8b^7c^5 + 40a^4c^3d^9 - 80a^5c^5$

11. $10x^2 - 21y + 11z^3$

12. $15a^3 + 14b^5 - 13c$

13. $-24x^8y^3 - 12x^7y^4 + 48x^5y^5 + 60x^4y^6$

14. $64y^9z^5 + 48y^8z^6 - 16y^7z^7 - 80y^4z^8$

15. $m(a+b) + n(a+b)$

16. $3a(a-2b) + 2(a-2b)$

17. $x(y+1) - (y+1)$

18. $2e(3e-f) - 3(3e-f)$

19. $5(x-y) - (x-y)^2$

20. $4(a+b) - (a+b)^2$

21. $8x(y^2+3z)^2 - 6x^4(y^2+3z)$

22. $12a^3(b-2c^5)^3 - 15a^2(b-2c^5)^2$

23. $5(x+y)^3(a+b)^5 + 15(x+y)^2(a+b)^6$

24. $14(s+t)^4(u+v)^7 + 7(s+t)^5(u+v)^6$

Set II Factor each of the following expressions or write "Not factorable."

1. $168s^4 - 126st^3u$

2. $-5x^3 + 10x^5y - 15x^3y^2$

3. $25y^3 - 15y^2 + 5y$

4. $17 + 34x^2 - 85xy$

5. $5a^2 - 2b + 3c$

6. $3x(x+y) - 6x^2(x+y)^2$

7. $7xy + 3yx - 5z + z$

8. $7xy + 49xy^2 - 14x^2y$

9. $-36x^3y^7 - 18x^4y^5 + 27x^5y^3 - 9x^6y^2$

10. $5x^4 - 7y^3 + 3z$

11. $25s^2 - 28t^3 + 9u$

12. $6a(b-c)^3 - 9a^2(b-c)^2$

13. $10a^2b^4 - 15a^3c^3 - 25a^5c^7$

14. $3a^3 + 17b - 8c^2$

15. $a(a+b) + 3b(a+b)$

16. $15x^4y^3 - 18x^5y^6 + 27y^4$

17. $3x(2y-5) - 2(2y-5)$

18. $3x^3(y+3z)^2 - 6x^4(y+3z)^3$

19. $6(a-b)^2 - (a-b)$

20. $12(x + y)^3 - 3(x + y)^4$

21. $6s^4(t^3 + 2u)^3 - 9s^3(t^3 + 2u)^2$

22. $24x^3(a - 5b)^5 - 32x^4(a - 5b)^4$

23. $21(a + b)^4(c + d)^7 + 3(a + b)^5(c + d)^6$

24. $54s^4(u^2 - v^3)^3 - 36s^3(u^2 - v^3)^4$

5.2 Factoring by Grouping

In Section 4.3A, we did a few problems in which the product of two factors was a polynomial that contained four terms. In this section, we "undo" such problems. That is, we start with a polynomial that contains four terms, assume that it is the product of two factors, and attempt to find the factors that multiplied together would give such a product.

We begin by considering those four-term polynomials that can be factored by *grouping* the four terms into two groups of two terms each. In all the problems in this section, when we factor by grouping there will be a *common factor* after we've grouped the terms and factored each group. (In later sections of this chapter, we will consider other methods of factoring by grouping.) To "undo" the multiplication, we can use the suggestions in the following box. They are based on the fact that we're asking ourselves a question such as "If we had had $ax + ay + bx + by$ *as the answer to a multiplication problem*, what would the original factors have looked like?" (See Example 1.) The suggestions in the box assume that any factor common to *all four* terms has already been factored out.

TO FACTOR AN EXPRESSION OF FOUR TERMS BY GROUPING TWO AND TWO

1. Arrange the four terms into two groups of two terms each. Each group of two terms should be factorable.
2. Factor out the GCF from each group. You will now have two terms. *The expression will not yet be factored.*
3. Factor the two-term expression resulting from step 2, if possible.
4. If the two terms resulting from step 2 do not have a GCF, try a different arrangement of the original four terms.

Example 1 Factor $ax + ay + bx + by$.

Solution

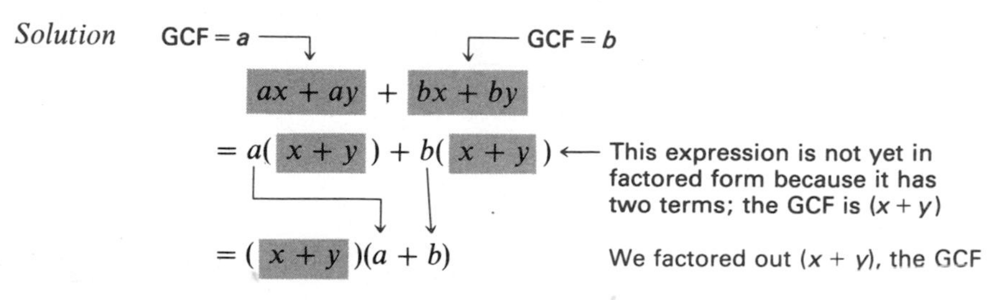

Therefore, $ax + ay + bx + by = (x + y)(a + b)$.

Check $(x + y)(a + b) = (x + y)a + (x + y)b$

$= ax + ay + bx + by$ ■

It is often possible to group terms differently and still be able to factor the expression. *The same factors are obtained no matter what grouping is used*, because factorization over the integers is unique.

Example 2 Factor $ab - b + a - 1$.

One Grouping

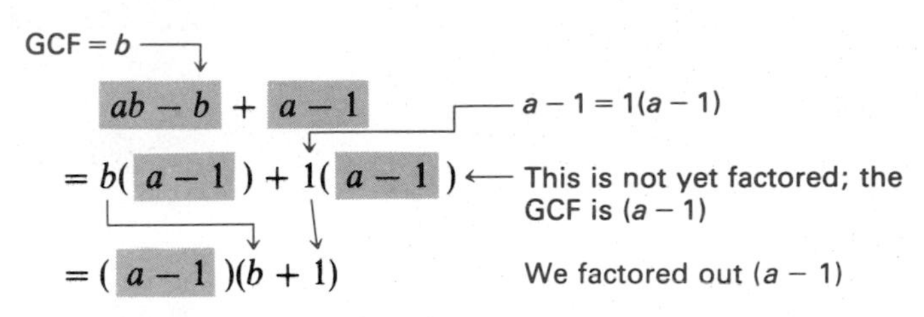

A Different Grouping If we rearrange the terms, we have

GCF $= a$ GCF $= -1$

$$ab + a - b - 1$$

$$= a(b + 1) - 1(b + 1)$$

Note that when we factor -1 from $-b - 1$, we must change the sign of each term that goes *into* the parentheses

$$= (b + 1)(a - 1)$$

Therefore, $ab - b + a - 1 = (a - 1)(b + 1) = (b + 1)(a - 1)$.

Check $(a - 1)(b + 1) = ab - b + a - 1$ ■

Example 3 Factor $2x^2 - 6xy + 3x - 9y$.

Solution

GCF $= 2x$ GCF $= 3$

$$2x^2 - 6xy + 3x - 9y$$

$$= 2x(x - 3y) + 3(x - 3y) \quad (x - 3y) \text{ is the GCF}$$

$$= (x - 3y)(2x + 3) \quad \text{We factored out } (x - 3y)$$

Therefore, $2x^2 - 6xy + 3x - 9y = (x - 3y)(2x + 3)$.

Check $(x - 3y)(2x + 3) = 2x^2 - 6xy + 3x - 9y$ ■

If we have an expression to factor that has six terms, we can try either three groups of two terms each *or* two groups of three terms each.

Example 4 Factor $ax^2 - ax + 5a + bx^2 - bx + 5b$.

Solution If we try two groups of three terms each, we have

$$ax^2 - ax + 5a + bx^2 - bx + 5b$$

$$= a(x^2 - x + 5) + b(x^2 - x + 5) \quad \text{There is a common } \textit{trinomial} \text{ factor}$$

$$= (x^2 - x + 5)(a + b)$$

If we try three groups of two terms each, we rearrange and have

$$ax^2 + bx^2 - ax - bx + 5a + 5b$$

$$= x^2(a + b) - x(a + b) + 5(a + b) \quad (a + b) \text{ is the GCF}$$

$$= (a + b)(x^2 - x + 5)$$

Check $(x^2 - x + 5)(a + b) = (a + b)(x^2 - x + 5)$
$= ax^2 + bx^2 - ax - bx + 5a + 5b$

Other arrangements are also possible. ■

A WORD OF CAUTION An expression is *not* factored until it has been written as a single term that is a product of factors. To illustrate this, consider Example 1 again:

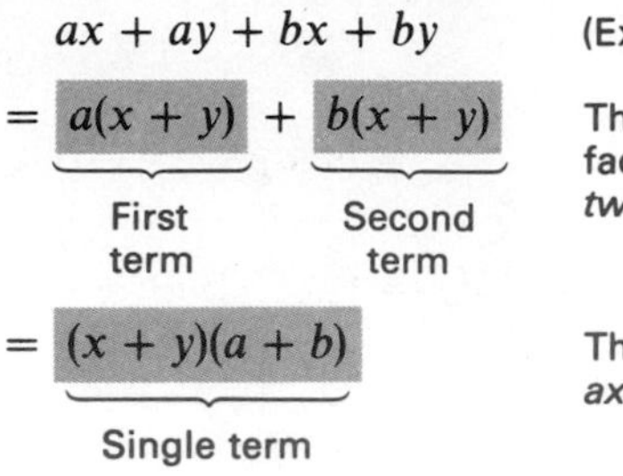

(Example 1)

This expression is *not* in factored form because it has *two* terms

The *factored form* of $ax + ay + bx + by$

✓

EXERCISES 5.2

Set I Factor each expression or write "Not factorable."

1. $mx - nx - my + ny$
2. $ah - ak - bh + bk$
3. $xy + x - y - 1$
4. $ad - d + a - 1$
5. $3a^2 - 6ab + 2a - 4b$
6. $2h^2 - 6hk + 5h - 15k$
7. $6e^2 - 2ef - 9e + 3f$
8. $8m^2 - 4mn - 6m + 3n$
9. $x^3 + 3x^2 - 2x - 6$
10. $a^3 - a^2 - 2a + 2$
11. $b^3 + 4b^2 + 5b - 20$
12. $6x^3 + 3x^2 - 2x + 1$
13. $2a^3 + 8a^2 - 3a - 12$
14. $5y^3 - 10y^2 + 2y - 4$
15. $acm + bcm + acn + bcn$
16. $cku + ckv + dku + dkv$
17. $a^2x + 2ax + 5x + a^2y + 2ay + 5y$
18. $ax^2 + 3ax + 7a + bx^2 + 3bx + 7b$
19. $s^2x - sx + 4x + s^2y - sy + 4y$
20. $at^2 - at + 3a + bt^2 - bt + 3b$
21. $ax^2 + ax + a - x^2 - x - 1$
22. $xy^2 + xy + 2x - y^2 - y - 2$

Set II Factor each expression or write "Not factorable."

1. $hw - kw - hz + kz$
2. $3a - 5b + 6ac - 10bc$
3. $xy - x + y - 1$
4. $3x^3 - 7x^2 + 3x - 7$
5. $12m^2 - 6mn + 14m - 7n$
6. $a^3m + 4a^2m - a^2n - 4an$
7. $6x^2 - 3xy - 4x + 2y$
8. $6x + 3y - 12ax - 6ay$
9. $x^3 + x^2 - 2x - 2$
10. $7xy - 3z - 14axy + 6az$
11. $4a^3 + 8a^2 - a + 2$
12. $8x^2 - 11xy - 16x + 22y$
13. $3t^3 - 12t^2 + 2t - 8$
14. $ab - ac - bd - cd$
15. $acx + acy + bcx + bcy$
16. $x^4 + x^3 + x^2 + x$
17. $ax^2 + 3ax + 8a + bx^2 + 3bx + 8b$
18. $x^3 + x^2a + x^2 + ax + 4x + 4a$
19. $a^2x - ax + 4x + a^2y - ay + 4y$
20. $2x^3 - 2ax^2 + x^2 - ax + x - a$
21. $b^2y + 2by + 3y - b^2 - 2b - 3$
22. $a^2x + 3ax + 4x - a^2 - 3a - 4$

5.3 Factoring the Difference of Two Squares

The expression $a^2 - b^2$ is called a **difference of two squares**. Because $(a + b)(a - b) = a^2 - b^2$ (we proved this in Section 4.4A), it must be true that $a^2 - b^2$ *factors into* $(a + b)(a - b)$.

> TO FACTOR THE DIFFERENCE OF TWO SQUARES
>
> **1.** Find the principal square root of each of the terms by inspection.
>
> **2.** The two binomial factors are the sum of and the difference of the square roots found in step 1.
>
> $$a^2 - b^2 = (a + b)(a - b)$$

Example 1 Factor $49a^6b^2 - 81c^4d^8$.

Solution

A difference of two squares

$$49a^6b^2 - 81c^4d^8 = (7a^3b)^2 - (9c^2d^4)^2$$

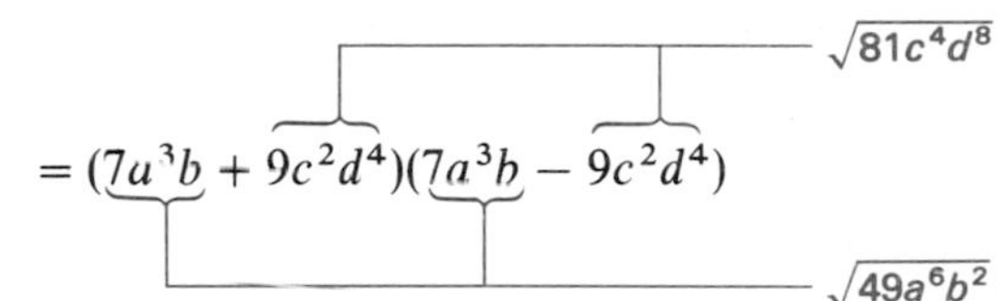

$$= (7a^3b + 9c^2d^4)(7a^3b - 9c^2d^4)$$

Check $(7a^3b + 9c^2d^4)(7a^3b - 9c^2d^4) = 49a^6b^2 - 81c^4d^8$ ■

A WORD OF CAUTION A *sum* of two squares is *not factorable* over the integers. (Exception: If a sum of two squares is of some degree that is a multiple of four, it *may* be factorable by completing the square, a method of factoring covered in Section 5.7.)

It is clear that there is no common factor for $a^2 + b^2$. We suggest that you verify that $a^2 + b^2$ is not factorable by assuming that it factors into a product of two binomials and attempting to find those binomials that multiplied together give $a^2 + b^2$. You will be unsuccessful. You might try this *again* after you've studied Section 5.4. ☑

We will consider an expression to be *completely factored* when all its factors are prime. By this we mean that no more factoring can be done by any method.

Example 2 Factor $a^4 - b^4$.
Solution

This factor can be factored again

$$a^4 - b^4 = (a^2 + b^2)(a^2 - b^2) = (a^2 + b^2)(a + b)(a - b)$$

This factor is a sum of two squares and cannot be factored

Check $(a^2 + b^2)(a + b)(a - b) = (a^2 + b^2)(a^2 - b^2) = a^4 - b^4$ ■

A WORD OF CAUTION Always remove the GCF first whenever possible. ☑

Example 3 Factor $27x^4 - 12y^2$.

Solution

$$27x^4 - 12y^2 = 3(\underbrace{9x^4 - 4y^2})$$ ← This has been factored, but this factor is not prime, so it factors further

$$= 3(3x^2 + 2y)(3x^2 - 2y)$$

(The checking is left to the student.) ■

Sometimes the quantities that are squares are not monomials. If the entire expression is in the form $a^2 - b^2$, we can still factor it (see Example 4).

Example 4 Factor $(x - y)^2 - (a + b)^2$.

Solution $(x - y)^2 - (a + b)^2$ A difference of two squares

$= [\quad + \quad][\quad - \quad]$ Blank outline

$= [(x - y) + (a + b)][(x - y) - (a + b)]$

$= (x - y + a + b)(x - y - a - b)$

Check $(x - y + a + b)(x - y - a - b)$

$= ([x - y] + [a + b])([x - y] - [a + b])$

$= [x - y]^2 - [a + b]^2$ ■

Example 5 Factor $a^2 - b^2 + a - b$.

Solution If we group the terms into two groups of two terms each, we have

$a^2 - b^2 + a - b$ ← Difference of two squares ($a^2 - b^2$)

$= (a + b)(a - b) + (a - b)$ $a - b$ is a common binomial factor

$= (a + b)(a - b) + 1(a - b)$ We write the coefficient 1 for $(a - b)$

$= (a - b)([a + b] + 1)$ We factor out $(a - b)$, the GCF

$= (a - b)(a + b + 1)$

Check $(a - b)(a + b + 1) = a^2 - b^2 + a - b$ ■

EXERCISES 5.3

Set I Factor completely or write "Not factorable."

1. $2x^2 - 8y^2$
2. $3x^2 - 27y^2$
3. $98u^4 - 72v^4$
4. $243m^6 - 300n^4$
5. $x^4 - y^4$
6. $a^4 - 16$
7. $4c^2 + 1$
8. $16d^2 + 1$
9. $4h^4k^4 - 1$
10. $9x^4 - 1$
11. $25a^4b^2 - a^2b^4$
12. $x^2y^4 - 100x^4y^2$

13. $16x^2 + 8x$ **14.** $25y^2 + 5y$ **15.** $9x^2 + 7y$

16. $36a^2 + 5b$ **17.** $(x + y)^2 - 4$ **18.** $(a + b)^2 - 9$

19. $(a + b)x^2 - (a + b)y^2$ **20.** $(x + y)a^2 - (x + y)b^2$

21. $9x^2 + y^2$ **22.** $25x^2 + y^2$

23. $x^2 - 9y^2 + x - 3y$ **24.** $x^2 - y^2 + x - y$

25. $x + 2y + x^2 - 4y^2$ **26.** $a + b + a^2 - b^2$

Set II Factor completely or write "Not factorable."

1. $6m^2 - 54n^4$ **2.** $3x^2 - 27x$ **3.** $50a^2 - 8b^2$

4. $4x^2 - 9$ **5.** $x^4 - 81$ **6.** $x^2 + 16$

7. $16y^4z^4 + 1$ **8.** $1 - 25a^2b^2$ **9.** $16y^4z^4 - 1$

10. $81x^2 - y^2$ **11.** $3b^2x^4 - 12b^2y^2$ **12.** $y^4 - 256$

13. $4x^2 + 8x$ **14.** $28a^2b^2 - 7a^2$

15. $25x^2 + 3y$ **16.** $49a^2 - 7a$

17. $(s + t)^2 - 16$ **18.** $(s + t)a^2 - (s + t)b^2$

19. $(s + t)x^2 - (s + t)y^2$ **20.** $(a - b)^2 - 1$

21. $4 + x^2$ **22.** $9x^2 - 3x$

23. $s^2 - t^2 + s - t$ **24.** $4x + 4y + x^2 - y^2$

25. $4x + y + 16x^2 - y^2$ **26.** $49x^2 - a^2 + 14x - 2a$

5.4 Factoring Trinomials

5.4A Factoring a Trinomial with a Leading Coefficient of 1

In Section 4.3A, we did many multiplication problems in which the product of two binomials was a trinomial. In this section, we "undo" such problems. That is, we start with a trinomial, assume that it is the product of two binomial factors, and attempt to find those factors. (Though most of the trinomials seen in this chapter *are* factorable, *most* trinomials are *not* factorable.) Careful examination of the following multiplication problems leads us to the *Rule of Signs for Factoring Trinomials* that follows.

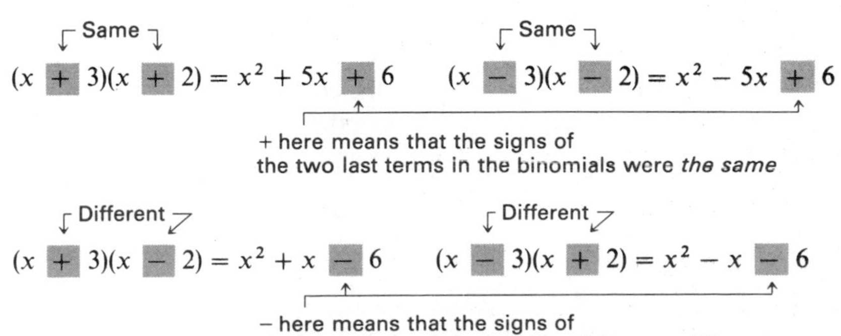

RULE OF SIGNS FOR FACTORING TRINOMIALS

Arrange the trinomial in descending powers of one variable.

1. If the sign of the last term of the trinomial is $+$, the signs of the two last terms in the binomials will be the same, and:

a. If the sign of the middle term of the trinomial is $+$, the signs of the two last terms in the binomials will both be $+$.

b. If the sign of the middle term of the trinomial is $-$, the signs of the two last terms in the binomials will both be $-$.

2. If the sign of the last term of the trinomial is $-$, the signs of the two last terms in the binomials will be different (one $+$, one $-$), and it will be more convenient to put the signs in last.

The method of factoring a trinomial with a leading coefficient of 1 is summarized as follows:

TO FACTOR A TRINOMIAL WITH A LEADING COEFFICIENT OF 1

(It is assumed that like terms have been combined.)

1. Factor out the GCF.

2. Arrange the trinomial in descending powers of the variable.

3. Make a blank outline. We *assume* that the trinomial will factor into the product of two binomials. If the Rule of Signs indicates that the signs will both be positive or both be negative, insert those signs.

4. The first term inside each set of parentheses will be the square root of the first term of the trinomial. If the third term of the trinomial has a variable, the square root of *that* variable must be a factor of the second term in each binomial.

5. To find the last term of each binomial.*

a. List (mentally, at least) *all* pairs of integral factors of the coefficient of the last term of the trinomial.

b. Select the particular pair of factors that has a sum equal to the coefficient of the middle term of the trinomial. If no such pair of factors exists, the trinomial is *not factorable.*

6. Check your result by multiplying the binomials together.

* In step 5a, you can, if you wish, list only the *positive* factors of the coefficient of the last term of the trinomial. Then in step 5b, *if the signs in both binomials will be the same,* select the pair of factors that has a *sum* equal to the absolute value of the coefficient of the middle term of the trinomial; *if the signs in both binomials will be different,* select the pair of factors that has a *difference* equal to the absolute value of the coefficient of the middle term of the trinomial.

Example 1 Factor $z^2 + 8zw + 12w^2$.

Solution

$$z^2 + 8zw + 12w^2$$

+ here tells us that the signs in the binomials will be the same (the + before $12w^2$)

+ here tells us they will both be + (the + before $8zw$)

$$(\ \ +\ \ )(\ \ +\ \ )$$ ← Blank outline

$\sqrt{z^2}$ is the first term of each binomial

$$(z + \square w)(z + \square w)$$

$\sqrt{w^2}$ is a factor of the second term in each binomial

The pairs of integral factors of $+12$ are $(1)(12), (-1)(-12), (2)(6), (-2)(-6), (3)(4)$, and $(-3)(-4)$. Because the coefficient of the middle term of the trinomial was 8, we must select the pair whose sum is $+8$—the pair 2 and 6. We then have

$$z^2 + 8zw + 12w^2 = (z + 2w)(z + 6w) \text{ or } (z + 6w)(z + 2w)$$

Check $(z + 2w)(z + 6w) = (z + 6w)(z + 2w) = z^2 + 8zw + 12w^2$ ■

A WORD OF CAUTION *Both sets of parentheses are essential.* Note:

$$(z + 2w)z + 6w = z^2 + 2zw + 6w \neq z^2 + 8zw + 12w^2$$

$$z + 2w(z + 6w) = z + 2zw + 12w^2 \neq z^2 + 8zw + 12w^2$$

$$z + 2w \cdot z + 6w = z + 2zw + 6w \neq z^2 + 8zw + 12w^2$$

☑

Example 2 Factor $w^2 + 8w - 20$.

Solution

$$w^2 + 8w - 20$$

− here tells us the signs in the binomials will be *different*

We must find two integers whose product is -20 and whose sum is $+8$. The integers are $+10$ and -2. Therefore,

$$w^2 + 8w - 20 = (w - 2)(w + 10) \text{ or } (w + 10)(w - 2)$$

The order of the factors is unimportant

Check $(w - 2)(w + 10) = (w + 10)(w - 2) = w^2 + 8w - 20$ ■

Example 3 Factor $x^2 + 3x - 5$.

Solution We must find two integers whose product is -5 and whose sum is 3. The only pairs of integers whose product is -5 are $(-1)(5)$ and $(1)(-5)$. The sum of neither pair is $+3$. Therefore, this trinomial is *not factorable* over the integers; it is a prime polynomial. ■

Example 4 Factor $2ax^2 - 14axy - 60ay^2$.

Solution All three terms have a common factor. The GCF is $2a$. When we factor out the GCF, we have

$$2ax^2 - 14axy - 60ay^2 = 2a(x^2 - 7xy - 30y^2)$$

The trinomial has been factored, but not completely. We must now factor $x^2 - 7xy - 30y^2$. To do this, we must find two integers whose product is -30 and whose sum is -7. The only such pair of integers is (3) and (-10). Therefore,

$$2ax^2 - 14axy - 60ay^2 = 2a(x + 3y)(x - 10y)$$

Check $2a(x + 3y)(x - 10y) = 2a(x^2 - 7xy - 30y^2)$
$= 2ax^2 - 14axy - 60ay^2$ ■

Sometimes the terms of the trinomial are not monomials (see Example 5).

Example 5 Factor $(x - y)^2 - 2(x - y) - 8$.

Solution This problem can be solved by using a substitution. If we let $(x - y) = a$, we have

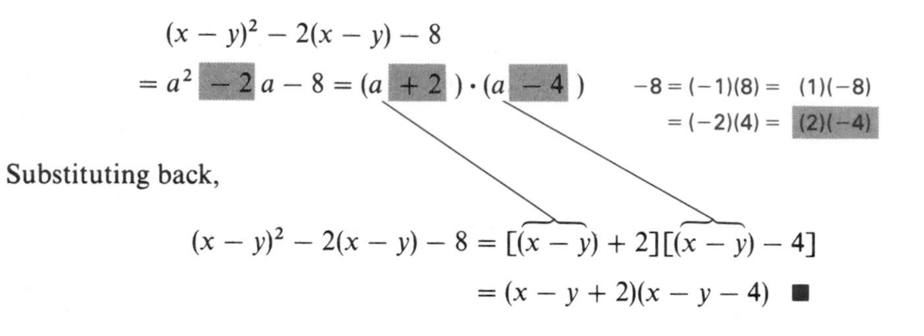

Substituting back,

$$(x - y)^2 - 2(x - y) - 8 = [(x - y) + 2][(x - y) - 4]$$
$$= (x - y + 2)(x - y - 4) \quad ■$$

EXERCISES 5.4A

Set I Factor completely or write "Not factorable."

1. $t^2 + 7t - 30$ **2.** $m^2 - 13m - 30$ **3.** $m^2 + 13m + 12$

4. $x^2 + 15x + 14$ **5.** $x^2 + 7x + 1$ **6.** $x^2 + 5x + 1$

7. $x^2 - 7x + 18$ **8.** $x^2 - 3x + 18$ **9.** $x^2 + 4x + 6x$

10. $y^2 + 7y + 3y$ **11.** $u^4 - 15u^2 + 14$ **12.** $y^4 - 16y^2 + 15$

13. $u^2 + 12u - 64$ **14.** $v^2 - 30v - 64$ **15.** $3x + x^2 + 2$

16. $7a + a^2 + 10$ **17.** $x^4 - 6x^3 + 2x^2$ **18.** $y^6 - 2y^4 + 2y^2$

19. $x^2 + xy - 2y^2$ **20.** $x^2 + 6xy + 9y^2$

21. $(a + b)^2 + 6(a + b) + 8$ **22.** $(m + n)^2 + 9(m + n) + 8$

23. $(x + y)^2 - 13(x + y) - 30$ **24.** $(x + y)^2 - 10(x + y) - 24$

Set II Factor completely or write "Not factorable."

1. $z^2 - 9z + 20$ **2.** $x^3 - 6x^2 - 16x$ **3.** $u^2 + 11u + 10$

4. $x^2 - 12x + 35$ **5.** $x^2 + 4x + 1$ **6.** $x^2 + x - 56$

7. $x^2 + 9x + 10$ **8.** $x^2 + 6x + 6$ **9.** $x^2 + 5x + 3x$

10. $x^2 + 12x - 13$ **11.** $x^4 + 2x^2 - 3$ **12.** $x^4 + 2x^3 - 3x^2$

13. $f^2 - 7f - 18$ **14.** $x^2 + 8x + 7x$ **15.** $10v + v^2 + 16$

16. $15 - 8x + x^2$ **17.** $z^4 + 3z^3 + 5z^2$ **18.** $x^4 - 21 - 4x^2$

19. $a^2 + 4ab + 3b^2$ **20.** $x^2 + 5x - 24$

21. $(a + b)^2 + 9(a + b) + 14$

22. $(x + y)^2 - (x + y) - 12$

23. $(x + y)^2 - 2(x + y) - 15$

24. $(a - b)^2 - 6(a - b) + 5$

5.4B Factoring a Trinomial with a Leading Coefficient Unequal to 1

Two methods will be shown for factoring a trinomial in which the leading coefficient is not 1. The first is a trial-and-error method; the second, the Master Product Method, makes use of factoring by grouping.

The Trial Method

The rules that follow assume that like terms have been combined and that any common factors have been factored out. It is also assumed that the trinomial has been arranged in descending powers of the variable and that the leading coefficient is positive. If the leading coefficient is negative, factor out -1 before proceeding, or, if the *third* term is positive, arrange the trinomial in ascending powers.* We will continue to use the Rule of Signs for Factoring Trinomials from Section 5.4A.

TO FACTOR A TRINOMIAL WITH A LEADING COEFFICIENT UNEQUAL TO 1

1. Make a blank outline and fill in the *literal parts* of each term of each binomial.† Fill in the signs in each binomial if they are both positive or both negative.
2. List (mentally, at least) all pairs of factors of the coefficient of the first term of the trinomial and of the last term of the trinomial.
3. By trial and error, select the pairs of factors (if they exist) from step 2 that make the sum of the inner and outer products of the binomials equal to the middle term of the trinomial. If no such pairs exist, the trinomial is *not factorable.*
4. Check the result.

Example 6 Factor $6x^2 - 89xy - 15y^2$.

Solution Blank outline

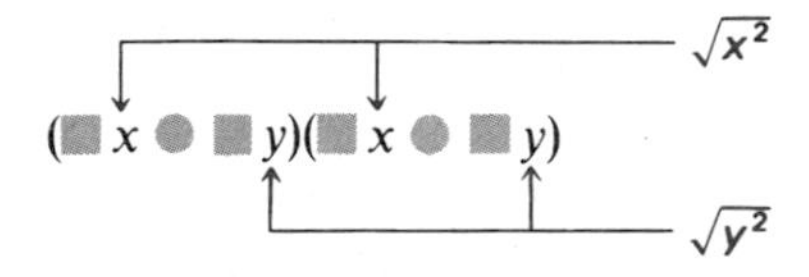

The variables are in place; the signs will be put in last, since they are different from each other

There is no common factor for all three terms. The pairs of positive factors of 6 are (2)(3) and (1)(6). The pairs of factors of -15 are $(3)(-5)$, $(-3)(5)$, $(15)(-1)$, and

* A polynomial is in *ascending powers* if its exponents *increase* as we read from left to right.
† See step 4, p. 200.

$(-15)(1)$. Thus, the *possible* factorizations are

$(2x - 3y)(3x + 5y)$	$(2x + 3y)(3x - 5y)$
$\star(2x - 5y)(3x + 3y)$	$\star(2x + 5y)(3x - 3y)$
$\star(2x - y)(3x + 15y)$	$\star(2x + y)(3x - 15y)$
$(2x - 15y)(3x + y)$	$(2x + 15y)(3x - y)$
$\star(6x - 3y)(x + 5y)$	$\star(6x + 3y)(x - 5y)$
$(6x - 5y)(x + 3y)$	$(6x + 5y)(x - 3y)$
$\star(6x - 15y)(x + y)$	$\star(6x + 15y)(x - y)$
$(6x - y)(x + 15y)$	$(6x + y)(x - 15y)$

This is the *only* combination in which the sum of the inner and outer products is $-89xy$

Therefore, $6x^2 - 89xy - 15y^2 = (6x + y)(x - 15y)$.

Check $(6x + y)(x - 15y) = 6x^2 - 89xy - 15y^2$

The starred combinations all have a common factor of 3 in one of the binomials. Since the original trinomial had no common factor, these combinations need not be given any consideration. ■

Example 7 Factor $3x^2 - x + 7$.

Solution Possible factorizations:

$$(3x - 7)(x - 1)$$
$$(3x - 1)(x - 7)$$

In both of these factorizations, the product of the two first terms is $3x^2$ and the product of the two last terms is $+7$. However, in neither one is the sum of the inner and outer products $-x$. Therefore, $3x^2 - x + 7$ is *not factorable.* ■

Example 8 Factor $12 + 7x - 10x^2$.

Solution If we arranged this trinomial in *descending* powers of x, the leading coefficient would be negative, so we leave it in *ascending* powers. Note that 12, 7, and 10 have no common factor. The pairs of positive factors of 12 are (3)(4), (2)(6), and (1)(12). The pairs of factors of -10 are $(2)(-5)$, $(-2)(5)$, $(1)(-10)$, and $(-1)(10)$.

This time, rather than write *all* possible combinations of binomials, we will stop when we find a combination that gives us $12 + 7x - 10x^2$. In each starred combination, one of the binomials has a common factor of 2; therefore, these combinations need not be considered.

$$\star(3 + 5x)(4 - 2x)$$
$$(3 + 2x)(4 - 5x)$$
$$\star(3 - 5x)(4 + 2x)$$
$$(3 - 2x)(4 + 5x) \longleftarrow \text{The sum of the inner and outer products is } +7x$$

Therefore, $12 + 7x - 10x^2 = (3 - 2x)(4 + 5x)$.

Check $(3 - 2x)(4 + 5x) = 12 + 7x - 10x^2$

NOTE We could have factored out -1 from the original trinomial before factoring. The answer would have been $(-1)(2x - 3)(5x + 4)$, which is equivalent to $(3 - 2x)(4 + 5x)$. ■

Example 9 Factor $15x^2 - 25xy - 10y^2$.

Solution $5(3x^2 - 5xy - 2y^2)$ First, remove the common factor, 5

$5(x - 2y)\quad(3x + 1y)$ Then factor the remaining trinomial

$$\begin{array}{r} -6xy \\ +1xy \\ \hline -5xy \end{array}$$

Therefore, $15x^2 - 25xy - 10y^2 = 5(x - 2y)(3x + y)$.

Check $5(x - 2y)(3x + y) = 5(3x^2 - 5xy - 2y^2) = 15x^2 - 25xy - 10y^2$ ■

A WORD OF CAUTION If the 5 isn't factored out first in Example 9, we will get

$15x^2 - 25xy - 10y^2 = (5x - 10y)(3x + y)$ Not factored completely

or $15x^2 - 25xy - 10y^2 = (x - 2y)(15x + 5y)$ Not factored completely

Neither of these answers is *completely* factored. ✓

Example 10 Factor $5(2y - z)^2 + 12(2y - z) + 4$.
Solution We can use a substitution here. Let $(2y - z) = a$. Then

$$5(2y - z)^2 + 12(2y - z) + 4 = 5a^2 + 12a + 4 = (1a + 2)\quad(5a + 2)$$

$5 = 1 \cdot 5 \quad | \quad 4 = 1 \cdot 4,\ = 2 \cdot 2$

$$\begin{array}{r} +10a \\ +2a \\ \hline +12a \end{array}$$

Therefore, $5a^2 + 12a + 4 = (a + 2)(5a + 2)$. Substituting back $2y - z$ for a, we have

$$\begin{aligned} 5(2y - z)^2 + 12(2y - z) + 4 &= [(2y - z) + 2][5(2y - z) + 2] \\ &= (2y - z + 2)(10y - 5z + 2) \end{aligned}$$

Check $([2y - z] + 2)(5[2y - z] + 2) = 5(2y - z)^2 + 12(2y - z) + 4$ ■

Example 11 Factor $2x^2 + 5xy - 3y^2 + 8x - 4y$.
Solution Since this expression has more than three terms, we will have to factor by grouping. We'll try grouping the first three terms and the last two terms (if this doesn't work, we'll try a different grouping).

$$2x^2 + 5xy - 3y^2 + 8x - 4y$$

Factoring each group, we have

$$(2x - y)(x + 3y) + 4(2x - y)$$

Note that we still have two terms. Therefore, the expression has not yet been factored. However, we now have a common *binomial* factor, namely, $2x - y$. When we factor this out we have

$$(2x - y)[(x + 3y) + 4]$$

Therefore, $2x^2 + 5xy - 3y^2 + 8x - 4y = (2x - y)(x + 3y + 4)$.

Check $(2x - y)(x + 3y + 4) = 2x^2 + 5xy - 3y^2 + 8x - 4y$ ■

The Master Product Method

The **Master Product Method**, which makes use of factoring by grouping, is a method of factoring trinomials that eliminates some of the guesswork. It can also be used to determine whether a trinomial is factorable or not. The rules that follow assume that any common factors have been factored out.

> TO FACTOR A TRINOMIAL BY THE MASTER PRODUCT METHOD
>
> Arrange the trinomial in descending powers of one variable.
>
> $$ax^2 + bxy + cy^2$$
>
> 1. Find the Master Product (MP) by multiplying the first and last coefficients of the trinomial being factored (MP $= a \cdot c$).
> 2. List the pairs of factors of the Master Product (MP).
> 3. Choose the pair of factors whose sum is the coefficient of the middle term (b).
> 4. Rewrite the given trinomial, replacing the middle term by the sum of two terms whose coefficients are the pair of factors found in step 3.
> 5. Factor the step 4 expression by grouping.
>
> Check your factoring by multiplying the binomial factors to see if their product is the given trinomial.

Example 12 Factor $5x + 2x^2 + 3$.

Solution $2x^2 + 5x + 3$ Arranged in descending powers

Master Product $= (2)(+3) = +6$

$6 = (+1)(+6) = (-1)(-6)$ We must find the factors of 6 whose sum equals 5

$= (+2)(+3) = (-2)(-3)$

$(+2) + (+3) = 5 \leftarrow$ The middle coefficient

$2x^2 + 5x + 3 = \underbrace{2x^2 + 2x} + \underbrace{3x + 3} \leftarrow$ Factor by grouping

$= 2x(x + 1) + 3(x + 1)$ The GCF is $(x + 1)$

$= (x + 1)(2x + 3)$.

Check $(x + 1)(2x + 3) = 2x^2 + 5x + 3$ ■

Example 13 Factor $3m^2 - 2mn - 8n^2$.

Solution Master Product $= (3)(-8) = -24$

$-24 = (-1)(24) = (1)(-24)$ We must find the factors of -24 whose sum is -2

$= (-2)(12) = (2)(-12)$

$= (-3)(8) = (3)(-8)$

$= (-4)(6) = (4)(-6)$

$$(+4) + (-6) = -2 \leftarrow \text{The middle coefficient}$$

$$3m^2 - 2mn - 8n^2 = \underbrace{3m^2 + 4mn} - \underbrace{6mn - 8n^2} \leftarrow \text{Factor by grouping}$$

$$= m(3m + 4n) - 2n(3m + 4n) \quad \text{The GCF is } (3m + 4n)$$

$$= (3m + 4n)(m - 2n)$$

(The checking is left to the student.) ■

Example 14 Factor $12x^2 - 5x + 10$.

Solution Master Product $= 12(10) = 120$

$$\begin{aligned} 120 &= 1(120) = (-1)(-120) \\ &= 2(60) = (-2)(-60) \\ &= 3(40) = (-3)(-40) \\ &= 4(30) = (-4)(-30) \\ &= 5(24) = (-5)(-24) \\ &= 6(20) = (-6)(-20) \\ &= 8(15) = (-8)(-15) \\ &= 10(12) = (-10)(-12) \end{aligned}$$

None of the sums of these pairs is -5. Therefore, the trinomial is not factorable. ■

Factoring Trinomials That Are Perfect Squares

We learned in Section 4.4B that

$$(a + b)^2 = a^2 + 2ab + b^2$$

and

$$(a - b)^2 = a^2 - 2ab + b^2$$

Trinomials that fit the patterns $a^2 + 2ab + b^2$ and $a^2 - 2ab + b^2$ are called **perfect squares**.

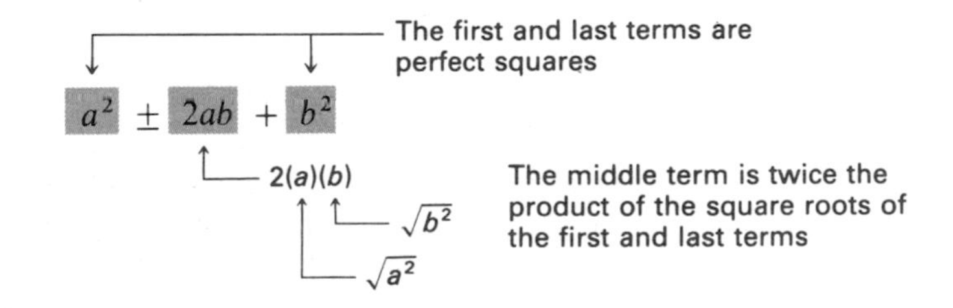

Therefore, $a^2 \pm 2ab + b^2 = (\sqrt{a^2} \pm \sqrt{b^2})^2 = (a \pm b)^2$.

> TO FACTOR A TRINOMIAL THAT IS A PERFECT SQUARE
>
> 1. Find a, the square root of the first term, and b, the square root of the last term.
> 2. The factors of the trinomial are
>
> $(a + b)^2$ if the middle term is $+2ab$
>
> $(a - b)^2$ if the middle term is $-2ab$

NOTE Such trinomials can *also* be factored by the trial method or by the Master Product Method. ☑

Example 15 Factor $4x^2 - 12x + 9$.

Solution

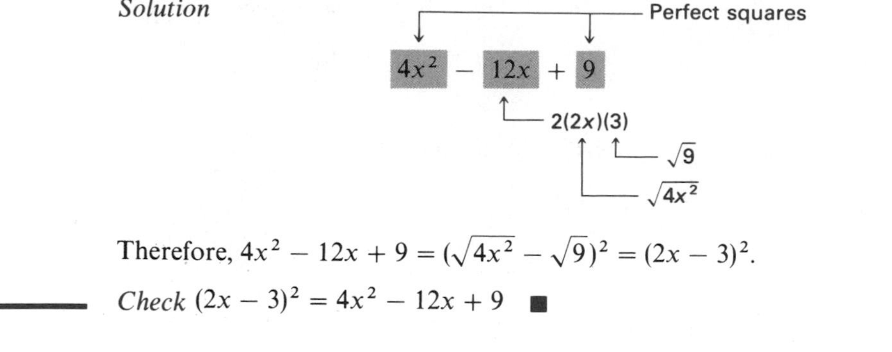

Therefore, $4x^2 - 12x + 9 = (\sqrt{4x^2} - \sqrt{9})^2 = (2x - 3)^2$.

Check $(2x - 3)^2 = 4x^2 - 12x + 9$ ■

Example 16 Factor $36x^4 - 13x^2 + 1$.

Solution $36x^4 - 13x^2 + 1$ *might* be a perfect square, since $36x^4 = (6x^2)^2$ and $1 = 1^2$. We check:

$$(6x^2 - 1)^2 \stackrel{?}{=} 36x^4 - 13x^2 + 1$$
$$\text{No; } (6x^2 - 1)^2 = 36x^4 - 12x^2 + 1$$

However, the expression still might be factorable. We try

$$(9x^2 - 1)(4x^2 - 1) \stackrel{?}{=} 36x^4 - 13x^2 + 1$$

This *does* check, and the expression has been factored. However, neither factor is prime (both factors are a difference of two squares), so we must factor again.

$$36x^4 - 13x^2 + 1 = (3x - 1)(3x + 1)(2x - 1)(2x + 1)$$

Check $(3x - 1)(3x + 1)(2x - 1)(2x + 1) = (9x^2 - 1)(4x^2 - 1)$
$= 36x^4 - 13x^2 + 1$ ■

In Section 5.2, we discussed factoring by grouping; in that section, we always had a common factor after factoring each group. We can *also* complete the factoring after we've grouped if we have a difference of two squares (see Example 17) or a factorable trinomial (see Example 18).

Example 17 Factor $x^2 + 4xy + 4y^2 - 9$.

Solution If we grouped the first two terms and the last two terms, we would *not* have a common binomial factor. (Try it!) Let's try grouping the first *three* terms together.

$$(x^2 + 4xy + 4y^2) - (9)$$
$$= (x + 2y)^2 - (3)^2 \leftarrow \text{This is a difference of two squares}$$
$$= ([x + 2y] + 3)([x + 2y] - 3)$$
$$= (x + 2y + 3)(x + 2y - 3)$$

Therefore, $x^2 + 4xy + 4y^2 - 9 = (x + 2y + 3)(x + 2y - 3)$.

Check $(x + 2y + 3)(x + 2y - 3) = (x + 2y)^2 - 3^2 = x^2 + 4xy + 4y^2 - 9$ ■

The type of factoring seen in Example 17 is used in Section 5.7 on factoring by completing the square.

Example 18 Factor $5x^2 - 30xy + 45y^2 - 14x + 42y - 3$.

Solution Since the polynomial has more than three terms, we must factor by grouping. Let's look at the first three terms only:

$$5x^2 - 30xy + 45y^2 = 5(x^2 - 6xy + 9y^2) = 5(x - 3y)^2$$

That result leads us to believe that we may have a factorable trinomial. Let's try the following grouping:

$$\boxed{5x^2 - 30xy + 45y^2} - \boxed{14x + 42y} - \boxed{3}$$
$$= 5(x - 3y)^2 - 14(x - 3y) - 3$$

Double check: If we removed the parentheses, would we get back the original polynomial? Yes. Continuing, we let $a = x - 3y$. We then have

$$5a^2 - 14a - 3 = (5a + 1)(a - 3)$$

Substituting back,

$$5x^2 - 30xy + 45y^2 - 14x + 42y - 3$$
$$= (5[x - 3y] + 1)([x - 3y] - 3)$$
$$= (5x - 15y + 1)(x - 3y - 3)$$

(The checking is left to the student.) ■

EXERCISES 5.4B

Set I Factor completely or write "Not factorable."

1. $5x^2 + 9x + 4$ **2.** $5x^2 + 12x + 4$ **3.** $7 - 22b + 3b^2$

4. $7 - 10u + 3u^2$ **5.** $x + 3x^2 + 1$ **6.** $x + 11x^2 + 1$

7. $3n^2 + 14n - 5$ **8.** $3n^2 + 2n - 5$ **9.** $3t^2 - 17tz - 6z^2$

10. $3x^2 - 7xy - 6y^2$ **11.** $-7 + 4x^2 + x$ **12.** $-13 + 6x^2 + x$

13. $8 + 12z - 8z^2$ **14.** $9 + 21z - 18z^2$ **15.** $1 + 4a^2 + 4a$

16. $1 + 9b^2 + 6b$ **17.** $2x^2 - 18$ **18.** $5y^2 - 80$

19. $4 + 7h^2 - 11h$ **20.** $4 + 7h^2 - 16h$ **21.** $24xy + 72x^2 + 2y^2$

22. $16xy + 32x^2 + 2y^2$ **23.** $x^2 - 7xy + 49y^2$ **24.** $x^2 - 5xy + 25y^2$

25. $2x^2y + 8xy^2 + 8y^3$ **26.** $3x^3 + 6x^2y + 3xy^2$ **27.** $6e^4 - 7e^2 - 20$

28. $10f^4 - 29f^2 - 21$ **29.** $12x^4 - 75x^2y^2$ **30.** $36x^4 - 16x^2y^2$

31. $a^4 + 2a^2b^2 + b^4$ **32.** $x^4 + 6x^2y^2 + 9y^4$

33. $2(a + b)^2 + 7(a + b) + 3$ **34.** $3(a - b)^2 + 7(a - b) + 2$

35. $4(x - y)^2 - 8(x - y) - 5$ **36.** $4(x + y)^2 - 4(x + y) - 3$

37. $5x^2 + 10xy + 5y^2 - 21x - 21y + 4$

38. $5x^2 - 10xy + 5y^2 - 12x + 12y + 4$

39. $(2x - y)^2 - (3a + b)^2$ **40.** $(2x + 3y)^2 - (a - b)^2$

41. $x^2 + 10xy + 25y^2 - 9$ **42.** $x^2 + 8xy + 16y^2 - 25$

43. $a^2 - 4x^2 - 4xy - y^2$
44. $x^2 - 9a^2 - 6ab - b^2$
45. $4x^4 - 13x^2 + 9$
46. $9x^4 - 13x^2 + 4$
47. $3x^2 - 7xy - 6y^2 - x + 3y$
48. $3t^2 - 17tz - 6z^2 - 3t - z$
49. $3n^2 + 2mn - 5m^2 + 3n + 5m$
50. $3n^2 + 14mn - 5m^2 + 3n - m$

Set II Factor completely or write "Not factorable."

1. $3x^2 + 7x + 4$
2. $7x^2 + 13x - 2$
3. $2 - 9x + 4x^2$
4. $3x^2 - 5x - 8$
5. $x + 7x^2 + 1$
6. $3x^2 + 6x - 6$
7. $5k^2 + 2k - 7$
8. $13a^2 - 18a + 5$
9. $3w^2 + 7wx - 6x^2$
10. $2x^2 - 2x - 12$
11. $-5 + 8x^2 + x$
12. $29x - 5 + 6x^2$
13. $8 - 14v - 4v^2$
14. $64c^2 - 16c + 1$
15. $1 + 16m^2 + 8m$
16. $5x^2 + 5x + 1$
17. $3z^2 - 75$
18. $5k + 3k^2 - 2$
19. $6 + 7x^2 - 23x$
20. $21x^2 + 22x - 8$
21. $128x^2 + 32xy + 2y^2$
22. $2y^2 - 72$
23. $x^2 - 2xy + 4y^2$
24. $11x^2 + 32xy - 3y^2$
25. $18mn^2 - 24m^2n + 8m^3$
26. $5x^2 + 7xy + 3x$
27. $2e^4 + 11e^2 - 6$
28. $5x^4 - 17x^2 + 6$
29. $36x^4 - 81x^2y^2$
30. $-24 - 6x + 45x^2$
31. $w^4 + 8w^2y^2 + 16y^4$
32. $17u - 12 + 5u^2$
33. $3(a - b)^2 + 16(a - b) + 5$
34. $4x^2 + 8xy + 4y^2 + 11x + 11y + 7$
35. $6(x + y)^2 - 7(x + y) - 3$
36. $6a^2 + 12ab + 6b^2 + 11a + 11b + 5$
37. $5a^2 + 10ab + 5b^2 - 16a - 16b + 3$
38. $4a^2 - 8ab + 4b^2 + 9a - 9b + 5$
39. $(3x + y)^2 - (a - 2b)^2$
40. $-8 + 14v + 4v^2$
41. $x^2 + 6xy + 9y^2 - 16$
42. $5z^3 + 40z^2 + 15z$
43. $y^2 - 25x^2 - 10xz - z^2$
44. $8x^3 + 24x^2 + 8x$
45. $9x^4 - 37x^2 + 4$
46. $7y^3 + 28y^2 + 14y$
47. $3w^2 + 7wx - 6x^2 - w - 3x$
48. $5(a + b) - 2 + 12(a + b)^2$
49. $5k^2 + 2km - 7m^2 + 5k + 7m$
50. $-8 + 10(x - y) + 12(x - y)^2$

5.5 Review: 5.1–5.4

Methods of Factoring

5.1 *First, check for a common factor.* If there is a common factor, factor it out.

5.3 *If the expression to be factored has two terms*: Is it a *difference* of two *squares*?

If the expression to be factored has three terms:

5.4A 1. Is the leading coefficient 1?

5.4B 2. Are the first and last terms perfect squares? If so, is the trinomial a perfect square?

5.4B 3. Is the leading coefficient unequal to 1?

5.2 *If the expression to be factored has four or more terms*: Can it be factored by grouping?

Check to see if any factor can be factored again.

Check your solution.

Review Exercises 5.5 Set I

Factor completely or write "Not factorable."

1. $65x^2y^3 - 39xy^4 - 13xy$ **2.** $3xyz - 5a + 3b$ **3.** $3x^3 + 9x^2 - 12x$

4. $x^3y^6 + 5$ **5.** $x^2 + 13x + 40$ **6.** $x^2 + x - 20$

7. $8 + x^2 + 8x$ **8.** $x^2 - 11x + 18$ **9.** $x^2 - 256$

10. $x^2 - 14x + 13$ **11.** $3x^2 - x - 14$ **12.** $2x^2 - 11x - 40$

13. $3xy + 12x + 2y + 8$ **14.** $x^3 + 5x^2 - x - 5$ **15.** $5ag - 7x + 10cd - 14$

16. $x^2 - x + 12$ **17.** $8x^3 - 2x$ **18.** $x^2 + 9$

19. $4x^2 + 11x - 3$ **20.** $11a - 10 + 8a^2$

Review Exercises 5.5 Set II

NAME

Factor completely or write "Not factorable."

1. $14ac + 56ab - 7a$
2. $x^2 - 144$
3. $x^3 - 7x^2 + x - 4$
4. $x^2 - 14x + 48$
5. $7x^2 + 5x - 2$
6. $x^2 + 16$
7. $5x^2 + 11x - 12$
8. $6 - 15x + 6x^2$
9. $2xy + 5y + 8x + 20$
10. $4x^2 - y^2 + 2x - y$
11. $6x + 5x^2 - 8$
12. $x^3 + 8x^2 - 4x - 32$
13. $2 + 15x + 7x^2$
14. $3x^3 - x^2 + 3x - 1$
15. $5x^2 + x + 1$
16. $8x^2 + 30x - 8$

ANSWERS

1. ______
2. ______
3. ______
4. ______
5. ______
6. ______
7. ______
8. ______
9. ______
10. ______
11. ______
12. ______
13. ______
14. ______
15. ______
16. ______

17. $3x^2y^2 + 9xy^3 - 21xy^2$

18. $x^3 - x^2 - 8x + 8$

19. $11x^2 + 31x - 6$

20. $5ax - 10a + 3bx - 6b - x + 2$

ANSWERS

17. ______________________

18. ______________________

19. ______________________

20. ______________________

5.6 Factoring the Sum or Difference of Two Cubes

A **sum of two cubes** is a binomial that fits the pattern $a^3 + b^3$; for example, $(2x)^3 + (3y)^3$ is a sum of two cubes. A **difference of two cubes** is a binomial that fits the pattern $a^3 - b^3$.

The factoring formulas for factoring $a^3 + b^3$ and $a^3 - b^3$ are based on two special products. Consider the products $(a + b)(a^2 - ab + b^2)$ and $(a - b)(a^2 + ab + b^2)$.

$$\begin{array}{rrrr} & a^2 & -ab & +b^2 \\ & & a & +b \\ \hline & a^2b & -ab^2 & +b^3 \\ a^3 & -a^2b & +ab^2 & \\ \hline a^3 & & & +b^3 \end{array} \qquad \begin{array}{rrrr} & a^2 & +ab & +b^2 \\ & & a & -b \\ \hline & -a^2b & -ab^2 & -b^3 \\ a^3 & +a^2b & +ab^2 & \\ \hline a^3 & & & -b^3 \end{array}$$

Since $(a + b)(a^2 - ab + b^2) = a^3 + b^3$, $a^3 + b^3$ factors as follows:

Sum of two cubes

$$a^3 + b^3 = (a + b)(a^2 - ab + b^2)$$

Same sign

Also, since $(a - b)(a^2 + ab + b^2) = a^3 - b^3$, $a^3 - b^3$ factors as follows:

Difference of two cubes

$$a^3 - b^3 = (a - b)(a^2 + ab + b^2)$$

Same sign

Note that each product of factors contains exactly one negative sign. When we're factoring a *difference* of two cubes, the negative sign is in the *binomial factor*. When we're factoring a *sum* of two cubes, the negative sign is the sign of the *second term* of the *trinomial factor*.

A WORD OF CAUTION $a^3 + b^3 \neq (a + b)^3$

Remember that

$$\begin{aligned}(a + b)^3 &= (a + b)(a + b)(a + b) \\ &= a^3 + 3a^2b + 3ab^2 + b^3\end{aligned}$$

whereas

$$a^3 + b^3 = (a + b)(a^2 - ab + b^2)$$ ✓

In order to use the factoring formulas, you must be able to find the cubic root of a term. The cubic root of the numerical coefficient is found by using the methods learned in Section 1.7C. The cubic root of the *variable* is found by dividing each exponent by 3.

Example 1 Find $\sqrt[3]{27a^6b^3}$.

Solution

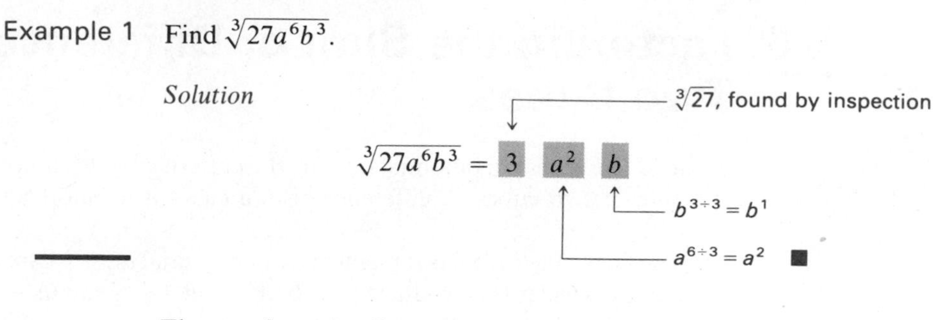

The new factoring formulas are summarized as follows:

> TO FACTOR $a^3 + b^3$ OR $a^3 - b^3$ (THE SUM OR DIFFERENCE OF TWO CUBES)
>
> **1.** Find a, the cubic root of the first term, and b, the cubic root of the second term. (Express numerical coefficients in prime factored form if necessary.)
>
> **2.** Substitute into the following formulas:
>
> $$a^3 + b^3 = (a + b)(a^2 - ab + b^2)$$
> $$a^3 - b^3 = (a - b)(a^2 + ab + b^2)$$

A WORD OF CAUTION A mistake students often make is to think that the middle term of the trinomial factor is $2ab$ instead of ab.

$$a^3 + b^3 = (a + b)(a^2 - ab + b^2)$$

↑ Not $2ab$ ☑

NOTE The expression $a^2 - ab + b^2$ does *not* factor further. You should verify this by attempting to find two binomials whose product is $a^2 - ab + b^2$; you might also verify that $a^2 + ab + b^2$ does not factor further. ☑

Example 2 Factor $x^3 + 8$.

Solution

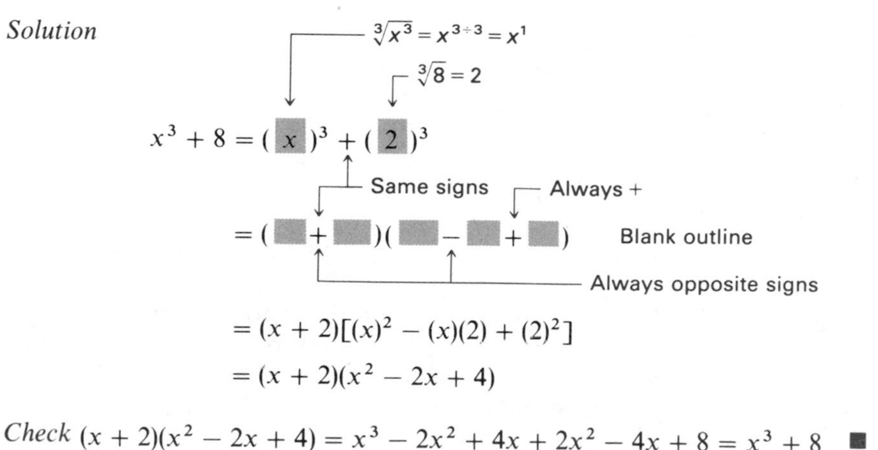

$$= (x + 2)[(x)^2 - (x)(2) + (2)^2]$$
$$= (x + 2)(x^2 - 2x + 4)$$

Check $(x + 2)(x^2 - 2x + 4) = x^3 - 2x^2 + 4x + 2x^2 - 4x + 8 = x^3 + 8$ ■

Example 3 Factor $27x^3 - 64y^3$.

Solution $27x^3 - 64y^3 = (3x)^3 - (4y)^3$

Same signs — Always +

$= (\blacksquare - \blacksquare)(\blacksquare + \blacksquare + \blacksquare)$ Blank outline

Always opposite signs

$= (3x - 4y)[(3x)^2 + (3x)(4y) + (4y)^2]$

$= (3x - 4y)(9x^2 + 12xy + 16y^2)$

(The checking is left to the student.) ■

Example 4 Factor $(x + 2)^3 - (y - 2z)^3$.

Solution Let $(x + 2) = a$ and $(y - 2z) = b$. Then

$$(x + 2)^3 - (y - 2z)^3 = a^3 - b^3 = (a - b)(a^2 + ab + b^2)$$

Since $a = x + 2$ and $b = y - 2z$, substituting back gives

$$(x + 2)^3 - (y - 2z)^3$$
$$= [(x + 2) - (y - 2z)][(x + 2)^2 + (x + 2)(y - 2z) + (y - 2z)^2]$$
$$= [x + 2 - y + 2z][x^2 + 4x + 4 + xy + 2y - 2xz - 4z + y^2 - 4yz + 4z^2]$$

(The checking is left to the student.) ■

EXERCISES 5.6

Set I Factor completely or write "Not factorable."

1. $x^3 - 8$ **2.** $x^3 - 27$ **3.** $64 + a^3$

4. $8 + b^3$ **5.** $125 - x^3$ **6.** $1 - a^3$

7. $8x^3 - 2x$ **8.** $27x^3 - 3x$ **9.** $c^3 - 27a^3b^3$

10. $c^3 - 64a^3b^3$ **11.** $8x^3y^6 + 27$ **12.** $64a^6b^3 + 125$

13. $125x^6y^4 - 1$ **14.** $64a^6b^3 - 9$ **15.** $a^4 + ab^3$

16. $x^3y + y^4$ **17.** $81 - 3x^3$ **18.** $40 - 5b^3$

19. $(x + y)^3 + 1$ **20.** $1 + (x - y)^3$ **21.** $64x^3 - y^6$

22. $125w^3 - v^6$ **23.** $4a^3b^3 + 108c^6$ **24.** $5x^3y^6 + 40z^9$

25. $x^6 - 729$ **26.** $y^6 - 64$ **27.** $(x + 1)^3 - (y - z)^3$

28. $(x - y)^3 - (a + b)^3$

Set II Factor completely or write "Not factorable."

1. $a^3 - 125$ **2.** $b^3 + 1$ **3.** $27 + c^3$

4. $16 + 2x^3$ **5.** $8 - b^3$ **6.** $x^3y^3 + 64$

7. $64a^3 - 4a$ **8.** $8a^3 + 27$ **9.** $x^3 - 64y^3z^3$

10. $x^3 - xy^2$ **11.** $8a^6b^3 + 1$ **12.** $250 - 2c^3$

13. $8x^9y^5 - 27$

14. $x^3 + 4x$

15. $x^4y^3 + x$

16. $a^4 + 8a$

17. $16 - 2m^3$

18. $3x^3 - 81$

19. $(a + b)^3 - (x - y)^3$

20. $(x + y)^3 + (u + v)^3$

21. $27a^3 - b^6$

22. $(s - 2)^3 + (t - 3v)^3$

23. $2x^3y^3 + 16z^6$

24. $x^6 + 64y^6$

25. $z^6 - 1$

26. $6x^4 + 48x$

27. $(a - 2)^3 - (b + c)^3$

28. $(a + x)^3 - 8(b - z)^3$

5.7 Factoring by Completing the Square of a Polynomial

Some binomials and trinomials are not factorable by any method discussed so far but *may* be factorable by a method called **completing the square**. An expression cannot be factored by completing the square unless it satisfies the following conditions:

> **1.** The first and last terms of the expression to be factored must be perfect squares.
>
> **2.** The degree of any literal factor in the first and last terms must be a multiple of *four*.

If these conditions are satisfied, we then try to find a *positive perfect square term* to add to the given binomial or trinomial to make it a polynomial that is a perfect square. Of course, *when a term is added, the same term must also be subtracted* so that the value of the original expression is unchanged. We will then have an expression that is a difference of two squares, and a difference of two squares is *always* factorable. If you need to, you can use steps 1–5 below to help you form this difference of two squares.

A complete procedure for factoring a polynomial that is in the form $ax^4 + bx^2y^2 + cy^4$, where a and c are perfect squares, follows:

1. Form the binomials $\sqrt{ax^4} + \sqrt{cy^4}$ and $\sqrt{ax^4} - \sqrt{cy^4}$.
2. Square one of the binomials found in step 1.
3. Subtract the polynomial being factored from the trinomial found in step 2.
4. If the difference found in step 3 is a positive term that is a perfect square, *add it to* and *subtract it from* the polynomial being factored, and continue with step 5. If it is not, repeat steps 2 and 3, using the other binomial from step 1. If the difference found in step 3 is still not a positive term that is a perfect square, the polynomial cannot be factored over the integers.
5. Regroup the terms to form a difference of two squares.
6. Factor the difference of two squares.
7. Determine whether any factor can be factored further.
8. Check the result.

NOTE The polynomials found in steps 1, 2, and 3 do *not* equal the polynomial that is being factored, and steps 1, 2, and 3 need not be shown. ✓

Example 1 Factor $x^4 + x^2 + 1$.

Step 1. The binomials are $\sqrt{x^4} + \sqrt{1} = x^2 + 1$ and $\sqrt{x^4} - \sqrt{1} = x^2 - 1$.

Step 2. We'll try $x^2 + 1$:

$$(x^2 + 1)^2 = x^4 + 2x^2 + 1$$

The difference is positive and is a perfect square

Step 3. $(x^4 + 2x^2 + 1) - (x^4 + x^2 + 1) = x^2$

$x^2 - x^2 = 0$

Step 4. $x^4 + x^2 + 1 = x^4 + x^2 + 1 + x^2 - x^2$

Step 5. $= (x^4 + x^2 + 1 + x^2) - x^2$

$= (x^4 + 2x^2 + 1) - x^2$

$= (x^2 + 1)^2 - x^2$ A difference of two squares

Step 6. $= [(x^2 + 1) + x][(x^2 + 1) - x]$

Step 7. $= (x^2 + x + 1)(x^2 - x + 1)$ Both factors are prime

Step 8. $(x^2 + x + 1)(x^2 - x + 1)$

$= x^2(x^2 - x + 1) + x(x^2 - x + 1) + 1(x^2 - x + 1)$

$= x^4 - x^3 + x^2 + x^3 - x^2 + x + x^2 - x + 1$

$= x^4 + x^2 + 1$ ■

Example 2 Factor $h^4 - 14h^2k^2 + 25k^4$.

Step 1. The binomials are $\sqrt{h^4} + \sqrt{25k^4} = h^2 + 5k^2$ and $\sqrt{h^4} - \sqrt{25k^4} = h^2 - 5k^2$.

Step 2. We'll try $h^2 + 5k^2$:

$$(h^2 + 5k^2)^2 = h^4 + 10h^2k^2 + 25k^4$$

Step 3. $(h^4 + 10h^2k^2 + 25k^4) - (h^4 - 14h^2k^2 + 25k^4) = 24h^2k^2$

$24h^2k^2$ is not a perfect square. We will repeat steps 2 and 3, using the binomial $h^2 - 5k^2$.

Step 2. $(h^2 - 5k^2)^2 = h^4 - 10h^2k^2 + 25k^4$

Step 3. $(h^4 - 10h^2k^2 + 25k^4) - (h^4 - 14h^2k^2 + 25k^4) = 4h^2k^2$

$4h^2k^2$ is positive and is a perfect square.

$4h^2k^2 - 4h^2k^2 = 0$

Step 4. $h^4 - 14h^2k^2 + 25k^4 = h^4 - 14h^2k^2 + 25k^4 + 4h^2k^2 - 4h^2k^2$

Step 5. $= (h^4 - 14h^2k^2 + 25k^4 + 4h^2k^2) - 4h^2k^2$

$= (h^4 - 10h^2k^2 + 25k^4) - 4h^2k^2$

$= (h^2 - 5k^2)^2 - (2hk)^2$

Step 6. $= [(h^2 - 5k^2) + 2hk][(h^2 - 5k^2) - 2hk]$

Step 7. $= (h^2 + 2hk - 5k^2)(h^2 - 2hk - 5k^2)$

Both factors are prime.

Step 8. The checking is left to the student. ■

Example 3 Factor $a^4 + 4$.

Step 1. The binomials are $\sqrt{a^4} + \sqrt{4} = a^2 + 2$ and $\sqrt{a^4} - \sqrt{4} = a^2 - 2$.

Step 2. We'll try $a^2 + 2$:

$$(a^2 + 2)^2 = a^4 + 4a^2 + 4$$

Step 3. $(a^4 + 4a^2 + 4) - (a^4 + 4) = 4a^2 \longleftarrow$ A positive, perfect square

$4a^2 - 4a^2 = 0$

Step 4. $a^4 + 4 = a^4 + 4 + 4a^2 - 4a^2$

Step 5. $= (a^4 + 4a^2 + 4) - 4a^2$

$= (a^2 + 2)^2 - (2a)^2$

Step 6. $= [(a^2 + 2) + 2a][(a^2 + 2) - 2a]$

Step 7. $= (a^2 + 2a + 2)(a^2 - 2a + 2)$

Both factors are prime.

Step 8. The checking is left to the student. ■

Example 4 Factor $h^4 - 6h^2k^2 + 25k^4$.

Step 1. The binomials are $\sqrt{h^4} + \sqrt{25k^4} = h^2 + 5k^2$ and $\sqrt{h^4} - \sqrt{25k^4} = h^2 - 5k^2$.

Step 2. We'll try $h^2 - 5k^2$:

$$(h^2 - 5k^2)^2 = h^4 - 10h^2k^2 + 25k^4$$

Step 3. $(h^4 - 10h^2k^2 + 25k^4) - (h^4 - 6h^2k^2 + 25k^4) = -4h^2k^2$

The difference is negative, not positive. We repeat steps 2 and 3, trying the binomial $h^2 + 5k^2$.

Step 2. $(h^2 + 5k^2)^2 = h^4 + 10h^2k^2 + 25k^4$

Step 3. $(h^4 + 10h^2k^2 + 25k^4) - (h^4 - 6h^2k^2 + 25k^4) = 16h^2k^2$

$16h^2k^2 - 16h^2k^2 = 0$

Step 4. $h^4 - 6h^2k^2 + 25k^4 = h^4 - 6h^2k^2 + 25k^4 + 16h^2k^2 - 16h^2k^2$

Step 5. $= (h^4 - 6h^2k^2 + 25k^4 + 16h^2k^2) - 16h^2k^2$

$= (h^4 + 10h^2k^2 + 25k^4) - 16h^2k^2$

$= (h^2 + 5k^2)^2 - (4hk)^2$

Step 6. $= [(h^2 + 5k^2) + 4hk][(h^2 + 5k^2) - 4hk]$

Step 7. $= (h^2 + 4hk + 5k^2)(h^2 - 4hk + 5k^2)$

Both factors are prime.

Step 8. The checking is left to the student. ■

Example 5 Factor $x^4 - 2x^2 + 9$.

Step 1. The binomials are $\sqrt{x^4} + \sqrt{9} = x^2 + 3$ and $\sqrt{x^4} - \sqrt{9} = x^2 - 3$.

Step 2. We'll try $x^2 + 3$:

$$(x^2 + 3)^2 = x^4 + 6x^2 + 9$$

Step 3. $(x^4 + 6x^2 + 9) - (x^4 - 2x^2 + 9) = 8x^2$

$8x^2$ is not a perfect square. We repeat steps 2 and 3, using the binomial $x^2 - 3$.

Step 2. $(x^2 - 3)^2 = x^4 - 6x^2 + 9$

Step 3. $(x^4 - 6x^2 + 9) - (x^4 - 2x^2 + 9) = -4x^2$

The difference is negative, not positive. We have tried both binomials from step 1, and neither one gave us a difference that was a positive, perfect square. Therefore, $x^4 - 2x^2 + 9$ is not factorable over the integers. ■

EXERCISES 5.7

Set I Factor completely or write "Not factorable."

1. $x^4 + 3x^2 + 4$ **2.** $x^4 + 5x^2 + 9$ **3.** $4m^4 + 3m^2 + 1$

4. $9u^4 + 5u^2 + 1$ **5.** $64a^4 + b^4$ **6.** $a^4 + 4b^4$

7. $x^4 - 3x^2 + 9$ **8.** $x^4 - x^2 + 16$ **9.** $a^4 - 17a^2b^2 + 16b^4$

10. $a^4 - 37a^2 + 36$ **11.** $x^2 + x + 1$ **12.** $x^2 + 5x + 9$

13. $a^4 - 3a^2b^2 + 9b^4$ **14.** $a^4 - 15a^2b^2 + 25b^4$ **15.** $a^4 + 9$

16. $x^4 + 25$ **17.** $50x^4 - 12x^2y^2 + 2y^4$ **18.** $32x^4 - 2x^2y^2 + 2y^4$

19. $8m^4n + 2n^5$ **20.** $3m^5 + 12mn^4$

21. $50x^4y + 32x^2y^3 + 8y^5$ **22.** $48x^5 + 21x^3y^2 + 3xy^4$

Set II Factor completely or write "Not factorable."

1. $u^4 + 4u^2 + 16$ **2.** $x^4 + 7x^2 + 16$ **3.** $y^4 - 9y^2 + 16$

4. $16m^4 + 8m^2 + 1$ **5.** $4a^4 + 1$ **6.** $x^4 + 10x^2 + 49$

7. $x^4 - 21x^2 + 4$ **8.** $4x^4 - 76x^2 + 9$ **9.** $x^4 - 10x^2 + 9$

10. $x^4 - 3x^2 + 1$ **11.** $x^2 + 3x + 4$ **12.** $2x^4 - 38x^2 + 18$

13. $z^4 + z^2w^2 + 25w^4$ **14.** $x^4 - 14x^2 + 25$ **15.** $y^4 + 1$

16. $9x^4 - 7x^2y^2 + y^4$ **17.** $2x^8y - 6x^4y + 18y$ **18.** $x^2 + 7x + 16$

19. $2a^5 + 128a$ **20.** $x^4 - 17x^2y^2 + 64y^4$ **21.** $27x^4 + 6x^2y^2 + 3y^4$

22. $8x^4 - 80x^2 + 2$

5.8 Factoring by Using Synthetic Division

Consider the factored polynomial $(x - r_1)(x - r_2)(x - r_3)$. If these factors are multiplied, the *constant* term in their product must be $r_1r_2r_3$, the product of the constant terms in each factor. For example,

$$(x - 5)(x + 2)(x - 3) = x^3 - 6x^2 - x + 30$$

$$(-5)(+2)(-3) = 30$$

If a polynomial *does* have a factor of the form $x - a$ or $x + a$, then a must be a factor of the constant term of the polynomial. We know that $x - 7$, for example, cannot be a factor of $x^3 - 6x^2 - x + 30$, because 7 is not a factor of 30. We recall that if the remainder in a division problem is zero, the divisor and quotient are factors of the dividend.

The rules for factoring a polynomial using synthetic division are summarized below. They assume that any common factors have been factored out and that the polynomial has been arranged in descending powers of the variable.

TO FACTOR A POLYNOMIAL BY SYNTHETIC DIVISION WHEN THE LEADING COEFFICIENT IS 1

1. Find all the positive and negative factors of the constant term of the polynomial.
2. Begin dividing the polynomial synthetically by each of the factors found in step 1. Stop when a remainder of zero is obtained.
3. If a remainder of zero is obtained for a, then $(x - a)$ and the quotient are factors of the polynomial.
4. Apply *any* method of factoring to see if the quotient can be factored further.
5. Check.

Example 1 Factor $x^3 - x^2 + 2x - 8$.

Step 1. Factors of the constant term 8 are ± 1, ± 2, ± 4, ± 8.

Step 2. Divide synthetically by ± 1, ± 2, ± 4, ± 8 until a remainder of zero is obtained.

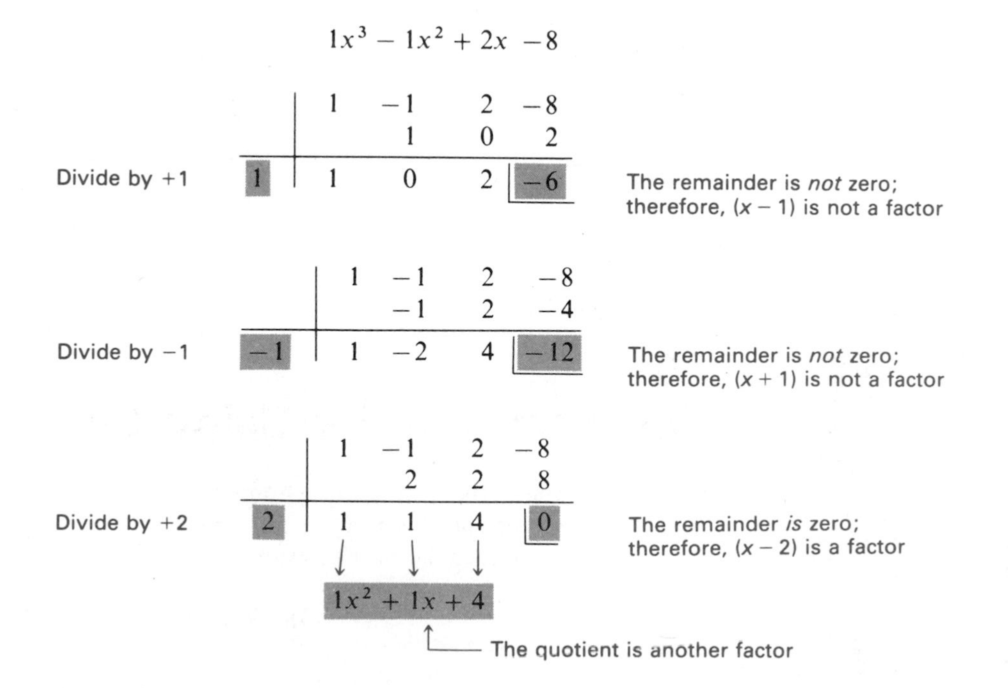

Step 3. $x^3 - x^2 + 2x - 8 = (x - 2)(x^2 + x + 4)$

Step 4. Applying the methods for factoring trinomials, we see that the quotient $x^2 + x + 4$ cannot be factored further, so it is an *irreducible* quadratic factor.

Step 5. *Check*

$$\begin{aligned} &(x - 2)(x^2 + x + 4) \\ &= x^3 + x^2 + 4x - 2x^2 - 2x - 8 \\ &= x^3 - x^2 + 2x - 8 \quad \blacksquare \end{aligned}$$

Example 2 Factor $x^5 - 3x^4 + 2x^3 - x^2 + 3x - 2$.

Step 1. Integral factors of the constant term 2 are ± 1, ± 2.

Step 2. Divide synthetically by ± 1 and ± 2 until a remainder of zero is obtained.

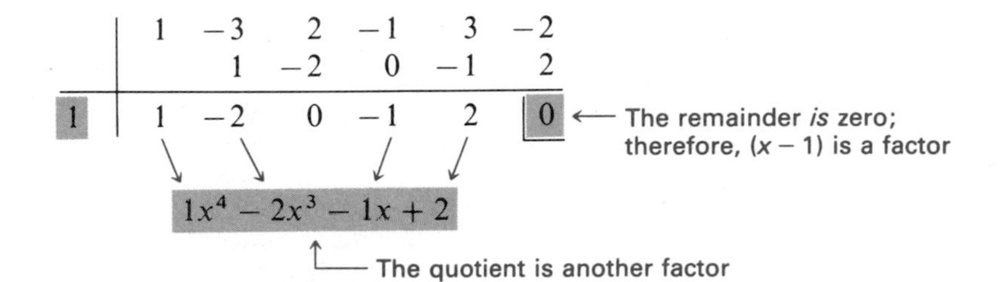

Step 3. $x^5 - 3x^4 + 2x^3 - x^2 + 3x - 2 = (x - 1)(x^4 - 2x^3 - x + 2)$

Step 4. To see if the quotient $x^4 - 2x^3 - x + 2$ can be factored further, we will use synthetic division again, even though this particular quotient can be factored by other methods.

Step 1. Factors of the constant term 2 are ± 1 and ± 2. *It is important to try the factor* 1 *again.*

Step 2.

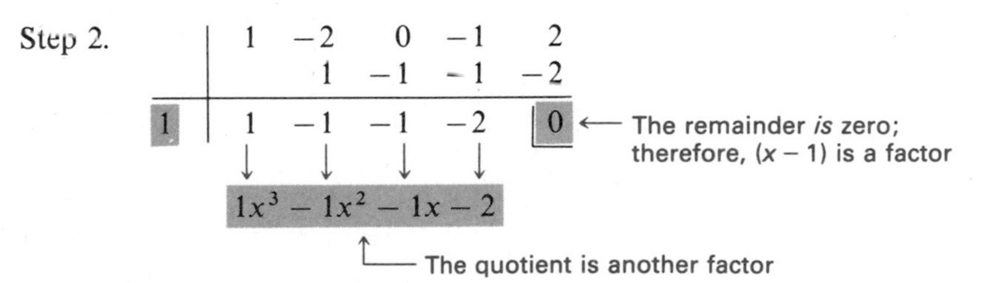

Step 3.

$$\begin{aligned} &x^5 - 3x^4 + 2x^3 - x^2 + 3x - 2 \\ &= (x - 1)(x^4 - 2x^3 - x + 2) \\ &= (x - 1)(x - 1)(x^3 - x^2 - x - 2) \end{aligned}$$

Step 4. To see if the quotient $x^3 - x^2 - x - 2$ can be factored further, we'll use synthetic division *again.*

Step 1. Factors of the constant term -2 are ± 1, ± 2.

Step 2.

$$\begin{array}{c|cccc} & 1 & -1 & -1 & -2 \\ & & 1 & 0 & -1 \\ \hline 1 & 1 & 0 & -1 & \boxed{-3} \end{array}$$

← The remainder is *not* zero

$(x - 1)$ *is not* a factor of $x^3 - x^2 - x - 2$.

We'll try $+2$:

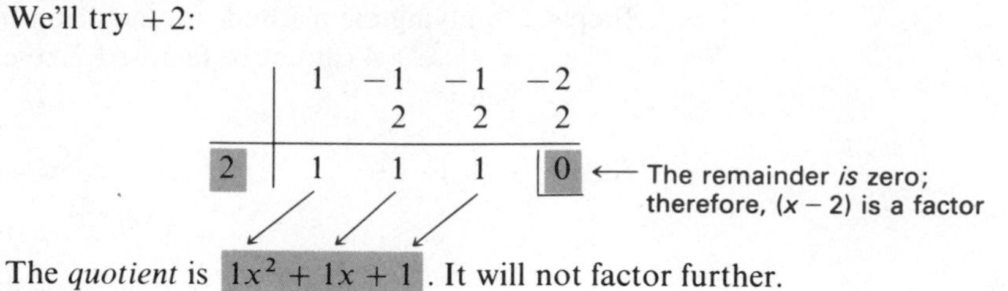

The *quotient* is $1x^2 + 1x + 1$. It will not factor further.

Therefore,

$$\begin{aligned} & x^5 - 3x^4 + 2x^3 - x^2 + 3x - 2 \\ &= (x-1)(x^4 - 2x^3 - x + 2) \\ &= (x-1)(x-1)(x^3 - x^2 - x - 2) \\ &= (x-1)^2(x-2)(x^2 + x + 1) \end{aligned}$$

Step 5. *Check*

$$\begin{aligned} & (x-1)^2(x-2)(x^2 + x + 1) \\ &= (x^2 - 2x + 1)(x-2)(x^2 + x + 1) \\ &= (x^3 - 4x^2 + 5x - 2)(x^2 + x + 1) \\ &= x^5 - 3x^4 + 2x^3 - x^2 + 3x - 2 \quad ■ \end{aligned}$$

When the leading coefficient is *not* 1, we may need to try all integer factors of the constant *divided by* all factors of the leading coefficient in addition to all integral factors of the constant (see Example 3).

Example 3 Factor $2x^4 + x^3 - x + 1$.

Solution The integral factors of 1 are ± 1. Because $2x^4 + x^3 - x + 1 = 2(x^4 + \frac{1}{2}x^3 - \frac{1}{2}x + \frac{1}{2})$, we may also need to try $\pm\frac{1}{2}$, but we will try the integers first.

$$2x^4 + x^3 - x + 1 = 2x^4 + x^3 + 0x^2 - x + 1$$

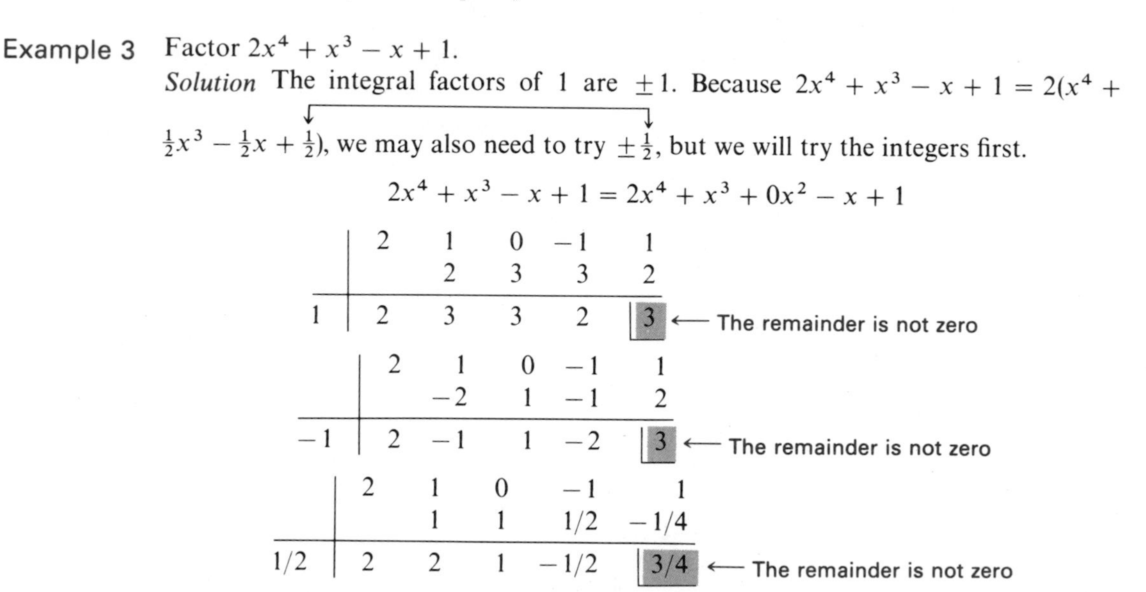

The polynomial is not factorable; it is prime, or irreducible. ■

Example 4 Factor $x^5 + 1$.

Solution Factors of the constant term $+1$ are ± 1.

$$1x^5 + 0x^4 + 0x^3 + 0x^2 + 0x + 1$$

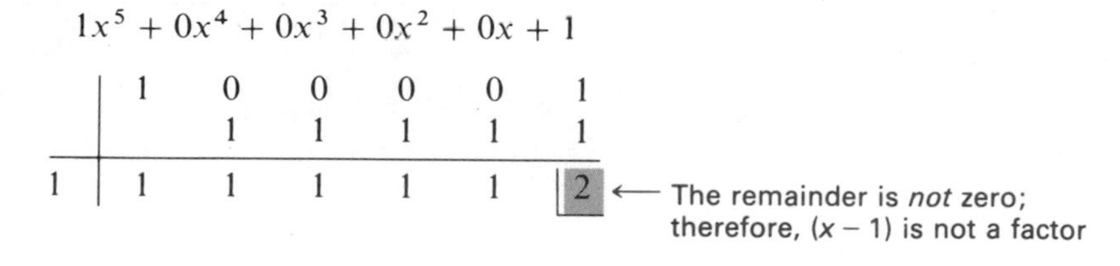

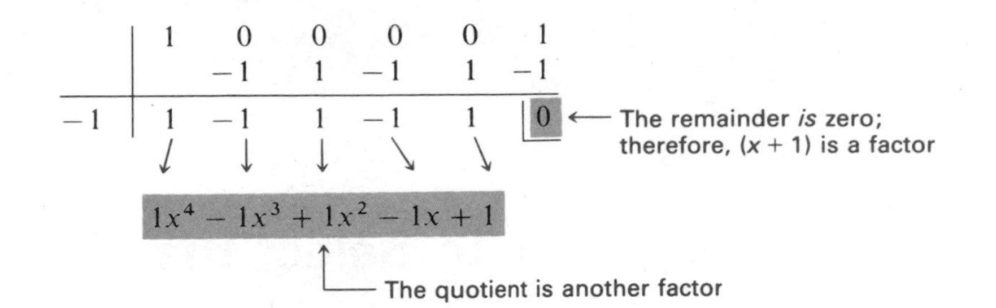

See if the quotient can be factored further.

$1x^4 - 1x^3 + 1x^2 - 1x + 1$

	1	−1	1	−1	1
		−1	2	−3	4
−1	1	−2	3	−4	5

±1 are the only factors of the constant term 1; since we already know that $(x - 1)$ is not a factor, the only factor we need to try is $(x + 1)$

The remainder is *not* zero

Therefore, $x^4 - x^3 + x^2 - x + 1$ is a prime polynomial, and $x^5 + 1 = (x + 1)(x^4 - x^3 + x^2 - x + 1)$.

Check $(x + 1)(x^4 - x^3 + x^2 - x + 1)$

$$= x^5 - \cancel{x^4} + \cancel{x^3} - \cancel{x^2} + \cancel{x} + \cancel{x^4} - \cancel{x^3} + \cancel{x^2} - \cancel{x} + 1$$

$$= x^5 + 1 \quad \blacksquare$$

EXERCISES 5.8

Set I Factor completely or write "Not factorable."

1. $x^3 + x^2 + x - 3$

2. $x^3 + x^2 - 5x - 2$

3. $x^3 - 3x^2 - 4x + 12$

4. $x^3 - 2x^2 - 5x + 6$

5. $2x^3 - 8x^2 + 2x + 12$

6. $2x^3 + 12x^2 + 22x + 12$

7. $6x^3 - 13x^2 + 4$

8. $x^3 - 7x - 6$

9. $x^4 - 3x^2 + 4x + 4$

10. $x^4 - x^3 - 5x^2 + 5x + 6$

11. $x^4 + 2x^3 - 3x^2 - 8x - 4$

12. $x^4 + 4x^3 + 3x^2 - 4x - 4$

13. $3x^4 - 4x^3 - 1$

14. $2x^4 + 3x^3 + 2x^2 + x + 1$

15. $x^4 - 4x^3 - 7x^2 + 34x - 24$

16. $x^4 + 6x^3 + 3x^2 - 26x - 24$

17. $6x^3 + x^2 - 11x - 6$

18. $6x^3 - 19x^2 + x + 6$

Set II Factor completely or write "Not factorable."

1. $x^3 + 2x^2 + 2x - 5$

2. $x^3 + 7x^2 + 11x + 2$

3. $x^3 + 2x^2 - 5x - 6$

4. $x^3 - 11x^2 - 25x + 3$

5. $2x^3 - 12x^2 + 22x - 12$

6. $x^3 - 2x^2 + 2x - 1$

7. $x^3 - 7x + 6$

8. $x^4 + x^3 - 3x^2 - 4x - 4$

9. $x^4 + 3x^3 + 3x^2 + x - 2$

10. $2x^3 + 8x^2 + 8x$

11. $x^4 + 6x^3 + 8x^2 - 6x - 9$

12. $2x^3 + x^2 + x + 1$

13. $2x^4 + 2x^3 + 3x^2 + x - 1$

14. $x^3 - 3x^2 - 8x - 10$

15. $x^4 + 2x^3 - 13x^2 - 38x - 24$

16. $x^4 + 3x^3 - 3x^2 - 7x + 6$

17. $6x^3 + 7x^2 - 7x - 6$

18. $x^7 + 1$

5.9 How to Select the Method of Factoring

With so many different kinds of factoring to choose from, a student is often confused as to which method of factoring to try first. The following is a procedure you can use to select the correct method for factoring a particular algebraic expression.

First, check for a common factor, no matter how many terms the expression has. If there is a common factor, factor it out. This simplifies the remaining polynomial factor.

If the expression to be factored has two terms:

1. Is it a *difference* of two *squares*? (Section 5.3)
2. Is it a *sum* of two *squares*? If so, it is *not factorable,* unless its degree is a multiple of four and it can be factored by completing the square. (Section 5.7)
3. Is it a *sum* or *difference* of two *cubes*? (Section 5.6)
4. Can it be factored by completing the square? (Section 5.7)
5. Can it be factored by synthetic division? (Section 5.8)

If the expression to be factored has three terms:

1. Is the leading coefficient 1? (Section 5.4A)
2. Are the first and last terms perfect squares? If so, is the trinomial a perfect square? (Section 5.4B)
3. Is the leading coefficient unequal to 1? (Section 5.4B)
4. Can it be factored by completing the square? (Section 5.7)
5. Can it be factored by synthetic division? (Section 5.8)

If the expression to be factored has four or more terms:

1. Can it be factored by grouping? (Section 5.2)
2. Can it be factored by synthetic division? (Section 5.8)

Check to see if any factor can be factored again. When the expression is *completely factored,* the same factors are obtained no matter what method is used.

Check the result by multiplying the factors together.

Example 1 Examples of selecting the method of factoring:

	Method
a. $6x^2y - 12xy + 4y$	Common factor
$= 2y(3x^2 - 6x + 2)$	$3x^2 - 6x + 2$ is not factorable
b. $3x^3 - 27xy^2$	Common factor
$= 3x(x^2 - 9y^2)$	Difference of two squares
$= 3x(x + 3y)(x - 3y)$	

	Method
c. $2ac - 3ad + 10bc - 15bd$ $= a(2c - 3d) + 5b(2c - 3d)$ $= (2c - 3d)(a + 5b)$	Grouping
d. $2a^3b + 16b$ $= 2b(a^3 + 8)$ $= 2b(a + 2)(a^2 - 2a + 4)$	Common factor Sum of two cubes
e. $6x^2 + 9x - 10$	Master Product Method shows this is not factorable; you verify this
f. $12x^2 - 13x - 4$ $= (3x - 4)(4x + 1)$	Trinomial with a leading coefficient $\neq 1$
g. $2xy^3 - 14xy^2 + 24xy$ $= 2xy(y^2 - 7y + 12)$ $= 2xy(y - 3)(y - 4)$	Common factor Trinomial with a leading coefficient of 1
h. $16x^2 - 24xy + 9y^2$ $= (4x - 3y)^2$	Trinomial that is a perfect square
i. $x^2 - y^2 + 2x - 2y$ $= (x + y)(x - y) + 2(x - y)$ $= (x - y)(x + y + 2)$ ■	Grouping

EXERCISES 5.9

Set I Factor completely or write "Not factorable."

1. $12e^2 + 13e - 35$

2. $30f^2 + 17f - 21$

3. $6ac - 6bd + 6bc - 6ad$

4. $10cy - 6cz + 5dy - 3dz$

5. $2xy^3 - 4xy^2 - 30xy$

6. $3yz^3 - 6yz^2 - 24yz$

7. $3x^3 + 24h^3$

8. $54f^3 - 2g^3$

9. $9e^2 - 30ef + 25f^2$

10. $16m^2 + 56mp + 49p^2$

11. $x^3 + 3x^2 - 4x - 12$

12. $a^3 - 2a^2 - 9a + 18$

13. $a^2 - b^2 - a + b$

14. $x^2 - y^2 - x - y$

15. $x^2 + x - 10$

16. $x^2 + x + 14$

17. $x^3 - 8y^3 + x^2 - 4y^2$

18. $a^3 - b^3 + a^2 - b^2$

19. $10x^2 + 2x - 21$

20. $10x^2 + 10x - 21$

21. $x^2 - 4xy + 4y^2 - 5x + 10y + 6$

22. $x^2 - 6xy + 9y^2 - 8x + 24y + 15$

23. $x^2 - 6xy + 9y^2 - 25$

24. $a^2 - 8ab + 16b^2 - 1$

Set II Factor completely or write "Not factorable."

1. $5x^2 + 12x + 4$

2. $2a^2mn - 18b^2mn$

3. $4ac + 4bc - 8ad - 8bd$

4. $15x^2 - 13x - 2$

5. $6x^3y + 4x^2y - 10xy$

6. $3x^2 - 18x + 15$

7. $4x^3 + 32y^3$

8. $4x^2 + 4x + 1 - y^2$

9. $16x^2 - 24xy + 9y^2$

10. $x^2 + 100$

11. $b + b^2 - 16b - 16$

12. $1 + 4a^2$

13. $x^2 - 4y^2 - x + 2y$

14. $4x^2 - y^2 + 6x + 3y$

15. $x^2 + x + 3$

16. $27 + 12x^2 + 36x$

17. $x^3 + y^3 + x^2 - y^2$

18. $2x^3 - x^2 - 18x + 9$

19. $10x^2 + 28x - 21$

20. $6x^2 + 7xy - 10y^2$

21. $x^2 - 2xy + y^2 - 10x + 10y + 21$

22. $9x^2 + 6xy + y^2 - z^2 - 2z - 1$

23. $x^2 - 10xy + 25y^2 - 16$

24. $25x^2 - 10xy + y^2 + 35x - 7y + 12$

5.10 Solving Equations by Factoring

In Chapter 2, we discussed solving first-degree equations. We are now ready to solve quadratic (second-degree) and higher-degree equations. One method of solving such equations is based on Rule 5.1, which is stated without proof.

RULE 5.1

If the product of two factors is zero, then one or both of the factors must be zero.

$$\text{If } a \cdot b = 0, \text{ then } \begin{cases} a = 0 \\ \text{or } b = 0 \\ \text{or both } a \text{ and } b = 0 \end{cases}$$

Rule 5.1 can be extended to include more than two factors; that is, if a product of factors is zero, at least one of the factors must be zero.

Many higher-degree equations can be solved by factoring. The method is summarized below.

TO SOLVE AN EQUATION BY FACTORING

1. Get all nonzero terms to one side of the equation by adding the same expression to both sides. *Only zero must remain on the other side.* Then arrange the polynomial in descending powers.
2. Factor the nonzero polynomial completely.
3. Applying Rule 5.1, set each factor equal to zero.*
4. Solve each resulting first-degree equation.
5. Check apparent solutions in the original equation.

* If any of the factors are *not* first-degree polynomials, we cannot solve the equation at this time.

Example 1 Solve $4x^2 - 16x = 0$.
Solution The polynomial must be factored first.

$$4x^2 - 16x = 0$$

$$4x(x - 4) = 0 \quad \text{This is in the form } ab = 0$$

$$4x = 0 \quad \Big| \quad x - 4 = 0 \quad \text{Applying Rule 5.1}$$

$$x = 0 \quad \Big| \quad x = 4 \quad \text{Solving each equation}$$

Check for $x = 0$ $\quad 4(0^2) - 16(0) = 0$

Check for $x = 4$ $\quad 4(4^2) - 16(4) = 64 - 64 = 0$

The solution set is $\{0, 4\}$. ■

Example 2 Solve $x^2 - x = 6$.

Solution

$$x^2 - x = 6$$

$$x^2 - x - 6 = 0 \quad \text{Adding } -6 \text{ to both sides}$$

$$(x + 2)(x - 3) = 0 \quad \text{Factoring}$$

$$x + 2 = 0 \quad \Big| \quad x - 3 = 0 \quad \text{Applying Rule 5.1}$$

$$x = -2 \quad \Big| \quad x = 3$$

Check for $x = -2$ $\quad (-2)^2 - (-2) = 4 + 2 = 6$

Check for $x = 3$ $\quad (3)^2 - (3) = 9 - 3 = 6$

The solution set is $\{-2, 3\}$. ■

Sometimes, it is convenient to get all terms to the right side of the equation, leaving only zero on the left side (see Example 3).

Example 3 Solve $4 - x = 3x^2$.
Solution We add $(-4 + x)$ to both sides and arrange the terms in descending powers.

$$4 - x = 3x^2$$

$$0 = 3x^2 + x - 4 \quad \text{Adding } (-4 + x) \text{ to both sides}$$

$$0 = (x - 1)(3x + 4) \quad \text{Factoring}$$

$$x - 1 = 0 \quad \Big| \quad 3x + 4 = 0 \quad \text{Applying Rule 5.1}$$

$$x = 1 \quad \Big| \quad 3x = -4$$

$$\Big| \quad x = -\frac{4}{3}$$

The solution set is $\left\{1, -\frac{4}{3}\right\}$. (The checking is left to the student.) ■

A WORD OF CAUTION The product must equal *zero*, or no conclusions can be drawn about the factors.

Suppose $(x-1)(x-3) = \boxed{8}$.

↑ No conclusion can be drawn because the product $\neq 0$

Students sometimes think that if $(x-1)(x-3) = 8$, then ~~$x - 1 = 8$ *or* $x - 3 = 8$~~. Both these assumptions are incorrect. Consider the following:

If $x - 1 = 8$ then $x = 9$.

Check for $x = 9$

$$(x-1)(x-3) = 8$$
$$(9-1)(9-3) \stackrel{?}{=} 8$$
$$8 \cdot 6 \stackrel{?}{=} 8$$
$$48 \neq 8$$

If $x - 3 = 8$ then $x = 11$.

Check for $x = 11$

$$(x-1)(x-3) = 8$$
$$(11-1)(11-3) \stackrel{?}{=} 8$$
$$10 \cdot 8 \stackrel{?}{=} 8$$
$$80 \neq 8$$

If the product of two numbers is 8, we *don't* know what the factors are; one factor could be 2 and the other 4, one factor could be 16 and the other 1/2, one factor could be -1 and the other -8, and so forth.

The correct solution is

$$(x-1)(x-3) = 8$$
$$x^2 - 4x + 3 = 8$$
$$x^2 - 4x - 5 = 0 \quad \text{Adding } -8 \text{ to both sides}$$
$$(x-5)(x+1) = 0 \quad \text{Factoring}$$
$$x - 5 = 0 \quad \Big| \quad x + 1 = 0 \quad \text{Applying Rule 5.1}$$
$$x = 5 \quad \Big| \quad x = -1$$

(The checking is left to the student.) The solution set is $\{5, -1\}$. ☑

Example 4 Solve $(x+2)^3 = x^3 + 56$.

Solution We first remove the grouping symbols, using the binomial theorem. We then have

$$x^3 + 3x^2(2) + 3x(2)^2 + (2)^3 = x^3 + 56$$
$$x^3 + 6x^2 + 12x + 8 - x^3 - 56 = 0 \quad \text{Moving all terms to the left side}$$
$$6x^2 + 12x - 48 = 0$$
$$6(x^2 + 2x - 8) = 0$$
$$(x+4)(x-2) = 0 \quad \text{Dividing both sides by 6 and factoring}$$
$$x + 4 = 0 \quad \Big| \quad x - 2 = 0$$
$$x = -4 \quad \Big| \quad x = 2$$

Check for $x = -4$

$$(-4+2)^3 \stackrel{?}{=} (-4)^3 + 56$$
$$(-2)^3 \stackrel{?}{=} -64 + 56$$
$$-8 = -8$$

Check for $x = 2$

$$(2+2)^3 \stackrel{?}{=} 2^3 + 56$$
$$4^3 \stackrel{?}{=} 8 + 56$$
$$64 = 64$$

The solution set is $\{-4, 2\}$. ■

EXERCISES 5.10

Set I Solve each of the following equations.

1. $3x(x - 4) = 0$
2. $5x(x + 6) = 0$
3. $4x^2 = 12x$
4. $6a^2 = 9a$
5. $x(x - 4) = 12$
6. $x(x - 2) = 15$
7. $2x^3 + x^2 = 3x$
8. $4x^3 = 10x - 18x^2$
9. $2x^2 - 15 = -7x$
10. $3x^2 - 10 = -13x$
11. $4x^2 + 9 = 12x$
12. $25x^2 + 4 = 20x$
13. $21x^2 + 60x = 18x^3$
14. $68x^2 = 30x^3 + 30x$
15. $4x(2x - 1)(3x + 7) = 0$
16. $5x(4x - 3)(7x - 6) = 0$
17. $x^3 + 3x^2 - 4x = 12$
18. $x^3 + x^2 - 9x = 9$
19. $x^4 - 10x^2 + 9 = 0$
20. $x^4 - 13x^2 + 36 = 0$
21. $(x + 3)^3 = x^3 + 63$
22. $(x + 1)^3 = x^3 + 37$
23. $(x + 3)^4 = x^4 + 108x + 81$
24. $(x + 2)^4 = x^4 + 32x + 16$

Set II Solve each of the following equations.

1. $4x(x - 2) = 0$
2. $16x^2 = 5x$
3. $x(3x - 1) = 2$
4. $x^2 + 9 = 6x$
5. $x(x + 1) = 12$
6. $(x + 2)^3 = x^3 + 8$
7. $6x^3 + 10x^2 = 4x$
8. $4x^3 + 10x^2 = 6x$
9. $3x^2 - 2 = x$
10. $4x^4 - 5x^2 + 1 = 0$
11. $6x^2 + 12 = 17x$
12. $2x^3 + 3x^2 = 2x + 3$
13. $30x - 3x^2 = 9x^3$
14. $x^3 - 5x^2 - 4x + 20 = 0$
15. $3x(3x - 1)(5x + 7) = 0$
16. $4x^3 = 20x^2 + 9x - 45$
17. $z^3 - 2z^2 + 2 = z$
18. $14x^4 + 41x^3 = 3x^2$
19. $x^4 - 5x^2 + 4 = 0$
20. $(x - 1)^4 = x^4 - 2x + 1$
21. $(x + 2)^3 = x^3 + 26$
22. $(x - 2)^3 = x^3 - 2$
23. $(x + 1)^4 = x^4 + 4x + 1$
24. $7x^2 + 4 = 29x$

5.11 Word Problems Solved by Factoring

Some word problems and some problems about geometric figures lead to equations that can be solved by factoring. Several formulas relating to geometric figures were given in Section 3.2. You might want to review them at this time. It is often helpful to sketch the figures and label them.

Example 1 The width of a rectangle is 5 cm less than its length. Its area is 10 more (numerically) than its perimeter. What are the dimensions of the rectangle?

Think The domain is the set of positive real numbers, since the length of a rectangle cannot be negative or zero.

Step 1. Let ℓ = length (in cm)
$\ell - 5$ = width (in cm)

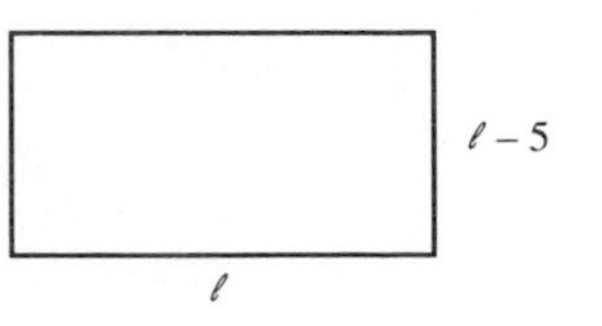

Think Area = (length) × (width) = $\ell(\ell - 5)$

Perimeter = 2(length) + 2(width) = $2\ell + 2(\ell - 5)$

Reread | Its area | is | 10 more than | its perimeter |

Step 2. $\ell(\ell - 5) = 10 + 2\ell + 2(\ell - 5)$

Step 3. $\ell^2 - 5\ell = 10 + 2\ell + 2\ell - 10$

$\ell^2 - 9\ell = 0$

$\ell(\ell - 9) = 0$

Step 4. Not in the domain → ~~$\ell = 0$~~ | $\ell - 9 = 0$

$\ell = 9$ The length

$\ell - 5 = 4$ The width

Step 5. *Check* Area = (9 cm)(4 cm) = 36 sq. cm

Perimeter = 2(9 cm) + 2(4 cm) = 18 cm + 8 cm = 26 cm

36 is 10 more than 26

Step 6. Therefore, the rectangle has a length of 9 cm and a width of 4 cm. ■

Example 2 The base of a triangle is 5 cm more than the height. If the area is 33 sq. cm, find the height and the base of the triangle.

Think The domain is the set of positive real numbers, since neither the height nor the base of a triangle can be negative or zero.

Step 1. Let h = height of triangle (in cm)

$5 + h$ = base of triangle (in cm)

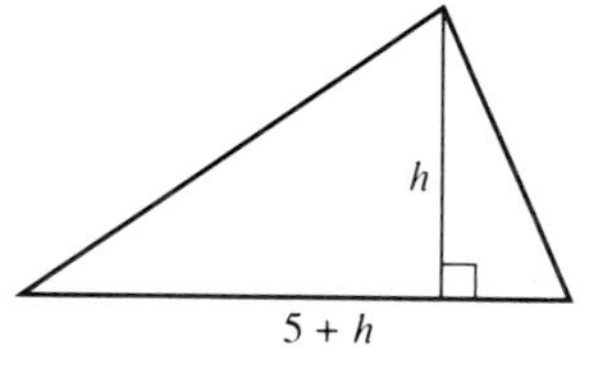

Think The area of a triangle is $\frac{1}{2}$(base)(height).

Reread | Area of triangle | is | 33 |

Step 2. $\frac{1}{2}h(5 + h) = 33$

Step 3. $5h + h^2 = 66$

$h^2 + 5h - 66 = 0$

$(h - 6)(h + 11) = 0$

$h - 6 = 0$ | $h + 11 = 0$

Step 4. $h = 6$ | ~~$h = -11$~~ −11 is not in the domain, since the height can't be negative

$5 + h = 11$

Step 5. *Check* The area is $\frac{1}{2}$(6 cm)(11 cm) = 33 sq. cm.

Step 6. Therefore, the height is 6 cm and the base is 11 cm. ■

Figure 5.11.1 shows some more geometric relationships often used in word problems. The *domain* in such problems is the set of *positive* real numbers, since the length of an edge of a cube or box cannot be negative or zero.

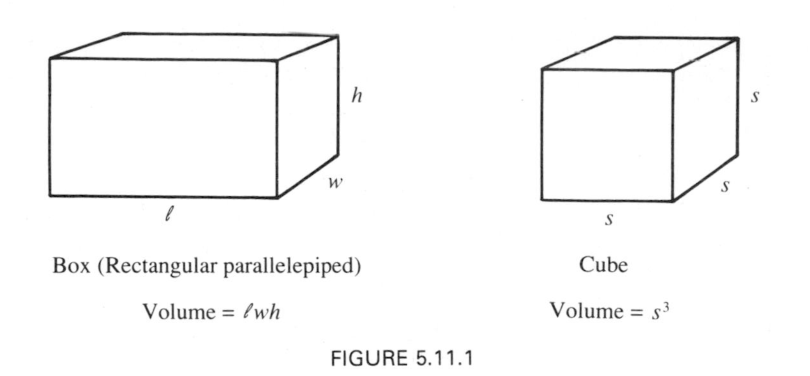

FIGURE 5.11.1

Example 3 One cube has a side 2 cm longer than the side of a second cube. If the volume of the larger cube is 56 cc (cubic centimeters) more than the volume of the smaller cube, find the length of the side of each cube.

Think The domain is the set of all positive real numbers.

Step 1. Let x = length of the side of the smaller cube (in cm)
$x + 2$ = length of the side of the larger cube (in cm)

Think The volume of a cube is s^3.

Reread [Volume of larger cube] [is] [56 more than] [volume of smaller cube]

Step 2. $(x + 2)^3 = 56 + x^3$

Step 3.
$$x^3 + 6x^2 + 12x + 8 = 56 + x^3$$
$$6x^2 + 12x - 48 = 0$$
$$6(x^2 + 2x - 8) = 0$$
$$(x - 2)(x + 4) = 0$$

$x - 2 = 0$ | $x + 4 = 0$

Step 4. $x = 2$ | ~~$x = -4$~~ -4 is not in the domain

$x + 2 = 4$

Step 5. *Check* Volume of smaller cube: $(2\text{ cm})^3 = 8\text{ cm}^3$ or 8 cc
Volume of larger cube: $(4\text{ cm})^3 = 64\text{ cm}^3$ or 64 cc
64 cc = 56 cc + 8 cc

Step 6. Therefore, the length of a side of the smaller cube is 2 cm and the length of a side of the larger cube is 4 cm. ■

EXERCISES 5.11

Set I Set up algebraically and solve. Be sure to state what your variables represent.

1. Find three consecutive even integers such that the product of the first two is 38 more than the third number.

2. Find three consecutive odd integers such that the product of the first two is 52 more than the third number.

3. The length and width of a rectangle are in the ratio of 3 to 2. The area is 150. Find the length and width.

4. The length and width of a rectangle are in the ratio of 4 to 3. The area is 192. Find the length and width.

5. The height of a rectangular box equals the length of a side of a certain cube. The width of the box is 3 more than its height, and the length of the box is 4 times its height. The volume of the box is 8 times the volume of the cube. Find the *dimensions* of the box, and find the *volume* of the cube.

6. The height of a rectangular box equals the length of a side of a certain cube. The width of the box is twice its height, and the length of the box is 10 more than its height. The volume of the box is 6 times the volume of the cube. Find the *dimensions* of the box, and find the *volume* of the cube.

7. The width of a rectangle is 5 m less than its length. The area is 46 more (numerically) than its perimeter. What are the dimensions of the rectangle?

8. The width of a rectangle is 7 m less than its length. The area is 4 more (numerically) than its perimeter. What are the dimensions of the rectangle?

9. The base of a triangle is 7 cm more than the height. If the area is 39 sq. cm, find the height and the base of the triangle.

10. The base of a triangle is 4 m more than the height. If the area is 48 sq. m, find the height and the base of the triangle.

11. One square has a side 6 cm shorter than the side of a second square. The area of the larger square is 9 times as great as the area of the smaller square. Find the length of the side of each square.

12. One square has a side 4 cm shorter than the side of a second square. The area of the larger square is 4 times as great as the area of the smaller square. Find the length of the side of each square.

13. One cube has a side 3 cm longer than the side of a second cube. If the volume of the larger cube is 63 cc more than the volume of the smaller one, find the length of a side of each cube.

14. One cube has a side 1 cm longer than the side of a second cube. If the volume of the larger cube is 37 cc more than the volume of the smaller one, find the length of a side of each cube.

15. A box with no top is to be formed from a rectangular sheet of metal by cutting 2-in. squares from the corners and folding up the sides. The length of the box is to be 3 in. more than its width, and its volume is to be 80 cu. in.

    a. Find the dimensions of the original sheet of metal.

    b. Find the dimensions of the box.

16. A box with no top is to be formed from a rectangular sheet of metal by cutting 3-in. squares from the corners and folding up the sides. The length of the box is to be 3 in. more than its width, and its volume is to be 30 cu. in.

    a. Find the dimensions of the original sheet of metal.

    b. Find the dimensions of the box.

17. Thirty-five square yards of carpet are laid in a rectangular room. The length is 3 yd less than two times the width. Find the dimensions of the room.

18. Fifty-four square yards of carpet are laid in a rectangular room. The length is 3 yd less than two times the width. Find the dimensions of the room.

19. David is working on a square needlepoint picture, and the yarn costs 30¢ per square inch. If the picture were 3 in. longer and 2 in. narrower, the yarn would cost $45. What are the dimensions of the picture?

20. Chang's family room is square. If the room were 1 yd narrower and 1 yd longer, the total cost of covering it with carpeting that costs $24 per square yard, installed, would be $1,920. What are the dimensions of the room?

Set II Set up algebraically and solve. Be sure to state what your variables represent.

1. Find three consecutive even integers such that the product of the first two is 16 more than the third number.

2. Find three consecutive odd integers such that the product of the first and third is 13 more than 16 times the second.

3. The length and width of a rectangle are in the ratio of 5 to 3. The area is 735. Find the length and width.

4. The base and height of a triangle are in the ratio 3:2. If the area of the triangle is 48 sq. cm, find its base and height.

5. The width of a rectangular box equals the length of a side of a certain cube. The height of the box is 2 more than its width, and the length of the box is 3 times its width. The volume of the box is 6 times the volume of the cube. Find the *dimensions* of the box, and find the *volume* of the cube.

6. A 2-cm wide mat surrounds a picture. The area of the picture itself is 286 sq. cm. If the length of the outside of the mat is twice its width, what are the dimensions of the outside of the mat? What are the dimensions of the picture?

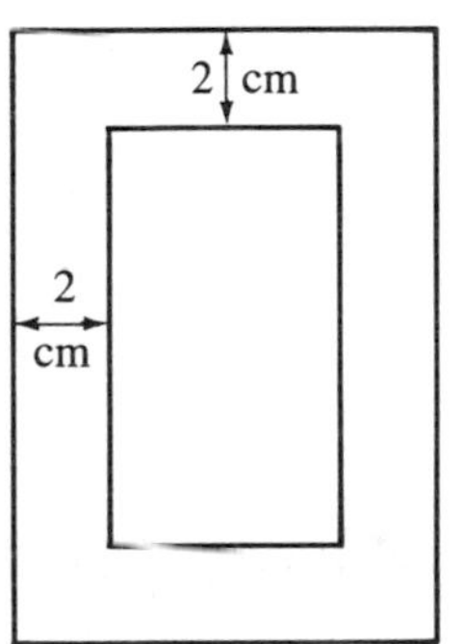

7. The width of a rectangle is 6 yd less than its length. The area is 23 more (numerically) than its perimeter. What are the dimensions of the rectangle?

8. Patricia is 3 years older than Sandra. The product of their ages is 154. How old is each?

9. The base of a triangle is 6 cm more than the height. If the area is 20 sq. cm, find the height and the base of the triangle.

10. Find three consecutive odd integers such that the product of the first and third is 9 more than 12 times the second.

11. One square has a side 2 m longer than the side of a second square. The area of the larger square is 16 times the area of the smaller square. Find the length of the side of each square.

12. The length of a rectangular box is twice its width, and its height is 2 cm less than its width. The volume of the box is 150 cc. Find the dimensions of the box.

13. One cube has a side 2 cm longer than the side of a second cube. If the volume of the larger cube is 26 cc more than the volume of the smaller one, find the length of a side of each cube.

14. Find three consecutive odd integers such that the product of the first and third is 11 more than 14 times the second.

15. A box with no top is formed from a square sheet of metal by cutting 3-in. squares from the corners and folding up the sides. The box is to have a volume of 12 cu. in.

a. Find the dimensions of the sheet of metal.

b. Find the dimensions of the box.

16. One cube has a side 2 cm longer than the side of a second cube. If the volume of the larger cube is 152 cc more than the volume of the smaller one, find the length of a side of each cube.

17. Forty square yards of carpet are laid in a rectangular room. The length is 2 yd less than three times the width. Find the dimensions of the room.

18. The base of a triangle is 5 m more than the height. The area is 12 sq. m. Find the base and the height.

19. Mr. Christ's kitchen is square, and he plans to have the floor covered with linoleum that costs $18 per square yard, installed. If the kitchen were 1 yd narrower and 2 yd longer, the cost of the linoleum would be $324. What are the dimensions of the kitchen?

20. A box with no top is to be formed from a rectangular piece of cardboard by cutting squares from the corners and folding up the sides. The width of the box is to be 2 in. less than the length, the height is to be 3 in., and the volume is to be 105 cu. in.

a. What size squares should be cut from the corners?

b. What are the dimensions of the box?

c. What are the dimensions of the original piece of cardboard?

5.12 Review: 5.6–5.11

Methods of Factoring

5.1 *First, check for a common factor.* If there is a common factor, factor it out.

If the expression to be factored has two terms:

5.3 1. Is it a *difference* of two *squares*?

2. Is it a *sum* of two *squares*? (It is not factorable unless it can be factored by completing the square.)

5.6 3. Is it a *sum* or *difference* of two *cubes*?

5.7 4. Can it be factored by completing the square?

5.8 5. Can it be factored by synthetic division?

If the expression to be factored has three terms:

5.4A 1. Is the leading coefficient 1?

5.4B 2. Are the first and last terms perfect squares? If so, is the trinomial a perfect square?

5.4B 3. Is the leading coefficient unequal to 1?

5.7 4. Can it be factored by completing the square?

5.8 5. Can it be factored by synthetic division?

If the expression to be factored has four or more terms:

5.2 1. Can it be factored by grouping?

5.8 2. Can it be factored by synthetic division?

Check to see if any factor can be factored again.

Check your solution.

To Solve an Equation by Factoring 5.10

1. Get *all* nonzero terms to one side of the equation by adding the same expression to both sides. *Only zero must remain on the other side.* Then arrange the polynomial in descending powers.
2. Factor the polynomial.
3. Set each factor equal to zero and solve for the variable.

Word Problems about Geometric Figures 5.11

In solving word problems about geometric figures, make a drawing of the figures and label them as an aid in writing the equation.

Review Exercises 5.12 Set I

In Exercises 1–29, factor completely or write "Not factorable."

1. $15u^2v - 3uv$ **2.** $3n^2 + 16n + 5$ **3.** $4x^2y - 8xy^2 + 4xy$

4. $5x^2 + 11x + 2$ **5.** $17u - 12 + 5u^2$ **6.** $4 + 21x + 5x^2$

7. $6u^3v^2 - 9uv^3 - 12uv$ **8.** $15a^2 + 15ab - 30b^2$ **9.** $81 - 9m^2$

10. $x^2 + 25$ **11.** $10x^2 - xy - 24y^2$ **12.** $28x^2 - 13xy - 6y^2$

13. $8a^3 - 27b^3$ **14.** $64h^3 - 125k^3$ **15.** $1 + 4y + 4y^2$

16. $4 + 12x + 9x^2$ **17.** $x^3 + 8y^3$ **18.** $27x^3 + y^3$

19. $x^2 + x - y - y^2$ **20.** $x^2 - x + y - y^2$ **21.** $8a^2 - 8ab + 2b^2$

22. $18h^2 - 12hk + 2k^2$ **23.** $x^3 - 4x^2 - 4x + 16$

24. $x^2 + x + 9$ **25.** $x^3 - 2x^2 - 9x + 18$

26. $x^2 + 8x + 16 - 25y^2$ **27.** $(2x + 3y)^2 + (2x + 3y) - 6$

28. $a^2 + 6ab + 9b^2 - 7a - 21b + 12$ **29.** $x^2 + 4xy + 4y^2 - 5x - 10y + 6$

In Exercises 30–38, solve each equation.

30. $5z^2 - 12z + 7 = 0$ **31.** $x^2 = 3(6 + x)$

32. $x^3 = 36x^2$ **33.** $6x^2 = 13x + 5$

34. $x^3 - 2x^2 - 9x + 18 = 0$ **35.** $(x + 3)^3 = x^3 + 27$

36. $4 + 5x^2 + 12x = 0$ **37.** $(x + 3)^4 = x^4 + 12x^3 + 81$

38. $(x + 2)^4 = x^4 + 8x^3 + 16$

In Exercises 39–44, set up algebraically and solve. Be sure to state what your variables represent.

39. Find three consecutive odd integers such that the product of the first and third is 5 more than 8 times the second.

40. The height and base of a triangle are in the ratio of 4 to 5. The area is 40. Find the height and the base.

41. The length and width of a rectangle are in the ratio of 5 to 3. The area is 240. Find the length and the width.

42. The length of a rectangle is 7 more than the width. The area is 40 more (numerically) than its perimeter. What are the dimensions of the rectangle?

43. One cube has a side 4 cm longer than the side of a second cube. If the volume of the larger cube is 316 cc more than the volume of the smaller one, find the length of a side of each cube.

44. One square has a side 3 cm longer than the side of a second square. The area of the larger square is 16 times the area of the smaller square. Find the length of a side of each square.

Review Exercises 5.12 Set II

NAME

In Exercises 1–29, factor completely or write "Not factorable."

1. $2a^2 - 9a + 10$
2. $a^2 - 2ab - 15b^2$
3. $5x^2 + x - 1$
4. $3x^2 + 13x - 10$
5. $6x^2 + xy - 15y^2$
6. $5a^2 + 13ab - 6b^2$
7. $2ac + bc - 3bd - 6ad$
8. $7x^3 + 40x^2 - 12x$
9. $64x^3 + 27y^3$
10. $6x^2 + 29x + 20$
11. $x^2 + 100$
12. $3x^3 - 12xy^2$
13. $x^2 + x + 17$
14. $4y^2 - 5y - 6$
15. $5a^4 - 40ab^3$
16. $x^3 + x^2 - 9x - 9$

ANSWERS

1. ____________
2. ____________
3. ____________
4. ____________
5. ____________
6. ____________
7. ____________
8. ____________
9. ____________
10. ____________
11. ____________
12. ____________
13. ____________
14. ____________
15. ____________
16. ____________

17. $x^2 - 10xy + 25y^2 - 9$

18. $3x^3 - 81$

19. $3n^2 - 8n - 5$

20. $x^3 - 3x^2 - 4x + 12$

21. $y^2 + 36$

22. $6x^2 + x - 2$

23. $x^2 + x + 15$

24. $y^2 + 10y + 25 - 4x^2$

25. $(5x - 2)^2 + 8(5x - 2)y + 15y^2$

26. $a^3 + a^2b - b^2a - b^3$

27. $x^2 + 6xy + 9y^2 - 2x - 6y - 24$

28. $(3x - 2y)^2 - (3x - 2y) - 6$

29. $9x^2 + 12xy + 4y^2 - 4a^2 - 4ab - b^2$

In Exercises 30–38, solve each equation.

30. $x(x - 2) = 15$

31. $x^3 = 16x^2$

32. $x^2 = 4(5 + 2x)$

33. $8x^2 = 10x + 12$

ANSWERS

17. ____________________

18. ____________________

19. ____________________

20. ____________________

21. ____________________

22. ____________________

23. ____________________

24. ____________________

25. ____________________

26. ____________________

27. ____________________

28. ____________________

29. ____________________

30. ____________________

31. ____________________

32. ____________________

33. ____________________

34. $x^3 = 9x^2$

35. $x^3 + 4x^2 - 25x = 100$

36. $x^3 - 5x^2 - 4x + 20 = 0$

37. $(x + 4)^3 = x^3 + 64$

38. $(x - 2)^3 = x^3 - 8$

ANSWERS

34. ______

35. ______

36. ______

37. ______

38. ______

39. ______

40. ______

41. ______

In Exercises 39–44, set up algebraically and solve. Be sure to state what your variables represent.

39. Find three consecutive odd integers such that the product of the first and third is 4 less than 25 times the second.

40. The length and width of a rectangle are in the ratio of 7 to 5. The area is 560. Find the length and the width.

41. One cube has a side 1 cm longer than the side of a second cube. If the volume of the larger cube is 91 cc more than the volume of the smaller one, find the length of a side of each cube.

42. Find three consecutive odd integers such that the product of the first and third is 7 more than 10 times the second.

43. The area of a square is 3 times its perimeter (numerically). What is the length of its side?

44. The height of a retangular box is 2 cm less than the width, and the length is 7 times the width. The volume is 525 cc. Find the dimensions of the box.

ANSWERS

42. ______________

43. ______________

44. ______________

Chapter 5 Diagnostic Test

The purpose of this test is to see how well you understand factoring and solving equations by factoring. We recommend that you work this diagnostic test *before* your instructor tests you on this chapter. Allow yourself about 50 minutes.

Complete solutions for all the problems on this test, together with section references, are given in the answer section at the end of the book. For the problems you do incorrectly, study the sections cited.

In Problems 1–5, factor the expressions completely.

1. a. $4x - 16x^3$

b. $43 + 7x^2 + 6$

c. $7x^2 + 23x + 6$

d. $x^2 + 81$

2. a. $2x^3 + 4x^2 + 16x$

b. $6x^2 - 5x - 6$

3. $y^3 - 1$

4. $3ac + 6bc - 5ad - 10bd$

5. $4x^2 + 4x + 1 - y^2$

In Problems 6–8, solve the equations.

6. a. $2x^2 + x - 15 = 0$

b. $8y^2 = 4y$

7. $6x^2 - 15 = 27x$

8. $(x + 1)^3 = x^3 + 1$

In Problems 9 and 10, set up algebraically and solve. Be sure to state what your variables represent.

9. Find three consecutive even integers such that the product of the first two is 68 more than the third number.

10. The base of a triangle is 8 cm more than the height. If the area is 64 sq. cm, find the height and the base.

Cumulative Review Exercises: Chapters 1–5

1. Evaluate $18 \div 2\sqrt{9} - 4^2 \cdot 3$.

2. Given the formula $h = 48t - 16t^2$, find h when $t = 2$.

3. Solve $2(x - 3) - 5 = 6 - 3(x + 4)$.

4. Solve $-6 < 4x - 2 < 10$.

5. Solve $|2x - 3| \geq 7$.

6. State which of the following are real numbers: $\frac{2}{3}$, 4.7958315 . . . , 0.

7. Solve $2x^2 = 9x + 5$

In Exercises 8 and 9, factor each expression completely or write "Not factorable."

8. $3x^3 + 81$

9. $x^3 + 5x^2 - x - 5$

In Exercises 10–14, perform the indicated operations and simplify.

10. $(2x^2 - 3x + 4) - (x^2 - 5x - 6) + (3 - x + 5x^2)$

11. $(a - 4)(a^2 - 2a + 5)$

12. $(3x + 5)^2$

13. $(2x - 1)^5$

14. $(x^3 - 2x^2 - 6x + 8) \div (x + 2)$

In Exercises 15–22, set up the problem algebraically, solve, and check.

15. How many quarts of pure antifreeze must be added to 10 qt of a 20% solution of antifreeze to make a 50% solution?

16. A motorboat crosses a lake traveling 6 mph and returns traveling 4 mph. If it takes $\frac{1}{2}$ hr longer for the return trip, what is the distance across the lake?

17. Yang has 27 coins. He has 4 times as many nickels as quarters. The product of the number of quarters and twice the number of dimes equals the number of nickels. How many coins of each kind does he have?

18. The three sides of a triangle are in the ratio 8:9:10. If the perimeter of the triangle is 189 cm, what are the lengths of the sides of the triangle?

19. A container is in the shape of a right circular cylinder, and the radius is 5 cm. At present, it is one-fourth full of water. If 300 cc of water are added, it will be one-third full. What is the volume of the container?

20. Two cars leave a certain point at the same time, traveling in opposite directions along a straight road. One is traveling 8 mph faster than the other. Fifteen minutes later, the cars are 26 mi apart. What is the speed of the slower car?

21. Consuelo invested $13,500, putting part of the money into an account with an annual interest rate of 7.2% and the rest into an account with an annual interest rate of 7.4%. If she earned $986.60 in interest for the first year, how much did she invest at each rate?

22. The sum of three consecutive odd integers is 58. What are the integers?

6 Rational Expressions and Equations

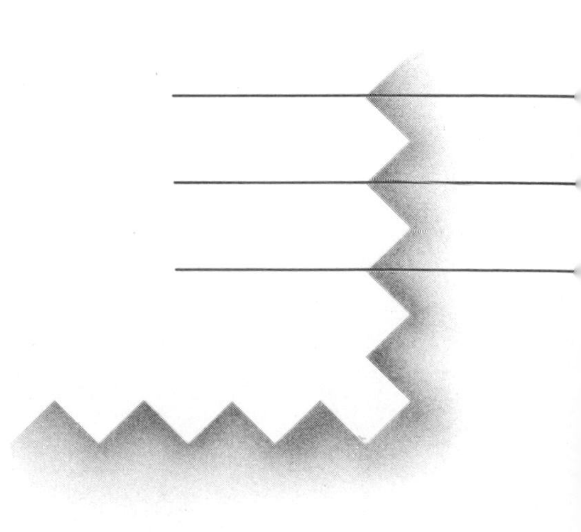

In this chapter, we define rational expressions and discuss how to perform necessary operations on them and how to solve equations and word problems involving them. A knowledge of factoring and of *arithmetic* fractions will help you in your work with rational expressions.

6.1 Rational Expressions

Rational Expressions

A simple **rational expression**, sometimes called an *algebraic fraction* or simply a *fraction,* is an algebraic expression of the form

$$\frac{P}{Q}$$

where P and Q are polynomials. We call P and Q the **terms** of the rational expression; P is called the **numerator**, and Q is called the **denominator**. We learned in Chapter 2 that the *domain of a variable* is the set of all the numbers that can be used in place of that variable, and we learned in Chapter 1 that we cannot divide by zero. Consequently, the domain of the variable in a rational expression is the set of all real numbers except those that would make the denominator, Q, equal to zero.

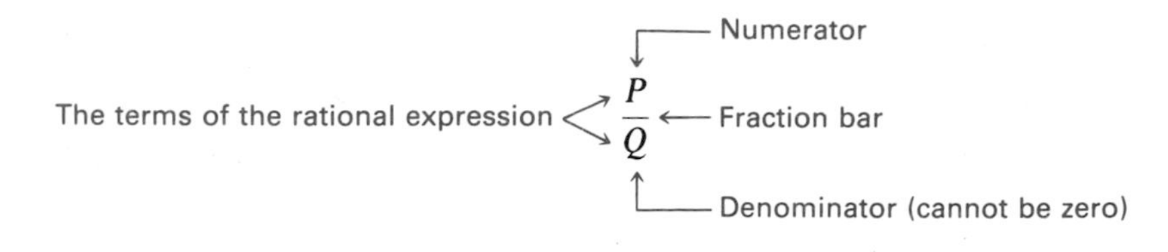

Example 1 Examples of rational expressions and their domains:

a. $\frac{x}{3}$

The domain is the set of all real numbers, since no value for x can make the denominator zero.

b. $\frac{5}{x}$

The domain is the set of all real numbers except zero.

c. $\frac{2x-5}{x-1}$

The domain is the set of all real numbers except 1.

d. $\frac{x^2+2}{x^2-3x-4} = \frac{x^2+2}{(x-4)(x+1)}$

The domain is the set of all real numbers except 4 and -1.

e. $\frac{2}{3}$

The domain is the set of all real numbers. Arithmetic fractions are also rational expressions; here, 2 and 3 are polynomials of degree 0. ■

NOTE After this section, whenever a rational expression is written, it will be understood that the value(s) of the variable(s) that make the denominator zero are excluded. ☑

A WORD OF CAUTION If you are accustomed to writing fractions with a *slanted bar* (/), you are *strongly urged* to break the habit! If you insist on using the slanted bar, be sure to put parentheses around any numerator or denominator that contains more than one term. For example, note that

$$(x + 2)/(x - 5) = \frac{x + 2}{x - 5}$$

but

$$x + 2/x - 5 = x + \frac{2}{x} - 5$$

☑

Equivalent Rational Expressions and the Fundamental Property of Rational Expressions

Equivalent rational expressions (equivalent fractions) are rational expressions (fractions) that have the same value. The fundamental property of rational expressions, which follows, allows us to reduce fractions and also to "build up" fractions in order to add and subtract them.

> THE FUNDAMENTAL PROPERTY OF RATIONAL EXPRESSIONS
>
> If P, Q, and C are polynomials and if $Q \neq 0$ and $C \neq 0$, then
>
> $$\frac{P \cdot C}{Q \cdot C} = \frac{P}{Q}$$

The fundamental property of rational expressions permits us to do the following:

1. Multiply both numerator and denominator by the same nonzero number; that is, $\frac{P}{Q} = \frac{P \cdot C}{Q \cdot C}$, $C \neq 0$.
2. Divide both numerator and denominator by the same nonzero number; that is, $\frac{P}{Q} = \frac{P \div C}{Q \div C}$, $C \neq 0$.

A WORD OF CAUTION When we use the fundamental property of rational expressions and multiply (or divide) both numerator and denominator by the same expression, we do not change the value of the original expression. We do *not* get a fraction equivalent to the one we started with if we *add* the same expression to or *subtract* the same expression from both the numerator and the denominator. For example, while $\frac{2}{3} = \frac{2 \cdot 4}{3 \cdot 4}$, it is *not* true that $\frac{2}{3} = \frac{2 + 4}{3 + 4}$, and while $\frac{6}{9} = \frac{6 \div 3}{9 \div 3}$, it is *not* true that $\frac{6}{9} = \frac{6 - 3}{9 - 3}$. ☑

Example 2 Determine whether the pairs of fractions are equivalent.

a. $\dfrac{x+3}{x+6}, \dfrac{2(x+3)}{2(x+6)}$

Solution Yes; the second rational expression can be obtained from the first by multiplying both numerator and denominator by 2.

b. $\dfrac{3+x}{5+x}, \dfrac{3}{5}$

Solution No; the second rational expression cannot be obtained from the first by multiplying or dividing both numerator and denominator by the same number. ■

The Three Signs of a Fraction

Every fraction has three signs associated with it, even if those signs are not visible: the sign of the fraction, the sign of the numerator, and the sign of the denominator.

The sign of the numerator

The sign of the fraction ⟶ $+\dfrac{+8}{+4}$

The sign of the denominator

Let us compare three other fractions with $+\dfrac{+8}{+4}$, which equals 2.

$$+\frac{-8}{-4} = +\left(\frac{-8}{-4}\right) = +(+2) = 2$$

The sign of the numerator and the sign of the denominator are different from $+\dfrac{+8}{+4}$

$$-\frac{-8}{+4} = -\left(\frac{-8}{+4}\right) = -(-2) = 2$$

The sign of the fraction and the sign of the numerator are different from $+\dfrac{+8}{+4}$

$$-\frac{+8}{-4} = -\left(\frac{+8}{-4}\right) = -(-2) = 2$$

The sign of the fraction and the sign of the denominator are different from $+\dfrac{+8}{+4}$

Since each of these fractions equals the same number, the fractions must all equal each other. Therefore,

$$+\frac{+8}{+4} = +\frac{-8}{-4} = -\frac{-8}{+4} = -\frac{+8}{-4} = 2$$

It can also be shown that

$$-\frac{+8}{+4} = +\frac{-8}{+4} = +\frac{+8}{-4} = -\frac{-8}{-4} = -2$$

> **RULE OF SIGNS FOR RATIONAL EXPRESSIONS**
>
> If any *two* of the three signs of a rational expression are changed, the value of the rational expression is unchanged; that is, the new rational expression is equivalent to the original one.

This rule of signs is sometimes helpful when we are reducing rational expressions or performing operations (addition, multiplication, and so forth) on them.

Example 3 Find the missing term.

a. $-\dfrac{-5}{xy} = \dfrac{?}{xy}$

Solution Since the signs of the *denominators* are both understood to be + and the signs of the rational expressions are different, the signs of the numerators must be different. Therefore,

$$-\frac{-5}{xy} = \frac{5}{xy}$$

b. $\dfrac{8}{-x} = \dfrac{-8}{?}$

Solution Since the signs of the rational expressions are both understood to be + and the signs of the *numerators* are different, the signs of the denominators must be different. Therefore,

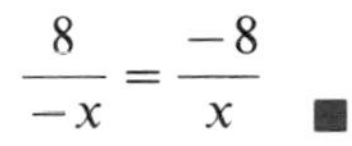

$$\frac{8}{-x} = \frac{-8}{x} \quad ■$$

Recall from Section 1.14C that

$$-(b - a) = -1(b - a) = -b + a = a - b$$

Consequently, $a - b$ can always be substituted for $-(b - a)$, and $-(b - a)$ can always be substituted for $a - b$.

Example 4 Find the missing term.

a. $-\dfrac{1}{2 - x} = \dfrac{1}{?}$

Solution

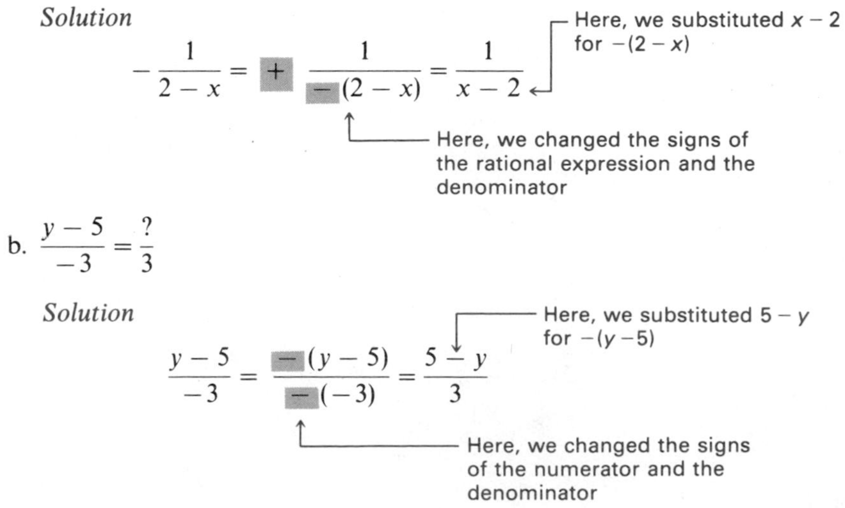

$$-\frac{1}{2 - x} = +\frac{1}{-(2 - x)} = \frac{1}{x - 2}$$

b. $\dfrac{y - 5}{-3} = \dfrac{?}{3}$

Solution

$$\frac{y - 5}{-3} = \frac{-(y - 5)}{-(-3)} = \frac{5 - y}{3}$$

c. $\dfrac{x-y}{(a+b)(c-d)} = \dfrac{?}{(a+b)(d-c)}$

Solution

$$\frac{x-y}{(a+b)(c-d)} = \frac{-(x-y)}{(a+b)(-[c-d])}$$

Here, we changed the signs of the numerator and the denominator

Here, we substituted $y-x$ for $-(x-y)$

$$= \frac{y-x}{(a+b)(d-c)}$$

Here, we substituted $d-c$ for $-[c-d]$ ■

EXERCISES 6.1

Set I In Exercises 1–8, find the domain of the variable.

1. $\dfrac{10-7y}{y+4}$ **2.** $\dfrac{3z+2}{5-z}$ **3.** $\dfrac{5x}{9}$

4. $\dfrac{7}{2x}$ **5.** $\dfrac{a^2+1}{a^2-25}$ **6.** $\dfrac{h^2+5}{h^2-h-6}$

7. $\dfrac{4c+3}{c^4-13c^2+36}$ **8.** $\dfrac{2x-5}{x^3-5x^2-9x+45}$

In Exercises 9–14, determine whether the pairs of rational expressions are equivalent.

9. $\dfrac{3}{6y}, \dfrac{1}{2y}$ **10.** $\dfrac{8}{4x}, \dfrac{2}{x}$ **11.** $\dfrac{x}{5}, \dfrac{x+5}{10}$

12. $\dfrac{3}{z}, \dfrac{9}{z+6}$ **13.** $\dfrac{2+x}{3-y}, \dfrac{8(2+x)}{8(3-y)}$ **14.** $\dfrac{4-x}{y+2}, \dfrac{3(4-x)}{3(y+2)}$

In Exercises 15–24, find the missing term.

15. $-\dfrac{5}{8} = \dfrac{5}{?}$ **16.** $-\dfrac{6}{-k} = \dfrac{?}{k}$ **17.** $\dfrac{-x}{5} = \dfrac{x}{?}$

18. $\dfrac{6}{-k} = \dfrac{?}{k}$ **19.** $\dfrac{8-y}{4y-7} = \dfrac{y-8}{?}$ **20.** $\dfrac{w-2}{5-w} = \dfrac{?}{w-5}$

21. $\dfrac{u-v}{a-b} = \dfrac{v-u}{?}$ **22.** $\dfrac{2-x}{y-5} = \dfrac{?}{5-y}$

23. $\dfrac{a-b}{(3a+2b)(a-5b)} = \dfrac{?}{(3a+2b)(5b-a)}$

24. $\dfrac{(e+4f)(7e-3f)}{2e-f} = \dfrac{(e+4f)(3f-7e)}{?}$

Set II In Exercises 1–8, find the domain of the variable.

1. $\dfrac{4y + 5}{2y - 9}$ **2.** $\dfrac{x + 2}{5}$ **3.** $\dfrac{11}{6d}$

4. $\dfrac{3 - x}{x + 2}$ **5.** $\dfrac{2a^2 + 13}{a^2 + 3a - 28}$ **6.** $\dfrac{3x^2 - 13x + 12}{9x^2 - 24x + 16}$

7. $\dfrac{u^2 + 10u + 1}{u^3 - 16u + 2u^2 - 32}$ **8.** $\dfrac{5}{v^3 + 3v^2 - v - 3}$

In Exercises 9–14, determine whether the pairs of rational expressions are equivalent.

9. $\dfrac{12}{3y}, \dfrac{4}{y}$ **10.** $\dfrac{12}{3 + y}, \dfrac{4}{y}$ **11.** $\dfrac{z}{2}, \dfrac{z + 6}{8}$

12. $\dfrac{6 + 3y}{8 + 4y}, \dfrac{3}{4}$ **13.** $\dfrac{9 - y}{x + 2}, \dfrac{2(9 - y)}{2(x + 2)}$ **14.** $\dfrac{y + 2}{y + 4}, \dfrac{1}{2}$

In Exercises 15–24, find the missing term.

15. $-\dfrac{2}{7} = \dfrac{2}{?}$ **16.** $\dfrac{-8}{x} = -\dfrac{?}{x}$ **17.** $\dfrac{-14}{3m} = \dfrac{14}{?}$

18. $\dfrac{4 - x}{x + 3} = -\dfrac{?}{x + 3}$ **19.** $\dfrac{5 - x}{3y - 2} = \dfrac{x - 5}{?}$ **20.** $\dfrac{?}{x + 5} = \dfrac{x + 10}{-(x + 5)}$

21. $\dfrac{x - y}{c - d} = \dfrac{?}{d - c}$ **22.** $\dfrac{y - 2}{?} = \dfrac{2 - y}{y + 7}$

23. $\dfrac{5t - 2u}{(t - 4u)(3t + u)} = \dfrac{?}{(4u - t)(3t + u)}$ **24.** $\dfrac{(8 - x)(8 + x)}{?} = \dfrac{(x - 8)(x + 8)}{(3 + x)(5 + x)}$

6.2 Reducing Rational Expressions to Lowest Terms

A rational expression is in *lowest terms* if the greatest common factor (GCF) of its numerator and denominator is 1. In this section and in the remainder of the book, it is understood that all rational expressions are to be reduced to lowest terms unless otherwise indicated.

One way to reduce a rational expression to lowest terms is as follows:

> TO REDUCE A RATIONAL EXPRESSION TO LOWEST TERMS
>
> **1.** Factor the numerator and denominator completely.
>
> **2.** Using the fundamental property of rational expressions, divide both numerator and denominator by any common factor.

If the numerator and denominator of a rational expression do not have a common factor (other than 1), we cannot reduce the expression; it is in lowest terms.

Example 1 Examples of reducing rational expressions to lowest terms:

a. $\dfrac{4x^2y}{2xy} = \dfrac{\overset{2}{\not{4}}x^2y}{\underset{1}{\not{2}}xy} = 2x$

Here, the literal parts of the rational expression are already factored; the common factors are 2, x, and y.

b. $\dfrac{3x^2 - 5xy - 2y^2}{6x^3y + 2x^2y^2} = \dfrac{(x - 2y)\overset{1}{\cancel{(3x + y)}}}{2x^2y\underset{1}{\cancel{(3x + y)}}} = \dfrac{x - 2y}{2x^2y}$

c. $\dfrac{2b^2 + ab - 3a^2}{4a^2 - 9ab + 5b^2}$

$(-1)(b - a) = (a - b)$

$$= \frac{(b - a)(2b + 3a)}{(a - b)(4a - 5b)} = \frac{(-1)(b - a)(2b + 3a)}{(-1)(a - b)(4a - 5b)} = \frac{\overset{1}{\cancel{(a - b)}}(2b + 3a)}{(-1)\underset{1}{\cancel{(a - b)}}(4a - 5b)}$$

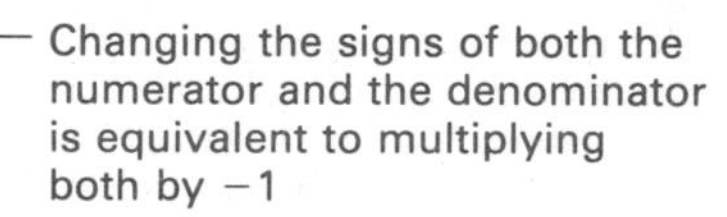

$$= \frac{(2b + 3a)}{(-1)(4a - 5b)} = \frac{2b + 3a}{5b - 4a}$$

or $\dfrac{(b - a)(2b + 3a)}{(a - b)(4a - 5b)} = \dfrac{\overset{1}{\cancel{(b - a)}}(2b + 3a)}{-\underset{1}{\cancel{(b - a)}}(4a - 5b)} = \dfrac{2b + 3a}{-(4a - 5b)} = \dfrac{2b + 3a}{5b - 4a}$

Substituting $-(b - a)$ for $a - b$

Difference of 2 cubes

d. $\dfrac{x^3 - 8}{x^3 - 2x^2 + 4x - 8} = \dfrac{\overset{1}{\cancel{(x - 2)}}(x^2 + 2x + 4)}{\underset{1}{\cancel{(x - 2)}}(x^2 + 4)} = \dfrac{x^2 + 2x + 4}{x^2 + 4}$

Factor by grouping

A WORD OF CAUTION An error often made in reducing rational expressions is to forget that the expression the numerator and denominator are divided by *must* be a *factor* of *both*. For example,

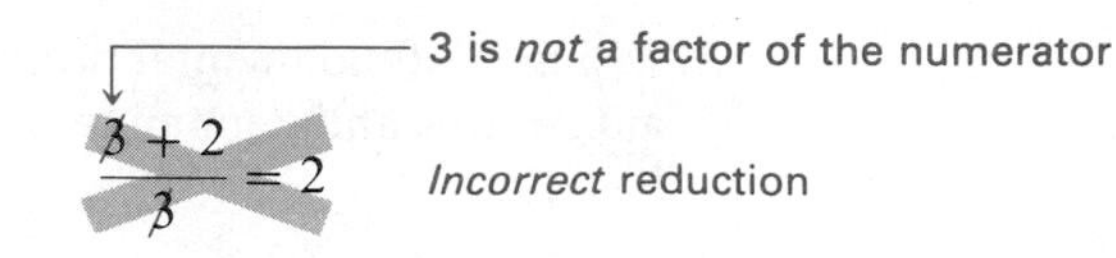

The above reduction is incorrect because

$$\frac{3 + 2}{3} = \frac{5}{3} \neq 2$$

Also see Example 1e. ✓

e. $\dfrac{x+3}{x+6}$

This rational expression cannot be reduced, since neither x nor 3 is a *factor* of the numerator or the denominator. ■

When a rational expression has only one term in the denominator but several terms in the numerator, it can be reduced as a single rational expression or it can be rewritten as a sum of several expressions, as was done in Section 4.5A; each resulting expression can then be reduced (see Example 2).

Example 2 Examples of reducing rational expressions to lowest terms:

a. $\dfrac{x+y}{x}$ ← x is *not* a factor of the numerator

This rational expression cannot be reduced. ← This is an acceptable answer

If we write $\dfrac{x+y}{x}$ as a sum of rational expressions, we have the following:

$$\frac{x+y}{x} = \frac{x}{x} + \frac{y}{x} = 1 + \frac{y}{x}$$ ← This is an acceptable answer

b. $$\frac{7pqr - 21p^2r + 28qr^3}{7pqr} = \frac{7r(pq - 3p^2 + 4qr^2)}{7pqr} = \frac{pq - 3p^2 + 4qr^2}{pq}$$ ← This is an acceptable answer

The same problem can also be done as follows:

$$\frac{7pqr - 21p^2r + 28qr^3}{7pqr} = \frac{7pqr}{7pqr} - \frac{21p^2r}{7pqr} + \frac{28qr^3}{7pqr} = 1 - \frac{3p}{q} + \frac{4r^2}{p}$$ ← This is an acceptable answer

Notice that

$$\frac{pq - 3p^2 + 4qr^2}{pq} = \frac{pq}{pq} - \frac{3p^2}{pq} + \frac{4qr^2}{pq} = 1 - \frac{3p}{q} + \frac{4r^2}{p}$$ ■

EXERCISES 6.2

Set I Reduce each rational expression to lowest terms.

1. $\dfrac{12m^3n}{4mn}$ **2.** $\dfrac{-6hk^4}{24hk}$ **3.** $\dfrac{15a^4b^3c^2}{-35ab^5c}$ **4.** $\dfrac{40e^2f^2g}{16e^4fg^3}$

5. $\dfrac{40x - 8x^2}{5x^2 + 10x}$ **6.** $\dfrac{16y^4 - 16y^3}{24y^2 - 24y}$ **7.** $\dfrac{c^2 - 4}{4}$ **8.** $\dfrac{9 + d^2}{9}$

9. $\dfrac{24w^2x^3 - 16wx^4}{18w^3x - 12w^2x^2}$ **10.** $\dfrac{18c^5d + 45c^4d^2}{12c^2d^3 + 30cd^4}$ **11.** $\dfrac{x^2 - 16}{x^2 - 9x + 20}$

12. $\dfrac{x^2 - 2x - 15}{x^2 - 9}$ **13.** $\dfrac{2x^2 - xy - y^2}{y^2 - 4xy + 3x^2}$ **14.** $\dfrac{2k^2 + 3hk - 5h^2}{3h^2 - 7hk + 4k^2}$

15. $\frac{2x^2 - 3x - 9}{12 - 7x + x^2}$
16. $\frac{15 + 7y - 2y^2}{4y^2 - 21y + 5}$
17. $\frac{2y^2 + xy - 6x^2}{3x^2 + xy - 2y^2}$
18. $\frac{10y^2 + 11xy - 6x^2}{4x^2 - 4xy - 15y^2}$
19. $\frac{2x^2 - 9x - 5}{2x^2 + 5x + 3}$
20. $\frac{3x^2 - 11x - 4}{3x^2 + 5x + 2}$
21. $\frac{a^3 - 1}{1 - a^2}$
22. $\frac{x^3 + y^3}{y^2 - x^2}$
23. $\frac{x^2 + 4}{x^2 + 4x + 4}$
24. $\frac{9 + y^2}{9 + 6y + y^2}$
25. $\frac{13x^3y^2 - 26xy^3 + 39xy}{13x^2y^2}$
26. $\frac{21m^2n^3 - 35m^3n^2 - 14mn}{7m^2n^2}$
27. $\frac{6a^2bc^2 - 4ab^2c^2 + 12bc}{6abc}$
28. $\frac{8a^3b^2c - 4a^2bc - 10ac}{4abc}$

Set II Reduce each rational expression to lowest terms.

1. $\frac{15c^2d^5}{12c^2d^3}$
2. $\frac{24x^4y^2}{48x^6y^3}$
3. $\frac{14e^5f^2g^3}{-42e^3f^4g}$
4. $\frac{x^2 + 9}{x^2 - 9}$
5. $\frac{32m^2 - 24m^3}{80m^3 - 60m^4}$
6. $\frac{x^2 + x - 6}{x^2 + 6x + 9}$
7. $\frac{6}{6 - x}$
8. $\frac{2x^2 - 7x - 15}{2x^2 + 7x - 15}$
9. $\frac{12a^4b - 30a^3b^2}{18a^2b^2 - 45ab^3}$
10. $\frac{x - 4}{x^2 - 16}$
11. $\frac{h^2 - 2h - 24}{h^2 - 36}$
12. $\frac{5x^2 - 11x + 2}{5x^2 + 7x + 2}$
13. $\frac{24x^2 + 14xy - 5y^2}{3y^2 - 14xy + 8x^2}$
14. $\frac{8x^3 + 1}{8x^3 + 4x^2 + 2x + 1}$
15. $\frac{6x^2 + 5x - 4}{4 - 3x - 10x^2}$
16. $\frac{x + 4}{3x^2 + 9x - 12}$
17. $\frac{21m^2 - mn - 2n^2}{4n^2 - 9nm - 9m^2}$
18. $\frac{2x^3 - x^2 - 8x + 4}{2 - 3x - 2x^2}$
19. $\frac{5x^2 - 4x - 1}{5x^2 + 7x + 2}$
20. $\frac{6x^2 - x - 2}{8x - 10x^2 - 3x^3}$
21. $\frac{4x^2 - w^2}{w^3 - 8x^3}$
22. $\frac{5 - x}{9x^2 - 39x - 30}$
23. $\frac{x^2 + 16}{x^2 + 8x + 16}$
24. $\frac{x^4 + 4}{x^4 - 4}$
25. $\frac{-30m^4n^2 + 60m^2n^3 - 45m^3n}{15m^3n^2}$
26. $\frac{15xyz^3 - 12xz^2 + 10y^2z}{6y^2z}$
27. $\frac{24abc + 18b^2c - 8bc}{12ac}$
28. $\frac{32x^3yz - 12xy^2 + 48xy^2z}{-16xy^2z}$

6.3 Multiplying and Dividing Rational Expressions

Multiplying Rational Expressions

Multiplication of rational expressions is defined as follows:

$$\frac{P}{Q} \cdot \frac{R}{S} = \frac{P \cdot R}{Q \cdot S}$$

where P, Q, R, and S are polynomials and where $Q \neq 0$ and $S \neq 0$.

In practice, however, we can often reduce the resulting rational expression. Therefore, we give the following suggestions for multiplying rational expressions:

TO MULTIPLY RATIONAL EXPRESSIONS

1. Factor any numerators or denominators that contain more than one term.
2. Divide the numerators and denominators by any factor common to both. (The common factors can be, but do not have to be, in the same rational expression.)
3. The numerator of the answer is the product of the factors remaining in the numerator; the denominator of the answer is the product of the factors remaining in the denominator. A factor of 1 will always remain in both numerator and denominator (see Examples 1a and 2b).

We can write the product as a single rational expression either *after* we divide by the common factors (see Examples 1a, 1c, and 1d) or *before* we divide by the common factors (see Example 1b).

Example 1 Examples of multiplying rational expressions:

a. $$\frac{a}{3}\cdot\frac{1}{a^2}=\frac{\overset{1}{\cancel{a}}}{3}\cdot\frac{1}{\underset{a}{\cancel{a^2}}}=\frac{1}{3a}$$

b. $$\frac{2y^3}{3x^2}\cdot\frac{12x}{5y^2}=\frac{2y^3\cdot\overset{4}{\cancel{12}}x}{\underset{1}{\cancel{3}}x^2\cdot 5y^2}=\frac{8y}{5x}$$

c. $$\frac{10xy^3}{x^2-y^2}\cdot\frac{2x^2+xy-y^2}{15x^2y}=\frac{\overset{2\cdot 1\,y^2}{\cancel{10xy^3}}}{\underset{1}{\cancel{(x+y)}}(x-y)}\cdot\frac{\overset{1}{\cancel{(x+y)}}(2x-y)}{\underset{3\,x\,1}{\cancel{15x^2y}}}$$

$$=\frac{2y^2(2x-y)}{3x(x-y)}$$

d. $$\frac{a^3-b^3}{a^2-b^2-a+b}\cdot\frac{8a^2b}{4a^3+4a^2b+4ab^2}$$

Factor by grouping: $= (a+b)(a-b)-1(a-b) = (a-b)(a+b-1)$

$$=\frac{\overset{1}{\cancel{(a-b)}}\overset{1}{\cancel{(a^2+ab+b^2)}}}{\underset{1}{\cancel{(a-b)}}(a+b-1)}\cdot\frac{\overset{2\cdot 1}{\cancel{8}\cancel{a}ab}}{\underset{1\cdot 1}{\cancel{4a}}\underset{1}{\cancel{(a^2+ab+b^2)}}}$$

$$=\frac{2ab}{a+b-1}\ \blacksquare$$

Now that we have discussed multiplication of rational expressions, we can give an alternate method for reducing rational expressions based on the definition of multiplication and the identity property of multiplication. We can reduce the rational expression $\dfrac{P \cdot C}{Q \cdot C}$ as follows (assuming $Q \neq 0$ and $C \neq 0$):

$$\frac{P \cdot C}{Q \cdot C} = \frac{P}{Q} \cdot \frac{C}{C} = \frac{P}{Q} \cdot 1 = \frac{P}{Q}$$

Using the definition of multiplication

$\frac{C}{C} = 1$

$\frac{P}{Q} \cdot 1 = \frac{P}{Q}$

Dividing Rational Expressions

Recall from Section 1.4 that two numbers are the *multiplicative inverses* (or *reciprocals*) of each other if their product is 1. This implies that the multiplicative inverse of $\dfrac{a}{b}$ is $\dfrac{b}{a}$, since $\dfrac{a}{b} \cdot \dfrac{b}{a} = 1$. We use these facts in *dividing* rational expressions.

The rule for dividing rational expressions is as follows:

> TO DIVIDE RATIONAL EXPRESSIONS
>
> Multiply the dividend by the multiplicative inverse (the reciprocal) of the divisor.
>
> $$\frac{P}{Q} \div \frac{S}{T} = \frac{P}{Q} \cdot \frac{T}{S}$$
>
> Dividend ↑ ↑ Divisor
>
> where P, Q, S, and T are polynomials and $Q \neq 0$, $S \neq 0$, and $T \neq 0$.

This method works (in algebra *and* in arithmetic) because a division problem is equivalent to a rational expression $\left(\text{that is, } P \div Q = \dfrac{P}{Q}\right)$ and because of the multiplicative identity and multiplicative inverse properties. That is,

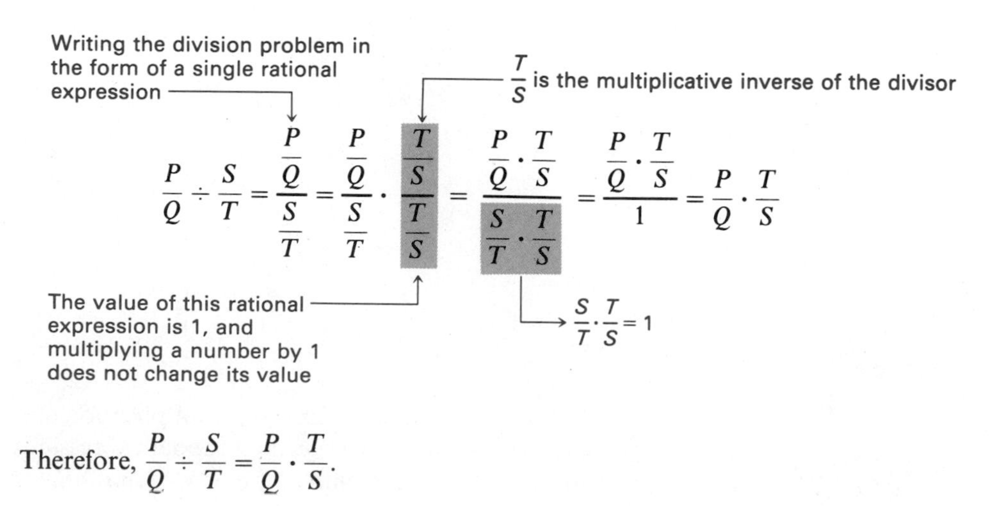

Therefore, $\dfrac{P}{Q} \div \dfrac{S}{T} = \dfrac{P}{Q} \cdot \dfrac{T}{S}$.

Example 2 Examples of dividing rational expressions:

a. $$\frac{3y^3 - 3y^2}{16y^5 + 8y^4} \div \frac{y^2 + 2y - 3}{4y + 12} = \frac{3y^3 - 3y^2}{16y^5 + 8y^4} \cdot \frac{4y + 12}{y^2 + 2y - 3}$$

$$= \frac{\overset{1}{\cancel{3y^2}}\overset{1}{\cancel{(y-1)}}}{\underset{2y^2}{\cancel{8y^4}}(2y + 1)} \cdot \frac{\overset{1}{\cancel{4}}\overset{1}{\cancel{(y+3)}}}{\underset{1}{\cancel{(y-1)}}\underset{1}{\cancel{(y+3)}}} = \frac{3}{2y^2(2y + 1)}$$

b. $$\frac{x^2 + x - 2}{x + 2} \div \frac{x^2 + 2x - 3}{x + 3} = \frac{x^2 + x - 2}{x + 2} \cdot \frac{x + 3}{x^2 + 2x - 3}$$

$$= \frac{\overset{1}{\cancel{(x+2)}}\overset{1}{\cancel{(x-1)}}}{\underset{1}{\cancel{x+2}}} \cdot \frac{\overset{1}{\cancel{x+3}}}{\underset{1}{\cancel{(x+3)}}\underset{1}{\cancel{(x-1)}}} = 1$$

c. $$\frac{y^2 - x^2}{4xy - 2y^2} \div \frac{2x - 2y}{2x^2 + xy - y^2} = \frac{y^2 - x^2}{4xy - 2y^2} \cdot \frac{2x^2 + xy - y^2}{2x - 2y}$$

$$= \frac{(y + x)(y - x)}{2y\underset{1}{\cancel{(2x - y)}}} \cdot \frac{(x + y)\overset{1}{\cancel{(2x - y)}}}{2(x - y)}$$

$$= \frac{(x + y)^2 (y - x)}{4y(x - y)} = \frac{(x + y)^2 (-1)\overset{1}{\cancel{(x - y)}}}{4y\underset{1}{\cancel{(x - y)}}}$$

$$= \frac{(-1)(x + y)^2}{4y} = -\frac{(x + y)^2}{4y}$$ ■

EXERCISES 6.3

Set I Perform the indicated operations.

1. $\frac{27x^4y^3}{22x^5yz} \cdot \frac{55x^2z^2}{9y^3z}$

2. $\frac{13b^2c^4}{42a^4b^3} \cdot \frac{35a^3bc^2}{39ac^5}$

3. $\frac{mn^3}{18n^2} \div \frac{5m^4}{24m^3n}$

4. $\frac{27k^3}{h^5k} \div \frac{15hk^2}{-4h^4}$

5. $\frac{15u - 6u^2}{10u^2} \cdot \frac{15u^3}{35 - 14u}$

6. $\frac{-22v^2}{63v + 84} \cdot \frac{42v^3 + 56v^2}{55v^3}$

7. $\frac{-15c^4}{40c^3 - 24c^2} \div \frac{35c}{35c^2 - 21c}$

8. $\frac{40d - 30d^2}{d^2} \div \frac{24d^2 - 18d^3}{12d^3}$

9. $\frac{d^2e^2 - d^3e}{12e^2d} \div \frac{d^2e^2 - de^3}{3e^2d + 3e^3}$

10. $\frac{9m^2n + 3mn^2}{16mn^2} \div \frac{2mn^2 - m^2n}{10mn^2 - 20n^3}$

11. $\frac{w^2 - 2w - 8}{6w - 24} \cdot \frac{5w^2}{w^2 - 3w - 10}$

12. $\frac{-15k^3}{8k + 32} \div \frac{15k^2 - 5k^3}{k^2 + k - 12}$

13. $\frac{4a^2 + 8ab + 4b^2}{a^2 - b^2} \div \frac{6ab + 6b^2}{b - a}$

14. $\frac{u^2 - v^2}{7u^2 - 14uv + 7v^2} \div \frac{2u^2 + 2uv}{14v - 14u}$

15. $\frac{4 - 2a}{2a + 2} \div \frac{2a^3 - 16}{a^2 + 2a + 1}$

16. $\frac{18 + 6a}{4 - 2a} \div \frac{2a^3 + 54}{4a - 8}$

17. $\dfrac{x^3+y^3}{2x-2y} \div \dfrac{x^2-xy+y^2}{x^2-y^2}$

18. $\dfrac{x^3-y^3}{3x+3y} \div \dfrac{x^2+xy+y^2}{x^2-y^2}$

19. $\dfrac{e^2+10ef+25f^2}{e^2-25f^2} \cdot \dfrac{3e-3f}{f-e} \div \dfrac{e+5f}{5f-e}$

20. $\dfrac{3x-y}{y+x} \div \dfrac{y^2+yx-12x^2}{4x^2+5xy+y^2} \cdot \dfrac{10x^2-8xy}{8xy-10x^2}$

21. $\dfrac{(x+y)^2+x+y}{(x-y)^2-x+y} \cdot \dfrac{x^2-2xy+y^2}{x^2+2xy+y^2} \div \dfrac{x-y}{x+y}$

22. $\dfrac{ac+bc+ad+bd}{ac-ad-bc+bd} \cdot \dfrac{c^2-2cd+d^2}{a^2+2ab+b^2} \div \dfrac{c+d}{a-b}$

Set II Perform the indicated operations.

1. $\dfrac{11m^4p^2}{30mn^2} \cdot \dfrac{21n^5p}{22m^3np^4}$

2. $\dfrac{65x^5y^3}{78x^2y^4} \cdot \dfrac{4y^2}{x^4}$

3. $\dfrac{5m^2n^3}{18mn^2} \div \dfrac{15n^3}{36m^2n}$

4. $\dfrac{x+3}{6} \cdot \dfrac{1}{3x+9}$

5. $\dfrac{7b}{24a^2b^2} \cdot \dfrac{32a^2b^3}{7-21a}$

6. $\dfrac{34a^3b^3}{7ab^2} \cdot \dfrac{21a^3b^2}{17a^2b}$

7. $\dfrac{-7f^2}{40f+16f^2} \div \dfrac{28f^3}{30f^3+12f^4}$

8. $\dfrac{12e^2-18e}{25e^4} \cdot \dfrac{15e^2}{18e^2-27e}$

9. $\dfrac{6x^2y-2xy^2}{9xy+18y^2} \div \dfrac{4x^2y^2-12x^3y}{3x^2+6xy}$

10. $\dfrac{9x^2-30x+25}{9x^2-25} \div \dfrac{3x^3+3x^2}{3x^2-2x-5}$

11. $\dfrac{6v^2-36v}{25-5v} \cdot \dfrac{v^2-11v+30}{-12v^2}$

12. $\dfrac{x^3+27}{x^2-9} \div \dfrac{x^2-3x+9}{x^2+9}$

13. $\dfrac{5h^2k-h^3}{9hk+18k^2} \div \dfrac{h^2-7hk+10k^2}{3h^2-12k^2}$

14. $\dfrac{x^2-2x-3}{x^4-10x^2+9} \cdot \dfrac{x-1}{x+3}$

15. $\dfrac{32+4b^3}{b^2-b-6} \div \dfrac{b^2-2b+4}{15b-5b^2}$

16. $\dfrac{2a^2-7a+6}{a^2+a-6} \cdot \dfrac{2a^2+13a+15}{15a-7a^2-2a^3}$

17. $\dfrac{5y^2+15yz+45z^2}{9z^2+3zy} \div \dfrac{2y^3-54z^3}{9z^2-y^2}$

18. $\dfrac{27x^3+8}{9x^2-4} \div (9x^2-6x+4)$

19. $\dfrac{6uv-6v^2+6v}{10uv} \cdot \dfrac{(u+v)-(u+v)^2}{(u-v)^2+(u-v)} \div \dfrac{3v^2+3vu}{5uv-5u^2}$

20. $\dfrac{x^2+6x+9-y^2}{5x-5y+15} \cdot \dfrac{15x-15y}{9x^2-9y^2} \cdot \dfrac{1}{6+2x+2y}$

21. $\dfrac{10nm-15n^2}{3n+2m} \div \dfrac{9mn-6m^2}{4m^2+12mn+9n^2} \div \dfrac{4m^2-9n^2}{27n-18m}$

22. $\dfrac{xs-ys+xt-yt}{sx+tx+sy+ty} \cdot \dfrac{sx+sy-ty-xt}{xs-ys-tx+yt}$

6.4 Least Common Denominator (LCD)

We ordinarily use the **least common denominator (LCD)** when we add or subtract unlike rational expressions (unlike rational expressions are defined in Section 6.5). The LCD is the least common multiple (LCM) of all the denominators (see Section 1.9 for review); that is, it is the *smallest* expression that all the denominators divide into exactly.

TO FIND THE LCD

1. Factor each denominator completely. Repeated factors should be expressed as powers.
2. Write down each different base that appears in any denominator.
3. Raise each base to the highest power to which it occurs in *any* denominator.
4. The LCD is the product of all the expressions found in step 3.

NOTE Reducing all the rational expressions to lowest terms before proceeding with step 1 is often helpful. ☑

Example 1 Find the LCD for $\frac{2}{x} + \frac{x}{x+2}$.

Step 1. The denominators are already factored.

Step 2. $x, (x + 2)$ All the different bases

Step 3. $x^1, (x + 2)^1$ The highest powers of the bases

Step 4. $\text{LCD} = x(x + 2)$ ■

A WORD OF CAUTION In Example 1, you might think that the LCD is $(x + 2)$ because the other denominator, x, is a *part* of $(x + 2)$. However, x is only a *term* of $x + 2$, *not* a factor of it, and so we need *both* x and $x + 2$ as *factors* of the LCD. An arithmetic example will confirm this. Suppose that in Example 1, $x = 3$. Then $\frac{2}{x} + \frac{x}{x+2} = \frac{2}{3} + \frac{3}{5}$, and the LCD for this problem is clearly $3 \cdot 5$, or 15, *not* 5. ☑

Example 2 Find the LCD for $\frac{2x-3}{x^2+10x+25} - \frac{5}{4x^2+20x} + \frac{4x-3}{x^2+2x-15}$.

Step 1.
$$\left.\begin{aligned} x^2 + 10x + 25 &= (x+5)^2 \\ 4x^2 + 20x &= 2^2 \cdot x \cdot (x+5) \\ x^2 + 2x - 15 &= (x+5)(x-3) \end{aligned}\right\}$$ The denominators in factored form

Step 2. $2, \; x, \; (x + 5), (x - 3)$ All the different bases

Step 3. $2^2, \; x^1, (x + 5)^2, (x - 3)^1$ The highest powers of the bases

Step 4. $\text{LCD} = 4x(x + 5)^2(x - 3)$ ■

Example 3 Find the LCD for $\frac{x+2}{x^3+8} + \frac{5x}{x^2-2x+4} - \frac{3x^2}{x^3-2x^2}$.

Solution It happens that two of these rational expressions are reducible:

$$\frac{x+2}{x^3+8} = \frac{\overset{1}{\cancel{x+2}}}{\underset{1}{\cancel{(x+2)}}(x^2-2x+4)} = \frac{1}{x^2-2x+4}$$

and

$$\frac{3x^2}{x^3-2x^2} = \frac{3\overset{1}{\cancel{x^2}}}{\underset{1}{\cancel{x^2}}(x-2)} = \frac{3}{x-2}$$

Now the problem is reduced to $\dfrac{1}{x^2 - 2x + 4} + \dfrac{5x}{x^2 - 2x + 4} - \dfrac{3}{x - 2}$, and so the denominators are $x^2 - 2x + 4$, $x^2 - 2x + 4$, and $x - 2$.

$$\text{LCD} = (x - 2)^1(x^2 - 2x + 4)^1 = (x - 2)(x^2 - 2x + 4) \quad \blacksquare$$

EXERCISES 6.4

Set I Find the LCD in each exercise. *Do not add the rational expressions.*

1. $\dfrac{9}{25a^3} + \dfrac{7}{15a}$
2. $\dfrac{13}{18b^2} + \dfrac{11}{12b^4}$
3. $\dfrac{49}{60hk^3} + \dfrac{71}{90h^2k^4}$
4. $\dfrac{44}{42x^2y^2} - \dfrac{45}{49x^3y}$
5. $\dfrac{11}{2w - 10} - \dfrac{15}{4w}$
6. $\dfrac{27}{2m^2} + \dfrac{19}{8m - 48}$
7. $\dfrac{15b}{9b^2 - c^2} + \dfrac{12c}{(3b - c)^2}$
8. $\dfrac{14e}{(5f - 2e)^2} + \dfrac{27f}{4e^2 - 25f^2}$
9. $\dfrac{5}{2g^3} - \dfrac{3g - 9}{g^2 - 6g + 9} + \dfrac{12g}{4g^2 - 12g}$
10. $\dfrac{5y - 30}{y^2 - 12y + 36} + \dfrac{7}{9y^2} - \dfrac{15y}{3y^2 - 18y}$
11. $\dfrac{2x - 5}{2x^2 - 16x + 32} + \dfrac{4x + 7}{x^2 + x - 20}$
12. $\dfrac{8k - 1}{5k^2 - 30k + 45} + \dfrac{3k - 4}{k^2 + 4k - 21}$
13. $\dfrac{35}{3e^2} - \dfrac{2e}{e^2 - 9} - \dfrac{13}{4e - 12}$
14. $\dfrac{3}{8u^3} - \dfrac{5u - 1}{6u^2 + 18u} - \dfrac{6u + 7}{u^2 - 5u - 24}$
15. $\dfrac{x^2 + 1}{12x^3 + 24x^2} - \dfrac{4x + 3}{x^2 - 4x + 4} - \dfrac{1}{x^2 - 4}$
16. $\dfrac{2y + 5}{y^2 + 6y + 9} + \dfrac{7y}{y^2 - 9} - \dfrac{11}{8y^2 - 24y}$

Set II Find the LCD in each exercise. *Do not add the rational expressions.*

1. $\dfrac{17}{50c} + \dfrac{23}{40c^2}$
2. $\dfrac{83}{12x^2} + \dfrac{5}{18x^4}$
3. $\dfrac{35}{24m^2n} + \dfrac{25}{63mn^3}$
4. $\dfrac{5}{28b^3c^2} - \dfrac{8}{49bc^5}$
5. $\dfrac{50}{9z - 63} - \dfrac{16}{3z}$
6. $\dfrac{18}{5 - 25x} + \dfrac{3}{10 - 50x}$
7. $\dfrac{6uv}{9v^2 - 16u^2} + \dfrac{12u^2v}{(4u - 3v)^2}$
8. $\dfrac{13x}{x^2 - 4} - \dfrac{15x^2}{x^2 + 4}$
9. $\dfrac{3}{25h} - \dfrac{20h^3}{5h^4 - 25h^3} + \dfrac{4h - 20}{h^2 - 10h + 25}$
10. $\dfrac{9}{4x^2 - 9} - \dfrac{7x}{6x + 9} + \dfrac{5}{2x^3 - x^2 - 3x}$
11. $\dfrac{5a - 3}{3a^2 - 30a + 72} + \dfrac{7a + 2}{a^2 - 12a + 36}$

12. $\dfrac{3x-7}{15x^2-55x-20}+\dfrac{5x-20}{6x^2+20x+6}$

13. $\dfrac{15}{4m^2}-\dfrac{11m}{2m^2-50}+\dfrac{2m-5}{m^2-2m-35}$

14. $\dfrac{5}{6x^2+30x}-\dfrac{7x}{2x^2-50}+\dfrac{8x+40}{3x^2-9x-30}$

15. $\dfrac{17}{2z^2-98}-\dfrac{9z+11}{4z^3-28z^2}-\dfrac{13z}{z^2-14z+49}$

16. $\dfrac{x-5}{2x^3+7x^2-15x}-\dfrac{4x^2+9}{4x^3-12x^2+9x}+\dfrac{9}{4x^4-9x^2}$

6.5 Adding and Subtracting Rational Expressions

Like Rational Expressions **Like rational expressions** are rational expressions that have the same denominator.

Example 1 Examples of like rational expressions:

a. $\dfrac{2}{x}, \dfrac{8}{x}, \dfrac{a}{x}, \dfrac{x+1}{x}$

b. $\dfrac{2}{a-b}, \dfrac{5a}{a-b}, \dfrac{3ab}{a-b}, \dfrac{a+b}{a-b}$ ■

Adding and Subtracting Like Rational Expressions

TO ADD OR SUBTRACT LIKE RATIONAL EXPRESSIONS

1. Write the sum or difference of the numerators over the denominator of the like rational expressions.

$$\frac{P}{Q}+\frac{R}{Q}=\frac{P+R}{Q} \quad \text{and} \quad \frac{P}{Q}-\frac{R}{Q}=\frac{P-R}{Q}, \qquad Q \neq 0$$

2. Reduce the resulting rational expression to lowest terms.

Example 2 Examples of adding and subtracting like rational expressions:

a. $\dfrac{3}{4a}-\dfrac{5}{4a}=\dfrac{3-5}{4a}=\dfrac{-2}{4a}=-\dfrac{\overset{1}{\cancel{2}}}{\underset{2}{\cancel{4}a}}=-\dfrac{1}{2a}$

b. $\dfrac{4x}{2x-y}-\dfrac{2y}{2x-y}=\dfrac{4x-2y}{2x-y}=\dfrac{2(2x-y)}{(2x-y)}=\dfrac{2\overset{1}{\cancel{(2x-y)}}}{\underset{1}{\cancel{(2x-y)}}}=2$

c. $\dfrac{15}{d-5} + \dfrac{-3d}{d-5} = \dfrac{15-3d}{d-5} = \dfrac{3(5-d)}{d-5}$

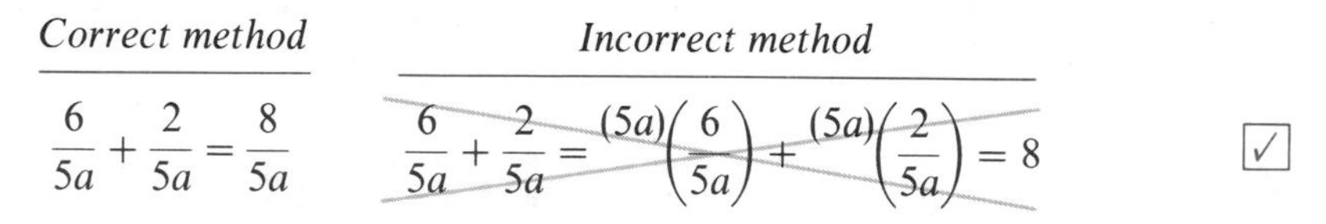

$= -\dfrac{3(d-5)}{(d-5)} = -\dfrac{3\cancel{(d-5)}}{\cancel{(d-5)}} = -3$

Changing the signs of the rational expression and the numerator ■

A WORD OF CAUTION Students often confuse *addition of rational expressions* with *solving equations*, and they multiply both rational expressions by the same number. This is *incorrect*.

Correct method	*Incorrect method*
$\dfrac{6}{5a} + \dfrac{2}{5a} = \dfrac{8}{5a}$	$\dfrac{6}{5a} + \dfrac{2}{5a} = (5a)\left(\dfrac{6}{5a}\right) + (5a)\left(\dfrac{2}{5a}\right) = 8$

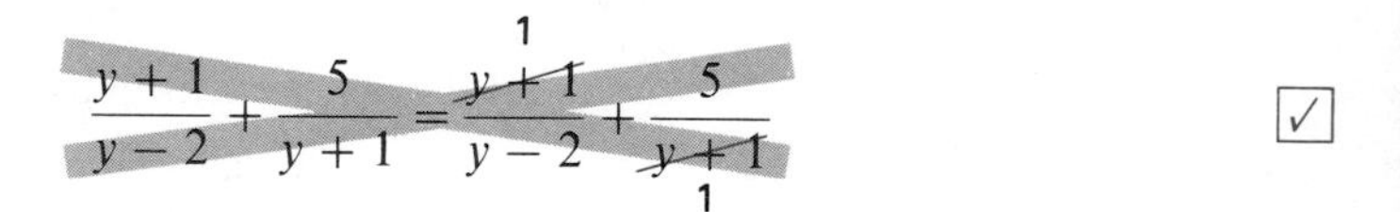

A WORD OF CAUTION It is *incorrect* to cancel a numerator of one rational expression against a denominator of a *different* rational expression when adding or subtracting. This is *incorrect*.

$$\frac{y+1}{y-2} + \frac{5}{y+1} = \frac{\cancel{y+1}}{y-2} + \frac{5}{\cancel{y+1}}$$

In a subtraction problem, if the numerator of the rational expression being subtracted has more than one term, you *must* put parentheses around that numerator when you rewrite the problem as a single rational expression (see Example 3).

Example 3 Subtract $\dfrac{3}{x+5} - \dfrac{x+2}{x+5}$.

Solution

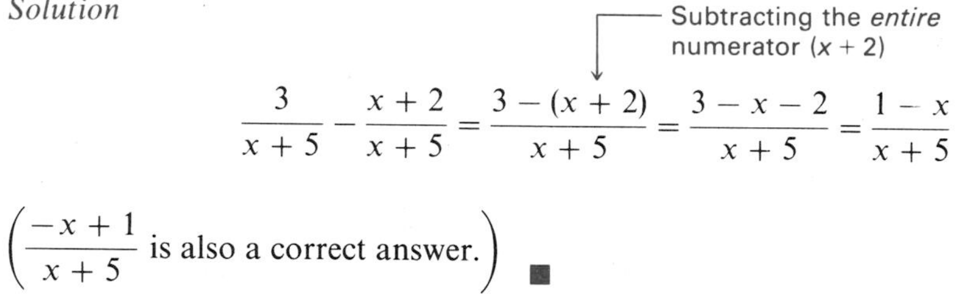

Subtracting the *entire* numerator $(x+2)$

$$\frac{3}{x+5} - \frac{x+2}{x+5} = \frac{3-(x+2)}{x+5} = \frac{3-x-2}{x+5} = \frac{1-x}{x+5}$$

$\left(\dfrac{-x+1}{x+5} \text{ is also a correct answer.}\right)$ ■

Unlike Rational Expressions **Unlike rational expressions** are rational expressions that have *different* denominators.

Example 4 Examples of unlike rational expressions:

a. $\dfrac{2}{x}, \dfrac{6}{x^2}, \dfrac{x-1}{5x}$

b. $\dfrac{5}{x+2}, \dfrac{2x}{x-1}$ ■

Adding and Subtracting Unlike Rational Expressions

Because the definition of addition of rational expressions is $\frac{P}{Q}+\frac{R}{Q}=\frac{P+R}{Q}$, we cannot add and subtract *unlike* rational expressions until they have been converted to *like* ones.

When two denominators are the negatives of each other, we can convert the rational expressions to *like* ones by changing two of the three signs of *one* of the expressions (see Example 5).

A WORD OF CAUTION If a numerator or denominator has more than one term in it, you *must* put parentheses around that numerator or denominator when you change its sign.

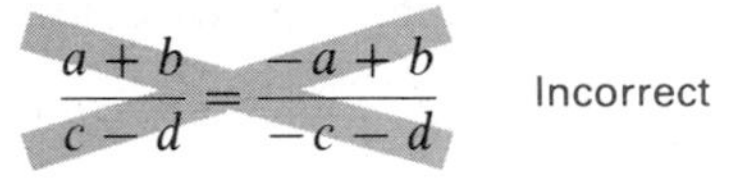

Incorrect

Rather,

$$\frac{a+b}{c-d}=\frac{-(a+b)}{-(c-d)}=\frac{-a-b}{d-c}$$ ☑

Example 5 Add: $\frac{9}{x-2}+\frac{5+x}{2-x}$

Solution

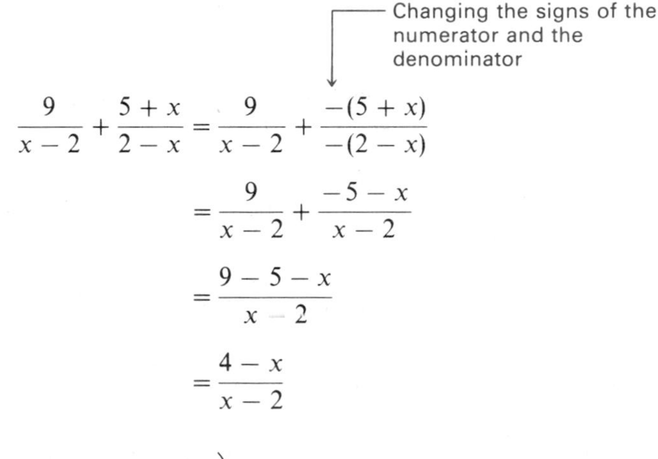

Changing the signs of the numerator and the denominator

$$\frac{9}{x-2}+\frac{5+x}{2-x}=\frac{9}{x-2}+\frac{-(5+x)}{-(2-x)}$$
$$=\frac{9}{x-2}+\frac{-5-x}{x-2}$$
$$=\frac{9-5-x}{x-2}$$
$$=\frac{4-x}{x-2}$$

$\left(\frac{x-4}{2-x}\text{ is also a correct answer.}\right)$

This problem could also be done by substituting $-1(x-2)$ for $2-x$:

$$\frac{9}{x-2}+\frac{5+x}{2-x}=\frac{9}{x-2}+\frac{5+x}{-1(x-2)}$$

The expressions do not yet have like denominators. We can then change the sign of the second rational expression and the sign of its denominator. The problem now becomes

$$\frac{9}{x-2}-\frac{5+x}{x-2}$$

The remainder of the solution is left to the student.

We could instead change the signs of the numerator and denominator of the second rational expression:

$$\frac{9}{x-2}+\frac{-(5+x)}{x-2}$$

Again, the remainder of the solution is left to the student. ■

When denominators are *not* the negatives of each other, we have two ways of converting unlike rational expressions to like ones (this procedure is often called "building fractions"). We can build fractions by using the identity property of multiplication (that is, by multiplying by a rational expression whose numerator and denominator are equal; this produces a rational equivalent to the original one, since $\frac{x}{x}=1$, $\frac{x+2}{x+2}=1$, and so forth), or we can build fractions by using the fundamental property of rational expressions (Section 6.1). Both methods are shown below.

Using the identity property of multiplication	*Using the fundamental property of rational expressions*
$\frac{x}{x+2}=\frac{x}{x+2}\cdot\frac{x}{x}=\frac{x^2}{x(x+2)}$	$\frac{x}{x+2}=\frac{x\cdot x}{(x+2)\cdot x}=\frac{x^2}{x(x+2)}$
$\frac{2}{x}=\frac{2}{x}\cdot\frac{x+2}{x+2}=\frac{2x+4}{x(x+2)}$	$\frac{2}{x}=\frac{2\cdot(x+2)}{x\cdot(x+2)}=\frac{2x+4}{x(x+2)}$

The procedure for adding or subtracting unlike rational expressions can be summarized as follows:

TO ADD OR SUBTRACT UNLIKE RATIONAL EXPRESSIONS

1. Find the LCD.
2. Convert each rational expression to an equivalent expression that has the LCD as its denominator by using either the identity property of multiplication or the fundamental property of rational expressions. (Dividing the LCD by the denominator will give you the factor to multiply that numerator and denominator by.)
3. Add or subtract the resulting *like* rational expressions according to the method given earlier in this section.
4. Reduce the sum or difference to lowest terms.

NOTE Reducing the given rational expressions to lowest terms before proceeding with step 1 is often helpful. ☑

A WORD OF CAUTION Students frequently *reduce* rational expressions just after they have converted them to equivalent expressions with the LCD as the denominator. The addition then cannot be done, because the rational expressions no longer have the same denominator. For example, in arithmetic we sometimes see the following:

$$\frac{1}{2}+\frac{1}{3}=\frac{1\cdot 3}{2\cdot 3}+\frac{1\cdot 2}{3\cdot 2}=\frac{\overset{1}{\cancel{3}}}{\underset{2}{\cancel{6}}}+\frac{\overset{1}{\cancel{2}}}{\underset{3}{\cancel{6}}}=\frac{1}{2}+\frac{1}{3}$$

← We can't do this addition because the denominators are different

Example 6 Add: $\frac{7}{18x^2y}+\frac{5}{8xy^4}$

Step 1. $18 = 2\cdot 3^2$; $8 = 2^3$; $\text{LCD} = 2^3\cdot 3^2x^2y^4 = 72x^2y^4$

$4y^3 = \frac{72x^2y^4}{18x^2y} = \frac{\text{LCD}}{\text{The denominator of the rational expression}}$

Step 2. $\frac{7}{18x^2y}=\frac{7\,(4y^3)}{18x^2y\,(4y^3)}=\frac{28y^3}{72x^2y^4}$

$\frac{5}{8xy^4}=\frac{5\,(9x)}{8xy^4\,(9x)}=\frac{45x}{72x^2y^4}$

$9x = \frac{72x^2y^4}{8xy^4} = \frac{\text{LCD}}{\text{The denominator of the rational expression}}$

Step 3. $\frac{7}{18x^2y}+\frac{5}{8xy^4}=\frac{28y^3}{72x^2y^4}+\frac{45x}{72x^2y^4}=\frac{28y^3+45x}{72x^2y^4}$

Step 4. You should verify that $\frac{28y^3+45x}{72x^2y^4}$ does not reduce. ■

Example 7 Subtract: $3-\frac{2a}{a+2}$

Step 1. $\text{LCD} = a+2$

Note the parentheses

Step 2. $3=\frac{3}{1}\cdot\frac{a+2}{a+2}=\frac{3(a+2)}{1(a+2)}=\frac{3a+6}{a+2}$

Step 3. $3-\frac{2a}{a+2}=\frac{3a+6}{a+2}-\frac{2a}{a+2}=\frac{3a+6-2a}{a+2}=\frac{a+6}{a+2}$

Step 4. $\frac{a+6}{a+2}$ cannot be reduced. ■

Example 8 Subtract: $\frac{z+1}{z+2}-\frac{z-1}{z-2}$

Step 1. $\text{LCD} = (z+2)(z-2)$

Note the parentheses

Step 2. $\frac{z+1}{z+2}=\frac{z+1}{z+2}\cdot\frac{z-2}{z-2}=\frac{(z+1)(z-2)}{(z+2)(z-2)}=\frac{z^2-z-2}{(z+2)(z-2)}$

$\frac{z-1}{z-2}=\frac{z-1}{z-2}\cdot\frac{z+2}{z+2}=\frac{(z-1)(z+2)}{(z-2)(z+2)}=\frac{z^2+z-2}{(z-2)(z+2)}$

Step 3. $\frac{z+1}{z+2} - \frac{z-1}{z-2} = \frac{z^2 - z - 2}{(z+2)(z-2)} - \frac{z^2 + z - 2}{(z-2)(z+2)}$

$$= \frac{z^2 - z - 2 - (z^2 + z - 2)}{(z+2)(z-2)}$$

$$= \frac{z^2 - z - 2 - z^2 - z + 2}{(z+2)(z-2)}$$

$$= \frac{-2z}{(z+2)(z-2)} = -\frac{2z}{z^2 - 4}$$

Changing the signs of the rational expression and the numerator

Step 4. $-\frac{2z}{z^2 - 4}$ cannot be reduced. ■

Example 9 Add: $\frac{3x + 9}{x^2 + 7x + 10} + \frac{14}{x^2 + 3x - 10}$

Step 1. $x^2 + 7x + 10 = (x+2)(x+5)$

$x^2 + 3x - 10 = (x+5)(x-2)$

$\text{LCD} = (x+2)(x+5)(x-2)$

Step 2. $\frac{3x+9}{x^2 + 7x + 10} = \frac{(3x+9)(x-2)}{(x+2)(x+5)(x-2)}$

You should verify that $\frac{3x + 9}{x^2 + 7x + 10}$ can't be reduced

$$\frac{14}{x^2 + 3x - 10} = \frac{14(x+2)}{(x+5)(x-2)(x+2)}$$

Step 3. $\frac{3x + 9}{x^2 + 7x + 10} + \frac{14}{x^2 + 3x - 10}$

$$= \frac{(3x+9)(x-2)}{(x+2)(x+5)(x-2)} + \frac{14(x+2)}{(x-2)(x+5)(x+2)}$$

$$= \frac{3x^2 + 3x - 18 + 14x + 28}{(x+2)(x+5)(x-2)}$$

$$= \frac{3x^2 + 17x + 10}{(x+2)(x+5)(x-2)}$$

Step 4. $= \frac{(3x+2)\overset{1}{\cancel{(x+5)}}}{(x+2)\underset{1}{\cancel{(x+5)}}(x-2)}$

Factoring the numerator and then dividing both numerator and denominator by $(x + 5)$

$$= \frac{3x + 2}{(x+2)(x-2)} \quad ■$$

When several operations are indicated, we must be sure to follow the correct order of operations (see Example 10).

Example 10 Perform the indicated operations: $\frac{x+3}{x+2}+\frac{1}{x-2}\div\frac{x+3}{x^2-4}$

Solution The division must be done *before* the addition.

$$\frac{x+3}{x+2}+\frac{1}{x-2}\div\frac{x+3}{x^2-4}=\frac{x+3}{x+2}+\frac{1}{x-2}\cdot\frac{x^2-4}{x+3}$$ Changing the division to multiplication

$$=\frac{x+3}{x+2}+\frac{1}{\cancel{x-2}}\cdot\frac{(x+2)\cancel{(x-2)}}{x+3}$$ Factoring x^2-4

$$=\frac{x+3}{x+2}+\frac{x+2}{x+3}$$ Simplifying; now the LCD is $(x+2)(x+3)$

$$=\frac{(x+3)(x+3)}{(x+2)(x+3)}+\frac{(x+2)(x+2)}{(x+3)(x+2)}$$ Using the fundamental property of rational expressions

$$=\frac{x^2+6x+9+x^2+4x+4}{(x+3)(x+2)}$$ Adding the rational expressions

$$=\frac{2x^2+10x+13}{(x+3)(x+2)}$$ ■

EXERCISES 6.5

Set I Perform the indicated operations.

1. $\frac{5a}{a+2}+\frac{10}{a+2}$

2. $\frac{6b}{b-3}-\frac{18}{b-3}$

3. $\frac{8m}{2m-3n}-\frac{12n}{2m-3n}$

4. $\frac{21k}{4h+3k}+\frac{28h}{4h+3k}$

5. $\frac{-15w}{1-5w}-\frac{3}{5w-1}$

6. $\frac{-35}{6w-7}-\frac{30w}{7-6w}$

7. $\frac{7z}{8z-4}+\frac{6-5z}{4-8z}$

8. $\frac{8x}{6x-5}+\frac{10-4x}{5-6x}$

9. $\frac{12x-31}{12x-28}-\frac{18x-39}{28-12x}$

10. $\frac{13-30w}{15-10w}-\frac{10w+17}{10w-15}$

11. $\frac{9}{25a^3}+\frac{7}{15a}$

12. $\frac{13}{18b^2}+\frac{11}{12b^4}$

13. $\frac{49}{60h^2k^2}-\frac{71}{90hk^4}$

14. $\frac{44}{42x^2y^2}-\frac{45}{49x^3y}$

15. $\frac{5}{t}+\frac{2t}{t-4}$

16. $\frac{6r}{r-8}-\frac{11}{r}$

17. $\frac{3k}{8k-4}-\frac{7}{6k}$

18. $\frac{2}{9j}+\frac{4j}{18j+12}$

19. $x^2-\frac{3}{x}+\frac{5}{x-3}$

20. $y^2-\frac{2}{y}+\frac{3}{y-5}$

21. $\frac{2a+3b}{b}+\frac{b}{2a-3b}$

22. $\frac{3x-5y}{y}+\frac{y}{3x+5y}$

23. $\dfrac{2}{a+3} - \dfrac{4}{a-1}$

24. $\dfrac{5}{b-2} - \dfrac{3}{b+4}$

25. $\dfrac{x+2}{x-3} - \dfrac{x+3}{x-2}$

26. $\dfrac{x-4}{x+6} - \dfrac{x-6}{x+4}$

27. $\dfrac{x+2}{x^2+x-2} + \dfrac{3}{x^2-1}$

28. $\dfrac{5}{x^2-4} + \dfrac{x+1}{x^2-x-2}$

29. $\dfrac{2x}{x-3} - \dfrac{2x}{x+3} + \dfrac{36}{x^2-9}$

30. $\dfrac{m}{m+6} - \dfrac{m}{m-6} - \dfrac{72}{m^2-36}$

31. $\dfrac{x-2}{x^2+4x+4} - \dfrac{x+1}{x^2-4}$

32. $\dfrac{x-2}{x^2-1} - \dfrac{x+1}{x^2-2x+1}$

33. $\dfrac{4}{x^2+2x+4} + \dfrac{x-2}{x+2}$

34. $\dfrac{x+3}{x-9} + \dfrac{3}{x^2-3x+9}$

35. $\dfrac{5}{2g^3} - \dfrac{3g-9}{g^2-6g+9} + \dfrac{12g}{4g^2-12g}$

36. $\dfrac{5y-30}{y^2-12y+36} + \dfrac{7}{9y^2} - \dfrac{15y}{3y^2-18y}$

37. $\dfrac{2x-5}{2x^2-16x+32} + \dfrac{4x+7}{x^2+x-20}$

38. $\dfrac{8k-1}{5k^2-30k+45} + \dfrac{3k-4}{k^2+4k-21}$

39. $\dfrac{35}{3e^2} - \dfrac{2e}{e^2-9} - \dfrac{3}{4e-12}$

40. $\dfrac{3}{8u^3} - \dfrac{5u-1}{6u^2+18u} - \dfrac{6u+7}{u^2-5u-24}$

41. $\dfrac{x^2+1}{12x^3+24x^2} - \dfrac{4x+3}{x^2-4x+4} - \dfrac{1}{x^2-4}$

42. $\dfrac{2y+5}{y^2+6y+9} + \dfrac{7y}{y^2-9} - \dfrac{11}{8y^2-24y}$

43. $\dfrac{7}{3y^3-12y^2+48y} + \dfrac{y^2+4}{y^3+64} - \dfrac{y}{y^2+8y+16}$

44. $\dfrac{x^2+9}{x^3+27} + \dfrac{5}{2x^3-6x^2+18x} - \dfrac{x}{x^2+6x+9}$

45. $\dfrac{x-1}{x^3+x^2-9x-9} - \dfrac{x+3}{x^3-3x^2-x+3}$

46. $\dfrac{x+2}{x^3-2x^2-x+2} - \dfrac{x-1}{x^3+x^2-4x-4}$

47. $\dfrac{x+6}{x-5} + \dfrac{1}{x+4} \div \dfrac{x+6}{x^2-x-20}$

48. $\dfrac{2x-3}{x+4} + \dfrac{1}{x-1} \div \dfrac{2x-3}{x^2+3x-4}$

Set II Perform the indicated operations.

1. $\dfrac{24}{8+c} + \dfrac{3c}{8+c}$

2. $\dfrac{35}{2-x} + \dfrac{15}{2-x}$

3. $\dfrac{30c}{4b-5c} - \dfrac{24b}{4b-5c}$

4. $\dfrac{16y}{4y+5} + \dfrac{20}{4y+5}$

5. $\dfrac{-35t}{3-7t} - \dfrac{15}{7t-3}$

6. $\dfrac{x+3}{3x-5} - \dfrac{2x+4}{5-3x}$

7. $\dfrac{5u}{6u-15} + \dfrac{30-7u}{15-6u}$

8. $\dfrac{5+a}{2a-5} + \dfrac{3a-2}{5-2a}$

9. $\dfrac{12x-5}{18x-6} - \dfrac{15x-4}{6-18x}$

10. $\frac{16 - 3x}{2x - 7} + \frac{3x - 5}{7 - 2x}$

11. $\frac{17}{50c} + \frac{23}{40c^2}$

12. $\frac{18}{25x^2} + \frac{7}{30x^3}$

13. $\frac{35}{27m^2n} - \frac{25}{63mn^3}$

14. $\frac{13}{15x^3y} - \frac{41}{40xy^3}$

15. $\frac{6}{e} + \frac{3e}{5 - e}$

16. $\frac{8}{x + 5} + x$

17. $\frac{9}{12d} - \frac{4}{18 - 27d}$

18. $\frac{6}{3x + 6} - \frac{5}{9x + 18}$

19. $\frac{5}{n} - n - \frac{8}{n - 6}$

20. $\frac{3 - x}{x^2} - \frac{x}{4x + x^2} - \frac{1}{x + 4}$

21. $\frac{k}{h - k} + \frac{k + h}{k}$

22. $\frac{c - d}{c} - \frac{d}{d - c}$

23. $\frac{7}{x - 5} - \frac{13}{x + 4}$

24. $\frac{8}{4 + a} + \frac{8}{2 + a}$

25. $\frac{z - 3}{z - 6} - \frac{z + 6}{z + 3}$

26. $\frac{x + 5}{x + 1} - \frac{x - 1}{x - 5}$

27. $\frac{16 - 4a}{a^2 + a - 12} + \frac{3}{a^2 - 16}$

28. $\frac{2x + 4}{x^2 + 2x - 8} + \frac{3}{5x^2 - 7x - 6}$

29. $\frac{3c}{c + 5} - \frac{3c}{c - 5} + \frac{150}{c^2 - 25}$

30. $\frac{5x}{x + 3} - \frac{5x}{x - 3} + \frac{90}{x^2 - 9}$

31. $\frac{m + 1}{m^2 - 9} - \frac{m - 2}{m^2 + 4m - 21}$

32. $x - \frac{x}{2x^2 + x - 3} + \frac{4}{3x^2 - x - 2}$

33. $\frac{u - 4}{u + 8} + \frac{8}{u^2 + 4u + 16}$

34. $\frac{x}{x^3 - 8} - \frac{2}{x^2 - 4} + \frac{5}{x + 2}$

35. $\frac{3}{25h} - \frac{20h^3}{5h^4 - 25h^3} + \frac{4h - 20}{h^2 - 10h + 25}$

36. $\frac{8 - x}{4 - x} - \frac{x}{x - 4} + \frac{4x}{3x^2 - 48}$

37. $\frac{5a - 3}{3a^2 - 30a + 72} + \frac{7a + 2}{a^2 - 12a + 36}$

38. $\frac{x + 3}{6x^2 - 7x - 6} - \frac{3x - 2}{2x^2 + 5x - 3}$

39. $\frac{15}{4m^2} - \frac{11m}{2m^2 - 50} + \frac{2m - 5}{m^2 - 2m - 35}$

40. $\frac{8}{3x^3 + 6x^2 - 9x} - \frac{4}{9x^3 - 9x^2} - \frac{x + 2}{5x^2 + 15x}$

41. $\frac{17}{2z^2 - 98} - \frac{9z + 11}{4z^3 - 28z^2} - \frac{13z}{z^2 - 14z + 49}$

42. $\frac{5}{4x^2 + 12x + 9} - \frac{8}{4x^2 - 9} - \frac{1}{6x^2 + 9x}$

43. $\frac{x^2 + 1}{x^3 + 1} + \frac{3}{5x^3 - 5x^2 + 5x} - \frac{x}{x^2 + 2x + 1}$

44. $\frac{x}{8x + 4} - \frac{3}{8x^3 + 1} + \frac{5}{8x^2 - 4x + 2}$

45. $\frac{x + 5}{x^3 - 5x^2 - x + 5} - \frac{x + 1}{x^3 - x^2 - 25x + 25}$

46. $\frac{x - 4}{x^3 - 5x^2 - 4x + 20} - \frac{x + 5}{x^3 - 2x^2 - 25x + 50}$

47. $\dfrac{x+7}{x-2} + \dfrac{1}{x-3} \div \dfrac{x+7}{x^2-5x+6}$

48. $\dfrac{3x-5}{x-4} - (x+1) \div \dfrac{3x^2-2x-5}{x-4}$

6.6 Simplifying Complex Fractions

If the numerator and/or the denominator of a fraction is itself a rational expression, we call the fraction a **complex fraction**. The following are examples of complex fractions:

$$\frac{\frac{2}{x}}{3}, \quad \frac{a}{\frac{1}{c}}, \quad \frac{\frac{3}{z}}{\frac{5}{z}}, \quad \frac{\frac{3}{x}-\frac{2}{y}}{\frac{5}{x}+\frac{3}{y}}$$

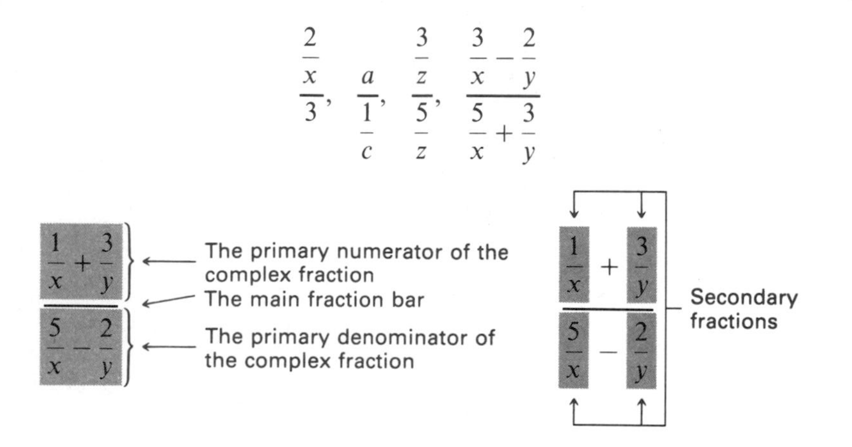

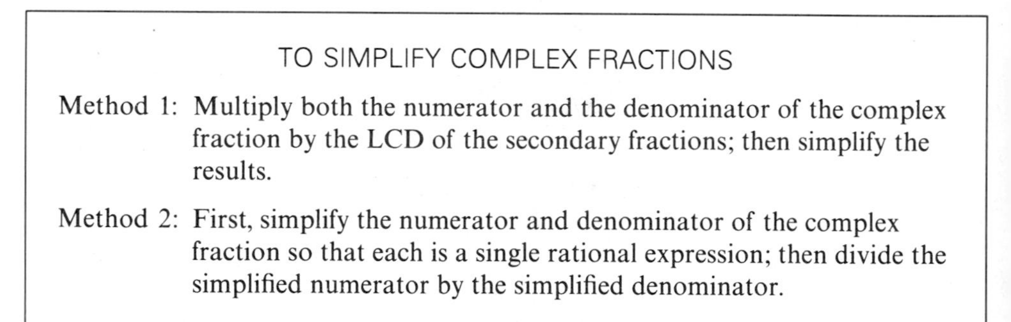

TO SIMPLIFY COMPLEX FRACTIONS

Method 1: Multiply both the numerator and the denominator of the complex fraction by the LCD of the secondary fractions; then simplify the results.

Method 2: First, simplify the numerator and denominator of the complex fraction so that each is a single rational expression; then divide the simplified numerator by the simplified denominator.

Note that in some of the examples below, the solution by method 1 is easier than that by method 2. In others, the opposite is true.

Example 1 Simplify $\dfrac{\frac{2}{x}-\frac{3}{x^2}}{5+\frac{1}{x}}$.

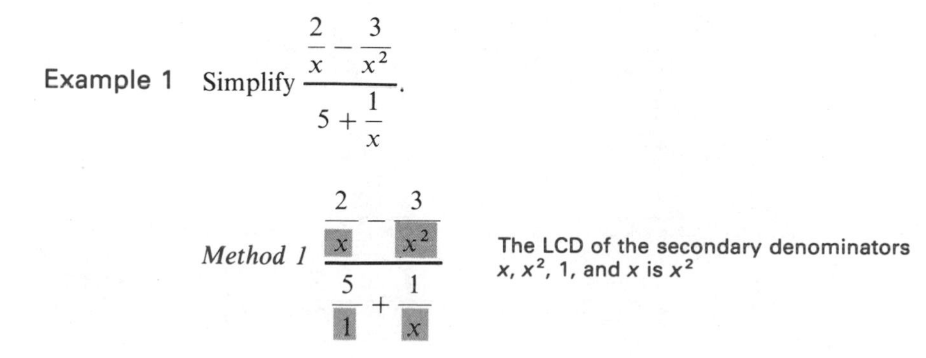

$$\frac{\frac{2}{x}-\frac{3}{x^2}}{\frac{5}{1}+\frac{1}{x}}$$

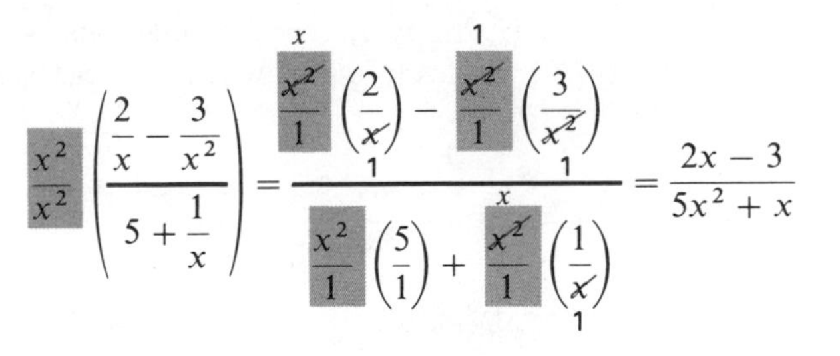

$$\frac{x^2}{x^2}\left(\frac{\frac{2}{x}-\frac{3}{x^2}}{5+\frac{1}{x}}\right)=\frac{\frac{x^2}{1}\left(\frac{2}{x}\right)-\frac{x^2}{1}\left(\frac{3}{x^2}\right)}{\frac{x^2}{1}\left(\frac{5}{1}\right)+\frac{x^2}{1}\left(\frac{1}{x}\right)}=\frac{2x-3}{5x^2+x}$$

This method clears all secondary denominators, "collapsing" the complex fraction into an ordinary rational expression.

Method 2 $$\frac{\frac{2}{x}-\frac{3}{x^2}}{\frac{5}{1}+\frac{1}{x}}=\left(\frac{2}{x}-\frac{3}{x^2}\right)\div\left(\frac{5}{1}+\frac{1}{x}\right)$$

$$=\left(\frac{2x}{x^2}-\frac{3}{x^2}\right)\div\left(\frac{5x}{x}+\frac{1}{x}\right)$$

$$=\frac{2x-3}{x^2}\div\frac{5x+1}{x}$$

$$=\frac{2x-3}{x^2}\cdot\frac{x}{5x+1}=\frac{2x-3}{5x^2+x}$$

This method converts the complex fraction into a division problem. ■

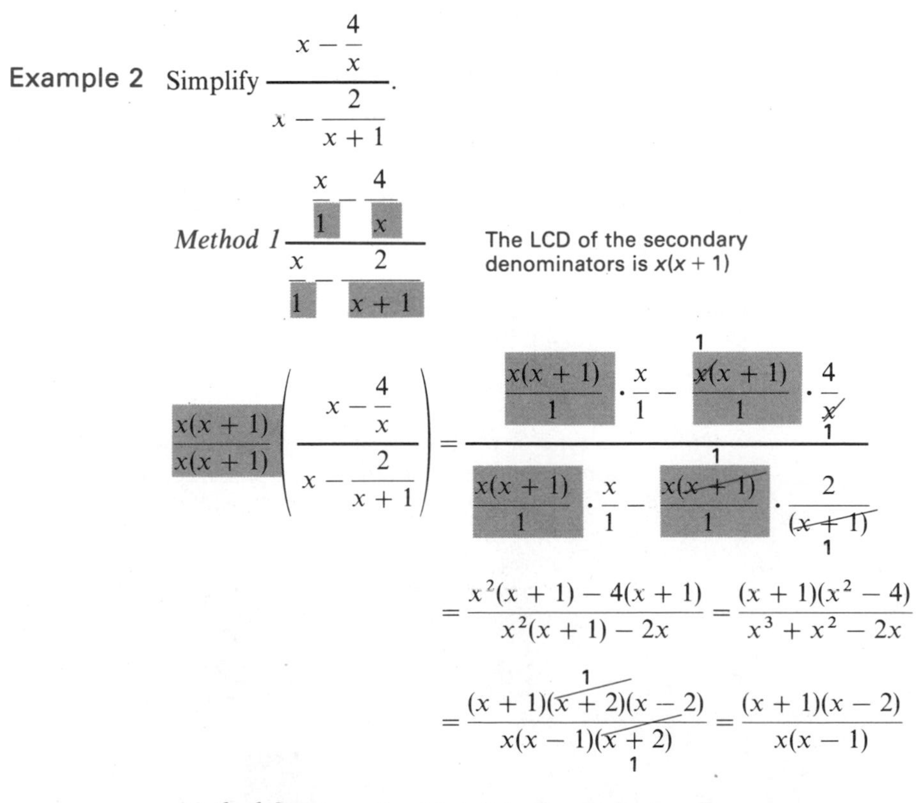

Example 2 Simplify $\dfrac{x-\frac{4}{x}}{x-\frac{2}{x+1}}$.

Method 1 $$\frac{\frac{x}{1}-\frac{4}{x}}{\frac{x}{1}-\frac{2}{x+1}}$$ The LCD of the secondary denominators is $x(x+1)$

$$\frac{x(x+1)}{x(x+1)}\left(\frac{x-\frac{4}{x}}{x-\frac{2}{x+1}}\right)=\frac{\frac{x(x+1)}{1}\cdot\frac{x}{1}-\frac{x(x+1)}{1}\cdot\frac{4}{x}}{\frac{x(x+1)}{1}\cdot\frac{x}{1}-\frac{x(x+1)}{1}\cdot\frac{2}{(x+1)}}$$

$$=\frac{x^2(x+1)-4(x+1)}{x^2(x+1)-2x}=\frac{(x+1)(x^2-4)}{x^3+x^2-2x}$$

$$=\frac{(x+1)(x+2)(x-2)}{x(x-1)(x+2)}=\frac{(x+1)(x-2)}{x(x-1)}$$

Method 2 This method is left to the student. ■

Many rational expressions with negative exponents on the variables become complex fractions when the negative exponents are removed (see Example 3).

Example 3 Simplify $\dfrac{16x^{-2} - y^{-2}}{4x^{-1} - y^{-1}}$.

Solution

Rewriting with positive exponents

$$\frac{16x^{-2} - y^{-2}}{4x^{-1} - y^{-1}} = \frac{\dfrac{16}{x^2} - \dfrac{1}{y^2}}{\dfrac{4}{x} - \dfrac{1}{y}} \qquad \text{The LCD of the secondary denominators is } x^2y^2$$

$$= \frac{x^2y^2}{x^2y^2}\left(\frac{\dfrac{16}{x^2} - \dfrac{1}{y^2}}{\dfrac{4}{x} - \dfrac{1}{y}}\right) \qquad \text{Using method 1}$$

$$= \frac{\dfrac{x^2y^2}{1}\left(\dfrac{16}{x^2}\right) - \dfrac{x^2y^2}{1}\left(\dfrac{1}{y^2}\right)}{\dfrac{x^2y^2}{1}\left(\dfrac{4}{x}\right) - \dfrac{x^2y^2}{1}\left(\dfrac{1}{y}\right)}$$

$$= \frac{16y^2 - x^2}{4xy^2 - x^2y}$$

$$= \frac{(4y - x)(4y + x)}{xy(4y - x)}$$

$$= \frac{4y + x}{xy} \qquad \blacksquare$$

Example 4 Simplify $\dfrac{3}{x + \dfrac{2}{x + \dfrac{1}{5x}}}$.

Solution This type of complex fraction is sometimes called a *continued fraction.* The primary denominator itself contains a complex fraction; we will simplify this complex fraction first:

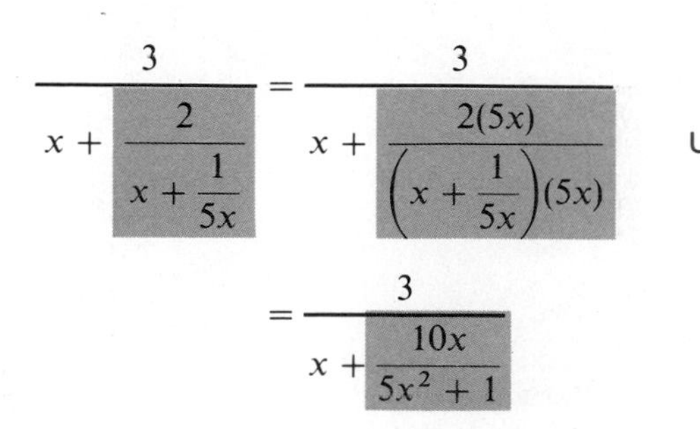

$$\frac{3}{x + \dfrac{2}{x + \dfrac{1}{5x}}} = \frac{3}{x + \dfrac{2(5x)}{\left(x + \dfrac{1}{5x}\right)(5x)}} \qquad \text{Using method 1}$$

$$= \frac{3}{x + \dfrac{10x}{5x^2 + 1}}$$

We now simplify the main complex fraction:

$$\frac{3}{x+\dfrac{10x}{5x^2+1}} = \frac{3(5x^2+1)}{\left(\dfrac{x}{1}+\dfrac{10x}{5x^2+1}\right)(5x^2+1)}$$

$$= \frac{3(5x^2+1)}{x(5x^2+1)+10x}$$

$$= \frac{15x^2+3}{5x^3+11x} \quad ■$$

EXERCISES 6.6

Set I Simplify each complex fraction.

1. $\dfrac{\dfrac{21m^3n}{14mn^2}}{\dfrac{20m^2n^2}{8mn^3}}$

2. $\dfrac{\dfrac{10a^2b}{12a^4b^3}}{\dfrac{5ab^2}{16a^2b^3}}$

3. $\dfrac{\dfrac{15h-6}{18h}}{\dfrac{30h^2-12h}{8h}}$

4. $\dfrac{\dfrac{9k^4}{20k^2-35k^3}}{\dfrac{12k}{16-28k}}$

5. $\dfrac{\dfrac{c}{d}+2}{\dfrac{c^2}{d^2}-4}$

6. $\dfrac{\dfrac{x^2}{y^2}-1}{\dfrac{x}{y}-1}$

7. $\dfrac{a+2-\dfrac{9}{a+2}}{a+1+\dfrac{a-7}{a+2}}$

8. $\dfrac{x-3+\dfrac{x-3}{x+2}}{x+4-\dfrac{4x+23}{x+2}}$

9. $\dfrac{\dfrac{x+y}{y}+\dfrac{y}{x-y}}{\dfrac{y}{x-y}}$

10. $\dfrac{\dfrac{a-b}{a}-\dfrac{a}{a+b}}{\dfrac{b^2}{a+b}}$

11. $\dfrac{\dfrac{x}{x+1}+\dfrac{4}{x}}{\dfrac{x}{x+1}-2}$

12. $\dfrac{\dfrac{4x}{4x+1}+\dfrac{1}{x}}{\dfrac{2}{4x+1}+2}$

13. $\dfrac{\dfrac{x+4}{x}-\dfrac{3}{x-1}}{x+1+\dfrac{2x+1}{x-1}}$

14. $\dfrac{\dfrac{2x-8}{x^2-6x}+\dfrac{x}{x-6}}{x-\dfrac{16}{x}}$

15. $\dfrac{4x^{-2}-y^{-2}}{2x^{-1}+y^{-1}}$

16. $\dfrac{x^{-2}-9y^{-2}}{x^{-1}-3y^{-1}}$

17. $\dfrac{\dfrac{x-2}{x+2}-\dfrac{x+2}{x-2}}{\dfrac{x-2}{x+2}+\dfrac{x+2}{x-2}}$

18. $\dfrac{\dfrac{m+3}{m-3}+\dfrac{m-3}{m+3}}{\dfrac{m+3}{m-3}-\dfrac{m-3}{m+3}}$

19. $\dfrac{\dfrac{2x+y}{x}+\dfrac{3x+y}{y-x}}{\dfrac{x+y}{y}+\dfrac{2(x+y)}{x-y}}$

20. $\dfrac{\dfrac{a+b}{b}+\dfrac{2(a+b)}{a-b}}{\dfrac{2a-b}{a}-\dfrac{5b-a}{b-a}}$

21. $\dfrac{1}{x+\dfrac{1}{x+\dfrac{1}{x+x}}}$

22. $\dfrac{2}{y + \dfrac{2}{y + \dfrac{2}{y + y}}}$

23. $\dfrac{x + \dfrac{1}{2 + \dfrac{x}{3}}}{x - \dfrac{3}{4 + \dfrac{x}{2}}}$

24. $\dfrac{x + \dfrac{5}{1 + \dfrac{x}{2}}}{x - \dfrac{4}{2 + \dfrac{x}{3}}}$

Set II Simplify each complex fraction.

1. $\dfrac{\dfrac{15e^3f^3}{36e^2f^4}}{\dfrac{5e^2f}{33e^3f^2}}$

2. $\dfrac{\dfrac{8x^2}{y^3}}{\dfrac{4x^3}{5y^4}}$

3. $\dfrac{\dfrac{30h^2 - 40hk}{2hk^2}}{\dfrac{54hk^2 - 72k^3}{6k^3}}$

4. $\dfrac{\dfrac{8x - 16}{5x^2 - 5x - 60}}{\dfrac{4x - 8}{5x + 15}}$

5. $\dfrac{\dfrac{9a^2}{5b^2} - 5}{\dfrac{3a}{b} + 5}$

6. $\dfrac{\dfrac{x^2}{y^2} - 4}{\dfrac{x + 2y}{y^3}}$

7. $\dfrac{z + 5 + \dfrac{z + 5}{z - 3}}{z - 4 - \dfrac{2}{z - 3}}$

8. $\dfrac{x - 3 - \dfrac{16}{x + 3}}{x - 4 - \dfrac{9}{x + 4}}$

9. $\dfrac{\dfrac{5m}{n - m}}{\dfrac{m + n}{n} + \dfrac{n}{m - n}}$

10. $\dfrac{\dfrac{a + 2b}{a} - \dfrac{3b}{a - 2b}}{a - b - \dfrac{6b^2}{a - 2b}}$

11. $\dfrac{\dfrac{1}{w} + \dfrac{9w}{6w + 1}}{\dfrac{5}{6w + 1} + 5}$

12. $\dfrac{\dfrac{1}{x} - \dfrac{3}{x + 2}}{\dfrac{5}{x + 2} - \dfrac{1}{x^2}}$

13. $\dfrac{2x - 5 - \dfrac{3}{x}}{\dfrac{1}{x} + \dfrac{x}{x - 12}}$

14. $\dfrac{3x + \dfrac{9x}{x - 1}}{x + 8 + \dfrac{18}{x - 1}}$

15. $\dfrac{5c^{-1} + 2d^{-1}}{25c^{-2} - 4d^{-2}}$

16. $\dfrac{25a^{-2} - b^{-2}}{5a^{-1} + b^{-1}}$

17. $\dfrac{\dfrac{2t - 1}{2t + 1} - \dfrac{2t + 1}{2t - 1}}{\dfrac{2t - 1}{2t + 1} + \dfrac{2t + 1}{2t - 1}}$

18. $\dfrac{\dfrac{x + 1}{x - 3} - \dfrac{x - 2}{x + 2}}{\dfrac{x + 3}{x + 2} - \dfrac{x + 3}{x - 3}}$

19. $\dfrac{\dfrac{u - 3v}{u} + \dfrac{3u - 4v}{u - v}}{\dfrac{2u - 7v}{v} + \dfrac{2u - v}{u - v}}$

20. $\dfrac{\dfrac{x}{x^2 - 1} + \dfrac{3x + 3}{x - 1}}{\dfrac{2x - 1}{x - 1} + \dfrac{x}{1 - x}}$

21. $\dfrac{1}{2z + \dfrac{1}{2z + \dfrac{1}{2z}}}$

22. $\dfrac{5}{1 + \dfrac{1}{x + \dfrac{1}{x}}}$

23. $\dfrac{x - \dfrac{1}{2 + \dfrac{x}{3}}}{x + \dfrac{1}{1 + \dfrac{x}{5}}}$

24. $\dfrac{\dfrac{1}{3 + \dfrac{x}{2}} + x}{x - \dfrac{1}{2 + \dfrac{x}{3}}}$

6.7 Review: 6.1–6.6

The Domain of a Rational Expression 6.1

The domain of the variable in a rational expression is the set of all real numbers except those that would make the denominator equal to zero.

The Fundamental Property of Rational Expressions 6.1

If P, Q, and C are polynomials and if $Q \neq 0$ and $C \neq 0$, then

$$\frac{P \cdot C}{Q \cdot C} = \frac{P}{Q}$$

The Rule of Signs for Rational Expressions 6.1

If any two of the three signs of a rational expression are changed, the value of the rational expression is unchanged.

To Reduce a Rational Expression to Lowest Terms 6.2

1. Factor the numerator and denominator completely.
2. Divide both numerator and denominator by any common factor.

To Multiply Rational Expressions 6.3

1. Factor any numerators or denominators that contain more than one term.
2. Divide the numerators and denominators by any factor common to both. (The common factors can be, but do not have to be, in the same rational expression.)
3. The numerator of the answer is the product of the factors remaining in the numerator; the denominator of the answer is the product of the factors remaining in the denominator. A factor of 1 will always remain in both numerator and denominator.

To Divide Rational Expressions 6.3

Multiply the dividend by the multiplicative inverse of the divisor.

$$\underset{\text{Dividend}}{\frac{P}{Q}} \div \underset{\text{Divisor}}{\frac{S}{T}} = \frac{P}{Q} \cdot \frac{T}{S}$$

where P, Q, S, and T are polynomials and $Q \neq 0$, $S \neq 0$, and $T \neq 0$.

To Find the LCD 6.4

1. Factor each denominator completely. Repeated factors should be expressed as powers.
2. Write down each different base that appears in any denominator.
3. Raise each base to the highest power to which it occurs in *any* denominator.
4. The LCD is the product of all the expressions found in step 3.

To Add or Subtract Like Rational Expressions 6.5

1. Write the sum or difference of the numerators over the denominator of the like rational expressions.

$$\frac{P}{Q} + \frac{R}{Q} = \frac{P + R}{Q} \quad \text{and} \quad \frac{P}{Q} - \frac{R}{Q} = \frac{P - R}{Q}, \qquad Q \neq 0$$

2. Reduce the resulting rational expression to lowest terms.

To Add or Subtract Unlike Rational Expressions 6.5

1. Find the LCD.
2. Convert each rational expression to an equivalent expression that has the LCD as its denominator.
3. Add or subtract the resulting *like* rational expressions.
4. Reduce the resulting sum or difference to lowest terms.

To Simplify Complex Fractions 6.6

Method 1: Multiply both the numerator and the denominator of the complex fraction by the LCD of the secondary fractions; then simplify the results.

Method 2: First, simplify the numerator and denominator of the complex fraction; then divide the simplified numerator by the simplified denominator.

Review Exercises 6.7 Set I

In Exercises 1 and 2, find the domain of the variable.

1. $\dfrac{15}{20 - 45a^2}$

2. $\dfrac{7m - 10}{6m^2 - 21m - 90}$

In Exercises 3–6, reduce each rational expression to lowest terms.

3. $\dfrac{4z^3 + 4z^2 - 24z}{2z^2 + 4z - 6}$

4. $\dfrac{6k^3 - 12k^2 - 18k}{3k^2 + 3k - 36}$

5. $\dfrac{a^3 - 27b^3}{a^2 - 3ab + 2a - 6b}$

6. $\dfrac{2x + 1}{8x^3 + 1}$

In Exercises 7–20, perform the indicated operations.

7. $\dfrac{-35mn^2p^2}{14m^3p^3} \cdot \dfrac{13m^4n}{52n^3p}$

8. $\dfrac{10b^2c}{6ab^4} \div \dfrac{15abc^2}{-12ac^3}$

9. $\dfrac{z^2 + 3z + 2}{z^2 + 2z + 1} \div \dfrac{z^2 + 2z - 3}{z^2 - 1}$

10. $\dfrac{a^2 - a - 2}{a^2 - 4a + 4} \cdot \dfrac{a^2 - 4}{a^2 + 3a + 2}$

11. $\dfrac{x^3 + y^3}{3x^2 - 3xy + 3y^2} \div \dfrac{x^2 - y^2}{x^2 + xy - 2y^2} \cdot \dfrac{15x^2y}{5x^2y + 10xy^2}$

12. $\dfrac{a^2 - 4b^2}{a^2 + 2ab + b^2} \cdot \dfrac{a^2 + 2ab + 4b^2}{a + 2b} \div \dfrac{a^3 - 8b^3}{2a^2 + 4ab + 2b^2}$

13. $\dfrac{20y - 7}{6y - 8} + \dfrac{17 + 2y}{8 - 6y}$

14. $\dfrac{23 - 22z}{30z - 24} + \dfrac{38z - 25}{24 - 30z}$

15. $\dfrac{11}{30e^3f} - \dfrac{7}{45e^2f^2}$

16. $\dfrac{3}{28u^2v^2} - \dfrac{7}{40u^4v}$

17. $\dfrac{a + 1}{a^2 - a - 2} - \dfrac{a - 2}{a^2 + a - 6}$

18. $\dfrac{x + 2}{x^2 + x - 2} - \dfrac{x - 3}{x^2 - x - 6}$

19. $\dfrac{15x}{5x^2 + 20x} - \dfrac{7}{3x^2} - \dfrac{3x + 12}{x^2 + 8x + 16}$

20. $\dfrac{4y - 20}{y^2 - 10y + 25} - \dfrac{5}{11y^3} + \dfrac{24y}{30y - 6y^2}$

In Exercises 21–24, simplify each complex fraction.

21. $\dfrac{\dfrac{x}{y+1}+2}{\dfrac{x}{y+1}-2}$

22. $\dfrac{3-\dfrac{a}{b+2}}{2+\dfrac{a}{b+2}}$

23. $\dfrac{8R^{-3}+T^{-3}}{4R^{-2}-T^{-2}}$

24. $\dfrac{m^{-2}-16n^{-2}}{m^{-3}-64n^{-3}}$

Review Exercises 6.7 Set II

NAME

ANSWERS

1.
2.
3.
4.
5.
6.
7.
8.
9.
10.
11.

In Exercises 1 and 2, find the domain of the variable.

1. $\dfrac{11y - 6}{6y^2 + 8y - 64}$

2. $\dfrac{3x + 9}{2x^2 + 9x - 5}$

In Exercises 3–6, reduce each rational expression to lowest terms.

3. $\dfrac{x^2 + 5x}{x^3 - 2x^2 - 35x}$

4. $\dfrac{2x^2 + 3x - 20}{x^3 + 64}$

5. $\dfrac{m^3 + 8n^3}{m^2 - 6n + 2mn - 3m}$

6. $\dfrac{6a^2 + 11ab - 2b^2}{ac + 2bc - ad - 2bd}$

In Exercises 7–20, perform the indicated operations.

7. $\dfrac{55f^5g^3}{-8e^2f^2} \cdot \dfrac{4e^3g}{15ef^2g^2}$

8. $\dfrac{39a^3b^5}{5ac^2} \cdot \dfrac{30c^3d^2}{65a^2b}$

9. $\dfrac{a^2 + 4a - 21}{6a^3 + 42a^2} \div \dfrac{3a^2 - 27}{36a^3}$

10. $\dfrac{2x^2 - 11x - 21}{2x^2 + 11x + 12} \div \dfrac{3x^2 + 13x + 4}{3x^3 - 20x^2 - 7x}$

11. $\dfrac{4h^2 + 8hk + 16k^2}{15hk^2 + 30k^3} \div \dfrac{3h^3 - 24k^3}{18kh^3} \div \dfrac{16h^2 - 8hk}{5h^2 - 20k^2}$

12. $\dfrac{7x^2 + 19x - 6}{2x^2 - 11x + 15} \cdot \dfrac{5x^2 - 16x + 3}{x^3 + 2x^2 - 3x} \cdot \dfrac{2x^2 - 7x + 5}{70x^2 - 34x + 4}$

13. $\dfrac{18 - 27u}{15 - 25u} + \dfrac{23u - 12}{25u - 15}$

14. $\dfrac{4x + 2}{3x - 5} + \dfrac{6x - 1}{5 - 3x}$

15. $\dfrac{7}{20a^3b^2} - \dfrac{13}{30ab^4}$

16. $\dfrac{5}{8x^2y} - \dfrac{3}{14xy^3}$

17. $\dfrac{3w - 15}{w^2 - 3w - 10} - \dfrac{5w + 30}{w^2 + w - 30}$

18. $\dfrac{x + 1}{x^2 - x - 6} - \dfrac{x - 3}{x^2 + 3x + 2}$

19. $\dfrac{12y^2 + 18y}{4y^2 - 9} + \dfrac{18y - 12y^2}{4y^2 - 12y + 9} - \dfrac{9}{8y^2}$

20. $\dfrac{7}{5y^2} - \dfrac{y + 1}{5y^2 - 9y - 2} - \dfrac{2y - 1}{2y^2 - y - 6}$

ANSWERS

12. __________

13. __________

14. __________

15. __________

16. __________

17. __________

18. __________

19. __________

20. __________

21. __________

22. __________

23. __________

24. __________

In Exercises 21–24, simplify each complex fraction.

21. $\dfrac{\dfrac{3}{w} + \dfrac{4w}{4w + 3}}{\dfrac{6}{4w + 3} + 2}$

22. $\dfrac{3 + \dfrac{7}{x - 1}}{\dfrac{5}{x + 2} + 1}$

23. $\dfrac{8y^{-3} - 27x^{-3}}{4y^{-2} - 9x^{-2}}$

24. $\dfrac{64x^{-3} + 1}{16x^{-2} - 1}$

6.8 Solving Rational Equations

A **rational equation** is an equation that contains one or more rational expressions. We solve such equations by first "clearing fractions"—that is, by multiplying both sides of the equation by the least common multiple of *all* the denominators. This procedure, however, can lead to an equation that is *not equivalent to* the original equation. (In Chapter 2, we learned that *equivalent equations* are equations with identical solution sets.) After we have "cleared fractions," we will have an equation that will look like the equations we solved in earlier chapters.

Equivalent equations are obtained when:

1. The same number or expression is added to or subtracted from both sides of the given equation.
2. Both sides of the given equation are multiplied or divided by the same *nonzero* constant.

Nonequivalent equations may be obtained when:

1. Both sides of the given equation are *multiplied* by an expression containing the variable. In this case, we may obtain an *apparent* solution to the equation that is *not in the domain of the variable*; such a solution is an **extraneous root** and must be rejected. (Any value of the variable that when checked gives a false statement, such as $3 = 0$, is also an extraneous root; we will see roots of this kind in Chapter 7.)
2. Both sides of the given equation are *divided* by an expression containing the variable. In this case, roots of the given equation may be lost. For example, the equation $(x - 5)(x + 4) = 0$ has the solution set $\{5, -4\}$. If, however, we divide both sides of that equation by $x - 5$, we obtain the equation $x + 4 = 0$, and *its* solution set is $\{-4\}$. Since the two equations do not have the same solution set, they are *not equivalent*. For this reason, *do not divide both sides of an equation by an expression containing the variable*.

TO SOLVE A RATIONAL EQUATION

1. Find the domain of the variable and find the LCD of all the denominators.

2. Remove denominators by multiplying *both sides of the equation* (that is, by multiplying *every term*) by the LCD.

3. a. Remove all grouping symbols.

b. Collect and combine like terms.

First-degree equations	*Second-degree equations*
4. a. Get all terms with the variable on one side of the equation and all other terms on the other side.	**4.** a. Get *all* nonzero terms on one side of the equation by adding the same expression to both sides. *Only zero must remain on the other side*. Then arrange the terms in descending powers.

Step 4 continues

First-degree equations	*Second-degree equations*
b. Divide both sides of the equation by the coefficient of the variable, or multiply both sides of the equation by the reciprocal of that coefficient.	b. Factor the polynomial.* c. Set each factor equal to zero and solve each equation.

5. Reject any apparent solutions that are not in the domain.

6. To find any possible errors, check any other apparent solutions in the original equation.

Example 1 Solve $\frac{3}{x+1} - \frac{2}{x} = \frac{5}{2x}$.

Step 1. The domain of the variable is the set of all real numbers except -1 and 0. The LCD is $2x(x+1)$.

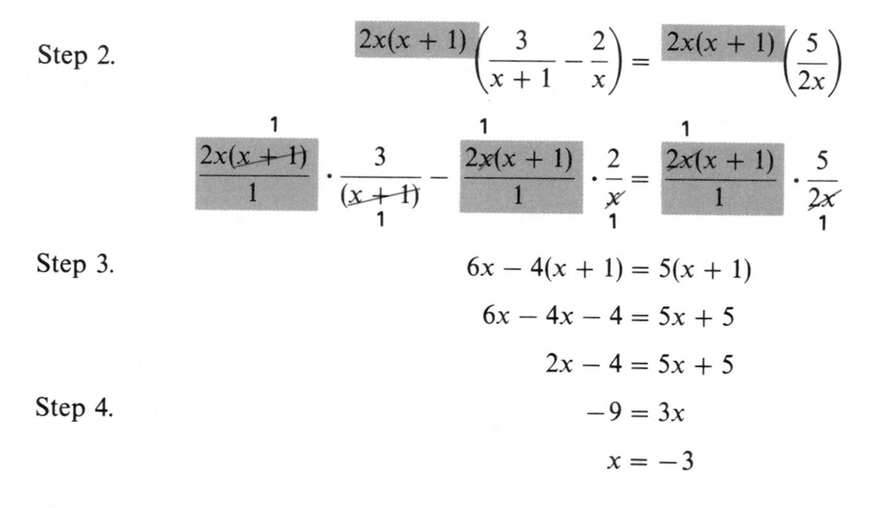

Step 2.

$$2x(x+1)\left(\frac{3}{x+1} - \frac{2}{x}\right) = 2x(x+1)\left(\frac{5}{2x}\right)$$

$$\frac{\overset{1}{2x(\cancel{x+1})}}{1} \cdot \frac{3}{\underset{1}{(\cancel{x+1})}} - \frac{\overset{1}{2\cancel{x}(x+1)}}{1} \cdot \frac{2}{\underset{1}{\cancel{x}}} = \frac{\overset{1}{\cancel{2x}(x+1)}}{1} \cdot \frac{5}{\underset{1}{\cancel{2x}}}$$

Step 3.

$$6x - 4(x+1) = 5(x+1)$$
$$6x - 4x - 4 = 5x + 5$$
$$2x - 4 = 5x + 5$$

Step 4.

$$-9 = 3x$$
$$x = -3$$

The solution -3 *is* in the domain of the variable.

Step 6. *Check*

$$\frac{3}{x+1} - \frac{2}{x} = \frac{5}{2x}$$

$$\frac{3}{-3+1} - \frac{2}{-3} \stackrel{?}{=} \frac{5}{2(-3)}$$

$$\frac{3}{-2} - \frac{2}{-3} \stackrel{?}{=} \frac{5}{-6}$$

$$-\frac{9}{6} + \frac{4}{6} \stackrel{?}{=} -\frac{5}{6}$$

$$-\frac{5}{6} = -\frac{5}{6} \quad \text{A true statement}$$

Therefore, the solution is -3, and the solution set is $\{-3\}$. ■

* If the polynomial can't be factored, the equation cannot be solved at this time.

Example 2 Solve $\dfrac{x}{x-3} = \dfrac{3}{x-3} + 4$.

Step 1. The domain is the set of all real numbers except 3. LCD $= x - 3$.

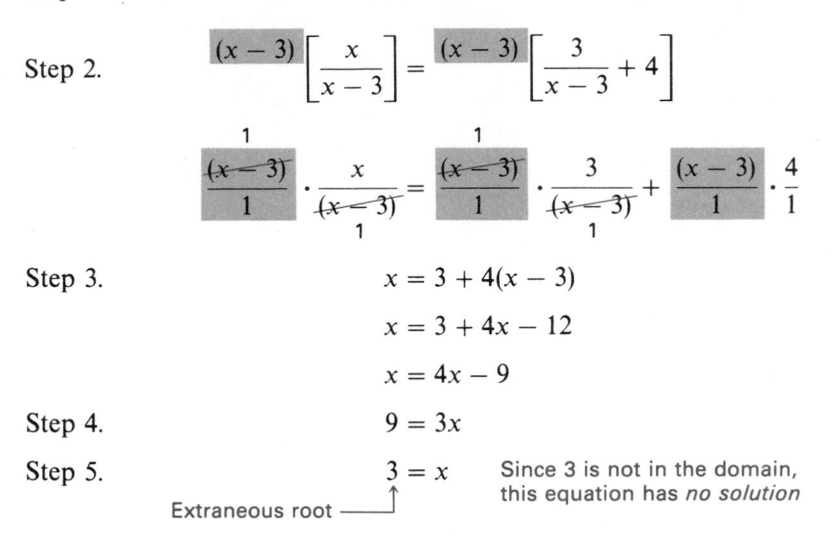

Step 2. $$(x-3)\left[\frac{x}{x-3}\right] = (x-3)\left[\frac{3}{x-3} + 4\right]$$

$$\frac{\overset{1}{\cancel{(x-3)}}}{1} \cdot \frac{x}{\underset{1}{\cancel{(x-3)}}} = \frac{\overset{1}{\cancel{(x-3)}}}{1} \cdot \frac{3}{\underset{1}{\cancel{(x-3)}}} + \frac{(x-3)}{1} \cdot \frac{4}{1}$$

Step 3. $$x = 3 + 4(x-3)$$

$$x = 3 + 4x - 12$$

$$x = 4x - 9$$

Step 4. $$9 = 3x$$

Step 5. $$3 = x$$

Extraneous root ↑ — Since 3 is not in the domain, this equation has *no solution*

If we try to check the value 3 in the original equation, we obtain expressions that are not defined:

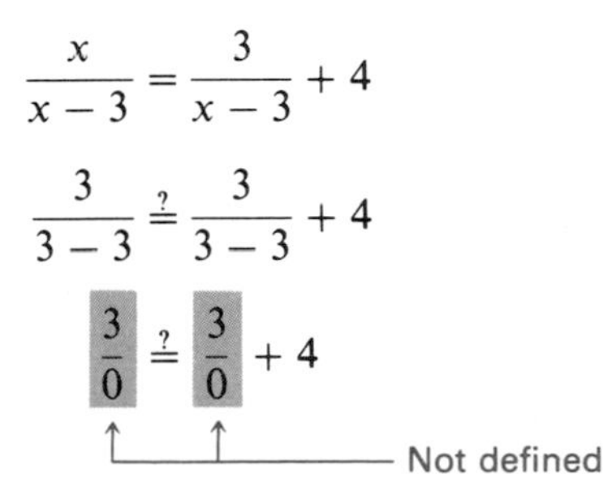

$$\frac{x}{x-3} = \frac{3}{x-3} + 4$$

$$\frac{3}{3-3} \stackrel{?}{=} \frac{3}{3-3} + 4$$

$$\frac{3}{0} \stackrel{?}{=} \frac{3}{0} + 4$$

Not defined

Therefore, 3 is not a root of the given equation. This equation has no roots. The solution set of the equation is { }. ■

Example 3 Solve $\dfrac{x+2}{x-2} = \dfrac{14}{x+1}$.

Step 1. The domain is the set of all real numbers except 2 and -1.
LCD $= (x-2)(x+1)$.

Step 2. $$\overset{1}{\cancel{(x-2)}}(x+1)\left(\frac{x+2}{\underset{1}{\cancel{x-2}}}\right) = (x-2)\overset{1}{\cancel{(x+1)}}\left(\frac{14}{\underset{1}{\cancel{x+1}}}\right)$$

$$(x+2)(x+1) = 14(x-2)$$

Step 3. $$x^2 + 3x + 2 = 14x - 28$$

Because there is a second-degree term, we will move all nonzero terms to the left side

Step 4. $$x^2 - 11x + 30 = 0$$

$$(x-5)(x-6) = 0$$

$$x - 5 = 0 \quad\Big|\quad x - 6 = 0$$

$$x = 5 \quad\Big|\quad x = 6$$

The numbers 5 and 6 *are* in the domain of the variable.

Step 6.

Check for $x = 5$

$$\frac{x+2}{x-2} = \frac{14}{x+1}$$

$$\frac{5+2}{5-2} \stackrel{?}{=} \frac{14}{5+1}$$

$$\frac{7}{3} \stackrel{?}{=} \frac{14}{6}$$

$$\frac{7}{3} = \frac{7}{3}$$

Check for $x = 6$

$$\frac{x+2}{x-2} = \frac{14}{x+1}$$

$$\frac{6+2}{6-2} \stackrel{?}{=} \frac{14}{6+1}$$

$$\frac{8}{4} \stackrel{?}{=} \frac{14}{7}$$

$$2 = 2$$

The solution set is $\{5, 6\}$. ■

Example 4 Solve $\frac{2}{x} + \frac{3}{x^2} = 1$.

Step 1. The domain is the set of all real numbers except zero. LCD $= x^2$.

Step 2.

$$x^2\left[\frac{2}{x} + \frac{3}{x^2}\right] = x^2\,[1]$$

$$\frac{\overset{x}{\cancel{x^2}}}{1}\left(\frac{2}{\underset{1}{\cancel{x}}}\right) + \frac{\overset{1}{\cancel{x^2}}}{1}\left(\frac{3}{\underset{1}{\cancel{x^2}}}\right) = \frac{x^2}{1}\left(\frac{1}{1}\right)$$

Step 3.

$$2x + 3 = x^2 \longleftarrow \text{Second-degree term}$$

Step 4.

$$0 = x^2 - 2x - 3$$

$$0 = (x - 3)(x + 1)$$

$$x - 3 = 0 \quad \bigg| \quad x + 1 = 0$$

$$x = 3 \quad \bigg| \quad x = -1$$

The numbers 3 and -1 *are* in the domain of the variable.

Step 6.

Check for $x = 3$

$$\frac{2}{x} + \frac{3}{x^2} = 1$$

$$\frac{2}{3} + \frac{3}{3^2} \stackrel{?}{=} 1$$

$$\frac{2}{3} + \frac{1}{3} \stackrel{?}{=} 1$$

$$1 = 1$$

Check for $x = -1$

$$\frac{2}{x} + \frac{3}{x^2} = 1$$

$$\frac{2}{-1} + \frac{3}{(-1)^2} \stackrel{?}{=} 1$$

$$-2 + 3 \stackrel{?}{=} 1$$

$$1 = 1$$

The solution set is $\{3, -1\}$. ■

A WORD OF CAUTION A mistake students often make is to confuse an equation such as $\frac{2}{x} + \frac{3}{x^2} = 1$ with an expression such as $\frac{2}{x} + \frac{3}{x^2}$.

The equation

Both *sides* are multiplied by the LCD to remove fractions.

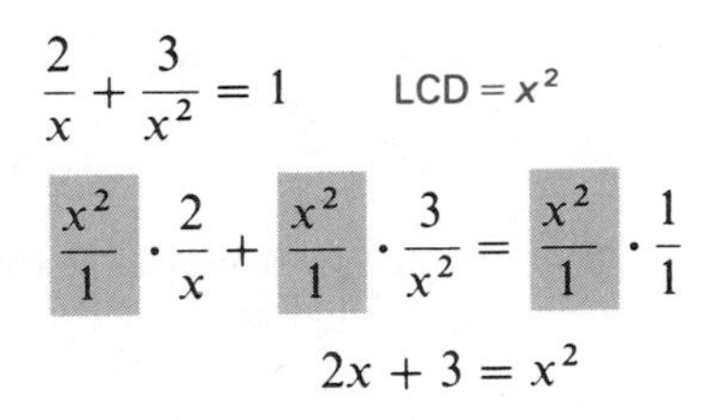

This equation is then solved by factoring (see Example 4). Here, the result is two numbers (-1 and 3) that make both sides of the given equation equal.

Multiplying both *sides* of an *equation* by a nonzero expression maintains equality.

The expression

Each fraction is changed into an equivalent fraction that has the LCD for a denominator.

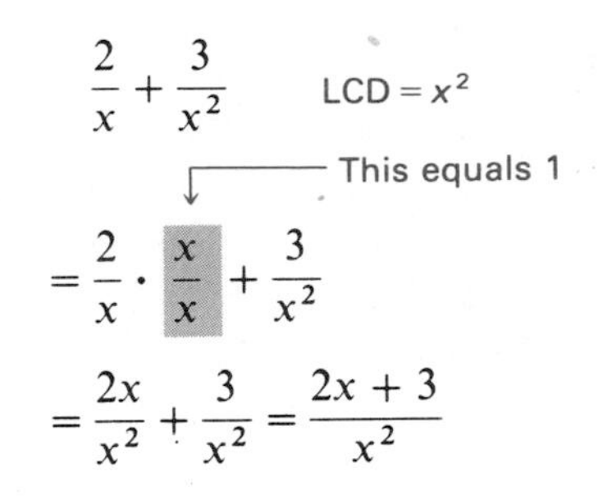

An *expression* can never be "solved." Here, the result is a fraction that represents the sum of the given fractions.

Common error

The usual mistake made is to multiply both terms of *the sum* by the LCD.

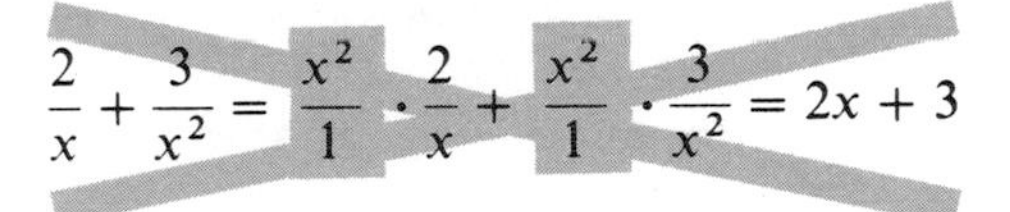

Multiplying the *terms* of an *expression* by an expression that does not equal 1 *changes* the value of the original expression. ☑

Example 5 Solve $\frac{x}{x - 3} - \frac{3x}{x^2 - x - 6} = \frac{4x^2 - 4x - 18}{x^2 - x - 6}$.

Step 1. Since $x^2 - x - 6 = (x + 2)(x - 3)$, the domain is the set of all real numbers except -2, and 3. The LCD is $(x + 2)(x - 3)$.

Step 2.

$$(x + 2)(x - 3)\left[\frac{x}{x - 3} - \frac{3x}{x^2 - x - 6}\right] = \overset{1}{\cancel{(x + 2)(x - 3)}}\left[\frac{4x^2 - 4x - 18}{\underset{1}{\cancel{x^2 - x - 6}}}\right]$$

$$\overset{1}{(x + 2)\cancel{(x - 3)}}\left[\frac{x}{\underset{1}{\cancel{x - 3}}}\right] - \overset{1}{\cancel{(x + 2)}}\overset{1}{\cancel{(x - 3)}}\left[\frac{3x}{\underset{1}{\cancel{(x + 2)}}\underset{1}{\cancel{(x - 3)}}}\right] = 4x^2 - 4x - 18$$

Step 3.

$$x(x + 2) - 3x = 4x^2 - 4x - 18$$
$$x^2 + 2x - 3x = 4x^2 - 4x - 18$$
$$x^2 - x = 4x^2 - 4x - 18$$

Step 4.
$$0 = 3x^2 - 3x - 18$$
$$0 = 3(x^2 - x - 6)$$
$$0 = 3(x + 2)(x - 3)$$
$$x + 2 = 0 \quad | \quad x - 3 = 0$$
$$x = -2 \quad | \quad x = 3$$

Step 5. Neither -2 nor 3 is in the domain of the variable. Therefore, there is no solution. The solution set is { }. ■

Example 6 Solve $\dfrac{5}{x-7} - \dfrac{1}{2x} = \dfrac{9x+7}{2x^2 - 14x}$.

Step 1. Since $2x^2 - 14x = 2x(x - 7)$, the domain is all real numbers except 0 and 7. The LCD is $2x(x - 7)$.

Step 2.

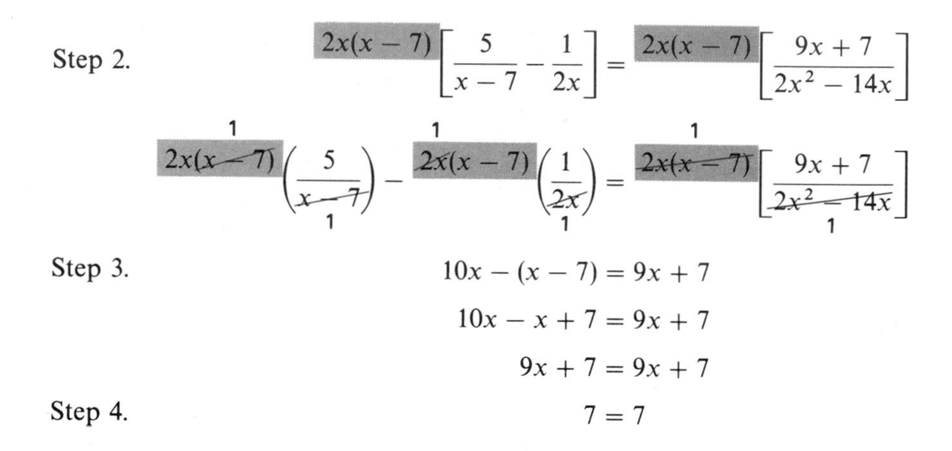

Step 3.
$$10x - (x - 7) = 9x + 7$$
$$10x - x + 7 = 9x + 7$$
$$9x + 7 = 9x + 7$$

Step 4.
$$7 = 7$$

This is an identity; however, the solution set is *not* the set of all real numbers, since 0 and 7 are not in the domain of the variable. The solution set, therefore, is the set of all real numbers *except* 0 and 7. ■

EXERCISES 6.8

Set I Find the solution set for each equation.

1. $\dfrac{2}{k-5} - \dfrac{5}{k} = \dfrac{3}{4k}$

2. $\dfrac{4}{h} - \dfrac{6}{h-7} = \dfrac{2}{3h}$

3. $\dfrac{x}{x-2} = \dfrac{2}{x-2} + 5$

4. $\dfrac{x}{x+5} = 4 - \dfrac{5}{x+5}$

5. $\dfrac{12m}{2m-3} = 6 + \dfrac{18}{2m-3}$

6. $\dfrac{40w}{5w+6} = 8 - \dfrac{48}{5w+6}$

7. $\dfrac{2y}{7y+5} = \dfrac{1}{3y}$

8. $\dfrac{2b}{3-4b} = \dfrac{1}{2b}$

9. $\dfrac{3e-5}{4e} = \dfrac{e}{2e+3}$

10. $\dfrac{2x-1}{3x} = \dfrac{3}{2x+7}$

11. $\dfrac{1}{2} - \dfrac{1}{x} = \dfrac{4}{x^2}$

12. $\dfrac{7}{4} - \dfrac{17}{4x} = \dfrac{3}{x^2}$

13. $\frac{4}{x+1} = \frac{3}{x} + \frac{1}{15}$

14. $\frac{1}{x+1} = \frac{3}{x} + \frac{1}{2}$

15. $\frac{6}{x+4} = \frac{5}{x+3} + \frac{4}{x}$

16. $\frac{7}{x+5} = \frac{3}{x-1} - \frac{4}{x}$

17. $\frac{6}{x^2-9} + \frac{1}{5} = \frac{1}{x-3}$

18. $\frac{6-x}{x^2-4} - 2 = \frac{x}{x+2}$

19. $\frac{x+2}{x-2} - \frac{x-2}{x+2} = \frac{16}{x^2-4}$

20. $\frac{x-5}{x+5} - \frac{x+5}{x-5} = \frac{100}{x^2-25}$

21. $\frac{1}{2x^2-11x+15} + \frac{x-1}{2x^2+x-15} = \frac{-4}{x^2-9}$

22. $\frac{1}{3x^2-10x+8} + \frac{x+8}{3x^2+2x-8} = \frac{6}{x^2-4}$

23. $\frac{8}{x^3+64} + \frac{3}{x^2-16} = \frac{-1}{x^2-4x+16}$

24. $\frac{12}{x^3+27} + \frac{1}{x^2-9} = \frac{1}{x^2-3x+9}$

Set II Find the solution set for each equation.

1. $\frac{7}{2t-3} - \frac{4}{5t} = \frac{3}{t}$

2. $\frac{6}{x} = \frac{7}{x-3} - \frac{1}{2}$

3. $\frac{8}{6+5e} = 3 - \frac{15e}{6+5e}$

4. $5 = \frac{3}{x-2} - \frac{8}{2-x}$

5. $\frac{36}{4-3d} = 9 + \frac{27d}{4-3d}$

6. $\frac{13}{3-x} = 8 - \frac{5x}{3-x}$

7. $\frac{4a}{33-5a} = \frac{2}{a}$

8. $\frac{9}{x} = \frac{24-3x}{5}$

9. $\frac{3}{1-2x} + \frac{5}{2-x} = 2$

10. $\frac{x}{8} = \frac{5}{6} - \frac{1}{x-1}$

11. $\frac{1}{2} - \frac{1}{6x} = \frac{7}{3x^2}$

12. $\frac{1}{5x} + \frac{1}{x^2} = \frac{2}{4x+5}$

13. $\frac{4}{3x-1} = \frac{2}{x} + 1$

14. $\frac{3}{1-2x} = \frac{2x+1}{x-2}$

15. $\frac{11}{2-3x} = \frac{6}{x+6} + \frac{3}{x}$

16. $\frac{x-5}{3+2x} = \frac{2x-1}{x-9}$

17. $\frac{x+25}{x^2-25} = 1 - \frac{12x}{x-5}$

18. $\frac{x+4}{x^2-16} + 1 = \frac{-7}{x-4}$

19. $\frac{x-4}{x+4} - \frac{x+4}{x-4} = \frac{64}{x^2-16}$

20. $\frac{4}{x^2-x-6} - \frac{2}{x^2-2x-3} = \frac{-1}{x+1}$

21. $\frac{1}{4x^2+11x-3} + \frac{24x-1}{4x^2-13x+3} = \frac{6}{x^2-9}$

22. $\frac{6}{x^2 + x - 6} - \frac{5}{x^2 - x - 2} = \frac{-1}{x + 1}$

23. $\frac{12}{x^3 + 8} - \frac{1}{x^2 - 4} = \frac{1}{x^2 - 2x + 4}$

24. $\frac{14}{x^3 + 1} + \frac{4}{x^2 - 1} = \frac{7}{x^2 - x + 1}$

6.9 Solving Literal Equations

Literal equations are equations that contain more than one variable. Such equations are also called *equations in two (or more) variables.* **Formulas** are literal equations that have applications in real-life situations. Sometimes literal equations have already been solved for one of the variables.

Example 1 Examples of literal equations:

a. $A = P(1 + rt)$

This is an equation in four variables; it is also a formula, because it has applications in business. It has already been solved for A; you might be asked to solve it for P, for r, or for t.

b. $\frac{4ab}{d} = 15$

This is an equation in three variables; it is not a formula. It has not been solved for any of its variables; you might be asked to solve for a, for b, or for d. ■

When we solve a literal equation for one of its variables, the solution will contain the other variables as well as constants. We must isolate the variable we are solving for; that is, *it must appear only once all by itself on one side of the equal sign.* All other variables and all constants must be on the other side of the equal sign. The suggestions in the following box are based on the addition, subtraction, multiplication, and division properties of equality.

TO SOLVE A LITERAL EQUATION

1. *Remove rational expressions* (if there are any) by multiplying both sides of the equation by the LCD.
2. *Remove grouping symbols* (if there are any).
3. *Collect and combine like terms,* with all terms containing the variable you are solving for on one side, all other terms on the other side.
4. *Factor out the variable you are solving for* (if it appears in more than one term).
5. *Divide both sides of the equation by the coefficient of the variable you are solving for* (or multiply both sides of the equation by the multiplicative inverse of that variable).

Example 2 Solve $A = P(1 + rt)$ for t.

Solution $A = P(1 + rt)$

Step 2. $A = P + Prt$ — Removing parentheses by using the distributive property

Step 3. $A - P = Prt$ — Collecting all terms with the variable we are solving for (t) on one side and all other terms on the other side

Step 5. $\dfrac{A - P}{Pr} = \dfrac{\overset{1}{\cancel{Pr}}t}{\underset{1}{\cancel{Pr}}}$ — Dividing both sides by Pr, the coefficient of t

$t = \dfrac{A - P}{Pr}$ — Solution ■

Example 3 Solve $\dfrac{1}{F} = \dfrac{1}{u} + \dfrac{1}{v}$ for u.

LCD $= Fuv$

Step 1. 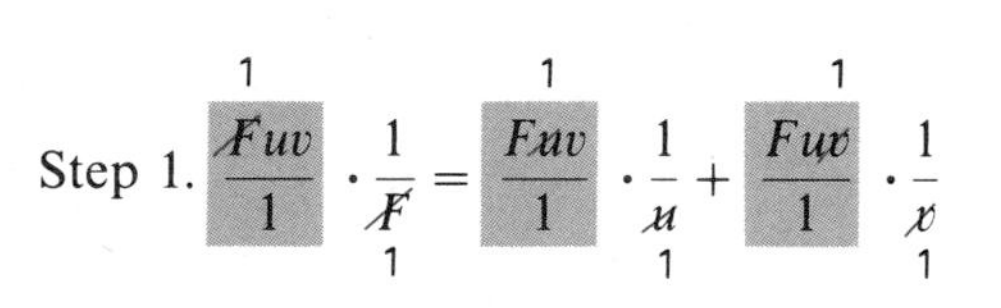 — Removing fractions by multiplying both sides by the LCD

$$\frac{\overset{1}{\cancel{F}}uv}{1} \cdot \frac{1}{\underset{1}{\cancel{F}}} = \frac{F\overset{1}{\cancel{u}}v}{1} \cdot \frac{1}{\underset{1}{\cancel{u}}} + \frac{Fu\overset{1}{\cancel{v}}}{1} \cdot \frac{1}{\underset{1}{\cancel{v}}}$$

$uv = Fv + Fu$

Step 3. $uv - Fu = Fv$ — Collecting all terms with the variable we are solving for (u) on one side and all other terms on the other side

Step 4. $u(v - F) = Fv$ — Factoring u from the left side

Step 5. $\dfrac{u\overset{1}{\cancel{(v - F)}}}{\underset{1}{\cancel{(v - F)}}} = \dfrac{Fv}{(v - F)}$ — Dividing both sides by $(v - F)$, the coefficient of u

$u = \dfrac{Fv}{v - F}$ — Solution ■

Example 4 Solve $I = \dfrac{nE}{R + nr}$ for n.

Step 1. $(R + nr)\,I = \overset{1}{\cancel{(R + nr)}}\,\dfrac{nE}{\underset{1}{\cancel{R + nr}}}$ — Multiplying both sides by the LCD

Step 2. $IR + Inr = nE$

Step 3. $IR = nE - Inr$ — Collecting all terms with the variable we are solving for (n) on one side and all other terms on the other side

Step 4. $IR = n(E - Ir)$ — Factoring n from the right side

Step 5. $n = \dfrac{IR}{E - Ir}$ — Dividing both sides by $(E - Ir)$, the coefficient of n ■

EXERCISES 6.9

Set I Solve for the variable listed after each equation.

1. $2(3x - y) = xy - 12; y$
2. $2(3x - y) = xy - 12; x$
3. $z = \frac{x - m}{s}; m$
4. $s^2 = \frac{N - n}{N - 1}; N$
5. $\frac{2x}{5yz} = z + x; x$
6. $\frac{xy}{z} = x + y; y$
7. $C = \frac{5}{9}(F - 32); F$
8. $A = \frac{h}{2}(B + b); B$
9. $s = c\left(1 + \frac{a}{c}\right); c$
10. $Z = \frac{Rr}{R + r}; R$
11. $A = P(1 + rt); r$
12. $S = \frac{1}{2}g(2t - 1); t$
13. $v^2 = \frac{2}{r} - \frac{1}{a}; a$
14. $\frac{1}{p} = 1 + \frac{1}{s}; s$
15. $S = \frac{a}{1 - r}; r$
16. $I = \frac{E}{R + r}; R$
17. $\frac{1}{F} = \frac{1}{u} + \frac{1}{v}; F$
18. $\frac{1}{c} = \frac{1}{a} + \frac{1}{b}; b$
19. $L = a + (n - 1)d; n$
20. $A = 2\pi rh + 2\pi r^2; h$
21. $C = \frac{a}{1 + \frac{a}{\pi A}}; a$
22. $R = \frac{V + v}{1 + \frac{vV}{c^2}}; v$

Set II Solve for the variable listed after each equation.

1. $5(2x + y) = 2xy + 25; x$
2. $\frac{mn}{m + n} = 1; m$
3. $3ab - 5 = 4(2a + b); b$
4. $\frac{x - y}{x} = x + y; y$
5. $\frac{1}{f} = \frac{1}{a} + \frac{1}{b}; a$
6. $A = \frac{h}{2}(B + b); b$
7. $A = \frac{h}{2}(B + b); h$
8. $\frac{1}{f} = \frac{1}{a} + \frac{1}{b}; f$
9. $a = b\left[1 + \frac{c}{b}\right]; b$
10. $PV = nRT; V$
11. $\frac{2}{15ab - 8a^2} = \frac{7}{20a^2 - 12ab}; b$
12. $x = \frac{y}{z + 4}; z$
13. $v^2 = \frac{2}{r} - \frac{1}{a}; r$
14. $\frac{2}{x} = \frac{y}{3} + \frac{1}{y}; x$
15. $I = \frac{E}{R + r}; r$
16. $A = P + Prt; P$

17. $\frac{1}{R} = \frac{1}{r_1} + \frac{1}{r_2}; r_1$

18. $y = mx + b; x$

19. $C = \frac{a}{1 + \frac{a}{\pi A}}; A$

20. $\frac{x}{a} + \frac{y}{b} = 1; b$

21. $E = \frac{k(1 - A)\pi R^2}{r^2}; A$

22. $S = \frac{n}{2}(A + L); A$

6.10 Solving Word Problems That Involve Rational Expressions

All the kinds of word problems discussed in previous chapters can lead to equations that contain rational expressions; for examples of how to solve word problems of those types, refer to Chapter 3.

In this section, we introduce a new type of word problem called a *work problem.* One basic relationship used to solve work problems is the following:

Rate × Time = Amount of work done

For example, suppose Jose can assemble a radio in 3 days.

$$\text{Jose's } rate = \frac{1 \text{ radio}}{3 \text{ days}} = \frac{1}{3} \text{ radio per day}$$

If his working *time* is 5 days, then

$$\boxed{\text{Rate}} \cdot \boxed{\text{Time}} = \boxed{\text{Amount of work done}}$$

$$\frac{1 \text{ radio}}{3 \ \cancel{\text{day}}} \cdot 5 \ \cancel{\text{days}} = \frac{5}{3} \text{ radios assembled}$$

The other basic relationship used to solve work problems is the following:

$$\begin{pmatrix}\text{Amount A}\\ \text{does in}\\ \text{time } x\end{pmatrix} + \begin{pmatrix}\text{Amount B}\\ \text{does in}\\ \text{time } x\end{pmatrix} = \begin{pmatrix}\text{Amount done}\\ \text{together in}\\ \text{time } x\end{pmatrix}$$

Example 1 Albert can paint a house in 6 days. Ben can do the same job in 8 days. How long would it take them to paint the same house if they work together?

Think

$$\text{Albert's rate} = \frac{1 \text{ house}}{6 \text{ days}} = \frac{1}{6} \text{ house per day}$$

$$\text{Ben's rate} = \frac{1 \text{ house}}{8 \text{ days}} = \frac{1}{8} \text{ house per day}$$

Step 1. Let x = number of days for Albert and Ben together to paint the house.

Reread

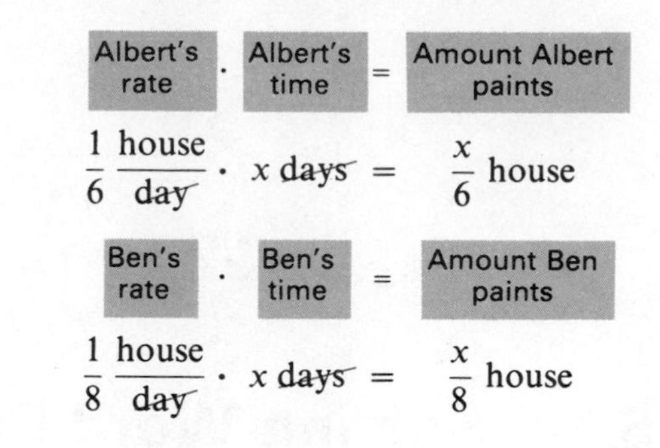

$$\boxed{\text{Albert's rate}} \cdot \boxed{\text{Albert's time}} = \boxed{\text{Amount Albert paints}}$$

$$\frac{1}{6}\frac{\text{house}}{\text{day}} \cdot x \text{ days} = \frac{x}{6} \text{ house}$$

$$\boxed{\text{Ben's rate}} \cdot \boxed{\text{Ben's time}} = \boxed{\text{Amount Ben paints}}$$

$$\frac{1}{8}\frac{\text{house}}{\text{day}} \cdot x \text{ days} = \frac{x}{8} \text{ house}$$

Therefore,

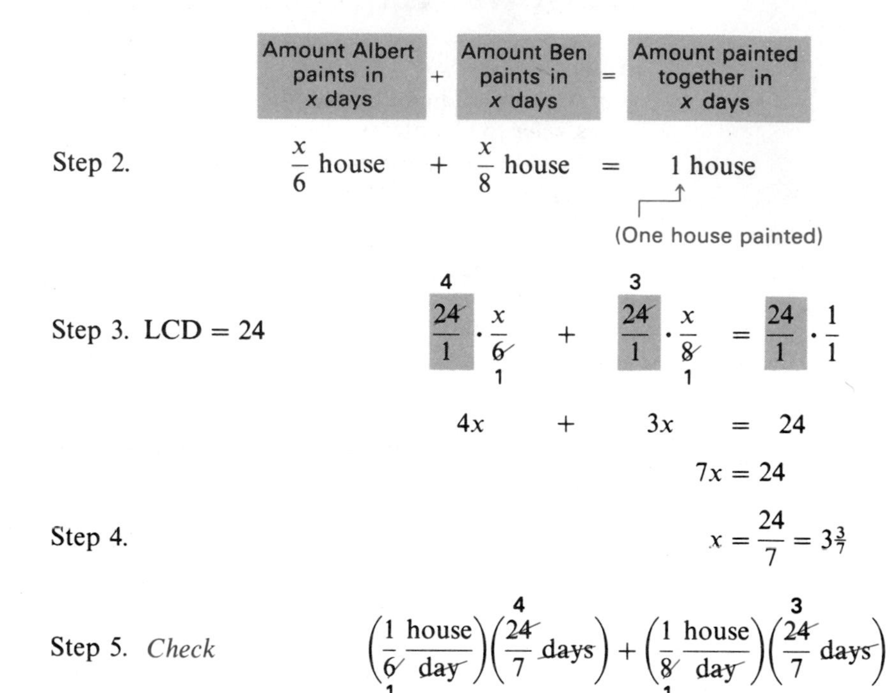

$$\boxed{\text{Amount Albert paints in } x \text{ days}} + \boxed{\text{Amount Ben paints in } x \text{ days}} = \boxed{\text{Amount painted together in } x \text{ days}}$$

Step 2. $\frac{x}{6}$ house $+$ $\frac{x}{8}$ house $=$ 1 house (One house painted)

Step 3. LCD = 24

$$\frac{24}{1} \cdot \frac{x}{6} + \frac{24}{1} \cdot \frac{x}{8} = \frac{24}{1} \cdot \frac{1}{1}$$

$$4x + 3x = 24$$

$$7x = 24$$

Step 4.

$$x = \frac{24}{7} = 3\tfrac{3}{7}$$

Step 5. *Check*

$$\left(\frac{1}{6}\frac{\text{house}}{\text{day}}\right)\left(\frac{24}{7}\text{ days}\right) + \left(\frac{1}{8}\frac{\text{house}}{\text{day}}\right)\left(\frac{24}{7}\text{ days}\right)$$

$$= \frac{4}{7} \text{ house} + \frac{3}{7} \text{ house} = 1 \text{ house}$$

Step 6. Therefore, it takes Albert and Ben $3\frac{3}{7}$ days to paint the house when they work together. ■

Example 2 Machine A can make 200 brackets in 6 hr. How long does it take machine B to make 100 brackets if the two machines working together can make 200 brackets in 5 hr?

Step 1. Let x = number of hours for machine B to make 100 brackets.

Think

$$\text{A's rate} = \frac{200 \text{ brackets}}{6 \text{ hr}} = \frac{100}{3} \text{ brackets per hour}$$

$$\text{B's rate} = \frac{100 \text{ brackets}}{x \text{ hr}} = \frac{100}{x} \text{ brackets per hour}$$

The two machines working together can make 200 brackets if each machine runs for 5 hours. Therefore,

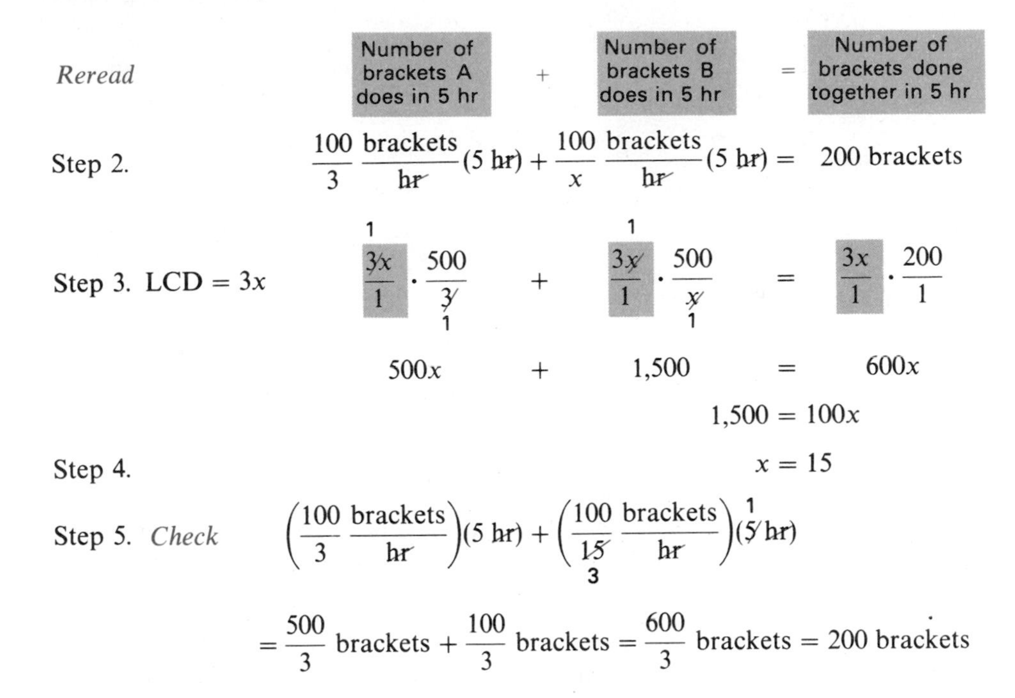

Reread [Number of brackets A does in 5 hr] + [Number of brackets B does in 5 hr] = [Number of brackets done together in 5 hr]

Step 2. $\frac{100}{3}\frac{\text{brackets}}{\text{hr}}(5\text{ hr}) + \frac{100}{x}\frac{\text{brackets}}{\text{hr}}(5\text{ hr}) = 200 \text{ brackets}$

Step 3. LCD = $3x$

$$\frac{3x}{1}\cdot\frac{500}{3} + \frac{3x}{1}\cdot\frac{500}{x} = \frac{3x}{1}\cdot\frac{200}{1}$$

$$500x + 1{,}500 = 600x$$

$$1{,}500 = 100x$$

Step 4. $x = 15$

Step 5. *Check* $\left(\frac{100}{3}\frac{\text{brackets}}{\text{hr}}\right)(5\text{ hr}) + \left(\frac{100}{15}\frac{\text{brackets}}{\text{hr}}\right)(5\text{ hr})$

$$= \frac{500}{3}\text{ brackets} + \frac{100}{3}\text{ brackets} = \frac{600}{3}\text{ brackets} = 200 \text{ brackets}$$

Step 6. Therefore, it takes machine B 15 hr to make 100 brackets. ■

Example 3 It takes pipe 1 twelve minutes to fill a particular tank. It takes pipe 2 only eight minutes to fill the same tank. Pipe 3 takes six minutes to *empty* the same tank. How long does it take to fill the tank when all three pipes are open?

Think

$$\text{Pipe 1's rate} = \frac{1\text{ tank}}{12\text{ min}} = \frac{1}{12}\text{ tank per minute}$$

$$\text{Pipe 2's rate} = \frac{1\text{ tank}}{8\text{ min}} = \frac{1}{8}\text{ tank per minute}$$

$$\text{Pipe 3's rate} = \frac{1\text{ tank}}{6\text{ min}} = \frac{1}{6}\text{ tank per minute}$$

Step 1. Let x = number of minutes for all three together to fill the tank

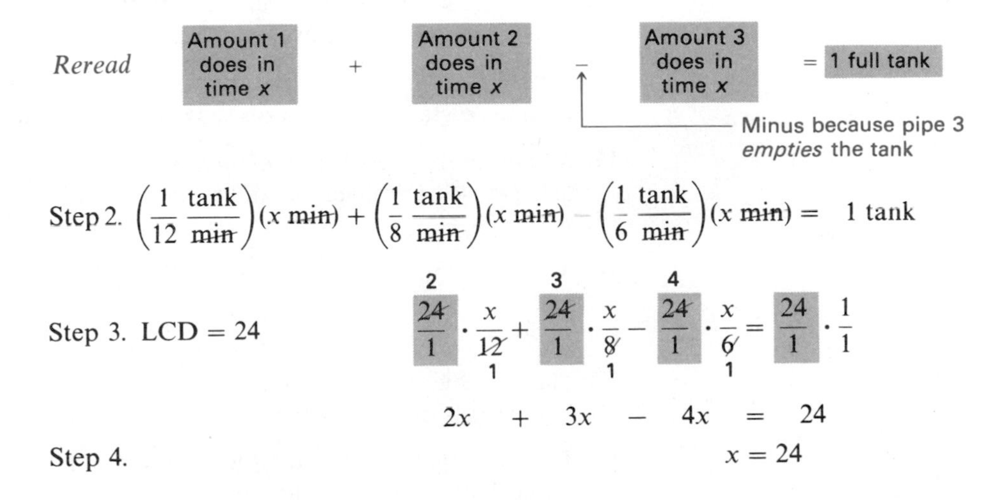

Reread [Amount 1 does in time x] + [Amount 2 does in time x] − [Amount 3 does in time x] = [1 full tank]

Minus because pipe 3 *empties* the tank

Step 2. $\left(\frac{1}{12}\frac{\text{tank}}{\text{min}}\right)(x\text{ min}) + \left(\frac{1}{8}\frac{\text{tank}}{\text{min}}\right)(x\text{ min}) - \left(\frac{1}{6}\frac{\text{tank}}{\text{min}}\right)(x\text{ min}) = 1\text{ tank}$

Step 3. LCD = 24

$$\frac{24}{1}\cdot\frac{x}{12} + \frac{24}{1}\cdot\frac{x}{8} - \frac{24}{1}\cdot\frac{x}{6} = \frac{24}{1}\cdot\frac{1}{1}$$

$$2x + 3x - 4x = 24$$

Step 4. $x = 24$

Step 5. *Check* $\left(\frac{1}{\cancel{12}_1}\frac{\text{tank}}{\cancel{\text{min}}}\right)(\overset{2}{\cancel{24}}\ \cancel{\text{min}}) + \left(\frac{1}{\cancel{8}_1}\frac{\text{tank}}{\cancel{\text{min}}}\right)(\overset{3}{\cancel{24}}\ \cancel{\text{min}}) - \left(\frac{1}{\cancel{6}_1}\frac{\text{tank}}{\cancel{\text{min}}}\right)(\overset{4}{\cancel{24}}\ \cancel{\text{min}})$

$= 2 \text{ tanks} + 3 \text{ tanks} - 4 \text{ tanks} = 1 \text{ tank}$

Step 6. Therefore, it takes 24 min to fill the tank when all three pipes are open. ■

EXERCISES 6.10

Set I

Set up each problem algebraically, solve, and check. Be sure to state what your variables represent.

1. Henry can paint a house in 5 days and Teri can do the same work in 4 days. If they work together, how many days will it take them to paint the house?
2. Jill can paint a house in 6 days and Karla can do it in 9 days. How many days will it take them to paint the house if they work together?
3. Trisha can type 100 pages of manuscript in 3 hr. How long does it take David to type 80 pages if he and Trisha working together can type a 500-page manuscript in 10 hr?
4. Carlos can wrap 200 newspapers in 55 min. How long does it take Heather to wrap 150 papers if she and Carlos working together can wrap 400 papers in 1 hr?
5. Machine A takes 36 hr to do a job that machine B does in 24 hr. If machine A runs for 12 hr before machine B is turned on, how long will it take both machines running together to finish the job?
6. Machine A takes 18 hr to do a job that machine B does in 15 hr. If machine A runs for 3 hr before machine B is turned on, how long will it take both machines running together to finish the job?
7. Two numbers differ by 8. One-fourth the larger is 1 more than one-third the smaller. Find the numbers.
8. Two numbers differ by 6. One-fifth the smaller exceeds one-half the larger by 3. Find the numbers.
9. It takes one plane $\frac{1}{2}$ hr longer than another to fly a certain distance. Find the distance if one plane travels 500 mph and the other 400 mph.
10. It takes Barbara 10 min longer than Ed to ride her bicycle a certain distance. Find the distance if Barbara's speed is 9 mph and Ed's is 12 mph.
11. An automobile radiator contains 14 qt of a 45% solution of antifreeze. How much must be drained out and replaced by pure antifreeze to make a 50% solution?
12. An automobile radiator contains 12 qt of a 30% solution of antifreeze. How much must be drained out and replaced with pure antifreeze to make a 50% solution?
13. The tens digit of a two-digit number is 1 more than its units digit. The product of the digits divided by the sum of the digits is $\frac{6}{5}$. Find the number.
14. The units digit of a two-digit number is 1 more than the tens digit. The product of the digits divided by the sum of the digits is $\frac{12}{7}$. Find the number.

15. Pipe 2 takes 1 hr longer to fill a particular tank than pipe 1. Pipe 3 can drain the tank in 2 hr. If it takes 3 hr to fill the tank when all three pipes are open, how long does it take pipe 1 to fill the tank alone?

16. Pipe 2 takes 2 hr more time to fill a particular tank than pipe 1. Pipe 3 can drain the tank in 4 hr. If it takes 2 hr to fill the tank when all three pipes are open, how long does it take pipe 1 to fill the tank alone?

17. Ruth can proofread 230 pages of a deposition in 4 hr. How long does it take Sandra to proofread 60 pages if, when they work together, they can proofread 525 pages in 6 hr?

18. Alice can crochet 3 sweaters in 48 hr. How long does it take Willa to crochet 2 sweaters if together they can crochet 15 sweaters in 112 hr?

19. In a film-processing lab, machine A can process 5,700 ft of film in 60 min. How long does it take machine B to process 4,300 ft of film if the two machines running together can process 15,500 ft of film in 50 min?

20. In a film-processing lab, machine C can process 3,225 ft of film in 15 min. How long does it take machine D to process 12,750 ft of film if the two machines running together can process 28,800 ft of film in 45 min?

Set II Set up each problem algebraically, solve, and check. Be sure to state what your variables represent.

1. Merwin can wax all his floors in 4 hr. His wife, June, can do the same job in 3 hr. How long will it take them if they work together?

2. The denominator of a fraction exceeds the numerator by 6. The value of the fraction is $\frac{3}{5}$. Find the fraction.

3. It takes Kevin 8 hr to clean the yard and mow the lawn. He and his brother Jason can do the same job in 3 hr working together. How long does it take Jason working alone?

4. The rate of Tom's boat in still water is 30 mph. If it can travel 132 miles *with* the current in the same time that it can travel 108 miles *against* the current, find the speed of the current.

5. Machine A takes 20 hr to do a job that machine B does in 28 hr. If machine A runs for 8 hr before machine B is turned on, how long will it take both machines running together to finish the job?

6. The sum of a fraction and its reciprocal is $\frac{73}{24}$. Find the fraction and its reciprocal.

7. Two numbers differ by 22. One-eighth the larger exceeds one-sixth the smaller by 2. Find the numbers.

8. Sherma can type 15 words per minute faster than Karen. If Sherma types 3,850 words in the same time that Karen types 3,100 words, find the rate of each.

9. It takes Mina 10 min longer than Susan to jog eight laps around the school soccer field. Find the distance around the soccer field if Mina's speed is 3 mph and Susan's is 4 mph.

10. Machine A can make 400 bushings in 8 hr. How long does it take machine B to make 50 bushings if the two machines working together can make 800 bushings in 12 hr?

11. One-fifth of the 10% antifreeze solution in a car radiator is drained and replaced by an antifreeze solution of unknown concentration. If the resulting mixture is 25% antifreeze, what was the unknown concentration?

12. The units digit of a two-digit number is 2 more than the tens digit. The product of the digits divided by the sum of the digits is $\frac{3}{4}$. Find the two-digit number.

13. The tens digit of a two-digit number is 4 more than its units digit. The product of the digits divided by the sum of the digits is $\frac{3}{2}$. Find the number.

14. The speed of a plane in still air is 480 mph. If it can fly 3,030 miles *with* the wind in the same time that it can fly 2,730 miles *against* the wind, what is the speed of the wind?

15. A refinery tank has one fill pipe and two drain pipes. Pipe 1 can fill the tank in 3 hr. Pipe 2 takes 6 hr longer to drain the tank than pipe 3. If it takes 12 hr to fill the tank when all three pipes are open, how long does it take pipe 3 to drain the tank alone?

16. It takes Matt 30 min longer than Rebecca to walk a certain distance. Find the distance if Matt's speed is 3 mph and Rebecca's is 4 mph.

17. Lori can knit 4 scarves in 24 hr. How long does it take Anita to knit one scarf if together they can knit 21 scarves in 72 hr?

18. The speed of the current in a river is 6 mph. If a boat can travel 222 miles *with* the current in the same time it can travel 150 miles *against* the current, what is the speed of the boat in still water?

19. In a film-processing lab, machine A can process 4,275 ft of film in 45 min. How long does it take machine B to process 10,625 ft of film if the two machines running together can process 41,600 ft of film in 80 min?

20. The tens digit of a two-digit number is 2 more than the units digit. If the product of the digits divided by the sum of the digits is $\frac{12}{5}$, find the number.

6.11 Review: 6.8–6.10

To Solve a Rational Equation 6.8

1. Find the domain of the variable and find the LCD of all the denominators.
2. Remove denominators by multiplying *both sides of the equation* by the LCD.
3. a. Remove all grouping symbols.
   b. Collect and combine like terms.

First-degree equations	*Second-degree equations*
4. a. Get all terms with the variable on one side of the equation and all other terms on the other side. b. Divide both sides of the equation by the coefficient of the variable, or multiply both sides of the equation by the reciprocal of that coefficient.	4. a. Get *all* nonzero terms on one side of the equation by adding the same expression to both sides. *Only zero must remain on the other side.* Then arrange the terms in descending powers. b. Factor the polynomial. c. Set each factor equal to zero and solve the equations.

5. Reject any apparent solutions that are not in the domain.
6. Check any other apparent solutions in the original equation.

Literal Equations 6.9

Literal equations are equations that have more than one variable.

To Solve a Literal Equation 6.9 Proceed in the same way used to solve an equation with one variable. The solution will be expressed in terms of the other variables given in the literal equation, as well as numbers.

Review Exercises 6.11 Set I

In Exercises 1–10, find the solution set for each equation.

1. $\dfrac{7}{2z+5} = \dfrac{13}{3z}$

2. $\dfrac{11}{6c-7} = \dfrac{8}{5c}$

3. $\dfrac{5a-4}{6a} = \dfrac{-2}{3a+10}$

4. $\dfrac{3b}{5b-4} = \dfrac{7-2b}{3}$

5. $\dfrac{3}{x} - \dfrac{8}{x^2} = \dfrac{1}{4}$

6. $\dfrac{4}{x^2} - \dfrac{3}{x} = \dfrac{5}{2}$

7. $\dfrac{9}{x+1} = \dfrac{4}{x} + \dfrac{1}{x-1}$

8. $\dfrac{4}{3x-1} - \dfrac{2}{x+1} = \dfrac{1}{x}$

9. $\dfrac{x}{x+5} = 8 - \dfrac{5}{x+5}$

10. $\dfrac{3}{x-3} - \dfrac{1}{3x} = \dfrac{8x+3}{3x^2-9x}$

In Exercises 11–14, solve for the variable listed after each equation.

11. $5(x-2y) = 14 + 3(2x-y)$; y

12. $2(5-2x) = 22 - 2(y-5x)$; x

13. $R = \dfrac{R_1R_2}{R_1+R_2}$; R_1

14. $F = \dfrac{1}{\dfrac{1}{u}+\dfrac{1}{v}}$; v

In Exercises 15–18, set up the problem algebraically and solve. Be sure to state what your variables represent.

15. Using tractor A, a man can cultivate his corn in 20 hr. With tractor B, the same corn can be cultivated in 15 hr. After one man cultivates with tractor A for 5 hr, he is joined by a second man using tractor B. How long will it take both men and tractors working together to finish the job?

16. Section gang A can lay a rail in 16 min. Section gang B can lay a rail in 12 min. After gang A has been working 4 min, how long will it take both gangs working together to finish laying one rail?

17. The denominator of a fraction is 10 more than its numerator. If 1 is added to both numerator and denominator, the value of the new fraction is $\frac{2}{3}$. Find the original fraction.

18. Pipes 1 and 2 run into a tank, while pipe 3 runs out the bottom. Pipe 1 can fill the tank in 4 hr; pipe 2 can fill it in 6 hr. When pipes 1, 2, and 3 are open, the tank is filled in 5 hr. How long does it take pipe 3 to drain the tank by itself?

Review Exercises 6.11 Set II

NAME

In Exercises 1–10, find the solution set for each equation.

1. $\dfrac{3}{5f-2} = \dfrac{7}{12f}$

2. $\dfrac{8}{x+5} = \dfrac{1}{13-2x}$

3. $\dfrac{4u+1}{5u} = \dfrac{20}{3u+6}$

4. $\dfrac{2x+1}{4x+2} = \dfrac{3x-2}{2}$

5. $\dfrac{17}{6x} + \dfrac{5}{2x^2} = \dfrac{2}{3}$

6. $\dfrac{3}{x} - \dfrac{1}{x^2} = \dfrac{11}{16}$

7. $\dfrac{9}{2x-1} = \dfrac{4}{x} - \dfrac{1}{x-3}$

8. $\dfrac{6}{6+x} - \dfrac{1}{x} = \dfrac{13}{4}$

9. $\dfrac{4}{x-4} - \dfrac{1}{2x} = \dfrac{7x+4}{2x^2-8x}$

10. $\dfrac{x}{x-8} = 7 + \dfrac{8}{x-8}$

In Exercises 11–14, solve for the variable listed after each equation.

11. $4(2y + x) = xy + 32$; x

12. $ab = \dfrac{c}{a} - \dfrac{b}{a}$; b

ANSWERS

1. ______

2. ______

3. ______

4. ______

5. ______

6. ______

7. ______

8. ______

9. ______

10. ______

11. ______

12. ______

13. $\frac{1}{p} = 1 + \frac{1}{s}$; p

14. $\frac{a}{b-3} = \frac{1}{a} + \frac{1}{4}$; b

ANSWERS

13. ______________

14. ______________

15. ______________

16. ______________

17. ______________

18. ______________

In Exercises 15–18, set up the problem algebraically and solve. Be sure to state what your variables represent.

15. A tank is filled through pipes 1 and 2 and drained through pipe 3. Pipe 1 can fill the tank in 5 hr and pipe 2 can fill it in 3 hr. When pipes 1, 2, and 3 are open, the tank is filled in 6 hr. How long does it take pipe 3 to drain the tank by itself?

16. Machine A can produce 25 sprocket wheels in 30 min. How many sprocket wheels can machine B produce in 1 hr if both machines working together can produce 315 sprocket wheels in $3\frac{1}{2}$ hours?

17. The numerator of a fraction is 32 less than its denominator. If the numerator and denominator are each increased by 3, the new fraction has a value of $\frac{1}{2}$. Find the original fraction.

18. Machine A can produce 25 sprocket wheels in 30 min. How long does it take machine B to produce 112 sprocket wheels if both machines working together can produce 477 sprocket wheels in $4\frac{1}{2}$ hr?

Chapter 6 Diagnostic Test

The purpose of this test is to see how well you understand operations with rational expressions. We recommend that you work this diagnostic test *before* your instructor tests you on this chapter. Allow yourself about 50 minutes.

Complete solutions for all the prolems on this test, together with section references, are given in the answer section at the end of the book. For the problems you do incorrectly, study the sections cited.

In Problems 1 and 2, find the domain of the variable.

1. $\dfrac{2x+3}{x^2-4x}$

2. $\dfrac{y+2}{3y^2-y-10}$

In Problems 3 and 4, find the missing term.

3. $\dfrac{4}{-h}=\dfrac{-4}{?}$

4. $\dfrac{-3}{k-2}=\dfrac{?}{2-k}$

5. Are $\dfrac{3x+4}{2y+4}$ and $\dfrac{3x}{2y}$ equivalent rational expressions?

In Problems 6 and 7, reduce each rational expression to lowest terms.

6. $\dfrac{f^2+5f+6}{f^2-9}$

7. $\dfrac{x^4-2x^3+5x^2-10x}{x^3-8}$

In Problems 8–13, perform the indicated operations.

8. $\dfrac{z}{2z^2-5z-3}\cdot\dfrac{6z^2-9z-6}{6z^2}$

9. $\dfrac{x^2-2x-24}{x^2-36}\cdot\dfrac{x^2+7x+6}{x^2+x-12}\div(x^3+1)$

10. $\dfrac{3m+3n}{m^3-n^3}\div\dfrac{m^2-n^2}{m^2+mn+n^2}$

11. $\dfrac{20a+27b}{12a-20b}+\dfrac{44a-13b}{20b-12a}$

12. $\dfrac{x}{x+4}-\dfrac{x}{x-4}-\dfrac{32}{x^2-16}$

13. $\dfrac{3}{x^2+x-6}-\dfrac{2}{x^2-4}-\dfrac{3}{x^2+5x+6}$

In Problems 14 and 15, simplify the complex fractions.

14. $\dfrac{\dfrac{8h^4}{5k}}{\dfrac{4h^2}{15k^3}}$

15. $\dfrac{6-\dfrac{4}{w}}{\dfrac{3w}{w-2}+\dfrac{1}{w}}$

In Problems 16–18, find the solution set.

16. $\dfrac{2}{3a+5}-\dfrac{6}{a-2}=3$

17. $\dfrac{x}{x+7}=3-\dfrac{7}{x+7}$

18. $\dfrac{3}{2z}+\dfrac{3}{z^2}=-\dfrac{1}{6}$

19. Solve for r in terms of the other variables: $I=\dfrac{E}{R+r}$

20. Set up algebraically, solve, and check: Rachelle can make 24 bushings on a lathe in 8 hr. How long does it take Ruben to make 8 bushings if he and Rachelle can make 14 bushings in 4 hr when they work together?

Cumulative Review Exercises: Chapters 1–6

1. Simplify. Write your answer using only positive exponents.

$$\left(\frac{x^4y^{-3}}{x^{-2}}\right)^{-2}$$

2. Solve $\frac{x-2}{4} = \frac{x+3}{2} + \frac{1}{4}$.

3. Solve $9 - 2(m - 6) \leq 3(4 - m) + 5$.

4. Is 0 a rational number?

5. Is 17 a real number?

In Exercises 6–11, perform the indicated operations and simplify.

6. $(2x - 5)(x^2 - 4x + 3)$

7. $(8a^2 + 6a + 1) \div (2a + 3)$

8. $(x - 2)^6$

9. $\frac{x+3}{2x^2+3x-2} \cdot \frac{2x^2-9x+4}{3x^2+10x+3} \div \frac{x-4}{x+2}$

10. $\frac{3x}{2x^2+15x+7} + \frac{4}{2x^2-5x-3}$

11. $\frac{a-3}{a^3+27} - \frac{1}{a^2-9}$

In Exercises 12–15, factor completely.

12. $10x^2 + 3xy - 18y^2$

13. $x^4 - x^2 - 12$

14. $x^3 - 8$

15. $6ax + 3bx - 2ay - by$

In Exercises 16–21, set up algebraically, solve, and check.

16. The length of a rectangle is 7 in. more than the width. The area is 60 sq. in. What are its dimensions?

17. Mrs. Kishinami spent $9.70 for 50 stamps. If she bought only 18¢ and 20¢ stamps, how many of each kind did she buy?

18. The tens digit of a three-digit number is twice the hundreds digit, and the sum of the digits is 17. If the digits are reversed, the new number is 99 more than the original number. Find the original number.

19. Four women plan to buy some investment property, sharing the cost equally. If they allow a fifth person to join them, with all five investing equal amounts, the share for each of the original four women will be reduced by $4,250. What is the cost of the property?

20. Forty-five square yards of carpet are laid in a rectangular room. The length is 1 yd less than twice the width. Find the dimensions of the room.

21. Alberto leaves Ft. Lauderdale at 3 A.M., traveling at 45 mph. Two hours later, Carlos leaves from the same point, traveling along the same route at 55 mph. How long will it take Carlos to overtake Alberto?

7 Exponents and Radicals

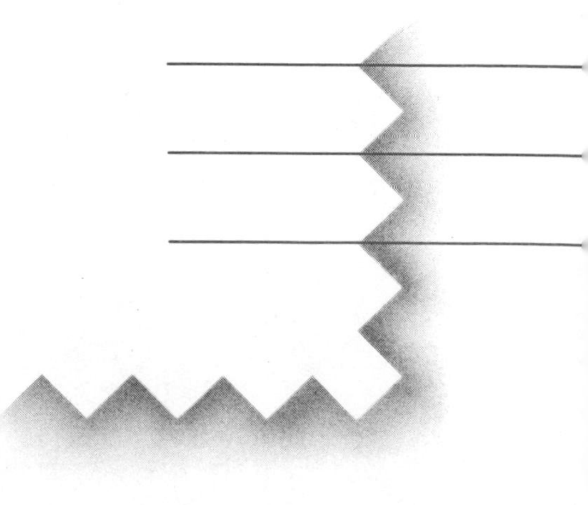

In this chapter, we extend the properties of exponents to include rational exponents, and we define the relation between exponents and radicals. Radicals are more fully discussed in this chapter than in Chapter 1, and *complex numbers* are introduced.

7.1 Rational Exponents

Let's review the eight basic properties of integral exponents from Section 1.11 (it is understood that none of the variables can have a value that makes a denominator zero):

The properties of exponents

1.1. $x^m \cdot x^n = x^{m+n}$	1.2. $(x^m)^n = x^{mn}$
1.3. $(xy)^n = x^n y^n$	1.4. $\dfrac{x^m}{x^n} = x^{m-n}$
1.5. $\left(\dfrac{x}{y}\right)^n = \dfrac{x^n}{y^n}$	1.6. $x^0 = 1$
1.7a. $x^{-n} = \dfrac{1}{x^n}$	1.8. $\left(\dfrac{x^a y^b}{z^c}\right)^n = \dfrac{x^{an} y^{bn}}{z^{cn}}$

In this section, we define additional properties of exponents that make it possible to work with exponents that are rational numbers.

7.1A The Meaning of Rational Exponents

The following discussion is not a proof; it is included to help you understand why we define rational exponents the way we do.

We know that for all $x \ge 0$, $\sqrt{x}$ is a number such that $(\sqrt{x})^2 = x$. Suppose we let $\sqrt{x} = x^p$, where p is some rational number to be determined. Then

$$(\sqrt{x})^2 = x$$

$$(x^p)^2 = x \qquad \text{Substituting } x^p \text{ for } \sqrt{x}$$

$$x^{2p} = x^1 \qquad \text{Property 1.2}$$

If $x^{2p} = x^1$, it must be true that $2p = 1$, or

$$p = \frac{1}{2}$$

That is, $x^p = x^{1/2}$. Therefore, we want $x^{1/2}$ to equal $\sqrt{x}$.

Similarly, we know that $\sqrt[3]{x}$ is a number such that $(\sqrt[3]{x})^3 = x$. Suppose we let $\sqrt[3]{x} = x^r$, where r is some rational number to be determined. Then

$$(\sqrt[3]{x})^3 = x$$

$$(x^r)^3 = x \qquad \text{Substituting } x^r \text{ for } \sqrt[3]{x}$$

$$x^{3r} = x^1 \qquad \text{Property 1.2}$$

If $x^{3r} = x^1$, it must be true that $3r = 1$, or

$$r = \frac{1}{3}$$

That is, $x^r = x^{1/3}$. Therefore, we want $x^{1/3}$ to equal $\sqrt[3]{x}$. Consequently, we make the following definitions:

> PROPERTY 7.1
>
> $\sqrt[n]{x} = x^{1/n}$ if $x \geq 0$ when n is even.
>
> $\sqrt[n]{x} = x^{1/n}$ for all x when n is odd.

A WORD OF CAUTION When you write rational exponents, *be sure your exponents look like exponents*! That is, be sure $2^{1/4}$ doesn't look like $2\frac{1}{4}$. ☑

Before we use Property 7.1, let's review the rules about principal roots from Chapter 1.

> PRINCIPAL ROOTS
>
> The symbol $\sqrt[n]{p}$ always represents the *principal* nth root of p.
>
> If the *radicand is positive*, the principal root is positive.
>
> If the *radicand is negative*:
>
> **1.** When the index is odd, the principal root is negative.
>
> **2.** When the index is even, the principal root is *not a real number*.*

Example 1 Examples of using Property 7.1 to change from radical form to exponential form:

a. $\sqrt[4]{z} = z^{1/4}$, if $z \geq 0$

b. $\sqrt[5]{y} = y^{1/5}$

c. $\sqrt[3]{-10} = -\sqrt[3]{10} = -10^{1/3}$

d. $\sqrt{14} = 14^{1/2}$

The rules about principal roots remind us that an *odd* root of a *negative* number is *negative* ■

Example 2 Examples of using Property 7.1 to change from exponential form to radical form:

a. $a^{1/4} = \sqrt[4]{a}$, if $a \geq 0$

b. $x^{1/7} = \sqrt[7]{x}$ ■

Example 3 Examples of evaluating expressions with rational exponents:

a. $27^{1/3} = \sqrt[3]{27} = 3$

b. $100^{1/2} = \sqrt{100} = 10$

c. $(-8)^{1/3} = \sqrt[3]{(-8)} = -2$

d. $32^{1/5} = \sqrt[5]{32} = 2$ ■

Example 4 Examples of using Property 7.1 to change from radical form to exponential form:

a. $\sqrt[3]{x^2} = (x^2)^{\frac{1}{3}}$ Property 7.1

$= x^{(2)\left(\frac{1}{3}\right)} = x^{\frac{2}{3}}$ Property 1.2

b. $(\sqrt[3]{x})^2 = \left(x^{\frac{1}{3}}\right)^2 = x^{\left(\frac{1}{3}\right)(2)} = x^{\frac{2}{3}}$

Therefore, $\sqrt[3]{x^2} = (\sqrt[3]{x})^2 = x^{\frac{2}{3}}$ ■

* Numbers that are not real numbers are discussed in Section 7.8.

The results of Examples 4a and 4b are generalized in the following property:

> PROPERTY 7.2
>
> $$\sqrt[b]{x^a} = (\sqrt[b]{x})^a = x^{a/b} \text{ if } x \geq 0 \text{ when } b \text{ is even}$$
> $$= x^{a/b} \text{ for all } x \text{ when } b \text{ is odd.}$$

Example 5 Examples of applying Property 7.2 (assume that all variables represent nonnegative numbers):

a. $\sqrt[5]{m^3} = m^{3/5}$

b. $(\sqrt[4]{z})^3 = (z^{1/4})^3 = z^{3/4}$

c. $d^{4/5} = \sqrt[5]{d^4}$ or $(\sqrt[5]{d})^4$

d. $a^{2/3} = (\sqrt[3]{a})^2$

e. $8^{2/3} = (\sqrt[3]{8})^2 = (\sqrt[3]{2^3})^2 = (2)^2 = 4$ ■

EXERCISES 7.1A

Set I

In Exercises 1–10, replace each radical by an equivalent exponential expression; assume that all variables represent nonnegative numbers.

1. $\sqrt{5}$ **2.** $\sqrt{7}$ **3.** $\sqrt[3]{z}$ **4.** $\sqrt[4]{x}$ **5.** $\sqrt[4]{x^3}$

6. $\sqrt[5]{y^2}$ **7.** $(\sqrt[3]{x^2})^2$ **8.** $(\sqrt[4]{x^3})^2$ **9.** $\sqrt[n]{x^{2n}}$ **10.** $\sqrt[n]{y^{5n}}$

In Exercises 11–16, rewrite each expression in radical form; assume that all variables represent nonnegative numbers.

11. $7^{1/2}$ **12.** $5^{1/3}$ **13.** $a^{3/5}$ **14.** $b^{2/3}$ **15.** $x^{m/n}$

16. $x^{(a+b)/a}$

In Exercises 17–24, evaluate each expression.

17. $8^{1/3}$ **18.** $27^{1/3}$ **19.** $(-27)^{2/3}$ **20.** $(-8)^{2/3}$

21. $4^{3/2}$ **22.** $9^{3/2}$ **23.** $(-16)^{3/4}$ **24.** $(-1)^{1/4}$

Set II

In Exercises 1–10, replace each radical by an equivalent exponential expression; assume that all variables represent nonnegative numbers.

1. $\sqrt{17}$ **2.** $\sqrt{31}$ **3.** $\sqrt[3]{y}$ **4.** $\sqrt[5]{x^2}$ **5.** $\sqrt[4]{x^7}$

6. $\sqrt[3]{z^4}$ **7.** $(\sqrt[3]{x})^2$ **8.** $(\sqrt{y})^3$ **9.** $\sqrt[n]{x^{3n}}$ **10.** $\sqrt[n]{x^{7n}}$

In Exercises 11–16, rewrite each expression in radical form; assume that the variables represent nonnegative numbers.

11. $13^{1/3}$ **12.** $7^{1/4}$ **13.** $x^{4/7}$ **14.** $c^{2/5}$ **15.** $y^{a/b}$

16. $z^{(a-b)/b}$

In Exercises 17–24, evaluate each expression.

17. $64^{2/3}$ **18.** $16^{3/4}$ **19.** $8^{2/3}$ **20.** $(-1)^{1/5}$ **21.** $27^{4/3}$

22. $1^{7/8}$ **23.** $25^{3/2}$ **24.** $(-64)^{2/3}$

7.1B Using the Properties of Exponents When the Exponents Are Rational Numbers

All of the properties of exponents reviewed at the beginning of this section can be used when the exponents are *any* rational numbers, as long as the radicand is nonnegative whenever the denominator of an exponent is an even number.

Example 6 Examples of using the properties of exponents to simplify expressions with rational exponents (assume that all variables represent nonnegative numbers):

a. $a^{1/2}a^{-1/4} = a^{1/2+(-1/4)} = a^{1/4}$ Property 1.1

b. $(8)^{2/3} = (2^3)^{2/3} = 2^{3 \cdot \frac{2}{3}}$ For an alternate solution, see Example 5e

$= 2^2 = 4$

c. $x^{-1/3} = \dfrac{1}{x^{1/3}}$ Property 1.7

d. $(z^{-1/2})^4 = z^{-\frac{1}{2} \cdot 4} = z^{-2} = \dfrac{1}{z^2}$ Properties 1.2 and 1.7

e. $\dfrac{y^{2/3}}{y^{1/2}} = y^{2/3-1/2} = y^{\frac{4}{6}-\frac{3}{6}} = y^{1/6}$ Property 1.4

f. $x^{-1/2}xx^{2/5} = x^{-1/2+1+2/5}$

$= x^{-5/10+10/10+4/10} = x^{9/10}$ Property 1.1

g. $\dfrac{b^{3/2}b^{-2/3}}{b^{5/6}} = b^{3/2-2/3-5/6}$

$= b^{9/6-4/6-5/6} = b^0 = 1$ Properties 1.1, 1.4, and 1.6

h. $(9h^{-2/5}k^{4/3})^{-3/2} = (3^2h^{-2/5}k^{4/3})^{-3/2}$

$= 3^{2\left(-\frac{3}{2}\right)}h^{-\frac{2}{5}\left(-\frac{3}{2}\right)}k^{\frac{4}{3}\left(-\frac{3}{2}\right)}$

$= 3^{-3}h^{3/5}k^{-2} = \dfrac{h^{3/5}}{27k^2}$ Properties 1.7 and 1.8

i. $\left(\dfrac{xy^{-2/3}}{z^{-4}}\right)^{-3/4} = \dfrac{x^{1\left(-\frac{3}{4}\right)}y^{-\frac{2}{3}\left(-\frac{3}{4}\right)}}{z^{-4\left(-\frac{3}{4}\right)}}$

$= \dfrac{x^{-3/4}y^{1/2}}{z^3} = \dfrac{y^{1/2}}{x^{3/4}z^3}$ Properties 1.7 and 1.8

j. $\left(\dfrac{c^{5/2}d^{-3}}{c^{-1/2}}\right)^{2/3} = (c^{5/2+1/2}d^{-3})^{2/3} = \left(\dfrac{c^3}{d^3}\right)^{2/3}$

$= \dfrac{c^{3\left(\frac{2}{3}\right)}}{d^{3\left(\frac{2}{3}\right)}} = \dfrac{c^2}{d^2}$ Properties 1.4 and 1.8 ■

It is often necessary in calculus to factor algebraic expressions that are *not* polynomials. We may use the techniques of factoring learned in Chapter 5 to factor algebraic expressions that contain rational exponents (see Example 7) and/or negative exponents (see Examples 8 and 9).

Example 7 Factor $6x^{4/3}y^{1/5} - 12x^{1/3}y^{6/5}$.

Solution Recall from Chapter 5 that we must use the *smaller* exponent on each variable for the greatest common factor. The smaller exponent on the x is $1/3$, and on the y it is $1/5$; therefore, the GCF is $6x^{1/3}y^{1/5}$. Also recall that we find the "polynomial factor" by dividing each term of the expression to be factored by the GCF. Therefore, we have

$$\frac{x^{4/3}}{x^{1/3}} = x^{4/3-1/3} = x^1$$

$$6x^{4/3}y^{1/5} - 12x^{1/3}y^{6/5} = 6x^{1/3}y^{1/5}\left(\frac{6x^{4/3}y^{1/5}}{6x^{1/3}y^{1/5}} - \frac{12x^{1/3}y^{6/5}}{6x^{1/3}y^{1/5}}\right) \longleftarrow \text{This step need not be written}$$

$$\frac{y^{6/5}}{y^{1/5}} = y^{6/5-1/5} = y^1$$

$$= 6x^{1/3}y^{1/5}(x - 2y) \longleftarrow \text{Factored form}$$

Check
$$6x^{1/3}y^{1/5}(x - 2y) = 6x^{1/3}y^{1/5}(x) - 6x^{1/3}y^{1/5}(2y)$$
$$= 6x^{4/3}y^{1/5} - 12x^{1/3}y^{6/5} \quad \blacksquare$$

Example 8 Factor $x^{-4} - x^{-2} - 6$.

Solution $x^{-4} - x^{-2} - 6$ is *quadratic in form*; that is, it can be expressed in the form $X^{2} + X^{1} + C$, where X represents some algebraic expression. In this case, the expression can be written as $(x^{-2})^{2} - (x^{-2})^{1} - 6$; we factor it as we would factor any trinomial.

$$x^{-4} - x^{-2} - 6 = (x^{-2})^2 - (x^{-2})^1 - 6$$
$$= (x^{-2} - 3)(x^{-2} + 2)$$
$$= \left(\frac{1}{x^2} - 3\right)\left(\frac{1}{x^2} + 2\right)$$

Check $(x^{-2} - 3)(x^{-2} + 2) = x^{-4} - x^{-2} - 6 \quad \blacksquare$

Example 9 Factor $8x^{-1/3}y^{3/5}z^{-5/4} + 48x^{2/3}y^{-2/5}z^{-1/4}$.

Solution The greatest common factor of 8 and 48 is 8; the smaller exponent on the x is $-1/3$, the smaller exponent on the y is $-2/5$, and the smaller exponent on the z is $-5/4$. Therefore, the GCF is $8x^{-1/3}y^{-2/5}z^{-5/4}$. Then

$$8x^{-1/3}y^{3/5}z^{-5/4} + 48x^{2/3}y^{-2/5}z^{-1/4}$$

$$\frac{y^{3/5}}{y^{-2/5}} = y^{3/5-(-2/5)} = y^1$$

$$= 8x^{-1/3}y^{-2/5}z^{-5/4}\left(\frac{8x^{-1/3}y^{3/5}z^{-5/4}}{8x^{-1/3}y^{-2/5}z^{-5/4}} + \frac{48x^{2/3}y^{-2/5}z^{-1/4}}{8x^{-1/3}y^{-2/5}z^{-5/4}}\right) \longleftarrow \text{This step need not be written}$$

$$\frac{z^{-1/4}}{z^{-5/4}} = z^{-1/4-(-5/4)} = z^1$$

$$\frac{x^{2/3}}{x^{-1/3}} = x^{2/3-(-1/3)} = x^1$$

$$= 8x^{-1/3}y^{-2/5}z^{-5/4}(y + 6xz) \quad \text{or} \quad \frac{8}{x^{1/3}y^{2/5}z^{5/4}}(y + 6xz)$$

Check

$$8x^{-1/3}y^{-2/5}z^{-5/4}(y+6xz)$$
$$= 8x^{-1/3}y^{-2/5}z^{-5/4}(y) + 8x^{-1/3}y^{-2/5}z^{-5/4}(6xz)$$
$$= 8x^{-1/3}y^{3/5}z^{-5/4} + 48x^{2/3}y^{-2/5}z^{-1/4} \quad \blacksquare$$

EXERCISES 7.1B

Set I In Exercises 1–30, perform the indicated operations. Express answers in exponential form, using positive exponents; assume that all variables represent nonnegative numbers.

1. $x^{1/2}x^{3/2}$ **2.** $y^{5/4}y^{3/4}$ **3.** $a^{3/4}a^{-1/2}$

4. $b^{5/6}b^{-1/3}$ **5.** $z^{-1/2}z^{2/3}$ **6.** $N^{-1/3}N^{3/4}$

7. $(H^{3/4})^2$ **8.** $(s^{5/6})^3$ **9.** $(x^{-3/4})^{1/3}$

10. $(y^{-2/3})^{1/2}$ **11.** $\dfrac{a^{3/4}}{a^{1/2}}$ **12.** $\dfrac{b^{2/3}}{b^{1/6}}$

13. $\dfrac{x^{1/2}}{x^{-1/3}}$ **14.** $\dfrac{z^{1/4}}{z^{-1/3}}$ **15.** $x^{2/3}xx^{-1/2}$

16. $x^{3/4}xx^{-1/3}$ **17.** $(x^{-1/2})^3(x^{2/3})^2$ **18.** $(x^{3/4})^2(x^{-2/3})^3$

19. $\dfrac{u^{1/2}v^{-2/3}}{u^{-1/4}v^{-1}}$ **20.** $\dfrac{u^{2/3}v^{-3/5}}{u^{-1/3}v^{-1}}$ **21.** $(16x^{-2/5}y^{4/9})^{3/2}$

22. $(8x^{9/8}y^{-3/4})^{2/3}$ **23.** $\left(\dfrac{a^6d^0}{b^{-9}c^3}\right)^{7/3}$ **24.** $\left(\dfrac{y^{10}z^{-5}}{x^0w^{-15}}\right)^{7/5}$

25. $\left(\dfrac{x^3y^0z^{-1}}{32x^{-1}z^2}\right)^{2/5}$ **26.** $\left(\dfrac{ab^{-1}c^0}{8a^{-4}b^4}\right)^{2/3}$ **27.** $\left(\dfrac{x^{-1}y^{2/3}}{z^{-5}}\right)^{-3/5}$

28. $\left(\dfrac{a^{3/4}b^{-1}}{c^{-2}}\right)^{-2/3}$ **29.** $\left(\dfrac{9x^{-2/3}y^{2/9}}{x^{-2}}\right)^{-3/2}$ **30.** $\left(\dfrac{27R^{3/5}S^{-5/2}}{S^{-4}}\right)^{-2/3}$

In Exercises 31–36, evaluate each expression.

31. $(8)^{-2/3}$ **32.** $(27)^{-2/3}$ **33.** $(4)^{1/2}(9)^{-3/2}$

34. $(8)^{1/3}(16)^{-3/2}$ **35.** $(100)^{-1/2}(-8)^{1/3}$ **36.** $(10^4)^{-1/2}(-27)^{1/3}$

In Exercises 37–48, factor each expression completely. Negative exponents may be left in your answers.

37. $3x^{2/3}y^{7/6} - 9x^{5/3}y^{1/6}$ **38.** $12a^{7/4}b^{3/8} - 8a^{3/4}b^{11/8}$

39. $18x^{5/2}y^{1/3} + 21x^{1/2}y^{4/3}$ **40.** $36x^{3/5}y^{5/4} + 30x^{8/5}y^{1/4}$

41. $5v^{-2} - 17v^{-1} + 6$ **42.** $5v^{-2} - 11v^{-1} + 6$

43. $35k^{-4} - 12k^{-2} + 1$ **44.** $18k^{-4} - 9k^{-2} + 1$

45. $2a^{1/4}b^{-8/3} - a^{-3/4}b^{1/3}$ **46.** $3x^{-2/5}y^{2/3} - 5x^{3/5}y^{-1/3}$

47. $8c^{-3/4}d^{-2/3}e^{1/5} + 12c^{1/4}d^{1/3}e^{-4/5}$ **48.** $15x^{1/2}y^{-1/3} + 10x^{-1/2}y^{2/3}$

Set II In Exercises 1–30, perform the indicated operations. Express answers in exponential form, using positive exponents; assume that all variables represent nonnegative numbers.

1. $a^{1/4}a^{3/4}$ **2.** $b^{3/5}b^{1/4}$ **3.** $x^{5/6}x^{-1/3}$ **4.** $c^{2/3}c^{-1/4}$

5. $z^{-1/5}z^{2/3}$ **6.** $y^{7/8}y^{-1/3}$ **7.** $(x^{3/5})^2$ **8.** $(a^{2/5})^{1/3}$

9. $(x^{-3/7})^{1/3}$ **10.** $(a^{-1/4})^{-8/9}$ **11.** $\frac{x^{7/8}}{x^{1/4}}$ **12.** $\frac{y^{5/6}}{y^{1/3}}$

13. $\frac{p^{3/5}}{p^{-1/2}}$ **14.** $\frac{a^{-2/5}}{a^{-7/10}}$ **15.** $z^{2/5}z^{-3/5}z^{7/10}$

16. $x^{-1/3}x^{2/5}x^{3/4}$ **17.** $(x^{-1/3})^4(x^{2/3})^5$ **18.** $(a^{-3/4})^3(a^{5/6})^2$

19. $\frac{x^{1/3}y^{-4/5}}{x^{-1/4}y^{-1}}$ **20.** $\frac{a^{-2/5}b^{2/3}}{a^{1/2}b^{1/2}}$ **21.** $(36R^{-4/3}S^{2/5})^{1/2}$

22. $(49x^{-1/5}y^{3/4})^{1/2}$ **23.** $\left(\frac{s^3t^0}{u^{-6}v^9}\right)^{4/3}$ **24.** $\left(\frac{a^0b^4}{c^8d^{-12}}\right)^{1/4}$

25. $\left(\frac{y^{2/3}z^{-2/3}}{4z^{-3}}\right)^{3/2}$ **26.** $\left(\frac{8a^{-12}}{b^9c^3}\right)^{2/3}$ **27.** $\left(\frac{R^{-3/2}S^{1/2}}{8R^{-2}}\right)^{2/3}$

28. $\left(\frac{16x^8y^4}{x^{12}}\right)^{3/4}$ **29.** $\left(\frac{x^2y^{-4}z^6}{32x^{-2}y^0z}\right)^{3/5}$ **30.** $\left(\frac{64a^6b^8}{a^{18}b^2}\right)^{5/6}$

In Exercises 31–36, evaluate each expression.

31. $(64)^{-2/3}$ **32.** $16^{-3/4}$ **33.** $9^{1/2}(25)^{3/2}$

34. $(10^6)^{-1/2}(125)^{1/3}$ **35.** $\frac{(10^4)^{1/2}(10^{-2})^{3/2}}{(10^5)^{-1/5}}$ **36.** $\frac{(2^4)^{1/4}(2^{-8})^{1/2}}{(2^{-6})^{-1/3}}$

In Exercises 37–48, factor each expression completely. Negative exponents may be left in your answers.

37. $8a^{1/6}b^{5/4} - 4a^{7/6}b^{1/4}$ **38.** $3x^{7/6}y^{1/4} - 6x^{1/6}y^{5/4}$

39. $9u^{3/8}v^{8/3} + 6u^{11/8}v^{2/3}$ **40.** $5z^{-2} - 19z^{-1} + 12$

41. $3t^{-2} + 17t^{-1} - 6$ **42.** $7x^{-4/5}y^{2/3}z^{1/4} + 14x^{1/5}y^{-1/3}z^{-3/4}$

43. $3t^{-2} + 14t^{-1} - 5$ **44.** $3t^{-2} + 7t^{-1} - 6$

45. $5a^{3/4}b^{-1/3} - a^{-1/4}b^{2/3}$ **46.** $2x^{-4} - 9x^{-2} + 9$

47. $6x^{-3/4}y^{1/3}z^{-1/2} + 8x^{1/4}y^{-2/3}z^{1/2}$ **48.** $5x^{3/5}y^{-3/5} - 25x^{-2/5}y^{2/5}$

7.2 Simplifying Radicals

In this section, we will simplify radicals of three kinds: (1) radicals that do not involve rational expressions, (2) radicals that do involve rational expressions, and (3) radicals in which the order of the radical can be reduced.

7.2A Radicals That Do Not Involve Rational Expressions

Recall from Section 1.14A that the rule for simplifying a square root with factors that are positive variables is to divide each exponent by 2. We can now show why this is so. If $x \geq 0$,

$$\sqrt{x^n} = (x^n)^{1/2} = x^{n/2}$$

Note that in the expression $\sqrt{x^n}$, we can consider x^n to be a *factor* of the radicand. The other factor is an understood 1.

A WORD OF CAUTION If the square root contains *two or more terms* with even exponents, we *cannot* simplify the radical by dividing each exponent by 2. That is,

$$\sqrt{x^2 + y^2} \neq x + y$$

Proof Suppose $x = 3$ and $y = 4$. Then $\sqrt{3^2 + 4^2} = \sqrt{9 + 16} = \sqrt{25} = 5$, but $3 + 4 = 7$. Therefore, $\sqrt{3^2 + 4^2} \neq 3 + 4$.

$\sqrt{x^2} = |x|$

As we mentioned in Section 1.14A, the *principal* square root of x^2 is $|x|$. If x is *assumed* to be positive, we may use $\sqrt{x^2} = x$, as we did in Chapter 1. If x is *not* assumed to be positive, we must use $\sqrt{x^2} = |x|$. To see why, let $x = -5$: Then "$\sqrt{(-5)^2} = -5$" is a *false* statement, since $\sqrt{(-5)^2} = \sqrt{25} = 5$, and $5 \neq -5$. Therefore, if $x = -5$, $\sqrt{x^2} \neq x$. However, "$\sqrt{(-5)^2} = |-5|$" is a *true* statement and is equivalent to $\sqrt{x^2} = |x|$.

Example 1 Simplify each radical.

a. $\sqrt{x^6 y^8}$

Solution $\sqrt{x^6 y^8} = |x^{6/2} y^{8/2}| = |x^3 y^4|$ or $|x^3| y^4$. We do not need absolute value symbols around y^4 because y^4 is positive when y is *any* real number, positive or negative.

b. $\sqrt{(a + b)^2}$. Assume that a and b are positive.

Solution $\sqrt{(a + b)^2} = (a + b)^{2/2} = (a + b)^1 = a + b$. We do not need absolute value symbols because a and b were *assumed* positive, and the sum of two positive numbers is always positive. ■

Higher-Order Radicals

Before we consider simplifying *higher*-order radicals, a new definition will be helpful.

Perfect Power A factor of a radicand raised to an exponent that is exactly divisible by the index of the radical is called a **perfect power**.

> TO SIMPLIFY A RADICAL THAT INVOLVES ONLY PRODUCTS OF PERFECT POWERS
>
> To simplify a radical when the radicand is a *product* of perfect powers, first express any numerical coefficients in prime factored form; then divide the exponent of each factor by the index of the radical. If the index is an *even* number, you may need to place absolute value symbols around the result.

Because the principal root is negative if the radicand is negative when the index is odd, cubic roots of negative numbers will be negative, fifth roots of negative numbers will be negative, and so forth. This fact suggests that a negative sign can be "moved across" the radical sign *if and only if* the index of the radical is an odd number. You can verify that $\sqrt[3]{-8} = -\sqrt[3]{8}$, that $\sqrt[5]{-32} = -\sqrt[5]{32}$, and so forth.

A radical is not considered to be in *simplest radical form* if the radicand is negative.

A WORD OF CAUTION $\sqrt{-16} \neq -\sqrt{16}$. Remember, $-\sqrt{16} = -4$, but $\sqrt{-16}$ is *not a real number*. (Numbers such as $\sqrt{-16}$ will be discussed in Section 7.8.) ☑

Example 2 Simplify each radical.

a. $\sqrt[4]{2^4}$

Solution $\sqrt[4]{2^4} = 2^{4/4} = 2^1 = 2$. No absolute value symbols are necessary because 2 is a positive number.

b. $\sqrt[4]{z^{12}}$

Solution $\sqrt[4]{z^{12}} = |z^{12/4}| = |z^3|$. Absolute value symbols are necessary because the index of the radical is even.

c. $\sqrt[3]{-b^6}$

Solution (The cubic root of a negative number is negative)

$\sqrt[3]{-b^6} = -\sqrt[3]{b^6} = -b^{6/3} = -b^2$. Note that $-b^2$ is a *negative* number. Let's verify that this is correct by substituting 2 for b: $\sqrt[3]{-2^6} = \sqrt[3]{-64} = -4 = -2^2$, *and* $-\sqrt[3]{64} = -4 = -2^2$.

d. $\sqrt[7]{-2^7y^7}$

Solution $\sqrt[7]{-2^7y^7} = -\sqrt[7]{2^7y^7} = -(2^7y^7)^{1/7} = -2^1y^1 = -2y$. Absolute value signs would be *incorrect* here because the index is odd. $-2y$ is a positive number if y is negative and a negative number if y is positive, as it should be. ■

A basic property of radicals is the following:

PROPERTY 7.3

$$\sqrt[n]{ab} = \sqrt[n]{a}\sqrt[n]{b}$$

if a and b are not *both* negative when n is even.

This property is easily proved if we change the radicals to exponential form (for simplicity in the proof, we assume that $a \geq 0$ and $b \geq 0$):

$$\sqrt[n]{ab} = (ab)^{1/n} = a^{1/n}b^{1/n} = \sqrt[n]{a}\sqrt[n]{b}$$

It is *also* true, of course, that $\sqrt[n]{a}\sqrt[n]{b} = \sqrt[n]{ab}$.

In this section we will use Property 7.3 for *simplifying* radicals, and in Section 7.4 we will use it for *multiplying* radicals.

Example 3 Examples verifying that Property 7.3 is true:

a. $\sqrt{36} \stackrel{?}{=} \sqrt{4}\sqrt{9}$

$6 \stackrel{?}{=} 2 \cdot 3$

$6 = 6$

b. $\sqrt[3]{8}\sqrt[3]{27} \stackrel{?}{=} \sqrt[3]{216}$

$2 \cdot 3 \stackrel{?}{=} 6$

$6 = 6$ ■

NOTE We now have an alternate way of approaching a problem such as $\sqrt[3]{-b^6}$. We can think of it this way:

$$\sqrt[3]{-b^6} = \sqrt[3]{(-1)b^6} = \sqrt[3]{-1}\sqrt[3]{b^6} = (-1)(b^{6/3}) = -b^2 \quad ✓$$

Simplified Form of a Radical

A number of conditions must be satisfied in order for a radical to be in *simplest radical form.* We will explore these conditions in the following sections. The *first* of these conditions is that the radicand must be positive. The *second* is that when the radicand is expressed in prime factored form, none of the factors can have an exponent that is greater than or equal to the index of the radical.

When any exponent *is* greater than the index of the radical, we simplify the radical as follows:

> TO SIMPLIFY A RADICAL THAT DOES NOT INVOLVE RATIONAL EXPRESSIONS
>
> 1. Express the radicand in prime factored form.
> 2. If the exponent of any factor of the radicand is greater than the index, break up that factor into the product of a perfect power and a nonperfect power *whose exponent is less than the index.*
> 3. Use Property 7.3 to rewrite the radical as a product of radicals.
> 4. Remove perfect powers by dividing their exponents by the index. All remaining exponents should be less than the index; factors with such exponents remain under the radical sign.
> 5. The simplified radical is the *product* of all factors from step 4. (Absolute value symbols may be necessary around the answer.)
> 6. It may be possible to reduce the order of the radical (see Section 7.2C).

Example 4 Examples of simplifying radicals:

a. $\sqrt{27} = \sqrt{3^3}$ Prime factored form of 27

$= \sqrt{3^2 \cdot 3}$ 3^2 is a perfect power

$= \sqrt{3^2}\sqrt{3}$

$= 3\sqrt{3}$

b. $\sqrt{48} = \sqrt{2^4 \cdot 3}$ ← Prime factored form of 48

$= \sqrt{2^4}\sqrt{3}$

$= 2^2\sqrt{3}$

$= 4\sqrt{3}$

2	48
2	24
2	12
2	6
	3

$48 = 2^4 \cdot 3$

c. $\sqrt[3]{54} = \sqrt[3]{2 \cdot 3^3}$ ← Prime factored form of 54

$= \sqrt[3]{2}\sqrt[3]{3^3}$

$= \sqrt[3]{2}(3)$

$= 3\sqrt[3]{2}$ ← It is customary to write rational factors before irrational ones

2	54
3	27
3	9
	3

$54 = 2 \cdot 3^3$

d. $\sqrt{360} = \sqrt{2^3 \cdot 3^2 \cdot 5}$

$= \sqrt{2^2 \cdot 2 \cdot 3^2 \cdot 5}$

$= 2 \cdot 3 \cdot \sqrt{2 \cdot 5} = 6\sqrt{10}$ Simplified form

e. $\sqrt{12x^4y^3} = \sqrt{2^2 \cdot \boxed{3} \cdot x^4 \cdot y^2 \cdot \boxed{y}}$

$= 2 \cdot x^2 \cdot |y| \sqrt{\boxed{3y}} = 2x^2|y|\sqrt{3y}$ Simplified form

We don't need absolute value symbols around x^2 since x^2 is always positive

f. $\sqrt[3]{-16a^5b^7} = \sqrt[3]{-2^4a^5b^7}$

$= \sqrt[3]{-2^3 \cdot \boxed{2} \cdot a^3 \cdot \boxed{a^2} \cdot b^6 \cdot \boxed{b}}$

$= -2 \cdot a \cdot b^2 \sqrt[3]{\boxed{2 \cdot a^2 \cdot b}} = -2ab^2\sqrt[3]{2a^2b}$

g. $\sqrt[4]{96h^{11}k^3m^8} = \sqrt[4]{2^5 \cdot 3h^{11}k^3m^8}$

$= \sqrt[4]{2^4 \cdot \boxed{2} \cdot \boxed{3} \cdot h^8 \cdot \boxed{h^3} \cdot \boxed{k^3} \cdot m^8}$

$= 2 \cdot h^2 \cdot m^2 \sqrt[4]{\boxed{2 \cdot 3 \cdot h^3 \cdot k^3}}$

$= 2h^2m^2\sqrt[4]{6h^3k^3}$

h. $5xy^2\sqrt{28x^4y^9z^3} = 5xy^2\sqrt{2^2 \cdot \boxed{7} \cdot x^4 \cdot y^8 \cdot \boxed{y} \cdot z^2 \cdot \boxed{z}}$

$= 5xy^2 \cdot 2 \cdot x^2 \cdot y^4 \cdot |z|\sqrt{\boxed{7 \cdot y \cdot z}}$

$= 10x^3y^6|z|\sqrt{7yz}$ ■

Sometimes you can find the root of an integer easily if by inspection you can see that the number has a factor that is a perfect power (see Example 5).

Example 5 Examples of simplifying radicals with factors that are perfect powers:

a. $\sqrt{50} = \sqrt{25 \cdot 2} = \sqrt{25}\sqrt{2} = 5\sqrt{2}$

b. $\sqrt[3]{24} = \sqrt[3]{8 \cdot 3} = \sqrt[3]{8}\sqrt[3]{3} = 2\sqrt[3]{3}$

c. $\sqrt[4]{32} = \sqrt[4]{16 \cdot 2} = \sqrt[4]{16}\sqrt[4]{2} = 2\sqrt[4]{2}$ ■

EXERCISES 7.2A

Set I Simplify each radical. Assume $a \neq 0$, $b \neq 0$, and $c \neq 0$.

1. $\sqrt{4x^2}$ **2.** $\sqrt{9y^2}$ **3.** $\sqrt[3]{8x^3}$ **4.** $\sqrt[3]{27y^3}$

5. $\sqrt[4]{16x^4y^8}$ **6.** $\sqrt[4]{81u^8v^4}$ **7.** $\sqrt{2^5}$ **8.** $\sqrt{3^3}$

9. $\sqrt{(-2)^2}$ **10.** $\sqrt{(-3)^2}$ **11.** $\sqrt[3]{-3^5}$ **12.** $\sqrt[3]{-2^8}$

13. $\sqrt[4]{32}$ **14.** $\sqrt[4]{48}$ **15.** $\sqrt[5]{-x^7}$ **16.** $\sqrt[5]{-z^8}$

17. $\sqrt{8a^4b^2}$ **18.** $\sqrt{20m^8u^2}$ **19.** $\sqrt{18m^3n^5}$ **20.** $\sqrt{50h^5k^3}$

21. $5\sqrt[3]{-24a^5b^2}$ **22.** $6\sqrt[3]{-54c^4d}$ **23.** $\sqrt[5]{64m^{11}p^{15}u}$

24. $\sqrt[5]{128u^4v^{10}w^{16}}$ **25.** $\sqrt[3]{8(a+b)^3}$ **26.** $\sqrt[3]{27(x-y)^6}$

27. $\dfrac{3}{2abc}\sqrt[3]{2^5a^8b^9c^{10}}$ **28.** $\dfrac{7}{2bc}\sqrt[3]{2^6a^5b^6c^7}$

Set II Simplify each radical. Assume $a \neq 0$, $b \neq 0$, $x \neq 0$, and $y \neq 0$.

1. $\sqrt{25x^2}$ **2.** $\sqrt{64a^4}$ **3.** $\sqrt[6]{64y^6}$ **4.** $\sqrt[5]{32b^8}$

5. $\sqrt[4]{256a^8b^{12}}$ **6.** $\sqrt[4]{81x^4y^7}$ **7.** $\sqrt{2^7}$ **8.** $\sqrt[3]{5^4}$

9. $\sqrt{(-7)^2}$ **10.** $\sqrt[5]{-3^5}$ **11.** $\sqrt[3]{-2^5}$ **12.** $\sqrt{(-3)^4}$

13. $\sqrt{40}$ **14.** $\sqrt[5]{128}$ **15.** $\sqrt[5]{-x^7}$ **16.** $\sqrt[3]{a^8}$

17. $\sqrt{75x^5y^4}$ **18.** $\sqrt{48a^3b^5c}$ **19.** $\sqrt[3]{27xy^8z^5}$ **20.** $\sqrt[3]{-16x^3y^2}$

21. $3\sqrt{32x^5y^9}$ **22.** $\sqrt[3]{-27ab^2c^3}$ **23.** $\sqrt[7]{256x^9y^{15}}$ **24.** $\sqrt{21x^4y^7z^{10}}$

25. $\sqrt[3]{(a+b)^6}$ **26.** $-8\sqrt[3]{-8s^2t^7}$ **27.** $\frac{3}{4xy}\sqrt[3]{2^6x^5y^6z^5}$ **28.** $\frac{5}{2ab^2}\sqrt[3]{8a^5b^3}$

7.2B Radicals Involving Rational Expressions

We now give another basic property of radicals:

> PROPERTY 7.4
>
> $$\sqrt[n]{\frac{a}{b}} = \frac{\sqrt[n]{a}}{\sqrt[n]{b}} \qquad (b \neq 0)$$
>
> if a and b are not *both* negative when n is even.

Proof (For simplicity in the proof, we assume $a \geq 0$ and $b > 0$.) If we change the radicals to exponential form, we have

$$\sqrt[n]{\frac{a}{b}} = \left(\frac{a}{b}\right)^{1/n} = \frac{a^{1/n}}{b^{1/n}} = \frac{\sqrt[n]{a}}{\sqrt[n]{b}}$$

It is also true, of course, that $\frac{\sqrt[n]{a}}{\sqrt[n]{b}} = \sqrt[n]{\frac{a}{b}}$.

Property 7.4 can be used in simplifying radicals in which the radicand is a rational expression (see Example 6), and in Section 7.4 we will use it in *dividing* radicals.

A *third* condition that must be satisfied in order for a radical to be in *simplest radical form* is that the radicand cannot be a rational expression.

Example 6 Examples of simplifying radicals involving rational expressions (assume $y \neq 0$ and $k \neq 0$):

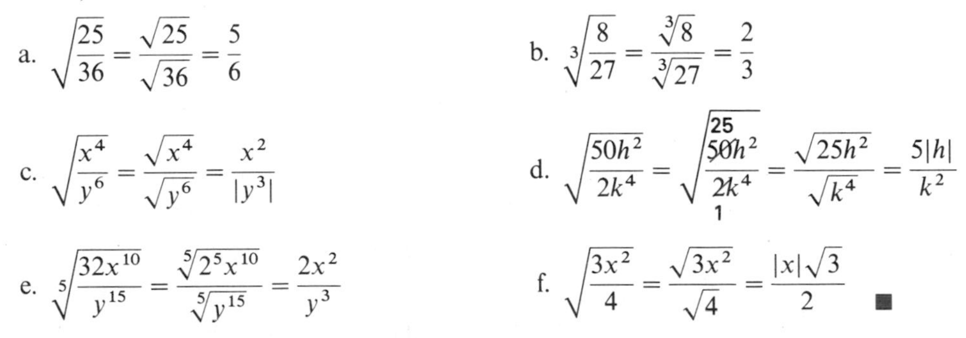

a. $\sqrt{\frac{25}{36}} = \frac{\sqrt{25}}{\sqrt{36}} = \frac{5}{6}$

b. $\sqrt[3]{\frac{8}{27}} = \frac{\sqrt[3]{8}}{\sqrt[3]{27}} = \frac{2}{3}$

c. $\sqrt{\frac{x^4}{y^6}} = \frac{\sqrt{x^4}}{\sqrt{y^6}} = \frac{x^2}{|y^3|}$

d. $\sqrt{\frac{50h^2}{2k^4}} = \sqrt{\frac{25h^2}{k^4}} = \frac{\sqrt{25h^2}}{\sqrt{k^4}} = \frac{5|h|}{k^2}$

e. $\sqrt[5]{\frac{32x^{10}}{y^{15}}} = \frac{\sqrt[5]{2^5x^{10}}}{\sqrt[5]{y^{15}}} = \frac{2x^2}{y^3}$

f. $\sqrt{\frac{3x^2}{4}} = \frac{\sqrt{3x^2}}{\sqrt{4}} = \frac{|x|\sqrt{3}}{2}$ ■

A *fourth* condition that must be satisfied if a radical is to be in *simplest radical form* is that no denominator can contain a radical. (In Example 6, each denominator happened to be a perfect power, so no radicals were left in the denominators, but that will not always be the case.)

If there is a rational expression under the radical sign or if the denominator contains a radical, we may need to multiply the numerator and denominator of the rational expression by some factor that will change the denominator into a rational number. This procedure is called **rationalizing the denominator** (see Example 7).

Example 7 Rationalize the denominator in each expression.

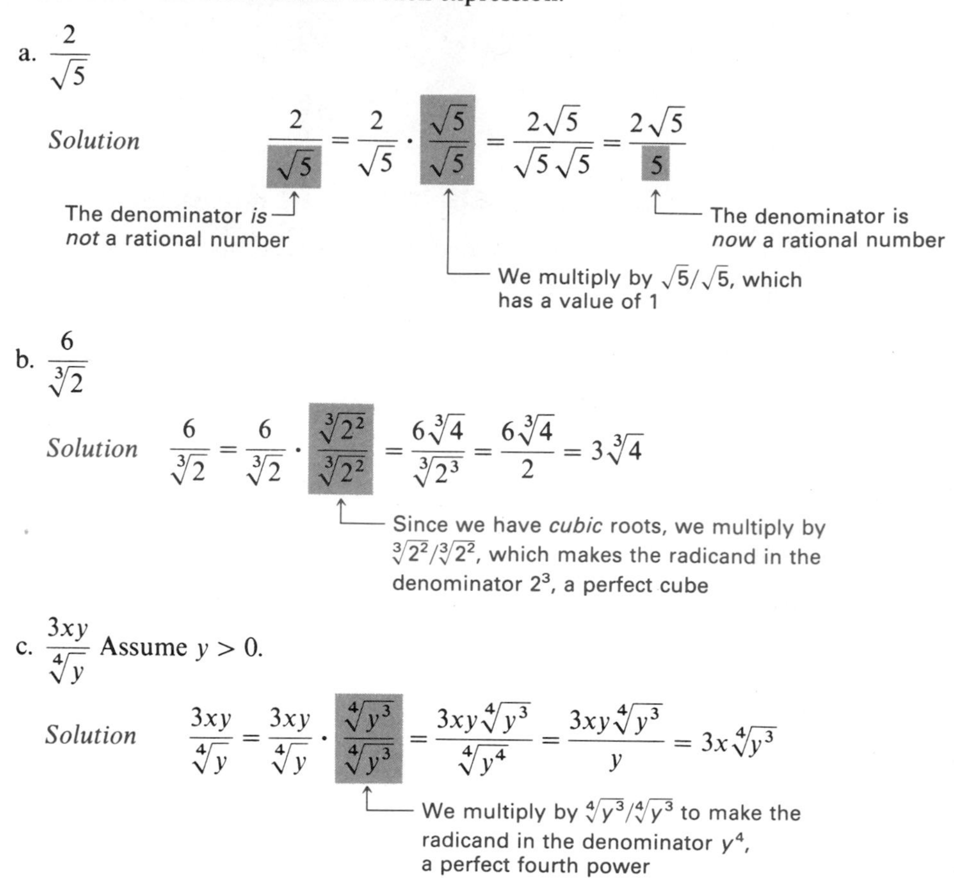

a. $\dfrac{2}{\sqrt{5}}$

Solution $\dfrac{2}{\sqrt{5}} = \dfrac{2}{\sqrt{5}} \cdot \dfrac{\sqrt{5}}{\sqrt{5}} = \dfrac{2\sqrt{5}}{\sqrt{5}\sqrt{5}} = \dfrac{2\sqrt{5}}{5}$

The denominator *is not* a rational number

We multiply by $\sqrt{5}/\sqrt{5}$, which has a value of 1

The denominator is *now* a rational number

b. $\dfrac{6}{\sqrt[3]{2}}$

Solution $\dfrac{6}{\sqrt[3]{2}} = \dfrac{6}{\sqrt[3]{2}} \cdot \dfrac{\sqrt[3]{2^2}}{\sqrt[3]{2^2}} = \dfrac{6\sqrt[3]{4}}{\sqrt[3]{2^3}} = \dfrac{6\sqrt[3]{4}}{2} = 3\sqrt[3]{4}$

Since we have *cubic* roots, we multiply by $\sqrt[3]{2^2}/\sqrt[3]{2^2}$, which makes the radicand in the denominator 2^3, a perfect cube

c. $\dfrac{3xy}{\sqrt[4]{y}}$ Assume $y > 0$.

Solution $\dfrac{3xy}{\sqrt[4]{y}} = \dfrac{3xy}{\sqrt[4]{y}} \cdot \dfrac{\sqrt[4]{y^3}}{\sqrt[4]{y^3}} = \dfrac{3xy\sqrt[4]{y^3}}{\sqrt[4]{y^4}} = \dfrac{3xy\sqrt[4]{y^3}}{y} = 3x\sqrt[4]{y^3}$

We multiply by $\sqrt[4]{y^3}/\sqrt[4]{y^3}$ to make the radicand in the denominator y^4, a perfect fourth power

A WORD OF CAUTION Be sure that the index is written so that it appears *inside* the "arm" of the radical symbol. Otherwise, the expression $3x\sqrt[4]{y^3}$ may look like $3x^4\sqrt{y^3}$. You may even want to write the expression as $3x \cdot \sqrt[4]{y^3}$ or as $3x(\sqrt[4]{y^3})$. ■

TO SIMPLIFY A RADICAL INVOLVING RATIONAL EXPRESSIONS

1. Reduce the rational expression to lowest terms.
2. Write the numerator and denominator in prime factored form.
3. Remove perfect powers by dividing their exponents by the index. (The root of a perfect power in the numerator of the radicand becomes a factor of the numerator outside the radical. The root of a perfect power in the denominator becomes a factor of the denominator outside the radical.)
4. Multiply the numerator and denominator of the remaining radicand by an expression that makes the denominator a perfect power.
5. Evaluate the perfect power in the denominator formed in step 4.
6. The simplified radical is the rational expression formed by those factors that were removed (reduced to lowest terms) multiplied by the radical whose radicand is the product of all factors that were not removed.

Rationalizing the denominator can often be combined with simplifying a radical, as Example 8 shows.

Example 8 Simplify the radicals.

a. $\sqrt{\dfrac{4}{5}}$

Solution We can do this problem in either of two ways:

$$\sqrt{\frac{4}{5}} = \frac{\sqrt{4}}{\sqrt{5}} = \frac{2}{\sqrt{5}} = \frac{2\sqrt{5}}{\sqrt{5}\sqrt{5}} = \frac{2\sqrt{5}}{5}$$

See Example 7a

or

$$\sqrt{\frac{4}{5}} = \sqrt{\frac{4}{5} \cdot \frac{5}{5}} = \frac{2\sqrt{5}}{\sqrt{25}} = \frac{2\sqrt{5}}{5}$$

Under the radical sign, we multiply by $\frac{5}{5}$ in order to make the denominator a perfect square

b. $\sqrt{\dfrac{x^3}{8}}$

Solution

$$\sqrt{\frac{x^3}{8}} = \sqrt{\frac{x^3}{8} \cdot \frac{2}{2}} = \sqrt{\frac{2x^3}{16}} = \sqrt{\frac{2 \cdot x^2 \cdot x}{16}} = \frac{|x|}{4}\sqrt{2x} \quad \text{Simplified form}$$

Under the radical sign, we multiply by $\frac{2}{2}$ in order to make the denominator a perfect square

c. $\sqrt[3]{\dfrac{a^4b^3}{-2c}}$ Assume $c \neq 0$.

Solution

$$\sqrt[3]{\frac{a^4b^3}{-2c}} = \sqrt[3]{\frac{a^4b^3}{-2c} \cdot \frac{2^2c^2}{2^2c^2}} = \frac{\sqrt[3]{a^3ab^32^2c^2}}{\sqrt[3]{-2^3c^3}} = \frac{ab\sqrt[3]{4ac^2}}{-2c} = -\frac{ab\sqrt[3]{4ac^2}}{2c}$$

Under the radical sign, we multiply by $2^2c^2/2^2c^2$ in order to make the denominator a perfect cube

d. $\sqrt[5]{\dfrac{24x^2y^6}{64x^4y}}$ Assume $x \neq 0$, $y \neq 0$.

We reduced the radicand first

Solution

$$\sqrt[5]{\frac{24x^2y^6}{64x^4y}} = \sqrt[5]{\frac{\overset{3}{\cancel{24}}x^2y^6}{\underset{8}{\cancel{64}}x^4y}} = \sqrt[5]{\frac{3y^5}{2^3x^2} \cdot \frac{2^2x^3}{2^2x^3}} = \frac{\sqrt[5]{12y^5x^3}}{\sqrt[5]{2^5x^5}} = \frac{y\sqrt[5]{12x^3}}{2x}$$

Under the radical sign, we multiply by $2^2x^3/2^2x^3$ in order to make the denominator a perfect fifth power

e. $\dfrac{8y}{x^4}\sqrt{\dfrac{5x^5y^2}{2xy^3}}$ Assume $x \neq 0$, $y > 0$.

Solution

$$\frac{8y}{x^4}\sqrt{\frac{5x^5y^2}{2xy^3}} = \frac{8y}{x^4}\sqrt{\frac{5x^4}{2y}} = \frac{8y}{x^4} \cdot \frac{x^2}{1}\sqrt{\frac{5}{2y} \cdot \frac{2y}{2y}}$$

$$= \frac{\overset{4}{\cancel{8}y}}{\underset{x^2}{\cancel{x^4}}} \cdot \frac{\overset{1}{\cancel{x^2}}}{\underset{1}{\cancel{2y}}}\sqrt{10y} = \frac{4}{x^2}\sqrt{10y}$$

■

EXERCISES 7.2B

Set I Simplify each radical; assume that all variables in denominators are nonzero.

1. $\sqrt{\frac{16}{25}}$ 2. $\sqrt{\frac{64}{100}}$ 3. $\sqrt[3]{\frac{-27}{64}}$ 4. $\sqrt[3]{\frac{-8}{125}}$

5. $\sqrt[4]{\frac{a^4b^8}{16c^0}}$ 6. $\sqrt[4]{\frac{c^8d^{12}}{81e^0}}$ 7. $\sqrt{\frac{4x^3y}{xy^3}}$ 8. $\sqrt{\frac{x^5y}{9xy^3}}$

9. $\frac{10}{\sqrt{5}}$ 10. $\frac{14}{\sqrt{2}}$ 11. $\frac{5}{\sqrt{3}}$ 12. $\frac{8}{\sqrt{7}}$

13. $\frac{9}{\sqrt[3]{3}}$ 14. $\frac{10}{\sqrt[3]{5}}$ 15. $\frac{8}{\sqrt[5]{4}}$ 16. $\frac{5}{\sqrt[4]{3}}$

17. $\sqrt[3]{\frac{m^5}{-3}}$ 18. $\sqrt[3]{\frac{k^4}{-5}}$ 19. $\frac{n}{2m}\sqrt[3]{\frac{8m^2n}{2n^3}}$

20. $\frac{x}{3y}\sqrt[3]{\frac{18xy^2}{2x^3}}$ 21. $\sqrt[4]{\frac{3m^7}{4m^3p^2}}$ 22. $\sqrt[4]{\frac{5a^9}{2a^5b^3}}$

23. $\sqrt[5]{\frac{15x^4y^7}{24x^6y^2}}$ 24. $\sqrt[5]{\frac{30mp^8}{48m^4p^3}}$ 25. $\frac{4x^2}{5y^3}\sqrt[3]{\frac{3x^2y^3}{8x^5y}}$

26. $\frac{3x}{4y^2}\sqrt[3]{\frac{2xy^5}{3x^3y}}$ 27. $\frac{2x^3}{y}\sqrt[3]{\frac{5y^0z}{16x^7z^8}}$ 28. $\frac{5c}{b}\sqrt[3]{\frac{b^4cd^2}{25a^0c^5}}$

Set II Simplify each radical; assume that all variables in denominators are nonzero.

1. $\sqrt{\frac{36}{81}}$ 2. $\sqrt{\frac{100}{49}}$ 3. $\sqrt[3]{\frac{-1}{27}}$ 4. $\sqrt[4]{\frac{16}{81}}$

5. $\sqrt{\frac{x^6y^0}{v^8}}$ 6. $\sqrt[4]{\frac{5^0}{a^4b^8}}$ 7. $\sqrt[5]{\frac{x^{15}y^{10}}{-243}}$ 8. $\sqrt[3]{\frac{6^0st^3}{s^3t^2}}$

9. $\frac{18}{\sqrt{3}}$ 10. $\frac{6}{\sqrt{15}}$ 11. $\frac{3}{\sqrt[5]{8}}$ 12. $\frac{16}{\sqrt[3]{12}}$

13. $\frac{8}{\sqrt[7]{4}}$ 14. $\frac{1}{\sqrt[3]{16}}$ 15. $\frac{2}{\sqrt[3]{3}}$ 16. $\frac{3}{\sqrt[3]{81}}$

17. $\sqrt[3]{\frac{x^7}{-2}}$ 18. $\sqrt[5]{\frac{-y^3}{4}}$ 19. $\frac{4}{a}\sqrt[3]{\frac{3a^2b}{4b^3}}$ 20. $\frac{x}{3}\sqrt[3]{\frac{75x^8}{3x}}$

21. $\sqrt[3]{\frac{8a^4b^5}{81ab^2}}$ 22. $\sqrt[3]{\frac{48x^3y^5}{8x^6y^2}}$ 23. $\sqrt[7]{\frac{2xy^5}{12x^2y^2}}$ 24. $\sqrt[3]{\frac{y^5zw^2}{4x^0z^5}}$

25. $\frac{3x^4}{2y^5}\sqrt[3]{\frac{96xy^4}{2x^2y}}$ 26. $\frac{5x}{2y^4}\sqrt[3]{\frac{8xy^6}{5x^3y}}$ 27. $\frac{7xz}{y}\sqrt[3]{\frac{3x}{4^0yz^2}}$ 28. $\frac{s^3}{t^2}\sqrt[3]{\frac{2st^4}{8s^2t^5}}$

7.2C Radicals in Which the Order of the Radical Can Be Reduced

In this section, we will assume that all variables represent nonnegative numbers. If we change $\sqrt[4]{x^2}$ to exponential form, we have $x^{2/4}$. But since $\frac{2}{4} = \frac{1}{2}$, $x^{2/4} = x^{1/2}$. Also, $x^{1/2} = \sqrt{x}$. Therefore, $\sqrt[4]{x^2} = \sqrt{x}$. Rewriting a radical as an *equivalent* radical with a *smaller index* is called **reducing the order** of the radical.

Whenever there is some number that is a factor of the index of the radical and *also* a factor of every exponent under the radical, the order of the radical can be reduced. The easiest way to do this is to change the radical to exponential form and reduce the rational exponents; then change the expression back to radical form.

Example 9 Examples of reducing the order of radicals:

a. $\sqrt[4]{x^2} = x^{2/4} = x^{1/2} = \sqrt{x}$

b. $\sqrt[6]{8b^3} = (2^3b^3)^{1/6} = 2^{3/6}b^{3/6} = 2^{1/2}b^{1/2} = (2b)^{1/2} = \sqrt{2b}$

c. $\sqrt[8]{16x^4y^4z^4} = (2^4x^4y^4z^4)^{1/8} = ((2xyz)^4)^{1/8} = (2xyz)^{1/2} = \sqrt{2xyz}$ ■

A *fifth* condition that must be satisfied if a radical is to be in simplest form is that the *order of the radical* must be as small as possible.

Let's summarize the rules we've had so far for expressing a radical in *simplest radical form*:

A RADICAL IS IN SIMPLEST FORM IF:

1. The radicand is a positive number.
2. No prime factor of a radicand has an exponent equal to or greater than the index.
3. No radicand contains a rational expression.
4. No denominator contains a radical.
5. The order of the radical cannot be reduced.

Example 10 Express the radicals in simplest radical form (assume that all variables represent positive numbers).

a. $\sqrt[8]{16x^{12}y^4z^{20}}$

Solution $\sqrt[8]{16x^{12}y^4z^{20}} = (2^4x^{12}y^4z^{20})^{1/8} = 2^{4/8}x^{12/8}y^{4/8}z^{20/8}$

$= 2^{1/2}x^{3/2}y^{1/2}z^{5/2} = (2x^3yz^5)^{1/2} = \sqrt{2x^3yz^5} = xz^2\sqrt{2xyz}$

b. $\sqrt[6]{\dfrac{8y^9}{x^3}}$

Solution $\sqrt[6]{\dfrac{8y^9}{x^3}} = \sqrt[6]{\dfrac{8y^9 \cdot x^3}{x^3 \cdot x^3}} = \sqrt[6]{\dfrac{2^3y^9x^3}{x^6}} = \left(\dfrac{2^3y^9x^3}{x^6}\right)^{1/6}$

$= \dfrac{2^{3/6}y^{9/6}x^{3/6}}{x^{6/6}} = \dfrac{2^{1/2}y^{3/2}x^{1/2}}{x} = \dfrac{(2xy^3)^{1/2}}{x}$

$= \dfrac{\sqrt{2xy^3}}{x} = \dfrac{y\sqrt{2xy}}{x}$ *or* $\dfrac{y}{x}\sqrt{2xy}$ ■

EXERCISES 7.2C

Set I Express the radicals in simplest radical form; assume that all variables represent nonnegative numbers.

1. $\sqrt[6]{x^3}$ **2.** $\sqrt[6]{x^2}$ **3.** $\sqrt[8]{a^6}$ **4.** $\sqrt[8]{a^2}$

5. $\sqrt[6]{27b^3}$ **6.** $\sqrt[6]{4b^4}$ **7.** $\sqrt[6]{49a^2}$ **8.** $\sqrt[4]{144x^2}$

9. $\sqrt[8]{81x^4y^0z^{12}}$ **10.** $\sqrt[6]{27x^9y^3z^0}$ **11.** $\sqrt[6]{256x^8y^4z^{10}}$ **12.** $\sqrt[4]{64x^8y^2z^6}$

13. $\sqrt[6]{\dfrac{x^3}{27}}$ **14.** $\sqrt[6]{\dfrac{x^2}{4}}$ **15.** $\dfrac{1}{\sqrt[6]{a^3}}, a \neq 0$ **16.** $\dfrac{1}{\sqrt[4]{x^2}}, x \neq 0$

Set II Express the radicals in simplest radical form; assume that all variables represent nonnegative numbers.

1. $\sqrt[6]{x^4}$ **2.** $\sqrt[10]{y^8}$ **3.** $\sqrt[6]{z^9}$ **4.** $\sqrt[8]{a^{10}}$

5. $\sqrt[6]{81x^4}$ **6.** $\sqrt[8]{16y^6}$ **7.** $\sqrt[4]{25y^2}$ **8.** $\sqrt[6]{64x^6y^9}$

9. $\sqrt[6]{16x^4y^0z^8w^{12}}$ **10.** $\sqrt[4]{81a^6b^8c^2}$ **11.** $\sqrt[6]{81a^6b^8c^{10}}$ **12.** $\sqrt[6]{5^0x^3y^9z^6}$

13. $\sqrt[6]{\dfrac{a^3}{64}}$ **14.** $\sqrt[6]{\dfrac{16x}{x^5}}, x \neq 0$ **15.** $\dfrac{1}{\sqrt[8]{y^4}}, y \neq 0$ **16.** $\dfrac{3}{\sqrt[4]{a^6}}, a \neq 0$

7.3 Combining Radicals

Like Radicals **Like radicals** are radicals that have the *same index* and the *same radicand.* The *coefficients* of the radicals can be different.

Example 1 Examples of like radicals:

a. $3\sqrt{5}, 2\sqrt{5}, -7\sqrt{5}$ Index = 2; radicand = 5

b. $2\sqrt[3]{x}, -9\sqrt[3]{x}, 11\sqrt[3]{x}$ Index = 3; radicand = x

c. $\frac{2}{3}\sqrt[4]{5ab}, -\frac{1}{2}\sqrt[4]{5ab}$ Index = 4; radicand = $5ab$ ■

Unlike Radicals **Unlike radicals** are radicals that have different indices or different radicands or both.

Example 2 Examples of unlike radicals:

a. $\sqrt{7}, \sqrt{5}$ Different radicands

b. $\sqrt[3]{x}, \sqrt{x}$ Different indices

c. $\sqrt[5]{2y}, \sqrt[3]{2}$ Different indices and different radicands ■

When two or more *unlike* radicals are connected with addition or subtraction symbols, we usually cannot express the sum or difference with just one radical sign.

A WORD OF CAUTION $\sqrt{a} + \sqrt{b} \neq \sqrt{a + b}$. To verify this, let $a = 9$ and $b = 16$. Then $\sqrt{9} + \sqrt{16} = 3 + 4 = 7$, but $\sqrt{9 + 16} = \sqrt{25} = 5$. Therefore, $\sqrt{9} + \sqrt{16} \neq \sqrt{9 + 16}$. ☑

Combining Like Radicals

We can combine like radicals, as we can combine any like terms, by using the distributive property.

Example 3 Examples of combining like radicals:

a. $5\sqrt{2} + 3\sqrt{2} = (5 + 3)\sqrt{2} = 8\sqrt{2}$

b. $6\sqrt[3]{x} - 4\sqrt[3]{x} = (6 - 4)\sqrt[3]{x} = 2\sqrt[3]{x}$

c. $2x\sqrt[4]{5x} - x\sqrt[4]{5x} = (2x - x)\sqrt[4]{5x} = x\sqrt[4]{5x}$

d. $\frac{2}{3}\sqrt[5]{4xy^2} - 6\sqrt[5]{4xy^2} = \left(\frac{2}{3} - 6\right)\sqrt[5]{4xy^2} = -\frac{16}{3}\sqrt[5]{4xy^2}$ ■

We can see from Example 3 that the coefficient of the answer is the sum of the coefficients of the like radicals, and the radical part of the answer is the same as the radical part of any *one* of the terms.

Combining Unlike Radicals

While most unlike radicals cannot be combined, some *can* be combined. Sometimes, terms that are not like radicals *become* like radicals after they have been expressed in simplest radical form; they can then be combined.

> TO COMBINE UNLIKE RADICALS
>
> **1.** Simplify the radical in each term.
>
> **2.** Then combine any like radicals.

A radical expression with two or more terms is not in *simplest radical form* unless each term is in simplest radical form *and* all like terms have been combined. This condition must be added to the list of rules in Section 7.2C for expressing a radical in simplest terms.

Example 4 Express in simplest radical form; assume that all variables represent positive numbers.

a. $\sqrt{8} + \sqrt{18}$

Solution $\sqrt{8} + \sqrt{18} = \sqrt{4 \cdot 2} + \sqrt{9 \cdot 2} = 2\sqrt{2} + 3\sqrt{2} = 5\sqrt{2}$

b. $\sqrt{12} + \sqrt{50}$

Solution $\sqrt{12} + \sqrt{50} = \sqrt{4 \cdot 3} + \sqrt{25 \cdot 2} = 2\sqrt{3} + 5\sqrt{2}$

This expression cannot be simplified further; the radicals are not like radicals. We could find an *approximation* of the sum by using a calculator.

c. $\sqrt[4]{4} + \sqrt{2}$

Solution $\sqrt[4]{4} + \sqrt{2} = \sqrt{2} + \sqrt{2}$ $\qquad \sqrt[4]{4} = 4^{1/4} = (2^2)^{1/4} = 2^{1/2} = \sqrt{2}$

$= (1 + 1)\sqrt{2} = 2\sqrt{2}$

d. $\sqrt{12x} - \sqrt{27x} + 5\sqrt{3x}$

Solution $\sqrt{12x} - \sqrt{27x} + 5\sqrt{3x} = \sqrt{4 \cdot 3 \cdot x} - \sqrt{9 \cdot 3 \cdot x} + 5\sqrt{3x}$

$= 2\sqrt{3x} - 3\sqrt{3x} + 5\sqrt{3x}$

$= (2 - 3 + 5)\sqrt{3x} = 4\sqrt{3x}$

e. $\sqrt[3]{24a^2} - 5\sqrt[3]{3a^5} + 2\sqrt[3]{81a^2}$

Solution $\sqrt[3]{24a^2} - 5\sqrt[3]{3a^5} + 2\sqrt[3]{81a^2} = \sqrt[3]{2^3 \cdot 3 \cdot a^2} - 5\sqrt[3]{3 \cdot a^3 \cdot a^2} + 2\sqrt[3]{3^3 \cdot 3 \cdot a^2}$

$$= 2\sqrt[3]{3a^2} - 5a\sqrt[3]{3a^2} + 2 \cdot 3\sqrt[3]{3a^2}$$

$$= (2 - 5a + 6)\sqrt[3]{3a^2} = (8 - 5a)\sqrt[3]{3a^2}$$

f. $2\sqrt{\frac{1}{2}} - 6\sqrt{\frac{1}{8}} - 10\sqrt{\frac{4}{5}}$

Solution $2\sqrt{\frac{1}{2}} - 6\sqrt{\frac{1}{8}} - 10\sqrt{\frac{4}{5}}$

$$= 2\sqrt{\frac{1}{2} \cdot \frac{2}{2}} - 6\sqrt{\frac{1}{2^3} \cdot \frac{2}{2}} - 10\sqrt{\frac{4}{5} \cdot \frac{5}{5}}$$ Rationalizing the denominators

$$= 2\left(\sqrt{\frac{2}{4}}\right) - 6\left(\sqrt{\frac{2}{2^4}}\right) - 10\left(\sqrt{\frac{4 \cdot 5}{25}}\right)$$

$$= \overset{1}{\cancel{2}}\left(\frac{\sqrt{2}}{\underset{1}{\cancel{2}}}\right) - \overset{3}{\cancel{6}}\left(\frac{\sqrt{2}}{\underset{2}{\cancel{4}}}\right) - \overset{2}{\cancel{10}}\left(\frac{2\sqrt{5}}{\underset{1}{\cancel{5}}}\right)$$

$$= \sqrt{2} - \frac{3}{2}\sqrt{2} - 4\sqrt{5}$$

$$= \left(1 - \frac{3}{2}\right)\sqrt{2} - 4\sqrt{5} = -\frac{1}{2}\sqrt{2} - 4\sqrt{5}$$

g. $5\sqrt[4]{\frac{3x}{8}} + \frac{2x}{3}\sqrt[4]{\frac{1}{6^3x^3}}$

Solution $5\sqrt[4]{\frac{3x}{8}} + \frac{2x}{3}\sqrt[4]{\frac{1}{6^3x^3}}$

$$= 5\sqrt[4]{\frac{3x}{2^3} \cdot \frac{2}{2}} + \frac{2x}{3}\sqrt[4]{\frac{1}{6^3x^3} \cdot \frac{6x}{6x}}$$ Rationalizing the denominators

$$= 5\left(\sqrt[4]{\frac{6x}{2^4}}\right) + \frac{2x}{3}\left(\sqrt[4]{\frac{6x}{6^4x^4}}\right)$$

$$= 5\left(\frac{\sqrt[4]{6x}}{2}\right) + \frac{\overset{1}{\cancel{2x}}}{3}\left(\frac{\sqrt[4]{6x}}{\underset{3}{\cancel{6x}}}\right)$$

$$= \frac{5}{2}\sqrt[4]{6x} + \frac{1}{9}\sqrt[4]{6x}$$

$$= \left(\frac{5}{2} + \frac{1}{9}\right)\sqrt[4]{6x} = \left(\frac{45}{18} + \frac{2}{18}\right)\sqrt[4]{6x} = \frac{47}{18}\sqrt[4]{6x} \quad \blacksquare$$

EXERCISES 7.3

Set I Express in simplest radical form; assume that all variables represent positive numbers.

1. $8\sqrt{2} + 3\sqrt{2}$ **2.** $12\sqrt{5} + 6\sqrt{5}$ **3.** $3\sqrt{6} + \sqrt{6}$

4. $\sqrt{7} + 5\sqrt{7}$ **5.** $\sqrt{15} + \sqrt{10}$ **6.** $\sqrt{2} + \sqrt{14}$

7. $3\sqrt{5} - \sqrt{5}$ **8.** $4\sqrt{3} - \sqrt{3}$ **9.** $\sqrt{9} + \sqrt{12}$

10. $\sqrt{45} + \sqrt{25}$ **11.** $\sqrt{12} - \sqrt{8}$ **12.** $\sqrt{28} - \sqrt{3}$

13. $\sqrt{18} + \sqrt[4]{4}$ **14.** $\sqrt{3} + \sqrt[4]{9}$ **15.** $5\sqrt[3]{xy} + 2\sqrt[3]{xy}$

16. $7\sqrt[4]{ab} + 3\sqrt[4]{ab}$ **17.** $2\sqrt{50} - \sqrt{32}$ **18.** $3\sqrt{24} - \sqrt{54}$

19. $3\sqrt{32x} - \sqrt{8x}$ **20.** $4\sqrt{27y} - 3\sqrt{12y}$

21. $\sqrt{125M} + \sqrt{20M} - \sqrt{45M}$ **22.** $\sqrt{75P} - \sqrt{48P} + \sqrt{27P}$

23. $\sqrt[3]{27x} + \frac{1}{2}\sqrt[3]{8x}$ **24.** $\frac{3}{4}\sqrt[3]{64a} + \sqrt[3]{27a}$

25. $\sqrt[3]{a^4} + 2a\sqrt[3]{8a}$ **26.** $H\sqrt[3]{8H^2} + 3\sqrt[3]{H^5}$

27. $\sqrt[5]{x^2y^6} + \sqrt[5]{x^7y}$ **28.** $\sqrt[5]{a^3b^8} + \sqrt[5]{a^8b^3}$

29. $3\sqrt{\frac{1}{6}} + \sqrt{12} - 5\sqrt{\frac{3}{2}}$ **30.** $3\sqrt{\frac{5}{2}} + \sqrt{20} - 5\sqrt{\frac{1}{10}}$

31. $10\sqrt{\frac{5b}{4}} - \frac{3b}{2}\sqrt{\frac{4}{5b}}$ **32.** $12\sqrt[3]{\frac{x^3}{16}} + x\sqrt[3]{\frac{1}{2}}$

33. $2k\sqrt[4]{\frac{3}{8k}} - \frac{1}{k}\sqrt[4]{\frac{2k^3}{27}} + 5k^2\sqrt[4]{\frac{6}{k^2}}$ **34.** $6\sqrt[3]{\frac{a^4}{54}} + 2a\sqrt[3]{\frac{a}{2}}$

35. $\sqrt{4x^2 + 4x + 1}$ **36.** $\sqrt{16a^2 + 8a + 1}$

Set II Express in simplest radical form; assume that all variables represent positive numbers.

1. $5\sqrt{7} + 10\sqrt{7}$ **2.** $8\sqrt{5} + 2\sqrt{5}$ **3.** $9\sqrt{11} + \sqrt{11}$

4. $\sqrt{6} + \sqrt{6}$ **5.** $\sqrt{7} + \sqrt{2}$ **6.** $10\sqrt{3} + \sqrt{3}$

7. $5\sqrt{13} - 3\sqrt{13}$ **8.** $\sqrt{6} - \sqrt{2}$ **9.** $\sqrt{49} + \sqrt{28}$

10. $\sqrt{7} + \sqrt{18}$ **11.** $5\sqrt{50} - \sqrt{2}$ **12.** $3\sqrt{16} - \sqrt{2}$

13. $\sqrt{5} + \sqrt[4]{25}$ **14.** $\sqrt{50} + \sqrt{18}$ **15.** $8\sqrt[5]{5a} + \sqrt[5]{5a}$

16. $3\sqrt[3]{6x^2} - 5\sqrt[3]{6x^2}$ **17.** $7\sqrt{98} - 3\sqrt{50}$ **18.** $2\sqrt[3]{54a^2} + 4\sqrt[3]{16a^2}$

19. $8\sqrt{75x^5} + 2x\sqrt{3x^3}$ **20.** $3\sqrt{25} + 2\sqrt{72} - \sqrt{16}$

21. $7\sqrt{14x} - 2\sqrt{5x} + \sqrt{3x}$ **22.** $4\sqrt{27a^5} - a\sqrt{12a^3}$

23. $\sqrt[3]{64x} + \frac{1}{5}\sqrt[3]{8x}$ **24.** $\frac{1}{3}x\sqrt[4]{4x^2} + x\sqrt{8x}$

25. $\sqrt[3]{x^7} - 5x^2\sqrt[3]{x}$ **26.** $\sqrt{x^5y^7} + 2xy\sqrt{16xy^3} - \sqrt{xy^3}$

27. $\sqrt[6]{a^6b^3} + 6a\sqrt{a^2b^3}$ **28.** $\sqrt[3]{8x^5} + x\sqrt[6]{x^4} - \sqrt[3]{x}$

29. $5\sqrt{\frac{1}{8}} - 7\sqrt{\frac{1}{18}} - 5\sqrt{\frac{1}{50}}$ **30.** $8\sqrt[3]{\frac{a^3}{32}} + a\sqrt[3]{54}$

31. $12\sqrt{\frac{7z}{9}} - \frac{4z}{5}\sqrt{\frac{25}{7z}}$ **32.** $12\sqrt[3]{\frac{x^5}{24}} + 6x\sqrt[3]{\frac{x^2}{3}}$

33. $3h^2\sqrt[4]{\frac{4}{27h}} + 4h^2\sqrt[4]{\frac{5}{8}} - 2h^2\sqrt[4]{\frac{12}{h}}$ **34.** $\sqrt{\frac{x^3}{3y^3}} + xy\sqrt{\frac{x}{3y^5}} - \frac{y^2}{3}\sqrt{\frac{3x^3}{y^7}}$

35. $\sqrt{a^2 + 6a + 9}$ **36.** $\sqrt{25x^2 - 10x + 1}$

7.4 Multiplying and Dividing Radicals

7.4A Multiplying Radicals When the Indices Are the Same

To multiply two or more radicals together when the *indices* are the same, we use Property 7.3.

> TO MULTIPLY RADICALS
>
> We assume that a and b are not both negative if n is even.
>
> **1.** Replace $\sqrt[n]{a}\sqrt[n]{b}$ by $\sqrt[n]{a \cdot b}$ (Property 7.3).
>
> **2.** Simplify the resulting radical (Section 7.2A).

Example 1 Examples of multiplying radicals (assume that all variables represent nonnegative numbers):

a. $\sqrt{3y}\sqrt{12y^3} = \sqrt{3y \cdot 12y^3} = \sqrt{36y^4} = 6y^2$

b. $2\sqrt[3]{2x^2} \cdot 3\sqrt[3]{4x} = 2 \cdot 3\sqrt[3]{8x^3} = 6 \cdot 2x = 12x$

c. $\sqrt[4]{8ab^3}\sqrt[4]{4a^3b^2} = \sqrt[4]{2^3ab^3 \cdot 2^2a^3b^2} = \sqrt[4]{2^5a^4b^5}$

$= \sqrt[4]{2^4 \cdot 2 \cdot a^4 \cdot b^4 \cdot b} = 2ab\sqrt[4]{2b}$

d. $(4\sqrt{3})^2 = 4\sqrt{3} \cdot 4\sqrt{3} = 16\sqrt{3 \cdot 3} = 16 \cdot 3 = 48$ ■

We now have a final condition to be added to our list for simplest radical form:

> AN ALGEBRAIC EXPRESSION IS IN SIMPLEST RADICAL FORM IF:
>
> **1.** The radicand is a positive number.
>
> **2.** No prime factor of a radicand has an exponent equal to or greater than the index.
>
> **3.** No radicand contains a rational expression.
>
> **4.** No denominator contains a radical.
>
> **5.** The order of the radical cannot be reduced.
>
> **6.** All like radicals have been combined.
>
> **7.** No term of the expression contains more than one radical sign.

The distributive property is used to multiply algebraic expressions containing two or more terms with radicals by another algebraic expression.

Example 2 Find $\sqrt{8x}(\sqrt{8x} - 3\sqrt{2})$ and simplify. Assume $x > 0$.

Solution

$$\begin{aligned}\sqrt{8x}(\sqrt{8x} - 3\sqrt{2}) &= \sqrt{8x} \cdot \sqrt{8x} - \sqrt{8x}(3\sqrt{2}) \\ &= \sqrt{64x^2} - 3\sqrt{16x} \\ &= 8x - 3 \cdot 4\sqrt{x} \\ &= 8x - 12\sqrt{x}\end{aligned}$$ ■

If we need to multiply two algebraic expressions containing two terms each, we can use the FOIL method, even when the expressions are not polynomials.

Example 3 Find $(\sqrt{3} - 5)(\sqrt{3} + 2)$ and simplify.

Solution Using the FOIL method, we have

$$\begin{aligned}(\sqrt{3} - 5)(\sqrt{3} + 2) &= \sqrt{3} \cdot \sqrt{3} + 2\sqrt{3} - 5\sqrt{3} - 5 \cdot 2 \\ &= 3 - 3\sqrt{3} - 10 \\ &= -7 - 3\sqrt{3} \quad \blacksquare\end{aligned}$$

Example 4 Find $(3\sqrt{5} - 4)^2$ and simplify.

Solution We square this as we would square a binomial

$$\begin{aligned}(3\sqrt{5} - 4)^2 &= (3\sqrt{5})^2 - 2(3\sqrt{5})(4) + 4^2 \\ &= 9 \cdot 5 - 24\sqrt{5} + 16 \\ &= 45 - 24\sqrt{5} + 16 \\ &= 61 - 24\sqrt{5} \quad \blacksquare\end{aligned}$$

Example 5 Find $(\sqrt[4]{x} - 1)^4$ and simplify; assume $x \ge 0$.

Solution We use the binomial theorem here, even though $\sqrt[4]{x} - 1$ is not a binomial. (Do you know why $\sqrt[4]{x} - 1$ is not a binomial?) The coefficients are 1, 4, 6, 4, and 1.

$$\begin{aligned}&(\sqrt[4]{x} - 1)^4 \\ &= 1(\sqrt[4]{x})^4 - 4(\sqrt[4]{x})^3(1) + 6(\sqrt[4]{x})^2(1)^2 - 4(\sqrt[4]{x})^1(1)^3 + 1(1)^4\end{aligned}$$

$(\sqrt[4]{x})^2 = \sqrt[4]{x^2} = \sqrt{x}$

$$= x - 4\sqrt[4]{x^3} + 6\sqrt{x} - 4\sqrt[4]{x} + 1 \quad \blacksquare$$

Example 6 Find $(x^{1/3} + y^{1/3})^3$ and simplify.

Solution We use the binomial theorem, even though $x^{1/3} + y^{1/3}$ is not a binomial. (Do you know why $x^{1/3} + y^{1/3}$ is not a binomial?) The coefficients are 1, 3, 3, and 1.

$$\begin{aligned}&(x^{1/3} + y^{1/3})^3 \\ &= 1(x^{1/3})^3 + 3(x^{1/3})^2(y^{1/3}) + 3(x^{1/3})(y^{1/3})^2 + 1(y^{1/3})^3 \\ &= x + 3x^{2/3}y^{1/3} + 3x^{1/3}y^{2/3} + y \quad \blacksquare\end{aligned}$$

EXERCISES 7.4A

Set I In Exercises 1–38, simplify each expression. Assume that all variables represent nonnegative numbers.

1. $\sqrt{3}\sqrt{3}$

2. $\sqrt{7}\sqrt{7}$

3. $\sqrt[3]{3}\sqrt[3]{9}$

4. $\sqrt[3]{4}\sqrt[3]{16}$

5. $\sqrt[4]{9}\sqrt[4]{9}$

6. $\sqrt[4]{25}\sqrt[4]{25}$

7. $\sqrt{5ab^2}\sqrt{20ab}$

8. $\sqrt{3x^2y}\sqrt{27xy}$

9. $3\sqrt[5]{2a^3b}(2\sqrt[5]{16a^2b})$

10. $4\sqrt[5]{4cb^4}(5\sqrt[5]{8c^2b})$

11. $(5\sqrt{7})^2$

12. $(4\sqrt{6})^2$

13. $\sqrt{2}(\sqrt{2} + 1)$

14. $\sqrt{3}(\sqrt{3} + 1)$

15. $\sqrt{x}(\sqrt{x} - 3)$

16. $\sqrt{y}(4 - \sqrt{y})$
17. $\sqrt{3}(2\sqrt{3} + 1)$
18. $\sqrt{5}(3\sqrt{5} + 1)$
19. $\sqrt{3x}(\sqrt{3x} - 4\sqrt{12})$
20. $\sqrt{5a}(\sqrt{10} + 3\sqrt{5a})$
21. $(\sqrt{7} + 2)(\sqrt{7} + 3)$
22. $(\sqrt{3} + 2)(\sqrt{3} + 4)$
23. $(5 + \sqrt{3})(5 - \sqrt{3})$
24. $(\sqrt{5} + \sqrt{3})(\sqrt{5} - \sqrt{3})$
25. $(2\sqrt{3} - 5)^2$
26. $(5\sqrt{2} - 3)^2$
27. $2\sqrt{7x^3y^3}(5\sqrt{3xy})(2\sqrt{7x^3y})$
28. $3\sqrt{5xy}(4\sqrt{2x^5y^3})(5\sqrt{5x^3y^5})$
29. $(3\sqrt{2x + 5})^2$
30. $(4\sqrt{3x - 2})^2$
31. $(\sqrt{2x} + 3)(\sqrt{2x} - 3)$
32. $(\sqrt{5x} + 7)(\sqrt{5x} - 7)$
33. $(\sqrt{xy} - 6\sqrt{y})^2$
34. $(\sqrt{ab} + 2\sqrt{a})^2$
35. $(\sqrt[5]{x} + 1)^5$
36. $(1 + \sqrt[3]{y})^3$
37. $(x^{1/4} + 1)^4$
38. $(x^{1/2} + y^{1/2})^2$
39. Find the value of $x^4 - 1$ if $x = 1 - \sqrt{2}$.
40. Find the value of $x^3 + 1$ if $x = 1 + \sqrt{2}$.

Set II In Exercises 1–38, simplify each expression. Assume that all variables represent nonnegative numbers.

1. $\sqrt{13}\sqrt{13}$
2. $\sqrt{2}\sqrt{32}$
3. $\sqrt[3]{5}\sqrt[3]{25}$
4. $\sqrt[5]{48}\sqrt[5]{4}$
5. $\sqrt[4]{49}\sqrt[4]{49}$
6. $\sqrt[4]{2}\sqrt[4]{8}$
7. $\sqrt[5]{4x^2y^3}\sqrt[5]{8x^3y^2}$
8. $\sqrt[3]{4x^2y^2}\sqrt[3]{2x^2y}$
9. $5\sqrt[5]{x^8y^6}(4\sqrt[5]{x^2y^4})$
10. $6\sqrt[3]{24a^4b^2}(5a\sqrt[3]{3ab^2})$
11. $(3\sqrt{2})^2$
12. $(2\sqrt{5})^3$
13. $\sqrt{5}(\sqrt{5} + 1)$
14. $\sqrt{8}(\sqrt{2} - 4)$
15. $\sqrt{z}(\sqrt{z - 5})$
16. $\sqrt{5x}(\sqrt{5x} - 2)$
17. $\sqrt{7}(3\sqrt{7} + 2)$
18. $\sqrt{27}(4\sqrt{3} + 5)$
19. $\sqrt{2y}(10\sqrt{10y} + 2\sqrt{6})$
20. $\sqrt{3a}(4\sqrt{3a} + 2)$
21. $(\sqrt{11} + 2)(\sqrt{11} + 5)$
22. $(\sqrt{7} + 3)(\sqrt{7} - 3)$
23. $(\sqrt{6} - \sqrt{3})(\sqrt{6} + \sqrt{3})$
24. $(2 - \sqrt{6})(2 + \sqrt{6})$
25. $(2\sqrt{5} - 4)^2$
26. $(\sqrt{5} + \sqrt{3})^2$
27. $3\sqrt{5x^4y^3}(2\sqrt{15xy^5})(5\sqrt{2x^2y})$
28. $8\sqrt[4]{6a^3b^5}(3\sqrt[4]{2ab^3})(2\sqrt[4]{8a^2b^4})$
29. $(5\sqrt{3x - 4})^2$
30. $(4\sqrt{5 - 2a})^2$
31. $(\sqrt{7x} + 3)(\sqrt{7x} - 3)$
32. $(8 - \sqrt{7x})(8 + \sqrt{7x})$
33. $(\sqrt{x} + 3\sqrt{y})^2$
34. $(3\sqrt{a} - \sqrt{2b})^2$
35. $(\sqrt[4]{x} + 2)^4$
36. $(\sqrt[4]{x} - \sqrt[4]{y})^4$
37. $(x^{1/3} - 1)^3$
38. $(x^{1/5} + 2)^5$
39. Find the value of $y^2 - 2y - 5$ if $y = 1 - \sqrt{6}$.
40. Find the value of $x^4 + 1$ if $x = 1 + \sqrt{2}$.

7.4B Dividing Radicals When the Indices Are the Same

To divide radicals when the indices are the same, we use Property 7.4.

> TO DIVIDE RADICALS
>
> We assume that a and b are not both negative if n is even.
>
> **1.** Replace $\dfrac{\sqrt[n]{a}}{\sqrt[n]{b}}$ by $\sqrt[n]{\dfrac{a}{b}}$ (Property 7.4).
>
> **2.** Simplify the resulting radical (Section 7.2B).

Example 7 Examples of dividing radicals (assume that all variables represent positive numbers):

a. $\dfrac{12\sqrt[3]{a^5}}{2\sqrt[3]{a^2}} = \dfrac{6}{1}\sqrt[3]{\dfrac{a^5}{a^2}} = 6\sqrt[3]{a^3} = 6a$

b. $\dfrac{5\sqrt[4]{28xy^6}}{10\sqrt[4]{7xy}} = \dfrac{1}{2}\sqrt[4]{\dfrac{28xy^6}{7xy}} = \dfrac{1}{2}\sqrt[4]{4y^5} = \dfrac{1}{2}\sqrt[4]{4y^4y} = \dfrac{y}{2}\sqrt[4]{4y}$

c. $\dfrac{\sqrt{5x}}{\sqrt{10x^2}} = \sqrt{\dfrac{5x}{10x^2}} = \sqrt{\dfrac{1}{2x}} = \sqrt{\dfrac{1}{2x}\cdot\dfrac{2x}{2x}} = \dfrac{1}{2x}\sqrt{2x}$ ■

When the dividend contains more than one term but the divisor contains only one term, we divide *each term* of the dividend by the divisor (see Example 8). This procedure is similar to that used in dividing a polynomial by a monomial.

Example 8 Examples of dividing radicals when the dividend contains more than one term:

a. $\dfrac{3\sqrt{14} - \sqrt{8}}{\sqrt{2}} = \dfrac{3\sqrt{14}}{\sqrt{2}} - \dfrac{\sqrt{8}}{\sqrt{2}} = 3\sqrt{\dfrac{14}{2}} - \sqrt{\dfrac{8}{2}} = 3\sqrt{7} - \sqrt{4} = 3\sqrt{7} - 2$

b. $\dfrac{3\sqrt[5]{64a} - 9\sqrt[5]{4a^3}}{3\sqrt[5]{2a}} = \dfrac{3\sqrt[5]{64a}}{3\sqrt[5]{2a}} - \dfrac{9\sqrt[5]{4a^3}}{3\sqrt[5]{2a}} = \sqrt[5]{\dfrac{64a}{2a}} - 3\sqrt[5]{\dfrac{4a^3}{2a}}$

$= \sqrt[5]{32} - 3\sqrt[5]{2a^2} = 2 - 3\sqrt[5]{2a^2}$ ■

Rationalizing a Denominator That Contains Two Terms

In order to discuss division problems in which the divisor contains radicals and has two terms, we need a new definition.

> CONJUGATE
>
> The **conjugate** of the algebraic expression $a + b$ is the algebraic expression $a - b$.

Example 9 Examples of conjugates:

a. The conjugate of $1 + \sqrt{3}$ is $1 - \sqrt{3}$.

b. The conjugate of $\sqrt{5} - \sqrt{3}$ is $\sqrt{5} + \sqrt{3}$.

c. The conjugate of $\sqrt{x} - \sqrt{y}$ is $\sqrt{x} + \sqrt{y}$. ■

If an algebraic expression contains *square roots* and has two terms, then the product of the expression and its conjugate will always be a rational number. For example, the conjugate of $1 + \sqrt{3}$ is $1 - \sqrt{3}$; the product of these numbers is:

$$(1 + \sqrt{3})(1 - \sqrt{3}) = (1)^2 - (\sqrt{3})^2$$
$$= 1 - 3 = -2 \quad \text{A rational number}$$

Because of this fact, the following procedure should be used when a denominator with two terms contains square roots.

> TO RATIONALIZE A DENOMINATOR THAT CONTAINS SQUARE ROOTS AND HAS TWO TERMS
>
> Multiply the numerator and the denominator by the conjugate of the denominator.

Example 10 Rationalize the denominator and simplify.

a. $\dfrac{2}{1 + \sqrt{3}}$

Solution

$$\frac{2}{1 + \sqrt{3}} = \frac{2}{(1 + \sqrt{3})}\frac{(1 - \sqrt{3})}{(1 - \sqrt{3})} = \frac{2(1 - \sqrt{3})}{1 - 3} = \frac{\overset{1}{\cancel{2}}(1 - \sqrt{3})}{\underset{-1}{\cancel{-2}}} = -1 + \sqrt{3} \text{ or } \sqrt{3} - 1$$

Multiplying numerator and denominator by $1 - \sqrt{3}$ (the conjugate of the denominator $1 + \sqrt{3}$)

b. $\dfrac{\sqrt{5} + \sqrt{3}}{\sqrt{5} - \sqrt{3}}$

Solution

$$\frac{\sqrt{5} + \sqrt{3}}{\sqrt{5} - \sqrt{3}} = \frac{(\sqrt{5} + \sqrt{3})}{(\sqrt{5} - \sqrt{3})}\frac{(\sqrt{5} + \sqrt{3})}{(\sqrt{5} + \sqrt{3})} = \frac{5 + 2\sqrt{15} + 3}{5 - 3} = \frac{8 + 2\sqrt{15}}{2}$$

$$= \frac{\overset{1}{\cancel{2}}(4 + \sqrt{15})}{\underset{1}{\cancel{2}}} = 4 + \sqrt{15}$$

Multiplying numerator and denominator by $\sqrt{5} + \sqrt{3}$ (the conjugate of the denominator $\sqrt{5} - \sqrt{3}$) ■

EXERCISES 7.4B

Set I Simplify each expression; assume that all variables represent positive numbers.

1. $\dfrac{\sqrt{32}}{\sqrt{2}}$ **2.** $\dfrac{\sqrt{98}}{\sqrt{2}}$ **3.** $\dfrac{\sqrt[3]{5}}{\sqrt[3]{4}}$

4. $\dfrac{\sqrt[3]{7}}{\sqrt[3]{2}}$ 5. $\dfrac{12\sqrt[4]{15x}}{4\sqrt[4]{5x}}$ 6. $\dfrac{15\sqrt[4]{18y}}{5\sqrt[4]{3y}}$

7. $\dfrac{\sqrt[5]{128z^7}}{\sqrt[5]{2z}}$ 8. $\dfrac{\sqrt[5]{3^7b^8}}{\sqrt[5]{3b^2}}$ 9. $\dfrac{\sqrt{72x^3y^2}}{\sqrt{2xy^2}}$

10. $\dfrac{\sqrt{27x^2y^3}}{\sqrt{3x^2y}}$ 11. $\dfrac{\sqrt{20}+5\sqrt{10}}{\sqrt{5}}$ 12. $\dfrac{2\sqrt{6}+\sqrt{18}}{\sqrt{6}}$

13. $\dfrac{6\sqrt[4]{2^5m^2}}{2\sqrt[4]{2m^3}}$ 14. $\dfrac{7\sqrt[4]{3^5H^3}}{14\sqrt[4]{3H^4}}$ 15. $\dfrac{4\sqrt[3]{8x}+6\sqrt[3]{32x^4}}{2\sqrt[3]{4x}}$

16. $\dfrac{10\sqrt[3]{81a^7}+15\sqrt[3]{6a}}{5\sqrt[3]{3a}}$ 17. $\dfrac{6}{\sqrt{3}-1}$ 18. $\dfrac{10}{\sqrt{3}+1}$

19. $\dfrac{\sqrt{2}}{\sqrt{3}+\sqrt{2}}$ 20. $\dfrac{\sqrt{7}}{\sqrt{7}-\sqrt{2}}$ 21. $\dfrac{\sqrt{7}+\sqrt{3}}{\sqrt{7}-\sqrt{3}}$

22. $\dfrac{\sqrt{11}-\sqrt{5}}{\sqrt{11}+\sqrt{5}}$ 23. $\dfrac{4\sqrt{3}-\sqrt{2}}{4\sqrt{3}+\sqrt{2}}$ 24. $\dfrac{\sqrt{x+1}-\sqrt{x}}{\sqrt{x+1}+\sqrt{x}}$

25. $\sqrt{\dfrac{a^2+2a-3}{a^2+4a+3}},\ a>1$ 26. $\sqrt{\dfrac{m^2-m-2}{m^2-3m+2}},\ m>2$

Set II Simplify each expression; assume that all variables represent positive numbers.

1. $\dfrac{\sqrt{72}}{\sqrt{2}}$ 2. $\dfrac{\sqrt{75}}{\sqrt{3}}$ 3. $\dfrac{\sqrt[5]{3}}{\sqrt[5]{4}}$

4. $\dfrac{\sqrt[3]{x^8y}}{\sqrt[3]{5xy^2}}$ 5. $\dfrac{30\sqrt[4]{64x}}{6\sqrt[4]{4x}}$ 6. $\dfrac{35\sqrt[5]{96x^5}}{7\sqrt[5]{6x^3}}$

7. $\dfrac{\sqrt{x^4y}}{\sqrt{5y}}$ 8. $\dfrac{\sqrt{m^6n}}{\sqrt{3n}}$ 9. $\dfrac{\sqrt{300a^5b^2}}{\sqrt{3ab^2}}$

10. $\dfrac{\sqrt[4]{2xy^3}}{\sqrt[4]{27x^2y}}$ 11. $\dfrac{\sqrt{34}+2\sqrt{6}}{\sqrt{2}}$ 12. $\dfrac{3\sqrt{27}-\sqrt{2}}{\sqrt{3}}$

13. $\dfrac{6\sqrt[3]{4B^2}}{9\sqrt[3]{8B^4}}$ 14. $\dfrac{5\sqrt[3]{3K^3}}{15\sqrt[3]{9K^5}}$ 15. $\dfrac{30\sqrt[4]{32a^6}-6\sqrt[4]{24a^2}}{3\sqrt[4]{2a^2}}$

16. $\dfrac{8\sqrt[3]{2x}-4\sqrt[3]{3x^4}}{2\sqrt[3]{5x^2}}$ 17. $\dfrac{8}{\sqrt{5}-1}$ 18. $\dfrac{3}{\sqrt{6}+1}$

19. $\dfrac{\sqrt{7}}{\sqrt{3}+\sqrt{7}}$ 20. $\dfrac{21}{3\sqrt{5}+2\sqrt{6}}$ 21. $\dfrac{\sqrt{5}+\sqrt{2}}{\sqrt{5}-\sqrt{2}}$

22. $\dfrac{3\sqrt{2}+\sqrt{6}}{3\sqrt{2}-\sqrt{6}}$ 23. $\dfrac{5\sqrt{7}+\sqrt{3}}{5\sqrt{7}-\sqrt{3}}$ 24. $\dfrac{x-4}{\sqrt{x}+2}$

25. $\sqrt{\dfrac{a^2+2ab+b^2}{a^2-b^2}},\ a>b$ 26. $\sqrt{\dfrac{x^2-y^2}{x^2-2xy+y^2}},\ x>y$

7.4C Multiplying and Dividing Radicals When the Indices Are Not the Same

In this section, we will assume that all variables represent positive numbers.

If all radicands can be expressed as powers of the same base, it is possible to multiply and divide radicals even when the indices are not the same. The easiest way to do this is to (1) write any numerical coefficients in prime factored form, (2) convert the radicals to exponential form, (3) perform the indicated multiplication or division, and (4) express the answer in simplest radical form.

Example 11 Examples of multiplying and dividing radicals with different indices:

a. $\sqrt{x}\sqrt[3]{x} = x^{1/2}x^{1/3} = x^{3/6}x^{2/6} = x^{5/6} = \sqrt[6]{x^5}$

b. $\sqrt{2}\sqrt[3]{-32} = 2^{1/2}(-2^5)^{1/3} = -2^{1/2+5/3} = -2^{3/6+10/6} = -2^{13/6}$

$= -2^{2+1/6} = -2^2 \cdot 2^{1/6} = -4\sqrt[6]{2}$

c. $\sqrt[4]{a^3}\sqrt{a}\sqrt[3]{a^2} = a^{3/4}a^{1/2}a^{2/3} = a^{9/12}a^{6/12}a^{8/12} = a^{23/12} = a^{1+11/12}$

$= a \cdot a^{11/12} = a\sqrt[12]{a^{11}}$

d. $\dfrac{\sqrt[3]{-d^2}}{\sqrt[4]{d^3}} = \dfrac{-d^{2/3}}{d^{3/4}} = -d^{2/3-3/4} = -d^{8/12-9/12} = -d^{-1/12} = -\dfrac{1}{\sqrt[12]{d}}$

We can use rational exponents to rationalize the denominator as follows:

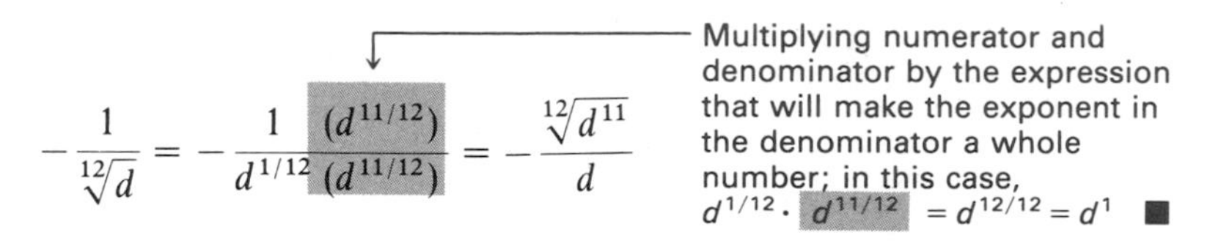

■

EXERCISES 7.4C

Set I In the exercises below, perform the indicated operations. Express the answers in simplest radical form. Assume that all variables represent positive numbers.

1. $\sqrt{a}\sqrt[4]{a}$

2. $\sqrt{b}\sqrt[3]{b}$

3. $\sqrt{8}\sqrt[3]{16}$

4. $\sqrt{27}\sqrt[3]{81}$

5. $\sqrt[3]{x^2}\sqrt[4]{x^3}\sqrt{x}$

6. $\sqrt{y}\sqrt[3]{y}\sqrt[4]{y^3}$

7. $\sqrt[3]{-8z^2}\sqrt[3]{-z}\sqrt[4]{16z^3}$

8. $\sqrt[3]{-27w}\sqrt[3]{-w^2}\sqrt[4]{16w^3}$

9. $\dfrac{\sqrt[4]{G^3}}{\sqrt[3]{G^2}}$

10. $\dfrac{\sqrt[5]{H^4}}{\sqrt{H}}$

11. $\dfrac{\sqrt[3]{-x^2}}{\sqrt[6]{x^5}}$

12. $\dfrac{\sqrt[6]{y^3}}{\sqrt[3]{-y^2}}$

Set II In the exercises below, perform the indicated operations. Express the answers in simplest radical form. Assume that all variables represent positive numbers.

1. $\sqrt[3]{a^2}\sqrt[6]{a^2}$

2. $\sqrt[2]{x^3}\sqrt[5]{x^4}$

3. $\sqrt{2}\sqrt[6]{8}$

4. $\sqrt[5]{3}\sqrt[5]{81}$

5. $\sqrt{x}\sqrt[4]{x^3}\sqrt[8]{x^6}$

6. $\sqrt[3]{a}\sqrt[5]{a^2}\sqrt[6]{a^4}$

7. $\sqrt[3]{-5x^2}\sqrt[4]{x^5}\sqrt[3]{-25x^3}$

8. $\sqrt[3]{x}\sqrt[5]{x^3}\sqrt[4]{x^6}$

9. $\dfrac{\sqrt[4]{a^3}}{\sqrt[3]{a^5}}$

10. $\dfrac{\sqrt[3]{x^2}}{\sqrt[5]{x^6}}$

11. $\dfrac{\sqrt[3]{-x^4}}{\sqrt{x^3}}$

12. $\dfrac{\sqrt[4]{32x^3}}{\sqrt{2x}}$

7.5 Review: 7.1–7.4

Relations between Exponents and Radicals 7.1

Property 7.1 $\sqrt[n]{x} = x^{1/n}$ if $x \geq 0$ when n is even
$= x^{1/n}$ for all x when n is odd.

Property 7.2 $\sqrt[b]{x^a} = (\sqrt[b]{x})^a = x^{a/b}$ if $x \geq 0$ when b is even
$= x^{a/b}$ for all x when b is odd.

Simplifying Radicals 7.2

Factor the radicand and use the following properties:

Property 7.3 $\sqrt[n]{ab} = \sqrt[n]{a}\,\sqrt[n]{b}$

Property 7.4 $\sqrt[n]{\dfrac{a}{b}} = \dfrac{\sqrt[n]{a}}{\sqrt[n]{b}}$

if a and b are not *both* negative when n is even.

Simplified Form of a Radical Expression 7.2, 7.3, 7.4

An algebraic expression is in simplest radical form if:

1. The radicand is a positive number.
2. No prime factor of a radicand has an exponent equal to or greater than the index.
3. No radicand contains a rational expression.
4. No denominator contains a radical.
5. The order of the radical cannot be reduced.
6. All like radicals have been combined.
7. No term of the expression contains more than one radical sign.

Rationalizing the Denominator 7.2B

Denominator with a single term: Multiply the numerator and denominator by an expression that will make the exponent of every factor in the radicand of the denominator exactly divisible by the index.

7.4B

Denominator with two terms: Multiply the numerator and denominator by the *conjugate* of the denominator.

Addition of Radicals 7.3

Like radicals: The coefficient of the sum is the sum of the coefficients. The radical part of the sum is the same as the radical part of any *one* of the terms.

Unlike radicals:

1. Simplify the radical in each term.
2. Then combine any like radicals.

Multiplication of Radicals 7.4A, 7.4C

Use the property $\sqrt[n]{a}\,\sqrt[n]{b} = \sqrt[n]{ab}$ (if a and b are not both negative when n is even); then simplify.

Division of Radicals 7.4B, 7.4C

Use the property $\dfrac{\sqrt[n]{a}}{\sqrt[n]{b}} = \sqrt[n]{\dfrac{a}{b}}$ (if a and b are not both negative when n is even); then simplify.

Review Exercises 7.5 Set I

In Exercises 1–3, rewrite each expression in an equivalent radical form. Assume $a \geq 0$ and $y \geq 0$.

1. $a^{3/4}$ **2.** $(3y)^{3/4}$ **3.** $(2x^2)^{2/5}$

In Exercises 4–6, replace each radical with an equivalent exponential expression. Assume $b \geq 0$.

4. $\sqrt[4]{b^3}$ **5.** $\sqrt[5]{8x^4}$ **6.** $\sqrt[5]{27x^3}$

In Exercises 7 and 8, evaluate each expression.

7. $(-64)^{2/3}$ **8.** $(-27)^{2/3}$

In Exercises 9 and 10, perform the indicated operations. Express the answers in exponential form, using positive exponents. Assume $R \geq 0$, $a > 0$, and $b > 0$.

9. $(P^{2/3}R^{3/4})^{2/3}$ **10.** $\left(\dfrac{27a^{1/3}b^{2/3}}{3b^{-1}}\right)^{-3/2}$

In Exercises 11–16, simplify each expression. Assume $y > 0$.

11. $\sqrt[3]{32}$ **12.** $\sqrt[3]{-125x^3}$ **13.** $\sqrt[4]{16y^8}$

14. $\dfrac{1}{3}\sqrt[5]{3^5m^6p}$ **15.** $\sqrt{\dfrac{20x^3}{5y}}$ **16.** $\dfrac{15}{\sqrt{3y}}$

In Exercises 17–28, perform the indicated operations and give your answers in simplest radical form. Assume that all variables represent positive numbers.

17. $\sqrt[3]{-2x}\sqrt[3]{-4x^4}$ **18.** $(5\sqrt{3x})^2$

19. $\dfrac{\sqrt{3}+\sqrt{7}}{\sqrt{3}-\sqrt{7}}$ **20.** $\dfrac{8}{\sqrt{5}+1}$

21. $(\sqrt{2}+3)(4\sqrt{2}-1)$ **22.** $(\sqrt{13}+\sqrt{3})^2$

23. $\sqrt[3]{16x^5}+x\sqrt[3]{54x^2}$ **24.** $(\sqrt{13}-2)(\sqrt{13}+2)$

25. $\sqrt[3]{-z^2}\sqrt[4]{z^2}$ **26.** $(\sqrt[4]{2}-x)^4$

27. $\dfrac{\sqrt[4]{G^2}}{\sqrt[5]{G}}$ **28.** $3\sqrt[4]{\dfrac{2ab}{32a^3}}-\dfrac{1}{2}\sqrt[4]{\dfrac{b}{a^2}}$

In Exercises 29 and 30, factor each expresssion completely.

29. $14x^{-2/3}y^{1/4}z^{1/3}-7x^{1/3}y^{-3/4}z^{-2/3}$ **30.** $4x^{-2}-x^{-1}-3$

Review Exercises 7.5 Set II

NAME

In Exercises 1–3, rewrite each expression in an equivalent radical form.

1. $R^{4/5}$ **2.** $(4x^2)^{3/5}$ **3.** $B^{2/3}$

In Exercises 4–6, replace each radical with an equivalent exponential expression. Assume $x \geq 0$.

4. $\sqrt[3]{8y^2}$ **5.** $(\sqrt[4]{x^3})^2$ **6.** $\sqrt[5]{a^2}$

In Exercises 7 and 8, evaluate each expression.

7. $(-32)^{2/5}$ **8.** $(-125)^{2/3}$

In Exercises 9 and 10, perform the indicated operations. Express the answers in exponential form, using positive exponents. Assume $x > 0$ and $y \geq 0$.

9. $(x^{3/2}y^{3/4})^{2/3}$ **10.** $\left(\dfrac{27x^{-1}y^{2/3}}{3x^{1/3}}\right)^{1/2}$

In Exercises 11–16, simplify each expression. Assume $b > 0$ and $a \geq 0$.

11. $\sqrt[3]{81}$ **12.** $\sqrt[5]{32x^{10}}$ **13.** $\sqrt[3]{36x^4}$

14. $3b\sqrt[4]{\dfrac{a^3}{27b}}$ **15.** $\dfrac{1}{2}\sqrt[3]{16m^4n^7}$ **16.** $\dfrac{b}{3}\sqrt[3]{\dfrac{-27a}{b^2}}$

ANSWERS

1. ______
2. ______
3. ______
4. ______
5. ______
6. ______
7. ______
8. ______
9. ______
10. ______
11. ______
12. ______
13. ______
14. ______
15. ______
16. ______

In Exercises 17–28, perform the indicated operations and give your answers in simplest radical form. Assume that all variables represent positive numbers.

17. $\sqrt[3]{-3a^2}\,\sqrt[3]{-9a^2}$

18. $\dfrac{3+\sqrt{5}}{1-\sqrt{5}}$

19. $(4+\sqrt{2x})^2$

20. $z\sqrt[3]{32z}+\sqrt[3]{4z^4}$

21. $\dfrac{3\sqrt{2}-1}{2\sqrt{2}+3}$

22. $2\sqrt[4]{\dfrac{xy}{81x^3}}-\dfrac{1}{3}\sqrt[4]{\dfrac{y^2}{x^2y}}$

23. $(\sqrt{7z}-\sqrt{3})(\sqrt{7z}+\sqrt{3})$

24. $(\sqrt[4]{x}+2)^4$

25. $(2\sqrt{3z})^3$

26. $\sqrt[3]{-M^2}\,\sqrt[6]{M^4}$

27. $\dfrac{\sqrt[6]{H^4}}{\sqrt[5]{H^3}}$

28. $(x-\sqrt{3})^2$

In Exercises 29 and 30, factor each expression completely.

29. $4x^{-2}-12x^{-1}+5$

30. $18a^{2/3}b^{-2/3}c^{4/5}-10a^{-1/3}b^{1/3}c^{-1/5}$

ANSWERS

17. ____________

18. ____________

19. ____________

20. ____________

21. ____________

22. ____________

23. ____________

24. ____________

25. ____________

26. ____________

27. ____________

28. ____________

29. ____________

30. ____________

7.6 Solving Radical Equations

Radical Equations A **radical equation** is an equation in which the variable appears in a radicand. Recall that even roots of negative numbers are not real numbers. Therefore, if the index of the radical is an *even* number, the domain of the variable is the set of all real numbers that will make the radicand greater than or equal to zero. If the index is an *odd* number, the domain is the set of all real numbers.

Example 1 Find the domain in each of these radical equations (*do not solve the equations*).

a. $\sqrt{x+2} = 3$

Solution $x + 2 \geq 0$ The *radicand* must be ≥ 0

$x \geq -2$

The domain is $\{x \mid x \geq -2\}$.

b. $\sqrt[3]{x-1} = -4$

Solution The domain is the set of all real numbers, because the index of the radical is not an even number.

c. $\sqrt{4x+5} - \sqrt{x-1} = \sqrt{14-x}$

Solution $4x + 5 \geq 0$ *and* $x - 1 \geq 0$ *and* $14 - x \geq 0$

$x \geq -\frac{5}{4}$ *and* $x \geq 1$ *and* $x \leq 14$

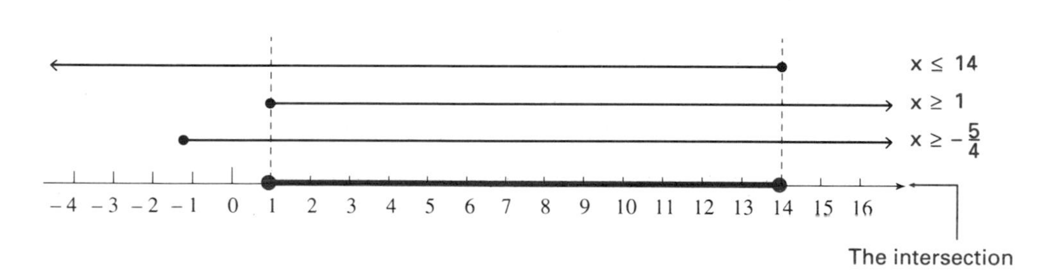

The numbers that satisfy *all three* conditions are the real numbers between 1 and 14. Therefore, the domain is $\{x \mid 1 \leq x \leq 14\}$. ■

Solving Radical Equations

In order to solve a radical equation, we must eliminate the radical signs by raising both sides of the equation to some power. *When we do this, we may introduce extraneous roots*; therefore, all apparent solutions to the equation must be checked in the original equation.

> TO SOLVE A RADICAL EQUATION
>
> **1.** Find the domain of the variable.
>
> **2.** Arrange the terms so that one term with a radical is by itself on one side of the equation.
>
> **3.** Raise each side of the equation to a power equal to the index of the radical.
>
> **4.** Simplify each side of the equation.

If a radical still remains, repeat steps 2, 3, and 4.

5. Solve the resulting equation for the variable.

6. Check apparent solutions in the original equation.

Example 2 Find the solution set for each equation.

a. $\sqrt{x+2} = 3$.

Step 1. The domain is $\{x \mid x \geq -2\}$.

Step 3.	$(\sqrt{x+2})^2 = 3^2$	Squaring both sides
Step 4.	$x + 2 = 9$	
Step 5.	$x = 7$	7 is in the domain

Step 6. *Check* $\sqrt{7+2} = \sqrt{9} = 3$. The solution set is $\{7\}$.

b. $\sqrt[3]{x-1} + 4 = 0$.

Step 1. The domain is the set of all real numbers.

	$\sqrt[3]{x-1} + 4 = 0$	
Step 2.	$\sqrt[3]{x-1} = -4$	Getting the radical by itself on one side
Step 3.	$(\sqrt[3]{x-1})^3 = (-4)^3$	Cubing both sides
Step 4.	$x - 1 = -64$	
Step 5.	$x = -63$	-63 is in the domain

Step 6. *Check*

$$\sqrt[3]{x-1} + 4 = 0$$
$$\sqrt[3]{-63-1} + 4 \stackrel{?}{=} 0$$
$$\sqrt[3]{-64} + 4 \stackrel{?}{=} 0$$
$$-4 + 4 \stackrel{?}{=} 0$$
$$0 = 0$$

The solution set is $\{-63\}$.

c. $\sqrt[4]{2x+1} = -2$.

Step 1. For the domain, $2x + 1 \geq 0$, or $x \geq -\frac{1}{2}$. We know the principal root of an even-index radical must be *positive*, and no positive number can equal -2. Therefore, this equation does not have a solution that is a real number. If you did not notice this and followed the procedure outlined for solving radical equations, the same result would be obtained. Following the procedure outlined, we have

	$\sqrt[4]{2x+1} = -2$	
Step 3.	$(\sqrt[4]{2x+1})^4 = (-2)^4$	Raising both sides to the fourth power
Step 4.	$2x + 1 = 16$	
	$2x = 15$	
Step 5.	$x = \frac{15}{2}$	$\frac{15}{2}$ is in the domain

Step 6. *Check* $\sqrt[4]{2x+1} = -2$

$$\sqrt[4]{2\left(\frac{15}{2}\right)+1} \stackrel{?}{=} -2$$

$$\sqrt[4]{15+1} \stackrel{?}{=} -2$$

$$\sqrt[4]{16} \stackrel{?}{=} -2$$

$$2 \neq -2$$

↑ Principal root

Therefore, $\frac{15}{2}$ is not a solution. The solution set is $\{\ \}$.

A WORD OF CAUTION If one side of the equation has more than one term, squaring each term is *not* the same as squaring both sides of the equation. That is, if $a = b + c$, then $a^2 \neq b^2 + c^2$; rather, $a^2 = (b+c)^2$, and $(b+c)^2 = b^2 + 2bc + c^2$. The squaring of $(b + c)$ will *always* result in a trinomial. ☑

d. $\sqrt{2x+1} = x - 1$.

Step 1. The domain is $\{x \mid x \geq -1/2\}$.

Step 3. $(\sqrt{2x+1})^2 = (x-1)^2$ — Squaring both sides

When squaring $(x - 1)$, do not forget this middle term

Step 4. $2x + 1 = x^2 - 2x + 1$

Step 5. $0 = x^2 - 4x$

$$0 = x(x-4)$$

$$x = 0 \quad or \quad x - 4 = 0$$

$$x = 4$$

Both apparent solutions are in the domain

Step 6. *Check for* $x = 0$

$$\sqrt{2x+1} = x - 1$$

$$\sqrt{2(0)+1} \stackrel{?}{=} (0) - 1$$

$$\sqrt{1} \stackrel{?}{=} -1$$

$$1 \neq -1$$

The symbol $\sqrt{1}$ *always* stands for the *principal* square root of 1, which is 1 (*not* -1)

Check for $x = 4$

$$\sqrt{2x+1} = x - 1$$

$$\sqrt{2(4)+1} \stackrel{?}{=} (4) - 1$$

$$\sqrt{9} \stackrel{?}{=} 3$$

$$3 = 3$$

Therefore, 0 *is not a solution*, while 4 *is* a solution. The solution set is $\{4\}$.

e. $\sqrt{4x+5} - \sqrt{x-1} = \sqrt{14-x}$

Step 1. The domain is $\{x \mid 1 \leq x \leq 14\}$ (see Example 1c).

$$\sqrt{4x+5} - \sqrt{x-1} = \sqrt{14-x}$$

One radical is by itself on one side of the equal sign

Step 3. $(\sqrt{4x+5} - \sqrt{x-1})^2 = (\sqrt{14-x})^2$ Squaring both sides

When squaring, don't forget this term

Step 4. $4x + 5 - 2\sqrt{(4x+5)(x-1)} + x - 1 = 14 - x$ Simplifying; a radical remains

$5x + 4 - 2\sqrt{(4x+5)(x-1)} = 14 - x$ Combining like terms

Step 2. $-2\sqrt{(4x+5)(x-1)} = 10 - 6x$ Adding $-4 - 5x$ to both sides

$\sqrt{(4x+5)(x-1)} = 3x - 5$ Dividing both sides by -2

Step 3. $(\sqrt{4x^2 + x - 5})^2 = (3x-5)^2$ Squaring both sides again

Step 4. $4x^2 + x - 5 = 9x^2 - 30x + 25$

Step 5. $0 = 5x^2 - 31x + 30$ Getting all terms on one side

$0 = (5x-6)(x-5)$ Factoring

$5x - 6 = 0$ *or* $x - 5 = 0$

$x = \frac{6}{5}$ *or* $x = 5$ Both apparent solutions are in the domain

Step 6.

Check for $x = \frac{6}{5}$

$$\sqrt{4x+5} - \sqrt{x-1} = \sqrt{14-x}$$

$$\sqrt{4\left(\frac{6}{5}\right) + 5} - \sqrt{\frac{6}{5} - 1} \stackrel{?}{=} \sqrt{14 - \frac{6}{5}}$$

$$\sqrt{\frac{24}{5} + \frac{25}{5}} - \sqrt{\frac{6}{5} - \frac{5}{5}} \stackrel{?}{=} \sqrt{\frac{70}{5} - \frac{6}{5}}$$

$$\sqrt{\frac{49}{5}} - \sqrt{\frac{1}{5}} \stackrel{?}{=} \sqrt{\frac{64}{5}}$$

$$\frac{7}{\sqrt{5}} - \frac{1}{\sqrt{5}} \stackrel{?}{=} \frac{8}{\sqrt{5}}$$

$$\frac{6}{\sqrt{5}} \neq \frac{8}{\sqrt{5}}$$

Check for $x = 5$

$$\sqrt{4x+5} - \sqrt{x-1} = \sqrt{14-x}$$

$$\sqrt{4(5)+5} - \sqrt{5-1} \stackrel{?}{=} \sqrt{14-5}$$

$$\sqrt{20+5} - \sqrt{4} \stackrel{?}{=} \sqrt{9}$$

$$\sqrt{25} - 2 \stackrel{?}{=} 3$$

$$5 - 2 \stackrel{?}{=} 3$$

$$3 = 3$$

Therefore, $\frac{6}{5}$ is *not* a solution, while 5 *is* a solution; the solution set is $\{5\}$. ■

We can use a similar method in solving equations that have rational exponents *if the numerators of those exponents are odd numbers* (see Example 3). We can also solve such equations by first changing terms with rational exponents to radical form. (In Section 10.2, we consider solving equations that have rational exponents when the numerators of the exponents are even numbers.)

Example 3 Find the solution set for $x^{-1/4} = 2$.

Solution The power to which we raise both sides must be the multiplicative inverse of $-1/4$ so that the *product of the two exponents* will be 1. Therefore, we must raise both sides to the negative fourth power.

Raising both sides to the negative fourth power

$$(x^{-1/4})^{-4} = 2^{-4}$$

$$x = 2^{-4} = \frac{1}{2^4} = \frac{1}{16}$$

(You should verify that you get the same result if you change the equation to radical form and solve.)

Check

$$\left(\frac{1}{16}\right)^{-1/4} = (16)^{1/4} = \sqrt[4]{16} = 2$$

The solution set is $\{1/16\}$. ■

Word Problems Involving Radical Equations

There are many applications of word problems involving radical equations in the sciences, business, engineering, and so forth.

Example 4 Find the amount of power, P, consumed if an appliance has a resistance of 5 ohms and draws 10 amps (amperes) of current, using this formula from electricity:

$I = \sqrt{\dfrac{P}{R}}$, where I, the current, is measured in amps (amperes); P, the power, is measured in watts; and R, the resistance, is measured in ohms.

Solution We must find the power, P, when $R = 5$ and $I = 10$.

$$I = \sqrt{\frac{P}{R}}$$

$$10 = \sqrt{\frac{P}{5}}$$

$$(10)^2 = \left(\sqrt{\frac{P}{5}}\right)^2$$

$$100 = \frac{P}{5}$$

$$P = 500$$

Check

$$I = \sqrt{\frac{P}{R}}$$

$$10 \stackrel{?}{=} \sqrt{\frac{500}{5}}$$

$$10 \stackrel{?}{=} \sqrt{100}$$

$$10 = 10$$

Therefore, 500 watts of power is consumed. ■

EXERCISES 7.6

Set I In Exercises 1–20, find the solution set for each equation.

1. $\sqrt{3x+1}=5$

2. $\sqrt{7x+8}=6$

3. $\sqrt{x+1}=\sqrt{2x-7}$

4. $\sqrt{3x-2}=\sqrt{x+4}$

5. $\sqrt[4]{4x-11}-1=0$

6. $\sqrt[4]{3x+1}-2=0$

7. $\sqrt{4x-1}=2x$

8. $\sqrt{6x-1}=3x$

9. $\sqrt{x+7}=2x-1$

10. $\sqrt{2x}+1=\sqrt{2x-3}$

11. $\sqrt{3x+4}-\sqrt{2x-4}=2$

12. $\sqrt{2x-1}+2=\sqrt{3x+10}$

13. $\sqrt[3]{2x+3}-2=0$

14. $\sqrt[3]{4x-3}-3=0$

15. $\sqrt{4u+1}-\sqrt{u-2}=\sqrt{u+3}$

16. $\sqrt{2-v}+\sqrt{v+3}=\sqrt{7+2v}$

17. $x^{1/2}=5$

18. $x^{1/3}=3$

19. $2x^{-5/3}=64$

20. $5x^{-3/2}=40$

In Exercises 21 and 22, solve each problem for the required unknown, using the formula $I=\sqrt{\dfrac{P}{R}}$, as given in Example 4.

21. Find the amount of power, P, consumed if an appliance has a resistance of 16 ohms and draws 5 amps of current.

22. Find the amount of power, P, consumed if an appliance has a resistance of 9 ohms and draws 8 amps of current.

In Exercises 23 and 24, use this formula from statistics: $\sigma=\sqrt{npq}$, where σ (*sigma*) is the *standard deviation*, n is the number of trials, p is the probability of success, and q is the probability of failure.

23. Find the probability of success, p, if the standard deviation is $3\frac{1}{3}$, the probability of failure is $\frac{2}{3}$, and the number of trials is 50.

24. Find the probability of success, p, if the standard deviation is 12, the probability of failure is $\frac{3}{5}$, and the number of trials is 600.

Set II In Exercises 1–20, find the solution set for each equation.

1. $\sqrt{3x-2}=x$

2. $x=\sqrt{10-3x}$

3. $\sqrt{3x-2}=\sqrt{5x+4}$

4. $\sqrt{5x-6}=x$

5. $\sqrt[5]{7x+4}-2=0$

6. $\sqrt{x+4}-\sqrt{2}=\sqrt{x-6}$

7. $x=\sqrt{12x-36}$

8. $\sqrt{3x+1}=\sqrt{1-x}$

9. $\sqrt{x-3}=x-5$

10. $\sqrt{4x+5}+5=2x$

11. $\sqrt{v+7}-\sqrt{v-2}=5$

12. $\sqrt{x}=\sqrt{x+16}-2$

13. $\sqrt[3]{x-1}=2$

14. $\sqrt{2+x}=1$

15. $\sqrt[5]{3x-14}-1=0$

16. $\sqrt{2x-9}-\sqrt{4x}+3=0$

17. $x^{1/4}=1$

18. $x^{1/5}=2$

19. $x^{3/5}=8$

20. $\sqrt{4v+1}=\sqrt{v+4}+\sqrt{v-3}$

In Exercises 21 and 22, solve each problem for the required unknown, using the formula $I = \sqrt{\dfrac{P}{R}}$, as given in Example 4.

21. Find the amount of power, P, consumed if an appliance has a resistance of 25 ohms and draws 4 amps of current.

22. Find R, the resistance in ohms, for an electrical system that consumes 600 watts and draws 5 amps of current.

In Exercises 23 and 24, use this formula from statistics: $\sigma = \sqrt{npq}$, where σ (*sigma*) is the *standard deviation*, n is the number of trials, p is the probability of success, and q is the probability of failure.

23. Find the probability of success, p, if the standard deviation is 8, the probability of failure is $\frac{4}{5}$, and the number of trials is 400.

24. Find the probability of success, p, if the standard deviation is 20, the probability of failure is $\frac{5}{6}$, and the number of trials is 2,880.

7.7 The Pythagorean Theorem

Right Triangles

A triangle that has a *right angle* (90°) is called a **right triangle**. The diagonal of a rectangle divides the rectangle into two right triangles. The parts of a right triangle and a rectangle and one of its diagonals are shown in Figure 7.7.1.

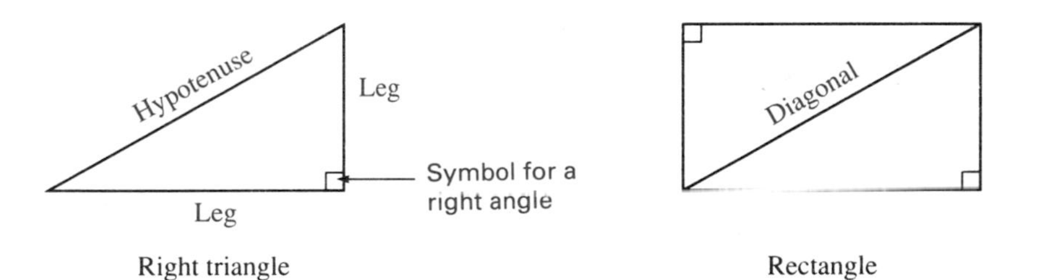

FIGURE 7.7.1.

The following theorem is proved in geometry:

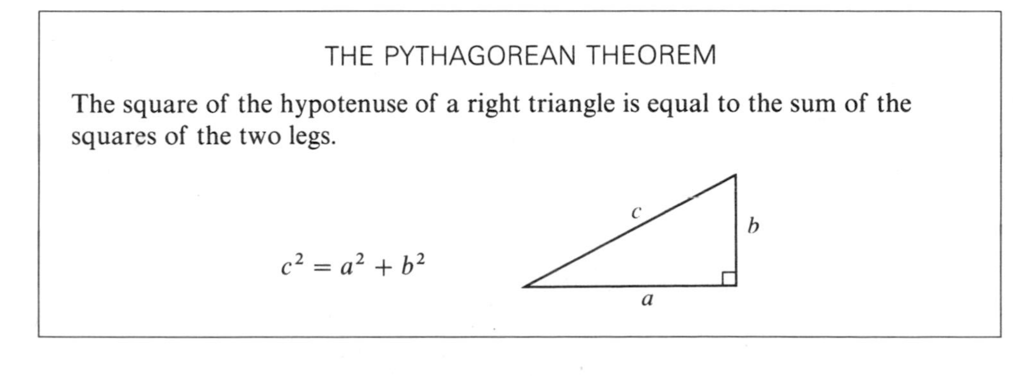

THE PYTHAGOREAN THEOREM

The square of the hypotenuse of a right triangle is equal to the sum of the squares of the two legs.

$$c^2 = a^2 + b^2$$

NOTE The Pythagorean theorem applies only to *right triangles*.

The Pythagorean theorem can be used to find one side of a right triangle when the other two sides are known (see Example 1).

Example 1 Find x, using the Pythagorean theorem.

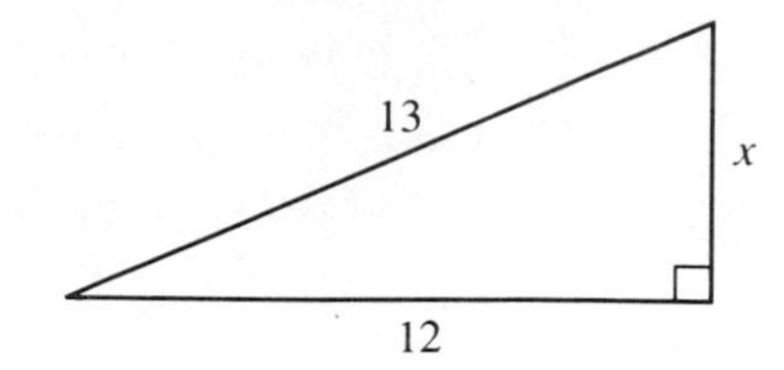

Solution The domain is the set of all *positive* numbers, since the length of a side of a triangle must be positive.

$$c^2 = a^2 + b^2 \quad \text{The Pythagorean theorem}$$
$$13^2 = 12^2 + x^2$$
$$169 = 144 + x^2$$
$$0 = x^2 - 25 \quad \text{Adding } -169 \text{ to both sides}$$
$$0 = (x + 5)(x - 5)$$
$$x + 5 = 0 \quad or \quad x - 5 = 0$$
$$x = -5 \quad or \quad x = 5$$

5 and -5 are both solutions of the original equation; however, -5 is not in the domain of the variable. Therefore, the only solution is 5. ■

Example 2 Find the hypotenuse of a right triangle with legs that are 8 and 6 units long.
Solution
The domain is the set of all positive numbers.
Let $x =$ the length of the hypotenuse.

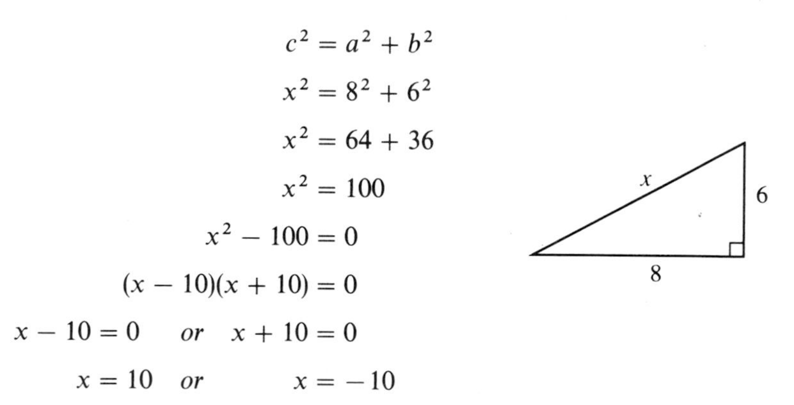

$$c^2 = a^2 + b^2$$
$$x^2 = 8^2 + 6^2$$
$$x^2 = 64 + 36$$
$$x^2 = 100$$
$$x^2 - 100 = 0$$
$$(x - 10)(x + 10) = 0$$
$$x - 10 = 0 \quad or \quad x + 10 = 0$$
$$x = 10 \quad or \quad x = -10$$

10 and -10 are solutions of the original equation. Because -10 is not in the domain of the variable, the only solution is 10. ■

If $k \geq 0$, we can use the following rule for solving an equation of the form $x^2 = k$.

> If $x^2 = k$ and if $k \geq 0$, then $x = \sqrt{k}$ or $x = -\sqrt{k}$.

We'll see in Chapter 10 that there are two solutions for every quadratic (*second* degree) equation, although sometimes the two solutions are identical and sometimes they are not real; notice that there are two possible solutions to the quadratic equation $x^2 = k$. We can easily verify that $\sqrt{k}$ and $-\sqrt{k}$ are solutions of $x^2 = k$. If $k \geq 0$, then

$$\text{if } x = \sqrt{k}, \quad x^2 = (\sqrt{k})^2 = \sqrt{k}\sqrt{k} = k$$
$$\text{if } x = -\sqrt{k}, \quad x^2 = (-\sqrt{k})^2 = (-\sqrt{k})(-\sqrt{k}) = \sqrt{k}\sqrt{k} = k$$

We sometimes combine the solutions and write $x = \pm\sqrt{k}$.*

Example 3 Find x, using the Pythagorean theorem.

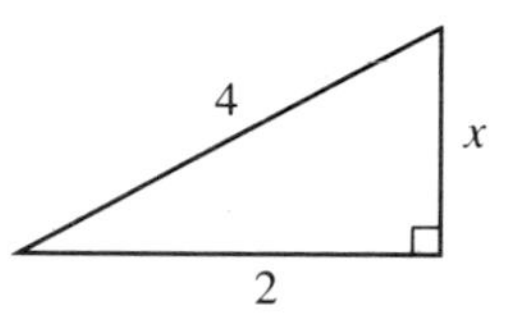

Solution The domain is the set of positive numbers.

$$c^2 = a^2 + b^2$$
$$4^2 = 2^2 + x^2$$
$$16 = 4 + x^2$$
$$x^2 = 12$$
$$x = \sqrt{12} \quad \text{or} \quad x = -\sqrt{12}$$
$$x = 2\sqrt{3} \quad \text{or} \quad x = -2\sqrt{3}$$

Because $-2\sqrt{3}$ is not in the domain, $2\sqrt{3}$ is the only solution. ■

Many word problems require the use of the Pythagorean theorem (see Examples 4 and 5). It is often helpful to draw and label a sketch of the figure.

Example 4 The length of a rectangle is 2 more than its width. If the length of its diagonal is 10, find the dimensions of the rectangle.

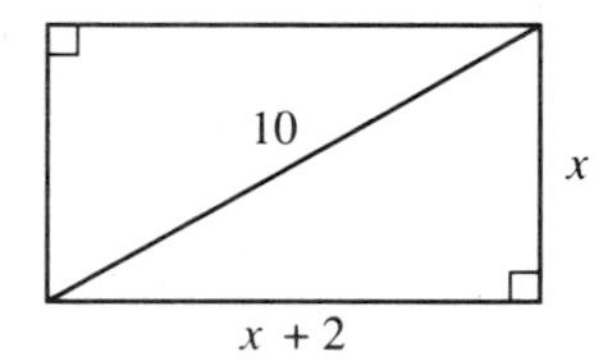

Think The domain is the set of all positive numbers.

Step 1. Let x = width
$x + 2$ = length

Step 2. $(10)^2 = (x)^2 + (x + 2)^2$ — The Pythagorean theorem

Step 3. $100 = x^2 + x^2 + 4x + 4$

$0 = 2x^2 + 4x - 96$

$0 = x^2 + 2x - 48$ — Dividing both sides by 2

$0 = (x + 8)(x - 6)$

* The symbol ± is read "plus or minus" (positive or negative). For example, ±2 is read "plus or minus 2," and $x = \pm 2$ is read "x equals plus or minus 2"; this means $x = +2$ or $x = -2$.

$$x + 8 = 0 \quad or \quad x - 6 = 0$$

Step 4. $x = -8$ *or* $x = 6$ The width

$\uparrow$ Not in the domain $\quad x + 2 = 8$ The length

Step 5. *Check* $6^2 + 8^2 = 36 + 64 = 100$, and $10^2 = 100$

Step 6. Therefore, the width of the rectangle is 6 and the length is 8. ■

Example 5 Two airplanes leave Denver at the same time. One plane is flying due east, and it's flying 50 mph faster than the other plane, which is flying due south. After 2 hr, the planes are 500 mi apart. What is the speed of the slower plane? (We assume that there is *no* wind blowing.)

Step 1. Let x = the speed of the slower plane
$x + 50$ = the speed of the faster plane

Think After 2 hr, the slower plane will have flown $\left(x \dfrac{\text{mi}}{\cancel{\text{hr}}}\right)(2\,\cancel{\text{hr}}) = 2x$ mi, and the faster plane will have flown

$$\left([x + 50]\frac{\text{mi}}{\cancel{\text{hr}}}\right)(2\,\cancel{\text{hr}}) = 2(x + 50) \text{ mi}$$

Sketch

Airport
$2(x + 50)$
East
$2x$
500
South

Step 2.	$(2x)^2 + [2(x + 50)]^2 = 500^2$	The Pythagorean theorem
Step 3.	$4x^2 + (2x + 100)^2 = 250{,}000$	Simplifying
	$4x^2 + 4x^2 + 400x + 10{,}000 = 250{,}000$	
	$8x^2 + 400x - 240{,}000 = 0$	
	$x^2 + 50x - 30{,}000 = 0$	Dividing both sides by 8
	$(x - 150)(x + 200) = 0$	Factoring the left side
	$x - 150 = 0$ *or* $x + 200 = 0$	Setting each factor equal to zero
Step 4.	$x = 150$ *or* $x = -200$	Solving each equation
	$x + 50 = 200$	We reject this solution ($x = -200$), since speed cannot be negative

Step 5. *Check* The faster plane is flying 50 mph faster than the slower plane. The slower plane flies $\left(150 \dfrac{\text{mi}}{\cancel{\text{hr}}}\right)(2\,\cancel{\text{hr}}) = 300$ mi, and the faster plane flies $\left(200 \dfrac{\text{mi}}{\cancel{\text{hr}}}\right)(2\,\cancel{\text{hr}}) = 400$ mi. The sum of the squares of the distances is $300^2 + 400^2$, or $90{,}000 + 160{,}000 = 250{,}000$, which is 500^2.

Step 6. Therefore, the slower plane is flying 150 mph. ■

EXERCISES 7.7

Set I In Exercises 1–6, use the Pythagorean theorem to find x in each figure.

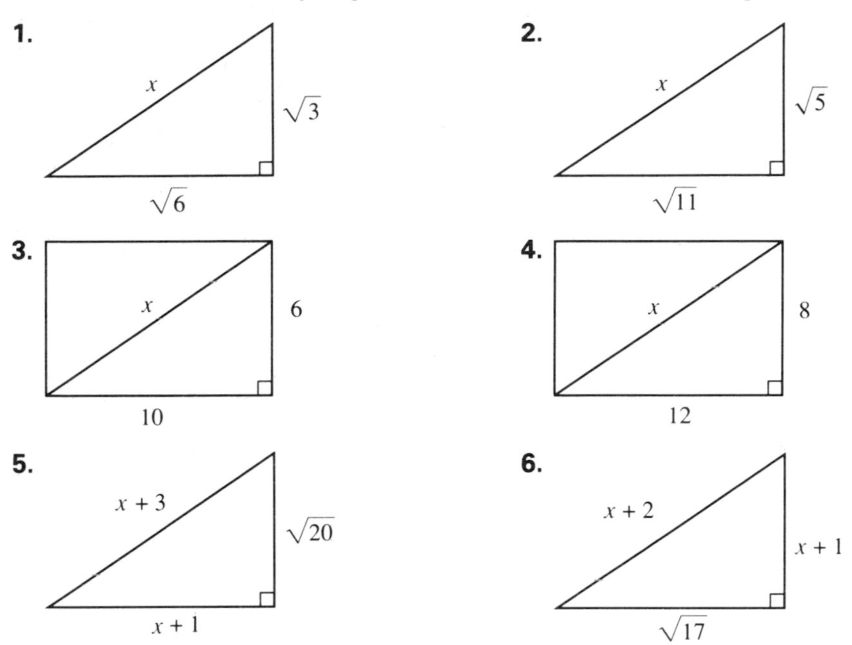

In Exercises 7–14, set up each problem algebraically, solve, and check.

7. Find the diagonal of a square with sides of length 4.

8. Find the diagonal of a square with sides of length 3.

9. Find the width of a rectangle that has a diagonal of 25 and a length of 24.

10. Find the width of a rectangle that has a diagonal of 41 and a length of 40.

11. One leg of a right triangle is 4 less than twice the other leg. If the hypotenuse is 10, how long are the two legs?

12. One leg of a right triangle is 4 more than twice the other leg. If the hypotenuse is $\sqrt{61}$, how long are the two legs?

13. Jaime and Nguyen leave a corner at the same time. Jaime is jogging due west; Nguyen is walking due north; Jaime's speed is 7 mph faster than Nguyen's. After 1 hr, the distance between the two men is 13 mi. How fast is Nguyen walking?

14. Rob and Julie start from the same point at the same time, walking at right angles to one another. Julie is walking 1 mph faster than Rob. After 3 hr, they are 15 mi apart. How fast is Julie walking?

Set II In Exercises 1–6, use the Pythagorean theorem to find x in each figure.

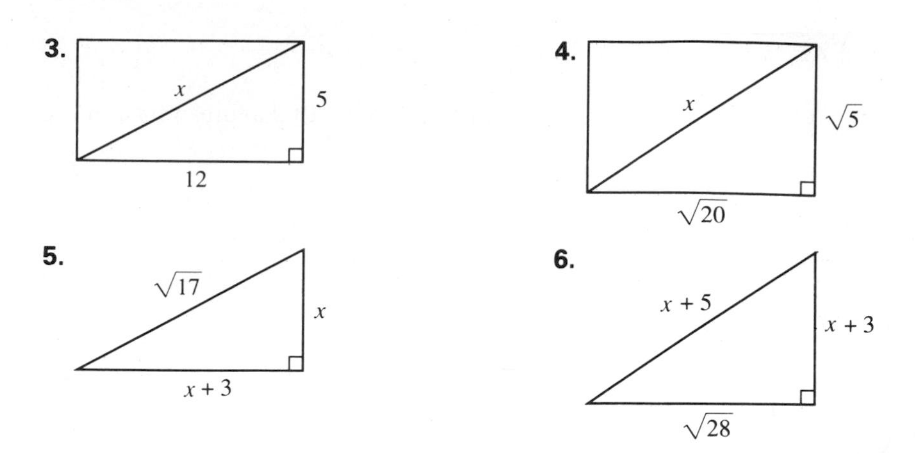

In Exercises 7–14, set up each problem algebraically, solve, and check.

7. Find the diagonal of a square with sides of length 5.

8. Find the diagonal of a rectangle if the width is 8 and the length is 11.

9. Find the width of a rectangle that has a diagonal of 17 and a length of 15.

10. One leg of a right triangle is 2 more than twice the other leg. If the hypotenuse is 2 less than three times the shorter leg, find the lengths of the legs and the hypotenuse of the triangle.

11. One leg of a right triangle is 3 more than twice the other leg. If the hypotenuse is $\sqrt{137}$, how long are the two legs?

12. The length of a rectangle is 3 more than its width. If the length of its diagonal is 15, find the dimensions of the rectangle.

13. Tran and Francisco leave a corner at the same time. Francisco is running due south, and Tran is walking due west; Francisco's speed is 7 mph faster than Tran's. After 1 hr, the distance between the two men is 17 mi. How fast is Francisco running?

14. Two cars leave an intersection at the same time, one going north and the other going east. The car going east is traveling 10 mph faster than the other car. After 30 min, the cars are 25 mi apart. How fast is the slower car going?

7.8 Complex Numbers

All the numbers discussed up to this point in the book have been *real* numbers. Recall that the set of real numbers is the union of the set of rational numbers with the set of irrational numbers. In this section, we discuss a new kind of number that is *not* a real number but is essential in many applications of mathematics.

7.8A Basic Definitions

In order for us to be able to solve equations such as $x^2 = -4$, we must invent, or *define*, a new set of numbers that will *not* be real numbers.

The basis for our new set of numbers is the number i, which is defined as follows:

$$i = \sqrt{-1}$$
$$i^2 = -1$$

Pure Imaginary Numbers

A **pure imaginary number** is a number that can be written as bi, where b is a real number and $i = \sqrt{-1}$. If b is an irrational number, the i is usually written first, as ib.

Example 1 Examples of pure imaginary numbers:

$$-5i, \qquad -2i\sqrt{3}, \qquad -\frac{2}{5}i, \qquad 0.63i. \quad ■$$

A WORD OF CAUTION Students often believe that $i = -1$. *This is not true*! $i = \sqrt{-1}$, not -1. Also, students often write a number such as $\sqrt{5}\,i$ as $\sqrt{5i}$. The i should *not* be under the radical sign. (Numbers such as $\sqrt{5i}$ are discussed in higher-level courses.) $\sqrt{5}\,i$ can be written as $i\sqrt{5}$. ☑

Complex Numbers

A **complex number** is a number that can be written in the form $a + bi$, where a and b are real numbers and $i = \sqrt{-1}$.

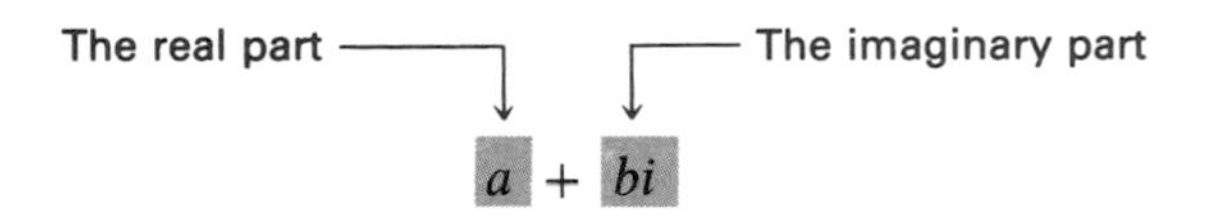

Example 2 Examples of complex numbers:

$$2 - 3i, \qquad \sqrt{5} - \frac{1}{2}i, \qquad -0.7 + 0.4i, \qquad 4 + 0i \quad ■$$

Because the set of complex numbers is a completely new set, we need definitions for equality, addition, subtraction, and so forth.

We should emphasize that the *ordering* property does not hold for complex numbers; that is, the relations "less than" and "greater than" are *not defined* for the set of complex numbers.

The definition of equality of complex numbers follows:

> DEFINITION OF EQUALITY OF TWO COMPLEX NUMBERS
>
> If $a + bi = c + di$, then $a = c$ (the real parts are equal) and $b = d$ (the coefficients of i are equal).

By definition,

$$0i = 0$$

When the *imaginary* part is zero, the complex number is a *real number.*

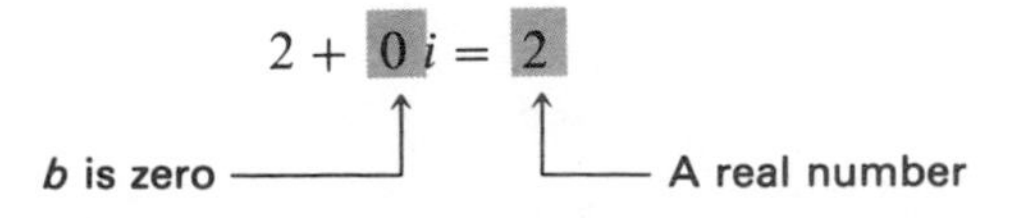

The set of real numbers is a subset of the set of complex numbers.

When the *real* part is zero, the complex number is a *pure imaginary number*.

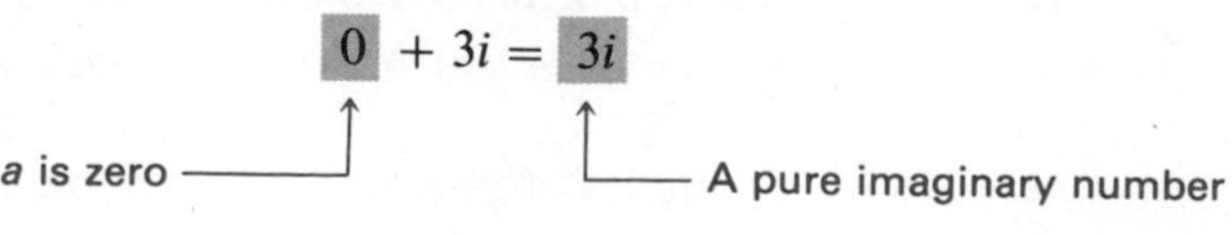

The set of pure imaginary numbers is a subset of the set of complex numbers.

NOTE Some authors call the number $a + bi$ an *imaginary* number (but not a *pure* imaginary number) if $b \neq 0$. We call such a number a complex number. ☑

Example 3 Examples of writing numbers in the form $a + bi$:

Property 7.3

a. $\sqrt{-9} = \sqrt{9(-1)} = \sqrt{9}\sqrt{-1} = 3i = 0 + 3i$

b. $\sqrt{-17} = \sqrt{17(-1)} = \sqrt{17}\sqrt{-1} = \sqrt{17}i = 0 + i\sqrt{17}$

c. $2 - \sqrt{-25} = 2 - \sqrt{25(-1)} = 2 - \sqrt{25}\sqrt{-1} = 2 - 5i$ ■

A WORD OF CAUTION In writing complex numbers in the form $a + bi$, we have used Property 7.3 from Section 7.2A:

$$\sqrt[n]{ab} = \sqrt[n]{a}\sqrt[n]{b}$$

This property does *not* apply when *both* a and b are negative and n is even.

$$\sqrt{(-4)(-9)} = \sqrt{36} = 6$$

but

$$\sqrt{-4}\sqrt{-9} = 2i \cdot 3i = 6i^2 = -6$$

Therefore, $\sqrt{(-4)(-9)} \neq \sqrt{-4}\sqrt{-9}$. Property 7.3 does not apply in this case. ☑

We can use the definition of equality of complex numbers in solving equations involving complex numbers. We simply set the two *real* parts equal to each other and solve that equation; then we set the coefficient of i from one side of the equation equal to the coefficient of i from the other side and solve *that* equation (see Example 4).

Example 4 Examples of solving equations involving complex numbers:

a. If $x - 3i = 5 + yi$, then

$x = 5$ Setting the real parts equal to each other

and $-3 = y$ Setting the coefficients of i equal to each other

$y = -3$

b. If $3x + 7yi = 10 - 2i$, then

$3x = 10$ Setting the real parts equal to each other

$x = \frac{10}{3}$

and $7y = -2$ Setting the coefficients of i equal to each other

$y = -\frac{2}{7}$ ■

EXERCISES 7.8A

Set I In Exercises 1–12, convert each expression to the form $a + bi$. Express all radicals in simplest radical form.

1. $3 + \sqrt{-16}$	**2.** $4 - \sqrt{-25}$	**3.** $\sqrt{-64}$
4. $\sqrt{-100}$	**5.** $5 + \sqrt{-32}$	**6.** $6 + \sqrt{-18}$
7. $\sqrt{-36} + \sqrt{4}$	**8.** $\sqrt{9} - \sqrt{-25}$	**9.** $2i - \sqrt{9}$
10. $3i - \sqrt{16}$	**11.** 14	**12.** -7

In Exercises 13–20, solve for x and y.

13. $3 - 4i = x + 2yi$	**14.** $3x + 5i = 6 + yi$
15. $5x - 3i = 6 - 7yi$	**16.** $-3 - yi = 2x + 3i$
17. $x\sqrt{3} - yi = 2 + i\sqrt{2}$	**18.** $3 + yi\sqrt{5} = x\sqrt{8} - i$
19. $\frac{3}{4}x - \frac{1}{3}yi = \frac{3}{5}x + \frac{1}{2}yi$	**20.** $\frac{2}{3}x - yi = \frac{1}{2}x + 3yi$

Set II In Exercises 1–12, convert each expression to the form $a + bi$. Express all radicals in simplest radical form.

1. $5 + \sqrt{-49}$	**2.** $\sqrt{-4} - 6$	**3.** $\sqrt{-81}$
4. -16	**5.** $3 + \sqrt{-8}$	**6.** $\sqrt{7} + \sqrt{-36}$
7. $\sqrt{-100} + \sqrt{16}$	**8.** 7	**9.** $5i - \sqrt{7}$
10. $\sqrt{8} - \sqrt{-18}$	**11.** $\sqrt{50}$	**12.** $\sqrt{-27}$

In Exercises 13–20, solve for x and y.

13. $5 - yi = x + 4i$	**14.** $8x - 6i = 12 + yi$
15. $3x + 7yi = 2 + 3i$	**16.** $2x - 3yi = 5 + 2i$
17. $4 + yi\sqrt{2} = x\sqrt{5} - 3i$	**18.** $x\sqrt{7} + 2i = \sqrt{2} + 4yi$
19. $\frac{1}{3}x - \frac{3}{4}i = \frac{1}{2} + \frac{1}{5}yi$	**20.** $\frac{3}{5}x - \frac{1}{2}yi = 3 - 4i$

7.8B Addition and Subtraction of Complex Numbers

We can add and subtract complex numbers by treating i as if it were a variable and performing the operations as usual. Addition and subtraction are formally defined as follows:

> ADDITION AND SUBTRACTION OF COMPLEX NUMBERS
>
> $$(a + bi) + (c + di) = (a + c) + (b + d)i$$
>
> $$(a + bi) - (c + di) = (a - c) + (b - d)i$$

Example 5 Examples of adding complex numbers:

a. $(2 + 3i) + (-4 + 5i) = \underline{2} + \underline{\underline{3i}} + \underline{(-4)} + \underline{\underline{5i}}$
$= (2 + [-4]) + (3 + 5)i = -2 + 8i$

b. $(7) + (-5 + 3i) = \underline{7} + \underline{(-5)} + \underline{\underline{3i}}$
$= (7 + [-5]) + 3i = 2 + 3i$

c. $(-7 + 4i) + (6 - 3i) = \underline{-7} + \underline{\underline{4i}} + \underline{6} - \underline{\underline{3i}} = -1 + i$

d. $(\underline{8} + \underline{\underline{7i}}) + (\underline{\underline{-5i}}) + (\underline{-13} + \underline{\underline{4i}}) = -5 + 6i$ ■

Example 6 Examples of subtracting complex numbers:

a. $(-5 + 2i) - (6 - 2i) = \underline{-5} + \underline{\underline{2i}} \underline{- 6} + \underline{\underline{2i}} = (-5 - 6) + (2 + 2)i = -11 + 4i$

b. $(-2) - (-9 - 4i) = \underline{-2} + \underline{9} + \underline{\underline{4i}} = 7 + 4i$

c. $(-13 + 8i) - (7 - 11i) = \underline{-13} + \underline{\underline{8i}} - \underline{7} + \underline{\underline{11i}} = -20 + 19i$

d. $(7 - i) - (6 - 10i) - (-4) = \underline{7} - \underline{\underline{i}} - \underline{6} + \underline{\underline{10i}} + \underline{4} = 5 + 9i$ ■

When complex numbers are not written in the form $a + bi$, you must change them into that form before you perform any operation on them.

Example 7 Add $(2 + \sqrt{-4}) + (3 - \sqrt{-9}) + (\sqrt{-16})$.

Solution $(\underline{2} + \underline{\underline{2i}}) + (\underline{3} - \underline{\underline{3i}}) + (\underline{\underline{4i}}) = 5 + 3i$ ■

EXERCISES 7.8B

Set I In Exercises 1–10, perform the indicated operations; write the answers in the form $a + bi$.

1. $(4 + 3i) + (5 - i)$

2. $(6 - 2i) + (-3 + 5i)$

3. $(7 - 4i) - (5 + 2i)$

4. $(8 - 3i) - (4 + i)$

5. $(2 + i) + (3i) - (2 - 4i)$

6. $(3 - i) + (2i) - (-3 + 5i)$

7. $(2 + 3i) - (x + yi)$

8. $(x - i) - (7 + yi)$

9. $(9 + \sqrt{-16}) + (2 + \sqrt{-25}) + (6 - \sqrt{-64})$

10. $(13 - \sqrt{-36}) - (10 - \sqrt{-49}) + (8 + \sqrt{-4})$

In Exercises 11–14, solve for x and y.

11. $(4 + 3i) - (5 - i) = (3x + 2yi) + (2x - 3yi)$

12. $(3 - 2i) - (x + i) = (2x - yi) - (x + 2yi)$

13. $(2 - 5i) - (5 + 3i) = (3x + 2yi) - (5x + 3yi)$

14. $(4x - 3yi) - (7x + 2yi) = (7 - 2i) - (3 + 4i)$

Set II In Exercises 1–10, perform the indicated operations; write the answers in the form $a + bi$.

1. $(8 - 5i) + (7 + 2i)$

2. $(5 - 2i) - (4) - (3 + 4i)$

3. $(9 - 12i) - (7 + i)$

4. $(7i) - (2 + i) + (-3 - 5i)$

5. $(7 + 3i) - (2i) - (2 - 5i)$

6. $(8 + 3i) - (2 - i) + (-9i)$

7. $(3 - xi) - (y - 5i)$

8. $(3x - 4i) + (x - yi) - (3 + yi)$

9. $(3 + \sqrt{-9}) - (4 - \sqrt{-81}) + (2 - \sqrt{-1})$

10. $(5 - \sqrt{-16}) + (\sqrt{-4} - 5) - (3 - \sqrt{-25})$

In Exercises 11–14, solve for x and y.

11. $(5 - i) + (-3 + 2i) = (4x - 5yi) - (7x - 2yi)$

12. $5 - yi = x + 4i$

13. $(3 - 2i) - (x + yi) = (2x + 5i) - (4 + i)$

14. $(5 + 2i) - (3i) - (x - 3i) = (x + yi) - (3 + 2yi)$

7.8C Multiplication of Complex Numbers

Multiplication is distributive over addition for the set of complex numbers, as it is for the set of real numbers. Because of this, and because $i^2 = -1$, the definition of the multiplication of two complex numbers is

$$(a + bi)(c + di) = (ac - bd) + (ad + bc)i$$

However, instead of memorizing this definition, you can apply the following technique.

TO MULTIPLY TWO COMPLEX NUMBERS

1. Multiply the numbers as you would multiply two binomials.
2. Replace i^2 by -1.
3. Collect and combine like terms and write the result in the form $a + bi$.

Example 8 Multiply $(4 + 3i)(-5 + 2i)$.

Solution

$$\begin{aligned}&(4 + 3i)(-5 + 2i)\\ &= -20 + 8i - 15i + 6i^2\\ &= -20 - 7i + 6(-1)\\ &= -20 - 7i - 6\\ &= -26 - 7i \quad \text{The product in the form } a + bi \quad \blacksquare\end{aligned}$$

Example 9 Find $(5 - 3i)^2$.

Solution

$$\begin{aligned}(5 - 3i)^2 &= 25 - 2(5)(3i) + 9i^2\\ &= 25 - 30i - 9\\ &= 16 - 30i \quad \blacksquare\end{aligned}$$

Because $\sqrt{a}\sqrt{b} \neq \sqrt{ab}$ if a and b are *both* negative, in Example 10 we *must* express $\sqrt{-25}$ and $\sqrt{-9}$ in simplest radical form before we multiply.

Example 10 Multiply $(2 - \sqrt{-25})(-3 + \sqrt{-9})$.

Solution

$$\left.\begin{aligned} 2 - \sqrt{-25} &= 2 - 5i \\ -3 + \sqrt{-9} &= -3 + 3i \end{aligned}\right\} \quad \text{First convert to the form } a + bi$$

Therefore,

$$\begin{aligned}(2 - \sqrt{-25})(-3 + \sqrt{-9}) &= (2 - 5i)(-3 + 3i) \\ &= -6 + 21i - 15i^2 \\ &= -6 + 21i + 15 \\ &= 9 + 21i \quad \blacksquare\end{aligned}$$

Simplifying Powers of i

Any power of i can be rewritten as ± 1 or $\pm i$. Therefore, an algebraic expression is not considered to be in *simplest form* if it contains any powers of i greater than 1.

Example 11 Examples of finding powers of i:

a. $i = \boxed{i}$

b. $i^2 = \boxed{-1}$

c. $i^3 = i^2 \cdot i = (-1)i = \boxed{-i}$

d. $i^4 = i^2 \cdot i^2 = (-1)(-1) = \boxed{1}$

(a–d: Each integral power of i must be one of these four numbers)

e. $i^{13} = (i^4)^3 \cdot i = (1)^3 \cdot i = 1 \cdot i = i$

f. $i^{51} = (i^4)^{12} \cdot i^3 = (1)^{12} \cdot i^3 = 1(-i) = -i$

g. $i^{100} = (i^4)^{25} = (1)^{25} = 1 \quad \blacksquare$

Example 12 Simplify $(1 + i)^6$.

Solution We will use the binomial theorem. The coefficients are 1, 6, 15, 20, 15, 6, and 1. Therefore,

$$\begin{aligned}(1 + i)^6 &= 1(1)^6 + 6(1)^5 i + 15(1)^4 i^2 + 20(1)^3 i^3 + 15(1)^2 i^4 + 6(1) i^5 + 1 i^6 \\ &= 1 + 6i + 15(-1) + 20(-i) + 15(1) + 6i + (-1) \\ &= 0 - 8i \text{ or } -8i \quad \blacksquare\end{aligned}$$

EXERCISES 7.8C

Set I Perform the indicated operations. Express your answers in the form $a + bi$.

1. $(1 + i)(1 - i)$
2. $(3 + 2i)(3 - 2i)$
3. $(4 - i)(3 + 2i)$
4. $(5 + 2i)(2 - 3i)$
5. $(6 - 2i)(2 - 3i)$
6. $(4 + 7i)(3 + 2i)$
7. $(\sqrt{5} + 2i)(\sqrt{5} - 2i)$
8. $(\sqrt{7} - 3i)(\sqrt{7} + 3i)$
9. $5i(i - 2)$
10. $6i(2i - 1)$
11. $(2 + 5i)^2$
12. $(3 - 4i)^2$
13. i^{10}
14. i^{23}
15. i^{87}
16. i^{73}
17. $(3i)^3$
18. $(2i)^3$
19. $(2i)^4$
20. $(3i)^4$
21. $(3 - \sqrt{-4})(4 + \sqrt{-25})$
22. $(5 + \sqrt{-64})(2 - \sqrt{-36})$

23. $(2 - \sqrt{-1})^2$ **24.** $(3 + \sqrt{-1})^2$ **25.** $[3 + i^6]^2$

26. $[4 - i^{10}]^2$ **27.** $i^{10}(i^{23})$ **28.** $i^{34}(i^{16})$

29. $i^{15} + i^7$ **30.** $i^{27} + i^{14}$ **31.** $[2 + (-i)^{11}]^2$

32. $[3 - (-i)^5]^2$ **33.** $(1 - i)^5$ **34.** $(1 + i)^4$

35. $(2 - i)^4$ **36.** $(4 - i)^3$

Set II Perform the indicated operations. Express your answers in the form $a + bi$.

1. $(8 + 5i)(3 - 4i)$ **2.** $(7 - i)(3 + 2i)$ **3.** $(6 - 3i)(4 + 5i)$

4. $(8 - i)(7 + 2i)$ **5.** $(7 + 2i)(1 - i)$ **6.** $(1 + 3i)(1 - 3i)$

7. $(4 - 2i)(4 + 2i)$ **8.** $(1 + i)(1 + i)$ **9.** $8i(2 - 3i)$

10. $-4i(3 - 2i)$ **11.** $(-4 + 3i)^2$ **12.** $(-7 - 2i)^2$

13. i^{13} **14.** i^{83} **15.** i^{91}

16. i^{74} **17.** $(2i)^5$ **18.** $(-3i)^3$

19. $(-5i)^2$ **20.** $(-3i)^4$

21. $(5 + \sqrt{-36})(3 - \sqrt{-100})$ **22.** $(2 - \sqrt{-49})(4 + \sqrt{-81})$

23. $(4 + \sqrt{-1})^2$ **24.** $(6 - \sqrt{-25})^2$ **25.** $(3 - i^{13})^2$

26. $(3 + [-i]^7)^2$ **27.** $i^{24}(i^{18})$ **28.** $i^{37}(i^{23})$

29. $i^{14} + i^{25}$ **30.** $i^{25} - i^{17}$ **31.** $[8 - (-i)^{13}]^2$

32. $[6 + 5(-i)^7]^2$ **33.** $(1 + i)^5$ **34.** $(1 + 2i)^4$

35. $(3 - i)^3$ **36.** $(1 - i)^6$

7.8D Division of Complex Numbers

The **conjugate of a complex number** is obtained by changing the sign of its imaginary part.

The conjugate of $a + bi$ is $a - bi$

The conjugate of $a - bi$ is $a + bi$

Example 13 Examples of conjugate complex numbers:

a. The conjugate of $3 - 2i$ is $3 + 2i$.

b. The conjugate of $-5 + 4i$ is $-5 - 4i$.

c. The conjugate of $7i$ is $-7i$, because $7i = 0 + 7i$ with conjugate $0 - 7i = -7i$.

d. The conjugate of 5 is 5, because $5 = 5 + 0i$ with conjugate $5 - 0i = 5$. ■

The product of a complex number and its conjugate is a positive real number.

Proof $(a + bi)(a - bi) = a^2 - b^2i^2 = a^2 + b^2$ A positive real number

In order to divide one complex number by another, we must convert the expression to the form $a + bi$; in other words, we must make the divisor (or denominator) a *real* number. Because the product of a complex number and its conjugate is a real number, we have the following rules for division of complex numbers:

TO DIVIDE ONE COMPLEX NUMBER BY ANOTHER

1. Write the division as a rational expression.
2. Multiply the numerator and denominator by the conjugate of the denominator.

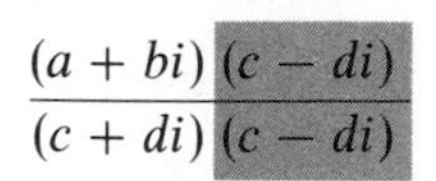

$$\frac{(a + bi)\,(c - di)}{(c + di)\,(c - di)}$$

3. Simplify, and write the result in the form $a + bi$.

This procedure is similar to the procedure for dividing radicals when the divisor (or denominator) contains two terms.

Example 14 Examples of dividing complex numbers:

a. $$\frac{10i}{1 - 3i} = \frac{10i}{(1 - 3i)}\frac{(1 + 3i)}{(1 + 3i)} = \frac{10i + 30i^2}{1 - 9i^2} = \frac{10i - 30}{1 + 9}$$

$$= \frac{\overset{1}{\cancel{10}}(i - 3)}{\underset{1}{\cancel{10}}} = i - 3 = -3 + i$$

The denominator has been converted into a real number

b. $$(2 + i) \div (3 - 2i) = \frac{2 + i}{3 - 2i} = \frac{(2 + i)}{(3 - 2i)}\frac{(3 + 2i)}{(3 + 2i)}$$

$$= \frac{6 + 7i + 2i^2}{9 - 4i^2} = \frac{6 + 7i - 2}{9 + 4} = \frac{4 + 7i}{13}$$

$$= \frac{4}{13} + \frac{7}{13}i$$ The quotient in the form $a + bi$

c. $$(3 + i) \div i = \frac{(3 + i)}{i}\frac{(-i)}{(-i)} = \frac{-3i - i^2}{-i^2} = \frac{-3i + 1}{1} = 1 - 3i$$

The conjugate of i is $-i$, because $i = 0 + i$ and its conjugate is $0 - i = -i$

d. $$5 \div (-2i) = \frac{5}{-2i} = \frac{5 \cdot i}{-2i \cdot i} = \frac{5i}{-2i^2} = \frac{5}{2}i = 0 + \frac{5}{2}i$$

Even though the conjugate of $-2i$ is $2i$, we were able to make the denominator a real number by multiplying by i instead of by $2i$. ■

EXERCISES 7.8D

Set I In Exercises 1–6, write the conjugate of each complex number.

1. $3 - 2i$ **2.** $5 + 4i$ **3.** $5i$ **4.** $-7i$ **5.** 10 **6.** -8

In Exercises 7–18, perform the indicated operations and write the answers in the form $a + bi$.

7. $\dfrac{10}{1 + 3i}$ **8.** $\dfrac{5}{1 + 2i}$ **9.** $\dfrac{1 + i}{1 - i}$ **10.** $\dfrac{1 - i}{1 + i}$

11. $\dfrac{8 + i}{i}$ **12.** $\dfrac{4 - i}{i}$ **13.** $\dfrac{3}{2i}$ **14.** $\dfrac{4}{5i}$

15. $\dfrac{15i}{1 - 2i}$ **16.** $\dfrac{20i}{1 - 3i}$ **17.** $\dfrac{4 + 3i}{2 - i}$ **18.** $\dfrac{3 + 2i}{4 + 2i}$

Set II In Exercises 1–6, write the conjugate of each complex number.

1. $5 - 3i$ **2.** $-2 + 3i$ **3.** $-3i$ **4.** 0 **5.** 7 **6.** $3i$

In Exercises 7–18, perform the indicated operations and write the answers in the form $a + bi$.

7. $\dfrac{8}{2 - 3i}$ **8.** $\dfrac{2 + 6i}{2 + i}$ **9.** $\dfrac{3 + i}{3 - i}$ **10.** $\dfrac{8 + i}{6i}$

11. $\dfrac{4 + 2i}{3i}$ **12.** $\dfrac{5 + 3i}{5 - 3i}$ **13.** $\dfrac{9}{4i}$ **14.** $\dfrac{8 - 3i}{8 + 3i}$

15. $\dfrac{29i}{2 + 5i}$ **16.** $\dfrac{3}{7i}$ **17.** $\dfrac{2i + 5}{3 + 2i}$ **18.** $\dfrac{6i - 1}{6i + 1}$

7.9 Review: 7.6–7.8

To Solve a Radical Equation 7.6

1. Find the domain of the variable.
2. Arrange the terms so that one term with a radical is by itself on one side of the equation.
3. Raise each side of the equation to a power equal to the index of the radical.
4. Simplify each side of the equation. If a radical still remains, repeat steps 1, 2, and 3.
5. Solve the resulting equation for the variable.
6. Check apparent solutions in the original equation.

The Pythagorean Theorem 7.7

The square of the hypotenuse of a right triangle is equal to the sum of the squares of the two legs.

$$c^2 = a^2 + b^2$$

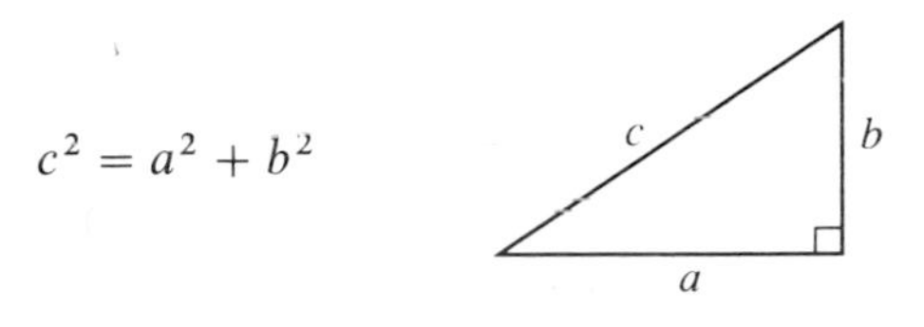

Complex Numbers 7.8

A **complex number** is a number of the form $a + bi$, where a and b are real numbers and $i = \sqrt{-1}$.

The set of complex numbers: $C = \{a + bi \mid a, b \in R, i = \sqrt{-1}\}$.

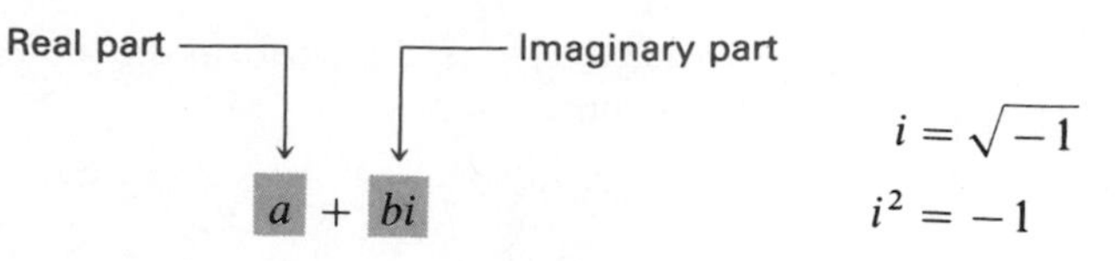

$i = \sqrt{-1}$

$i^2 = -1$

When the real part is zero, the complex number is a *pure imaginary number*.

When the imaginary part is zero, the complex number is a *real number*.

Equality of Complex Numbers 7.8A

If $a + bi = c + di$, then $a = c$ (the real parts are equal) and $b = d$ (the coefficients of i are equal).

Addition and Subtraction of Complex Numbers 7.8B

$(a + bi) + (c + di) = (a + c) + (b + d)i$

$(a + bi) - (c + di) = (a - c) + (b - d)i$

Perform the operations as usual, treating i as if it were a variable.

Multiplication of Two Complex Numbers 7.8C

1. Multiply the numbers as you would multiply two binomials.
2. Replace i^2 by -1.
3. Collect and combine like terms and write the result in the form $a + bi$.

Simplifying Powers of i 7.8C

Any power of i can be rewritten as ± 1 or $\pm i$.

Division of Complex Numbers 7.8D

1. Write the division as a rational expression.
2. Multiply the numerator and denominator by the conjugate of the denominator.

$$\frac{(a + bi)(c - di)}{(c + di)(c - di)}$$

3. Simplify, and write the result in the form $a + bi$.

Review Exercises 7.9 Set I

In Exercises 1–6, find the solution set for each equation.

1. $\sqrt{x - 5} = \sqrt{3x + 8}$

2. $\sqrt{3x - 5} + 3 = x$

3. $\sqrt{5x - 4} - \sqrt{2x + 1} = 1$

4. $\sqrt[5]{5x + 4} = -2$

5. $x^{5/6} = 32$

6. $x^{3/4} = 27$

In Exercises 7 and 8, solve for x and y.

7. $5 - yi = x + 6i$

8. $(2 - 5i) - (x + yi) = (x - 7i) + (4 - 2yi)$

In Exercises 9–13, perform the indicated operations and write the answers in the form $a + bi$.

9. $(3i + 2)(4 - 2i)$

10. $(3 - i)^4$

11. $\dfrac{2+i}{1+3i}$

12. $\dfrac{3+i}{1-2i}$

13. $(4+\sqrt{-27})+(2-\sqrt{-12})-(1-\sqrt{-3})$

In Exercises 14–17, set up the problem algebraically, solve, and check.

14. The length of a rectangle is 2 ft more than its width. Its diagonal is $\sqrt{34}$ ft. Find the dimensions of the rectangle.

15. The width of a rectangle is 3 m less than its length. Its diagonal is $\sqrt{45}$ m. Find the dimensions of the rectangle.

16. Using the formula $\sigma = \sqrt{npq}$, as given in Section 7.6, find p, the probability of success, if σ, the standard deviation, is $2\sqrt{6}$, q, the probability of failure, is $2/5$, and n, the number of trials, is 100.

17. Two cars leave a desert intersection at the same time, one traveling east and the other traveling south; twenty minutes later, the cars are 25 mi apart. The car traveling south is going 15 mph faster than the other car. What is the speed of the faster car?

Review Exercises 7.9 Set II

NAME ____________________

In Exercises 1–6, find the solution set for each equation.

1. $x = \sqrt{3x + 10}$

2. $\sqrt[3]{4x - 1} = -2$

3. $\sqrt{3x + 7} - \sqrt{x - 2} = 3$

4. $x^{3/5} = -8$

5. $x = 2\sqrt{x + 6} - 3$

6. $x^{1/3} = -2$

In Exercises 7 and 8, solve for x and y.

7. $3x - 3i = 8 - yi$

8. $(4 - 3i) - (x + 2yi) = (x + yi) - (4 - 5i)$

In Exercises 9–13, perform the indicated operations and write the answers in the form $a + bi$.

9. $(8 + 7i) - (3 - 2i) + (\sqrt{2} - 3i)$

10. $4i(6 - 2i)$

11. $(2 + i)^5$

ANSWERS

1. ____________________

2. ____________________

3. ____________________

4. ____________________

5. ____________________

6. ____________________

7. ____________________

8. ____________________

9. ____________________

10. ____________________

11. ____________________

12. $(3 - \sqrt{-50}) - (1 + \sqrt{-18}) + \sqrt{-32}$

13. $\dfrac{8 + 5i}{8 - 5i}$

In Exercises 14–17, set up the problem algebraically, solve, and check.

14. The length of a rectangle is 7 ft more than its width. If the length of its diagonal is 13 ft, find the dimensions of the rectangle.

15. The longer leg of a right triangle is 3 in. longer than the shorter leg, and the hypotenuse is 6 in. longer than the shorter leg. Find the lengths of the three sides.

16. Using the formula $I = \sqrt{\dfrac{P}{R}}$, as given in Section 7.6, find P, the amount of power consumed, if an appliance has a resistance, R, of 36 ohms and draws 3 amps of current, I.

17. Eduardo and Miguel leave a certain point at the same time, bicycling along paths that are at right angles to one another. Miguel, who's going downhill, is bicycling 6 mph faster than Eduardo. Two hours later, they are 60 mi apart. How fast is Eduardo bicycling?

ANSWERS

12. ______________

13. ______________

14. ______________

15. ______________

16. ______________

17. ______________

Chapter 7 Diagnostic Test

The purpose of this test is to see how well you understand operations with radicals, rational exponents, and complex numbers. We recommend that you work this diagnostic test *before* your instructor tests you on this chapter. Allow yourself about 50 minutes.

Complete solutions for all the problems on this test, together with section references, are given in the answer section at the end of the book. For the problems you do incorrectly, study the sections cited.

In Problems 1–5, perform the indicated operations. Express the answers in exponential form with positive exponents. Assume that the variables represent positive numbers.

1. $x^{1/2}x^{-1/4}$

2. $(R^{-4/3})^3$

3. $\dfrac{a^{5/6}}{a^{1/3}}$

4. $\left(\dfrac{x^{-2/3}y^{3/5}}{x^{1/3}y}\right)^{-5/2}$

5. $\dfrac{b^{2/3}}{b^{-1/5}}$

In Problems 6–10, write each expression in simplest radical form.

6. $\sqrt[3]{54x^6y^7}$

7. $\dfrac{4xy}{\sqrt{2x}}$, $x > 0$

8. $\sqrt[6]{a^3}$

9. $\sqrt{40} + \sqrt{9}$

10. $\sqrt{x}\sqrt[3]{x}$, $x \geq 0$

11. Evaluate the expression $(-27)^{2/3}$.

In Problems 12–18, perform the indicated operations. Give your answers in simplified form. Assume $x > 0$ and $y \geq 0$.

12. $4\sqrt{8y} + 3\sqrt{32y}$

13. $3\sqrt{\dfrac{5x^2}{2}} - 5\sqrt{\dfrac{x^2}{10}}$

14. $\sqrt{2x^4}\sqrt{8x^3}$

15. $\sqrt{2x}(\sqrt{8x} - 5\sqrt{2})$

16. $\dfrac{\sqrt{10x} + \sqrt{5x}}{\sqrt{5x}}$

17. $\dfrac{5}{\sqrt{7} + \sqrt{2}}$

18. $(1 - \sqrt[3]{x})^3$

In Problems 19–22, perform the indicated operations; write the answers in the form $a + bi$.

19. $(5 - \sqrt{-8}) - (3 - \sqrt{-18})$

20. $(3 + i)(2 - 5i)$

21. $\dfrac{10}{1 - 3i}$

22. $(2 - i)^3$

23. Solve and check: $x^{3/2} = 8$

24. Solve and check: $\sqrt{x - 3} + 5 = x$

25. Find the value of x shown in the right triangle.

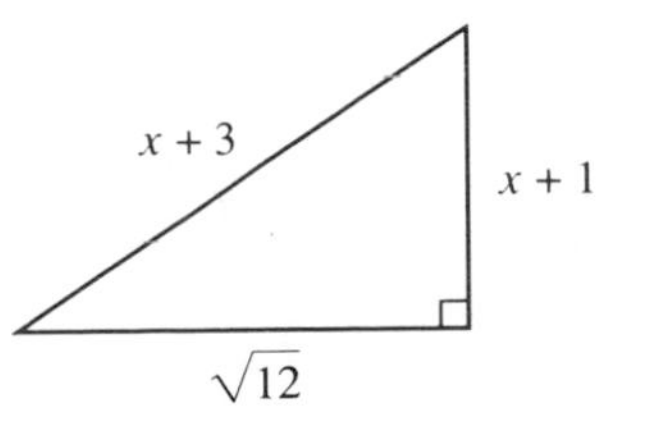

Cumulative Review Exercises: Chapters 1–7

In Exercises 1–6, perform the indicated operations and simplify. Assume that the value(s) of the variable(s) that make the denominator zero are excluded.

1. $\dfrac{6}{a^2-9} - \dfrac{2}{a^2-4a+3}$

2. $\dfrac{x^3-8}{2x^2+x-10} \div \dfrac{3x^3-x^2}{6x^2+13x-5}$

3. $\dfrac{\frac{4}{x}-\frac{8}{x^2}}{\frac{1}{x}-\frac{2}{x^2}}$

4. $\sqrt{5x}(\sqrt{5x}+2),\ x \geq 0$

5. $(\sqrt{26}-\sqrt{10})^2$

6. $\dfrac{6}{2-\sqrt{5}}$

In Exercises 7–11, find the solution set for each equation or inequality.

7. $3(4x-1)-(3-2x)=4(2x-3)$

8. $\dfrac{3x}{4}-\dfrac{5}{6} \leq \dfrac{7x}{12}$

9. $\sqrt{2x+2} = 1+\sqrt{3x-12}$

10. $\dfrac{3}{x+2}-\dfrac{11}{2x+4}=\dfrac{5}{2}$

11. $6x^2+11x=10$

12. Solve for a in terms of the other variables:

$$\frac{1}{a}+\frac{1}{b}=\frac{1}{c}$$

13. Find the solution set for $|2x-5|>3$.

In Exercises 14–18, set up each problem algebraically, solve, and check.

14. Virginia can do a job in 5 hr. Adolph can do the same job in 6 hr. How long will the job take if they work together?

15. The area of a rectangle is 54 sq. cm. Find its dimensions if its sides are in the ratio of 2 to 3.

16. Min invested \$26,500, placing some of the money in an account that earned 7.6% interest annually and the rest of the money in an account that earned 7.3% annually. If she earned \$1,977.10 in the first year, how much did she invest at each rate?

17. A boat's engine is running so that the boat would go 27 mph in still water. It can go 120 mi upstream (against the current) in the same amount of time that it could go 150 mi downstream. What is the speed of the current?

18. Amir's living room is square. If the room were 2 yd longer and 1 yd narrower, the cost of carpeting it with carpeting that costs \$35 per square yard would be \$3,080. What are the dimensions of his living room?

8 Graphing and Functions

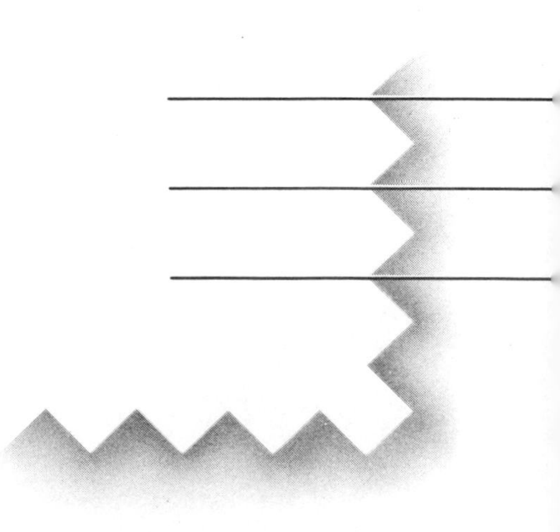

Graphs are mathematical pictures that help explain relationships between variables. In this chapter, we discuss graphs of ordered pairs and graphs of equations and inequalities in two variables. We also discuss relations, functions and their inverses, and variation.

8.1 The Rectangular Coordinate System and Mathematical Relations

Ordered Pairs Two **ordered pairs** are not equal to each other unless they contain the same elements *in the same order*. We enclose the elements of an ordered pair in *parentheses*. Thus,

$$\{1, 4\} = \{4, 1\} \qquad (1, 4) \neq (4, 1)$$

Sets Ordered pairs

The Rectangular Coordinate System

The **rectangular coordinate system** in the plane usually consists of a horizontal number line, called the **horizontal axis** or ***x*-axis**, and a vertical number line, called the **vertical axis** or ***y*-axis**. These lines intersect at a point we call the **origin**. The units of measure on these two lines are usually equal, but in some cases—for example, in graphing the ordered pair (3, 200)—it is best for them to be *unequal.*

The two axes divide the plane into four **quadrants**. Some of the terms commonly used with a rectangular coordinate system are shown in Figures 8.1.1 and 8.1.2.

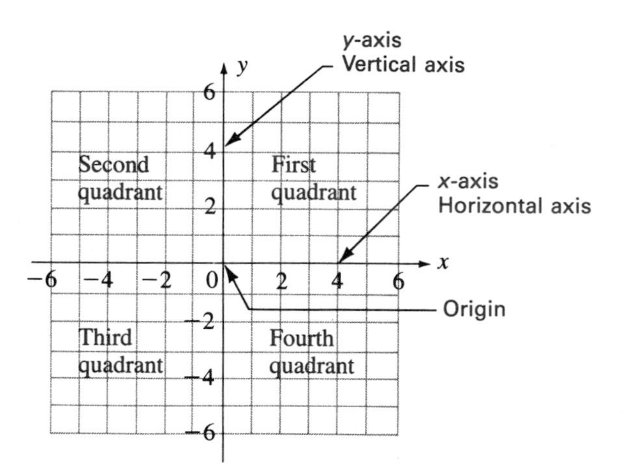

FIGURE 8.1.1 RECTANGULAR COORDINATE SYSTEM

The Graph of a Point There is exactly one point in the plane that corresponds to each ordered pair of real numbers. The origin corresponds to the ordered pair (0, 0).

The *first* number of an ordered pair tells us how far the point is from the y-axis. A *positive* first coordinate indicates that the point is to the *right* of the y-axis; a *negative* first coordinate indicates that the point is to the *left* of the y-axis.

The *second* number in the ordered pair tells us how far the point is from the x-axis. A *positive* second coordinate indicates that the point is *above* the x-axis; a *negative* second coordinate indicates that the point is *below* the x-axis.

The point (3, 2) is shown in Figure 8.1.2.

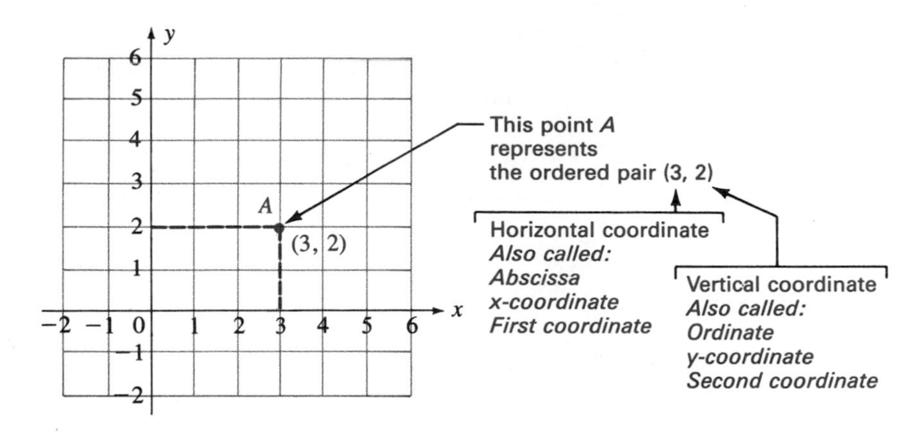

FIGURE 8.1.2 GRAPH OF AN ORDERED PAIR

The statement "plot the points" means the same as "graph the points."

Example 1 Graph (plot) the points $A(1, 4)$, $B(4, 1)$, $C(-1, 3)$, $D(3, -1)$, and $E(0, 2)$.

Solution

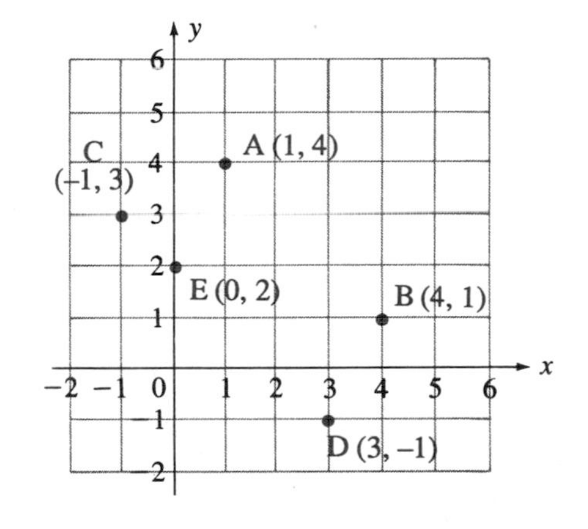

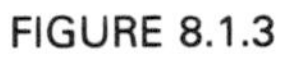

FIGURE 8.1.3 ■

NOTE When the order is changed in an ordered pair, we get a different point. We see in Figure 8.1.3 for Example 1 that points A and B are different from each other, as are points C and D. ☑

Subscripts

x_1 The small number written below and to the right of the variable is used to indicate a particular value of that variable; it is called a **subscript**. x_1 is read "x sub one."

x_2 A different subscript indicates a different value of that variable. x_2 is read "x sub two."

Example 2 Examples of subscripted variables:

a. y_1 and y_2 are different values of y. y_1 is read "y sub one"; y_2 is read "y sub two."

b. $P_1(x_1, y_1)$ and $P_2(x_2, y_2)$ represent two different points. ■

If two points have equal y-coordinates, they must lie on the same *horizontal* line (and vice versa); therefore, $P_1(x_1, y_1)$ and $P_2(x_2, y_1)$ lie on the same horizontal line. Similarly, if two points have equal x-coordinates, they must lie on the same *vertical* line (and vice versa); therefore, $P_1(x_1, y_1)$ and $P_3(x_1, y_2)$ lie on the same vertical line.

The Distance Between Two Points

We use the notation $|P_1P_2|$ to represent the **distance** between points P_1 and P_2.

> TO FIND THE DISTANCE BETWEEN POINTS P_1 AND P_2
>
> 1. If P_1 is the point (x_1, y_1) and P_2 is the point (x_2, y_1), the points lie on the same *horizontal* line and the distance between them is defined to be $|P_1P_2| = |x_2 - x_1|$.
> 2. If P_1 is the point (x_1, y_1) and P_2 is the point (x_1, y_2), the points lie on the same *vertical* line and the distance between them is defined to be $|P_1P_2| = |y_2 - y_1|$.
> 3. If P_1 is the point (x_1, y_1) and P_2 is the point (x_2, y_2), the points lie anywhere in the plane, and the distance, d, between them is
>
> $$d = |P_1P_2| = \sqrt{(x_2 - x_1)^2 + (y_2 - y_1)^2}$$

Proof of (3) Suppose $P_1(x_1, y_1)$ and $P_2(x_2, y_2)$ are two points that do not lie on the same horizontal or vertical line. Let's draw a horizontal line through $P_1(x_1, y_1)$ and a vertical line through $P_2(x_2, y_2)$. These two lines meet at a point we will call C (see Figure 8.1.4); C *must* be the point (x_2, y_1). The triangle formed by joining P_1, C, and P_2 is a right triangle with P_1P_2 as its hypotenuse.

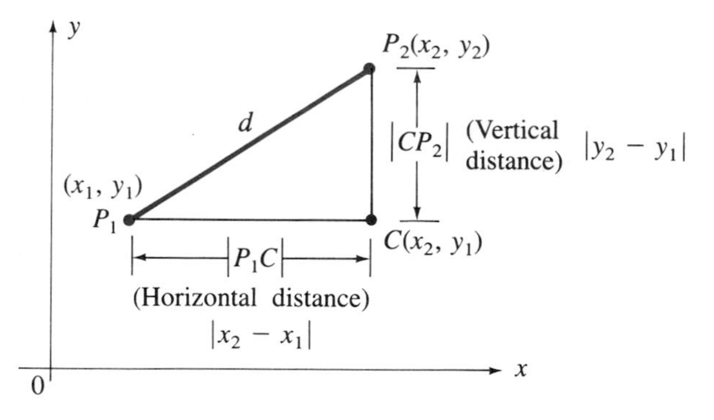

FIGURE 8.1.4

The horizontal distance $|P_1C| = |x_2 - x_1|$, and the vertical distance $|CP_2| = |y_2 - y_1|$. Because $|x_2 - x_1|^2 = (x_2 - x_1)^2$ and $|y_2 - y_1|^2 = (y_2 - y_1)^2$ (you should verify this), using the Pythagorean theorem to find d gives

$$\begin{aligned}(d)^2 &= (|x_2 - x_1|)^2 + (|y_2 - y_1|)^2 \\ &= (x_2 - x_1)^2 + (y_2 - y_1)^2\end{aligned}$$

Solving for d, we have

$$d = \pm\sqrt{(x_2 - x_1)^2 + (y_2 - y_1)^2}$$

However, since d represents the length of a line segment, we reject the negative answer. Therefore,

$$d = \sqrt{(x_2 - x_1)^2 + (y_2 - y_1)^2}$$

NOTE This formula *can* be used even if the points lie on the same vertical line or on the same horizontal line. ☑

Example 3 Find the distance between the points $(-6, 5)$ and $(6, -4)$.

Solution Let $P_1 = (-6, 5)$ and $P_2 = (6, -4)$.

Then
$$\begin{aligned} d &= \sqrt{(x_2 - x_1)^2 + (y_2 - y_1)^2} \\ &= \sqrt{[6 - (-6)]^2 + [-4 - 5]^2} \\ &= \sqrt{(12)^2 + (-9)^2} = \sqrt{144 + 81} \\ &= \sqrt{225} = 15 \end{aligned}$$

The distance is not changed if the points P_1 and P_2 are interchanged.

Let $P_1 = (6, -4)$ and $P_2 = (-6, 5)$.

Then
$$\begin{aligned} d &= \sqrt{[-6 - 6]^2 + [5 - (-4)]^2} \\ &= \sqrt{(-12)^2 + (9)^2} = \sqrt{144 + 81} \\ &= \sqrt{225} = 15 \quad \blacksquare \end{aligned}$$

Mathematical Relations

A **mathematical relation** is a set of ordered pairs (x, y).

Example 4 Examples of relations:

($\mathscr{R}$ represents the relation)

a. $\mathscr{R} = \{(-1, 2), (3, -4), (0, 5), (4, 3)\}$

b. $\mathscr{R} = \{(3, 5), (-5, 2), (-5, -4), (0, -3), (4, -6)\}$ ■

The **domain** *of a relation* is the set of all the first coordinates of the ordered pairs of that relation. We represent the domain of the relation $\mathscr{R}$ by the symbol $D_{\mathscr{R}}$.

The **range** *of a relation* is the set of all the second coordinates of the ordered pairs of that relation. We represent the range of the relation $\mathscr{R}$ by the symbol $R_{\mathscr{R}}$.

The **graph** *of a relation* is the graph of all the ordered pairs of that relation.

Example 5 Find the domain, range, and graph of the relation

$$\mathscr{R} = \{(-1, 2), (3, -4), (0, 5), (4, 3)\}$$

Solution

The domain of $\{(-1, 2), (3, -4), (0, 5), (4, 3)\}$ is $\{-1, 3, 0, 4\} = D_{\mathscr{R}}$.

The range of $\{(-1, 2), (3, -4), (0, 5), (4, 3)\}$ is $\{2, -4, 5, 3\} = R_{\mathscr{R}}$.

The graph of $\{(-1, 2), (3, -4), (0, 5), (4, 3)\}$ is shown in Figure 8.1.5.

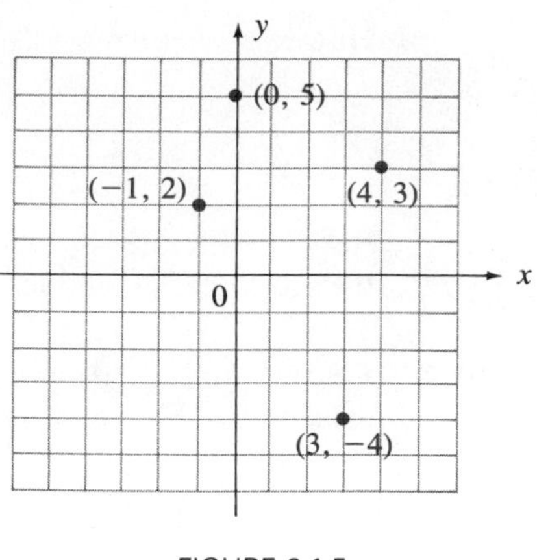

FIGURE 8.1.5

EXERCISES 8.1

Set I

1. Find the ordered pair corresponding to each of the following points as shown in Figure 8.1.6.

   a. R b. N c. U d. S

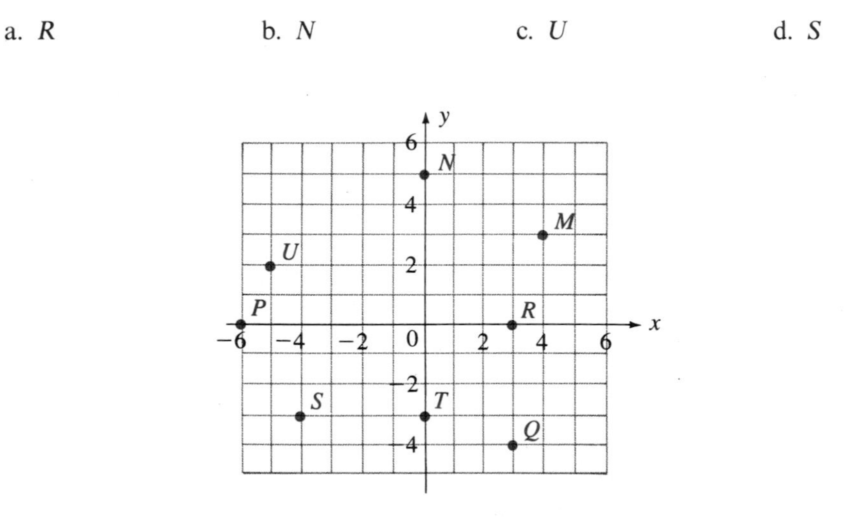

FIGURE 8.1.6

2. Find the ordered pair corresponding to each of the following points as shown in Figure 8.1.6.

   a. M b. P c. Q d. T

3. Draw the triangle that has vertices at the following points:

   $A(0, 0)$, $B(3, 2)$, $C(-4, 5)$

4. Draw the triangle that has vertices at the following points:

   $A(-2, -3)$, $B(-2, 4)$, $C(3, 5)$

5. Find the distance between the two given points.

   a. $(-2, -2)$ and $(2, 1)$ b. $(-3, 3)$ and $(3, -1)$

   c. $(5, 3)$ and $(-2, 3)$ d. $(2, -2)$ and $(2, -5)$

   e. $(4, 6)$ and $(0, 0)$

6. Find the distance between the two given points.

a. $(-4, -3)$ and $(8, 2)$
b. $(-3, 2)$ and $(4, -3)$
c. $(-3, -4)$ and $(-3, 2)$
d. $(-1, -2)$ and $(5, -2)$
e. $(-6, 9)$ and $(0, 0)$

7. Find the perimeter of the triangle that has vertices at $A(-2, 2)$, $B(4, 2)$, and $C(6, 8)$.

8. Find the perimeter of the triangle that has vertices at $A(0, 2)$, $B(11, 2)$, and $C(8, 6)$.

9. Use the distance formula to discover whether the triangle with vertices $A(-3, -2)$, $B(5, -1)$, and $C(3, 2)$ is or is not a right triangle.

10. Use the distance formula to discover whether the triangle with vertices $A(3, 2)$, $B(-5, 1)$, and $C(-3, -2)$ is or is not a right triangle.

11. Find the domain and range of the relation $\{(2, -1), (3, 4), (0, 2), (-3, -2)\}$ and graph it.

12. Find the domain and range of the relation $\{(-4, 0), (0, 0), (3, -2), (1, 5), (-3, -3)\}$ and graph it.

Set II

1. Find the ordered pair corresponding to each of the following points as shown in Figure 8.1.7.

a. F
b. G
c. K
d. L

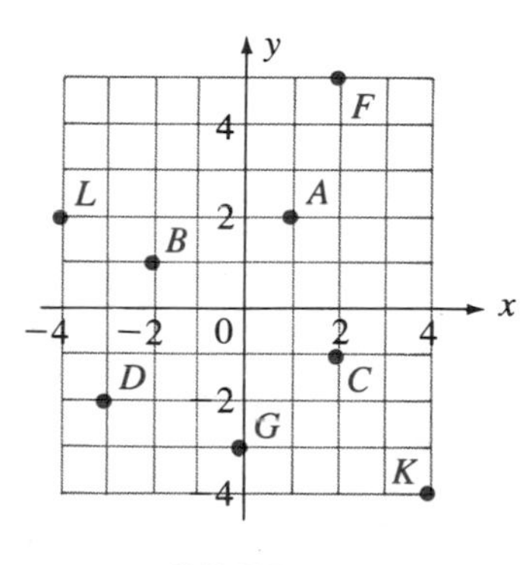

FIGURE 8.1.7

2. Find the ordered pair corresponding to each of the following points as shown in Figure 8.1.7.

a. A
b. B
c. C
d. D

3. Draw the triangle that has vertices at the following points:

$A(4, -4)$, $B(2, 3)$, $C(-4, 1)$

4. Draw the parallelogram that has vertices at the following points:

$A(-2, -5)$, $B(3, -5)$, $C(0, 4)$, $D(5, 4)$

5. Find the distance between the two given points.

a. $(7, -3)$ and $(15, 3)$
b. $(-14, 6)$ and $(-14, -13)$
c. $(9, -8)$ and $(-6, -5)$
d. $(-11, -18)$ and $(-17, -18)$
e. $(2, 12)$ and $(-6, 16)$

6. Find the distance between the two given points.

a. $(3, 0)$ and $(-1, 4)$ b. $(3, -1)$ and $(-3, 1)$

c. $(8, 3)$ and $(-2, 3)$ d. $(-4, 2)$ and $(3, -1)$

e. $(7, 2)$ and $(7, -5)$

7. Find the perimeter of the triangle that has vertices at $R(4, -7)$, $S(9, 5)$, and $T(-8, -2)$.

8. Use the distance formula to discover whether the triangle with vertices $A(-2, -5)$, $B(-4, -2)$, and $C(5, 0)$ is or is not a right triangle.

9. Use the distance formula to discover whether the triangle with vertices $A(-2, -3)$, $B(2, -2)$, and $C(0, 3)$ is or is not a right triangle.

10. Do the points $P(10, -6)$, $Q(-5, 3)$, and $R(-15, 9)$ lie in a straight line? Why or why not?

11. Find the domain and range of the relation $\{(-4, 3), (-2, 0), (-3, -5), (4, -2), (4, 4)\}$ and graph it.

12. Find the domain and range of the relation $\{(-6, 3), (2, -4), (0, 5), (-1, -1)\}$ and graph it.

8.2 Graphing Linear Equations

A first-degree equation is also called a **linear equation**. In this section, we will graph linear equations in one or two variables. The general form of a linear equation in two variables is

$$Ax + By + C = 0 \quad \text{where } A, B, \text{ and } C \text{ are real numbers and } A \text{ and } B \text{ are not both zero}$$

Before we discuss graphs of linear equations in more detail, we need some definitions.

A Solution of an Equation in x and y An ordered pair (a, b) is a **solution** of an equation in x and y if we get a *true statement* when we substitute a for x and b for y in that equation; if (a, b) is a solution of an equation, we often say that the point (a, b) *satisfies* the equation. The solution *set* of an equation is the set of all the ordered pairs that satisfy the equation; therefore, the solution set of an equation is a relation. There are always many ordered pairs that are solutions of an equation in two variables and many others that are not solutions.

The Graph of an Equation in Two Variables A point (a, b) lies on the **graph** of an equation in two variables if its coordinates satisfy that equation.

The graph of an equation in two variables must fulfill the following conditions:

1. It must contain *only* those points whose coordinates satisfy the equation.
2. It must include, within the limits of the size of the graph, *all* the points whose coordinates satisfy the equation. (Because there are infinitely many points whose coordinates satisfy an equation in one or two variables, we usually cannot graph *all* of these points; our graph would have to be infinitely large! However, we use arrowheads to indicate that the graph extends beyond the picture we have drawn.)

The Graph of a Linear Equation The following statement will not be proved, but *it must be memorized.*

> **THE GRAPH OF A LINEAR EQUATION IN TWO VARIABLES**
>
> The graph of any linear equation in one or two variables is the *straight line* that contains, within the limits of the size of the graph, all the ordered pairs, and only those ordered pairs, that satisfy the given equation.

Therefore, when we graph a first-degree equation in one or two variables, after we have graphed a few points that lie on the line (see the NOTE below), *we must connect those points with a straight line*, and *we must put arrowheads at each end of the line*. Only in this manner can we indicate that we're including *all* the ordered pairs that satisfy the equation.

NOTE A straight line can be drawn if we know any *two* points on that line. Although only two points are *necessary*, it is advisable to graph a third point as a *checkpoint*. If the equation is a first-degree equation and if the three points do *not* lie in a straight line, some mistake has been made either in the calculation of the coordinates of the points or in graphing the points. ☑

The Intercept Method of Graphing a Straight Line

To find a point on the line $Ax + By + C = 0$ when $B \neq 0$, we can let x have *any* value whatever and then solve the equation for y; when $A \neq 0$, we can, instead, let y have *any* value whatever and then solve the equation for x. Either way, we will have found a point on the line.

However, it's usually more convenient to find the points where the line crosses the coordinate axes. These points are called the **intercepts** of the line. The **x-intercept** $(x, 0)$ is the point where the line crosses the x-axis; we find it by letting $y = 0$. The **y-intercept** $(0, y)$ is the point where the line crosses the y-axis; we find it by letting $x = 0$. (See Figure 8.2.1 and Example 1.) Some lines have only an x-intercept (see Example 3), and some have only a y-intercept (see Example 4). We often enter the coordinates of the points we find into a "table of values" instead of writing them as ordered pairs of numbers.

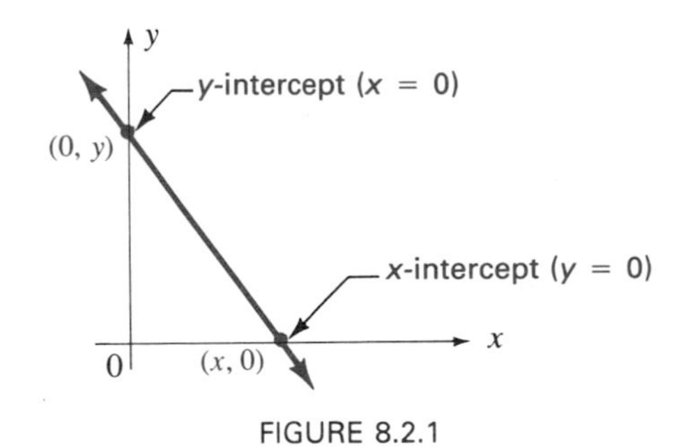

FIGURE 8.2.1

Example 1 Graph the relation $4x + 3y = 12$.

Solution *x-intercept* Set $y = 0$.

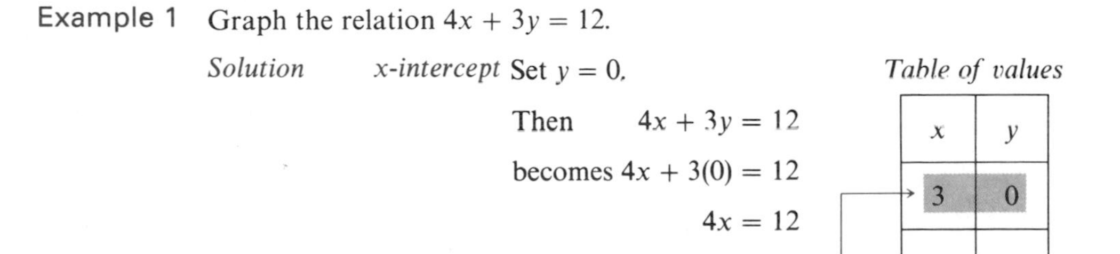

Then $4x + 3y = 12$

becomes $4x + 3(0) = 12$

$4x = 12$

$x = 3$

The x-intercept is $(3, 0)$

Table of values

x	y
3	0

We sometimes say "the x-intercept is 3" (naming the *x-coordinate* of the point where the line crosses the x-axis, instead of the point itself).

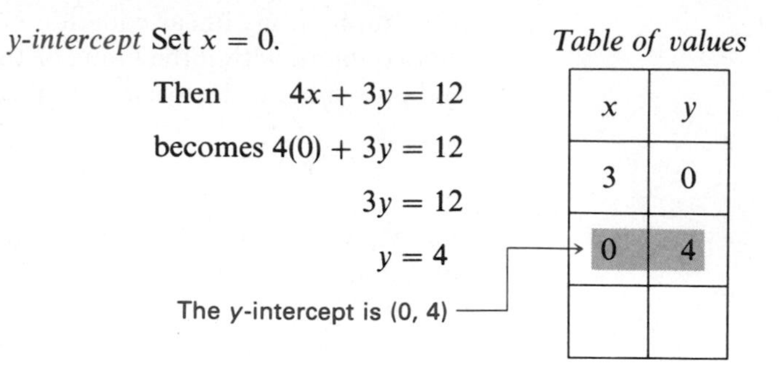

y-intercept Set $x = 0$.

Then $4x + 3y = 12$

becomes $4(0) + 3y = 12$

$3y = 12$

$y = 4$

The y-intercept is (0, 4)

Table of values

x	y
3	0
0	4

We sometimes say "the y-intercept is 4" (naming the *y-coordinate* of the point where the line crosses the y-axis, instead of the point itself).

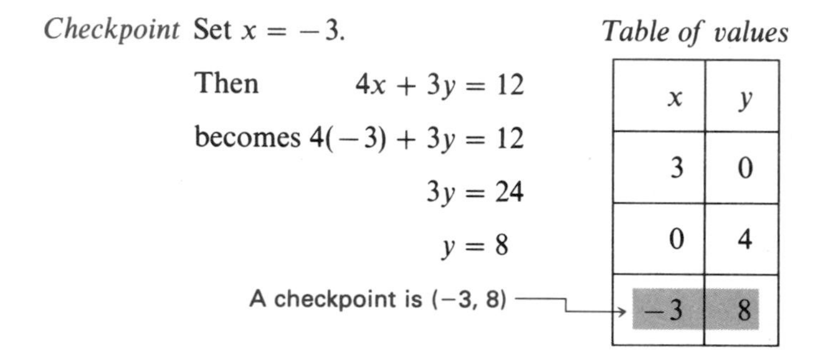

Checkpoint Set $x = -3$.

Then $4x + 3y = 12$

becomes $4(-3) + 3y = 12$

$3y = 24$

$y = 8$

A checkpoint is (−3, 8)

Table of values

x	y
3	0
0	4
−3	8

We then graph the x- and y-intercepts and the third point and note that these three points *do* appear to lie on the same straight line. See Figure 8.2.2.

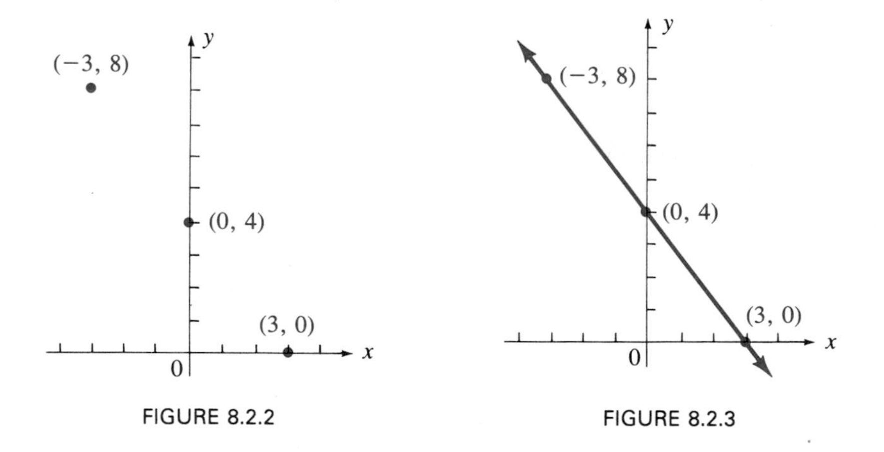

FIGURE 8.2.2

FIGURE 8.2.3

This is *not yet* the graph of $4x + 3y = 12$. There are *infinitely* many points whose coordinates satisfy the equation. Therefore, we must now connect these points with a *straight line*. The line actually extends infinitely far in both directions; to indicate this, we put arrowheads on both ends of the line. See Figure 8.2.3. ■

Example 2 Graph the relation $3x - 4y = 0$.

Solution

x-intercept Set $y = 0$. If $y = 0$, $x = 0$. (Verify this!)

Therefore, the x-intercept is (0, 0)—the origin. Since the line goes through the origin, the *y-intercept* is also (0, 0).

We have found only one point on the line: (0, 0). Therefore, *we must find another point on the line*. To find another point, we can set either variable equal to a nonzero number and then solve the equation for the other variable.

Second point Set $y = 3$. If $y = 3$, $x = 4$. (Verify this!)

This gives the point (4, 3).

Third point Set $x = -4$. If $x = -4$, $y = -3$. (Verify this!)

This gives the point $(-4, -3)$. We now plot the points (0, 0), (4, 3), and $(-4, -3)$ and connect them with a straight line. See Figure 8.2.4.

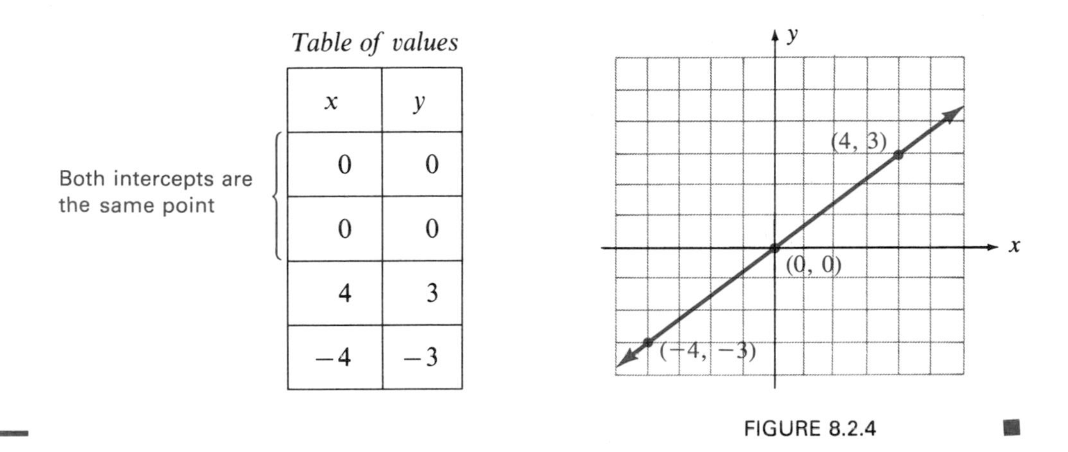

Table of values

x	y
0	0
0	0
4	3
-4	-3

FIGURE 8.2.4

Sometimes, the x- and y-intercepts are very close together and to draw the line through them accurately would be very difficult. In this case, find another point on the line far enough away from the intercepts so that drawing an accurate line is easy.

Some equations of a line have only one variable. Such equations have graphs that are either vertical or horizontal lines (see Examples 3 and 4).

Example 3 Graph the relation $x = 3$. (The equation $x = 3$ is equivalent to $x + 0 \cdot y = 3$ and to $x - 3 = 0$.)

Solution The domain of the relation is $\{3\}$, since 3 is the only value that x can have. While y is not mentioned in the equation, it is understood that $x = 3$ *for all y-values.* We could make a chart of values, but x would always equal 3. We will plot the points (3, 0), $(3, -2)$, and (3, 4) and then connect them with a straight line. See Figure 8.2.5.

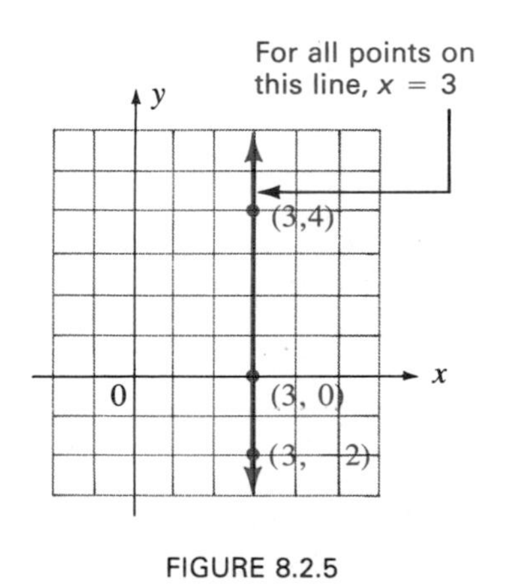

FIGURE 8.2.5

The graph of $\boldsymbol{x = k}$ is always a vertical line.

Example 4 Graph the relation $y + 4 = 0$. (The equation $y + 4 = 0$ is equivalent to $y = -4$ and to $0 \cdot x + y + 4 = 0$.)

Solution The domain of the relation is the set of all real numbers. While x is not mentioned in the equation, it is understood that $y + 4 = 0$ *for all* x. We will plot the points $(0, -4)$, $(3, -4)$, and $(-2, -4)$ and connect them with a straight line. The graph of the relation $y + 4 = 0$ is shown in Figure 8.2.6.

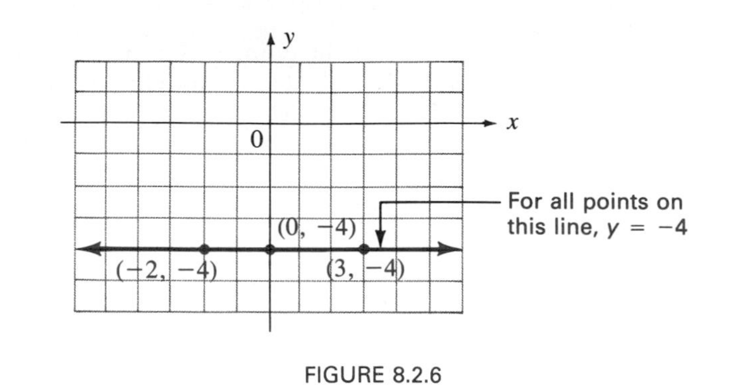

FIGURE 8.2.6 ■

The graph of $\boldsymbol{y = k}$ is always a horizontal line.

The following example illustrates the use of different scales on the two axes.

Example 5 Graph the relation $\{(x, y) \mid 50y - x = 100\}$.

Solution The statement in set-builder notation means that the relation consists of all of the ordered pairs (x, y) that satisfy the equation $50y - x = 100$. Since the domain of the relation is not specified, it is the set of all real numbers, and the graph is a straight line.

Table of values

x	y
0	2
−100	0
50	3

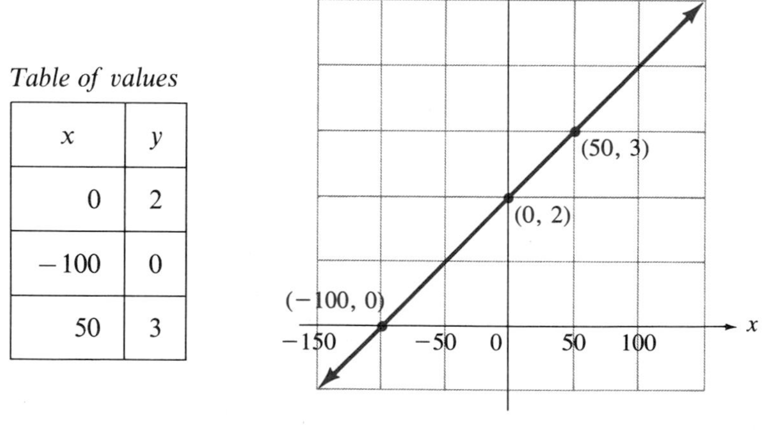

FIGURE 8.2.7

It would be impractical to have the units on the two axes be of equal lengths. The graph of the relation is shown in Figure 8.2.7. ■

Example 6 Find the domain and range of the relation $\{(x, y) \mid 2x + 3y = 6, x = -3, 0, 3, 6\}$ and graph the relation.

Solution The domain is *not* the set of real numbers; it is $\{-3, 0, 3, 6\}$.

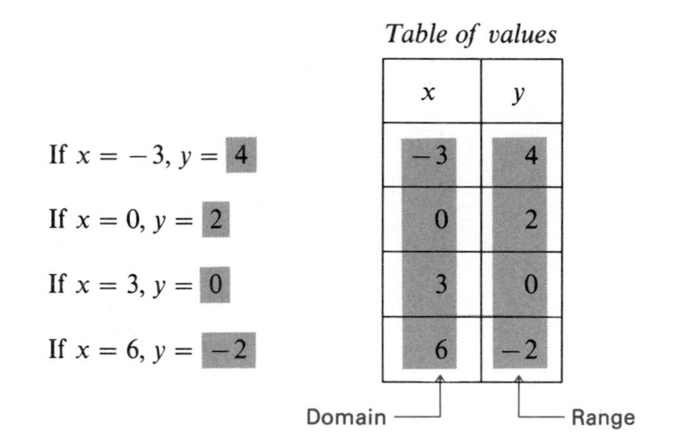

Table of values

If $x = -3$, $y = 4$

If $x = 0$, $y = 2$

If $x = 3$, $y = 0$

If $x = 6$, $y = -2$

x	y
-3	4
0	2
3	0
6	-2

Domain — Range

Therefore, the range $R_{\mathscr{R}} = \{4, 2, 0, -2\}$. The graph of the relation is shown in Figure 8.2.8.

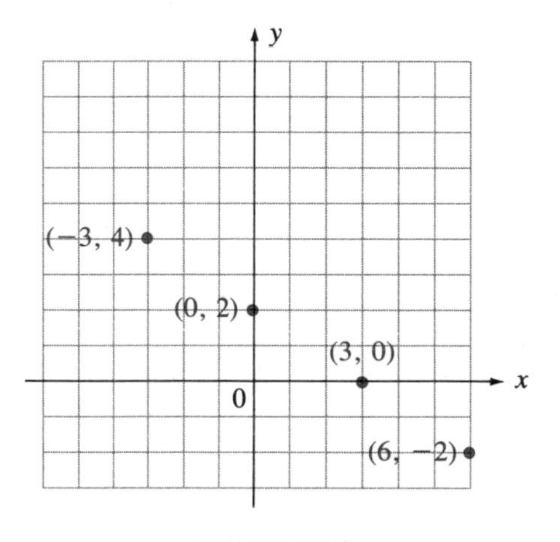

FIGURE 8.2.8

This relation can also be written $\{(-3, 4), (0, 2), (3, 0), (6, -2)\}$ ■

In Example 6, it would be *incorrect* to connect the four points with a straight line. However, if the domain of the relation were changed from $\{-3, 0, 3, 6\}$ to the set of *all* real numbers, then the graph would be the straight line through the four points shown in Figure 8.2.8.

The intercept method of graphing a straight line is summarized in the following box:

TO GRAPH A STRAIGHT LINE (INTERCEPT METHOD)

If the equation contains both x and y:

1. Find the x-intercept: Set $y = 0$; then solve for x.
2. Find the y-intercept: Set $x = 0$; then solve for y.

3. Find the coordinates of one other point. (If the line passes through the origin, find two points other than the origin.)
4. Plot the three points.
5. Draw a straight line through the points and put arrowheads at each end of the line.

If the equation contains only one variable:

The graph is a vertical line that passes through $(a, 0)$ if the equation is $x = a$.

The graph is a horizontal line that passes through $(0, b)$ if the equation is $y = b$.

EXERCISES 8.2

Set I In Exercises 1–20, sketch the graph of each relation.

1. $3x + 2y = 6$	**2.** $4x - 3y = 12$	**3.** $5x - 3y = 15$
4. $2x + 5y = 10$	**5.** $9x + 5y = 18$	**6.** $6x - 11y = 22$
7. $10x = 21 + 7y$	**8.** $13y = 40 - 8x$	**9.** $9y = 25 - 7x$
10. $17x = 31 + 6y$	**11.** $8x - 41 = 14y$	**12.** $5y - 33 = -15x$
13. $6x + 11y = 0$	**14.** $3x + 2y = 0$	**15.** $4y = -8x$
16. $9x = 3y$	**17.** $x + 5 = 0$	**18.** $x = 4$
19. $y = -3$	**20.** $y - 2 = 0$	

For Exercises 21–34, graph each given relation.

21. $\{(x, y) \mid 7x + 5y = 2\}$	**22.** $\{(x, y) \mid 3x + 8y = 4\}$
23. $\{(x, y) \mid y = \frac{1}{2}x - 1\}$	**24.** $\{(x, y) \mid y = \frac{1}{3}x + 2\}$
25. $\{(x, y) \mid 3(x - 5) = 7y\}$	**26.** $\{(x, y) \mid 4x = 3(y - 6)\}$
27. $\{(x, y) \mid 50x + y = -100\}$	**28.** $\{(x, y) \mid x - 30y = 90\}$
29. $\{(x, y) \mid x = 50y\}$	**30.** $\{(x, y) \mid y = 70x\}$
31. $\{(x, y) \mid y = -2x, x = 1, 3, 5\}$	**32.** $\{(x, y) \mid y = 3x, x = -1, 3, 4\}$
33. $\{(x, y) \mid x - y = 5, x = 1, 4\}$	**34.** $\{(x, y) \mid 2x - y = 3, x = 0, 2\}$

In Exercises 35 and 36, (a) graph the two relations for each exercise on the same set of axes and (b) find the ordered pair corresponding to the point where the two lines cross.

35. $\begin{cases} 5x - 7y = 18 \\ 2x + 3y = -16 \end{cases}$ **36.** $\begin{cases} 4x + 9y = 3 \\ 2x - 5y = 11 \end{cases}$

Set II In Exercises 1–20, sketch the graph of each relation.

1. $3x + 8y = 24$	**2.** $4x - 12y = 24$	**3.** $4x - 5y = 20$
4. $x = 5$	**5.** $15x + 8y = 32$	**6.** $y = -6$
7. $7x - 28 = 9y$	**8.** $-2x + 5y = 10$	**9.** $19y = 35 - 8x$
10. $y = 3x + 4$	**11.** $7y = 16x - 47$	**12.** $y = \frac{2}{3}x + 4$
13. $5x - 13y = 0$	**14.** $3x = 5y$	**15.** $2x = 8y$

16. $y = \frac{2}{3}x - 1$ **17.** $x + 4 = 0$ **18.** $y - 6 = 0$

19. $y = 4$ **20.** $x + y = 0$

For Exercises 21–34, graph each given relation.

21. $\{(x, y) \mid 11x - 4y = 2\}$

22. $\{(x, y) \mid x = 0\}$

23. $\{(x, y) \mid y = \frac{2}{5}x - 3\}$

24. $\{(x, y) \mid y = 0\}$

25. $\{(x, y) \mid 4(y + 7) = 9x\}$

26. $\{(x, y) \mid x = -4\}$

27. $\{(x, y) \mid 20y - x = 80\}$

28. $\{(x, y) \mid y = \frac{2}{3}x\}$

29. $\{(x, y) \mid y = 100x\}$

30. $\{(x, y) \mid x = \frac{2}{3}y\}$

31. $\{(x, y) \mid y = 4x, x = -2, 2\}$

32. $\{(x, y) \mid y = 5x - 2, x = 2, 4, 6\}$

33. $\{(x, y) \mid x - 2y = 4, x = 2, 4, 6\}$

34. $\{(x, y) \mid x = 5y, x = -5, 0, 5\}$

In Exercises 35 and 36, (a) graph the two relations for each exercise on the same set of axes and (b) find the ordered pair corresponding to the point where the two lines cross.

35. $\begin{cases} 7x + 10y = -1 \\ 3x - 8y = -25 \end{cases}$

36. $\begin{cases} 8x - y = 4 \\ x + 2y = 9 \end{cases}$

8.3 The Slope of a Line; Deriving Equations of Straight Lines

In Section 8.2, we discussed the graph of a straight line. In this section, we discuss the slope of a line and show how to write the equation of a line given certain facts about the line.

8.3A The Slope of a Line

If we imagine a line as representing a hill, then the slope of a line is a measure of the steepness of the hill. To measure the **slope** of a line, we choose any two points $P_1(x_1, y_1)$ and $P_2(x_2, y_2)$, on the line. (see Figure 8.3.1).

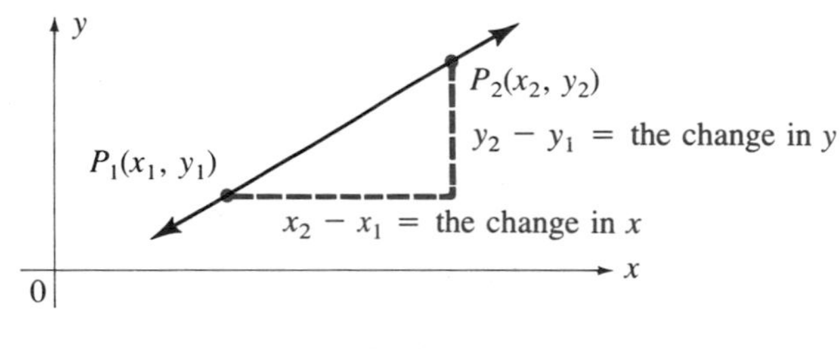

FIGURE 8.3.1

We use m to represent the slope of a line. The slope is defined as follows:

$$\text{Slope} = \frac{\text{The change in } y}{\text{The change in } x}$$

$$m = \frac{y_2 - y_1}{x_2 - x_1}$$

Example 1 Find the slope of the line through the points $(-3, 5)$ and $(6, -1)$.
Solution Let $P_1 = (-3, 5)$ and $P_2 = (6, -1)$.

$$\text{Then } m = \frac{y_2 - y_1}{x_2 - x_1} = \frac{-1 - 5}{6 - (-3)} = \frac{-6}{9} = -\frac{2}{3}$$

The slope is not changed if the points P_1 and P_2 are interchanged.

Let $P_1 = (6, -1)$ and $P_2 = (-3, 5)$

$$\text{Then } m = \frac{y_2 - y_1}{x_2 - x_1} = \frac{5 - (-1)}{-3 - 6} = \frac{6}{-9} = -\frac{2}{3}$$

The line is graphed in Figure 8.3.2.

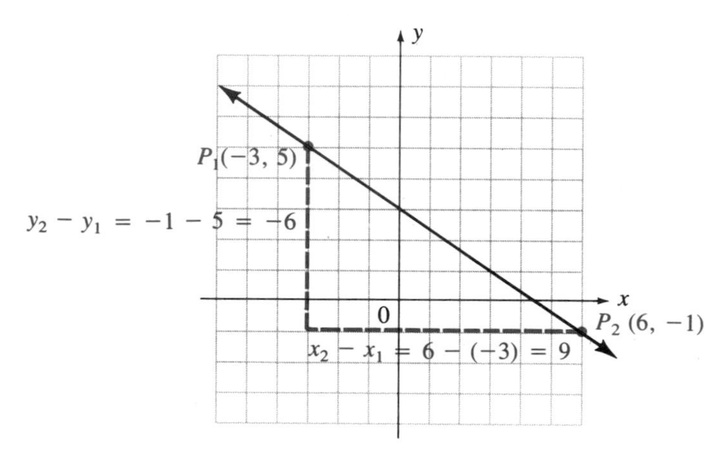

FIGURE 8.3.2

Notice in Example 1 that the slope of the line is *negative* and that a point moving along the line in the positive x-direction *falls*. Observe, in fact, that for every three units we move toward the right, we move two units downward.

Example 2 Find the slope of the line through the points $A(-2, -4)$ and $B(5, 1)$.

Solution

$$m = \frac{y_2 - y_1}{x_2 - x_1}$$

$$= \frac{1 - (-4)}{5 - (-2)} = \frac{5}{7}$$

The line is graphed in Figure 8.3.3.

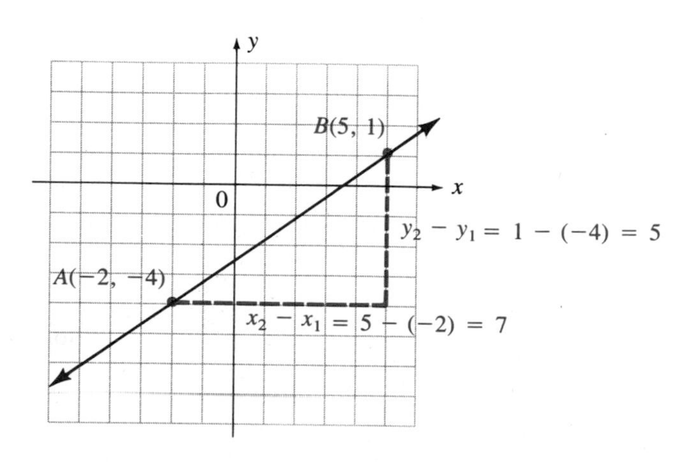

FIGURE 8.3.3

Notice in Example 2 that the slope of the line is *positive* and that a point moving along the line in the positive x-direction *rises*. Observe, in fact, that for every seven units we move toward the right, we move five units upward.

Example 3 Find the slope of the line through the points $E(-4, -3)$ and $F(2, -3)$.

Solution

$$m = \frac{y_2 - y_1}{x_2 - x_1}$$

$$= \frac{-3 - (-3)}{2 - (-4)} = \frac{0}{6} = 0$$

Whenever the slope is zero, the line is horizontal (see Figure 8.3.4).

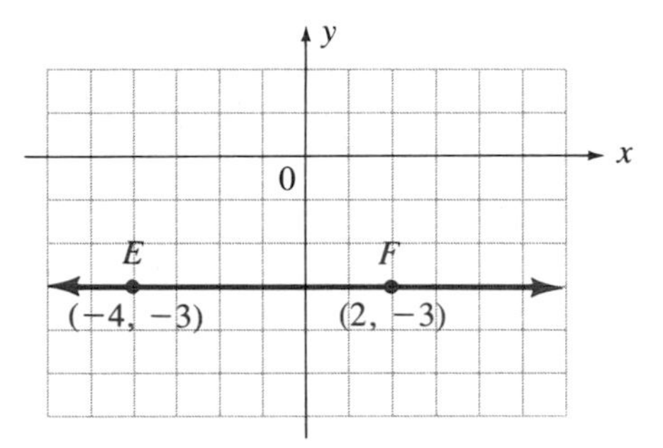

FIGURE 8.3.4

Example 4 Find the slope of the line through the points $R(4, 5)$ and $S(4, -2)$.

Solution

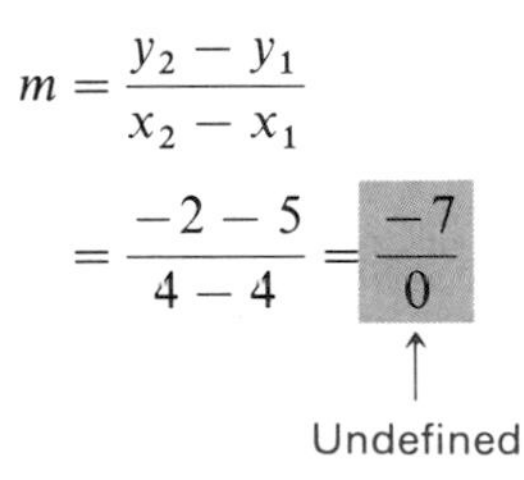

$$m = \frac{y_2 - y_1}{x_2 - x_1}$$

$$= \frac{-2 - 5}{4 - 4} = \frac{-7}{0}$$

Undefined

Note that $\frac{-7}{0}$ is not a real number. The slope *does not exist* when a line is vertical (see Figure 8.3.5).

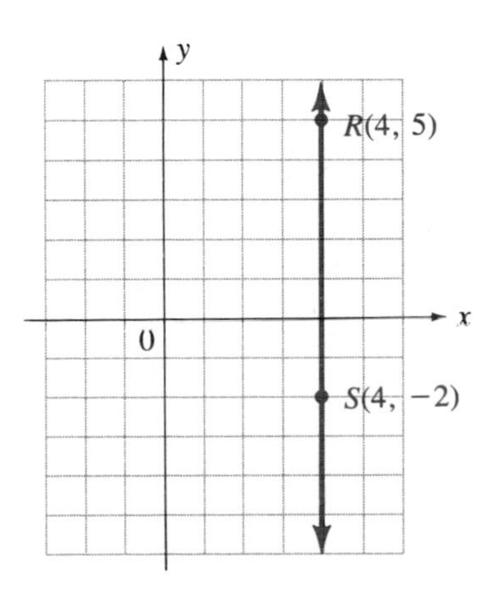

FIGURE 8.3.5

The facts about the slope of a line are summarized as follows:

The slope of a line is positive if a point moving along the line in the positive x-direction rises. (See Figure 8.3.3.)

The slope of a line is negative if a point moving along the line in the positive x-direction falls. (See Figure 8.3.2.)

The slope is zero if the line is horizontal. (See Figure 8.3.4.)

The slope does not exist if the line is vertical. (See Figure 8.3.5.)

Example 5 Graph each of the following lines and find each slope:

a. $y = -\frac{2}{3}x$

Solution

x	y
-3	2
0	0
3	-2

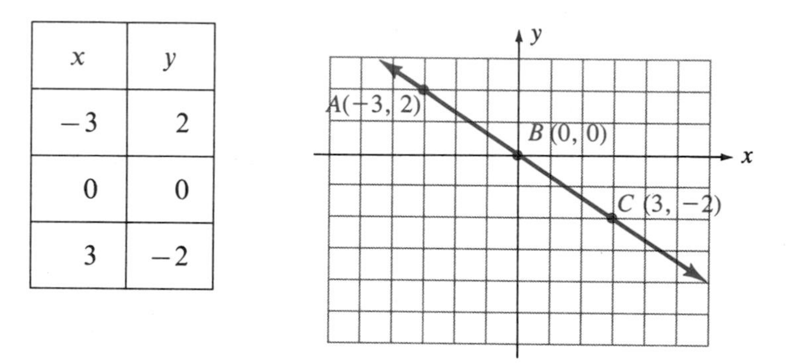

For the slope, if we use the points $A(-3, 2)$ and $B(0, 0)$, we have

$$m = \frac{0-2}{0-(-3)} = \frac{-2}{3} = -\frac{2}{3}$$

If we use the points $B(0, 0)$ and $C(3, -2)$, we have

$$m = \frac{-2-0}{3-0} = \frac{-2}{3} = -\frac{2}{3}$$

b. $y = -\frac{2}{3}x - 2$

Solution

x	y
-6	2
-3	0
0	-2

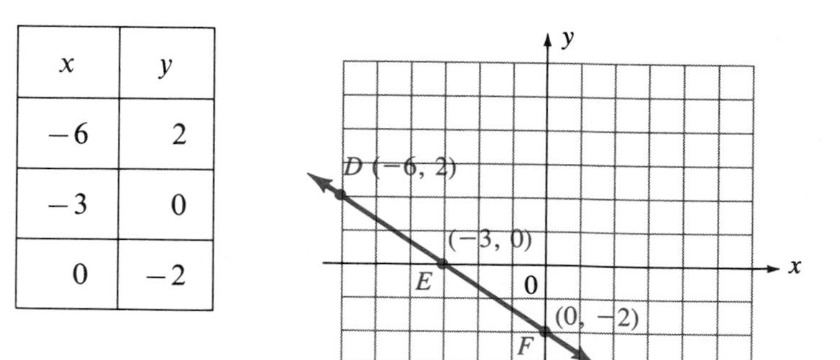

For the slope, if we use the points $D\,(-6, 2)$ and $E\,(-3, 0)$, we have

$$m = \frac{0 - 2}{-3 - (-6)} = \frac{-2}{3} = -\frac{2}{3}$$

If we use the points $E\,(-3, 0)$ and $F\,(0, -2)$, we have

$$m = \frac{-2 - 0}{0 - (-3)} = \frac{-2}{3} = -\frac{2}{3}$$

You can verify that for each of the lines, if any other points had been used, the slope would have remained unchanged.

c. $y = \frac{3}{2}x$

Solution

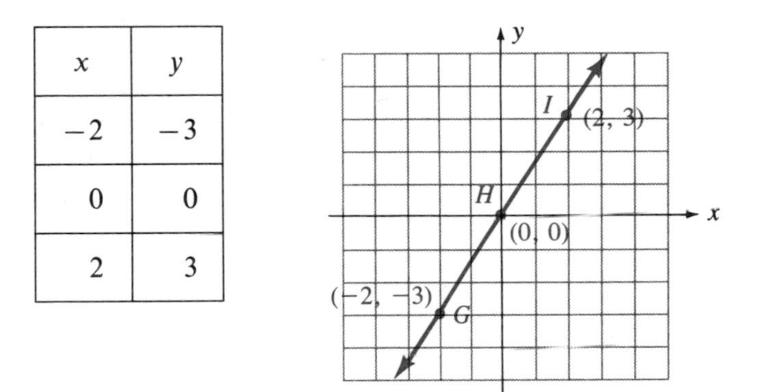

x	y
-2	-3
0	0
2	3

For the slope, if we use the points $G\,(-2, -3)$ and $H\,(0, 0)$, we have

$$m = \frac{0 - (-3)}{0 - (-2)} = \frac{3}{2}$$

If we use the points $H\,(0, 0)$ and $I\,(2, 3)$, we have

$$m = \frac{3 - 0}{2 - 0} = \frac{3}{2}$$

d. Graph the lines from Example 5a and Example 5b on the same axes.

Solution

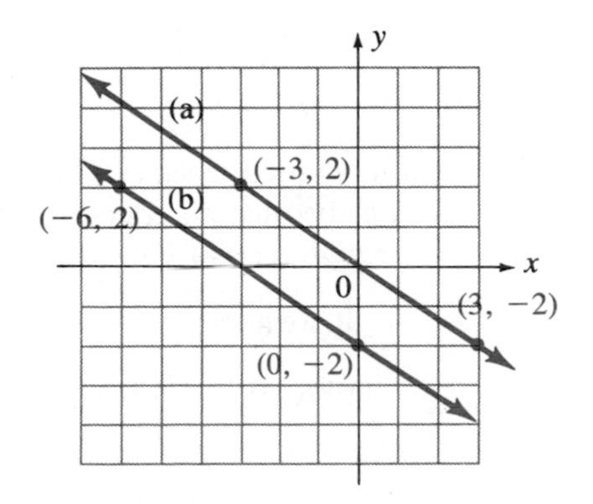

e. Graph the lines from Example 5a and Example 5c on the same axes.
Solution

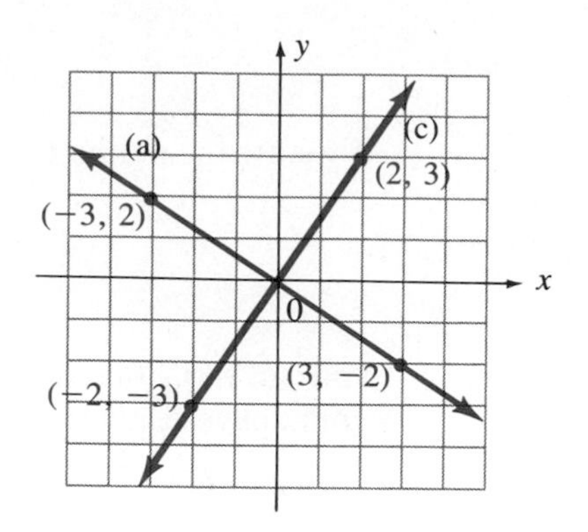

Note that $m = -\frac{2}{3}$ in Examples 5a and 5b and that $m = +\frac{3}{2}$, the negative reciprocal of $-\frac{2}{3}$, in Example 5c. ■

The following facts are stated without proof:

Parallel lines have the same slope; conversely, any lines that have the same slope are parallel.

Perpendicular lines (if neither is vertical) have slopes whose product is -1; equivalently, if one line having slope m_1 is perpendicular to another line having slope m_2, then

$$m_1 = -\frac{1}{m_2}$$

A vertical line is parallel to any other vertical line and is perpendicular to any horizontal line.

In Examples 5a and 5b, the lines are *parallel* to each other and the *slopes* of the two lines are equal. The lines of Examples 5a and 5c appear to be perpendicular to each other, and the *product* of the slopes of these lines is -1.

Example 6 Graph the line that passes through the point (3, 4) and has a slope of $-\frac{2}{3}$.

Solution If we interpret the slope as $\frac{-2}{3}$, we find the line falls 2 units for each 3 units it moves to the right. The line *falls* because the numerator is negative; it moves to the *right* because the denominator is *positive*. We know that one point the line passes through is (3, 4). Therefore, the line is as shown in Figure 8.3.6.

If we interpret the slope as $\frac{2}{-3}$, we find that the line rises 2 units for each 3 units it moves to the left. It *rises* because the numerator is *positive*; it moves to the *left* because the denominator is *negative*. We see that when we interpret the slope as $\frac{2}{-3}$, we simply get a different point on the same line.

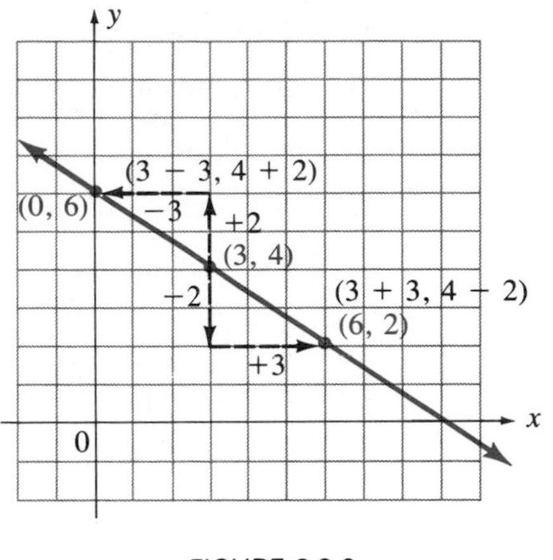

FIGURE 8.3.6

EXERCISES 8.3A

Set I In Exercises 1–8, find the slope of the line through the given pair of points.

1. (1, 4) and (10, 6)

2. (−3, −5) and (3, 0)

3. (−5, −5) and (1, −7)

4. (−1, 0) and (7, −4)

5. (−7, −5) and (2, −5)

6. (0, −2) and (5, −2)

7. (−4, 3) and (−4, −2)

8. (−5, 2) and (−5, 7)

In Exercises 9 and 10, graph each set of lines on the same axes and find the slope of each line.

9. a. $y = \frac{3}{4}x$ b. $y = \frac{3}{4}x + 5$ c. $y = \frac{3}{4}x - 2$

10. a. $y = x$ b. $y = x - 3$ c. $y = x + 4$

In Exercises 11–14, graph the line described.

11. Passes through (−1, 3) and has a slope of 3.

12. Passes through (2, −4) and has a slope of 5.

13. Passes through (0, 4) and has a slope of $-\frac{2}{5}$.

14. Passes through (1, −3) and has a slope of $-\frac{5}{3}$.

Set II In Exercises 1–8, find the slope of the line through the given pair of points.

1. (−5, −8) and (−9, 4)

2. (9, 15) and (−6, 3)

3. (−2, −11) and (−7, −11)

4. (14, −9) and (14, −21)

5. (−3, 2) and (4, −3)

6. (8, 1) and (8, 6)

7. (0, 5) and (−3, 5)

8. (4, −2) and (−2, 4)

In Exercises 9 and 10, graph each set of lines on the same axes and find the slope of each line.

9. a. $y = -\frac{1}{2}x$ b. $y = -\frac{1}{2}x + 2$ c. $y = -\frac{1}{2}x - 3$

10. a. $y = -4x$ b. $y = -4x + 3$ c. $y = -4x - 2$

In Exercises 11–14, graph the line described.

11. Passes through (5, 0) and has a slope of -3.

12. Passes through $(-2, 4)$ and has a slope of $\frac{3}{5}$.

13. Passes through (0, 0) and has a slope of 0.

14. Passes through $(3, -1)$ and has a slope of $-\frac{1}{4}$.

8.3B Deriving Equations of Lines

Three forms of the equation of a straight line are especially useful: the *general form*, the *point-slope form*, and the *slope-intercept form*.

The General Form of the Equation of a Line

The **general form** of the equation of a line is as follows:

> GENERAL FORM OF THE EQUATION OF A LINE
>
> $$Ax + By + C = 0$$
>
> where A, B, and C are constants and A and B are not both zero.

NOTE *In this text*, we prefer A, B, and C to be *integers*, with $A \geq 0$. ☑

To write the equation of a straight line in the *general form* as preferred in this text, clear fractions and move all terms to the left of the equal sign. If A is then negative, multiply both sides of the equation by -1.

Example 7 Write $-\frac{2}{3}x + \frac{1}{2}y = 1$ in the general form.

Solution

LCD = 6

$$\frac{6}{1}\left(-\frac{2}{3}x\right) + \frac{6}{1}\left(\frac{1}{2}y\right) = \frac{6}{1}\left(\frac{1}{1}\right)$$

$$-4x \quad + \quad 3y \quad = \quad 6$$

$$4x - 3y + 6 = 0 \quad \text{General form} \; ■$$

The Point-Slope Form of the Equation of a Line

Let $P_1(x_1, y_1)$ be a known point on the line with slope m. Let $P(x, y)$ represent *any* other point on the line (see Figure 8.3.7). Then

$$m = \frac{y - y_1}{x - x_1}$$

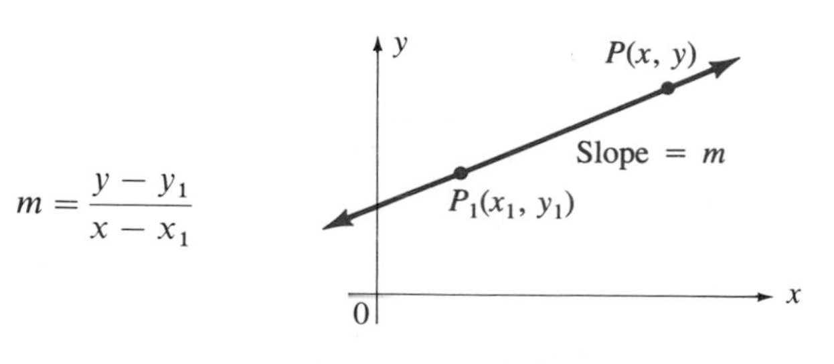

FIGURE 8.3.7

Therefore,

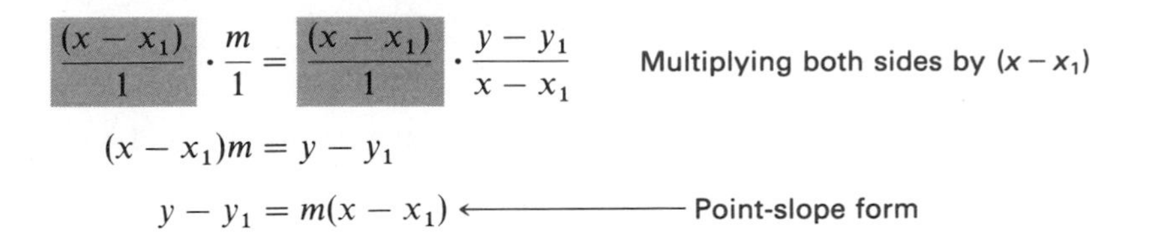

$$\frac{(x - x_1)}{1} \cdot \frac{m}{1} = \frac{(x - x_1)}{1} \cdot \frac{y - y_1}{x - x_1}$$ Multiplying both sides by $(x - x_1)$

$$(x - x_1)m = y - y_1$$

$$y - y_1 = m(x - x_1)$$ ← Point-slope form

POINT-SLOPE FORM OF THE EQUATION OF A LINE

$$y - y_1 = m(x - x_1)$$

where m is the slope of the line and $P_1(x_1, y_1)$ is a known point on the line.

When we are given the *slope* of a line and a *point* through which it passes, we will use the **point-slope form** of the equation of a line in order to write the equation of the line (see Example 8).

Example 8 Write the general form of the equation of the line that passes through $(-1, 4)$ and has a slope of $-\frac{2}{3}$.

Solution Into the point-slope form, we substitute -1 for x_1, 4 for y_1, and $-\frac{2}{3}$ for m.

$$y - y_1 = m(x - x_1)$$

$$y - 4 = -\frac{2}{3}[x - (-1)]$$ Point-slope form

$$3(y - 4) = -2(x + 1)$$

$$3y - 12 = -2x - 2$$

$$2x + 3y - 10 = 0$$ General form ■

The Slope-Intercept Form of the Equation of a Line

Let $(0, b)$ be the *y-intercept* of a line with slope m (see Figure 8.3.8). Then

$$y - y_1 = m(x - x_1)$$

$$y - b = m(x - 0)$$ Substituting b for y_1 and 0 for x_1

$$y - b = mx$$

$$y = mx + b$$ ← Slope-intercept form

Slope → m; b → y-intercept

FIGURE 8.3.8

SLOPE-INTERCEPT FORM OF THE EQUATION OF A LINE

$$y = mx + b$$

where m is the slope of the line, and b is the y-intercept of the line.

When we are given the *slope* of a line and its *y-intercept*, we can write the equation of the line by using the **slope-intercept form** of the equation of a line (see Example 9). The point-slope form could be used instead.

Example 9 Write the general form of the equation of the line that has a slope of $-\frac{3}{4}$ and a y-intercept of -2.

Solution Into the slope-intercept form, we substitute $-\frac{3}{4}$ for m and -2 for b.

$$y = mx + b$$

$$y = -\frac{3}{4}x - 2 \qquad \text{Slope-intercept form}$$

$$4y = -3x - 8$$

$$3x + 4y + 8 = 0 \qquad \text{General form}$$

You might verify that we obtain the same equation if we use the point-slope form of the equation, remembering that the ordered pair corresponding to the y-intercept is $(0, -2)$. ■

The Equation of a Horizontal Line

The following statement can be used in writing the equation of a horizontal line:

The equation of a horizontal line is of the form $y = k$ or $y - k = 0$, where k is the y-coordinate of every point on the line.

Example 10 Find the equation of the horizontal line that passes through $(2, -4)$.

Solution Because the y-coordinate of the given point is -4, the equation is

$$y = -4$$

$$y + 4 = 0 \qquad \text{General form}$$ ■

The Equation of a Vertical Line

Since a *vertical* line has no slope, we cannot use the point-slope form or the slope-intercept form when writing its equation. In fact, the only way to write the equation of a vertical line is to memorize the following statement:

The equation of a vertical line is of the form $x = k$ or $x - k = 0$, where k is the x-coordinate of every point on the line.

Example 11 Find the equation of the vertical line through $(3, -5)$.

Solution Since the line is vertical, we know its equation will be of the form $x = k$. Because the x-coordinate of the given point is 3, the equation is $x = 3$, or $x - 3 = 0$. ■

Example 12 illustrates the technique of finding the equation of a line when we know two points on the line.

Example 12 Find the equation of the line that passes through the points $(-15, -9)$ and $(-5, 3)$.

Step 1. Find the slope from the two given points.

$$m = \frac{3 - (-9)}{-5 - (-15)} = \frac{12}{10} = \frac{6}{5}$$

Step 2. Use this slope with *either* given point to find the equation of the line.

Using the point $(-5, 3)$	*Using the point* $(-15, -9)$
$y - y_1 = m(x - x_1)$	$y - y_1 = m(x - x_1)$
$y - 3 = \frac{6}{5}[x - (-5)]$	$y - (-9) = \frac{6}{5}[x - (-15)]$
$5y - 15 = 6(x + 5)$	$5(y + 9) = 6(x + 15)$
$5y - 15 = 6x + 30$	$5y + 45 = 6x + 90$
$0 = 6x - 5y + 45$	$0 = 6x - 5y + 45$

Note that the same equation is obtained no matter which of the two given points is used. ■

It is sometimes useful to change the general form of the equation of a line into the slope-intercept form (see Example 13).

Example 13 Uses of the slope-intercept form.

a. Write the general form of the equation of the line through $(-4, 7)$ and parallel to the line $2x + 3y + 6 = 0$.

Solution *Parallel lines have the same slope.* Therefore, the line we are trying to find has the same slope as $2x + 3y + 6 = 0$. To find the slope of $2x + 3y + 6 = 0$, write $2x + 3y + 6 = 0$ in the slope-intercept form by solving it for y, using the techniques discussed in Section 6.9.

$$2x + 3y + 6 = 0 \qquad \text{First, we will add } -2x - 6 \text{ to both sides}$$

$$3y = -2x - 6 \qquad \text{Now we will divide both sides by 3}$$

$$y = -\frac{2}{3}x - 2 \qquad \text{Slope-intercept form}$$

Slope → $-\frac{2}{3}$

Therefore, we must write the equation of the line that has a slope of $-\frac{2}{3}$ and passes through $(-4, 7)$.

$$y - y_1 = m(x - x_1)$$

$$y - 7 = -\tfrac{2}{3}[x - (-4)] \qquad \text{Point-slope form}$$

$$3(y - 7) = -2(x + 4)$$

$$3y - 21 = -2x - 8$$

$$2x + 3y - 13 = 0 \qquad \text{General form}$$

b. Find the y-intercept of the line $2x + 3y + 6 = 0$.

Solution In Example 13a, we found that the slope-intercept form of the equation was

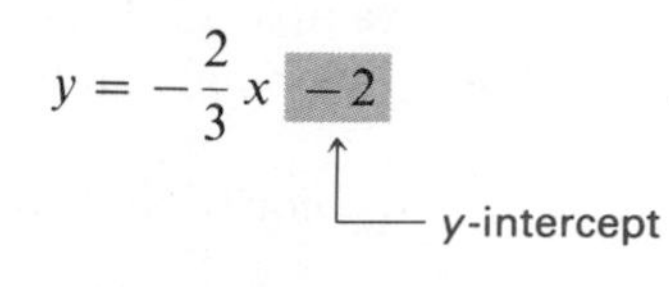

Therefore, the y-intercept is -2. ■

Example 14 Write the general form of the equation of the line through $(-4, 7)$ and perpendicular to the line $2x + 3y + 6 = 0$.

Solution In Example 13, we found that the slope of the line $2x + 3y + 6 = 0$ was $-\frac{2}{3}$. Since the required line is to be perpendicular to that one, its slope must be the negative reciprocal of $-\frac{2}{3}$, which is

$$-\frac{1}{-\frac{2}{3}} = \frac{3}{2}$$

Therefore, we must write the equation of the line that has a slope of $\frac{3}{2}$ and passes through $(-4, 7)$.

$$y - 7 = \tfrac{3}{2}(x - [-4]) \qquad \text{Point-slope form}$$
$$2(y - 7) = 3(x + 4)$$
$$2y - 14 = 3x + 12$$
$$3x - 2y + 26 = 0 \qquad \text{General form}$$ ■

EXERCISES 8.3B

Set I In Exercises 1–8, write each equation in the general form.

1. $5x = 3y - 7$ **2.** $4x = -9y + 3$ **3.** $\frac{x}{2} - \frac{y}{5} = 1$

4. $\frac{y}{6} - \frac{x}{7} = 1$ **5.** $y = -\frac{5}{3}x + 4$ **6.** $y = -\frac{3}{8}x - 5$

7. $4(x - y) = 11 - 2(x + 3y)$ **8.** $15 - 6(3x + y) = 7(x - 2y)$

In Exercises 9–14, write the general form of the equation of the line that passes through the given point and has the indicated slope.

9. $(4, -3)$, $m = \frac{1}{5}$ **10.** $(-2, -1)$, $m = -\frac{5}{6}$ **11.** $(-6, 5)$, $m = \frac{1}{4}$

12. $(-3, -2)$, $m = -\frac{4}{5}$ **13.** $(-1, 3)$, $m = -4$ **14.** $(3, 0)$, $m = 2$

In Exercises 15–20, write the general form of the equation of the line that has the indicated slope and y-intercept.

15. $m = \frac{5}{7}$, y-intercept $= -3$ **16.** $m = -\frac{1}{4}$, y-intercept $= -2$

17. $m = -\frac{4}{3}$, y-intercept $= \frac{1}{2}$ **18.** $m = -\frac{3}{5}$, y-intercept $= \frac{3}{4}$

19. $m = 0$, y-intercept $= 5$ **20.** $m = 0$, y-intercept $= 7$

21. Write the equation of the horizontal line that passes through the point $(-4, 3)$.

22. Write the equation of the horizontal line that passes through the point $(2, -5)$.

23. Write the equation of the vertical line that passes through the point $(7, -2)$.

24. Write the equation of the vertical line that passes through the point $(-6, 4)$.

In Exercises 25–28, (a) write the given equation in the slope-intercept form, (b) give the slope of the line, and (c) give the y-intercept of the line.

25. $4x - 5y + 20 = 0$

26. $8x + 3y - 24 = 0$

27. $\frac{2}{3}x + 3y + 5 = 0$

28. $3x - \frac{5}{6}y - 2 = 0$

In Exercises 29–34, find the general form of the equation of the line that passes through the given points.

29. $(8, -1)$ and $(6, 4)$

30. $(7, -2)$ and $(5, 1)$

31. $(10, 0)$ and $(7, 4)$

32. $(-4, 0)$ and $(-6, -5)$

33. $(-9, 3)$ and $(-3, -1)$

34. $(-11, 4)$ and $(-3, -2)$

35. Write the general form of the equation of the line through $(-4, 7)$ and parallel to the line $3x - 5y = 6$.

36. Write the general form of the equation of the line through $(8, -5)$ and parallel to the line $7x + 4y + 3 = 0$.

37. Write the general form of the equation of the line through $(6, 2)$ and perpendicular to the line $2x + 4y = 3$.

38. Write the general form of the equation of the line through $(5, -1)$ and perpendicular to the line $3x - 6y = 2$.

39. Write the general form of the equation of the line that has an x-intercept of 4 and is parallel to the line $3x + 5y - 12 = 0$.

40. Write the general form of the equation of the line that has an x-intercept of -3 and is parallel to the line $9x - 14y + 6 = 0$.

41. Write the general form of the equation of the line that has an x-intercept of 6 and a y-intercept of 4.

42. Write the general form of the equation of the line that has an x-intercept of 15 and a y-intercept of -12.

Set II In Exercises 1–8, write each equation in the general form.

1. $6y = 8 - 13x$

2. $8y = 2x - 4$

3. $\dfrac{x}{-3} - \dfrac{y}{6} = 1$

4. $5x = 2y$

5. $y = \frac{2}{9}x - 13$

6. $x = \frac{1}{5}y + 3$

7. $17 - 5(2x - y) = 8(x + 4y)$

8. $3x - 5(2y - 5x) = 4y + 7x - 1$

In Exercises 9–14, write the general form of the equation of the line that passes through the given point and has the indicated slope.

9. $(7, -4)$, $m = \frac{1}{6}$

10. $(8, 3)$, $m = -4$

11. $(-5, -8)$, $m = -\frac{3}{4}$

12. $(0, 0)$, $m = \frac{2}{5}$

13. $(2, -5)$, $m = -1$

14. $(-3, 0)$, $m = 8$

In Exercises 15–20, write the general form of the equation of the line that has the indicated slope and y-intercept.

15. $m = \frac{5}{6}$, y-intercept $= -4$

16. $m = -\frac{2}{9}$, y-intercept $= \frac{1}{3}$

17. $m = 0$, y-intercept $= -3$

18. $m = 3$, y-intercept $= 0$

19. $m = -\frac{1}{4}$, y-intercept $= 1$

20. $m = -5$, y-intercept $= -3$

21. Write the equation of the horizontal line that passes through the point $(-3, 7)$.

22. Write the equation of the horizontal line that passes through the point $(-9, -7)$.

23. Write the equation of the vertical line that passes through the point $(-5, -1)$.

24. Write the equation of the vertical line that passes through the point $(0, -5)$.

In Exercises 25–28, (a) write the given equation in the slope-intercept form, (b) give the slope of the line, and (c) give the y-intercept of the line.

25. $3x - 5y + 30 = 0$

26. $8x + 3y = 5$

27. $\frac{4}{5}x - 6y + 3 = 0$

28. $3x + \frac{4}{7}y - 5 = 0$

In Exercises 29–34, find the general form of the equation of the line that passes through the given points.

29. $(12, -7)$ and $(8, -9)$

30. $(-17, 3)$ and $(-5, -12)$

31. $(14, -6)$ and $(-1, 4)$

32. $(-3, 5)$ and $(-3, 2)$

33. $(4, 7)$ and $(7, 4)$

34. $(0, 4)$ and $(-3, 5)$

35. Write the general form of the equation of the line through $(-7, -13)$ and parallel to the line $6x - 8y = 15$.

36. Write the general form of the equation of the line through $(-7, -13)$ and parallel to the line $6y - 4x = 5$.

37. Write the general form of the equation of the line through $(0, 0)$ and perpendicular to the line $2x + 5y = 10$.

38. Write the general form of the equation of the line through $(0, 3)$ and perpendicular to the line $2y + 5x = 1$.

39. Write the general form of the equation of the line that has an x-intercept of -8 and is parallel to the line $12x - 9y - 7 = 0$.

40. Write the general form of the equation of the line that has a y-intercept of -5 and is parallel to the line $8x + 3y = 5$.

41. Write the general form of the equation of the line that has an x-intercept of -14 and a y-intercept of -6.

42. Write the general form of the equation of the line that has an x-intercept of 3 and a y-intercept of 5.

8.4 Graphing First-Degree Inequalities

Half-Planes Any line in a plane divides that plane into two **half-planes.** For example, in Figure 8.4.1, the line AB divides the plane into the two half-planes shown.

The following statement will not be proved, but *it must be memorized*:

> In the real plane, any first-degree inequality in one or two variables has a graph that is a half-plane.

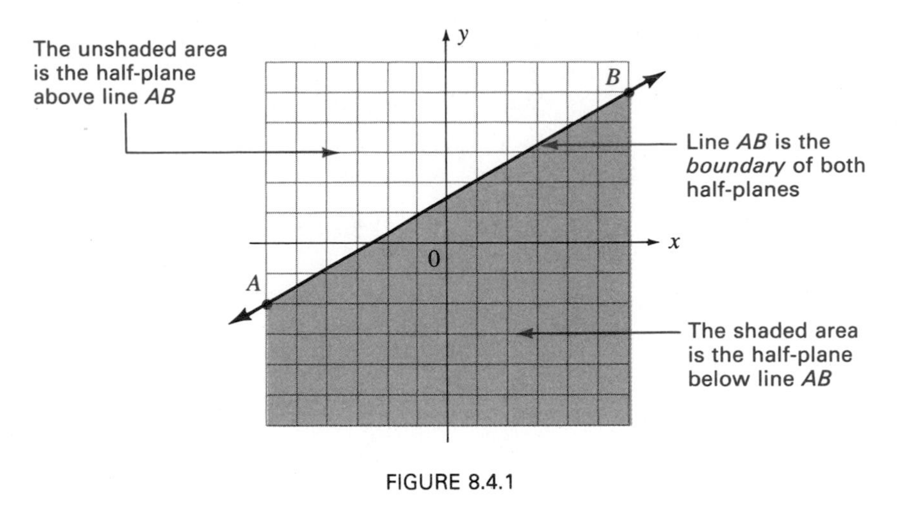

FIGURE 8.4.1

The Boundary Line The equation of the **boundary line** of the half-plane is obtained by replacing the inequality sign with an equal sign.

How to Determine Whether the Boundary Is a Dashed or a Solid Line The boundary line is drawn as a *solid* line and is part of the solution when the inequality symbol is $\geq$ or $\leq$. When the inequality symbol is $>$ or $<$, the boundary line is drawn as a dashed line and is not part of the solution.

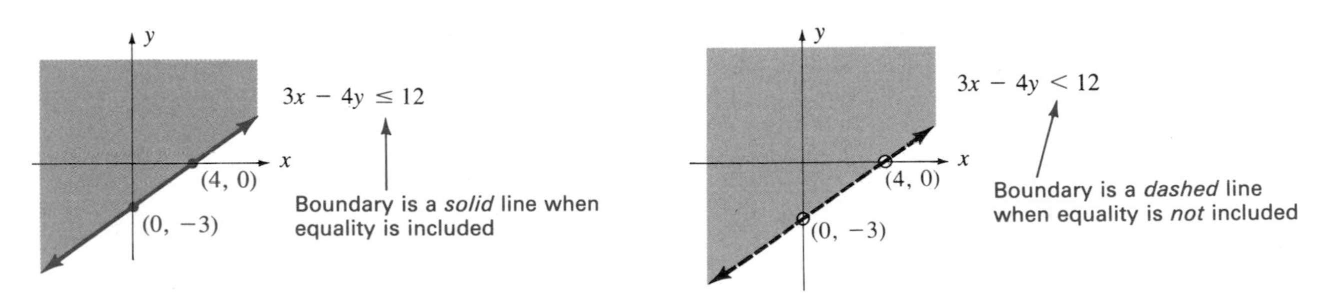

How to Determine the Correct Half-Plane Into the inequality, substitute the coordinates of *any* point that is *not on the boundary line.* (If the boundary line does not pass through the origin, the origin (0, 0) is the easiest point to use.)

If the resulting statement is *true,* the solution is the half-plane containing the point selected.

If the resulting statement is *false,* the solution is the half-plane *not* containing the point selected.

TO GRAPH A FIRST-DEGREE INEQUALITY (IN THE PLANE)

1. Determine whether the boundary line will be *solid* or *dashed.* It must be:

   a. *solid* if the inequality is $\leq$ or $\geq$ (equality is included);

   b. *dashed* if the inequality is $<$ or $>$ (equality is *not* included).

2. Find the equation of the boundary line by substituting an equal sign for the inequality symbol; graph the boundary line.

3. Select and shade the correct half-plane.

Example 1 Graph the inequality $2x - 3y < 6$.

Step 1. The boundary is a *dashed* line because equality is *not* included.

$$2x - 3y < 6$$

Step 2. To find the equation of the boundary line, change $<$ to $=$:

$$2x - 3y = 6 \qquad \text{Boundary line}$$

x	y
3	0
0	−2
−3	−4

Graph the boundary line $2x - 3y = 6$.

x-intercept Set $y = 0$. If $y = 0$, $x = 3$. (Verify this.)

y-intercept Set $x = 0$. If $x = 0$, $y = -2$. (Verify this.)

Checkpoint Set $x = -3$. If $x = -3$, $y = -4$. (Verify this.)

Therefore, the boundary line goes through (3, 0), (0, −2), and (−3, −4). We graph these points as *hollow dots*, because they are *not* in the solution set of $2x - 3y < 6$. (Equality was *not* included.) We graph the line as a dashed line.

Step 3. Select the correct half-plane. The solution of the inequality is only one of the two half-planes determined by the boundary line. Substituting the coordinates of the origin (0, 0) into the inequality:

$$2x - 3y < 6$$
$$2(0) - 3(0) < 6$$
$$0 < 6 \qquad \text{True}$$

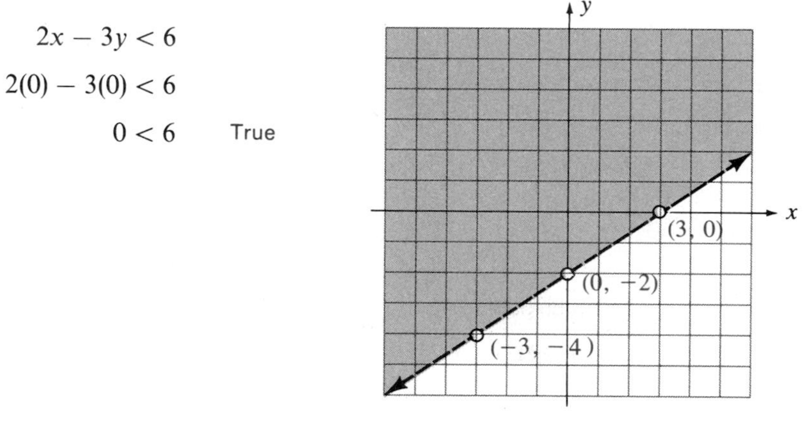

FIGURE 8.4.2

Therefore, the half-plane containing the origin is the solution. The solution is the shaded area in Figure 8.4.2. ■

Example 2 Graph the inequality $3x + 4y \leq -12$.

Step 1. The boundary is a *solid* line because equality is included.

$$3x + 4y \leq -12$$

x	y
−4	0
0	−3
4	−6

Step 2. Graph the boundary line $3x + 4y = -12$.

x-intercept Set $y = 0$. If $y = 0$, $x = -4$. (Verify this.)

y-intercept Set $x = 0$. If $x = 0$, $y = -3$. (Verify this.)

Checkpoint Set $x = 4$. If $x = 4$, $y = -6$. (Verify this.)

We graph the points $(-4, 0)$, $(0, -3)$, and $(4, -6)$ as *solid dots*, because they *are* in the solution set of $3x + 4y \leq 12$. (Equality *was* included.) We graph the line as a solid line.

Step 3. Select the correct half-plane. Substituting the coordinates of the origin (0, 0) into the inequality:

$$3x + 4y \leq -12$$
$$3(0) + 4(0) \leq -12$$
$$0 \leq -12 \quad \text{False}$$

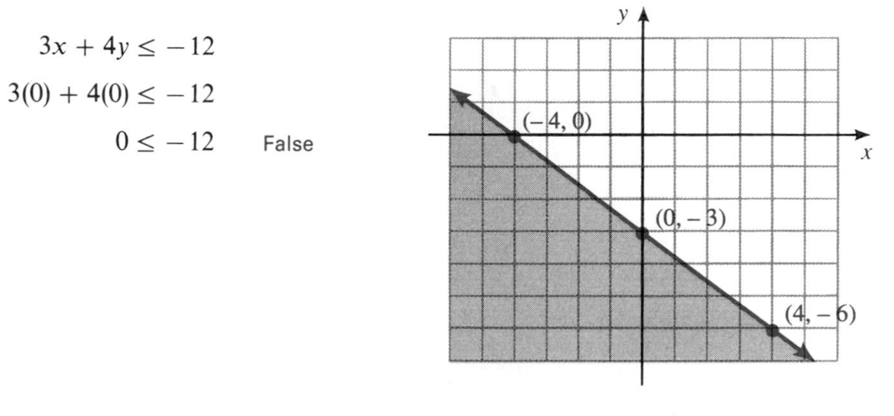

FIGURE 8.4.3

Therefore, the solution is the half-plane *not* containing (0, 0). (See the shaded area in Figure 8.4.3.) ■

Some inequalities have equations with only one variable. Such inequalities have graphs whose boundaries are either vertical or horizontal lines (see Example 3).

Example 3 Graph the inequality $x + 4 < 0$.

Step 1. The boundary is a *dashed* line because equality is *not* included.

$$x + 4 < 0$$

Step 2. Graph the boundary line $x + 4 = 0$, or $x = -4$ as a dashed line.

Step 3. Select the correct half-plane. Substituting the coordinates of the origin (0, 0) into $x + 4 < 0$:

$$x + 4 < 0$$
$$0 + 4 < 0$$
$$4 < 0 \quad \text{False}$$

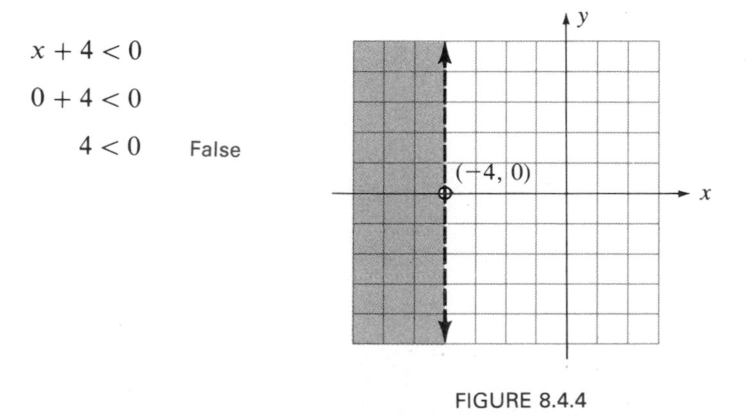

FIGURE 8.4.4

Therefore, the solution is the half-plane *not* containing (0, 0). (See the shaded area in Figure 8.4.4.) ■

Example 4 Graph the inequality $2y - 5x \geq 0$.

Step 1. The boundary is a *solid* line because equality is included.

$$2y - 5x \geq 0$$

Step 2. Graph the boundary line $2y - 5x = 0$.

x-intercept Set $y = 0$. If $y = 0$, $x = 0$. (Verify this.)

Therefore, the y-intercept is also 0. We should find two other points.

Set $x = 2$. If $x = 2$, $y = 5$. (Verify this.)

Set $x = -2$. If $x = -2$, $y = -5$. (Verify this.)

x	y
0	0
2	5
−2	−5

We graph the points (0, 0), (2, 5), and (−2, −5) as *solid dots*, because they *are* in the solution set of $2y - 5x \geq 0$. (Equality *was* included.) We graph the line as a solid line.

Step 3. Select the correct half-plane. Since the boundary goes through the origin (0, 0), we select a point *not* on the boundary, say (1, 0). Substituting the coordinates of (1, 0) into $2y - 5x \geq 0$:

$$2(0) - 5(1) \geq 0$$

$$-5 \geq 0 \quad \text{False}$$

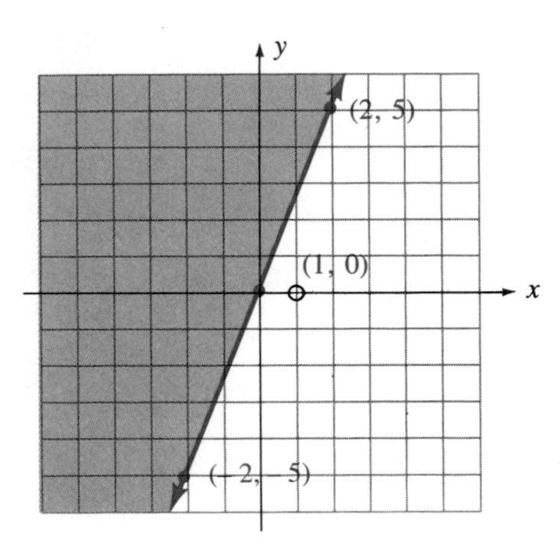

FIGURE 8.4.5

Therefore, the solution is the half-plane *not* containing (1, 0). (See the shaded area in Figure 8.4.5.) ■

EXERCISES 8.4

Set I Graph each of the following inequalities.

1. $4x + 5y < 20$ **2.** $5x - 3y > 15$ **3.** $3x - 8y > -16$

4. $6x + 5y < -18$ **5.** $9x + 7y \leq -27$ **6.** $5x - 14y \geq -28$

7. $x \geq -1$ **8.** $y \leq -4$ **9.** $6x - 13y > 0$

10. $4x + 9y < 0$ **11.** $14x + 3y \leq 17$ **12.** $10x - 4y \geq 23$

13. $\frac{x}{4} - \frac{y}{2} > 1$ **14.** $\frac{x}{3} + \frac{y}{5} < 1$

15. $4(x + 2) + 7 \le 3(5 - 2x)$

16. $3(y + 4) - 8 \ge 4(1 - 2y)$

17. $\dfrac{2x + y}{3} - \dfrac{x - y}{2} \ge \dfrac{5}{6}$

18. $\dfrac{4x - 3y}{5} - \dfrac{2x - y}{2} \ge \dfrac{2}{5}$

Set II Graph each of the following inequalities.

1. $3x - 4y > 12$

2. $7x - 2y < 14$

3. $8x + 11y < -24$

4. $x - 5y \ge 0$

5. $12x - 17y \ge -36$

6. $x > 3$

7. $y \le 3$

8. $x + y < 0$

9. $5x + 8y < 0$

10. $x \le -2$

11. $12x - 7y \ge 19$

12. $4y + 2x > 4$

13. $\dfrac{x}{2} + \dfrac{y}{6} < -1$

14. $\dfrac{y}{3} - \dfrac{x}{2} \ge 2$

15. $6(4 - y) + 11 \ge 5(6y + 7)$

16. $8(3 + x) + 2 < 3(x - 4)$

17. $\dfrac{3x - 5y}{3} - \dfrac{4x + 7y}{12} < \dfrac{3}{4}$

18. $\dfrac{2}{3} - \dfrac{4x - y}{2} > y$

8.5 Review: 8.1–8.4

Ordered Pair 8.1

An **ordered pair** of numbers is used to represent a point in the plane.

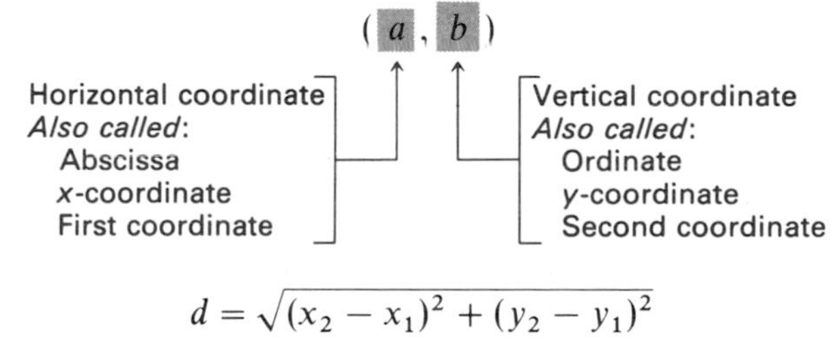

Distance between Two Points $P_1(x_1, y_1)$ and $P_2(x_2, y_2)$ 8.1

$$d = \sqrt{(x_2 - x_1)^2 + (y_2 - y_1)^2}$$

Relations 8.1

A **mathematical relation** is a set of ordered pairs (x, y).

The domain of a relation is the set of all the first coordinates of the ordered pairs of that relation.

The range of a relation is the set of all the second coordinates of the ordered pairs of that relation.

The graph of a relation is the graph of all the ordered pairs of that relation.

To Graph a Straight Line (Intercept Method) 8.2

If the equation contains both x and y:

1. Find the x-intercept: Set $y = 0$; then solve for x.
2. Find the y-intercept: Set $x = 0$; then solve for y.
3. Find the coordinates of one other point. (If the line passes through the origin, find two points other than the origin.)

4. Plot the three points.
5. Draw a straight line through the points and put arrowheads at each end of the line.

If the equation contains only one variable:

The graph is a vertical line that passes through $(a, 0)$ if the equation is $x = a$.

The graph is a horizontal line that passes through $(0, b)$ if the equation is $y = b$.

Slope of the Line through Points $P_1(x_1, y_1)$ and $P_2(x_2, y_2)$ 8.3A

$$m = \frac{y_2 - y_1}{x_2 - x_1}$$

Parallel lines have the same slope. If two lines are perpendicular (and neither is a vertical line), the product of their slopes is -1.

Equations of a Line 8.3B

General form: $Ax + By + C = 0$, where A and B are not both 0.

Point-slope form: $y - y_1 = m(x - x_1)$, where (x_1, y_1) is a known point on the line and m is the slope of the line.

Slope-intercept form: $y = mx + b$, where m is the slope and b is the y-intercept of the line.

To Graph a First-Degree Inequality in the Plane 8.4

1. The boundary line is *solid* if equality *is* included $(\leq, \geq)$.

   The boundary line is *dashed* if equality is *not* included $(<, >)$.
2. Find the equation of the boundary line by substituting an equal sign for the inequality symbol; graph the boundary line.
3. Select and shade the correct half-plane.

Review Exercises 8.5 Set I

1. Find the distance between each given pair of points.

a. $(-3, -4)$ and $(2, -4)$ b. $(-2, -3)$ and $(4, 1)$

2. Find the domain and range of the relation $\{(0, 5), (-2, 3), (3, -4), (0, 0)\}$.

3. Find the domain and range of $\{(x, y) \mid 3x - 2y = 6, x = -2, 0, 6, 8\}$ and graph the relation.

In Exercises 4–7, graph each relation.

4. $3y + 2x = -12$

5. $x + 2y = 0$

6. $y = 2x$

7. $\dfrac{2x + 3y}{5} - \dfrac{x - 3y}{4} = \dfrac{9}{10}$

In Exercises 8 and 9, find the slope of the line through each given pair of points.

8. $(3, -5)$ and $(-2, 4)$

9. $(-6, 2)$ and $(3, 2)$

10. Write the equation of the line that passes through $(-8, 4)$ and has a slope of $-\frac{3}{4}$.

11. Write the equation of the line that has a slope of $-\frac{1}{2}$ and a y-intercept of 6.

12. Write the equation of the line that passes through the points $(-2, -4)$ and $(1, 3)$.

13. Find the slope and y-intercept of the line $\frac{2x}{5} - \frac{3y}{2} = 3$.

In Exercises 14–16, graph each inequality in the plane.

14. $y > 3$

15. $2x - 5y > 10$

16. $3x - 2y \geq 0$

Review Exercises 8.5 Set II

NAME

ANSWERS

1. Find the distance between each given pair of points.

   a. $(8, -6)$ and $(-12, -2)$ b. $(-7, 5)$ and $(7, -13)$

1a. ______

b. ______

2. Find the domain and range of the relation $\{(10, 6), (-5, -2), (0, 7), (-9, -6)\}$.

2. ______

______

3. Find the domain and range of $\{(x, y) | 4x + 3y = 12, x = -3, 0, 3\}$ and graph the relation.

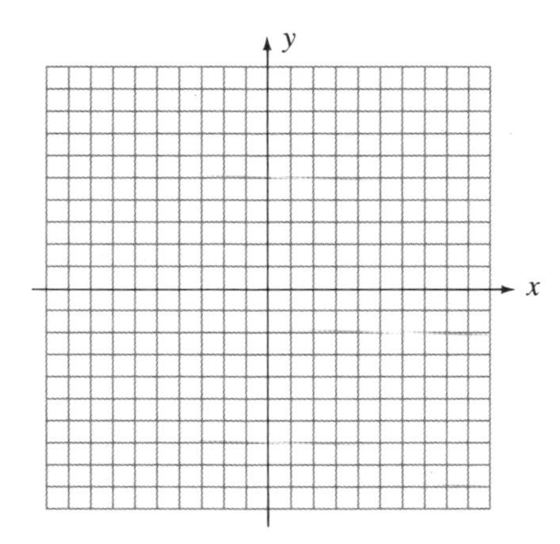

3. ______

______

Use the grid.

4. Use the grid.

5. Use the grid.

In Exercises 4–7, graph each relation.

4. $6x - 3y = 6$

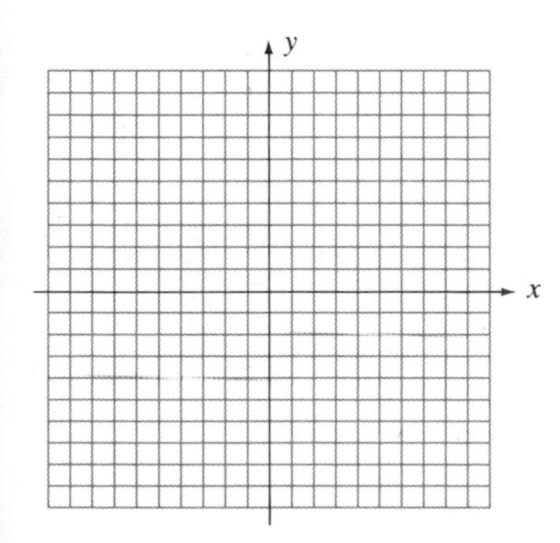

5. $5x - 3y = 0$

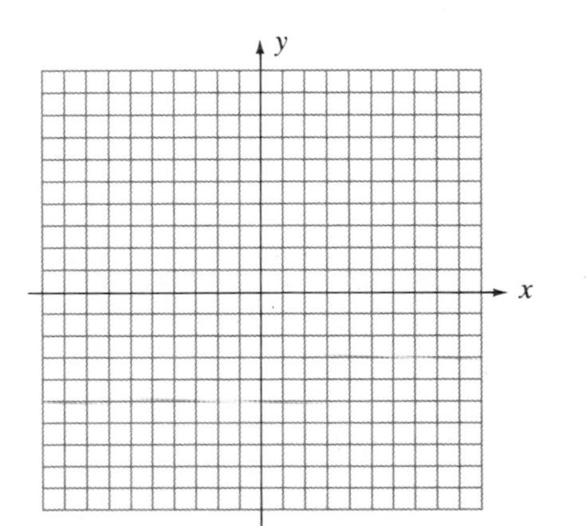

6. $\frac{x-4y}{7} - \frac{3x-7y}{6} = \frac{20}{21}$

7. $x = 5$

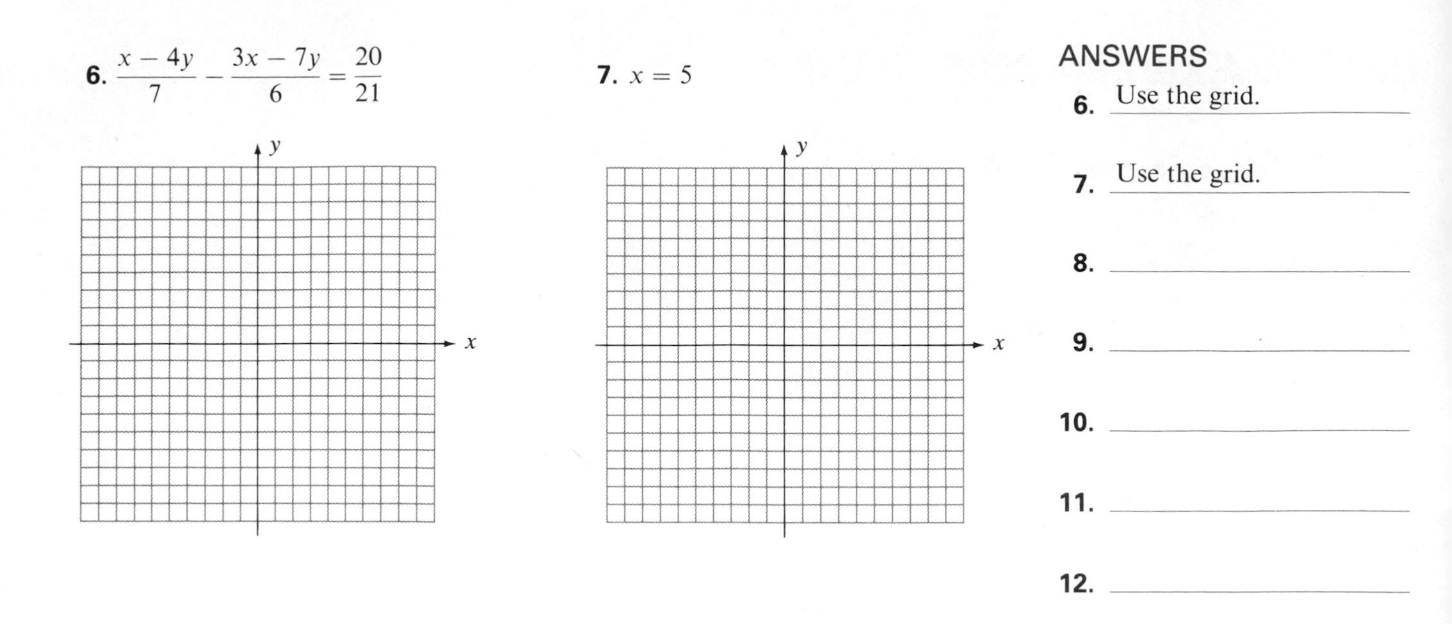

ANSWERS

6. Use the grid.

7. Use the grid.

8. ______

9. ______

10. ______

11. ______

12. ______

In Exercises 8 and 9, find the slope of the line through each given pair of points.

8. $(-12, 11)$ and $(-25, -28)$

9. $(-17, -14)$ and $(35, -22)$

10. Write the equation of the line that passes through $(-7, 12)$ and has a slope of $-\frac{4}{5}$.

11. Write the equation of the line that has a slope of $-\frac{7}{4}$ and a y-intercept of -6.

12. Write the equation of the line through the points $(-16, 15)$ and $(-38, -25)$.

13. Find the slope and y-intercept of the line $\dfrac{4x}{7} - \dfrac{6y}{5} = 2$.

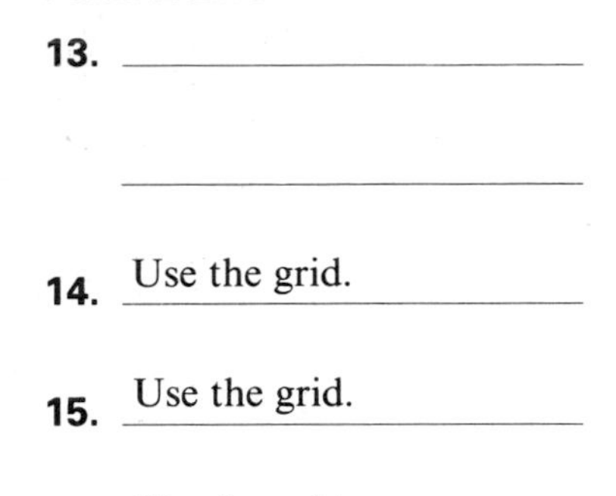

ANSWERS

13. ____________________

____

14. Use the grid.

15. Use the grid.

16. Use the grid.

In Exercises 14–16, graph each inequality in the plane.

14. $x + 4y \geq 0$

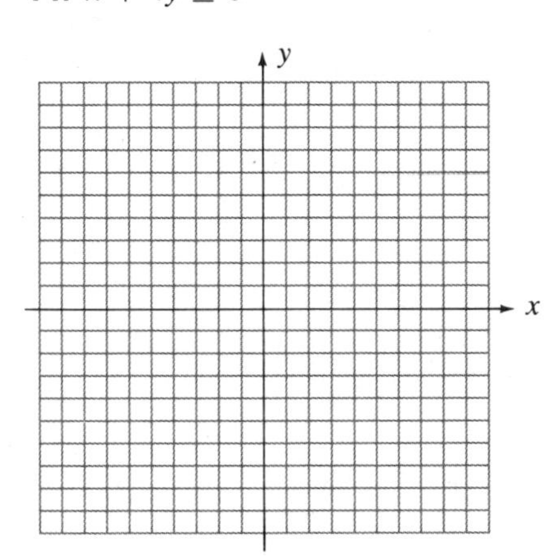

15. $y \leq -1$

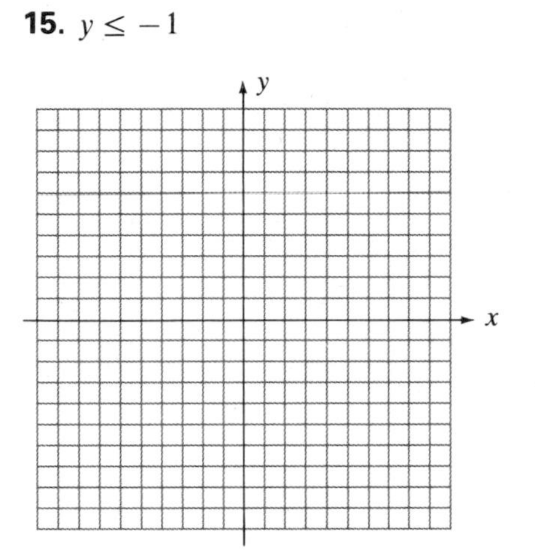

16. $3x - 9y > 9$

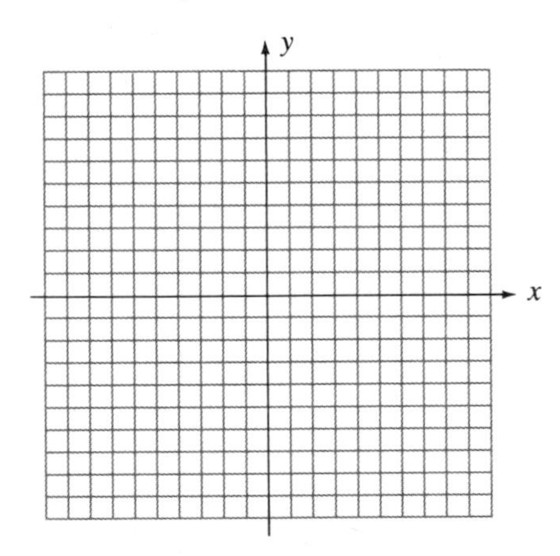

8.6 Functions

We live in a world of functions. For example, if you are paid \$5.65 an hour, your weekly salary is $S = 5.65h$, where h is the number of hours you work in the week. Your salary is a *function* of the number of hours you work. When you rent a car, you usually pay a flat rate plus mileage. If the flat rate is \$12 a day and you pay 15¢ for every mile you drive, your daily car rental cost C is $C = 12 + 0.15m$, where m is the number of miles you drive in a day. Therefore, the cost is a *function* of the number of miles driven.

8.6A Definition of a Function

A **function** is a special relation in which no two ordered pairs have the same first coordinate and different second coordinates. A function can be described in a number of different ways: by a table of values, by a set of ordered pairs, by a written statement, by an equation, or by a graph. We can think of a function as a *rule* that assigns *one and only one* value to y for each value of x in the domain of the variable.

No vertical line can meet the graph of a function in more than one point.

Example 1 Determine from each graph whether the relation is a function.

a.

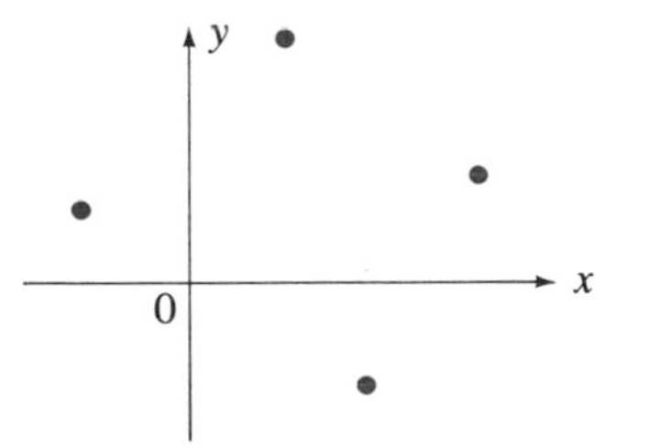

Solution This *is* the graph of a function, because any vertical line meets the graph in no more than one point.

b.

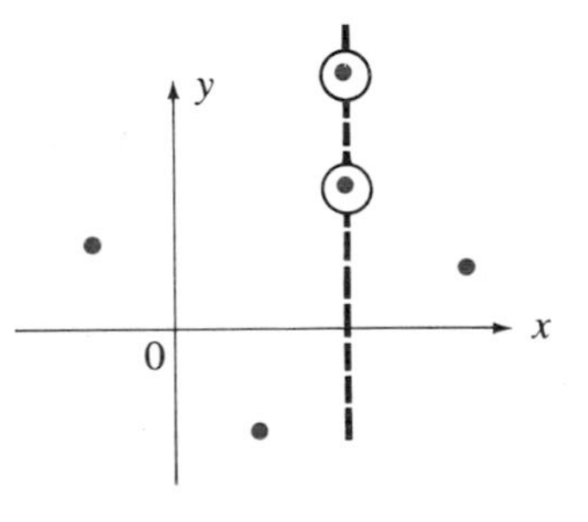

Solution This *is not* the graph of a function, because one vertical line meets the graph in two points.

c.

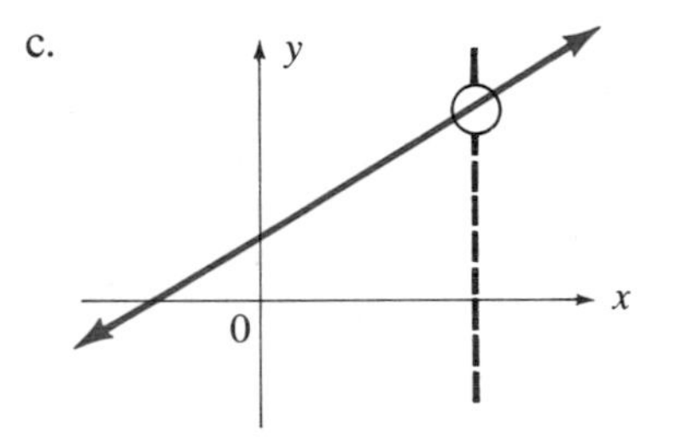

Solution This *is* the graph of a function, because any vertical line meets the graph in no more than one point. Straight lines *other* than vertical ones are graphs of linear functions. A **linear function** is a function with an equation of the form $y = mx + b$, where m and b are constants (see Section 8.3).

d.

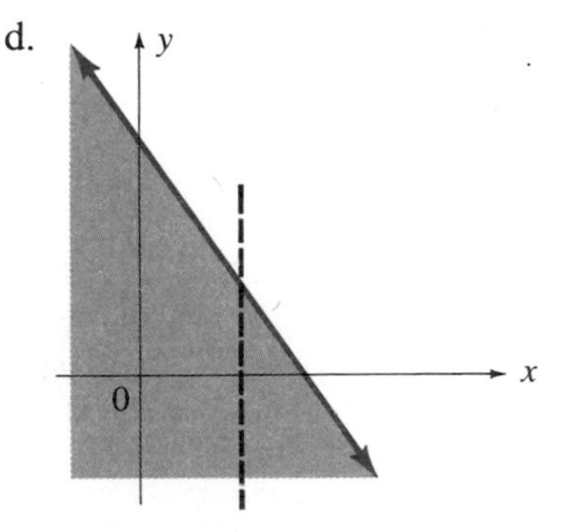

Solution This *is not* the graph of a function, because any vertical line meets the graph in an infinite number of points. Linear inequalities are *relations* (see Section 8.1) but not functions.

e.

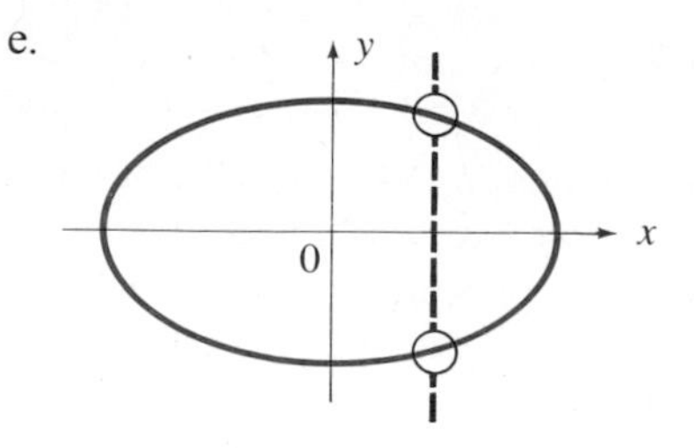

Solution This *is not* the graph of a function, because a vertical line can meet the graph in two points. ■

Example 2 Determine whether the relation is a function.

a. $\{(1, 7), (3, 6), (4, 7)\}$

Solution Since no two ordered pairs have the same first coordinate, the relation *is* a function.

b. $\{(2, 8), (2, 9), (3, 7)\}$

Solution Because (2, 8) and (2, 9) have the same first coordinate but different second coordinates, the relation is *not* a function. ■

Domain, Range, and Graph of a Function

Since a function is a relation, the meanings of its domain, range, and graph are the same as for a relation.

The **domain** of a function is the set of all permitted x-values. When an equation for a function is given, the domain will be the set of real numbers *unless* one of the following occurs:

1. The domain is restricted by some statement accompanying the equation.
2. There are variables in a denominator (or, equivalently, variable expressions with negative exponents).
3. Variables occur under a radical sign when the index of the radical is an even number.
4. Variable expressions with rational exponents occur where the denominator of the exponent is an even number.

To determine the domain of a function by looking at its graph, we look "right and left" to see how much of the plane is "used up."

The **range** of a function is the set of all the y-values. Because the value of y depends on the value of x, we often call y the **dependent variable** and x the **independent variable**. It is often more difficult to find the range of a function than to find its domain. In fact, in some cases the range cannot be found without using techniques beyond the scope of this course. If the function can be graphed, then the range can be found by examining the graph. To do this, we look "up and down" to see how much of the plane is "used up."

Example 3 Find the domain and the range for each of the following functions.

a.

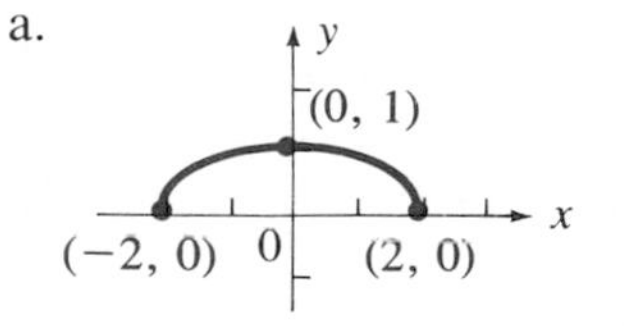

Solution Because no part of the graph is to the *left* of $x = -2$ and no part is to the *right* of $x = 2$, the domain is $\{x \mid -2 \le x \le 2\}$. Because no part of the graph is *above* $y = 1$ and no part is below $y = 0$, the range is $\{y \mid 0 \le y \le 1\}$.

b.

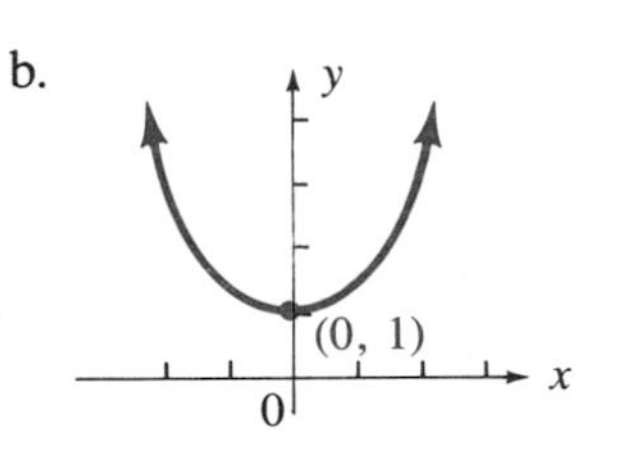

Solution Because the graph extends infinitely far left and right, the domain is $\{x \mid x \in R\}$. Because no part of the graph is *below* $y = 1$, the range is $\{y \mid y \geq 1\}$.

c.

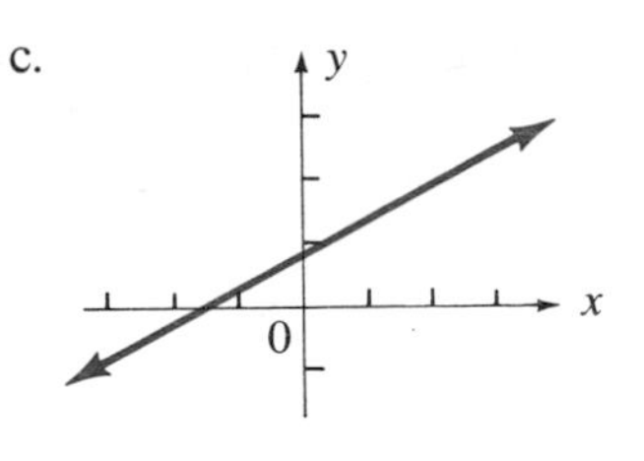

Solution Because the graph extends infinitely far left and right, the domain is $\{x \mid x \in R\}$. Because the graph extends infinitely far up and down, the range is $\{y \mid y \in R\}$.

d.

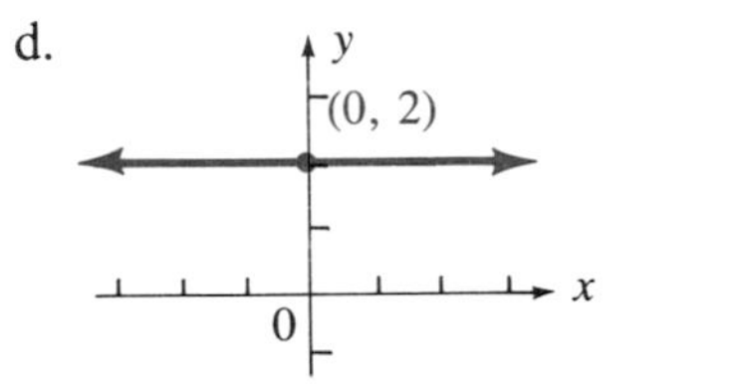

Solution Because the graph extends infinitely far left and right, the domain is $\{x \mid x \in R\}$. Because the only y-value that is "used up" is $y = 2$, the range is $\{2\}$.

e.

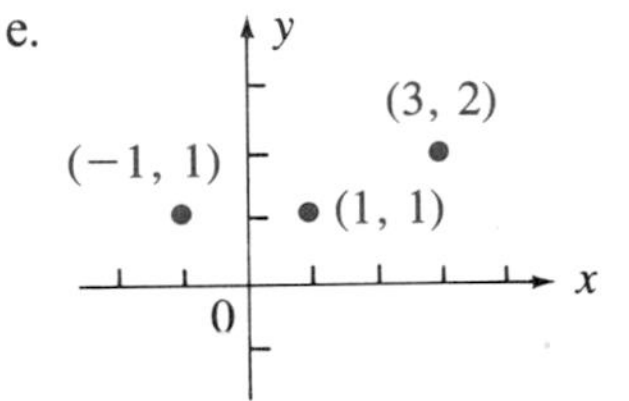

Solution Because the only x-values that are "used up" are -1, 1, and 3, the domain is $\{-1, 1, 3\}$. Because the only y-values that are "used up" are 1 and 2, the range is $\{1, 2\}$. ■

Example 4 Find the domain and range of each function.

a. $y = 3x + 2$

Solution Since there are no denominators, no radical signs, no negative exponents, no rational exponents, and no other restrictions on the domain, the domain is the set of all real numbers. We recognize the equation as the equation of a straight line that is *not* a horizontal line; therefore, the range is also the set of all real numbers.

b. $\{(x, y) \mid y = 2x, x = -1, 4\}$

Solution The domain is restricted to $\{-1, 4\}$. From the equation, we see that when $x = -1$, $y = -2$, and when $x = 4$, $y = 8$. Therefore, the range is $\{-2, 8\}$.

c. $y = 7$

Solution The equation is that of a horizontal line. Therefore, the domain is $\{x \mid x \in R\}$, and the range is $\{7\}$.

d. $y = \sqrt{2 + x}$

Solution Because we have a square root, the radicand must be nonnegative. Therefore, $2 + x \geq 0$, or $x \geq -2$. The domain is $\{x \mid x \geq -2\}$. Finding the range is more difficult. We know that y can never be negative, because the principal square root is never negative. If $x = -2$, $y = \sqrt{0} = 0$; therefore, y *can* equal 0. We can see, too, that as x gets larger and larger, y gets larger and larger. Therefore, the range is $\{y \mid y \geq 0\}$. ■

EXERCISES 8.6A

Set I

1. Which of the following are graphs of functions?

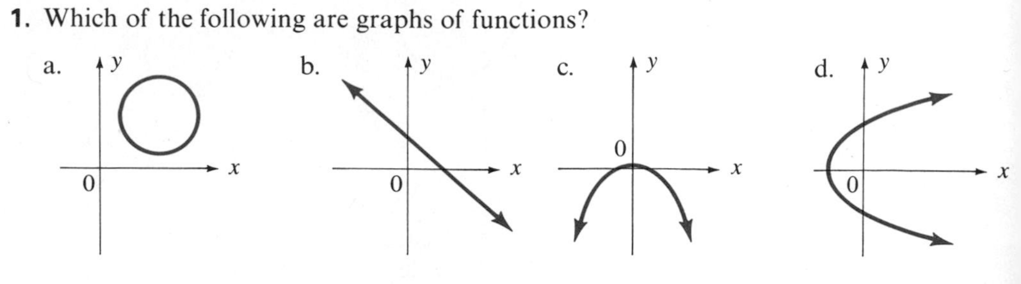

2. Which of the following are graphs of functions?

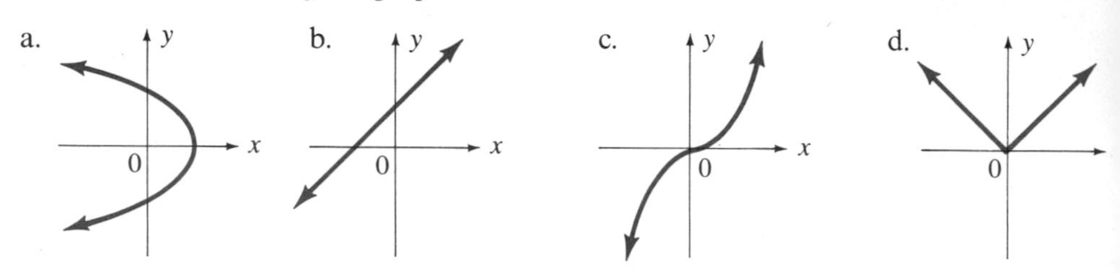

3. Find the range of the function $\{(x, y) \mid 2x + 5y = 10, x = -5, -1, 0, 2\}$ and graph the function.

4. Find the range of the function $\{(x, y) \mid 3x - 6 = 2y, x = -2, 0, 2, 4\}$ and graph the function.

5. Graph the function $\{(x, y) \mid y = 2x - 3\}$ with domain the set of real numbers.

6. Graph the function $\left\{(x, y) \middle| \frac{x}{2} - \frac{y}{3} = 1\right\}$ with domain the set of real numbers.

In Exercises 7 and 8, determine whether each relation is a function.

7. a. $\{(2, 3), (5, 3), (1, 6)\}$ b. $\{(-1, 6), (-1, 8)\}$

c. $\{(x, y) \mid x = 7\}$ d. $\{(x, y) \mid y = 3\}$

8. a. $\{(-3, 6), (-3, -8), (4, 2)\}$ b. $\{(2, -8), (5, -8)\}$

c. $\{(x, y) \mid y = 3x + 4\}$ d. $\{(x, y) \mid x = 0\}$

In Exercises 9 and 10, find the domain and range of each function.

9. a. The graph is

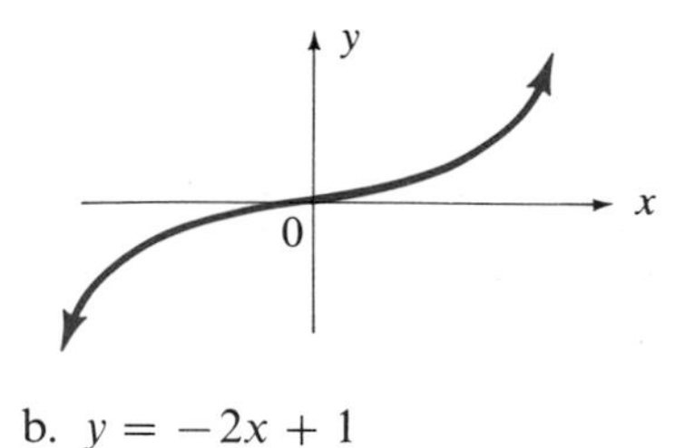

b. $y = -2x + 1$

c. $\{(x, y) \mid y = \sqrt{x - 5}\}$

d. $\{(x, y) \mid y = -x + 1, x = 1, 4, 7\}$

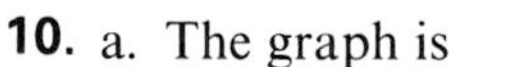

10. a. The graph is

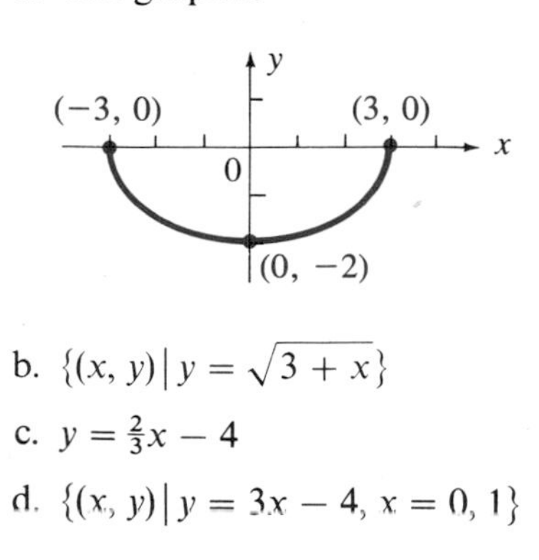

b. $\{(x, y) \mid y = \sqrt{3 + x}\}$

c. $y = \frac{2}{3}x - 4$

d. $\{(x, y) \mid y = 3x - 4, x = 0, 1\}$

Set II **1.** Which of the following are graphs of functions?

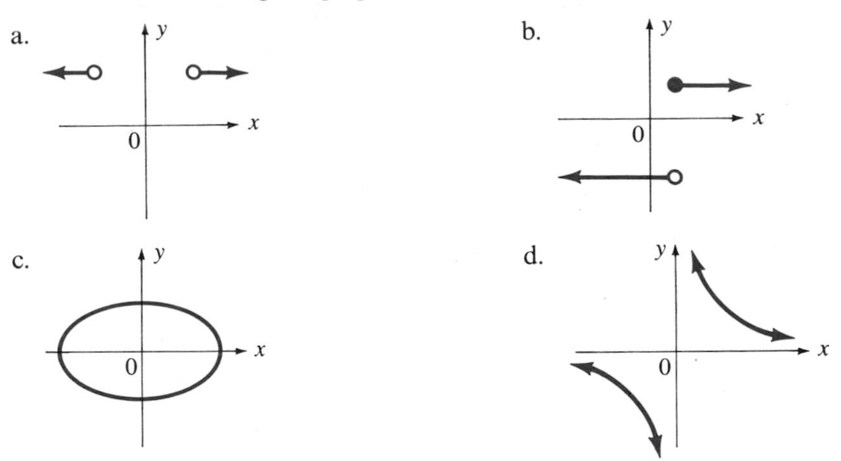

2. Which of the following are graphs of functions?

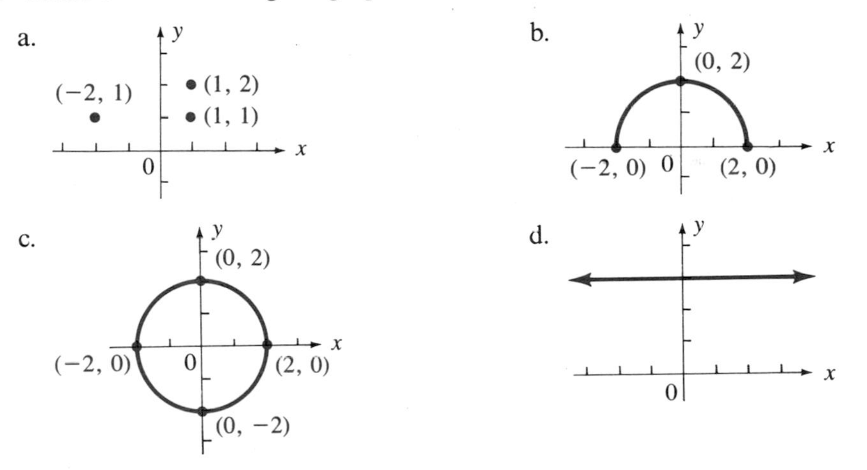

3. Find the range of the function $\{(x, y) \mid 10x - 7y = 25, x = -1, 2, 4, 6\}$ and graph the function.

4. Find the range of the function $\{(x, y) \mid 3x - 9y = 6, x = -1, 0, 1, 2\}$ and graph the function.

5. Graph the function $\{(x, y) \mid 4x - 12y = 12\}$ with domain the set of real numbers.

6. Graph the function $\left\{(x, y) \middle| \frac{x}{2} + \frac{y}{5} = -1\right\}$ with domain the set of real numbers.

In Exercises 7 and 8, determine whether each relation is a function.

7. a. $\{(0, 3), (0, 7), (1, 2)\}$
b. $\{(1, 6), (2, 6)\}$
c. $\{(x, y) \mid y = x\}$
d. $\{(x, y) \mid x = 6\}$

8. a. $\{(3, 5), (4, 5), (7, 5)\}$
b. $\{(1, 0), (1, 1)\}$
c. $\{(x, y) \mid x = 3\}$
d. $\{(x, y) \mid y = -1\}$

In Exercises 9 and 10, find the domain and range of each function.

9. a. The graph is

10. a. The graph is

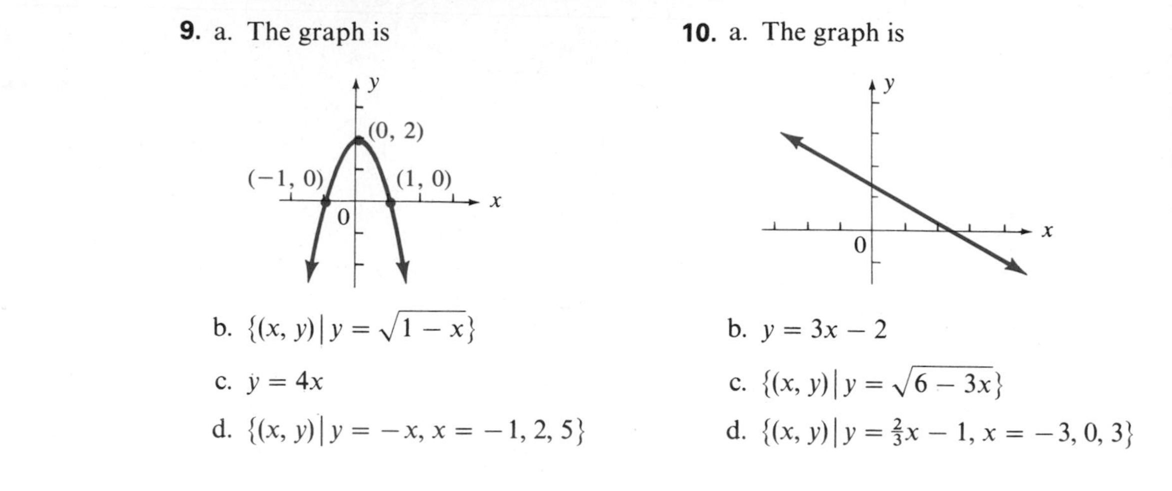

9.
b. $\{(x, y) \mid y = \sqrt{1 - x}\}$
c. $y = 4x$
d. $\{(x, y) \mid y = -x, x = -1, 2, 5\}$

10.
b. $y = 3x - 2$
c. $\{(x, y) \mid y = \sqrt{6 - 3x}\}$
d. $\{(x, y) \mid y = \frac{2}{3}x - 1, x = -3, 0, 3\}$

8.6B Functional Notation

If a rule or formula relating y and x is known and if there is no more than one value of y for each value of x, then y is said to be a **function of x**. This is written $y = f(x)$ and is read as "y equals f of x." (See Figure 8.6.1.)

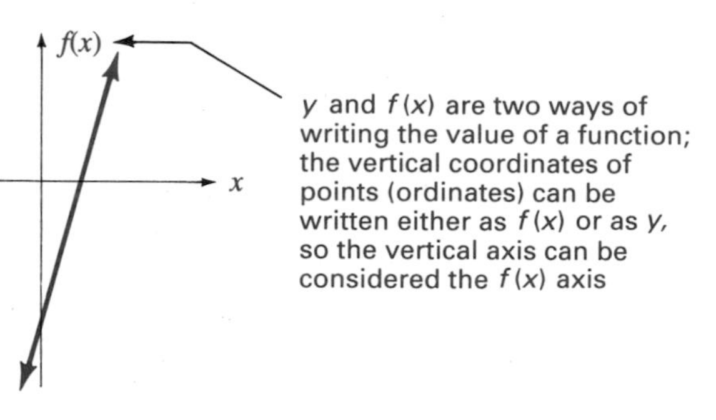

FIGURE 8.6.1

A WORD OF CAUTION The notation $f(x)$ does *not* mean f times x. It is simply an alternate way of writing y when y is a function of x. ✓

The equations $y = 5x - 7$ and $f(x) = 5x - 7$ mean the same thing, and we often combine the equations as follows:

$$y = f(x) = 5x - 7$$

Evaluating a Function

To *evaluate a function* means to determine what value y or $f(x)$ has for a particular x-value. If a function $f(x)$ is to be evaluated at $x = a$, we denote this by $f(a)$, where $f(a)$ is read as "f of a."

Example 5 If $f(x) = 5x - 7$, find $f(2)$.
Solution Since $f(x) = 5x - 7$,

$$f(2) = 5(2) - 7 \quad \text{Substituting 2 for } x$$
$$f(2) = 10 - 7 = 3$$

The statement $f(2) = 3$ means that $y = 3$ when $x = 2$. ■

Example 6 Evaluate the function $f(x) = \dfrac{2x^2 - 7x}{5x - 3}$ when $x = -2, 0, h$, and $a + b$.

Solution

$$f(x) = \frac{2(x)^2 - 7(x)}{5(x) - 3}$$

a. $f(-2) = \dfrac{2(-2)^2 - 7(-2)}{5(-2) - 3} = \dfrac{8 + 14}{-10 - 3} = -\dfrac{22}{13}$ Substituting -2 for every x

b. $f(0) = \dfrac{2(0)^2 - 7(0)}{5(0) - 3} = \dfrac{0 - 0}{0 - 3} = 0$ Substituting 0 for every x

c. $f(h) = \dfrac{2(h)^2 - 7(h)}{5(h) - 3} = \dfrac{2h^2 - 7h}{5h - 3}$ Substituting h for every x

d. $f(a + b) = \dfrac{2(a + b)^2 - 7(a + b)}{5(a + b) - 3}$ Substituting $(a + b)$ for every x

$$= \frac{2a^2 + 4ab + 2b^2 - 7a - 7b}{5a + 5b - 3} \quad ■$$

Example 7 Given $f(x) = 2x + 1$, evaluate $\dfrac{f(-5) - f(2)}{4}$.

Solution

$$f(x) = 2(x) + 1$$
$$f(-5) = 2(-5) + 1 = -10 + 1 = -9$$
$$f(2) = 2(2) + 1 = 4 + 1 = 5$$

Therefore, $\dfrac{f(-5) - f(2)}{4} = \dfrac{-9 - 5}{4} = \dfrac{-14}{4} = -\dfrac{7}{2}$ ■

Example 8 Given $f(x) - x^5$, find and simplify $\dfrac{f(x + h) - f(x)}{h}$.

Solution If $f(x) = x^5$, then $f(x + h) = (x + h)^5$. Using the binomial theorem, we have

$$(x + h)^5 = x^5 + 5x^4h + 10x^3h^2 + 10x^2h^3 + 5xh^4 + h^5$$

Therefore,

$$f(x + h) - f(x) = \overbrace{x^5 + 5x^4h + 10x^3h^2 + 10x^2h^3 + 5xh^4 + h^5}^{f(x+h)} - (\overset{f(x)}{x^5})$$

and

$$\frac{f(x + h) - f(x)}{h} = \frac{\cancel{x^5} + 5x^4h + 10x^3h^2 + 10x^2h^3 + 5xh^4 + h^5 - \cancel{x^5}}{h}$$

$$= \frac{5x^4h + 10x^3h^2 + 10x^2h^3 + 5xh^4 + h^5}{h}$$

$$= \frac{\overset{1}{\cancel{h}}(5x^4 + 10x^3h + 10x^2h^2 + 5xh^3 + h^4)}{\underset{1}{\cancel{h}}}$$

$$= 5x^4 + 10x^3h + 10x^2h^2 + 5xh^3 + h^4 \quad ■$$

Letters other than f can be used to name functions.

Variables other than x can be used for the independent variable.

Variables other than y can be used for the dependent variable.

Example 9 Examples of using other letters to represent functions and variables:

a. $g(x)$, $h(x)$, $F(x)$, $G(x)$

b. $f(z)$, $g(y)$, $G(t)$, $H(s)$ ■

Example 10 Examples of graphs of functions using letters other than x, y, and f:

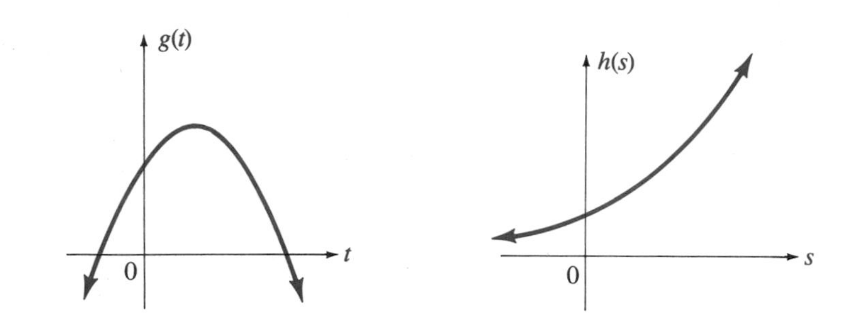

■

Example 11 Given $h(r) = 5r^2$ and $g(s) = 4s - 1$, find $h(a + 1) - g(3a)$.

Solution

$$h(r) = 5(r)^2$$

$$h(a + 1) = 5(a + 1)^2 = 5(a^2 + 2a + 1) = 5a^2 + 10a + 5$$

$$g(s) = 4(s) - 1$$

$$g(3a) = 4(3a) - 1 = 12a - 1$$

Therefore,

$$\begin{aligned} h(a + 1) - g(3a) &= (5a^2 + 10a + 5) - (12a - 1) \\ &= 5a^2 + 10a + 5 - 12a + 1 \\ &= 5a^2 - 2a + 6 \end{aligned}$$ ■

Word Problems Involving Functions

There are many applications of functions and functional notation in business, statistics, the sciences, and so on.

Example 12 If h is measured in feet and t in seconds, and if the formula for finding the height of an arrow that is shot upward from ground level is $h(t) = 64t - 16t^2$, $0 \leq t \leq 4$, find the height of the arrow 1 sec after it is shot into the air.

Solution We need to find h when $t = 1$. Therefore, we must find $h(1)$. $h(1) = 64(1) - 16(1^2) = 64 - 16 = 48$. The arrow is 48 ft above the ground 1 sec after it is shot upward. ■

Functions and Relations of More Than One Variable

All the functions and relations discussed so far in this section have been functions of *one* variable. Many functions and relations have more than one variable.

Example 13 Example of a function of more than one variable:
If you put \$100 (P) in a bank that pays $\frac{1}{2}\%$ a month interest (i) and leave it there for 24 months (n), the amount of money (A) in your account after 24 months is found by using the formula

$$A = P(1 + i)^n$$

This is a function of three variables: $A = f(P, i, n)$, read as "A is a function of P, i, and n." In this case, the *dependent variable A* is a function of *three independent variables P, i,* and *n*, and $f(100, 0.005, 24) = 100(1 + 0.005)^{24} \approx 112.72$. ■

EXERCISES 8.6B

Set I

1. Given $f(x) = 3x - 1$, find the following.

a. $f(2)$ b. $f(0)$ c. $f(a - 2)$ d. $f(x + 2)$

2. Given $f(x) = 4x + 1$, find the following.

a. $f(3)$ b. $f(-5)$ c. $f(0)$ d. $f(x - 2)$

3. Given $f(x) = 2x^2 - 3$, evaluate $\dfrac{f(5) - f(2)}{6}$.

4. Given $f(x) = 5x^2 - 2$, evaluate $\dfrac{f(3) - f(1)}{5}$.

5. Given $f(x) = 3x^2 - 2x + 4$, find $2f(3) + 4f(1) - 3f(0)$.

6. Given $f(x) = (x + 1)(x^2 - x + 1)$, find $3f(0) + 5f(2) - f(1)$.

7. If $f(x) = x^3$ and $g(x) = \dfrac{1}{x}$, find $f(-3) - 6g(2)$.

8. If $f(x) = x^2$ and $g(x) = \dfrac{1}{x}$, find $2f(-4) - 9g(3)$.

9. If $H(x) = 3x^2 - 2x + 4$ and $K(x) = x - x^2$, find $2H(2) - 3K(3)$.

10. If $P(a) = a^2 - 4$ and $Q(a) = 2 - a$, find $3P(2) + 2Q(4)$.

11. Find the domain and range of the function $f(x) = \sqrt{x - 4}$.

12. Find the domain and range of the function $f(x) = -\sqrt{9x - 16}$.

In Exercises 13–16, find and simplify $\dfrac{f(x + h) - f(x)}{h}$ for the given functions.

13. $f(x) = x^2 - x$ **14.** $f(x) = 3x^2$ **15.** $f(x) = x^4$ **16.** $f(x) = x^3$

17. If $A(r) = \pi r^2$ and $C(r) = 2\pi r$, find $\dfrac{3A(r) - 2C(r)}{\pi r}$.

18. If $A(r) = \pi r^2$ and $C(r) = 2\pi r$, find $\dfrac{5A(r) + 3C(r)}{\pi r}$.

19. If the formula for finding the height of an arrow that was shot upward from ground level is $h(t) = 64t - 16t^2$, $0 \leq t \leq 4$, find the height of the arrow 2 sec after it is shot into the air.

20. If the formula for finding the height of an arrow that was shot upward from ground level is $h(t) = 128t - 16t^2, 0 \le t \le 8$, find the height of the arrow 1 sec after it is shot into the air.

21. If the formula for finding the cost of manufacturing x number of bushings per day is $C(x) = 500 + 20x - 0.1x^2$, find the cost of manufacturing 100 bushings per day.

22. If the formula for finding the cost of manufacturing x number of brackets per day is $C(x) = 400 + 10x - 0.1x^2$, find the cost of manufacturing 100 brackets per day.

23. Evaluate the function $z = g(x, y) = 5x^2 - 2y^2 + 7x - 4y$ when $x = 3$ and $y = -4$.

24. Evaluate the function $z = h(x, y) = 2x^2 - 6xy + 5y^2$ when $x = -1$ and $y = -2$.

25. Given that $A = f(P, i, n) = P(1 + i)^n$, use a calculator to approximate $f(100, 0.08, 12)$.

26. Given that $F = f(A, v) = 58.6Av^2$, use a calculator to find $f(126.5, 634)$.

Set II

1. Given $f(x) = 13 - 5x$, find the following.

a. $f(9)$ b. $f(-7)$ c. $f(0)$ d. $f(x - 6)$

2. Given $f(x) = 8 + 2x$, find the following.

a. $f(0)$ b. $f(-1)$ c. $f(x + 2)$ d. $f(x) + 2$

3. Given $f(x) = 8 - 3x^2$, evaluate $\dfrac{f(4) - f(-6)}{12}$.

4. Given $f(x) = x^2 - 2$, evaluate $\dfrac{f(0) + f(2)}{3}$.

5. Given $f(x) = (3x - 5)(x^2 - 9)$, find $6f(8) - 13f(0) - f(-5)$.

6. Given $f(x) = (x - 1)(2x + 1)(x + 1)$, find $3f(0) - 7f(-1) + 2f(4)$.

7. If $f(x) = 2x^2$ and $g(x) = \dfrac{1}{x^2}$, find $\dfrac{8}{f(-6)} - 12g(4)$.

8. If $f(x) = x + 3$ and $g(x) = \sqrt{x - 1}$, find $\dfrac{1}{f(0)} + g(5)$.

9. If $h(x) = \dfrac{x}{x - 2}$ and $k(x) = x^2 + 4$, find $h(0) + 3h(3) - k(1)$.

10. If $F(e) = \dfrac{3e - 1}{e^2}$ and $G(w) = \dfrac{1}{5 - w}$, find $\dfrac{4F(-3)}{5G(4)}$.

11. Find the domain and range of the function $f(x) = \sqrt{15 - 4x}$.

12. Find the domain and range of the function $f(x) = \sqrt{5x - 2}$.

In Exercises 13–16, find and simplify $\dfrac{f(x + h) - f(x)}{h}$ for the given functions.

13. $f(x) = 5x^2 - 3$

14. $f(x) = 1 - x^2$

15. $f(x) = -2x^3$

16. $f(x) = -x^3$

17. If $h(x) = 3x^2$ and $k(x) = x^2 - 2x$, find $\dfrac{h(x) - k(x)}{x}$.

18. If $A(r) = \pi r^2$ and $C(r) = 2\pi r$, find $\dfrac{6A(r) + 3C(r)}{12\pi r}$.

19. If the formula for finding the height of an arrow that was shot upward from ground level is $h(t) = 128t - 16t^2, 0 \le t \le 8$, find the height of the arrow 3 sec after it is shot into the air.

20. If the formula for finding the height of an arrow that was shot upward from ground level is $h(t) = 64t - 16t^2, 0 \le t \le 4$, find the height of the arrow 3 sec after it is shot into the air.

21. If the formula for finding the cost of manufacturing x number of handles per day is $C(x) = 600 + 30x - 0.1x^2$, find the cost of manufacturing 100 handles per day.

22. If the formula for finding the cost of manufacturing x number of faucets per day is $C(x) = 1{,}000 + 20x - 0.1x^2$, find the cost of manufacturing 100 faucets per day.

23. Evaluate the function $w = f(x, y, z) = \dfrac{5xyz - x^2}{2z}$ when $x = -3$, $y = 4$, and $z = -15$.

24. Evaluate the function $w = f(x, y, z) = \dfrac{3x + y^2}{z}$ when $x = -2$, $y = -3$, and $z = 5$.

25. Given that $A = f(P, i, n) = P(1 + i)^n$, use a calculator to approximate $f(500, 0.11, 20)$.

26. Given that $w = f(x, y, z) = \dfrac{x - z}{2y}$, find $f(3, -2, -4)$.

8.6C Graphing $y = |x|$, and Comparing $y = f(x) + h$ and $y = f(x + h)$ with $y = f(x)$

Example 14 Find the domain and range of the function $f(x) = |x|$ and graph the function.

Solution There are no restrictions on x; therefore, the domain is $\{x \mid x \in R\}$. Since $y = |x|$, we know that y can never be negative. The range is $\{y \mid y \ge 0\}$.

Because $f(0) = 0$, one point on the graph will be the origin, $(0, 0)$.

By the definition of absolute value, if $x > 0$, $|x| = x$. Therefore, to the *right* of the y-axis (that is, where $x > 0$), we will graph the line $y = x$.

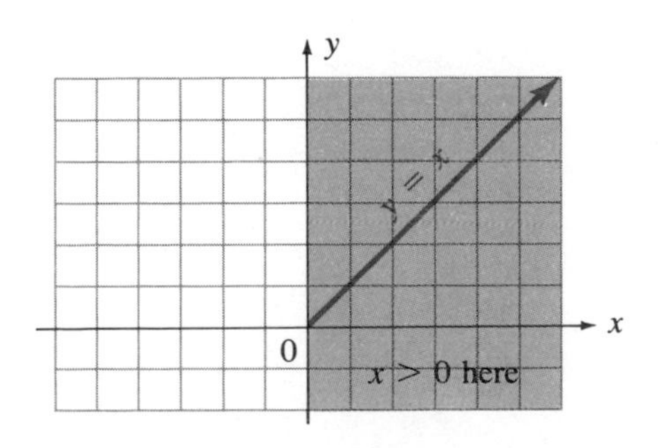

By the definition of absolute value, if $x < 0$, $|x| = -x$. To the *left* of the y-axis (that is, where $x < 0$), we will graph the line $y = -x$.

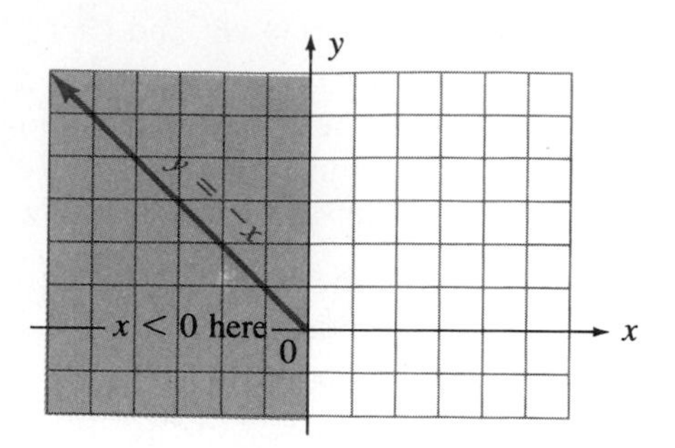

Putting these two graphs together with the point (0, 0), we have the graph of $y = f(x) = |x|$. It is always shaped like a **v**.

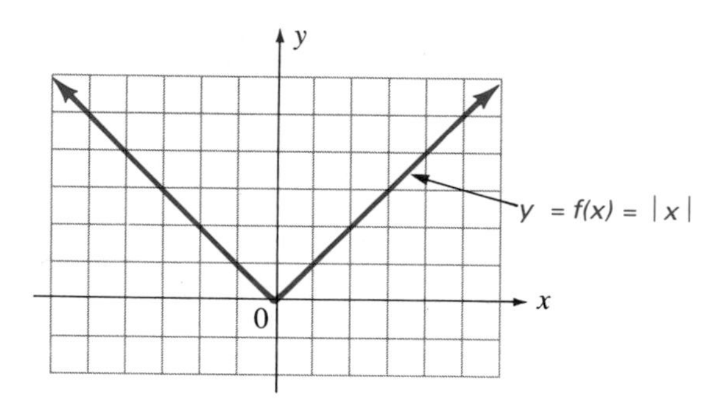

Graphing $y = f(x) + h$ Once we find the graph of $y = f(x)$, we can find the graph of $y = f(x) + h$ by shifting $y = f(x)$ *upward* $|h|$ units if h is positive (see Example 15) and *downward* $|h|$ units if h is negative.

Example 15 Graph $y = |x| + 3$.

Solution We can graph $y = |x| + 3$ by shifting the graph of $y = |x|$ *upward* three units. Verify that the graph is correct by examining the table of values.

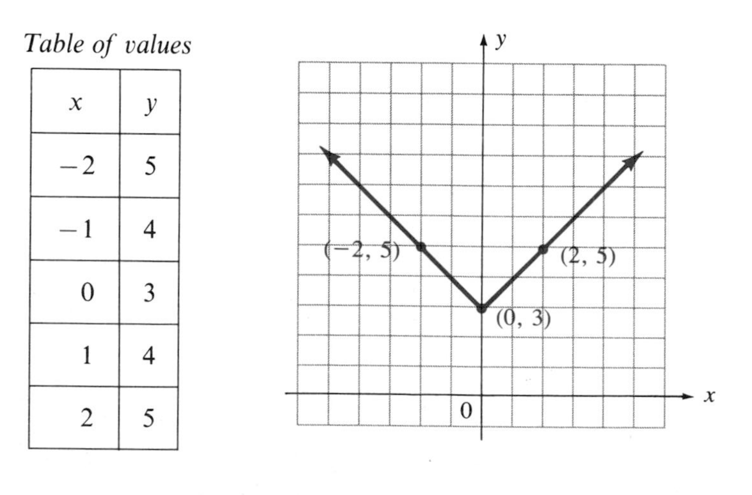

Table of values

x	y
-2	5
-1	4
0	3
1	4
2	5

Graphing $y = f(x + h)$ The graph of $y = f(x + h)$ can be found from the graph of $y = f(x)$ by shifting $y = f(x)$ to the right or left $|h|$ units. We move to the *left* if h is positive (see Example 16) and to the *right* if h is negative.

Example 16 Graph $y = |x + 3|$.

Solution We can graph $y = |x + 3|$ by shifting the graph of $y = |x|$ three units to the *left*. (We shift to the left because *h is positive.*) Verify that the graph is correct by examining the table of values.

Table of values

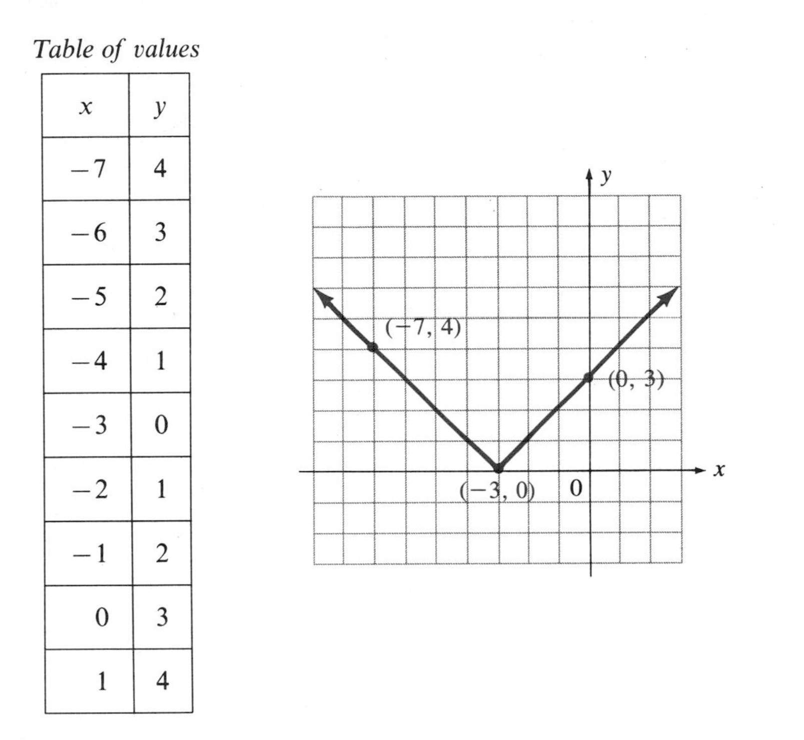

x	y
-7	4
-6	3
-5	2
-4	1
-3	0
-2	1
-1	2
0	3
1	4

■

NOTE The graph of the function $f(x) = |ax + b|$ is shaped like a **v**. The graph of the function $f(x) = -|ax + b|$ is shaped like an upside-down **v**. ☑

EXERCISES 8.6C

Set I Sketch the graph of each function.

1. a. $f(x) = |x| - 2$ b. $f(x) = |x - 2|$ c. $f(x) = -|x - 2|$

2. a. $f(x) = |x| + 5$ b. $f(x) = |x + 5|$ c. $f(x) = -|x + 5|$

3. a. $f(x) = -3x$ b. $f(x) = -3x + 2$ c. $f(x) = -3(x + 2)$

4. a. $f(x) = 2x$ b. $f(x) = 2x - 3$ c. $f(x) = 2(x - 3)$

Set II Sketch the graph of each function.

1. a. $f(x) = |x| + 4$ b. $f(x) = |x + 4|$ c. $f(x) = -|x + 4|$

2. a. $f(x) = |x| - 3$ b. $f(x) = |x - 3|$ c. $f(x) = -|x - 3|$

3. a. $f(x) = 4x$ b. $f(x) = 4x + 2$ c. $f(x) = 4(x + 2)$

4. a. $f(x) = -3x$ b. $f(x) = -3x + 2$ c. $f(x) = -3(x + 2)$

8.7 Inverse Relations and Functions

Inverse Relations and Their Graphs

A relation $\mathscr{R}$ was defined as a set of ordered pairs (Section 8.1). The **inverse relation of** $\mathscr{R}$ is defined to be the set of ordered pairs obtained by interchanging the first and second coordinates in each ordered pair of $\mathscr{R}$.

$$\mathscr{R} = \{(3, 1), (3, -2), (2, -4)\}$$

$$\mathscr{R}^{-1} = \{(1, 3), (-2, 3), (-4, 2)\}$$

$\mathscr{R}^{-1}$ represents the inverse relation of $\mathscr{R}$

NOTE $\mathscr{R}^{-1} \neq \dfrac{1}{\mathscr{R}}$ ☑

It follows from this definition of the inverse relation that the *domain* of $\mathscr{R}^{-1}$ is the *range* of $\mathscr{R}$ (that is, $D_{\mathscr{R}^{-1}} = R_{\mathscr{R}}$) and the *range of* $\mathscr{R}^{-1}$ is the *domain* of $\mathscr{R}$ (that is, $R_{\mathscr{R}^{-1}} = D_{\mathscr{R}}$).

The points of $\mathscr{R}$ and the points of $\mathscr{R}^{-1}$ are graphed in Figure 8.7.1.

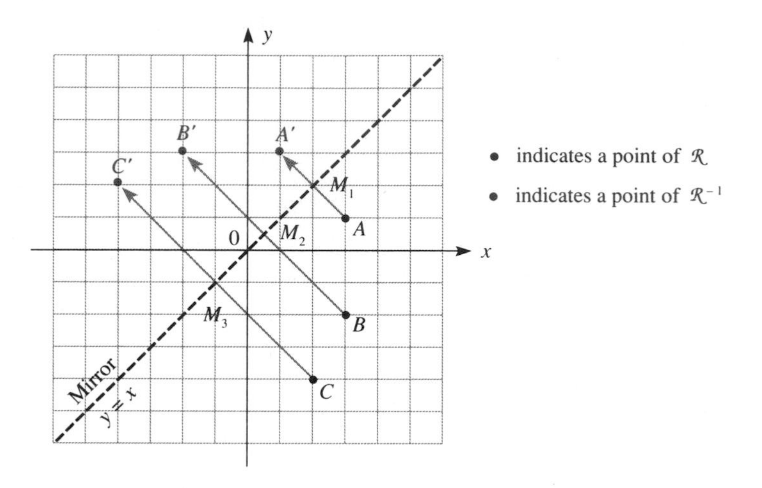

FIGURE 8.7.1

The points of $\mathscr{R}$		*The points of* $\mathscr{R}^{-1}$	
$A(3, 1)$	$\longrightarrow$	$A'(1, 3)$	A' is the *image* of A
$B(3, -2)$	$\longrightarrow$	$B'(-2, 3)$	B' is the *image* of B
$C(2, -4)$	$\longrightarrow$	$C'(-4, 2)$	C' is the *image* of C

Each point of the inverse relation $\mathscr{R}^{-1}$ is the *mirror image*, with respect to the line $y = x$, of the corresponding point of the relation $\mathscr{R}$. The mirror image of a point is the same distance behind the "mirror" (in this case, the line $y = x$) as the actual point is in front of the mirror. Therefore, in Figure 8.7.1, $|AM_1| = |M_1A'|$, $|BM_2| = |M_2B'|$, and $|CM_3| = |M_3C'|$.

Example 1 Given $\mathscr{R} = \{(2, -3), (-3, -1), (-4, 5), (4, 0)\}$, find $\mathscr{R}^{-1}$, $D_{\mathscr{R}^{-1}}$, and $R_{\mathscr{R}^{-1}}$, and graph $\mathscr{R}$ and $\mathscr{R}^{-1}$.

Solution

$$\mathscr{R} = \{(2, -3), (-3, -1), (-4, 5), (4, 0)\} \qquad \mathscr{R} \text{ is a } \textit{function}$$

$$\mathscr{R}^{-1} = \{(-3, 2), (-1, -3), (5, -4), (0, 4)\} \qquad \mathscr{R}^{-1} \text{ is a } \textit{function}$$

$$D_{\mathscr{R}^{-1}} = \{-3, -1, 5, 0\} = R_{\mathscr{R}}$$

$$R_{\mathscr{R}^{-1}} = \{2, -3, -4, 4\} = D_{\mathscr{R}}$$

The graph of the relation and its inverse is shown in Figure 8.7.2.

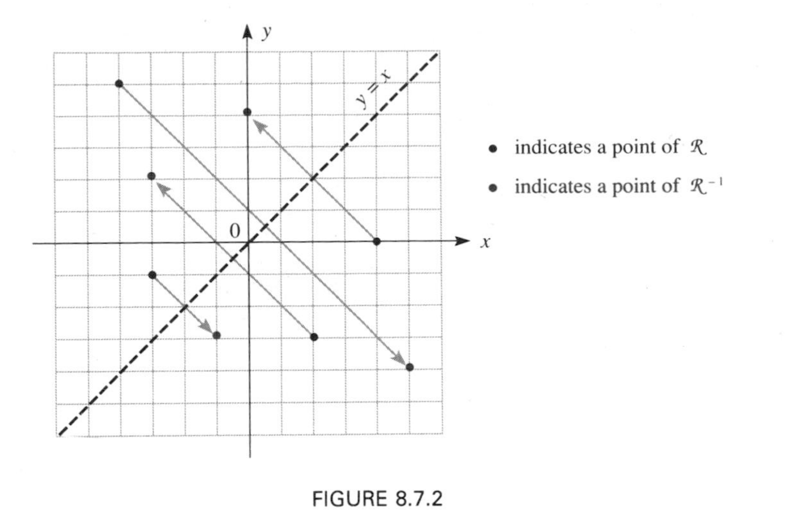

FIGURE 8.7.2

■

Inverse Functions and Their Graphs

A function will have an inverse if for each y-value there is only one x-value. We call such a function **invertible**. Graphically, this means that a horizontal line will not touch the graph in more than one point. If $f(x)$ is invertible, the inverse of $y = f(x)$ will *also* be a function, and we will write $y = f^{-1}(x)$.

A WORD OF CAUTION In the notation $f^{-1}(x)$, the -1 is *not* an exponent, and $f^{-1}(x) \neq \dfrac{1}{f(x)}$. ☑

Finding the Inverse of a Function When Its Equation Is Given To find the **inverse of a function** whose equation is given, substitute y for $f(x)$ if necessary, interchange x and y in the equation, and then solve the new equation for y.

The *domain* of $f(x)$ equals the *range* of $f^{-1}(x)$, and the *range* of $f(x)$ equals the *domain* of $f^{-1}(x)$, as was true for relations.

The graphs of $y = f(x)$ and $y = f^{-1}(x)$ will be the mirror images of each other with respect to the line $y = x$.

Example 2 Find the inverse function for $y = f(x) = 2x - 3$, and graph $y = f(x)$ and its inverse.

Solution

$$y = 2x - 3$$

$$x = 2y - 3 \qquad \text{Interchanging } x \text{ and } y$$

$$\left.\begin{aligned} x + 3 &= 2y \\ \frac{x+3}{2} &= y \\ y = f^{-1}(x) &= \frac{x+3}{2} \end{aligned}\right\} \text{Solving for } y$$

$f^{-1}(x) = \dfrac{x+3}{2}$ is the inverse function for $f(x) = 2x - 3$

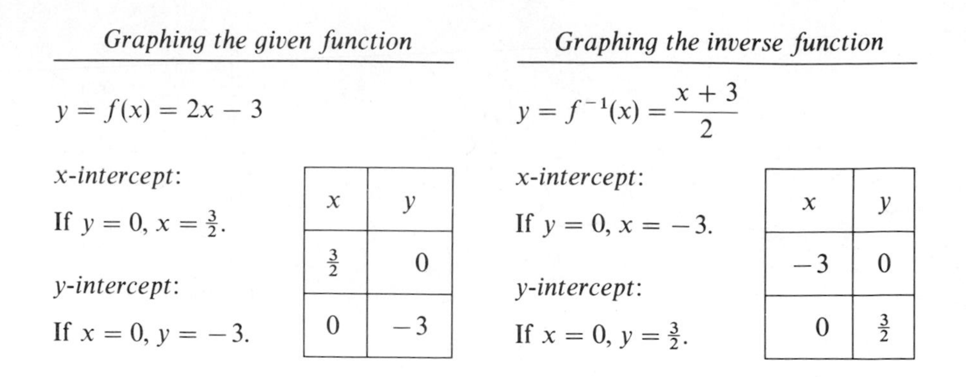

Graphing the given function	*Graphing the inverse function*
$y = f(x) = 2x - 3$	$y = f^{-1}(x) = \frac{x+3}{2}$
x-intercept: If $y = 0$, $x = \frac{3}{2}$.	*x-intercept*: If $y = 0$, $x = -3$.
y-intercept: If $x = 0$, $y = -3$.	*y-intercept*: If $x = 0$, $y = \frac{3}{2}$.

x	y
$\frac{3}{2}$	0
0	-3

x	y
-3	0
0	$\frac{3}{2}$

We did not have to calculate the entries of the inverse table. We could have used the entries from the function table, but entered the x-values in the y-column and the y-values in the x-column. The graphs are shown in Figure 8.7.3.

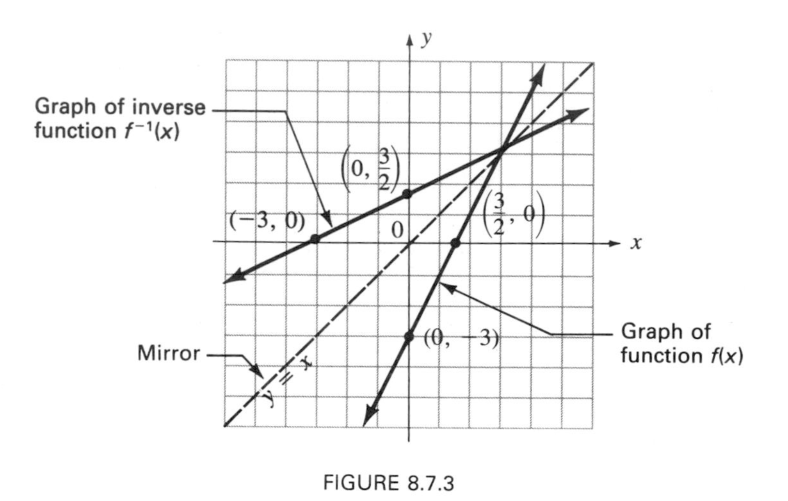

FIGURE 8.7.3

In Example 2, it is apparent from the graphs that $f(x)$ and $f^{-1}(x)$ are both functions (Section 8.6).

Example 3 Find the inverse function for $y = f(x) = \frac{5x - 1}{x}$ and find the domain and range for both functions.

Solution

$$y = \frac{5x-1}{x}$$

$$x = \frac{5y-1}{y} \qquad \text{Interchanging } x \text{ and } y$$

$$\left.\begin{aligned} xy &= 5y - 1 \\ 1 &= 5y - xy \\ 1 &= y(5 - x) \\ y &= \frac{1}{5-x} = f^{-1}(x) \end{aligned}\right\} \text{Solving for } y$$

$f^{-1}(x) = \frac{1}{5-x}$ is the inverse function for $f(x) = \frac{5x-1}{x}$

The domain of $f(x)$ is the set of all real numbers except 0, since the denominator cannot equal 0. Therefore, the range of $f^{-1}(x)$ must be the set of all real numbers except 0, since the domain of $f(x)$ equals the range of $f^{-1}(x)$. We cannot determine the range of $f(x)$ by looking at its equation; however, we *can* determine that the domain of $f^{-1}(x) = \dfrac{1}{5-x}$ is the set of all real numbers except 5. Because the range of $f(x)$ equals the domain of $f^{-1}(x)$, we now know that the range of $f(x)$ must be the set of all real numbers except 5.

In summary,

$$\begin{aligned} \text{domain of } f(x) &= \{x \mid x \in R, x \neq 0\} \\ \text{range of } f(x) &= \{y \mid y \in R, y \neq 5\} \\ \text{domain of } f^{-1}(x) &= \{x \mid x \in R, x \neq 5\} \\ \text{range of } f^{-1}(x) &= \{y \mid y \in R, y \neq 0\} \end{aligned}$$ ■

TO FIND THE INVERSE RELATION OR FUNCTION

1. If a relation, $\mathscr{R}$, or function, $f(x)$, is given as a set of ordered pairs:
   a. Interchange the first and second coordinates of each ordered pair.
   b. The inverse relation, $\mathscr{R}^{-1}$, or function, $f^{-1}(x)$, is the set of ordered pairs obtained in step 1a.
2. If an equation is given and if the relation is a function that is invertible:
   a. Substitute y for $f(x)$ if necessary.
   b. Interchange x and y in the equation.
   c. Solve the resulting equation for y.
   d. The inverse function, $f^{-1}(x)$, is found by substituting $f^{-1}(x)$ for y in the equation obtained in step 2c.

EXERCISES 8.7

Set I

1. Given $\mathscr{R} = \{(-10, 7), (3, -8), (-5, -4), (3, 9)\}$, find $\mathscr{R}^{-1}$, $D_{\mathscr{R}^{-1}}$, and $R_{\mathscr{R}^{-1}}$, and graph $\mathscr{R}$ and $\mathscr{R}^{-1}$. Is $\mathscr{R}$ a function? Is $\mathscr{R}^{-1}$ a function?

2. Given $\mathscr{R} = \{(9, -6), (0, 11), (3, 8), (-2, -6), (10, -4)\}$, find $\mathscr{R}^{-1}$, $D_{\mathscr{R}^{-1}}$, and $R_{\mathscr{R}^{-1}}$, and graph $\mathscr{R}$ and $\mathscr{R}^{-1}$. Is $\mathscr{R}$ a function? Is $\mathscr{R}^{-1}$ a function?

3. Find the inverse function for $y = f(x) = 5 - 2x$. Graph $y = f(x)$ and its inverse.

4. Find the inverse function for $y = f(x) = 3x - 10$. Graph $y = f(x)$ and its inverse.

5. Find the inverse function for $y = f(x) = \dfrac{4x - 3}{5}$.

6. Find the inverse function for $y = f(x) = \dfrac{2x - 7}{3}$.

7. Find the inverse function for $y = f(x) = \dfrac{5}{x+2}$.

8. Find the inverse function for $y = f(x) = \dfrac{10}{2x-1}$.

Set II

1. Given $\mathscr{R} = \{(-10, 2), (6, 5), (-11, -9), (1, -3), (-7, 5)\}$, find $\mathscr{R}^{-1}$, $D_{\mathscr{R}^{-1}}$, and $R_{\mathscr{R}^{-1}}$, and graph $\mathscr{R}$ and $\mathscr{R}^{-1}$. Is $\mathscr{R}$ a function? Is $\mathscr{R}^{-1}$ a function?

2. Given $\mathscr{R} = \{(-3, 6), (5, 6), (2, 3)\}$, find $\mathscr{R}^{-1}$, $D_{\mathscr{R}^{-1}}$, and $R_{\mathscr{R}^{-1}}$, and graph $\mathscr{R}$ and $\mathscr{R}^{-1}$. Is $\mathscr{R}$ a function? Is $\mathscr{R}^{-1}$ a function?

3. Find the inverse function for $y = f(x) = \frac{1}{5}x + 2$. Graph $y = f(x)$ and its inverse.

4. Find the inverse function for $y = f(x) = \frac{2}{3}x - 4$. Graph $y = f(x)$ and its inverse.

5. Find the inverse function for $y = f(x) = \dfrac{3x-8}{4}$.

6. Find the inverse function for $y = f(x) = \dfrac{3}{x-2}$.

7. Find the inverse function for $y = f(x) = \dfrac{6}{4-3x}$.

8. Find the inverse function for $y = f(x) = \dfrac{8}{4-x}$.

8.8 Graphing Polynomial Functions

It is shown in calculus that the graph of an equation that can be expressed in the form

$$y = a_n x^n + a_{n-1}x^{n-1} + \cdots + a_2x^2 + a_1x + a_0$$

is always a *continuous, smooth curve*; that is, it has *no holes, no jumps,* and *no sharp corners*. Because the right side of the above equation is a polynomial, we call the function a **polynomial function**.

Example 1 Examples of polynomial functions:

a. $f(x) = 3x - 5$ — A *linear* polynomial function

b. $f(x) = x^2 - x - 2$ — A *quadratic* polynomial function

c. $f(x) = x^3 - 4x$ — A *cubic* polynomial function

d. $f(x) = 7x^5 - 2x^3 + 6$ — A *fifth-degree* polynomial function ■

In Section 8.2, we showed how to graph straight lines (linear polynomial functions). In this section, we show how to graph polynomial functions that are not linear. Their graphs are curves, not straight lines.

Two points are all that we need to draw a straight line. To draw a curve, however, we must find more than two points.

Example 2 Graph the function $y = f(x) = x^2$.

Solution

x	y
-3	9
-2	4
-1	1
$-\frac{1}{2}$	$\frac{1}{4}$
$-\frac{1}{4}$	$\frac{1}{16}$
0	0
$\frac{1}{4}$	$\frac{1}{16}$
$\frac{1}{2}$	$\frac{1}{4}$
1	1
2	4
3	9

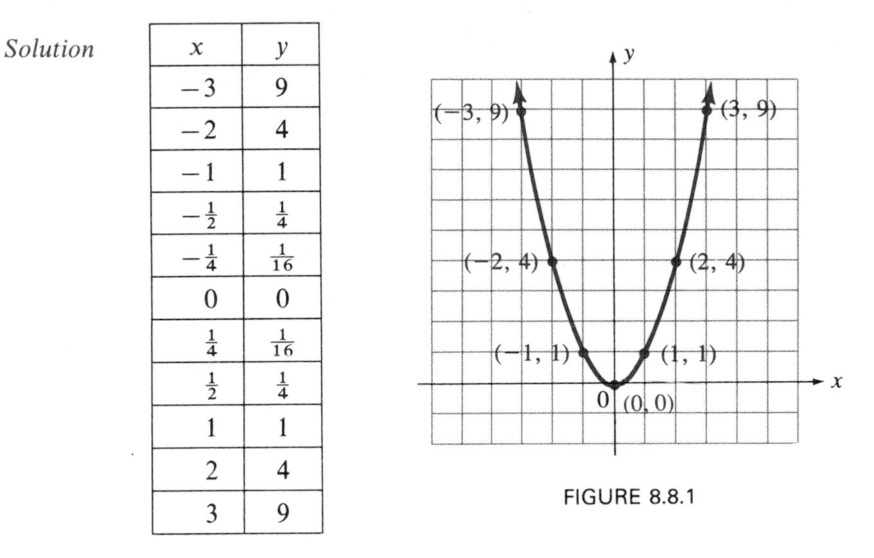

FIGURE 8.8.1

In drawing the smooth curve, we start with the point in the table of values having the smallest x-value, and we draw to the point having the next larger value of x. We continue in this way, drawing a smooth curve through all the points. The graph of the equation $y = x^2$ is called a **parabola** and is shown in Figure 8.8.1. ■

A parabola is one of a family of curves of quadratic polynomial functions called *conic sections*; they are discussed further in Sections 10.7 and 10.8.

A WORD OF CAUTION The graph in Example 2 has *no sharp corners* and thus does *not* look like this:

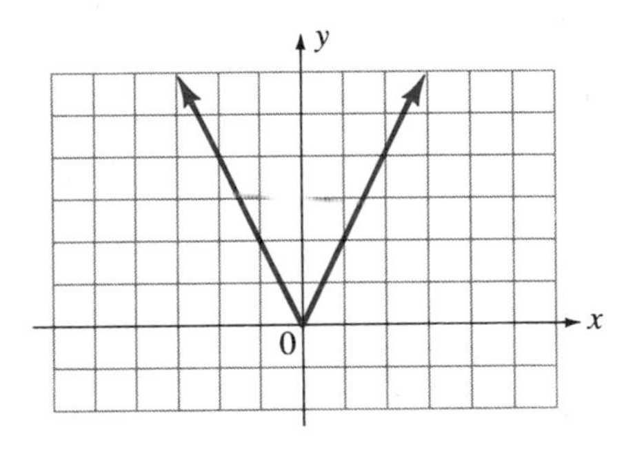

Example 3 Graph the function $y = g(x) = x^2 - 2$.

Solution Using the technique discussed in Section 8.6C, we can graph $y = x^2 - 2$ by shifting the graph of Figure 8.8.1, $y = f(x) = x^2$, two units *down* (see Figure 8.8.2). You might want to make a table of values to verify that the points shown on the graph in Figure 8.8.2 are, indeed, points on the graph of $y = x^2 - 2$.

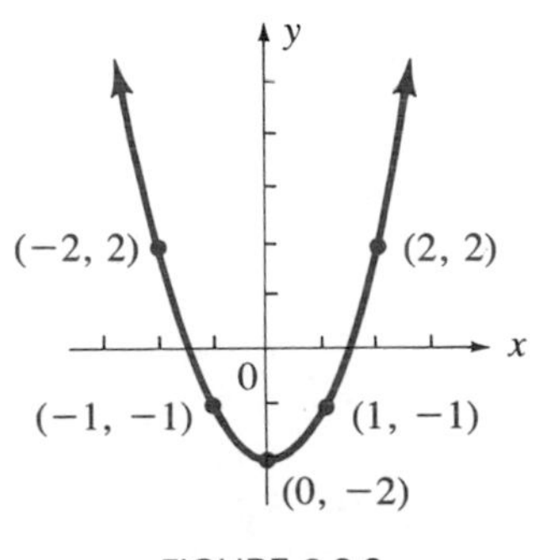

FIGURE 8.8.2

■

Example 4 Graph the function $y = h(x) = (x - 2)^2$.

Solution Using the technique discussed in Section 8.6C, we can graph $y = (x - 2)^2$ by shifting the graph of Figure 8.8.1, $y = f(x) = x^2$, two units *to the right* (see Figure 8.8.3). You might want to make a table of values to verify that the points shown on the graph in Figure 8.8.3 are points on the graph of $y = (x - 2)^2$.

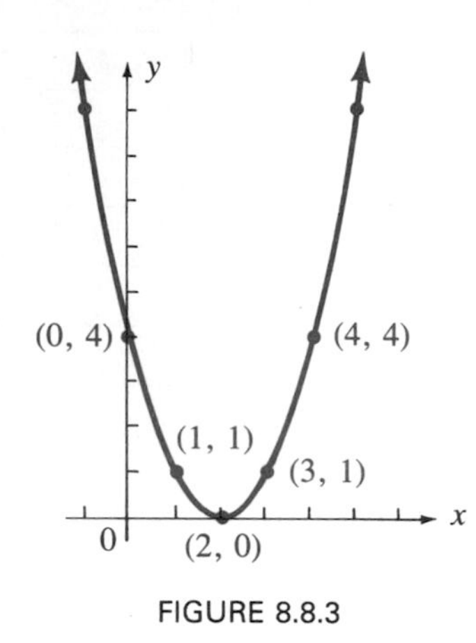

FIGURE 8.8.3

Example 5 Graph the function $y = x^2 - x - 2$.

Solution Make a table of values by substituting values of x in the equation and finding the corresponding values for y. Because the domain of the variable is the set of all real numbers, we must be sure to include some negative values for x, some positive values, and zero. (We could also, of course, have included some noninteger values for x.)

If $x = -2$, then $y = (-2)^2 - (-2) - 2 = 4 + 2 - 2 = 4$

If $x = -1$, then $y = (-1)^2 - (-1) - 2 = 1 + 1 - 2 = 0$

If $x = 0$, then $y = (0)^2 - (0) - 2 = -2$

If $x = 1$, then $y = (1)^2 - (1) - 2 = 1 - 1 - 2 = -2$

If $x = 2$, then $y = (2)^2 - (2) - 2 = 4 - 2 - 2 = 0$

If $x = 3$, then $y = (3)^2 - (3) - 2 = 9 - 3 - 2 = 4$

x	y
-2	4
-1	0
0	-2
1	-2
2	0
3	4

In Figure 8.8.4, we graph these points and draw a smooth curve through them.

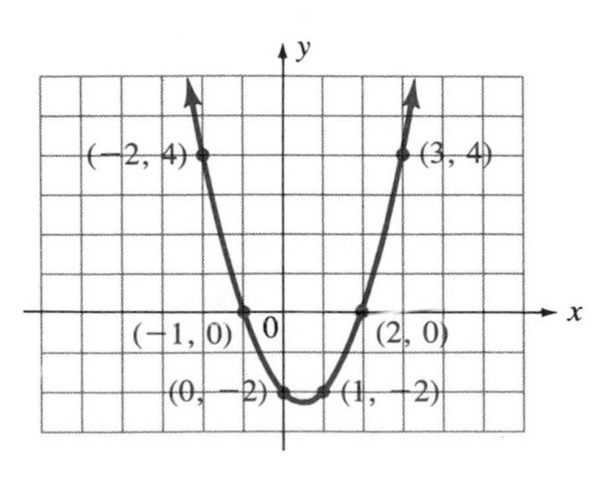

FIGURE 8.8.4

Example 6 Graph the function $y = x^3 - 4x$.

Solution As we mentioned at the beginning of this section, it is shown in calculus that the graph of any polynomial function *is a smooth curve*. Since $y = x^3 - 4x$ is a polynomial function, its graph must be a smooth curve. We'll find seven points on the curve by making a table of values, letting x have integral values from -3 to 3. (We could also, of course, let x have noninteger values.)

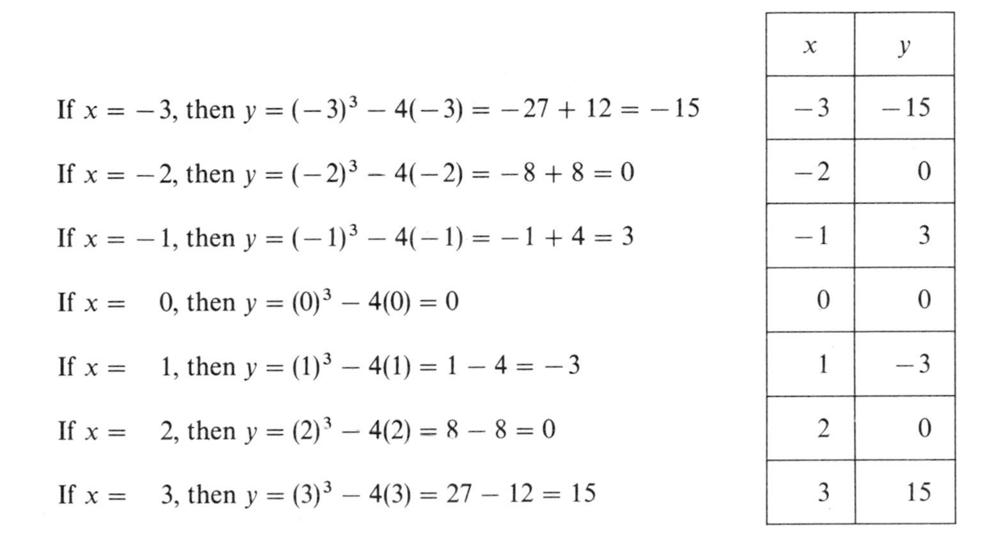

If $x = -3$, then $y = (-3)^3 - 4(-3) = -27 + 12 = -15$

If $x = -2$, then $y = (-2)^3 - 4(-2) = -8 + 8 = 0$

If $x = -1$, then $y = (-1)^3 - 4(-1) = -1 + 4 = 3$

If $x = \ \ 0$, then $y = (0)^3 - 4(0) = 0$

If $x = \ \ 1$, then $y = (1)^3 - 4(1) = 1 - 4 = -3$

If $x = \ \ 2$, then $y = (2)^3 - 4(2) = 8 - 8 = 0$

If $x = \ \ 3$, then $y = (3)^3 - 4(3) = 27 - 12 = 15$

x	y
-3	-15
-2	0
-1	3
0	0
1	-3
2	0
3	15

In Figure 8.8.5, we graph these points and then draw a smooth curve through them, taking them in order from left to right.

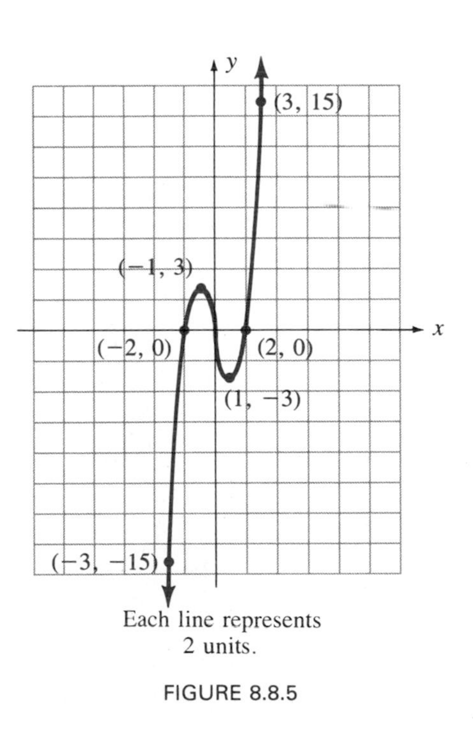

FIGURE 8.8.5

Finding the Table of Values by Using Synthetic Division (Optional)

Graphing polynomial functions by using synthetic division to find the y-values is often simpler than finding the y-values by direct substitution. To find y-values by synthetic division, we use the remainder theorem, which is stated below and proved in Appendix A.

> **THE REMAINDER THEOREM**
>
> If we divide a polynomial in x by $x - a$, the remainder is equal to the value of the polynomial when $x = a$. In symbols, if the remainder is R and the function is $y = f(x)$, then $f(a) = R$.

Example 7 Given $f(x) = 2x^4 - 11x^3 + 7x^2 - 12x + 14$, find $f(5)$.
First method Using substitution,

$$f(x) = 2x^4 - 11x^3 + 7x^2 - 12x + 14$$
$$f(5) = 2(5)^4 - 11(5)^3 + 7(5)^2 - 12(5) + 14$$
$$= 1250 - 1375 + 175 - 60 + 14 = 4$$

Second method Using synthetic division and the remainder theorem, we divide $f(x)$ by $x - 5$ because we want to find $f(5)$.

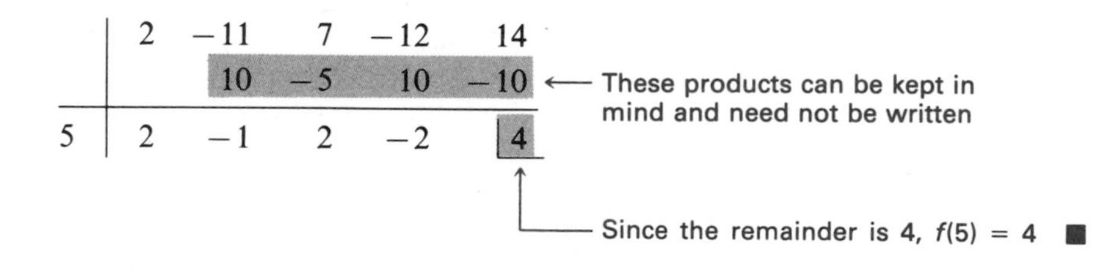

Example 8 Use synthetic division to find the table of values for the function given in Example 6.

Solution $y = f(x) = x^3 - 4x$
$$= 1x^3 + 0x^2 - 4x + 0$$

x	1	0	−4	0 y
−3	1	−3	5	−15
−2	1	−2	0	0
−1	1	−1	−3	3
0	1	0	−4	0
1	1	1	−3	−3
2	1	2	0	0
3	1	3	5	15

To simplify the writing of this table, the *products* are not written ■

EXERCISES 8.8

Set I In Exercises 1 and 2, use integer values of x from -3 to $+3$ to make a table of values for each equation. Graph the points and draw a smooth curve through them.

1. $y = x^2 + 3$ **2.** $y = x^2 + 1$

3. Use integer values of x from -4 to $+2$ to make a table of values for the equation $y = (x + 1)^2$. Graph the points and draw a smooth curve through them.

4. Use integer values of x from -6 to 0 to make a table of values for the equation $y = (x + 3)^2$. Graph the points and draw a smooth curve through them.

5. Use integer values of x from -2 to $+4$ to make a table of values for the equation $y = x^2 - 2x$. Graph the points and draw a smooth curve through them.

6. Use integer values of x from -2 to $+4$ to make a table of values for the equation $y = 3x - x^2$. Graph the points and draw a smooth curve through them.

7. Use integer values of x from -2 to $+4$ to make a table of values for the equation $y = 2x - x^2$. Graph the points and draw a smooth curve through them.

8. Use integer values of x from -3 to $+1$ to make a table of values for the equation $y = 2x + x^2$. Graph the points and draw a smooth curve through them.

9. Use integer values of x from -2 to $+2$ to make a table of values for the equation $y = x^3$. Graph the points and draw a smooth curve through them.

10. Use integer values of x from -2 to $+2$ to make a table of values for the equation $y = x^3 + 2$. Graph the points and draw a smooth curve through them.

11. Use integer values of x from -3 to $+1$ to make a table of values for the equation $y = (x + 1)^3$. Graph the points and draw a smooth curve through them.

12. Use integer values of x from -1 to $+3$ to make a table of values for the equation $y = (x - 1)^3$. Graph the points and draw a smooth curve through them.

13. Find the range of the function $f(x) = x^3 - 2x^2 - 13x + 20$ if the domain is $\{-4, -3, -2, 0, 2, 3, 4, 5\}$.

14. Find the range of the function $f(x) = x^3 - 3x^2 - 20x + 12$ if the domain is $\{-3, -2, 0, 2, 3, 4, 5, 6\}$.

Set II

1. Use integer values of x from -3 to $+3$ to make a table of values for the equation $y = x^2 - 1$. Graph the points and draw a smooth curve through them.

2. Use integer values of x from -2 to $+4$ to make a table of values for the equation $y = (x - 1)^2$. Graph the points and draw a smooth curve through them.

3. Use integer values of x from -2 to $+4$ to make a table of values for the equation $y = x - \frac{1}{2}x^2$. Graph the points and draw a smooth curve through them.

4. Use integer values of x from -3 to $+3$ to make a table of values for the equation $y = 2x^2$. Graph the points and draw a smooth curve through them.

5. Use integer values of x from -3 to $+3$ to make a table of values for the equation $y = 2x^2 + 3$. Graph the points and draw a smooth curve through them.

6. Use integer values of x from -2 to $+4$ to make a table of values for the equation $y = 2x^2 - 4x$. Graph the points and draw a smooth curve through them.

7. Use integer values of x from -1 to $+4$ to make a table of values for the equation $y = 3x - x^2$. Graph the points and draw a smooth curve through them.

8. Use integer values of x from -4 to $+1$ to make a table of values for the equation $y = 3x + x^2$. Graph the points and draw a smooth curve through them.

9. Use integer values of x from -2 to $+2$ to make a table of values for the equation $y = x^3 - 1$. Graph the points and draw a smooth curve through them.

10. Use integer values of x from -2 to $+2$ to make a table of values for the equation $y = x^4$. Graph the points and draw a smooth curve through them.

11. Use integer values of x from -2 to $+2$ to make a table of values for the equation $y = x^4 - 1$. Graph the points and draw a smooth curve through them.

12. Use integer values of x from -2 to $+2$ to make a table of values for the equation $y = x^3 - 3x + 4$. Graph the points and draw a smooth curve through them.

13. Find the range of the function $f(x) = x^3 + 3x^2 - x - 3$ if the domain is $\{-3, -2, -1, 0, 1, 2\}$.

14. Find the range of the function $f(x) = x^3 + x^2 - 4x - 4$ if the domain is $\{-2, -1, 0, 1, 2, 3\}$.

8.9 Variation and Proportion

In this section, we will discuss direct variation and proportion, as well as inverse variation, joint variation, and combined variation.

8.9A Direct Variation and Proportion

Direct Variation

If a function can be represented by an equation of the form $y = kx^n$, where k is some constant and n is some positive number, then that function is called a **direct variation**. The constant k is called the **constant of proportionality**. In a direct variation, as one of the variables increases, the other increases also; as one of the variables decreases, the other decreases also.

Many examples of direct variation exist:

If the number of miles per gallon of gasoline obtained by a certain car is a constant, then the number of miles driven in that car is directly proportional to the number of gallons of gasoline used.

If someone is working at a certain wage per hour (k), then as the number of hours (h) changes, the salary (s) changes according to the formula $s = kh$. Therefore, s is directly proportional to h, and s is a function of h.

In circles, as the diameter (D) changes, the circumference (C) changes according to the formula $C = \pi D$, where $\pi \approx 3.14$. In this case, π is the constant of proportionality. C is a function of D, and we can say that C varies as D, or that C is directly proportional to D.

DIRECT VARIATION

The following statements all translate into the equation

$$y = kx^n, \text{ where } n > 0:$$

Constant of proportionality (k)

y varies directly with x^n.

y varies directly as x^n.

y is directly proportional to x^n.

y is proportional to x^n.

Notice that the relationship between x and y in a direct variation need not be linear; that is, x can be raised to any positive power. Thus, $y = kx^2$ and $y = kx^3$ are examples of direct variation.

Example 1 Given: y varies directly with x. If $y = 10$ when $x = 2$, find k (the constant of proportionality) and find y when $x = 3$.

Solution Since y varies directly with x, our equation is $y = kx$. Substituting 10 for y and 2 for x, we have $10 = k(2)$. Therefore, $k = 5$.

The equation is now $y = 5x$. Substituting 3 for x, we have $y = 5(3) = 15$. ■

Subscripts make it possible to indicate *corresponding values* of variables. In Example 1, we consider $y = 10$ and $x = 2$ to be a pair of corresponding values, because we are told that $y = 10$ *when* $x = 2$. To indicate by subscripted variables that these are corresponding values, we use the *same* subscript. That is,

$$\text{Condition 1:} \quad x_1 = 2 \text{ and } y_1 = 10$$

In Example 1, we also consider $y = 15$ and $x = 3$ to be a pair of corresponding values. Therefore,

$$\text{Condition 2:} \quad x_2 = 3 \text{ and } y_2 = 15$$

In general, for the direct variation whose equation is $y = kx$,

$$\text{Condition 1} \begin{Bmatrix} \text{If} & y = y_1 \\ \text{when} & x = x_1 \end{Bmatrix} \text{ then } y_1 = kx_1 \qquad (1)$$

$$\text{Condition 2} \begin{Bmatrix} \text{If} & y = y_2 \\ \text{when} & x = x_2 \end{Bmatrix} \text{ then } y_2 = kx_2 \qquad (2)$$

Proportion

When an equation can be expressed in the form $\frac{a}{b} = \frac{c}{d}$, it is called a **proportion**. (Stated differently, a proportion is a statement that two ratios are equal.)

RELATED PROPORTIONS FOR THE DIRECT VARIATION WHOSE EQUATION IS $y = kx^n$

$$\frac{y_1}{y_2} = \frac{x_1^n}{x_2^n} \quad \text{and} \quad \frac{y_1}{x_1^n} = \frac{y_2}{x_2^n}$$

The proportions in the box on page 427 are derived as follows: If the equation $y_1 = kx_1^n$ is divided by the equation $y_2 = kx_2^n$, we have

$$\frac{y_1}{y_2} = \frac{\overset{1}{\cancel{k}}x_1^n}{\underset{1}{\cancel{k}}x_2^n} = \frac{x_1^n}{x_2^n}$$

Alternately, if the equation $y_1 = kx_1^n$ is solved for k, we have $k = \dfrac{y_1}{x_1^n}$, and if the equation $y_2 = kx_2^n$ is solved for k, we have $k = \dfrac{y_2}{x_2^n}$. Then, since $\dfrac{y_1}{x_1^n}$ and $\dfrac{y_2}{x_2^n}$ both equal k, they must equal each other:

$$\frac{y_1}{x_1^n} = \frac{y_2}{x_2^n}$$

Because a proportion is simply a rational equation, it can be solved by the techniques learned in Chapter 6. An alternate method of solution is based on the following fact: When we multiply both sides of the equation $\dfrac{a}{b} = \dfrac{c}{d}$ by the LCD, we obtain the new equation $ad = bc$.

Therefore, the equation $ad = bc$ is equivalent to the equation $\dfrac{a}{b} = \dfrac{c}{d}$ *except* that in the equation $\dfrac{a}{b} = \dfrac{c}{d}$, neither b nor d can equal zero. When we rewrite $\dfrac{a}{b} = \dfrac{c}{d}$ as $ad = bc$, we say we are **cross-multiplying**.

We can, then, cross-multiply as the first step in solving a proportion. However, since we may introduce extraneous roots when we do this, we must be sure to reject any apparent solutions that aren't in the domain of the variable.

In solving variation problems, we can, if we wish, solve the related proportion without first finding k. This method will be designated as the alternate method in most of the examples in this section.

PROPORTION METHOD FOR SOLVING WORD PROBLEMS INVOLVING DIRECT VARIATION

1. Represent the unknown quantity by a variable.
2. Write a proportion, putting the units of measure next to the numbers when writing the proportion. Be sure the same units occupy corresponding positions in the two ratios of the proportion.

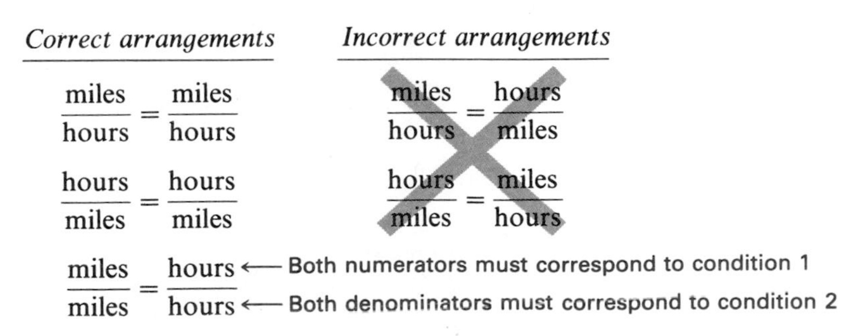

3. Once the numbers have been correctly entered in the proportion by using the units as a guide, drop the units and then solve for the unknown.

Example 2 The number of miles driven is proportional to the number of gallons of gasoline used. Sherma knows she can drive 2,220 mi on 60 gal of gasoline. Find k, the constant of proportionality, and find how many miles she can expect to drive on 17 gal of gasoline.
Solution Because the number of miles (m) driven is proportional to the number of gallons of gasoline (g), the equation of the variation is $m = kg$.

Let m_2 = the number of miles on 17 gal of gasoline.

$$\text{Condition 1} \begin{cases} m_1 = 2{,}220 \text{ mi} \\ g_1 = 60 \text{ gal} \end{cases} \qquad \text{Condition 2} \begin{cases} m_2 = ? \\ g_2 = 17 \text{ gal} \end{cases}$$

From condition 1,

$$m_1 = kg_1$$
$$2{,}220 = k(60)$$
$$k = \frac{2{,}220}{60} = 37$$

From condition 2, $m_2 = kg_2 = 37(17) = 629$

Alternate method Using a proportion:

$$\frac{m_1}{g_1} = \frac{m_2}{g_2}$$
$$\frac{2{,}220 \text{ mi}}{60 \text{ gal}} = \frac{m_2}{17 \text{ gal}}$$
$$2{,}220(17) = 60m_2$$
$$m_2 = \frac{2{,}220(17)}{60} = 629$$

Sherma can expect to drive 629 mi on 17 gal of gasoline. ■

The following theorem from geometry leads to equations involving direct variation: If two triangles are *similar*, their sides are directly proportional. (Recall that two triangles are similar if their corresponding angles are equal.) Example 3 uses this theorem and shows the alternate method only.

Example 3 If a 6-ft man casts a $4\frac{1}{2}$-ft shadow, how tall is a tree that casts a 30-ft shadow?
Solution We use the fact that the length of a shadow is determined by the angle of a ray of the sun. (We assume that the tree and the man are both vertical; you should verify that the triangles are similar.)

NOTE The ground does not have to be horizontal. ☑

Let x = height of tree (in feet).

A 6-ft man casts a $4\frac{1}{2}$-ft shadow	How tall is a tree that casts a 30-ft shadow?
Man	Tree

$$\text{Height} \longrightarrow \frac{6 \text{ ft}}{4\frac{1}{2} \text{ ft}} = \frac{x \text{ ft}}{30 \text{ ft}} \longleftarrow \text{Height}$$
$$\text{Shadow} \longrightarrow \qquad\qquad \longleftarrow \text{Shadow}$$

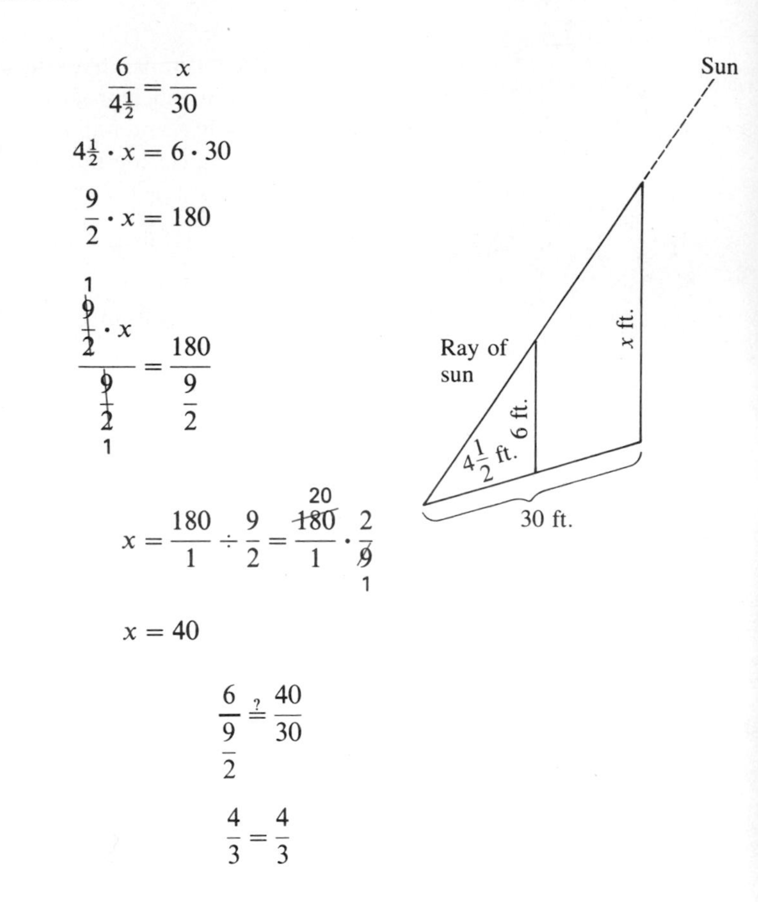

$$\frac{6}{4\frac{1}{2}} = \frac{x}{30}$$

$$4\tfrac{1}{2} \cdot x = 6 \cdot 30$$

$$\frac{9}{2} \cdot x = 180$$

$$\frac{\frac{\cancel{9}^{1}}{\cancel{2}_{1}} \cdot x}{\frac{\cancel{9}}{\cancel{2}}_{1}} = \frac{180}{\frac{9}{2}}$$

$$x = \frac{180}{1} \div \frac{9}{2} = \frac{\cancel{180}^{20}}{1} \cdot \frac{2}{\cancel{9}_{1}}$$

$$x = 40$$

Check

$$\frac{6}{\frac{9}{2}} \stackrel{?}{=} \frac{40}{30}$$

$$\frac{4}{3} = \frac{4}{3}$$

The tree is 40 ft tall. ■

Example 4 Given: The area (A) of a circle varies directly with the square of the radius (r). If $A = 200.96$ when $r = 8$, find k, the constant of proportionality, and find A when $r = 5$.

Solution Because A varies directly as r^2, the equation for the variation is $A = kr^2$.

Condition 1 $\begin{cases} A_1 = 200.96 \\ r_1 = 8 \end{cases}$ Condition 2 $\begin{cases} A_2 = ? \\ r_2 = 5 \end{cases}$

From condition 1,

$$A_1 = kr_1^2$$

$$200.96 = k(8^2)$$

$$k = \frac{200.96}{64} = 3.14$$

The constant of proportionality is 3.14

From condition 2, $A_2 = kr_2^2 = 3.14(5^2) = 78.5$

Alternate method Using a proportion:

$$\frac{A_1}{r_1^2} = \frac{A_2}{r_2^2}$$

$$\frac{200.96}{8^2} = \frac{A_2}{5^2}$$

$$A_2 = \frac{200.96}{8^2}(5^2) = 78.5$$

The area is 78.5 when the radius is 5. ■

EXERCISES 8.9A

Set I

1. Given: y varies directly with x. If $y = 12$ when $x = 3$, find k, the constant of proportionality, and find y when $x = 5$.

2. Given: y varies directly with x. If $y = 10$ when $x = 4$, find k, the constant of proportionality, and find y when $x = -6$.

For Exercises 3 and 4, assume that the number of miles driven is proportional to the number of gallons of gasoline used.

3. Leon knows he can drive his motorhome 161 mi on 23 gal of gasoline. Find k, the constant of proportionality. If his tanks are full and together hold 50 gal, how far can he expect to drive before stopping for more gasoline?

4. Nick knows he can drive his motorhome 85 mi on 17 gal of gasoline. Find k, the constant of proportionality. If his tanks are full and together hold 75 gal, how far can he expect to drive before stopping for more gasoline?

5. A 5-ft woman casts a 3-ft shadow when a tree casts a 27-ft shadow. How tall is the tree?

6. A 6-ft man casts a 4-ft shadow when a tree casts a 22-ft shadow. How tall is the tree?

7. The circumference (C) of a circle varies directly with the radius (r). If $C = 47.1$ when $r = 7.5$, find k, and find C when $r = 4.5$.

8. The pressure (P) in water varies directly with the depth (d). If $P = 4.33$ when $d = 10$, find k, and find P when $d = 18$.

9. The area (A) of a circle varies directly with the square of the radius (r). If $A = 28.26$ when $r = 3$, find k, and find A when $r = 6$.

10. The surface area (S) of a sphere varies directly with the square of the radius (r). If $S = 50.24$ when $r = 2$, find k, and find S when $r = 4$.

11. The amount of sediment a stream will carry is directly proportional to the sixth power of its speed. If a stream carries 1 unit of sediment when the speed of the current is 2 mph, how many units of sediment will it carry when the current is 4 mph?

12. The salary John earns is directly proportional to the number of hours he works. One week he worked 18 hours, and his salary was \$153.54. The next week he worked 25 hours. How much did he earn the second week?

Set II

1. Given: y varies directly with x. If $y = 3$ when $x = -6$, find k, the constant of proportionality, and find y when $x = 8$.

2. Given: y varies directly as x. If $y = 4$ when $x = -2$, find k, the constant of proportionality, and find y when $x = 7$.

For Exercises 3 and 4, assume that the number of miles driven is proportional to the number of gallons of gasoline used.

3. Ted knows he can drive his motorhome 117 mi on 13 gal of gasoline. Find k, the constant of proportionality. If his tank is full and holds 50 gal, how far can he expect to drive before stopping for more gasoline?

4. Jo knows she can drive her car 494 mi on 13 gal of gasoline. Find k, the constant of proportionality. If her tank is full and holds 15 gal, how far can she expect to drive before stopping for more gasoline?

5. A 6-ft man casts a 4-ft shadow when a building casts a 24-ft shadow. How tall is the building?

6. Ruth drives 1,008 miles in $3\frac{1}{2}$ days. At this rate, how far should she be able to drive in 5 days? Assume that the number of miles driven varies directly with the number of hours.

7. The circumference (C) of a circle varies directly with the diameter (D). If $C = 9.42$ when $D = 3$, find k, and find C when $D = 15$.

8. The pressure (P) in water varies directly with the depth (d). If $P = 8.66$ when $d = 20$, find k, and find P when $d = 50$.

9. Given: y varies directly as the square of x. If $y = 20$ when $x = 10$, find k, and find y when $x = 15$.

10. The area (A) of a circle varies directly with the square of the radius (r). If $A = 13.8474$ when $r = 2.1$, find k, and find A when $r = 1.2$.

11. The distance d in miles that a person can see to the horizon from a point h ft above the surface of the earth varies approximately as the square root of the height h. If, for a height of 600 ft, the horizon is 30 miles distant, how far is the horizon from a point that is 1,174 ft high?

12. Given: y is directly proportional to the cube of x. If $y = 72$ when $x = 6$, find k, and find y when $x = -3$.

8.9B Inverse Variation

If a function can be represented by an equation of the form $y = \frac{k}{x^n}$, where k is some constant, $x \neq 0$, and n is some positive number, then that function is called an **inverse variation**. The constant k is called the **constant of proportionality**. In an inverse variation, as one of the variables increases, the other decreases.

Two examples of inverse variation follow:

If the amount of work to be done is held constant, then the amount of time required to do that work is inversely proportional to the rate of speed of the worker or machine.

If the area of a rectangle is held constant, then the length of the rectangle varies inversely as its width.

INVERSE VARIATION

The following statements all translate into the equation

Constant of proportionality (k)

$$y = \frac{k}{x^n}, \text{ where } n > 0,\ x \neq 0:$$

y varies inversely with x^n.

y varies inversely as x^n.

y is inversely proportional to x^n.

Example 5 The amount of time necessary to complete a job is inversely proportional to the rate of speed of the worker. Ruth types at the rate of $5\frac{1}{2}$ pages per hour. If she spends 8 hours on a certain job, find k, the constant of proportionality. How long would it take Mike, who types at the rate of $3\frac{2}{3}$ pages per hour, to do the same job?

Solution Because *time* is inversely proportional to *rate*, the equation of the variation is $t = \dfrac{k}{r}$. Let $t_2 =$ Mike's time.

$$\text{Condition 1}\ \begin{cases} r_1 = 5\frac{1}{2} \\ t_1 = 8 \end{cases} \qquad \text{Condition 2}\ \begin{cases} r_2 = 3\frac{2}{3} \\ t_2 = ? \end{cases}$$

From condition 1,

$$t_1 = \frac{k}{r_1}$$

$$8 = \frac{k}{5.5}$$

$$k = 8(5.5) = 44$$

From condition 2,

$$t_2 = \frac{k}{r_2} = \frac{44}{3\frac{2}{3}} = 12$$

It would take Mike 12 hr to do the job. ■

An alternate way to solve inverse variation problems is based on the following facts:

$$\text{Condition 1}\ \begin{Bmatrix} \text{If} & y = y_1 \\ \text{when} & x = x_1 \end{Bmatrix} \text{ then } y_1 = \frac{k}{x_1^n}$$

If we solve this equation for k, we have $k = x_1^n y_1$.

$$\text{Condition 2}\ \begin{Bmatrix} \text{If} & y = y_2 \\ \text{when} & x = x_2 \end{Bmatrix} \text{ then } y_2 = \frac{k}{x_2^n}$$

If we solve this equation for k, we have $k = x_2^n y_2$. Then, since $x_1^n y_1$ and $x_2^n y_2$ both equal k, they must equal each other:

$$x_1^n y_1 = x_2^n y_2$$

We can, if we wish, use the equation $x_1^n y_1 = x_2^n y_2$ in solving problems dealing with inverse proportions. This method will be shown in the examples in this section as the alternate method.

Example 6 Under certain conditions, the pressure (P) of a gas varies inversely with the volume (V). If $P = 30$ when $V = 500$, find k, the constant of proportionality, and find P when $V = 200$.

Solution

$$\text{Condition 1}\ \begin{cases} P_1 = 30 \\ V_1 = 500 \end{cases} \qquad \text{Condition 2}\ \begin{cases} P_2 = ? \\ V_2 = 200 \end{cases}$$

Since P varies *inversely* with V, our equation is $P = \dfrac{k}{V}$. From condition 1, we have

$$P_1 = \frac{k}{V_1}$$

$$30 = \frac{k}{500}$$

$$k = 30(500) = 15{,}000$$

The constant of proportionality is 15,000

We can then find P when $V = 200$:

$$P_2 = \frac{k}{V_2} = \frac{15{,}000}{200} = 75$$

Alternate method The equation $P_1V_1 = P_2V_2$ can be used:

$$P_1V_1 = P_2V_2$$
$$30(500) = P_2(200)$$
$$P_2 = \frac{30(500)}{200} = 75$$

The pressure is 75 when the volume is 200. ■

Example 7 Given: y varies inversely with the square of x. If $y = -3$ when $x = 4$, find k, and find y when $x = -6$.

Solution Condition 1 $\begin{cases} x_1 = 4 \\ y_1 = -3 \end{cases}$ Condition 2 $\begin{cases} x_2 = -6 \\ y_2 = ? \end{cases}$

Since y varies inversely with x^2, the equation of the variation is $y = \frac{k}{x^2}$. From condition 1, we have

$$y_1 = \frac{k}{x_1^2}$$
$$-3 = \frac{k}{4^2}$$
$$k = -3(4^2) = -48$$

From condition 2, $y_2 = \frac{k}{x_2^2} = \frac{-48}{(-6)^2} = \frac{-48}{36} = -\frac{4}{3}$

Alternate method We use the equation $x_1^2y_1 = x_2^2y_2$. Therefore,

$$x_1^2y_1 = x_2^2y_2$$
$$4^2(-3) = (-6)^2y_2$$
$$y_2 = \frac{4^2(-3)}{(-6)^2} = \frac{-48}{36} = -\frac{4}{3}$$ ■

EXERCISES 8.9B

Set I

1. Given: y varies inversely with x. If $y = 7$ when $x = 2$, find k, the constant of proportionality, and find y when $x = -7$.
2. Given: z varies inversely with w. If $z = 4$ when $w = -12$, find k, and find z when $w = 8$.
3. Given: Pressure (P) varies inversely with volume (V). If $P = 18$ when $V = 15$, find k, and find P when $V = 10$.
4. Given: s varies inversely with t. If $s = 8$ when $t = 5$, find k, and find s when $t = 4$.

For Exercises 5 and 6, assume that time is inversely proportional to rate.

5. Machine A works for 3 hr to complete one order. It makes a certain part at the rate of 375 parts per hour. Find k, and find how long it would take Machine B, which makes the parts at the rate of 225 parts per hour, to complete the same order.

6. Machine C works for 9 hr to complete one order. It makes a certain part at the rate of 275 parts per hour. Find k, and find how long it would take Machine D, which makes the parts at the rate of 330 parts per hour, to complete the same order.

7. Given: y varies inversely with the square of x. If $y = 9$ when $x = 4$, find k, and find y when $x = 3$.

8. Given: C varies inversely with the square of v. If $C = 8$ when $v = 3$, find k, and find C when $v = 6$.

9. Given: F varies inversely with the square of d. If $F = 3$ when $d = 4$, find k, and find F when $d = 2$.

10. The intensity (I) of light received from a light source varies inversely with the square of the distance (d) from the source. If the light intensity is 15 foot-candles at a distance of 10 ft from the light source, what is the light intensity at a distance of 15 ft?

11. a. Write the equation for the following statement: When the area of a rectangle is held constant, the length of the rectangle varies inversely as its width.

    b. If the area of a rectangle is held constant and if the length is 15 cm when the width is 3 cm, find k, the constant of proportionality, and find the length when the width is 5 cm.

12. a. Write the equation for the following statement: When the area of a parallelogram is held constant, the altitude of the parallelogram varies inversely as the length of its base.

    b. If the area of a parallelogram is held constant and if the altitude is 12 cm when the base is 5 cm, find k, the constant of proportionality, and find the altitude when the base is 10 cm.

Set II

1. Given: y varies inversely with x. If $y = 5$ when $x = 4$, find k, the constant of proportionality, and find y when $x = 2$.

2. Given: u varies inversely with v. If $u = 15$ when $v = 7$, find k, and find u when $v = 3$.

3. Given: V varies inversely with P. If $V = 340$ when $P = 30$, find k, and find V when $P = 85$.

4. Given: w varies inversely with x. If $w = 38$ when $x = 8$, find k, and find w when $x = 19$.

For Exercises 5 and 6, assume that time is inversely proportional to rate.

5. Machine A works for 4 hr to complete one order. It makes a certain part at the rate of 130 parts per hour. Find k, and find how long it would take Machine B, which makes the parts at the rate of 120 parts per hour, to complete the same order.

6. Machine A, which prints negatives at the rate of 290 ft per minute, works for 3.5 hr to complete one job. Find k, the constant of proportionality. How long would it take Machine B, which prints negatives at the rate of 350 ft per minute, to do the same job?

7. Given: L varies inversely with the square of r. If $L = 9$ when $r = 5$, find k, and find L when $r = 3$.

8. Given: y varies inversely with the cube of x. If $y = 960$ when $x = 4$, find k, and find y when $x = 2$.

9. Given: x varies inversely with the cube of y. If $x = \frac{1}{4}$ when $y = 2$, find k, and find x when $y = 4$.

10. The gravitational attraction (F) between two bodies varies inversely with the square of the distance (d) separating them. If the attraction measures 36 when the distance is 4 cm, find the attraction when the distance is 80 cm.

11. a. Write the equation for the following statement: When the area of a rectangle is held constant, the width of the rectangle varies inversely as its length.

    b. If the area of a rectangle is held constant and if the width is 15 m when the length is 16 m, find k, the constant of proportionality, and find the width when the length is 12 m.

12. a. Write the equation for the following statement: When the area of a triangle is held constant, the altitude of the triangle varies inversely as the length of its base.

    b. If the area of a triangle is held constant and if the altitude is 18 in. when the base is 12 in., find k, the constant of proportionality, and find the altitude when the base is 36 in.

8.9C Joint Variation and Combined Variation

While direct and inverse variation are functions of *one* variable, variations also exist that are functions of *several* variables. We will now discuss variations of this kind.

Joint Variation

Joint variation is a type of variation that relates one variable to the *product* of two (or more) other variables.

> JOINT VARIATION
>
> The following statements translate into the equation
>
> $$z = kx^n y^m, \text{ where } n > 0 \text{ and } m > 0:$$
>
> (k: Constant of proportionality)
>
> z varies jointly with x^n and y^m.
>
> z is jointly proportional to x^n and y^m.

Example 8 Examples of joint variation:
Given: z varies jointly with x and y according to the formula $z = 2xy$:

a. If $x = 3$ and $y = 5$,
then $z = 2(3)(5) = 30$

b. If $x = 4$ and $y = 5$,
then $z = 2(4)(5) = 40$

c. If $x = 3$ and $y = 4$,
then $z = 2(3)(4) = 24$

The value of z depends on the product xy

Note that as x *increased* from 3 to 4 while y was held constant at 5, z *increased* from 30 to 40. Also, as y *decreased* from 5 to 4 while x was held constant at 3, z *decreased* from 30 to 24. Also, when the product xy *increased* from $(3)(5) = 15$ to $(4)(5) = 20$, z *increased* from 30 to 40. ■

Example 9 If the principal is held constant, then the amount of interest (I) earned in time (t) varies jointly with t and the interest rate (r). If a certain amount of money earns \$440 in 2 years when it is invested at 5.5% interest (per year), find k, the constant of proportionality, and find I when the money is invested at 5% interest for 3 years.

Solution

$$\text{Condition } \begin{cases} I_1 = 440 \\ r_1 = 0.055 \\ t_1 = 2 \end{cases} \qquad \text{Condition 2 } \begin{cases} I_2 = ? \\ r_2 = 0.05 \\ t_2 = 3 \end{cases}$$

Because I varies jointly with t and r, the equation of the variation is $I = krt$.

From condition 1

$$I_1 = kr_1t_1$$
$$440 = k(0.055)(2)$$
$$k = \frac{440}{0.055(2)} = 4{,}000$$

From condition 2, $\quad I_2 = kr_2t_2 = 4{,}000(0.05)(3) = 600$

Therefore, the amount of money earned in 3 years at 5% is \$600. In this example, k is the principal, \$4,000. ■

Example 10 The heat (H) generated by an electric heater is jointly proportional to R and the square of I. If $H = 1{,}200$ when $I = 8$ and $R = 15$, find k, and find H when $I = 5.5$ and $R = 20$.

Solution

$$\text{Condition 1 } \begin{cases} I_1 = 8 \\ R_1 = 15 \\ H_1 = 1{,}200 \end{cases} \qquad \text{Condition 2 } \begin{cases} I_2 = 5.5 \\ R_2 = 20 \\ H_2 = ? \end{cases}$$

Since H is jointly proportional to R and I^2, our formula is $H = kRI^2$.

From condition 1,

$$H_1 = kR_1I_1^2$$
$$1{,}200 = k(15)(8^2)$$
$$k = \frac{1{,}200}{(15)(8^2)} = 1.25$$

From condition 2, $\quad H_2 = kR_2I_2^2 = 1.25(20)(5.5)^2 = 756.25$ ■

Combined Variation

A variable can be *directly* or *jointly* proportional to some variables at the same time that it is *inversely* proportional to others. The equations resulting from such circumstances are called **combined variations**.

Example 11 Examples of combined variation, with k the constant of proportionality in each equation:

a. $w = \dfrac{kx}{y}$

The variable w is directly proportional to x and inversely proportional to y.

b. $R = \dfrac{kL}{d^2}$

The variable R is directly proportional to L and inversely proportional to d^2.

c. $F = \dfrac{kMm}{d^2}$

The variable F is jointly proportional to M and m and inversely proportional to d^2. ■

Example 12 The strength (S) of a rectangular beam is jointly proportional to b and the square of d, and inversely proportional to L, where b is the breadth, d is the depth, and L is the length of the beam. If $S = 2{,}000$ when $b = 2$, $d = 10$, and $L = 15$, find k, and find S when $b = 4$, $d = 8$, and $L = 12$.

Solution

$$\text{Condition 1} \begin{cases} b_1 = 2 \\ d_1 = 10 \\ L_1 = 15 \\ S_1 = 2{,}000 \end{cases} \qquad \text{Condition 2} \begin{cases} b_2 = 4 \\ d_2 = 8 \\ L_2 = 12 \\ S_2 = ? \end{cases}$$

Since we have a combined variation, the formula is $S = \dfrac{kbd^2}{L}$.

From condition 1,

$$S_1 = \frac{kb_1d_1^2}{L_1}$$

$$2{,}000 = \frac{k(2)(10^2)}{15}$$

$$k = \frac{2{,}000(15)}{(2)(10^2)} = 150$$

When $b = 4$, $d = 8$, and $L = 12$,

$$S_2 = \frac{kb_2d_2^2}{L_2} = \frac{150(4)(8^2)}{12} = 3{,}200$$ ■

EXERCISES 8.9C

Set I

1. Given: z varies jointly with x and y. If $z = -36$ when $x = -3$ and $y = 2$, find k, the constant of proportionality, and find z when $x = 4$ and $y = 3$.
2. Given: A varies jointly with L and W. If $A = 120$ when $L = 6$ and $W = 5$, find k, and find A when $L = 7$ and $W = 3$.
3. The simple interest I earned in a given time t varies jointly as the principal P and the interest rate r. If $I = 115.50$ when $P = 880$ and $r = 0.0875$, find k, and find I when $P = 760$ and $r = 0.0925$.
4. If I varies jointly with P and r, and $I = 240.50$ when $P = 860$ and $r = 0.0775$, find k, and find I when $P = 1{,}250$ and $r = 0.0950$. (Round off I to two decimal places.)
5. The wind force F on a vertical surface varies jointly as the area A of the surface

and as the square of the wind velocity V. When the wind is blowing 20 mph, the force on 1 sq. ft of surface is 1.8 lb. Find the force exerted on a surface of 2 sq. ft when the wind velocity is 60 mph.

6. The pressure P in a liquid varies jointly with the depth h and density D of the liquid. If $P = 204$ when $h = 163.2$ and $D = 1.25$, find P when $h = 182.5$ and $D = 13.56$.

7. Given: z is directly proportional to x and inversely proportional to y. If $z = 12$ when $x = 6$ and $y = 2$, find k, and find z when $x = -8$ and $y = -4$.

8. The electrical resistance R of a wire is directly proportional to L and inversely proportional to the square of d. If $R = 2$ when $L = 8$ and $d = 4$, find k, and find R when $L = 10$ and $d = 5$.

9. The elongation e of a wire is jointly proportional to P and L and inversely proportional to A. If $e = 3$ when $L = 45$, $P = 2.4$, and $A = 0.9$, find k, and find e when $L = 40$, $P = 1.5$, and $A = 0.75$.

10. When a horizontal beam with rectangular cross-section is supported at both ends, its strength S varies jointly as the breadth b and the square of the depth d and inversely as the length L. A 2- by 4-in. beam 8 ft long resting on the 2-in. side will safely support 600 lb. What is the safe load when the beam is resting on the 4-in. side?

Set II

1. Given: V varies jointly with L and H. If $V = 144$ when $L = 3$ and $H = 8$, find k, the constant of proportionality, and find V when $L = 2$ and $H = 5$.

2. Given: C varies jointly with L and W. If $C = 7{,}500$ when $L = 25$ and $W = 20$, find k, and find C when $L = 18$ and $W = 23$.

3. The simple interest I earned in a given time t varies jointly as the principal P and the interest rate r. If $I = 255$ when $P = 1{,}250$ and $r = 0.0975$, find k, and find I when $P = 1{,}500$ and $r = 0.0825$.

4. On a certain truck line, it costs $56.80 to send 5 tons of goods 8 mi. How much will it cost to send 14 tons a distance of 15 mi if the cost varies jointly with the weight and the distance?

5. Given: H varies jointly with R and the square of I. If $H = 1{,}458$ when $R = 24$ and $I = 4.5$, find k, and find H when $R = 22$ and $I = 5.5$.

6. Given: y varies jointly with x and the square of z. If $y = 72$ when $x = 2$ and $z = 3$, find k, and find y when $x = 5$ and $z = 2$.

7. Given: z is directly proportional to x and inversely proportional to y. If $z = 2$ when $x = 4$ and $y = 10$, find k, and find z when $x = 3$ and $y = 6$.

8. Given: P is directly proportional to T and inversely proportional to V. If $P = 10$ when $T = 250$ and $V = 400$, find k, and find P when $T = 280$ and $V = 350$.

9. Given: W is jointly proportional to x and y and inversely proportional to the square of z. If $W = 1{,}200$ when $x = 8$, $y = 6$, and $z = 2$, find k, and find W when $x = 5.6$, $y = 3.8$, and $z = 1.5$.

10. The gravitational attraction F between two masses is jointly proportional to M and m and inversely proportional to the square of d. If $F = 1{,}000$ when $d = 100$, $m = 50$, and $M = 2{,}000$, find k, and find F when $d = 66$, $m = 125$, and $M = 1{,}450$.

8.10 Review: 8.6–8.9

Functions 8.6

A **function** is a relation such that no two of its ordered pairs have the *same first coordinate* and *different second coordinates.* This means that no vertical line can meet the graph of a function in more than one point.

Linear function: $f(x) = mx + b$; graph is a *straight line*

Quadratic function: $f(x) = ax^2 + bx + c$; graph is a *parabola*

Cubic function: $f(x) = ax^3 + bx^2 + cx + d$

The value of a function: $f(a)$, read "f of a," is the value of $f(x)$ when x is replaced by a.

The Graphs of $y = f(x) + h$ and $y = f(x + h)$ 8.6C

The graph of $y = f(x) + h$ can be found by shifting the graph of $y = f(x)$ upward $|h|$ units if h is positive and downward $|h|$ units if h is negative.

The graph of $y = f(x + h)$ can be found by shifting the graph of $y = f(x)$ to the left $|h|$ units if h is positive and to the right $|h|$ units if h is negative.

To Find the Inverse Relation $\mathscr{R}^{-1}$ of a Relation $\mathscr{R}$ 8.7

1. If the relation is given as a set of ordered pairs, $\mathscr{R}^{-1}$ is the set of ordered pairs found by interchanging the first and second coordinates of each ordered pair in $\mathscr{R}$.
2. If the relation is given by an equation, $\mathscr{R}^{-1}$ is found by interchanging x and y in the equation and then solving the new equation for y.

The graph of the inverse relation $\mathscr{R}^{-1}$ is the mirror image of the relation $\mathscr{R}$, with the mirror being the line $y = x$.

The Inverse of a Function 8.7

If a function is invertible, that is, if for each y-value there is only one x-value, then $f^{-1}(x)$ is found by substituting y for $f(x)$ (if necessary), interchanging x and y, and then solving the new equation for y.

The graph of the inverse function $y = f^{-1}(x)$ is the mirror image of the graph of the function $y = f(x)$, with the mirror being the line $y = x$.

To Graph a Curve (Polynomial Functions Other Than Linear) 8.8

1. Use the equation to make a table of values.
2. Plot the points from the table of values.
3. Draw a smooth curve through the points, taking them in order from left to right.

Variation 8.9

The equation for "y varies directly as x^n" is $y = kx^n$.

The equation for "y varies inversely as x^n" is $y = \dfrac{k}{x^n}$.

The equation for "y varies jointly with s^n and t^m" is $y = ks^n t^m$.

The equation for "y varies directly as x^n and inversely as z^m" is $y = \dfrac{kx^n}{z^m}$.

Proportion 8.9

An equation that can be expressed in the form $\dfrac{a}{b} = \dfrac{c}{d}$ is called a **proportion**.

Review Exercises 8.10 Set I

1. Which of the following are graphs of functions?

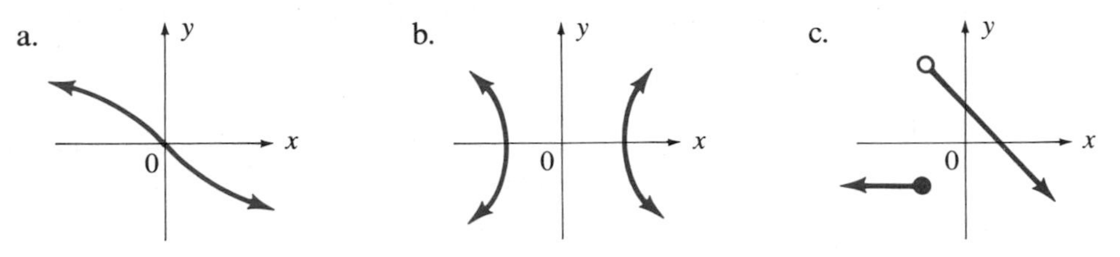

2. Which of the following are graphs of functions?

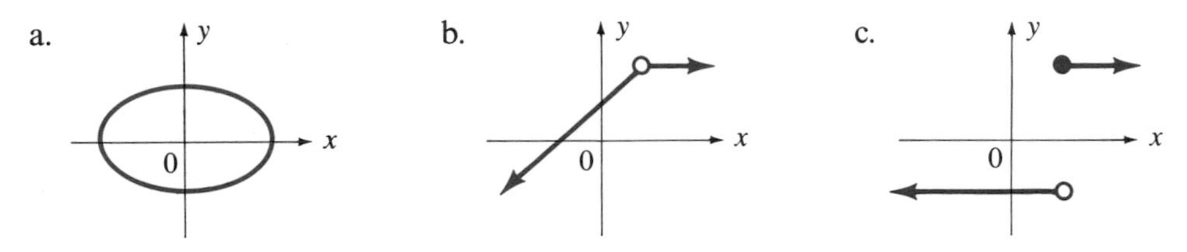

3. Given $f(x) = 3x^2 - 5x + 4$, find the following.

a. $f(2)$ b. $f(0)$ c. $f(x - 1)$

4. If $F(x) = x^3 - 2x^2 + 6$ and $g(x) = 2x^2 - 7x$, find $\frac{1}{3}F(2) - 6g(4)$.

In Exercises 5–7, graph each function.

5. $y = f(x) = |x| + 1$ **6.** $y = f(x) = |x| - 1$ **7.** $y = f(x) = |x + 1|$

8. Given $\mathscr{R} = \{(-1, -2), (-4, 4), (5, 0), (4, -2), (-5, -3)\}$, find $\mathscr{R}^{-1}$, $D_{\mathscr{R}^{-1}}$, and $R_{\mathscr{R}^{-1}}$, and graph $\mathscr{R}$ and $\mathscr{R}^{-1}$. Is $\mathscr{R}$ a function? Is $\mathscr{R}^{-1}$ a function?

In Exercises 9 and 10, find the inverse function for each function. Graph each function and its inverse. Find the domain and range for $y = f(x)$ and $y = f^{-1}(x)$.

9. $y = f(x) = \dfrac{x + 6}{3}$ **10.** $y = f(x) = -\dfrac{5}{2}x + 1$

In Exercises 11 and 12, find the inverse function for each function.

11. $y = f(x) = \dfrac{9}{2 - 7x}$ **12.** $y = f(x) = \dfrac{11}{2(4x - 5)}$

13. Use x-values of $-4, -2, -1, 0, 1, 2$, and 4 to make a table of values for the equation $y = \dfrac{x^2}{2}$. Graph the points and draw a smooth curve through them.

14. Use x-values of $-4, -2, -1, 0, 1, 2$, and 4 to make a table of values for the equation $y = \dfrac{x^2}{2} + 1$. Graph the points and draw a smooth curve through them.

15. Use integer values of x from -3 to $+3$ to make a table of values for the equation $y = x^3 - 3x$. Graph the points and draw a smooth curve through them.

16. The variable c varies jointly with p and q and inversely with the square of t. If $c = 30$ when $p = 3$, $q = 5$, and $t = 4$, find k, and find c when $p = 7$, $q = 13$, and $t = 2$.

17. The surface area, S, of a sphere varies directly with the square of the radius, r. If $S = 615.44$ when $r = 7$, find k, and find S when $r = 10$.

18. (For this exercise, assume that time is inversely proportional to rate.) Machine A works for 4 hr to complete one order, making a certain part at the rate of 125 parts per hour. Find k, and find how long it would take Machine B, which makes the parts at the rate of 75 parts per hour, to complete the same order.

Review Exercises 8.10 Set II

NAME

ANSWERS

1a. ______

b. ______

c. ______

2a. ______

b. ______

c. ______

3. ______

4. ______

5. Use the grid.

6. Use the grid.

1. Which of the following are graphs of functions?

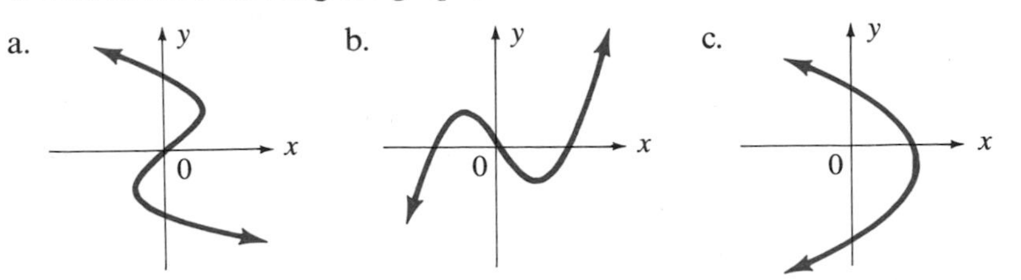

2. Given $f(x) = 5x^2 - 12x + 16$, find the following.

a. $f(-1)$ b. $f(4)$ c. $f(2a - 1)$

3. If $H(x) = 5x^3 + 7$ and $g(y) = 4y - 2y^2$, find $\frac{5}{11}H(-2) - \frac{1}{6}g(5)$.

4. If $F(x) = 3x - x^2$ and $G(w) = 4 - w^2$, find $\dfrac{F(1 - c)}{G(c)}$.

In Exercises 5 and 6, graph each function.

5. $y = f(x) = |x| - 4$

6. $y = f(x) = |x - 4|$

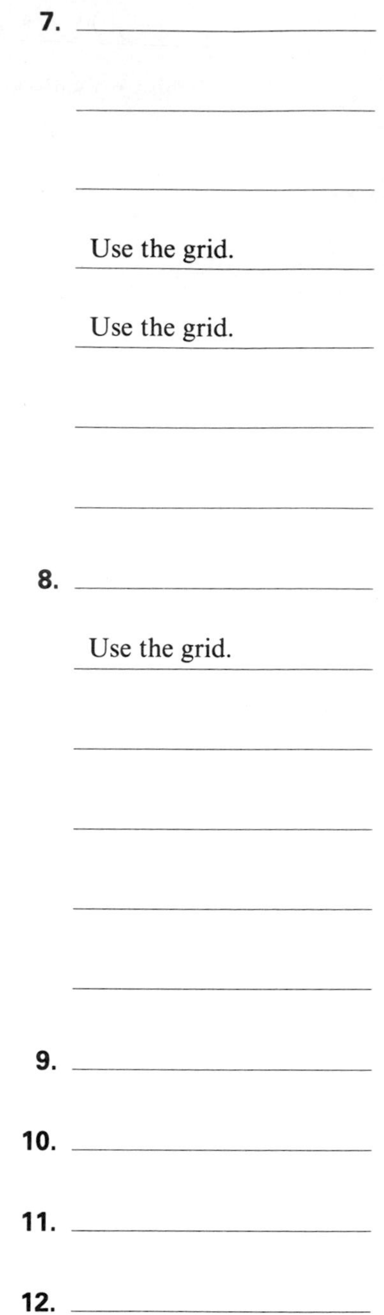

7. Given: $\mathscr{R} = \{(-2, 4), (-1, 2), (3, -1), (0, 4)\}$. Find $\mathscr{R}^{-1}$, $D_{\mathscr{R}^{-1}}$, and $R_{\mathscr{R}^{-1}}$ and graph $\mathscr{R}$ and $\mathscr{R}^{-1}$. Is $\mathscr{R}$ a function? Is $\mathscr{R}^{-1}$ a function?

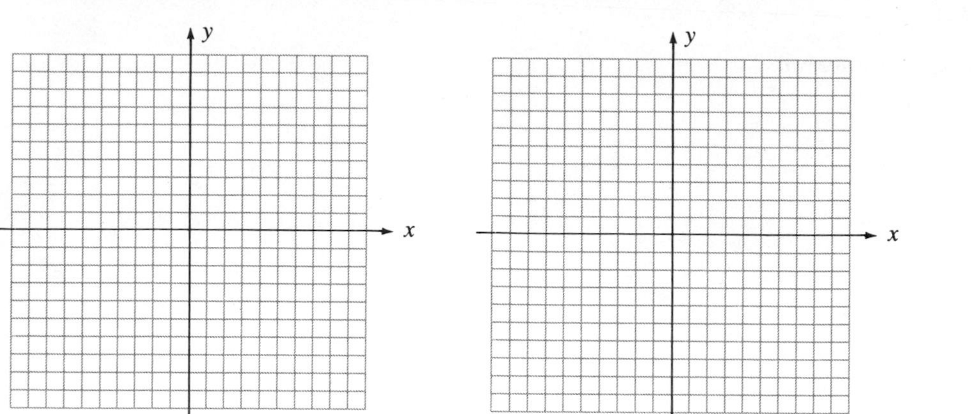

8. For $y = f(x) = \frac{4}{7}x - 2$, find the inverse function, $f^{-1}(x)$. Graph the function and its inverse on the same axes. Find the domain and range for $y = f(x)$ and $y = f^{-1}(x)$.

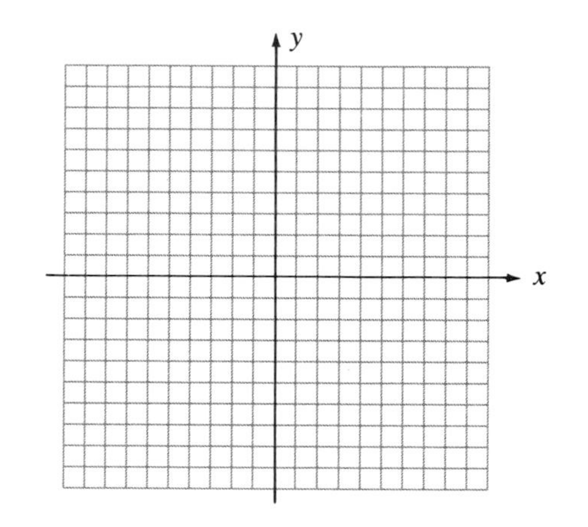

For Exercises 9–12, find the inverse function, $y = f^{-1}(x)$.

9. $y = f(x) = x - 2$

10. $y = f(x) = \dfrac{x - 2}{5}$

11. $y = f(x) = \dfrac{1}{x}$

12. $y = f(x) = \dfrac{13}{3(2x + 3)}$

13. Use integer values of x from -4 to $+2$ to make a table of values for the equation $y = \frac{(x + 1)^2}{2}$. Graph the points and draw a smooth curve through them.

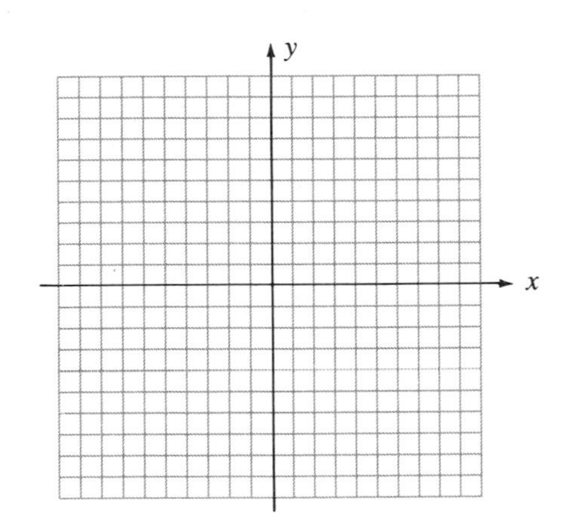

14. Use integer values of x from -3 to $+3$ to make a table of values for the equation $y = 5x - x^3$. Graph the points and draw a smooth curve through them.

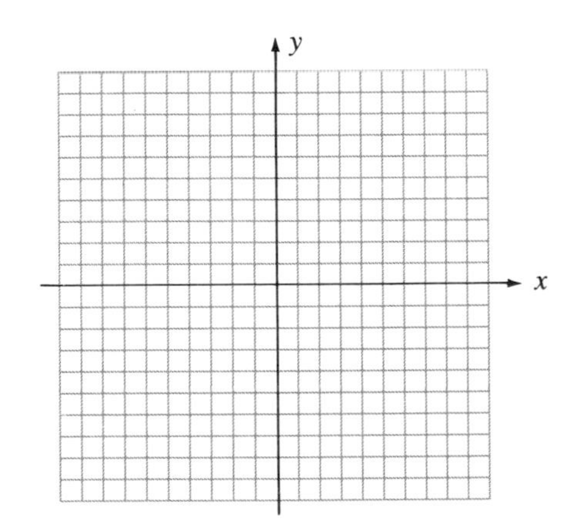

15. The variable d varies directly with e and inversely with the cube of f. If $d = 60$ when $e = 32$ and $f = 2$, find k, and find d when $e = 9$ and $f = 3$.

ANSWERS

13. Use the grid.

14. Use the grid.

15. ____________

____

16. The variable y varies directly with the square of x and inversely with the cube of z. If $y = \frac{4}{9}$ when $x = 2$ and $z = 3$, find k.

17. The pressure, P, in water varies directly with the depth, d. If $P = 4.33$ when $d = 10$, find k, and find P when $d = 25$.

18. a. Write the equation for the following statement: When the area of a rectangle is held constant, the width of the rectangle varies inversely as its length. Use the equation from (a) in (b) and (c).

b. If the width is 8 m when the length is 12 m, find k.

c. Find the width when the length is 16 m.

ANSWERS

16. ______________

17. ______________

____

18a. ______________

b. ______________

c. ______________

Chapter 8 Diagnostic Test

The purpose of this test is to see how well you understand functions and graphing. We recommend that you work this diagnostic test *before* your instructor tests you on this chapter. Allow yourself about 50 minutes.

Complete solutions for all the problems on this test, together with section references, are given in the answer section at the end of the book. For the problems you do incorrectly, study the sections cited.

1. a. Draw the triangle with vertices at $A(-4, 2)$, $B(1, -3)$, and $C(5, 3)$.

b. Find the length of the side AB.

c. Find the slope of the line through B and C.

d. Write the general form of the equation of the line through A and B.

e. Find the x-intercept of the line through A and B.

f. Find the y-intercept of the line through A and B.

2. Find the general form of the equation of the line that has a slope of $\frac{6}{5}$ and a y-intercept of -4.

3. Graph the relation $x - 2y = 6$.

4. Use integer values of x from -2 to 3 and graph the relation $y = 1 + x - x^2$.

5. Graph the inequality $4x - 3y \leq -12$.

6. Given the relation $\{(-4, -5), (2, 4), (4, -2), (-2, 3), (2, -1)\}$:

a. Find its domain.

b. Find its range.

c. Draw the graph of the relation.

d. Is this relation a function?

7. Given $f(x) = 3x^2 - 5$. Find the following.

a. $f(-2)$ b. $f(4)$ c. $\dfrac{f(4) - f(-2)}{6}$

8. Given $\mathscr{R} = \{(1, 4), (-5, -3), (4, -2), (-5, 2)\}$, find $\mathscr{R}^{-1}$, $D_{\mathscr{R}^{-1}}$, and $R_{\mathscr{R}^{-1}}$, and graph $\mathscr{R}$ and $\mathscr{R}^{-1}$. Is $\mathscr{R}$ a function? Is $\mathscr{R}^{-1}$ a function?

9. a. Find the inverse function for $y = f(x) = -\frac{3}{2}x + 1$. Graph $y = f(x)$ and $y = f^{-1}(x)$ on the same axes.

b. Find the inverse function for $y = f(x) = \dfrac{15}{4(2 - 5x)}$. Find the domain and range for $y = f(x)$ and $y = f^{-1}(x)$.

10. The variable w varies jointly with x and y and inversely with the square of z. If $w = 20$ when $x = 8$, $y = 6$, and $z = 12$, find k, the constant of proportionality, and find w when $x = 6$, $y = 10$, and $z = 5$.

Cumulative Review Exercises: Chapters 1–8

In Exercises 1 and 2, factor completely.

1. $2x^4 - 24x^2 + 54$

2. $8a^4b - 27ab^4$

In Exercises 3 and 4, perform the indicated operations and express your answers in simplest radical form.

3. $\sqrt[3]{-4a^4} \cdot \sqrt[3]{2a}$

4. $4\sqrt{18x^4} + \sqrt{32x^5} - \sqrt{50x^4}$

In Exercises 5–7, perform the indicated operations and write the answers in the form $a + bi$.

5. $\dfrac{3 + 2i}{2 - i}$

6. $(7 - \sqrt{-16}) - (5 - \sqrt{-36})$

7. $(3 + i)^4$

8. Solve for r: $S = \dfrac{a}{1 - r}$

9. Find the solution set for $x^{-1/3} = 3$.

10. Subtract $(2x^2 - 3x + 4)$ from the sum of $(9x^2 + x - 7)$ and $(8 - 6x - 4x^2)$.

11. If $f(x) = 3x^2 - 2x + 7$, find $f(0)$ and $f(-2)$.

12. Graph $2x - y = 4$.

13. Write the equation of the straight line that passes through the points $(-3, 4)$ and $(1, -2)$.

14. The variable y varies inversely with the cube of x. If $y = 3$ when $x = -2$, find k, and find y when $x = 3$.

In Exercises 15–18, set up each problem algebraically, solve, and check.

15. A grocer makes up a 60-lb mixture of cashews and peanuts. If the cashews cost $7.40 a pound and the peanuts cost $2.80 a pound, how many pounds of each kind of nut must be used in order for the mixture to cost $4.18 a pound?

16. Dorothy can drive from A to B at 55 mph, but in driving from B to C, she can travel only 50 mph. B is 8 mi further from A than it is from C. If it takes her the same length of time to get from A to B as it takes her to get from B to C, what is the distance from A to B?

17. The longer leg of a right triangle is 7 cm more than the shorter leg, and the hypotenuse is 9 cm longer than the shorter leg. What are the lengths of the three sides?

18. The surface area of a cube varies directly with the square of the length of one side. If the surface area is 54 cc when the length of one side is 3 cm, find k, and find the surface area when the length of a side is 5 cm.

9 Exponential and Logarithmic Functions

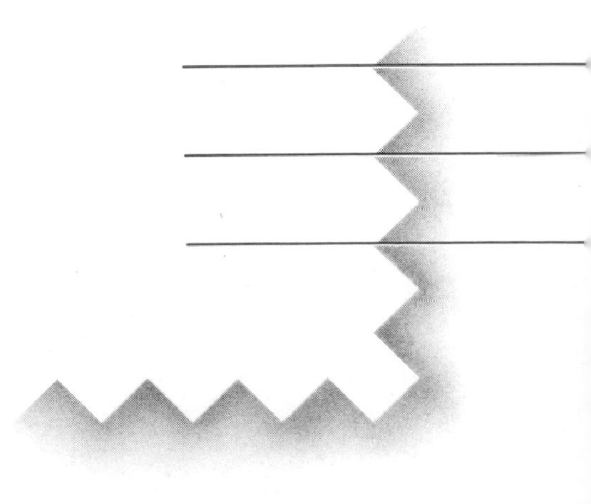

In this chapter, we discuss exponential and logarithmic functions and some of their applications. Exponential functions are very important in business and in the sciences, as we will see in Section 9.6. Logarithms have many applications in mathematics and science; for example, they are used to measure the magnitude of earthquakes and the pH factor of solutions.*

9.1 Basic Definitions

9.1A The Exponential Function and Its Graph

Exponential Functions

An **exponential function** is a function of the form $y = f(x) = b^x$, where b is any real number greater than zero except 1 and x is any real number.

The *domain* of the exponential function is the set of all real numbers. It is proved in higher mathematics courses that the graph of the exponential function is a smooth curve, that the *range* of the exponential function $y = b^x$ is the set of all real numbers greater than zero, and that there is only one x-value for each y-value. The technique of graphing an exponential function is demonstrated in Example 1.

Example 1 Graph the exponential function $y = 2^x$ and discuss its range.

Solution We calculate the values of the function for integral values of x from -2 to $+3$.

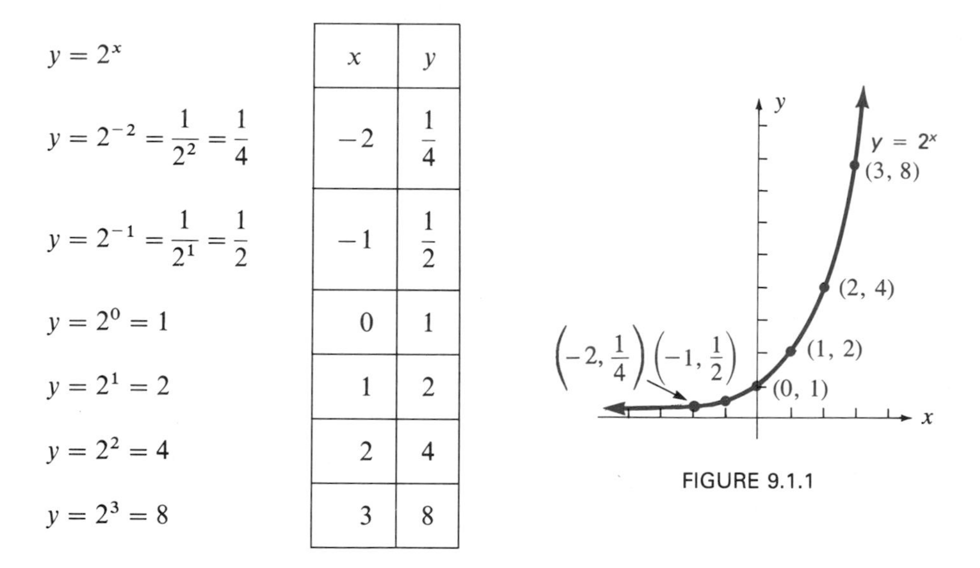

$y = 2^x$	x	y
$y = 2^{-2} = \frac{1}{2^2} = \frac{1}{4}$	-2	$\frac{1}{4}$
$y = 2^{-1} = \frac{1}{2^1} = \frac{1}{2}$	-1	$\frac{1}{2}$
$y = 2^0 = 1$	0	1
$y = 2^1 = 2$	1	2
$y = 2^2 = 4$	2	4
$y = 2^3 = 8$	3	8

FIGURE 9.1.1

Next, we graph the points listed in the table of values and draw a *smooth curve* through them, taking the points in order from left to right. The graph of $y = 2^x$ is shown in Figure 9.1.1. We see from the graph that the curve never touches the x-axis and is never below it. Thus, the range of the function $y = 2^x$ is, indeed, the set of all real numbers greater than 0. We also see that no horizontal line touches the graph in more than one point. ■

Example 2 Graph $y = \left(\frac{1}{3}\right)^x$ and discuss its range.

---

* Note to the Instructor: In this chapter, logarithms are found by using calculators. If you prefer the use of tables, you may substitute Appendix B for Section 9.4. A section on calculations using logarithms, including an exercise set, is also included in Appendix B.

Solution

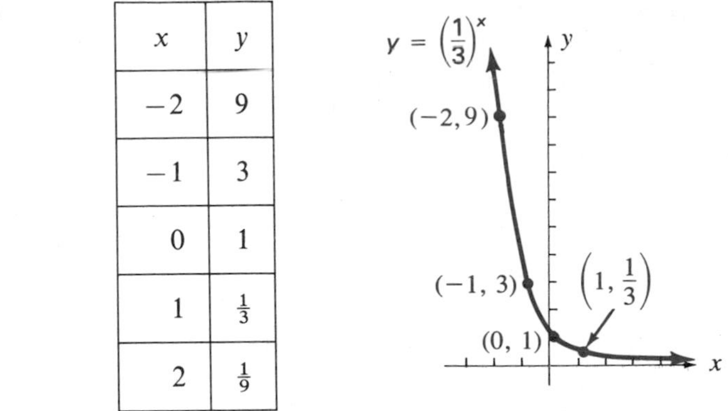

x	y
−2	9
−1	3
0	1
1	$\frac{1}{3}$
2	$\frac{1}{9}$

We see again that the graph *never* touches the x-axis and is never below it, so the *range* is the set of all real numbers greater than zero. ■

If $b > 1$, the graph of $y = f(x) = b^x$ will always look like the graph in Figure 9.1.2. The curve *rises* as we move toward the right and lies entirely above the x-axis. It passes through the point (0, 1).

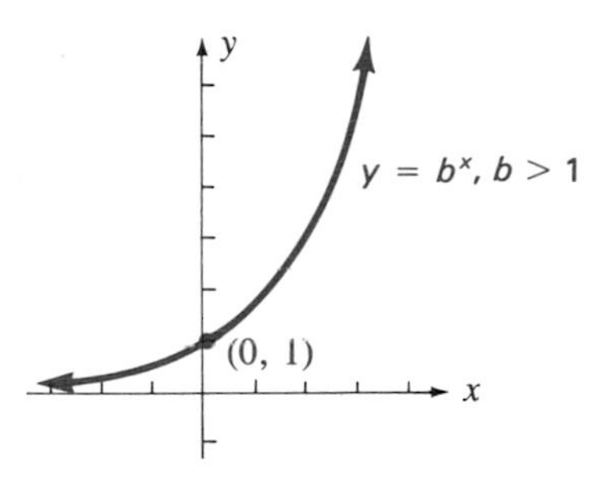

FIGURE 9.1.2

The Number *e*

An *irrational number*, designated by e, is particularly important as a base for exponential functions.* It is so important, in fact, that special tables are made up for the function $y = f(x) = e^x$ (see Appendix C). On most scientific calculators, you find the values for exponential functions with the base e by using the key labeled [ln x] or [ln] (on many calculators, that key has "e^x" written above it); for example, to find the decimal approximation for e^1, press these keys, in order: [1], [INV], [ln x]. If you do this, you will find that e^1 is approximately 2.7. (As usual, you must consult the manual for your calculator for details.) To find the decimal approximation for e^{-5}, you'll probably press [5], [+/−], [INV], [ln x], and so forth.

We state here (without proof) two other properties of exponential functions; see Examples 4, 6, 7, 8, and 9 in Section 9.2B for examples of the use of these rules.

> RULE 9.1
>
> For $b > 0$ but $b \neq 1$, if $b^x = b^y$, then $x = y$.
>
> RULE 9.2
>
> For $a > 0$ and $b > 0$, if $a^x = b^x$, then $a = b$.

* An explanation of how e is calculated is beyond the scope of this text.

EXERCISES 9.1A

Set I In Exercises 1–4, find the decimal approximation for each expression; use a calculator and round off the answers to four decimal places, or use Table I of Appendix C.

1. e^3 **2.** $e^{0.5}$ **3.** e^{-3} **4.** $e^{-0.5}$

In Exercises 5–9, graph each of the given functions.

5. $y = 4^x$ **6.** $y = 5^x$

7. $y = e^x$ (Use a calculator or Table I, Appendix C, for the y-values.)

8. $y = (\frac{1}{2})^x$ **9.** $y = (\frac{3}{2})^x$

Set II In Exercises 1–4, find the decimal approximation for each expression; use a calculator and round off the answers to four decimal places, or use Table I of Appendix C.

1. $e^{2.2}$ **2.** $e^{-4.6}$ **3.** $e^{-0.03}$ **4.** $e^{0.24}$

In Exercises 5–9, graph each of the given functions.

5. $y = 6^x$ **6.** $y = -3^x$

7. $y = -e^x$ (Use a calculator or Table I, Appendix C, for the y-values.)

8. $y = (\frac{5}{2})^x$ **9.** $y = (\frac{1}{4})^x$

9.1B The Logarithmic Function and Its Graph

Because there is only one x-value corresponding to each y-value for the function $y = f(x) = b^x$, that function must have an *inverse*. We even know what the graph of the inverse function looks like—its graph is the mirror image, with respect to the line $y = x$, of the graph of $y = b^x$ (see Figure 9.1.3). The domain of the inverse function must be the set of all real numbers greater than zero, because the domain of the inverse of a function equals the range of the original function. The range of the inverse function must be the set of all real numbers, because the range of an inverse function equals the domain of the original function.

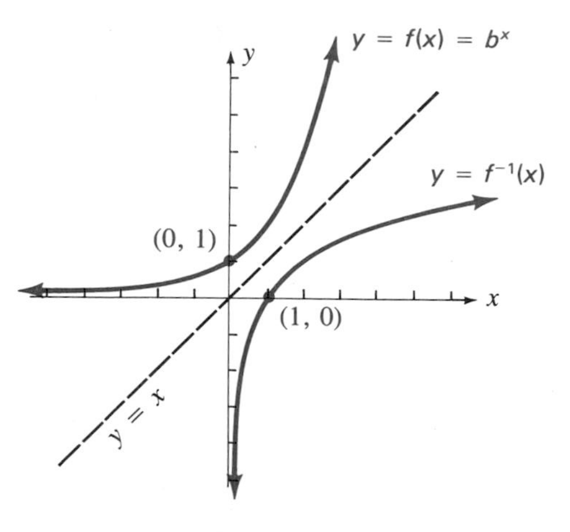

FIGURE 9.1.3

Notice that the graph of the *inverse* function never touches the y-axis and is never to the left of the y-axis.

To find the inverse of $y = f(x) = b^x$, we use the method shown in Section 8.7 for finding the inverse of a function; that is, we first interchange x and y and write $x = b^y$. Then we try to solve this new equation for y. However, we need new notation and a definition before we can solve the equation.

Logarithms

The definition of a logarithm is as follows:

> The **logarithm** of a positive* number x is the exponent y to which the base b ($b > 0$, $b \neq 1$) must be raised to give x. In symbols,
>
> $$y = \log_b x \quad \text{if and only if } b^y = x$$
>
> where x is any real number greater than zero.

NOTE $y = \log_b x$ is read "y is the logarithm of x to the base b." ☑

The Argument In the expression $\log_b x$, x is called the **argument** of the logarithmic function. The argument can be any expression, such as $(x + 1)$, $\dfrac{2x - 1}{5 - x}$, or $(x^2 - 3x + 2)$. Because the argument must *always* be positive, however, the domain is generally some subset of the set of real numbers. We discuss the argument further in Section 9.5.

Example 3 Graph the logarithmic function $y = \log_2 x$.

Solution We rewrite $y = \log_2 x$ in the form $x = 2^y$ and let y be the independent variable; we'll choose some negative values for y as well as some nonnegative ones.

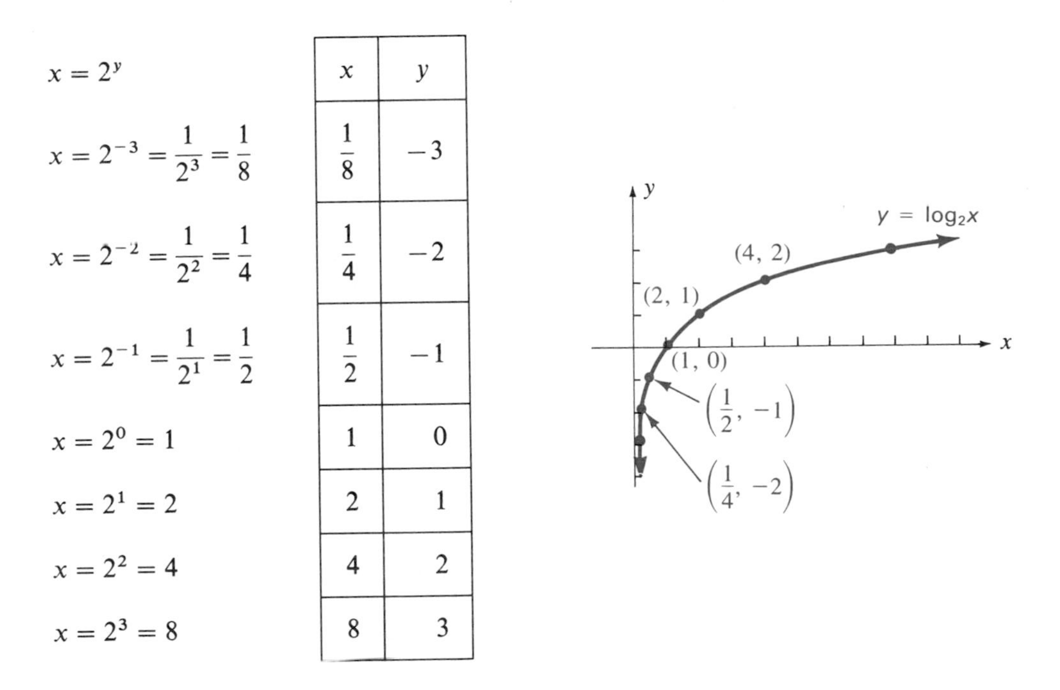

$x = 2^y$

$x = 2^{-3} = \dfrac{1}{2^3} = \dfrac{1}{8}$

$x = 2^{-2} = \dfrac{1}{2^2} = \dfrac{1}{4}$

$x = 2^{-1} = \dfrac{1}{2^1} = \dfrac{1}{2}$

$x = 2^0 = 1$

$x = 2^1 = 2$

$x = 2^2 = 4$

$x = 2^3 = 8$

x	y
$\frac{1}{8}$	-3
$\frac{1}{4}$	-2
$\frac{1}{2}$	-1
1	0
2	1
4	2
8	3

Notice that the *domain* is the set of all real numbers greater than zero and the *range* is the set of all real numbers. ■

* The logarithm of a negative number exists, but it is a complex number. We will not discuss the logarithm of a negative number in this book.

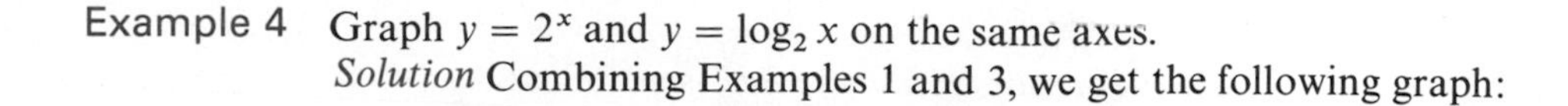

Example 4 Graph $y = 2^x$ and $y = \log_2 x$ on the same axes.
Solution Combining Examples 1 and 3, we get the following graph:

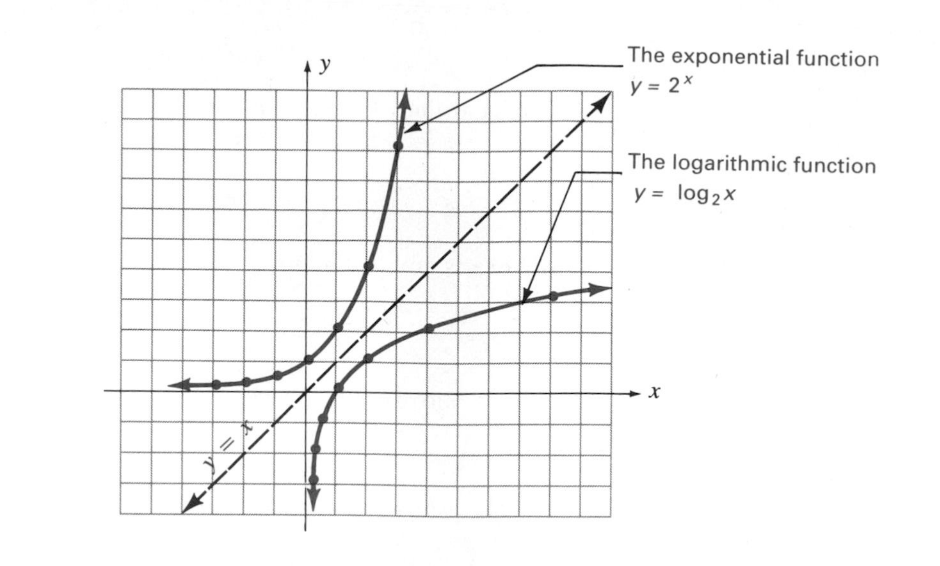

In general, the graph of any logarithmic function with a base greater than 1 has the appearance and characteristics of the curve shown in Figure 9.1.4.

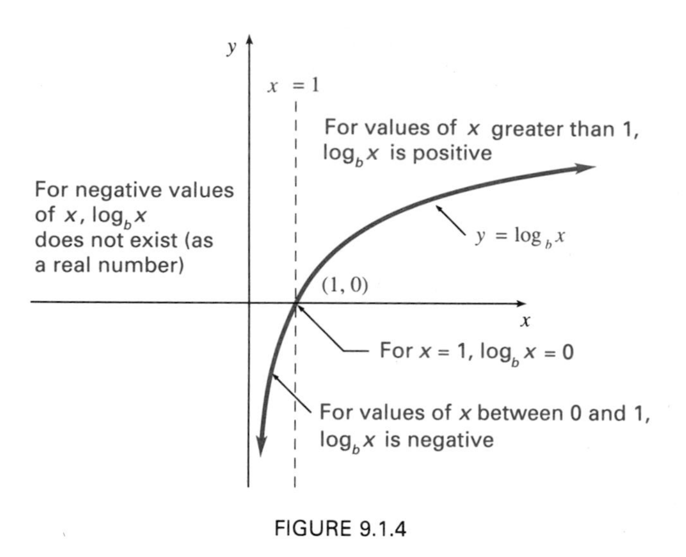

FIGURE 9.1.4

The following three properties of logarithms are stated without proof.

1. The logarithmic function is an *increasing function.* This means that as x gets larger, $\log_b x$ also gets larger.
2. If two numbers are equal, then their logarithms (taken to the same base) must be equal, because $\log_b x$ is a *function.* This fact is used in solving exponential equations (Section 9.5) and is stated in symbols in Rule 9.3.

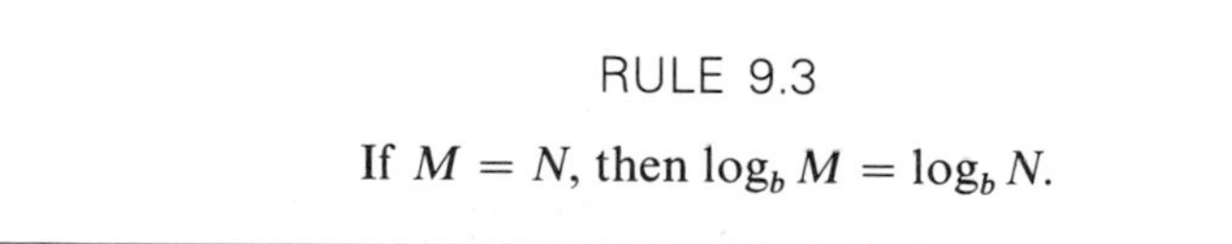

RULE 9.3

If $M = N$, then $\log_b M = \log_b N$.

3. If the logarithms of two numbers are equal (when the bases are the same), then the numbers must be equal. This fact is used in solving logarithmic equations and is stated in symbols in Rule 9.4.

> RULE 9.4
>
> If $\log_b M = \log_b N$, then $M = N$.

EXERCISES 9.1B

Set I

1. Graph $y = \log_2 x$ and $y = \log_{10} x$ on the same set of axes.
2. Graph the exponential function $y = 3^x$ and its inverse function, $y = \log_3 x$, on the same set of axes.
3. Graph $y = e^x$ and its inverse function, $y = \log_e x$, on the same set of axes.

Set II

1. Graph $y = \log_2 x$ and $y = \log_3 x$ on the same set of axes.
2. Graph $y = 4^x$ and its inverse logarithmic function on the same set of axes.
3. Graph $y = 5^x$ and its inverse function on the same set of axes.

9.2 Working with Exponential and Logarithmic Functions

9.2A Exponential and Logarithmic Forms

Because of the definition of logarithms, *logarithms are exponents.* Consequently, every exponential equation can be written in logarithmic form, and every logarithmic equation can be written in exponential form. In the following box, the symbol ⇔ means "is equivalent to."

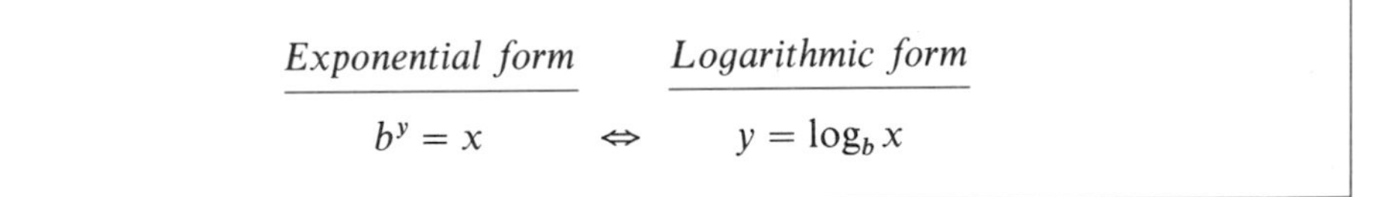

Example 1 Change $10^2 = 100$ to logarithmic form.

Solution

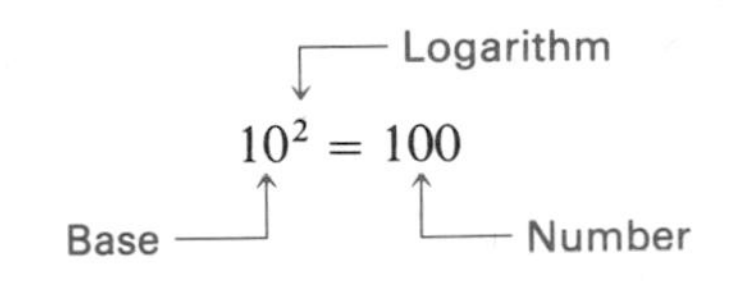

The *logarithm* is the *exponent* (2) to which the base (10) must be raised to give the number (100). This can be written $\log_{10} 100 = 2$ and is read "the logarithm of 100 to the base 10 equals 2."

Exponential form		*Logarithmic form*
$10^2 = 100$	⇔	$\log_{10} 100 = 2$ ■

Example 2 Examples of changing exponential equations to logarithmic form:

	Exponential form		*Logarithmic form*
a.	$5^2 = 25$	$\Leftrightarrow$	$\log_5 25 = 2$
b.	$2^3 = 8$	$\Leftrightarrow$	$\log_2 8 = 3$ ■

Example 3 Examples of changing logarithmic equations to exponential form:

	Logarithmic form		*Exponential form*
a.	$\log_4 16 = 2$	$\Leftrightarrow$	$4^2 = 16$
b.	$\log_{25} 5 = \frac{1}{2}$	$\Leftrightarrow$	$25^{1/2} = 5$
c.	$\log_{10} 10{,}000 = 4$	$\Leftrightarrow$	$10^4 = 10{,}000$
d.	$\log_{10} 0.01 = -2$	$\Leftrightarrow$	$10^{-2} = 0.01$ ■

EXERCISES 9.2A

Set I In Exercises 1–12, write each equation in logarithmic form.

1. $3^2 = 9$ **2.** $4^3 = 64$ **3.** $10^3 = 1{,}000$

4. $10^5 = 100{,}000$ **5.** $2^4 = 16$ **6.** $4^2 = 16$

7. $3^{-2} = \dfrac{1}{9}$ **8.** $2^{-3} = \dfrac{1}{8}$ **9.** $12^0 = 1$

10. $8^0 = 1$ **11.** $16^{1/2} = 4$ **12.** $8^{1/3} = 2$

In Exercises 13–24, write each equation in exponential form.

13. $\log_8 64 = 2$ **14.** $\log_2 32 = 5$ **15.** $\log_7 49 = 2$

16. $\log_4 64 = 3$ **17.** $\log_5 1 = 0$ **18.** $\log_6 1 = 0$

19. $\log_9 3 = \frac{1}{2}$ **20.** $\log_8 2 = \frac{1}{3}$ **21.** $\log_{10} 100 = 2$

22. $\log_{10} 1{,}000 = 3$ **23.** $\log_{10} 0.001 = -3$ **24.** $\log_{10} 0.01 = -2$

Set II In Exercises 1–12, write each equation in logarithmic form.

1. $2^3 = 8$ **2.** $10^4 = 10{,}000$ **3.** $4^{-2} = \frac{1}{16}$

4. $5^0 = 1$ **5.** $4^{1/2} = 2$ **6.** $3^4 = 81$

7. $5^2 = 25$ **8.** $3^{-3} = \frac{1}{27}$ **9.** $6^{-3} = \frac{1}{216}$

10. $17^0 = 1$ **11.** $64^{1/3} = 4$ **12.** $2^5 = 32$

In Exercises 13–24, write each equation in exponential form.

13. $\log_4 64 = 3$ **14.** $\log_{16} 4 = \frac{1}{2}$ **15.** $\log_3 1 = 0$

16. $\log_7 7 = 1$ **17.** $\log_{10} 100{,}000 = 5$ **18.** $\log_b a = c$

19. $\log_2 64 = 6$ **20.** $\log_7 1 = 0$ **21.** $\log_6 36 = 2$

22. $\log_{13} 13 = 1$ **23.** $\log_{10} 0.0001 = -4$ **24.** $\log_{16} 2 = \dfrac{1}{4}$

9.2B Solving Simple Logarithmic Equations and Evaluating Logarithms

We can *sometimes* solve a logarithmic equation by rewriting the equation in exponential form and then solving the exponential equation.

Example 4 Solve for b: $\log_b 64 = 3$.

Solution

Logarithmic form		*Exponential form*
$\log_b 64 = 3$	$\Leftrightarrow$	$b^3 = 64$

Because $64 = 4^3$, we can write $b^3 = 64 = 4^3$. Then by Rule 9.2, if $b^3 = 4^3$, $b = 4$. ■

Example 5 Solve for N: $\log_2 N = 4$.

Solution

Logarithmic form		*Exponential form*
$\log_2 N = 4$	$\Leftrightarrow$	$2^4 = N$

If $N = 2^4$, then $N = 16$. ■

Example 6 Solve for x: $\log_2 32 = x$.

Solution

Logarithmic form		*Exponential form*
$\log_2 32 = x$	$\Leftrightarrow$	$2^x = 32$

We know that $32 = 2^5$. If $2^x = 2^5$, then by Rule 9.1, $x = 5$. ■

It is also possible to *evaluate* certain logarithms as follows: (1) let the logarithmic expression equal x, (2) rewrite the logarithmic equation in exponential form, and (3) solve the exponential equation for x.

Example 7 Find $\log_{10} 1{,}000$.

Solution Let $x = \log_{10} 1{,}000$.

	Logarithmic form		*Exponential form*	
Then	$x = \log_{10} 1{,}000$	$\Leftrightarrow$	$10^x = 1{,}000 = 10^3$	
			$x = 3$	Rule 9.1

Therefore, $\log_{10} 1{,}000 = 3$. ■

Example 8 Find $\log_8 4$.

Solution Let $x = \log_8 4$.

	Logarithmic form		*Exponential form*	
Then	$x = \log_8 4$	$\Leftrightarrow$	$8^x = 4$	
			$(2^3)^x = 2^2$	Writing 8 and 4 as powers of 2
			$2^{3x} = 2^2$	
			$3x = 2$	Rule 9.1
			$x = \frac{2}{3}$	

Therefore, $\log_8 4 = \frac{2}{3}$. ■

Example 9 Find $\log_7 \frac{1}{49}$.

Solution Let $x = \log_7 \frac{1}{49}$.

	Logarithmic form		*Exponential form*	
Then	$x = \log_7 \frac{1}{49}$	$\Leftrightarrow$	$7^{x} = \frac{1}{49} = \frac{1}{7^2} = 7^{-2}$	
			$x = -2$	Rule 9.1

Therefore, $\log_7 \frac{1}{49} = -2$. ■

NOTE The logarithm of a positive number can be negative, but, as was mentioned previously, the logarithm of a *negative* number is *complex* and is not discussed in this book. ✓

Example 10 Find $\log_{10} 0$.
Solution Let $x = \log_{10} 0$.

	Logarithmic form		*Exponential form*
Then	$x = \log_{10} 0$	$\Leftrightarrow$	$10^x = 0$

There is *no solution* because no value of x makes $10^x = 0$

Therefore, $\log_{10} 0$ does not exist. ■

The result in Example 10 can be generalized as follows:

> The logarithm of 0 to any base b ($b > 0$, $b \neq 1$) does not exist.

EXERCISES 9.2B

Set I In Exercises 1–18, find the value of the unknown b, N, or x.

1. $\log_5 N = 2$
2. $\log_2 N = 5$
3. $\log_3 9 = x$
4. $\log_2 16 = x$
5. $\log_b 27 = 3$
6. $\log_b 81 = 4$
7. $\log_5 125 = x$
8. $\log_3 81 = x$
9. $\log_{10} 10^{-4} = x$
10. $\log_{10} 10^{-3} = x$
11. $\log_{3/2} N = 2$
12. $\log_{5/3} N = 3$
13. $\log_9 \frac{1}{3} = x$
14. $\log_8 \frac{1}{64} = x$
15. $\log_b 8 = 1.5$
16. $\log_b 125 = 1.5$
17. $\log_2 N = -2$
18. $\log_{10} N = -2$

In Exercises 19–30, find the value of each logarithm.

19. $\log_5 25$
20. $\log_2 16$
21. $\log_{10} 10{,}000$
22. $\log_{10} 100{,}000$
23. $\log_4 8$
24. $\log_{27} 9$
25. $\log_3 3^4$
26. $\log_2 2^5$
27. $\log_{16} 16$
28. $\log_{20} 20$
29. $\log_8 1$
30. $\log_7 1$

Set II In Exercises 1–18, find the value of the unknown b, N, or x.

1. $\log_{25} N = 1.5$ **2.** $\log_b 9 = \frac{2}{3}$ **3.** $\log_{16} 32 = x$

4. $\log_{16} N = \frac{5}{4}$ **5.** $\log_4 8 = x$ **6.** $\log_{625} N = 0.5$

7. $\log_b \frac{1}{2} = -\frac{1}{3}$ **8.** $\log_b \frac{27}{8} = 3$ **9.** $\log_b 32 = \frac{5}{3}$

10. $\log_8 4 = x$ **11.** $\log_{1/2} N = 2$ **12.** $\log_b 32 = 5$

13. $\log_{25} \frac{1}{5} = x$ **14.** $\log_{3/4} N = -1$ **15.** $\log_b \frac{1}{6} = -1$

16. $\log_{27} 3 = x$ **17.** $\log_4 N = 0$ **18.** $\log_{25} 5 = x$

In Exercises 19–30, find the value of each logarithm.

19. $\log_4 64$ **20.** $\log_{13} 13$ **21.** $\log_4 1$ **22.** $\log_{64} 8$

23. $\log_9 27$ **24.** $\log_5 1$ **25.** $\log_{81} 3$ **26.** $\log_7 7^8$

27. $\log_{16} 4$ **28.** $\log_{16} 8$ **29.** $\log_3 1$ **30.** $\log_{49} 7$

9.3 Rules of Logarithms

Since logarithms are exponents, we can derive the rules of logarithms from the properties of exponents. We will develop several rules of logarithms by using the techniques shown in Section 9.2.

> RULE 9.5
>
> If $b > 0$ and $b \neq 1$, then the logarithm of b to the base b equals 1.
>
> $$\log_b b = 1$$

Proof We know that $b = b^1$. If we rewrite this exponential equation in logarithmic form, we have $\log_b b = 1$.

Example 1 Find $\log_6 6$.

Solution By rule 9.5, $\log_6 6 = 1$. ■

> RULE 9.6
>
> The logarithm of 1 to any base b ($b > 0$, $b \neq 1$) equals 0.
>
> $$\log_b 1 = 0$$

Proof By definition, $b^0 = 1$. If we rewrite this exponential equation in logarithmic form, we have $\log_b 1 = 0$.

Example 2 Find $\log_5 1$.

Solution By Rule 9.6, $\log_5 1 = 0$. ■

RULE 9.7

$$\log_b MN = \log_b M + \log_b N$$

Proof Let $x = \log_b M$
$y = \log_b N$

Then $x + y = \log_b M + \log_b N$

Also, $x = \log_b M \Leftrightarrow b^x = M$

$y = \log_b N \Leftrightarrow b^y = N$

and $MN = b^x b^y = b^{x+y} \Leftrightarrow \log_b MN = x + y$

Therefore, since $\log_b MN$ and $\log_b M + \log_b N$ both equal $x + y$, $\log_b MN = \log_b M + \log_b N$.

NOTE $\log_b MN = \log_b(MN)$, *not* $(\log_b M)N$. ✓

A WORD OF CAUTION A mistake students frequently make is to think that

$$\log_b(M + N) = \log_b M + \log_b N \quad \text{Incorrect}$$

Since $M + N \neq MN$, $\log_b(M + N) \neq \log_b MN$. Then, since $\log_b MN = \log_b M + \log_b N$, $\log_b(M + N) \neq \log_b M + \log_b N$.

WARNING: $\log_b(M + N)$ *cannot* be rewritten as a sum of terms. ✓

RULE 9.8

$$\log_b \frac{M}{N} = \log_b M - \log_b N$$

Proof Let $x = \log_b M$
$y = \log_b N$

Then $x - y = \log_b M - \log_b N$

Also, $x = \log_b M \Leftrightarrow b^x = M$

$y = \log_b N \Leftrightarrow b^y = N$

and $\frac{M}{N} = \frac{b^x}{b^y} = b^{x-y} \Leftrightarrow \log_b \frac{M}{N} = x - y$

Therefore, since $\log_b \frac{M}{N}$ and $\log_b M - \log_b N$ both equal $x - y$, $\log_b \frac{M}{N} = \log_b M - \log_b N$.

A WORD OF CAUTION $\log_b(M - N) \neq \log_b M - \log_b N$.
WARNING: $\log_b(M - N)$ *cannot* be rewritten as a difference of two terms. ☑

A WORD OF CAUTION $\log_b \frac{M}{N} \neq \frac{\log_b M}{\log_b N}$. The rule is $\log_b \frac{M}{N} = \log_b M - \log_b N$. ☑

RULE 9.9

$$\log_b N^p = p \log_b N$$

Proof Let $y = \log_b N$.

Then $py = p \log_b N$ — Multiplying both sides by p

Also, $y = \log_b N \iff b^y = N$ ⟵ Next, we will raise both sides of $b^y = N$ to the pth power

$N^p = (b^y)^p = b^{yp} \iff \log_b N^p = yp = py$

Therefore, since $\log_b N^p$ and $p \log_b N$ both equal py, $\log_b N^p = p \log_b N$.

The rules of logarithms can be summarized as follows:

THE RULES OF LOGARITHMS

9.3 If $M = N$, then $\log_b M = \log_b N$.

9.4 If $\log_b M = \log_b N$, then $M = N$.

9.5 $\log_b b = 1$

9.6 $\log_b 1 = 0$

9.7 $\log_b MN = \log_b M + \log_b N$

9.8 $\log_b \frac{M}{N} = \log_b M - \log_b N$

9.9 $\log_b N^p = p \log_b N$

where $\begin{cases} M \text{ and } N \text{ are positive real numbers} \\ b > 0, b \neq 1 \\ p \text{ is any real number} \end{cases}$

A WORD OF CAUTION You must be careful about the use of parentheses, as usual; $\log_b(M + N) \neq \log_b M + N$. ☑

Example 3 Examples of transforming logarithmic expressions by using the rules of logarithms:

a. $\log_{10}(56)(107) = \log_{10} 56 + \log_{10} 107$ — Rule 9.7

b. $\log_{10} \frac{275}{89} = \log_{10} 275 - \log_{10} 89$ — Rule 9.8

c. $\log_{10}(37)^2 = 2\log_{10} 37$ Rule 9.9

d. $\log_{10}\sqrt{5} = \log_{10} 5^{1/2} = \frac{1}{2}\log_{10} 5$ Rule 9.9

e. $\log_{10} \frac{(57)(23)}{101} = \log_{10}(57)(23) - \log_{10} 101$ Rule 9.8

$= \log_{10} 57 + \log_{10} 23 - \log_{10} 101$ Rule 9.7

f. $\log_{10} \frac{(47)(19)^3}{(1.04)^7} = \log_{10}(47)(19)^3 - \log_{10}(1.04)^7$ Rule 9.8

$= \log_{10} 47 + \log_{10}(19)^3 - \log_{10}(1.04)^7$ Rule 9.7

$= \log_{10} 47 + 3\log_{10} 19 - 7\log_{10} 1.04$ Rule 9.9

g. $\log_{10} \frac{\sqrt[5]{21.4}}{(3.5)^4} = \log_{10}(21.4)^{1/5} - \log_{10}(3.5)^4$ Rule 9.8

$= \frac{1}{5}\log_{10} 21.4 - 4\log_{10} 3.5$ Rule 9.9 ■

Example 4 Given $\begin{Bmatrix} \log_{10} 2 \approx 0.301 \\ \log_{10} 3 \approx 0.477 \end{Bmatrix}^*$, approximate the following logarithms.

a. $\log_{10} 6 = \log_{10}(2)(3) = \log_{10} 2 + \log_{10} 3$ Rule 9.7

$\approx 0.301 + 0.477 = 0.778$

b. $\log_{10} 1.5 = \log_{10} \frac{3}{2} = \log_{10} 3 - \log_{10} 2$ Rule 9.8

$\approx 0.477 - 0.301 = 0.176$

c. $\log_{10} 8 = \log_{10} 2^3 = 3\log_{10} 2$ Rule 9.9

$\approx 3(0.301) = 0.903$

d. $\log_{10}\sqrt{3} = \log_{10} 3^{1/2} = \frac{1}{2}\log_{10} 3$ Rule 9.9

$\approx \frac{1}{2}(0.477) = 0.2385$

e. $\log_{10} 5 = \log_{10} \frac{10}{2} = \log_{10} 10 - \log_{10} 2$ Rule 9.8

$\approx 1 - 0.301 = 0.699$ ■

The rules of logarithms can be used to simplify some logarithmic expressions, as in Example 5.

Example 5 Examples of simplifying logarithmic expressions using the rules of logarithms:

a. $\log_b 5x + 2\log_b x = \log_b 5x + \log_b x^2$ Rule 9.9

$= \log_b[(5x)(x^2)]$ Rule 9.7

$= \log_b 5x^3$

b. $\frac{1}{2}\log_b x - 4\log_b y = \log_b x^{1/2} - \log_b y^4$ Rule 9.9

$= \log_b \frac{\sqrt{x}}{y^4}$ Rule 9.8 ■

* The methods for finding that $\log_{10} 2 \approx 0.301$ and $\log_{10} 3 \approx 0.477$ are shown in Section 9.4 and in Appendix B.

EXERCISES 9.3

Set I

In Exercises 1–10, transform each expression by using the rules of logarithms, as was done in Example 3.

1. $\log_{10}(31)(7)$
2. $\log_{10}(17)(29)$
3. $\log_{10}\dfrac{41}{13}$
4. $\log_{10}\dfrac{19}{23}$
5. $\log_{10}(19)^3$
6. $\log_{10}(7)^4$
7. $\log_{10}\sqrt[5]{75}$
8. $\log_{10}\sqrt[4]{38}$
9. $\log_{10}\dfrac{35\sqrt{73}}{(1.06)^8}$
10. $\log_{10}\dfrac{27\sqrt{31}}{(1.03)^{10}}$

In Exercises 11–20, approximate the value of each logarithm, given that $\log_{10} 2 \approx 0.301$, $\log_{10} 3 \approx 0.477$, and $\log_{10} 7 \approx 0.845$. (Remember, $\log_{10} 10 = 1$.)

11. $\log_{10} 14$
12. $\log_{10} 21$
13. $\log_{10} \frac{9}{7}$
14. $\log_{10} \frac{7}{4}$
15. $\log_{10}\sqrt{27}$
16. $\log_{10}\sqrt{8}$
17. $\log_{10}(36)^2$
18. $\log_{10}(98)^2$
19. $\log_{10} 6{,}000$
20. $\log_{10} 1{,}400$

In Exercises 21–30, write each expression as a single logarithm and simplify.

21. $\log_b x + \log_b y$
22. $4 \log_b x + 2 \log_b y$
23. $2 \log_b x - 3 \log_b y$
24. $\frac{1}{2} \log_b x^4$
25. $3(\log_b x - 2 \log_b y)$
26. $\frac{1}{3} \log_b y^3$
27. $\log_b(x^2 - y^2) - 3 \log_b(x + y)$
28. $\log_b(x^2 - z^2) - \log_b(x - z)$
29. $2 \log_b 2xy - \log_b 3xy^2 + \log_b 3x$
30. $2 \log_b 3xy - \log_b 6x^2y^2 + \log_b 2y^2$

Set II

In Exercises 1–10, transform each expression by using the rules of logarithms, as was done in Example 3.

1. $\log_{10}(27)(11)$
2. $\log_{10}(8)(12)(4)$
3. $\log_{10} \frac{5}{14}$
4. $\log_{10} \frac{83}{7}$
5. $\log_{10}(24)^5$
6. $\log_{10}(18)^4$
7. $\log_{10}\dfrac{(17)(31)}{29}$
8. $\log_{10}\dfrac{(7)(11)}{13}$
9. $\log_{10}\dfrac{53}{(11)(19)^2}$
10. $\log_{10}\dfrac{29}{(31)^3(47)}$

In Exercises 11–20, approximate the value of each logarithm, given that $\log_{10} 2 \approx 0.301$, $\log_{10} 3 \approx 0.477$, and $\log_{10} 7 \approx 0.845$. (Remember, $\log_{10} 10 = 1$.)

11. $\log_{10} 40$
12. $\log_{10} 90$
13. $\log_{10}\sqrt[4]{3}$
14. $\log_{10}\sqrt[5]{7}$
15. $\log_{10} 7^3$
16. $\log_{10} 32$
17. $\log_{10}\sqrt{14}$
18. $\log_{10} \frac{3}{7}$
19. $\log_{10} \frac{20}{3}$
20. $\log_{10} 8^5$

In Exercises 21–30, write each expression as a single logarithm and simplify.

21. $\log_b x - \log_b y$
22. $2(3 \log_b x - \log_b y)$
23. $\frac{1}{2} \log_b(x - a) - \frac{1}{2} \log_b(x + a)$
24. $3 \log_b v + 2 \log_b v^2 - \log_b v$
25. $\log_b \frac{6}{7} - \log_b \frac{27}{4} + \log_b \frac{21}{16}$
26. $\log_b x - \log_b y - \log_b z$
27. $5 \log_b xy^3 - \log_b x^2y$
28. $2 \log_b x - 3 \log_b y + \log_b z$
29. $2 \log_b x + 2 \log_b y$
30. $\log_b(x^3 + y^3) - \log_b(x + y)$

9.4 Logarithms and the Calculator*

9.4A Finding Logarithms Using a Calculator

There are two systems of logarithms in widespread use:

1. Common logarithms (base 10)
2. Natural logarithms (base e, where $e \approx 2.71828$)

When the logarithm of a number x is written as **$\log x$**, *without* a base being specified, the base of the logarithm is understood to be 10. For example, log 8 is understood to mean $\log_{10} 8$. The special abbreviation **$\ln x$** is reserved for *natural* logarithms. For example, ln 6 means $\log_e 6$.

Most scientific calculators use symbols that reflect these abbreviations. Most have a key marked [log] or [log x] that is used for common logarithms and a key marked [ln] or [ln x] for natural logarithms.

Nearly all the logarithms found by using a calculator or by using tables are approximations. Nevertheless, we commonly say "Find the logarithm" rather than "Approximate the logarithm."

Example 1 Using a calculator, find the logarithm of each given number. Round off answers to four decimal places.

a. log 2

Solution Press these keys in this order: [2] [log]. When we round off the answer, we see that $\log 2 \approx 0.3010$.

b. log 20

Solution Press these keys in this order: [2] [0] [log]. When we round off the answer, we see that $\log 20 \approx 1.3010$.

c. log 200

Solution Press these keys in this order: [2] [0] [0] [log]; $\log 200 \approx 2.3010$.

d. log 0.002

Solution Press these keys in this order: [·] [0] [0] [2] [log]; $\log 0.002 \approx -2.6990$.

e. $3 + \log 0.002$

Solution $3 + \log 0.002 \approx 3 + (-2.6990) = 0.3010$.

f. log 0

Solution When we press [0] [log], we get an error message. This is to be expected, since $\log_{10} 0$ does not exist.

g. $\log(-3)$

Solution When we press [3] [+/−] [log], we get an error message, because $\log_{10}(-3)$ is not a real number. ■

* Note to the Instructor: If you prefer your students to use tables, use Appendix B in place of this section.

NOTE In Example 1, observe that the answers to parts b, c, and d are consistent with what we know about the properties of logarithms. That is, because $20 = 10 \times 2$,

$$\log 20 = \log(10 \times 2) = \log 10 + \log 2 = 1 + \log 2 \approx 1 + 0.3010 = 1.3010$$

Similarly,

$$\begin{aligned}\log 200 &= \log(10^2 \times 2) = \log 10^2 + \log 2 = 2 \log 10 + \log 2 = 2(1) + \log 2 \\ &= 2 + \log 2 \approx 2 + 0.3010 = 2.3010\end{aligned}$$

and

$$\begin{aligned}\log 0.002 &= \log(10^{-3} \times 2) = \log 10^{-3} + \log 2 = -3 \log 10 + \log 2 \\ &= -3(1) + \log 2 \approx -3 + 0.3010 = -2.6990\end{aligned}$$

For further discussion, see Appendix B. ☑

Example 2 Using a calculator, find the logarithm of each given number. Round off answers to four decimal places.

a. ln 2

Solution Press these keys in this order: [2] [ln]. When we round off, we see that $\ln 2 \approx 0.6931$.

b. ln 20

Solution Press these keys in this order: [2] [0] [ln]; $\ln 20 \approx 2.9957$.

c. ln 0.002

Solution When we press [·] [0] [0] [2] [ln], we see that $\ln 0.002 \approx -6.2146$.

d. ln 0

Solution When we press [0] [ln], we get an error message, because ln 0 does not exist.

e. $\ln(-5)$

Solution When we press [5] [+/−] [ln], we get an error message, because $\ln(-5)$ is not a real number. ■

In Examples 2a, b, and c, notice that *natural* logarithms do not exhibit the consistent decimal part we saw for *common* logarithms in Examples 1a, b, and c.

EXERCISES 9.4A

Set I Find the logarithm of each given number, using a calculator. Round off all answers to four decimal places.

1. a. log 3 b. log 3,000 c. log 30
d. ln 3 e. ln 3,000 f. ln 30

2. a. log 10 b. log 100 c. log 100,000
d. ln 10 e. ln 100 f. ln 100,000

3. a. log 0.1 b. log 0.01 c. log 0.001
d. ln 0.1 e. ln 0.01 f. ln 0.001

4. a. log 0.3 b. log 0.0003 c. log 0.000003
d. ln 0.3 e. ln 0.0003 f. ln 0.000003

Set II Find the logarithm of each given number, using a calculator. Round off all answers to four decimal places.

1. a. log 5 b. log 5,000,000 c. log 500
d. ln 5 e. ln 5,000,000 f. ln 500

2. a. log 0.5 b. log 0.000005 c. log 0.005
d. ln 0.5 e. ln 0.000005 f. ln 0.005

3. a. log 12 b. log 120,000 c. log 0.00012
d. ln 12 e. ln 120,000 f. ln 0.00012

4. a. log 0.16 b. log 0.0016 c. log 0.000016
d. ln 0.16 e. ln 0.0016 f. ln 0.000016

9.4B Finding Antilogarithms Using a Calculator

Significant Digits

When we use a calculator to find a logarithm or to find a number whose logarithm is known, we almost never get an exact answer. For that reason, we will give here a brief explanation of how to find the number of **significant digits** in a number.

To determine the number of significant digits in a number, read from left to right and start counting from the *first nonzero* digit. If the number does not have a decimal point, or if it has a decimal point but no digits to the right of the decimal point, *stop counting* when all the remaining digits are zeros. If the number has a decimal point and one or more digits to the right of the decimal point, start counting with the first nonzero digit and continue counting to the end of the number.

Example 3 Find the number of significant digits in each number.

a. 6,080,000

Solution

Start here
6,080,000
End here—this is the last nonzero digit

6,080,000 has three significant digits.

b. 300

Solution

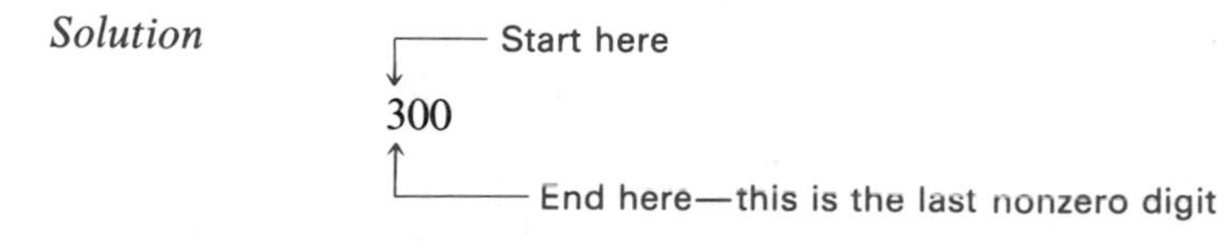

300 has one significant digit.

c. 300.00

Solution

Start here

300.00

End here; we count to the last digit of the number, because there are digits to the right of the decimal point

300.00 has five significant digits.

d. 0.00030

Solution

Start here—this is the first nonzero digit

0.00030

End here

0.00030 has two significant digits. ■

Antilogarithms

In Section 9.4A, we discussed finding the logarithm of a given number by using a calculator. In this section, we discuss the inverse operation: finding the *number* when its logarithm is known.

If the logarithm of N is L, then N is the **antilogarithm** of L. In symbols,

If	$\log N = L$
then	$N = \text{antilog } L$

If a base is not mentioned, it is understood to be 10. Problems in which we need to find an antilogarithm can be worded in either of two ways:

Find antilog 3.6263.
Find N if $\log N = 3.6263$.
} Both statements mean the same thing

Example 4 Find the (common) antilogs and round off all answers to three significant digits:

a. antilog 3.6263

Solution Press these keys in this order:

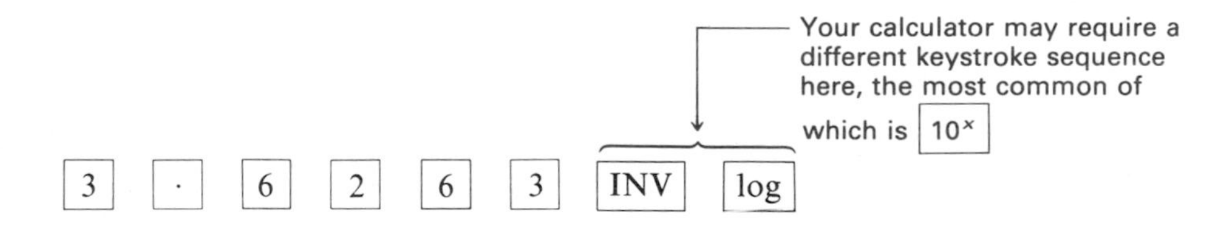

The display shows 4229.6068. When we round this off to three significant digits, we get

$$\text{antilog } 3.6263 \approx 4{,}230$$

NOTE The statement "antilog 3.6263 ≈ 4,230" is equivalent to the statement "log 4,230 ≈ 3.6263." ☑

b. Find N if $\log N = -1.1864$.

Solution Press these keys in this order:

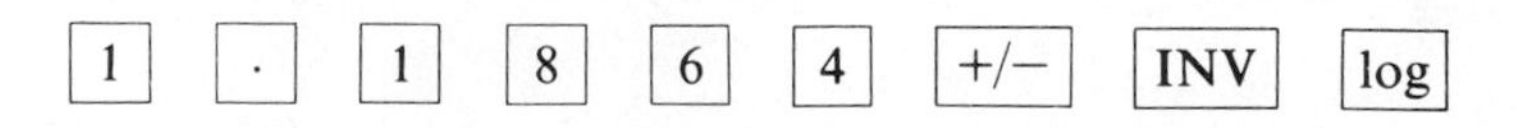

The display is 0.0651028, which rounds off to 0.0651. Therefore, $N \approx 0.0651$. (The problem could have been stated this way: Find antilog (-1.1864).) ■

Example 5 Find N if $\ln N = 0.4886$. Round off the answer to three significant digits.

Solution Note that this time we've been given the *natural* log of N, rather than the common log of N. We press these keys in this order:

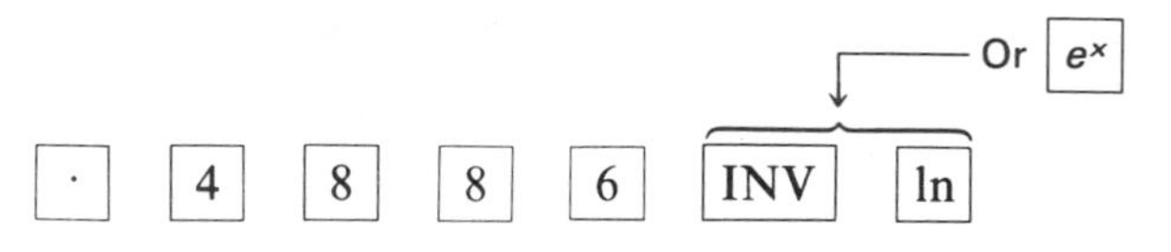

The display is 1.6300326, which rounds off to 1.63. Therefore, $N \approx 1.63$. ■

EXERCISES 9.4B

Set I In Exercises 1–4, find the common antilogarithms. Round off answers to three significant digits.

1. antilog 0.3711

2. antilog 0.6085

3. antilog(-0.0701)

4. antilog(-0.1221)

In Exercises 5–12, find N. Round off answers to three significant digits.

5. $\log N = 3.4082$

6. $\log N = 1.8136$

7. $\log N = -2.5143$

8. $\log N = -3.3883$

9. $\ln N = 2.7183$

10. $\ln N = 5.4366$

11. $\ln N = -2.5849$

12. $\ln N = -5.1534$

Set II In Exercises 1–4, find the common antilogarithms. Round off answers to three significant digits.

1. antilog 1.6990

2. antilog(-1.6990)

3. antilog 4.7275

4. antilog(-2.2381)

In Exercises 5–12, find N. Round off answers to three significant digits.

5. $\log N = 2.5119$

6. $\log N = 4.5465$

7. $\log N = -2.6271$

8. $\log N = -3.2652$

9. $\ln N = 3.0204$

10. $\ln N = 5.8435$

11. $\ln N = -2.7790$

12. $\ln N = -7.0436$

9.5 Exponential and Logarithmic Equations

In this section, we discuss the application of logarithms and using the rules of logarithms in the solution of exponential and logarithmic equations. Although calculating with logarithms is being replaced by the use of hand calculators and computers, the use of logarithms will continue to be important. Natural logarithms are used extensively in higher-level mathematics, especially in calculus, so it is important that you understand the rules of logarithms, and when and how to apply them.

Exponential Equations

An **exponential equation** is an equation in which the variable appears in one or more exponents.

Example 1 Examples of exponential equations:

a. $3^x = 17$ b. $(5.26)^{x+1} = 75.4$ ■

It is possible to solve some exponential equations by expressing both sides as powers of the same base (see Example 2).

Example 2 Solve $16^x = \frac{1}{4}$.

Solution

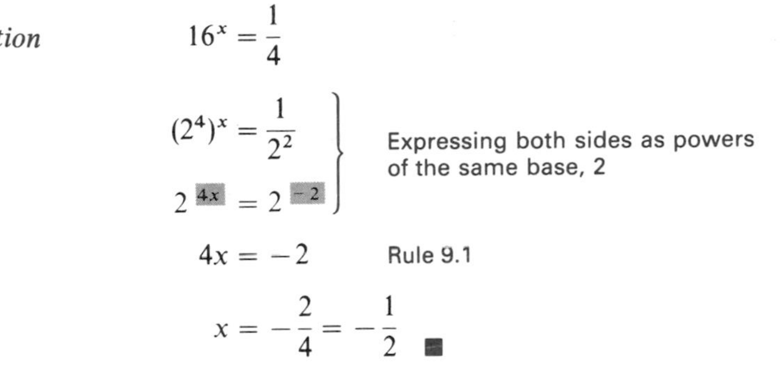

$$16^x = \frac{1}{4}$$

$$(2^4)^x = \frac{1}{2^2}$$

$$2^{4x} = 2^{-2}$$

Expressing both sides as powers of the same base, 2

$$4x = -2 \qquad \text{Rule 9.1}$$

$$x = -\frac{2}{4} = -\frac{1}{2} \quad ■$$

When both sides of an exponential equation *cannot* be expressed as powers of the same base, we solve the equation by taking logarithms of both sides. Rule 9.3 justifies our doing this. Most exponential equations are of this type. We can take either the *common* logarithm or the *natural* logarithm of both sides (see Example 3).

Example 3 Solve $3^x = 17$. Round off the answer to four significant digits.

Method 1

$$3^x = 17$$

$$\log 3^x = \log 17 \qquad \text{Rule 9.3 (taking the } common \text{ logarithm of both sides)}$$

$$x \log 3 = \log 17 \qquad \text{Rule 9.9}$$

$$x = \frac{\log 17}{\log 3} \qquad \text{Dividing both sides of the equation by log 3}$$

$$x \approx \frac{1.2304}{0.4771} \approx 2.579$$

A WORD OF CAUTION A mistake students often make is to think that

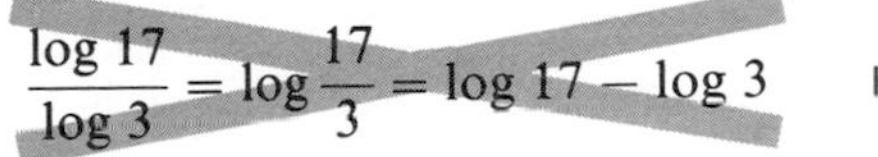

$$\frac{\log 17}{\log 3} = \log \frac{17}{3} = \log 17 - \log 3 \qquad \text{Incorrect}$$

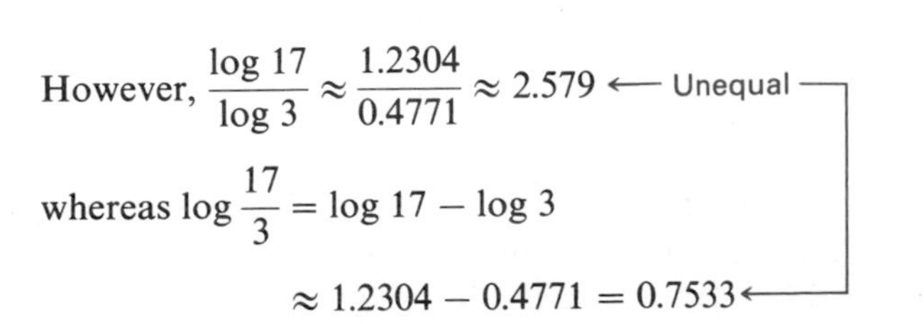

However, $\frac{\log 17}{\log 3} \approx \frac{1.2304}{0.4771} \approx 2.579$ ← Unequal

whereas $\log \frac{17}{3} = \log 17 - \log 3$

$\approx 1.2304 - 0.4771 = 0.7533$ ←

☑

Method 2 $3^x = 17$

$\ln 3^x = \ln 17$ Rule 9.3 (taking the *natural* logarithm of both sides)

$x \ln 3 = \ln 17$ Rule 9.9

$x = \dfrac{\ln 17}{\ln 3}$ Dividing both sides of the equation by ln 3

$$x \approx \frac{2.8332}{1.0986} \approx 2.579$$

Note that we obtained the same answer whether we took the *natural* logarithm or the *common* logarithm of both sides. ■

Example 4 Solve $e^{4x} = 7$. Round off the answer to four significant digits.

Solution We could take the common logarithm of both sides of the equation, but because the base of the exponential function is e, we will instead take the natural logarithm of both sides:

$e^{4x} = 7$

$\ln e^{4x} = \ln 7$ Rule 9.3 (taking the natural logarithm of both sides)

$4x \ln e = \ln 7$ Rule 9.9

$4x(1) = \ln 7$ Rule 9.5 ($\ln e = \log_e e = 1$)

$4x = \ln 7$

$$x = \frac{\ln 7}{4} \approx \frac{1.9459101}{4} \approx 0.4865$$

A completely different method can be used. We can rewrite the given exponential equation in logarithmic form:

$$e^{4x} = 7 \Leftrightarrow \log_e 7 = 4x$$

$\ln 7 = 4x$ Writing $\log_e 7$ as $\ln 7$

$$x = \frac{\ln 7}{4} \approx 0.4865 \quad ■$$

Logarithmic Equations

Recall from Section 9.1B that in the expression $\log_b x$, x is called the *argument* of the logarithmic function. A **logarithmic equation** is an equation in which the variable appears in the argument of a logarithm.

Example 5 Examples of logarithmic equations:

a. $\ln(2x + 1) + \ln 5 = \ln(x + 6)$

b. $\log(x - 1) + \log(x + 2) = 1$ ■

If every term contains a logarithm, as in Example 5a, we can solve the equation by simplifying both sides of the equation and then using Rule 9.4: If $\log_b M = \log_b N$, then $M = N$ (see Example 6). If any term of the equation contains no logarithms, as in Example 5b, we can solve it by either of two methods: The equation can be simplified to the form $\log_b x = k$, rewritten in exponential form, and then solved (see Example 7, Method 1), *or* the term that doesn't contain a logarithm can be rewritten as a logarithm (see Example 7, Method 2, and Example 8).

Apparent solutions to logarithmic equations *must* be checked, because we often obtain extraneous roots when we solve logarithmic equations (see Example 7). Any solution that would result in zero or a negative number as the *argument* of a logarithm must be rejected.

Example 6 Solve $\ln(2x + 1) + \ln 5 = \ln(x + 6)$.

Solution

$$\ln(2x + 1) + \ln 5 = \ln(x + 6)$$

$$\ln(2x + 1)(5) = \ln(x + 6) \qquad \text{Rule 9.7}$$

$$(2x + 1)(5) = x + 6 \qquad \text{Rule 9.4 (If } \log_b M = \log_b N\text{, then } M = N\text{)}$$

$$10x + 5 = x + 6$$

$$9x = 1$$

$$x = \frac{1}{9}$$

Check

$$\ln(2x + 1) + \ln 5 = \ln(x + 6)$$

$$\ln[2(\tfrac{1}{9}) + 1] + \ln 5 \stackrel{?}{=} \ln(\tfrac{1}{9} + 6)$$

$$\ln(\tfrac{11}{9}) + \ln 5 \stackrel{?}{=} \ln(\tfrac{55}{9})$$

$$\ln(\tfrac{11}{9})(5) \stackrel{?}{=} \ln(\tfrac{55}{9})$$

$$\ln(\tfrac{55}{9}) = \ln(\tfrac{55}{9}) \qquad \text{True}$$

Therefore, $\frac{1}{9}$ is the solution. ■

Example 7 Solve $\log(x - 1) + \log(x + 2) = 1$.

Method 1

$$\log(x - 1) + \log(x + 2) = 1$$

$$\log_{10}(x - 1)(x + 2) = 1 \qquad \text{Rule 9.7}$$

$$(x - 1)(x + 2) = 10^1 \qquad \text{Rewriting the equation in exponential form}$$

$$x^2 + x - 2 = 10$$

$$x^2 + x - 12 = 0$$

$$(x - 3)(x + 4) = 0$$

$$x - 3 = 0 \quad \text{or} \quad x + 4 = 0$$

$$x = 3 \quad \text{or} \quad x = -4$$

Method 2

$$\log(x - 1) + \log(x + 2) = 1 \qquad \text{Rewriting 1 as log 10}$$

$$\log(x - 1)(x + 2) = \log 10 \qquad \text{Rule 9.7}$$

$$(x - 1)(x + 2) = 10 \qquad \text{Rule 9.4}$$

(See Method 1 for the remainder of the solution.)

Check for $x = 3$

$$\log(x - 1) + \log(x + 2) = 1$$

$$\log(3 - 1) + \log(3 + 2) \stackrel{?}{=} 1$$

$$\log 2 + \log 5 \stackrel{?}{=} 1$$

$$\log(2)(5) \stackrel{?}{=} 1$$

$$\log 10 = 1 \qquad \text{True}$$

Check for $x = -4$

$$\log(x - 1) + \log(x + 2) = 1$$

$$\log(-4 - 1) + \log(-4 + 2) \stackrel{?}{=} 1$$

$$\log(-5) + \log(-2) \stackrel{?}{=} 1$$

Logarithms of negative numbers are not real numbers

Therefore, 3 is a solution, while -4 is *not* a solution; it is an extraneous root. The only solution is 3. ■

Example 8 Solve $\log(2x + 3) - \log(x - 1) = 0.73$.

Solution We can first write 0.73 as a logarithm; that is, we can find antilog 0.73. Antilog $0.73 \approx 5.370$. Therefore, $0.73 \approx \log 5.370$. We then have

$$\log(2x + 3) - \log(x - 1) \approx \log 5.370$$
$$\log\frac{2x+3}{x-1} \approx \log 5.370$$
$$\frac{2x+3}{x-1} \approx 5.370 \qquad \text{Rule 9.4}$$
$$2x + 3 \approx 5.370(x - 1)$$
$$2x + 3 \approx 5.370x - 5.370$$
$$8.370 \approx 3.370x$$
$$\frac{8.370}{3.370} \approx x$$
$$x \approx 2.484$$

(We can also solve the equation by writing the equation $\log\left(\frac{2x+3}{x-1}\right) = 0.73$ in exponential form—that is, in the form $10^{0.73} = \frac{2x+3}{x-1}$—and then solving for x.)

Check for $x \approx 2.484$

$$\log(2x + 3) - \log(x - 1) \approx 0.73$$
$$\log[2(2.484) + 3] - \log[2.484 - 1] \overset{?}{\approx} 0.73$$
$$\log 7.968 - \log 1.484 \overset{?}{\approx} 0.73$$
$$0.9013 - 0.1714 \overset{?}{\approx} 0.73$$
$$0.7299 \approx 0.73 \qquad \text{True}$$

Therefore, $x \approx 2.484$. ■

EXERCISES 9.5

Set I In Exercises 1–6, solve each equation by expressing each side as a power of the same base.

1. $27^x = \frac{1}{9}$ **2.** $8^x = \frac{1}{16}$ **3.** $4^x = \frac{1}{8}$

4. $125^x = \frac{1}{25}$ **5.** $25^{2x+3} = 5^{x-1}$ **6.** $27^{3x-1} = 9^{x+2}$

In Exercises 7–16, solve each equation by taking the logarithm of both sides of the equation. Round off the answers to three significant digits.

7. $2^x = 3$ **8.** $5^x = 4$

9. $e^x = 8$ **10.** $e^x = 20$

11. $(7.43)^{x+1} = 9.55$ **12.** $(5.14)^{x-1} = 7.08$

13. $(8.71)^{2x+1} = 8.57$ **14.** $(9.55)^{3x-1} = 3.09$

15. $e^{3x+4} = 5$ **16.** $3^{2x-1} = 23$

In Exercises 17–30, solve and check each logarithmic equation.

17. $\log(3x - 1) + \log 4 = \log(9x + 2)$

18. $\log(2x - 1) + \log 3 = \log(4x + 1)$

19. $\ln(x + 4) - \ln 3 = \ln(x - 2)$

20. $\ln(2x + 1) - \ln 5 = \ln(x - 1)$

21. $\log(5x + 2) - \log(x - 1) = 0.7782$

22. $\log(8x + 11) - \log(x + 1) = 0.9542$

23. $\log x + \log(7 - x) = \log 10$

24. $\log x + \log(11 - x) = \log 10$

25. $\ln x + \ln(x - 3) = \ln 4$

26. $\ln x + \ln(x + 2) = \ln 8$

27. $\log(x + 1) + \log(x - 2) = 1$

28. $\log(x + 6) + \log(x - 3) = 1$

29. $\log 10x - \log(x - 450) = 2$

30. $\log(x + 48) - \log(x - 6) = 1$

Set II In Exercises 1–6, solve each equation by expressing each side as a power of the same base.

1. $2^{-3x} = \frac{1}{8}$

2. $9^x = \frac{1}{3^{-2}}$

3. $5^{3x-2} = 25^x$

4. $4^{3x} = 8$

5. $27^{5x} = 9^2$

6. $25^x = \frac{1}{125}$

In Exercises 7–16, solve each equation by taking the logarithm of both sides of the equation. Round off the answers to three significant digits.

7. $3^x = 50$

8. $8^x = 17$

9. $e^x = 29$

10. $e^x = 14$

11. $(4.6)^{x+1} = 100$

12. $(34.7)^{2x} = (12.5)^{3x-2}$

13. $(13.5)^{4x-2} = 7.12$

14. $(2.03)^{2x-1} = 142$

15. $e^{2x-3} = 60$

16. $3^{3x+5} = 25$

In Exercises 17–30, solve and check each logarithmic equation.

17. $\log(x + 4) - \log 10 = \log 6 - \log x$

18. $\log(x - 2) - \log 5 = \log 3 - \log x$

19. $\ln(5x - 7) = \ln(2x - 3) + \ln 3$

20. $\ln x + \ln(x + 4) = \ln 21$

21. $\log(2x + 1) = \log 1 + \log(x + 2)$

22. $2 \log(x + 3) = \log(7x + 1) + \log 2$

23. $\log x = \log(7x + 12) - \log(x + 3)$

24. $\log 2x + \log(x + 2) = \log(12 - x)$

25. $\ln(2x + 3) - \ln(x - 2) = \ln 5$

26. $\ln x + \ln(3x + 8) = \ln 3$

27. $\log(x + 4) + \log(x + 1) = 1$

28. $\log(x + 13) + \log(x - 8) = 2$

29. $\log 25x - \log(x - 60) = 2$

30. $\log(x + 28) - \log(x - 44) = 1$

9.6 Word Problems Involving Exponential and Logarithmic Functions

Many applications of exponential and logarithmic functions exist in the real world. The formulas for calculating bacterial growth and radioactive decay contain exponential functions, as do the formulas for calculating compound interest in business. Many formulas from the sciences, including the social sciences, earth science, and astronomy, require the ability to solve exponential and logarithmic equations.

We have selected just a few applications for this section. Listed below are the formulas we will use. In formulas 1–3, P is the amount originally invested, r is the interest rate in decimal form, t is the number of years, A is the amount in the account at the end of t years, and k is the number of times per year the interest is compounded.

1. $A = P(1 + r)^t$ Interest compounded annually
2. $A = P\left(1 + \frac{r}{k}\right)^{kt}$ Interest compounded k times a year
3. $A = Pe^{rt}$ Interest compounded continuously ($e \approx 2.71828$)
4. $y = Ce^{kt}$ Exponential growth of bacteria

where

C = the amount initially present
t = the time
k = some positive constant (it varies from one bacteria to another)
y = the amount present after time t

5. $y = Ce^{-kt}$ Radioactive decay

where

C = the amount initially present
t = the time
k = some positive constant
y = the amount present after time t

Example 1 $1,000 is invested at $5\frac{1}{4}\%$ interest. Find the amount that will be in the account after 10 years if the interest is compounded:

a. annually b. monthly c. daily d. continuously

(Round off answers to the nearest cent.)

Solutions

a. We will use the formula $A = P(1 + r)^t$, letting $P = \$1{,}000$, $r = 0.0525$ ($5\frac{1}{4}\%$ changed to a decimal), and $t = 10$.

$$A = \$1{,}000(1 + 0.0525)^{10}$$
$$A = \$1{,}000(1.0525)^{10}$$
On the calculator, $(1.0525)^{10}$ is [1.0525] [x^y] [10]
$$A \approx \$1{,}000(1.6680960)$$
$$A \approx \$1{,}668.10 \text{ (to the nearest cent)}$$

b. We will use the formula $A = P\left(1 + \frac{r}{k}\right)^{kt}$, letting $P = \$1{,}000$, $r = 0.0525$, $k = 12$, and $t = 10$.

$$A = \$1{,}000\left(1 + \frac{0.0525}{12}\right)^{(12)(10)}$$
$$A = \$1{,}000(1 + 0.004375)^{120}$$
$$A = \$1{,}000(1.004375)^{120}$$
On the calculator, $(1.004375)^{120}$ is [1.004375] [x^y] [120]
$$A \approx \$1{,}000(1.6885242)$$
$$A \approx \$1{,}688.52$$

c. We will use the same formula as for Example 1b and the same values for P, r, and t; k will be 365.

$$A = \$1{,}000\left(1 + \frac{0.0525}{365}\right)^{(365)(10)}$$

$$A = \$1{,}000(1 + 0.000143836)^{3650}$$

$$A = \$1{,}000(1.000143836)^{3650}$$

$$A \approx \$1{,}000(1.6903949)$$

$$A \approx \$1{,}690.39$$

d. We will use the formula $A = Pe^{rt}$, with P, r, and t the same as in Examples 1a, 1b, and 1c. Therefore,

$$A = \$1{,}000e^{0.0525(10)}$$

$$A = \$1{,}000e^{0.525}$$ On the calculator, $e^{0.525}$ is .525 INV ln x

$$A \approx \$1{,}000(1.6904588)$$

$$A \approx \$1{,}690.46$$

Therefore, the amount in the account at the end of ten years will be \$1,668.10 if the interest is compounded annually, \$1,688.52 if it is compounded monthly, \$1,690.39 if it is compounded daily, and \$1,690.46 if it is compounded continuously. ■

Example 2 \$1,000 is invested at $5\frac{1}{4}\%$ annual interest, compounded continuously. How long will it take for the money to double? Round off the answer to three significant digits.

Solution We will use the formula $A = Pe^{rt}$, with $P = \$1{,}000$, $r = 0.0525$, and $A = \$2{,}000$. We must find t. Therefore,

$$\$2{,}000 = \$1{,}000e^{0.0525t}$$

$$2 = e^{0.0525t}$$ Dividing both sides by \$1,000

We show two ways to finish the problem:

Method 1

We can take the natural logarithm of both sides of $2 = e^{0.0525t}$:

$$\ln 2 = \ln e^{0.0525t}$$

$$0.693147 \approx 0.0525t(\ln e)$$

$$0.693147 \approx 0.0525t$$

Method 2

We can rewrite the equation $2 = e^{0.0525t}$ in logarithmic form:

$$\log_e 2 = 0.0525t$$

$$\ln 2 = 0.0525t$$

$$0.693147 \approx 0.0525t$$

Then $$\frac{0.693147}{0.0525} \approx t$$

or $$t \approx 13.2$$

It will take about 13.2 years for the money to double. ■

Example 3 Suppose a certain culture of bacteria increases according to the formula $y = Ce^{0.04t}$. How many hours will it take for the bacteria to grow from 1,000 to 4,000? Round off the answer to four significant digits.

Solution We will substitute $C = 1{,}000$ and $y = 4{,}000$ into the formula $y = Ce^{0.04t}$. We must find t.

$$4{,}000 = 1{,}000e^{0.04t}$$

$$4 = e^{0.04t} \qquad \text{Dividing both sides by 1,000}$$

Method 1	*Method 2*
$\ln 4 = \ln e^{0.04t}$	$\log_e 4 = 0.04t$
$1.3862944 \approx 0.04t(\ln e)$	$1.3862944 \approx 0.04t$
$1.3862944 \approx 0.04t$	

Then $$t \approx \frac{1.3862944}{0.04}$$

or $$t \approx 34.66$$

It will take about 34.66 hours. ■

EXERCISES 9.6

Set I In the exercises below, round off each answer to the nearest cent if it is an amount of money. Otherwise, round it off to four significant digits.

1. \$1,250 is invested at $5\frac{1}{2}\%$ annual interest. Find the amount that will be in the account after 20 years if the interest is compounded:

a. annually b. monthly c. daily d. continuously

2. \$2,500 is invested at $5\frac{3}{4}\%$ annual interest. Find the amount that will be in the account after 10 years if the interest is compounded:

a. annually b. monthly c. daily d. continuously

3. \$1,500 is invested at $5\frac{3}{4}\%$ annual interest. How long will it take for the money to grow to \$2,000 if the interest is compounded continuously?

4. \$2,500 is invested at $5\frac{1}{2}\%$ annual interest. How long will it take for the money to grow to \$4,000 if the interest is compounded continuously?

5. It is known that a certain type of bacteria grows according to the formula $y = Ce^{0.035t}$.

a. How many bacteria will there be after 3 hr if 500 bacteria were present initially?

b. How many hours will it take for the bacteria count to grow from 500 to 800?

6. It is known that a certain type of bacteria grows according to the formula $y = Ce^{0.025t}$.

a. How many bacteria will there be after 2 hr if 900 bacteria were present initially?

b. How many hours will it take for the bacteria count to grow from 900 to 1,600?

7. A certain radioactive material decomposes according to the formula $y = Ce^{-0.3t}$. How many years will it take for 100 g to decompose to 80 g?

8. A certain radioactive material decomposes according to the formula $y = Ce^{-0.4t}$. How many years will it take for 150 g to decompose to 110 g?

Set II In the exercises below, round off each answer to the nearest cent if it is an amount of money. Otherwise, round it off to four significant digits.

1. \$2,000 is invested at $5\frac{1}{4}\%$ annual interest. Find the amount that will be in the account after 8 years if the interest is compounded:

a. annually b. monthly c. daily d. continuously

2. The formula from physics for measuring sound intensity, N, in decibels is $N = 10 \log \frac{I}{I_0}$, where I_0 is a constant and I is the power of the sound being measured. Find the number of decibels in a sound whose power is 3×10^{-10}, if $I_0 = 10^{-16}$.

3. \$2,000 is invested at $5\frac{1}{2}\%$ annual interest. How long will it take for the money to grow to \$2,800 if the interest is compounded continuously?

4. In chemistry, the number pH is a measure of the acidity or alkalinity of a solution. If $(H+)$ is the hydronium ion concentration measured in moles per liter, then $pH = -\log(H+)$. Find the pH of a solution with a hydronium ion concentration of 4.0×10^{-3}.

5. It is known that a certain type of bacteria grows according to the formula $y = Ce^{0.04t}$.

a. How many bacteria will there be after 4 hr if 800 bacteria were present initially?

b. How many hours will it take for the bacteria count to grow from 800 to 1,100?

6. The magnitude, M, of an earthquake, as measured on the Richter scale, is calculated as follows: $M = \log \frac{a}{a_0}$, where a_0 is a constant and a is the amplitude of the seismic wave. If $a_0 = 10^{-3}$, find:

a. the amplitude of the seismic wave if the magnitude of the earthquake is 6

b. the amplitude of the seismic wave if the magnitude of the earthquake is 5

7. A certain radioactive material decomposes according to the formula $y = Ce^{-0.3t}$. How many years will it take for 120 g to decompose to 100 g?

8. The formula for finding the monthly payment on a homeowner's mortgage is

$$R = \frac{Ai(1 + i)^n}{(1 + i)^n - 1}$$

where R = monthly payment
i = interest rate per month expressed as a decimal
n = number of months
A = original amount of mortgage

Find the monthly payment on a 25-year, \$40,000 loan at 12% interest.

9.7 Change of Base

In some applications, logarithms to a base different from 10 are given. To find such logarithms, we must be able to *change the base* of a logarithm. The formula for changing bases is derived as follows:

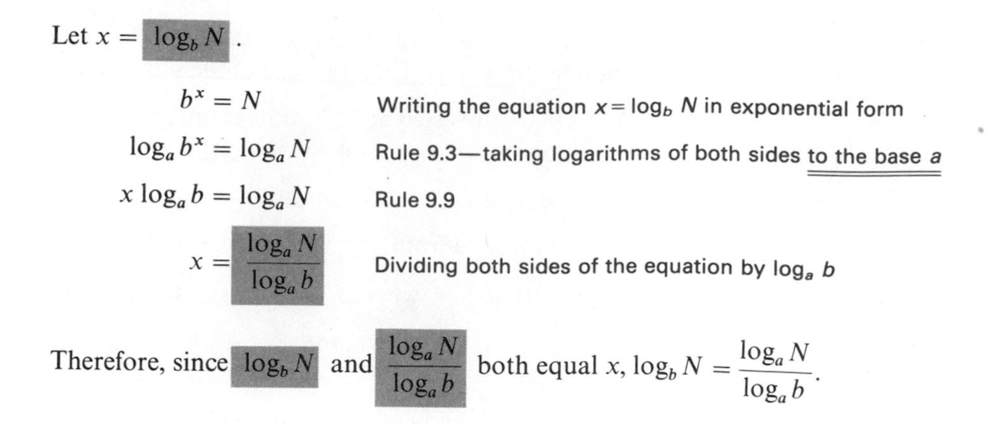

Let $x = \log_b N$.

$b^x = N$	Writing the equation $x = \log_b N$ in exponential form
$\log_a b^x = \log_a N$	Rule 9.3—taking logarithms of both sides to the base a
$x \log_a b = \log_a N$	Rule 9.9
$x = \dfrac{\log_a N}{\log_a b}$	Dividing both sides of the equation by $\log_a b$

Therefore, since $\log_b N$ and $\dfrac{\log_a N}{\log_a b}$ both equal x, $\log_b N = \dfrac{\log_a N}{\log_a b}$.

> RULE 9.10
>
> To find a logarithm with base b when we are given logarithms with base a, we use the formula
>
> $$\log_b N = \frac{\log_a N}{\log_a b}$$

Rule 9.10 makes it possible to use any table of logarithms available to find the logarithm of a number to a different base.

Example 1 Find $\log_5 51.7$.

Solution

$\log_b N = \dfrac{\log_a N}{\log_a b}$	Rule 9.10
$\log_5 51.7 = \dfrac{\log_{10} 51.7}{\log_{10} 5}$	By setting $\begin{cases} b = 5 \\ a = 10 \\ N = 51.7 \end{cases}$
$\log_5 51.7 \approx \dfrac{1.7135}{0.6990} \approx 2.451$ ■	

Example 2 Find $\log_e 51.7$ ($e \approx 2.7183$).

Solution

$\log_b N = \dfrac{\log_a N}{\log_a b}$	Rule 9.10
$\log_e 51.7 \approx \dfrac{\log_{10} 51.7}{\log_{10} 2.7183} \approx \dfrac{1.7135}{0.4343} \approx 3.945$ ■	

EXERCISES 9.7

Set I Find each logarithm. Round off each answer to four significant digits.

1. $\log_2 156$ **2.** $\log_3 231$ **3.** $\log_{12} 7.54$ **4.** $\log_{20} 9.75$

5. $\log_e 3.04$ **6.** $\log_e 4.08$ **7.** $\log_{6.8} 0.507$ **8.** $\log_{8.3} 0.0304$

Set II Find each logarithm. Round off each answer to four significant digits.

1. $\log_5 29.8$ **2.** $\log_e 53.7$ **3.** $\log_{14} 0.842$ **4.** $\log_{5.2} 0.926$

5. $\log_e 16.1$ **6.** $\log_e 0.076$ **7.** $\log_8 12$ **8.** $\log_3 0.333$

9.8 Review: 9.1–9.7

Exponential Functions 9.1A

An **exponential function** is a function of the form $y = f(x) = b^x$, where $b > 0, b \neq 1$, and $x =$ any real number.

Rules of exponents:

9.1 For $b > 0$ but $b \neq 1$, if $b^x = b^y$, then $x = y$.

9.2 For $a > 0$ and $b > 0$, if $a^x = b^x$, then $a = b$.

Logarithmic Functions 9.1B

A **logarithmic function** is a function of the form $y = f(x) = \log_b x$, where $b > 0, b \neq 1$, and $x > 0$.

The logarithm of a number x is the exponent y to which the base b ($b > 0$, $b \neq 1$) must be raised to give x. The logarithm of a number can be found either by using a calculator (Section 9.4) or by using tables (Appendix B).

Exponential and Logarithmic Forms 9.2

Logarithmic form	*Exponential form*
$y = \log_b x$	$x = b^y$

Rules of Logarithms 9.3

9.3 If $M = N$, then $\log_b M = \log_b N$.

9.4 If $\log_b M = \log_b N$, then $M = N$.

9.5 $\log_b b = 1$

9.6 $\log_b 1 = 0$

9.7 $\log_b MN = \log_b M + \log_b N$

9.8 $\log_b \dfrac{M}{N} = \log_b M - \log_b N$

9.9 $\log_b N^p = p \log_b N$

9.7 9.10 $\log_b N = \dfrac{\log_a N}{\log_a b}$ Change of base

where M and N are positive real numbers; $b > 0, b \neq 1$; p is any real number; $a > 0, a \neq 1$

Logarithms and the Calculator 9.4

Use the key marked [log] or [log x] to find common logarithms and the key marked [ln] or [ln x] to find natural logarithms.

Exponential Equations 9.5

An **exponential equation** is an equation in which the variable appears in one or more exponents.

To solve an exponential equation:

When both sides of the equation can be expressed as powers of the same base:

1. Express both sides as powers of the same base.
2. Use Rule 9.1 (if $b^x = b^y$, then $x = y$).
3. Solve the resulting equation for the variable.

When both sides of the equation cannot be expressed as powers of the same base:

1. Take the logarithms (to the same base) of both sides.

2. Use Rule 9.9 to rewrite the equation with no exponents.
3. Solve the resulting equation for the variable.

Logarithmic Equations 9.5

A **logarithmic equation** is an equation in which the variable appears in the argument of a logarithm.

To solve a logarithmic equation when every term contains a logarithm:

1. Use the rules of logarithms to write each side as a single logarithm.
2. Use Rule 9.4 (if $\log_b M = \log_b N$, then $M = N$).
3. Solve the resulting equation for the variable.
4. Check apparent solutions in the given logarithmic equation.

To solve a logarithmic equation when one term doesn't contain a logarithm:

Method 1 The equation can be simplified to the form $\log_b x = k$, rewritten in exponential form, and then solved.

Method 2 The term that doesn't contain a logarithm can be rewritten as a logarithm. Then follow the procedure for solving a logarithmic equation when every term contains a logarithm.

Review Exercises 9.8 Set I

1. Write $3^4 = 81$ in logarithmic form.

2. Write $\log_4 0.0625 = -2$ in exponential form.

In Exercises 3–7, find the value of the unknown b, N, or x.

3. $\log_{10} 1{,}000 = x$

4. $\log_{10} 0.01 = x$

5. $\log_9 N = \frac{3}{2}$

6. $\log_b \frac{1}{8} = -3$

7. $\log_{10} 145.6 = x$ (Round off the answer to four decimal places.)

In Exercises 8–10, write each expression as a single logarithm and simplify.

8. $\frac{1}{5} \log x^5 + 3 \log x^4$

9. $\log \frac{3}{5} + \log \frac{5}{3}$

10. $\log x^4y^4 - \log 6x + \log 3 - 4 \log xy$

In Exercises 11–14, find each logarithm. Round off each answer to four decimal places.

11. $\log 25.48$

12. $\log 0.0008005$

13. $\ln 0.0342$

14. $\ln 5{,}300$

In Exercises 15 and 16, find each antilogarithm. Round off each answer to four significant digits.

15. antilog 3.4072

16. Find N if $\ln N = 2.4849$.

In Exercises 17 and 18, solve each equation.

17. $81^{x-1} = \frac{1}{9}$

18. $\log 2 + \log(3x - 1) = \log(4x + 1)$

19. Find $\log_4 75$. Round off the answer to three decimal places.

20. Graph $y = 6^x$ and its inverse logarithmic function on the same set of axes.

21. $2,000 is invested at 5.8% interest. How long will it take for the money to grow to $2,500 if the interest is compounded continuously? Round off the answer to two significant digits.

Review Exercises 9.8 Set II

NAME ______________________

ANSWERS

1. ______________________
2. ______________________
3. ______________________
4. ______________________
5. ______________________
6. ______________________
7. ______________________
8. ______________________
9. ______________________
10. ______________________

1. Write $4^2 = 16$ in logarithmic form.

2. Write $\log_{1.3} 1.69 = 2$ in exponential form.

In Exercises 3–7, find the value of the unknown b, N, or x.

3. $\log_{10} 0.001 = x$

4. $\log_{27} N = \frac{2}{3}$

5. $\log_b \frac{1}{16} = -4$

6. $\log_7 N = 0$

7. $\log_b \frac{1}{16} = -2$

In Exercises 8–10, write each expression as a single logarithm and simplify.

8. $\frac{1}{4} \log x^4 + 2 \log x^2$

9. $\log \frac{14}{3} - \log \frac{7}{3}$

10. $\log(x^2 - x - 12) - \log(x - 4)$

In Exercises 11–14, find each logarithm correct to four decimal places.

11. $\log 28.25$

12. $\log 0.0003684$

13. $\ln 0.00235$

14. $\ln 12$

In Exercises 15 and 16, find each antilogarithm correct to two decimal places.

15. antilog 2.6551

16. Find N if $\ln N = 2.70805$.

In Exercises 17 and 18, solve each equation.

17. $(4.55)^{x+1} = 8.45$
Round off to three decimal places.

18. $\log(x + 3) + \log(x - 2) = \log 6$

19. Find $\log_6 148$ correct to three decimal places.

20. Graph $y = 3^x$ and its inverse logarithmic function on the same set of axes.

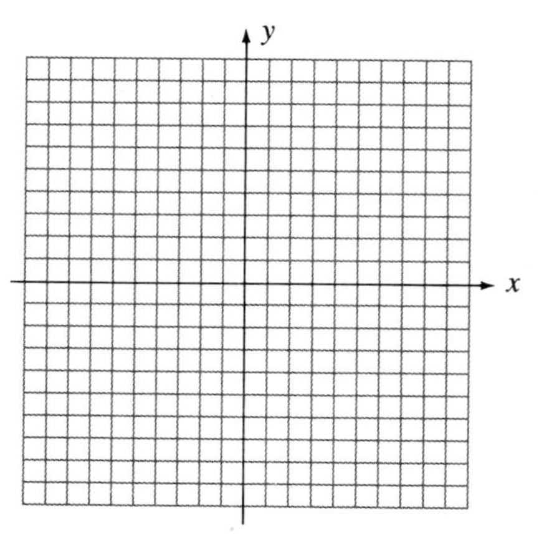

21. $1,600 is invested at 5.7% interest. How long will it take for the money to grow to $2,000 if the interest is compounded continuously? Round off the answer to two significant digits.

ANSWERS

11. ______

12. ______

13. ______

14. ______

15. ______

16. ______

17. ______

18. ______

19. ______

20. Use the grid.

21. ______

Chapter 9 Diagnostic Test

The purpose of this test is to see how well you understand exponential and logarithmic functions. We recommend that you work this diagnostic test *before* your instructor tests you on this chapter. Allow yourself about 50 minutes.

Complete solutions for all the problems on this test, together with section references, are given in the answer section at the end of the book. For the problems you do incorrectly, study the sections cited.

1. a. Write $2^4 = 16$ in logarithmic form.

b. Write $\log_{2.5} 6.25 = 2$ in exponential form.

2. Find the value of the unknown b, N, or x.

a. $\log_4 N = 3$

b. $\log_{10} 10^{-2} = x$

c. $\log_b 6 = 1$

d. $\log_5 1 = x$

e. $\log_{0.5} N = -2$

In Problems 3 and 4, write each expression as a single logarithm.

3. a. $\log x + \log y - \log z$

b. $\frac{1}{2} \log x^4 + 2 \log x$

4. $\log(x^2 - 9) - \log(x - 3)$

5. Use the rules of logarithms to solve $\log(3x + 5) - \log 7 = \log(x - 1)$.

6. Use the rules of logarithms to solve $\log(x + 8) + \log(x - 2) = \log 11$.

In Problems 7 and 8, express each answer correct to three significant digits.

7. Use logarithms to solve $e^{3x-4} = 8$.

8. Find $\log_2 718$.

9. Graph $y = 7^x$ and its inverse logarithmic function on the same set of axes.

10. Suppose a certain culture of bacteria increases according to the formula $y = Ce^{0.05t}$. If there were initially 1,500 bacteria, find:

a. the number of bacteria present after 2 hr (round off your answer to four significant digits).

b. the number of hours it will take for the bacteria to triple (round off your answer to two significant digits).

Cumulative Review Exercises: Chapters 1–9

In Exercises 1 and 2, perform the indicated operations and simplify.

1. $\dfrac{7}{x^2 - x - 12} - \dfrac{3}{x^2 - 5x + 4}$

2. $\dfrac{\dfrac{2}{x} - \dfrac{2}{y}}{\dfrac{y^2 - x^2}{xy}}$

In Exercises 3 and 4, graph the solution set for each equation.

3. $2x - 7y = 7$

4. $y = x^2 - 2x - 3$

In Exercises 5–8, find the solution set for each equation or inequality.

5. $\dfrac{2x}{3} - 1 \le \dfrac{x + 2}{5}$

6. $\sqrt{1 - 4x} = x + 5$

7. $\log(5x + 2) + \log 3 = \log(12x + 15)$

8. $e^{3x} = 5$ (Round off the answer to three significant digits.)

9. List any prime numbers greater than 7 and less than 31 that yield a remainder of 1 when divided by 5.

10. Find the length of the diagonal of a rectangle that is 2 m long and 1 m wide.

11. Find the distance between $(3, -4)$ and $(1, 2)$.

12. Write the general form of the equation of the line that has an x-intercept of -14 and a y-intercept of -6.

13. Write the general form of the equation of the line that passes through $(9, -13)$ and is parallel to the line $10x - 6y + 15 = 0$.

14. What is the multiplicative identity element?

15. What is the multiplicative inverse of -3?

16. Write $2^7 = 128$ in logarithmic form.

17. Find log 0.0312. Round off the answer to four decimal places.

18. Find N if $\ln N = 4.276666$. Round off the answer to three significant digits.

In Exercises 19–21, set up each problem algebraically and solve.

19. \$3,000 is invested at 5.6% interest. How long will it take for the money to grow to \$5,000 if the interest is compounded continuously? (Round off the answer to three significant digits.)

20. How many liters of water must be added to 5 ℓ of a 60% solution of alcohol to make a 40% solution?

21. It takes Claudio 2 hr longer to do a certain job than it does Maria. After Maria has worked on the job for 2 hr, Claudio joins her, and together they finish the job in $3\frac{1}{3}$ hr. How long would it take each of them to do the entire job working alone?

10 Nonlinear Equations and Inequalities

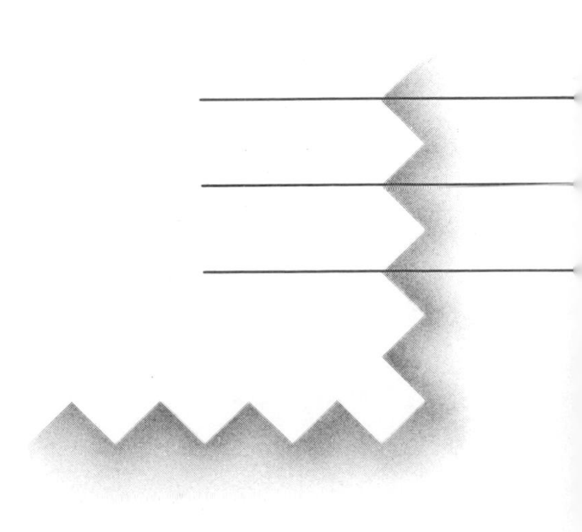

In this chapter, we discuss several methods of solving quadratic equations and inequalities. We also discuss graphing quadratic functions and conic sections. (We have already solved quadratic equations by *factoring* in Chapters 5, 6, and 7.)

10.1 Basic Definitions

Quadratic Equations

A **quadratic equation** is a polynomial equation with a *second-degree* term as its highest-degree term. Such equations are also called **second-degree equations**. The number of roots of a polynomial equation is at *most* equal to the degree of the polynomial. For this reason, we expect to get two roots when we solve quadratic equations. (Refer to the footnote on page 502.)

Example 1 Examples of quadratic equations:

a. $3x^2 + 7x + 2 = 0$ b. $\frac{1}{2}x^2 = \frac{2}{3}x - 4$

c. $x^2 - 4 = 0$ d. $5x^2 - 15x = 0$ ■

General Form (Standard Form) The *general form* of a quadratic equation is as follows:

THE GENERAL FORM OF A QUADRATIC EQUATION

$$ax^2 + bx + c = 0$$

where a, b, and c are any real numbers, except that a cannot be zero (see the NOTE below).

NOTE *In this text*, when we write the general form of a quadratic equation, a, b, and c will be *integers*, with $a > 0$. ✓

In Section 5.10, we changed quadratic equations into the general form just before we factored one side of the equation; we use the same procedure now. That is, we remove any denominators by multiplying both sides of the equation by the LCD, remove any grouping symbols, combine any like terms, get *all* terms to one side of the equation with zero on the other side, and finally, arrange the terms in descending powers.

We can then identify a, b, and c: In the equation $ax^2 + bx + c$, a is the coefficient of x^2, b is the coefficient of x, and c is the constant.

Example 2 Change each of the following quadratic equations into the general form, and then identify a, b, and c.

a. $7x = 5 - 2x^2$

Solution $2x^2 + 7x - 5 = 0 \longleftarrow$ General form $\begin{cases} a = 2 \\ b = 7 \\ c = -5 \end{cases}$

b. $5x^2 = 3$

Solution $5x^2 + 0x - 3 = 0 \longleftarrow$ General form $\begin{cases} a = 5 \\ b = 0 \\ c = -3 \end{cases}$

c. $6x = 11x^2$

Solution $0 = 11x^2 - 6x + 0$

$11x^2 - 6x + 0 = 0 \leftarrow$ General form $\begin{cases} a = 11 \\ b = -6 \\ c = 0 \end{cases}$

d. $\frac{2}{3}x^2 - 5x = \frac{1}{2}$

Solution LCD = 6

$$\frac{6}{1} \cdot \frac{2}{3}x^2 + \frac{6}{1} \cdot (-5x) = \frac{6}{1} \cdot \frac{1}{2}$$

$4x^2 - 30x = 3$

$4x^2 - 30x - 3 = 0 \leftarrow$ General form $\begin{cases} a = 4 \\ b = -30 \\ c = -3 \end{cases}$

e. $(x + 2)(2x - 3) = 3x - 7$

Solution $2x^2 + x - 6 = 3x - 7$

$2x^2 - 2x + 1 = 0 \leftarrow$ General form $\begin{cases} a = 2 \\ b = -2 \\ c = 1 \end{cases}$ ■

EXERCISES 10.1

Set I Write each of the quadratic equations in the general form; then identify *a*, *b*, and *c*.

1. $3x^2 + 5x = 2$ **2.** $3x + 5 = 2x^2$ **3.** $3x^2 = 4$

4. $16 = x^2$ **5.** $\frac{4x}{3} = 4 + x^2$ **6.** $\frac{3x}{2} - 5 = x^2$

7. $2 - 4x = 3x^2$ **8.** $6x - 3 = 5x^2$ **9.** $x(x + 2) = 4$

10. $4(x - 5) = x^2$ **11.** $3x(x - 2) = (x + 1)(x - 5)$

12. $7x(2x + 3) = (x - 3)(x + 4)$

Set II Write each of the quadratic equations in the general form; then identify *a*, *b*, and *c*.

1. $2x - 5 = 6x^2$ **2.** $6 - 3x = 5x^2$ **3.** $8 = 2x^2$

4. $x^2 - 1 = 3x$ **5.** $\frac{x^2}{2} - 3 = 7x$ **6.** $\frac{3 - x^2}{4} = x$

7. $8x - x^2 = -2$ **8.** $4 - \frac{x}{2} = \frac{x^2}{4}$ **9.** $2x(3 - x) = 5$

10. $x(x - 4) = 3$ **11.** $5x(x + 3) = (4x - 5)(6x + 3)$

12. $(2x - 1)(3x + 5) = 2x - 1$

10.2 Incomplete Quadratic Equations

An **incomplete quadratic equation** is one in which *b* or *c* (or both) is zero. The only constant that *cannot* be zero is *a*. If *a* were zero, the equation would not be quadratic.

Example 1 Examples of incomplete quadratic equations:

a. $12x^2 + 5 = 0$ $(b = 0)$ b. $7x^2 - 2x = 0$ $(c = 0)$

c. $3x^2 = 0$ $(b \text{ and } c = 0)$ ■

An incomplete quadratic equation in which $c = 0$ can be solved by factoring (see Example 2).

Example 2 Solve $12x^2 = 3x$.

Solution

$$12x^2 - 3x = 0 \quad \text{General form}$$

$$3x(4x - 1) = 0 \quad \text{GCF} = 3x$$

$$3x = 0 \quad | \quad 4x - 1 = 0 \quad \text{Setting each factor equal to 0}$$

$$x = 0 \quad | \quad 4x = 1$$

$$x = \tfrac{1}{4}$$

The check is left to the student. The solution set is $\{0, \frac{1}{4}\}$. ■

A WORD OF CAUTION In Example 2, a mistake students frequently make is to divide both sides of the equation by x.

$$12x^2 = 3x$$

$$\cancel{12x = 3} \quad \text{Dividing both sides by } x$$

$$x = \tfrac{1}{4}$$

Using this method, we found only the solution $\frac{1}{4}$. The equations $12x^2 = 3x$ and $12x = 3$ are *not* equivalent equations. By dividing both sides of the equation by x, we lost the solution 0.

Do not divide both sides of an equation by an expression containing the variable because you may lose solutions. ☑

In Section 7.7, the following method was shown for solving an incomplete quadratic equation in which $b = 0$ *and* $k > 0$:

RULE 10.1

If $x^2 = k$, then $x = \sqrt{k}$ or $x = -\sqrt{k}$.

It can be shown that this rule holds even if k is negative; therefore, we can solve an equation of the form $ax^2 + c = 0$ by the following method:

TO SOLVE THE QUADRATIC EQUATION $ax^2 + c = 0$

1. Add $-c$ to both sides of the equation.
2. Divide both sides of the equation by a, the coefficient of x^2.
3. Use Rule 10.1: If $x^2 = k$, then $x = \pm\sqrt{k}$. When we do this, we often say we are taking the square root of both sides of the equation.
4. Express the square roots in simplest form. When the radicand is:

   positive or zero, the square roots are real numbers;

   negative, the square roots are complex numbers.

NOTE In this section, when the solutions are irrational, we will express them in simplest radical form; we will not find decimal approximations to the solutions. ☑

Example 3 Solve $3x^2 - 5 = 0$.

Solution

$$3x^2 = 5 \quad \text{Adding 5 to both sides}$$

$$x^2 = \frac{5}{3} \quad \text{Dividing both sides by 3}$$

$$x = \pm\sqrt{\frac{5}{3}} \quad \text{Using Rule 10.1}$$

$$x = \pm\sqrt{\frac{5 \cdot 3}{3 \cdot 3}} = \pm\frac{\sqrt{15}}{3} \quad \text{Simplifying the square roots}$$

The check is left to the student. The solution set is

$$\left\{\frac{\sqrt{15}}{3}, -\frac{\sqrt{15}}{3}\right\} \quad ■$$

Example 4 Solve $x^2 + 25 = 0$.

Solution

$$x^2 = -25$$

$$x = \pm\sqrt{-25} \quad \text{Using Rule 10.1}$$

$$x = \pm 5i$$

Check

$$x^2 + 25 = 0$$

$$(\pm 5i)^2 + 25 \stackrel{?}{=} 0$$

$$25i^2 + 25 \stackrel{?}{=} 0$$

$$-25 + 25 = 0$$

The solution set is $\{5i, -5i\}$. ■

In Section 7.6, we solved equations with rational exponents by raising both sides of the equation to the same power. However, the problems in that section were carefully selected to ensure that the power was never a fraction with an even-number denominator. Raising both sides of an equation to a fractional power in which the *denominator* of the exponent is an *even* number is equivalent to taking an *even root* of both sides of the equation. Remember that when we do take an even root of both sides of an equation there will be both a positive and a negative root, so we must put a $\pm$ sign in front of one side of the equation (usually in front of the constant). Therefore, in Example 5, we will put a $\pm$ sign in front of the constant when we raise both sides to the $-\frac{3}{2}$ power.

Example 5 Solve $x^{-2/3} = 4$.

Solution

$$x^{-2/3} = 2^2$$

$$(x^{-2/3})^{-3/2} = \pm(2^2)^{-3/2}$$

We raise both sides to the $-\frac{3}{2}$ power in order to make the exponent of x equal to 1: $(-\frac{2}{3})(-\frac{3}{2}) = 1$

$$x^{(-2/3)(-3/2)} = \pm 2^{2(-3/2)}$$

$$x = \pm 2^{-3} = \pm\tfrac{1}{8}$$

Check for $x = \frac{1}{8}$	*Check for* $x = -\frac{1}{8}$
$x^{-2/3} = 4$	$x^{-2/3} = 4$
$(\frac{1}{8})^{-2/3} \stackrel{?}{=} 4$	$(-\frac{1}{8})^{-2/3} \stackrel{?}{=} 4$
$(8)^{2/3} \stackrel{?}{=} 4$	$(-8)^{2/3} \stackrel{?}{=} 4$
$(\sqrt[3]{8})^2 \stackrel{?}{=} 4$	$\sqrt[3]{(-8)^2} \stackrel{?}{=} 4$
$2^2 \stackrel{?}{=} 4$	$\sqrt[3]{64} \stackrel{?}{=} 4$
$4 = 4$	$4 = 4$

The solution set is $\{-\frac{1}{8}, \frac{1}{8}\}$. ■

EXERCISES 10.2

Set I In Exercises 1–20, find the solution set for each equation.

1. $x^2 - 27 = 0$ **2.** $x^2 - 8 = 0$ **3.** $x^2 + 16 = 0$

4. $x^2 + 81 = 0$ **5.** $x^2 = 7$ **6.** $x^2 = 13$

7. $x^2 = -12$ **8.** $x^2 = -75$ **9.** $12x = 8x^3$

10. $9x = 12x^3$ **11.** $5x^2 + 4 = 0$ **12.** $3x^2 + 25 = 0$

13. $\frac{2x^2}{3} = 4x$ **14.** $\frac{3x}{5} = 6x^2$

15. $x(x - 2) = (2x + 3)x$ **16.** $x(x - 3) = x(2x - 8)$

17. $\frac{x + 2}{3x} = \frac{x + 1}{x}$ **18.** $\frac{3x - 2}{4x} = \frac{x + 1}{3x}$

19. $3x^{-2/3} = 48$ **20.** $x^{2/5} = 4$

In Exercises 21–24, set up each problem algebraically and solve. In Exercises 22–24, express each answer in simplest radical form.

21. The length of the diagonal of a square is $\sqrt{32}$. What is the length of a side?

22. The length of the diagonal of a square is 18. What is the length of a side?

23. A rectangle is 7 cm wide and 10 cm long. Find the length of its diagonal.

24. A rectangle is 12 cm long and 8 cm wide. Find the length of its diagonal.

Set II In Exercises 1–20, find the solution set for each equation.

1. $x^2 - 50 = 0$ **2.** $x^2 - 48 = 0$ **3.** $x^2 + 144 = 0$

4. $x^2 = -5$ **5.** $x^2 = 3$ **6.** $x^2 - 12 = 0$

7. $x^2 = -20$ **8.** $x^2 = 3x$ **9.** $8x = 10x^3$

10. $5x = 3x^2$ **11.** $3x^2 + 7 = 0$ **12.** $2x^2 = 11$

13. $\frac{5x}{2} = 15x^2$ **14.** $3x = \frac{2x^2}{5}$

15. $2x(x - 1) = 3x(2x + 1)$ **16.** $4x(3x + 2) = x(x - 5)$

17. $\frac{2x - 1}{3x} = \frac{x - 3}{x}$ **18.** $\frac{4x - 3}{5x} = \frac{2x - 1}{x}$

19. $x^{4/5} = 16$ **20.** $x^{4/3} = 81$

In Exercises 21–24, set up each problem algebraically and solve. In Exercises 21–23, express each answer in simplest radical form.

21. The length of the diagonal of a square is 100. What is the length of a side?

22. The area of a square is 75. What is the length of a side?

23. A rectangle is 12 m long and 4 m wide. Find the length of its diagonal.

24. The area of a certain square is numerically equal to its perimeter. Find the length of a side of the square.

10.3 Solving Equations That Are Quadratic in Form

Sometimes, equations that are not quadratics can be solved like quadratics after an appropriate substitution is made. If an equation can be written in the form $aX^{2} + bX^{1} + c = 0$, where X represents any algebraic expression, we say the equation is **quadratic in form** (see Example 1).

Example 1 Examples of recognizing equations that are quadratic in form:

a. $9 + 5y^{-8} - 19y^{-4} = 0$ can be written as $5(y^{-4})^{2} - 19(y^{-4})^{1} + 9 = 0$; therefore, it is quadratic in form.

b. $h^{-2/3} - h^{-1/3} = 0$ can be written as $(h^{-1/3})^{2} - (h^{-1/3})^{1} = 0$; therefore, it is quadratic in form.

c. $12 + (x^2 - 2x)^2 - 7(x^2 - 2x) = 0$ can be written as $(x^2 - 2x)^{2} - 7(x^2 - 2x)^{1} + 12 = 0$; therefore, it is quadratic in form. ■

USING SUBSTITUTION TO SOLVE AN EQUATION THAT IS WRITTEN AS $a(X)^2 + b(X)^1 + c = 0$

1. Let some different variable equal X.
2. Substitute that variable for X.
3. Solve the resulting quadratic equation.
4. Set each solution of the quadratic equation equal to X.
5. Solve each resulting equation for its variable.
6. Check all answers in the *original* equation, because there may be extraneous roots.

Example 2 Solve $x^4 - 29x^2 + 100 = 0$.

Solution (Because this is a fourth-degree equation, we expect to find four roots.) $x^4 - 29x^2 + 100 = 0$ can be written as $(x^2)^2 - 29(x^2)^1 + 100 = 0$. Let $z = x^2$; then $z^2 = x^4$.

Therefore, $x^4 - 29x^2 + 100 = 0$

becomes $z^2 - 29z + 100 = 0$ This is quadratic in z

$(z - 4)(z - 25) = 0$

$z - 4 = 0$	$z - 25 = 0$	Setting each factor equal to 0
$z = 4$	$z = 25$	
$x^2 = 4$	$x^2 = 25$	Replacing z with x^2
$x = \pm 2$	$x = \pm 5$	

Check for $x = \pm 2$	*Check for* $x = \pm 5$
$(\pm 2)^4 - 29(\pm 2)^2 + 100 \stackrel{?}{=} 0$	$(\pm 5)^4 - 29(\pm 5)^2 + 100 \stackrel{?}{=} 0$
$16 - 29(4) + 100 \stackrel{?}{=} 0$	$625 - 29(25) + 100 \stackrel{?}{=} 0$
$16 - 116 + 100 \stackrel{?}{=} 0$	$625 - 725 + 100 \stackrel{?}{=} 0$
$0 = 0$	$0 = 0$

The solution set is $\{\pm 2, \pm 5\}$. ■

Example 3 Solve $h^{-2/3} - h^{-1/3} = 0$.

Solution $h^{-2/3} - h^{-1/3} = 0$ can be written as $(h^{-1/3})^2 - (h^{-1/3})^1 = 0$. Let $z = h^{-1/3}$; then $z^2 = h^{-2/3}$.

Therefore, $$h^{-2/3} - h^{-1/3} = 0$$

becomes $$z^2 - z = 0$$

$$z(z - 1) = 0$$

$z = 0$	$z - 1 = 0$	
	$z = 1$	
$h^{-1/3} = 0$	$h^{-1/3} = 1$	Replacing z with $h^{-1/3}$
$(h^{-1/3})^{-3} = (0)^{-3}$ ← Does not exist	$(h^{-1/3})^{-3} = (1)^{-3}$	
	$h = 1$	

The solution 1 checks. (We leave the check to the student.) The solution set is $\{1\}$. ■

Example 4 Solve $(x^2 - 2x)^2 - 11(x^2 - 2x) + 24 = 0$.

Solution Let $z = x^2 - 2x$; then $z^2 = (x^2 - 2x)^2$.

Therefore, $$(x^2 - 2x)^2 - 11(x^2 - 2x) + 24 = 0$$

becomes $$z^2 - 11z + 24 = 0$$

$$(z - 3)(z - 8) = 0$$

$z - 3 = 0$	$z - 8 = 0$	
$z = 3$	$z = 8$	
$x^2 - 2x = 3$	$x^2 - 2x = 8$	Replacing z with $x^2 - 2x$
$x^2 - 2x - 3 = 0$	$x^2 - 2x - 8 = 0$	
$(x + 1)(x - 3) = 0$	$(x + 2)(x - 4) = 0$	

$x + 1 = 0$	$x - 3 = 0$	$x + 2 = 0$	$x - 4 = 0$
$x = -1$	$x = 3$	$x = -2$	$x = 4$

All these solutions check. (We leave the checks to the student.) The solution set is $\{-1, 3, -2, 4\}$. ■

EXERCISES 10.3

Set I In Exercises 1–12, find the solution set for each equation by factoring, after making appropriate substitutions.

1. $x^4 - 37x^2 + 36 = 0$

2. $y^4 - 13y^2 + 36 = 0$

3. $z^{-4} - 10z^{-2} + 9 = 0$

4. $x^{-4} - 5x^{-2} + 4 = 0$

5. $y^{2/3} - 5y^{1/3} = -4$

6. $x^{2/3} - 10x^{1/3} = -9$

7. $z^{-4} - 4z^{-2} = 0$

8. $R^{-4} - 9R^{-2} = 0$

9. $K^{-2/3} + 2K^{-1/3} + 1 = 0$

10. $M^{-1} - 2M^{-1/2} + 1 = 0$

11. $(x^2 - 4x)^2 - (x^2 - 4x) - 20 = 0$

12. $(x^2 - 2x)^2 - 2(x^2 - 2x) - 3 = 0$

In Exercises 13–16, set up each problem algebraically and solve.

13. The length of a rectangle is twice its width. If the numerical sum of its area and perimeter is 80, find the length and width.

14. The length of a rectangle is three times its width. If the numerical sum of its area and perimeter is 80, find its dimensions.

15. Bruce drives from Los Angeles to the Mexican border and back to Los Angeles, a total distance of 240 mi. His average speed returning to Los Angeles was 20 mph faster than his average speed going to Mexico. If his total driving time was 5 hr, what was his average speed driving from Los Angeles to Mexico?

16. Ruth drives from Creston to Des Moines, a distance of 90 mi. Then she continues on from Des Moines to Omaha, a distance of 120 mi. Her average speed was 10 mph faster on the second part of the journey than on the first part. If the total driving time was 6 hr, what was her average speed on the first leg of the journey?

Set II In Exercises 1–12, find the solution set for each equation by factoring, after making appropriate substitutions.

1. $y^4 - 26y^2 + 25 = 0$

2. $x^4 + 49 = 50x^2$

3. $x^{-4} - 17x^{-2} = -16$

4. $x^{-4} - 29x^{-2} + 100 = 0$

5. $x^{2/3} - 4x^{1/3} + 4 = 0$

6. $y^{2/3} - 2y^{1/3} = 3$

7. $x^{-4} - 16x^{-2} = 0$

8. $x^4 = 16x^2$

9. $x^{-2/3} - 6x^{-1/3} + 9 = 0$

10. $x^{-2/3} + x^{-1/3} = 2$

11. $(x^2 - 6x)^2 + 17(x^2 - 6x) + 72 = 0$

12. $(x^2 + 2x)^2 - 7(x^2 + 2x) = 8$

In Exercises 13–16, set up each problem algebraically and solve.

13. The length of a rectangle is 2 more than twice its width. If its diagonal is 3 more than twice its width, what are its dimensions?

14. The tens digit of a two-digit number is 4 more than the units digit. If the product of the units digit and tens digit is 21, find the number.

15. Jeff jogged from his home to a park 15 mi away, and then he walked back home. He jogged 2 mph faster than he walked. If his total traveling time was 8 hr, how fast did he jog?

16. If the product of two consecutive even integers is increased by 4, the result is 84. Find the integers.

10.4 Completing the Square of a Quadratic Equation and the Quadratic Formula

In Section 5.7, we used *completing the square* as a method of factoring. In this section, we use completing the square to solve quadratic equations and to derive the quadratic formula. The methods we have used in this chapter so far can be used only to solve *some* quadratic equations. The quadratic formula (or completing the square) can be used to solve *any* quadratic equation.

Solving a Quadratic Equation by Completing the Square

The technique we use here for completing the square is different from the one shown in Chapter 5, because in this chapter we are working with *equations*. In Chapter 5, we had to add an amount *to* and then subtract it *from* a polynomial, since we were not working with an equation. Here, we will add the same number to both sides of the equation. The method is as follows:

TO SOLVE A QUADRATIC EQUATION (IN x) BY COMPLETING THE SQUARE

1. Get all the terms containing x on one side of the equal sign and the constant term on the other side.
2. Divide both sides of the equation by a, the coefficient of x^2.
3. Find the quantity that, when added to the left side, makes the left side a trinomial square. Do this by squaring one-half of the coefficient of x^1.
4. Add the number found in step 3 to both sides of the equation.
5. Factor the left side.
6. Use Rule 10.1: If $x^2 = k$, then $x = \pm\sqrt{k}$.
7. Solve the resulting first-degree (linear) equation for x.
8. Check your solutions.

Example 1 Solve $x^2 - 4x + 1 = 0$.

Solution

Step 3.

To find the quantity that, when added to both sides, makes the left side a trinomial square:

Take $\frac{1}{2}$ of -4: $\frac{1}{2}(-4) = -2$

Then square the result: $(-2)^2 = 4$

Step 1. $x^2 - 4x = -1$

Step 4. $x^2 - 4x + 4 = -1 + 4$ Adding 4 to both sides to make the left side a trinomial square

Step 5. $(x - 2)^2 = 3$ Factoring the left side and simplifying the right side

Step 6. $x - 2 = \pm\sqrt{3}$ Using Rule 10.1

Step 7. $x = 2 \pm \sqrt{3}$ Adding 2 to both sides

Step 8. *Check for* $x = 2 + \sqrt{3}$

$$x^2 - 4x + 1 = 0$$
$$(2 + \sqrt{3})^2 - 4(2 + \sqrt{3}) + 1 \stackrel{?}{=} 0$$
$$4 + 4\sqrt{3} + 3 - 8 - 4\sqrt{3} + 1 \stackrel{?}{=} 0$$
$$0 = 0$$

We leave the check for $x = 2 - \sqrt{3}$ to the student. ■

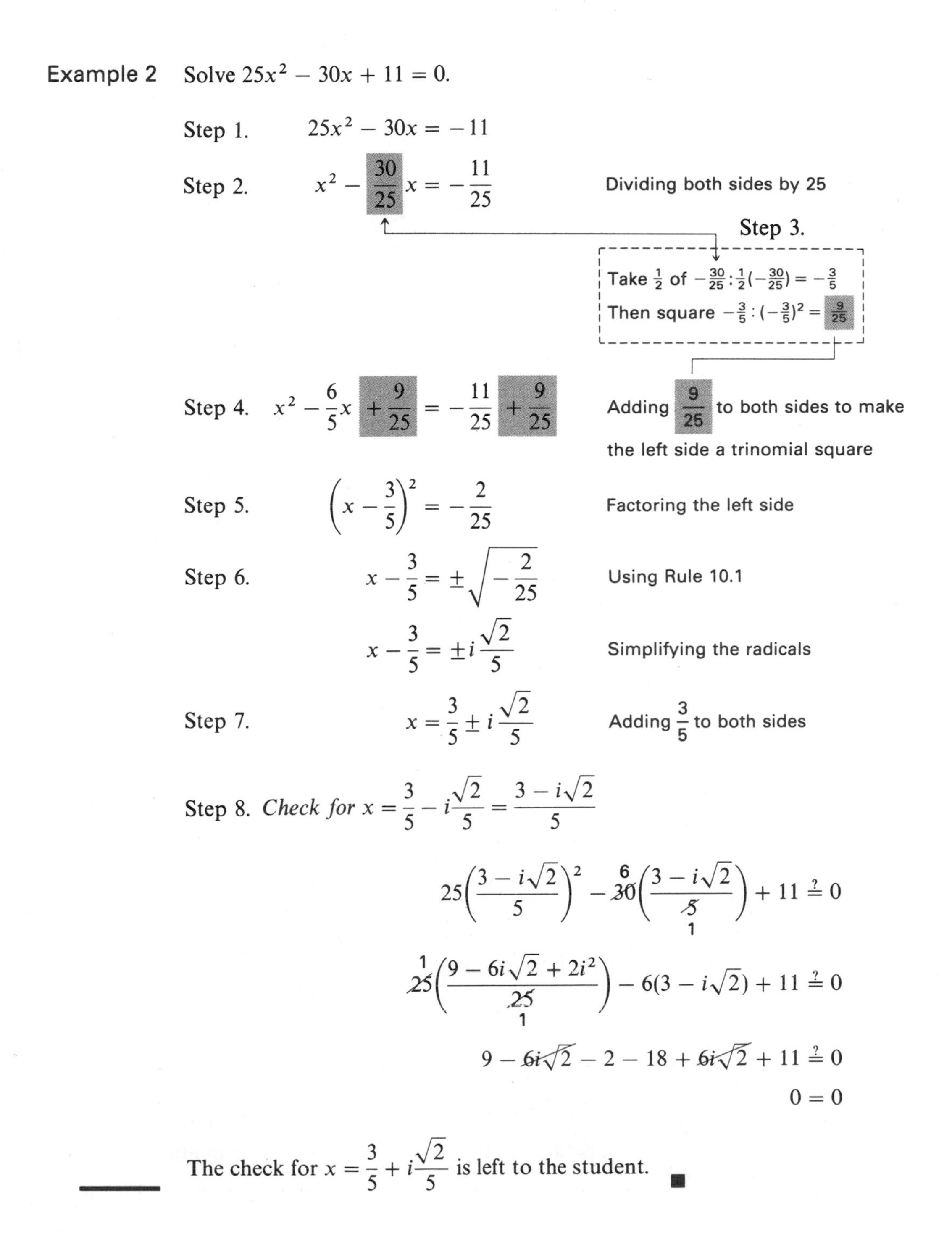

Example 2 Solve $25x^2 - 30x + 11 = 0$.

Step 1. $25x^2 - 30x = -11$

Step 2. $x^2 - \frac{30}{25}x = -\frac{11}{25}$ — Dividing both sides by 25

Step 3.

Take $\frac{1}{2}$ of $-\frac{30}{25}$: $\frac{1}{2}\left(-\frac{30}{25}\right) = -\frac{3}{5}$

Then square $-\frac{3}{5}$: $\left(-\frac{3}{5}\right)^2 = \frac{9}{25}$

Step 4. $x^2 - \frac{6}{5}x + \frac{9}{25} = -\frac{11}{25} + \frac{9}{25}$ — Adding $\frac{9}{25}$ to both sides to make the left side a trinomial square

Step 5. $\left(x - \frac{3}{5}\right)^2 = -\frac{2}{25}$ — Factoring the left side

Step 6. $x - \frac{3}{5} = \pm\sqrt{-\frac{2}{25}}$ — Using Rule 10.1

$x - \frac{3}{5} = \pm i\frac{\sqrt{2}}{5}$ — Simplifying the radicals

Step 7. $x = \frac{3}{5} \pm i\frac{\sqrt{2}}{5}$ — Adding $\frac{3}{5}$ to both sides

Step 8. *Check for* $x = \frac{3}{5} - i\frac{\sqrt{2}}{5} = \frac{3 - i\sqrt{2}}{5}$

$$25\left(\frac{3 - i\sqrt{2}}{5}\right)^2 - \overset{6}{\cancel{30}}\left(\frac{3 - i\sqrt{2}}{\underset{1}{\cancel{5}}}\right) + 11 \stackrel{?}{=} 0$$
$$\overset{1}{\cancel{25}}\left(\frac{9 - 6i\sqrt{2} + 2i^2}{\underset{1}{\cancel{25}}}\right) - 6(3 - i\sqrt{2}) + 11 \stackrel{?}{=} 0$$
$$9 - \cancel{6i\sqrt{2}} - 2 - 18 + \cancel{6i\sqrt{2}} + 11 \stackrel{?}{=} 0$$
$$0 = 0$$

The check for $x = \frac{3}{5} + i\frac{\sqrt{2}}{5}$ is left to the student. ■

Solving a Quadratic Equation by Using the Quadratic Formula

The method of completing the square can be used to solve *any* quadratic equation. We now use it to solve the *general form* of the quadratic equation and in this way, we derive the *quadratic formula.*

	$ax^2 + bx + c = 0$	The general form
Step 1.	$ax^2 + bx = 0 - c$	Subtracting c from both sides
Step 2.	$x^2 + \frac{b}{a}x = -\frac{c}{a}$	Dividing both sides by a
Step 3.	Take $\frac{1}{2}$ of $\frac{b}{a}$: $\frac{1}{2}\left(\frac{b}{a}\right) = \frac{b}{2a}$ Then $\left(\frac{b}{2a}\right)^2 = \frac{b^2}{4a^2}$	
Step 4.	$x^2 + \frac{b}{a}x + \frac{b^2}{4a^2} = \frac{b^2}{4a^2} - \frac{c}{a}$	Adding $\frac{b^2}{4a^2}$ to both sides to make the left side a trinomial square
Step 5.	$\left(x + \frac{b}{2a}\right)^2 = \frac{b^2 - 4ac}{4a^2}$	Factoring the left side and adding the fractions on the right side
Step 6.	$x + \frac{b}{2a} = \pm\sqrt{\frac{b^2 - 4ac}{4a^2}}$	Using Rule 10.1
	$x + \frac{b}{2a} = \pm\frac{\sqrt{b^2 - 4ac}}{\sqrt{4a^2}} = \pm\frac{\sqrt{b^2 - 4ac}}{2a}$	Simplifying the radicals
Step 7.	$x = -\frac{b}{2a} \pm \frac{\sqrt{b^2 - 4ac}}{2a}$	Adding $-\frac{b}{2a}$ to both sides
Therefore,	$x = \frac{-b \pm \sqrt{b^2 - 4ac}}{2a}$	The quadratic formula

The procedure for using the quadratic formula can be summarized as follows:

TO SOLVE A QUADRATIC EQUATION BY THE QUADRATIC FORMULA

1. Arrange the equation in the general form.

$$ax^2 + bx + c = 0$$

and identify a, b, and c.

2. Substitute the values of a, b, and c into the quadratic formula.

$$x = \frac{-b \pm \sqrt{b^2 - 4ac}}{2a} \qquad (a \neq 0)$$

3. Simplify your answers.

4. Check your answers by substituting them in the original equation.

Example 3 Solve $x^2 - 5x + 6 = 0$ by the quadratic formula.

Solution Substitute $\begin{Bmatrix} a = 1 \\ b = -5 \\ c = 6 \end{Bmatrix}$ in the formula $x = \dfrac{-b \pm \sqrt{b^2 - 4ac}}{2a}$.

$$x = \frac{-(-5) \pm \sqrt{(-5)^2 - 4(1)(6)}}{2(1)}$$

$$x = \frac{5 \pm \sqrt{25 - 24}}{2} = \frac{5 \pm \sqrt{1}}{2}$$

$$x = \frac{5 \pm 1}{2} = \begin{cases} \dfrac{5+1}{2} = \dfrac{6}{2} = 3 \\ \dfrac{5-1}{2} = \dfrac{4}{2} = 2 \end{cases}$$

This equation can also be solved by factoring.

$$x^2 - 5x + 6 = 0$$

$$(x - 2)(x - 3) = 0$$

$$x - 2 = 0 \quad \big| \quad x - 3 = 0$$

$$x = 2 \quad \big| \quad x = 3$$

The check is left to the student. ■

Solving a quadratic equation by factoring is ordinarily faster than using the quadratic formula. Therefore, first check to see if the equation can be factored easily by any of the methods discussed in Chapter 5. If it cannot, use the quadratic formula or complete the square.

Example 4 Solve $\frac{1}{4}x^2 = 1 - x$.

Solution LCD = 4. We first change the equation to the general form.

$$\frac{4}{1} \cdot \frac{1}{4}x^2 = 4 \cdot (1 - x)$$

$$x^2 = 4 - 4x$$

$$x^2 + 4x - 4 = 0$$

Substitute $\begin{Bmatrix} a = 1 \\ b = 4 \\ c = -4 \end{Bmatrix}$ into $x = \dfrac{-b \pm \sqrt{b^2 - 4ac}}{2a}$.

$$x = \frac{-(4) \pm \sqrt{(4)^2 - 4(1)(-4)}}{2(1)}$$

$$= \frac{-4 \pm \sqrt{16 + 16}}{2} = \frac{-4 \pm \sqrt{32}}{2}$$

$$= \frac{-4 \pm 4\sqrt{2}}{2} = -\frac{4}{2} \pm \frac{4\sqrt{2}}{2} = -2 \pm 2\sqrt{2}$$

Check for $x = -2 - 2\sqrt{2}$

$$\frac{1}{4}x^2 = 1 - x$$

$$\frac{1}{4}(-2 - 2\sqrt{2})^2 \stackrel{?}{=} 1 - (-2 - 2\sqrt{2})$$

$$\frac{1}{4}(4 + 8\sqrt{2} + 8) \stackrel{?}{=} 1 + 2 + 2\sqrt{2}$$

$$\frac{1}{4}(12 + 8\sqrt{2}) \stackrel{?}{=} 3 + 2\sqrt{2}$$

$$3 + 2\sqrt{2} = 3 + 2\sqrt{2}$$

The check for $x = -2 + 2\sqrt{2}$ is left to the student. ■

Example 5 Solve $x^2 - 6x + 13 = 0$.

Solution Substitute $\begin{Bmatrix} a = 1 \\ b = -6 \\ c = 13 \end{Bmatrix}$ into $x = \dfrac{-b \pm \sqrt{b^2 - 4ac}}{2a}$.

$$x = \frac{-(-6) \pm \sqrt{(-6)^2 - 4(1)(13)}}{2(1)}$$

$$= \frac{6 \pm \sqrt{36 - 52}}{2} = \frac{6 \pm \sqrt{-16}}{2}$$

$$= \frac{6 \pm 4i}{2} = \frac{6}{2} \pm \frac{4i}{2} = 3 \pm 2i$$

Check for $x = 3 + 2i$

$$x^2 \quad - \quad 6x \quad + 13 = 0$$

$$(3 + 2i)^2 - 6(3 + 2i) + 13 \stackrel{?}{=} 0$$

$$9 + 12i + 4i^2 - 18 - 12i + 13 \stackrel{?}{=} 0$$

$$9 + 12i - 4 - 18 - 12i + 13 \stackrel{?}{=} 0$$

$$0 = 0$$

We leave the check for $x = 3 - 2i$ to the student. ■

Example 6 Solve $x^2 - 5x + 3 = 0$. Express the answers in simplest radical form *and* approximate the answers correct to two decimal places. Do not check the solutions.

Solution Substitute $\begin{Bmatrix} a = 1 \\ b = -5 \\ c = 3 \end{Bmatrix}$ into $x = \dfrac{-b \pm \sqrt{b^2 - 4ac}}{2a}$.

$$x = \frac{-(-5) \pm \sqrt{(-5)^2 - 4(1)(3)}}{2(1)}$$

$$= \frac{5 \pm \sqrt{25 - 12}}{2} = \frac{5 \pm \sqrt{13}}{2} \qquad \sqrt{13} \approx 3.606$$

$$\approx \frac{5 \pm 3.606}{2} = \begin{cases} \dfrac{5 + 3.606}{2} = \dfrac{8.606}{2} = 4.303 \approx 4.30 \\ \dfrac{5 - 3.606}{2} = \dfrac{1.394}{2} = 0.697 \approx 0.70 \end{cases}$$

■

EXERCISES 10.4

Set I In Exercises 1–6, use the method of completing the square to find the solution set for each equation.

1. $x^2 = 6x + 11$ **2.** $x^2 = 10x - 13$ **3.** $x^2 - 13 = 4x$

4. $x^2 + 20 = 8x$ **5.** $x^2 - 2x - 2 = 0$ **6.** $4x^2 - 8x + 1 = 0$

In Exercises 7–24, use the quadratic formula to find the solution set for each equation.

7. $3x^2 - x - 2 = 0$ **8.** $2x^2 + 3x - 2 = 0$ **9.** $x^2 - 4x + 1 = 0$

10. $x^2 - 4x - 1 = 0$ **11.** $x^2 - 4x + 2 = 0$ **12.** $x^2 - 2x - 2 = 0$

13. $x^2 + x + 5 = 0$ **14.** $x^2 + x + 7 = 0$ **15.** $3x^2 + 2x + 1 = 0$

16. $4x^2 + 3x + 2 = 0$ **17.** $2x^2 = 8x - 9$ **18.** $3x^2 = 6x - 4$

19. $x + \frac{1}{3} = \frac{-1}{3x}$ **20.** $x + \frac{1}{4} = \frac{-1}{4x}$ **21.** $2x^2 - 5x = -7$

22. $3x^2 - 5x = -6$

In Exercises 23 and 24, show the check.

23. $x^2 - 4x + 5 = 0$ **24.** $x^2 - 6x + 10 = 0$

In Exercises 25–28, set up each problem algebraically and solve. Express answers in simplest radical form *and* use a calculator to approximate the answers correct to two decimal places.

25. The length of a rectangle is 2 more than its width. If its area is 2, find its dimensions.

26. The length of a rectangle is 4 more than its width. If its area is 6, find its dimensions.

27. The perimeter of a square is numerically 4 more than its area. Find the length of its side.

28. The area of a square is numerically 2 more than its perimeter. Find the length of its side.

Set II In Exercises 1–6, use the method of completing the square to find the solution set for each equation.

1. $x^2 = 4x + 10$ **2.** $x^2 - 6x - 5 = 0$ **3.** $x^2 - 3 = 8x$

4. $x^2 + x = 5$ **5.** $x^2 - 3x - 3 = 0$ **6.** $3x^2 - 6x + 1 = 0$

In Exercises 7–24, use the quadratic formula to find the solution set for each equation.

7. $4x^2 = 12x - 7$ **8.** $3x^2 + 2x = -5$ **9.** $3x^2 = 4x + 1$

10. $5x^2 = 1 - 2x$ **11.** $2x^2 = 3 - 5x$ **12.** $3x = 1 - x^2$

13. $x^2 + x + 4 = 0$ **14.** $x^2 - x + 4 = 0$ **15.** $4x^2 + 3x + 1 = 0$

16. $6x^2 = 2 - x$ **17.** $x^2 + 6 = 2x$ **18.** $x^2 + 3x = -4$

19. $\frac{x}{2} + \frac{6}{x} = \frac{5}{2}$ **20.** $\frac{x}{3} + \frac{3}{x} = 2$ **21.** $4x^2 - 5x = -2$

22. $5x^2 - x + 3 = 0$

In Exercises 23 and 24, show the check.

23. $x^2 - x + 1 = 0$ **24.** $x^2 + x + 1 = 0$

In Exercises 25–28, set up each problem algebraically and solve. Express the answers in simplest radical form *and* use a calculator to approximate the answers correct to two decimal places.

25. The length of a rectangle is 4 more than its width. If its area is 1, find its dimensions.

26. Find the length of the diagonal of a square with side equal to 4.

27. The area of a square is numerically 12 more than its perimeter. Find the length of its side.

28. The diagonal of a rectangle is 4, and the length of one side of the rectangle is 2. Find the length of the other side of the rectangle.

10.5 The Nature of Quadratic Roots

Of the quadratic equations solved so far, some have had real roots, some complex roots, some rational roots, and some irrational roots. Some have had equal roots, and some have had unequal roots. In this section, we will show how to determine what kinds of roots a quadratic equation has *without actually solving the equation.*

The Quadratic Discriminant

We know that the roots of the quadratic equation $ax^2 + bx + c = 0$ are

$$x = \frac{-b \pm \sqrt{b^2 - 4ac}}{2a}$$

If, for some equation, $\sqrt{b^2 - 4ac} = \sqrt{-24}$, then the roots of that equation are *complex conjugates.* If, for another equation, $\sqrt{b^2 - 4ac} = \sqrt{13}$, then its roots must be *real, irrational conjugates.* If, for still another equation, $\sqrt{b^2 - 4ac} = \sqrt{25} = 5$, then its roots must be *real* and *rational.* Since the value of $b^2 - 4ac$ determines the *nature* of the roots, $b^2 - 4ac$ is called the **quadratic discriminant**.

The information showing the relationship between the quadratic discriminant and the roots of an equation is summarized in the following box:

THE RELATION BETWEEN THE QUADRATIC DISCRIMINANT AND ROOTS

If $b^2 - 4ac$ *is:*	*The roots will be:*
Positive and a perfect square	Real, rational, and unequal
Positive and not a perfect square	Real, unequal, irrational conjugates
Zero	One real, rational root *of multiplicity two**
Negative	Complex conjugates

* When $b^2 - 4ac = 0$, the one real root *of multiplicity two* is sometimes considered as two equal roots. This is done so that *all* quadratic equations can be considered to have two roots.

NOTE When a, b, and c are integers, any *real, irrational* roots occur in *conjugate pairs*. This means that if $2 + \sqrt{3}$ is a root, then $2 - \sqrt{3}$ must also be a root. Also, any *complex* roots occur in conjugate pairs, so that if $2 - i$ is a root, $2 + i$ must also be a root. ☑

Example 1 Determine the nature of the roots of each of the following equations without solving the equation.

a. $2x^2 + 5x - 12 = 0$

Solution

$$\left\{\begin{array}{l} a = 2 \\ b = 5 \\ c = -12 \end{array}\right\} \quad b^2 - 4ac = (5)^2 - 4(2)(-12) = 25 + 96 = 121 \quad \left\{\begin{array}{l}\text{Perfect} \\ \text{square:} \\ 11^2 = 121\end{array}\right.$$

Therefore, the two roots are real, rational, and unequal.

b. $x^2 - 2x - 2 = 0$

Solution

$$\left\{\begin{array}{l} a = 1 \\ b = -2 \\ c = -2 \end{array}\right\} \quad b^2 - 4ac = (-2)^2 - 4(1)(-2) = 4 + 8 = 12 \quad \left\{\begin{array}{l}\textit{Not}\text{ a} \\ \text{perfect} \\ \text{square}\end{array}\right.$$

Therefore, the two roots are real, unequal, irrational conjugates.

c. $9x^2 - 6x + 1 = 0$

Solution

$$\left\{\begin{array}{l} a = 9 \\ b = -6 \\ c = 1 \end{array}\right\} \quad b^2 - 4ac = (-6)^2 - 4(9)(1) = 36 - 36 = 0$$

Therefore, there is one real, rational root of multiplicity two.

d. $x^2 - 6x + 11 = 0$

Solution

$$\left\{\begin{array}{l} a = 1 \\ b = -6 \\ c = 11 \end{array}\right\} \quad b^2 - 4ac = (-6)^2 - 4(1)(11) = 36 - 44 = -8$$

Therefore, the two roots are complex conjugates. ■

Using Roots to Find the Equation

Sometimes we want to find an equation that has a given set of numbers as its roots. We do this by *reversing* the procedure used to solve an equation by factoring.

Example 2 Find a quadratic equation that has the roots -3 and 5.

Solution If the roots are -3 and 5, then $x = -3$ and $x = 5$, so $x + 3 = 0$ and $x - 5 = 0$. Then the *product* of $x + 3$ and $x - 5$ must equal zero. Therefore, $(x + 3)(x - 5) = 0$. Removing the parentheses, we get $x^2 - 2x - 15 = 0$. Therefore, a quadratic equation that has roots -3 and 5 is $x^2 - 2x - 15 = 0$. ■

Example 3 Find an equation of lowest degree that has integral coefficients and that has $1 + \sqrt{2}$ as a root.

Solution Since the coefficients are integers, we know that the *conjugate* of $1 + \sqrt{2}$ must also be a root.

$x = 1 + \sqrt{2}$	$x = 1 - \sqrt{2}$	Irrational conjugate roots
$x - 1 - \sqrt{2} = 0$	$x - 1 + \sqrt{2} = 0$	Rewriting each equation with zero on one side of the equal sign

$$(x - 1 - \sqrt{2}) \cdot (x - 1 + \sqrt{2}) = 0$$ Using the fact that if $a = 0$ or $b = 0$, then $ab = 0$

$$[(x - 1) - \sqrt{2}][(x - 1) + \sqrt{2}] = 0$$ Enclosing $(x - 1)$ in parentheses to show that the left side is in the form $[a - b][a + b]$

$$(x - 1)^2 - (\sqrt{2})^2 = 0$$

$$x^2 - 2x + 1 - 2 = 0$$

$$x^2 - 2x - 1 = 0$$ This equation has the given roots ■

Example 4 Find an equation of lowest degree that has integral coefficients and that has $2 - 3i$ as a root.

Solution Since the coefficients are integers, complex roots occur in conjugate pairs; therefore, $2 + 3i$ must also be a root.

$x = 2 + 3i$	$x = 2 - 3i$	Complex conjugate roots
$x - 2 - 3i = 0$	$x - 2 + 3i = 0$	

$$(x - 2 - 3i) \cdot (x - 2 + 3i) = 0$$

$$[(x - 2) - 3i][(x - 2) + 3i] = 0$$ The left side is in the form $[a - b][a + b]$

$$(x - 2)^2 - (3i)^2 = 0$$

$$x^2 - 4x + 4 - 9i^2 = 0$$

$$x^2 - 4x + 13 = 0$$ ■

Example 5 Find a *cubic* equation that has roots $\frac{1}{2}$, -3, and $\frac{2}{3}$.

Solution

$x = \frac{1}{2}$	$x = -3$	$x = \frac{2}{3}$
$2x - 1 = 0$	$x + 3 = 0$	$3x - 2 = 0$

$$(2x - 1) \cdot (x + 3) \cdot (3x - 2) = 0$$

$$(2x^2 + 5x - 3)(3x - 2) = 0$$

$$6x^3 + 11x^2 - 19x + 6 = 0$$ ■

EXERCISES 10.5

Set I In Exercises 1–8, use the quadratic discriminant to determine the nature of the roots without solving the equation.

1. $x^2 - x - 12 = 0$

2. $x^2 + 3x - 10 = 0$

3. $6x^2 - 7x = 2$

4. $10x^2 - 11x = 5$

5. $x^2 - 4x = -4$

6. $x^2 + 8x = -16$

7. $9x^2 + 2 = 6x$

8. $2x^2 + 6x + 5 = 0$

In Exercises 9–18, find a quadratic equation that has the given roots.

9. 4 and -2

10. -3 and 2

11. 0 and 5

12. 6 and 0

13. $2 + \sqrt{3}$ and $2 - \sqrt{3}$

14. $3 + \sqrt{5}$ and $3 - \sqrt{5}$

15. $\frac{1}{2}$ and $\frac{2}{3}$

16. $\frac{3}{5}$ and $\frac{2}{3}$

17. $\dfrac{1 + i\sqrt{3}}{2}$ and $\dfrac{1 - i\sqrt{3}}{2}$

18. $\dfrac{1 - i\sqrt{2}}{3}$ and $\dfrac{1 + i\sqrt{2}}{3}$

In Exercises 19–22, find a cubic equation that has the given roots.

19. 1, 3, and 4

20. 2, 1, and 5

21. 3, $-2i$, and $+2i$

22. 2, $-3i$, and $+3i$

In Exercises 23–28, find and simplify the equation of lowest degree that has integral coefficients and that has the given roots.

23. $5 - i$

24. $2 + 3i$

25. $1 - 2\sqrt{5}$

26. $3 + 5\sqrt{2}$

27. 2 and $-3i$

28. -3 and $6i$

Set II

In Exercises 1–8, use the quadratic discriminant to determine the nature of the roots without solving the equation.

1. $x^2 + 25 = 10x$

2. $2x^2 + 3x = 2$

3. $x^2 - 2x = 2$

4. $x^2 - x + 1 = 0$

5. $x^2 + 4x = 0$

6. $x^2 - 8x = -16$

7. $3x^2 + 5x = 2$

8. $x^2 = 8$

In Exercises 9–18, find a quadratic equation that has the given roots.

9. 3 and -4

10. $3\sqrt{2}$ and $-3\sqrt{2}$

11. -8 and 0

12. $3 - \sqrt{5}$ and $3 + \sqrt{5}$

13. $4 - \sqrt{7}$ and $4 + \sqrt{7}$

14. $2i$ and $-2i$

15. $\frac{3}{4}$ and $\frac{1}{8}$

16. 3 and $\frac{2}{5}$

17. $\dfrac{3 + 5i}{3}$ and $\dfrac{3 - 5i}{3}$

18. $3 - i\sqrt{5}$ and $3 + i\sqrt{5}$

In Exercises 19–22, find a cubic equation that has the given roots.

19. 0, 2, and 5

20. 2, $1 + \sqrt{2}$, and $1 - \sqrt{2}$

21. 4, $5i$, and $-5i$

22. 0, $2 - 3i$, and $2 + 3i$

In Exercises 23–28, find and simplify the equation of lowest degree that has integral coefficients and that has the given roots.

23. $6 + i$

24. $2 - 4i$

25. $5 + 3\sqrt{7}$

26. 3 and $\sqrt{5}$

27. 3 and $7i$

28. 0 and i

10.6 Review: 10.1–10.5

Quadratic Equations 10.1

A **quadratic equation** is a polynomial equation that has a second-degree term as its highest-degree term.

The General Form 10.1

The general form of a quadratic equation is $ax^2 + bx + c = 0$, where a, b, and c are real numbers ($a \neq 0$). In this book, we will not consider the quadratic equation to be in the general form unless a, b, and c are *integers* and a is positive.

Solving Incomplete Quadratics 10.2

When $c = 0$, find the greatest common factor (GCF); then solve by factoring.

When $b = 0$, to solve the equation $ax^2 + c = 0$:

1. Add $-c$ to both sides of the equation.
2. Divide both sides by the coefficient of x^2.
3. Use Rule 10.1: If $x^2 = k$, then $x = \pm\sqrt{k}$.

Using Substitution in Solving an Equation That Is Written As $a(X)^2 + b(X)^1 + c = 0$ 10.3

1. Let some different variable equal X.
2. Substitute that variable for X.
3. Solve the resulting quadratic equation.
4. Set each solution of the quadratic equation equal to X.
5. Solve each resulting equation for its variable.
6. Check all answers in the original equation, because there may be extraneous roots.

Solving Quadratic Equations by the Quadratic Formula 10.4

1. Arrange the equation in the general form and identify a, b, and c.
2. Substitute the values of a, b, and c into

$$x = \frac{-b \pm \sqrt{b^2 - 4ac}}{2a}$$

3. Simplify the answers.

The Relation between the Quadratic Discriminant and Roots 10.5

If $b^2 - 4ac > 0$ and is a perfect square	The two roots are real, rational, and unequal
If $b^2 - 4ac > 0$ and is not a perfect square	The two roots are real, unequal, irrational conjugates
If $b^2 - 4ac = 0$	There is one real, rational root of multiplicity two
If $b^2 - 4ac < 0$	The two roots are complex conjugates

Using Roots to Find the Equation 10.5

Reverse the procedure used to solve an equation by factoring.

Review Exercises 10.6 Set I

In Exercises 1–16, find the solution set for each equation by any convenient method.

1. $x^2 + x = 6$

2. $x^2 = 3x + 10$

3. $x^2 - 2x - 4 = 0$

4. $x^2 - 4x + 2 = 0$

5. $x^2 - 2x + 5 = 0$

6. $x^2 - 4x + 7 = 0$

7. $\dfrac{2x}{3} = \dfrac{3}{8x}$

8. $\dfrac{3x}{5} = \dfrac{5}{12x}$

9. $\dfrac{x+2}{3} = \dfrac{1}{x-2} + \dfrac{2}{3}$

10. $\dfrac{x+2}{4} = \dfrac{2}{x+2} + \dfrac{1}{2}$

11. $(x + 5)(x - 2) = x(3 - 2x) + 2$

12. $(2x - 1)(3x + 5) = x(x + 7) + 4$

13. $(x^2 - 4x)^2 + 5(x^2 - 4x) + 4 = 0$

14. $(x^2 + 6x)^2 + 13(x^2 + 6x) + 36 = 0$

15. $x^4 - 65x^2 + 64 = 0$

16. $x^4 - 82x^2 + 81 = 0$

In Exercises 17 and 18, use the quadratic discriminant to determine the nature of the roots without solving the equation.

17. $x^2 - 6x + 7 = 0$

18. $x^2 - 8x + 13 = 0$

In Exercises 19–22, set up each problem algebraically and solve. If answers are not rational, express in simplest radical form *and* give decimal approximations rounded off to two decimal places.

19. If the product of two consecutive odd integers is decreased by 14, the result is 85. Find the integers.

20. If the product of two consecutive integers is increased by 4, the result is 60. Find the integers.

21. The length of a rectangle is 3 more than its width. If its area is 8, find its dimensions.

22. The length of one leg of a right triangle is 2 more than the other leg. If its hypotenuse is 3, find the length of each leg.

23. Find a quadratic equation that has the roots $1 - \sqrt{7}$ and $1 + \sqrt{7}$.

24. Find an equation of lowest degree that has integral coefficients and that has the roots 2 and $-i\sqrt{6}$.

Review Exercises 10.6 Set II

NAME ____________________

In Exercises 1–16, find the solution set for each equation by any convenient method.

1. $x^2 + x = 20$
2. $x^2 - 13 = 0$
3. $x^2 - 2x - 15 = 0$
4. $4x^2 + 3x = 0$
5. $x^2 - x + 7 = 0$
6. $x^2 - x = 30$
7. $\dfrac{3x}{5} = \dfrac{5}{3x}$
8. $\dfrac{5x}{2} = \dfrac{1}{x - 2}$
9. $\dfrac{3}{x} - \dfrac{x}{x + 2} = 2$
10. $\dfrac{2}{x} + \dfrac{x}{x + 1} = 5$
11. $(x + 3)(x - 4) = x(3 - x) + 3$
12. $(3x - 4)(x - 1) = x(2 - 4x) - 2$
13. $(x^2 - 6x)^2 + 2(x^2 - 6x) - 63 = 0$
14. $(x^2 - 3x)^2 - 14(x^2 - 3x) + 40 = 0$
15. $y^4 - 17y^2 + 16 = 0$
16. $x^{2/3} - 3x^{1/3} + 2 = 0$

ANSWERS

1. ____________________
2. ____________________
3. ____________________
4. ____________________
5. ____________________
6. ____________________
7. ____________________
8. ____________________
9. ____________________
10. ____________________
11. ____________________
12. ____________________
13. ____________________
14. ____________________
15. ____________________
16. ____________________

In Exercises 17 and 18, use the quadratic discriminant to determine the nature of the roots without solving the equation.

ANSWERS

17. ______

18. ______

19. ______

______

20. ______

21. ______

22. ______

______

23. ______

24. ______

17. $x^2 + x + 5 = 0$

18. $3x^2 - 5 = 0$

In Exercises 19–22, set up each problem algebraically and solve. If answers are not rational, express in simplest radical form *and* give decimal approximations rounded off to two decimal places.

19. If the product of two consecutive even integers is decreased by 8, the result is 40. Find the integers.

20. The area of a square is 5. Find the length of a side of the square.

21. The length of a rectangle is 4 more than its width. The area is 3. Find the width of the rectangle.

22. One leg of a right triangle is 3 units shorter than the other. The hypotenuse is $\sqrt{11}$. Find the length of each leg.

23. Find a quadratic equation that has the roots $-2 - 3i$ and $-2 + 3i$.

24. Find the equation of lowest degree that has integral coefficients and that has the roots $2i$ and $-\sqrt{5}$.

10.7 Graphing Quadratic Functions of One Variable

In Section 8.8, we introduced the graphing of quadratic functions of one variable. In this section, we include additional information about quadratic functions of one variable that will simplify the drawing of their graphs.

Recall that a *quadratic function* is a function that can be written in the form $y = f(x) = ax^2 + bx + c$ and that its graph is a curve called a *parabola.* For $y = f(x) = ax^2 + bx + c$, if $a > 0$, the parabola opens *upward*; if $a < 0$, the parabola opens *downward.* (See Figure 10.7.1.)

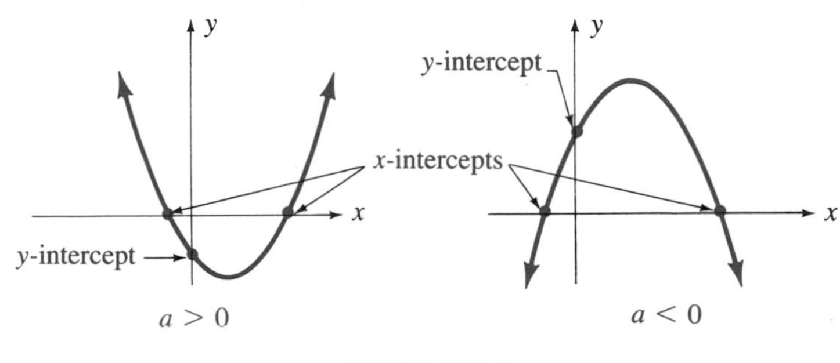

FIGURE 10.7.1

y-intercept

The *y-intercept* is the point at which the parabola crosses the y-axis; it is found by letting $x = 0$. For the parabola $y = f(x) = ax^2 + bx + c$, there is only one y-intercept—the point $(0, c)$.

$$y = f(x) = ax^2 + bx + c$$
$$y = f(0) = a(0)^2 + b(0) + c$$
$$y = c$$

The y-coordinate of the y-intercept of a quadratic function of one variable is *always equal to the constant term.*

x-intercepts

The *x-intercepts* are the points at which the parabola crosses the x-axis. There may be two x-intercepts, one, or none. They are found by setting $y = 0$ and then solving the resulting equation for x.

$$y = f(x) = ax^2 + bx + c = 0$$

If we do not get *real* values for x, then there are no x-intercepts; the graph does not cross the x-axis.

Example 1 Determine whether each parabola opens upward or downward and find all the intercepts.

a. $y = f(x) = 6 - x - x^2$

Solution $a = -1 < 0$. Therefore, the parabola opens downward. Because $c = 6$, the y-intercept is $(0, 6)$. Since we know that the x-coordinate of the y-intercept will *always* be zero, we sometimes simply say that the y-intercept is 6.

If $y = 0$, we have

$$0 = 6 - x - x^2$$
$$0 = (2 - x)(3 + x)$$
$$2 - x = 0 \quad \Big| \quad 3 + x = 0$$
$$x = 2 \quad \Big| \quad x = -3$$

The x-intercepts are $(2, 0)$ and $(-3, 0)$. Since we know that the y-coordinates of the x-intercepts are *always* zero, we sometimes say that the x-intercepts are 2 and -3.

b. $y = f(x) = x^2 - 2x - 1$

Solution $a = 1 > 0$. Therefore, the parabola opens upward. Because $c = -1$, the y-intercept is $(0, -1)$.

Set $y = 0$ to find the x-intercepts.

$$y = f(x) = x^2 - 2x - 1 = 0$$

Substitute $\begin{Bmatrix} a = 1 \\ b = -2 \\ c = -1 \end{Bmatrix}$ into $x = \dfrac{-b \pm \sqrt{b^2 - 4ac}}{2a}$.

$$x = \frac{-(-2) \pm \sqrt{(-2)^2 - 4(1)(-1)}}{2(1)}$$
$$= \frac{2 \pm \sqrt{4 + 4}}{2} = \frac{2 \pm \sqrt{8}}{2} = \frac{2 \pm 2\sqrt{2}}{2} = 1 \pm \sqrt{2}$$

The x-coordinates of the x-intercepts are $1 + \sqrt{2}$ and $1 - \sqrt{2}$. These can be approximated by using a calculator to find $\sqrt{2} \approx 1.4$.

$$\text{Therefore } \begin{cases} 1 + \sqrt{2} \approx 1 + 1.4 = 2.4 \\ 1 - \sqrt{2} \approx 1 - 1.4 = -0.4 \end{cases}$$

The x-intercepts are approximately $(2.4, 0)$ and $(-0.4, 0)$.

c. $y = f(x) = x^2 - 4x + 5$

Solution $a = 1 > 0$; the parabola opens upward. Because $c = 5$, the y-intercept is $(0, 5)$.

Set $y = 0$ to find the x-intercepts.

$$y = f(x) = x^2 - 4x + 5 = 0$$

Substitute $\begin{Bmatrix} a = 1 \\ b = -4 \\ c = 5 \end{Bmatrix}$ into $x = \dfrac{-b \pm \sqrt{b^2 - 4ac}}{2a}$.

$$x = \frac{-(-4) \pm \sqrt{(-4)^2 - 4(1)(5)}}{2(1)}$$
$$= \frac{4 \pm \sqrt{16 - 20}}{2} = \frac{4 \pm \sqrt{-4}}{2}$$
$$= \frac{4 \pm 2i}{2} = 2 \pm i$$

Since $2 \pm i$ are not real numbers, *the parabola does not cross the x-axis*. There are no x-intercepts in this case. ■

NOTE The graphs of the parabolas of Example 1 are shown in Example 4 on pages 515 and 516. ☑

Symmetrical Points and the Axis of Symmetry

If two points are symmetric with respect to a line, that line is the perpendicular bisector of the line segment that joins the two points, and the line is called the **axis of symmetry** (see Figure 10.7.2).

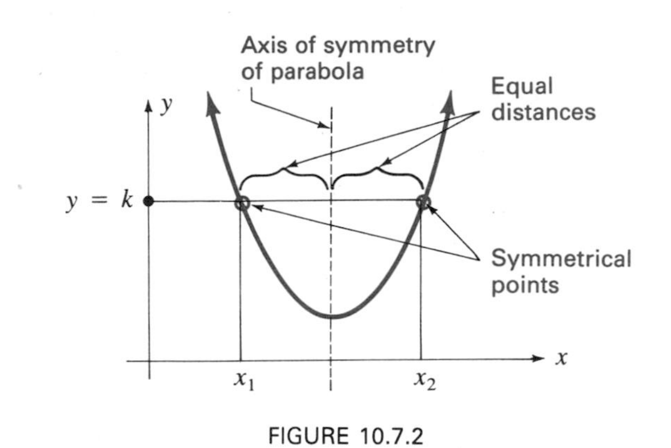

FIGURE 10.7.2

The axis of symmetry of the parabola $y = f(x) = ax^2 + bx + c$ is a *vertical* line midway between any pair of symmetrical points on that parabola. It is shown in Appendix A that the equation of the axis of symmetry for $y = f(x) = ax^2 + bx + c$ can be found as follows:

> The equation of the axis of symmetry for the parabola $y = f(x) = ax^2 + bx + c$ is
>
> $$x = -\frac{b}{2a}$$

Example 2 Find the equation of the axis of symmetry for the graph of the quadratic function $y = f(x) = 2x^2 + 7x - 4$.

Solution

$$x = -\frac{b}{2a} = -\frac{7}{2(2)} = -\frac{7}{4}$$

Therefore, $x = -\frac{7}{4}$ is the equation of the axis of symmetry. ■

A WORD OF CAUTION Students often say that the equation of the axis of symmetry is $-\frac{b}{2a}$. *This is incorrect;* $-\frac{b}{2a}$ is not an equation at all! An equation must have an equal sign in it. ☑

Vertex

The point at which the parabola crosses its axis of symmetry is called the **vertex** of the parabola. The vertex is the point

$$\left(-\frac{b}{2a}, f\left(-\frac{b}{2a}\right)\right)$$

While this notation may look difficult and confusing, it is simply telling you how to find the coordinates of the vertex: First find the x-coordinate of the vertex by using the formula $x = -\frac{b}{2a}$, and then use that value of x in the original equation to find its corresponding y-value (see Example 3).

If the parabola opens upward, the vertex will be the *lowest* point on the parabola, and y will have its *minimum* value there. If the parabola opens downward, the vertex will be the *highest* point on the parabola, and y will have its *maximum* value there. (See Figure 10.7.3.)

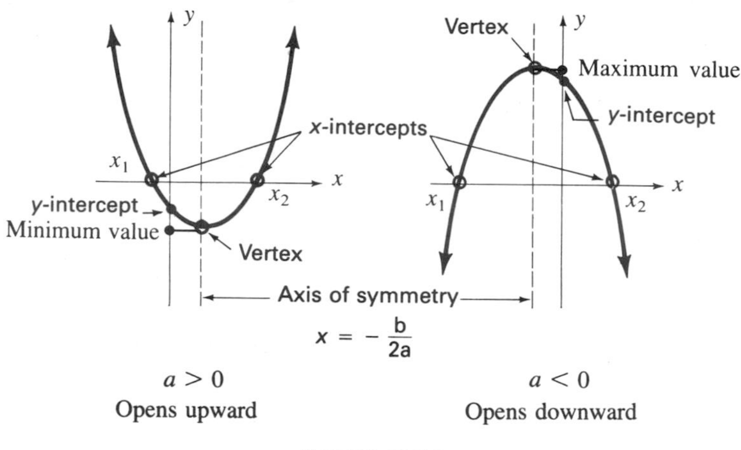

FIGURE 10.7.3

Example 3 Find the equation of the axis of symmetry and the coordinates of the vertex for each parabola of Example 1.

a. $y = f(x) = 6 - x - x^2$

Solution $a = -1, b = -1, c = 6$.

The equation of the axis of symmetry is $x = -\frac{-1}{2(-1)}$, or $x = -\frac{1}{2}$.

$$f(-\tfrac{1}{2}) = 6 - (-\tfrac{1}{2}) - (-\tfrac{1}{2})^2 = 6 + \tfrac{1}{2} - \tfrac{1}{4} = 6\tfrac{1}{4}$$

Therefore, the vertex is $(-\frac{1}{2}, 6\frac{1}{4})$.

A WORD OF CAUTION Students often think that they can rewrite $y = 6 - x - x^2$ as $y = x^2 + x - 6$ because they remember that $0 = 6 - x - x^2$ is equivalent to $0 = x^2 + x - 6$. This is true, however, only because one side of each equation is *zero*. Multiplying both sides of $y = 6 - x - x^2$ by -1 gives $-y = x^2 + x - 6$, *not* $y = x^2 + x - 6$. The equations $y = 6 - x - x^2$ and $y = x^2 + x - 6$ are *not equivalent* and their graphs are not the same. ✓

b. $y = f(x) = x^2 - 2x - 1$

Solution $a = 1, b = -2, c = -1$.

The equation of the axis of symmetry is $x = -\frac{-2}{2(1)}$, or $x = 1$.

$$f(1) = 1^2 - 2(1) - 1 = -2$$

Therefore, the vertex is $(1, -2)$.

c. $y = f(x) = x^2 - 4x + 5$

Solution $a = 1, b = -4, c = 5$.

The equation of the axis of symmetry is $x = -\dfrac{-4}{2(1)}$, or $x = 2$.

$$f(2) = (2)^2 - 4(2) + 5 = 4 - 8 + 5 = 1$$

Therefore, the vertex is (2, 1). ■

Graphing Quadratic Functions of One Variable

In graphing a parabola that has two x-intercepts, we can draw a fairly accurate graph by plotting the x- and y-intercepts and the vertex and then connecting these points with a smooth curve. If the parabola has fewer than two x-intercepts, we need to find additional points on the curve. The procedure is summarized below:

> TO GRAPH A QUADRATIC FUNCTION OF ONE VARIABLE
>
> $$y = f(x) = ax^2 + bx + c$$
>
> 1. Determine whether the parabola opens upward or downward.
> 2. The y-intercept is $(0, c)$.
> 3. Find the x-intercepts: Set $y = 0$ and solve the resulting equation for x.
> 4. Find the equation of the axis of symmetry: $x = \dfrac{-b}{2a}$.
> 5. Find the vertex: $\left(-\dfrac{b}{2a}, f\left(-\dfrac{b}{2a}\right)\right)$.
> 6. Plot the points found in steps 2, 3, and 5. Then plot the points symmetrical to those points with respect to the axis of symmetry.
> 7. If necessary, plot some additional points. Draw a smooth curve through all the points, taking the points in order from left to right.

Example 4 Graph the parabolas from Example 1; label the intercepts and the vertex, and sketch the axis of symmetry.

a. $y = 6 - x - x^2$

Solution

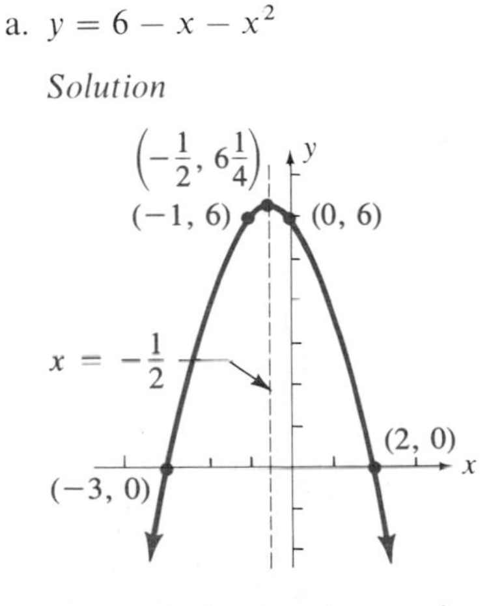

Parabola opens downward

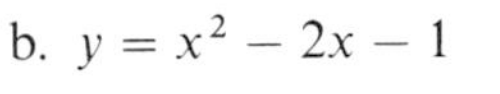

b. $y = x^2 - 2x - 1$

Solution

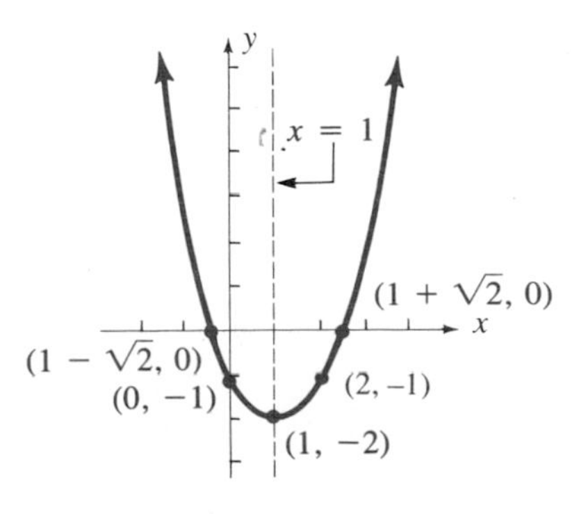

Parabola opens upward

c. $y = x^2 - 4x + 5$

Solution The point symmetrical to (0, 5) is (4, 5).

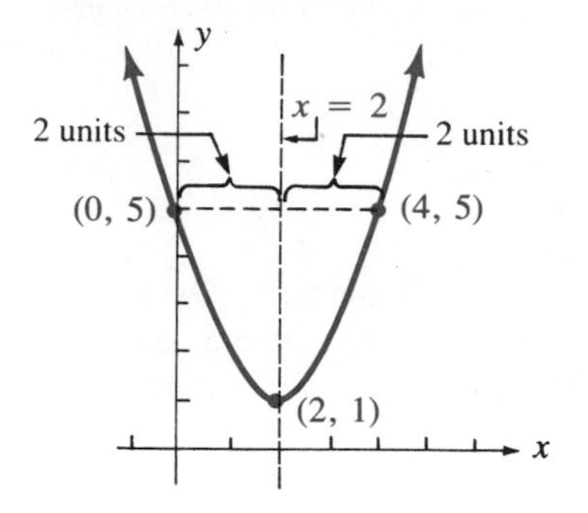

■

Example 5 Discuss and graph $y = f(x) = 9x^2 - 6x + 1$.

Step 1. The parabola opens upward, since $a = 9 > 0$.

Step 2. The y-intercept is (0, 1), since $c = 1$.

Step 3.

$$y = f(x) = 9x^2 - 6x + 1 = 0 \quad \text{Setting } y = 0$$

$$(3x - 1)(3x - 1) = 0$$

$$3x - 1 = 0 \quad \Big| \quad 3x - 1 = 0$$

$$x = \tfrac{1}{3} \quad \Big| \quad x = \tfrac{1}{3} \quad \tfrac{1}{3} \text{ is a root of multiplicity two}$$

The x-intercepts are both the same, which means that the parabola touches the x-axis in only one point, $x = \frac{1}{3}$ (see Figure 10.7.4); that is, the parabola is *tangent* to the x-axis.

Step 4. The equation of the axis of symmetry is $x = -\dfrac{-6}{2(9)}$, or $x = \frac{1}{3}$.

Step 5. The vertex is $(\frac{1}{3}, 0)$.

Step 6. Plot the points found so far (see Figure 10.7.4). The point symmetrical to (0, 1) is $(\frac{2}{3}, 1)$.

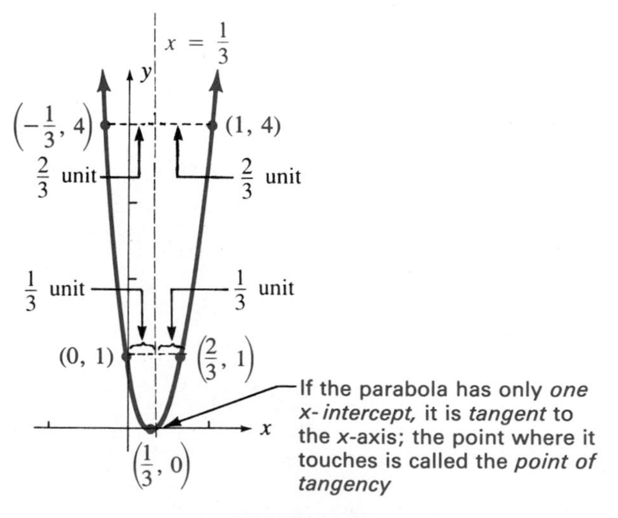

FIGURE 10.7.4.

Step 7. Because the three points found so far are quite close together, we might find one other point and its symmetric point. Let $x = 1$. Then $f(1) = 9(1)^2 \quad 6(1) + 1 = 4$. Onc other point, then, is $(1, 4)$, and its symmetric point is $(-\frac{1}{3}, 4)$. We draw a smooth curve through all the points mentioned in steps 6 and 7, taking them in order from left to right (Figure 10.7.4). ■

Example 6 Graph the function $f(x) = x^2 - x - 2$.

Step 1. The parabola opens upward, since $a = 1 > 0$.

Step 2. The y-intercept is $(0, -2)$ since $c = -2$.

Step 3. x-intercepts

$$y = f(x) = x^2 - x - 2 = 0$$

$$(x + 1)(x - 2) = 0$$

$$x + 1 = 0 \quad \Big| \quad x - 2 = 0$$

$$x = -1 \quad \Big| \quad x = 2$$

Step 4. Axis of symmetry: $x = \dfrac{-b}{2a} = \dfrac{-(-1)}{2(1)} = \dfrac{1}{2}$

Step 5. Vertex:

$$x = \frac{-b}{2a} = \frac{1}{2}$$

$$y = f\left(\frac{-b}{2a}\right) = f\left(\frac{1}{2}\right) = \left(\frac{1}{2}\right)^2 - \left(\frac{1}{2}\right) - 2 = -2\frac{1}{4}$$

The vertex is $(\frac{1}{2}, -2\frac{1}{4})$.

Step 6. Plot the points found so far (see Figure 10.7.5). The point symmetrical to $(0, -2)$ is $(1, -2)$.

Step 7. Find an additional point to help draw the graph:
Choose $x = 3$. Then $f(3) = (3)^2 - (3) - 2 = 9 - 3 - 2 = 4$. Therefore an additional point is $(3, 4)$; its symmetrical point is $(-2, 4)$.

Draw a smooth curve through all points mentioned in steps 6 and 7, taking them in order from left to right (Figure 10.7.5). Optionally, a table of values can be found by using synthetic division (see Section 8.8) as follows.

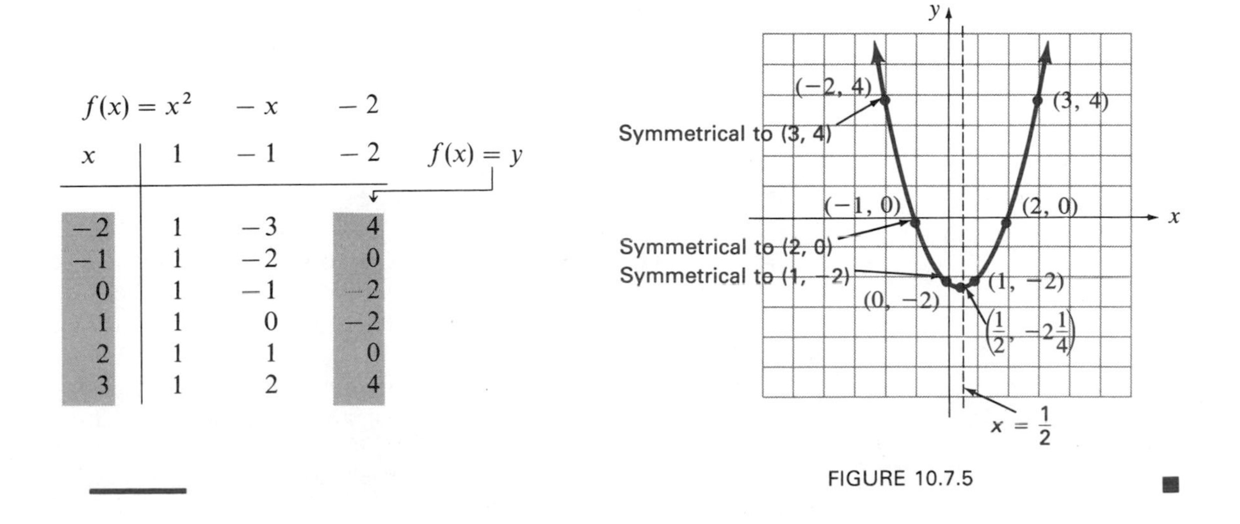

$f(x) = x^2$		$-x$	-2	
x	1	-1	-2	$f(x) = y$
-2	1	-3	4	
-1	1	-2	0	
0	1	-1	-2	
1	1	0	-2	
2	1	1	0	
3	1	2	4	

FIGURE 10.7.5 ■

Recall from Section 8.6C that the graph of $y = f(x) + h$ can be obtained from the graph of $y = f(x)$ by shifting the graph of $y = f(x)$ up or down, and that the graph of $y = f(x + h)$ can be obtained from the graph of $y = f(x)$ by shifting the graph of $y = f(x)$ to the right or left. Therefore, the graph of $y - k = (x - h)^2$ has the exact same *size* and *shape* as the graph of $y = x^2$; however, the *vertex* of $y - k = (x - h)^2$ is at the point (h, k) instead of at $(0, 0)$. The equation of the axis of symmetry is $x = h$.

Example 7 Graph $y - 2 = (x - 3)^2$.

Solution Since $y - 2 = (x - 3)^2$ fits the pattern $y - k = (x - h)^2$ if $k = 2$ and $h = 3$, the vertex of the parabola will be at $(3, 2)$, and the parabola will open upward. The equation of the axis of symmetry is $x = 3$. The y-intercept is $(0, 11)$, and its symmetric point is $(6, 11)$. The curve is as follows:

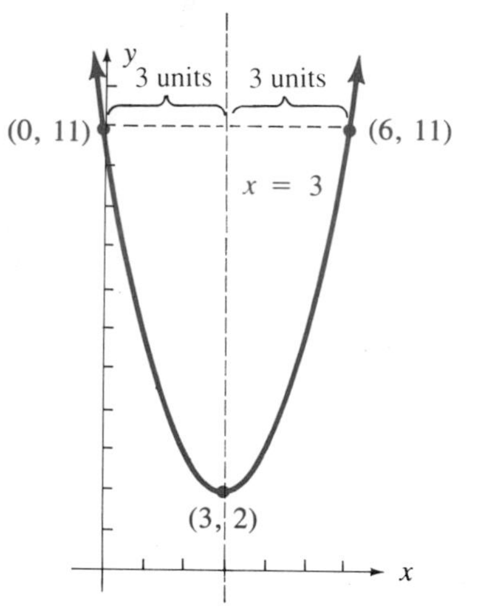

Example 8 Graph $y = (x - 4)^2 - 1$.

Solution To change the given equation into the form $y - k = (x - h)^2$, we add 1 to both sides of the equation, obtaining $y + 1 = (x - 4)^2$. Therefore, $k = -1$ and $h = 4$. The vertex is the point $(4, -1)$, and the equation of the axis of symmetry is $x = 4$. The curve is as follows:

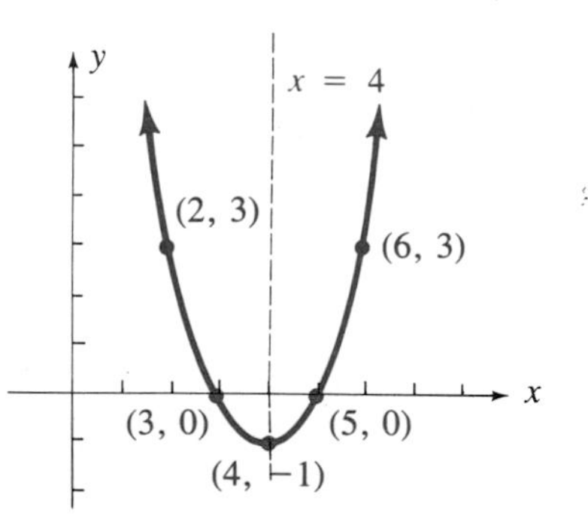

Minimum and Maximum Values

If a parabola opens upward, then the vertex will be the lowest point on the curve. In this case, the y-coordinate of the vertex is known as the **minimum value** of the function.

If a parabola opens downward, then the vertex will be the highest point on the curve, in which case the y-coordinate of the vertex is known as the **maximum value** of the function.

In other words, for the quadratic function $y = f(x) = ax^2 + bx + c$, if a is positive, the function $f(x)$ has a *minimum*, and if a is negative, $f(x)$ has a *maximum*. The minimum or maximum is $f\left(-\frac{b}{2a}\right)$.

Example 9 Find the maximum or minimum value of $f(x) = 3x^2 - 6x + 2$.
Solution a is *positive*; therefore, $f(x)$ has a *minimum* value.

$$\text{Minimum value} = f\left(-\frac{b}{2a}\right) = f\left(-\frac{-6}{2(3)}\right) = f(1)$$

$$= 3(1)^2 - 6(1) + 2 = 3 - 6 + 2 = -1 \quad \blacksquare$$

Example 10 Find the maximum or minimum value of $f(x) = 5 - 2x - 4x^2$.
Solution In the general form, $f(x) = -4x^2 - 2x + 5$. a is *negative*; therefore, $f(x)$ has a *maximum* value.

$$\text{Maximum value} = f\left(-\frac{b}{2a}\right) = f\left(-\frac{-2}{2(-4)}\right) = f\left(-\frac{1}{4}\right)$$

$$= -4\left(-\frac{1}{4}\right)^2 - 2\left(-\frac{1}{4}\right) + 5 = -\frac{1}{4} + \frac{2}{4} + 5 = 5\frac{1}{4} = \frac{21}{4} \quad \blacksquare$$

EXERCISES 10.7

Set I In Exercises 1–18, analyze and graph the functions.

1. $y = f(x) = x^2 - 2x - 3$

2. $y = f(x) = x^2 - 2x - 15$

3. $y = f(x) = x^2 - 2x - 13$

4. $y = f(x) = x^2 - 4x - 8$

5. $y = f(x) = 3 + x^2 - 4x$

6. $y = f(x) = 2x - 8 + x^2$

7. $y = f(x) = x^2 + 3x - 10$

8. $y = f(x) = 2x^2 - 7x - 4$

9. $f(x) = x^2 - 8x + 12$

10. $f(x) = x^2 - 4x - 6$

11. $f(x) = 2x - x^2 + 3$

12. $f(x) = 5 - 4x - x^2$

13. $y = f(x) = x^2 - 6x + 10$

14. $y = f(x) = x^2 - 6x + 11$

15. $y = f(x) = 4x^2 - 9$

16. $y = f(x) = 9x^2 - 4$

17. $f(x) = 6 + x^2 - 4x$

18. $f(x) = 2x - x^2 - 2$

In Exercises 19–24, find the maximum or minimum value and the vertex of each quadratic function.

19. $f(x) = x^2 - 6x + 7$

20. $f(x) = x^2 - 4x + 5$

21. $f(x) = 8x - 2x^2 - 3$

22. $f(x) = 4 + 6x - 3x^2$

23. $f(x) = -\frac{1}{2}x^2 + x + \frac{3}{2}$

24. $f(x) = -\frac{2}{3}x^2 - \frac{8}{3}x + \frac{1}{3}$

In Exercises 25–32, graph the functions.

25. $y = x^2 - 6$

26. $y = x^2 - 3$

27. $y - 3 = (x + 2)^2$

28. $y - 1 = (x + 3)^2$

29. $y + 4 = (x - 1)^2$

30. $y + 2 = (x - 1)^2$

31. $y = (x - 1)^2 + 3$

32. $y = (x - 3)^2 + 1$

Set II In Exercises 1–18, analyze and graph the functions.

1. $y = f(x) = x^2 + x - 2$

2. $y = f(x) = 2 + x - x^2$

3. $y = f(x) = 2x^2 - x - 6$

4. $y = f(x) = 3x^2 - 3x - 6$

5. $y = f(x) = 3x^2 - 10x + 6$

6. $y = f(x) = 3 + 5x - 2x^2$

7. $y = f(x) = x^2 - 2x - 8$

8. $y = f(x) = x^2 + 4x + 4$

9. $y = f(x) = 6x^2 - 7x - 3$

10. $y = f(x) = x^2 + 2x + 4$

11. $y = f(x) = 2x^2 + 7x - 4$

12. $y = f(x) = 2x - x^2 - 1$

13. $y = f(x) = 30x - 9x^2 - 25$

14. $y = f(x) = 3 + 2x - x^2$

15. $y = f(x) = x^2 - 1$

16. $y = f(x) = 4 - x^2$

17. $f(x) = 2x^2 - 4x + 1$

18. $f(x) = 3x^2 - 6x + 1$

In Exercises 19–24, find the maximum or minimum value and the vertex of each quadratic function.

19. $y = f(x) = 4x^2 - 8x + 7$

20. $y = f(x) = x^2$

21. $y = f(x) = 5x^2 + 10x - 3$

22. $y = f(x) = 3x^2 - x$

23. $y = f(x) = x^2 + 2x - 8$

24. $y = f(x) = 5x^2 - 1$

In Exercises 25–32, graph the function.

25. $y = x^2 - 2$

26. $y = (x - 2)^2$

27. $y - 2 = (x + 1)^2$

28. $y = (x + 2)^2 - 2$

29. $y + 3 = (x - 2)^2$

30. $y = (x + 1)^2 - 4$

31. $y = (x - 2)^2 + 5$

32. $y = (x + 5)^2 - 1$

10.8 Conic Sections

Conic sections are curves formed when a plane cuts through a cone. (See Figures 10.8.1, 10.8.2, 10.8.3, and 10.8.4.) Parabolas, circles, ellipses, and hyperbolas are all conic sections. Circles, ellipses, and hyperbolas are never functions; a parabola is a function only if its axis of symmetry is a vertical line.

In this book, we will consider only those conic sections that have a vertical or horizontal axis of symmetry and only those ellipses and hyperbolas that have their centers at the origin.

The Equations of the Conic Sections

GENERAL EQUATIONS OF PARABOLAS

Vertical axis of symmetry: $y = f(x) = ax^2 + bx + c$ ← The axis is $x = -\dfrac{b}{2a}$

or $y - k = a(x - h)^2$ ← The vertex is at (h, k)

The parabola opens upward if $a > 0$ and downward if $a < 0$.

Horizontal axis of symmetry: $x = ay^2 + by + c$ ← The axis is $y = -\dfrac{b}{2a}$

or $x - h = a(y - k)^2$ ← The vertex is at (h, k)

The parabola opens to the right if $a > 0$ and to the left if $a < 0$.

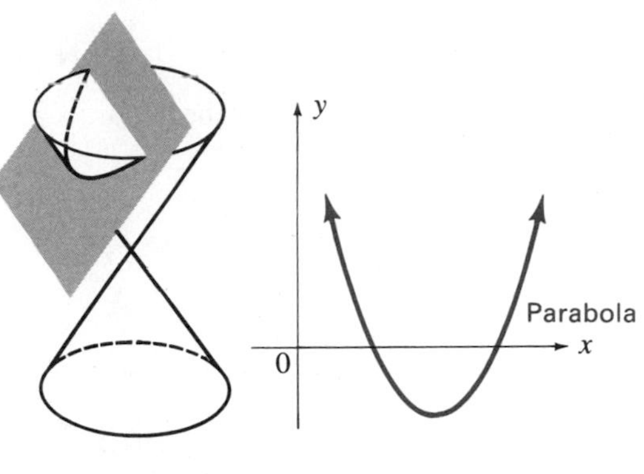

FIGURE 10.8.1

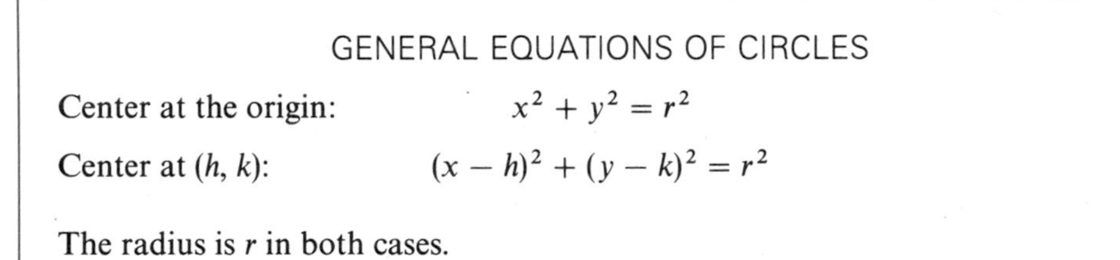

GENERAL EQUATIONS OF CIRCLES

Center at the origin: $x^2 + y^2 = r^2$

Center at (h, k): $(x - h)^2 + (y - k)^2 = r^2$

The radius is r in both cases.

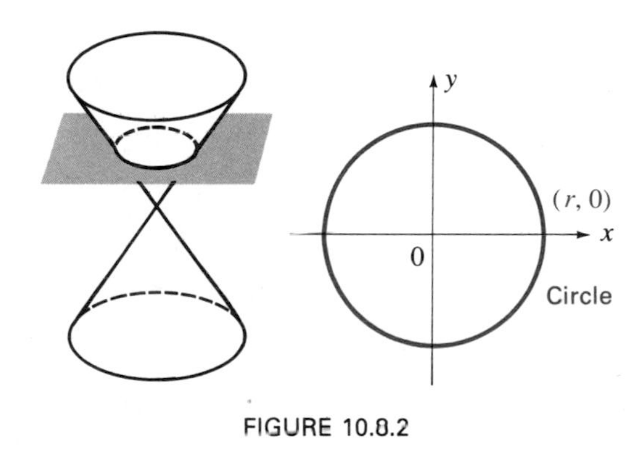

FIGURE 10.8.2

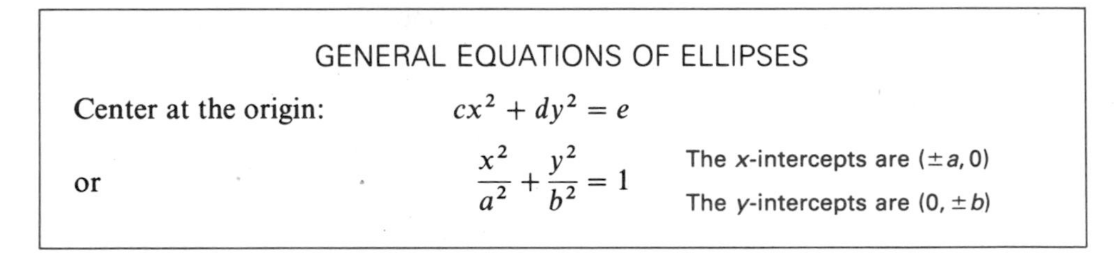

GENERAL EQUATIONS OF ELLIPSES

Center at the origin: $cx^2 + dy^2 = e$

or $\dfrac{x^2}{a^2} + \dfrac{y^2}{b^2} = 1$ The x-intercepts are $(\pm a, 0)$
The y-intercepts are $(0, \pm b)$

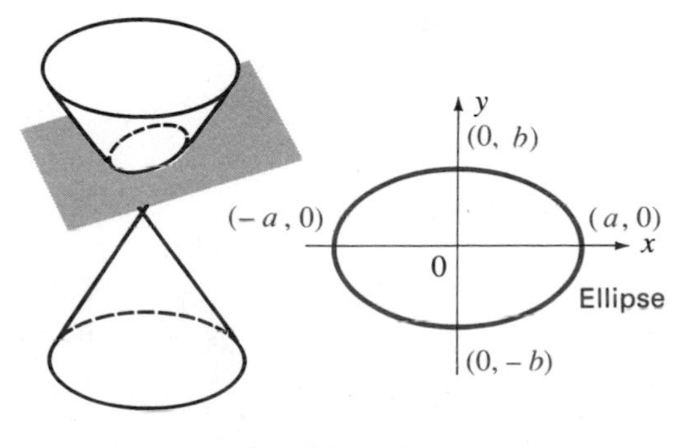

FIGURE 10.8.3

GENERAL EQUATIONS OF HYPERBOLAS

Center at the origin:	$ex^2 - fy^2 = g$	
or	$\frac{x^2}{a^2} - \frac{y^2}{b^2} = 1$	The x-intercepts are $(\pm a, 0)$ (There are no y-intercepts)
or	$\frac{y^2}{c^2} - \frac{x^2}{d^2} = 1$	The y-intercepts are $(0, \pm c)$ (There are no x-intercepts)

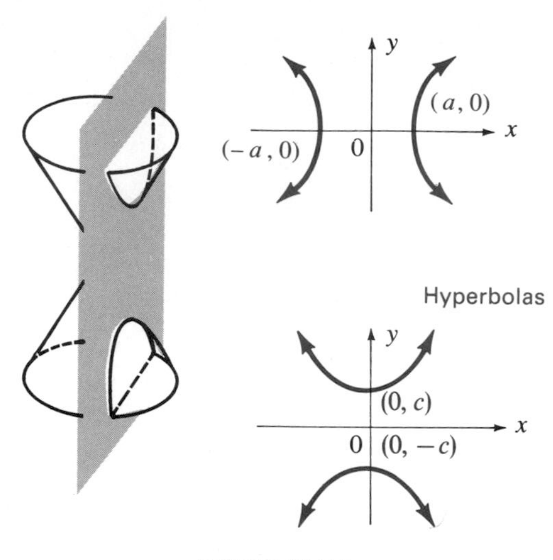

FIGURE 10.8.4

How to Determine Whether an Equation Is That of a Parabola, a Circle, an Ellipse, or a Hyperbola Notice that in the equation of a parabola, either x is squared or y is squared, *but not both.* In the equations of the other conic sections, x and y are *both* squared. Both equations of the hyperbola have a *negative* sign between the x^2 term and the y^2 term, while the circle and the ellipse both have a *plus* sign there. How can you tell the equation of a circle from that of an ellipse? In the equation for a circle, the coefficients of x^2 and y^2 are equal; in the equation for an ellipse, the coefficients are *not* equal.

Example 1 Graph $9x^2 + 4y^2 = 36$.

Solution The equation fits the pattern $cx^2 + dy^2 = e$; therefore, the graph will be an *ellipse* with its center at the origin. If we divide both sides of the equation by 36, we have

$$\frac{9x^2}{36} + \frac{4y^2}{36} = \frac{36}{36}$$

or

$$\frac{x^2}{4} + \frac{y^2}{9} = 1 \quad \text{This is in the form } \frac{x^2}{a^2} + \frac{y^2}{b^2} = 1$$

Then $a^2 = 4$ and $b^2 = 9$, so the x-intercepts are $(2, 0)$ and $(-2, 0)$ and the y-intercepts are $(0, 3)$ and $(0, -3)$. (The intercepts could also be found, of course, by letting $y = 0$ and solving for x and then letting $x = 0$ and solving for y.) The graph is as follows:

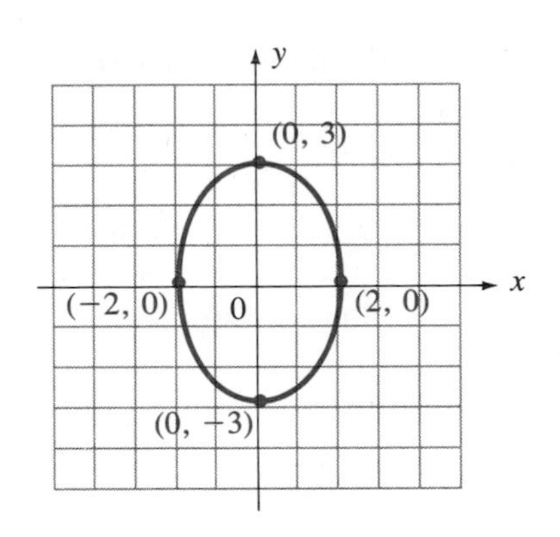

Example 2 Name and graph $2y^2 - 4y + x = 0$.

Solution Because y is squared and x is not, we know that the curve is a *parabola* with a horizontal axis of symmetry. We want to get the equation into the form

$$x - h = a(y - k)^2$$

Therefore, we will *complete the square* (as we did in Section 10.4) for the y-terms. We start by getting a *positive* x on onc side of the equation and all other terms on the other side.

$$2y^2 - 4y + x = 0$$

$$x = -2y^2 + 4y$$ Adding $-2y^2 + 4y$ to both sides

$$x = -2(y^2 - 2y + \quad)$$ Factoring out the coefficient of y^2

We now want to complete the square for $y^2 - 2y$; therefore, we must add $(\frac{1}{2}[-2])^2$, or $+1$, *inside* the parentheses (Because of the -2 *outside* the parentheses, we're really adding -2 to the right side and must add -2 to the left side also)

$$x - 2 = -2(y^2 - 2y + 1)$$

$$x - 2 = -2(y - 1)^2$$

This equation fits the pattern $x - h = a(y - k)^2$, where $h = 2$, $k = 1$, and $a = -2$. Therefore, the curve is a parabola that has its vertex at (2, 1); since a is negative, it opens to the left. The line of symmetry is the line $y = \dfrac{-4}{2(-2)} = 1$. The parabola passes through the point (0, 0), since $x = 0$ if $y = 0$. (Verify this!) By symmetry, we know that one other point must be (0, 2). The graph is at the right.

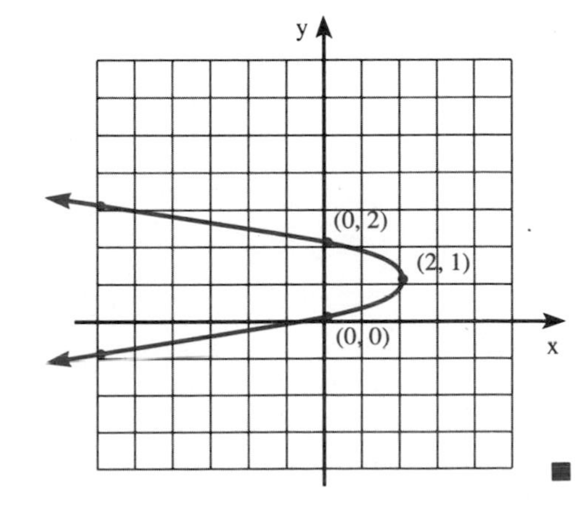

Example 3 Name and graph $x^2 + y^2 - 6x + 2y - 6 = 0$.

Solution Because x and y are both squared and their coefficients are equal, this conic section must be a *circle*. We want to get the equation into the form

$$(x - h)^2 + (y - k)^2 = r^2$$

Therefore, we will complete the squares for both x and y. We start by getting the x-term next to the x^2-term, the y-term next to the y^2-term, and the constant on the right side of the equation.

$$x^2 + y^2 - 6x + 2y - 6 = 0$$

$$(x^2 - 6x + \square) + (y^2 + 2y + \square) = 6 \quad \longleftarrow \text{Rearranging terms and adding 6 to both sides}$$

$$\left(\tfrac{1}{2}[-6]\right)^2 = 9$$

$$(x^2 - 6x + 9) + (y^2 + 2y + 1) = 6 + 9 + 1 \quad \text{Completing the squares}$$

$$\left(\tfrac{1}{2}[2]\right)^2 = 1$$

$$(x - 3)^2 + (y + 1)^2 = 16 = 4^2$$

This equation fits the pattern $(x - h)^2 + (y - k)^2 = r^2$, where $h = 3$, $k = -1$, and $r = 4$. Therefore, the graph will be a circle of radius 4 with its center (indicated on the graph by the hollow dot) at the point $(3, -1)$, as follows:

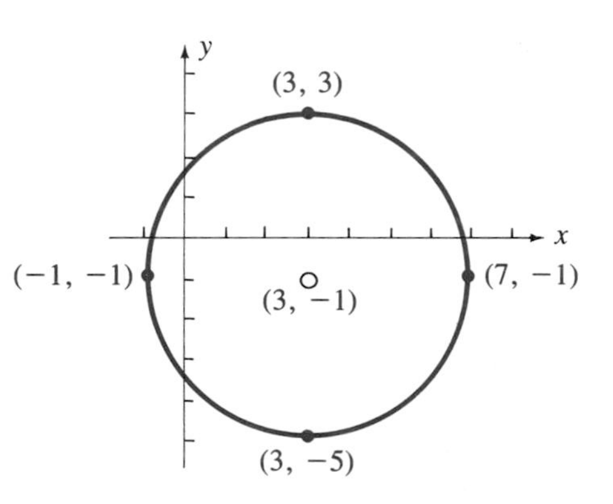

■

Asymptotes of a Hyperbola Associated with each hyperbola are two straight lines called the **asymptotes** of the hyperbola. The hyperbola gets closer to these straight lines as the points of the graph get further from the center of the hyperbola, but the curve never touches the asymptotes. The asymptotes are helpful in sketching the graph of the hyperbola, but they are *not* part of the graph itself.

If the equation of a hyperbola can be put into the form $\frac{x^2}{a^2} - \frac{y^2}{b^2} = 1$, one asymptote passes through the origin and through the point (a, b), while the other passes through the origin and through the point $(-a, b)$. The rectangle with vertices at (a, b), $(-a, b)$, $(-a, -b)$, and $(a, -b)$ is called the *rectangle of reference* for the hyperbola $\frac{x^2}{a^2} - \frac{y^2}{b^2} = 1$; the asymptotes of the hyperbola are the diagonals of this rectangle.

If the equation of a hyperbola can be put into the form $\frac{y^2}{c^2} - \frac{x^2}{d^2} = 1$, one asymptote passes through the origin and through the point (d, c), while the other passes through the origin and through the point $(-d, c)$. The rectangle with vertices at (d, c), $(-d, c)$, $(-d, -c)$, and $(d, -c)$ is called the rectangle of reference for the hyperbola $\frac{y^2}{c^2} - \frac{x^2}{d^2} = 1$.

Example 4 Graph $3y^2 - 4x^2 = 12$.

Solution In this equation, x and y are both squared, but there is a *negative* sign between the two squared terms; therefore, the graph will be a *hyperbola.* If we divide both sides of the equation by 12, we have

$$\frac{3y^2}{12} - \frac{4x^2}{12} = \frac{12}{12}$$

or

$$\frac{y^2}{4} - \frac{x^2}{3} = 1 \qquad \text{This is in the form } \frac{y^2}{c^2} - \frac{x^2}{d^2} = 1$$

Therefore, $c^2 = 4$, and the y-intercepts are $(0, 2)$ and $(0, -2)$. (The y-intercepts could be found, of course, by letting $x = 0$ and solving for y.) There are *no* x-intercepts. (If we try to find the x-intercepts by letting $y = 0$, we obtain the solutions $x = \pm i\sqrt{3}$; since these are imaginary numbers, they cannot be graphed in a coordinate system in which points on both axes represent only real numbers.) This means that the graph does not intersect the x-axis.

We know that $c = \pm 2$, but in order to sketch the asymptotes, we need to identify d also. Comparing our equation to the general equation again, we see that $d^2 = 3$, or $d = \pm\sqrt{3}$. Therefore, the vertices of the rectangle of reference are $(\sqrt{3}, 2)$, $(-\sqrt{3}, 2)$, $(-\sqrt{3}, -2)$, and $(\sqrt{3}, -2)$. We can now sketch the asymptotes of the hyperbola; they are the diagonals of this rectangle (see Figure 10.8.5).

A WORD OF CAUTION The hyperbola does *not* pass through the points $(\sqrt{3}, 2)$, $(-\sqrt{3}, 2)$, $(-\sqrt{3}, -2)$, and $(\sqrt{3}, -2)$. ☑

Let's find some other points on the graph. You can verify that if $x = 3$, $y = \pm 4$; therefore, the graph passes through the points $(3, 4)$ and $(3, -4)$. If $x = -3$, $y = \pm 4$; therefore, two more points on the curve are $(-3, 4)$ and $(-3, -4)$. Because we know what a hyperbola should look like (see Figure 10.8.4), we can draw the graph as shown in Figure 10.8.6.

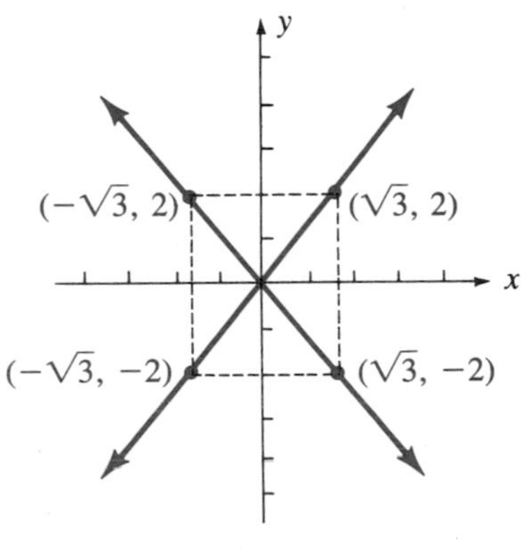

FIGURE 10.8.5

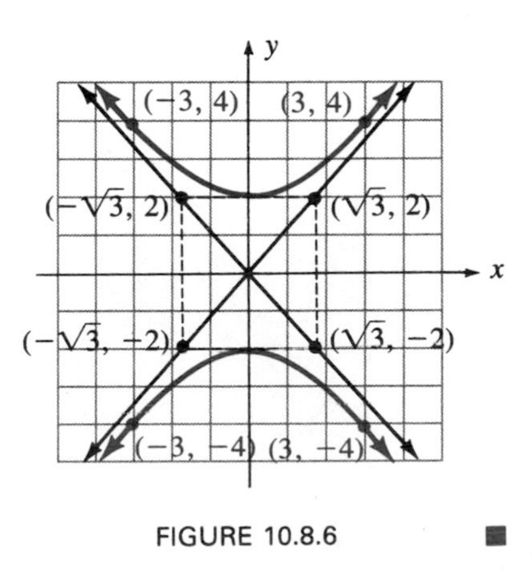

FIGURE 10.8.6 ■

EXERCISES 10.8

Set I Sketch and identify (name) each conic section.

1. $x^2 + y^2 = 9$
2. $x^2 + y^2 = 4$
3. $y = x^2 + 2x - 3$
4. $y = 6 - x - x^2$
5. $3x^2 + 3y^2 = 21$
6. $3x^2 + 4y^2 = 12$
7. $4x^2 + 9y^2 = 36$
8. $16x^2 + y^2 = 16$
9. $9x^2 - 4y^2 = 36$
10. $x^2 - 4y^2 = 4$
11. $y^2 = 8x$
12. $y^2 = -4x$
13. $x^2 + y^2 + 2x - 4y + 1 = 0$
14. $x^2 + y^2 - 4x + 6y + 12 = 0$
15. $y^2 + 2y + x = 0$
16. $2y^2 - 4y - x = 0$

Set II Sketch and identify (name) each conic section.

1. $x^2 + y^2 = 20$
2. $\dfrac{x^2}{9} + \dfrac{y^2}{4} = 1$
3. $y = 5 + 4x - x^2$
4. $x^2 - \dfrac{y^2}{4} = 1$
5. $2x^2 + 2y^2 = 8$
6. $4x^2 + y^2 = 4$
7. $9x^2 + 16y^2 = 144$
8. $x^2 = 6y$
9. $4y^2 - 16x^2 = 64$
10. $x^2 + y^2 = 9$
11. $y^2 = 4x$
12. $x^2 - y^2 = 9$
13. $x^2 + y^2 + 10x - 4y + 20 = 0$
14. $x^2 + y^2 - 4x + 2y - 11 = 0$
15. $y^2 - 4y - x = 0$
16. $y^2 + 4y - 2x = 0$

10.9 Solving Nonlinear Inequalities

10.9A Solving Nonlinear Inequalities in One Variable

In Chapter 2, we discussed solving first-degree inequalities with one variable and graphing their solution sets on the real number line. In this section, we discuss the solution of *nonlinear* inequalities with one variable and we graph those solution sets on the real number line.

We will show two different methods of solving nonlinear inequalities in one variable. The solutions can be graphed on the real number line, and they can also be written in set-builder notation or in interval notation.

TO SOLVE A NONLINEAR INEQUALITY IN ONE VARIABLE

1. For *either* method, temporarily substitute an equal sign for the inequality symbol.
2. Solve the resulting equation. (For *method 1*, the solutions will be the x-intercepts of the curve that will be sketched. For *method 2*, the solutions will be the *critical points* that separate the real number line into several intervals.)
3. Proceed with method 1 or method 2. (Both methods are explained in the following examples.)

Example 1 Solve $x^2 + x - 2 > 0$ and graph its solution set on the number line.

Step 1. $x^2 + x - 2 = 0$ Substituting = for >

Step 2.

$$x^2 + x - 2 = 0$$

$$(x + 2)(x - 1) = 0$$

$$x + 2 = 0 \quad | \quad x - 1 = 0$$

$$x = -2 \quad | \quad x = 1$$

Step 3. We will show both methods of solution.

Method 1 We will sketch the graph of the related quadratic function $x^2 + x - 2 = y$ obtained by temporarily substituting $= y$ for > 0.

We know from Section 10.7 that this is the equation of a parabola that opens upward, since $a > 0$. We found in step 2 that the x-intercepts are -2 and 1. To find the y-intercept, set $x = 0$.

$$y = f(0) = 0^2 + 0 - 2 = -2$$

The intercepts $(-2, 0)$, $(1, 0)$, and $(0, -2)$ are all we need to make a rough sketch of the parabola (Figure 10.9.1).

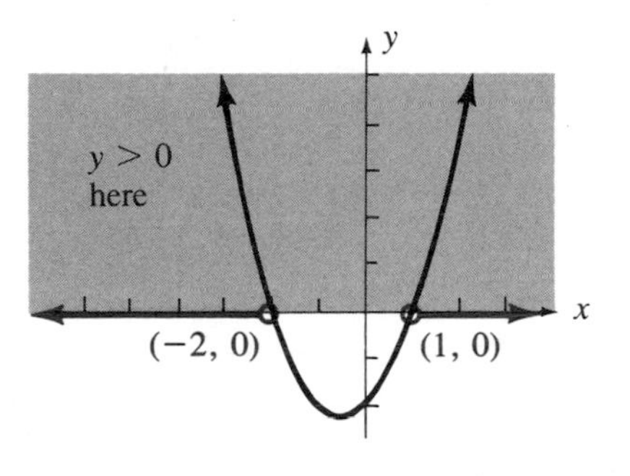

FIGURE 10.9.1

Since we are solving the inequality $y = x^2 + x - 2 > 0$, we need only indicate which x-values make $y > 0$. That is, we must determine those x-values for which the graph is *above* the x axis. These values are shown by the heavy red arrows in Figure 10.9.1.

Method 2 We found the *critical points* (-2 and 1) in step 2. These are the points where the sign of the function can change and we plot them on the real number line; because our inequality was $>$ and not $\geq$, we use *hollow* circles. We see that these two points separate the number line into three intervals: the set of numbers less than -2, the set of numbers between -2 and 1, and the set of numbers greater than 1 (see Figure 10.9.2).

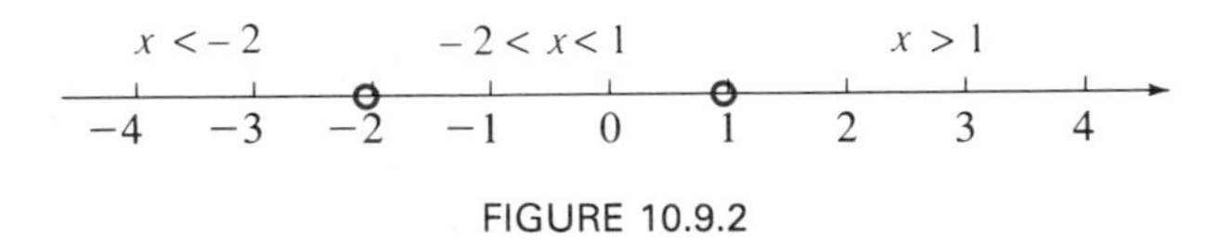

FIGURE 10.9.2

We now look at the sign of each of the factors of $x^2 + x - 2$. The factor $x + 2$ is negative for all $x < -2$ and positive for all $x > -2$. The factor $x - 1$ is negative for all $x < 1$ and positive for all $x > 1$. We indicate these facts on the number line as follows:

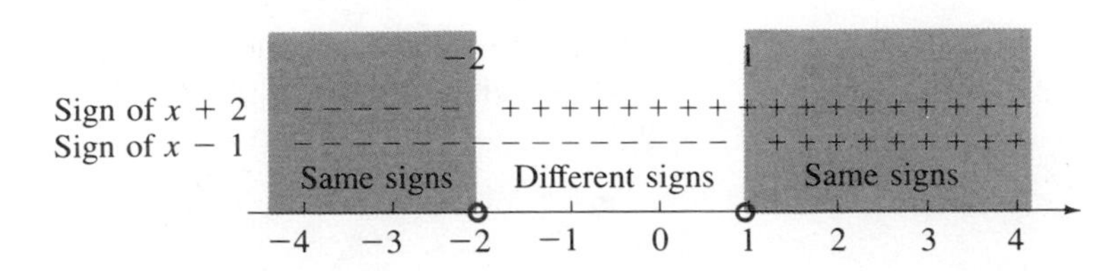

Because we're interested in the interval(s) on which the *product* of $x + 2$ and $x - 1$ is *positive*, we select the interval(s) where both signs are the *same*.

For *either* method, the final solution set is $\{x \mid x < -2\} \cup \{x \mid x > 1\}$, or, in interval notation, $(-\infty, -2) \cup (1, +\infty)$. This means that the solution set consists of all the values of $x < -2$, *as well as* all the values of $x > 1$. The graph of the solution set is as follows:

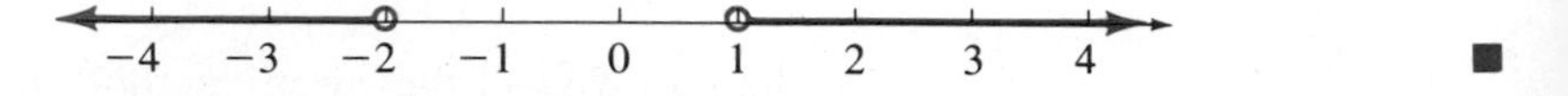

■

Example 2 Find the domain of the function $y = f(x) = \sqrt{8 - 10x - 3x^2}$.

Solution We know that the radicand must be greater than or equal to 0. Therefore, we must solve the inequality $8 - 10x - 3x^2 \geq 0$.

Step 1. $8 - 10x - 3x^2 = 0$ Substituting $=$ for $\geq$

Step 2.

$$(2 - 3x)(4 + x) = 0$$

$$2 - 3x = 0 \quad\Big|\quad 4 + x = 0$$

$$x = \tfrac{2}{3} \quad\Big|\quad x = -4$$

Step 3.

Method 1 The parabola $y = 8 - 10x - 3x^2$ opens downward, since $a < 0$, and its x-intercepts are $\frac{2}{3}$ and -4. Its y-intercept is 8. This means that the parabola must go through (0, 8). The heavy red line segment in Figure 10.9.3 is the graph of the solution set, $\{x \mid -4 \leq x \leq \frac{2}{3}\}$. These are the x-values that make $y \geq 0$.

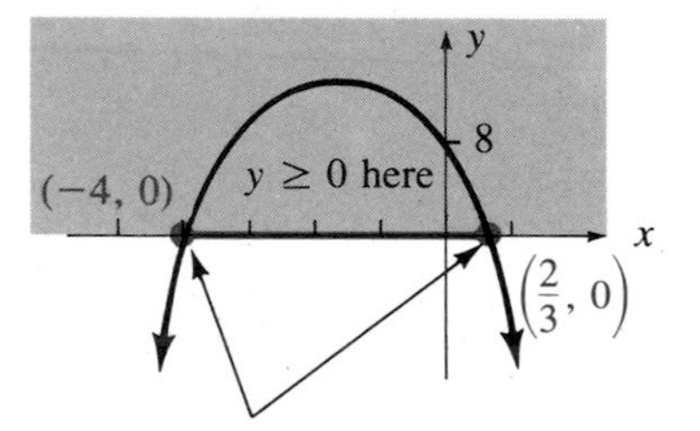

x-intercepts *are* included because the inequality is $\geq$

FIGURE 10.9.3

Method 2 We found the *critical points* (-4 and $\frac{2}{3}$) in step 2. We plot those points as *solid* dots because our inequality symbol ($\geq$) includes the equal sign.

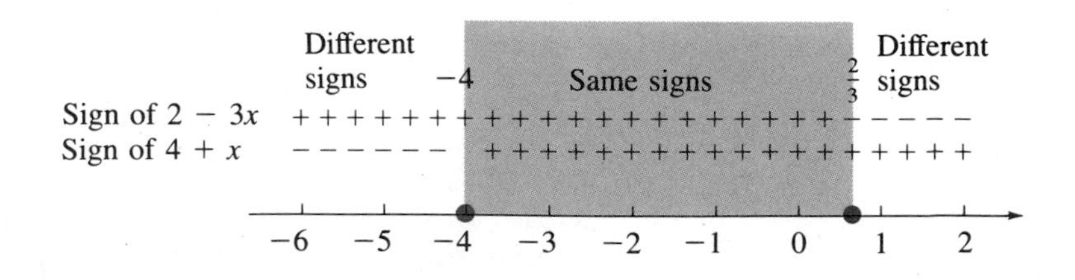

Because we want to find the interval in which the function is *positive*, we select the interval in which the signs of both factors are the *same*—the interval between -4 and $\frac{2}{3}$.

For either method, the final solution set for the inequality is $\{x \mid -4 \le x \le \frac{2}{3}\}$, or $[-4, \frac{2}{3}]$. The graph is as follows:

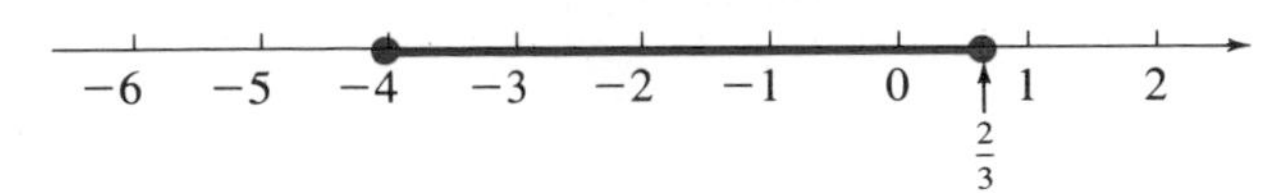

Therefore, the domain of the function $y = \sqrt{8 - 10x - 3x^2}$ is $\{x \mid -4 \le x \le \frac{2}{3}\}$. ■

When a quadratic function cannot be factored, use the *quadratic formula* to find the x-intercepts for method 1 or the critical points for method 2.

Example 3 Solve $x^2 - 2x - 4 < 0$, and graph the solution set on the real number line.

Step 1. $x^2 - 2x - 4 = 0$ Substituting = for <

Step 2.

$$x^2 - 2x - 4 = 0 \begin{cases} a = 1 \\ b = -2 \\ c = -4 \end{cases}$$

$$x = \frac{-b \pm \sqrt{b^2 - 4ac}}{2a} = \frac{-(-2) \pm \sqrt{(-2)^2 - 4(1)(-4)}}{2(1)}$$

$$= \frac{2 \pm \sqrt{4 + 16}}{2} = \frac{2 \pm \sqrt{20}}{2} = \frac{2 \pm 2\sqrt{5}}{2} = 1 \pm \sqrt{5}$$

$$x_1 = 1 - \sqrt{5} \approx 1 - 2.2 = -1.2 \qquad \sqrt{5} \approx 2.236$$

$$x_2 = 1 + \sqrt{5} \approx 1 + 2.2 = 3.2$$

Step 3. We will show only *method 1* for this example. The parabola opens upward, since $a > 0$, and the x-intercepts are approximately -1.2 and 3.2. The solution set is chosen to make $y < 0$, because we're solving $x^2 - 2x - 4 < 0$.

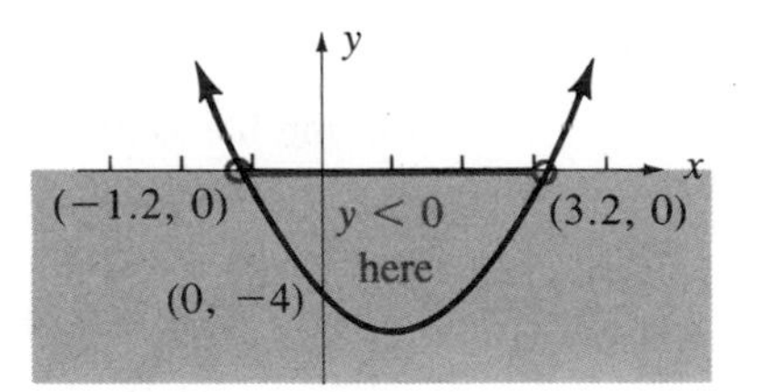

The solution set is $\{x \mid 1 - \sqrt{5} < x < 1 + \sqrt{5}\}$, or $(1 - \sqrt{5}, 1 + \sqrt{5})$. The graph of the solution set is as follows:

■

Example 4 Solve $(x - 2)^2(x + 3) \ge 0$, and graph the solution set on the real number line.

Step 1. $(x - 2)^2(x + 3) = 0$ Substituting = for ≥

Step 2.

$$(x - 2)^2(x + 3) = 0$$

$x - 2 = 0$	$x + 3 = 0$
$x = 2$	$x = -3$

Step 3. We will show only *method 2* for this example. The critical points are -3 and 2. We graph them with *solid* dots, since our inequality ($\geq$) includes the equal sign. Of course, $(x-2)^2$ is never negative.

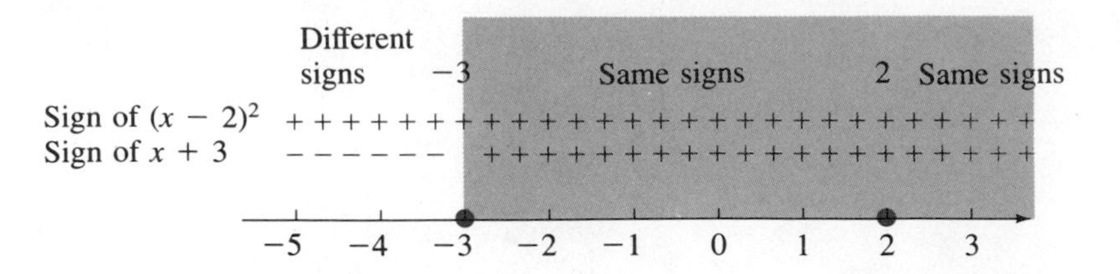

Because our inequality symbol was $\geq$, we select the interval where both signs are *the same*. The solution set, then, is $\{x \mid x \geq -3\}$, or $[-3, +\infty)$. The graph of the solution is as follows:

$-5 \quad -4 \quad -3 \quad -2 \quad -1 \quad 0 \quad 1 \quad 2 \quad 3$ ■

In Examples 5 and 6, the inequality contains a denominator with variables. When the denominator contains variables, *the factors of the denominator*, as well as those of the numerator, *must be set equal to zero* in order to find critical points.

Example 5 Solve $\dfrac{x+5}{x-3} \leq 0$, and graph the solution set on the real number line.

Step 1. $\dfrac{x+5}{x-3} = 0$ Substituting $=$ for $<$

Step 2. $\dfrac{x+5}{x-3} = 0$ Note that 3 is not in the domain

$$x + 5 = 0$$

$$x = -5$$

The equation has only one solution. However, when we set the denominator equal to zero, we find that 3 is a *critical point*.

Step 3. We will not use method 1, because we cannot easily graph $y = \dfrac{x+5}{x-3}$. Using *method 2*, we graph the critical point -5 with a *solid* dot, because the inequality symbol includes the equal sign. However, because 3 *is not in the domain* of the variable, we must graph 3 with a hollow circle.

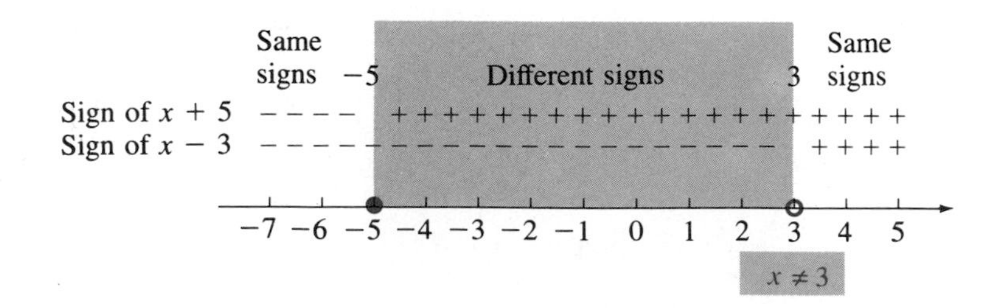

Since a fraction is negative if the numerator and denominator have different signs, we select the interval on which the signs are *different*. The solution set is $\{x \mid -5 \leq x < 3\}$, or $[-5, 3)$, and the graph of the solution set is as follows:

$-7 \quad -6 \quad -5 \quad -4 \quad -3 \quad -2 \quad -1 \quad 0 \quad 1 \quad 2 \quad 3 \quad 4 \quad 5$

A WORD OF CAUTION The solution set is *not* $\{x \mid -5 \le x \le 3\}$. The 3 *cannot* be included, since $x - 3$ is in the denominator. ■

Example 6 Solve $x \le \dfrac{3}{x+2}$, and graph the solution set on the real number line.

Solution We can solve this inequality by moving $\dfrac{3}{x+2}$ to the left, so that we have zero on one side of the inequality symbol. We then proceed in a manner similar to the method used in Example 5.

$$x - \frac{3}{x+2} \le 0$$

$$\frac{x(x+2) - 3}{x+2} \le 0 \qquad \text{Rewriting the left side as a single fraction}$$

$$\frac{x^2 + 2x - 3}{x+2} \le 0$$

$$\frac{(x+3)(x-1)}{x+2} \le 0 \qquad \text{Note that } -2 \text{ is not in the domain}$$

We set each factor of the numerator *and* of the denominator equal to zero to find the critical points:

$x + 3 = 0$	$x - 1 = 0$	$x + 2 = 0$
$x = -3$	$x = 1$	$x = -2$

The critical points are -3, 1, and -2, and these three points separate the number line into four intervals. We must graph -2 as a hollow circle because -2 is *not in the domain* of the variable.

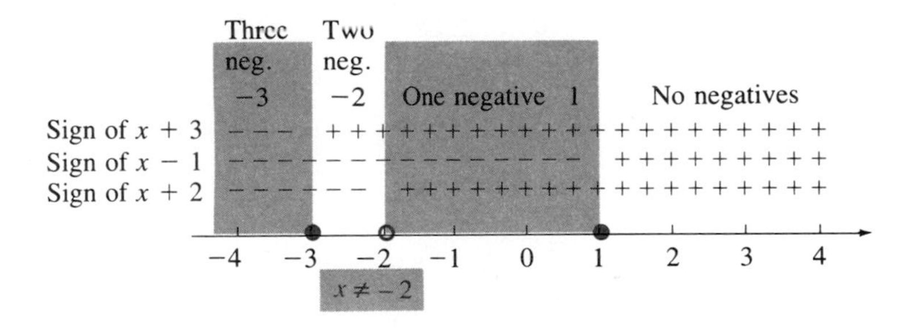

Because the fraction is to be negative or 0, we must have an *odd* number of negative signs. There is an odd number of negative signs to the left of -3 and between -2 and 1.

The solution set is $\{x \mid x \le -3\} \cup \{x \mid -2 < x \le 1\}$, or $(-\infty, -3] \cup (-2, 1]$. This means that the solution set consists of all the values of $x \le -3$, *as well as* all values of x between -2 and 1, including 1. The graph of the solution set is as follows:

$-4 \quad -3 \quad -2 \quad -1 \quad 0 \quad 1 \quad 2 \quad 3 \quad 4$

A different method of solution is shown below. To remove fractions in equations, we usually multiply both sides by the LCD. Since we're solving an *inequality*, we must multiply by $(x + 2)^2$ to make sure both sides are being multiplied by a *positive* number so that the *sense of the inequality does not change.*

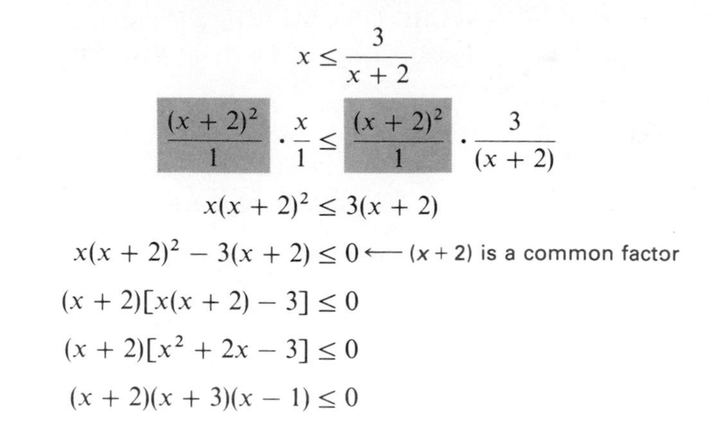

$$x \le \frac{3}{x + 2}$$

$$\frac{(x + 2)^2}{1} \cdot \frac{x}{1} \le \frac{(x + 2)^2}{1} \cdot \frac{3}{(x + 2)}$$

$$x(x + 2)^2 \le 3(x + 2)$$

$$x(x + 2)^2 - 3(x + 2) \le 0 \longleftarrow (x + 2) \text{ is a common factor}$$

$$(x + 2)[x(x + 2) - 3] \le 0$$

$$(x + 2)[x^2 + 2x - 3] \le 0$$

$$(x + 2)(x + 3)(x - 1) \le 0$$

-2, -3, and 1 are the critical points.

The problem can then be finished using method 2; the results will, of course, be the same. ■

EXERCISES 10.9A

Set I In Exercises 1–20, solve each inequality, writing the solution in set-builder notation and in interval notation, and graph the solution set on the real number line.

1. $(x + 1)(x - 2) < 0$ **2.** $(x + 1)(x - 2) > 0$ **3.** $3 - 2x - x^2 \ge 0$

4. $2 - x - x^2 \le 0$ **5.** $x^2 - 3x - 4 < 0$ **6.** $x^2 + 4x - 5 > 0$

7. $x^2 + 7 > 6x$ **8.** $x^2 + 13 < 8x$ **9.** $x^2 \le 5x$

10. $x^2 \ge 3x$ **11.** $3x - x^2 > 0$ **12.** $7x - x^2 < 0$

13. $x^3 - 3x^2 - x + 3 > 0$ **14.** $x^3 + 2x^2 - x - 2 > 0$

15. $(x + 1)(x - 2)^2 < 0$ **16.** $(x - 1)(x + 2)^2 < 0$

17. $\dfrac{x - 2}{x + 3} \ge 0$ **18.** $\dfrac{x + 4}{x - 2} \le 0$

19. $x < \dfrac{2}{x + 1}$ **20.** $x > \dfrac{4}{x + 3}$

In Exercises 21–24, find the domain of the given function.

21. $y = f(x) = \sqrt{x^2 - 4x - 12}$ **22.** $y = f(x) = \sqrt{x^2 + 3x - 10}$

23. $y = f(x) = \sqrt{4 + 3x - x^2}$ **24.** $y = f(x) = \sqrt{5 - 4x - x^2}$

Set II In Exercises 1–20, solve each inequality, writing the solution in set-builder notation and in interval notation, and graph the solution set on the real number line.

1. $(x + 4)(x - 1) < 0$ **2.** $(2x - 5)(x + 2) > 0$ **3.** $8 - 2x - x^2 \ge 0$

4. $x^2 - 2x \le 3$ **5.** $x^2 - 5x - 6 < 0$ **6.** $(x - 1)^2(x - 5) < 0$

7. $x^2 + 8 > 7x$ **8.** $x^2 < 9$ **9.** $x^2 \ge x$

10. $(x+1)^2(x-4) < 0$ **11.** $3x - x^2 < 0$ **12.** $1 + 4x + 3x^2 \geq 0$

13. $x^3 + x^2 - 4x - 4 > 0$ **14.** $(x-1)^2(x-3) < 0$

15. $(x-2)(x+1)(x+3) < 0$ **16.** $x^2 + 9 < 6x$

17. $\dfrac{x+2}{x-4} \geq 0$ **18.** $\dfrac{3x-1}{2-x} \leq 0$

19. $x < \dfrac{6}{x+5}$ **20.** $x \geq \dfrac{5}{6-x}$

In Exercises 21–24, find the domain of the given function.

21. $y = f(x) = \sqrt{x^2 + 6x - 7}$ **22.** $y = f(x) = \sqrt{x^2 + 4x + 3}$

23. $y = f(x) = \sqrt{8 + 2x - x^2}$ **24.** $y = f(x) = \sqrt{10 - x^2 + 3x}$

10.9B Solving Quadratic Inequalities in Two Variables by Graphing

A parabola separates the plane into two regions: the region "inside" the parabola and the region "outside" the parabola. Similarly, circles and ellipses separate the plane into the regions *inside* them and the regions *outside* them. In Figures 10.9.4. and 10.9.5, the regions inside a parabola and inside an ellipse are shaded. A hyperbola separates the plane into *three* regions (see Figure 10.9.6).

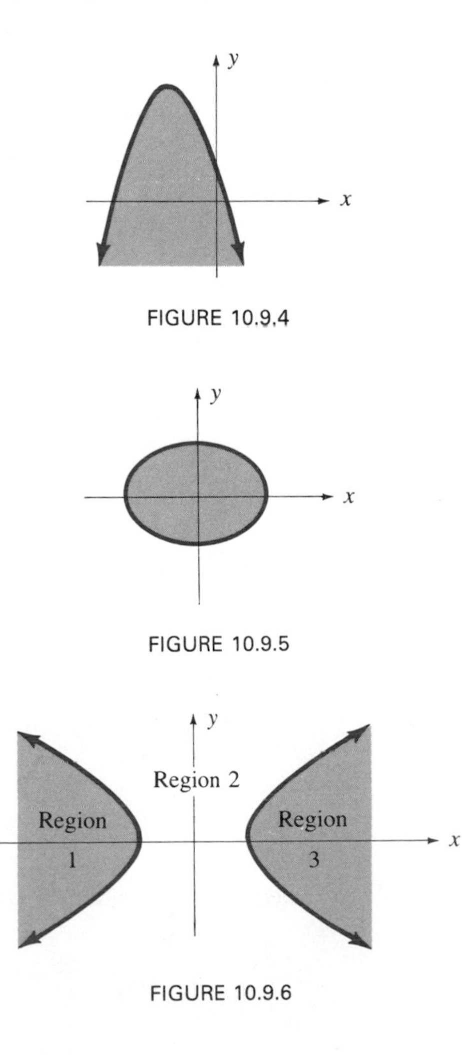

FIGURE 10.9.4

FIGURE 10.9.5

FIGURE 10.9.6

Quadratic inequalities in two variables can be solved *only* by graphing. Since we need to include regions of the xy-plane (rather than just portions of the real number line), it is not possible to write these solutions in either set-builder notation or interval notation.

TO SOLVE A QUADRATIC INEQUALITY IN TWO VARIABLES GRAPHICALLY

1. Temporarily substitute an equal sign for the inequality symbol and graph the resulting conic section.
2. Select any point *not* on the conic section and substitute its coordinates into the original inequality. If the statement is
   a. *true*, shade the region of the plane that contains the point selected;
   b. *false*, shade the region of the plane that does *not* contain the point selected.

   Any point that lies in the shaded region(s) is in the graphic solution; the coordinates of any such point will satisfy the inequality.

Example 7 Solve $y \geq x^2 - 5x - 6$.

Solution We first sketch the graph of the equation $y = x^2 - 5x - 6$. It is a parabola that opens upward; the x-intercepts are 6 and -1, and the y-intercept is -6. We draw the graph as a *solid curve* because the equal sign is included in the original inequality.

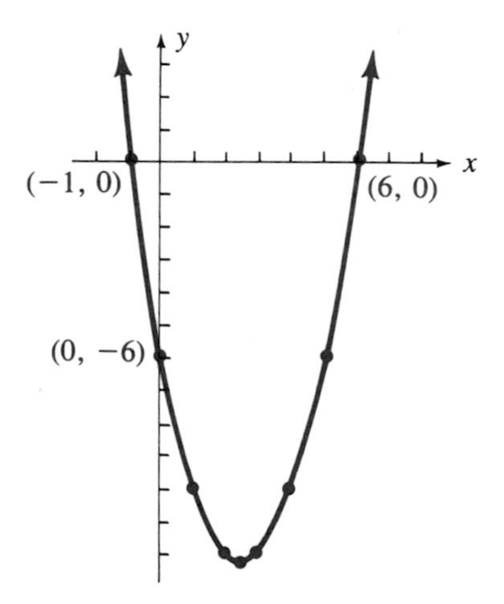

Now we must select a point *not* on the parabola. Let's use (0, 0). Substituting 0 for x and 0 for y into the original inequality, we have

$$0 \geq 0^2 - 5(0) - 6$$

$$0 \geq -6 \quad \text{A } \textit{true} \text{ statement}$$

Therefore, we shade the region that contains the point (0, 0).

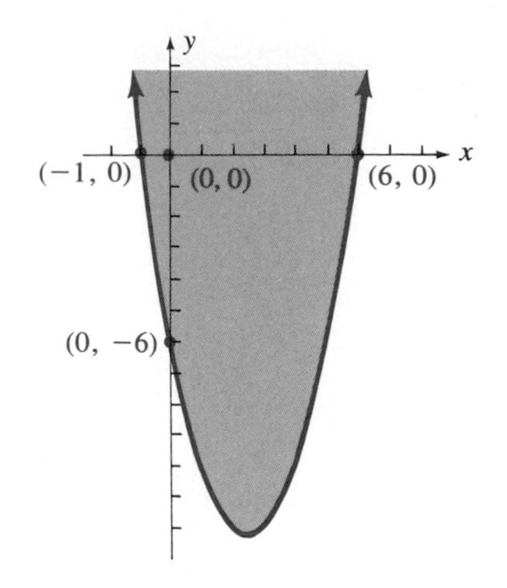

Any point that lies in the shaded region is in the graphic solution. ■

Example 8 Solve $(x + 1)^2 + (y - 3)^2 > 4$.

Solution We temporarily replace $>$ with $=$, giving $(x + 1)^2 + (y - 3)^2 = 4$. This is the equation of a circle with its center at the point $(-1, 3)$ and radius 2. Because the equal sign was *not* included in the original inequality, we sketch the circle with a *dotted line.*

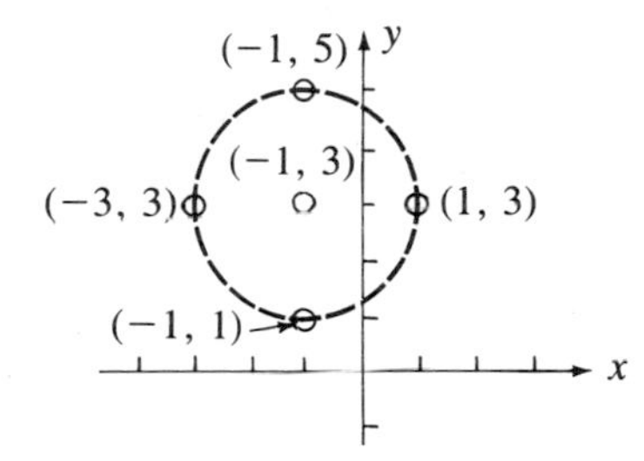

We now select any point not *on* the circle. Again, let's choose (0, 0). Substituting 0 for x and 0 for y into the inequality, we have

$$(0 + 1)^2 + (0 - 3)^2 > 4$$
$$1 + 9 > 4$$
$$10 > 4 \quad \text{A } \textit{true} \text{ statement}$$

Therefore, we shade the region that contains the point (0, 0).

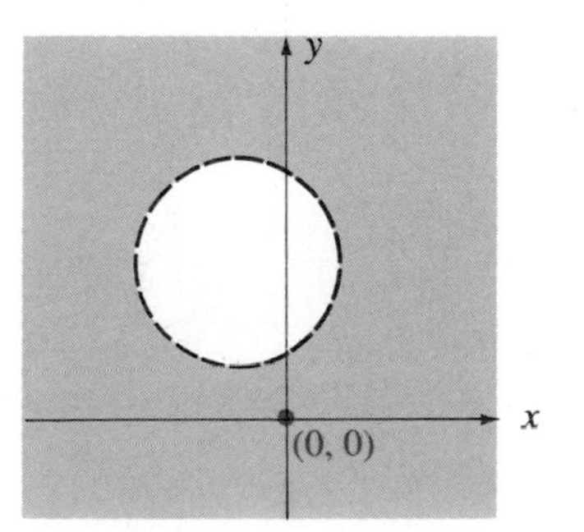

Any point in the shaded region is in the graphic solution. ■

Example 9 Solve $x^2 - y^2 > 1$.

Solution We temporarily replace $>$ with $=$, giving $x^2 - y^2 = 1$. This is a *hyperbola.* If $y = 0$, $x^2 = 1$. Therefore, the x-intercepts are $(1, 0)$ and $(-1, 0)$. We can see that the hyperbola separates the plane into three regions.

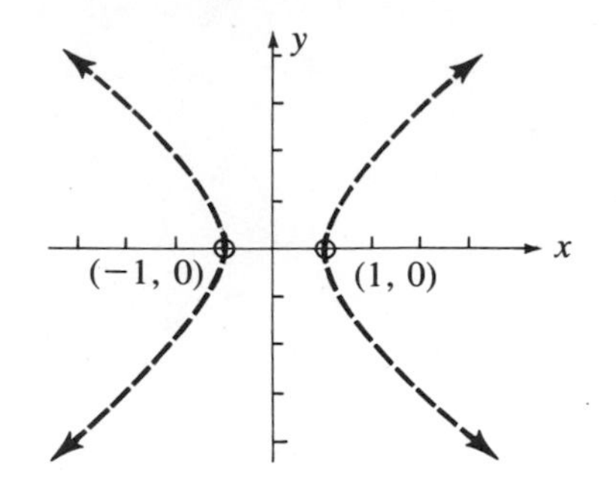

Again, let's choose $(0, 0)$ as the point to try in the inequality.

$$0^2 - 0^2 > 1$$

$$0 > 1 \quad \text{A false statement}$$

Therefore, we shade the portions of the plane that do *not* contain $(0, 0)$.

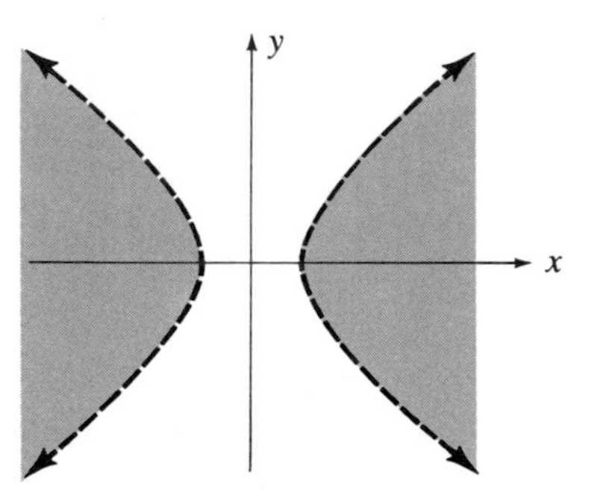

Any point in either shaded region is in the graphic solution. (You should verify that the coordinates of several points in both shaded regions do satisfy the inequality.) ■

EXERCISES 10.9B

Set I Solve each of the following inequalities graphically.

1. $y \leq x^2 + 5x + 6$

2. $y \leq x^2 + 4x + 3$

3. $x^2 + 4y^2 > 4$

4. $9x^2 + y^2 > 9$

5. $(x + 2)^2 + (y - 3)^2 \leq 9$

6. $(x - 1)^2 + (y + 2)^2 \leq 4$

7. $4x^2 - y^2 < 4$

8. $9x^2 - y^2 < 9$

Set II Solve each of the following inequalities graphically.

1. $y \leq x^2 + 3x + 2$

2. $(x + 1)^2 + (y + 1)^2 \leq 1$

3. $16x^2 + y^2 > 16$

4. $y < x^2 - 3x + 2$

5. $(x - 3)^2 + (y - 2)^2 \leq 16$

6. $x^2 - 9y^2 > 9$

7. $16x^2 - y^2 < 16$

8. $x^2 + 9y^2 \geq 9$

10.10 Review: 10.7–10.9

Quadratic Functions 10.7

A **quadratic function** is a second-degree function.

$$y = f(x) = ax^2 + bx + c \qquad \text{General form of a quadratic function}$$

Its graph is called a *parabola.*

Its *x-intercepts* are found by setting the function equal to zero and then solving the resulting quadratic equation for x.

The *y-intercept* is found by setting $x = 0$ and then solving the resulting equation for y. The y-coordinate of the y-intercept of a quadratic function of one variable is always equal to the constant term.

The parabola opens *downward* and a *maximum value* occurs if $a < 0$.

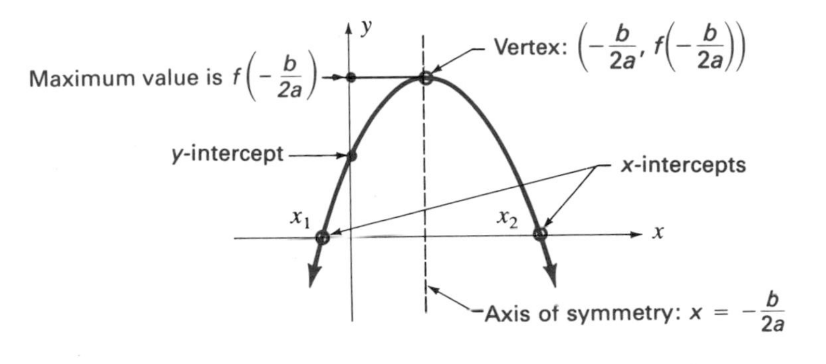

The parabola opens *upward* and a *minimum value* occurs if $a > 0$.

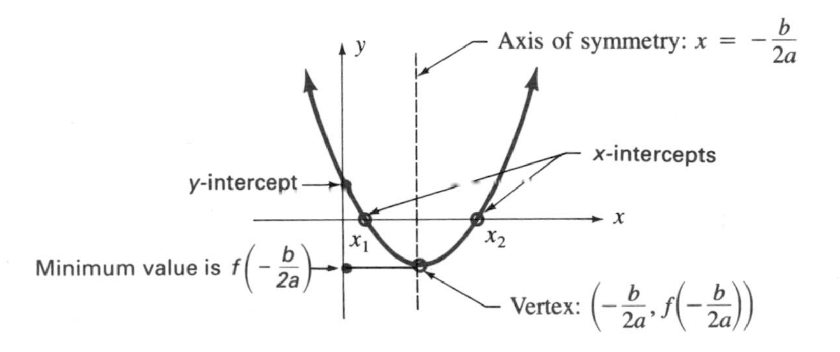

Conic Sections 10.8

Conic sections are curves formed when a plane cuts through a cone. They are the parabola, circle, ellipse, and hyperbola. A quick sketch of the conics covered in this section can usually be made by using the *intercept method.*

To Solve a Quadratic Inequality in One Variable 10.9A

1. Substitute an equal sign for the inequality symbol.
2. Solve the resulting equation. The solutions will be the x-intercepts if method 1 is used, and they will be the critical points if method 2 is used.
3. a. *Method 1:* Sketch the graph of the related function that is obtained by substituting y for 0 in the inequality. Read the solution set from the graph.

   b. *Method 2:* Separate the real number line into intervals by graphing the critical points. Determine the sign of the function in each of the intervals.

To Solve a Quadratic Inequality in Two Variables 10.9B

1. Substitute an equal sign for the inequality symbol and graph the resulting conic section.
2. Select any point *not on* the conic section, and substitute its coordinates into the *original inequality*. If the statement is
   a. *true*, shade the region of the plane that contained the point selected;
   b. *false*, shade the region of the plane that did *not* contain the point selected.

Review Exercises 10.10 Set I

In Exercises 1 and 2, (a) find the x-intercepts, if they exist; (b) find the y-intercept; (c) find the coordinates of the vertex; and (d) graph the function.

1. $f(x) = x^2 - 2x - 8$ **2.** $f(x) = 6x - 9 - x^2$

In Exercises 3–5, solve each inequality, writing the solution in set-builder notation and in interval notation.

3. $(4 - x)(2 + x) > 0$ **4.** $x + \dfrac{2}{x - 3} < 0$ **5.** $x > \dfrac{3}{x + 2}$

6. Find the domain of the function $y = f(x) = \sqrt{6 + 5x - x^2}$.

In Exercises 7–9, sketch and identify each conic section.

7. $3x^2 + 3y^2 = 27$ **8.** $4y^2 - x^2 = 16$ **9.** $9x^2 + 4y^2 = 36$

10. Solve $y \geq x^2 + 2x - 3$ graphically.

Review Exercises 10.10 Set II

NAME ______________________

In Exercises 1 and 2, graph the function.

1. $f(x) = 2x + 3 - x^2$

2. $f(x) = x^2 + 4x + 4$

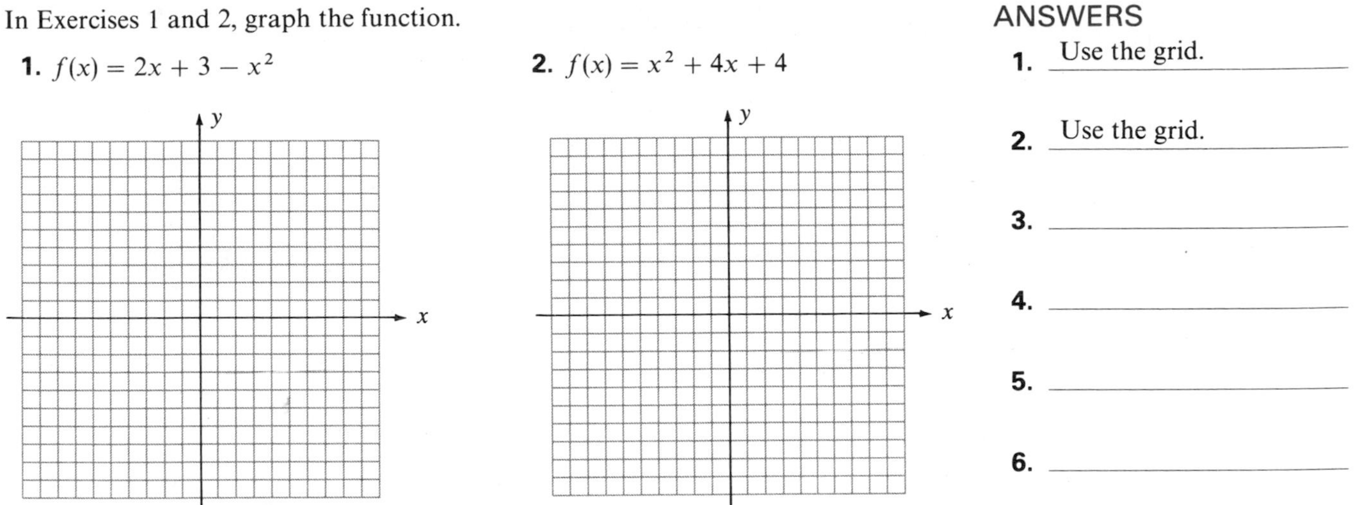

ANSWERS

1. Use the grid.

2. Use the grid.

3. ______________________

4. ______________________

5. ______________________

6. ______________________

In Exercises 3–5, solve each inequality, writing the solution in set-builder notation and in interval notation.

3. $x^2 \leq 3x + 10$

4. $x^2 \geq 1 - 2x$

5. $x < \dfrac{4}{x + 3}$

6. Find the domain of the function $y = f(x) = \sqrt{x^2 + 7x - 8}$.

In Exercises 7–9, sketch and identify each conic section.

7. $y = x^2 + 2x - 3$

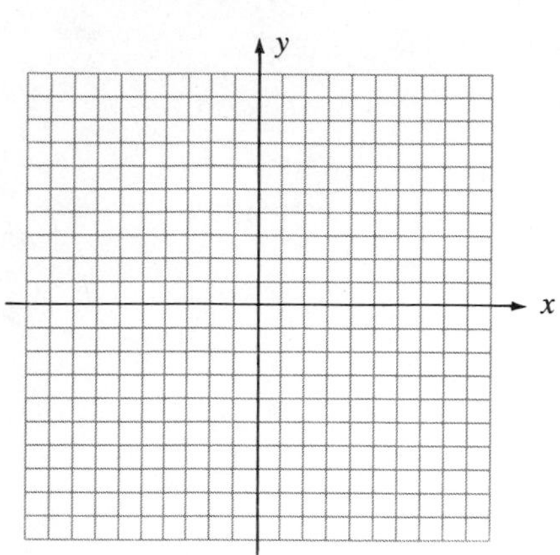

8. $x^2 + 16y^2 = 16$

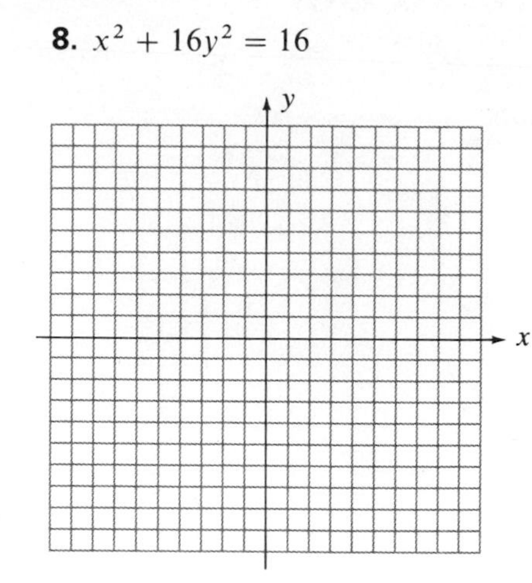

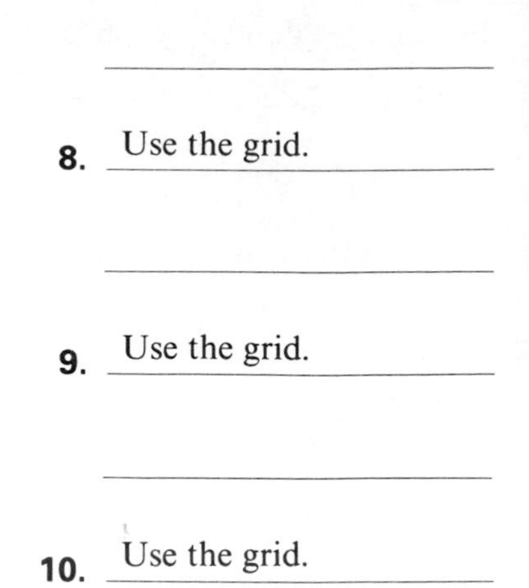

9. $x^2 - 4y^2 = 4$

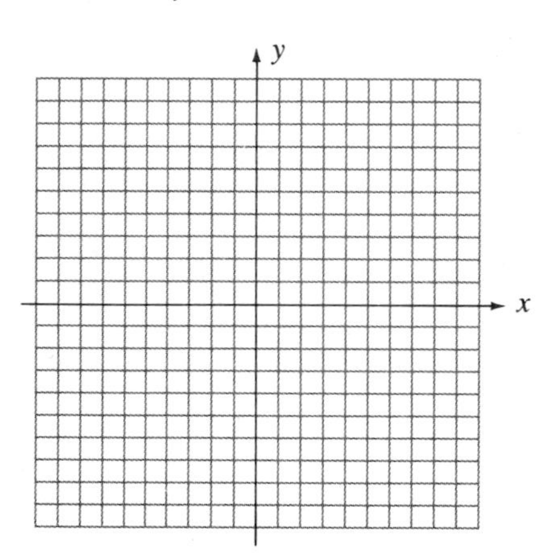

10. Solve $y \geq x^2 + 5x + 4$ graphically.

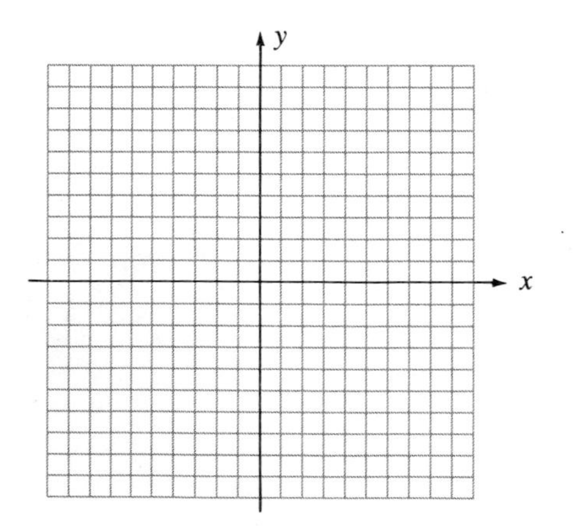

Chapter 10 Diagnostic Test

The purpose of this test is to see how well you understand nonlinear equations and inequalities. We recommend that you work this diagnostic test *before* your instructor tests you on this chapter. Allow yourself about 50 minutes.

Complete solutions for all the problems on this test, together with section references, are given in the answer section at the end of the book. For the problems you do incorrectly, study the sections cited.

In Problems 1–5, find the solution set for each equation by any convenient method.

1. a. $2x^2 = 6x$

b. $2x^2 = 18$

2. $\dfrac{x-1}{2} + \dfrac{4}{x+1} = 2$

3. $2x^{2/3} + 3x^{1/3} = 2$

4. a. $x^2 = 6x - 7$

b. $3x^2 + 7x = 1$

5. $x^2 + 6x + 10 = 0$

6. a. Use the quadratic discriminant to determine the nature of the roots of the equation $25x^2 - 20x + 7 = 0$. Do *not* solve for the roots of the equation.

b. Write an equation (with integral coefficients) of least degree that has roots 3 and $2 + i\sqrt{3}$.

7. Identify and sketch the graph of the conic section whose equation is $25x^2 + 16y^2 = 400$.

8. a. Find the solution set of the inequality $x^2 + 5 < 6x$.

b. Find the solution set of the inequality $x^2 + 2x \geq 8$.

9. Given the quadratic function $f(x) = x^2 - 4x$, find:

a. the equation of the axis of symmetry

b. the coordinates of the vertex

c. the x-intercepts (if they exist)

Draw the graph.

Set up the following problem algebraically and solve.

10. It takes Jan 2 hr longer to do a job than it does Oscar. After Jan works on the job for 1 hr, Oscar joins her. Together they finish the job in 3 hr. How long would it take each of them working alone to do the entire job?

Cumulative Review Exercises: Chapters 1–10

1. Find the slope of the line through $(4, -1)$ and $(2, 5)$.

2. Find the general form of the equation of the line with slope $-\frac{1}{2}$ that passes through $(-2, 3)$.

3. Solve $|4 - 3x| \geq 10$.

4. Find the solution set for $\sqrt{2 - x} - 4 = x$.

5. Graph $4x - 2y < 8$.

6. Find the inverse relation for $y = f(x) = 2x - 3$. Graph $f(x)$ and its inverse.

In Exercises 7–9, perform the indicated operations and express the answers in exponential form with positive exponents or no exponents.

7. $\left(\dfrac{x^{-3}y^4}{x^2y}\right)^{-2}$

8. $(8x^{-3/5})^{2/3}$

9. $a^{1/3} \cdot a^{-1/2}$

10. Find the solution set for $x^2 - x + 7 = 0$.

11. Solve $3x^2 + x = 3$.

12. Solve $7^x = 45$. (Round off the answer to three decimal places.)

13. Solve $\log(x + 11) - \log(x + 1) = \log 6$.

In Exercises 14 and 15, find the value of the variable.

14. $\log_b 2 = \dfrac{1}{4}$

15. $\log_7 \dfrac{1}{7} = x$

16. Graph $9x^2 + 16y^2 = 144$.

17. Find the domain for $y = f(x) = \sqrt{4 - 3x - x^2}$.

In Exercises 18–21, set up each problem algebraically and solve.

18. The length of a rectangle is 2 more than its width. If the length of its diagonal is 10, find the dimensions of the rectangle.

19. It takes Mina 3 hr longer to do a job than it does Merwin. After Mina has worked on the job for 5 hr, Merwin joins her. Together they finish the job in 3 hr. How long would it take each of them to do the entire job working alone?

20. Eric and Colin left an intersection at the same time, Eric walking east and Colin walking south. Eric was walking 3 mph faster than Colin, and at the end of an hour the two men were 5 mi apart. How fast was Colin walking? (Express your answer in simplest radical form and also give a decimal approximation, rounded off to one decimal place.)

21. \$1,000 is invested at $6\frac{1}{4}\%$ annual interest. How long will it take for the money to double if the interest is compounded continuously? Round off the answer to the nearest year.

11 Systems of Equations and Inequalities

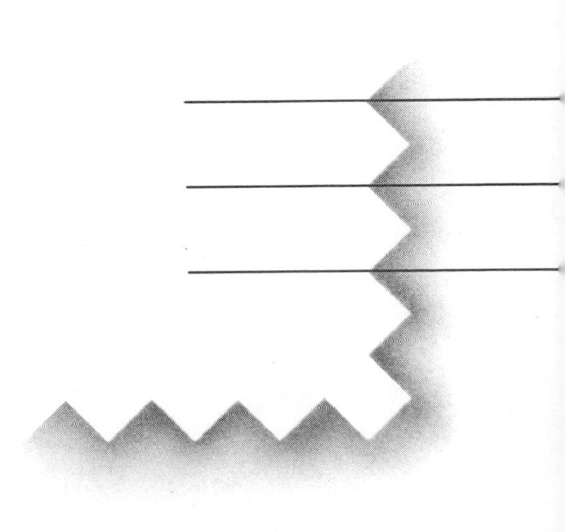

In previous chapters, we showed how to solve a single equation for a single variable. In this chapter, we show how to solve systems of two or more equations in two or more variables. We also include brief discussions of matrices, determinants, and systems of inequalities.

11.1 Basic Definitions

Two Equations in Two Variables

$\begin{Bmatrix} x + y = 6 \\ x - y = 2 \end{Bmatrix}$ is an example of a **system of two equations in two variables.**

A *solution* of a system of two equations in two variables is an ordered pair that satisfies both equations.

Example 1 Verify that (4, 2) is a solution for the system $\begin{Bmatrix} x + y = 6 \\ x - y = 2 \end{Bmatrix}$.

Solution Substituting (4, 2) into each equation, we have

First equation	*Second equation*
$x + y = 6$	$x - y = 2$
$4 + 2 = 6$	$4 - 2 = 2$
True	True

Therefore, (4, 2) satisfies both equations and is a solution for the system $\begin{Bmatrix} x + y = 6 \\ x - y = 2 \end{Bmatrix}$. ■

Three Equations in Three Variables

To be able to solve a system of equations in three variables, we need the following definition.

Ordered Triple Two *ordered triples* are not equal unless they contain the same three elements *in the same order*. We enclose the elements of an ordered triple in *parentheses*:

$$(1, 4, 7) \neq (4, 1, 7)$$

Ordered triples

$\begin{Bmatrix} 2x - 3y + z = 1 \\ x + 2y + z = -1 \\ 3x - y + 3z = 4 \end{Bmatrix}$ is an example of a **system of three equations in three variables.**

A *solution* of a system of three equations in three variables is an ordered triple that satisfies all three equations.

Example 2 Verify that the ordered triple $(-3, -1, 4)$ is a solution for the system

$$\begin{Bmatrix} 2x - 3y + z = 1 \\ x + 2y + z = -1 \\ 3x - y + 3z = 4 \end{Bmatrix}$$

Solution Substituting $(-3, -1, 4)$ into each equation, we have

First equation	*Second equation*	*Third equation*
$2x - 3y + z = 1$	$x + 2y + z = -1$	$3x - y + 3z = 4$
$2(-3) - 3(-1) + 4 \stackrel{?}{=} 1$	$-3 + 2(-1) + 4 \stackrel{?}{=} -1$	$3(-3) - (-1) + 3(4) \stackrel{?}{=} 4$
$-6 + 3 + 4 = 1$	$-3 - 2 + 4 = -1$	$-9 + 1 + 12 = 4$
True	True	True

Therefore, the ordered triple $(-3, -1, 4)$ satisfies all three equations and is a solution of the given system of equations. ■

Systems of Equations and Inequalities

Linear System If each equation of a system is a first-degree equation, the system is called a **linear system**.

Quadratic System If the highest-degree equation of a system is second degree, then the system is called a **quadratic system**.

System of Inequalities If a system contains inequalities rather than equations, it is called a **system of inequalities**.

Example 3 Examples of systems of equations and inequalities:

a. $\begin{Bmatrix} 2x - 3y + z = 1 \\ x + 2y + z = -1 \\ 3x - y + 3z = 4 \end{Bmatrix}$ is a system of linear equations (a linear system).

b. $\begin{Bmatrix} x^2 + y^2 = 25 \\ x - y = 4 \end{Bmatrix}$ is a quadratic system.

c. $\begin{Bmatrix} 3x - 2y > 6 \\ x - 2y < 4 \end{Bmatrix}$ is a system of linear *inequalities*. ■

Solving a System of Equations To *solve* a system of two equations in two variables means to find an ordered pair (if one exists) that satisfies *both* equations. To solve a system of three equations in three variables means to find an ordered triple (if one exists) that satisfies *all three* equations. In this text, we will consider solving only those systems with the same number of equations as variables.

11.2 Graphical Method for Solving a Linear System of Two Equations in Two Variables

The graphical method is not an exact method of solution for two equations in two variables, but it can sometimes be used successfully in solving such systems.

TO SOLVE A LINEAR SYSTEM OF TWO EQUATIONS BY THE GRAPHICAL METHOD

1. Graph each equation of the system on the same set of axes.

2. Three outcomes are possible:

a. *The lines intersect at one point.* The solution is the ordered pair representing the point of intersection. (See Figure 11.2.1.)

b. *The lines never cross* (they are parallel). There is no solution. (See Figure 11.2.2.)

c. *Both equations have the same line for their graph.* For the line $ax + by = c$, the solution set is the set of all ordered pairs of the form $\left(t, \dfrac{c - at}{b}\right)$*, where t represents some real number. (See Figure 11.2.3.)

3. If a unique solution was found in step 2, check it in *both* equations.

When the lines intersect in exactly one point, we say:

1. There is a *unique* solution for the system.
2. The system of equations is *consistent* and *independent*.

When the lines are parallel, we say:

1. There is *no* solution for the system.
2. The system of equations is *inconsistent* and *independent*.

When both equations have the same line for their graph (that is, when the lines coincide), we say:

1. There are *infinitely* many solutions for the system.
2. The system of equations is *consistent* and *dependent*.

Example 1 Solve the system $\begin{cases} x + y = 6 \\ x - y = 2 \end{cases}$ graphically.

Solution We draw the graph of each equation on the same set of axes (see Figure 11.2.1).

Line 1 $x + y = 6$ has intercepts (6, 0) and (0, 6).

Line 2 $x - y = 2$ has intercepts (2, 0) and (0, −2).

The lines appear to intersect at the point (4, 2). We can check to see whether they do in fact intersect there by substituting 4 for x and 2 for y in *both* equations:

$$x + y = 6 \qquad\qquad x - y = 2$$

$$4 + 2 = 6 \quad \text{True} \qquad 4 - 2 = 2 \quad \text{True}$$

Therefore, the solution (4, 2) is correct, and we can say that the lines intersect in a *unique* point, that the system of equations is *consistent* and *independent*, and that the *solution* of the system is (4, 2).

* If $x = t$, then $at + by = c$; solving for y gives $y = \dfrac{c - at}{b}$.

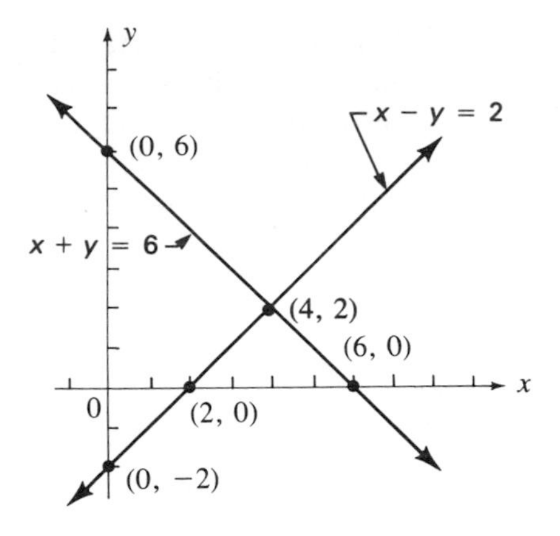

FIGURE 11.2.1 ■

Example 2 Solve the system $\begin{Bmatrix} 2x - 3y = 6 \\ 6x - 9y = 36 \end{Bmatrix}$ graphically.

Solution We draw the graph of each equation on the same set of axes (Figure 11.2.2).

Line 1 $2x - 3y = 6$ has intercepts $(3, 0)$ and $(0, -2)$.

Line 2 $6x - 9y = 36$ has intercepts $(6, 0)$ and $(0, -4)$.

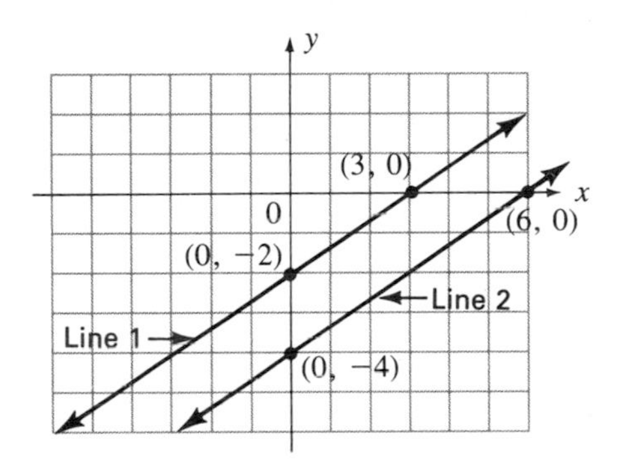

FIGURE 11.2.2

The lines *appear* to be parallel. The slopes of both lines are $\frac{2}{3}$ (you should verify this); therefore, the lines *are* parallel, the system of equations is *inconsistent* and *independent*, and there is *no solution* for the system of equations. ■

Example 3 Solve the system $\begin{Bmatrix} 3x + 5y = 15 \\ 6x + 10y = 30 \end{Bmatrix}$ graphically.

Solution We draw the graph of each equation on the same set of axes (Figure 11.2.3).

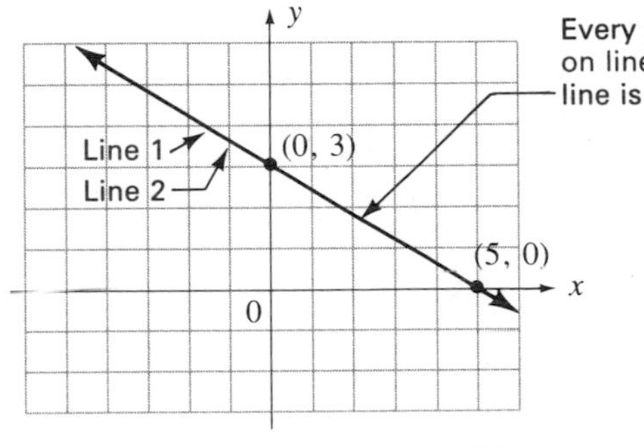

FIGURE 11.2.3

Line 1 $3x + 5y = 15$ has intercepts $(5, 0)$ and $(0, 3)$.

Line 2 $6x + 10y = 30$ has intercepts $(5, 0)$ and $(0, 3)$.

Since both lines go through the same two points, they must be the same line. We can say that the lines *coincide*, the system of equations is *consistent* and *dependent*, and there are *infinitely many solutions* for the system of equations.

We find the solution set by substituting some variable, such as t, for x in one of the equations and then solving for y:

$$3t + 5y = 15$$
$$5y = 15 - 3t$$
$$y = \frac{15 - 3t}{5}$$

The solution set is the set of ordered pairs $\left(t, \frac{15 - 3t}{5}\right)$, where t is any real number. If we substitute s for y instead of t for x, we find that the solution set can also be expressed as the set of ordered pairs $\left(\frac{15 - 5s}{3}, s\right)$, where s is any real number. (Both answers are correct.) ■

EXERCISES 11.2

Set I Find the solution of each system graphically (there may be no solution).

1. $\begin{cases} 2x + y = 6 \\ x - y = 0 \end{cases}$
2. $\begin{cases} 2x - y = -4 \\ x + y = -2 \end{cases}$
3. $\begin{cases} x + 2y = 3 \\ 3x - y = -5 \end{cases}$
4. $\begin{cases} 5x + 4y = 12 \\ x + 3y = -2 \end{cases}$
5. $\begin{cases} x + 2y = 0 \\ 2x - y = 0 \end{cases}$
6. $\begin{cases} 2x + y = 0 \\ x - 3y = 0 \end{cases}$
7. $\begin{cases} 8x - 5y = 15 \\ 10y - 16x = 16 \end{cases}$
8. $\begin{cases} 3x + 9y = 18 \\ 2x + 6y = -24 \end{cases}$
9. $\begin{cases} 8x - 10y = 16 \\ 15y - 12x = -24 \end{cases}$
10. $\begin{cases} 14x + 30y = -70 \\ 15y + 7x = -35 \end{cases}$

Set II Find the solution of each system graphically (there may be no solution).

1. $\begin{cases} 8x - 5y = 30 \\ 2x - 7y = -4 \end{cases}$
2. $\begin{cases} x + 2y = -5 \\ x - y = 4 \end{cases}$
3. $\begin{cases} 3x + 2y = -8 \\ 5x - 2y = -8 \end{cases}$
4. $\begin{cases} 3x - 2y = 0 \\ 2x + 5y = 0 \end{cases}$
5. $\begin{cases} 2x + 3y = -12 \\ x - y = -1 \end{cases}$
6. $\begin{cases} 3x + 8y = 9 \\ 9x + 5y = -30 \end{cases}$
7. $\begin{cases} 4x + 8y = -20 \\ 10y + 5x = 10 \end{cases}$
8. $\begin{cases} 5x + 8y = -22 \\ 4x - 3y = 20 \end{cases}$
9. $\begin{cases} 15x + 7y = -35 \\ 14y + 30x = -70 \end{cases}$
10. $\begin{cases} 5x - 8y = 14 \\ 9x - 2y = -12 \end{cases}$

11.3 Addition Method for Solving a Linear System of Two Equations in Two Variables

The graphical method for solving a system of equations has two disadvantages: (1) it is slow, and (2) it is not an exact method of solution. The method we discuss in this section has neither of these disadvantages. We need one definition before we proceed.

> EQUIVALENT SYSTEMS OF EQUATIONS
>
> Two systems of equations are said to be **equivalent systems** if they both have the same solution set.

The operations in the following box always give us a system of equations equivalent to the system we started with; we usually use the second and third operations when we use the addition method for solving a system of equations.

> OPERATIONS THAT YIELD EQUIVALENT SYSTEMS OF EQUATIONS
>
> **1.** Interchanging the equations
>
> **2.** Rewriting any of the equations by using the multiplication property of equality
>
> **3.** Adding any two of the equations together; that is, using the following property of equality:
>
> If $a = b$ and $c = d$, then $a + c = b + d$.

In solving a system of two equations in two variables algebraically, our objective is to find an *equivalent* system in which one of the equations is in the form $x = a$ and the other is in the form $y = b$. That is, we hope to find an equivalent system in which each equation has only one variable.

We will be able to find such an equivalent system if the system of equations is consistent and independent. In the other two cases, *both* variables will drop out. We will be left with a *false* statement (such as $0 = 5$) if the system of equations is inconsistent and with a *true* statement (such as $0 = 0$) if the system is dependent.

We must begin by *eliminating one of the variables* from the equations; the following procedure for solving a linear system of equations in two variables enables us to do this. (In this procedure, it is understood that both cquations have already been simplified.)

The complete procedure is as follows:

> TO SOLVE A LINEAR SYSTEM OF TWO EQUATIONS BY THE ADDITION METHOD
>
> **1.** If necessary, multiply one or both equations by numbers that make the coefficients of one of the variables equal numerically but opposite in sign. Write the equations one under the other with the equal signs aligned and *with like terms aligned.*
>
> **2.** Add the equations from step 1 together. (This will eliminate at least one of the variables.)
>
> **3.** Three outcomes are possible:
>
> a. *One variable remains.* Solve *the resulting equation for that variable.* Then substitute this value into either equation to find the value of the other variable.
>
> b. *Both variables are eliminated and a false statement results.* There is no solution.
>
> c. *Both variables are eliminated and a true statement results.* There are infinitely many solutions. For the line $ax + by = c$, the solution set is the set of all ordered pairs of the form $\left(t, \dfrac{c - at}{b}\right)$.
>
> **4.** If a unique solution is found in step 3, check it in *both* equations.

We will give examples of each type of system.

Systems That Have Only One Solution (Consistent, Independent Systems)

Example 1 Solve the system $\begin{Bmatrix} 3x + 2y = 13 \\ 3x - 4y = 1 \end{Bmatrix}$.

Solution If we multiply the second equation by -1, the coefficients of x will be equal numerically and opposite in sign:

$$\begin{aligned} 3x + 2y &= 13 \\ -3x + 4y &= -1 \\ \hline 6y &= 12 \\ y &= 2 \end{aligned}$$

We add *both* sides of both equations

Don't drop the equal sign

Substituting 2 for y in the second equation, we have

$$\begin{aligned} 3x - 4y &= 1 \\ 3x - 4(2) &= 1 \\ 3x &= 9 \\ x &= 3 \end{aligned}$$

Check

$$\begin{aligned} 3x + 2y &= 13 \\ 3(3) + 2(2) &\stackrel{?}{=} 13 \\ 9 + 4 &\stackrel{?}{=} 13 \\ 13 &= 13 \quad \text{True} \end{aligned} \qquad \begin{aligned} 3x - 4y &= 1 \\ 3(3) - 4(2) &\stackrel{?}{=} 1 \\ 9 - 8 &\stackrel{?}{=} 1 \\ 1 &= 1 \quad \text{True} \end{aligned}$$

Therefore, the solution of the system is (3, 2). The system $\begin{Bmatrix} x = 3 \\ y = 2 \end{Bmatrix}$ is equivalent to the system $\begin{Bmatrix} 3x + 2y = 13 \\ 3x - 4y = 1 \end{Bmatrix}$, since both systems have the same solution set. ■

When we have found the value of one variable, that value may be substituted into either of the equations to find the value of the other variable. Often one equation is easier to work with than the other.

Example 2 Solve the system $\begin{matrix} (1) \\ (2) \end{matrix} \begin{Bmatrix} 3x + 4y = 6 \\ 2x + 3y = 5 \end{Bmatrix}$ (a) by eliminating the x's and (b) by eliminating the y's.

a. *Solution* If Equation 1 is multiplied by -2 and Equation 2 is multiplied by 3, the coefficients of x will be equal numerically but opposite in sign.

This symbol means the equation is to be multiplied by -2

$$\begin{array}{lll} (1) & -2\,] & 3x + 4y = 6 \Rightarrow {}^{*}-6x - 8y = -12 \\ (2) & 3\,] & 2x + 3y = 5 \Rightarrow \underline{6x + 9y = 15} \\ & & \qquad\qquad\qquad\qquad\quad y = 3 \end{array}$$

We add *both* sides of both equations

This pair of numbers is found by interchanging the coefficients of x and making one of them negative

Substituting 3 for y in Equation 2, we have

$$\begin{aligned} (2) \quad 2x + 3y &= 5 \\ 2x + 3(3) &= 5 \\ 2x + 9 &= 5 \\ 2x &= -4 \\ x &= -2 \end{aligned}$$

Therefore, the solution of the system is $(-2, 3)$.

---

* The symbol $\Rightarrow$, read "implies," means that the second statement is true if the first statement is true. For example,

$$3x + 4y = 6 \Rightarrow -6x - 8y = -12$$

is read "$3x + 4y = 6$ implies $-6x - 8y = -12$"

and means that $-6x - 8y = -12$ is true *if* $3x + 4y = 6$ is true.

b. *Solution*

This pair is found by interchanging the coefficients of y and making one of them negative

$$\begin{array}{lll} (1) & 3\,] & 3x + 4y = 6 \Rightarrow \quad 9x + 12y = 18 \\ (2) & -4\,] & 2x + 3y = 5 \Rightarrow -8x - 12y = -20 \\ & & \hline \qquad\qquad\qquad\qquad\quad x \qquad\quad = -\ 2 \end{array}$$

We add *both* sides of both equations

Substituting -2 for x in Equation 1, we have

$$\begin{aligned} (1) \qquad 3x + 4y &= 6 \\ 3(-2) + 4y &= 6 \\ -6 + 4y &= 6 \\ 4y &= 12 \\ y &= 3 \end{aligned}$$

Therefore, we get the same solution as before: $(-2, 3)$. (The check will not be shown.) ■

In Examples 1 and 2, we can say that the lines intersect in exactly one point and that the system of equations is consistent and independent.

We can always make the coefficients of one of the variables equal numerically by simply interchanging the coefficients. Sometimes, however, smaller numbers can be used (see Example 3).

Example 3 Consider the system $\begin{Bmatrix} 10x - 9y = 5 \\ 15x + 6y = 4 \end{Bmatrix}$.

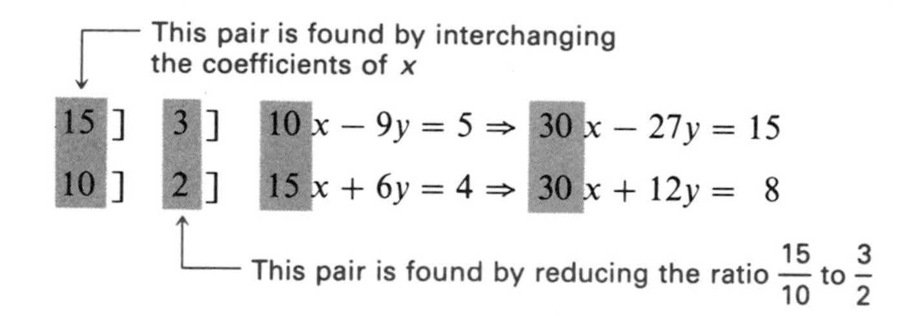

Notice that 30 is the least common multiple of 10 and 15.

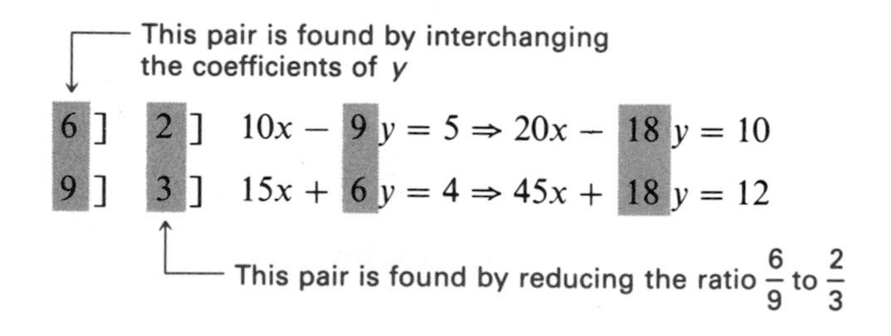

Notice that 18 is the least common multiple of 9 and 6. ■

Systems That Have No Solution (Inconsistent, Independent Systems)

In Section 11.2, we found that a system whose graphs are parallel lines has no solution (see Section 11.2, Example 2). Here, we show how to identify systems that have no solution when we're using the *addition method*.

When we attempt to solve a system of equations algebraically and *both* variables drop out and a *false statement* results, there is no solution for the system.

Example 4 Solve the system $\begin{Bmatrix} 2x - 3y = 6 \\ 6x - 9y = 36 \end{Bmatrix}$.

Solution

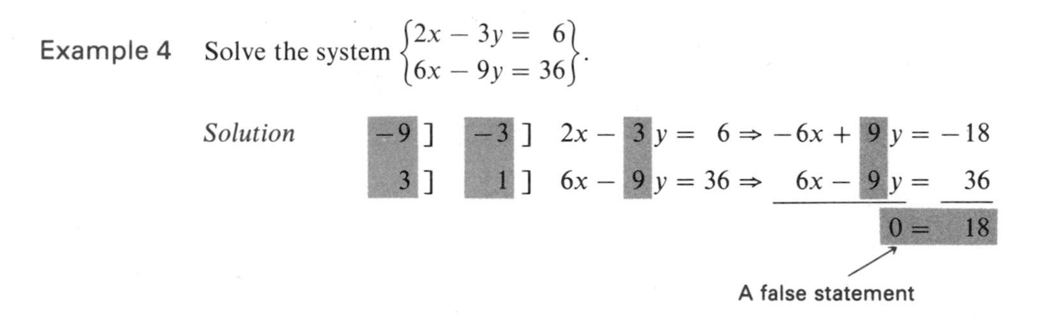

$$\begin{aligned} -9]\ \ -3]\ \ 2x - 3y &= 6 \Rightarrow -6x + 9y = -18 \\ 3]\ \ \ \ 1]\ \ 6x - 9y &= 36 \Rightarrow \ \ \ 6x - 9y = \ \ 36 \\ & \qquad\qquad\qquad\qquad 0 = \ \ 18 \end{aligned}$$

A false statement

You can verify that the slopes of the lines are the same; therefore, the lines *are* parallel, and there is no solution for this system of equations. The system is inconsistent and independent. ■

Systems That Have More Than One Solution (Consistent, Dependent Systems)

In Section 11.2, we found that a system whose equations have the same graph has an infinite number of solutions (see Section 11.2, Example 3). Here, we show how to identify such systems when we're using the *addition method.*

When we attempt to solve a system of equations algebraically and *both* variables drop out and a *true* statement results, there are infinitely many solutions for the system.

Example 5 Solve the system $\begin{Bmatrix} 4x + 6y = 4 \\ 6x + 9y = 6 \end{Bmatrix}$.

Solution

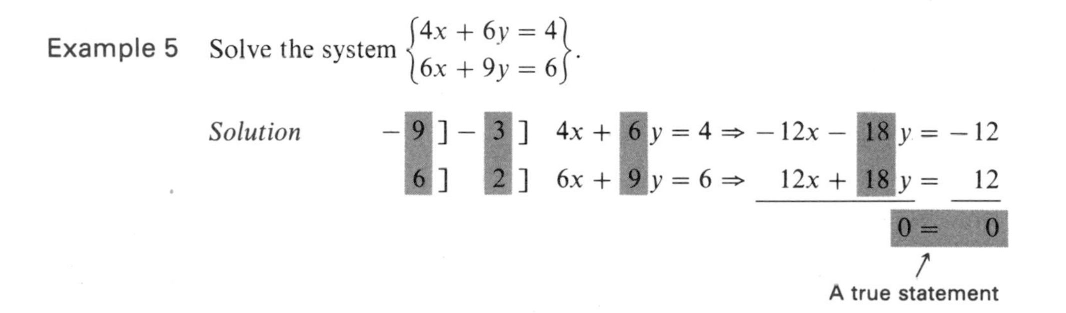

$$\begin{aligned} -9]\ -3]\ \ 4x + 6y &= 4 \Rightarrow -12x - 18y = -12 \\ 6]\ \ \ \ 2]\ \ 6x + 9y &= 6 \Rightarrow \ \ \ 12x + 18y = \ \ 12 \\ & \qquad\qquad\qquad\qquad\quad 0 = \ \ 0 \end{aligned}$$

A true statement

You can verify that both equations simplify to $2x + 3y = 2$; therefore, the lines coincide and the system of equations is consistent and dependent. We can find the solution set for the system by letting $x = t$ in either equation and solving for y:

$$4t + 6y = 4$$

$$6y = 4 - 4t$$

$$y = \frac{4 - 4t}{6} = \frac{\overset{1}{\cancel{2}}(2 - 2t)}{\underset{3}{\cancel{6}}} = \frac{2 - 2t}{3}$$

Therefore, the solution set is the set of ordered pairs $\left(t, \frac{2 - 2t}{3}\right)$, where t is any real number. If we had let $y = s$, the solution set would have been $\left(\frac{2 - 3s}{2}, s\right)$. ■

NOTE Whenever both of the original equations can be written as the *same* equation (as in Example 5), the system of equations is dependent. ☑

EXERCISES 11.3

Set I Find the solution of each system by the addition method. Write "inconsistent" if no solution exists. Write "dependent" if an infinite number of solutions exist.

1. $\left\{\begin{aligned} 3x - y &= 11 \\ 3x + 2y &= -4 \end{aligned}\right\}$

2. $\left\{\begin{aligned} 6x + 5y &= 2 \\ 2x - 5y &= -26 \end{aligned}\right\}$

3. $\left\{\begin{aligned} 8x + 15y &= 11 \\ 4x - y &= 31 \end{aligned}\right\}$

4. $\left\{\begin{aligned} x + 6y &= 24 \\ 5x - 3y &= 21 \end{aligned}\right\}$

5. $\left\{\begin{aligned} 7x - 3y &= 3 \\ 20x - 9y &= 12 \end{aligned}\right\}$

6. $\left\{\begin{aligned} 10x + 7y &= -1 \\ 2x + y &= 5 \end{aligned}\right\}$

7. $\left\{\begin{aligned} 6x + 5y &= 0 \\ 4x - 3y &= 38 \end{aligned}\right\}$

8. $\left\{\begin{aligned} 7x + 4y &= 12 \\ 2x - 3y &= -38 \end{aligned}\right\}$

9. $\left\{\begin{aligned} 4x + 6y &= 5 \\ 8x + 12y &= 7 \end{aligned}\right\}$

10. $\left\{\begin{aligned} 5x - 2y &= 3 \\ 15x - 6y &= 4 \end{aligned}\right\}$

11. $\left\{\begin{aligned} 7x - 3y &= 5 \\ 14x - 6y &= 10 \end{aligned}\right\}$

12. $\left\{\begin{aligned} 8x - 12y &= 16 \\ 2x - 3y &= 4 \end{aligned}\right\}$

13. $\left\{\begin{aligned} 9x + 4y &= -4 \\ 15x - 6y &= 25 \end{aligned}\right\}$

14. $\left\{\begin{aligned} 16x - 25y &= -38 \\ 8x + 5y &= -12 \end{aligned}\right\}$

15. $\left\{\begin{aligned} 9x + 10y &= -3 \\ 14y &= 7 - 15x \end{aligned}\right\}$

16. $\left\{\begin{aligned} 35x + 18y &= 30 \\ 7x &= 24y - 17 \end{aligned}\right\}$

Set II Find the solution of each system by the addition method. Write "inconsistent" if no solution exists. Write "dependent" if an infinite number of solutions exist.

1. $\left\{\begin{aligned} 2x - 7y &= 6 \\ 4x + 7y &= -30 \end{aligned}\right\}$

2. $\left\{\begin{aligned} x - 2y &= -32 \\ 7x + 8y &= -4 \end{aligned}\right\}$

3. $\left\{\begin{aligned} 12x + 17y &= 30 \\ 3x - y &= 39 \end{aligned}\right\}$

4. $\left\{\begin{aligned} 3x - 2y &= -51 \\ 2x + 3y &= -21 \end{aligned}\right\}$

5. $\left\{\begin{aligned} 3x - 5y &= -2 \\ 2x + y &= 16 \end{aligned}\right\}$

6. $\left\{\begin{aligned} 3x - y &= -14 \\ x + 2y &= 7 \end{aligned}\right\}$

7. $\left\{\begin{aligned} 2x - y &= 0 \\ 3x + 4y &= -22 \end{aligned}\right\}$

8. $\left\{\begin{aligned} 2x - 5y &= 15 \\ 3x + 2y &= 13 \end{aligned}\right\}$

9. $\left\{\begin{aligned} 12x - 9y &= 28 \\ 20x - 15y &= 35 \end{aligned}\right\}$

10. $\left\{\begin{aligned} 7x - 2y &= 4 \\ 14x - 4y &= 8 \end{aligned}\right\}$

11. $\left\{\begin{aligned} 21x + 35y &= 28 \\ 12x + 20y &= 16 \end{aligned}\right\}$

12. $\left\{\begin{aligned} 8x - 5y &= 3 \\ 24x - 15y &= 6 \end{aligned}\right\}$

13. $\left\{\begin{aligned} 3x + 10y &= 4 \\ 9x - 20y &= -18 \end{aligned}\right\}$

14. $\left\{\begin{aligned} 7x + 2y &= 4 \\ 2x + 3y &= -11 \end{aligned}\right\}$

15. $\left\{\begin{aligned} 3x + 4y &= 2 \\ 4y - 3x &= 0 \end{aligned}\right\}$

16. $\left\{\begin{aligned} 22x - 15y &= -29 \\ 33x &= 10y - 11 \end{aligned}\right\}$

11.4 Substitution Method for Solving a Linear System of Two Equations In Two Variables

All linear systems of two equations in two variables can be solved by the addition method shown in Section 11.3. However, if one equation has already been solved for one variable, the substitution method is easier to use than the addition method. Also, you should learn the *substitution method* of solution at this time, so that you can apply it later to solving quadratic systems (Section 11.9).

TO SOLVE A LINEAR SYSTEM OF TWO EQUATIONS BY THE SUBSTITUTION METHOD

1. Solve one equation for one of the variables in terms of the other by using the method of solving literal equations that was given in Section 6.9.

2. Substitute the expression obtained in step 1 into the *other* equation, and simplify both sides of the equation.

3. Three outcomes are possible:

a. *One variable remains.* Solve the equation resulting from step 2 for its variable, and substitute that value into either equation to find the value of the other variable. There is a unique solution for the system.

b. *Both variables are eliminated, and a false statement results.* There is no solution for the system.

c. *Both variables are eliminated, and a true statement results.* There are many solutions for the system; the solution set can be found by using the method described in Section 11.3.

4. If a unique solution is found, check it in *both* equations.

Examples 1 through 3 show how to decide which equation to use and which variable to solve for in step 1 of the substitution method.

Example 1 An example in which onc of the equations is already solved for a variable:

$$\begin{Bmatrix} 2x - 5y = 4 \\ \boxed{y} = 3x + 7 \end{Bmatrix}$$

↑ Already solved for y

Use $\boxed{y} = 3x + 7$. ■

Example 2 An example in which one of the variables has a coefficient of 1:

$$\begin{Bmatrix} 2x + 6y = 3 \\ \boxed{x} - 4y = 2 \end{Bmatrix} \Rightarrow \boxed{x} = 4y + 2$$

↑ x has a coefficient of 1

Use $\boxed{x} = 4y + 2$. ■

Example 3 An example of choosing the variable with the smallest coefficient:

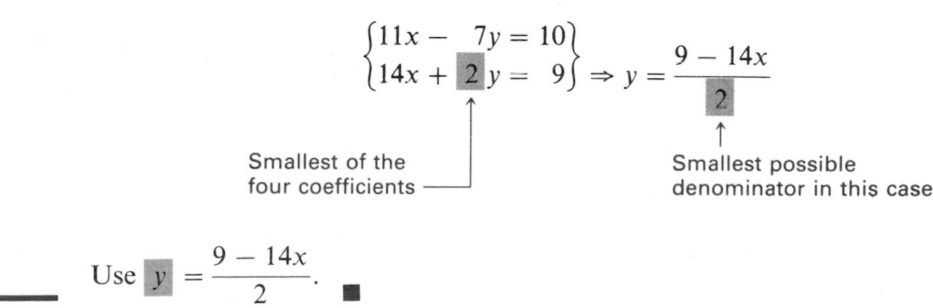

$$\begin{Bmatrix} 11x - 7y = 10 \\ 14x + \boxed{2}y = 9 \end{Bmatrix} \Rightarrow y = \frac{9 - 14x}{\boxed{2}}$$

Use $\boxed{y} = \dfrac{9 - 14x}{2}$. ■

Systems That Have a Unique Solution (Consistent, Independent Systems)

We now give two examples of using the substitution method with systems that have a unique solution.

Example 4 Solve the system $\left\{\begin{array}{l} 2x - 5y = 4 \\ y = 3x + 7 \end{array}\right\}$.

Step 1. The second equation has already been solved for y.

Step 2. Substitute $3x + 7$ for y in the first equation:

$$\begin{aligned} 2x - 5y &= 4 \\ 2x - 5(3x + 7) &= 4 \\ 2x - 15x - 35 &= 4 \\ -13x &= 39 \\ x &= -3 \end{aligned}$$

Step 3. Substitute -3 for x in $y = 3x + 7$:

$$y = 3(-3) + 7 = -9 + 7 = -2$$

Therefore, $(-3, -2)$ is the solution for this system.

Step 4. *Check*

$$\begin{aligned} 2x - 5y &= 4 \\ 2(-3) - 5(-2) &\stackrel{?}{=} 4 \\ -6 + 10 &\stackrel{?}{=} 4 \\ 4 &= 4 \quad \text{True} \end{aligned} \qquad \begin{aligned} y &= 3x + 7 \\ -2 &\stackrel{?}{=} 3(-3) + 7 \\ -2 &\stackrel{?}{=} -9 + 7 \\ -2 &= -2 \quad \text{True} \end{aligned}$$

■

Example 5 Solve the system $\left\{\begin{array}{l} x - 2y = 11 \\ 3x + 5y = -11 \end{array}\right\}$.

Step 1. Solve the first equation for x:

$$\begin{aligned} x - 2y &= 11 \\ x &= 11 + 2y \end{aligned}$$

Step 2. Substitute $11 + 2y$ for x in the second equation:

$$\begin{aligned} 3x + 5y &= -11 \\ 3(11 + 2y) + 5y &= -11 \\ 33 + 6y + 5y &= -11 \\ 11y &= -44 \\ y &= -4 \end{aligned}$$

Step 3. Substitute -4 for y in $x = 11 + 2y$:

$$\begin{aligned} x &= 11 + 2(-4) \\ x &= 11 - 8 \\ x &= 3 \end{aligned}$$

Step 4. *Check*

$$\begin{aligned} x - 2y &= 11 \\ 3 - 2(-4) &\stackrel{?}{=} 11 \\ 3 + 8 &\stackrel{?}{=} 11 \\ 11 &= 11 \quad \text{True} \end{aligned} \qquad \begin{aligned} 3x + 5y &= -11 \\ 3(3) + 5(-4) &\stackrel{?}{=} -11 \\ 9 - 20 &\stackrel{?}{=} -11 \\ -11 &= -11 \quad \text{True} \end{aligned}$$

Therefore, $(3, -4)$ is the solution. ■

Systems That Have No Solution (Inconsistent, Independent Systems)

We now show how to identify systems that have no solution when we're using the *substitution method*.

Example 6 Solve the system $\begin{cases} 2x - 3y = 6 \\ 6x - 9y = 36 \end{cases}$. (Smallest coefficient: 2)

Step 1. Solve the first equation for x:

$$2x - 3y = 6 \Rightarrow 2x = 3y + 6$$
$$x = \frac{3y + 6}{2}$$

Step 2. Substitute $\frac{3y+6}{2}$ for x in the second equation:

$$6x - 9y = 36$$
$$\overset{3}{\cancel{6}}\left(\frac{3y+6}{\underset{1}{\cancel{2}}}\right) - 9y = 36$$
$$3(3y + 6) - 9y = 36$$
$$9y + 18 - 9y = 36$$
$$18 = 36$$

A false statement

Step 3. *No* values for x and y can make $18 = 36$. Therefore, there is *no solution* for this system of equations. ■

Systems That Have More Than One Solution (Consistent, Dependent Systems)

We now show how to identify systems that have more than one solution when we're using the *substitution method*.

Example 7 Solve the system $\begin{cases} 9x + 6y = 6 \\ 6x + 4y = 4 \end{cases}$. (Smallest coefficient: 4)

Step 1. Solve the second equation for y:

$$6x + 4y = 4 \Rightarrow 4y = 4 - 6x$$

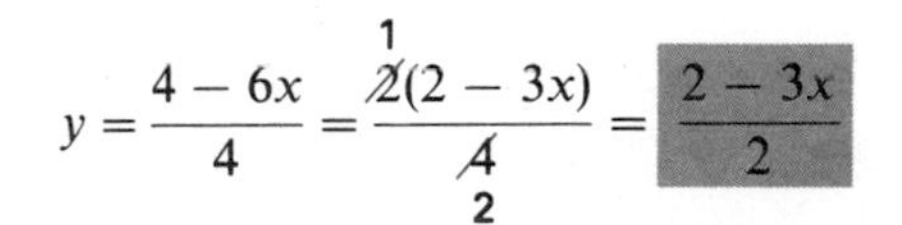

$$y = \frac{4 - 6x}{4} = \frac{\overset{1}{\cancel{2}}(2 - 3x)}{\underset{2}{\cancel{4}}} = \frac{2 - 3x}{2}$$

Step 2. Substitute $\frac{2-3x}{2}$ for y in the first equation:

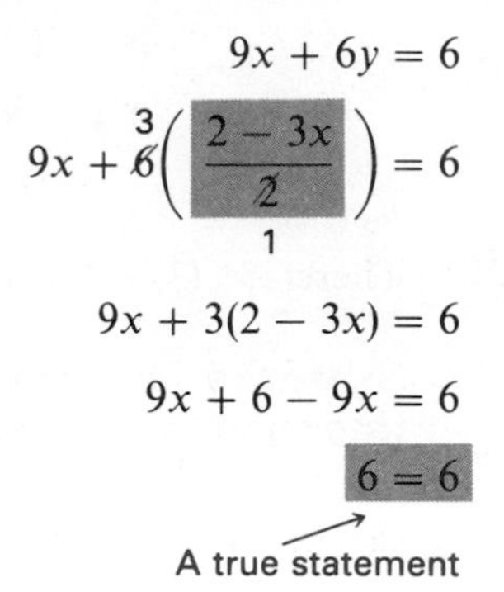

$$9x + 6y = 6$$

$$9x + \overset{3}{\cancel{6}}\left(\frac{2-3x}{\underset{1}{\cancel{2}}}\right) = 6$$

$$9x + 3(2 - 3x) = 6$$

$$9x + 6 - 9x = 6$$

$$6 = 6$$

A true statement

Step 3. To find the solution set, we let $x = t$. Then $y = \frac{2-3t}{2}$. Therefore, the solution set is the set of all ordered pairs of the form $\left(t, \frac{2-3t}{2}\right)$. ■

EXERCISES 11.4

Set I Find the solution set of each system. Write "inconsistent" if no solution exists. Write "dependent" if an infinite number of solutions exist.

In Exercises 1–10, use the substitution method.

1. $\begin{Bmatrix} 7x + 4y = 4 \\ y = 6 - 3x \end{Bmatrix}$

2. $\begin{Bmatrix} 2x + 3y = -5 \\ x = y - 10 \end{Bmatrix}$

3. $\begin{Bmatrix} 5x - 4y = -1 \\ 3x + y = -38 \end{Bmatrix}$

4. $\begin{Bmatrix} 3x - 5y = 5 \\ x - 6y = 19 \end{Bmatrix}$

5. $\begin{Bmatrix} 8x - 5y = 4 \\ x - 2y = -16 \end{Bmatrix}$

6. $\begin{Bmatrix} 12x - 6y = 24 \\ 3x - 2y = -2 \end{Bmatrix}$

7. $\begin{Bmatrix} 15x + 5y = 8 \\ 6x + 2y = -10 \end{Bmatrix}$

8. $\begin{Bmatrix} 15x - 5y = 30 \\ 12x - 4y = 11 \end{Bmatrix}$

9. $\begin{Bmatrix} 20x - 10y = 70 \\ 6x - 3y = 21 \end{Bmatrix}$

10. $\begin{Bmatrix} 2x - 10y = 18 \\ 5x - 25y = 45 \end{Bmatrix}$

In Exercises 11–14, solve by any convenient method.

11. $\begin{Bmatrix} 8x + 4y = 7 \\ 3x + 6y = 6 \end{Bmatrix}$

12. $\begin{Bmatrix} 5x - 4y = 2 \\ 15x + 12y = 12 \end{Bmatrix}$

13. $\begin{Bmatrix} 4x + 4y = 3 \\ 6x + 12y = -6 \end{Bmatrix}$

14. $\begin{Bmatrix} 4x + 9y = -11 \\ 10x + 6y = 11 \end{Bmatrix}$

Set II Find the solution set of each system. Write "inconsistent" if no solution exists. Write "dependent" if an infinite number of solutions exist.

In Exercises 1–10, use the substitution method.

1. $\begin{Bmatrix} 4x + 3y = -7 \\ y = 2x - 9 \end{Bmatrix}$

2. $\begin{Bmatrix} 13y - 7x = 17 \\ 2x - y = -13 \end{Bmatrix}$

3. $\begin{cases} 12x - 16y = -3 \\ 8x - 4y = 8 \end{cases}$

4. $\begin{cases} y = x + 8 \\ 2y - 3x = 18 \end{cases}$

5. $\begin{cases} x + 4y = 1 \\ 2x + 9y = 1 \end{cases}$

6. $\begin{cases} 3x + y = 5 \\ -6x - 2y = -10 \end{cases}$

7. $\begin{cases} 3x - 9y = 15 \\ 4x - 12y = 7 \end{cases}$

8. $\begin{cases} x + 4y = 8 \\ 3y - 2x = -5 \end{cases}$

9. $\begin{cases} 8x - 2y = 26 \\ 16x - 4y = 52 \end{cases}$

10. $\begin{cases} x - 4y = 3 \\ 4y - x = 3 \end{cases}$

In Exercises 11–14, solve by any convenient method.

11. $\begin{cases} -19x + 10y = 26 \\ 12x - 5y = 2 \end{cases}$

12. $\begin{cases} 7x - 4y = 5 \\ 4y - 7x = 5 \end{cases}$

13. $\begin{cases} 3x + 2y = -1 \\ 15x + 14y = -23 \end{cases}$

14. $\begin{cases} 3x - 2y = 3 \\ 8x + 5y = 8 \end{cases}$

11.5 Higher-Order Systems

So far we have considered only systems of two equations in two variables. If a system has more than two equations and more than two variables, it is usually called a **higher-order system**. A system with three equations and three variables is called a *third-order system*, one with four equations and four variables is a *fourth-order system*, and so on.

Third-Order Systems

In solving third-order systems, we want to find, if it exists, an *equivalent* system of equations in which one of the equations is in the form $x = a$, another is in the form $y = b$, and another is in the form $z = c$. The solution set will be the ordered triple (a, b, c).

The addition method can be used to solve third-order systems, as outlined below.

> TO SOLVE A THIRD-ORDER LINEAR SYSTEM BY ADDITION
>
> 1. Eliminate one variable from a pair of equations by addition.
> 2. Eliminate the *same* variable from a *different* pair of equations by addition.
> 3. The equation in two variables obtained in step 1 and the equation in the same two variables obtained in step 2 form a second-order system that can then be solved.

Example 1 Solve the system
$$\begin{matrix} (1) \\ (2) \\ (3) \end{matrix} \begin{cases} 2x - 3y + z = 1 \\ x + 2y + z = -1 \\ 3x - y + 3z = 4 \end{cases}.$$

Solution First, eliminate z from Equations 1 and 2. (While any one of the variables can be eliminated, we choose z in this case.)

$$\begin{array}{lrl} (1) & 1] & 2x - 3y + z = 1 \Rightarrow 2x - 3y + z = 1 \\ (2) & -1] & x + 2y + z = -1 \Rightarrow \underline{-x - 2y - z = 1} \\ & & (4) \qquad x - 5y = 2 \end{array}$$

Add the equations

Next, eliminate z from Equations 1 and 3:

$$\begin{array}{llrcl} (1) & 3\,] & 2x - 3y + z = 1 & \Rightarrow & 6x - 9y + 3z = 3 \\ (3) & -1\,] & 3x - y + 3z = 4 & \Rightarrow & -3x + y - 3z = -4 \\ & & & (5) & 3x - 8y = -1 \end{array}$$

Add the equations

(We could have eliminated z from Equations 2 and 3, instead.)
Equations 4 and 5 form a second-order system that we solve by addition. (The system *could* be solved by substitution.)

$$\begin{array}{llrcl} (4) & 3\,] & x - 5y = 2 & \Rightarrow & 3x - 15y = 6 \\ (5) & -1\,] & 3x - 8y = -1 & \Rightarrow & -3x + 8y = 1 \\ & & & & -7y = 7 \\ & & & & y = -1 \end{array}$$

$$\begin{aligned} (4)\quad x - 5y &= 2 \\ x - 5(-1) &= 2 \quad \text{Substituting } -1 \text{ for } y \text{ in Equation 4} \\ x + 5 &= 2 \\ x &= -3 \end{aligned}$$

The two values found are substituted into any one of the original equations to find the value of the third variable.

$$\begin{aligned} (2)\quad x + 2y + z &= -1 \\ (-3) + 2(-1) + z &= -1 \quad \text{Substituting } -3 \text{ for } x \text{ and } -1 \text{ for } y \text{ in Equation 2} \\ -3 - 2 + z &= -1 \\ z &= 4 \end{aligned}$$

Check The check for this solution is shown in Example 2 of Section 11.1. Therefore, the solution of the system is $(-3, -1, 4)$. ■

Higher-Order Systems

The method for solving a third-order system can be extended to a fourth-order system as follows:

1. Eliminate one variable from a pair of equations. This gives one equation in three variables.
2. Eliminate the *same* variable from a different pair of equations. This gives a second equation in three variables.
3. Eliminate the *same* variable from a third different pair of equations. This gives a third equation in three variables. (Note that each equation of the system must be used in at least one pair of equations.)
4. The three equations obtained in steps 1, 2, and 3 form a third-order system that is then solved by the method given in the box on page 559 and shown in Example 1.

This same method can be extended to solve systems of *any* order. However, because of the amount of work involved in solving higher-order systems, solutions are usually carried out by computer.

Some higher-order systems have no solution (inconsistent, independent systems) and others have many solutions (consistent, dependent systems), as was true for second-order systems. In this section we consider only higher-order systems that have a single solution (consistent, independent systems).

The graph of a linear equation in three variables is a *plane* in three-dimensional space. Therefore, graphical solutions of third-order linear systems are too complicated for practical use. Graphical solutions for *higher* than third-order linear systems are not possible, as we cannot visualize (nor represent on paper) four-dimensional space, five-dimensional space, and so on.

EXERCISES 11.5

Set I

Solve each system.

1. $\begin{Bmatrix} 2x + y + z = 4 \\ x - y + 3z = -2 \\ x + y + 2z = 1 \end{Bmatrix}$

2. $\begin{Bmatrix} x + y + z = 1 \\ 2x + y - 2z = -4 \\ x + y + 2z = 3 \end{Bmatrix}$

3. $\begin{Bmatrix} x + 2y + 2z = 0 \\ 2x - y + z = -3 \\ 4x + 2y + 3z = 2 \end{Bmatrix}$

4. $\begin{Bmatrix} 2x + y - z = 0 \\ 3x + 2y + z = 3 \\ x - 3y - 5z = 5 \end{Bmatrix}$

5. $\begin{Bmatrix} x + 2z = 7 \\ 2x - y = 5 \\ 2y + z = 4 \end{Bmatrix}$

6. $\begin{Bmatrix} x - 2y = 4 \\ y + 3z = 8 \\ 2x - z = 1 \end{Bmatrix}$

7. $\begin{Bmatrix} 2x + 3y + z = 7 \\ 4x - 2z = -6 \\ 6y - z = 0 \end{Bmatrix}$

8. $\begin{Bmatrix} 4x + 5y + z = 4 \\ 10y - 2z = 6 \\ 8x + 3z = 3 \end{Bmatrix}$

9. $\begin{Bmatrix} x + y + z + w = 5 \\ 2x - y + 2z - w = -2 \\ x + 2y - z - 2w = -1 \\ -x + 3y + 3z + w = 1 \end{Bmatrix}$

10. $\begin{Bmatrix} x + y + z + w = 4 \\ x - 2y - z - 2w = -2 \\ 3x + 2y + z + 3w = 4 \\ 2x + y - 2z - w = 0 \end{Bmatrix}$

11. $\begin{Bmatrix} 6x + 4y + 9z + 5w = -3 \\ 2x + 8y - 6z + 15w = 8 \\ 4x - 4y + 3z - 10w = -3 \\ 2x - 4y + 3z - 5w = -1 \end{Bmatrix}$

12. $\begin{Bmatrix} 12x + 9y + 4z - 8w = 1 \\ 6x + 15y + 2z + 4w = 2 \\ 3x + 6y + 4z + 2w = 5 \\ 4x + 4y + 4z + 4w = 3 \end{Bmatrix}$

Set II

Solve each system.

1. $\begin{Bmatrix} x + y - z = 0 \\ 2x - y + 3z = 1 \\ 3x + y + z = 2 \end{Bmatrix}$

2. $\begin{Bmatrix} 2x + y + z = -1 \\ 3x + 5y + z = 0 \\ 7x - y + 2z = 1 \end{Bmatrix}$

3. $\begin{Bmatrix} x + 2y - 3z = 5 \\ x + y + z = 0 \\ 3x + 4y + 2z = -1 \end{Bmatrix}$

4. $\begin{Bmatrix} x + 2y + z = 0 \\ 2x + 3y - 5z = 1 \\ -3x + y + 4z = -7 \end{Bmatrix}$

5. $\begin{Bmatrix} 3x + 2z = 0 \\ 5y - z = 6 \\ 2x - 3y = 8 \end{Bmatrix}$

6. $\begin{Bmatrix} x + 2y = 0 \\ y - 2z = 0 \\ x - 4z = 0 \end{Bmatrix}$

7. $\begin{Bmatrix} 5x + y + 6z = -2 \\ 2y - 3z = 3 \\ 5x + 6z = -4 \end{Bmatrix}$

8. $\begin{Bmatrix} x + y + 2z = 7 \\ 5x - y + z = 2 \\ 3x + 3y - z = 0 \end{Bmatrix}$

9. $\begin{Bmatrix} x + y + z + w = 5 \\ 3x - y + 2z - w = 0 \\ 2x + y - z + 2w = 4 \\ 2x + y + z + 2w = 2 \end{Bmatrix}$

10. $\begin{Bmatrix} 2x + 5y + 3z - 4w = 0 \\ -x + y + 6w = -1 \\ 3x - z - 2w = 1 \\ 4x + 2y - z = 3 \end{Bmatrix}$

11. $\begin{Bmatrix} x + 2y + 2z + w = 0 \\ -x + y + 3z + w = 2 \\ 2x + 3y + 2z - w = -5 \\ 3x - y - 7z + w = 2 \end{Bmatrix}$

12. $\begin{Bmatrix} 3x + 2y + 4z + 5w = 5 \\ 9x - 8z + 10w = -4 \\ 6y + 12z + 5w = 5 \\ -6x - 4y + 15w = -1 \end{Bmatrix}$

11.6 Review: 11.1–11.5

Systems of Equations 11.1

In a linear system of equations, each equation is first degree.

In a quadratic system of equations, the highest-degree equation is second degree.

A solution of a system of two equations in two variables is an *ordered pair* that satisfies both equations.

A solution of a system of three equations in three variables is an *ordered triple* that satisfies all three equations.

Methods of Solving a System of Linear Equations 11.2 11.3 11.4

Second-order system

1. Graphical method
2. Addition method
3. Substitution method

When solving a system of linear equations, three outcomes are possible:

1. *There is only one solution.*
   a. Graphical method: The lines intersect at one point.
   b. Algebraic method:
      Addition / Substitution } The equations can be solved for a single ordered pair.
2. *There is no solution.*
   a. Graphical method: The lines are parallel.
   b. Algebraic method:
      Addition / Substitution } Both variables drop out and a false statement results.
3. *There are many solutions.*
   a. Graphical method: Both equations have the same line for a graph.
   b. Algebraic method:
      Addition / Substitution } Both variables drop out and a true statement results.

11.5 *Higher-order systems* can be solved by the addition method.

Review Exercises 11.6 Set I

In Exercises 1–3, solve each system graphically (there may be no solution).

1. $\begin{Bmatrix} 4x + 5y = 22 \\ 3x - 2y = 5 \end{Bmatrix}$

2. $\begin{Bmatrix} 9x - 12y = 3 \\ 12x - 16y = 4 \end{Bmatrix}$

3. $\begin{Bmatrix} 2x - 3y = 3 \\ 3y - 2x = 6 \end{Bmatrix}$

In Exercises 4–10, solve each system by any convenient method. Write "inconsistent" if no solution exists. Write "dependent" if an infinite number of solutions exist.

4. $\begin{Bmatrix} 6x + 4y = 13 \\ 8x + 10y = 1 \end{Bmatrix}$

5. $\begin{Bmatrix} 4x - 8y = 4 \\ 3x - 6y = 3 \end{Bmatrix}$

6. $\begin{Bmatrix} 4x - 7y = 28 \\ 7y - 4x = 20 \end{Bmatrix}$

7. $\begin{Bmatrix} x = y + 2 \\ 4x - 5y = 3 \end{Bmatrix}$

8. $\begin{Bmatrix} 7x - 3y = 1 \\ y = x + 5 \end{Bmatrix}$

9. $\begin{Bmatrix} 2x + y - z = 1 \\ 3x - y + 2z = 3 \\ x + 2y + 3z = -6 \end{Bmatrix}$

10. $\begin{Bmatrix} x - 2y + z = -3 \\ 3x + 4y + 2z = 4 \\ 2x - 4y - z = 3 \end{Bmatrix}$

Review Exercises 11.6 Set II

NAME

In Exercises 1–3, solve each system graphically (there may be no solution).

1. $\begin{cases} 2x + 3y = -5 \\ 4x - 5y = 23 \end{cases}$

2. $\begin{cases} 5y - 3x = 10 \\ 3x - 5y = 15 \end{cases}$

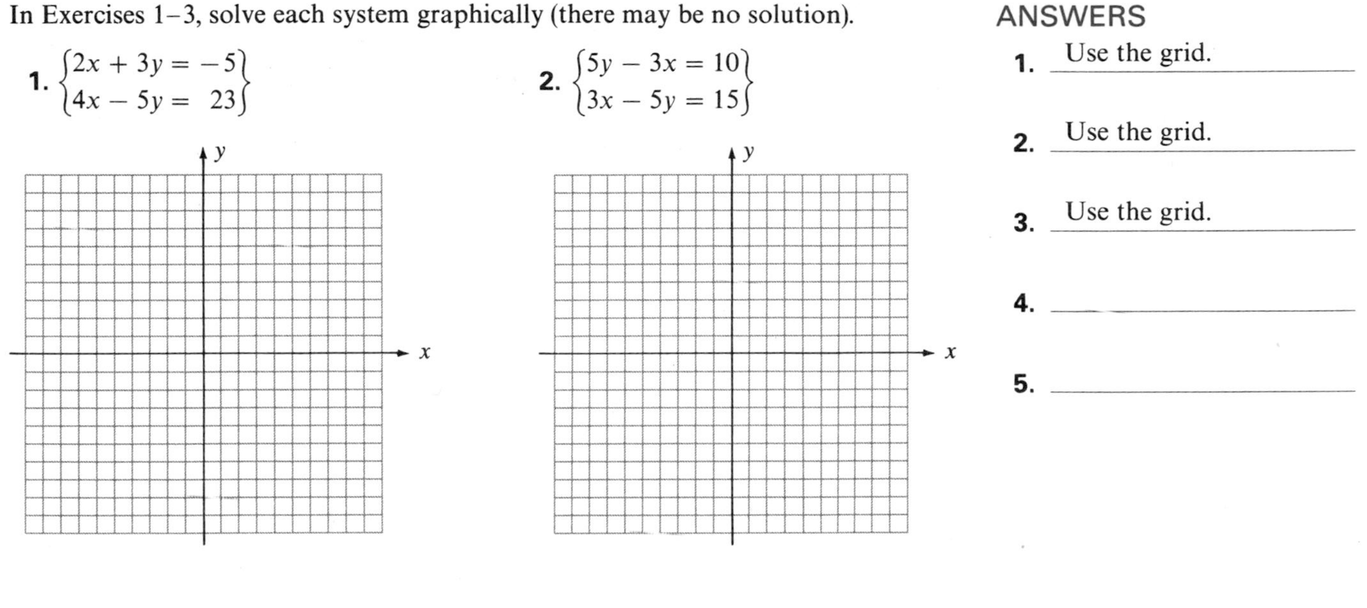

3. $\begin{cases} 7x - 8y = -9 \\ 5x + 6y = 17 \end{cases}$

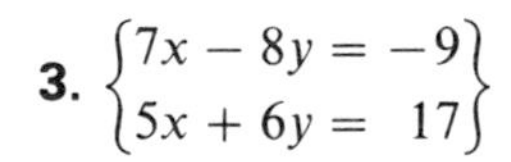

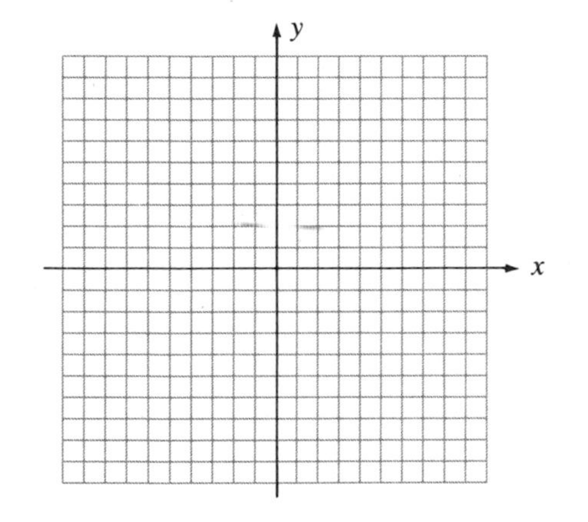

In Exercises 4–10, solve each system by any convenient method. Write "inconsistent" if no solution exists. Write "dependent" if an infinite number of solutions exist.

4. $\begin{cases} 6x + 4y = -1 \\ 4x + 6y = -9 \end{cases}$

5. $\begin{cases} 8x - 4y = 12 \\ 6x - 3y = 9 \end{cases}$

ANSWERS

1. Use the grid.
2. Use the grid.
3. Use the grid.
4. ________
5. ________

6. $\begin{cases} 3x + 2y = 1 \\ 7x + 3y = 9 \end{cases}$

7. $\begin{cases} 4x + 6y = -9 \\ 2x - 8y = 23 \end{cases}$

8. $\begin{cases} 3x - 2y = 10 \\ 5x + 4y = 24 \end{cases}$

9. $\begin{cases} 3x + 2y + z = 5 \\ 2x - 3y - 3z = 1 \\ 4x + 3y + 2z = 8 \end{cases}$

10. $\begin{cases} 4x + 3y + 2z = 6 \\ 8x + 3y - 3z = -1 \\ -4x + 6y + 5z = 5 \end{cases}$

ANSWERS

6. ______________________

7. ______________________

8. ______________________

9. ______________________

10. ______________________

11.7 Alternate Methods for Solving Systems of Linear Equations

11.7A A Matrix Method for Solving Systems of Linear Equations

In this section, we discuss a **matrix** method for solving a system of linear equations. The method can be used for solving any system of linear equations, and it is based on those operations that can be performed on a system of equations and yield an equivalent system—the operations listed in Section 11.3.

Matrix

A **matrix** (plural: *matrices*) is a rectangular array of numbers; we will use brackets around a matrix (some authors use parentheses). A **row** of a matrix is a *horizontal* line of its elements. A **column** of a matrix is a *vertical* line of its elements. An $\boldsymbol{m \times n}$ (read "m by n") **matrix** has m rows and n columns. The following are examples of matrices:

3×4 *matrix*

$$\begin{bmatrix} 2 & -3 & 1 & 1 \\ 1 & 2 & 1 & -1 \\ 3 & -1 & 3 & 4 \end{bmatrix}$$

(3 rows, 4 columns)

4×5 *matrix*

$$\begin{bmatrix} 2 & -1 & 3 & 1 & -4 \\ 1 & 2 & -1 & 2 & -4 \\ 3 & 1 & -1 & -2 & 4 \\ -2 & 3 & 1 & 3 & 0 \end{bmatrix}$$

(4 rows, 5 columns)

Main Diagonal The **main diagonal** of a matrix is the diagonal of numbers that goes from the upper left-hand corner downward toward the right.

Triangular Form We say that a matrix is in **triangular form** if the numbers *below* the main diagonal are all zeros; that is, a matrix is in triangular form if it is in the following form:

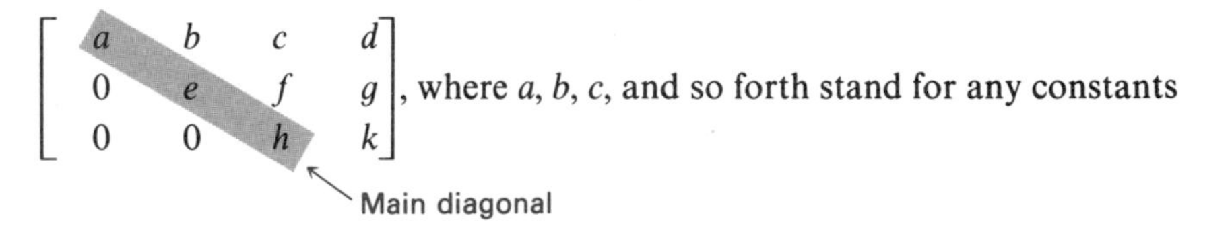

$$\begin{bmatrix} a & b & c & d \\ 0 & e & f & g \\ 0 & 0 & h & k \end{bmatrix}$$, where a, b, c, and so forth stand for any constants

Row-Echelon Form If a matrix is in triangular form *and* has only 1's down the main diagonal, we say that it is in **row-echelon form**.

NOTE The actual definition of the row-echelon form of a matrix is slightly more complicated than we've indicated here; however, our definition of row-echelon form will suffice for the problems in this text. ☑

The Augmented Matrix for a System of Equations

An **augmented matrix** is associated with each system of linear equations. To write the augmented matrix for a system of equations, it's helpful to first write the equations as follows:

1. Write the equations one under the other with all the terms containing variables on the left side of each equation and with *like terms aligned*; write the constant on the right side of each equation.

2. Write 0 for any missing terms, and *write* 1 as the coefficient of any term that has an understood coefficient of 1.

The augmented matrix is then formed from that system of equations as follows:

1. Drop (delete) all the variables, all the plus signs, *and* all the equal signs. (Do not drop any negative signs.)
2. Put brackets around the resulting array of numbers.

Thus, if we're solving a system of four equations in x, y, z, and w, for example, and if the variables are listed in that order, then the first *column* of the augmented matrix will be the coefficients of x, the second column will be the coefficients of y, and so forth. The rightmost column will be the column of constants.

The augmented matrix for a system of n equations in n unknowns will always be an $n \times (n + 1)$ matrix.

NOTE When writing an augmented matrix, some authors insert a solid vertical line or a dashed vertical line before the column of constants; we do not follow that practice. ☑

Example 1 Write the augmented matrix for each system of equations.

a. $\begin{cases} 2x - 3y + 1z = 1 \\ 1x + 2y + 1z = -1 \\ 3x - 1y + 3z = 4 \end{cases}$ Three equations in three unknowns

Solution There are no 0's or 1's to write; all like terms are aligned, and the constants are on the right.

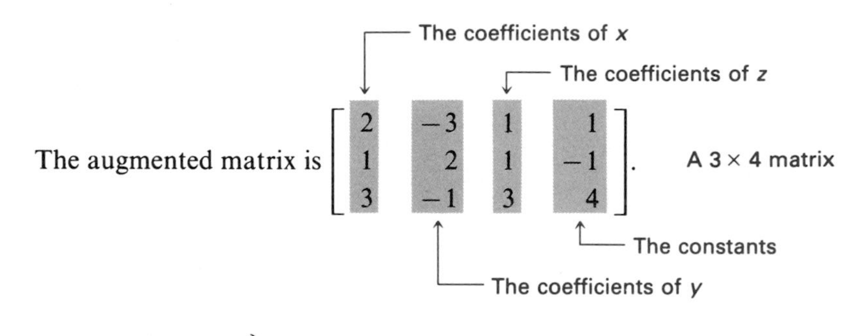

b. $\begin{cases} x + y + 2z = 5 \\ 2x = z - 1 \\ 3y = -2z \end{cases}$

Solution Writing all the terms with variables on the left, aligning like terms, and writing in the 0's and 1's, we have

$$\begin{cases} 1x + 1y + 2z = 5 \\ 2x + 0y - 1z = -1 \\ 0x + 3y + 2z = 0 \end{cases}$$ Three equations in three unknowns

The augmented matrix is $\begin{bmatrix} 1 & 1 & 2 & 5 \\ 2 & 0 & -1 & -1 \\ 0 & 3 & 2 & 0 \end{bmatrix}$. A 3 × 4 matrix ■

Example 2 Assume that each of the following matrices is the augmented matrix for a system of linear equations. Find a system of linear equations associated with each matrix.

a. $\begin{bmatrix} 1 & 3 & 2 & 4 \\ 2 & -5 & -4 & 0 \\ 3 & 3 & 2 & 3 \end{bmatrix}$

Solution The system of equations is $\begin{Bmatrix} x + 3y + 2z = 4 \\ 2x - 5y - 4z = 0 \\ 3x + 3y + 2z = 3 \end{Bmatrix}$.

b. $\begin{bmatrix} 5 & 1 & 0 & 2 & 7 \\ -3 & 4 & 2 & 6 & 4 \\ 1 & 0 & 1 & 0 & -3 \\ 5 & 2 & -7 & 1 & 0 \end{bmatrix}$

Solution The system of equations is $\begin{Bmatrix} 5x + y + 2w = 7 \\ -3x + 4y + 2z + 6w = 4 \\ x + z = -3 \\ 5x + 2y - 7z + w = 0 \end{Bmatrix}$. ■

Suppose that a matrix representing a system of three equations in three variables is in row-echelon form; that is, suppose it is in the following form:

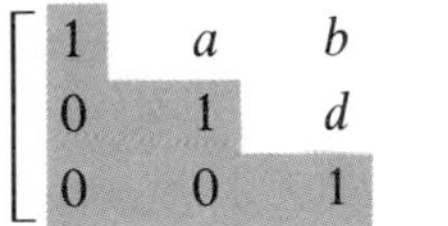

$\begin{bmatrix} 1 & a & b & c \\ 0 & 1 & d & e \\ 0 & 0 & 1 & k \end{bmatrix}$, where a, b, c, d, e, and k are any constants

Then the last row represents the equation $0x + 0y + 1z = k$, or $z = k$, and the system has been solved for z. The second row represents an equation in y and z, so we can find y. After finding y, we can find x by substitution, and the system is solved.

We see, then, that it is easy to solve a system of equations if the augmented matrix for the system is in row-echelon form.

Row-Equivalent Matrices and the Elementary Row Operations

If we perform any of the following three operations on a matrix, we say that the new matrix is **row-equivalent** to the given matrix.

> THE ELEMENTARY ROW OPERATIONS
>
> 1. Interchanging any two rows
> 2. Multiplying each element in any row by a nonzero constant
> 3. Adding a (nonzero) multiple of the elements of one row to the corresponding elements of another row

We use the symbol $\sim$ between the two matrices to indicate that we have performed one or more of the elementary row operations. We do *not* use an equal sign! The two matrices *are not equal*; they are row-equivalent.

Performing any of these operations on the augmented matrix for a system of equations is equivalent to performing any of the operations listed in Section 11.3 on a system of equations. For instance, interchanging two rows of the matrix is equivalent to interchanging two equations in the system of equations.

Converting a Matrix to Row-Echelon Form

There are several different ways to convert a matrix to row-echelon form. We discuss only one of the methods in this text.

CONVERTING A MATRIX TO ROW-ECHELON FORM

In this order, use the elementary row operations to:

1. Move any row that consists of all zeros to the bottom of the matrix.
2. Get a 1 in the upper left-hand corner, usually by interchanging two rows or by multiplying the elements of row 1 by the multiplicative inverse of the element in row 1, column 1—assuming that that element is not zero.
3. Get 0's beneath that 1, usually by using the *third* elementary row operation.

Without performing any operation that will change the first column:

4. Get a 1 in row 2, column 2, if possible.* (See step 2.)
5. Get 0's beneath that 1. (See step 3.)

Without performing any operation that will change the first two columns:

6. Get a 1 in row 3, column 3, if possible. (See step 2.)
7. Get 0's beneath that 1. (See step 3.)

Continue in this manner, trying to get 1's down the main diagonal and 0's beneath that diagonal, to the last row of the matrix.

In Example 3, we show how to use this procedure.

NOTE When you're using the third elementary row operation and multiplying one row by some constant, *that row must remain unchanged* in the new matrix; the row you're adding *to* changes. ✓

Example 3 Convert the matrix from Example 1b to row-echelon form.

Solution There is already a 1 in the upper left-hand corner; we want 0's beneath that 1.

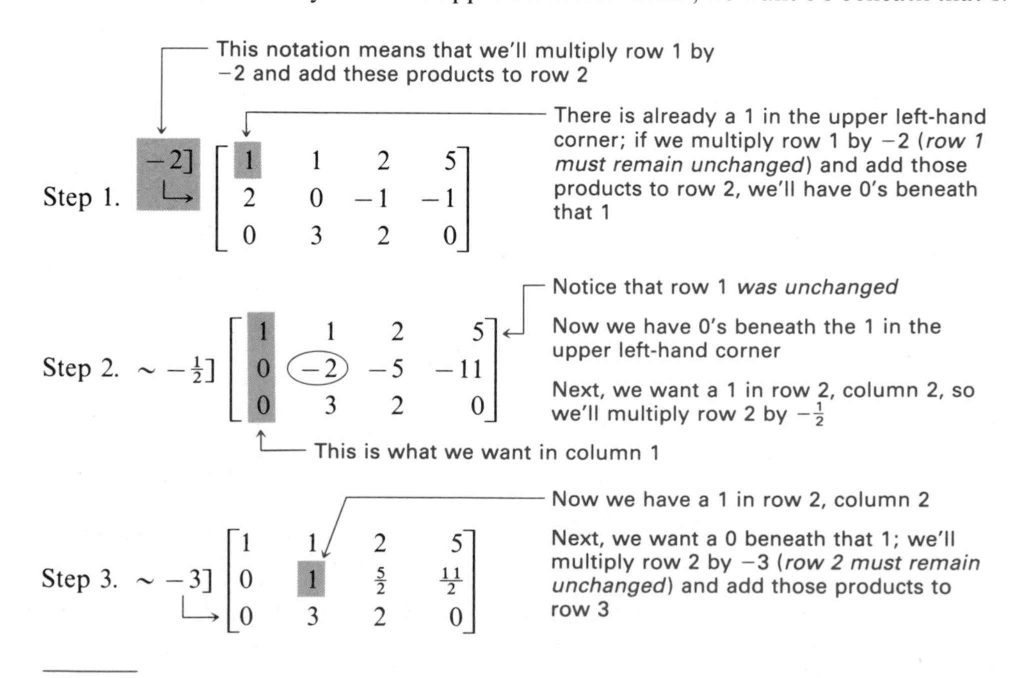

* If you can't get a 1 in row 2, column 2, try to get a 1 in row 2, column 3; if not there, then in row 2, column 4, and so forth.

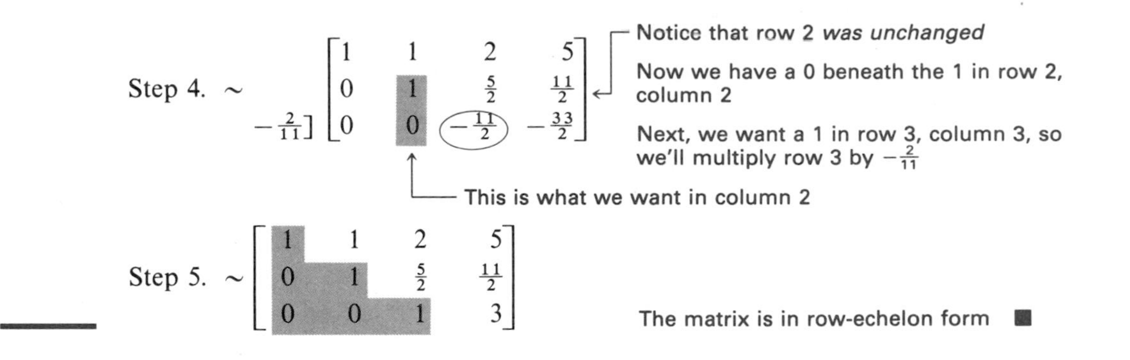

The Gaussian Elimination Method for Solving a System of Equations

We can solve a system of equations by forming the augmented matrix, reducing it to row-echelon form, and then writing the system of equations that corresponds to the final matrix. If no errors have been made, that system will be equivalent to the original system. This method is called the **Gaussian elimination method**.

If there is the same number of equations as variables (as there will be in this text), then three possibilities exist:

1. If the next-to-last number of the last row of the augmented matrix is 1, the system is consistent and independent.
2. If any row of the augmented matrix is of the form $0\ 0 \cdots 0\ k$, where k is some nonzero number, the system is inconsistent and independent.
3. If no row of the augmented matrix is of the form $0\ 0 \cdots 0\ k$, where k is some nonzero number, and if the last row of the matrix consists of all zeros, the system is consistent and dependent.

To see that statement 2 is true, consider this case: Suppose we're solving a system of four equations in four variables and suppose that some row of the augmented matrix is

$$0 \quad 0 \quad 0 \quad 0 \quad 5$$

This is equivalent to the equation $0x + 0y + 0z + 0w = 5$, or $0 = 5$, which can't be true for *any* values of x, y, z, and w. Therefore, the system is inconsistent.

To see that statement 3 is true, consider this case: Suppose we're solving a system of four equations in four variables and suppose that no row is of the form $0\ 0 \cdots 0\ k$, where k is some nonzero number, and that the last row of the augmented matrix is

$$0 \quad 0 \quad 0 \quad 0 \quad 0$$

This is equivalent to the equation $0x + 0y + 0z + 0w = 0$, or $0 = 0$. Since this statement is true for all x, y, z, and w, there are infinitely many solutions for the system of equations.

If the system is consistent and independent, the last equation will already be solved for one of the variables.

Example 4 Solve the following system of equations by the matrix method described in this section.

$$\begin{cases} 2x - y + 3z + w = -4 \\ x + 2y - z + 2w = -4 \\ 3x + y - z - 2w = 4 \\ -2x + 3y + z + 3w = 0 \end{cases}$$

Solution The augmented matrix for this system of equations is

$$\begin{bmatrix} 2 & -1 & 3 & 1 & -4 \\ 1 & 2 & -1 & 2 & -4 \\ 3 & 1 & -1 & -2 & 4 \\ -2 & 3 & 1 & 3 & 0 \end{bmatrix}$$

We reduce this matrix to row-echelon form as follows:

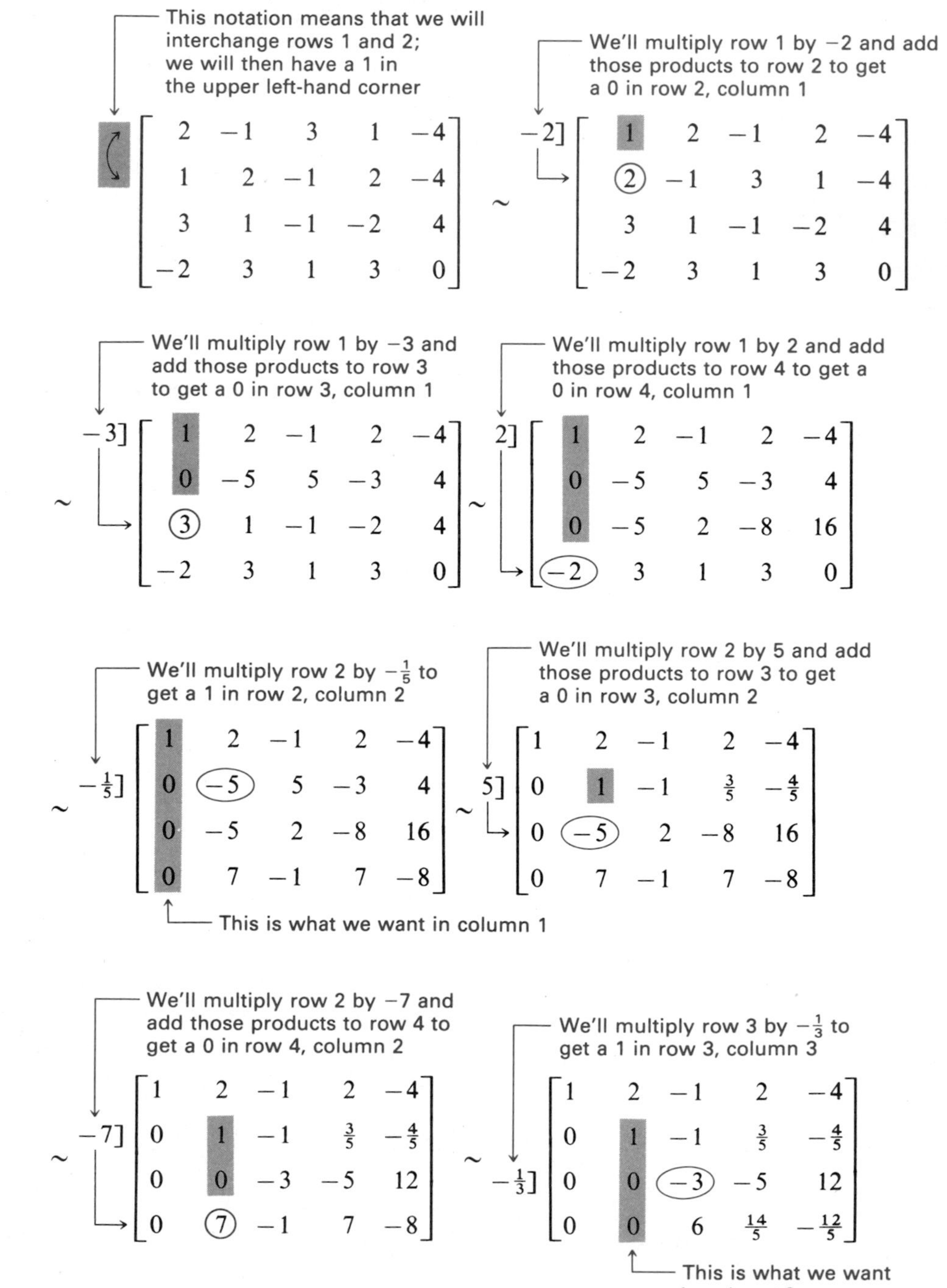

$$\begin{bmatrix} 2 & -1 & 3 & 1 & -4 \\ 1 & 2 & -1 & 2 & -4 \\ 3 & 1 & -1 & -2 & 4 \\ -2 & 3 & 1 & 3 & 0 \end{bmatrix} \sim \begin{bmatrix} 1 & 2 & -1 & 2 & -4 \\ 2 & -1 & 3 & 1 & -4 \\ 3 & 1 & -1 & -2 & 4 \\ -2 & 3 & 1 & 3 & 0 \end{bmatrix}$$

$$\sim \begin{bmatrix} 1 & 2 & -1 & 2 & -4 \\ 0 & -5 & 5 & -3 & 4 \\ 3 & 1 & -1 & -2 & 4 \\ -2 & 3 & 1 & 3 & 0 \end{bmatrix} \sim \begin{bmatrix} 1 & 2 & -1 & 2 & -4 \\ 0 & -5 & 5 & -3 & 4 \\ 0 & -5 & 2 & -8 & 16 \\ -2 & 3 & 1 & 3 & 0 \end{bmatrix}$$

$$\sim \begin{bmatrix} 1 & 2 & -1 & 2 & -4 \\ 0 & -5 & 5 & -3 & 4 \\ 0 & -5 & 2 & -8 & 16 \\ 0 & 7 & -1 & 7 & -8 \end{bmatrix} \sim \begin{bmatrix} 1 & 2 & -1 & 2 & -4 \\ 0 & 1 & -1 & \frac{3}{5} & -\frac{4}{5} \\ 0 & -5 & 2 & -8 & 16 \\ 0 & 7 & -1 & 7 & -8 \end{bmatrix}$$

$$\sim \begin{bmatrix} 1 & 2 & -1 & 2 & -4 \\ 0 & 1 & -1 & \frac{3}{5} & -\frac{4}{5} \\ 0 & 0 & -3 & -5 & 12 \\ 0 & 7 & -1 & 7 & -8 \end{bmatrix} \sim \begin{bmatrix} 1 & 2 & -1 & 2 & -4 \\ 0 & 1 & -1 & \frac{3}{5} & -\frac{4}{5} \\ 0 & 0 & -3 & -5 & 12 \\ 0 & 0 & 6 & \frac{14}{5} & -\frac{12}{5} \end{bmatrix}$$

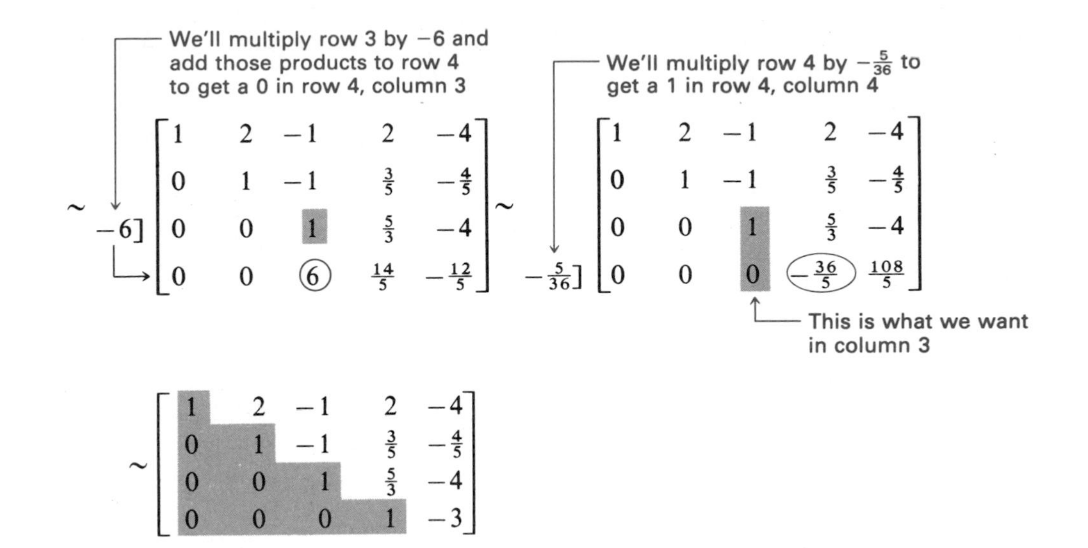

The matrix is in row-echelon form. The last row is equivalent to the equation

$$0x + 0y + 0z + 1w = -3$$

Therefore, $w = -3$.

The third row is equivalent to the equation

$$0x + 0y + 1z + \tfrac{5}{3}w = -4$$

Substituting -3 for w, we have

$$z + \tfrac{5}{3}(-3) = -4$$

or

$$z = 1$$

The second row is equivalent to the equation

$$0x + 1y - 1z + \tfrac{3}{5}w = -\tfrac{4}{5}$$

Substituting -3 for w and 1 for z, we have

$$y - 1(1) + \tfrac{3}{5}(-3) = -\tfrac{4}{5}$$

or

$$y = 2$$

The first row is equivalent to the equation

$$1x + 2y - 1z + 2w = -4$$

Substituting -3 for w, 1 for z, and 2 for y, we have

$$1x + 2(2) - 1(1) + 2(-3) = -4$$

or

$$x = -1$$

Therefore, the solution *appears* to be $(-1, 2, 1, -3)$.

Check (We will not write the original equations in this check.)

$2(-1) - 2 + 3(1) + (-3) \stackrel{?}{=} -4$	$-4 = -4$	True
$-1 + 2(2) - 1 + 2(-3) \stackrel{?}{=} -4$	$-4 = -4$	True
$3(-1) + 2 - 1 - 2(-3) \stackrel{?}{=} 4$	$4 = 4$	True
$-2(-1) + 3(2) + 1 + 3(-3) \stackrel{?}{=} 0$	$0 = 0$	True

Therefore, the solution *is* $(-1, 2, 1, -3)$. ■

NOTE This matrix method *can* be used for two equations in two unknowns. However, the methods discussed in Sections 11.3 and 11.4 are probably easier to use in solving such systems. ☑

EXERCISES 11.7A

Set I Solve each of the following systems of equations by using the matrix method discussed in this section.

1. $\begin{Bmatrix} 2x + y + z = 4 \\ x - y + 3z = -2 \\ x + y + 2z = 1 \end{Bmatrix}$

2. $\begin{Bmatrix} x + y + z = 1 \\ 2x + y - 2z = -4 \\ x + y + 2z = 3 \end{Bmatrix}$

3. $\begin{Bmatrix} 2x + 3y + z = 7 \\ 4x - 2z = -6 \\ 6y - z = 0 \end{Bmatrix}$

4. $\begin{Bmatrix} x - 2y = 4 \\ y + 3z = 8 \\ 2x - z = 1 \end{Bmatrix}$

5. $\begin{Bmatrix} x + y + z + w = 5 \\ 2x - y + 2z - w = -2 \\ x + 2y - z - 2w = -1 \\ -x + 3y + 3z + w = 1 \end{Bmatrix}$

6. $\begin{Bmatrix} x + y + z + w = 4 \\ x - 2y - z - 2w = -2 \\ 3x + 2y + z + 3w = 4 \\ 2x + y - 2z - w = 0 \end{Bmatrix}$

Set II Solve each of the following systems of equations by using the matrix method discussed in this section.

1. $\begin{Bmatrix} x + 2z = 7 \\ 2x - y = 5 \\ 2y + z = 4 \end{Bmatrix}$

2. $\begin{Bmatrix} 4x + 5y + z = 4 \\ 10y - 2z = 6 \\ 8x + 3z = 3 \end{Bmatrix}$

3. $\begin{Bmatrix} 2x + y - 3z = 3 \\ x + 2y = 0 \\ -2x - 2y + 5z = -2 \end{Bmatrix}$

4. $\begin{Bmatrix} 3x - y + 2z = 4 \\ x - 2y - 3z = -4 \\ x + y = 2 \end{Bmatrix}$

5. $\begin{Bmatrix} x + y + z + w = 5 \\ 3x - y + 2z - w = 0 \\ 2x + y - z + 2w = 4 \\ 2x + y + z + 2w = 2 \end{Bmatrix}$

6. $\begin{Bmatrix} 2x + 5y + 3z - 4w = 0 \\ -x + y + 6w = -1 \\ 3x - z - 2w = 1 \\ 4x + 2y - z = 3 \end{Bmatrix}$

11.7B Determinants

In Section 11.7C, we will discuss using Cramer's rule for solving systems of linear equations. In order to do this, we must first discuss *determinants*. We will begin with a few definitions that apply to determinants of any size.

A **determinant** is a *real number* that is represented as a square array of numbers enclosed between *two vertical bars* (see Figure 11.7.1). (Students who have studied

Section 11.7A should note the difference in notation between a *matrix* and a *determinant*: A *matrix* is enclosed within *brackets*, while a *determinant* is enclosed between *vertical bars*.)

The *elements* of a determinant are the numbers in its array.

The *principal diagonal* of a determinant is the line of its elements from the upper left corner to the lower right corner.

The *secondary diagonal* of a determinant is the line of elements from the lower left corner to the upper right corner.

Second-Order Determinants

A **second-order determinant** is a determinant that has two rows and two columns of elements (Figure 11.7.1).

The *value of a second-order determinant* is the product of the elements in its principal diagonal minus the product of the elements in its secondary diagonal.

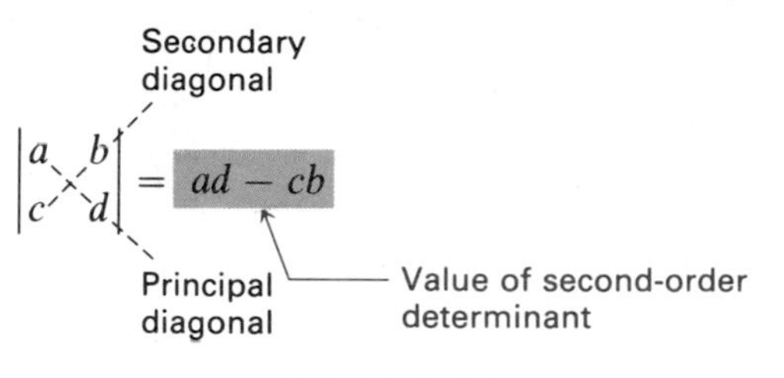

FIGURE 11.7.1
SECOND-ORDER DETERMINANT

Example 5 Examples of finding the value of second-order determinants:

a. $\begin{vmatrix} -5 & -6 \\ 2 & 3 \end{vmatrix} = (-5)(3) - (2)(-6) = -15 + 12 = -3$

b. $\begin{vmatrix} 6 & -7 \\ 4 & 0 \end{vmatrix} = (6)(0) - (4)(-7) = 0 + 28 = 28$ ■

Third-Order Determinants

A **third-order determinant** is a determinant that has three rows and three columns (see Figure 11.7.2).

	Column 1	Column 2	Column 3
Row 1	a_1	b_1	c_1
Row 2	a_2	b_2	c_2
Row 3	a_3	b_3	c_3

FIGURE 11.7.2
THIRD-ORDER DETERMINANT

Minor The **minor** of an element of a determinant is the determinant that remains after striking out the row and column in which that element appears. For example, in Figure 11.7.2, the minor of element a_2 is the determinant $\begin{vmatrix} b_1 & c_1 \\ b_3 & c_3 \end{vmatrix}$.

Cofactor The **cofactor** of an element of a determinant is the *signed* minor of that element; in particular, the cofactor of an element in the ith row and the jth column of a determinant is $(-1)^{i+j}$ times the minor of that element. In Figure 11.7.2, the cofactor of element a_2 is

$$(-1)^{2+1}\begin{vmatrix} b_1 & c_1 \\ b_3 & c_3 \end{vmatrix} \quad \text{or} \quad (-1)\begin{vmatrix} b_1 & c_1 \\ b_3 & c_3 \end{vmatrix}$$

Example 6 Find the required minors and cofactors, given the determinant $\begin{vmatrix} 2 & 0 & -1 \\ 5 & -4 & 6 \\ -3 & 1 & 7 \end{vmatrix}$.

a. Find the minor of 1.

Solution Striking out the row and column that contain 1, we have $\begin{vmatrix} 2 & 0 & -1 \\ 5 & -4 & 6 \\ -3 & ① & 7 \end{vmatrix}$.

Therefore, the minor of 1 is $\begin{vmatrix} 2 & -1 \\ 5 & 6 \end{vmatrix}$.

b. Find the cofactor of 1.

Solution Because 1 is in the *third* row and the *second* column, the sign of its cofactor is $(-1)^{3+2}$ or -1. Therefore, the cofactor of 1 is $(-1)\begin{vmatrix} 2 & -1 \\ 5 & 6 \end{vmatrix}$.

c. Find the minor of 7.

Solution Striking out the row and column that contain 7, we have $\begin{vmatrix} 2 & 0 & -1 \\ 5 & -4 & 6 \\ -3 & 1 & ⑦ \end{vmatrix}$.

Therefore, the minor of 7 is $\begin{vmatrix} 2 & 0 \\ 5 & -4 \end{vmatrix}$.

d. Find the cofactor of 7.

Solution Because 7 is in the *third* row and the *third* column, the sign of its cofactor is $(-1)^{3+3}$ or $+1$. Therefore, the cofactor of 7 is $\begin{vmatrix} 2 & 0 \\ 5 & -4 \end{vmatrix}$. ■

An easier way to find the sign of the cofactor of any element is to memorize the pattern of signs shown in Figure 11.7.3. (The pattern can be extended to a determinant of any size.)

$$\begin{vmatrix} + & - & + \\ - & + & - \\ + & - & + \end{vmatrix}$$

FIGURE 11.7.3

The *value of the third-order determinant*

$$\begin{vmatrix} a_1 & b_1 & c_1 \\ a_2 & b_2 & c_2 \\ a_3 & b_3 & c_3 \end{vmatrix}$$

is defined to be

$$a_1b_2c_3 - a_1b_3c_2 - a_2b_1c_3 + a_2b_3c_1 + a_3b_1c_2 - a_3b_2c_1$$

You should verify that this value could be rewritten as

$$a_1(b_2c_3 - b_3c_2) - a_2(b_1c_3 - b_3c_1) + a_3(b_1c_2 - b_2c_1)$$

or

$$a_1\begin{vmatrix} b_2 & c_2 \\ b_3 & c_3 \end{vmatrix} - a_2\begin{vmatrix} b_1 & c_1 \\ b_3 & c_3 \end{vmatrix} + a_3\begin{vmatrix} b_1 & c_1 \\ b_2 & c_2 \end{vmatrix}$$

Notice the negative sign, and note that the cofactor of a_2 is negative

When we rewrite the value of a third-order determinant in this last way, we say we are *expanding by cofactors of the elements of the first column.*

We can expand a determinant by cofactors of *any row* or *any* column. To expand by cofactors of a *row*, we multiply each element in that row by its cofactor and then add the products. To expand by cofactors of a *column*, we multiply each element in that column by its cofactor and then add the products. *The same value is obtained no matter what row or column is used.*

Example 7 Find the value of $\begin{vmatrix} 1 & 3 & 2 \\ 2 & -1 & 1 \\ -4 & 1 & -3 \end{vmatrix}$ by (a) expansion by column 1 and (b) expansion by row 2.

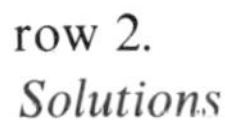

Solutions

a.

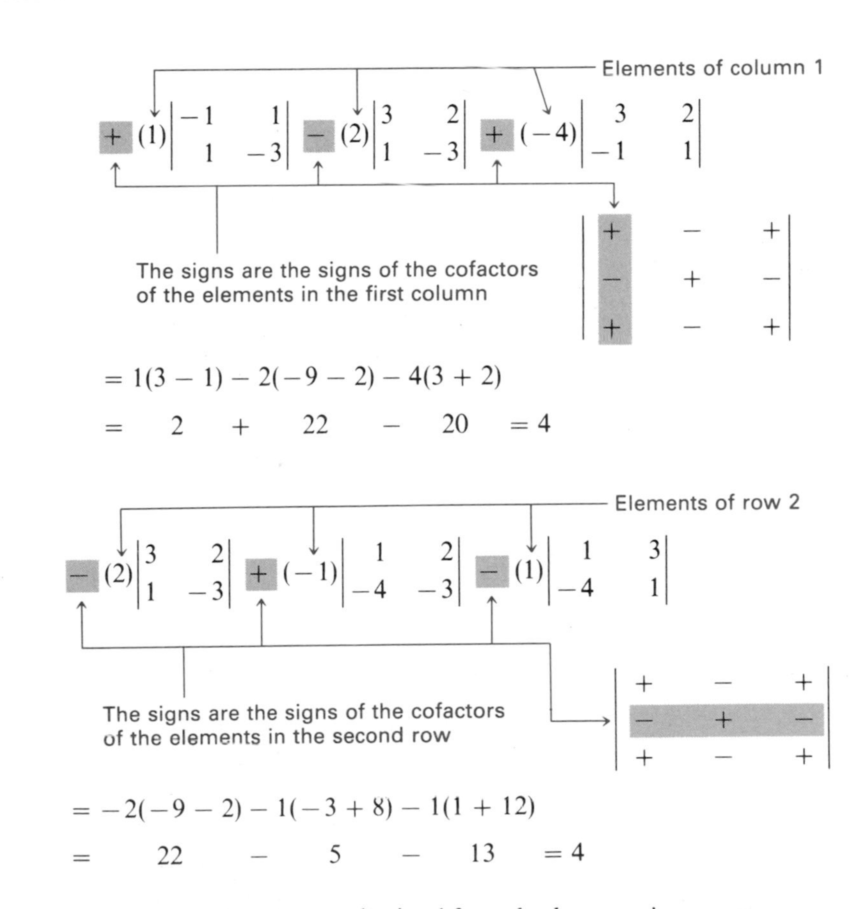

$$+(1)\begin{vmatrix} -1 & 1 \\ 1 & -3 \end{vmatrix} - (2)\begin{vmatrix} 3 & 2 \\ 1 & -3 \end{vmatrix} + (-4)\begin{vmatrix} 3 & 2 \\ -1 & 1 \end{vmatrix}$$

$$= 1(3 - 1) - 2(-9 - 2) - 4(3 + 2)$$

$$= 2 + 22 - 20 = 4$$

b.

$$-(2)\begin{vmatrix} 3 & 2 \\ 1 & -3 \end{vmatrix} + (-1)\begin{vmatrix} 1 & 2 \\ -4 & -3 \end{vmatrix} - (1)\begin{vmatrix} 1 & 3 \\ -4 & 1 \end{vmatrix}$$

$$= -2(-9 - 2) - 1(-3 + 8) - 1(1 + 12)$$

$$= 22 - 5 - 13 = 4$$

Note that the same value, 4, was obtained from *both* expansions. ■

If zeros appear anywhere in a determinant, expand the determinant by the row or column containing the most zeros. This minimizes the numerical work that needs to be done in evaluating the determinant.

Example 8 Examples of selecting the best row or column to expand the determinant by:

a. $\begin{vmatrix} -2 & 1 & 0 \\ 4 & -2 & 1 \\ 3 & -1 & 5 \end{vmatrix}$ Expand by row 1 (or column 3) because it contains a zero

Expanding by row 1:

$$= (-2)\begin{vmatrix} -2 & 1 \\ -1 & 5 \end{vmatrix} - (1)\begin{vmatrix} 4 & 1 \\ 3 & 5 \end{vmatrix} + (0)\begin{vmatrix} 4 & -2 \\ 3 & -1 \end{vmatrix}$$

$$= (-2)(-10 + 1) - 1(20 - 3) + 0 = 1$$

b. $\begin{vmatrix} -4 & 0 & -1 \\ 1 & 2 & -3 \\ -2 & 0 & 5 \end{vmatrix}$ Expand by column 2 because it contains two zeros

Expanding by column 2:

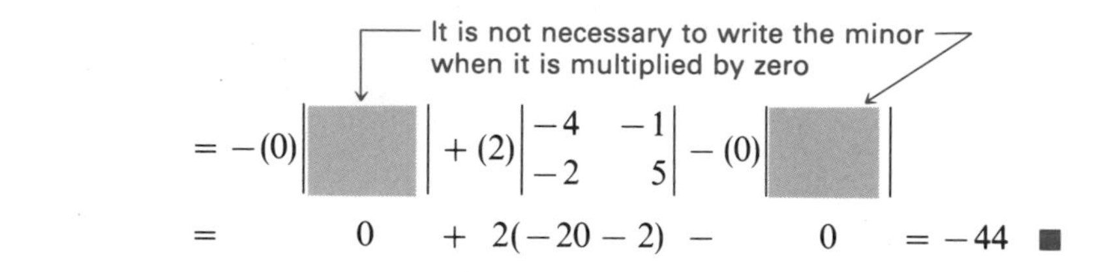

$$= -(0)\begin{vmatrix} \; \end{vmatrix} + (2)\begin{vmatrix} -4 & -1 \\ -2 & 5 \end{vmatrix} - (0)\begin{vmatrix} \; \end{vmatrix}$$

$$= \quad 0 \quad + \; 2(-20 - 2) \; - \quad 0 \quad = -44 \; \blacksquare$$

EXERCISES 11.7B

Set I In Exercises 1–6, find the value of each second-order determinant.

1. $\begin{vmatrix} 3 & 4 \\ 2 & 5 \end{vmatrix}$ **2.** $\begin{vmatrix} 4 & 3 \\ 2 & 7 \end{vmatrix}$ **3.** $\begin{vmatrix} 2 & -4 \\ 5 & -3 \end{vmatrix}$

4. $\begin{vmatrix} 5 & 0 \\ -9 & 8 \end{vmatrix}$ **5.** $\begin{vmatrix} -7 & -3 \\ 5 & 8 \end{vmatrix}$ **6.** $\begin{vmatrix} 2 & -4 \\ -3 & 6 \end{vmatrix}$

In Exercises 7 and 8, solve for x.

7. $\begin{vmatrix} 2 & -4 \\ 3 & x \end{vmatrix} = 20$ **8.** $\begin{vmatrix} 3 & x \\ -4 & 5 \end{vmatrix} = 27$

9. For the determinant $\begin{vmatrix} 1 & 2 & 3 \\ 4 & 5 & -1 \\ -3 & -5 & 0 \end{vmatrix}$, find the following.

a. the minor of 2 b. the cofactor of 2

c. the minor of -5 d. the cofactor of -5

e. the minor of -3 f. the cofactor of -3

10. For the determinant $\begin{vmatrix} 2 & 0 & 1 \\ 4 & 1 & 5 \\ -3 & -2 & 4 \end{vmatrix}$, find the following.

a. the minor of -3 b. the cofactor of -3

c. the minor of 5 d. the cofactor of 5

e. the minor of -2 f. the cofactor of -2

In Exercises 11–14, find the value of each determinant by expanding by cofactors of the indicated row or column.

11. $\begin{vmatrix} 1 & 2 & 1 \\ 3 & 1 & 2 \\ 4 & 2 & 0 \end{vmatrix}$ column 3

12. $\begin{vmatrix} 2 & 1 & 1 \\ 1 & 3 & 1 \\ 0 & 2 & 4 \end{vmatrix}$ column 1

13. $\begin{vmatrix} 1 & 3 & -2 \\ -1 & 2 & -3 \\ 0 & 4 & 1 \end{vmatrix}$ row 3

14. $\begin{vmatrix} 2 & -1 & 1 \\ 0 & 5 & -3 \\ 1 & 2 & -2 \end{vmatrix}$ row 2

In Exercises 15–18, expand by any row or column.

15. $\begin{vmatrix} 1 & -2 & 3 \\ -3 & 4 & 0 \\ 2 & 6 & 5 \end{vmatrix}$

16. $\begin{vmatrix} 2 & -4 & -1 \\ 5 & 0 & -6 \\ -3 & 4 & -2 \end{vmatrix}$

17. $\begin{vmatrix} 6 & 7 & 8 \\ -6 & 7 & -9 \\ 0 & 0 & -2 \end{vmatrix}$

18. $\begin{vmatrix} 0 & 0 & -3 \\ 5 & 6 & 9 \\ -5 & 6 & -8 \end{vmatrix}$

In Exercises 19 and 20, solve for x.

19. $\begin{vmatrix} x & 0 & 1 \\ 0 & 2 & 3 \\ 4 & -1 & -2 \end{vmatrix} = 6$

20. $\begin{vmatrix} 0 & x & 1 \\ -3 & 2 & 4 \\ -2 & 1 & 0 \end{vmatrix} = 17$

Set II In Exercises 1–6, find the value of each second-order determinant.

1. $\begin{vmatrix} 6 & 7 \\ -5 & -2 \end{vmatrix}$

2. $\begin{vmatrix} -9 & 8 \\ -3 & 2 \end{vmatrix}$

3. $\begin{vmatrix} 1 & -3 \\ 2 & -6 \end{vmatrix}$

4. $\begin{vmatrix} 3 & 4 \\ 1 & 5 \end{vmatrix}$

5. $\begin{vmatrix} -1 & 4 \\ 3 & 2 \end{vmatrix}$

6. $\begin{vmatrix} 5 & -1 \\ 6 & -2 \end{vmatrix}$

In Exercises 7 and 8, solve for x.

7. $\begin{vmatrix} -2 & 3 \\ x & 4 \end{vmatrix} = 10$

8. $\begin{vmatrix} 2x & 5 \\ 2 & 3x \end{vmatrix} = -11x$

9. For the determinant $\begin{vmatrix} -3 & 1 & 2 \\ 0 & 4 & -1 \\ 5 & -2 & 6 \end{vmatrix}$, find the following.

a. the minor of 4

b. the cofactor of 4

c. the minor of 2

d. the cofactor of 2

e. the minor of 1

f. the cofactor of 1

10. For the determinant $\begin{vmatrix} 2 & 5 & 6 \\ 1 & 8 & 4 \\ 3 & 7 & 9 \end{vmatrix}$, find the following.

a. the minor of 2

b. the cofactor of 2

c. the minor of 1

d. the cofactor of 1

e. the minor of 7

f. the cofactor of 7

In Exercises 11–14, find the value of each determinant by expanding by cofactors of the indicated row or column.

11. $\begin{vmatrix} 2 & -1 & -3 \\ 4 & 0 & -2 \\ -5 & 2 & 3 \end{vmatrix}$ column 2

12. $\begin{vmatrix} 0 & 5 & -2 \\ -3 & 4 & -1 \\ 2 & -4 & 6 \end{vmatrix}$ row 1

13. $\begin{vmatrix} 3 & -2 & 4 \\ 5 & 1 & 0 \\ 0 & -1 & 2 \end{vmatrix}$ row 3

14. $\begin{vmatrix} 1 & 5 & -1 \\ 2 & 0 & -2 \\ 3 & 1 & -1 \end{vmatrix}$ column 2

In Exercises 15–18, expand by any row or column.

15. $\begin{vmatrix} -1 & 3 & -2 \\ -4 & 2 & 5 \\ 0 & 1 & -3 \end{vmatrix}$

16. $\begin{vmatrix} 4 & -5 & 2 \\ -2 & 0 & -3 \\ 6 & 0 & -1 \end{vmatrix}$

17. $\begin{vmatrix} 3 & -2 & 0 \\ -2 & 1 & -2 \\ 2 & -1 & 2 \end{vmatrix}$

18. $\begin{vmatrix} 3 & 0 & 5 \\ -1 & -1 & 3 \\ 2 & -3 & 1 \end{vmatrix}$

In Exercises 19 and 20, solve for x.

19. $\begin{vmatrix} 0 & -4 & 3 \\ x & 2 & 0 \\ -1 & 5 & x \end{vmatrix} = 31$

20. $\begin{vmatrix} 2 & x & 1 \\ -1 & x & 2 \\ x & 0 & 1 \end{vmatrix} = 18$

11.7C Using Cramer's Rule for Solving Second- and Third-Order Systems

Terms Associated with Linear Systems

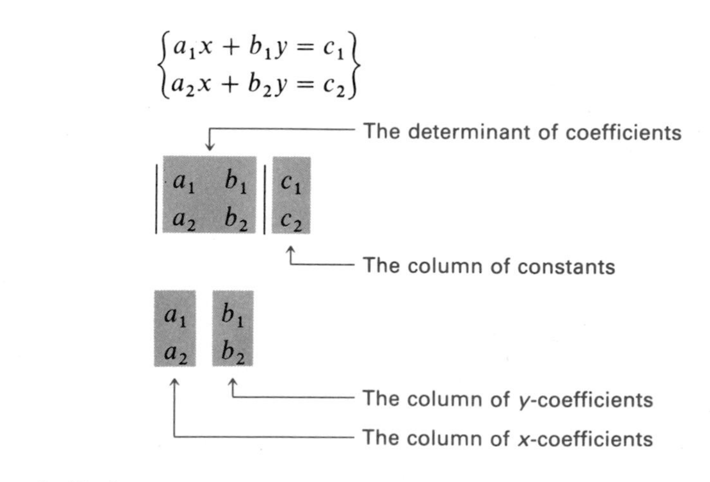

Cramer's Rule

Consider the linear system $\begin{cases} a_1x + b_1y = c_1 \\ a_2x + b_2y = c_2 \end{cases}$. Solving for x by addition, we have

$$\begin{aligned} b_2\,]\quad a_1x + b_1y = c_1 &\Rightarrow \quad a_1b_2x + b_1b_2y = c_1b_2 \\ -b_1\,]\quad a_2x + b_2y = c_2 &\Rightarrow -a_2b_1x - b_1b_2y = -c_2b_1 \end{aligned}$$

Add the equations

$$(a_1b_2 - a_2b_1)x = c_1b_2 - c_2b_1$$

$$x = \frac{c_1b_2 - c_2b_1}{a_1b_2 - a_2b_1}$$

This can be written

$$x = \frac{c_1b_2 - c_2b_1}{a_1b_2 - a_2b_1} = \frac{\begin{vmatrix} c_1 & b_1 \\ c_2 & b_2 \end{vmatrix}}{\begin{vmatrix} a_1 & b_1 \\ a_2 & b_2 \end{vmatrix}}$$

Solving for y by addition (the work is left to the student), we find that

$$y = \frac{a_1c_2 - a_2c_1}{a_1b_2 - a_2b_1}$$

This can be written

$$y = \frac{a_1c_2 - a_2c_1}{a_1b_2 - a_2b_1} = \frac{\begin{vmatrix} a_1 & c_1 \\ a_2 & c_2 \end{vmatrix}}{\begin{vmatrix} a_1 & b_1 \\ a_2 & b_2 \end{vmatrix}}$$

The formulas derived above make up Cramer's rule.

CRAMER'S RULE

In the solution of a linear system of equations, each variable is equal to the ratio of two determinants. The denominator for *every* variable is the determinant of the coefficients. The numerator for each variable is the determinant of the coefficients with the column of coefficients for that variable replaced by the column of constants.

TO SOLVE $\begin{Bmatrix} a_1x + b_1y = c_1 \\ a_2x + b_2y = c_2 \end{Bmatrix}$ BY USING CRAMER'S RULE

The solution is (x, y), where

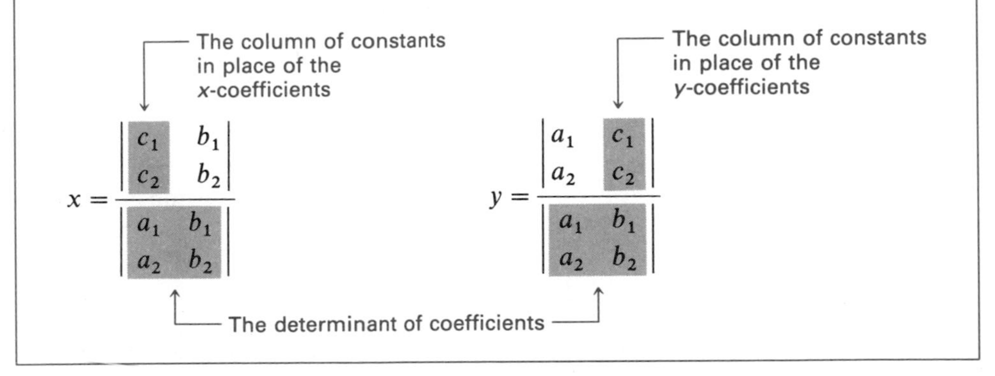

When a system of equations (any order) is solved by Cramer's rule, three outcomes are possible:

1. *Only one solution* (*consistent, independent system*): The determinant of the coefficients is not equal to zero.
2. *No solution* (*inconsistent, independent system*): The determinant of the coefficients is equal to zero, *and* the determinants serving as numerators for the variables are not equal to zero.

3. *Many solutions* (*consistent, dependent system*): The determinant of the coefficients is equal to zero, *and* the determinants serving as numerators for the variables are also equal to zero.

   In a second-order system, the determinant of coefficients equals zero if $a_1b_2 = a_2b_1$.

Example 9 Solve $\begin{cases} x + 4y = 7 \\ 3x + 5y = 0 \end{cases}$ by using Cramer's rule.

Solution

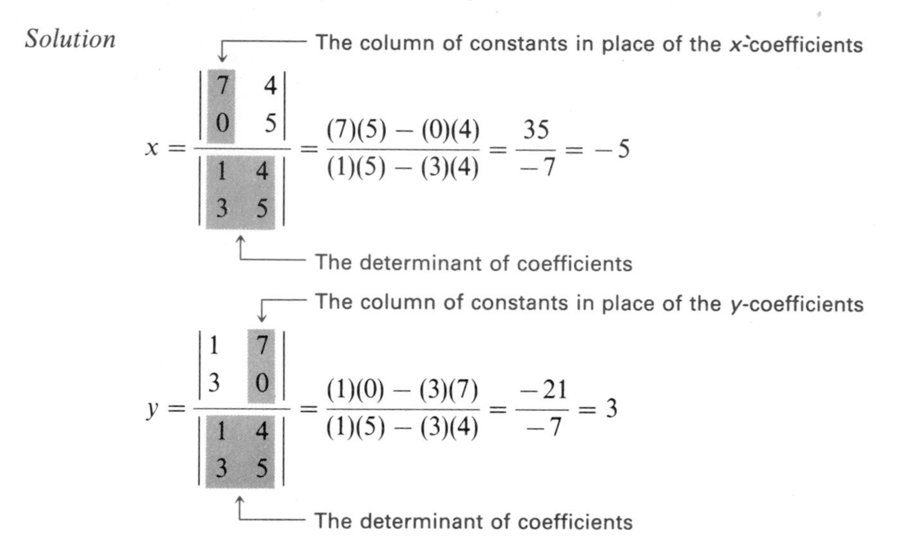

Therefore, the solution is $(-5, 3)$. (The check will not be shown.) This is a consistent, independent system. ■

Example 10 Solve $\begin{cases} 2x - 3y = 6 \\ 6x - 9y = 36 \end{cases}$ by using Cramer's rule.

Solution

$$x = \frac{\begin{vmatrix} 6 & -3 \\ 36 & -9 \end{vmatrix}}{\begin{vmatrix} 2 & -3 \\ 6 & -9 \end{vmatrix}} = \frac{(6)(-9) - (36)(-3)}{(2)(-9) - (6)(-3)} = \frac{-54 + 108}{-18 + 18} = \frac{54}{0} \quad \text{Not defined}$$

The denominator was zero and the numerator nonzero; we need not try to solve for y. This system has *no solution*. It is an inconsistent, independent system (see Example 2, Section 11.2). ■

Example 11 Solve $\begin{cases} 3x + 5y = 15 \\ 6x + 10y = 30 \end{cases}$ by using Cramer's rule.

Solution

$$x = \frac{\begin{vmatrix} 15 & 5 \\ 30 & 10 \end{vmatrix}}{\begin{vmatrix} 3 & 5 \\ 6 & 10 \end{vmatrix}} = \frac{(15)(10) - (30)(5)}{(3)(10) - (6)(5)} = \frac{150 - 150}{30 - 30} = \frac{0}{0} \quad \text{Cannot be determined}$$

The denominator and numerator were both zero (we need not try to solve for y). This is a consistent, dependent system. There are many solutions (see Example 3, Section 11.2). The solutions cannot be found by Cramer's rule. ■

The method for solving a third-order linear system by using Cramer's rule can be summarized as follows (the rule is stated without proof):

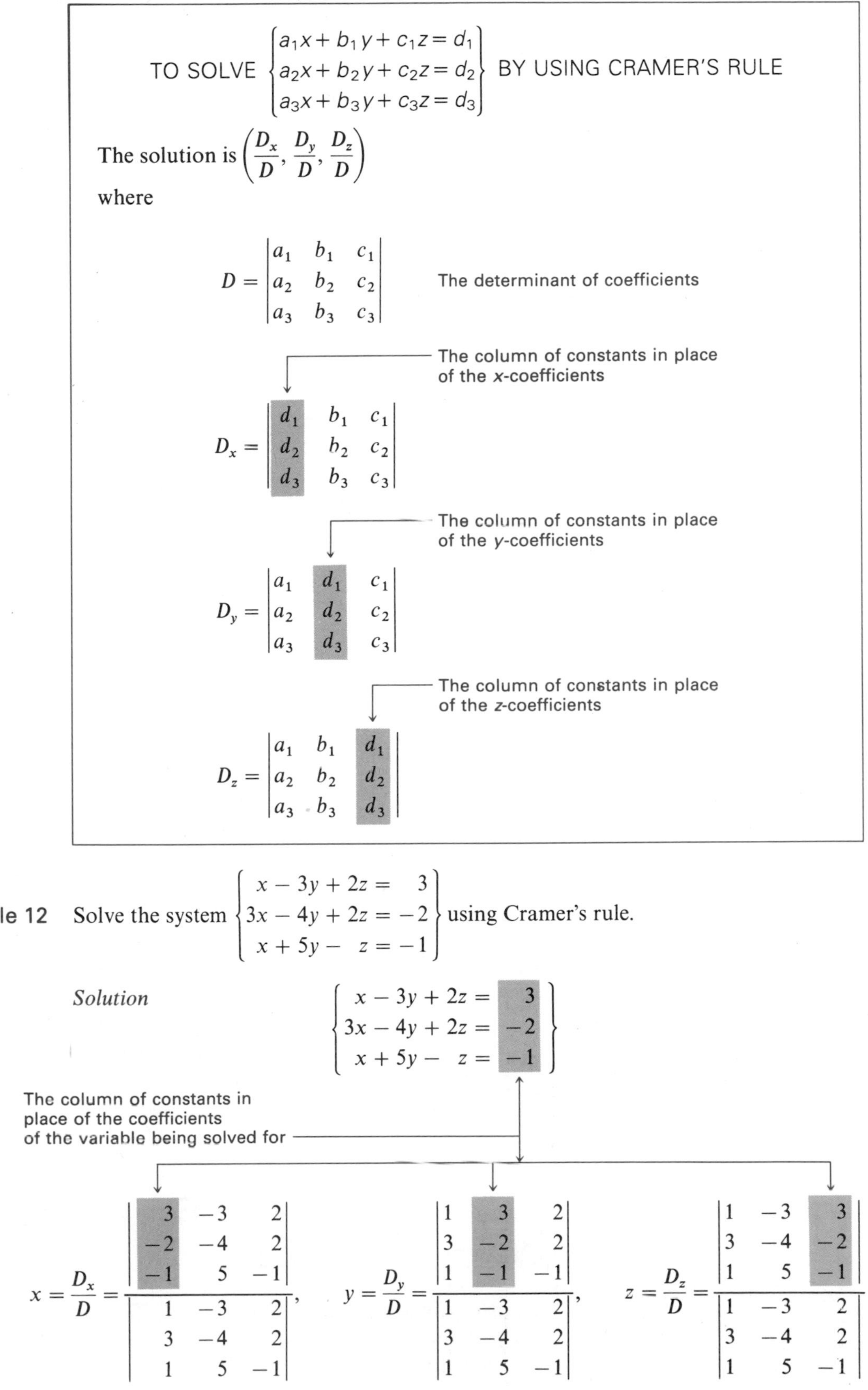

TO SOLVE $\begin{cases} a_1x + b_1y + c_1z = d_1 \\ a_2x + b_2y + c_2z = d_2 \\ a_3x + b_3y + c_3z = d_3 \end{cases}$ BY USING CRAMER'S RULE

The solution is $\left(\dfrac{D_x}{D}, \dfrac{D_y}{D}, \dfrac{D_z}{D}\right)$

where

$$D = \begin{vmatrix} a_1 & b_1 & c_1 \\ a_2 & b_2 & c_2 \\ a_3 & b_3 & c_3 \end{vmatrix}$$ The determinant of coefficients

The column of constants in place of the x-coefficients

$$D_x = \begin{vmatrix} d_1 & b_1 & c_1 \\ d_2 & b_2 & c_2 \\ d_3 & b_3 & c_3 \end{vmatrix}$$

The column of constants in place of the y-coefficients

$$D_y = \begin{vmatrix} a_1 & d_1 & c_1 \\ a_2 & d_2 & c_2 \\ a_3 & d_3 & c_3 \end{vmatrix}$$

The column of constants in place of the z-coefficients

$$D_z = \begin{vmatrix} a_1 & b_1 & d_1 \\ a_2 & b_2 & d_2 \\ a_3 & b_3 & d_3 \end{vmatrix}$$

Example 12 Solve the system $\begin{cases} x - 3y + 2z = 3 \\ 3x - 4y + 2z = -2 \\ x + 5y - z = -1 \end{cases}$ using Cramer's rule.

Solution $\begin{cases} x - 3y + 2z = 3 \\ 3x - 4y + 2z = -2 \\ x + 5y - z = -1 \end{cases}$

The column of constants in place of the coefficients of the variable being solved for

$$x = \frac{D_x}{D} = \frac{\begin{vmatrix} 3 & -3 & 2 \\ -2 & -4 & 2 \\ -1 & 5 & -1 \end{vmatrix}}{\begin{vmatrix} 1 & -3 & 2 \\ 3 & -4 & 2 \\ 1 & 5 & -1 \end{vmatrix}}, \quad y = \frac{D_y}{D} = \frac{\begin{vmatrix} 1 & 3 & 2 \\ 3 & -2 & 2 \\ 1 & -1 & -1 \end{vmatrix}}{\begin{vmatrix} 1 & -3 & 2 \\ 3 & -4 & 2 \\ 1 & 5 & -1 \end{vmatrix}}, \quad z = \frac{D_z}{D} = \frac{\begin{vmatrix} 1 & -3 & 3 \\ 3 & -4 & -2 \\ 1 & 5 & -1 \end{vmatrix}}{\begin{vmatrix} 1 & -3 & 2 \\ 3 & -4 & 2 \\ 1 & 5 & -1 \end{vmatrix}}$$

(In evaluating D, D_x, D_y, and D_z, we will shade the row or column we're expanding by.)

$$D = \begin{vmatrix} 1 & -3 & 2 \\ 3 & -4 & 2 \\ 1 & 5 & -1 \end{vmatrix} = +(1)\begin{vmatrix} -4 & 2 \\ 5 & -1 \end{vmatrix} - (-3)\begin{vmatrix} 3 & 2 \\ 1 & -1 \end{vmatrix} + (2)\begin{vmatrix} 3 & -4 \\ 1 & 5 \end{vmatrix}$$

$$= 1(4 - 10) + 3(-3 - 2) + 2(15 + 4) = 17$$

$$D_x = \begin{vmatrix} 3 & -3 & 2 \\ -2 & -4 & 2 \\ -1 & 5 & -1 \end{vmatrix} = +(3)\begin{vmatrix} -4 & 2 \\ 5 & -1 \end{vmatrix} - (-2)\begin{vmatrix} -3 & 2 \\ 5 & -1 \end{vmatrix} + (-1)\begin{vmatrix} -3 & 2 \\ -4 & 2 \end{vmatrix}$$

$$= 3(4 - 10) + 2(3 - 10) - 1(-6 + 8) = -34$$

$$D_y = \begin{vmatrix} 1 & 3 & 2 \\ 3 & -2 & 2 \\ 1 & -1 & -1 \end{vmatrix} = -(3)\begin{vmatrix} 3 & 2 \\ 1 & -1 \end{vmatrix} + (-2)\begin{vmatrix} 1 & 2 \\ 1 & -1 \end{vmatrix} - (-1)\begin{vmatrix} 1 & 2 \\ 3 & 2 \end{vmatrix}$$

$$= -3(-3 - 2) - 2(-1 - 2) + 1(2 - 6) = 17$$

$$D_z = \begin{vmatrix} 1 & -3 & 3 \\ 3 & -4 & -2 \\ 1 & 5 & -1 \end{vmatrix} = +(3)\begin{vmatrix} 3 & -4 \\ 1 & 5 \end{vmatrix} - (-2)\begin{vmatrix} 1 & -3 \\ 1 & 5 \end{vmatrix} + (-1)\begin{vmatrix} 1 & -3 \\ 3 & -4 \end{vmatrix}$$

$$= 3(15 + 4) + 2(5 + 3) - 1(-4 + 9) = 68$$

Therefore, $x = \dfrac{D_x}{D} = \dfrac{-34}{17} = -2$, $y = \dfrac{D_y}{D} = \dfrac{17}{17} = 1$, and $z = \dfrac{D_z}{D} = \dfrac{68}{17} = 4$, and the solution is $(-2, 1, 4)$. ■

Cramer's rule *can* be used to solve linear systems whenever there is the same number of equations as variables.

EXERCISES 11.7C

Set I For Exercises 1–12, use Cramer's rule to work Exercises 1–12 from Set I, Exercises 11.3, page 554.

For Exercises 13–16, use Cramer's rule to work Exercises 1–4 from Set I, Exercises 11.7A, page 574.

Set II For Exercises 1–12, use Cramer's rule to work Exercises 1–12 from Set II, Exercises 11.3, page 554.

For Exercises 13–16, use Cramer's rule to work Exercises 1–4 from Set II, Exercises 11.7A, page 574.

11.8 Using Systems of Equations to Solve Word Problems

In solving word problems that involve more than one unknown, it is sometimes difficult to represent each unknown in terms of a single variable. In this section, we eliminate that difficulty by using a different variable for each unknown.

TO SOLVE A WORD PROBLEM USING A SYSTEM OF EQUATIONS

Read	First read the problem very carefully.
Think	Determine what *type* of problem it is, if possible. Determine what is unknown.
Sketch	Draw a sketch *with labels*, if possible.
Step 1	Represent each unknown number by a *different* variable.
Reread	Reread the word problem, breaking it up into small pieces.
Step 2	Translate each English phrase into an algebraic expression; fit these expressions together into *two or more different equations. There must be as many equations as variables.*
Step 3	Solve the *system* of equations for *all* the variables, using one of the following: a. the addition method (Section 11.3) b. the substitution method (Section 11.4). c. the matrix method or Cramer's rule (Section 11.7) d. the graphical method (Section 11.2)
Step 4	Be sure you've answered *all* the questions asked.
Step 5	Check the solution(s) *in the word statements.*
Step 6	State your results clearly.

Example 1 Doris has 17 coins in her purse with a total value of \$1.15. If she has only nickels and dimes, how many of each are there?

Step 1. Let D = number of dimes
N = number of nickels

Reread Doris has 17 coins — She has only nickels and dimes

$$(1) \qquad 17 = N + D$$

The coins in her purse have a total value of \$1.15 (115¢).

Amount of money in dimes + Amount of money in nickels = Total amount of money

$$(2) \qquad 10D + 5N = 115$$

Step 2. $\begin{matrix}(1)\\(2)\end{matrix} \begin{Bmatrix} N + D = 17 \\ 5N + 10D = 115 \end{Bmatrix}$ We solve this system by addition

Step 3.
$$\begin{array}{lrl} (1) & -5] & N + D = 17 \Rightarrow -5N - 5D = -85 \\ (2) & 1] & 5N + 10D = 115 \Rightarrow \underline{5N + 10D = 115} \\ & & 5D = 30 \end{array}$$

Step 4. $D = 6$ Number of dimes

$N + 6 = 17$ Substituting into Equation 1

$N = 11$ Number of nickels

Step 5. *Check* If there are 6 dimes and 11 nickels, there are 17 coins altogether. The value of 6 dimes is $0.60, and the value of 11 nickels is $0.55; the total value is $1.15.

Step 6. Therefore, Doris has 6 dimes and 11 nickels in her purse. ■

Example 2 Jeff left Riverside at 5 A.M. traveling toward Stockton. His friend George left Riverside at 5:40 A.M., also traveling toward Stockton. George drove 10 mph faster than Jeff and overtook Jeff at 9 A.M. Find the average speed of each car and the total distance traveled before they met.

Solution Note that Jeff traveled 4 hr and George traveled 3 hr, 20 min before they met and that 3 hr, 20 min can be expressed as $3\frac{1}{3}$ hr. or $\frac{10}{3}$ hr.

Step 1. Let x = Jeff's average speed (in mph)
y = George's average speed (in mph)

Then $4x$ = Jeff's distance $\left(x\frac{\text{mi}}{\cancel{\text{hr}}}\right)(4\ \cancel{\text{hr}}) = 4x$ mi

$\frac{10}{3}y$ = George's distance $\left(y\frac{\text{mi}}{\cancel{\text{hr}}}\right)\left(\frac{10}{3}\cancel{\text{hr}}\right) = \frac{10}{3}y$ mi

Step 2.
$$\begin{cases} (1)\quad 4x = \frac{10}{3}y & \text{The distances are equal} \\ (2)\quad y = x + 10 & \text{George's speed was 10 mph faster than Jeff's} \end{cases}$$

Because Equation 2 has already been solved for y, we'll use substitution. Clearing fractions in Equation 1, we have

Step 3. $12x = 10y$

$12x = 10(x + 10)$ Substituting $x + 10$ for y in $12x = 10y$

$12x = 10x + 100$

$2x = 100$

Step 4. $x = 50$ Jeff's average speed in $\frac{\text{mi}}{\text{hr}}$

$y = x + 10 = 60$ George's average speed in $\frac{\text{mi}}{\text{hr}}$.

$4x = 200$ The number of miles Jeff traveled before they met

Step 5. *Check* 60 mph = 50 mph + 10 mph George's speed is 10 mph more than Jeff's

$\left(50\frac{\text{mi}}{\cancel{\text{hr}}}\right)(4\ \cancel{\text{hr}}) = 200$ mi Jeff's distance

$\left(60\frac{\text{mi}}{\cancel{\text{hr}}}\right)\left(\frac{10}{3}\cancel{\text{hr}}\right) = 200$ mi George's distance

George's distance equals Jeff's distance.

Step 6. Therefore, Jeff's average speed is 50 mph, George's average speed is 60 mph, and they traveled 200 mi before they met. ■

Example 3 If crew A works 9 hr, crew B works 4 hr, and crew C works 10 hr, then 7 cars are completed. If crew A works 3 hr and crew B works 8 hr, then 5 cars are completed. If crew B works 6 hr and crew C works 5 hr, then 4 cars are completed. How long does it take each crew, working alone, to complete 1 car?

Step 1. Let crew A's rate $= a \dfrac{\text{cars}}{\text{hr}}$

crew B's rate $= b \dfrac{\text{cars}}{\text{hr}}$

crew C's rate $= c \dfrac{\text{cars}}{\text{hr}}$

The basic relationship used in solving work problems is as follows:

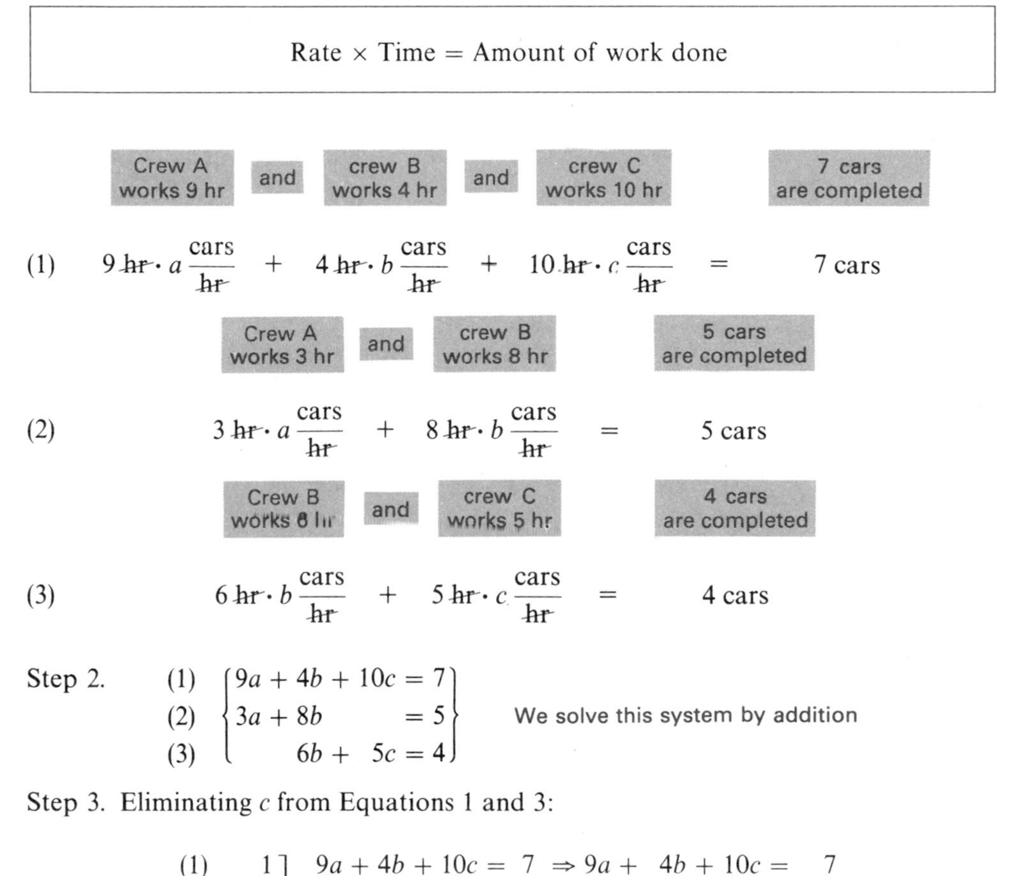

(1) $9\,\text{hr} \cdot a \dfrac{\text{cars}}{\text{hr}} + 4\,\text{hr} \cdot b \dfrac{\text{cars}}{\text{hr}} + 10\,\text{hr} \cdot c \dfrac{\text{cars}}{\text{hr}} = 7 \text{ cars}$

(2) $3\,\text{hr} \cdot a \dfrac{\text{cars}}{\text{hr}} + 8\,\text{hr} \cdot b \dfrac{\text{cars}}{\text{hr}} = 5 \text{ cars}$

(3) $6\,\text{hr} \cdot b \dfrac{\text{cars}}{\text{hr}} + 5\,\text{hr} \cdot c \dfrac{\text{cars}}{\text{hr}} = 4 \text{ cars}$

Step 2.

$$\begin{array}{ll} (1) & \\ (2) & \\ (3) & \end{array} \left\{\begin{array}{l} 9a + 4b + 10c = 7 \\ 3a + 8b \qquad\;\;\; = 5 \\ \qquad 6b + 5c = 4 \end{array}\right\}$$

We solve this system by addition

Step 3. Eliminating c from Equations 1 and 3:

$$\begin{array}{llrl} (1) & 1] & 9a + 4b + 10c = 7 & \Rightarrow 9a + 4b + 10c = 7 \\ (3) & -2] & 6b + 5c = 4 & \Rightarrow \quad -12b - 10c = -8 \\ & & (4) & \quad 9a - 8b \qquad = -1 \end{array}$$

$$\begin{array}{lrl} (2) & 3a + 8b = & 5 \\ (4) & 9a - 8b = & -1 \\ \hline & 12a \qquad = & 4 \end{array}$$

Add Equations 2 and 4

Step 4. $$a = \frac{4}{12} = \frac{1}{3}$$

$$(2) \qquad 3a + 8b = 5$$

$$3\left(\frac{1}{3}\right) + 8b = 5 \qquad \text{Substituting } \frac{1}{3} \text{ for } a \text{ in Equation 2}$$

$$1 + 8b = 5$$

$$8b = 4$$

$$b = \frac{4}{8} = \frac{1}{2}$$

$$(3) \qquad 6b + 5c = 4$$

$$6\left(\frac{1}{2}\right) + 5c = 4 \qquad \text{Substituting } \frac{1}{2} \text{ for } b \text{ in Equation 3}$$

$$3 + 5c = 4$$

$$5c = 1$$

$$c = \frac{1}{5}$$

Step 5. The check is left to the student.

Step 6. Crew A's rate is $\frac{1}{3}\frac{\text{car}}{\text{hr}}$, or $\frac{1 \text{ car}}{3 \text{ hr}}$; therefore, crew A completes 1 car in 3 hr. Crew B's rate is $\frac{1}{2}\frac{\text{car}}{\text{hr}}$; therefore, crew B completes 1 car in 2 hr. Crew C's rate is $\frac{1}{5}\frac{\text{car}}{\text{hr}}$; therefore, crew C completes 1 car in 5 hr. ■

EXERCISES 11.8

Set I Set up each word problem algebraically, *using at least two variables*, solve, and check. Be sure to state what your variables represent.

1. The sum of two numbers is 30. Their difference is 12. What are the numbers?

2. The sum of two angles is 180°. Their difference is 70°. Find the angles.

3. Beatrice has 15 coins with a total value of \$1.75. If the coins are nickels and quarters, how many of each kind are there?

4. Raul has 22 coins with a total value of \$5.00. If the coins are dimes and half-dollars, how many of each kind are there?

5. A fraction has a value of two-thirds. If 10 is added to its numerator and 5 is subtracted from its denominator, the value of the fraction becomes 1. What was the original fraction?

6. A fraction has a value of one-half. If 8 is added to its numerator and 6 is added to its denominator, the value of the fraction becomes two-thirds. What was the original fraction?

7. The sum of the digits of a three-digit number is 20. The tens digit is 3 more than the units digit. The sum of the hundreds digit and the tens digit is 15. Find the number.

8. The sum of the digits of a three-digit number is 21. The units digit is 1 less than the tens digit. Twice the hundreds digit plus the tens digit is 17. Find the number.

9. Albert, Bill, and Carlos working together can do a job in 2 hr. Bill and Carlos together can do the job in 3 hr. Albert and Bill together can do the job in 4 hr. How long would it take each man to do the entire job working alone?

10. Crews A and C working together can assemble 1 machine in 3 hr. Crews A and B together take 8 hr to assemble 3 machines. Crews B and C together take 24 hr to assemble 5 machines. How long would it take each crew, working alone, to assemble 1 machine?

11. A pilot takes $5\frac{1}{2}$ hr to fly 2,750 mi against the wind and only 5 hr to return with the wind. Find the average speed of the plane in still air and the average speed of the wind.

12. A pilot takes $2\frac{1}{2}$ hr to fly 1,200 mi against the wind and only 2 hr to return with the wind. Find the average speed of the plane in still air and the average speed of the wind.

13. A 90-lb mixture of two different grades of coffee costs \$338.90. If grade A costs \$3.85 a pound and grade B costs \$3.65 a pound, how many pounds of each grade were used?

14. An 80-lb mixture of two different grades of coffee costs \$305.25. If grade A costs \$3.95 a pound and grade B costs \$3.70 a pound, how many pounds of each grade were used?

15. The sum of the digits of a two-digit number is 12. If the digits are reversed, the new number is 54 less than the original number. Find the original number.

16. The sum of the digits of a two-digit number is 14. If the digits are reversed, the new number is 18 less than the original number. Find the original number.

17. Tom spent \$7.21 on 29 stamps, buying 18¢, 22¢, and 45¢ stamps. He bought twice as many 22¢ stamps as 18¢ stamps. How many of each kind did he buy?

18. Sherma spent \$5.07 on 21 stamps, buying 18¢, 22¢, and 45¢ stamps. She bought twice as many 18¢ stamps as 45¢ stamps. How many of each kind did she buy?

19. A tie and a pin cost \$1.10. The tie costs \$1.00 more than the pin. What is the cost of each?

20. A number of birds are resting on two limbs of a tree. One limb is above the other. A bird on the lower limb says to the birds on the upper limb, "If one of you will come down here, we will have an equal number on each limb." A bird from above replies, "If one of you will come up here, we will have twice as many up here as you have down there." How many birds are sitting on each limb?

Set II Set up each word problem algebraically, *using at least two variables*, solve, and check. Be sure to state what your variables represent.

1. The sum of two numbers is 50. Their difference is 22. What are the numbers?

2. The sum of two angles is 90°. Their difference is 40°. Find the angles.

3. Carol has 18 coins with a total value of \$3.45. If the coins are dimes and quarters, how many of each kind are there?

4. Don spent \$3.70 for 22 stamps. If he bought only 15¢ stamps and 20¢ stamps, how many of each kind did he buy?

5. A fraction has a value of three-fourths. If 4 is subtracted from the numerator and 2 is added to the denominator, the value of the new fraction is one-half. What was the original fraction?

6. Several families went to a movie together. They spent \$24.75 for 8 tickets. If adults' tickets cost \$4.50 and children's tickets cost \$2.25, how many of each kind of ticket were bought?

7. The sum of the digits of a three-digit number is 15. The tens digit is 5 more than the units digit. The sum of the hundreds digit and the units digit is 9. Find the number.

8. One-third the sum of two numbers is 12. Twice their difference is 12. Find the numbers.

9. A refinery tank has one fill pipe and two drain pipes. Pipe 1 can fill the tank in 3 hr. Pipe 2 takes 6 hr longer to drain the tank than pipe 3. If it takes 12 hr to fill the tank when all three pipes are open, how long does it take pipe 3 to drain the tank alone?

10. The sum of the digits of a three-digit number is 11. The tens digit is twice the units digit. If the digits are reversed, the new number is 297 less than the original number. Find the original number.

11. Jerry takes 6 hr to ride his bicycle 30 mi against the wind. He takes 2 hr to return with the wind. Find Jerry's average riding speed in still air and the average speed of the wind.

12. A fraction has a value of two-thirds. If 5 is added to the numerator and 4 is added to the denominator, the value of the new fraction is three-fourths. What was the original fraction?

13. A 60-lb mixture of two different grades of coffee costs \$218.50. If grade A costs \$3.80 a pound and grade B costs \$3.55 a pound, how many pounds of each grade were used?

14. Find two numbers such that 5 times the larger plus 3 times the smaller is 47, and 4 times the larger minus twice the smaller is 20.

15. The sum of the digits of a two-digit number is 11. If the digits are reversed, the new number is 27 less than the original number. Find the original number.

16. The sum of the digits of a three-digit number is 13. The tens digit is 4 times the units digit. If the digits are reversed, the new number is 99 less than the original number. Find the original number.

17. Gloria spent \$6.31 on 26 stamps, buying 18¢, 22¢, and 45¢ stamps. She bought twice as many 18¢ stamps as 22¢ stamps. How many of each kind did she buy?

18. \$26,000 was divided and placed in three different accounts. Twice as much was put into an account earning 7% per year as was put into an account earning 6% per year. The rest was put into an account earning 5% per year. The interest earned in one year by all three accounts together was \$1,650. How much was invested at each rate?

19. Al, Chet, and Muriel are two brothers and a sister. Ten years ago, Al was twice as old as Chet was then. Three years ago, Muriel was three-fourths Chet's age at that time. In 15 years, Al will be 8 years older than Muriel is then. Find their ages now.

20. Tickets for a certain concert sold for \$6, \$8, and \$12 each. There were 280 more \$6 tickets sold than \$8 and \$12 tickets together, and the total revenue from 840 tickets was \$6,080. How many tickets were sold at each price?

11.9 Quadratic Systems

All the systems studied in this chapter so far have been *linear systems* of equations. In this section, we discuss the solution of *quadratic systems*. A quadratic system is one that has a second-degree equation as its highest-degree equation.

Most of the quadratic equations that appear in the quadratic systems of this section are the conic sections discussed and graphed in Section 10.8.

Systems That Have One Quadratic Equation and One Linear Equation

When the system has one quadratic equation and one linear equation, there are three possible cases:

1. Two points of intersection (two real solutions)

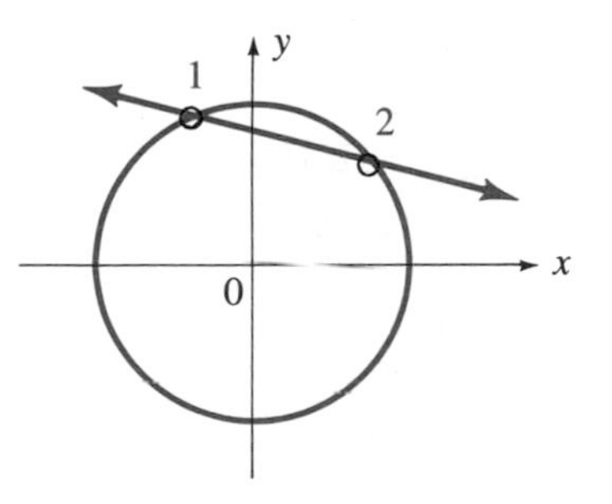

2. One point of intersection (one real solution)

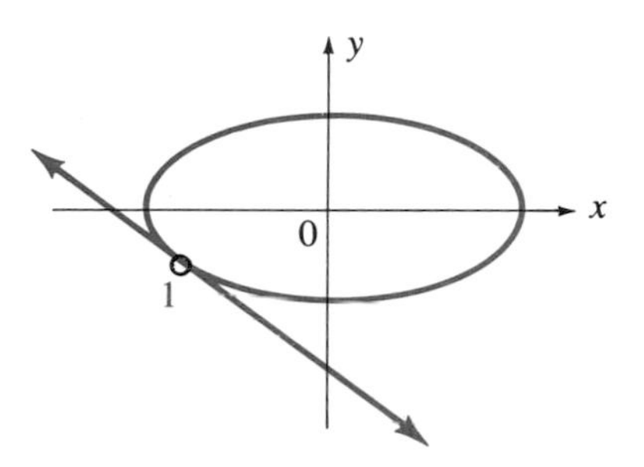

3. No point of intersection (two *complex* solutions)

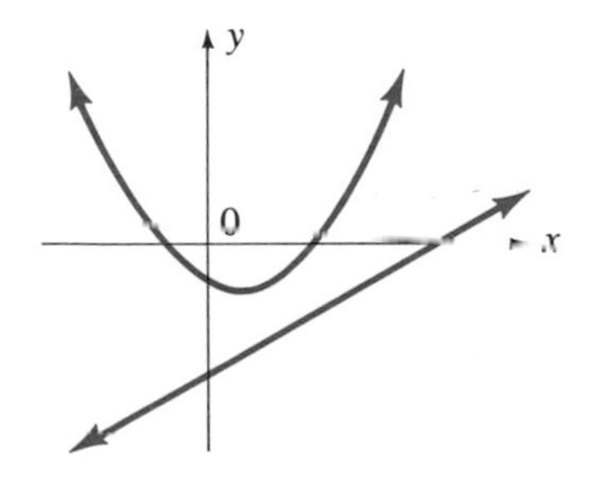

> TO SOLVE A SYSTEM THAT HAS ONE QUADRATIC AND ONE LINEAR EQUATION
>
> **1.** Solve the linear equation for one variable.
>
> **2.** Substitute the expression obtained in step 1 into the quadratic equation of the system; then solve the resulting equation.
>
> **3.** Substitute the solutions obtained in step 2 into the expression obtained in step 1 to solve for the remaining variable.

Example 1 Solve the quadratic system $\begin{matrix}(1)\\(2)\end{matrix} \left\{\begin{aligned} x - 2y &= -4 \\ x^2 &= 4y \end{aligned}\right\}.$

Step 1. Solve Equation 1 for x: (1) $x - 2y = -4$

$$x = 2y - 4$$

Step 2. Substitute $2y - 4$ for x in Equation 2:

$$
\begin{aligned}
(2) \qquad x^2 &= 4y \\
(2y - 4)^2 &= 4y \\
4y^2 - 16y + 16 &= 4y \\
4y^2 - 20y + 16 &= 0 \\
y^2 - 5y + 4 &= 0 \\
(y - 1)(y - 4) &= 0
\end{aligned}
$$

$$
\begin{array}{c|c}
y - 1 = 0 & y - 4 = 0 \\
y = 1 & y = 4
\end{array}
$$

Step 3. If $y = 1$, then $x = 2y - 4 = 2(1) - 4 = -2$. Therefore, one solution is $(-2, 1)$.

If $y = 4$, then $x = 2y - 4 = 2(4) - 4 = 8 - 4 = 4$. Therefore, the second solution is $(4, 4)$.

When the graphs of Equations 1 and 2 are drawn on the same set of axes, we see that the two solutions, $(-2, 1)$ and $(4, 4)$, are the points where the graphs *intersect* (Figure 11.9.1). Refer to Section 8.2 for graphing $x - 2y = -4$. Refer to Section 10.8 for graphing $x^2 = 4y$.

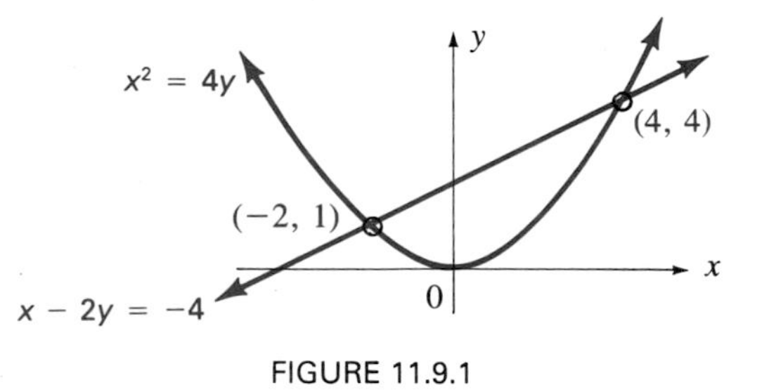

FIGURE 11.9.1

■

Systems That Have Two Quadratic Equations

When the system has two quadratic equations, there are five possible cases:

1. No points of intersection (*complex* solutions)

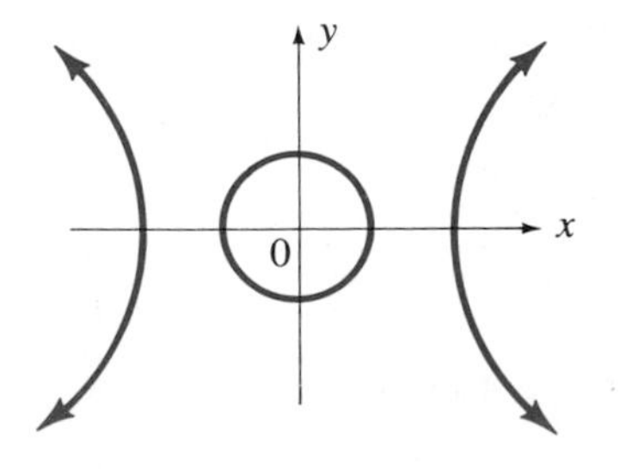

2. One point of intersection (one real solution)

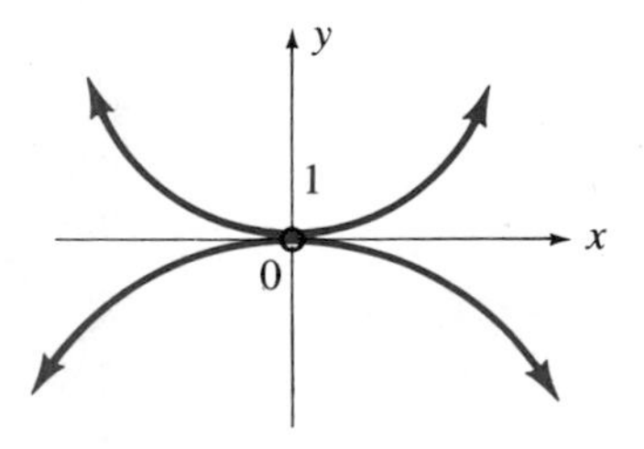

3. Two points of intersection (two real solutions)

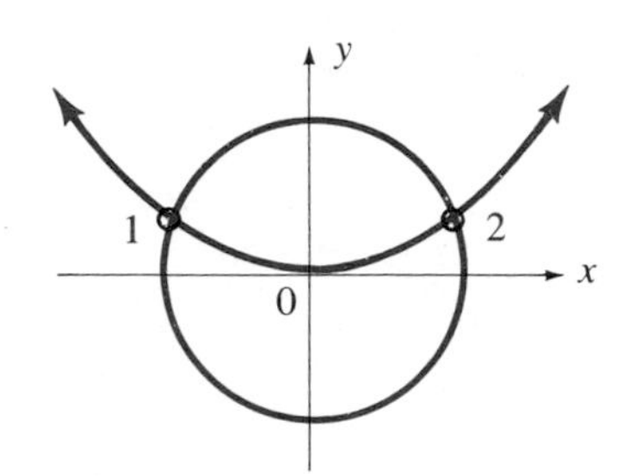

4. Three points of intersection (three real solutions)

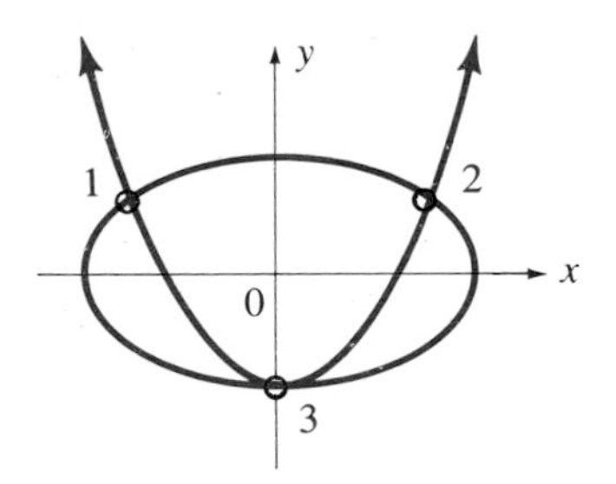

5. Four points of intersection (four real solutions)

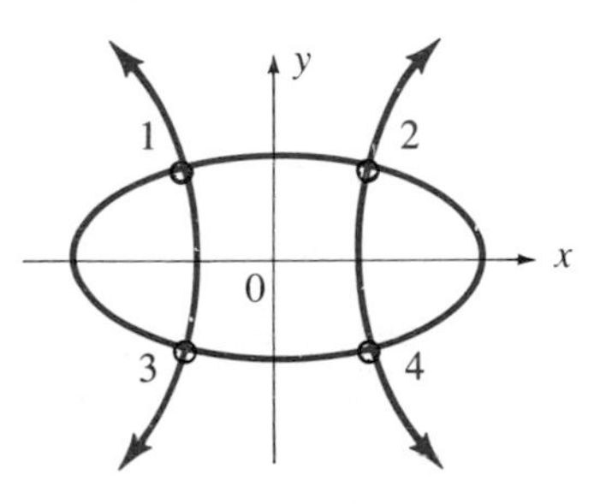

We now give two examples of solving some systems that have two quadratic equations.

Example 2 Solve the quadratic system $\begin{matrix}(1)\\(2)\end{matrix}\ \begin{Bmatrix} x^2 + y^2 = 13 \\ 2x^2 + 3y^2 = 30 \end{Bmatrix}$.

Solution We can use the addition method because the two equations have like terms.

$$\begin{array}{lrlrl}(1) & -2] & x^2 + y^2 = 13 \Rightarrow & -2x^2 - 2y^2 = & -26 \\ (2) & 1] & 2x^2 + 3y^2 = 30 \Rightarrow & 2x^2 + 3y^2 = & 30 \\ \hline & & & y^2 = & 4 \\ & & & y = \pm 2 & \end{array}$$

Add the equations

We find the corresponding x-values by substitution as follows:

$$\begin{aligned} (1)\quad x^2 + y^2 &= 13 \\ x^2 + (2)^2 &= 13 \qquad \text{Substituting 2 for } y \text{ in Equation 1} \\ x^2 + 4 &= 13 \\ x^2 &= 9 \\ x &= \pm 3 \end{aligned}$$

Therefore, two solutions are (3, 2) and (−3, 2).

$$\begin{aligned} (1)\quad x^2 + y^2 &= 13 \\ x^2 + (-2)^2 &= 13 \qquad \text{Substituting } -2 \text{ for } y \text{ in Equation 1} \\ x^2 + 4 &= 13 \\ x^2 &= 9 \\ x &= \pm 3 \end{aligned}$$

Therefore, two more solutions are (3, −2) and (−3, −2).

When the graphs of Equations 1 and 2 are drawn on the same set of axes, we see that the four solutions, (3, 2), (3, −2), (−3, 2) and (−3, −2), are the points where the two graphs *intersect* (Figure 11.9.2). The solutions should *not* be written as (±3, ±2); this notation does not make it clear that there are four distinct solutions.

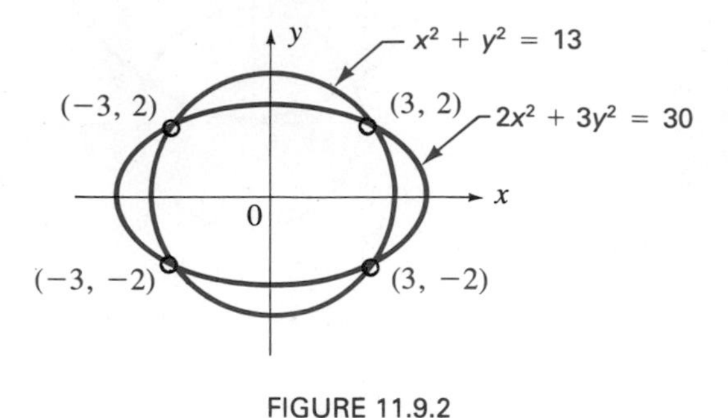

FIGURE 11.9.2

Example 3 Solve the system $\begin{matrix}(1)\\(2)\end{matrix}\left\{\begin{aligned} xy &= 1\\ x^2 + y^2 &= 2\end{aligned}\right\}$.

Solution We have not previously discussed an equation such as $xy = 1$. It is a quadratic equation, because it is of degree 2. (Remember that to find the degree of the term xy, we add the exponents on the variables: $xy = x^1y^1$, and $1 + 1 = 2$.) It happens to be a *hyperbola* with the x- and y-axes as its asymptotes.

We must use the *substitution method* in solving this system of equations.

We first solve Equation 1 for y: (1) $xy = 1$

$$y = \frac{1}{x}$$

Next, we substitute $\frac{1}{x}$ for y in Equation 2:

$$\begin{aligned}
(2) \qquad x^2 + y^2 &= 2\\
x^2 + \left(\frac{1}{x}\right)^2 &= 2\\
x^2 + \frac{1}{x^2} &= 2\\
x^4 + 1 &= 2x^2\\
x^4 - 2x^2 + 1 &= 0\\
(x^2 - 1)^2 &= 0\\
x^2 - 1 &= 0\\
x^2 &= 1\\
x &= \pm 1
\end{aligned}$$

If $x = 1$, then $y = \frac{1}{x} = \frac{1}{1} = 1$. Therefore, one solution is (1, 1).

If $x = -1$, then $y = \frac{1}{x} = \frac{1}{-1} = -1$. Therefore, the second solution is (−1, −1). The graph is as follows:

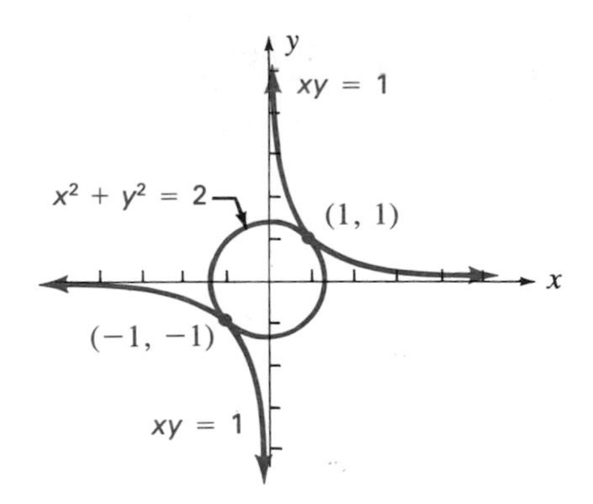

Verify that if we substitute 1 for x in Equation 2, we obtain the solutions $y = \pm 1$, and if we substitute -1 for x in Equation 2, we obtain the solutions $y = \pm 1$. Therefore, we might think that $(1, 1)$, $(1, -1)$, $(-1, 1)$, and $(-1, -1)$ are all points of intersection. However, checking these solutions in Equation 1 (*or* examining the graph) shows that $(-1, 1)$ and $(1, -1)$ are extraneous solutions. Note that the solutions should *not* be written as $(\pm 1, \pm 1)$. ■

NOTE When a quadratic equation contains an xy-term, we cannot determine which of the conic sections it is by simply looking at the equation. ☑

EXERCISES 11.9

Set I Solve each system of equations.

1. $\begin{cases} x^2 = 2y \\ x - y = -4 \end{cases}$ **2.** $\begin{cases} x^2 = 4y \\ x + 2y = 4 \end{cases}$ **3.** $\begin{cases} x^2 = 4y \\ x - y = 1 \end{cases}$

4. $\begin{cases} x^2 + y^2 = 25 \\ x - 3y = -5 \end{cases}$ **5.** $\begin{cases} xy = 4 \\ x - 2y = 2 \end{cases}$ **6.** $\begin{cases} xy = 3 \\ x + y = 0 \end{cases}$

7. $\begin{cases} x^2 + y^2 = 61 \\ x^2 - y^2 = 11 \end{cases}$ **8.** $\begin{cases} x^2 + 4y^2 = 4 \\ x^2 - 4y = 4 \end{cases}$

9. $\begin{cases} 2x^2 + 3y^2 = 21 \\ x^2 + 2y^2 = 12 \end{cases}$ **10.** $\begin{cases} 4x^2 - 5y^2 = 62 \\ 5x^2 + 8y^2 = 106 \end{cases}$

Set II Solve each system of equations.

1. $\begin{cases} x^2 = 2y \\ 2x + y = -2 \end{cases}$ **2.** $\begin{cases} xy = 6 \\ x - y = 1 \end{cases}$

3. $\begin{cases} x^2 + y^2 = 25 \\ 3x + y = -5 \end{cases}$ **4.** $\begin{cases} 4x^2 + 9y^2 = 36 \\ 2x^2 - 9y = 18 \end{cases}$

5. $\begin{cases} 3x^2 + 4y^2 = 35 \\ 2x^2 + 5y^2 = 42 \end{cases}$ **6.** $\begin{cases} y = 3 - 2x - x^2 \\ y = x^2 + 2x + 3 \end{cases}$

7. $\begin{cases} 9x^2 + 16y^2 = 144 \\ 3x + 4y = 12 \end{cases}$ **8.** $\begin{cases} x^2 - y^2 = 4 \\ x^2 + y^2 = 4 \end{cases}$

9. $\begin{cases} x + y = 4 \\ y = x^2 - 4x + 4 \end{cases}$ **10.** $\begin{cases} 9x^2 + y^2 = 9 \\ 3x - y = -3 \end{cases}$

11.10 Solving a System of Inequalities in Two Variables by Graphing

11.10A Solving a System of Linear Inequalities by Graphing

So far we have discussed only systems of *equations*. In this section, we discuss solving systems of linear *inequalities*. You may wish to refer back to Section 8.4, where drawing the graph of a linear inequality was discussed in detail.

TO SOLVE A SYSTEM OF TWO LINEAR INEQUALITIES GRAPHICALLY

1. Graph the first inequality, shading the half-plane that represents its solution (Section 8.4).
2. Graph the second inequality on the same set of axes. Use a different type of shading for the half-plane that represents its solution.
3. Heavily shade the region that contains *both* types of shading if such a region exists.
4. Any point that lies in the heavily shaded region is in the graphic solution; the coordinates of any such point will satisfy *both* inequalities.

Example 1 Solve the system $\begin{matrix}(1)\\(2)\end{matrix}\begin{Bmatrix}2x - 3y < 6\\3x + 4y \leq -12\end{Bmatrix}$ graphically.

Step 1. Graph inequality 1: $2x - 3y < 6$

The boundary, $2x - 3y = 6$, is a *dashed* line because equality is *not* included

$$2x - 3y < 6$$

The boundary line has intercepts (3, 0) and (0, −2). We graph these points as *hollow dots* and graph the boundary line as a dashed line. We then substitute the coordinates of the origin, (0, 0), into inequality 1:

$$\begin{aligned} (1) \qquad 2x - 3y &< 6 \\ 2(0) - 3(0) &< 6 \\ 0 &< 6 \qquad \text{True} \end{aligned}$$

Therefore, the half-plane containing (0, 0) is the solution. We shade this region with black diagonal lines (see Figure 11.10.1).

Step 2. Graph inequality 2: $3x + 4y \leq -12$

The boundary, $3x + 4y = -12$, is a *solid* line because equality *is* included

$$3x + 4y \leq -12$$

The boundary line has intercepts $(-4, 0)$ and $0, -3)$. We graph these points as *solid dots* and graph the boundary line as a solid line. We then substitute the coordinates of the origin, (0, 0), into inequality 2:

$$(2) \qquad 3x + 4y \le -12$$
$$3(0) + 4(0) \le -12$$
$$0 \le -12 \qquad \text{False}$$

Therefore, the solution is the half-plane *not* containing (0, 0). We shade this region with red diagonal lines (see Figure 11.10.1).

Step 3. We heavily shade the region that contains both types of shading (see Figure 11.10.1).

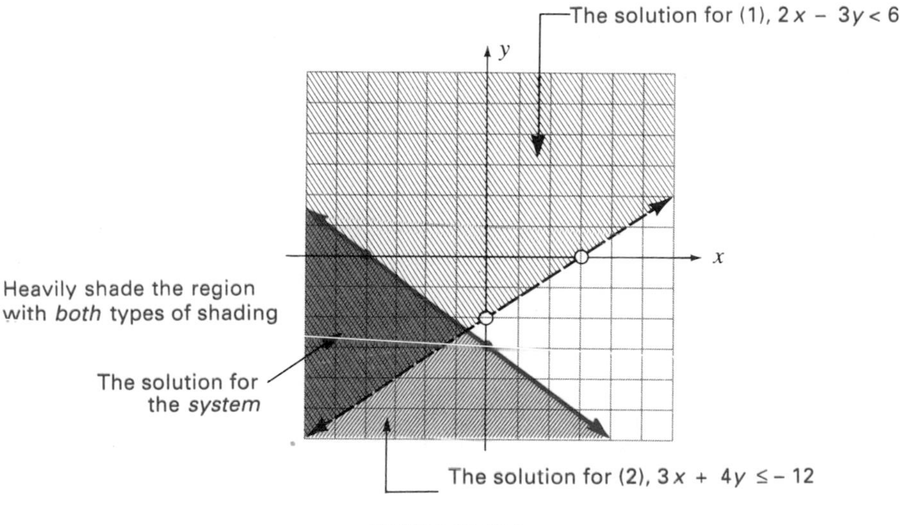

FIGURE 11.10.1

Step 4. Any point that lies in the heavily shaded region is in the graphic solution; its coordinates will satisfy both inequalities. ■

EXERCISES 11.10A

Set I Solve each system of linear inequalities graphically.

1. $\begin{cases} 4x - 3y > -12 \\ y > 2 \end{cases}$ **2.** $\begin{cases} x - y > -2 \\ 2x + y > 2 \end{cases}$ **3.** $\begin{cases} 2x - y \le 2 \\ x + y \ge 5 \end{cases}$

4. $\begin{cases} x - y \le 4 \\ x + y \ge -2 \end{cases}$ **5.** $\begin{cases} 2x + y < 0 \\ x - y \ge -3 \end{cases}$ **6.** $\begin{cases} 3x > 6 - y \\ y + 3x \le 0 \end{cases}$

7. $\begin{cases} 3x - 2y < 6 \\ x + 2y \le 4 \\ 6x + y > -6 \end{cases}$ **8.** $\begin{cases} 3x + 4y \le 15 \\ 2x + y \le 2 \\ -4 < x < 1 \\ -2 < y < 6 \end{cases}$

Set II Solve each system of linear inequalities graphically.

1. $\begin{cases} 3x - 5y < 15 \\ x < 3 \end{cases}$ **2.** $\begin{cases} 2x + 3y < 6 \\ x - 3y < 3 \end{cases}$ **3.** $\begin{cases} 3x - y \ge 0 \\ 3x + 2y < 6 \end{cases}$

4. $\begin{Bmatrix} 2x \geq y - 4 \\ y - 2x > 0 \end{Bmatrix}$

5. $\begin{Bmatrix} 3x - 2y < 6 \\ x + 2y \geq 2 \end{Bmatrix}$

6. $\begin{Bmatrix} 4x + y \leq 4 \\ 2x - y > 2 \end{Bmatrix}$

7. $\begin{Bmatrix} 4x + 3y \geq 12 \\ y - x > 1 \\ 9x - 28y \geq 42 \end{Bmatrix}$

8. $\begin{Bmatrix} 4x - 3y \geq 12 \\ x < 3 \end{Bmatrix}$

11.10B Solving a System of Quadratic Inequalities by Graphing

TO SOLVE A SYSTEM OF QUADRATIC INEQUALITIES GRAPHICALLY

1. Graph the first inequality, shading the region that represents its solution (Section 10.9B).
2. Graph the second inequality on the same set of axes. Use a different type of shading for the region that represents its solution.
3. Heavily shade the region that contains *both* types of shading.
4. Any point that lies in the heavily shaded region is in the graphic solution; the coordinates of any such point will satisfy both inequalities.

This technique also applies when one of the inequalities is linear and the other is quadratic (see Example 3).

Example 2 Solve the system $\begin{Bmatrix} y \geq x^2 - 2x - 3 \\ x^2 + 4y^2 \leq 4 \end{Bmatrix}$ graphically.

Solution First, we graph the equation $y = x^2 - 2x - 3$. It is a parabola that opens upward; the intercepts are $(3, 0)$, $(-1, 0)$, and $(0, -3)$, and the vertex is at $(1, -4)$. A *true* statement is obtained if we substitute $x = 0$ and $y = 0$ into the inequality. Therefore, we shade the region that contains $(0, 0)$, using *red diagonal* lines (see Figure 11.10.2).

Next, we graph the equation $x^2 + 4y^2 = 4$. It is an ellipse; the intercepts are $(2, 0)$, $(-2, 0)$, $(0, 1)$, and $(0, -1)$. *A true* statement is obtained if we substitute $x = 0$ and $y = 0$ into the inequality. Therefore, we shade the region that contains $(0, 0)$ with *black vertical* lines (see Figure 11.10.2).

We now heavily shade that portion of the plane that has both diagonal *and* vertical lines in it (see Figure 11.10.2).

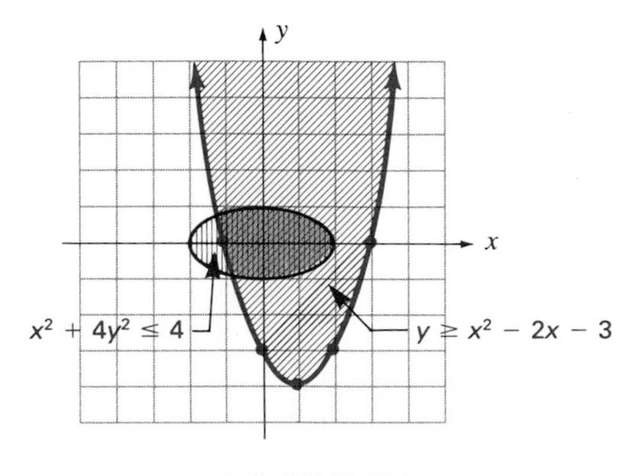

FIGURE 11.10.2

Any point that lies in the heavily shaded region is in the solution set for the system. ■

Example 3 Solve the system $\begin{Bmatrix} y < 4 - x^2 \\ y \le x + 3 \end{Bmatrix}$ graphically.

Solution We will graph the inequality $y < 4 - x^2$. $y = 4 - x^2$ is a parabola that opens downward; the intercepts are (2, 0), (−2, 0), and (0, 4), and the vertex is at (0, 4). We graph the parabola with a dashed curve, since the inequality is $<$. Because (0, 0) satisfies the inequality, we shade, with black diagonal lines, the region inside the parabola (see Figure 11.10.3).

We next graph the inequality $y \le x + 3$. The equation $y = x + 3$ is a *linear* equation. We graph that line and then find that (0, 0) satisfies the inequality $y \le x + 3$. Therefore, we shade, with *red vertical* lines, the half-plane that contains the origin (see Figure 11.10.3).

We heavily shade that portion of the plane that contains both diagonal *and* vertical lines (see Figure 11.10.3).

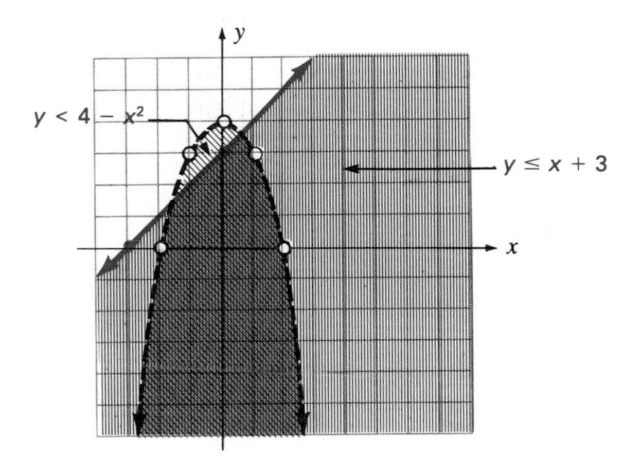

FIGURE 11.10.3

Any point that lies in the heavily shaded region is in the solution set of the system of inequalities. ■

EXERCISES 11.10B

Set I Solve each system of inequalities graphically.

1. $\begin{Bmatrix} \dfrac{x^2}{9} + \dfrac{y^2}{4} < 1 \\ \dfrac{x^2}{4} + \dfrac{y^2}{9} < 1 \end{Bmatrix}$

2. $\begin{Bmatrix} \dfrac{x^2}{4} + y^2 \le 1 \\ x + y < 1 \end{Bmatrix}$

3. $\begin{Bmatrix} y > 1 - x^2 \\ x^2 + y^2 < 4 \end{Bmatrix}$

4. $\begin{Bmatrix} x^2 + y^2 \ge 1 \\ x^2 + y^2 \le 9 \end{Bmatrix}$

5. $\begin{Bmatrix} x^2 + y^2 \le 9 \\ y \ge x - 2 \end{Bmatrix}$

Set II Solve each system of inequalities graphically.

1. $\begin{Bmatrix} x^2 - \dfrac{y^2}{9} \le 1 \\ \dfrac{x^2}{9} + \dfrac{y^2}{4} \le 1 \end{Bmatrix}$

2. $\begin{Bmatrix} y \le x^2 - 1 \\ 4x^2 + y^2 \le 4 \end{Bmatrix}$

3. $\begin{Bmatrix} x^2 + \dfrac{y^2}{9} \le 1 \\ y \le 1 - x^2 \end{Bmatrix}$

4. $\begin{Bmatrix} \dfrac{x^2}{9} + y^2 \le 1 \\ x^2 + y^2 \ge 4 \end{Bmatrix}$

5. $\begin{Bmatrix} x^2 + y^2 \le 16 \\ x + y > 2 \end{Bmatrix}$

11.11 Review: 11.7–11.10

A Matrix 11.7A

A **matrix** (plural: **matrices**) is a rectangular array of numbers; we use brackets around a matrix. A *row* of a matrix is a *horizontal* line of its elements. A *column* of a matrix is a *vertical* line of its elements. An $m \times n$ (read "m by n") matrix has m rows and n columns.

Augmented Matrix for a System of Equations 11.7A

An **augmented matrix** is associated with each system of linear equations. The augmented matrix for a system of n equations in n unknowns is always an $n \times (n + 1)$ matrix.

Row-Echelon Form 11.7A

If a matrix is in the form $\begin{bmatrix} 1 & a & b & c \\ 0 & 1 & d & e \\ 0 & 0 & 1 & k \end{bmatrix}$, where a, b, c, d, e, and k are any constants, we say it is in **row-echelon form**.

Elementary Row Operations 11.7A

If any of the following three operations are performed on a matrix, we say that the new matrix is **row-equivalent** to the given matrix.

1. Interchanging any two rows
2. Multiplying each element in any row by a nonzero constant
3. Adding a (nonzero) multiple of the elements of one row to the corresponding elements of another row

Gaussian Elimination 11.7A

To use **Gaussian elimination** to solve a system of linear equations, we form the augmented matrix, reduce it to row-echelon form, and then write the system of equations that corresponds to the final matrix.

Determinants 11.7B

A **determinant** is a *real number* that is represented as a square array of numbers enclosed between two vertical bars.

The value of the second-order determinant $\begin{vmatrix} a & b \\ c & d \end{vmatrix}$ is $ad - cb$.

To find the value of a third-order determinant, see Section 11.7B.

Cramer's Rule 11.7C

Cramer's rule can be used to solve linear systems of equations if the system contains the same number of equations as variables; however, if the system is dependent, Cramer's rule does not give the solutions.

To Solve a Word Problem Using a System of Equations 11.8

Read First read the problem very carefully.

Think Determine what *type* of problem it is, if possible. Determine what is unknown.

Sketch Draw a sketch *with labels*, if possible.

Step 1 Represent each unknown number by a *different* variable.

Reread Reread the word problem, breaking it up into small pieces.

Step 2 Translate each English phrase into an algebraic expression; fit these expressions together into *two or more different equations. There must be as many equations as variables.*

Step 3 Solve the *system* of equations for *all* the variables, using one of the following:

11.3 a. the addition method

11.4 b. the substitution method

11.7 c. the matrix method or Cramer's rule

11.2 d. the graphical method

Step 4 Be sure you've answered *all* the questions asked.

Step 5 Check the solution(s) *in the word statements.*

Step 6 State your results clearly.

Methods of Solving a Quadratic System of Equations 11.9

1. *One quadratic and one linear equation*: Solve the linear equation for one variable; then substitute the expression obtained into the quadratic equation.
2. *Two quadratic equations* (with like terms): Use addition.

To Solve a System of Inequalities 11.10

1. Graph the first inequality, shading the region that represents its solution.
2. Graph the second inequality, using a different type of shading.
3. Heavily shade the region that contains *both* types of shading.
4. Any point that lies in the heavily shaded region is in the solution set of both inequalities.

Review Exercises 11.11 Set I

In Exercises 1–3, use the matrix method or Cramer's rule to solve each system.

1. $\begin{cases} 3x - 2y = 8 \\ 2x + 5y = -1 \end{cases}$

2. $\begin{cases} 3x + y + 2z = 4 \\ 2x - 3y + 3z = -5 \\ 5x + 2y + 3z = 7 \end{cases}$

3. $\begin{cases} 5x + 2y = 1 \\ 7x - 6y = 8 \end{cases}$

In Exercises 4–8, solve each system by any convenient method. Write "inconsistent" if no solution exists. Write "dependent" if an infinite number of solutions exist.

4. $\begin{cases} 5x - 4y = -7 \\ -6x + 8y = 2 \end{cases}$

5. $\begin{cases} x - 3y = 15 \\ 5x + 7y = -13 \end{cases}$

6. $\begin{cases} 2x + 3y - 4z = 4 \\ 3x + 2y - 5z = 0 \\ 4x + 5y + 2z = -4 \end{cases}$

7. $\begin{cases} x - 2y = -1 \\ 2x^2 - 3y^2 = 6 \end{cases}$

8. $\begin{cases} x + 3y = -2 \\ x^2 - 2y^2 = 8 \end{cases}$

In Exercises 9–11, solve each system of inequalities graphically.

9. $\begin{Bmatrix} x + 4y \le 4 \\ 3x + 2y > 2 \end{Bmatrix}$

10. $\begin{Bmatrix} 3x + 6 \ge 2y \\ 3x + y < 3 \end{Bmatrix}$

11. $\begin{Bmatrix} y \le 12 + 4x - x^2 \\ x^2 + y^2 \le 16 \end{Bmatrix}$

In Exercises 12–16, set up algebraically and solve, using at least two variables.

12. The sum of two numbers is 6. Their difference is 40. What are the numbers?

13. A mail-order office paid $730 for a total of 80 rolls of stamps in two denominations. If one kind cost $10 per roll and the other kind cost $8 per roll, how many rolls of each kind were bought?

14. Jennifer motors 64 mi upstream in 4 hr and motors back in half the time. Find the average speed of her boat in still water and the average speed of the river.

15. The sum of the digits of a three-digit number is 12. The hundreds digit is 3 times the units digit. The tens digit is 2 more than the units digit. Find the number.

16. The difference of two numbers is -7. The difference of their squares is 7. Find the numbers.

Review Exercises 11.11 Set II

NAME ______________________

In Exercises 1–3, use the matrix method or Cramer's rule to solve each system.

1. $\begin{Bmatrix} 3x - 6y = -4 \\ -5x + 8y = 6 \end{Bmatrix}$

2. $\begin{Bmatrix} 3x - 2y = 10 \\ 5x + 4y = 24 \end{Bmatrix}$

3. $\begin{Bmatrix} x + 2y + z = 3 \\ 3x + y - 2z = -1 \\ 2x - 3y - z = 4 \end{Bmatrix}$

ANSWERS

1. ______________________

2. ______________________

3. ______________________

4. ______________________

5. ______________________

6. ______________________

7. ______________________

8. ______________________

In Exercises 4–8, solve each system by any convenient method. Write "inconsistent" if no solution exists. Write "dependent" if an infinite number of solutions exist.

4. $\begin{Bmatrix} 7x + 3y = -1 \\ 5x + y = -5 \end{Bmatrix}$

5. $\begin{Bmatrix} 3x + 2y + z = 5 \\ 2x - 3y - 3z = 1 \\ 4x + 3y + 2z = 8 \end{Bmatrix}$

6. $\begin{Bmatrix} 4x + 3y + 2z = 6 \\ 8x + 3y - 3z = -1 \\ -4x + 6y + 5z = 5 \end{Bmatrix}$

7. $\begin{Bmatrix} x - 3y = 1 \\ x^2 = 6y - 8 \end{Bmatrix}$

8. $\begin{Bmatrix} x^2 + y^2 = 5 \\ x^2 - 3y^2 = 1 \end{Bmatrix}$

In Exercises 9–11, solve each system of inequalities graphically.

9. $\begin{cases} 5x + 2y \geq 0 \\ 5y - 6x > 15 \end{cases}$

10. $\begin{cases} y \leq x^2 - 4x + 3 \\ x^2 + 4y^2 \leq 4 \end{cases}$

11. $\begin{cases} x^2 + y^2 \geq 4 \\ x^2 + 9y^2 \leq 9 \end{cases}$

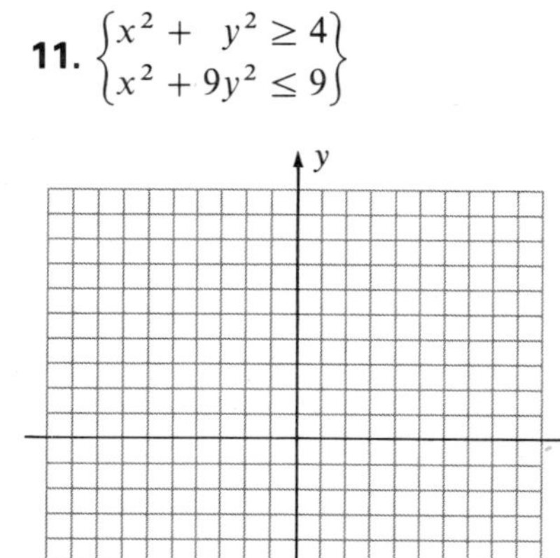

ANSWERS

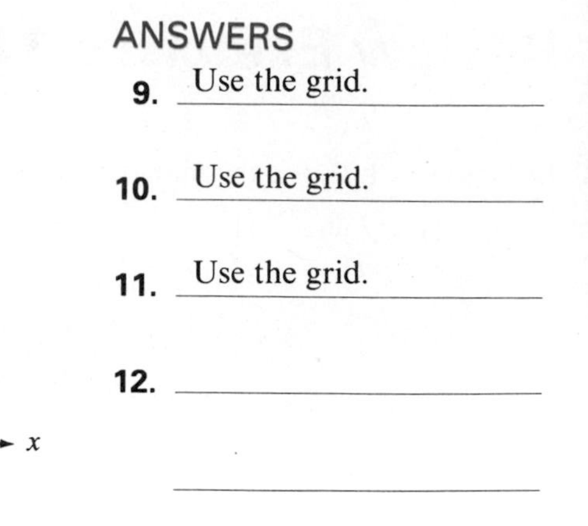

9. Use the grid.

10. Use the grid.

11. Use the grid.

12. ______

______

In Exercises 12–16, set up algebraically and solve, using at least two variables.

12. Half the sum of two numbers is 15. Half their difference is 8. Find the numbers.

13. Brian worked at two jobs during the week for a total of 32 hr. For this he received a total of \$86.60. If he was paid \$2.20 an hour as a tutor and \$3.10 an hour as a waiter, how many hours did he work at each job?

14. Leticia worked at two jobs during the week for a total of 26 hr. For this she received a total of \$118.25. If she was paid \$4.75 an hour as a lab assistant and \$4.40 an hour as a clerk-typist, how many hours did she work at each job?

15. The sum of the digits of a three-digit number is 11. The units digit is 3 more than the tens digit. The hundreds digit is 1 less than the units digit. Find the number.

16. Fausto motors 21 mi upstream in 3 hr and motors back in one-third the time. Find the average speed of the boat in still water and the average speed of the river.

ANSWERS

13. ____________

____

14. ____________

____

15. ____________

16. ____________

____

Chapter 11 Diagnostic Test

The purpose of this test is to see how well you understand systems of equations and systems of inequalities. We recommend that you work this diagnostic test *before* your instructor tests you on this chapter. Allow yourself about 50 minutes.

Complete solutions for all the problems on this test, together with section references, are given in the answer section at the end of the book. For the problems you do incorrectly, study the sections cited.

1. Solve the system $\begin{Bmatrix} 3x + 2y = 4 \\ x - y = 3 \end{Bmatrix}$ graphically.

In Problems 2–7, solve each system by any convenient method. Write "inconsistent" if no solution exists. Write "dependent" if an infinite number of solutions exist.

2. $\begin{Bmatrix} 4x - 3y = 13 \\ 5x - 2y = 4 \end{Bmatrix}$

3. $\begin{Bmatrix} 5x + 4y = 23 \\ 3x + 2y = 9 \end{Bmatrix}$

4. $\begin{Bmatrix} 15x + 8y = -18 \\ 9x + 16y = -8 \end{Bmatrix}$

5. $\begin{Bmatrix} 35y - 10x = -18 \\ 4x - 14y = 8 \end{Bmatrix}$

6. $\begin{Bmatrix} x + y + z = 0 \\ 2x - 3z = 5 \\ 3y + 4z = 3 \end{Bmatrix}$

7. $\begin{Bmatrix} y^2 = 8x \\ 3x + y = 2 \end{Bmatrix}$

8. Solve the system $\begin{Bmatrix} 2x + 3y \leq 6 \\ y - 2x < 2 \end{Bmatrix}$ graphically.

9. Solve the system $\begin{Bmatrix} x - y - z = 0 \\ x + 3y + z = 4 \\ 7x - 2y - 5z = 2 \end{Bmatrix}$ by the matrix method, or by Cramer's rule.

Set up the following problem algebraically, *using two variables*, and solve.

10. Barney motors 30 mi upstream in 2 hr and motors back in half the time. Find the average speed of his boat in still water and the average speed of the river.

Cumulative Review Exercises: Chapters 1–11

1. Add $\dfrac{x}{xy - y^2} + \dfrac{y}{xy - x^2}$.

2. Divide $(x^3 - 5x^2 + 9x + 20)$ by $(x - 3)$.

In Exercises 3 and 4, simplify each radical.

3. $\sqrt[6]{4x^2}$

4. $\sqrt{\dfrac{3a^3b^7}{8a^5b^4}}$

5. Write the general form of the equation of the line through $(-1, 2)$ and $(2, 4)$.

6. Find the solution set for the inequality $x^2 - 8x + 12 < 0$.

In Exercises 7 and 8, find the solution set for each equation.

7. $4x^2 - 25 = 0$

8. $\dfrac{x+2}{3} + \dfrac{5}{x-2} = 4$

9. Use the quadratic formula to find the solution set for $x^2 - 4x + 5 = 0$.

10. Is division associative?

11. What is the additive inverse of $-\frac{3}{5}$?

12. What is the multiplicative inverse of $-\frac{3}{5}$?

In Exercises 13–15, solve each system by any convenient method. Write "inconsistent" if no solution exists. Write "dependent" if an infinite number of solutions exist.

13. $\begin{Bmatrix} 4x + 3y = 8 \\ 8x + 7y = 12 \end{Bmatrix}$

14. $\begin{Bmatrix} 2x - 3y = 3 \\ 3y - 2x = 6 \end{Bmatrix}$

15. $\begin{Bmatrix} 2x + 3y + z = 4 \\ x + 4y - z = 0 \\ 3x + y - z = -5 \end{Bmatrix}$

16. Solve the system $\begin{Bmatrix} 9x^2 + 4y^2 \geq 36 \\ 3x + 2y \geq 6 \end{Bmatrix}$ graphically.

17. Find the domain for the function $y = f(x) = \sqrt{12 - 4x - x^2}$.

In Exercises 18–21, set up the problem algebraically and solve.

18. The length of a rectangle is 7 more than its width. If the area is 60, find its dimensions.

19. It takes Darryl 2 hr longer to do a job than it does Jeannie. After Darryl has worked on the job for 1 hr, Jeannie joins him. Together they finish the job in 3 hr. How long would it take each of them to do the entire job working alone?

20. On a bicycle tour, it took Barbara 1 hr longer to ride 45 mi than it took Pat to ride 48 mi. If Pat averages 3 mph more than Barbara, find the average speed of each.

21. How much pure alcohol should be mixed with 3 ℓ of a 50% solution of alcohol to make a 70% solution?

12 Sequences and Series

In this chapter, we introduce *sequences* and *series.* There are many applications of sequences and series in the sciences and in the mathematics of finance. For example, formulas for the calculation of interest, annuities, and mortgage loans are derived using series. The numbers in the logarithm tables in this book were calculated using series. This chapter is just a brief introduction to a very extensive and important part of mathematics.

12.1 Basic Definitions

Sequences

When you look at the following set of numbers, do you know what number comes next?

$$30, 40, 50, 60, 70, \ldots$$

The 3 dots (called an *ellipsis*) indicate that the sequence never ends

What number comes next in the following set of numbers?

$$7, 10, 13, 16, \ldots$$

These sets of numbers are examples of sequences of numbers. A **sequence** of numbers is a set of numbers in which the numbers follow a definite pattern and are arranged in a definite order. The numbers that make up the sequence are called the **terms** of the sequence.

A sequence is usually written

$$a_1, a_2, a_3, \ldots, a_n, \ldots$$

The ellipsis here means that some terms between a_3 and a_n were omitted

where

$$a_1 = \text{first term}$$
$$a_2 = \text{second term}$$
$$a_3 = \text{third term}$$
$$\vdots$$
$$a_n = n\text{th term}$$ This is also called the *general term* of the sequence
$$\vdots$$

The *subscript* of each term represents the *term number*. The symbol a_n represents the nth term of *any* sequence.

If, in counting the terms of a sequence, the counting comes to an end, the sequence is called a *finite* sequence. The last term of a finite sequence with n terms is represented by the symbol a_n. (Some authors represent the last term by ℓ.)

If, in counting the terms of a sequence, the counting never comes to an end, the sequence is called an *infinite* sequence.

Example 1 Examples of finite and infinite sequences:

a. 0, 1, 2, 3, 4, 5, 6, 7, 8, 9

This set, the set of digits, is a *finite* sequence. (We know that 9 is the last number in the sequence because there is no ellipsis after the 9.) Each succeeding term is found by adding 1 to the preceding term. This sequence could also be written 0, 1, 2, . . . , 9.

b. 0, 1, 2, 3, . . .

This set, the set of whole numbers, is an *infinite* sequence. Each term is found by adding 1 to the preceding term.

c. 15, 10, 5, 0, −5, −10

This is a *finite* sequence. Each term is found by adding −5 to the preceding term. ■

In the sequences discussed so far, it is possible to discover each succeeding term *by inspection.* For many sequences, this is not the case.

Actually, each term of a sequence is a function of n, where n, the term number, is any *natural number.*

$$a_n = f(n)$$

This means that a sequence can be thought of as a *function* with domain the set of natural numbers.

Example 2 Examples of sequences written in functional form:

a. If $a_n = f(n) = \frac{n}{2}$,

then $a_1 = \frac{(1)}{2} = \frac{1}{2}$

$a_2 = \frac{(2)}{2} = 1$

$a_3 = \frac{(3)}{2} = \frac{3}{2}$

⋮

b. $a_n = f(n) = \frac{n+1}{4}$

$a_1 = \frac{(1)+1}{4} = \frac{2}{4} = \frac{1}{2}$

$a_2 = \frac{(2)+1}{4} = \frac{3}{4}$

⋮

$a_{11} = \frac{(11)+1}{4} = \frac{12}{4} = 3$

⋮ ■

Series

A **series** is the indicated *sum* of a finite or infinite sequence of terms. It is a *finite* or an *infinite series* according to whether the number of terms is finite or infinite. An infinite series is usually written

$$a_1 + a_2 + a_3 + \cdots + a_n + \cdots$$

A *partial sum of a series* is the sum of a *finite* number of consecutive terms of the series, beginning with the first term.

$S_1 = a_1$ — *First* partial sum

$S_2 = a_1 + a_2$ — *Second* partial sum

$S_3 = a_1 + a_2 + a_3$ — *Third* partial sum

⋮

$S_n = a_1 + a_2 + \cdots + a_n$ — nth partial sum

⋮

Example 3 Consider the infinite *sequence*, $f(n) = \dfrac{1}{n}$:

The *sequence is* $\dfrac{1}{1}, \dfrac{1}{2}, \dfrac{1}{3}, \ldots$

A *sequence* is a *list* of numbers

The *series for this sequence* is: $1 + \dfrac{1}{2} + \dfrac{1}{3} + \cdots$

A *series* is a *sum*—the sum of the numbers of the sequence

The first three *partial sums* are

$$S_1 = 1 \qquad\qquad = 1$$

$$S_2 = 1 + \frac{1}{2} \qquad = \frac{3}{2}$$

$$S_3 = 1 + \frac{1}{2} + \frac{1}{3} = \frac{11}{6} \quad ■$$

Sigma Notation (or **Summation Notation**) We often use the symbol $\sum$, the Greek letter (capital) sigma, to indicate the sum of a series, and we call this notation **sigma notation** or **summation notation**. Sigma notation for the finite series $f(1) + f(2) + f(3) + \cdots + f(n)$ is

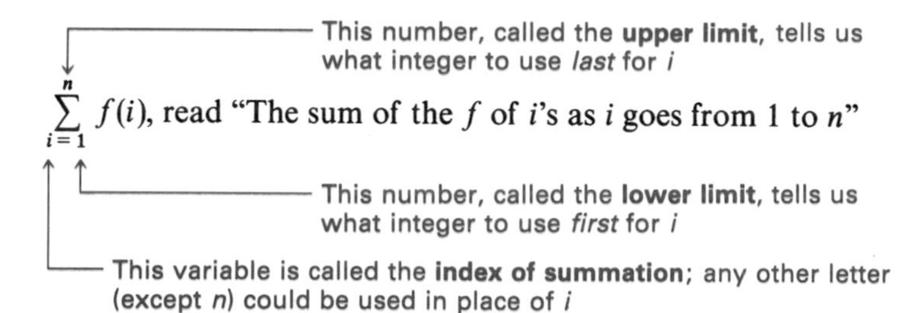

Sigma notation for the infinite series $g(1) + g(2) + g(3) + \cdots$ is

$$\sum_{k=1}^{\infty} g(k), \text{ read "The sum of the } g \text{ of } k\text{'s as } k \text{ goes from 1 to infinity"}$$

Example 4 Write $\sum_{i=2}^{6} i^2$ in expanded form.

Solution The function is i^2; we first substitute 2 for i, getting 2^2. Then we write a plus sign (remember, $\sum$ stands for *sum*), substitute 3 for i, then write a plus sign, and so forth.

The *last* number we substitute for i is 6

$$\sum_{i=2}^{6} i^2 = 2^2 + 3^2 + 4^2 + 5^2 + 6^2$$

The *first* number we substitute for i is 2 ■

Example 5 Write $\sum_{k=3}^{\infty} \dfrac{1}{k}$ in expanded form.

Solution Because there is no last term, we will write an ellipsis ($\cdots$) after we've written a few terms of the series.

$$\sum_{k=3}^{\infty} \frac{1}{k} = \frac{1}{3} + \frac{1}{4} + \frac{1}{5} + \frac{1}{6} + \frac{1}{7} + \cdots$$

3 is the *first* number we substitute for k ■

EXERCISES 12.1

Set I In Exercises 1–4, for each given sequence determine the next three terms by inspection.

1. 10, 15, 20, __, __, __, . . .

2. 8, 11, 14, __, __, __, . . .

3. 15, 13, 11, __, __, __, . . .

4. $1, \frac{4}{3}, \frac{5}{3}$, __, __, __, . . .

In Exercises 5–10, use the given general term to write the terms specified.

5. $a_n = f(n) = n + 4$; first three terms

6. $a_n = f(n) = 2n + 1$; first four terms

7. $a_n = \frac{1-n}{n}$; first four terms

8. $a_n = \frac{n(n-1)}{2}$; first three terms

9. $a_n = n^2 - 1$; first three terms

10. $a_n = n^3 + 1$; first four terms

In Exercises 11–14, find the indicated partial sum by using the given general term.

11. Given $a_n = 2n - 3$, find S_4.

12. Given $a_n = \frac{2n-1}{5-n}$, find S_4.

13. Given $a_n = \frac{n-1}{n+1}$, find S_3.

14. Given $a_n = 3^n + 2$, find S_3.

In Exercises 15–18, write each series in expanded form.

15. $\sum_{i=1}^{5} \frac{i}{7}$

16. $\sum_{k=2}^{6} \frac{k-1}{11}$

17. $\sum_{i=1}^{\infty} (2i+1)^2$

18. $\sum_{k=3}^{\infty} k^3$

Set II In Exercises 1–4, for each given sequence determine the next three terms by inspection.

1. 20, 16, 12, __, __, __, . . .

2. $\frac{1}{2}, \frac{3}{4}, 1$, __, __, __, . . .

3. 17, 19, 21, __, __, __, . . .

4. $4\frac{2}{3}, 4, 3\frac{1}{3}$, __, __, __, . . .

In Exercises 5–10, use the given general term to write the terms specified.

5. $a_n = f(n) = 1 - 2n$; first four terms

6. $a_n = \frac{n}{1-2n}$; first three terms

7. $a_n = f(n) = 3n - 1$; first four terms

8. $a_n = f(n) = 1 - n$; first four terms

9. $a_n = f(n) = n^2 + 1$; first four terms

10. $a_n = \frac{2-n}{1+n^2}$; first three terms

In Exercises 11–14, find the indicated partial sum by using the given general term.

11. Given $a_n = 4 - n$, find S_5.

12. Given $a_n = \dfrac{1 - 2^n}{3}$, find S_4.

13. Given $a_n = \dfrac{2n - 1}{n}$, find S_3.

14. Given $a_n = \dfrac{n^2 - 1}{2^n}$, find S_3.

In Exercises 15–18, write each series in expanded form.

15. $\sum_{k=1}^{7} 3k$

16. $\sum_{i=1}^{5} \dfrac{2i - 1}{13}$

17. $\sum_{i=2}^{\infty} (5i + 1)^3$

18. $\sum_{k=0}^{\infty} \dfrac{1}{k + 1}$

12.2 Arithmetic Sequences and Series

Arithmetic Sequences

An **arithmetic sequence (arithmetic progression)** is a sequence in which each term after the first is found by *adding* the same fixed number to the preceding term. The fixed number added is called the *common difference, d. Arithmetic progression* is abbreviated AP.

Example 1 Write the first five terms of the infinite arithmetic progression having first term $a_1 = 3$ and common difference $d = 2$.

Solution

$a_1 = 3$

$a_2 = 3 + 2 = 5$ Adding the common difference (2) to the first term (3)

$a_3 = 5 + 2 = 7$ Adding the common difference (2) to the second term (5)

$a_4 = 7 + 2 = 9$ And so on

$a_5 = 9 + 2 = 11$

Therefore, the first five terms are 3, 5, 7, 9, and 11. (The AP is 3, 5, 7, 9, 11, . . .) ■

Example 2 Write a six-term arithmetic progression having first term $a_1 = 15$ and common difference $d = -7$.

Solution

$$a_1 = 15$$
$$a_2 = 15 + (-7) = 8$$
$$a_3 = 8 + (-7) = 1$$
$$a_4 = 1 + (-7) = -6$$
$$a_5 = -6 + (-7) = -13$$
$$a_6 = -13 + (-7) = -20$$

Therefore, the AP is 15, 8, 1, -6, -13, -20. ■

Example 3 Determine which of the sequences below are arithmetic progressions.

Method Subtract each term from the following term. If every difference found this way is the same, the sequence is an AP.

a. $-8, -3, 2, 7, 12$

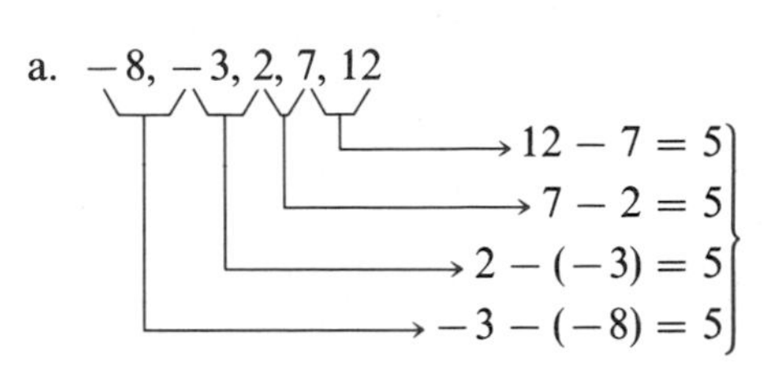

All these differences are the same (5)

The sequence *is* an AP.

b. $1, 5, 9, 12, 16$

$16 - 12 = 4$
$12 - 9 = 3$

These differences are *not* the same; we need not find the remaining differences

The sequence is *not* an AP.

c. $(5x - 3), (7x - 4), (9x - 5), \ldots$

$(9x - 5) - (7x - 4) = 2x - 1$
$(7x - 4) - (5x - 3) = 2x - 1$

The differences are the same $(2x - 1)$

The sequence *is* an AP. ■

In general, an arithmetic progression with first term a_1 and common difference d has the following terms:

$$a_1 = a_1$$
$$a_2 = a_1 + 1d$$
$$a_3 = (a_1 + d) + d = a_1 + 2d$$
$$a_4 = (a_1 + 2d) + d = a_1 + 3d$$
$$\vdots$$
$$a_n = ?$$

We can discover a formula for finding a_n by examining a few terms of an arithmetic progression:

Term number:	1	2	3	4	…	n
Term:	a_1,	$a_1 + 1d$,	$a_1 + 2d$,	$a_1 + 3d$,	…,	$a_1 + (n - 1)d$

(The coefficient of d is always 1 less than the term number.) Therefore,

THE nth TERM OF AN AP

$$a_n = a_1 + (n - 1)d$$

Example 4 Find the twenty-first term of the AP having $a_1 = 23$ and $d = -2$.
Solution

$$a_n = a_1 + (n - 1)d$$

$$a_{21} = 23 + (21 - 1)(-2) \quad \text{Substituting 23 for } a_1\text{, 21 for } n\text{, and } -2 \text{ for } d$$

$$a_{21} = 23 + (-40) = -17$$

Therefore, the twenty-first term is -17. ■

Example 5 Given an AP with $a_7 = -10$ and $a_{12} = 5$, find a_1 and d.
Solution When $n = 7$, $a_n = -10$; therefore, $-10 = a_1 + 6d$. When $n = 12$, $a_n = 5$; therefore, $5 = a_1 + 11d$. We now multiply both sides of the first equation by -1 and then add the two equations:

$$\begin{aligned} a_1 + 6d = -10 \Rightarrow -a_1 - 6d &= 10 \\ a_1 + 11d &= 5 \\ \hline 5d &= 15 \\ d &= 3 \end{aligned}$$

To find a_1, we substitute 3 for d in the first equation:

$$a_1 + 6d = -10$$

$$a_1 + 6(3) = -10$$

$$a_1 = -28$$

Therefore, $a_1 = -28$ and $d = 3$. ■

Arithmetic Series

An **arithmetic series** is the sum of the terms of an arithmetic progression. An infinite arithmetic series can be written

$$a_1 + (a_1 + d) + (a_1 + 2d) + \cdots + a_n + \cdots$$

For a finite arithmetic series of n terms, the *sum* of those terms, S_n, is

$$S_n = a_1 + (a_1 + d) + (a_1 + 2d) + \cdots + (a_n - 2d) + (a_n - d) + a_n$$

Subtract d from one term to get the *preceding* term

Example 6 Find the sum of the first 100 integers.
Solution The story is told that the famous German mathematician, Carl Friedrich Gauss, at the age of ten very quickly solved this problem when it was first presented in his arithmetic class.

We are to find

$$\sum_{i=1}^{100} i = S_{100} = \underset{\text{1st term}}{1} + \underset{\text{2nd term}}{2} + \underset{\text{3rd term}}{3} + \cdots + 98 + 99 + \underset{\text{100th term}}{100}$$

If we choose to, we can, of course, write the equation as follows:

$$S_{100} = 100 + 99 + 98 + \cdots + 3 + 2 + 1$$

Now let us add the two equations together, adding the right sides term by term. We then have

$$S_{100} + S_{100} = \underbrace{\overset{\substack{\text{1st}\\\text{term}\\\downarrow}}{101} + \overset{\substack{\text{2nd}\\\text{term}\\\downarrow}}{101} + \overset{\substack{\text{3rd}\\\text{term}\\\downarrow}}{101} + \cdots + 101 + 101 + \overset{\substack{\text{100th}\\\text{term}\\\downarrow}}{101}}_{\text{100 terms}}$$

Therefore, $2S_{100} = 101(100)$, and

$$\sum_{i=1}^{100} i = S_{100} = \frac{101(100)}{2} = 5{,}050 \quad \blacksquare$$

Using the same reasoning as in Example 6, let's attempt to find a general formula for the sum of an arithmetic series. In the general case we have

$$\begin{aligned}
&(1)\quad S_n = \overset{\substack{\text{1st}\\\text{term}\\\downarrow}}{a_1} + \overset{\substack{\text{2nd}\\\text{term}\\\downarrow}}{(a_1 + d)} + \overset{\substack{\text{3rd}\\\text{term}\\\downarrow}}{(a_1 + 2d)} + \cdots + \overset{\substack{(n-2)\text{nd}\\\text{term}\\\downarrow}}{(a_n - 2d)} + \overset{\substack{(n-1)\text{st}\\\text{term}\\\downarrow}}{(a_n - d)} + \overset{\substack{n\text{th}\\\text{term}\\\downarrow}}{a_n}\\
&(2)\quad S_n = a_n + (a_n - d) + (a_n - 2d) + \cdots + (a_1 + 2d) + (a_1 + d) + a_1
\end{aligned}$$

(Observe that the terms on the right side of Equation 2 are identical to the terms on the right side of Equation 1; they are just in the reverse order.) Adding the two equations term by term, we have

$$2S_n = \underbrace{(a_1 + a_n) + (a_1 + a_n) + (a_1 + a_n) + \cdots + (a_1 + a_n) + (a_1 + a_n) + (a_1 + a_n)}_{n \text{ terms}}$$

The right side of the last equation has n terms of $(a_1 + a_n)$; therefore,

$$2S_n = n(a_1 + a_n)$$

or

> THE SUM OF THE FIRST n TERMS OF AN AP
>
> $$S_n = \frac{n(a_1 + a_n)}{2}$$

You should verify that the sum could also be written $S_n = \dfrac{n[2a_1 + (n - 1)d]}{2}$.

We recommend that you find $\sum_{i=1}^{100} i$ by using these formulas and compare your answers with the answer for Example 6.

Note that for $\sum_{i=1}^{100} i$, if we try to find n by subtracting 1 (the lower limit of the sum) from 100 (the upper limit of the sum), we get 99. Yet if we expand the sum to $1 + 2 + 3 + \cdots + 100$, we see that there are 100 terms, not 99 terms; therefore, n is not 99, it is 100. To use the lower and upper limits of the sum to find n, we have to *add 1* to the difference between those two limits (see Example 7).

Example 7 Find $\sum_{i=7}^{92} 4i$ by using the formula for S_n.

Solution The first term, a_1, is 4(7), or 28, and the last term, a_n, is 4(92), or 368; n is (92 − 7 + 1), or 86. Substituting into the formula $S_n = \frac{n(a_1 + a_n)}{2}$, we have

$$\sum_{i=7}^{92} 4i = S_{86} = \frac{86(28 + 368)}{2} = 43(396) = 17{,}028 \quad \blacksquare$$

Example 8 Given an AP having $a_1 = -8$, $a_n = 20$, and $S_n = 30$, find d and n.

Solution When we substitute the given values for a_1, a_n, and S_n into the formula for S_n, we will be able to solve for n.

$$(1) \quad S_n = \frac{n(a_1 + a_n)}{2} \Rightarrow 30 = \frac{n(-8 + 20)}{2} \Rightarrow 30 = \frac{n(12)}{2} \Rightarrow 30 = 6n$$

$$5 = n$$

$$(2) \quad a_n = a_1 + (n - 1)d \Rightarrow 20 = -8 + (n - 1)d$$

We substitute 5 for n in Equation 2:

$$20 = -8 + (n - 1)d$$
$$20 = -8 + (5 - 1)d$$
$$20 = -8 + 4d$$
$$28 = 4d$$
$$7 = d$$

Therefore, $n = 5$ and $d = 7$. ■

EXERCISES 12.2

Set I In Exercises 1–8, determine whether each sequence is an AP. If it is, find the common difference.

1. 3, 8, 13, 18

2. 7, 11, 15, 19

3. 7, 4, 1, −2, . . .

4. 9, 4, −1, −6, . . .

5. 4, $5\frac{1}{2}$, 7, 9

6. 3, $4\frac{1}{4}$, $5\frac{1}{2}$, $6\frac{1}{2}$

7. $2x - 1, x, 1, -x + 2$

8. $3 - 2x, 2 - x, 1, x$

9. Write the first four terms of the AP for which $a_1 = 5$ and $d = -7$.

10. Write the first five terms of the AP for which $a_1 = 4$ and $d = -5$.

11. Write the AP with five terms for which $a_1 = 7$ and $a_5 = 31$.

12. Write the AP with six terms for which $a_1 = 6$ and $a_6 = 51$.

13. Write the thirty-first term of this AP: −8, −2, 4, ...

14. Write the forty-first term of this AP: −5, −1, 3, ...

15. Write the eleventh term of this AP: $x, 2x + 1, 3x + 2, \ldots$.

16. Write the ninth term of this AP: $2z + 1, 3z, 4z - 1, \ldots$.

17. Find the sum of the even integers from 2 to 100, inclusive.

18. Find the sum of the odd integers from 1 to 99, inclusive.

19. Find $\sum_{i=1}^{100} 2i$ by using the formula for S_n.

20. Find $\sum_{i=1}^{50} (2i - 1)$ by using the formula for S_n.

21. Find $\sum_{i=3}^{8} 5i$ by using the formula for S_n.

22. Find $\sum_{j=5}^{10} 6j$ by using the formula for S_n.

23. Given an AP with $a_6 = 15$ and $a_{12} = 39$, find a_1 and d.

24. Given an AP with $a_5 = 12$ and $a_{14} = 57$, find a_1 and d.

25. Given an AP with $a_1 = -5$, $d = 3$, and $a_n = 16$, find n and S_n.

26. Given an AP with $a_1 = -7$, $d = 4$, and $a_n = 25$, find n and S_n.

27. Given an AP with $a_1 = 5$, $a_n = 17$, and $S_n = 44$, find d and n.

28. Given an AP with $a_1 = 3$, $a_n = 42$, and $S_n = 180$, find d and n.

29. Given an AP with $d = \frac{3}{2}$, $n = 9$, and $S_n = -\frac{9}{4}$, find a_1 and a_n.

30. Given an AP with $d = \frac{3}{4}$, $n = 7$, and $S_n = \frac{21}{4}$, find a_1 and a_n.

31. A rock dislodged by a mountain climber falls approximately 16 ft during the first second, 48 ft during the second second, 80 ft during the third second, and so on. Find the distance it falls during the tenth second and the total distance it falls during the first 12 sec.

32. A college student's young son saves 10¢ on the first of May, 12¢ on the second, 14¢ on the third, and so on. If he continues saving in this manner, how much does he save during the month of May?

Set II

In Exercises 1–8, determine whether each sequence is an AP. If it is, find the common difference.

1. $-41, -24, -7, 10$

2. $13, 5, -3, -11, \ldots$

3. $-1, \frac{1}{2}, 1, 2\frac{1}{2}$

4. $-2 + x, -1 - x, -3x, 1 - 5x, \ldots$

5. $8, 4, 2, 1, \ldots$

6. $17, 14, 11, 8, \ldots$

7. $-9, -5, -1, 3, \ldots$

8. $\frac{13}{16}, \frac{1}{2}, \frac{3}{16}, -\frac{1}{8}, \ldots$

9. Write the first four terms of the AP for which $a_1 = -17$ and $d = -3$.

10. Write the first four terms of the AP for which $a_1 = 19$ and $d = -24$.

11. Write the first four terms of the AP for which $a_1 = 83$ and $a_6 = 48$.

12. Write the first five terms of the AP for which $a_1 = 7$ and $a_7 = -11$.

13. Write the twenty-ninth term of this AP: $-26, -22, -18, \ldots$.

14. Write the eighty-third term of this AP: $28, 34, 40, \ldots$.

15. Write the seventh term of this AP: $-x, x + 3, 3x + 6, \ldots$.

16. Write the eighth term of this AP: $-5x + 1, -3x - 1, -x - 3, \ldots$.

17. Find the sum of the even integers from 100 to 200, inclusive.

18. Find the sum of the odd integers from 101 to 201, inclusive.

19. Find $\sum_{i=1}^{40} 3i$ by using the formula for S_n.

20. Find $\sum_{i=1}^{35} (2i + 1)$ by using the formula for S_n.

21. Find $\sum_{i=7}^{22} 2i$ by using the formula for S_n.

22. Find $\sum_{k=2}^{13} 8k$ by using the formula for S_n.

23. Given an AP with $a_5 = 46$ and $a_9 = 74$, find a_1 and d.

24. Given an AP with $a_7 = 11$ and $a_{13} = 29$, find a_1 and d.

25. Given an AP with $a_1 = 18$, $d = -3$, and $a_n = -6$, find n and S_n.

26. Given an AP with $a_1 = -37$, $d = 4$, and $a_n = 7$, find n and S_n.

27. Given an AP with $a_1 = 13$, $a_n = 49$, and $S_n = 217$, find d and n.

28. Given an AP with $a_1 = 17$, $d = -12$, and $a_n = -103$, find n and S_n.

29. Given an AP with $a_1 = 7$, $a_n = -83$, and $S_n = -722$, find d and n.

30. Given an AP with $d = \frac{2}{3}$, $n = 15$, and $S_n = 35$, find a_1 and a_n.

31. If we put one penny on the first square of a chessboard, three pennies on the second square, five pennies on the third square, and continue in this way until all the squares are covered, how much money will there be on the board? Note that a chessboard has 64 squares.

32. Jason saves \$1 the first week of the year, \$2 the second week of the year, \$3 the third week of the year, and so on. If he continues saving in this manner, how much does he save during one year?

12.3 Geometric Sequences and Series

Geometric Sequences

A **geometric sequence (geometric progression)** is a sequence in which each term after the first is found by *multiplying* the preceding term by the same fixed number, called the *common ratio, r. Geometric progression* is abbreviated GP.

Example 1 Write the first four terms of the geometric progression having first term $a_1 = 5$ and common ratio $r = 2$.

Solution

$a_1 = 5$

$a_2 = 5(2) = 10$ Multiplying the first term (5) by the common ratio (2)

$a_3 = 10(2) = 20$ Multiplying the second term (10) by the common ratio (2)

$a_4 = 20(2) = 40$ Multiplying the third term (20) by the common ratio (2)

Therefore, the first four terms of the GP are 5, 10, 20, and 40. ■

Example 2 Determine which of the sequences below are geometric progressions.

Method *Divide* each term by the preceding term. If every ratio found this way is the same, the sequence is a GP.

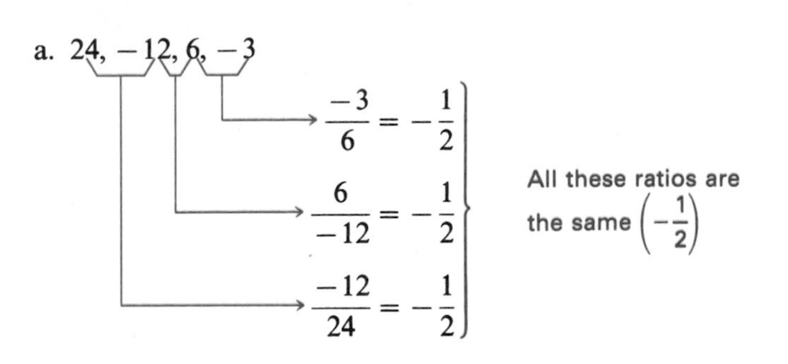

The sequence *is* a GP.

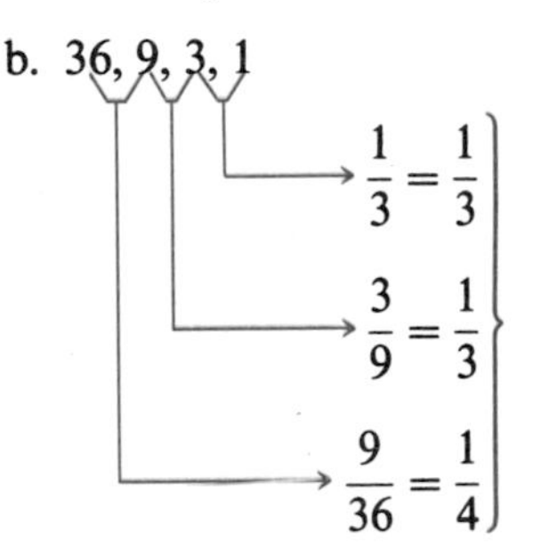

These ratios are *not* all the same

The sequence is *not* a GP.

c. $-\frac{3x}{y^2}, \frac{9x^2}{y}, -27x^3, \ldots$

$$-27x^3 \div \frac{9x^2}{y} = \frac{-27x^3}{1} \cdot \frac{y}{9x^2} = -3xy$$

$$\frac{9x^2}{y} \div \left(-\frac{3x}{y^2}\right) = \frac{9x^2}{y} \cdot \left(-\frac{y^2}{3x}\right) = -3xy$$

These ratios are the same $(-3xy)$

The sequence *is* a *GP*. ■

In general, a geometric progression with first term a_1 and common ratio r has the following terms:

$$a_1 = a_1$$
$$a_2 = a_1 r$$
$$a_3 = (a_1 r)r = a_1 r^2$$
$$a_4 = (a_1 r^2)r = a_1 r^3$$
$$\vdots$$
$$a_n = ?$$

We can discover a formula for finding a_n by examining a few terms of a geometric progression:

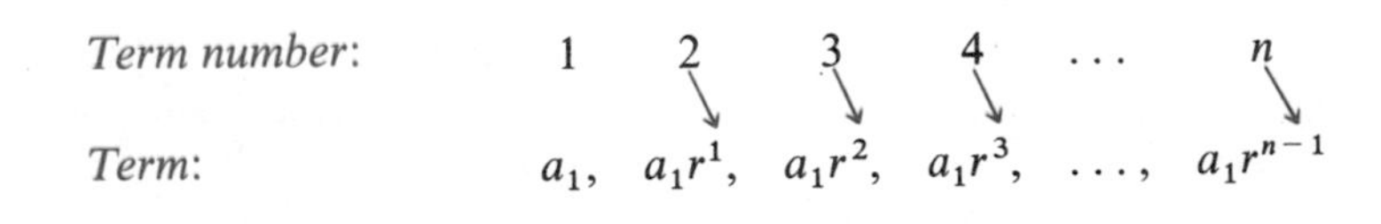

(The exponent of r is always 1 less than the term number.) Therefore,

> THE nth TERM OF A GP
>
> $$a_n = a_1 r^{n-1}$$

Example 3 Find the fifth term of the GP having $a_1 = 18$ and $r = -\dfrac{1}{3}$.

Solution

$$a_n = a_1 r^{n-1}$$

$$a_5 = 18\left(-\frac{1}{3}\right)^4 \qquad \text{Substituting 5 for } n\text{, 18 for } a_1\text{, and } -\tfrac{1}{3} \text{ for } r$$

$$a_5 = \frac{18}{1} \cdot \frac{1}{81} = \frac{2}{9}$$

Therefore, the fifth term is $\dfrac{2}{9}$. ■

Example 4 Given a GP with $a_2 = 12$ and $a_5 = 96$, find a_1 and r.

Solution $a_n = a_1 r^{n-1}$

$$\left.\begin{array}{ll}(1) & a_5 = 96 = a_1 r^4 \\ (2) & a_2 = 12 = a_1 r^1\end{array}\right\} \text{ Therefore, } \frac{96}{12} = \frac{a_1 r^4}{a_1 r} \qquad \text{Dividing Equation 1 by Equation 2}$$

$$8 = r^3$$

$$2 = r$$

To find a_1, we substitute 2 for r in Equation 2: $12 = a_1 r^1$

$$12 = a_1(2)$$

$$6 = a_1$$

Therefore, $a_1 = 6$ and $r = 2$. ■

Geometric Series

A **geometric series** is the sum of the terms of a geometric progression. An infinite geometric series can be written

$$\sum_{i=1}^{\infty} a_1 r^{i-1} = a_1 + a_1 r^1 + a_1 r^2 + \cdots + a_1 r^{n-1} + \cdots$$

For a finite geometric series of n terms, the *sum* of those terms, S_n, is

$$(1) \quad \sum_{i=1}^{n} a_1 r^{i-1} = S_n = a_1 + a_1 r^1 + a_1 r^2 + \cdots + a_1 r^{n-1}$$

Another formula for S_n can be found as follows: Multiply both sides of Equation 1 by r; then subtract the resulting equation from Equation 1.

$$
\begin{array}{lrl}
(1) & S_n = & a_1 + a_1r^1 + a_1r^2 + \cdots + a_1r^{n-1} \\
(2) & rS_n = & \qquad a_1r^1 + a_1r^2 + \cdots + a_1r^{n-1} + a_1r^n \\
\hline
 & S_n - rS_n = & a_1 \qquad\qquad\qquad\qquad\qquad\qquad - a_1r^n
\end{array}
$$

Multiplying both sides of Equation 1 by r

Subtracting Equation 2 from Equation 1

$$(1-r)S_n = a_1(1-r^n) \qquad \text{Factoring both sides}$$

$$S_n = \frac{a_1(1-r^n)}{1-r} \qquad \text{Dividing both sides by } 1-r$$

Therefore,

THE SUM OF THE FIRST n TERMS OF A GP

$$S_n = \frac{a_1(1-r^n)}{1-r}, \quad r \neq 1$$

Example 5 Evaluate $\sum_{i=1}^{6} 24\left(\frac{1}{2}\right)^{i-1}$.

Solution You should write out the first three terms of this series; you'll see that it is a geometric series, that the first term, a_1, is 24, and that r is $\frac{1}{2}$; n is 6. Substituting into the formula

$$S_n = \frac{a_1(1-r^n)}{1-r}$$

we have

$$S_6 = \frac{24\left[1-\left(\frac{1}{2}\right)^6\right]}{1-\frac{1}{2}} = \frac{24\left[1-\left(\frac{1}{64}\right)\right]}{\frac{1}{2}} = \frac{24\left[\frac{63}{64}\right]}{\frac{1}{2}} = \frac{189}{4}$$

■

You might verify that the answer for Example 5 is correct by expanding the series and finding its sum arithmetically.

Example 6 Given a GP having $r = -2$, $a_n = 80$, and $S_n = 55$, find a_1 and n.

Solution We will need to use the formula for a_n *and* the formula for S_n. Let's start with the formula for S_n, substituting 55 for S_n and -2 for r.

$$S_n = \frac{a_1(1-r^n)}{1-r}$$

$$55 = \frac{a_1(1-[-2]^n)}{1-[-2]}$$

$$(1) \qquad 55 = \frac{a_1 - a_1[-2]^n}{3}$$

To find a_1, we need to know what $a_1[-2]^n$ equals. Let's turn to the formula for a_n, substituting 80 for a_n and -2 for r.

$$a_n = a_1r^{n-1}$$

$$(2) \qquad 80 = a_1(-2)^{n-1}$$

If we multiply $(-2)^{n-1}$ by -2, we will have $(-2)^n$, which is what we need to use in Equation 1. Therefore, we will multiply both sides of Equation 2 by -2:

$$(-2)(80) = a_1(-2)^{n-1}(-2)$$

$$(3) \qquad -160 = a_1(-2)^n$$

We will now substitute -160 for $a_1(-2)^n$ in Equation 1:

$$55 = \frac{a_1 - [-160]}{3}$$

$$165 = a_1 + 160$$

$$a_1 = 5$$

To find n, we substitute 5 for a_1 in Equation 3:

$$-160 = 5(-2)^n$$

$$-32 = (-2)^n$$

$$(-2)^5 = (-2)^n$$

$$5 = n$$

Therefore, $a_1 = 5$ and $n = 5$. ■

EXERCISES 12.3

Set I In Exercises 1–8, determine whether each sequence is a GP. If it is, find the common ratio.

1. 4, 12, 36, 108

2. 7, 14, 28, 56

3. $-5, 15, -45, 135, \ldots$

4. $-6, 24, -96, 384, \ldots$

5. $2, \frac{1}{2}, \frac{1}{8}, \frac{1}{16}$

6. $3, \frac{1}{2}, \frac{1}{8}, \frac{1}{48}$

7. $5x, 10xy, 20xy^2, 40xy^3, \ldots$

8. $4xz, 12xz^2, 36xz^3, 108xz^4, \ldots$

9. Write the first five terms of the GP for which $a_1 = 12$ and $r = \frac{1}{3}$.

10. Write the first four terms of the GP for which $a_1 = 8$ and $r = \frac{3}{2}$.

11. Write the GP with five terms for which $a_1 = \frac{2}{3}$ and $a_5 = 54$. You will get two answers.

12. Write the GP with five terms for which $a_1 = \frac{3}{25}$ and $a_5 = 75$. You will get two answers.

13. Write the seventh term of this GP: $-9, -6, -4, \ldots$.

14. Write the eighth term of this GP: $-12, -18, -27, \ldots$.

15. Write the eighth term of this GP: $16x, 8xy, 4xy^2, \ldots$.

16. Write the seventh term of this GP: $27y, 9x^2y, 3x^4y, \ldots$.

17. Evaluate $\sum_{i=1}^{5} 3\left(\frac{1}{3}\right)^{i-1}$.

18. Evaluate $\sum_{k=1}^{7} 5\left(\frac{1}{2}\right)^{k-1}$.

19. Given a GP with $a_5 = 80$ and $r = \frac{2}{3}$, find a_1 and S_5.

20. Given a GP with $a_7 = 320$ and $r = 2$, find a_1 and S_5.

21. Given a GP with $a_3 = 28$ and $a_5 = \frac{112}{9}$, find a_1, r, and S_5. You will get two answers.

22. Given a GP with $a_2 = 384$ and $a_4 = 24$, find a_1, r, and S_4. You will get two answers.

23. Given a GP with $r = \frac{1}{2}$, $a_n = 3$, and $S_n = 189$, find a_1 and n.

24. Given a GP with $r = \frac{1}{3}$, $a_n = 4$, and $S_n = 160$, find a_1 and n.

25. A woman invested a certain amount of money that earned $1\frac{1}{5}$ times as much in the second year as in the first year, $1\frac{1}{5}$ times as much in the third year as in the second year, and so on. If the investment earned $22,750 in the first 3 years, how much would it earn in the fifth year?

26. A man invested a certain amount of money that earned $1\frac{1}{4}$ times as much in the second year as in the first year, $1\frac{1}{4}$ times as much in the third year as in the second year, and so on. If the investment earned $9,760 in the first 3 years, how much would it earn in the fifth year?

27. Suppose you took a job that pays 1¢ the first day, 2¢ the second day, and 4¢ the third day, with the pay continuing to increase in this manner for a month of 31 days.

a. How much would you make on the tenth day?

b. How much would you make on the thirty-first day?

c. What would be your total earnings for the month?

Set II In Exercises 1–8, determine whether each sequence is a GP. If it is, find the common ratio.

1. $-11, -44, -176, -704$

2. $-6, 30, -150, 750, \ldots$

3. $1, -\frac{1}{2}, -\frac{1}{4}, -\frac{1}{8}$

4. $20ab, -5a^3b, \frac{5}{4}a^5b, -\frac{5}{16}a^7b, \ldots$

5. $36, 18, 9, \frac{9}{2}, \ldots$

6. $-16, -8, 0, 8, \ldots$

7. $12x^5, 4x^3, \frac{4}{3}x, \frac{4}{9x}, \ldots$

8. $2x^4, -8x^3, 16x^2, -32x, \ldots$

9. Write the first five terms of the GP for which $a_1 = 4$ and $r = 3$.

10. Write the first four terms of the GP for which $a_1 = -9$ and $r = -\frac{1}{3}$.

11. Write the GP with five terms for which $a_1 = -\frac{3}{4}$ and $a_5 = -12$. You will get two answers.

12. Write the GP with five terms for which $a_1 = \frac{1}{4}$ and $a_5 = 4$. You will get two answers.

13. Write the seventh term of this GP: $-\frac{25}{54}, \frac{5}{18}, -\frac{1}{6}, \ldots$.

14. Write the ninth term of this GP: $-16, -8, -4, \ldots$.

15. Write the eighth term of this GP: $\frac{24}{hk}, -\frac{12k}{h}, \frac{6k^3}{h}, \ldots$.

16. Write the seventh term of this GP: $-32, -8x, -2x^2, \ldots$.

17. Evaluate $\sum_{i=1}^{4} 8\left(\frac{1}{5}\right)^{i-1}$.

18. Evaluate $\sum_{k=1}^{8} 2\left(\frac{1}{10}\right)^{k-1}$.

19. Given a GP with $a_5 = 40$ and $r = -\frac{2}{3}$, find a_1 and S_5.

20. Given a GP with $a_4 = -\frac{10}{3}$ and $r = -\frac{1}{3}$, find a_1 and S_5.

21. Given a GP with $a_3 = 16$ and $a_5 = 9$, find a_1, r, and S_5. You will get two answers.

22. Given a GP with $a_7 = 192$ and $a_4 = 24$, find a_1, r, and S_5.

23. Given a GP with $a_n = 972$, $r = 3$, and $S_n = 1{,}456$, find a_1 and n.

24. Given a GP with $r = -\frac{1}{2}$, $a_n = 5$, and $S_n = 55$, find a_1 and n.

25. If it takes 1 sec for a certain type of microbe to split into two microbes, how long will it take a colony of 1,500 such microbes to exceed 6 million?

26. If we put one penny on the first square of a chessboard, two pennies on the second square, four pennies on the third square, and continue in this way until all squares are covered, how much money will there be on the board? Recall that a chessboard has 64 squares.

27. Suppose you took a job that pays 1¢ the first day, 3¢ the second day, and 9¢ the third day, with the pay continuing to increase in this manner for a month of 31 days.

a. How much would you make on the eighth day?

b. How much would you make on the thirty-first day?

c. What would be your total earnings for the month?

12.4 Infinite Geometric Series

Consider the formula derived in Section 12.3 for the sum of n terms of a geometric series:

$$S_n = \frac{a_1(1 - r^n)}{1 - r}, \quad r \neq 1$$

If $|r| < 1$, then r^n gets smaller and smaller as n gets larger. For example, suppose $r = \frac{1}{2}$. Then

$$\left(\frac{1}{2}\right)^1 = \frac{1}{2} = 0.5$$
$$\left(\frac{1}{2}\right)^2 = \frac{1}{4} = 0.25$$
$$\left(\frac{1}{2}\right)^3 = \frac{1}{8} = 0.125$$
$$\vdots$$
$$\left(\frac{1}{2}\right)^{10} = \frac{1}{1{,}024} \approx 0.001$$
$$\vdots$$
$$\left(\frac{1}{2}\right)^{20} \approx 0.000001$$
$$\vdots$$
$$\left(\frac{1}{2}\right)^{100} \approx 8 \times 10^{-31} \approx 0$$

$\left(\frac{1}{2}\right)^n$ gets closer and closer to zero as n gets larger and larger

Therefore, $S_{100} \approx \frac{1}{2} + \frac{1}{4} + \frac{1}{8} + \cdots + \frac{1}{1{,}024} + \cdots + \underset{\left(\frac{1}{2}\right)^{98}}{0} + \underset{\left(\frac{1}{2}\right)^{99}}{0} + \underset{\left(\frac{1}{2}\right)^{100}}{0}$. That is, when $|r| < 1$, r^n is so close to zero that it contributes essentially nothing to the sum S_n when n is large.

Therefore, the formula for S_n becomes

This term contributes essentially nothing when n becomes infinitely large if $|r| < 1$

$$S_n = \frac{a_1(1 - r^n)}{1 - r} \approx \frac{a_1}{1 - r} \quad \text{if } |r| < 1 \text{ and } n \text{ is large}$$

The symbol S_∞ represents S_n when n becomes infinitely large. Therefore,

THE SUM OF AN INFINITE GEOMETRIC SERIES

$$S_\infty = \sum_{i=1}^{\infty} a_1 r^{i-1} = \frac{a_1}{1 - r} \quad \text{if} \quad |r| < 1$$

Notice that we use the *equal* sign in the formula for S_∞, not just the *approximates* sign ($\approx$).

Example 1 Evaluate the infinite geometric series $1 + \frac{1}{2} + \frac{1}{4} + \cdots$.

Solution

$$1, \frac{1}{2}, \frac{1}{4}, \ldots$$

$$\frac{1}{2} \div 1 = \frac{1}{2} = r$$

$$S_\infty = \frac{a_1}{1 - r} = \frac{1}{1 - \frac{1}{2}} = \frac{1}{\frac{1}{2}} = 2$$

Therefore, the sum is 2. ■

Example 2 Evaluate $\sum_{i=1}^{\infty} 6\left(-\frac{2}{3}\right)^{i-1}$.

Solution r is $-\frac{2}{3}$, which is between -1 and 1, and a_1 is 6. Therefore, we have

$$S_\infty = \frac{a_1}{1 - r} = \frac{6}{1 - \left(-\frac{2}{3}\right)} = \frac{6}{1 + \left(+\frac{2}{3}\right)} = \frac{6}{\frac{5}{3}} = \frac{18}{5}$$

Therefore, the sum is $\frac{18}{5}$. ■

Example 3 Write the repeating decimal $0.\overline{25}$ as a fraction.

Solution $0.\overline{25} = 0.252525\ldots = 0.25 + 0.0025 + 0.000025 + \cdots$

$$\frac{0.0025}{0.25} = 0.01 = r$$

$$S_\infty = \frac{a_1}{1 - r} = \frac{0.25}{1 - 0.01} = \frac{0.25}{0.99} = \frac{25}{99}$$

Therefore, $0.\overline{25} = \frac{25}{99}$. ■

The reason that $|r|$ must be less than 1 is that if $|r| > 1$, the absolute values of the succeeding terms in the geometric series become larger and larger. Therefore, S_∞ would be infinitely large.

EXERCISES 12.4

Set I In Exercises 1–6, find the sum of each geometric series.

1. $3 + 1 + \frac{1}{3} + \cdots$

2. $9 - 1 + \frac{1}{9} - \cdots$

3. $\frac{4}{3} + 1 + \frac{3}{4} + \cdots$

4. $10^{-1} + 10^{-2} + 10^{-3} + \cdots$

5. $-6 - 4 - \frac{8}{3} - \cdots$

6. $-49 - 35 - 25 - \cdots$

In Exercises 7–12, write each repeating decimal as a fraction reduced to lowest terms.

7. $0.\overline{2}$ **8.** $0.\overline{21}$ **9.** $0.0\overline{54}$

10. $0.0\overline{39}$ **11.** $8.6\overline{4}$ **12.** $5.2\overline{6}$

13. Evaluate $\sum_{i=1}^{\infty} 5\left(-\frac{3}{4}\right)^{i-1}$. **14.** Evaluate $\sum_{i=1}^{\infty} 2\left(-\frac{1}{5}\right)^{i-1}$.

15. A rubber ball is dropped from a height of 9 ft. Each time it strikes the floor, it rebounds to a height that is two-thirds the height from which it last fell. Find the total distance the ball travels *vertically* before coming to rest.

16. A ball bearing is dropped from a height of 10 ft. Each time it strikes the metal floor, it rebounds to a height that is three-fifths the height from which it last fell. Find the total distance the bearing travels *vertically* before coming to rest.

17. The first swing of a pendulum is 12 in. Each succeeding swing is nine-tenths as long as the preceding one. Find the total distance traveled by the pendulum before it comes to rest.

Set II In Exercises 1–6, find the sum of each geometric series.

1. $5 - 1 + \frac{1}{5} - \cdots$ **2.** $10^{-2} + 10^{-4} + 10^{-6} + \cdots$

3. $\frac{6}{5} + 1 + \frac{5}{6} + \cdots$ **4.** $\frac{1}{4} + \frac{1}{8} + \frac{1}{16} + \cdots$

5. $2 - \frac{2}{3} + \frac{2}{9} - \cdots$ **6.** $4 + 2 + 1 + \cdots$

In Exercises 7–12, write each repeating decimal as a fraction reduced to lowest terms.

7. $0.\overline{26}$ **8.** $0.0\overline{143}$ **9.** $0.\overline{70}$

10. $0.\overline{12}$ **11.** $0.00\overline{136}$ **12.** $3.\overline{765}$

13. Evaluate $\sum_{i=1}^{\infty} 3\left(-\frac{1}{8}\right)^{i-1}$. **14.** Evaluate $\sum_{i=1}^{\infty} 3\left(\frac{1}{2}\right)^{i-1}$.

15. When Bob stops his car suddenly, the 3-ft radio antenna oscillates back and forth. Each swing is three-fourths as great as the previous one. If the initial travel of the antenna tip is 4 in., how far will the tip travel before it comes to rest?

16. If a rabbit moves 10 yd in the first second, 5 yd in the second second, and continues to move one-half as far in each succeeding second as he did in the preceding second:

a. How many yards does the rabbit travel before he comes to rest?

b. What is the total time the rabbit is moving?

17. The first swing of a pendulum is 10 in. Each succeeding swing is eight-ninths as long as the preceding one. Find the total distance traveled by the pendulum before it comes to rest.

12.5 Review: 12.1–12.4

Sequences 12.1 A **sequence** of numbers is a set of numbers in which the numbers follow a definite pattern and are arranged in a definite order. The numbers that make up the sequence are called the *terms* of the sequence. If, in counting the terms of a sequence, the counting comes to an end, the sequence is a *finite sequence*; if the counting never ends, the sequence is an *infinite sequence*. The usual notation for the general term of a sequence is a_n.

Series 12.1

A **series** is the indicated sum of a finite or infinite sequence of terms. It is a finite or an infinite series according to whether the number of terms is finite or infinite. The *partial sum of a series* is the sum of a finite number of consecutive terms of the series, beginning with the first term.

Sigma Notation 12.1

The symbol $\sum$ is used to indicate the sum of a series. **Sigma notation** for the series $f(1) + f(2) + f(3) + \cdots + f(n)$ is $\sum_{i=1}^{n} f(i)$, and for the infinite series $g(1) + g(2) + g(3) + \cdots$, it is $\sum_{k=1}^{\infty} g(k)$.

Arithmetic Progressions 12.2

An **AP** is a sequence in which each term after the first is found by *adding* the same fixed number to the preceding term. The fixed number added is called the *common difference, d*; *d* can be found by subtracting any term from the term that follows it.

The nth term of an AP: $a_n = a_1 + (n - 1)d$

The sum of n terms of an AP: $S_n = \frac{n(a_1 + a_n)}{2} = \frac{n[2a_1 + (n-1)d]}{2}$

Geometric Progressions 12.3

A **GP** is a sequence in which each term after the first is found by *multiplying* the preceding term by the same fixed number, called the *common ratio, r*. The common ratio can be found by dividing any term by the term that precedes it.

The nth term of a GP: $a_n = a_1 r^{n-1}$

The sum of n terms of a GP: $S_n = \frac{a_1(1 - r^n)}{1 - r}, r \neq 1$

The Sum of an Infinite Geometric Series 12.4

$$S_\infty = \frac{a_1}{1 - r} \quad \text{if} \quad |r| < 1$$

Review Exercises 12.5 Set I

In Exercises 1–5, determine which of the sequences are arithmetic progressions, which are geometric progressions, and which are neither.

1. $7, 5, 3, \ldots$

2. $-2, -6, -18, \ldots$

3. $\frac{1}{2}, \frac{1}{4}, \frac{1}{6}, \ldots$

4. $\frac{3}{5}, -\frac{1}{5}, \frac{1}{15}, \ldots$

5. $3x - 2, 2x - 1, x, \ldots$

6. Write the first four terms of the AP for which $a_n = 2n - 6$. Then find d, a_{30}, and S_{30}.

7. Write the first three terms of the GP for which $a_n = \left(\frac{1}{2}\right)^n$. Then find S_5 and S_∞.

8. Given an AP with $a_1 = -5$, $a_n = 7$, and $S_n = 16$, find d and n.

9. Given an AP with $d = \frac{3}{2}$, $n = 7$, and $S_n = \frac{7}{2}$, find a_1 and a_n.

10. Given a GP with $a_3 = \frac{9}{4}$ and $r = -\frac{2}{3}$, find a_1 and S_6.

11. Given a GP with $a_3 = 8$ and $a_5 = \frac{32}{9}$, find a_1, r, and S_5. You will get two answers.

12. Write $3.2\overline{845}$ as a fraction reduced to lowest terms.

13. A rubber ball is dropped from a height of 8 ft. Each time it strikes the floor, it rebounds to a height that is three-fourths the height from which it last fell. Find the total distance the ball travels *vertically* before coming to rest.

14. Write the series $\sum_{i=2}^{6} 3\left(\frac{i+1}{11}\right)$ in expanded form.

15. Evaluate $\sum_{i=1}^{6} 2\left(\frac{1}{4}\right)^{i-1}$. (Use a calculator.)

Review Exercises 12.5 Set II

NAME

In Exercises 1–5, determine which of the sequences are arithmetic progressions, which are geometric progressions, and which are neither.

1. $23, 4, -15, \ldots$

2. $-5, -35, -245, \ldots$

3. $\frac{1}{15}, \frac{1}{20}, \frac{1}{25}, \ldots$

4. $\frac{10}{3}, -5, \frac{15}{2}, \ldots$

5. $6 - x, 3, x, \ldots$

6. Write the first four terms of the AP for which $a_n = 5n - 2$. Then find d, a_{25}, and S_{25}.

7. Write the first three terms of the GP for which $a_n = \left(\frac{2}{5}\right)^n$. Then find S_5 and S_∞.

8. Given an AP with $a_1 = 2$, $a_n = -1$, and $S_n = 2\frac{1}{2}$, find d and n.

ANSWERS

1. ______

2. ______

3. ______

4. ______

5. ______

6. ______

______

______

______

______

______

______

7. ______

______

______

______

______

8. ______

______

ANSWERS

9. ______

______

10. ______

______

11. ______

______

______

______

______

______

12. ______

13. ______

14. ______

15. ______

9. Given an AP with $d = -\frac{4}{3}$, $n = 7$, and $S_n = 7$, find a_1 and a_n.

10. Given a GP with $a_4 = 27$ and $r = -\frac{3}{2}$, find a_1 and S_6.

11. Given a GP with $a_3 = 18$ and $a_5 = 32$, find a_1, r, and S_5. You will get two answers.

12. Write $2.9\overline{048}$ as a fraction reduced to lowest terms.

13. A mine's output in the first month was \$10,000 in gold. If each succeeding monthly output is $\frac{9}{10}$ the previous month's output, what is the most this mine's total output can be?

14. Write the series $\sum_{i=1}^{4} 3\left(\frac{i^2}{2}\right)$ in expanded form.

15. Evaluate $\sum_{i=1}^{8} 5\left(\frac{1}{3}\right)^{i-1}$. (Use a calculator).

Chapter 12 Diagnostic Test

The purpose of this test is to see how well you understand sequences and series. We recommend that you work this diagnostic test *before* your instructor tests you on this chapter. Allow yourself about 50 minutes.

Complete solutions for all the problems on this test, together with section references, are given in the answer section at the end of the book. For the problems you do incorrectly, study the sections referred to.

1. Given $a_n = \dfrac{2n - 1}{n}$, find S_4.

2. Determine whether each of the following sequences is an AP, a GP, or neither. If an AP, give d. If a GP, give r.

a. $8, -20, 50, \ldots$

b. $\frac{1}{2}, \frac{3}{4}, 1, \frac{5}{4}, \frac{3}{2}, \ldots$

c. $2x - 1, 3x, 4x + 2, \ldots$

d. $\frac{c^4}{16}, -\frac{c^3}{8}, \frac{c^2}{4}, \ldots$

3. a. Write the first five terms of the AP for which $a_1 = x + 1$ and $d = x - 1$.

b. Write the AP with five terms for which $a_1 = 2$ and $a_5 = -2$.

c. Write the fifteenth term of this AP: $1 - 6h, 2 - 4h, 3 - 2h, \ldots$.

4. a. Given an AP with $a_3 = 1$ and $a_7 = 2$, find a_1 and d.

b. Given an AP with $a_1 = 10$, $a_n = -8$, and $S_n = 7$, find d and n.

5. Write the first five terms of the GP for which $a_1 = \dfrac{c^4}{16}$ and $r = -\dfrac{2}{c}$.

6. Given a GP with $a_2 = -18$ and $a_4 = -8$, find a_1, r, and S_4.

7. Given a GP with $r = -\frac{2}{3}$, $a_n = -16$, and $S_n = 26$, find a_1 and n.

8. Find the sum of $8, -4, 2, -1, \ldots$.

9. Write $3.0\overline{3}$ as a fraction reduced to lowest terms.

10. A ball bearing is dropped from a height of 6 ft. Each time it strikes the floor, it rebounds to a height that is two-thirds the height from which it last fell. Find the total distance the ball bearing travels *vertically* before it comes to rest.

Cumulative Review Exercises: Chapters 1–12

In Exercises 1 and 2, perform the indicated operations and write the answers using only positive exponents or no exponents.

1. $\left(\dfrac{x^3y^{-4}z^{-1}}{x^{-2}z^{-3}}\right)^{-2}$

2. $(32a^{-1/2} \cdot a^{1/3})^{-6/5}$

In Exercises 3 and 4, perform the indicated operations and write the answers in simplest radical form. (Assume $x > 0$.)

3. $\sqrt{27x^3y^4} + 3xy^2\sqrt{12x} - x^2y^2\sqrt{\dfrac{3}{x}}$

4. $\dfrac{15}{\sqrt{7} - \sqrt{2}}$

5. Perform the indicated operation and write the answer in the form $a + bi$: $\dfrac{6 - 5i}{6 + 5i}$.

6. Reduce $\dfrac{x^3 + 2x^2 - x - 2}{x^3 + 8}$ to lowest terms.

In Exercises 7–10, perform the indicated operations and write the answers in simplest form.

7. $\dfrac{x}{x - 3} - \dfrac{9}{x + 3} - \dfrac{1}{2x^2 - 18}$

8. $(x^4 + 3x^3 - 6x^2 - 2x + 1) \div (x^2 - x + 1)$

9. $\dfrac{x + 1}{3x^2 + 14x - 5} \div \dfrac{2x^2 - x - 3}{6x^2 - 11x + 3}$

10. $(x + 2)^5$

In Exercises 11–19, find the solution set for each equation or inequality and the solution for each system of equations.

11. $\dfrac{6 - x}{x^2 - 4} - \dfrac{x}{x + 2} = 2$

12. $\log(x + 3) + \log(x - 2) = \log 6$

13. $x = \sqrt{x + 9} + 3$

14. $x^2 + 2x - 15 \geq 0$

15. a. $\log_5 \frac{1}{25} = x$ b. $\log_b 4 = \frac{1}{2}$

16. $x^2 - 4x = 8$

17. $\begin{Bmatrix} 2x + 4y = 0 \\ 5x - 3y = 13 \end{Bmatrix}$

18. $\begin{Bmatrix} x + y + z = 1 \\ 2x + y - 2z = -4 \\ x + y + 2z = 3 \end{Bmatrix}$

19. $\left|\dfrac{x}{2} + 3\right| < 4$

20. Graph $y = x^2 - 2x - 3$.

21. Solve $\begin{Bmatrix} 2x - 5y \geq 10 \\ 5x + y < 5 \end{Bmatrix}$ graphically.

For Exercises 22 and 23, use the function $y = f(x) = \dfrac{2}{3x - 5}$.

22. a. Find the domain. b. Find $f(x + h)$.

23. Find the inverse function, $f^{-1}(x)$.

For Exercises 24 and 25, use the points $(3, -4)$ and $(1, 6)$.

24. a. Find the distance between the points.

b. Find the slope of the line through the points.

25. Write the equation of the line through the points.

26. Given an AP with $a_1 = -7$, $a_n = 11$, and $S_n = 10$, find d and n.

27. Given a GP with $a_4 = -1$ and $r = -\frac{2}{7}$, find a_1 and S_5. (Use a calculator.)

In Exercises 28–30, set up each problem algebraically, solve, and check.

28. Find three consecutive integers such that the product of the first two is 10 more than 4 times the third.

29. Coffee that costs \$3.50 per pound is to be mixed with coffee that costs \$4.25 per pound to obtain 30 lb of a blend worth \$4.00 per pound. How much of each type of coffee should be used?

30. Roberta and John leave an intersection at the same time; Roberta is jogging west, and John is jogging north. Roberta is jogging 3 mph faster than John. Twenty minutes later, they are 5 mi apart. What are their rates?

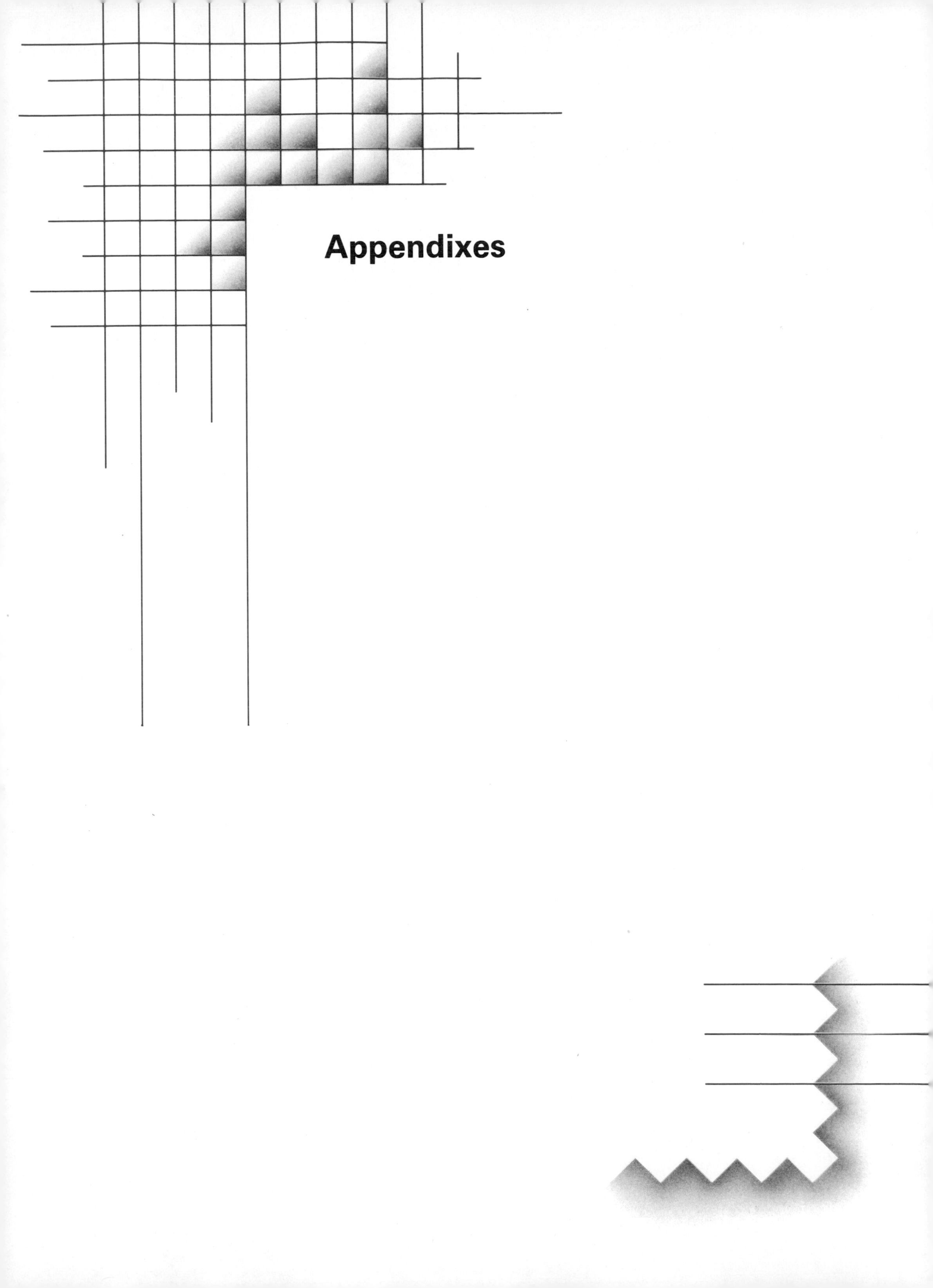

Appendixes

Appendix A Proofs

Property 1.7b $\dfrac{1}{x^{-n}} = x^n$ (page 55)

Proof $\dfrac{1}{x^{-n}} = \dfrac{1}{\frac{1}{x^n}} = 1 \div \dfrac{1}{x^n} = 1 \cdot \dfrac{x^n}{1} = x^n$ ■

Property 1.7c $\left(\dfrac{x}{y}\right)^{-n} = \left(\dfrac{y}{x}\right)^n$ (page 55)

Proof $\left(\dfrac{x}{y}\right)^{-n} = \dfrac{1}{\left(\frac{x}{y}\right)^n} = 1 \div \left(\dfrac{x}{y}\right)^n = 1 \div \dfrac{x^n}{y^n} = 1 \cdot \dfrac{y^n}{x^n} = \dfrac{y^n}{x^n} = \left(\dfrac{y}{x}\right)^n$ ■

Rule 2.1 If $a \geq 0$ and if $|N| = a$, then $N = a$ or $N = -a$, where N is any algebraic expression. (page 105)

Proof
Case I, $N \geq 0$.
If $N \geq 0$, $|N| = N$. Then $|N| = a \Rightarrow N = a$.

Case II, $N < 0$.
If $N < 0$, $|N| = -N$. Then $|N| = a \Rightarrow -N = a$. But if $-N = a$, $N = -a$.

The final answer is the *union* of the solution sets of cases I and II. Therefore, if $|N| = a$, then $N = a$ or $N = -a$. ■

Rule 2.2 If a is a positive real number,

1. if $|N| < a$, then $-a < N < a$, and
2. if $|N| \leq a$, then $-a \leq N \leq a$,

where N is any algebraic expression. (page 105)

Proof of 1
Case I, $N \geq 0$.
If $N \geq 0$, $|N| = N$. Therefore, $|N| < a \Rightarrow N < a$. We must find the intersection between $N \geq 0$ and $N < a$. Because $a > 0$, the intersection is $0 \leq N < a$. On the number line:

Case II, $N < 0$.
If $N < 0$, $|N| = -N$. Therefore, $|N| < a \Rightarrow -N < a$. If $-N < a$, then $N > -a$. We must find the intersection between $N < 0$ and $N > -a$. Because $a > 0$, the intersection is $-a < N < 0$. On the number line:

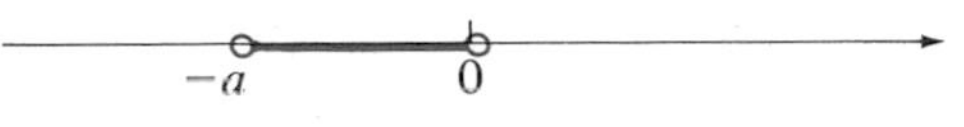

The final answer is the *union* of the solution sets of Cases I and II, which is $-a < N < a$. On the number line:

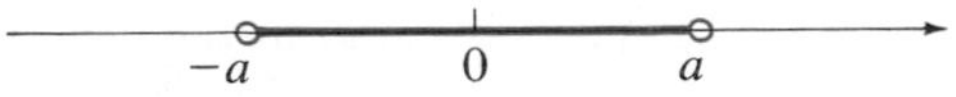

$-a$ $\quad$ 0 $\quad$ a ■

The proof of 2 is similar to the proof of 1 and will not be shown.

Rule 2.3 If a is a positive real number,

1. if $|N| > a$, then $N > a$ or $N < -a$, and
2. if $|N| \geq a$, then $N \geq a$ or $N \leq -a$,

where N is any algebraic expression. (page 107)

Proof of 1
Case I, $N \geq 0$.
If $N \geq 0$, $|N| = N$. Therefore, $|N| > a \Rightarrow N > a$. We must find the intersection between $N \geq 0$ and $N > a$. Since $a > 0$, the intersection is $N > a$. On the number line:

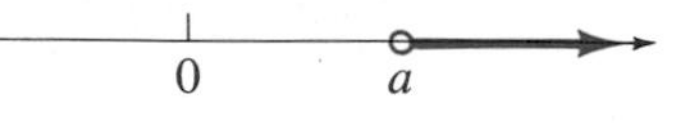

Case II, $N < 0$.
If $N < 0$, $|N| = -N$. Therefore, $|N| > a \Rightarrow -N > a$. If $-N > a$, then $N < -a$. We must find the intersection between $N < 0$ and $N < -a$. Because $a > 0$, the intersection is $N < -a$. On the number line:

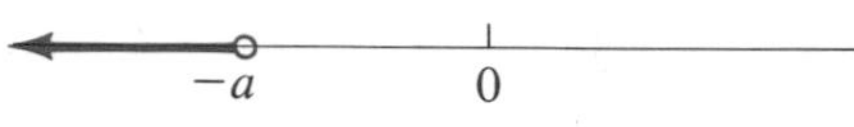

The final answer is the *union* between the solution sets for Cases I and II: $\{N > a\} \cup \{N < -a\}$ is $N > a$ or $N < -a$. On the number line:

■

The proof of 2 is similar to the proof of 1 and will not be shown.

The Remainder Theorem If $f(x)$ is divided by $(x - a)$, then the remainder equals $f(a)$. (page 424)

Proof Let us call the quotient $q(x)$ and the remainder R. Then if $f(x)$ is a polynomial of degree n, $q(x)$ is a polynomial of degree $n - 1$, and "$f(x) = (x - a)q(x) + R$" must be a true statement for all x. Let $x = a$. We then have

$$f(a) = (a - a)q(a) + R$$

But since $a - a = 0$, $f(a) = R$. Therefore, if $f(x)$ is divided by $x - a$, $R = f(a)$. ■

Axis of Symmetry The equation of the axis of symmetry of the parabola $y = ax^2 + bx + c$ is $x = -\frac{b}{2a}$. (page 513)

Proof For simplicity, let us assume that the parabola crosses the x-axis. The x-intercepts are $\frac{-b + \sqrt{b^2 - 4ac}}{2a}$ and $\frac{-b - \sqrt{b^2 - 4ac}}{2a}$. The axis of symmetry passes through a point whose x-coordinate is the average of these two values—that is, the x-coordinate of a point on the axis of symmetry is

$$\frac{\frac{-b + \sqrt{b^2 - 4ac}}{2a} + \frac{-b - \sqrt{b^2 - 4ac}}{2a}}{2} = \frac{\frac{-2b}{2a}}{2} = -\frac{b}{2a}$$

The equation of the vertical line that passes through the point $\left(-\frac{b}{2a}, 0\right)$ is $x = -\frac{b}{2a}$. ■

Appendix B The Use of Tables for Logarithms; Computations with Logarithms

B.1 Common Logarithms: The Characteristic

We first consider logarithms of powers of 10. *The logarithm of a number is the exponent to which the base must be raised to give that number.*

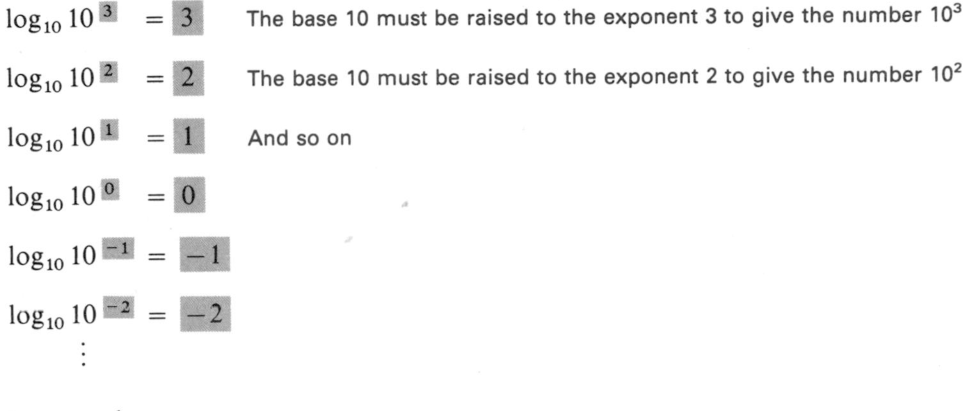

In general,

$$\log_{10} 10^k = k$$

The logarithm of a number that is *not* an integral power of 10 is made up of two parts:

1. *An integer part called the* **characteristic**. The characteristic of the logarithm of a number equals the exponent on the 10 when that number is written in scientific notation.
2. *A decimal part called the* **mantissa** (found in Table II, Appendix C). *The mantissas in the table are never negative.* We discuss finding the mantissa in Section B.2.

Finding the Characteristic

While the characteristic of the logarithm of a number is the exponent of 10 when the number is written in scientific notation, it is better for computational purposes to write *negative* characteristics as the *difference* of two numbers, as shown in Examples 1b, 1c, 1d, and 1f.

Example 1 Examples of writing the characteristic of the logarithm of a number:

Number	*Number in Scientific Notation*	*Characteristic*	*Written*
a. 76.3	7.63×10^1	1	1. ▇
b. 0.506	5.06×10^{-1}	-1	9. ▇ -10
c. 0.0932	9.32×10^{-2}	-2	8. ▇ -10
d. 0.00000479	4.79×10^{-6}	-6	4. ▇ -10
e. 1.83×10^4	1.83×10^4	4	4. ▇
f. 236×10^{-9}	2.36×10^{-7}	-7	3. ▇ -10

The mantissa goes here ↑ ■

The characteristic of the logarithm of a number that is written in standard decimal notation can also be found as follows:

If the number is greater than or equal to 1, the characteristic of its logarithm is *one less than* the *number of digits* to the *left* of the decimal point.

If the number is less than 1, the characteristic of its logarithm is negative, and the absolute value of the characteristic equals the *number of decimal places* between the actual decimal point and standard position for the decimal point.

EXERCISES B.1

Set I Write the characteristic of the logarithm of each number.

1. 386 **2.** 27 **3.** 5.67 **4.** 30.4

5. 0.516 **6.** 0.089 **7.** 93,000,000 **8.** 186,000

9. 0.0000806 **10.** 0.000777 **11.** 78,000 **12.** 1,400

13. 2.06×10^5 **14.** 3.55×10^4 **15.** 7.14×10^{-3} **16.** 8.96×10^{-5}

Set II Write the characteristic of the logarithm of each number.

1. 784 **2.** 8.99 **3.** 0.314 **4.** 0.000578

5. 2.56×10^4 **6.** 3.14×10^{-3} **7.** 7.0005 **8.** 0.000109

9. 3,480 **10.** 0.00437 **11.** 2,300,000 **12.** 8.32

13. 2.78×10^3 **14.** 8.62×10^6 **15.** 3.58×10^{-4} **16.** 9.51×10^{-2}

B.2 Common Logarithms: The Mantissa

In this section, we discuss common logarithms and the use of the Table of Common Logarithms (Table II, Appendix C). Table II gives the *mantissas* (the decimal parts) of the logarithms of the numbers from 1.00 to 9.99. Except for the first one (0.0000), the mantissas are all approximations. *The mantissas in the table are never negative.*

Tables of common logarithms come in different accuracies. Some tables have mantissas rounded off to four places, others to five places, and so on. Table II is a four-place table. If you use other than a four-place table to solve the problems in this chapter, you may get slightly different answers.

Finding the Mantissa

Hereafter, *when the notation* log *N is used, the base is understood to be* 10; that is, $\log N = \log_{10} N$.

Example 2 Find the mantissa for log 5.74. (Figure B.2.1 is part of Table II, Appendix C.)
Solution The number 5.74 is already in scientific notation, so we look up 5‸74 in the table. Notice that the *caret* shows us where the decimal point belongs.

The *first two digits* of 5.7 4

The *third digit* of 5.7 4

N	0	1	2	3	4	5	6	7	8	9
5‸5	0.7404	0.7412	0.7419	0.7427	0.7435	0.7443	0.7451	0.7459	0.7466	0.7474
5‸6	0.7482	0.7490	0.7497	0.7505	0.7513	0.7520	0.7528	0.7536	0.7543	0.7551
5‸7	0.7559	0.7566	0.7574	0.7582	0.7589	0.7597	0.7604	0.7612	0.7619	0.7627
5‸8	0.7634	0.7642	0.7649	0.7657	0.7664	0.7672	0.7679	0.7686	0.7694	0.7701
5‸9	0.7709	0.7716	0.7723	0.7731	0.7738	0.7745	0.7752	0.7760	0.7767	0.7774
6‸0	0.7782	0.7789	0.7796	0.7803	0.7810	0.7818	0.7825	0.7832	0.7839	0.7846
6‸1	0.7853	0.7860	0.7868	0.7875	0.7882	0.7889	0.7896	0.7903	0.7910	0.7917

0.7589 is the *mantissa* of log 5.74

FIGURE B.2.1

■

Example 3 Find log 8,360.
Solution We first write 8,360 in scientific notation.

$$\log 8{,}360 = \log(8.36 \times 10^3) = \log 8.36 + \log 10^3 = \log 8.36 + 3$$

We find log 8.36 in Table II, a portion of which is shown below.

N	0	1	2	3	4	5	6	7	8	9
7‸8	0.8921	0.8927	0.8932	0.8938	0.8943	0.8949	0.8954	0.8960	0.8965	0.8971
7‸9	0.8976	0.8982	0.8987	0.8993	0.8998	0.9004	0.9009	0.9015	0.9020	0.9025
8‸0	0.9031	0.9036	0.9042	0.9047	0.9053	0.9058	0.9063	0.9069	0.9074	0.9079
8‸1	0.9085	0.9090	0.9096	0.9101	0.9106	0.9112	0.9117	0.9122	0.9128	0.9133
8‸2	0.9138	0.9143	0.9149	0.9154	0.9159	0.9165	0.9170	0.9175	0.9180	0.9186
8‸3	0.9191	0.9196	0.9201	0.9206	0.9212	0.9217	0.9222	0.9227	0.9232	0.9238
8‸4	0.9234	0.9248	0.9253	0.9258	0.9263	0.9269	0.9274	0.9279	0.9284	0.9289

Mantissa

Therefore, $\log 8{,}360 \approx 3.9222$

Characteristic

(The characteristic is 3, because $8{,}360 = 8.36 \times 10^{3}$.) In practice, we usually determine and write down the characteristic and a decimal point and then look up the mantissa and write it after the decimal point. That is, we usually do *not* write all the steps found at the beginning of this example; we simply write

$$\log 8{,}360 \approx 3.9222 \quad ■$$

Example 4 Find log 727.
Solution The characteristic is 2, because $727 = 7.27 \times 10^2$. The mantissa is 0.8615 (from Table II). Therefore, $\log 727 \approx 2.8615$.

NOTE The characteristic and the mantissa may be positive, negative, or zero; however, the mantissas *in the table* are never negative. Therefore, if we were going to find a logarithm, perform further calculations using that logarithm, *and then use Table II* again, we would want the mantissa to be *positive*. ✓

Example 5 Find log 0.0438.
Solution

$$\log 0.0438 = \log(4.38 \times 10^{-2}) = \log 4.38 + \log 10^{-2} = \log 4.38 + (-2)$$

From Table II, we find that $\log 4.38 \approx 0.6415$. Therefore,

$$\log 0.0438 \approx 0.6415 - 2$$

but we usually write -2 as $8. \quad -10$ and write

$$\log 0.0438 \approx 8.6415 - 10$$

If we performed either of the subtractions, we would have $\log 0.0438 \approx -1.3585$, which would equal the rounded-off calculator display for the logarithm. As previously mentioned, if we were going to perform any further calculations using this logarithm *and then use Table II* again, we would want the mantissa to be *positive* and we would leave the answer as $0.6415 - 2$, or, more commonly, as $8.6415 - 10$. ■

Example 6 Find log 0.0000429.
Solution The characteristic is -5, which we will write as $5. \quad -10$. The mantissa is 0.6325. Therefore, $\log 0.0000429 \approx 5.6325 - 10$. ■

EXERCISES B.2

Set I Find each logarithm by using Table II.

1. log 754 **2.** log 186 **3.** log 17
4. log 29 **5.** log 3,350 **6.** log 4,610
7. log 7,000 **8.** log 200 **9.** log 0.0604
10. log 0.0186 **11.** $\log(5.64 \times 10^3)$ **12.** $\log(2.14 \times 10^{-4})$

Set II Find each logarithm by using Table II.

1. log 0.905 **2.** log 0.306 **3.** log 58.9
4. log 36.7 **5.** $\log(5.77 \times 10^{-4})$ **6.** $\log(3.96 \times 10^3)$
7. log 15,000 **8.** log 0.0123 **9.** log 0.0013
10. log 5 **11.** $\log(3.16 \times 10^5)$ **12.** $\log(4.1 \times 10^{-3})$

B.3 Natural Logarithms

A special abbreviation, ln, is used for logarithms to the base e. That is, $\ln x = \log_e x$.
A brief table of natural logarithms is given in Table III, Appendix C. Natural logarithms do not have a characteristic or a mantissa. To find the natural logarithm of

a number in Table III, we look for that number under the column headed n. If we find the desired number in that column, we read the value of its logarithm directly from the table, under the column headed $\log_e n$. If we do *not* find the desired number in the column headed n, we must use the laws of logarithms before we proceed (see Example 8). (Except for ln 1.0, all the logarithms in Table III are approximations.)

Example 7 Find ln 4.7.

Solution Figure B.3.1 is part of Table III, Appendix C. We *do* find 4.7 in the table under the column headed n. Therefore, we read $\ln 4.7 \approx 1.5476$.

n	$\log_e n$
4.5	1.5041
4.6	1.5261
→ 4.7	1.5476 ←
4.8	1.5686
4.9	1.5892

FIGURE B.3.1 ■

Example 8 Find ln 2,000.

Solution We do *not* find 2,000 in the column headed n. We can, however, rewrite 2,000 as 20(100). Then

$$\begin{aligned}\ln 2{,}000 &= \ln 20(100)\\ &= \ln 20 + \ln 100\\ &\approx 2.9957 + 4.6052\\ &= 7.6009\end{aligned}$$

We could, instead, write 2,000 as $2(10^3)$. The result is the same. ■

EXERCISES B.3

Set I Find the logarithms by using Table III.

1. ln 3.6 **2.** ln 5.2 **3.** ln 8.1 **4.** ln 7.5

5. ln 83 **6.** ln 62 **7.** ln 0.002 **8.** ln 0.006

Set II Find the logarithms by using Table III.

1. ln 2.8 **2.** ln 7.6 **3.** ln 1.6 **4.** ln 140

5. ln 78 **6.** ln 14,000 **7.** ln 0.008 **8.** ln 0.035

B.4 Interpolation

Sometimes we want the logarithm of a number that lies *between* two consecutive numbers in Table II. The process of finding such a number is called **interpolation**. When we interpolate to find a logarithm, the answer should not have more decimal places than the mantissas in the table have.

Example 9 Find log 29.38.

Solution $2_{\wedge}9.38$ is *between* $2_{\wedge}9.3$ and $2_{\wedge}9.4$, and the characteristic of log 29.38 is 1.

N	0	1	2	3	4	5	6	7	8	9
$2_{\wedge}8$	0.4472	0.4487	0.4502	0.4518	0.4533	0.4548	0.4564	0.4579	0.4594	0.4609
$2_{\wedge}9$	0.4624	0.4639	0.4654	0.4669	0.4683	0.4698	0.4713	0.4728	0.4742	0.4757
$3_{\wedge}0$	0.4771	0.4786	0.4800	0.4814	0.4829	0.4843	0.4857	0.4871	0.4886	0.4900
$3_{\wedge}1$	0.4914	0.4928	0.4942	0.4955	0.4969	0.4983	0.4997	0.5011	0.5024	0.5038
$3_{\wedge}2$	0.5051	0.5065	0.5079	0.5092	0.5105	0.5119	0.5132	0.5145	0.5159	0.5172
$3_{\wedge}3$	0.5185	0.5198	0.5211	0.5224	0.5237	0.5250	0.5263	0.5276	0.5289	0.5302
$3_{\wedge}4$	0.5315	0.5328	0.5340	0.5353	0.5366	0.5378	0.5391	0.5403	0.5416	0.5428

The mantissa for $2_{\wedge}9.38$ is a number *between* 0.4669 and 0.4683

We generally arrange the work as follows:

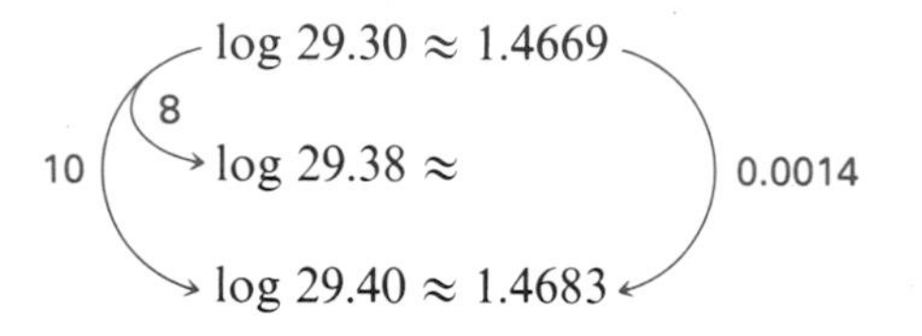

Since 29.38 is eight-tenths of the way from 29.30 to 29.40, we assume that log 29.38 is eight-tenths of the way from 1.4669 to 1.4683. We first subtract 1.4669 from 1.4683, getting 0.0014. Next we find $\frac{8}{10}$ of 0.0014, which is 0.00112. We round that answer off to four decimal places (to 0.0011) and then add 0.0011 to 1.4669 (the smaller number). It is customary to omit the decimal point and the zeros in the numbers 0.0014, 0.00112, and 0.0011 and to show the work as follows:

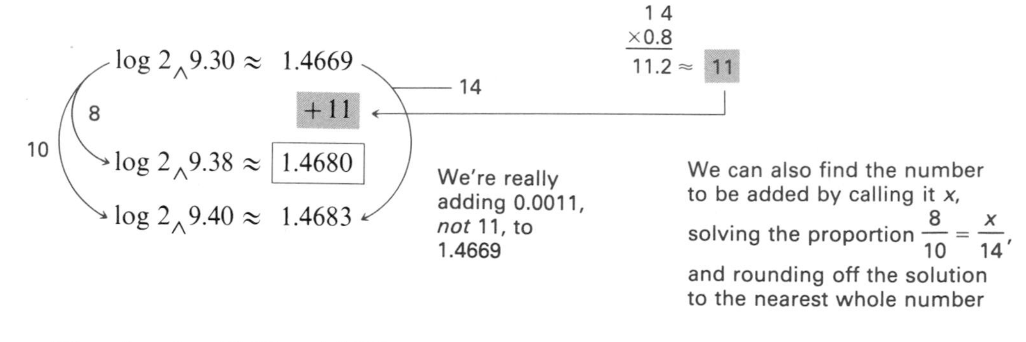

Therefore, log 29.38 ≈ 1.4680. ■

Example 10 Find log 0.002749.

Solution $2_{\wedge}749$ is *between* $2_{\wedge}740$ and $2_{\wedge}750$. The characteristic of log 0.002749 is -3, or $7 - 10$.

N	0	1	2	3	4	5	6	7	8	9
2‸0	0.3010	0.3032	0.3054	0.3075	0.3096	0.3118	0.3139	0.3160	0.3181	0.3201
2‸1	0.3222	0.3243	0.3263	0.3284	0.3304	0.3324	0.3345	0.3365	0.3385	0.3404
2‸2	0.3424	0.3444	0.3464	0.3483	0.3502	0.3522	0.3541	0.3560	0.3579	0.3598
2‸3	0.3617	0.3636	0.3655	0.3674	0.3692	0.3711	0.3729	0.3747	0.3766	0.3784
2‸4	0.3802	0.3820	0.3838	0.3856	0.3874	0.3892	0.3909	0.3927	0.3945	0.3962
2‸5	0.3979	0.3997	0.4014	0.4031	0.4048	0.4065	0.4082	0.4099	0.4116	0.4133
2‸6	0.4150	0.4166	0.4183	0.4200	0.4216	0.4232	0.4249	0.4265	0.4281	0.4298
2‸7	0.4314	0.4330	0.4346	0.4362	0.4378	0.4393	0.4409	0.4425	0.4440	0.4456

The mantissa for 2‸749 is a number *between* 0.4378 and 0.4393

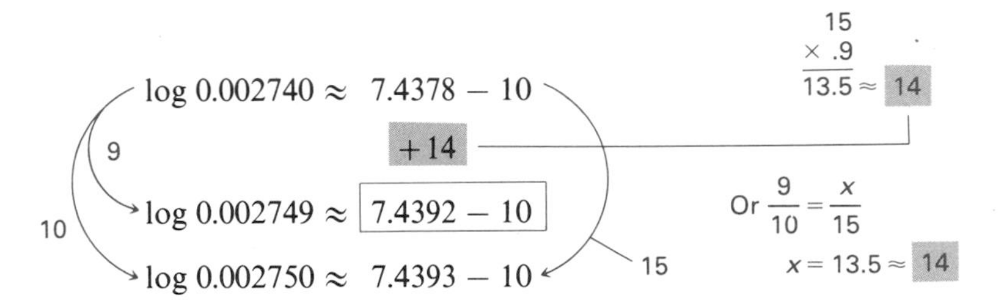

Therefore, $\log 0.002749 \approx 7.4392 - 10$. ■

The logarithms of numbers can be found by using a calculator with a log key as well as by using tables. If you have a calculator with a log key, we suggest that you use your calculator to check logarithms found by using the tables. One difference between logarithms obtained by calculator and those obtained from Table II occurs when the characteristic of the logarithm is negative. In Example 10,

$$\log 0.002749 \approx 7.4392 - 10$$

was found by using Table II. If this same logarithm were found by depressing the log key of a calculator, we would get

$$\log 0.002749 \approx -2.5608$$

But

$$7.4392 - 10 = -2.5608$$

EXERCISES B.4

Set I Find each logarithm by using Table II and check each answer by using a calculator.

1. $\log 23.35$ **2.** $\log 27.85$ **3.** $\log 3.062$

4. $\log 4.098$ **5.** $\log 0.06644$ **6.** $\log 0.5839$

7. $\log 150.7$ **8.** $\log 20.88$ **9.** $\log(8.375 \times 10^6)$

10. $\log(3.875 \times 10^{-5})$ **11.** $\log 324.38$ **12.** $\log 75.062$

Set II Find each logarithm by using Table II and check each answer by using a calculator.

1. log 186,300	**2.** log 92,840,000	**3.** log 0.8006
4. log 0.07093	**5.** log 0.004003	**6.** $\log(1.756 \times 10^{-6})$
7. log 62.34	**8.** log 0.005326	**9.** log 2,168
10. log 0.0003152	**11.** $\log(1.425 \times 10^{4})$	**12.** $\log(4.257 \times 10^{-3})$

B.5 Finding Common Antilogarithms

NOTE The term *antilog* is discussed in Section 9.4B, and finding N when ln N (its natural logarithm) is known is also discussed there. ☑

In the preceding sections, we discussed finding the logarithm of a given number. In this section, we discuss finding a number when its logarithm is known. The two statements that follow are equivalent:

$$\log N = L$$
$$N = \text{antilog } L$$

Example 11 If $\log N = 3.6263$, find N.
Solution

1. Locate the mantissa (0.6263) in the body of Table II, and find the number $4_\wedge 23$ by the method shown in the figure below.

The first two digits of N

The third digit of N

N	**0**	**1**	**2**	**3**	**4**	**5**	**6**	**7**	**8**	**9**
$4_\wedge 0$	0.6021	0.6031	0.6042	0.6053	0.6064	0.6075	0.6085	0.6096	0.6107	0.6117
$4_\wedge 1$	0.6128	0.6138	0.6149	0.6160	0.6170	0.6180	0.6191	0.6201	0.6212	0.6222
$4_\wedge 2$	0.6232	0.6243	0.6253	0.6263	0.6274	0.6284	0.6294	0.6304	0.6314	0.6325
$4_\wedge 3$	0.6335	0.6345	0.6355	0.6365	0.6375	0.6385	0.6395	0.6405	0.6415	0.6425
$4_\wedge 4$	0.6435	0.6444	0.6454	0.6464	0.6474	0.6484	0.6493	0.6503	0.6513	0.6522

The mantissa

2. Use 3, the characteristic in **3**.6263, to locate the actual decimal point (the decimal point must go three places to the *right* of standard position). Therefore,

$$N \approx 4.230 \times 10^3 = 4{,}230$$

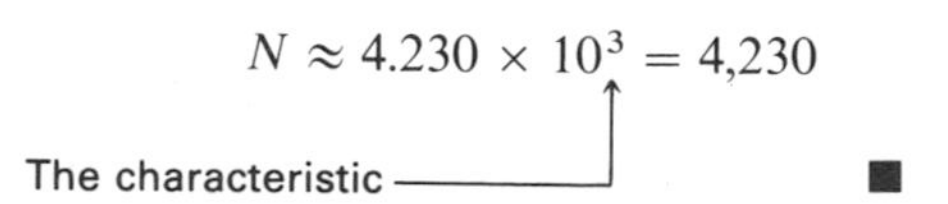

■

Example 12 If $N = \text{antilog}(7.8675 - 10)$, find N.

Solution $\log N = 7.8675 - 10$

1. Locate the mantissa (0.8675) in the body of Table II, and find the number $7_\wedge 37$ by the method shown in the figure at the top of page 651.

The first two digits of N — The third digit of N — The mantissa

N	0	1	2	3	4	5	6	7	8	9
$7_\wedge0$	0.8451	0.8457	0.8463	0.8470	0.8476	0.8482	0.8488	0.8494	0.8500	0.8506
$7_\wedge1$	0.8513	0.8519	0.8525	0.8531	0.8537	0.8543	0.8549	0.8555	0.8561	0.8567
$7_\wedge2$	0.8573	0.8579	0.8585	0.8591	0.8597	0.8603	0.8609	0.8615	0.8621	0.8627
$7_\wedge3$	0.8633	0.8639	0.8645	0.8651	0.8657	0.8663	0.8669	0.8675	0.8681	0.8686
$7_\wedge4$	0.8692	0.8698	0.8704	0.8710	0.8716	0.8722	0.8727	0.8733	0.8739	0.8745

2. Use -3, the characteristic in 7.8675 − 10, to locate the actual decimal point (the decimal point must go three places to the *left* of standard position). Therefore,

$$N \approx 0.007_\wedge37 = 0.00737$$

−3 ← Characteristic ■

When the mantissa of log N falls between two consecutive numbers in Table II, we must *interpolate* to find a fourth significant digit for N. (See Examples 13 and 14.)

Example 13 Find antilog 4.7129.
Solution Let N = antilog 4.7129. Then log N = 4.7129.

The first two digits of N — The third digit of N

N	0	1	2	3	4	5	6	7	8	9
$5_\wedge0$	0.6990	0.6998	0.7007	0.7016	0.7024	0.7033	0.7042	0.7050	0.7059	0.7067
$5_\wedge1$	0.7076	0.7084	0.7093	0.7101	0.7110	0.7118	0.7126	0.7135	0.7143	0.7152
$5_\wedge2$	0.7160	0.7168	0.7177	0.7185	0.7193	0.7202	0.7210	0.7218	0.7226	0.7235
$5_\wedge3$	0.7243	0.7251	0.7259	0.7267	0.7275	0.7284	0.7292	0.7300	0.7308	0.7316
$5_\wedge4$	0.7324	0.7332	0.7340	0.7348	0.7356	0.7364	0.7372	0.7380	0.7388	0.7396

The mantissa 0.7129 is *between* 0.7126 and 0.7135

The mantissa 0.7129 is *between* 0.7126 and 0.7135. The mantissa 0.7126 has $5_\wedge16$ as its antilog, and the mantissa 0.7135 has $5_\wedge17$ as its antilog; we attach a fourth digit, 0, to each of these numbers as shown below.

Again omitting the decimal points, we find that the difference 7,129 − 7,126 is 3 and that the difference 7,135 − 7,126 is 9.

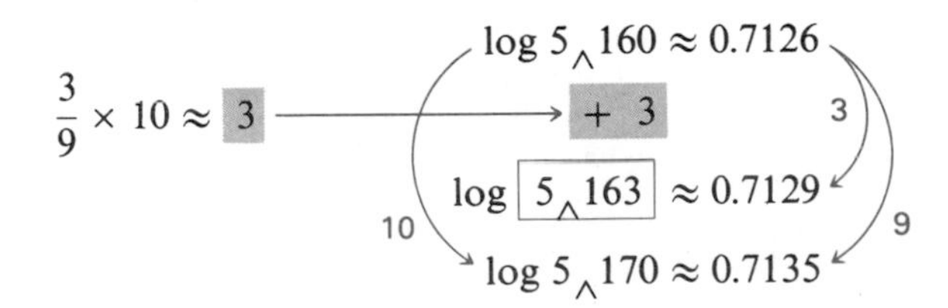

Since the mantissa 0.7129 is three-ninths of the way from 0.7126 to 0.7135, we assume that N is three-ninths of the way from $5_\wedge160$ to $5_\wedge170$. Therefore, we add $\frac{3}{9} \times 10 \approx 3$ to the *last* place of $5_\wedge160$ and get $5_\wedge163$. Then, since we want to find antilog 4.7129, we locate the actual decimal point by using the characteristic, 4; that is, the decimal point must be four places to the right of standard position.

$$N \approx 5_{\wedge}\underrightarrow{1630}. = 51{,}630$$

4 ← The characteristic

Therefore, antilog 4.7129 ≈ 51,630. ■

Example 14 Find antilog(8.7385 − 10).

Solution $\log N = 8.7385 - 10$

The mantissa 0.7385 is *between* 0.7380 and 0.7388. The mantissa 0.7380 has $5_{\wedge}47$ as its antilog, and the mantissa 0.7388 has $5_{\wedge}48$ as its antilog; we attach a fourth digit, 0, to each of these numbers. The difference 7,385 − 7,380 is 5 and the difference 7,388 − 7,380 is 8.

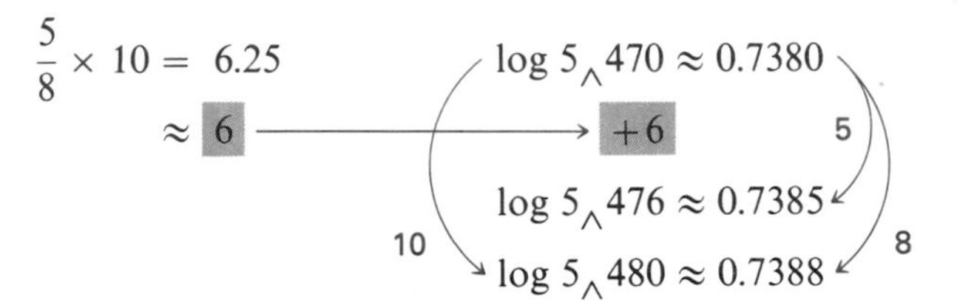

Then, since the characteristic of log N is 8 − 10, or −2, the decimal point must go two places to the left of standard position.

$$N \approx 0.\underleftarrow{05}_{\wedge}476 = 0.05476$$

−2 ← The characteristic = 8 − 10 = −2 ■

The antilogarithms of numbers can be found with a calculator by using the [INV] [log] keys or the [10^x] key.

EXERCISES B.5

Set I In Exercises 1–12, find each antilogarithm by using Table II.

1. antilog 3.5478

2. antilog 2.4409

3. antilog 0.9605

4. antilog 0.8848

5. antilog(9.2529 − 10)

6. antilog(8.1271 − 10)

7. antilog 3.5051

8. antilog 2.6335

9. antilog 4.0588

10. antilog 3.0846

11. antilog(6.9900 − 10)

12. antilog(7.9596 − 10)

In Exercises 13–16, find N by using Table II.

13. $\log N = 7.7168 - 10$

14. $\log N = 4.9410 - 10$

15. $\log N = 1.7120$

16. $\log N = 4.9873$

Set II In Exercises 1–12, find each antilogarithm by using Table II.

1. antilog(7.6117 − 10)

2. antilog(6.6010 − 10)

3. antilog 1.1685

4. antilog 2.5470

5. antilog(8.6908 − 10)

6. antilog(9.7995 − 10)

7. antilog 4.7388

8. antilog(7.6096 − 10)

9. antilog 2.0486

10. antilog(8.8136 − 10)

11. antilog(6.2963 − 10)

12. antilog 6.0925

In Exercises 13–16, find N by using Table II.

13. $\log N = 3.4084$

14. $\log N = 0.5011$

15. $\log N = 0.6860$

16. $\log N = 8.0367 - 10$

B.6 Calculating with Logarithms

Logarithms can be used to perform some arithmetic calculations. In this chapter, we assume that all numbers given in examples and exercises are *exact* numbers, and we round off all answers to four significant digits.

Example 15 Multiply (37.5)(0.00842).

Solution We let N equal the given expression *and then we take the logarithm of both sides of the resulting equation.*

Let $N = (37.5)(0.00842)$

$\log N = \log(37.5)(0.00842)$ Rule 9.3

$\log N = \log 37.5 + \log 0.00842$ Rule 9.7

$\log N \approx 1.5740 + 7.9253 - 10$

$$\begin{array}{r} 1.5740 \\ 7.9253 - 10 \\ \hline 9.4993 - 10 \end{array}$$

$\log N \approx 9.4993 - 10$ ← The characteristic is 9 − 10, or − 1

$N \approx 0.3_{\wedge}157 = 0.3157$ N is antilog (9.4993 − 10)

The decimal point goes one place to the left of standard position

Before any calculating is done, it is helpful to analyze the problem and make an outline of the procedure to be followed. ■

The method for calculating with logarithms can be summarized as follows:

TO CALCULATE WITH LOGARITHMS

1. Let N equal the given expression.
2. Take the logarithm of both sides of the resulting equation.
3. Analyze the problem and make a blank outline.
4. Write all the characteristics in the blank outline.
5. Use Table II to find all the mantissas, and then write them in the blank outline.
6. Carry out the calculations indicated in the outline.
7. Solve for N.

Example 16 Divide $\dfrac{6.74}{0.0391}$.

Solution Let

$$N = \frac{6.74}{0.0391}$$

$$\log N = \log \frac{6.74}{0.0391} \qquad \text{Rule 9.3}$$

$$\log N = \log 6.74 - \log 0.0391 \qquad \text{Rule 9.8}$$

The characteristic of the logarithm of 6.74, zero, is written as 10 − 10 to help with the subtraction

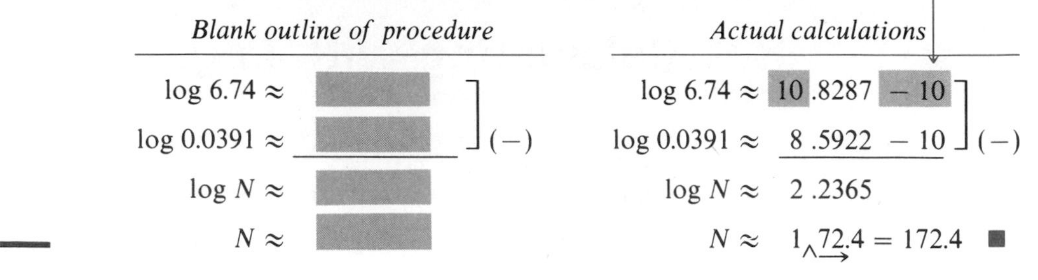

Example 17 Find $(1.05)^{10}$.

Solution Let

$$N = (1.05)^{10}$$

$$\log N = \log(1.05)^{10} \qquad \text{Rule 9.3}$$

$$\log N = 10 \log 1.05 \qquad \text{Rule 9.9}$$

Blank outline	*Actual calculations*
$\log 1.05 \approx$	$\log 1.05 \approx 0.0212$
$\log N = 10 \log 1.05 \approx$	$\log N = 10 \log 1.05 \approx 0.2120$
$N \approx$	$N \approx 1{,}629$ (with caret after 1)
	$N \approx 1.629$ ■

Example 18 Find $\sqrt[3]{0.506}$.

Solution

Let $N = \sqrt[3]{0.506} = (0.506)^{1/3}$ — Rewriting the cubic root as a rational exponent

$\log N = \log(0.506)^{1/3}$ — Rule 9.3

$\log N = \frac{1}{3} \log 0.506$ — Rule 9.9

The characteristic 9 − 10 is written as 29 − 30 so that the second term, −30, is exactly divisible by 3

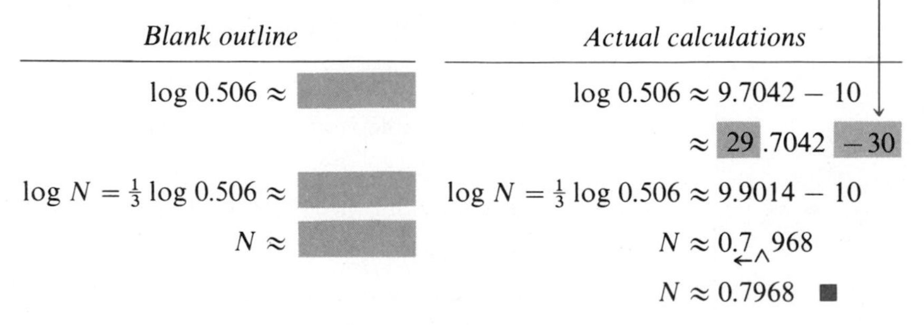

Example 19 Find N if $N = \dfrac{(1.16)^5(31.7)}{\sqrt{481}\,(0.629)}$, or $\dfrac{(1.16)^5(31.7)}{(481)^{1/2}(0.629)}$

Solution In the blank outline, we will represent the logarithm of the numerator by "log(num)" and the logarithm of the denominator by "log(den)."

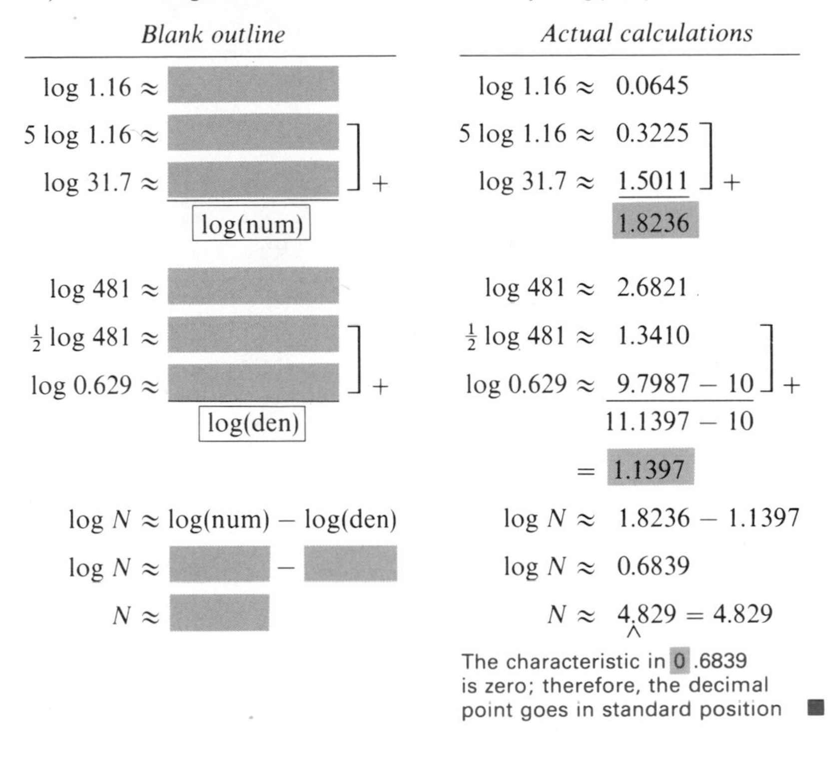

EXERCISES B.6

In the exercises (in both sets) that follow, use logarithms to perform the calculations. Assume that all the given numbers are *exact* numbers. Round off all answers to four significant digits. We suggest that you use a calculator to verify the results obtained by the logarithmic calculations. There may be a slight difference in the answers obtained by using a calculator and those found by using the table because of their different levels of accuracy.

Set I

1. 74.3×0.618

2. 0.314×14.9

3. $\dfrac{562}{21.4}$

4. $\dfrac{651}{30.6}$

5. $(1.09)^5$

6. $(3.4)^4$

7. $\sqrt[3]{0.444}$

8. $\sqrt[4]{0.897}$

9. $2.863 + \log 38.46$

10. $\dfrac{\log 7.86}{\log 38.4}$

11. $\sqrt[5]{\dfrac{(5.86)(17.4)}{\sqrt{450}}}$

12. $\dfrac{(5.65)\sqrt[6]{175}}{(2.4)^4}$

Set II

1. $\dfrac{(4.92)(25.7)}{388}$

2. $\dfrac{(2.04)^5}{(5.9)(0.66)}$

3. $\log 786.4 + 3.154$

4. $\sqrt[3]{0.564}$

5. $\dfrac{\log 58.4}{\log 2.50}$

6. $\sqrt[4]{\dfrac{(39.4)(7.86)}{\sqrt[3]{704}}}$

7. $\sqrt{0.00378}$

8. $\dfrac{1.26}{(0.0245)(0.00143)}$

9. $(2.87)^{12}$

10. $\sqrt{0.792}$

11. $\dfrac{\sqrt{4.63}}{\sqrt{0.0732}}$

12. $\dfrac{\log 1.25}{\log 1.05}$

Appendix C Tables

Table I Exponential Functions

x	e^x	e^{-x}	x	e^x	e^{-x}
0.00	1.0000	1.0000	1.5	4.4817	0.2231
0.01	1.0101	0.9901	1.6	4.9530	0.2019
0.02	1.0202	0.9802	1.7	5.4739	0.1827
0.03	1.0305	0.9705	1.8	6.0496	0.1653
0.04	1.0408	0.9608	1.9	6.6859	0.1496
0.05	1.0513	0.9512	2.0	7.3891	0.1353
0.06	1.0618	0.9418	2.1	8.1662	0.1225
0.07	1.0725	0.9324	2.2	9.0250	0.1108
0.08	1.0833	0.9231	2.3	9.9742	0.1003
0.09	1.0942	0.9139	2.4	11.023	0.0907
0.10	1.1052	0.9048	2.5	12.182	0.0821
0.11	1.1163	0.8958	2.6	13.464	0.0743
0.12	1.1275	0.8869	2.7	14.880	0.0672
0.13	1.1388	0.8781	2.8	16.445	0.0608
0.14	1.1503	0.8694	2.9	18.174	0.0550
0.15	1.1618	0.8607	3.0	20.086	0.0498
0.16	1.1735	0.8521	3.1	22.198	0.0450
0.17	1.1853	0.8437	3.2	24.533	0.0408
0.18	1.1972	0.8353	3.3	27.113	0.0369
0.19	1.2092	0.8270	3.4	29.964	0.0334
0.20	1.2214	0.8187	3.5	33.115	0.0302
0.21	1.2337	0.8106	3.6	36.598	0.0273
0.22	1.2461	0.8025	3.7	40.447	0.0247
0.23	1.2586	0.7945	3.8	44.701	0.0224
0.24	1.2712	0.7866	3.9	49.402	0.0202
0.25	1.2840	0.7788	4.0	54.598	0.0183
0.30	1.3499	0.7408	4.1	60.340	0.0166
0.35	1.4191	0.7047	4.2	66.686	0.0150
0.40	1.4918	0.6703	4.3	73.700	0.0136
0.45	1.5683	0.6376	4.4	81.451	0.0123
0.50	1.6487	0.6065	4.5	90.017	0.0111
0.55	1.7333	0.5769	4.6	99.484	0.0101
0.60	1.8221	0.5488	4.7	109.95	0.0091
0.65	1.9155	0.5220	4.8	121.51	0.0082
0.70	2.0138	0.4966	4.9	134.29	0.0074
0.75	2.1170	0.4724	5.0	148.41	0.0067
0.80	2.2255	0.4493	5.5	244.69	0.0041
0.85	2.3396	0.4274	6.0	403.43	0.0025
0.90	2.4596	0.4066	6.5	665.14	0.0015
0.95	2.5857	0.3867	7.0	1096.6	0.0009
1.0	2.7183	0.3679	7.5	1808.0	0.0006
1.1	3.0042	0.3329	8.0	2981.0	0.0003
1.2	3.3201	0.3012	8.5	4914.8	0.0002
1.3	3.6693	0.2725	9.0	8103.1	0.0001
1.4	4.0552	0.2466	10.0	22026.5	0.00005

Table II Common Logarithms

N	0	1	2	3	4	5	6	7	8	9
1‸0	0.0000	0.0043	0.0086	0.0128	0.0170	0.0212	0.0253	0.0294	0.0334	0.0374
1‸1	0.0414	0.0453	0.0492	0.0531	0.0569	0.0607	0.0645	0.0682	0.0719	0.0755
1‸2	0.0792	0.0828	0.0864	0.0899	0.0934	0.0969	0.1004	0.1038	0.1072	0.1106
1‸3	0.1139	0.1173	0.1206	0.1239	0.1271	0.1303	0.1335	0.1367	0.1399	0.1430
1‸4	0.1461	0.1492	0.1523	0.1553	0.1584	0.1614	0.1644	0.1673	0.1703	0.1732
1‸5	0.1761	0.1790	0.1818	0.1847	0.1875	0.1903	0.1931	0.1959	0.1987	0.2014
1‸6	0.2041	0.2068	0.2095	0.2122	0.2148	0.2175	0.2201	0.2227	0.2253	0.2279
1‸7	0.2304	0.2330	0.2355	0.2380	0.2405	0.2430	0.2455	0.2480	0.2504	0.2529
1‸8	0.2553	0.2577	0.2601	0.2625	0.2648	0.2672	0.2695	0.2718	0.2742	0.2765
1‸9	0.2788	0.2810	0.2833	0.2856	0.2878	0.2900	0.2923	0.2945	0.2967	0.2989
2‸0	0.3010	0.3032	0.3054	0.3075	0.3096	0.3118	0.3139	0.3160	0.3181	0.3201
2‸1	0.3222	0.3243	0.3263	0.3284	0.3304	0.3324	0.3345	0.3365	0.3385	0.3404
2‸2	0.3424	0.3444	0.3464	0.3483	0.3502	0.3522	0.3541	0.3560	0.3579	0.3598
2‸3	0.3617	0.3636	0.3655	0.3674	0.3692	0.3711	0.3729	0.3747	0.3766	0.3784
2‸4	0.3802	0.3820	0.3838	0.3856	0.3874	0.3892	0.3909	0.3927	0.3945	0.3962
2‸5	0.3979	0.3997	0.4014	0.4031	0.4048	0.4065	0.4082	0.4099	0.4116	0.4133
2‸6	0.4150	0.4166	0.4183	0.4200	0.4216	0.4232	0.4249	0.4265	0.4281	0.4298
2‸7	0.4314	0.4330	0.4346	0.4362	0.4378	0.4393	0.4409	0.4425	0.4440	0.4456
2‸8	0.4472	0.4487	0.4502	0.4518	0.4533	0.4548	0.4564	0.4579	0.4594	0.4609
2‸9	0.4624	0.4639	0.4654	0.4669	0.4683	0.4698	0.4713	0.4728	0.4742	0.4757
3‸0	0.4771	0.4786	0.4800	0.4814	0.4829	0.4843	0.4857	0.4871	0.4886	0.4900
3‸1	0.4914	0.4928	0.4942	0.4955	0.4969	0.4983	0.4997	0.5011	0.5024	0.5038
3‸2	0.5051	0.5065	0.5079	0.5092	0.5105	0.5119	0.5132	0.5145	0.5159	0.5172
3‸3	0.5185	0.5198	0.5211	0.5224	0.5237	0.5250	0.5263	0.5276	0.5289	0.5302
3‸4	0.5315	0.5328	0.5340	0.5353	0.5366	0.5378	0.5391	0.5403	0.5416	0.5428
3‸5	0.5441	0.5453	0.5465	0.5478	0.5490	0.5502	0.5514	0.5527	0.5539	0.5551
3‸6	0.5563	0.5575	0.5587	0.5599	0.5611	0.5623	0.5635	0.5647	0.5658	0.5670
3‸7	0.5682	0.5694	0.5705	0.5717	0.5729	0.5740	0.5752	0.5763	0.5775	0.5786
3‸8	0.5798	0.5809	0.5821	0.5832	0.5843	0.5855	0.5866	0.5877	0.5888	0.5899
3‸9	0.5911	0.5922	0.5933	0.5944	0.5955	0.5966	0.5977	0.5988	0.5999	0.6010
4‸0	0.6021	0.6031	0.6042	0.6053	0.6064	0.6075	0.6085	0.6096	0.6107	0.6117
4‸1	0.6128	0.6138	0.6149	0.6160	0.6170	0.6180	0.6191	0.6201	0.6212	0.6222
4‸2	0.6232	0.6243	0.6253	0.6263	0.6274	0.6284	0.6294	0.6304	0.6314	0.6325
4‸3	0.6335	0.6345	0.6355	0.6365	0.6375	0.6385	0.6395	0.6405	0.6415	0.6425
4‸4	0.6435	0.6444	0.6454	0.6464	0.6474	0.6484	0.6493	0.6503	0.6513	0.6522
4‸5	0.6532	0.6542	0.6551	0.6561	0.6571	0.6580	0.6590	0.6599	0.6609	0.6618
4‸6	0.6628	0.6637	0.6646	0.6656	0.6665	0.6675	0.6684	0.6693	0.6702	0.6712
4‸7	0.6721	0.6730	0.6739	0.6749	0.6758	0.6767	0.6776	0.6785	0.6794	0.6803
4‸8	0.6812	0.6821	0.6830	0.6839	0.6848	0.6857	0.6866	0.6875	0.6884	0.6893
4‸9	0.6902	0.6911	0.6920	0.6928	0.6937	0.6946	0.6955	0.6964	0.6972	0.6981
5‸0	0.6990	0.6998	0.7007	0.7016	0.7024	0.7033	0.7042	0.7050	0.7059	0.7067
5‸1	0.7076	0.7084	0.7093	0.7101	0.7110	0.7118	0.7126	0.7135	0.7143	0.7152
5‸2	0.7160	0.7168	0.7177	0.7185	0.7193	0.7202	0.7210	0.7218	0.7226	0.7235
5‸3	0.7243	0.7251	0.7259	0.7267	0.7275	0.7284	0.7292	0.7300	0.7308	0.7316
5‸4	0.7324	0.7332	0.7340	0.7348	0.7356	0.7364	0.7372	0.7380	0.7388	0.7396
N	0	1	2	3	4	5	6	7	8	9

Table II *Continued*

N	0	1	2	3	4	5	6	7	8	9
5‸5	0.7404	0.7412	0.7419	0.7427	0.7435	0.7443	0.7451	0.7459	0.7466	0.7474
5‸6	0.7482	0.7490	0.7497	0.7505	0.7513	0.7520	0.7528	0.7536	0.7543	0.7551
5‸7	0.7559	0.7566	0.7574	0.7582	0.7589	0.7597	0.7604	0.7612	0.7619	0.7627
5‸8	0.7634	0.7642	0.7649	0.7657	0.7664	0.7672	0.7679	0.7686	0.7694	0.7701
5‸9	0.7709	0.7716	0.7723	0.7731	0.7738	0.7745	0.7752	0.7760	0.7767	0.7774
6‸0	0.7782	0.7789	0.7796	0.7803	0.7810	0.7818	0.7825	0.7832	0.7839	0.7846
6‸1	0.7853	0.7860	0.7868	0.7875	0.7882	0.7889	0.7896	0.7903	0.7910	0.7917
6‸2	0.7924	0.7931	0.7938	0.7945	0.7952	0.7959	0.7966	0.7973	0.7980	0.7987
6‸3	0.7993	0.8000	0.8007	0.8014	0.8021	0.8028	0.8035	0.8041	0.8048	0.8055
6‸4	0.8062	0.8069	0.8075	0.8082	0.8089	0.8096	0.8102	0.8109	0.8116	0.8122
6‸5	0.8129	0.8136	0.8142	0.8149	0.8156	0.8162	0.8169	0.8176	0.8182	0.8189
6‸6	0.8195	0.8202	0.8209	0.8215	0.8222	0.8228	0.8235	0.8241	0.8248	0.8254
6‸7	0.8261	0.8267	0.8274	0.8280	0.8287	0.8293	0.8299	0.8306	0.8312	0.8319
6‸8	0.8325	0.8331	0.8338	0.8344	0.8351	0.8357	0.8363	0.8370	0.8376	0.8382
6‸9	0.8388	0.8395	0.8401	0.8407	0.8414	0.8420	0.8426	0.8432	0.8439	0.8445
7‸0	0.8451	0.8457	0.8463	0.8470	0.8476	0.8482	0.8488	0.8494	0.8500	0.8506
7‸1	0.8513	0.8519	0.8525	0.8531	0.8537	0.8543	0.8549	0.8555	0.8561	0.8567
7‸2	0.8573	0.8579	0.8585	0.8591	0.8597	0.8603	0.8609	0.8615	0.8621	0.8627
7‸3	0.8633	0.8639	0.8645	0.8651	0.8657	0.8663	0.8669	0.8675	0.8681	0.8686
7‸4	0.8692	0.8698	0.8704	0.8710	0.8716	0.8722	0.8727	0.8733	0.8739	0.8745
7‸5	0.8751	0.8756	0.8762	0.8768	0.8774	0.8779	0.8785	0.8791	0.8797	0.8802
7‸6	0.8808	0.8814	0.8820	0.8825	0.8831	0.8837	0.8842	0.8848	0.8854	0.8859
7‸7	0.8865	0.8871	0.8876	0.8882	0.8887	0.8893	0.8899	0.8904	0.8910	0.8915
7‸8	0.8921	0.8927	0.8932	0.8938	0.8943	0.8949	0.8954	0.8960	0.8965	0.8971
7‸9	0.8976	0.8982	0.8987	0.8993	0.8998	0.9004	0.9009	0.9015	0.9020	0.9025
8‸0	0.9031	0.9036	0.9042	0.9047	0.9053	0.9058	0.9063	0.9069	0.9074	0.9079
8‸1	0.9085	0.9090	0.9096	0.9101	0.9106	0.9112	0.9117	0.9122	0.9128	0.9133
8‸2	0.9138	0.9143	0.9149	0.9154	0.9159	0.9165	0.9170	0.9175	0.9180	0.9186
8‸3	0.9191	0.9196	0.9201	0.9206	0.9212	0.9217	0.9222	0.9227	0.9232	0.9238
8‸4	0.9243	0.9248	0.9253	0.9258	0.9263	0.9269	0.9274	0.9279	0.9284	0.9289
8‸5	0.9294	0.9299	0.9304	0.9309	0.9315	0.9320	0.9325	0.9330	0.9335	0.9340
8‸6	0.9345	0.9350	0.9355	0.9360	0.9365	0.9370	0.9375	0.9380	0.9385	0.9390
8‸7	0.9395	0.9400	0.9405	0.9410	0.9415	0.9420	0.9425	0.9430	0.9435	0.9440
8‸8	0.9445	0.9450	0.9455	0.9460	0.9465	0.9469	0.9474	0.9479	0.9484	0.9489
8‸9	0.9494	0.9499	0.9504	0.9509	0.9513	0.9518	0.9523	0.9528	0.9533	0.9538
9‸0	0.9542	0.9547	0.9552	0.9557	0.9562	0.9566	0.9571	0.9576	0.9581	0.9586
9‸1	0.9590	0.9595	0.9600	0.9605	0.9609	0.9614	0.9619	0.9624	0.9628	0.9633
9‸2	0.9638	0.9643	0.9647	0.9652	0.9657	0.9661	0.9666	0.9671	0.9675	0.9680
9‸3	0.9685	0.9689	0.9694	0.9699	0.9703	0.9708	0.9713	0.9717	0.9722	0.9727
9‸4	0.9731	0.9736	0.9741	0.9745	0.9750	0.9754	0.9759	0.9763	0.9768	0.9773
9‸5	0.9777	0.9782	0.9786	0.9791	0.9795	0.9800	0.9805	0.9809	0.9814	0.9818
9‸6	0.9823	0.9827	0.9832	0.9836	0.9841	0.9845	0.9850	0.9854	0.9859	0.9863
9‸7	0.9868	0.9872	0.9877	0.9881	0.9886	0.9890	0.9894	0.9899	0.9903	0.9908
9‸8	0.9912	0.9917	0.9921	0.9926	0.9930	0.9934	0.9939	0.9943	0.9948	0.9952
9‸9	0.9956	0.9961	0.9965	0.9969	0.9974	0.9978	0.9983	0.9987	0.9991	0.9996
N	0	1	2	3	4	5	6	7	8	9

Table III Natural Logarithms

n	$\log_e n$	n	$\log_e n$	n	$\log_e n$
	*	4.5	1.5041	9.0	2.1972
0.1	7.6974	4.6	1.5261	9.1	2.2083
0.2	8.3906	4.7	1.5476	9.2	2.2192
0.3	8.7960	4.8	1.5686	9.3	2.2300
0.4	9.0837	4.9	1.5892	9.4	2.2407
0.5	9.3069	5.0	1.6094	9.5	2.2513
0.6	9.4892	5.1	1.6292	9.6	2.2618
0.7	9.6433	5.2	1.6487	9.7	2.2721
0.8	9.7769	5.3	1.6677	9.8	2.2824
0.9	9.8946	5.4	1.6864	9.9	2.2925
1.0	0.0000	5.5	1.7047	10	2.3026
1.1	0.0953	5.6	1.7228	11	2.3979
1.2	0.1823	5.7	1.7405	12	2.4849
1.3	0.2624	5.8	1.7579	13	2.5649
1.4	0.3365	5.9	1.7750	14	2.6391
1.5	0.4055	6.0	1.7918	15	2.7081
1.6	0.4700	6.1	1.8083	16	2.7726
1.7	0.5306	6.2	1.8245	17	2.8332
1.8	0.5878	6.3	1.8405	18	2.8904
1.9	0.6419	6.4	1.8563	19	2.9444
2.0	0.6931	6.5	1.8718	20	2.9957
2.1	0.7419	6.6	1.8871	25	3.2189
2.2	0.7885	6.7	1.9021	30	3.4012
2.3	0.8329	6.8	1.9169	35	3.5553
2.4	0.8755	6.9	1.9315	40	3.6889
2.5	0.9163	7.0	1.9459	45	3.8067
2.6	0.9555	7.1	1.9601	50	3.9120
2.7	0.9933	7.2	1.9741	55	4.0073
2.8	1.0296	7.3	1.9879	60	4.0943
2.9	1.0647	7.4	2.0015	65	4.1744
3.0	1.0986	7.5	2.0149	70	4.2485
3.1	1.1314	7.6	2.0281	75	4.3175
3.2	1.1632	7.7	2.0412	80	4.3820
3.3	1.1939	7.8	2.0541	85	4.4427
3.4	1.2238	7.9	2.0669	90	4.4998
3.5	1.2528	8.0	2.0794	100	4.6052
3.6	1.2809	8.1	2.0919	110	4.7005
3.7	1.3083	8.2	2.1041	120	4.7875
3.8	1.3350	8.3	2.1163	130	4.8676
3.9	1.3610	8.4	2.1282	140	4.9416
4.0	1.3863	8.5	2.1401	150	5.0106
4.1	1.4110	8.6	2.1518	160	5.0752
4.2	1.4351	8.7	2.1633	170	5.1358
4.3	1.4586	8.8	2.1748	180	5.1930
4.4	1.4816	8.9	2.1861	190	5.2470

* Subtract 10 for $n < 1$. Thus, $\log_e 0.1 = 7.6974 - 10 = -2.3026$.

Answers

to Set I Exercises, Diagnostic Tests, and Cumulative Review Exercises

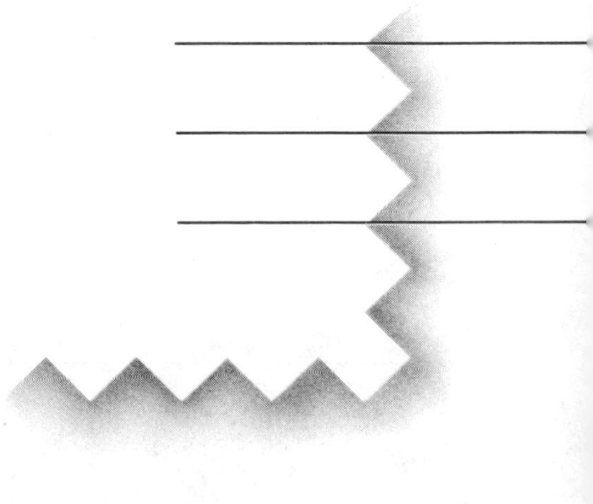

Exercises 1.1A (page 5)

1. a. True; a set is a collection of things
   b. False; $\{1, 2, 3\} \neq \{1, 2, 3, \ldots\}$ c. True
   d. False; "23" is not a set e. False; 0 is *not* a natural number
   f. False; $\{2\}$ is a *set*, not an element g. True
   h. True; $\{\ \}$ is a subset of every set i. False; 10 is not a digit
   j. False; the empty set has *no* elements k. True
   l. True; it is the empty set
2. a. True b. False c. True d. True e. False
   f. False g. True h. True i. True j. False
   k. True l. True
3. a. Infinite b. Finite c. Finite
4. a. Infinite b. Finite c. Finite
5. $\{\ \}, \{a\}, \{b\}, \{c\}, \{a, b\}, \{a, c\}, \{b, c\}, \{a, b, c\}$
6. $\{\ \}, \{1\}, \{2\}, \{1, 2\}$
7. a. False; $12 \in B$ but $12 \notin A$
   b. True, because the elements of C are also elements of B
   c. False, because all elements of C are also elements of A
8. a. True b. False c. True
9. a. $\{0, 2, 4, 6, 8\}$ b. $\{5, 10, 15, \ldots\}$
10. a. $\{1, 3, 5, \ldots\}$ b. $\{11, 22, 33, \ldots\}$
11. a. $\{x \mid x$ is one of the first 5 letters of the alphabet$\}$
    b. $\{x \mid x$ is a multiple of 4 and $x \in N$ and x is less than $13\}$
    c. $\{x \mid x$ is a natural number divisible by $10\}$
12. a. $\{x \mid x$ is a digit divisible by $2\}$
    b. $\{x \mid x$ is one of the last three letters of the alphabet$\}$
    c. $\{x \mid x$ is a natural number divisible by $5\}$

Exercises 1.1B (page 8)

1. a. $\{1, 2, 3, 4, 5\}$ b. $\{2, 4\}$ c. $\{1, 2, 3, 4, 5\}$ d. $\{2, 4\}$
2. a. $\{\ \}$ or $\varnothing$ b. $\{2, 5, 6, 7, 8, 9, 12\}$ c. $\{\ \}$ or $\varnothing$
   d. $\{2, 5, 6, 7, 8, 9, 12\}$
3. a. $\{5, 11\}$ b. $\{0, 3, 4, 5, 6, 7, 11, 13\}$ c. $\{6\}$
   d. $\{\ \}$ or $\varnothing$ e. $\{2, 5, 6, 7, 11, 13\}$
   f. $\{0, 3, 4, 5, 6, 7, 11, 13\}$ g. $\{5, 11\} \cap \{0, 3, 4, 6\} = \{\ \}$
   h. $\{2, 5, 6, 11\} \cap \{\ \} = \{\ \}$
   i. $\{2, 5, 6, 7, 11, 13\} \cup \{0, 3, 4, 6\} = \{0, 2, 3, 4, 5, 6, 7, 11, 13\}$
   j. $\{2, 5, 6, 11\} \cup \{0, 3, 4, 5, 6, 7, 11, 13\} = \{0, 2, 3, 4, 5, 6, 7, 11, 13\}$
4. a. $\{4, b\}$ b. $\{a, b, m, 4, 6, 7\}$ c. $\{a, b, n, t, 3, 4, 5, 7\}$
   d. $\{\ \}$ or $\varnothing$ e. $\{\ \}$ or $\varnothing$ f. $\{b, m, n, t, 3, 4, 5, 6\}$
   g. $\{\ \}$ or $\varnothing$ h. $\{\ \}$ or $\varnothing$ i. $\{a, b, m, n, t, 3, 4, 5, 6, 7\}$
   j. $\{a, b, m, n, t, 3, 4, 5, 6, 7\}$

Exercises 1.2 (page 12)

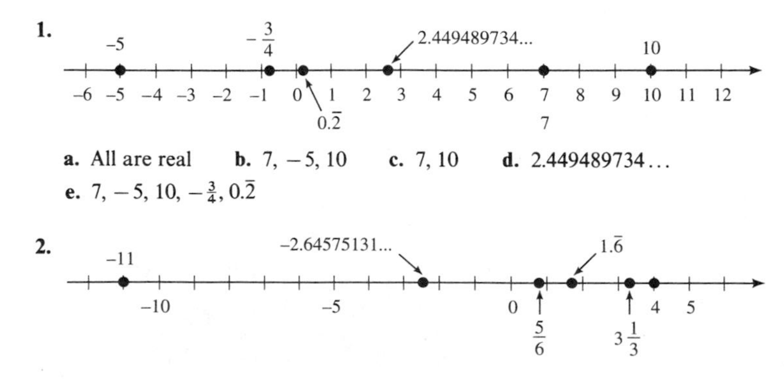

1. a. All are real b. $7, -5, 10$ c. $7, 10$ d. $2.449489734\ldots$
   e. $7, -5, 10, -\frac{3}{4}, 0.\bar{2}$
2. a. All b. $-11, 4$ c. 4 d. $-2.64575131\ldots$
   e. $-11, 1.\bar{6}, \frac{5}{6}, 3\frac{1}{3}, 4$
3. a. $\subseteq$ b. $\in$ c. $\subseteq$ d. $\in$ e. $\in$ f. $\subseteq$
4. a. $\subseteq$ b. $\subseteq$ c. $\in$ d. $\in$ e. $\subseteq$ f. $\subseteq$
5. a. False, because irrationals are real b. True, because $J \subseteq R$
   c. False; fractions are real numbers but not whole numbers
   d. True; 3 is an element of the set of integers
   e. False; 3 is not an irrational number
   f. True; the irrational numbers are a subset of the set of real numbers
   g. True; a nonrepeating, nonterminating decimal is irrational
   h. True, because a nonrepeating decimal is real
6. a. True b. False c. False d. True e. True
   f. True g. False h. True

Exercises 1.3A (page 15)

1. a. 3 b. x, y 2. a. $4, -2$ b. y, z
3. a. Three terms b. $5F$ c. E, 1
4. a. Three terms b. $2T$ c. R, 1
5. a. Two terms b. $-2(x + y)$ c. $(R + S)$, 1
6. a. Two terms b. $-5(W + V)$ c. $(A + 2B)$, 1
7. a. Two terms b. 4 c. $3, X, Y, Z$
8. a. Two terms b. $3x$ c. $4, a, b$
9. a. Two terms b. $\dfrac{3B - C}{DE}$ c. $2, A$
10. a. Two terms b. $\dfrac{w + z}{xyz}$ c. $3, s, t$ 11. a. 2 b. RT
12. a. 4 b. xy 13. a. -1 b. y 14. a. -1 b. b

Exercises 1.3B (page 19)

1. a. 7 b. $|-5| = -(-5) = 5$ c. $-|-12| = -(12) = -12$
2. a. 34 b. -5 c. -3 3. a. $<$ b. $<$ c. $>$ d. $<$
4. a. $>$ b. $<$ c. $>$ d. $<$
5. a. Valid ($3 < 5$, $5 < 10$, and $3 < 10$ are all true)
   b. Invalid ($-2 > 7$, $1 > 7$, and $-2 > 1$ are false statements)
   c. Invalid (one "$<$" and one "$>$")
   d. Valid ($8 > 3$, $3 > -1$, and $8 > -1$ are all true)
   e. Invalid (one "$<$" and one "$>$")
   f. Invalid ($0 > 8$, $3 > 8$, and $0 > 3$ are false statements)
6. a. Valid b. Invalid (one "$<$" and one "$>$")
   c. Invalid (one "$<$" and one "$>$") d. Valid
   e. Invalid ($4 > 7$, $7 > 11$, and $4 > 11$ are false)
   f. Invalid ($5 < 0$, $0 < -2$, and $5 < -2$ are false)
7. False 8. True

Exercises 1.4 (page 25)

1. $-(|-9| - |5|) = -(9 - 5) = -4$ **2.** -3

3. $-(|-12| + |-7|) = -(12 + 7) = -19$ **4.** $-\frac{5}{4}$

5. $-\left(\left|-\frac{7}{12}\right| + \left|-\frac{1}{6}\right|\right) = -\left(\frac{7}{12} + \frac{1}{6}\right) = -\frac{9}{12} = -\frac{3}{4}$

6. $-\frac{9}{8}$ or $-1\frac{1}{8}$ **7.** $\frac{21}{4} + \left(\frac{-5}{2}\right) = \frac{21}{4} + \left(\frac{-10}{4}\right) = \frac{11}{4}$ **8.** $1\frac{7}{10}$

9. $-13.5 + 8.06 = -5.44$ **10.** -0.37 **11.** $2.4 + 13 = 15.4$

12. 96.2 **13.** $\frac{1}{3} + \frac{1}{2} = \frac{5}{6}$ **14.** $26\frac{1}{6}$

15. $\left(-5\frac{3}{4}\right) + \left(-2\frac{2}{4}\right) = -7\frac{5}{4} = -8\frac{1}{4}$ **16.** -11.26

17.
$$\begin{array}{r} 16.71 \\ +\ 18.90 \\ \hline 35.61 \end{array}$$
18. 12.61 **19.** -3 **20.** 0

21. 260 (When the signs are the same, the product is positive.)

22. 7.7 **23.** $\frac{\cancel{3}^{\,1}}{\cancel{8}_{\,2}} \cdot \frac{-\cancel{4}^{\,-1}}{\cancel{9}_{\,3}} = -\frac{1}{6}$ (Unlike signs give a negative product.)

24. -15 **25.** $-\frac{1}{3}$ **26.** $\frac{5}{2}$ or $2\frac{1}{2}$ **27.** $\left(\frac{15}{2}\right) \cdot \left(\frac{-2}{1}\right) = -15$

28. Undefined

29. 0 (0 divided by a nonzero number *is* possible, and the answer is 0.)

30. Undefined **31.** $\left(-\frac{5}{4}\right)\left(-\frac{8}{3}\right) = \frac{10}{3}$ or $3\frac{1}{3}$ **32.** 250

33. $-8 - (-3) = -8 + 3 = -5$ **34.** -1.56 **35.** $5 + \frac{2}{3} = 5\frac{2}{3}$

36. -7 **37.** $10 + 25 = 35$ **38.** -7

39. $-3(4) = -12$ **40.** 11

Exercises 1.5 (page 29)

1. True; commutative property of addition
2. True; commutative property of addition
3. True; commutative properties of addition and multiplication
4. True; commutative properties of addition and multiplication
5. True; distributive property **6.** True; distributive property
7. False **8.** False **9.** False **10.** False **11.** False
12. False **13.** True; commutative property of multiplication
14. True; commutative properties of addition and multiplication
15. True; 0 is the additive identity
16. True; 0 is the additive identity
17. True; addition is associative **18.** False **19.** False
20. False **21.** True; distributive property
22. True; distributive property
23. True; 1 is the multiplicative identity
24. True; 1 is the multiplicative identity **25.** False
26. True; addition is commutative
27. True; additive inverse property
28. True; additive inverse property
29. True; commutative property of addition (twice)
30. True; commutative property of addition (twice)
31. True; multiplicative inverse property
32. True; additive inverse property
33. $-5(-4)$ **34.** $-3(8)$ **35.** $[7 + (-2)] + 8$
36. $(-9 + 12) + (-3)$ **37.** $-3(2) + (-3)(7)$ **38.** $5(-3) + 5(6)$
39. $8 + (-7)$ **40.** $-4 + (-5)$ **41.** $[9(-4)](7)$
42. $[9(-3)](8)$ **43.** $-8(6 + 3)$ **44.** $9(6 + 8)$

Exercises 1.6 (page 33)

1. $4 \cdot 4 \cdot 4 = 64$ **2.** 49 **3.** $(-3)(-3)(-3)(-3) = 81$ **4.** 16

5. $-2^4 = -(2 \cdot 2 \cdot 2 \cdot 2) = -16$ **6.** -81 **7.** 0 **8.** 0

9. -1; an odd power of a negative number is negative **10.** 1

11. $\left(\frac{1}{2}\right)\left(\frac{1}{2}\right)\left(\frac{1}{2}\right)\left(\frac{1}{2}\right) = \frac{1}{16}$ **12.** $\frac{49}{64}$

13. $(-0.1)(-0.1)(-0.1) = -0.001$ **14.** 0.0001

15. $\left(\frac{1}{10}\right)\left(\frac{1}{10}\right)\left(\frac{1}{10}\right) = \frac{1}{1{,}000}$ **16.** 299.29 **17.** 778.688

18. 5.0625 **19.** 39.0625 **20.** -148.877

Exercises 1.7A (page 34)

1. -6 **2.** 7 **3.** -5 **4.** 8 **5.** -10 **6.** -12 **7.** 9
8. 11 **9.** -16 **10.** -4

Exercises 1.7B (page 35)

1. 458 **2.** 624 **3.** 3.464 **4.** 4.123 **5.** 13.565
6. 13.820 **7.** 1.673 **8.** 3.063

Exercises 1.7C (page 36)

1. 2 **2.** 1 **3.** -3 **4.** -2 **5.** -4 **6.** -3 **7.** -2
8. -4 **9.** -10 **10.** -6 **11.** $-(\sqrt[5]{-32}) = -(-2) = 2$
12. 5

Exercises 1.8 (page 39)

1. $(16 - 9) - 4 = 7 - 4 = 3$ **2.** 9

3. $(12 \div 6) \div 2 = 2 \div 2 = 1$ **4.** 2

5. $(10 \div 2)(-5) = 5(-5) = -25$ **6.** -9 **7.** $3 \times 16 = 48$

8. 45 **9.** $8 + 30 = 38$ **10.** 11 **11.** $7 + \frac{5}{3} = 8\frac{2}{3}$ **12.** $6\frac{1}{2}$

13. $10 - 6 = 4$ **14.** 4 **15.** $10(225) - 64 = 2{,}250 - 64 = 2{,}186$

16. 32 **17.** $\frac{1}{2} - 20 = -19\frac{1}{2}$ **18.** $-3\frac{2}{3}$ or $-\frac{11}{3}$

19. $(100)(5)(4) = 2{,}000$ **20.** $1{,}500$

21. $2 + 300 \div 25 = 2 + 12 = 14$ **22.** 53 **23.** $28 + 2 = 30$

24. 42 **25.** $2(8 - 5)(3) = 2(3)(3) = 6 \times 3 = 18$ **26.** 240

27. $6(-6) = -36$ **28.** -8

29. $-1{,}000 - 5(100)(-3) = -1{,}000 + 1{,}500 = 500$ **30.** $10{,}100$

31. $20 - [5 - (-3)] = 20 - [8] = 12$ **32.** 3

33. $-\frac{5}{5} = -1$ **34.** -4

35. $8 - [5(-8) - 4] = 8 - [-40 - 4] = 8 - [-44] = 52$

36. -12

37. $(3 \times 25 - 5) \div (-7) = (75 - 5) \div (-7) = 70 \div (-7) = -10$

38. -20

39. $15 - \{4 - [2 - 3(2)]\} = 15 - \{4 - [2 - 6]\} = 15 - \{4 - [-4]\}$
$= 15 - \{4 + 4\} = 15 - 8 = 7$

40. 10

Exercises 1.9 (page 43)

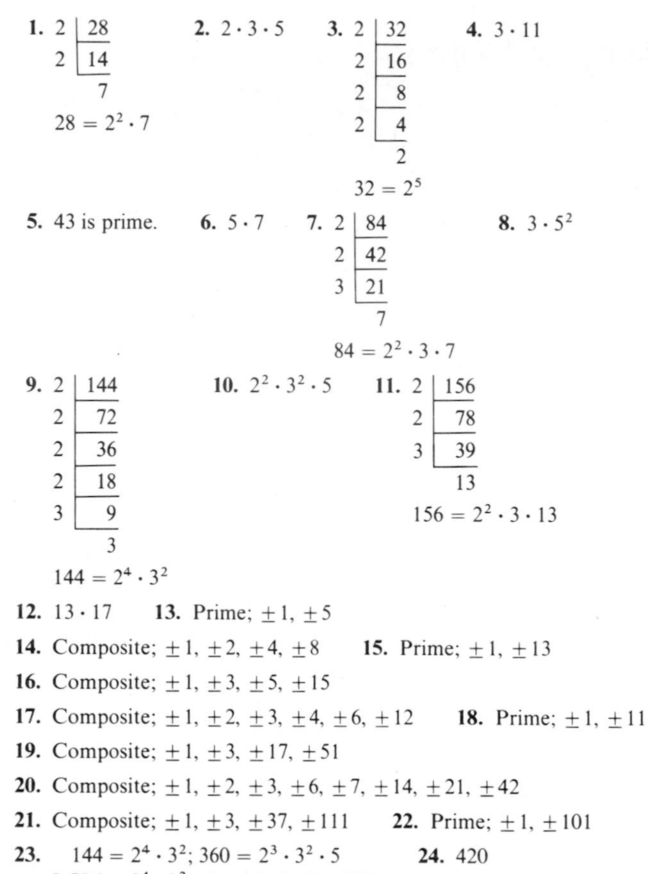

1. $2 \mid 28$, $2 \mid 14$, 7; $28 = 2^2 \cdot 7$ **2.** $2 \cdot 3 \cdot 5$ **3.** $2 \mid 32$, $2 \mid 16$, $2 \mid 8$, $2 \mid 4$, 2; $32 = 2^5$ **4.** $3 \cdot 11$

5. 43 is prime. **6.** $5 \cdot 7$ **7.** $2 \mid 84$, $2 \mid 42$, $3 \mid 21$, 7; $84 = 2^2 \cdot 3 \cdot 7$ **8.** $3 \cdot 5^2$

9. $2 \mid 144$, $2 \mid 72$, $2 \mid 36$, $2 \mid 18$, $3 \mid 9$, 3; $144 = 2^4 \cdot 3^2$ **10.** $2^2 \cdot 3^2 \cdot 5$ **11.** $2 \mid 156$, $2 \mid 78$, $3 \mid 39$, 13; $156 = 2^2 \cdot 3 \cdot 13$

12. $13 \cdot 17$ **13.** Prime; $\pm 1, \pm 5$

14. Composite; $\pm 1, \pm 2, \pm 4, \pm 8$ **15.** Prime; $\pm 1, \pm 13$

16. Composite; $\pm 1, \pm 3, \pm 5, \pm 15$

17. Composite; $\pm 1, \pm 2, \pm 3, \pm 4, \pm 6, \pm 12$ **18.** Prime; $\pm 1, \pm 11$

19. Composite; $\pm 1, \pm 3, \pm 17, \pm 51$

20. Composite; $\pm 1, \pm 2, \pm 3, \pm 6, \pm 7, \pm 14, \pm 21, \pm 42$

21. Composite; $\pm 1, \pm 3, \pm 37, \pm 111$ **22.** Prime; $\pm 1, \pm 101$

23. $144 = 2^4 \cdot 3^2$; $360 = 2^3 \cdot 3^2 \cdot 5$
LCM $= 2^4 \cdot 3^2 \cdot 5 = 16 \cdot 9 \cdot 5 = 720$ **24.** 420

25. $270 = 2 \cdot 3^3 \cdot 5$
$900 = 2^2 \cdot 3^2 \cdot 5^2$
$75 = 3 \cdot 5^2$
LCM $= 2^2 \cdot 3^3 \cdot 5^2 = 2{,}700$ **26.** 2,940

27. The prime numbers between 17 and 37 are 19, 23, 29, and 31; the required one is 19.

28. 19, 31, 37

Review Exercises 1.10 (page 47)

1. Infinite **2.** Finite **3.** $\{0, 1, 2, 3, 4\}$ **4.** No

5. a. $\{2, 5, 7, 8\}$ **b.** $\{5, 7\}$ **c.** $\{\ \}$ or $\varnothing$ **d.** $\{\ \}$ or $\varnothing$

6. a. All are real numbers. **b.** -2 and 0 are integers.
c. None are natural numbers.
d. 2.6457513... is an irrational number.
e. $-2, 4.53, 0.\overline{16}, \frac{2}{3}$, and 0 are rational numbers.

7. 4 **8.** 0 **9.** 10 **10.** 36 **11.** $-(5)(5) = -25$ **12.** 0

13. Undefined **14.** 0 **15.** $6 + 2 \cdot 4 - 8 = 6 + 8 - 8 = 6$

16. 34 **17.** $6 + 18 \div 6 \div 3 = 6 + 3 \div 3 = 6 + 1 = 7$

18. a. $\pm 1, \pm 2, \pm 4, \pm 7, \pm 14, \pm 28$ **b.** $2^2 \cdot 7$ **c.** $2^3 \cdot 3 \cdot 7$
d. $2^3 \cdot 3 \cdot 7 = 168$

19. Yes **20.** No; $1 > 5$, $5 > 7$, and $1 > 7$ are false

21. No; the inequalities are not both $>$ or both $<$

22. False **23.** True; multiplicative identity

24. True; associative property of addition

25. True; additive inverse **26.** False

27. True; additive identity

28. True; commutative property of addition **29.** False

30. True; distributive property

Exercises 1.11A (page 53)

1. $10^{2+4} = 10^6$ **2.** 2^5 **3.** $x^{2+5} = x^7$ **4.** y^9

5. Cannot be simplified **6.** Cannot be simplifed **7.** 2^{x+y}

8. 3^{m+n} **9.** $a^{8-3} = a^5$ **10.** x^3 **11.** Cannot be simplified

12. Cannot be simplified **13.** $10^{3 \cdot 2} = 10^6$ **14.** 5^8

15. $3^{a \cdot b} = 3^{ab}$ **16.** 2^{mn} **17.** x^5y^5 **18.** u^4v^4

19. 2^6x^6 or $64x^6$ **20.** 3^4x^4 **21.** Cannot be simplified

22. Cannot be simplified **23.** $3^{4x+7x} = 3^{11x}$ **24.** 5^{4x}

25. a^{x-y} **26.** x^{a-b}

27. Cannot be simplified because the bases are different

28. Cannot be simplified **29.** $\frac{x^3}{y^3}$ **30.** $\frac{u^7}{v^7}$

31. $\frac{3^4}{x^4}$ or $\frac{81}{x^4}$ **32.** $\frac{x^2}{25}$

Exercises 1.11B (page 58)

1. $\frac{1}{a^3}$ **2.** $\frac{1}{x^2}$ **3.** $\frac{1}{10^3}$ or $\frac{1}{1{,}000}$ **4.** $\frac{1}{10^5}$ **5.** $\frac{5}{b^7}$ **6.** $\frac{3}{y^2}$

7. $\frac{1}{(5b)^2} = \frac{1}{5^2b^2} = \frac{1}{25b^2}$ **8.** $\frac{1}{9y^2}$ **9.** $\frac{1}{x^3} \cdot \frac{y^2}{1} \cdot 1 = \frac{y^2}{x^3}$ **10.** $\frac{r^3}{s^4}$

11. $\frac{x}{1} \cdot \frac{1}{y^2} \cdot \frac{1}{z^3} \cdot 1 = \frac{x}{y^2z^3}$ **12.** $\frac{b}{z^4c^5}$ **13.** $a^3 \cdot \frac{b^4}{1} = a^3b^4$

14. c^4d^5 **15.** $\frac{1}{x^3} \cdot \frac{y^2}{1} = \frac{y^2}{x^3}$ **16.** $\frac{Q^4}{P^2}$ **17.** $x^{-7} = \frac{1}{x^7}$

18. $\frac{1}{y^5}$ **19.** $a^{-6} = \frac{1}{a^6}$ **20.** $\frac{1}{b^8}$ **21.** $x^{8+(-2)} = x^6$ **22.** a^2

23. $x^{-6a} = \frac{1}{x^{6a}}$ **24.** $\frac{1}{y^{6c}}$ **25.** $y^{3-(-2)} = y^5$ **26.** x^{12}

27. $x^{3a-(-a)} = x^{4a}$ **28.** a^{6x} **29.** 1 **30.** 1 **31.** $5 \cdot 1 = 5$

32. 2 **33.** $\frac{1}{x^3} + \frac{1}{x^5}$ **34.** $\frac{1}{y^2} + \frac{1}{y^6}$ **35.** $x^7 - \frac{1}{x^5}$

36. $y^{10} - \frac{1}{y^3}$ **37.** $1 + 1 = 2$ **38.** 3 **39.** 1 **40.** 1

41. $10^{5+(-2)} = 10^3 = 1{,}000$ **42.** 4 **43.** $3^{(-2)(-2)} = 3^4 = 81$

44. 1,000 **45.** $(1)^5 = 1$ **46.** 1

47. $10^{2+(-1)-(-3)} = 10^{2-1+3} = 10^4 = 10{,}000$ **48.** 8 **49.** $x^{-3}y$

50. xy^{-2} **51.** a^4x **52.** m^2n^3 **53.** $x^4y^{-3}z^2$ **54.** $a^{-1}b^3c^4$

Exercises 1.11C (page 61)

1. $\frac{1}{(5x)^3} = \frac{1}{5^3x^3} = \frac{1}{125x^3}$ **2.** $\frac{1}{16y^2}$ **3.** $\frac{7}{1} \cdot \frac{1}{x^2} = \frac{7}{x^2}$ **4.** $\frac{3}{y^4}$

5. $\frac{27}{x^3}$ **6.** $\frac{7^2}{y^2}$ or $\frac{49}{y^2}$ **7.** $\frac{8^2}{z}$ or $\frac{64}{z}$ **8.** $\frac{8}{x}$ **9.** $a^{2 \cdot 2}b^{3 \cdot 2} = a^4b^6$

10. $x^{12}y^{15}$ **11.** $m^{(-2)4}n^{1\cdot 4} = m^{-8}n^4 = \dfrac{n^4}{m^8}$ **12.** $\dfrac{r^5}{p^{15}}$

13. $x^{(-2)(-4)}y^{3(-4)} = x^8y^{-12} = \dfrac{x^8}{y^{12}}$ **14.** $\dfrac{w^6}{z^8}$ **15.** $k^{(-4)(-2)} = k^8$

16. z^{10} **17.** $2^{1\cdot 3}x^{2\cdot 3}y^{(-4)3} = 8x^6y^{-12} = \dfrac{8x^6}{y^{12}}$ **18.** $\dfrac{9b^{10}}{a^2}$

19. $5^{1(-2)}m^{(-3)(-2)}n^{5(-2)} = 5^{-2}m^6n^{-10} = \dfrac{m^6}{25n^{10}}$ **20.** $\dfrac{y^2}{8x^8}$

21. $\dfrac{x^{1\cdot 2}y^{4\cdot 2}}{z^{2\cdot 2}} = \dfrac{x^2y^8}{z^4}$ **22.** $\dfrac{a^9b^3}{c^6}$ **23.** $\dfrac{M^{(-2)4}}{N^{3\cdot 4}} = \dfrac{M^{-8}}{N^{12}} = \dfrac{1}{M^8N^{12}}$

24. $R^{15}S^{12}$ **25.** $\dfrac{x^{(-5)(-2)}}{y^{4(-2)}z^{(-3)(-2)}} = \dfrac{x^{10}}{y^{-8}z^6} = \dfrac{x^{10}y^8}{z^6}$ **26.** $\dfrac{a^{12}b^6}{c^{15}}$

27. 1 **28.** 1 **29.** $\dfrac{3^{1\cdot 2}x^{2\cdot 2}}{y^{3\cdot 2}} = \dfrac{9x^4}{y^6}$ **30.** $\dfrac{16a^{16}}{b^8}$

31. $\dfrac{4^{1(-1)}a^{(-2)(-1)}}{b^{3(-1)}} = \dfrac{4^{-1}a^2}{b^{-3}} = \dfrac{a^2b^3}{4}$ **32.** $25m^8n^6$

33. $(x^{-1-4}y^2)^{-2} = (x^{-5}y^2)^{-2} = x^{(-5)(-2)}y^{2(-2)} = x^{10}y^{-4} = \dfrac{x^{10}}{y^4}$

34. $u^{12}v^3$ **35.** $(x^{3n+2n-4n})^2 = (x^n)^2 = x^{2n}$ **36.** $x^{2n-2}y^{2n-4}$

37. $(2)(-3)(x^5 \cdot x^4)(y^2) = -6x^9y^2$ **38.** $-28a^2b^5$

39. $(-4)(-2)(x^{-2}xy^3y^{-1}) = 8x^{-1}y^2 = \dfrac{8y^2}{x}$ **40.** $\dfrac{10b}{a}$

41. $(2)(3)(-1)(s^3 \cdot s \cdot s)(1)(u^{-4} \cdot u^3) = -6s^{3+1+1}u^{-4+3}$
$= -6s^5u^{-1} = \dfrac{-6s^5}{u}$

42. $\dfrac{-8x^4}{z}$

Exercises 1.12 (page 66)

1. 2.856×10^1 **2.** 3.754×10^2 **3.** 6.184×10^{-2}
4. 3.056×10^{-3} **5.** 7.8×10^4 **6.** 1.4×10^3
7. 2.006×10^{-1} **8.** 9.5×10^{-5}
9. $(3.62 \times 10^{-1}) \times 10^{-2} = 3.62 \times 10^{-3}$ **10.** 6.311×10^{-4}
11. $(2.452 \times 10^2) \times 10^{-5} = 2.452 \times 10^{-3}$ **12.** 3.17×10^{-3}
13. $\dfrac{(6 \times 10^{-5}) \times (8 \times 10^8)}{(5 \times 10^7) \times (3 \times 10^{-4})} = \dfrac{6 \times 8}{5 \times 3} \times \dfrac{10^{-5} \times 10^8}{10^7 \times 10^{-4}} = \dfrac{16}{5} \times \dfrac{10^3}{10^3}$
$= 3.2 \times 10^0$, or 3.2
14. 2.5×10^4, or 25,000 **15.** 1.288×10^{10} **16.** 3×10^{-9}
17. 1.6×10^{-3} **18.** 9×10^{-4}
19. $\left(6.02 \times 10^{23} \dfrac{\text{molecules}}{\text{mole}}\right)(600 \text{ moles}) = 3{,}612 \times 10^{23}$ molecules
$= (3.612 \times 10^3) \times 10^{23}$ molecules $= 3.612 \times 10^{26}$ molecules
20. \$9,360,000,000
21. $\dfrac{4{,}400{,}000{,}000 \text{ mi}}{12 \text{ years}}\left(\dfrac{1 \text{ year}}{365 \text{ days}}\right)\left(\dfrac{1 \text{ day}}{24 \text{ hr}}\right) \approx 41{,}857 \dfrac{\text{mi}}{\text{hr}}$
22. 1.60704×10^{10} mi = 16,070,400,000 mi

Exercises 1.13A (page 69)

1. $2(-6)^2 + 3(5) = 2(36) + 15 = 72 + 15 = 87$ **2.** 5
3. $(-5) - 12\left(\dfrac{1}{3}\right)^2 = -5 - 12\left(\dfrac{1}{9}\right) = -5 - \dfrac{4}{3} = -\dfrac{19}{3} = -6\dfrac{1}{3}$
4. $-\dfrac{26}{3}$ or $-8\dfrac{2}{3}$ **5.** $(-5^2) - 4(5)(-6) = 25 - (-120) = 145$
6. 61 **7.** $(5 + [-6])^2 = (-1)^2 = 1$ **8.** 36
9. $(5)^2 + 2(5)(-6) + (-6)^2 = 25 - 60 + 36 = 1$ **10.** 36
11. $5^2 + (-6)^2 = 25 + 36 = 61$ **12.** 26
13. $\dfrac{3(0)}{5 + (-15)} = \dfrac{0}{-10} = 0$ **14.** 0
15. $2(-1) - [5 - (0 - 5(-15))] = -2 - [5 - (75)]$
$= -2 - [-70] = 68$
16. -32
17. $-(-1) - \sqrt{(-1)^2 - 4(-4)(5)} = 1 - \sqrt{1 + 80}$
$= 1 - \sqrt{81} = 1 - 9 = -8$
18. 12 **19.** $5b^2 - 4b + 8$ **20.** $4c + 3$
21. $2(x^2 - 4x)^2 - 3(x^2 - 4x) + 7$ **22.** $(y^4 + 2)^2 - 2(y^4 + 2)$

Exercises 1.13B (page 70)

1. $q = \dfrac{DQ}{H} = \dfrac{5(420)}{30} = 70$ **2.** 125
3. $A = P(1 + rt) = 500[1 + 0.09(2.5)]$
$= 500(1 + 0.225) = 500(1.225) = 612.50$
4. 498 **5.** $A = P(1 + i)^n = 600(1.085)^2 = 600(1.177225) \approx 706.34$
6. $808.9375 \approx 808.94$
7. $C = \dfrac{5}{9}(F - 32) = \dfrac{5}{9}(-10 - 32) = \dfrac{5}{9}(-42) = -23\dfrac{1}{3}$ **8.** $-21\dfrac{2}{3}$
9. $S = \dfrac{1}{2}gt^2 = \dfrac{1}{2}(32)\left(8\dfrac{1}{2}\right)^2 = (16)\left(\dfrac{17}{2}\right)^2 = 1{,}156$ **10.** 361
11. $Z = \dfrac{Rr}{R + r} = \dfrac{22(8)}{22 + 8} = \dfrac{176}{30} = 5\dfrac{13}{15} \approx 5.87$ **12.** $17\dfrac{3}{16}$
13. $S = 2\pi r^2 + 2\pi rh \approx 2(3.14)(3)^2 + 2(3.14)(3)(20)$
$= 2(3.14)(9) + 376.8 = 56.52 + 376.8 = 433.32$
14. ≈ 602.88

Exercises 1.14A (page 72)

1. $2x^{2/2} = 2x$ **2.** $3y$ **3.** $m^{4/2}n^{2/2} = m^2n$ **4.** u^5v^3
5. $5a^{4\div 2}b^{2\div 2} = 5a^2b$ **6.** $10b^2c$ **7.** $x^{10\div 2}y^{4\div 2} = x^5y^2$
8. x^6y^4 **9.** $10a^{10\div 2}y^{2\div 2} = 10a^5y$ **10.** $11a^{12}b^2$
11. $9m^{8\div 2}n^{16\div 2} = 9m^4n^8$ **12.** $7c^9d^5$

Exercises 1.14B (page 73)

1. $(3a)(6) + (3a)(x) = 18a + 3ax$ **2.** $35b + 5by$
3. $(x)(-4) + (-5)(-4) = -4x + 20$ **4.** $-5y + 10$
5. $(-3)(x) - (-3)(2y) + (-3)(2)$ **6.** $-2x + 6y - 8$
$= -3x - (-6y) + (-6)$
$= -3x + 6y - 6$
7. $(x)(xy) - (x)(3) = x^2y - 3x$ **8.** $a^2b - 4a$
9. $(3a)(ab) + (3a)(-2a^2) = 3a^2b - 6a^3$ **10.** $12x^2 - 8xy^2$
11. $(3x^3)(-2xy) + (-2x^2y)(-2xy) + (y^3)(-2xy)$
$= -6x^4y + 4x^3y^2 - 2xy^4$
12. $-8yz^4 + 2y^2z^3 + 2y^4z$
13. $(-2)(3)(6)(a \cdot a^2 \cdot a)(b \cdot b \cdot b)(c^3) = -36a^4b^3c^3$
14. $-30x^4y^5z^2$
15. $(4xy^2)(3x^3y^2) - (4xy^2)(2x^2y^3) + (4xy^2)(5xy^4)$
$= 12x^4y^4 - 8x^3y^5 + 20x^2y^6$
16. $-10x^6y^1 + 4x^5y^2 + 6x^3y^3$

17. $(3mn^2)(-2m^2n)(5m^2) + (3mn^2)(-2m^2n)(-n^2)$
$= (3)(-2)(5)m^{1+2+2}n^{2+1} + (3)(-2)(-1)m^{1+2}n^{2+1+2}$
$= -30m^5n^3 + 6m^3n^5$

18. $-36a^5b^3 + 18a^3b^5$

19. $-8x^3y^2z[3xy - 2xz + 5yz]$
$= (-8x^3y^2z)(3xy) - (-8x^3y^2z)(2xz) + (-8x^3y^2z)(5yz)$
$= -24x^4y^3z - (-16x^4y^2z^2) + (-40x^3y^3z^2)$
$= -24x^4y^3z + 16x^4y^2z^2 - 40x^3y^3z^2$

20. $21a^4b^3 - 14a^3b^4 - 7a^3b^3c$ 21. $7x + 3y - 2x^2$

22. $3a - b - 4c$

Exercises 1.14C (page 76)

1. $10 + 4x - y$ 2. $8 + 3a - b$
3. $7 - 1(-4R - S) = 7 + 4R + S$ 4. $9 + 3m + n$
5. $6 - 2a + 6b$ 6. $12 - 6R + 3S$ 7. $3 - 2x^2 + 8xy$
8. $2 - 10x^2 + 15xy$ 9. $-x + y + 2 - a$
10. $-a + b + x - 3$ 11. $2a - 2b - 6$ 12. $3x - 3y - 5$
13. $x - [a + y - b] = x - a - y + b$ 14. $y - m - x + n$
15. $5 - 3[a - 8x + 4y] = 5 - 3a + 24x - 12y$
16. $7 - 5x + 30a - 15b$
17. $2 - [a - 1(b - c)] = 2 - 1[a - b + c] = 2 - a + b - c$
18. $5 - x + y + z$
19. $9 - 2[-3a - 8x + 4y] = 9 + 6a + 16x - 8y$
20. $P - x + y - 4 + z$

Exercises 1.14D (page 77)

1. $-2x$ 2. $-a$ 3. $6x^2y$ 4. $7ab^2$ 5. $6xy^2 + 8x^2y$
6. $5a^2b - 4ab^2$ 7. $-2xy$ 8. $4mn$ 9. $5xyz^2 - 2x^2y^2z^2$
10. $3a^3b^3c^3 - 4abc^3$ 11. $0xyz^2 = 0$ 12. $1a^2bc$ or a^2bc
13. $3x^2y - 2xy^2$ 14. $2xy^2 - 5x^2y$ 15. $2ab - a + b$
16. $6xy - x - y$ 17. $2x^3 - 2x^2 - 2x$ 18. $-3y^3 + 5y^2 - 2y$
19. $2x - 9y + 11$ 20. $-7a - 2b - 5$
21. $-2a^2b + (-ab) + 7ab^2 = -2a^2b - ab + 7ab^2$
22. $-4x^2y + 4xy^2 + y$

Exercises 1.14E (page 79)

1. $6h^3 - 2hk - hk + 3k^4 = 6h^3 - 3hk + 3k^4$ 2. $11xy^2 - 14x^2$
3. $3x - 4 - 5x = -2x - 4$ 4. $-3x - 7$
5. $(-5x)(3x) - (-5x)(4) = -15x^2 + 20x$ 6. $-40x^2 + 56x$
7. $2 + 3x$ 8. $5 + 8y$
9. $3x - [5y - 2x + 4y] = 3x - [9y - 2x] = 3x - 9y + 2x$
$= 5x - 9y$
10. $5x - 9y$ 11. $-10[-6x + 10 + 17] - 4x$
$= -10[-6x + 27] - 4x$
$= 60x - 270 - 4x$
$= 56x - 270$
12. $115x - 640$
13. $8 - 2(x - y + 3x) = 8 - 2(4x - y) = 8 - 8x + 2y$
14. $9 - 12u + 4t$ 15. $8x + 10x^2 - 4x$
$= 4x + 10x^2$
16. $4y + 25y^2$
17. $3u - v - \{2u - 10 + v - 20\} - 8v$
$= 3u - v - \{2u + v - 30\} - 8v$
$= 3u - v - 2u - v + 30 - 8v = u - 10v + 30$

18. $-5x - 2y + 23$
19. $50 - \{-2t - [5t - 6 + 2t]\} + 7^0$
$= 50 - \{-2t - [7t - 6]\} + 1$
$= 50 - \{-2t - 7t + 6\} + 1$
$= 50 - \{-9t + 6\} + 1$
$= 50 + 9t - 6 + 1 = 45 + 9t$

20. $22 + 11x$
21. $100v - 3\{-4[8 + 2v - 5v]\}$
$= 100v - 3\{-4[8 - 3v]\}$
$= 100v - 3\{-32 + 12v\}$
$= 100v + 96 - 36v$
$= 64v + 96$

22. $72z + 96$ 23. $w^4 - 4w^2 + 4w^2 - 16 = w^4 - 16$
24. $x^4 - 81$ 25. $(3)(5)(4)(2)(x \cdot x^2 \cdot x^3) = 120x^6$ 26. $60y^8$
27. $5X^{-4+6} + 3(1) = 5X^2 + 3$ 28. $3Y^2 + 2$
29. $3(x + 2y) - 5(3x - y)$
$= 3x + 6y - 15x + 5y = -12x + 11y$
30. $-2x + 10y$
31. $2[(3)(x + 2y) - (3x - y)]$
$= 2[3x + 6y - 3x + y] = 2(7y) = 14y$
32. $42x - 21y$

Review Exercises 1.15 (page 81)

1. $x^{3+5} = x^8$ 2. $x^4 + x^2$ 3. $N^{2 \cdot 3} = N^6$ 4. $s^6 - s^2$
5. $a^{5-2} = a^3$ 6. $\dfrac{x^6}{y^4}$ 7. $\dfrac{2^3a^3}{b^{2 \cdot 3}} = \dfrac{8a^3}{b^6}$ 8. $\dfrac{x^4}{y^2}$
9. $\left(\dfrac{x^2y}{x^4}\right)^{-1} = \left(\dfrac{y}{x^2}\right)^{-1} = \dfrac{y^{-1}}{x^{-2}} = \dfrac{x^2}{y}$ 10. 2 11. 1 12. xy^4
13. $3c^4d^2 - 12c^3d^3$ 14. $8 - 6x + 2y$
15. $(-10)(-8)(-1)(x^2x^3x)(y^3y^2)(z^4) = -80x^6y^5z^4$
16. $-8x^4 - 4x^2 + 2xy$ 17. $5 - 2[3 - 5x + 5y + 4x - 6]$
$= 5 - 2[-3 - x + 5y]$
$= 5 + 6 + 2x - 10y$
$= 11 + 2x - 10y$
18. $4x^3 - y$ 19. 1.486×10^2 20. 0.00317
21. $A = P(1 + rt) = 550[1 + (0.09)(2.5)]$
$= 550(1 + 0.225)$
$= 550(1.225) = 673.75$
22. $C = 40$
23. $S = R\left[\dfrac{(1 + i)^n - 1}{i}\right] = 750\left[\dfrac{(1.09)^3 - 1}{0.09}\right] = 2{,}458.575$
24. 472.5

Chapter 1 Diagnostic Test (page 85)

Following each problem number is the textbook section number (in parentheses) where that kind of problem is discussed.

1. a. (1.1) True b. (1.2) True c. (1.2) True
d. (1.4) False; division by zero is undefined
2. a. (1.5) False; it illustrates the associative property of multiplication
b. (1.5) False
c. (1.1) False; 0 is a digit but not a natural number
d. (1.2) True
3. a. (1.3) True
b. (1.2) False; irrational numbers are real but not rational
c. (1.3) False; one symbol is "<" and the other is ">"
d. (1.5) False; $3 \cdot (7 \cdot 2) = 3 \cdot 14 = 42$, while $(3 \cdot 7)(3 \cdot 2) = 21(6) = 126$
4. (1.1) a. False; $y \notin A$ b. True
5. (1.2) a. All b. $-3, 0, 5$ c. 5 d. 2.8652916...
e. $-3, 2.4, 0, 5, \frac{1}{2}, 0.\overline{18}$

6. (1.1) **a.** $\{r, s, w, x, y, z\}$ **b.** $\{\ \}$ **c.** $\{x, w\}$
d. $B \cap C = \{y\}$; $A \cup (B \cap C) = \{x, z, w\} \cup \{y\} = \{w, x, y, z\}$

7. a. (1.3) $|-17| = 17$ **b.** (1.6) $(-5)^2 = (-5)(-5) = 25$
c. (1.4) -6 **d.** (1.4) $-35 + (-2) = -37$

8. a. (1.4) $-27 + (+17) = -10$ **b.** (1.4) 72
c. (1.4) 0 **d.** (1.4) 5

9. a. (1.4) -22 **b.** (1.6) $-(6)(6) = -36$ **c.** (1.11) 1
d. (1.3) $-(3) = -3$

10. a. (1.7) -3 **b.** (1.7) 2 **c.** (1.11) $3^{(-2)(-1)} = 3^2 = 9$
d. (1.11) $10^{-3+5} = 10^2 = 100$

11. a. (1.11) $2^{-4-(-7)} = 2^{-4+7} = 2^3 = 8$
b. (1.8) $(16 \div 4)(2) = 4 \cdot 2 = 8$
c. (1.8) $2 \cdot 3 - 5 = 6 - 5 = 1$ **d.** (1.7) 9

12. (1.12) $$\frac{(8.1 \times 10^7) \times (3 \times 10^{-8})}{(4 \times 10^{-5}) \times (6 \times 10^8)} = \frac{8.1 \times 3}{4 \times 6} \times \frac{10^7 \times 10^{-8}}{10^{-5} \times 10^8}$$
$$= 1.0125 \times \frac{10^{-1}}{10^3} = 1.0125 \times 10^{-4} = 0.00010125$$

13. (1.9) The prime numbers between 7 and 29 are 11, 13, 17, 19, and 23. The only one that yields a remainder of 2 when divided by 5 is 17.

14. (1.9) **a.** $2\,\underline{|\,78}$, $3\,\underline{|\,39}$, 13; $78 = 2 \cdot 3 \cdot 13$ **b.** $5\,\underline{|\,65}$, 13; $65 = 5 \cdot 13$

15. (1.9) LCM $= 2 \cdot 3 \cdot 5 \cdot 13 = 390$

16. (1.11) $x^{2+(-5)} = x^{-3} = \dfrac{1}{x^3}$ **17.** (1.11) $N^{2 \cdot 4} = N^8$

18. (1.11) $\dfrac{2^2X^6}{Y^2} = \dfrac{4X^6}{Y^2}$ **19.** (1.11) $\dfrac{x^{-1}y^2}{y^3} = \dfrac{1}{xy}$

20. (1.11) $\dfrac{1}{a^{-3}} = a^3$ (Property 1.7b)

21. (1.14) $7x - 10 + 2x + 9x = 18x - 10$

22. (1.14) $\underline{12x^2y^2} - \underline{18x^4} - \underline{6x^2y^2} + \underline{18x^4} = 6x^2y^2$

23. (1.14) $7x - 2\{6 - 3[8 - 2x + 6 - 12 + 2x]\}$
$= 7x - 2\{6 - 3[2]\}$
$= 7x - 2\{6 - 6\} = 7x - 2\{0\} = 7x - 0 = 7x$

24. (1.13)
$$S = \frac{a(1 - r^n)}{1 - r} = \frac{-8[1 - (3)^2]}{1 - (3)} = \frac{-8[1 - 9]}{1 - 3} = \frac{-8[-8]}{-2} = \frac{64}{-2} = -32$$

25. (1.12) $p = 3{,}400{,}000{,}000 = 3.4 \times 10^9$, $r = 0.075$, $t = 3$
$I = prt = (3.4 \times 10^9)(0.075)(3) = 765{,}000{,}000$; therefore, the interest is \$765,000,000.

Exercises 2.1 (page 92)

Checks will not always be shown.

1. $4x + 12 + 9x = -1$
$13x + 12 = -1$
$13x = -13$
$x = -1$

Check for $x = -1$
$4(-1) + 3(4 + 3[-1]) \stackrel{?}{=} -1$
$-4 + 3(4 + [-3]) \stackrel{?}{=} -1$
$-4 + 3(1) \stackrel{?}{=} -1$
$-1 = -1$

Solution set: $\{-1\}$; graph:

−3 −2 −1 0 1

2. $\{-2\}$; graph:

−3 −2 −1 0 1

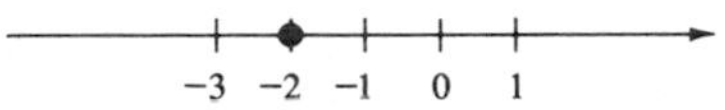

3. $7y - 10 - 8y = 8$
$-y = 18$
$y = -18$
Solution set: $\{-18\}$; graph:

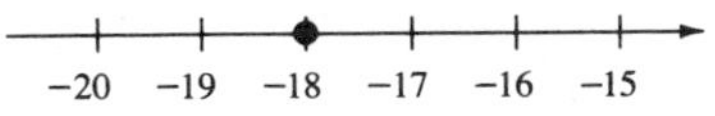

−20 −19 −18 −17 −16 −15

4. $\{-15\}$; graph:

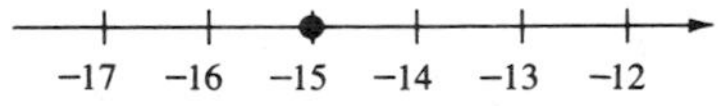

−17 −16 −15 −14 −13 −12

5. $4x + 12 = 12 + 4x$
$12 = 12 \leftarrow$ True
Identity; solution set is R

6. Identity; solution set is R

7. $3[5 - 10 + 2z] = 6z + 14$
$3[-5 + 2z] = 6x + 14$
$-15 + 6z = 6z + 14$
$-15 = 14 \leftarrow$ False
No solution; solution set is $\{\ \}$

8. No solution; solution set is $\{\ \}$

9. (LCM is 6)
$$(6)\left(\frac{x}{3} - \frac{x}{6}\right) = (6)(18)$$
$$\overset{2}{(\cancel{6})}\left(\frac{x}{\underset{1}{\cancel{3}}}\right) - \overset{1}{(\cancel{6})}\left(\frac{x}{\underset{1}{\cancel{6}}}\right) = 108$$
$2x - x = 108$
$x = 108$

Check for $x = 108$
$\dfrac{108}{3} - \dfrac{108}{6} \stackrel{?}{=} 18$
$36 - 18 \stackrel{?}{=} 18$
$18 = 18$

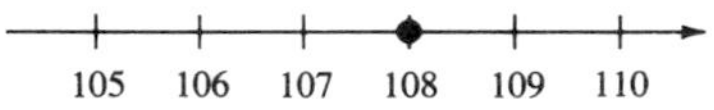

$\{108\}$; graph:

105 106 107 108 109 110

10. $\{128\}$; graph:

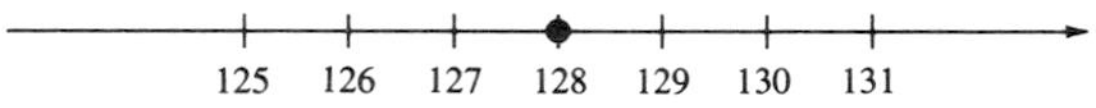

125 126 127 128 129 130 131

11. LCD $= 40$
$$\frac{\overset{5}{\cancel{40}}}{1}\left(\frac{y+3}{\underset{1}{\cancel{8}}}\right) - \frac{\overset{10}{\cancel{40}}}{1}\left(\frac{3}{\underset{1}{\cancel{4}}}\right) = \frac{\overset{4}{\cancel{40}}}{1}\left(\frac{y+6}{\underset{1}{\cancel{10}}}\right)$$
$5y + 15 - 30 = 4y + 24$
$5y - 15 = 4y + 24$
$y = 39$

$\{39\}$; graph:

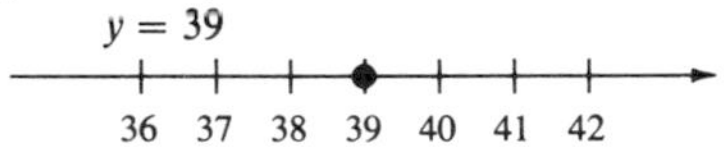

36 37 38 39 40 41 42

12. $\{-25\}$; graph:

−28 −27 −26 −25 −24 −23 −22

13. $5z - 6 - 9z = 6$
$-6 - 4z = 6$
$-4z = 12$
$z = -3$
$-3 \notin N$; no solution; $\{\ \}$

14. No solution; $\{\ \}$

15. $7x - 10 - 8x = 8$
$-x = 18$
$x = -18$
$x \in J$
$\{-18\}$; graph:

−20 −19 −18 −17 −16

16. $\{0\}$; graph:

−2 −1 0 1 2

17. $6x - 12 - 15x - 12 = 35x - 40$
$-9x - 24 = 35x - 40$
$16 = 44x$
$x = \frac{16}{44} = \frac{4}{11}$
$\frac{4}{11} \notin J$; no solution; { }

18. No solution; { }

19. LCD = 10

$$\frac{\overset{2}{\cancel{10}}}{1} \cdot \frac{2(y-3)}{\underset{1}{\cancel{5}}} - \frac{\overset{5}{\cancel{10}}}{1} \cdot \frac{3(y+2)}{\underset{1}{\cancel{2}}} = \frac{\overset{1}{\cancel{10}}}{1} \cdot \frac{7}{\underset{1}{\cancel{10}}}$$

$4y - 12 - 15y - 30 = 7$
$-11y = 49$
$y = \frac{-49}{11}$ or $-4\frac{5}{11}$
$\{-4\frac{5}{11}\}$; graph:
−6 −5 −4 −3 −2 −1

20. $\{\frac{81}{11}\}$ or $\{7\frac{4}{11}\}$; graph:
5 6 7 8 9 10

21. $6.23x + 2.5(3.08 - 8.2x) = -14.7$
$6.23x + 7.7 - 20.5x = -14.7$
$-14.27x + 7.7 = -14.7$
$-14.27x = -22.4$
$x \approx 1.57$
$\{\approx 1.57\}$; graph:
−1 0 1 2 3

22. $\{\approx 0.97\}$; graph:
−1 0 1 2

Exercises 2.2 (page 98)

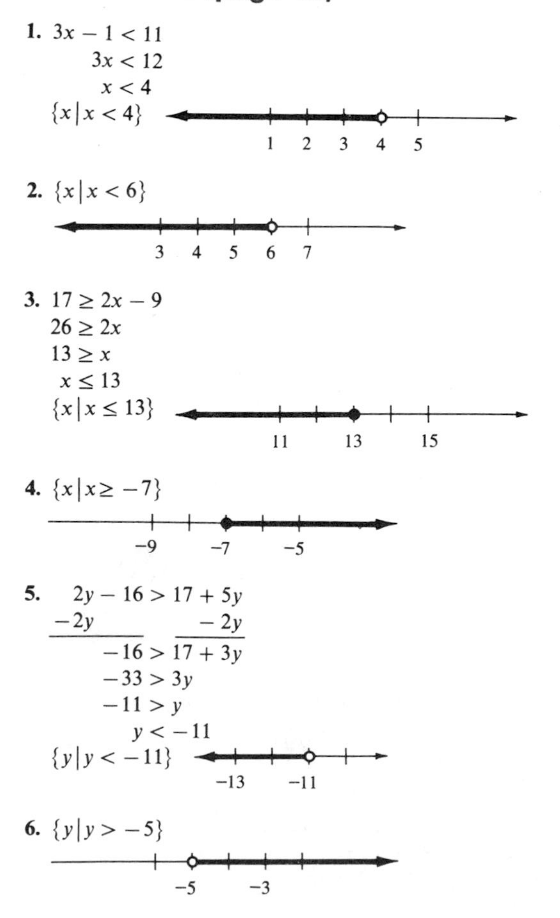

1. $3x - 1 < 11$
$3x < 12$
$x < 4$
$\{x \mid x < 4\}$
1 2 3 4 5

2. $\{x \mid x < 6\}$
3 4 5 6 7

3. $17 \geq 2x - 9$
$26 \geq 2x$
$13 \geq x$
$x \leq 13$
$\{x \mid x \leq 13\}$
11 13 15

4. $\{x \mid x \geq -7\}$
−9 −7 −5

5. $2y - 16 > 17 + 5y$
$\underline{-2y \qquad\qquad -2y}$
$-16 > 17 + 3y$
$-33 > 3y$
$-11 > y$
$y < -11$
$\{y \mid y < -11\}$
−13 −11

6. $\{y \mid y > -5\}$
−5 −3

7. $4z - 22 < 6z - 42$
$\underline{-4z \qquad\qquad -4z}$
$-22 < 2z - 42$
$20 < 2z$
$10 < z$ or $z > 10$
$\{z \mid z > 10\}$
9 10 11 12 13

8. $\{a \mid a < -2\}$
−4 −3 −2 −1 0

9. $18 - 45m - 4 \geq 13m + 24 - 56m$
$14 - 45m \geq 24 - 43m$
$\underline{+45m \qquad\qquad +45m}$
$14 \geq 24 + 2m$
$-10 \geq 2m$
$-5 \geq m$
$m \leq -5$
$\{m \mid m \leq -5\}$
−7 −6 −5 −4 −3

10. $\{k \mid k \leq 1\}$
−1 0 1 2 3

11. $10 - 5x > 2[3 - 5x + 20]$
$10 - 5x > 2[23 - 5x]$
$10 - 5x > 46 - 10x$
$10 + 5x > 46$
$5x > 36$
$x > \frac{36}{5}$ or $x > 7\frac{1}{5}$
$\{x \mid x > 7\frac{1}{5}\}$
5 6 7 8 9

12. $\{y \mid y < -6\}$
−8 −7 −6 −5 −4

13. LCD = 12

$$\frac{\overset{4}{\cancel{12}}}{1} \cdot \frac{z}{\underset{1}{\cancel{3}}} > \frac{12}{1} \cdot \frac{7}{1} - \frac{\overset{3}{\cancel{12}}}{1} \cdot \frac{z}{\underset{1}{\cancel{4}}}$$

$4z > 84 - 3z$
$7z > 84$
$z > 12$
$\{z \mid z > 12\}$
10 11 12 13 14

14. $\{t \mid t > 15\}$
13 14 15 16 17

15. LCD = 15

$$\frac{\overset{5}{\cancel{15}}}{1} \cdot \frac{1}{\underset{1}{\cancel{3}}} + \frac{\overset{3}{\cancel{15}}}{1} \cdot \frac{(w+2)}{\underset{1}{\cancel{5}}} \geq \frac{\overset{5}{\cancel{15}}(w-5)}{1 \quad \underset{1}{\cancel{3}}}$$

$5 + 3w + 6 \geq 5w - 25$
$11 \geq 2w - 25$
$36 \geq 2w$
$18 \geq w$ or $w \leq 18$
$\{w \mid w \leq 18\}$
16 18 20

16. $\{u \mid u \geq 6\}$
4 5 6 7 8

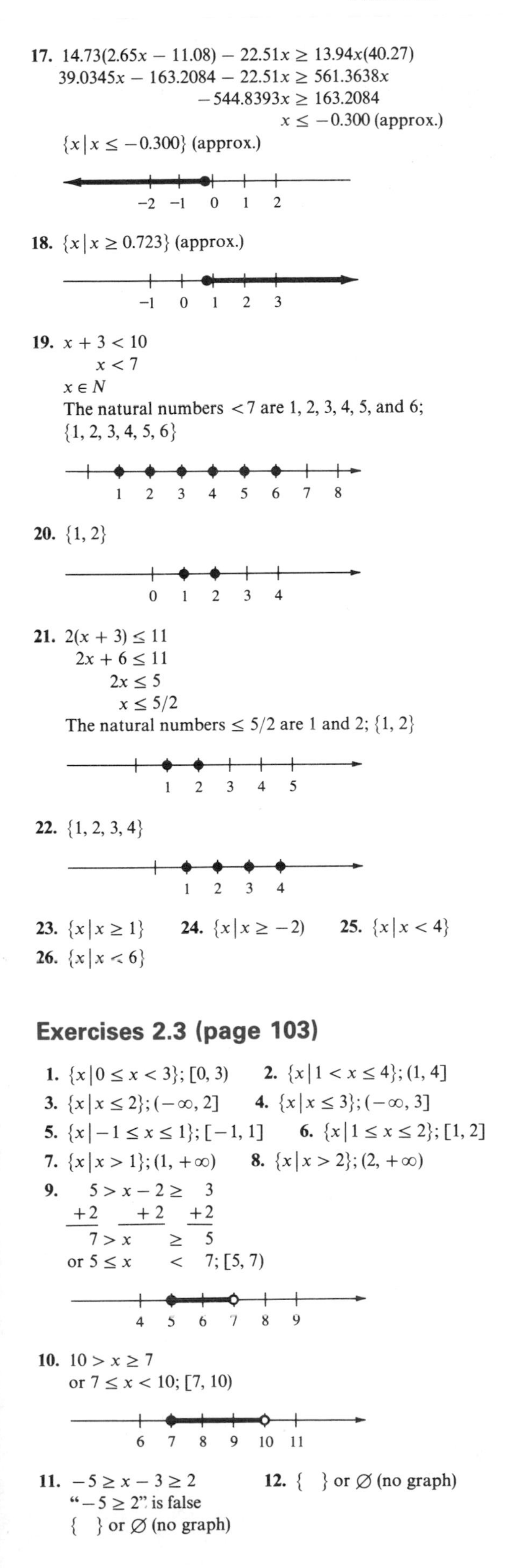

17. $14.73(2.65x - 11.08) - 22.51x \geq 13.94x(40.27)$
$39.0345x - 163.2084 - 22.51x \geq 561.3638x$
$-544.8393x \geq 163.2084$
$x \leq -0.300$ (approx.)
$\{x \mid x \leq -0.300\}$ (approx.)

18. $\{x \mid x \geq 0.723\}$ (approx.)

19. $x + 3 < 10$
$x < 7$
$x \in N$
The natural numbers < 7 are 1, 2, 3, 4, 5, and 6;
$\{1, 2, 3, 4, 5, 6\}$

20. $\{1, 2\}$

21. $2(x + 3) \leq 11$
$2x + 6 \leq 11$
$2x \leq 5$
$x \leq 5/2$
The natural numbers $\leq 5/2$ are 1 and 2; $\{1, 2\}$

22. $\{1, 2, 3, 4\}$

23. $\{x \mid x \geq 1\}$ **24.** $\{x \mid x \geq -2)$ **25.** $\{x \mid x < 4\}$
26. $\{x \mid x < 6\}$

Exercises 2.3 (page 103)

1. $\{x \mid 0 \leq x < 3\}$; $[0, 3)$ **2.** $\{x \mid 1 < x \leq 4\}$; $(1, 4]$
3. $\{x \mid x \leq 2\}$; $(-\infty, 2]$ **4.** $\{x \mid x \leq 3\}$; $(-\infty, 3]$
5. $\{x \mid -1 \leq x \leq 1\}$; $[-1, 1]$ **6.** $\{x \mid 1 \leq x \leq 2\}$; $[1, 2]$
7. $\{x \mid x > 1\}$; $(1, +\infty)$ **8.** $\{x \mid x > 2\}$; $(2, +\infty)$

9. $5 > x - 2 \geq 3$
$+2 \quad +2 \quad +2$
$7 > x \geq 5$
or $5 \leq x < 7$; $[5, 7)$

10. $10 > x \geq 7$
or $7 \leq x < 10$; $[7, 10)$

11. $-5 \geq x - 3 \geq 2$
"$-5 \geq 2$" is false
{ } or $\varnothing$ (no graph)

12. { } or $\varnothing$ (no graph)

13. $-4 < 3x - 1 \leq 7$
$+1 \quad +1 \quad +1$
$\frac{-3}{3} < \frac{3x}{3} \leq \frac{8}{3}$
$-1 < x \leq \frac{8}{3}$; $(-1, \frac{8}{3}]$

14. $-1 < x \leq \frac{7}{4}$; $(-1, \frac{7}{4}]$

15. $x - 1 > 3$ or $x - 1 < -3$
$+1 \quad +1 \qquad +1 \quad +1$
$x > 4$ or $x < -2$; $(-\infty, -2) \cup (4, +\infty)$

16. $x > 7$ or $x < -3$; $(-\infty, -3) \cup (7, +\infty)$

17. $2x + 1 \geq 3$ or $2x + 1 \leq -3$
$-1 \quad -1 \qquad -1 \quad -1$
$\frac{2x}{2} \geq \frac{2}{2}$ or $\frac{2x}{2} \leq \frac{-4}{2}$
$x \geq 1$ or $x \leq -2$; $(-\infty, -2] \cup [1, +\infty)$

18. $x \geq 7/3$ or $x \leq -1$; $(-\infty, -1] \cup [7/3, +\infty)$

19.

$x > 4$

$x \geq 2$

Final inequality: $x > 4$; $(4, +\infty)$

20. $x < -1$; $(-\infty, -1)$

21. $x > 4$ or $x \geq 2$

$x > 4$

$x \geq 2$

Final inequality: $x \geq 2$; $[2, +\infty)$

22. $x < 3; (-\infty, 3)$

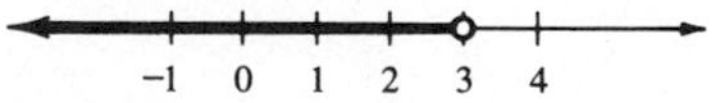

−1 0 1 2 3 4

23. $-5 \le x - 3 \le 2$
$\underline{+3} \quad \underline{+3} \quad \underline{+3}$
$-2 \le x \le 5$, but $x \in N$. Therefore, $\{1, 2, 3, 4, 5\}$.

0 1 2 3 4 5 6 7

24. $\{1, 2, 3, 4, 5, 6\}$

0 1 2 3 4 5 6 7

25. $4 \ge x - 3 > -5$
$\underline{+3} \quad \underline{+3} \quad \underline{+3}$
$7 \ge x > -2$, but $x \in J$.
Therefore, $\{-1, 0, 1, 2, 3, 4, 5, 6, 7\}$.

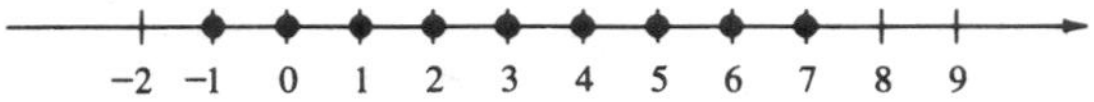

−2 −1 0 1 2 3 4 5 6 7 8 9

26. $\{-1, 0, 1, 2, 3, 4, 5, 6, 7, 8\}$

−2 −1 0 1 2 3 4 5 6 7 8 9

27. $-3 \le 2x + 1 \le 7$
$\underline{-1} \quad \underline{-1} \quad \underline{-1}$
$\frac{-4}{2} \le \frac{2x}{2} \le \frac{6}{2}$
$-2 \le x \le 3$, but $x \in N$. Therefore, $\{1, 2, 3\}$.

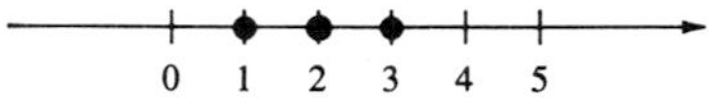

0 1 2 3 4 5

28. $\{1\}$

0 1 2 3

Exercises 2.4 (page 108)

1. If $|x| = 3$, $x = 3$ or $x = -3$.
$\{3, -3\}$

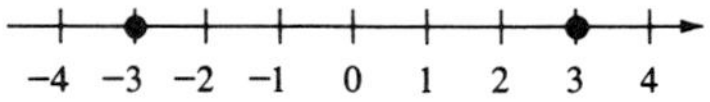

−4 −3 −2 −1 0 1 2 3 4

2. $\{5, -5\}$

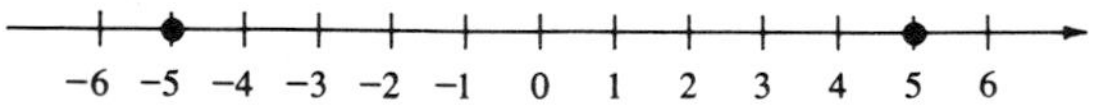

−6 −5 −4 −3 −2 −1 0 1 2 3 4 5 6

3. If $|3x| = 12$, $3x = 12$ or $3x = -12$; $x = 4$ or $x = -4$.
$\{4, -4\}$

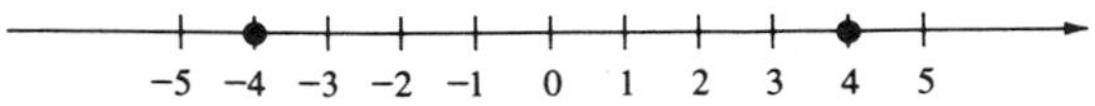

−5 −4 −3 −2 −1 0 1 2 3 4 5

4. $\{5, -5\}$

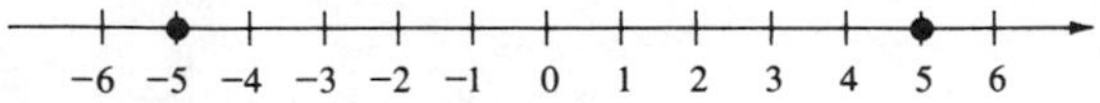

−6 −5 −4 −3 −2 −1 0 1 2 3 4 5 6

5. If $|x| < 2$, $-2 < x < 2$. $\{x \mid -2 < x < 2\}$

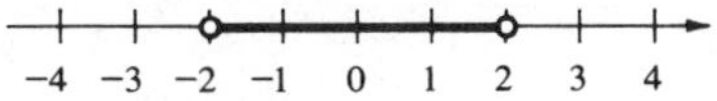

−4 −3 −2 −1 0 1 2 3 4

6. $\{x \mid -7 < x < 7\}$

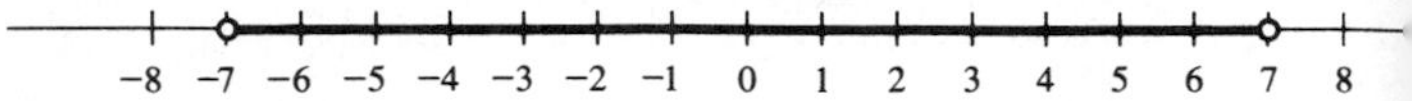

−8 −7 −6 −5 −4 −3 −2 −1 0 1 2 3 4 5 6 7 8

7. If $|4x| < 12$,
$-12 < 4x < 12$
$-3 < x < 3$
$\{x \mid -3 < x < 3\}$

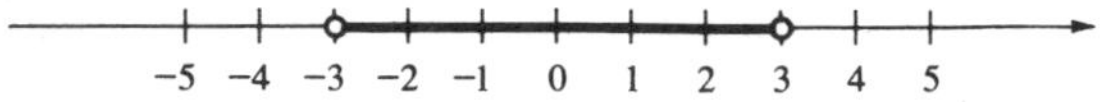

−5 −4 −3 −2 −1 0 1 2 3 4 5

8. $\{x \mid -3 < x < 3\}$

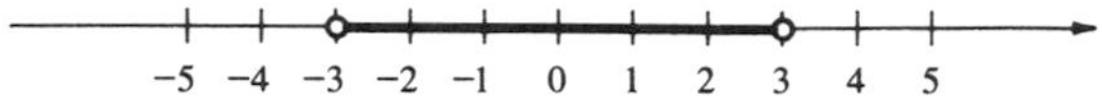

−5 −4 −3 −2 −1 0 1 2 3 4 5

9. If $|5x| \le 25$,
$-25 \le 5x \le 25$
$-5 \le x \le 5$ $\{x \mid -5 \le x \le 5\}$

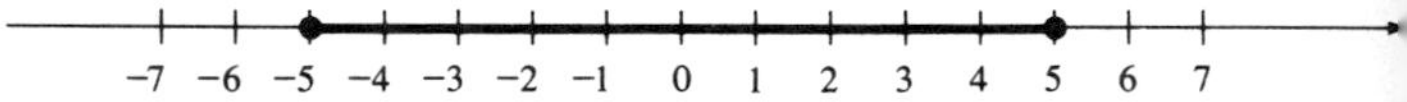

−7 −6 −5 −4 −3 −2 −1 0 1 2 3 4 5 6 7

10. $\{x \mid -1 \le x \le 1\}$

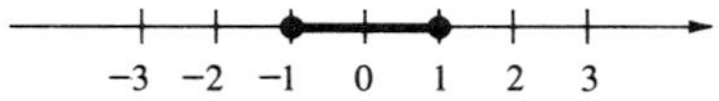

−3 −2 −1 0 1 2 3

11. If $|x| > 2$, $x > 2$ or $x < -2$.
$\{x \mid x > 2 \text{ or } x < -2\}$

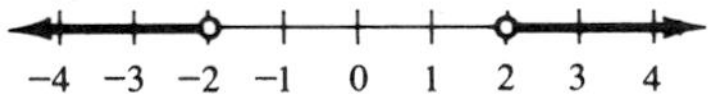

−4 −3 −2 −1 0 1 2 3 4

12. $\{x \mid x > 3 \text{ or } x < -3\}$

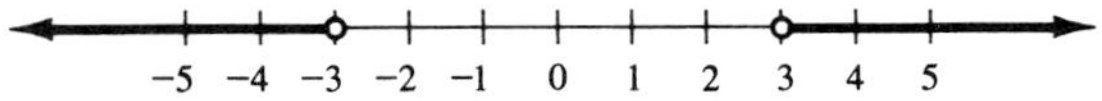

−5 −4 −3 −2 −1 0 1 2 3 4 5

13. Since $|3x|$ is always ≥ 0, it will always be > -3. Therefore, the solution set is the set of all real numbers.

−3 −2 −1 0 1 2 3 4 5

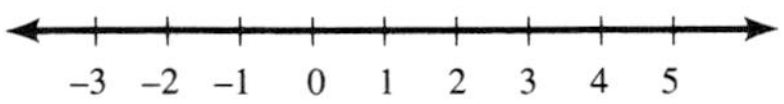

14. The solution set is the set of all real numbers

−3 −2 −1 0 1 2 3 4 5

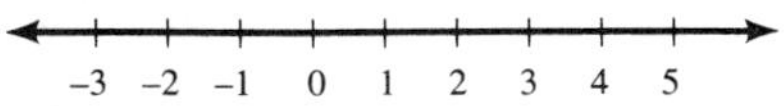

15. If $|x + 2| = 5$,
$x + 2 = 5$ or $x + 2 = -5$
$x = 3$ or $x = -7$
$\{3, -7\}$

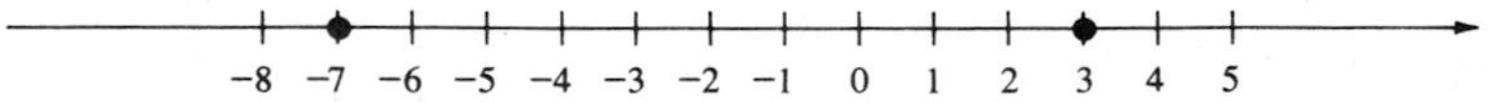

−8 −7 −6 −5 −4 −3 −2 −1 0 1 2 3 4 5

16. $\{4, -10\}$

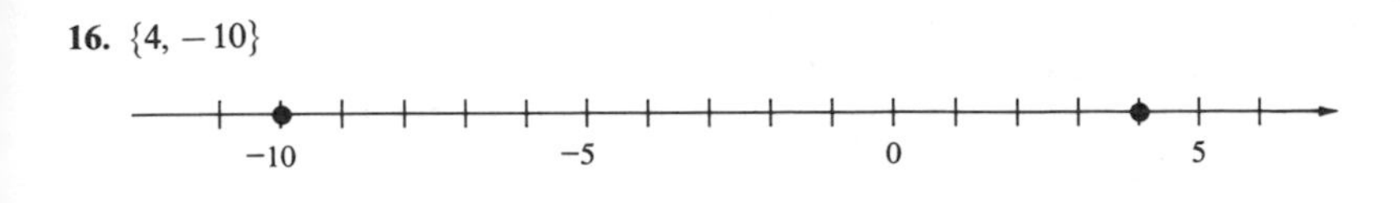

17. If $|x-3|<2$,
$$-2 < x-3 < 2$$
$$\underline{3} \qquad \underline{3} \quad \underline{3}$$
$$1 < x < 5$$
$\{x \mid 1 < x < 5\}$

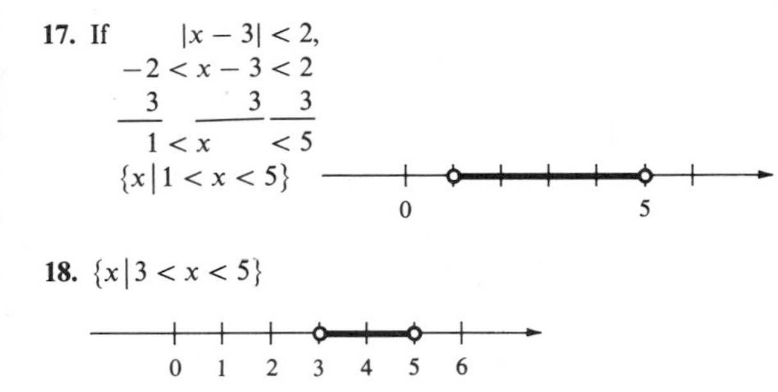

18. $\{x \mid 3 < x < 5\}$

19. Since $|x+4|$ is always ≥ 0, it will *never* be ≤ -3. Therefore, there is no solution; the solution set is the empty set.

20. No solution

21. If $|x+1| > 3$,
$$x+1 > 3 \text{ or } x+1 < -3$$
$$\underline{-1} \quad \underline{-1} \qquad \underline{-1} \quad \underline{-1}$$
$$x > 2 \text{ or } x < -4$$
$\{x \mid x > 2 \text{ or } x < -4\}$

22. $\{x \mid x > 6 \text{ or } x < -2\}$

23. If $|x+5| \geq 2$,
$$x+5 \geq 2 \text{ or } x+5 \leq -2$$
$$x \geq -3 \text{ or } \quad x \leq -7 \qquad \{x \mid x \leq -7 \text{ or } x \geq -3\}$$

24. $\{x \mid x \geq -3 \text{ or } x \leq -5\}$

25. If $|3x+4| = 3$,
$$3x+4 = 3 \text{ or } 3x+4 = -3$$
$$3x = -1 \text{ or } \quad 3x = -7$$
$$x = -\frac{1}{3} \text{ or } \quad x = -\frac{7}{3}$$
$\left\{-\frac{1}{3}, -\frac{7}{3}\right\}$

26. $\left\{-2, \frac{1}{2}\right\}$

27. If $|2x-3| < 4$,
$$-4 < 2x-3 < 4$$
$$\underline{3} \qquad \underline{3} \quad \underline{3}$$
$$-1 < 2x < 7$$
$$-\frac{1}{2} < x < \frac{7}{2} \qquad \left\{x \,\middle|\, -\frac{1}{2} < x < \frac{7}{2}\right\}$$

28. $\{x \mid -4/3 < x < 2\}$

29. If $|3x-5| \geq 6$,
$$3x-5 \geq 6 \text{ or } 3x-5 \leq -6$$
$$3x \geq 11 \text{ or } \quad 3x \leq -1$$
$$x \geq \frac{11}{3} \text{ or } \quad x \leq -\frac{1}{3}$$
$\{x \mid x \geq \frac{11}{3} \text{ or } x \leq -\frac{1}{3}\}$

30. $\{x \mid x \geq 1 \text{ or } x \leq -1/2\}$

31. If $|1-2x| \leq 5$,
$$-5 \leq 1-2x \leq 5$$
$$\underline{-1} \quad \underline{-1} \qquad -1$$
$$-6 \leq -2x \leq 4$$
$$3 \geq x \geq -2$$
$\{x \mid -2 \leq x \leq 3\}$

32. $\{x \mid -2/3 < x < 2\}$

33. If $|2-3x| > 4$,
$$2-3x > 4 \text{ or } 2-3x < -4$$
$$-3x > 2 \qquad -3x < -6$$
$$x < -\frac{2}{3} \text{ or } \quad x > 2 \qquad \{x \mid x < -\frac{2}{3} \text{ or } x > 2\}$$

34. $\{x \mid x \leq -1/2 \text{ or } x \geq 11/2\}$

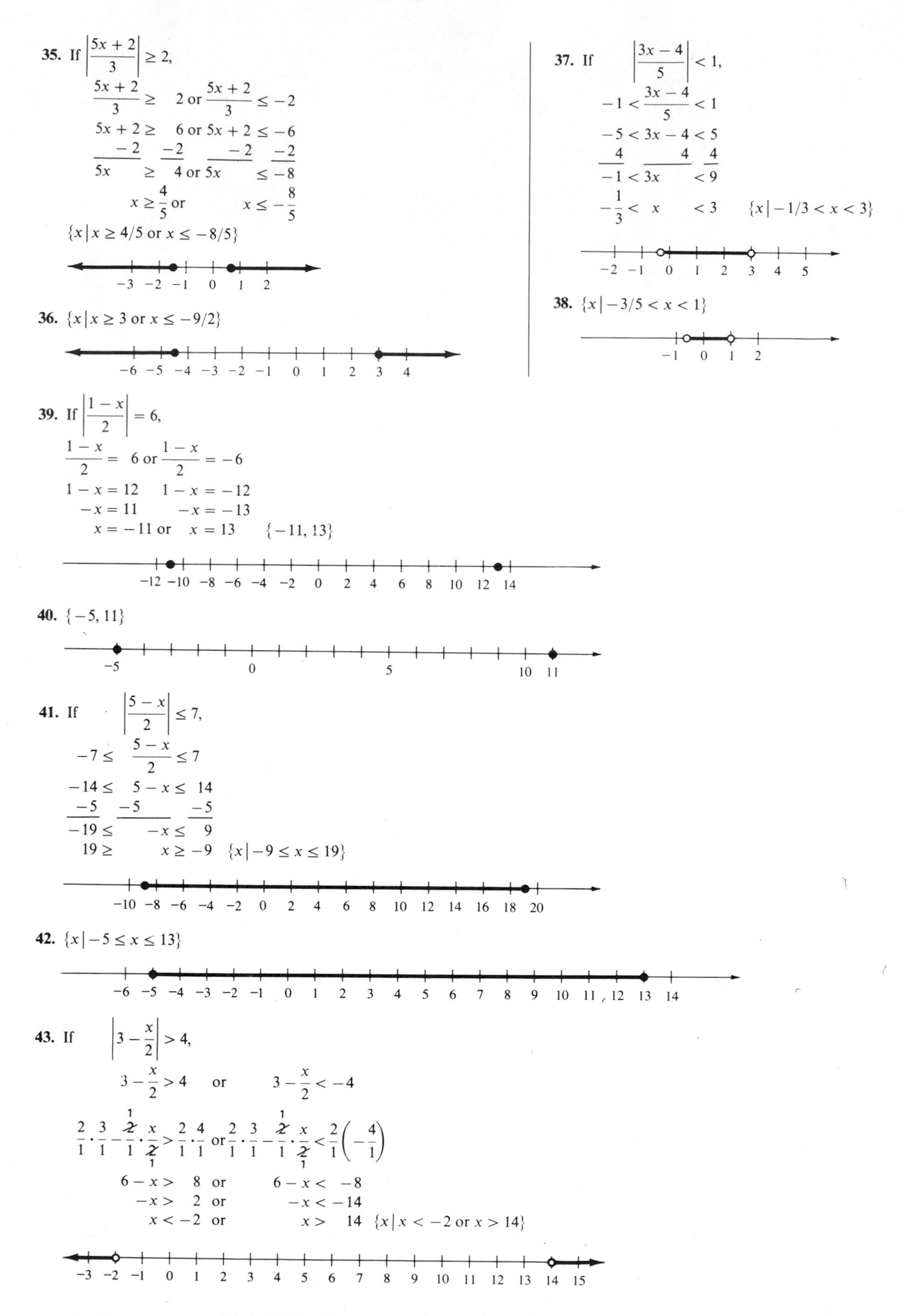

35. If $\left|\frac{5x+2}{3}\right| \geq 2$,

$$\frac{5x+2}{3} \geq 2 \text{ or } \frac{5x+2}{3} \leq -2$$
$$5x+2 \geq 6 \text{ or } 5x+2 \leq -6$$
$$\underline{-2 \quad -2} \qquad \underline{-2 \quad -2}$$
$$5x \geq 4 \text{ or } 5x \leq -8$$
$$x \geq \frac{4}{5} \text{ or } x \leq -\frac{8}{5}$$

$\{x \mid x \geq 4/5 \text{ or } x \leq -8/5\}$

36. $\{x \mid x \geq 3 \text{ or } x \leq -9/2\}$

37. If $\left|\frac{3x-4}{5}\right| < 1$,

$$-1 < \frac{3x-4}{5} < 1$$
$$-5 < 3x-4 < 5$$
$$\underline{\ 4 \qquad\quad 4 \quad 4\ }$$
$$-1 < 3x < 9$$
$$-\frac{1}{3} < x < 3 \qquad \{x \mid -1/3 < x < 3\}$$

38. $\{x \mid -3/5 < x < 1\}$

39. If $\left|\frac{1-x}{2}\right| = 6$,

$$\frac{1-x}{2} = 6 \text{ or } \frac{1-x}{2} = -6$$
$$1-x = 12 \qquad 1-x = -12$$
$$-x = 11 \qquad -x = -13$$
$$x = -11 \text{ or } x = 13 \qquad \{-11, 13\}$$

40. $\{-5, 11\}$

41. If $\left|\frac{5-x}{2}\right| \leq 7$,

$$-7 \leq \frac{5-x}{2} \leq 7$$
$$-14 \leq 5-x \leq 14$$
$$\underline{-5 \quad -5 \qquad\quad -5}$$
$$-19 \leq -x \leq 9$$
$$19 \geq x \geq -9 \qquad \{x \mid -9 \leq x \leq 19\}$$

42. $\{x \mid -5 \leq x \leq 13\}$

43. If $\left|3 - \frac{x}{2}\right| > 4$,

$$3 - \frac{x}{2} > 4 \quad \text{or} \quad 3 - \frac{x}{2} < -4$$
$$\frac{2}{1}\cdot\frac{3}{1} - \frac{\overset{1}{\cancel{2}}}{1}\cdot\frac{x}{\underset{1}{\cancel{2}}} > \frac{2}{1}\cdot\frac{4}{1} \text{ or } \frac{2}{1}\cdot\frac{3}{1} - \frac{\overset{1}{\cancel{2}}}{1}\cdot\frac{x}{\underset{1}{\cancel{2}}} < \frac{2}{1}\left(-\frac{4}{1}\right)$$
$$6 - x > 8 \text{ or } 6 - x < -8$$
$$-x > 2 \text{ or } -x < -14$$
$$x < -2 \text{ or } x > 14 \qquad \{x \mid x < -2 \text{ or } x > 14\}$$

44. $\{x \mid x < 9 \text{ or } x > 15\}$

45. If $\left|4 - \frac{x}{2}\right| < 3,$

$$-3 < 4 - \frac{x}{2} < 3$$
$$-6 < 8 - x < 6$$
$$\underline{-8} \quad \underline{-8} \quad \underline{-8}$$
$$-14 < -x < -2$$
$$14 > x > 2 \quad \{x \mid 2 < x < 14\}$$

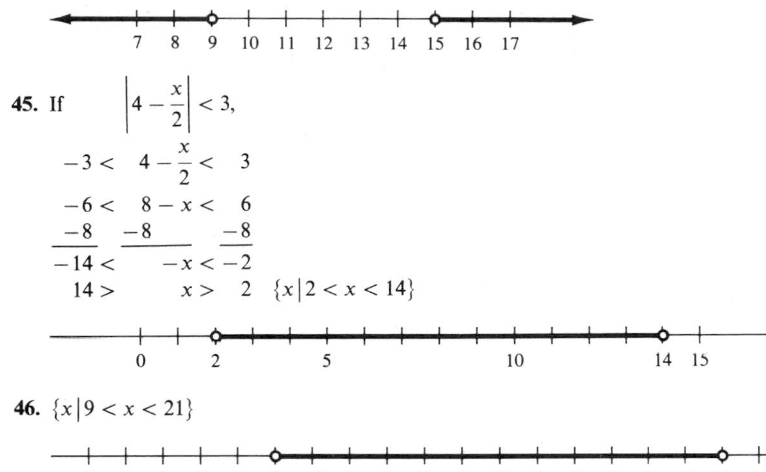

46. $\{x \mid 9 < x < 21\}$

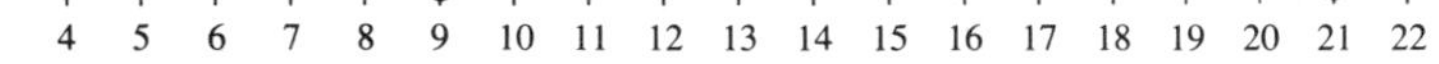

Review Exercises 2.5 (page 111)

1.
$$7 - 2M + 8 = 5$$
$$-2M + 15 = 5$$
$$\frac{-2M}{-2} = \frac{-10}{-2}$$
$$M = 5 \quad \{5\}$$

2. No solution; $\{\ \}$

3. LCD = 12

$$\frac{\overset{3}{\cancel{12}}}{1} \cdot \frac{3(x+3)}{\underset{1}{\cancel{4}}} - \frac{\overset{4}{\cancel{12}}}{1} \cdot \frac{2(x-3)}{\underset{1}{\cancel{3}}} = \frac{12}{1} \cdot 1$$
$$9(x+3) - 8(x-3) = 12$$
$$9x + 27 - 8x + 24 = 12$$
$$x = -39 \quad \{-39\}$$

4. Identity; $\{x \mid x \in R\}$

5.
$$2[-7y - 15 + 12y + 10] = 10y - 12$$
$$2[5y - 5] = 10y - 12$$
$$10y - 10 = 10y - 12$$
$$0 = -2 \leftarrow \text{False}$$
No solution $\{\ \}$

6. $\{x \mid x \le \frac{1}{3}\}$

7. LCD = 30

$$\frac{\overset{6}{\cancel{30}}}{1} \cdot \frac{3z}{\underset{1}{\cancel{5}}} + \frac{\overset{10}{\cancel{30}}}{1} \cdot \frac{-2z}{\underset{1}{\cancel{3}}} < \frac{\overset{15}{\cancel{30}}}{1} \cdot \frac{1}{\underset{1}{\cancel{2}}}$$
$$18z - 20z < 15$$
$$-2z < 15$$
$$z > -\tfrac{15}{2} \quad \{z \mid z > -\tfrac{15}{2}\}$$

8. $\{w \mid w > -3\frac{1}{2}\}$

9. LCD = 20

$$\frac{\overset{2}{\cancel{20}}}{1} \cdot \frac{2(x+6)}{\underset{1}{\cancel{10}}} + \frac{\overset{1}{\cancel{20}}}{1} \cdot \frac{3x}{\underset{1}{\cancel{20}}} < \frac{20}{1} \cdot \frac{3}{1}$$
$$4(x+6) + 3x < 60$$
$$4x + 24 + 3x < 60$$
$$7x < 36$$
$$x < 5\tfrac{1}{7} \quad \{x \mid x < 5\tfrac{1}{7}\}$$

10. $\left\{x \,\middle|\, x > -9\frac{1}{11}\right\}$

11.
$$3x + 2 = 11$$
$$3x = 9$$
$$x = 3 \quad \{3\}$$

12. $\{3, -3\}$

13. Since $|x|$ is always ≥ 0, it will *never* be ≤ -3. Therefore, there is no solution; the solution set is the empty set.

14. $\{x \mid x \ge 2\} \cup \{x \mid x \le -2\}$, or $(-\infty, -2] \cup [2, +\infty)$

15. If $|6 - 2x| = 10$,
$$6 - 2x = 10 \text{ or } 6 - 2x = -10$$
$$-2x = 4 \qquad -2x = -16$$
$$x = -2 \qquad x = 8$$
$\{-2, 8\}$

16. $\{x \mid 0 \le x \le 4\}$ or $[0, 4]$

17.
$$2|x - 3| \le 4$$
$$|x - 3| \le 2$$
$$-2 \le x - 3 \le 2$$
$$\underline{3} \quad \underline{3} \quad \underline{3}$$
$$1 \le x \le 5 \quad \{x \mid 1 \le x \le 5\} \text{ or } [1, 5].$$

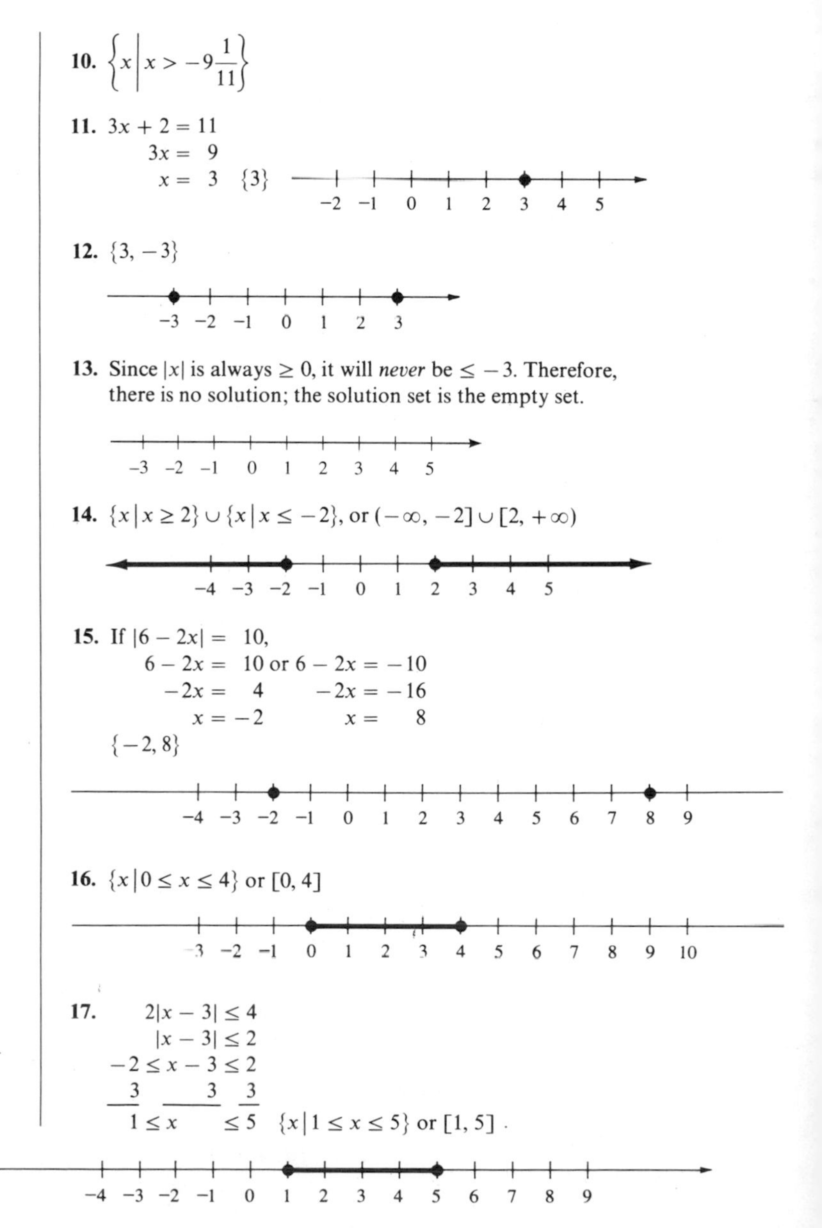

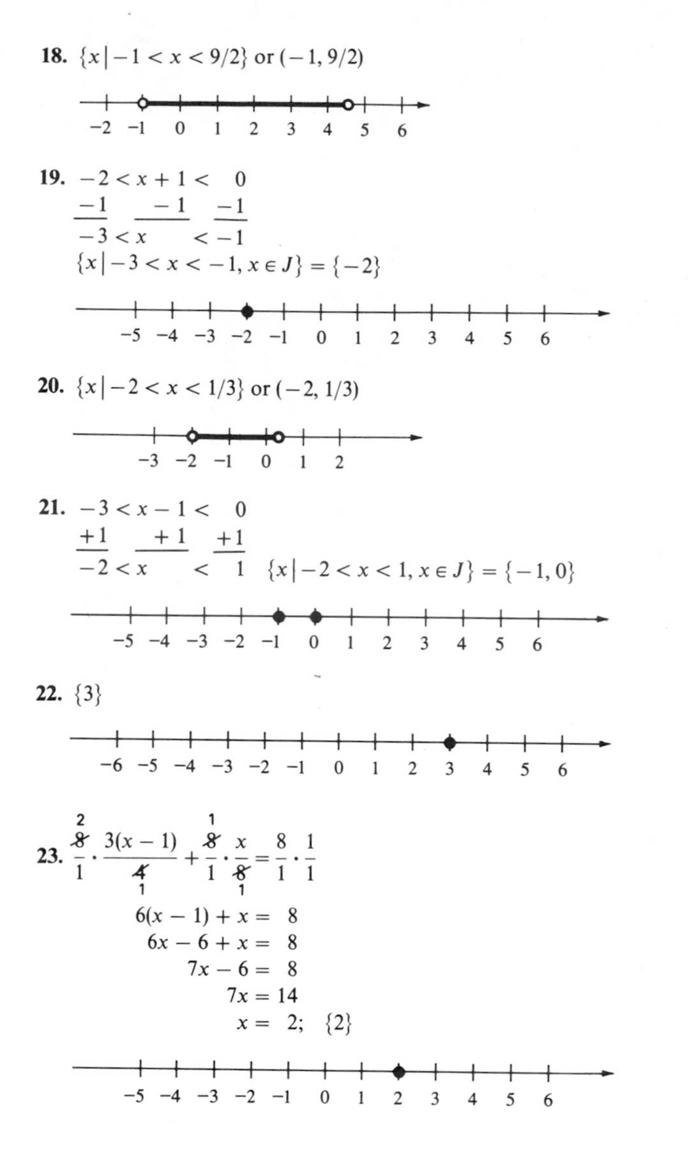

18. $\{x \mid -1 < x < 9/2\}$ or $(-1, 9/2)$

19.
$$\begin{aligned} -2 &< x + 1 < 0 \\ -1 &\quad\; -1 \quad -1 \\ -3 &< x \quad\;\; < -1 \end{aligned}$$
$\{x \mid -3 < x < -1, x \in J\} = \{-2\}$

20. $\{x \mid -2 < x < 1/3\}$ or $(-2, 1/3)$

21.
$$\begin{aligned} -3 &< x - 1 < 0 \\ +1 &\quad\; +1 \quad +1 \\ -2 &< x \quad\;\; < 1 \end{aligned}$$
$\{x \mid -2 < x < 1, x \in J\} = \{-1, 0\}$

22. $\{3\}$

23. $\frac{\overset{2}{\not{8}}}{1} \cdot \frac{3(x-1)}{\underset{1}{\not{4}}} + \frac{\overset{1}{\not{8}}}{1} \cdot \frac{x}{\underset{1}{\not{8}}} = \frac{8}{1} \cdot \frac{1}{1}$

$$\begin{aligned} 6(x-1) + x &= 8 \\ 6x - 6 + x &= 8 \\ 7x - 6 &= 8 \\ 7x &= 14 \\ x &= 2; \quad \{2\} \end{aligned}$$

24. $\{x \mid 3 \le x \le 9\}$ or $[3, 9]$

25. If $\left|\frac{2x-4}{3}\right| \ge 2$,

$$\frac{2x-4}{3} \ge 2 \quad \text{or} \quad \frac{2x-4}{3} \le -2$$

$$\frac{\overset{1}{\not{3}}}{1} \cdot \frac{2x-4}{\underset{1}{\not{3}}} \ge \frac{3}{1} \cdot \frac{2}{1} \qquad \frac{\overset{1}{\not{3}}}{1} \cdot \frac{2x-4}{\underset{1}{\not{3}}} \le \frac{3}{1}\left(\frac{-2}{1}\right)$$

$$\begin{aligned} 2x - 4 &\ge 6 & 2x - 4 &\le -6 \\ 2x &\ge 10 & 2x &\le -2 \\ x &\ge 5 & x &\le -1 \end{aligned}$$

$\{x \mid x \ge 5 \text{ or } x \le -1\}$ or $(-\infty, -1] \cup [5, +\infty)$

26. $\{x \mid x < 1\}$ **27.** $\{x \mid -3 < x \le 1\}$ **28.** $\{x \mid x \ge 4\}$

Chapter 2 Diagnostic Test (page 115)

Following each problem number is the textbook section number (in parentheses) where that kind of problem is discussed.

1. (2.1) LCD is 12.

$$12\left\{\frac{x}{6} - \frac{x+2}{4}\right\} = 12\left\{\frac{1}{3}\right\}$$

$$\begin{aligned} 2x - 3(x+2) &= 4 \\ 2x - 3x - 6 &= 4 \\ -x &= 10 \\ x &= -10 \end{aligned}$$

$\{-10\}$

Check $\frac{-10}{6} - \frac{-10+2}{4} \stackrel{?}{=} \frac{1}{3}$

$\frac{-5}{3} - \frac{-8}{4} \stackrel{?}{=} \frac{1}{3}$

$\frac{-5}{3} + 2 \stackrel{?}{=} \frac{1}{3}$

$\frac{1}{3} = \frac{1}{3}$

2. (2.1) **a.**
$$\begin{aligned} 3(x-6) &= 5(1+2x) - 7(x-4) \\ 3x - 18 &= 5 + 10x - 7x + 28 \\ 3x - 18 &= 3x + 33 \\ -18 &= 33 \leftarrow \text{False No solution} \end{aligned}$$

b.
$$\begin{aligned} 2[7x - 4(1+3x)] &= 5(3-2x) - 23 \\ 2[7x - 4 - 12x] &= 15 - 10x - 23 \\ 2[-4 - 5x] &= -10x - 8 \\ -8 - 10x &= -8 - 10x \\ 0 &= 0 \leftarrow \text{True Identity} \end{aligned}$$
$\{x \mid x \in R\}$

3. (2.2) **a.**
$$\begin{aligned} 5w + 2 &\le 10 - w \\ 6w &\le 8 \\ w &\le \tfrac{8}{6} \\ w &\le \tfrac{4}{3} \end{aligned}$$
$\{w \mid w \le \frac{4}{3}\}$

b.
$$\begin{aligned} 13h - 4(2+3h) &\ge 0 \\ 13h - 8 - 12h &\ge 0 \\ h &\ge 8 \end{aligned}$$
$\{h \mid h \ge 8\}$

4. (2.3) **a.** $\{x \mid -3 < x + 1 < 5, x \in J\}$

$$\begin{array}{rcl} -3 < x + 1 < & 5 \\ \underline{-1 \quad\; -1} & \underline{-1} \\ -4 < x \quad\; < & 4 \end{array}$$

Solution set $= \{x \mid -4 < x < 4, x \in J\}$
$= \{-3, -2, -1, 0, 1, 2, 3\}$

−4 −3 −2 −1 0 1 2 3 4

b. $\{x \mid 4 \geq 3x + 7 > -2, x \in R\}$

$$\begin{array}{rcl} 4 \geq 3x + 7 & > & -2 \\ \underline{-7 \quad\; -7} & & \underline{-7} \\ -3 \geq 3x \quad\; & > & -9 \\ -1 \geq x \quad\; & > & -3 \end{array}$$

Solution set $= \{x \mid -3 < x \leq -1, x \in R\}$ or $(-3, -1]$

−4 −3 −2 −1 0 1 2 3

5. (2.2) LCD is 12.

$$\frac{\overset{4}{\cancel{12}}}{1} \cdot \frac{5(x-2)}{\underset{1}{\cancel{3}}} + \frac{\overset{3}{\cancel{12}}}{1} \cdot \frac{x}{\cancel{4}} \leq \frac{12}{1} \cdot \frac{12}{1}$$

$$\begin{aligned} 20(x-2) + 3x &\leq 144 \\ 20x - 40 + 3x &\leq 144 \\ 23x &\leq 184 \\ x &\leq 8, x \in R \end{aligned}$$

$\{x \mid x \leq 8, x \in R\}$ or $(-\infty, 8]$

−2 −1 0 1 2 3 4 5 6 7 8 9 10

6. (2.4) If $\left|\dfrac{2x+3}{5}\right| = 1,$

$$\begin{array}{rcl} \dfrac{2x+3}{5} = 1 & \text{or} & \dfrac{2x+3}{5} = -1 \\ 2x + 3 = 5 & & 2x + 3 = -5 \\ 2x = 2 & & 2x = -8 \\ x = 1 & & x = -4 \end{array}$$

Solution set $= \{-4, 1\}$

−6 −5 −4 −3 −2 −1 0 1 2 3 4

7. (2.4) If $|3x - 1| > 2,$

$$\begin{array}{rcl} 3x - 1 > 2 & \text{or} & 3x - 1 < -2 \\ \underline{\quad\; 1 \quad 1} & & \underline{\quad\; 1 \quad\; 1} \\ 3x \quad > 3 & & 3x \quad < -1 \\ x > 1 & & x < -\frac{1}{3} \end{array}$$

Solution set $= \{x \mid x > 1\} \cup \{x \mid x < -\frac{1}{3}\}$
or $(-\infty, -1/3) \cup (1, +\infty)$

−2 −1 0 1 2 3

8. (2.4) If $|7 - 3x| \geq 6,$

$$\begin{array}{rcl} 7 - 3x \geq 6 & \text{or} & 7 - 3x \leq -6 \\ 7 \geq 3x + 6 & & 7 \leq 3x - 6 \\ 1 \geq 3x & & 13 \leq 3x \\ \frac{1}{3} \geq x & & \frac{13}{3} \leq x \end{array}$$

Solution set $= \{x \mid x \geq 4\frac{1}{3}\} \cup |x \mid x \leq \frac{1}{3}\}$
or $(-\infty, 1/3] \cup [4\frac{1}{3}, +\infty)$

−4 −3 −2 −1 0 1 2 3 4 5 6 7 8

9. (2.4) If $\left|\dfrac{5x+1}{2}\right| \leq 7,$

$$\begin{array}{rcl} -7 \leq \dfrac{5x+1}{2} & \leq & 7 \\ -14 \leq 5x + 1 & \leq & 14 \\ \underline{-1 \qquad\; -1} & & \underline{-1} \\ -15 \leq 5x \quad & \leq & 13 \\ -3 \leq x \quad & \leq & \frac{13}{5} \end{array}$$

Solution set $= \{x \mid -3 \leq x \leq \frac{13}{5}\}$
or $[-3, 13/5]$

−4 −3 −2 −1 0 1 2 3

10. (2.4) If $|2x - 5| < 11,$

$$\begin{array}{rcl} -11 < 2x - 5 & < & 11 \\ \underline{5 \qquad\; 5} & & \underline{5} \\ -6 < 2x \quad & < & 16 \\ -3 < x \quad & < & 8 \end{array}$$

Solution set $= \{x \mid -3 < x < 8\}$ or $(-3, 8)$

−4 −3 −2 −1 0 1 2 3 4 5 6 7 8 9 10

Cumulative Review Exercises: Chapters 1 and 2 (page 116)

1. $(-14) + (+22) = 8$ **2.** 48 **3.** −1 **4.** −14
5. $(-2)(-2)(-2)(-2) = 16$ **6.** 8
7. Not possible (not defined) **8.** 0 **9.** $\dfrac{20}{-5} = -4$ **10.** −25
11. −4 **12.** −36 **13.** −5 **14.** 0 **15.** −1 **16.** 1,000
17. 1 **18.** 8 **19.** $-3 - 4 \cdot 6 = -3 - 24 = -27$ **20.** 11
21. 4 **22.** −28 **23.** 0 **24.** 9 **25.** 3 **26.** 13
27. Cannot be determined (not defined) **28.** 1 **29.** 0 **30.** 4
31. $2 \cdot 3 \cdot 13$ **32.** 19, 23 **33.** 0 **34.** $(3 + 2) + 7$
35. $(17 \times 8) \times 6$ **36.** Yes
37. $y - 2x + 2y - 3 + 3y - 2y = 4y - 2x - 3$
38. $x^3 - 3x^2 + 4x$ **39.** $5^2x^2(3^3x^6) = 25 \cdot 27x^{2+6} = 675x^8$
40. $-3x^3 - 4y^2$ **41.**

$$\begin{aligned} 8x - 8 - 12x &= 12 \\ -4x - 8 &= 12 \\ \underline{+8} &= \underline{+8} \\ -4x &= 20 \\ x &= -5 \end{aligned}$$

$\{-5\}$

42. $\{y \mid y > 3\}$ **43.**

$$\begin{aligned} 18x - 30 + 7 &= 27 + 18x - 1 \\ 18x - 23 &= 18x + 26 \\ -23 &= 26 \leftarrow \text{False} \end{aligned}$$

No solution; { }

44. $\{x \mid x \geq -11/3\}$ **45.**

$$\begin{array}{rcl} 2x + 3 & = & 4x + 10 \\ \underline{-2x - 10} & & \underline{-2x - 10} \\ -7 & = & 2x \\ \dfrac{-7}{2} & = & x \end{array}$$

$\{-7/2\}$

46. $\{5\}$

47.

$$\begin{array}{rcl} -5 < x + 4 & \leq & 3 \\ \underline{-4 \qquad -4} & & \underline{-4} \\ -9 < x \quad & \leq & -1 \end{array}$$

$\{x \mid -9 < x \leq -1\}$

48. Identity; $\{x \mid x \in R\}$

49. If $|2x+3|>1$, $2x+3>1$ or $2x+3<-1$

$$2x>-2 \qquad 2x<-4$$
$$x>-1 \qquad x<-2$$

$\{x \mid x>-1 \text{ or } x<-2\}$

50. $\{x \mid 1 \le x \le 7\}$

Exercises 3.1 (page 121)

Note: The checks for these exercises will not be shown.

1. Let x = unknown number.

$$\begin{aligned} 2x+7&=23\\ 2x&=16\\ x&=8 \end{aligned}$$

The number is 8.

2. 9

3. Let x = unknown number.

$$\begin{aligned} 4x-7&=25\\ 4x&=32\\ x&=8 \end{aligned}$$

The number is 8.

4. 11

5. Let x = unknown number.

$$\begin{aligned} x-7&=\tfrac{1}{2}x\\ 2(x-7)&=2(\tfrac{1}{2}x)\\ 2x-14&=x\\ x&=14 \end{aligned}$$

The number is 14.

6. 24

7. Let x = number of centimeters in second piece of string
$3x$ = number of centimeters in first piece of string

$$\begin{aligned} x+3x&=12\\ 4x&=12\\ x&=3\\ 3x&=9 \end{aligned}$$

The first piece is 9 cm long and the second piece is 3 cm long.

8. 8 m, 4 m

9. Let x = first integer
$x+1$ = second integer
$x+2$ = third integer

$$\begin{aligned} x+(x+1)+(x+2)&=19\\ 3x+3&=19\\ 3x&=16\\ x&=\tfrac{16}{3} \quad \text{not an integer} \end{aligned}$$

No solution

10. No solution

11. Let x = length of second piece
$x+8$ = length of first piece

$$\begin{aligned} x+(x+8)&=42\\ 2x+8&=42\\ 2x&=34\\ x&=17\\ x+8&=25 \end{aligned}$$

The first piece is 25 cm long and the second piece is 17 cm long.

12. 28 m, 22 m

13. Let x = first integer
$x+1$ = second integer
$x+2$ = third integer

$$\begin{aligned} x+(x+1)-(x+2)&=10\\ x+x+1-x-2&=10\\ x-1&=10\\ x&=11\\ x+1&=12\\ x+2&=13 \end{aligned}$$

The integers are 11, 12, and 13.

14. 18, 19, and 20

15. Let x = first odd integer
$x+2$ = second odd integer
$x+4$ = third odd integer

$$\begin{aligned} 3[(x+2)+(x+4)]&=40+5(x)\\ 3(2x+6)&=40+5x\\ 6x+18&=40+5x\\ x&=22 \end{aligned}$$

22 is not an odd integer. No solution

16. No solution

17. Let x = number of cans of peas
$x+4$ = number of cans of corn
$3x$ = number of cans of green beans

$$\begin{aligned} x+(x+4)+(3x)&=24\\ 5x+4&=24\\ 5x&=20\\ x&=4\\ x+4&=8\\ 3x&=12 \end{aligned}$$

David bought 4 cans of peas, 8 cans of corn, and 12 cans of green beans.

18. 3 cans of pears, 9 cans of peaches, 9 cans of cherries

19. Let x = unknown number.

$$\begin{aligned} 3(8+x)&=2(x+7)\\ 24+3x&=2x+14\\ x&=-10 \end{aligned}$$

The number is -10.

20. 7

21. Let x = unknown number.

$$\begin{aligned} 8x-2(5+x)&=4(8+2x)\\ 8x-10-2x&=32+8x\\ -42&=2x\\ x&=-21 \end{aligned}$$

The number is -21.

22. 2

23. Let x = first even integer
$x+2$ = second even integer
$x+4$ = third even integer

$$\begin{aligned} 21 &< x+(x+2)+(x+4) < 45\\ 21 &< 3x+6 < 45\\ 15 &< 3x < 39\\ 5 &< x < 13 \end{aligned}$$

The even integers between 5 and 13 are 6, 8, 10, and 12. Therefore, the answers are 6, 8, and 10; 8, 10, and 12; 10, 12, and 14; and 12, 14, and 16.

24. 26, 28, and 30; and 28, 30, and 32

25. Let x = score needed.

$$x \le 200$$
$$\begin{aligned} 560 \le\ &396+x \le 640\\ -396\ \ &-396 \quad\ -396\\ 164 \le\ &x \le 244 \end{aligned}$$

but $x \le 200$, so

$$164 \le x \le 200$$

To receive a C, Clark must have a score between 164 and 200 on the final exam.

26. 136 to 200

Exercises 3.2 (page 126)

Note: The checks for these exercises will not always be shown.

1. Let $4x$ = smaller number
$5x$ = larger number

$$\begin{aligned} 4x+5x&=81\\ 9x&=81\\ x&=9\\ 4x&=4(9)=36\\ 5x&=5(9)=45 \end{aligned}$$

Check: $\frac{36}{45}=\frac{4}{5}$, $36+45=81$

The numbers are 36 and 45.

2. The numbers are 56 and 21.

3. Let $3x$ = first side
$4x$ = second side
$5x$ = third side

The perimeter is 108.

$$\begin{aligned} 3x+4x+5x&=108\\ 12x&=108\\ x&=9\\ 3x&=27\\ 4x&=36\\ 5x&=45 \end{aligned}$$

The lengths of the sides are 27, 36 and 45.

4. 32, 40, 48

5. Let $4x$ = study hours
$2x$ = class hours
$3x$ = work hours
$54 = 4x + 2x + 3x$
$54 = 9x$
$6 = x$
$4x = 4(6) = 24$
$2x = 2(6) = 12$
$3x = 3(6) = 18$
The student spends 24 hr studying, 12 hr in class, and 18 hr working.

6. Study hours = 16
Class hours = 8
Work hours = 24

7. Let $7x$ = length of rectangle
$6x$ = width of rectangle
$2(7x) + 2(6x) = 78$
$14x + 12x = 78$
$26x = 78$
$x = 3$
$7x = 21$
$6x = 18$
Check: Perimeter is $2(21) + 2(18) = 42 + 36 = 78$.
The length is 21 and the width is 18.

8. Width = 35, length = 63

9. Let $7x$ = one number
$6x$ = other number
$143 = 7x + 6x$ *Check:* $\frac{77}{66} = \frac{7}{6}$
$143 = 13x$ $77 + 66 = 143$
$x = 11$
$7x = 77$
$6x - 66$
One number is 77 and the other is 66.

10. 136, 85

11. Let $4x$ = shortest side
$5x$ = second side
$6x$ = third side
$0 < 4x + 5x + 6x < 60$
$0 < 15x < 60$
$0 < x < 4$
$0 < 4x < 16$
The shortest side must be between 0 and 16.

12. The shortest side must be between 0 and 24.

Exercises 3.3 (page 131)

Note: The checks for these exercises will not be shown.

1. Let x = speed of Malone car (in mph)
$x + 9$ = speed of King car (in mph)
$6x$ = distance for Malone car
$5(x + 9)$ = distance for King car
$6x = 5(x + 9)$ Distances are equal.
$6x = 5x + 45$
a. $x = 45$
$x + 9 = 45 + 9 = 54$
The speed of the Malone car was 45 mph and the speed of the King car was 54 mph.
b. $45 \times 6 = 54 \times 5$
$= 270$ The distance traveled was 270 mi.

2. **a.** Duran car—40 mph; Silva car—50 mph   **b.** 400 mi

3. Let x = hours required to return from lake
$x + 3$ = hours required to hike to lake
$2(x + 3)$ = distance going to lake
$5x$ = distance returning from lake
$5x = 2(x + 3)$ Distances are equal.
$5x = 2x + 6$
$3x = 6$
$x = 2$
a. $x + 3 = 5$ It took Eric 5 hr to hike to the lake.
b. The distance is $\left(2\frac{\text{mi}}{\cancel{\text{hr}}}\right)(5\,\cancel{\text{hr}}) = 10$ mi.

4. **a.** 3 hr   **b.** 6 mi

5. Let x = Fran's speed
$\frac{4}{5}x$ = Ron's speed
$3x$ = distance traveled by Fran
$3(\frac{4}{5}x)$ = distance traveled by Ron
$3x + 3(\frac{4}{5}x) = 54$ Sum of diatances is 54.
LCD = 5
$\frac{5}{1}(\frac{3}{1}x) + \frac{5}{1}(\frac{3}{1})(\frac{4}{5}x) = \frac{5}{1}(\frac{54}{1})$
$15x + 12x = 270$
$27x = 270$
$x = 10$
$\frac{4}{5}x = \frac{4}{5}(10)$
$= 8$
Fran's speed is 10 mph and Ron's speed is 8 mph.

6. Tran's rate is 9 mph; Atour's rate is 6 mph.

7. Let x = speed of boat in still water
$x + 2$ = speed of boat downstream
$x - 2$ = speed of boat upstream
$5(x - 2)$ = distance boat traveled upstream
$3(x + 2)$ = distance boat traveled downstream

Distance traveled downstream = 6 mi + Distance traveled upstream

$3(x + 2) = 6 + 5(x - 2)$
$3x + 6 = 6 + 5x - 10$
$10 = 2x$
$x = 5$
$3(x + 2) = 3(5 + 2) = 3(7) = 21$
The speed of the boat is 5 mph in still water, and Colin traveled 21 mi downstream.

8. 3 mph in still water; 16 mi downstream

9. Let t = time with a headwind
$10 - t$ = time with a tailwind
$80t$ = distance with a headwind
$(10 - t)(120)$ = distance with a tailwind
$80t = 120(10 - t)$ Distances are equal.
$80t = 1{,}200 - 120t$
$200t = 1{,}200$
$t = 6$
The distance is $\left(80\frac{\text{mi}}{\cancel{\text{hr}}}\right)(6\,\cancel{\text{hr}}) = 480$ mi.

10. 6 mi

Exercises 3.4 (page 136)

Note: The checks for these exercises will not be shown.

1. Let x = number of pounds of Colombian coffee
$100 - x$ = number of pounds of Brazilian coffee
$3.90x + 3.60(100 - x) = 3.72(100)$
$39x + 36(100 - x) = 37.2(100)$
$39x + 3{,}600 - 36x = 3{,}720$
$3x = 120$
$x = 40$
$100 - x = 60$
40 lb of Colombian coffee and 60 lb of Brazilian coffee should be used.

2. 16 lb of cashews, 34 lb of peanuts

3. Let x = number of pounds of almonds
$10 - x$ = number of pounds of walnuts
$$3.00x + 4.50(10 - x) = 39.72$$
$$300x + 450(10 - x) = 3{,}972$$
$$300x + 4{,}500 - 450x = 3{,}972$$
$$-150x = -528$$
$$x = 3.52$$
$$10 - x = 6.48$$
3.52 lb of almonds and 6.48 lb of walnuts should be used.

4. 1.5 lb of apple chunks, 4.5 lb of granola

5. Let D = number of dimes
$17 - D$ = number of nickels
Each dime is worth 10¢; therefore, D dimes are worth $10D$¢.
Each nickel is worth 5¢; therefore, $(17 - D)$ nickels are worth $5(17 - D)$¢.

Amount of money in dimes	+	Amount of money in nickels	=	115¢
$10D$	+	$5(17 - D)$	=	115

$$10D + 85 - 5D = 115$$
$$5D = 30$$
$$D = 6$$
$$17 - D = 17 - 6 = 11$$
Doris has 6 dimes and 11 nickels.

6. 11 quarters, 6 dimes

7. Let x = number of quarters
$x + 7$ = number of dimes
$3x$ = number of nickels
$$0.25x + 0.10(x + 7) + 0.05(3x) = 3.20$$
$$25x + 10(x + 7) + 5(3x) = 320$$
$$25x + 10x + 70 + 15x = 320$$
$$50x = 250$$
$$x = 5$$
$$x + 7 = 12$$
$$3x = 15$$
Dianne has 5 quarters, 12 dimes, and 15 nickels.

8. 6 dimes, 11 nickels, 12 quarters

9. Let x = number of pounds of peanut brittle.
$$6.60(15) + 4.20x = 5.10(15 + x)$$
$$66(15) + 42x = 51(15 + x)$$
$$990 + 42x = 765 + 51x$$
$$225 = 9x$$
$$x = 25$$
25 lb of peanut brittle should be used.

10. 16 lb

11. Let x = number of children's tickets
$23 - x$ = number of adults' tickets
Each child's ticket is worth \$3.50, so the total value = $3.50x$.
Each adult's ticket is worth \$5.25, so the total value = $5.25(23 - x)$.
$$3.50x + 5.25(23 - x) = 108.50$$
$$350x + 525(23 - x) = 10{,}850$$
$$350x + 12{,}075 - 525x = 10{,}850$$
$$-175x = -1{,}225$$
$$x = 7$$
$$23 - x = 16$$
There were 7 children's tickets and 16 adults' tickets.

12. Twenty-seven 22¢ stamps; eighteen 18¢ stamps

Exercises 3.5 (page 139)

Note: The checks for these exercises will not be shown.

1. Let x = number of milliliters of water.
$$0.40(500) + 0x = 0.25(500 + x)$$
$$40(500) = 25(500 + x)$$
$$20{,}000 = 12{,}500 + 25x$$
$$7{,}500 = 25x$$
$$x = 300$$
300 ml of water must be added.

2. 5 ℓ

3. Let x = number of liters of pure alcohol.
$$0.20(10) + x = 0.50(10 + x)$$
$$2(10) + 10x = 5(10 + x)$$
$$20 + 10x = 50 + 5x$$
$$5x = 30$$
$$x = 6$$
6 liters of pure alcohol should be added.

4. 500 ml

5. Let x = number of cc of 20% solution.
$$0.20x + 0.50(100) = 0.25(100 + x)$$
$$20x + 50(100) = 25(100 + x)$$
$$20x + 5{,}000 = 2{,}500 + 25x$$
$$2{,}500 = 5x$$
$$x = 500$$
500 cc of 20% solution should be added.

6. 20 pt

7. Let x = number of gallons of 30% solution
$100 - x$ = number of gallons of 90% solution
$$.30(x) + .90(100 - x) = 100(.75)$$
$$30x + 90(100 - x) = 100(75)$$
$$30x + 9{,}000 - 90x = 7{,}500$$
$$-60x = -1{,}500$$
$$x = 25$$
$$100 - x = 75$$
25 gal of the 30% solution and 75 gal of the 90% solution should be used.

8. 400 g of 65% alloy; 800 g of 20% alloy

9. Let x = number of liters of 40% solution
$10 - x$ = number of liters of 90% solution
$$.40x + .90(10 - x) = .50(10)$$
$$40x + 90(10 - x) = 50(10)$$
$$40x + 900 - 90x = 500$$
$$-50x = -400$$
$$x = 8$$
$$10 - x = 2$$
8 ℓ of the 40% solution and 2 ℓ of the 90% solution should be used.

10. 1,200 ml

Exercises 3.6 (page 141)

1. Let x = units digit
$13 - x$ = tens digit

Original number: $10(13 - x) + x$ *Number with digits reversed:* $10x + (13 - x)$
$$10(13 - x) + x + 27 = 10x + (13 - x)$$
$$130 - 10x + x + 27 = 10x + 13 - x$$
$$-9x + 157 = 9x + 13$$
$$144 = 18x$$
$$x = 8, \text{ units digit}$$
$$13 - x = 13 - 8 = 5, \text{ tens digit}$$
Therefore, the original number is apparently 58.

Check: Sum of digits = 5 + 8 = 13
Original number 58
Increased by 27
85 Number with digits reversed

2. 100 mi **3.** Let x = volume of the tank.

$$\frac{1}{3}x + 3{,}000 = \frac{3}{4}x$$
$$4x + 36{,}000 = 9x$$
$$36{,}000 = 5x$$
$$x = 7{,}200$$

Therefore, the volume of the tank is 7,200 cu. in. The check will not be shown.

4. 52

5. Let x = hundreds digit
$2x$ = units digit
$6 - x - 2x$ = tens digit (Sum of the digits is 6)

Original number: $100x + 10(6 - 3x) + 2x$ *Digits reversed:* $100(2x) + 10(6 - 3x) + x$

$$100(2x) + 10(6 - 3x) + x = 198 + 100x + 10(6 - 3x) + 2x$$
$$200x + 60 - 30x + x = 198 + 100x + 60 - 30x + 2x$$
$$171x + 60 = 72x + 258$$
$$99x = 198$$
$$x = 2, \text{ hundreds digit}$$
$$2x = 4, \text{ units digit}$$
$$6 - 3x = 0, \text{ tens digit}$$

The original number is apparently 204.

Check: Sum of digits = 2 + 0 + 4 = 6
Original number 204
Increased by 198
402 Number with digits reversed

Therefore, the original number *is* 204.

6. 45 min.

7. In the original 20 lb of alloy:
40% of 20 lb is (0.40)(20 lb) = 8 lb (amount of nickel)
60% of 20 lb is (0.60)(20 lb) = 12 lb (amount of copper)
Let x = number of pounds of pure copper added.

$$\frac{x + 12}{8} = \frac{3}{1} \quad \text{Required ratio is 3 to 1.}$$
$$x + 12 = 24$$
$$x = 12$$

12 lb of pure copper should be used. (The check will not be shown.)

8. 42 lb

9. Let x = amount invested at 5.75%
$13{,}000 - x$ = amount invested at 5.25%

$$0.0575x + 0.0525(13{,}000 - x) = 702.50$$
$$575x + 525(13{,}000 - x) = 7{,}025{,}000$$
$$575x + 6{,}825{,}000 - 525x = 7{,}025{,}000$$
$$50x = 200{,}000$$
$$x = 4{,}000$$
$$13{,}000 - x = 9{,}000$$

Check: \$4,000(0.0575) = \$230.00
\$9,000(0.0525) = \$472.50
Sum: \$702.50

\$4,000 was invested at 5.75% and \$9,000 was invested at 5.25%

10. 326

11. Let x = price of cheaper camera
$x + 48$ = price of expensive camera
$18x$ = total price paid for cheaper cameras
$10(x + 48)$ = total price paid for expensive cameras

$$18x = 10(x + 48)$$
$$18x = 10x + 480$$
$$8x = 480$$
$$x = 60$$
$$x + 48 = 108$$

Check: 18(\$60) = \$1,080; 10(\$108) = \$1,080 } Same expenditure

The cheaper cameras were \$60 each and the more expensive ones were \$108 each.

12. 300 children

Review Exercises 3.7 (page 144)

Note: The checks for these exercises will not be shown.

1. Let x = unknown number. **2.** No solution

$$\frac{28 - 5x}{3} = 3x$$
$$\frac{\not{3}}{1}\left(\frac{28 - 5x}{\not{3}}\right) = \frac{3}{1}(3x)$$
$$28 - 5x = 9x$$
$$28 = 14x$$
$$x = 2$$

The number is 2.

3. Let $3x$ = first side
$7x$ = second side
$8x$ = third side

The perimeter is less than 36.

$$0 < 3x + 7x + 8x < 36$$
$$0 < 18x < 36$$
$$0 < x < 2$$
$$0 < 3x < 6$$

The shortest side must be between 0 and 6.

4. 5 quarters; 8 dimes; 3 nickels

5. Let x = number of pounds at \$2.20
$15 - x$ = number of pounds at \$2.60

$$2.20x + 2.60(15 - x) = 2.36(15)$$
$$220x + 260(15 - x) = 236(15)$$
$$220x + 3{,}900 - 260x = 3{,}540$$
$$-40x = -360$$
$$x = 9 \text{ lb at } \$2.20$$
$$15 - x = 6 \text{ lb at } \$2.60$$

9 lb of the cheaper candy and 6 lb of the more expensive candy should be used.

6. 840 cc **7.** Let x = speed returning from work
$x + 10$ = speed going to work
$\frac{1}{2}(x + 10)$ = distance going to work
$\frac{3}{4}x$ = distance returning from work

$$\frac{1}{2}(x + 10) = \frac{3}{4}(x) \quad \text{Distances are equal.}$$
$$\frac{1}{2}x + 5 = \frac{3}{4}x$$
$$5 = \frac{1}{4}x$$
$$x = 20$$

The distance is (rate)(time) $= 20\frac{\text{mi}}{\not{\text{hr}}}\left(\frac{3}{4}\not{\text{hr}}\right) = 15$ mi.

8. 83 **9.** Let x = amount invested at 12%
$27{,}000 - x$ = amount invested at 8%

$$.12x + .08(27{,}000 - x) = 2{,}780$$
$$12x + 8(27{,}000 - x) = 278{,}000$$
$$12x + 216{,}000 - 8x = 278{,}000$$
$$4x = 62{,}000$$
$$x = 15{,}500$$
$$27{,}000 - x = 11{,}500$$

\$15,500 was invested at 12% and \$11,500 at 8%.

10. \$13,500

11. Let x = rate of slower plane
$x + 25$ = rate of faster plane
The sum of the distances is 570.

$$2x + 2(x + 25) = 570$$
$$2x + 2x + 50 = 570$$
$$4x = 520$$
$$x = 130$$

Therefore, the slower plane is going 130 mph and the faster one is going 155 mph.

Chapter 3 Diagnostic Test (page 147)

Following each problem number is the textbook section number (in parentheses) where that kind of problem is discussed. The checks will not be shown.

1. (3.1) Let x = unknown number.

When 23	is added to	four times an unknown number	the sum is	31.
23	+	$4 \cdot x$	=	31

$$23 + 4x = 31$$
$$4x = 8$$
$$x = 2$$

The unknown number is 2.

2. (3.1) Let x = first integer
$x + 1$ = second integer
Their sum is 55.

$$x + (x + 1) = 55$$
$$x + x + 1 = 55$$
$$2x = 54$$
$$x = 27, \text{ first integer}$$
$$x + 1 = 27 + 1 = 28, \text{ second integer}$$

The integers are 27 and 28.

3. (3.2) Let $5x$ = length
$4x$ = width

Perimeter	is	90.

$$2(5x) + 2(4x) = 90$$
$$10x + 8x = 90$$
$$18x = 90$$
$$x = 5$$
$$5x = 25, \text{ length}$$
$$4x = 20, \text{ width}$$

The length is 25 and the width is 20.

4. (3.4) Let x = number of dimes
$14 - x$ = number of quarters

Value of dimes	+	Value of quarters	=	Total value
$10x$	+	$25(14 - x)$	=	215

$$10x + 350 - 25x = 215$$
$$-15x = -135$$
$$x = 9$$
$$14 - x = 5$$

Linda has 9 dimes and 5 quarters.

5. (3.4) Let x = number of pounds of cashews
$60 - x$ = number of pounds of peanuts

$$7.40x + 2.80(60 - x) = 4.18(60)$$
$$740x + 280(60 - x) = 418(60)$$
$$740x + 16{,}800 - 280x = 25{,}080$$
$$460x = 8{,}280$$
$$x = 18$$
$$60 - x = 42$$

The grocer needs to mix 18 lb of cashews and 42 lb of peanuts.

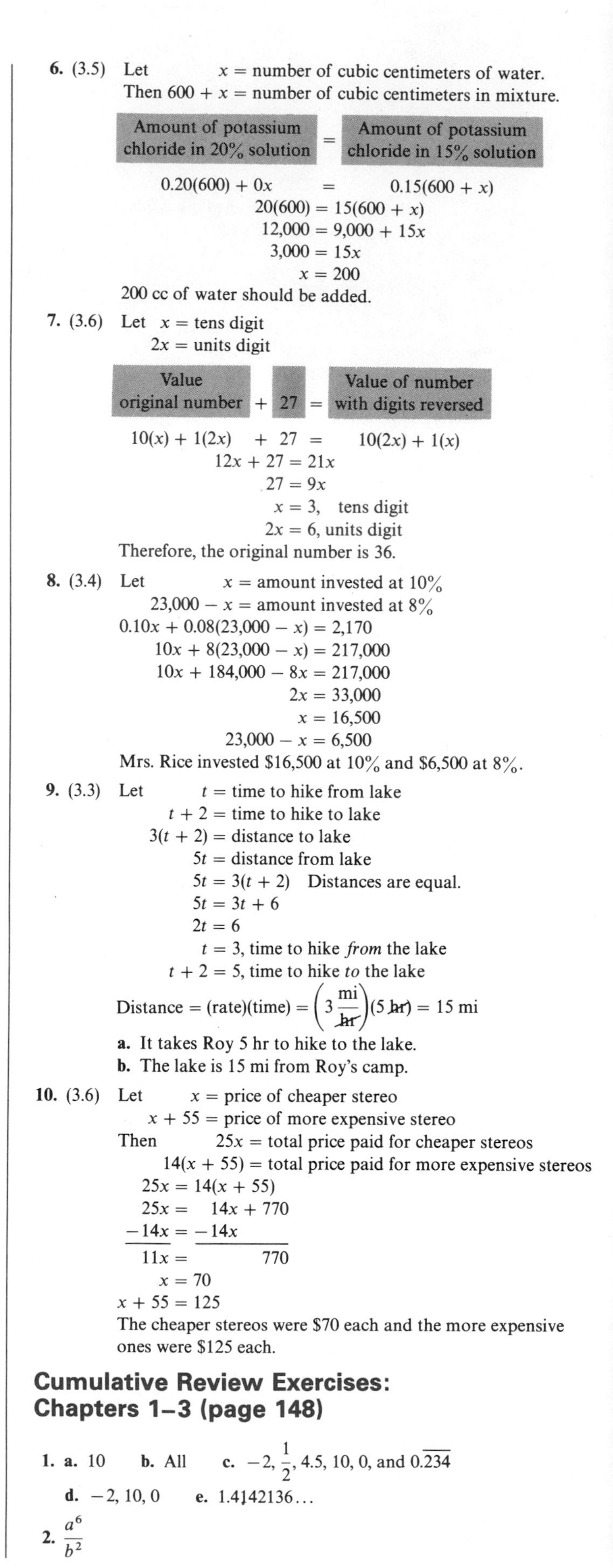

6. (3.5) Let x = number of cubic centimeters of water.
Then $600 + x$ = number of cubic centimeters in mixture.

Amount of potassium chloride in 20% solution	=	Amount of potassium chloride in 15% solution
$0.20(600) + 0x$	=	$0.15(600 + x)$

$$20(600) = 15(600 + x)$$
$$12{,}000 = 9{,}000 + 15x$$
$$3{,}000 = 15x$$
$$x = 200$$

200 cc of water should be added.

7. (3.6) Let x = tens digit
$2x$ = units digit

Value original number	+	27	=	Value of number with digits reversed
$10(x) + 1(2x)$	+	27	=	$10(2x) + 1(x)$

$$12x + 27 = 21x$$
$$27 = 9x$$
$$x = 3, \text{ tens digit}$$
$$2x = 6, \text{ units digit}$$

Therefore, the original number is 36.

8. (3.4) Let x = amount invested at 10%
$23{,}000 - x$ = amount invested at 8%

$$0.10x + 0.08(23{,}000 - x) = 2{,}170$$
$$10x + 8(23{,}000 - x) = 217{,}000$$
$$10x + 184{,}000 - 8x = 217{,}000$$
$$2x = 33{,}000$$
$$x = 16{,}500$$
$$23{,}000 - x = 6{,}500$$

Mrs. Rice invested \$16,500 at 10% and \$6,500 at 8%.

9. (3.3) Let t = time to hike from lake
$t + 2$ = time to hike to lake
$3(t + 2)$ = distance to lake
$5t$ = distance from lake

$$5t = 3(t + 2) \quad \text{Distances are equal.}$$
$$5t = 3t + 6$$
$$2t = 6$$
$$t = 3, \text{ time to hike } \textit{from} \text{ the lake}$$
$$t + 2 = 5, \text{ time to hike } \textit{to} \text{ the lake}$$

$$\text{Distance} = (\text{rate})(\text{time}) = \left(3\frac{\text{mi}}{\cancel{\text{hr}}}\right)(5\ \cancel{\text{hr}}) = 15 \text{ mi}$$

a. It takes Roy 5 hr to hike to the lake.
b. The lake is 15 mi from Roy's camp.

10. (3.6) Let x = price of cheaper stereo
$x + 55$ = price of more expensive stereo
Then $25x$ = total price paid for cheaper stereos
$14(x + 55)$ = total price paid for more expensive stereos

$$25x = 14(x + 55)$$
$$25x = 14x + 770$$
$$\underline{-14x = -14x}$$
$$11x = 770$$
$$x = 70$$
$$x + 55 = 125$$

The cheaper stereos were \$70 each and the more expensive ones were \$125 each.

Cumulative Review Exercises: Chapters 1–3 (page 148)

1. a. 10 **b.** All **c.** $-2, \frac{1}{2}, 4.5, 10, 0$, and $0.\overline{234}$
d. $-2, 10, 0$ **e.** $1.4142136\ldots$

2. $\dfrac{a^6}{b^2}$

3. $S = \frac{1}{2}gt^2$
$= \frac{1}{2}(32)(2)^2$
$= \frac{1}{2}(32)(4) = 64$

4. $2x^2y^2 - 10x^2y + 7xy$

5. $6 - \{4 - [3x - 10 + 6x]\}$
$= 6 - \{4 - [9x - 10]\}$
$= 6 - \{4 - 9x + 10\}$
$= 6 - \{14 - 9x\}$
$= 6 - 14 + 9x$
$= 9x - 8$

6. $\{-1\}$

7. $12 - 12x + 15 \geq 18 - 9x - 6$
$27 - 12x \geq 12 - 9x$
$-3x \geq -15$
$\frac{-3x}{-3} \leq \frac{-15}{-3}$
$x \leq 5$
$\{x \mid x \leq 5\}$

8. $\{x \mid -2 < x < 2\}$

9. If $\left|\frac{2x-6}{4}\right| = 1$,
$\frac{2x-6}{4} = 1$ or $\frac{2x-6}{4} = -1$
$2x - 6 = 4$ $\quad$ $2x - 6 = -4$
$2x = 10$ $\quad$ $2x = 2$
$x = 5$ $\quad$ $x = 1$
$\{5, 1\}$

10. $\{x \mid -2 < x < 3\}$

11. Let $7x$ = length of rectangle
$4x$ = width of rectangle
$2(7x) + 2(4x) = 66$
$14x + 8x = 66$
$22x = 66$
$x = 3$
$7x = 21$
$4x = 12$
(The check will not be shown.) The length is 21 in. and the width is 12 in.

12. \$7,500 at 7.4%; \$4,500 at 6.8%

13. Let x = volume of the tank (in cubic inches).
$\frac{1}{4}x + 80 = \frac{1}{3}x$
$3x + 960 = 4x$
$960 = x$
$x = 960$ (The check will not be shown.)
The volume of the tank is 960 cu. in.

14. 3 bags of perlite; 8 bags of potting soil

Exercises 4.1 (page 152)

1. 2nd degree **2.** 2nd degree
3. Not a polynomial because the exponents are not positive integers
4. Not a polynomial **5.** 0 degree **6.** 0 degree
7. Not a polynomial because the variable is in the denominator
8. Not a polynomial **9.** 1st degree **10.** 1st degree
11. Not a polynomial because the variable is under a radical sign
12. Not a polynomial
13. Not a polynomial because the variable is in the denominator
14. Not a polynomial **15.** 6th degree **16.** 6th degree
17. $8x^5 + 7x^3 - 4x - 5$; 8 **18.** $-3y^5 - 2y^3 + 4y^2 + 10$; -3
19. $-y^5 + y^3 + 3x^2y + 8x^3$; -1 **20.** $6y^3 - 4y^2 + y + 7x^2$; 6

Exercises 4.2 (page 154)

1. $-3x^4 - 2x^3 + 5 + 2x^4 + x^3 - 7x - 12 = -x^4 - x^3 - 7x - 7$
2. $-2y^3 + 3y^2 - 4y + 4$
3. $7 - 8v^3 + 9v^2 + 4v - 9v^3 - 6 + 8v^2 - 4v = -17v^3 + 17v^2 + 1$
4. $13x^3 - 4x^2 + 8x - 7$ **5.** $6x^3 + 2x^2 - 2$
6. $-2y^4 + 8y + 4$
7. $4x^3 + 6 + x - 6 - 3x^5 + 4x^2 = -3x^5 + 4x^3 + 4x^2 + x$
8. $4x^4 - 2x^3 - 3x$
9. $2y^4 + 4y^3 + 8 - 3y^4 - 2y^3 + 5y = -y^4 + 2y^3 + 5y + 8$
10. $5y^3 + 8y^2 - 5y - 4$ **11.** $-5x^4 - 7x^3 + x^2 + 4x - 2$
12. $-2x^4 + x^3 + 5x^2 - 2x + 6$ **13.** $7x^3 + 4x^2 + 7x - 11$
14. $7x^3 - 2x^2 - x - 12$
15. $6m^2n^2 - 8mn + 9 - 10m^2n^2 + 18mn - 11 + 3m^2n^2 - 2mn + 7$
$= -m^2n^2 + 8mn + 5$
16. $27u^2v + 4uv^2 + 17$
17. $[5 + xy^2 + x^3y - 6 - 3xy^2 + 4x^3y]$
$-[x^3y + 3xy^2 - 4 + 2x^3y - xy^2 + 5]$
$= [-1 - 2xy^2 + 5x^3y] - [3x^3y + 2xy^2 + 1]$
$= -1 - 2xy^2 + 5x^3y - 3x^3y - 2xy^2 - 1$
$= 2x^3y - 4xy^2 - 2$
18. $6m^2n + 6$
19. $8.586x^2 - 9.030x + 6.976 - 1.946x^2 + 41.45x + 7.468 - 3.914x^2$
$= 2.726x^2 + 32.42x + 14.444$
20. $-40.06x^2 - 117.32x + 88.55$.

Exercises 4.3A (page 157)

1. $x^2 + 9x + 20$ **2.** $y^2 + 10y + 21$ **3.** $y^2 - y - 72$
4. $z^2 + 7z - 30$ **5.** $6x^2 - 7x - 20$ **6.** $8y^2 + 14y - 15$
7. $16x^2 - 25y^2$ **8.** $s^2 - 4t^2$ **9.** $10x^2 - 17xy + 3y^2$
10. $18u^2 - 27uv + 4v^2$ **11.** $56wz + 21z - 16w - 6$
12. $36xz + 45z - 4x - 5$ **13.** $64x^2 - 81y^2$
14. $49w^2 - 4x^2$ **15.** $2s^3 + 2s^2 + 5s + 5$
16. $3u^3 + 3u^2 + 2u + 2$ **17.** $4x^2 - 4xy + y^2$
18. $25z^2 - 10wz + w^2$ **19.** $6xy - 8x + 9y - 12$
20. $12uv - 20u - 6v + 10$ **21.** $49x^4 - 42x^2 + 9$
22. $16x^4 - 40x^2 + 25$ **23.** $6x^3 - 8x^2 + 15x - 20$
24. $15y^3 - 6y^2 + 5y - 2$
25. $3x[2x^2 - 6x + 4] = 6x^3 - 18x^2 + 12x$
26. $14x^3 + 7x^2 - 105x$

Exercises 4.3B (page 159)

1.
$$\begin{array}{r} 4h^2 - 5h + 7 \\ 2h - 3 \\ \hline -12h^2 + 15h - 21 \\ 8h^3 - 10h^2 + 14h \quad\quad \\ \hline 8h^3 - 22h^2 + 29h - 21 \end{array}$$

2. $10k^3 + 23k^2 - 57k + 18$

3.
$$\begin{array}{r} a^4 + 3a^2 - 2a + 4 \\ a + 3 \\ \hline 3a^4 + 9a^2 - 6a + 12 \\ a^5 + 3a^3 - 2a^2 + 4a \quad\quad \\ \hline a^5 + 3a^4 + 3a^3 + 7a^2 - 2a + 12 \end{array}$$

4. $b^5 - 7b^4 + 10b^3 + 3b^2 - 20b + 25$

5.
$$\begin{array}{r} -\ 3z^3 + \ z^2 - \ 5z + \ 4 \\ -\ \ z + \ 4 \\ \hline -\ 12z^3 + 4z^2 - 20z + 16 \\ 3z^4 - \ \ z^3 + 5z^2 - \ 4z \qquad \\ \hline 3z^4 - 13z^3 + 9z^2 - 24z + 16 \end{array}$$

6. $v^4 - 4v^3 + 5v + 6$

7.
$$\begin{array}{r} 3u^2 - \ \ u + 5 \\ 2u^2 + \ 4u - 1 \\ \hline -\ 3u^2 + \ \ u - 5 \\ 12u^3 - \ 4u^2 + 20u \qquad \\ 6u^4 - \ 2u^3 + 10u^2 \qquad\qquad \\ \hline 6u^4 + 10u^3 + \ 3u^2 + 21u - 5 \end{array}$$

8. $10w^4 - w^3 - 40w^2 + 20w + 7$ **9.** $6x^3y^2 - 2x^2y^3 + 8xy^4$

10. $12x^2y^3 + 3x^3y^2 - 9x^4y$

11. $-5a^5b + 15a^4b^2 - 15a^3b^3 + 5a^2b^4$

12. $-4m^4p^2 + 12m^3p^3 - 12m^2p^4 + 4mp^5$

13.
$$\begin{array}{r} x^2 + \ 2x + 3 \\ x^2 + \ 2x + 3 \\ \hline 3x^2 + \ 6x + 9 \\ 2x^3 + \ 4x^2 + \ 6x \qquad \\ x^4 + 2x^3 + \ 3x^2 \qquad\qquad \\ \hline x^4 + 4x^3 + 10x^2 + 12x + 9 \end{array}$$

14. $z^4 - 6z^3 + z^2 + 24z + 16$

15.
$$\begin{array}{r|r} x^2 - xy + y^2 & x^2 + xy + y^2 \\ x + y & x - y \\ \hline x^2y - xy^2 + y^3 & -x^2y - xy^2 - y^3 \\ x^3 - x^2y + xy^2 \qquad & x^3 + x^2y + xy^2 \qquad \\ \hline x^3 \qquad\qquad + y^3 & x^3 \qquad\qquad - y^3 \end{array}$$

Then:

$[(x + y)(x^2 - xy + y^2)][(x - y)(x^2 + xy + y^2)]$
$= [x^3 + y^3][x^3 - y^3] = x^6 - y^6$

16. $a^6 - 1$

Exercises 4.4A (page 160)

1. $4u^2 - 25v^2$ **2.** $9m^2 - 49n^2$ **3.** $4x^4 - 81$ **4.** $100y^4 - 9$

5. $x^{10} - y^{12}$ **6.** $a^{14} - b^8$ **7.** $49m^2n^2 - 4r^2s^2$

8. $64h^2k^2 - 25e^2f^2$ **9.** $144x^8y^6 - u^{14}v^2$

10. $121a^{10}b^4 - 81c^6d^{12}$

11. $(a + b)^2 - (2)^2 = (a + b)(a + b) - 4 = a^2 + 2ab + b^2 - 4$

12. $x^2 + 2xy + y^2 - 25$

13. $([x^2 + y] + 5)([x^2 + y] - 5) = [x^2 + y]^2 - 5^2$
$= [x^2 + y][x^2 + y] - 25 = x^4 + 2x^2y + y^2 - 25$

14. $u^4 - 2u^2v + v^2 - 49$ **15.** $7x(x^2 - 1) = 7x^3 - 7x$

16. $12y^3 - 75y$

Exercises 4.4B (page 162)

1. $(2x)^2 + 2(2x)(3) + 3^2 = 4x^2 + 12x + 9$ **2.** $36x^2 + 60x + 25$

3. $(5x)^2 - 2(5x)(3) + 3^2 = 25x^2 - 30x + 9$ **4.** $81x^2 - 108x + 36$

5. $49x^2 - 140xy + 100y^2$ **6.** $16u^2 - 72uv + 81v^2$

7. $(u + v)^2 + 2(u + v)(7) + (7)^2 = u^2 + 2uv + v^2 + 14u + 14v + 49$

8. $s^2 + 2st + t^2 + 8s + 8t + 16$

9. $[2x - y]^2 - 2[2x - y][3] + 3^2$
$= (2x)^2 - 2(2x)(y) + y^2 - 12x + 6y + 9$
$= 4x^2 - 4xy + y^2 - 12x + 6y + 9$

10. $9z^2 - 6zw + w^2 - 42z + 14w + 49$

11. $x^2 + 2x(u + v) + (u + v)^2 = x^2 + 2xu + 2xv + u^2 + 2uv + v^2$

12. $s^2 + 2sy + 4s + y^2 + 4y + 4$

13. $y^2 - 2y[x - 2] + [x - 2]^2 = y^2 - 2xy + 4y + x^2 - 4x + 4$

14. $x^2 - 2xz + 10x + z^2 - 10z + 25$

15. $5[x^2 - 2(x)(3) + 3^2] = 5(x^2 - 6x + 9) = 5x^2 - 30x + 45$

16. $7x^2 + 14x + 7$

Exercises 4.4C (page 170)

1. $4! = 4 \cdot 3 \cdot 2 \cdot 1 = 24$ **2.** 720 **3.** $\binom{3}{2} = 3$ **4.** 35

5. $\binom{6}{4} = \dfrac{6!}{4!(6-4)!} = \dfrac{6 \cdot 5 \cdot \overset{1}{\cancel{4!}}}{\underset{1}{\cancel{4!}}(2!)} = \dfrac{6 \cdot 5}{2} = 15$ **6.** 1

7. The coefficient of the 6th term is

$\binom{9}{5} = \dfrac{9!}{5!(4!)} = \dfrac{9 \cdot \overset{2}{\cancel{8}} \cdot 7 \cdot \overset{1}{\cancel{6}} \cdot \overset{1}{\cancel{5!}}}{\underset{1}{\cancel{5!}}(\underset{1}{\cancel{4}} \cdot \cancel{3} \cdot \underset{1}{\cancel{2}} \cdot 1)} = 126.$

The 6th term is $126a^4b^5$.

8. $35a^4b^3$

9. $x^5 + 5x^4y + \dfrac{5!}{(2!)(3!)}x^3y^2 + \dfrac{5!}{(3!)(2!)}x^2y^3 + 5xy^4 + y^5$
$= x^5 + 5x^4y + 10x^3y^2 + 10x^2y^3 + 5xy^4 + y^5$

10. $r^4 + 4r^3s + 6r^2s^2 + 4rs^3 + s^4$

11. $x^5 - 5x^4(2) + \dfrac{5!}{(2!)(3!)}x^3(2)^2 - \dfrac{5!}{(3!)(2!)}x^2(2)^3 + 5x(2)^4 - (2)^5$
$= x^5 - 10x^4 + 40x^3 - 80x^2 + 80x - 32$

12. $y^4 - 12y^3 + 54y^2 - 108y + 81$

13. $(3r + s)^6 = (3r)^6 + 6(3r)^5s + \dfrac{6!}{(2!)(4!)}(3r)^4s^2 + \dfrac{6!}{(3!)(3!)}(3r)^3s^3$
$+ 15(3r)^2s^4 + 6(3r)s^5 + s^6$
$= 729r^6 + 1{,}458r^5s + 1{,}215r^4s^2 + 540r^3s^3$
$+ 135r^2s^4 + 18rs^5 + s^6$

14. $64x^6 + 192x^5y + 240x^4y^2 + 160x^3y^3 + 60x^2y^4 + 12xy^5 + y^6$

15. $(x + y^2)^4 = x^4 + 4x^3(y^2) + 6x^2(y^2)^2 + 4x(y^2)^3 + (y^2)^4$
$= x^4 + 4x^3y^2 + 6x^2y^4 + 4xy^6 + y^8$

16. $u^5 + 5u^4v^2 + 10u^3v^4 + 10u^2v^6 + 5uv^8 + v^{10}$

17. $(2x)^5 + 5(2x)^4(-\tfrac{1}{2}) + \dfrac{5!}{(2!)(3!)}(2x)^3(-\tfrac{1}{2})^2 + 10(2x)^2(-\tfrac{1}{2})^3$
$+ 5(2x)(-\tfrac{1}{2})^4 + (-\tfrac{1}{2})^5$
$= 32x^5 - 40x^4 + 20x^3 - 5x^2 + \tfrac{5}{8}x - \tfrac{1}{32}$

18. $81x^4 - 36x^3 + 6x^2 - \dfrac{4}{9}x + \dfrac{1}{81}$

19. $\left(\dfrac{x}{3}\right)^4 + 4\left(\dfrac{x}{3}\right)^3\left(\dfrac{3}{2}\right) + \dfrac{4!}{(2!)(2!)}\left(\dfrac{x}{3}\right)^2\left(\dfrac{3}{2}\right)^2 + 4\left(\dfrac{x}{3}\right)\left(\dfrac{3}{2}\right)^3 + \left(\dfrac{3}{2}\right)^4$
$= \tfrac{1}{81}x^4 + \tfrac{2}{9}x^3 + \tfrac{3}{2}x^2 + \tfrac{9}{2}x + \tfrac{81}{16}$

20. $\tfrac{1}{625}x^4 + \tfrac{2}{25}x^3 + \tfrac{3}{2}x^2 + \tfrac{25}{2}x + \tfrac{625}{16}$

21. $(4x^2 - 3y^2)^5 = (4x^2)^5 + 5(4x^2)^4(-3y^2) + 10(4x^2)^3(-3y^2)^2$
$+ 10(4x^2)^2(-3y^2)^3 + 5(4x^2)(-3y^2)^4 + (-3y^2)^5$
$= 1{,}024x^{10} - 3{,}840x^8y^2 + 5{,}760x^6y^4$
$- 4{,}320x^4y^6 + 1{,}620x^2y^8 - 243y^{10}$

22. $243x^{10} - 810x^8y^3 + 1{,}080x^6y^6 - 720x^4y^9 + 240x^2y^{12} - 32y^{15}$

23. $(x + x^{-1})^4 = (x)^4 + 4(x)^3(x^{-1}) + 6(x)^2(x^{-1})^2$
$+ 4(x)(x^{-1})^3 + (x^{-1})^4$
$= x^4 + 4x^2 + 6 + 4x^{-2} + x^{-4}$
$= x^4 + 4x^2 + 6 + \dfrac{4}{x^2} + \dfrac{1}{x^4}$

24. $\dfrac{1}{x^4} - \dfrac{4}{x^2} + 6 - 4x^2 + x^4$

25. $(x)^{10} + 10(x)^9(2y^2) + \frac{10!}{(2!)(8!)}(x)^8(2y^2)^2 + \frac{10!}{(3!)(7!)}(x)^7(2y^2)^3 + \cdots$

$= x^{10} + 20x^9y^2 + 180x^8y^4 + 960x^7y^6 + \cdots$

26. $x^8 + 24x^7y^2 + 252x^6y^4 + 1{,}512x^5y^6 + \cdots$

27. $(x - 3y^2)^{10} = (x)^{10} + 10(x)^9(-3y^2) + \frac{10!}{(2!)(8!)}(x)^8(-3y^2)^2$

$+ \frac{10!}{(3!)(7!)}(x)^7(-3y^2)^3 + \cdots$

$= x^{10} - 30x^9y^2 + 405x^8y^4 - 3{,}240x^7y^6 + \cdots$

28. $x^{11} - 22x^{10}y^3 + 220x^9y^6 - 1{,}320x^8y^9 + \cdots$

Exercises 4.5A (page 171)

1. $\frac{18x^5}{6x^2} + \frac{-24x^4}{6x^2} + \frac{-12x^3}{6x^2} = 3x^3 - 4x^2 - 2x$

2. $-4y^2 + 9y - 5$

3. $\frac{55a^4b^3}{-11ab} + \frac{-33ab^2}{-11ab} = -5a^3b^2 + 3b$

4. $-2mn^3 + 4m^2$

5. $\frac{-15x^2y^2z^2}{-5xyz} + \frac{-30xyz}{-5xyz} = 3xyz + 6$

6. $3abc + 2$

7. $\frac{13x^3y^2}{13x^2y^2} + \frac{-26x^5y^3}{13x^2y^2} + \frac{39x^4y^6}{13x^2y^2} = x - 2x^3y + 3x^2y^4$

8. $3n - 5m - 2mn^3$

Exercises 4.5B (page 176)

Note: The checks for these exercises will not be shown.

1.
$$\begin{array}{r} x+3 \\ x+7\,\overline{)\,x^2+10x-5} \\ \underline{x^2+7x} \\ 3x-5 \\ \underline{3x+21} \\ -26 \end{array}$$

Answer: $x + 3 - \frac{26}{x+7}$

2. $x + 5 - \frac{25}{x+4}$

3.
$$\begin{array}{r} 2z^2-z-3 \\ 3z-5\,\overline{)\,6z^3-13z^2-4z+15} \\ \underline{6z^3-10z^2} \\ -3z^2-4z \\ \underline{-3z^2+5z} \\ -9z+15 \\ \underline{-9z+15} \\ 0 \end{array}$$

Answer: $2z^2 - z - 3$

4. $3y^2 - y - 4$

5.
$$\begin{array}{r} v^3-4v^2-4v-5 \\ v+1\,\overline{)\,v^4-3v^3-8v^2-9v-5} \\ \underline{v^4+v^3} \\ -4v^3-8v^2 \\ \underline{-4v^3-4v^2} \\ -4v^2-9v \\ \underline{-4v^2-4v} \\ -5v-5 \\ \underline{-5v-5} \\ 0 \end{array}$$

Answer: $v^3 - 4v^2 - 4v - 5$

6. $w^3 - 4w^2 - 4w - 5$

7.
$$\begin{array}{r} 4x^2+8x+8 \\ -x+2\,\overline{)\,-4x^3+8x+10} \\ \underline{-4x^3+8x^2} \\ -8x^2+8x \\ \underline{-8x^2+16x} \\ -8x+10 \\ \underline{-8x+16} \\ -6 \end{array}$$

Answer: $4x^2 + 8x + 8 - \frac{6}{2-x}$

8. $x^2 + 3x - 3 + \frac{-6}{3-x}$

9.
$$\begin{array}{r} 4x^2-xy+5y^2 \\ 3x-2y\,\overline{)\,12x^3-11x^2y+17xy^2-10y^3} \\ \underline{12x^3-8x^2y} \\ -3x^2y+17xy^2 \\ \underline{-3x^2y+2xy^2} \\ 15xy^2-10y^3 \\ \underline{15xy^2-10y^3} \\ 0 \end{array}$$

Answer: $4x^2 - xy + 5y^2$

10. $5x^2 - xy - 4y^2$

11.
$$\begin{array}{r} x^3+2x^2+4x+1 \\ x-2\,\overline{)\,x^4+0x^3+0x^2-7x+6} \\ \underline{x^4-2x^3} \\ 2x^3+0x^2 \\ \underline{2x^3-4x^2} \\ 4x^2-7x \\ \underline{4x^2-8x} \\ x+6 \\ \underline{x-2} \\ 8 \end{array}$$

Answer: $x^3 + 2x^2 + 4x + 1 + \frac{8}{x-2}$

12. $x^3 + x^2 + x + 7 + \frac{10}{x-1}$

13.
$$\begin{array}{r} x^2-x+2 \\ 4x^2+x+1\,\overline{)\,4x^4-3x^3+8x^2+x+2} \\ \underline{4x^4+x^3+x^2} \\ -4x^3+7x^2+x \\ \underline{-4x^3-x^2-x} \\ 8x^2+2x+2 \\ \underline{8x^2+2x+2} \\ 0 \end{array}$$

Answer: $x^2 - x + 2$

14. $x^2 - x + 5$

15.
$$\begin{array}{r} 2x^2+0x+3 \\ x^2+2\,\overline{)\,2x^4+0x^3+7x^2+0x+5} \\ \underline{2x^4+4x^2} \\ 3x^2+5 \\ \underline{3x^2+6} \\ -1 \end{array}$$

Answer: $2x^2 + 3 - \frac{1}{x^2+2}$

16. $3x^2 + 2 + \frac{-5}{x^2+3}$

17.
$$\begin{array}{r} x^2-5 \\ 2x^3+3\,\overline{)\,2x^5+0x^4-10x^3+3x^2+0x-15} \\ \underline{2x^5+3x^2} \\ -10x^3-15 \\ \underline{-10x^3-15} \\ 0 \end{array}$$

Answer: $x^2 - 5$

18. $x^2 - 2$

19.
$$
\begin{array}{r}
x^2 + 5x + 2 \\
3x^2 - x + 1\overline{)3x^4 + 14x^3 + 2x^2 + 3x + 2} \\
\underline{3x^4 - x^3 + x^2} \\
15x^3 + x^2 + 3x \\
\underline{15x^3 - 5x^2 + 5x} \\
6x^2 - 2x + 2 \\
\underline{6x^2 - 2x + 2} \\
0
\end{array}
$$

Answer: $x^2 + 5x + 2$

20. $x^2 + 7x - 2$

Exercises 4.6 (page 180)

Note: The checks for these exercises will not be shown.

1.

	1	2	−18
		3	15
3	1	5	−3

Answer: $x + 5 - \dfrac{3}{x - 3}$

2. $x + 6 + \dfrac{2}{x - 2}$

3.

	1	3	−5	6
		−4	4	4
−4	1	−1	−1	10

Answer: $x^2 - x - 1 + \dfrac{10}{x + 4}$

4. $x^2 + x - 1 + \dfrac{-2}{x + 5}$

5.

	1	6	0	−1	−4
		−6	0	0	6
−6	1	0	0	−1	2

Answer: $x^3 - 1 + \dfrac{2}{x + 6}$

6. $2x^3 - x^2 + 3x + 1 + \dfrac{-5}{x + 3}$

7.

	1	0	0	0	−16
		2	4	8	16
2	1	2	4	8	0

Answer: $x^3 + 2x^2 + 4x + 8$

8. $x^6 + x^5 + x^4 + x^3 + x^2 + x + 1$

9.

	1	−3	0	0	−2	3	5
		3	0	0	0	−6	−9
3	1	0	0	0	−2	−3	−4

Answer: $x^5 - 2x - 3 + \dfrac{-4}{x - 3}$

10. $x^5 + 2x^4 + x^3 + 2x^2 + 4x + 1$

11.

	3	−1	9	0	−1
		1	0	3	1
$\frac{1}{3}$	3	0	9	3	0

Answer: $3x^3 + 9x + 3$

12. $3x^3 - 6x^2 + 3x - 3$

13.

	1	1	−45	−45	324	324
		−3	6	117	−216	−324
−3	1	−2	−39	72	108	0

Answer: $x^4 - 2x^3 - 39x^2 + 72x + 108$

14. $x^4 - 5x^3 - 15x^2 + 45x + 54$

15.

	4	0	−45	3	100
		8	16	−58	−110
2	4	8	−29	−55	−10

Answer: $4x^3 + 8x^2 - 29x - 55 + \dfrac{-10}{x - 2}$

16. $9x^3 + 9x^2 - 4x - 2 + \dfrac{4}{x - 1}$

17.

	2.6	0	1.8	−6.4
		3.9	5.85	11.475
1.5	2.6	3.9	7.65	5.075

Answer: $2.6x^2 + 3.9x + 7.65 + \dfrac{5.075}{x - 1.5}$

18. $3.8x^2 + 8.1x + 20.25 + \dfrac{26.725}{x - 2.5}$

19.

	2.7	0	−1.6	3.289
		3.24	3.888	2.7456
1.2	2.7	3.24	2.288	6.0346

Answer: $2.7x^2 + 3.24x + 2.288 + \dfrac{6.0346}{x - 1.2}$

20. $3x^2 + 6x + 9.6 + \dfrac{13.86}{x - 1.6}$

Review Exercises 4.7 (page 183)

1. 3rd degree **2.** Not a polynomial

3. Not a polynomial because the variable is under a radical sign

4. Not a polynomial **5.** $-2x^3 - 38x^2 + 24x - 3$

6. $9x^2y + xy^2 + 2y^2 - 12$ **7.** $-2x^3 - 2x^2 - 5x + 7$

8. $x^4 - x^3 + x^2 - x + 1$

9. $\underline{2k^3} - 7k + 11 + \underline{4k^3} + k^2 - 9k - 3k^2 + 5k + 6$
$= 6k^3 - 2k^2 - 11k + 17$

10. $-15a^3b^3 - 9a^2b^2 + 6ab^2c$ **11.** $35x^2 - 38xy + 8y^2$

12. $x^8 + 12x^6 + 54x^4 + 108x^2 + 81$

13.
$$
\begin{array}{rrrrr}
 & & 4x^2 & -\ 5x & +\ 1 \\
 & & 2x^2 & +\ x & -\ 3 \\
\hline
 & & -\ 12x^2 & +\ 15x & -\ 3 \\
 & 4x^3 & -\ 5x^2 & +\ x & \\
8x^4 & -\ 10x^3 & +\ 2x^2 & & \\
\hline
8x^4 & -\ 6x^3 & -\ 15x^2 & +\ 16x & -\ 3
\end{array}
$$

14. $49x^2 - 25$

15. $(2x^2)^5 + 5(2x^2)^4(\frac{1}{2})^1 + \dfrac{5!}{(2!)(3!)}(2x^2)^3(\frac{1}{2})^2 + 10(2x^2)^2(\frac{1}{2})^3 + 5(2x^2)^1(\frac{1}{2})^4 + (\frac{1}{2})^5$
$= 32x^{10} + 40x^8 + 20x^6 + 5x^4 + \frac{5}{8}x^2 + \frac{1}{32}$

16. $x^2 + 2xy + y^2 + 6x + 6y + 9$ **17.** $a^2 - 10a + 25$

18. $z^3 + 27$

19.
$$
\begin{array}{rrrr}
 & a + & b + & 4 \\
 & a + & b + & 4 \\
\hline
 & 4a + & 4b + & 16 \\
 & ab & +\ 4b & +\ b^2 \\
a^2 + & ab + & 4a & \\
\hline
\multicolumn{4}{r}{a^2 + 2ab + 8a + 8b + 16 + b^2}
\end{array}
$$

20. $\dfrac{1}{z^8} + \dfrac{4}{z^4} + 6 + 4z^4 + z^8$ **21.** $4x(x^4 - y^4) = 4x^5 - 4xy^4$

22. $\frac{1}{81} + \frac{4}{9}x^2 + 6x^4 + 36x^6 + 81x^8$

23. $\dfrac{-15a^2b^3}{-5ab} + \dfrac{20a^4b^2}{-5ab} + \dfrac{-10ab}{-5ab} = 3ab^2 - 4a^3b + 2$

24. $3x + \dfrac{10}{2x - 3}$

25.
$$
\begin{array}{r}
2a + 5b \\
5a - b\overline{)10a^2 + 23ab - 5b^2} \\
\underline{10a^2 - 2ab} \\
25ab - 5b^2 \\
\underline{25ab - 5b^2} \\
0
\end{array}
$$

Answer: $2a + 5b$

26. $x^2 - 1 + \dfrac{-10}{3x^2 - 2x + 5}$

27.
$$
\begin{array}{r}
x^3 - 3x^2 + 9x - 27 \\
x + 3\,\overline{)\,x^4 + 0x^3 + 0x^2 + 0x - 81} \\
\underline{x^4 + 3x^3} \qquad\qquad\qquad \\
-3x^3 + 0x^2 \qquad\qquad \\
\underline{-3x^3 - 9x^2} \qquad\qquad \\
9x^2 + 0x \qquad \\
\underline{9x^2 + 27x} \qquad \\
-27x - 81 \\
\underline{-27x - 81} \\
0
\end{array}
$$

Answer: $x^3 - 3x^2 + 9x - 27$

28. $3x^4 + x^3 - 2x^2 + \dfrac{4}{x+2}$

29.
$$
\begin{array}{c|cccccc}
 & 5 & -2 & 10 & -4 & 0 & 2 \\
 & & 2 & 0 & 4 & 0 & 0 \\
\hline
\frac{2}{5} & 5 & 0 & 10 & 0 & 0 & \boxed{2}
\end{array}
$$

Answer: $5x^4 + 10x^2 + \dfrac{2}{x - \frac{2}{5}}$

30. $2x^4 + 1 + \dfrac{2}{x+3}$

Chapter 4 Diagnostic Test (page 187)

Following each problem number is the textbook section number (in parentheses) where that kind of problem is discussed.

1. (4.1) **a.** Degree of $-\frac{1}{3}x^1y^1$ is $1 + 1 = 2$.
b. Degree of a polynomial is degree of highest-degree term: degree of $2x^2y^3$ is $2 + 3 = 5$. Therefore, the polynomial is of 5th degree.

2. (4.2) **a.** $-x^3 - x^2 + 4x + 3$ **b.** $-2x^4 - 2x^3 - x^2 + 4x + 3$

3. (4.2) **a.** $(-3z^2 - 6z + 8) - (8 - z + 4z^2)$
$= \underline{-3z^2} - \underline{6z} + 8 - 8 + \underline{z} - \underline{4z^2} = -7z^2 - 5z$

b. $\underline{3ab^2} - \underline{5ab} - a^3 - \underline{2ab} + \underline{4ab} - \underline{7ab^2}$
$= -4ab^2 - 3ab - a^3$

4. a. (4.3) $(-3ab)(6a^2) + (-3ab)(-2ab^2) + (-3ab)(5b)$
$= -18a^3b + 6a^2b^3 - 15ab^2$

b. (4.5A) $\dfrac{9z^3w}{3zw} + \dfrac{6z^2w^2}{3zw} - \dfrac{12zw^3}{3zw} = 3z^2 + 2zw - 4w^2$

5. a. (4.3)
$$
\begin{array}{r}
x^2 + 2x + 4 \\
\underline{x - 2} \\
-2x^2 - 4x - 8 \\
\underline{x^3 + 2x^2 + 4x \qquad} \\
x^3 \qquad\qquad\quad - 8
\end{array}
$$

b. (4.4A) $4x^8 - 9$

6. a. (4.3A) $15m^2 + 14m - 8$ **b.** (4.4B) $9R^4 - 30R^2 + 25$

7. a. (4.5B)
$$
\begin{array}{r}
4y - 4 \\
3y + 2\,\overline{)\,12y^2 - 4y + 1} \\
\underline{12y^2 + 8y} \qquad \\
-12y + 1 \\
\underline{-12y - 8} \\
9
\end{array}
$$

Answer: $4y - 4 + \dfrac{9}{3y+2}$

b. (4.6) $2x^4 + 3x^3 - 7x^2 + 0x - 5$
$$
\begin{array}{c|ccccc}
 & 2 & 3 & -7 & 0 & -5 \\
 & & -6 & 9 & -6 & 18 \\
\hline
-3 & 2 & -3 & 2 & -6 & \boxed{13}
\end{array}
$$

Answer: $2x^3 - 3x^2 + 2x - 6 + \dfrac{13}{x+3}$

8. (4.3)
$$
\begin{array}{r}
m^2 - 2m + 5 \\
\underline{m^2 - 2m + 5} \\
5m^2 - 10m + 25 \\
-2m^3 + 4m^2 - 10m \qquad \\
\underline{m^4 - 2m^3 + 5m^2 \qquad\qquad\quad} \\
m^4 - 4m^3 + 14m^2 - 20m + 25
\end{array}
$$

9. (4.5B)
$$
\begin{array}{r}
3x^2 + 0x - 5 \\
2x^2 - x + 4\,\overline{)\,6x^4 - 3x^3 + 2x^2 + 5x - 7} \\
\underline{6x^4 - 3x^3 + 12x^2} \qquad\qquad \\
-10x^2 + 5x - 7 \\
\underline{-10x^2 + 5x - 20} \\
13
\end{array}
$$

Answer: $3x^2 - 5 + \dfrac{13}{2x^2 - x + 4}$

10. (4.4C) $(2x+1)^5 = (2x)^5 + 5(2x)^4(1) + \dfrac{5!}{(2!)(3!)}(2x)^3(1)^2$
$+ \dfrac{5!}{(3!)(2!)}(2x)^2(1)^3 + 5(2x)(1)^4 + 1^5$
$= 32x^5 + 80x^4 + 80x^3 + 40x^2 + 10x + 1$

Cumulative Review Exercises: Chapters 1–4 (page 188)

1. $10 - (-3 - 25) = 10 - (-28) = 10 + 28 = 38$ **2.** $\dfrac{1}{x^2y^3}$

3. $108 = 2^2 \cdot 3^3$; $360 = 2^3 \cdot 3^2 \cdot 5$; LCM $= 2^3 \cdot 3^3 \cdot 5 = 1{,}080$

4. $6a - 3$ **5.** $4x - 12 = 4 - x - 6$
$5x = 10$
$x = 2$
$\{2\}$

6. $\{x \mid x \geq 3\}$

7. If $|5 - x| = 6$,
$5 - x = 6$ or $5 - x = -6$
$-x = 1$ $\quad -x = -11$
$x = -1$ $\quad x = 11$
$\{-1, 11\}$

8. $\{x \mid x > 4\} \cup \{x \mid x < -\frac{2}{3}\}$

9.
$|2x - 1| \leq 4$
$-4 \leq 2x - 1 \leq 4$
$\underline{\ \ 1 \qquad\quad\ 1 \quad 1}$
$-3 \leq 2x \leq 5$
$-\frac{3}{2} \leq x \leq \frac{5}{2}$
$\{x \mid -\frac{3}{2} \leq x \leq \frac{5}{2}\}$

10. $3x^5 + 6x^4 + 7x^3 + 12x^2 - 4x$

11. $x^6 - 6x^5 + \dfrac{6!}{(2!)(4!)}x^4 - \dfrac{6!}{(3!)(3!)}x^3 + 15x^2 - 6x + 1$
$= x^6 - 6x^5 + 15x^4 - 20x^3 + 15x^2 - 6x + 1$

12. $x^2 - 2x - 5 + \dfrac{4}{x+4}$

13. Let $2x$ = length of shortest side
$4x$ = length of middle side
$5x$ = length of longest side
$2x + 4x + 5x = 88$
$11x = 88$
$x = 8$
$2x = 16$, shortest side
$4x = 32$, middle side
$5x = 40$, longest side
The sides are 16, 32, and 40 units long.

14. 4.5 lb

15. Let x = speed of stream in mph. (Note that 40 min = $\frac{2}{3}$ hr.)
Then $30 + x$ = speed downstream in mph
$\frac{2}{3}(30 + x)$ mi = distance downstream
$30 - x$ = speed upstream in mph
$\frac{2}{3}(30 - x)$ mi = distance upstream
$\frac{2}{3}(30 + x) = \frac{2}{3}(30 - x) + 4$
$20 + \frac{2}{3}x = 20 - \frac{2}{3}x + 4$
$\frac{4}{3}x = 4$
$x = 3$ (The check will not be shown.)
The speed of the stream is 3 mph.

16. \$600 at 6.25%; \$800 at 5.75%

17. The boat will go (28 − 4) mph, or 24 mph, when going upstream and (28 + 4) mph, or 32 mph, when going downstream.
Let x = the *time* the boater will take to go upstream (in hours)
$3\frac{1}{2} - x$ = the *time* the boater will take to go downstream (in hours)
Then $\left(24\frac{\text{mi}}{\cancel{\text{hr}}}\right)(x\,\cancel{\text{hr}})$ = the distance the boater will go upstream
$\left(32\frac{\text{mi}}{\cancel{\text{hr}}}\right)\left\{\left(\frac{7}{2} - x\right)\cancel{\text{hr}}\right\}$ = the distance the boater will go downstream
The distances are equal.
$$24x = 32\left(\frac{7}{2} - x\right)$$
$$24x = 112 - 32x$$
$$56x = 112$$
$x = 2$, number of hours for boat to go upstream
$\left(24\frac{\text{mi}}{\cancel{\text{hr}}}\right)(2\,\cancel{\text{hr}}) = 48$ mi, distance boater will go upstream
(The check will not be shown.)
The boater can go 48 mi upstream.

18. 7.5 ℓ of 2% solution; 4.5 ℓ of 10% solution

Exercises 5.1 (page 193)

1. $18xy(3x^2z^4 - 4y^2)$ 2. $15ab^2c(15b^3 - 7a^2c^5)$
3. $4x(4x^2 - 2x + 1)$ 4. $3a(9a^3 - 3a + 1)$ 5. Not factorable
6. Not factorable 7. $3(2my + 5mz - 3n)$
8. $4(nx + 2ny + 3z)$ 9. $-5r^7s^5(7t^4 + 11rs^4u^4 - 8p^8r^2s^3)$ or $5r^7s^5(-7t^4 - 11rs^4u^4 + 8p^8r^2s^3)$
10. $-40a^4c^3(3a^4b^7c^2 - d^9 + 2ac^2)$ or $40a^4c^3(-3a^4b^7c^2 + d^9 - 2ac^2)$
11. Not factorable 12. Not factorable
13. $-12x^4y^3(2x^4 + x^3y - 4xy^2 - 5y^3)$
14. $16y^4z^5(4y^5 + 3y^4z - y^3z^2 - 5z^3)$
15. GCF = $(a + b)$
$m(a + b) + n(a + b) = (a + b)(m + n)$
16. $(a - 2b)(3a + 2)$ 17. GCF = $(y + 1)$
$x(y + 1) - (y + 1) = (y + 1)(x - 1)$
18. $(3e - f)(2e - 3)$
19. GCF = $(x - y)$
$5(x - y) - (x - y)^2 = (x - y)[5 - (x - y)]$
$= (x - y)(5 - x + y)$
20. $(a + b)(4 - a - b)$
21. $8x(y^2 + 3z)^2 - 6x^4(y^2 + 3z) = 2x(y^2 + 3z)[4(y^2 + 3z) - 3x^3]$
$= 2x(y^2 + 3z)(4y^2 + 12z - 3x^3)$
22. $3a^2(b - 2c^5)^2(4ab - 8ac^5 - 5)$
23. $5(x + y)^2(a + b)^5([x + y] + 3[a + b])$
$= 5(x + y)^2(a + b)^5(x + y + 3a + 3b)$
24. $7(s + t)^4(u + v)^6(2u + 2v + s + t)$

Exercises 5.2 (196)

1. $x(m - n) - y(m - n) = (m - n)(x - y)$ 2. $(h - k)(a - b)$
3. $x(y + 1) - 1(y + 1) = (y + 1)(x - 1)$ 4. $(a - 1)(d + 1)$
5. $3a(a - 2b) + 2(a - 2b) = (a - 2b)(3a + 2)$ 6. $(h - 3k)(2h + 5)$
7. $2e(3e - f) - 3(3e - f) = (3e - f)(2e - 3)$
8. $(2m - n)(4m - 3)$ 9. $x^2(x + 3) - 2(x + 3) = (x + 3)(x^2 - 2)$
10. $(a - 1)(a^2 - 2)$ 11. $b^2(b + 4) + 5(b - 4)$ Not factorable
12. Not factorable 13. $2a^2(a + 4) - 3(a + 4) = (a + 4)(2a^2 - 3)$
14. $(y - 2)(5y^2 + 2)$
15. $c[am + bm + an + bn] = c[m(a + b) + n(a + b)]$
$= c(a + b)(m + n)$
16. $k(u + v)(c + d)$
17. $x(a^2 + 2a + 5) + y(a^2 + 2a + 5) = (a^2 + 2a + 5)(x + y)$
18. $(x^2 + 3x + 7)(a + b)$
19. $x(s^2 - s + 4) + y(s^2 - s + 4) = (s^2 - s + 4)(x + y)$
20. $(t^2 - t + 3)(a + b)$
21. $a(x^2 + x + 1) - (x^2 + x + 1) = (x^2 + x + 1)(a - 1)$
22. $(y^2 + y + 2)(x - 1)$

Exercises 5.3 (page 198)

1. $2(x^2 - 4y^2) = 2(x - 2y)(x + 2y)$ 2. $3(x + 3y)(x - 3y)$
3. $2(49u^4 - 36v^4) = 2(7u^2 + 6v^2)(7u^2 - 6v^2)$
4. $3(9m^3 + 10n^2)(9m^3 - 10n^2)$
5. $(x^2 + y^2)(x^2 - y^2) = (x^2 + y^2)(x + y)(x - y)$
6. $(a^2 + 4)(a + 2)(a - 2)$ 7. Not factorable 8. Not factorable
9. $(2h^2k^2 + 1)(2h^2k^2 - 1)$ 10. $(3x^2 + 1)(3x^2 - 1)$
11. $a^2b^2(25a^2 - b^2) = a^2b^2(5a + b)(5a - b)$
12. $x^2y^2(y + 10x)(y - 10x)$ 13. $8x(2x + 1)$ 14. $5y(5y + 1)$
15. Not factorable 16. Not factorable
17. $[(x + y) + 2][(x + y) - 2] = (x + y + 2)(x + y - 2)$
18. $(a + b + 3)(a + b - 3)$
19. $(a + b)[x^2 - y^2] = (a + b)(x + y)(x - y)$
20. $(x + y)(a + b)(a - b)$ 21. Not factorable
22. Not factorable
23. $(x + 3y)(x - 3y) + (x - 3y) = (x - 3y)([x + 3y] + 1)$
$= (x - 3y)(x + 3y + 1)$
24. $(x - y)(x + y + 1)$
25. $(x + 2y) + (x + 2y)(x - 2y) = (x + 2y)(1 + x - 2y)$
26. $(a + b)(1 + a - b)$

Exercises 5.4A (page 202)

1. $(t - 3)(t + 10)$ 2. $(m + 2)(m - 15)$ 3. $(m + 1)(m + 12)$
4. $(x + 14)(x + 1)$ 5. Not factorable 6. Not factorable
7. Not factorable 8. Not factorable
9. $x^2 + 10x = x(x + 10)$ 10. $y(y + 10)$
11. $(u^2 - 14)(u^2 - 1) = (u^2 - 14)(u - 1)(u + 1)$
12. $(y - 1)(y + 1)(y^2 - 15)$ 13. $(u - 4)(u + 16)$

14. $(v + 2)(v - 32)$ **15.** $x^2 + 3x + 2 = (x + 2)(x + 1)$

16. $(a + 5)(a + 2)$ **17.** $x^2(x^2 - 6x + 2)$ **18.** $y^2(y^4 - 2y^2 + 2)$

19. $(x + 2y)(x - y)$ **20.** $(x + 3y)^2$

21. $[(a + b) + 2][(a + b) + 4] = (a + b + 2)(a + b + 4)$

22. $(m + n + 1)(m + n + 8)$

23. $[(x + y) + 2][(x + y) - 15] = (x + y + 2)(x + y - 15)$

24. $(x + y + 2)(x + y - 12)$

Exercises 5.4B (page 209)

1. $(x + 1)(5x + 4)$ **2.** $(5x + 2)(x + 2)$

3. $(7 - b)(1 - 3b)$ or $(3b - 1)(b - 7)$

4. $(1 - u)(7 - 3u)$ or $(3u - 7)(u - 1)$ **5.** Not factorable

6. Not factorable **7.** $(3n - 1)(n + 5)$ **8.** $(3n + 5)(n - 1)$

9. $(3t + z)(t - 6z)$ **10.** $(3x + 2y)(x - 3y)$ **11.** Not factorable

12. Not factorable **13.** $4(2 + 3z - 2z^2) = 4(1 + 2z)(2 - z)$

14. $3(1 + 3z)(3 - 2z)$ **15.** $(2a + 1)^2$ **16.** $(3b + 1)^2$

17. $2(x^2 - 9) = 2(x - 3)(x + 3)$ **18.** $5(y + 4)(y - 4)$

19. $7h^2 - 11h + 4 = (7h - 4)(h - 1)$ **20.** $(h - 2)(7h - 2)$

21. $2(36x^2 + 12xy + y^2) = 2(6x + y)^2$ **22.** $2(4x + y)^2$

23. Not factorable **24.** Not factorable

25. $2y(x^2 + 4xy + 4y^2) = 2y(x + 2y)^2$ **26.** $3x(x + y)^2$

27. $(3e^2 + 4)(2e^2 - 5)$ **28.** $(5f^2 + 3)(2f^2 - 7)$

29. $3x^2(4x^2 - 25y^2) = 3x^2(2x + 5y)(2x - 5y)$

30. $4x^2(3x + 2y)(3x - 2y)$ **31.** $(a^2 + b^2)^2$ **32.** $(x^2 + 3y^2)^2$

33. Let $(a + b) = x$.
Then $2(a + b)^2 + 7(a + b) + 3 = 2x^2 + 7x + 3$
$= (2x + 1)(x + 3)$
But since $x = (a + b)$
$2(a + b)^2 + 7(a + b) + 3 = [2(a + b) + 1][(a + b) + 3]$
$= (2a + 2b + 1)(a + b + 3)$

34. $(3a - 3b + 1)(a - b + 2)$

35. Let $(x - y) = a$.
Then $4(x - y)^2 - 8(x - y) - 5 = 4a^2 - 8a - 5$
$= (2a + 1)(2a - 5)$
But since $a = (x - y)$
$4(x - y)^2 - 8(x - y) - 5 = (2[x - y] + 1)(2[x - y] - 5)$
$= (2x - 2y + 1)(2x - 2y - 5)$

36. $(2x + 2y + 1)(2x + 2y - 3)$

37. $5x^2 + 10xy + 5y^2 - 21x - 21y + 4$
$= 5(x^2 + 2xy + y^2) - 21(x + y) + 4$
$= 5(x + y)^2 - 21(x + y) + 4$. Let $a = (x + y)$
Then $5(x + y)^2 - 21(x + y) + 4 = 5a^2 - 21a + 4$
$= (5a - 1)(a - 4)$
But since $a = (x + y)$
$5(x + y)^2 - 21(x + y) + 4 = (5[x + y] - 1)([x + y] - 4)$
$= (5x + 5y - 1)(x + y - 4)$

38. $(5x - 5y - 2)(x - y - 2)$

39. $(2x - y)^2 - (3a + b)^2 \leftarrow$ Difference of two squares
$= [(2x - y) + (3a + b)][(2x - y) - (3a + b)]$
$= (2x - y + 3a + b)(2x - y - 3a - b)$

40. $(2x + 3y + a - b)(2x + 3y - a + b)$

41. $\underbrace{x^2 + 10xy + 25y^2} - 9$
$= (x + 5y)^2 - 9$
$= (x + 5y - 3)(x + 5y + 3)$

42. $(x + 4y + 5)(x + 4y - 5)$

43. $a^2 - 4x^2 - 4xy - y^2 = a^2 - (4x^2 + 4xy + y^2)$
Difference of two squares
$= a^2 - (2x + y)^2$
$= (a + [2x + y])(a - [2x + y])$
$= (a + 2x + y)(a - 2x - y)$

44. $(x + 3a + b)(x - 3a - b)$

45. $(4x^2 - 9)(x^2 - 1) = (2x - 3)(2x + 3)(x - 1)(x + 1)$

46. $(3x - 2)(3x + 2)(x + 1)(x - 1)$

47. $\underbrace{3x^2 - 7xy - 6y^2} - \underbrace{x + 3y}$
$(3x + 2y)(x - 3y) - (x - 3y)$
$= (x - 3y)([3x + 2y] - 1) = (x - 3y)(3x + 2y - 1)$

48. $(3t + z)(t - 6z - 1)$ **49.** $\underbrace{3n^2 + 2mn - 5m^2} + \underbrace{3n + 5m}$
$= (3n + 5m)(n - m) + (3n + 5m)$
$= (3n + 5m)(n - m + 1)$

50. $(3n - m)(n + 5m + 1)$

Review Exercises 5.5 (page 211)

1. $13xy(5xy^2 - 3y^3 - 1)$ **2.** Not factorable

3. $3x(x^2 + 3x - 4) = 3x(x + 4)(x - 1)$ **4.** Not factorable

5. $(x + 5)(x + 8)$ **6.** $(x - 4)(x + 5)$ **7.** Not factorable

8. $(x - 9)(x - 2)$ **9.** $(x - 16)(x + 16)$ **10.** $(x - 13)(x - 1)$

11. $(3x - 7)(x + 2)$ **12.** $(2x + 5)(x - 8)$

13. $3x(y + 4) + 2(y + 4) = (y + 4)(3x + 2)$

14. $(x + 5)(x + 1)(x - 1)$ **15.** Not factorable

16. Not factorable **17.** $2x(4x^2 - 1) = 2x(2x - 1)(2x + 1)$

18. Not factorable **19.** $(4x - 1)(x + 3)$ **20.** $(8a - 5)(a + 2)$

Exercises 5.6 (page 217)

1. $(x)^3 - (2)^3 = (x - 2)[(x)^2 + (2)(x) + (2)^2]$
$= (x - 2)(x^2 + 2x + 4)$

2. $(x - 3)(x^2 + 3x + 9)$

3. $(4)^3 + (a)^3 = (4 + a)[(4)^2 - (4)(a) + (a)^2]$
$= (4 + a)(16 - 4a + a^2)$

4. $(2 + b)(4 - 2b + b^2)$

5. $(5)^3 - x^3 = (5 - x)[(5)^2 + (5)(x) + (x)^2]$
$= (5 - x)(25 + 5x + x^2)$

6. $(1 - a)(1 + a + a^2)$

7. $2x(4x^2 - 1) = 2x(2x - 1)(2x + 1)$

8. $3x(3x + 1)(3x - 1)$

9. $c^3 - (3ab)^3 = (c - 3ab)[(c)^2 + (c)(3ab) + (3ab)^2]$
$= (c - 3ab)(c^2 + 3abc + 9a^2b^2)$

10. $(c - 4ab)(c^2 + 4abc + 16a^2b^2)$

11. $(2xy^2)^3 + (3)^3$
$= (2xy^2 + 3)[(2xy^2)^2 - (2xy^2)(3) + (3)^2]$
$= (2xy^2 + 3)(4x^2y^4 - 6xy^2 + 9)$

12. $(4a^2b + 5)(16a^4b^2 - 20a^2b + 25)$

13. Not factorable **14.** Not factorable

15. $a[a^3 + b^3] = a(a + b)(a^2 - ab + b^2)$

16. $y(x + y)(x^2 - xy + y^2)$

17. $3(27 - x^3) = 3(3 - x)[(3)^2 + (3)(x) + (x)^2]$
$= 3(3 - x)(9 + 3x + x^2)$

18. $5(2 - b)(4 + 2b + b^2)$

19. $(x+y)^3+(1)^3 = [(x+y)+(1)][(x+y)^2-(x+y)(1)+(1)^2]$
$= (x+y+1)(x^2+2xy+y^2-x-y+1)$

20. $(1+x-y)(1-x+y+x^2-2xy+y^2)$

21. $(4x)^3-(y^2)^3 = (4x-y^2)[(4x)^2+(4x)(y^2)+(y^2)^2]$
$= (4x-y^2)(16x^2+4xy^2+y^4)$

22. $(5w-v^2)(25w^2+5wv^2+v^4)$

23. $4(a^3b^3+27c^6) = 4[(ab)^3+(3c^2)^3]$
$= 4[ab+3c^2][(ab)^2-(ab)(3c^2)+(3c^2)^2]$
$= 4(ab+3c^2)(a^2b^2-3abc^2+9c^4)$

24. $5(xy^2+2z^3)(x^2y^4-2xy^2z^3+4z^6)$

25. A binomial that is both the difference of two squares and the difference of two cubes should be treated *first* as the difference of two squares.
$x^6-729 = (x^3-27)(x^3+27)$
$= (x-3)(x^2+3x+9)(x+3)(x^2-3x+9)$

26. $(y-2)(y^2+2y+4)(y+2)(y^2-2y+4)$

27. $(x+1)^3-(y-z)^3$
$= [(x+1)-(y-z)][(x+1)^2+(x+1)(y-z)+(y-z)^2]$
$= (x+1-y+z)$
$\cdot(x^2+2x+1+xy-xz+y-z+y^2-2yz+z^2)$

28. $(x-y-a-b)$
$\cdot(x^2-2xy+y^2+ax+bx-ay-by+a^2+2ab+b^2)$

Exercises 5.7 (page 221)

Note: Steps 1–3 will not be shown.

1. Try $(x^2+2)^2$.
Add and subtract x^2.
$x^4+3x^2+x^2+4-x^2 = x^4+4x^2+4-x^2$
$= (x^2+2)^2-x^2$
$= (x^2+2-x)(x^2+2+x)$
$= (x^2-x+2)(x^2+x+2)$

2. $(x^2+x+3)(x^2-x+3)$

3. Try $(2m^2+1)^2$.
Add and subtract m^2.
$4m^4+3m^2+m^2+1-m^2 = (4m^4+4m^2+1)-m^2$
$= (2m^2+1)^2-m^2$
$= (2m^2+1-m)(2m^2+1+m)$
$= (2m^2-m+1)(2m^2+m+1)$

4. $(3u^2+u+1)(3u^2-u+1)$

5. Add and subtract $16a^2b^2$.
$64a^4+16a^2b^2+b^4-16a^2b^2$
$= [(8a^2+b^2)^2-16a^2b^2]$
$= (8a^2+b^2-4ab)(8a^2+b^2+4ab)$
$= (8a^2-4ab+b^2)(8a^2+4ab+b^2)$

6. $(a^2+2ab+2b^2)(a^2-2ab+2b^2)$

7. Add and subtract $9x^2$.
$x^4-3x^2+9x^2+9-9x^2 = x^4+6x^2+9-9x^2$
$= (x^2+3)^2-9x^2$
$= (x^2+3-3x)(x^2+3+3x)$
$= (x^2-3x+3)(x^2+3x+3)$

8. $(x^2+3x+4)(x^2-3x+4)$

9. Add and subtract $9a^2b^2$.
$a^4-17a^2b^2+9a^2b^2+16b^4-9a^2b^2$
$(a^4-8a^2b^2+16b^4)-9a^2b^2$
$= (a^2-4b^2)^2-9a^2b^2$
$= (a^2-4b^2-3ab)(a^2-4b^2+3ab)$
$= (a^2-3ab-4b^2)(a^2+3ab-4b^2)$
$= (a-4b)(a+b)(a+4b)(a-b)$

10. $(a+1)(a-1)(a+6)(a-6)$

11. Not factorable **12.** Not factorable

13. Add and subtract $9a^2b^2$.
$a^4-3a^2b^2+9a^2b^2+9b^4-9a^2b^2$
$= (a^4+6a^2b^2+9b^4)-9a^2b^2$
$= (a^2+3b^2)^2-9a^2b^2$
$= (a^2+3b^2-3ab)(a^2+3b^2+3ab)$
$= (a^2-3ab+3b^2)(a^2+3ab+3b^2)$

14. $(a^2+5ab+5b^2)(a^2-5ab+5b^2)$

15. Not factorable **16.** Not factorable

17. $2(25x^4-6x^2y^2+y^4)$
Add and subtract $16x^2y^2$.
$= 2[25x^4-6x^2y^2+16x^2y^2+y^4-16x^2y^2]$
$= 2[25x^4+10x^2y^2+y^4-16x^2y^2]$
$= 2[(5x^2+y^2)^2-16x^2y^2]$
$= 2[(5x^2+y^2-4xy)(5x^2+y^2+4xy)]$
$= 2(5x^2-4xy+y^2)(5x^2+4xy+y^2)$

18. $2(4x^2+3xy+y^2)(4x^2-3xy+y^2)$

19. $2n(4m^4+n^4)$
Add and subtract $4m^2n^2$.
$= 2n[(4m^4+4m^2n^2+n^4)-4m^2n^2]$
$= 2n[(2m^2+n^2)^2-4m^2n^2]$
$= 2n(2m^2+2mn+n^2)(2m^2-2mn+n^2)$

20. $3m(m^2+2mn+2n^2)(m^2-2mn+2n^2)$

21. $2y(25x^4+16x^2y^2+4y^4)$
Add and subtract $4x^2y^2$.
$= 2y(25x^4+16x^2y^2+4x^2y^2+4y^4-4x^2y^2)$
$= 2y(25x^4+20x^2y^2+4y^4-4x^2y^2)$
$= 2y[(5x^2+2y^2)^2-4x^2y^2]$
$= 2y(5x^2+2xy+2y^2)(5x^2-2xy+2y^2)$

22. $3x(4x^2+xy+y^2)(4x^2-xy+y^2)$

Exercises 5.8 (page 225)

1. Factors of the constant term are $\pm1, \pm3$.

	1	1	1	−3	
		1	2	3	Remainder is zero;
Divide by 1	1	2	3	0	therefore, $(x-1)$ is a factor
	x^2+2x+3				Will not factor

Therefore, $x^3+x^2+x-3 = (x-1)(x^2+2x+3)$.

2. $(x-2)(x^2+3x+1)$

3. Factors of the constant term are $\pm1, \pm2, \pm3, \pm4, \pm6, \pm12$.

	1	−3	−4	12	
		2	−2	−12	Remainder is zero;
Divide by 2	1	−1	−6	0	therefore, $(x-2)$ is a factor
	x^2-x-6				Quotient is another factor
	$(x-3)(x+2)$				Factors of quotient

Therefore, $x^3-3x^2-4x+12 = (x-2)(x-3)(x+2)$.

4. $(x-1)(x+2)(x-3)$

5. $2(x^3-4x^2+x+6)$
We now factor x^3-4x^2+x+6.
Factors of the constant term are $\pm1, \pm2, \pm3, \pm6$.

	1	−4	1	6	Remainder is not zero;
		1	−3	−2	therefore, $(x-1)$ is not
Divide by 1	1	−3	−2	4	a factor

	1	−4	1	6	Remainder is zero;
		−1	5	−6	therefore, $(x+1)$ is
Divide by −1	1	−5	6	0	a factor
	x^2-5x+6				Quotient is another factor
	$(x-2)(x-3)$				Factors of quotient

Therefore, $x^3-4x^2+x+6 = (x+1)(x-2)(x-3)$
and $2x^3-8x^2+2x+12 = 2(x+1)(x-2)(x-3)$.

6. $2(x + 1)(x + 2)(x + 3)$

7. Factors of the constant term are ± 1, ± 2, ± 4.

	6	−13	0	+4	
		12	−2	−4	Remainder is zero;
Divide by 2	6	−1	−2	0	therefore $(x - 2)$ is a factor

Therefore, $6x^3 - 13x^2 + 4 = (x - 2)(6x^2 - x - 2)$
$= (x - 2)(3x - 2)(2x + 1)$

8. $(x + 1)(x + 2)(x - 3)$

9. Factors of the constant term are ± 1, ± 2, ± 4.

	1	0	−3	4	4	Remainder is zero;
		−2	4	−2	−4	therefore, $(x + 2)$ is
Divide by −2	1	−2	1	2	0	a factor
	$x^3 - 2x^2 + x + 2$					Quotient is another factor; it does not factor

Therefore, $x^4 - 3x^2 + 4x + 4 = (x + 2)(x^3 - 2x^2 + x + 2)$.

10. $(x + 2)(x^3 - 3x^2 + x + 3)$

11. Factors of the constant term are ± 1, ± 2, ± 4.

	1	2	−3	−8	−4	Remainder is not
		1	3	0	−8	zero; therefore, $(x - 1)$
Divide by 1	1	3	0	−8	−12	is not a factor

	1	2	−3	−8	−4	Remainder is zero;
		2	8	10	4	therefore, $(x - 2)$
Divide by 2	1	4	5	2	0	is a factor

We *now* work with the coefficients 1, 4, 5, 2.
Factors of the constant term are ± 1 and ± 2. However, since $+1$ did not work for the original coefficients, it will not work for the new ones. We must try $+2$ a second time:

	1	4	5	2	Remainder is not zero;
		2	12	34	therefore, $(x - 2)$ is not a
Divide by 2	1	6	17	36	factor of $x^3 + 4x^2 + 5x + 2$

	1	4	5	2	Remainder is zero;
		−1	−3	−2	therefore, $(x + 1)$ is a
Divide by −1	1	3	2	0	factor

Therefore, $x^4 + 2x^3 - 3x^2 - 8x - 4$
$= (x - 2)(x + 1)(x^2 + 3x + 2)$
$= (x - 2)(x + 1)(x + 1)(x + 2)$ or
$= (x - 2)(x + 1)^2(x + 2)$

12. $(x - 1)(x + 1)(x + 2)^2$

13. Factors of the constant term are ± 1. We may also have to try $\pm 1/3$.

	3	−4	0	0	−1	Remainder is not zero;
		3	−1	−1	−1	therefore, $(x - 1)$ is
Divide by 1	3	−1	−1	−1	−2	not a factor

	3	−4	0	0	−1	Remainder is not
		−3	7	−7	7	zero; therefore,
Divide by −1	3	−7	7	−7	6	$(x + 1)$ is not a factor

	3	−4	0	0	−1	Remainder is
		1	−1	−1/3	−1/9	not zero;
Divide by 1/3	3	−3	−1	−1/3	−10/9	therefore, $(x - 1/3)$ is not a factor

	3	−4	0	0	−1	Remainder
		−1	5/3	−5/9	5/27	is not zero;
Divide by −1/3	3	−5	5/3	−5/9	−22/27	therefore, $(x + 1/3)$ is not a factor

Since none of the possible factors work, the expression is not factorable.

14. Not factorable

15. Factors of the constant term are ± 1, ± 2, ± 3, ± 4, ± 6, ± 8, ± 12, ± 24.

	1	−4	−7	34	−24	Remainder is zero;
		1	−3	−10	24	therefore, $(x - 1)$ is
Divide by 1	1	−3	−10	24	0	a factor
		2	−2	−24		Remainder is zero;
Divide by 2	1	−1	−12	0		therefore, $(x - 2)$ is a factor

Therefore, $x^4 - 4x^3 - 7x^2 + 34x - 24$
$= (x - 1)(x - 2)(x^2 - x - 12)$
$= (x - 1)(x - 2)(x - 4)(x + 3)$

16. $(x - 2)(x + 1)(x + 3)(x + 4)$

17. Factors of the constant term are ± 1, ± 2, ± 3, ± 6. (We may also have to try fractions whose denominators are factors of 6.)

	6	1	−11	−6	Remainder is not zero;
		6	7	−4	therefore, $(x - 1)$ is not
Divide by 1	6	7	−4	−10	a factor

	6	1	−11	−6	Remainder is not zero;
		12	26	30	therefore, $(x - 2)$ is not
Divide by 2	6	13	15	24	a factor

	6	1	−11	−6	
		−6	5	6	Remainder is zero;
Divide by −1	6	−5	−6	0	$(x + 1)$ is a factor

Therefore, $6x^3 + x^2 - 11x - 6 = (x + 1)(6x^2 - 5x - 6)$
$= (x + 1)(2x - 3)(3x + 2)$

18. $(x - 3)(3x - 2)(2x + 1)$

Exercises 5.9 (page 227)

1. $(4e - 5)(3e + 7)$ **2.** $(6f + 7)(5f - 3)$

3. $6(ac - bd + bc - ad) = 6[a(c - d) + b(c - d)]$
$= 6(a + b)(c - d)$

4. $(2c + d)(5y - 3z)$

5. $2xy(y^2 - 2y - 15) = 2xy(y + 3)(y - 5)$ **6.** $3yz(z - 4)(z + 2)$

7. $3(x^3 + 8h^3) = 3(x + 2h)(x^2 - 2xh + 4h^2)$

8. $2(3f - g)(9f^2 + 3fg + g^2)$ **9.** $(3e - 5f)^2$ **10.** $(4m + 7p)^2$

11. $x^2(x + 3) - 4(x + 3) = (x + 3)(x^2 - 4) = (x + 3)(x + 2)(x - 2)$

12. $(a - 2)(a + 3)(a - 3)$

13. $(a + b)(a - b) - 1(a - b) = (a - b)(a + b - 1)$

14. $(x + y)(x - y - 1)$

15. Not factorable **16.** Not factorable

17. $\overbrace{x^3 - 8y^3} + \overbrace{x^2 - 4y^2}$
$= (x - 2y)(x^2 + 2xy + 4y^2) + (x - 2y)(x + 2y)$
$= (x - 2y)([x^2 + 2xy + 4y^2] + [x + 2y])$
$= (x - 2y)(x^2 + 2xy + 4y^2 + x + 2y)$

18. $(a - b)(a^2 + ab + b^2 + a + b)$

19. Not factorable **20.** Not factorable

21. $x^2 - 4xy + 4y^2 - 5x + 10y + 6 = (x - 2y)^2 - 5(x - 2y) + 6$
Let $a = (x - 2y)$.
Then $(x - 2y)^2 - 5(x - 2y) + 6 = a^2 - 5a + 6 = (a - 2)(a - 3)$
But since $a = (x - 2y)$,
$(a - 2)(a - 3) = ([x - 2y] - 2)([x - 2y] - 3)$
Therefore, $x^2 - 4xy + 4y^2 - 5x + 10y + 6$
$= (x - 2y - 2)(x - 2y - 3)$

22. $(x - 3y - 3)(x - 3y - 5)$

23. $(x^2 - 6xy + 9y^2) - 25 = (x - 3y)^2 - 25$
$= (x - 3y - 5)(x - 3y + 5)$

24. $(a - 4b + 1)(a - 4b - 1)$

Exercises 5.10 (page 231)

1.
$$\begin{array}{c|c} 3x = 0 & x - 4 = 0 \\ x = 0 & x = 4 \end{array}$$
$\{0, 4\}$

2. $\{0, -6\}$

3. $4x^2 - 12x = 0$
$4x(x - 3) = 0$
$$\begin{array}{c|c} 4x = 0 & x - 3 = 0 \\ x = 0 & x = 3 \end{array}$$
$\{0, 3\}$

4. $\{0, \frac{3}{2}\}$

5. $x^2 - 4x = 12$
$x^2 - 4x - 12 = 0$
$(x - 6)(x + 2) = 0$
$$\begin{array}{c|c} x - 6 = 0 & x + 2 = 0 \\ x = 6 & x = -2 \end{array}$$
$\{6, -2\}$

6. $\{-3, 5\}$

7. $2x^3 + x^2 - 3x = 0$
$x(2x^2 + x - 3) = 0$
$x(x - 1)(2x + 3) = 0$
$$\begin{array}{c|c|c} x = 0 & x - 1 = 0 & 2x + 3 = 0 \\ & x = 1 & 2x = -3 \\ & & x = -\frac{3}{2} \end{array}$$
$\{0, 1, -\frac{3}{2}\}$

8. $\{0, -5, \frac{1}{2}\}$

9. $2x^2 + 7x - 15 = 0$
$(2x - 3)(x + 5) = 0$
$$\begin{array}{c|c} 2x - 3 = 0 & x + 5 = 0 \\ 2x = 3 & x = -5 \\ x = \frac{3}{2} & \end{array}$$
$\{\frac{3}{2}, -5\}$

10. $\{\frac{2}{3}, -5\}$

11. $4x^2 - 12x + 9 = 0$
$(2x - 3)(2x - 3) = 0$
$2x - 3 = 0$
$2x = 3$
$x = \frac{3}{2}$
$\{\frac{3}{2}\}$

12. $\{\frac{2}{5}\}$

13. $18x^3 - 21x^2 - 60x = 0$
$3x(6x^2 - 7x - 20) = 0$
$3x(2x - 5)(3x + 4) = 0$
$$\begin{array}{c|c|c} 3x = 0 & 2x - 5 = 0 & 3x + 4 = 0 \\ x = 0 & 2x = 5 & 3x = -4 \\ & x = \frac{5}{2} & x = -\frac{4}{3} \end{array}$$
$\{0, \frac{5}{2}, -\frac{4}{3}\}$

14. $\{0, \frac{3}{5}, \frac{5}{3}\}$

15.
$$\begin{array}{c|c|c} 4x = 0 & 2x - 1 = 0 & 3x + 7 = 0 \\ x = 0 & 2x = 1 & 3x = -7 \\ & x = \frac{1}{2} & x = -\frac{7}{3} \end{array}$$
$\{0, \frac{1}{2}, -\frac{7}{3}\}$

16. $\{0, \frac{3}{4}, \frac{6}{7}\}$

17. $x^3 + 3x^2 - 4x - 12 = 0$
$x^2(x + 3) - 4(x + 3) = 0$
$(x + 3)(x^2 - 4) = 0$
$(x + 3)(x + 2)(x - 2) = 0$
$$\begin{array}{c|c|c} x + 3 = 0 & x + 2 = 0 & x - 2 = 0 \\ x = -3 & x = -2 & x = 2 \end{array}$$
$\{-3, -2, 2\}$

18. $\{-3, 3, -1\}$

19. $(x^2 - 9)(x^2 - 1) = 0$
$(x - 3)(x + 3)(x - 1)(x + 1) = 0$
$$\begin{array}{c|c|c|c} x - 3 = 0 & x + 3 = 0 & x - 1 = 0 & x + 1 = 0 \\ x = 3 & x = -3 & x = 1 & x = -1 \end{array}$$
$\{3, -3, 1, -1\}$

20. $\{2, -2, 3, -3\}$

21. $x^3 + 3x^2(3) + 3x(3^2) + 3^3 = x^3 + 63$
$x^3 + 9x^2 + 27x + 27 = x^3 + 63$
$9x^2 + 27x - 36 = 0$
$9(x^2 + 3x - 4) = 0$
$9(x + 4)(x - 1) = 0$
$$\begin{array}{c|c|c} 9 \neq 0 & x + 4 = 0 & x - 1 = 0 \\ & x = -4 & x = 1 \end{array}$$
$\{-4, 1\}$

22. $\{3, -4\}$

23. $x^4 + 12x^3 + 54x^2 + 108x + 81 = x^4 + 108x + 81$
$12x^3 + 54x^2 = 0$
$6x^2(2x + 9) = 0$
$$\begin{array}{c|c} 6x^2 = 0 & 2x + 9 = 0 \\ x = 0 & x = -\frac{9}{2} \end{array}$$
$\{0, -\frac{9}{2}\}$

24. $\{0, -3\}$

Exercises 5.11 (page 233)

Note: The checks for these exercises usually will not be shown.

1. Let $x =$ first even integer
$x + 2 =$ second even integer
$x + 4 =$ third even integer
$x(x + 2) = 38 + (x + 4)$
$x^2 + 2x = 38 + x + 4$
$x^2 + x - 42 = 0$
$(x - 6)(x + 7) = 0$
$$\begin{array}{c|c} x - 6 = 0 & x + 7 = 0 \\ x = 6 & x = -7 \leftarrow \text{Not an even integer} \end{array}$$
$x + 2 = 8$
$x + 4 = 10$
The integers are 6, 8, and 10.

2. 7, 9, and 11

3. Let $3x =$ length
$2x =$ width
$A = \text{(length)(width)}$
$150 = (3x)(2x)$
$150 = 6x^2$
$25 = x^2$
$25 - x^2 = 0$
$(5 - x)(5 + x) = 0$
$$\begin{array}{c|c} 5 - x = 0 & 5 + x = 0 \\ 5 = x & x = -5 \leftarrow \text{Not in the domain} \end{array}$$
$3x = 15$
$2x = 10$
The length is 15 and the width is 10.

4. Width is 12; length is 16.

5. Let $x =$ length of a side of the cube.
Then $x =$ height of box
$x + 3 =$ width of box
$4x =$ length of box
$x^3 =$ volume of cube
$x(x + 3)(4x) =$ volume of box
$x(x + 3)(4x) = 8x^3$
$4x^3 + 12x^2 = 8x^3$
$0 = 4x^3 - 12x^2$
$0 = 4x^2(x - 3)$
$$\begin{array}{c|c|c} 4 \neq 0 & x^2 = 0 & x - 3 = 0 \\ & x = 0 & x = 3 \end{array}$$
($x = 0$: Not in the domain)

The height of the box is 3, the width is 6, and the length is 12. The volume of the cube is $3^3 = 27$.

6. Height of box is 5; width is 10; length is 15. Volume of cube is 125.

7. Let $x = \text{length}$
$x - 5 = \text{width}$
Area = Perimeter + 46
$x(x-5) = 4x - 10 + 46$
$x^2 - 5x = 4x + 36$
$x^2 - 9x - 36 = 0$
$(x-12)(x+3) = 0$
$x - 12 = 0 \mid x + 3 = 0$
$x = 12 \mid x = -3$
$x - 5 = 7$ ↑ Not in the domain
The length is 12 m and the width is 7 m.

8. Length is 10 m; width is 3 m.

9. Let $h = \text{height (altitude)}$
$7 + h = \text{base}$
$\frac{1}{2}h(7+h) = \text{area}$
$\frac{1}{2}h(7+h) = 39$
$h(7+h) = 78$
$h^2 + 7h - 78 = 0$
$(h+13)(h-6) = 0$
$h + 13 = 0 \mid h - 6 = 0$
$h = -13$ Not in the domain $\mid h = 6$, $7 + h = 13$
The height is 6 cm and the base is 13 cm.

10. Height is 8 m; base is 12 m.

11. Let $x = \text{side of shorter square}$
$x + 6 = \text{side of longer square}$
Area of smaller square $= x^2$
Area of larger square $= (x+6)^2$
$9x^2 = (x+6)^2$
$9x^2 = x^2 + 12x + 36$
$8x^2 - 12x - 36 = 0$
$4(2x^2 - 3x - 9) = 0$
$2x^2 - 3x - 9 = 0$
$(2x+3)(x-3) = 0$
$2x + 3 = 0 \mid x - 3 = 0$
$x = -\frac{3}{2} \mid x = 3$
↑ Not in the domain
The length of a side of the smaller square is 3 cm and the length of a side of the larger square is 9 cm.

12. Side of larger square is 8 cm; side of smaller square is 4 cm.

13. Let $x = \text{length of side of smaller cube}$
$x + 3 = \text{length of side of larger cube}$
$x^3 = \text{volume of smaller cube}$
$(x+3)^3 = \text{volume of larger cube}$
$(x+3)^3 = 63 + x^3$
$x^3 + 9x^2 + 27x + 27 = 63 + x^3$
$9x^2 + 27x - 36 = 0$
$9(x^2 + 3x - 4) = 0$
$9(x+4)(x-1) = 0$
$9 \neq 0 \mid x + 4 = 0 \mid x - 1 = 0$
$x = -4$ Not in the domain $\mid x = 1$, $x + 3 = 4$
The length of a side of the smaller cube is 1 cm and the length of a side of the larger cube is 4 cm.

14. Length of side of smaller cube is 3 cm; length of side of larger cube is 4 cm.

15. Let $x = \text{width}$.
(length)(width)(depth) = volume
$(x+3)(x)(2) = 80$
$(x+3)(x) = 40$
$x^2 + 3x - 40 = 0$
$(x+8)(x-5) = 0$
$x + 8 = 0 \mid x - 5 = 0$
$x = -8 \mid x = 5$, $x + 3 = 8$
↑ Not in the domain

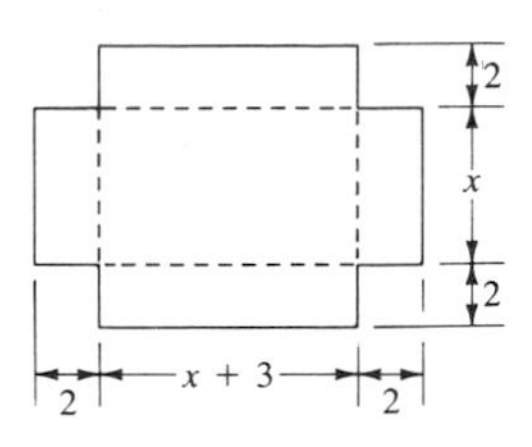

a. The dimensions of the metal sheet are 9 in. by 12 in.
b. The dimensions of the box are: depth = 2 in., width = 5 in., length = 8 in.

16. **a.** 8 in. by 11 in. **b.** 2 in. by 5 in. by 3 in.

17. Let $x = \text{width of rectangular room}$
$2x - 3 = \text{length of rectangular room}$
$x(2x-3) = 35$
$2x^2 - 3x = 35$
$2x^2 - 3x - 35 = 0$
$(2x+7)(x-5) = 0$
$2x + 7 = 0 \mid x - 5 = 0$
$x = -\frac{7}{2} \mid x = 5$, $2x - 3 = 7$
↑ Not in the domain
The width is 5 yd and the length is 7 yd.

18. 6 yd by 9 yd

19. Let x = the number of inches in a side of the square. If the picture were 3 in. longer and 2 in. narrower, its area would be $(x+3)(x-2)$ sq. in.; the cost would be $\$0.30(x+3)(x-2)$.
$0.30(x+3)(x-2) = 45$
$(x+3)(x-2) = 150$ Dividing both sides by 0.30
$x^2 + x - 6 = 150$
$x^2 + x - 156 = 0$
$(x-12)(x+13) = 0$
$x - 12 = 0 \mid x + 13 = 0$
$x = 12 \mid x = -13$ Reject
(The check will not be shown.) The picture is 12 in. by 12 in.

20. 9 yd by 9 yd

Review Exercises 5.12 (page 237)

1. $3uv(5u-1)$ **2.** $(3n+1)(n+5)$ **3.** $4xy(x-2y+1)$
4. $(5x+1)(x+2)$ **5.** $5u^2 + 17u - 12 = (5u-3)(u+4)$
6. $(1+5x)(4+x)$ **7.** $3uv(2u^2v - 3v^2 - 4)$
8. $15(a+2b)(a-b)$ **9.** $9(9-m^2) = 9(3-m)(3+m)$
10. Not factorable **11.** $(2x+3y)(5x-8y)$
12. $(4x-3y)(7x+2y)$
13. $(2a)^3 - (3b)^3 = (2a-3b)[(2a)^2 + (2a)(3b) + (3b)^2]$
$= (2a-3b)(4a^2 + 6ab + 9b^2)$
14. $(4h-5k)(16h^2 + 20hk + 25k^2)$ **15.** $(1+2y)^2$
16. $(2+3x)^2$
17. $(x)^3 + (2y)^3 = (x+2y)[(x)^2 - (x)(2y) + (2y)^2]$
$= (x+2y)(x^2 - 2xy + 4y^2)$
18. $(3x+y)(9x^2 - 3xy + y^2)$
19. $x^2 - y^2 + x - y = (x+y)(x-y) + 1(x-y)$
$= (x-y)(x+y+1)$
20. $(x-y)(x+y-1)$ **21.** $2(4a^2 - 4ab + b^2) = 2(2a-b)^2$
22. $2(3h-k)^2$

23. $x^2(x-4)-4(x-4)=(x-4)(x^2-4)$
$= (x-4)(x+2)(x-2)$

24. Not factorable

25. $x^2(x-2)-9(x-2)=(x-2)(x^2-9)$
$=(x-2)(x+3)(x-3)$

26. $(x+4+5y)(x+4-5y)$

27. Let $a=(2x+3y)$. Then
$(2x+3y)^2+(2x+3y)-6=a^2+a-6=(a+3)(a-2)$
But since $a=(2x+3y)$,
$(a+3)(a-2)=([2x+3y]+3)([2x+3y]-2)$
Therefore,
$(2x+3y)^2+(2x+3y)-6=(2x+3y+3)(2x+3y-2)$

28. $(a+3b-3)(a+3b-4)$

29. $x^2+4xy+4y^2-5x-10y+6=(x+2y)^2-5(x+2y)+6$
Let $a=(x+2y)$. Then
$(x+2y)^2-5(x+2y)+6=a^2-5a+6=(a-2)(a-3)$
But since $a=(x+2y)$,
$(a-2)(a-3)=([x+2y]-2)([x+2y]-3)$
Therefore,
$x^2+4xy+4y^2-5x-10y+6=(x+2y-2)(x+2y-3)$

30. $\{1, \frac{7}{5}\}$

31.
$x^2=18+3x$
$x^2-3x-18=0$
$(x-6)(x+3)=0$

$x-6=0$	$x+3=0$
$x=6$	$x=-3$

$\{6, -3\}$

32. $\{0, 36\}$

33. $6x^2-13x-5=0$
$(3x+1)(2x-5)=0$

$3x+1=0$	$2x-5=0$
$x=-\frac{1}{3}$	$x=\frac{5}{2}$

$\{-\frac{1}{3}, \frac{5}{2}\}$

34. $\{2, 3, -3\}$

35. $x^3+3x^2(3)+3x(3^2)+3^3=x^3+27$
$x^3+9x^2+27x+27=x^3+27$
$9x^2+27x=0$
$9x(x+3)=0$

$9\neq 0$	$x=0$	$x+3=0$
		$x=-3$

$\{0, -3\}$

36. $\{-2, -\frac{2}{5}\}$

37. $x^4+12x^3+54x^2+108x+81=x^4+12x^3+81$
$54x^2+108x=0$
$54x(x+2)=0$

$54\neq 0$	$x=0$	$x+2=0$
		$x=-2$

$\{0, -2\}$

38. $\{0, -\frac{4}{3}\}$

39. Let x = first odd integer
$x+2$ = second odd integer
$x+4$ = third odd integer
$x(x+4)=5+8(x+2)$
$x^2+4x=5+8x+16$
$x^2-4x-21=0$
$(x-7)(x+3)=0$

$x-7=0$	$x+3=0$
$x=7$	$x=-3$
$x+2=9$	$x+2=-1$
$x+4=11$	$x+4=1$

Therefore, one set of such numbers is 7, 9, and 11; another is -3, -1, and 1.

40. Height is 8, base is 10.

41. Let $5x$ = length
$3x$ = width
Area = (length)(width)
$240=5x(3x)$
$240=15x^2$
$0=15x^2-240$
$=15(x^2-16)$
$0=x^2-16$
$0=(x-4)(x+4)$

$x-4=0$	$x+4=0$
$x=4$	$x=-4$ Not in the domain

$5x=20$
$3x=12$
The length is 20 and the width is 12.

42. 6 by 13

43. Let x = length of side of smaller cube
$x+4$ = length of side of larger cube
x^3 = volume of smaller cube
$(x+4)^3$ = volume of larger cube
$(x+4)^3=316+x^3$
$x^3+12x^2+48x+64=316+x^3$
$12x^2+48x-252=0$
$12(x^2+4x-21)=0$
$12(x+7)(x-3)=0$

$12\neq 0$	$x+7=0$	$x-3=0$
	$x=-7$	$x=3$
	Not in the domain	$x+4=7$

The length of a side of the smaller cube is 3 cm, and the length of a side of the larger cube is 7 cm.

44. Length of side of smaller square is 1 cm; length of side of larger square is 4 cm.

Chapter 5 Diagnostic Test (page 243)

Following each problem number is the textbook section number (in parentheses) where that kind of problem is discussed.

1. **a.** (5.1, 5.3) $4x-16x^3=4x(1-4x^2)=4x(1-2x)(1+2x)$
b. (5.1) $43+7x^2+6=7x^2+49=7(x^2+7)$
c. (5.4B) $7x^2+23x+6=(7x+2)(x+3)$
d. (5.3) x^2+81 is a sum of two squares and cannot be factored.

2. **a.** (5.1) $2x^3+4x^2+16x=2x(x^2+2x+8)$
b. (5.4B) $6x^2-5x-6=(3x+2)(2x-3)$

3. (5.6) $y^3-1=(y)^3-(1)^3=(y-1)(y^2+y+1)$

4. (5.2) $3ac+6bc-5ad-10bd$
$=3c(a+2b)-5d(a+2b)$
$=(a+2b)(3c-5d)$

5. (5.3) $4x^2+4x+1-y^2$ Factor by grouping
$=(2x+1)^2-y^2$
$=([2x+1]+y)([2x+1]-y)$
$=(2x+1+y)(2x+1-y)$

6. **a.** (5.10) $(x+3)(2x-5)=0$

$x+3=0$	$2x-5=0$
$x=-3$	$x=\frac{5}{2}$

$\{-3, \frac{5}{2}\}$

b. (5.10) $8y^2-4y=0$
$4y(2y-1)=0$

$4\neq 0$	$y=0$	$2y-1=0$
		$y=\frac{1}{2}$

$\{0, \frac{1}{2}\}$

7. (5.10)
$$6x^2 - 27x - 15 = 0$$
$$3(2x^2 - 9x - 5) = 0$$
$$3(2x + 1)(x - 5) = 0$$
$3 \neq 0$ | $2x + 1 = 0$, $x = -\frac{1}{2}$ | $x - 5 = 0$, $x = 5$

$\{-\frac{1}{2}, 5\}$

8. (5.10)
$$(x + 1)^3 = x^3 + 1$$
$$x^3 + 3x^2 + 3x + 1 = x^3 + 1$$
$$3x^2 + 3x = 0$$
$$3x(x + 1) = 0$$
$3 \neq 0$ | $x = 0$ | $x + 1 = 0$, $x = -1$

$\{0, -1\}$

9. (5.11) Let x = first even integer
$x + 2$ = second even integer
$x + 4$ = third even integer
$$x(x + 2) = 68 + (x + 4)$$
$$x^2 + 2x = 68 + x + 4$$
$$x^2 + x - 72 = 0$$
$$(x + 9)(x - 8) = 0$$
$x + 9 = 0$, $x = -9$ Not an even integer | $x - 8 = 0$, $x = 8$, $x + 2 = 10$, $x + 4 = 12$

The integers are 8, 10, and 12.

10. (5.11) Let x = height
$x + 8$ = base
$\frac{1}{2}x(x + 8)$ = area
$$\frac{1}{2}x(x + 8) = 64$$
$$x(x + 8) = 128$$
$$x^2 + 8x - 128 = 0$$
$$(x + 16)(x - 8) = 0$$
$x + 16 = 0$, $x = -16$ Not in the domain | $x - 8 = 0$, $x = 8$, $x + 8 = 16$

The height is 8 cm and the base is 16 cm.

Cumulative Review Exercises: Chapters 1–5 (page 244)

1. $18 \div 2 \cdot 3 - 16 \cdot 3 = 9 \cdot 3 - 16 \cdot 3 = 27 - 48 = -21$ **2.** 32

3.
$$2x - 6 - 5 = 6 - 3x - 12$$
$$2x - 11 = -3x - 6$$
$$5x = 5$$
$$x = 1$$
$\{1\}$

4. $\{x \mid -1 < x < 3\}$

5. If $|2x - 3| \geq 7$
$2x - 3 \geq 7$ or $2x - 3 \leq -7$
$2x \geq 10$ or $2x \leq -4$
$x \geq 5$ or $x \leq -2$
$\{x \mid x \geq 5 \text{ or } x \leq -2\}$

6. All are real numbers.

7.
$$2x^2 - 9x - 5 = 0$$
$$(2x + 1)(x - 5) = 0$$
$2x + 1 = 0$ or $x - 5 = 0$
$x = -\frac{1}{2}$ or $x = 5$
$\{-\frac{1}{2}, 5\}$

8. $3(x + 3)(x^2 - 3x + 9)$

9.
$$x^3 + 5x^2 - x - 5 = x^2(x + 5) - (x + 5)$$
$$= (x + 5)(x^2 - 1)$$
$$= (x + 5)(x - 1)(x + 1)$$

10. $6x^2 + x + 13$

11.
$$(a - 4)(a^2 - 2a + 5)$$
$$a^3 - 2a^2 + 5a - 4a^2 + 8a - 20$$
$$a^3 - 6a^2 + 13a - 20$$

12. $9x^2 + 30x + 25$

13.
$$(2x - 1)^5 = (2x)^5 - 5(2x)^4(1) + \frac{5!}{(2!)(3!)}(2x)^3(1^2)$$
$$- 10(2x)^2(1^3) + 5(2x)(1^4) - (1)^5$$
$$= 32x^5 - 80x^4 + 80x^3 - 40x^2 + 10x - 1$$

14. $x^2 - 4x + 2 + \frac{4}{x + 2}$

15. Let x = number of quarts of antifreeze added
$$.20(10) + 1.00(x) = 0.50(10 + x)$$
$$20(10) + 100(x) = 50(10 + x)$$
$$200 + 100x = 500 + 50x$$
$$50x = 300$$
$$x = 6$$
6 qt of antifreeze should be added.

16. 6 mi

17. Let x = the number of quarters
$4x$ = the number of nickels
$27 - (x + 4x)$ = the number of dimes
$$x(2[27 - 5x]) = 4x$$
$$54x - 10x^2 = 4x$$
$$0 = 10x^2 - 50x$$
$$0 = 10x(x - 5)$$
$x - 5 = 0$, $x = 5$, $4x = 20$, $27 - 5x = 2$ | $10x = 0$, $x = 0$, $4x = 0$, $27 - 5x = 27$

(The checks will not be shown.) There are two answers: Yang has 5 quarters, 20 nickels, and 2 dimes, *or* he has 27 dimes, no nickels, and no quarters.

18. 56 cm, 63 cm, and 70 cm

19. Let x = volume of the tank.
$$\tfrac{1}{4}x + 300 = \tfrac{1}{3}x$$
$$3x + 3{,}600 = 4x$$
$$3{,}600 = x$$
$x = 3{,}600$ (The check will not be shown.)
The volume of the tank is 3,600 cu. cm.

20. 48 mph

21. Let x = amount invested at 7.2%
$13{,}500 - x$ = amount invested at 7.4%
$$0.072x + 0.074(13{,}500 - x) = 986.60$$
$$72x + 74(13{,}500 - x) = 986{,}600$$
$$72x + 999{,}000 - 74x = 986{,}600$$
$$999{,}000 - 2x = 986{,}600$$
$$2x = 12{,}400$$
$$x = 6{,}200$$
$$13{,}500 - x = 7{,}300$$
Consuelo invested \$6,200 at 7.2% and \$7,300 at 7.4%.

22. No solution

Exercises 6.1 (page 250)

1. All real numbers except -4 **2.** All real numbers except 5

3. No number can make the denominator zero. The domain is the set of all real numbers.

4. All real numbers except 0

5. All real numbers except 5 and -5

6. All real numbers except 3 and -2

7. Since $c^4 - 13c^2 + 36 = (c^2 - 4)(c^2 - 9)$
$= (c + 2)(c - 2)(c + 3)(c - 3),$
the domain is the set of all real numbers except −2, 2, −3, and 3.

8. All real numbers except −3, 3, and 5

9. Yes; $\frac{1}{2y} = \frac{3 \div 3}{6y \div 3}$ 10. Yes 11. No 12. No

13. Yes 14. Yes

15. $-\frac{+5}{+8} = \boxed{+}\frac{+5}{\boxed{-}8}$ The missing term is −8. 16. 6

17. $+\frac{-x}{+5} = +\frac{\boxed{+}x}{\boxed{-}5}$ The missing term is −5. 18. −6

19. $+\frac{+(8-y)}{+(4y-7)} = +\frac{\boxed{-}(8-y)}{\boxed{-}(4y-7)} = \frac{y-8}{7-4y}$ 20. $2 - w$
The missing term is $7 - 4y$.

21. $+\frac{+(u-v)}{+(a-b)} = +\frac{\boxed{-}(u-v)}{\boxed{-}(a-b)} = \frac{v-u}{b-a}$ The missing term is $b - a$.

22. $x - 2$

23. $+\frac{+(a-b)}{[+(3a+2b)][+(a-5b)]} = +\frac{\boxed{-}(a-b)}{[+(3a+2b)][\boxed{-}(a-5b)]}$
$= \frac{b-a}{(3a+2b)(5b-a)}$ The missing term is $b - a$.

24. $f - 2e$

Exercises 6.2 (page 253)

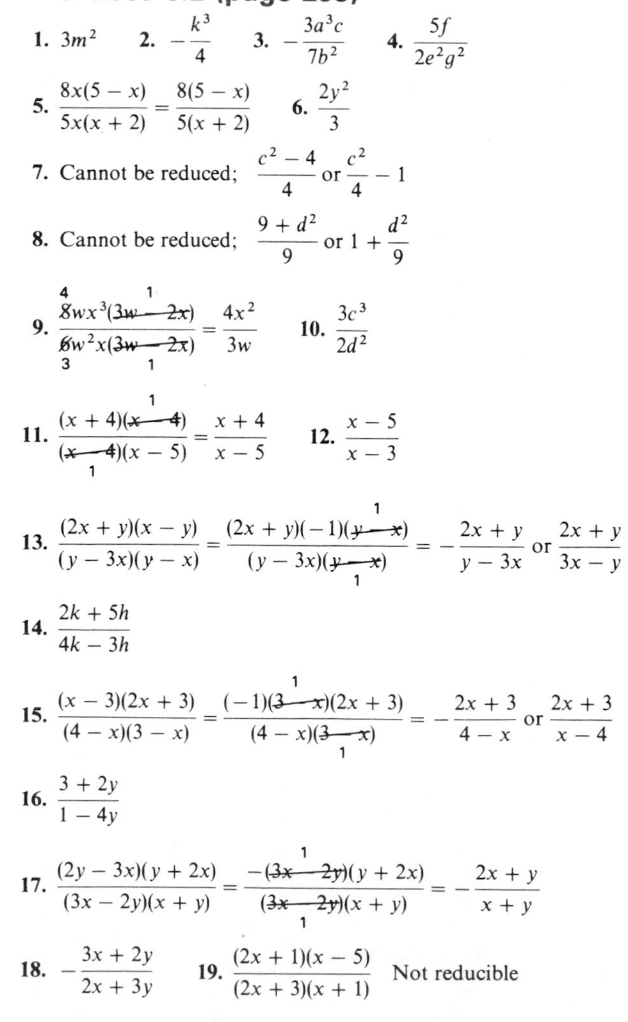

1. $3m^2$ 2. $-\frac{k^3}{4}$ 3. $-\frac{3a^3c}{7b^2}$ 4. $\frac{5f}{2e^2g^2}$

5. $\frac{8x(5-x)}{5x(x+2)} = \frac{8(5-x)}{5(x+2)}$ 6. $\frac{2y^2}{3}$

7. Cannot be reduced; $\frac{c^2-4}{4}$ or $\frac{c^2}{4} - 1$

8. Cannot be reduced; $\frac{9+d^2}{9}$ or $1 + \frac{d^2}{9}$

9. $\frac{\overset{4}{\cancel{8}}wx^3\overset{1}{\cancel{(3w-2x)}}}{\underset{3}{\cancel{6}}w^2x\underset{1}{\cancel{(3w-2x)}}} = \frac{4x^2}{3w}$ 10. $\frac{3c^3}{2d^2}$

11. $\frac{(x+4)\overset{1}{\cancel{(x-4)}}}{\underset{1}{\cancel{(x-4)}}(x-5)} = \frac{x+4}{x-5}$ 12. $\frac{x-5}{x-3}$

13. $\frac{(2x+y)(x-y)}{(y-3x)(y-x)} = \frac{(2x+y)(-1)\overset{1}{\cancel{(y-x)}}}{(y-3x)\underset{1}{\cancel{(y-x)}}} = -\frac{2x+y}{y-3x}$ or $\frac{2x+y}{3x-y}$

14. $\frac{2k+5h}{4k-3h}$

15. $\frac{(x-3)(2x+3)}{(4-x)(3-x)} = \frac{(-1)\overset{1}{\cancel{(3-x)}}(2x+3)}{(4-x)\underset{1}{\cancel{(3-x)}}} = -\frac{2x+3}{4-x}$ or $\frac{2x+3}{x-4}$

16. $\frac{3+2y}{1-4y}$

17. $\frac{(2y-3x)(y+2x)}{(3x-2y)(x+y)} = \frac{-\overset{1}{\cancel{(3x-2y)}}(y+2x)}{\underset{1}{\cancel{(3x-2y)}}(x+y)} = -\frac{2x+y}{x+y}$

18. $-\frac{3x+2y}{2x+3y}$ 19. $\frac{(2x+1)(x-5)}{(2x+3)(x+1)}$ Not reducible

20. Not reducible

21. $\frac{(a-1)(a^2+a+1)}{(1-a)(1+a)} = \frac{\overset{1}{\cancel{(a-1)}}(a^2+a+1)}{(-1)\underset{1}{\cancel{(a-1)}}(a+1)} = -\frac{a^2+a+1}{a+1}$

22. $\frac{x^2-xy+y^2}{y-x}$ 23. $\frac{x^2+4}{(x+2)^2}$ Not reducible

24. Not reducible

25. $\frac{13x^3y^2}{13x^2y^2} + \frac{-26xy^3}{13x^2y^2} + \frac{39xy}{13x^2y^2} = x - \frac{2y}{x} + \frac{3}{xy}$ or $\frac{x^2y - 2y^2 + 3}{xy}$

26. $3n - 5m - \frac{2}{mn}$ or $\frac{3mn^2 - 5m^2n - 2}{mn}$

27. $\frac{6a^2bc^2}{6abc} + \frac{-4ab^2c^2}{6abc} + \frac{12bc}{6abc} = ac - \frac{2bc}{3} + \frac{2}{a}$ or $\frac{3a^2c - 2abc + 6}{3a}$

28. $2a^2b - a - \frac{5}{2b}$ or $\frac{4a^2b^2 - 2ab - 5}{2b}$

Exercises 6.3 (page 257)

1. $\frac{\overset{3}{\cancel{27}}x^4y^3}{\underset{2}{\cancel{22}}x^5yz}\cdot\frac{\overset{5}{\cancel{55}}x^2z^2}{\underset{1}{\cancel{9}}y^3z} = \frac{15x}{2y}$ 2. $\frac{5c}{18a^2}$ 3. $\frac{mn^3}{\underset{3}{\cancel{18}}n^2}\cdot\frac{\overset{4}{\cancel{24}}m^3n}{5m^4} = \frac{4n^2}{15}$

4. $-\frac{36}{5h^2}$ 5. $\frac{3u\overset{1}{\cancel{(5-2u)}}}{\underset{2}{\cancel{10}}u^2}\cdot\frac{\overset{3}{\cancel{15}}u^3}{7\underset{1}{\cancel{(5-2u)}}} = \frac{9u^2}{14}$ 6. $-\frac{4v}{15}$

7. $\frac{-\overset{3}{\cancel{15}}c^4}{8c^2\underset{1}{\cancel{(5c-3)}}}\cdot\frac{\overset{1}{\cancel{7}}c\overset{1}{\cancel{(5c-3)}}}{\underset{\underset{1}{\cancel{5}}}{\cancel{35}}c} = -\frac{3c^2}{8}$ 8. 20

9. $\frac{d^2e\overset{-1}{\cancel{(e-d)}}}{\underset{4}{\cancel{12}}e^2d}\cdot\frac{\overset{1}{\cancel{3}}e^2(d+e)}{de^2\underset{1}{\cancel{(d-e)}}} = -\frac{d+e}{4e}$ 10. $-\frac{15(3m+n)}{8m}$

11. $\frac{\overset{1}{\cancel{(w-4)}}\overset{1}{\cancel{(w+2)}}}{6\underset{1}{\cancel{(w-4)}}}\cdot\frac{5w^2}{\underset{1}{\cancel{(w+2)}}(w-5)} = \frac{5w^2}{6(w-5)}$ 12. $\frac{3k}{8}$

13. $\frac{\overset{2}{\cancel{4}}\overset{1}{\cancel{(a+b)}}\overset{1}{\cancel{(a+b)}}}{\underset{1}{\cancel{(a+b)}}\underset{1}{\cancel{(a-b)}}}\cdot\frac{\overset{-1}{\cancel{(b-a)}}}{\underset{3}{\cancel{6}}b\underset{1}{\cancel{(a+b)}}} = -\frac{2}{3b}$ 14. $-\frac{1}{u}$

15. $\frac{\overset{1}{\cancel{2}}\overset{-1}{\cancel{(2-a)}}}{\underset{1}{\cancel{2}}\underset{1}{\cancel{(a+1)}}}\cdot\frac{\overset{1}{\cancel{(a+1)}}(a+1)}{2\underset{1}{\cancel{(a-2)}}(a^2+2a+4)} = -\frac{a+1}{2(a^2+2a+4)}$

16. $-\frac{6}{a^2-3a+9}$

17. $\frac{(x+y)\overset{1}{\cancel{(x^2-xy+y^2)}}}{2\underset{1}{\cancel{(x-y)}}}\cdot\frac{(x+y)\overset{1}{\cancel{(x-y)}}}{\underset{1}{\cancel{(x^2-xy+y^2)}}} = \frac{(x+y)^2}{2}$

18. $\frac{(x-y)^2}{3}$ 19. $\frac{\overset{1}{\cancel{(e+5f)}}\overset{1}{\cancel{(e+5f)}}}{\underset{1}{\cancel{(e+5f)}}\underset{-1}{\cancel{(e-5f)}}}\cdot\frac{3\overset{-1}{\cancel{(e-f)}}}{\underset{1}{\cancel{(f-e)}}}\cdot\frac{\overset{1}{\cancel{(5f-e)}}}{\underset{1}{\cancel{(e+5f)}}} = 3$

20. 1 21. $\frac{\overset{1}{\cancel{(x+y)}}(x+y+1)}{\underset{1}{\cancel{(x-y)}}(x-y-1)}\cdot\frac{\overset{1}{\cancel{(x-y)}}\overset{1}{\cancel{(x-y)}}}{\underset{1}{\cancel{(x+y)}}\underset{1}{\cancel{(x+y)}}}\cdot\frac{\overset{1}{\cancel{x+y}}}{\underset{1}{\cancel{x-y}}} = \frac{x+y+1}{x-y-1}$

22. $\frac{c-d}{a+b}$

Exercises 6.4 (page 260)

1. (1) $5^2 \cdot a^3$; $3 \cdot 5 \cdot a$ Denominators in factored form
(2) $3, 5, a$ All the different bases
(3) $3^1, 5^2, a^3$ Highest power of each base
(4) LCD $= 3^1 \cdot 5^2 \cdot a^3 = 75a^3$

2. $36b^4$

3. (1) $2^2 \cdot 3 \cdot 5 \cdot h \cdot k^3$; $2 \cdot 3^2 \cdot 5 \cdot h^2 \cdot k^4$
(2) $2, 3, 5, h, k$
(3) $2^2, 3^2, 5, h^2, k^4$
(4) LCD $= 2^2 \cdot 3^2 \cdot 5 \cdot h^2 \cdot k^4 = 180h^2k^4$

4. $294x^3y^2$ (or $147x^3y^2$ if fractions were reduced first)

5. (1) $2(w-5)$; 2^2w
(2) $2, w, (w-5)$
(3) $2^2, w^1, (w-5)^1$
(4) LCD $= 2^2 \cdot w^1 \cdot (w-5)^1 = 4w(w-5)$

6. $8m^2(m-6)$

7. (1) $(3b+c)(3b-c)$; $(3b-c)^2$
(2) $(3b+c), (3b-c)$
(3) $(3b+c)^1, (3b-c)^2$
(4) LCD $= (3b+c)(3b-c)^2$

8. $(2e+5f)(2e-5f)^2$

9. (1) $2 \cdot g^3$; $(g-3)^2$; $2^2 \cdot g \cdot (g-3)$
(2) $2, g, (g-3)$
(3) $2^2, g^3, (g-3)^2$
(4) LCD $= 2^2 \cdot g^3 \cdot (g-3)^2 = 4g^3(g-3)^2$
(The LCD is $2g^3(g-3)$ if fractions were reduced first.)

10. $9y^2(y-6)^2$ (LCD is $9y^2(y-6)$ if fractions were reduced first.)

11. (1) $2 \cdot (x-4)^2$; $(x-4) \cdot (x+5)$
(2) $2, (x-4), (x+5)$
(3) $2^1, (x-4)^2, (x+5)^1$
(4) LCD $= 2(x-4)^2(x+5)$

12. $5(k+7)(k-3)^2$

13. (1) $3 \cdot e^2$; $(e+3)(e-3)$; $2^2(e-3)$
(2) $2, 3, e, (e+3)(e-3)$
(3) $2^2, 3^1, e^2, (e+3)^1, (e-3)^1$
(4) LCD $= 2^2 \cdot 3^1 \cdot e^2 \cdot (e+3)(e-3) = 12e^2(e+3)(e-3)$

14. $24u^3(u+3)(u-8)$

15. (1) $2^2 \cdot 3 \cdot x^2 \cdot (x+2)$; $(x-2)^2$; $(x+2)(x-2)$
(2) $2, 3, x, (x+2), (x-2)$
(3) $2^2, 3^1, x^2, (x+2)^1, (x-2)^2$
(4) LCD $= 2^2 \cdot 3 \cdot x^2 \cdot (x+2) \cdot (x-2)^2 = 12x^2(x+2)(x-2)^2$

16. $8y(y+3)^7(y-3)$

Exercises 6.5 (page 267)

1. $\frac{5a+10}{a+2} = \frac{5\overset{1}{\cancel{(a+2)}}}{\underset{1}{\cancel{(a+2)}}} = 5$

2. 6

3. $\frac{8m-12n}{2m-3n} = \frac{4\overset{1}{\cancel{(2m-3n)}}}{\underset{1}{\cancel{(2m-3n)}}} = 4$

4. 7

5. $\frac{15w}{5w-1} - \frac{3}{5w-1} = \frac{15w-3}{5w-1} = \frac{3\overset{1}{\cancel{(5w-1)}}}{\underset{1}{\cancel{(5w-1)}}} = 3$

6. 5

7. $\frac{7z}{8z-4} + \frac{5z-6}{8z-4} = \frac{7z+5z-6}{8z-4} = \frac{12z-6}{8z-4} = \frac{\overset{3}{\cancel{6}}\overset{1}{\cancel{(2z-1)}}}{\underset{2}{\cancel{4}}\underset{1}{\cancel{(2z-1)}}} = \frac{3}{2}$

8. 2

9. $\frac{12x-31}{12x-28} + \frac{18x-39}{12x-28} = \frac{30x-70}{12x-28} = \frac{\overset{5}{\cancel{10}}\overset{1}{\cancel{(3x-7)}}}{\underset{2}{\cancel{4}}\underset{1}{\cancel{(3x-7)}}} = \frac{5}{2}$

10. 2

11. LCD $= 75a^3$

$$\frac{9}{25a^3} \cdot \frac{3}{3} + \frac{7}{15a} \cdot \frac{5a^2}{5a^2} = \frac{27}{75a^3} + \frac{35a^2}{75a^3} = \frac{27+35a^2}{75a^3}$$

12. $\frac{26b^2+33}{36b^4}$

13. LCD $= 180h^2k^4$

$$\frac{49}{60h^2k^2} \cdot \frac{3k^2}{3k^2} - \frac{71}{90hk^4} \cdot \frac{2h}{2h} = \frac{147k^2}{180h^2k^4} - \frac{142h}{180h^2k^4} = \frac{147k^2-142h}{180h^2k^4}$$

14. $\frac{154x-135y}{147x^3y^2}$

15. LCD $= t(t-4)$

$$\frac{5}{t} \cdot \frac{t-4}{t-4} + \frac{2t}{t-4} \cdot \frac{t}{t} = \frac{5t-20}{t(t-4)} + \frac{2t^2}{t(t-4)} = \frac{2t^2+5t-20}{t(t-4)}$$

16. $\frac{6r^2-11r+88}{r(r-8)}$

17. LCD $= 12k(2k-1)$

$$\frac{3k}{4(2k-1)} - \frac{7}{6k} = \frac{9k^2}{12k(2k-1)} - \frac{7 \cdot 2(2k-1)}{12k(2k-1)} = \frac{9k^2}{12k(2k-1)} - \frac{28k-14}{12k(2k-1)} = \frac{9k^2-28k+14}{12k(2k-1)}$$

18. $\frac{2(3j^2+3j+2)}{9j(3j+2)}$

19. LCD $= x(x-3)$

$$\frac{x^2(x)(x-3)}{1(x)(x-3)} - \frac{-3(x-3)}{x(x-3)} + \frac{5(x)}{(x-3)(x)} = \frac{x^4-3x^3-3x+9+5x}{x(x-3)} = \frac{x^4-3x^3+2x+9}{x(x-3)}$$

20. $\frac{y^4-5y^3+y+10}{y(y-5)}$

21. LCD $= b(2a-3b)$

$$\frac{(2a+3b)(2a-3b)}{b(2a-3b)} + \frac{b^2}{b(2a-3b)} = \frac{4a^2-9b^2}{b(2a-3b)} + \frac{b^2}{b(2a-3b)} = \frac{4a^2-8b^2}{b(2a-3b)} = \frac{4(a^2-2b^2)}{b(2a-3b)}$$

22. $\frac{9x^2-24y^2}{y(3x+5y)}$

23. LCD $= (a+3)(a-1)$

$$\frac{2(a-1)}{(a+3)(a-1)} + \frac{-4(a+3)}{(a-1)(a+3)} = \frac{2a-2-4a-12}{(a+3)(a-1)} = \frac{-2a-14}{(a+3)(a-1)} \text{ or } \frac{14+2a}{(3+a)(1-a)}$$

24. $\frac{2b+26}{(b-2)(b+4)}$

25. LCD $= (x-3)(x-2)$

$$\frac{(x+2)(x-2)}{(x-3)(x-2)} + \frac{-(x+3)(x-3)}{(x-2)(x-3)} = \frac{x^2-4}{(x-3)(x-2)} + \frac{-(x^2-9)}{(x-3)(x-2)} = \frac{x^2-4-x^2+9}{(x-3)(x-2)} = \frac{5}{(x-3)(x-2)}$$

26. $\dfrac{20}{(x+6)(x+4)}$

27. $\dfrac{\overset{1}{\cancel{(x+2)}}}{(x-1)\underset{1}{\cancel{(x+2)}}} + \dfrac{3}{(x+1)(x-1)} \qquad \text{LCD} = (x+1)(x-1)$

$$\frac{1(x+1)}{(x-1)(x+1)} + \frac{3}{(x+1)(x-1)} = \frac{x+1+3}{(x+1)(x-1)} = \frac{x+4}{x^2-1}$$

28. $\dfrac{x+7}{(x+2)(x-2)}$

29. LCD $= (x-3)(x+3)$

$$\frac{2x(x+3)}{(x-3)(x+3)} + \frac{-2x(x-3)}{(x+3)(x-3)} + \frac{36}{(x+3)(x-3)}$$

$$= \frac{2x^2+6x-2x^2+6x+36}{(x-3)(x+3)} = \frac{12x+36}{(x-3)(x+3)}$$

$$= \frac{12\overset{1}{\cancel{(x+3)}}}{(x-3)\underset{1}{\cancel{(x+3)}}} = \frac{12}{x-3}$$

30. $\dfrac{12}{6-m}$

31. LCD $= (x+2)^2(x-2)$

$$\frac{x-2}{(x+2)(x+2)} \cdot \frac{x-2}{x-2} - \frac{x+1}{(x+2)(x-2)} \cdot \frac{x+2}{x+2}$$

$$= \frac{x^2-4x+4}{(x+2)^2(x-2)} - \frac{x^2+3x+2}{(x+2)^2(x-2)}$$

$$= \frac{x^2-4x+4-x^2-3x-2}{(x+2)^2(x-2)} = \frac{2-7x}{(x+2)^2(x-2)}$$

32. $\dfrac{1-5x}{(x+1)(x-1)^2}$

33. LCD $= (x^2+2x+4)(x+2)$

$$\frac{4(x+2)}{(x^2+2x+4)(x+2)} + \frac{(x-2)(x^2+2x+4)}{(x+2)(x^2+2x+4)}$$

$$= \frac{4x+8}{(x^2+2x+4)(x+2)} + \frac{x^3-8}{(x^2+2x+4)(x+2)}$$

$$= \frac{4x+x^3}{(x^2+2x+4)(x+2)}$$

34. $\dfrac{x^3+3x}{(x-9)(x^2-3x+9)}$

35. $\dfrac{5}{2g^3} - \dfrac{3\overset{1}{\cancel{(g-3)}}}{(g-3)\underset{1}{\cancel{(g-3)}}} + \dfrac{\overset{3}{\cancel{12g}}}{\underset{1}{\cancel{4g}}(g-3)} = \dfrac{5}{2g^3} - \cancel{\dfrac{3}{g-3}} + \cancel{\dfrac{3}{g-3}} = \dfrac{5}{2g^3}$

36. $\dfrac{7}{9y^2}$

37. LCD $= 2(x-4)^2(x+5)$

$$\frac{(2x-5)(x+5)}{2(x-4)^2(x+5)} + \frac{(4x+7)(2)(x-4)}{(x+5)(x-4)(2)(x-4)}$$

$$= \frac{2x^2+5x-25+8x^2-18x-56}{2(x-4)^2(x+5)}$$

$$= \frac{10x^2-13x-81}{2(x-4)^2(x+5)}$$

38. $\dfrac{23k^2-10k+53}{5(k-3)^2(k+7)}$

39. LCD $= 12e^2(e+3)(e-3)$

$$\frac{35(4)(e+3)(e-3)}{3e^2(4)(e+3)(e-3)} - \frac{2e(12e^2)}{(e+3)(e-3)(12e^2)} - \frac{3(3e^2)(e+3)}{4(e-3)(3e^2)(e+3)}$$

$$= \frac{140(e^2-9)-2e(12e^2)-9e^2(e+3)}{12e^2(e+3)(e-3)}$$

$$= \frac{140e^2-1260-24e^3-9e^3-27e^2}{12e^2(e+3)(e-3)} = \frac{-33e^3+113e^2-1260}{12e^2(e+3)(e-3)}$$

$$= -\frac{33e^3-113e^2+1260}{12e^2(e+3)(e-3)}$$

40. $\dfrac{-164u^4-4u^3-23u^2-45u-216}{24u^3(u+3)(u-8)}$ or $\dfrac{164u^4+4u^3+23u^2+45u+216}{24u^3(u+3)(8-u)}$

41. LCD $= 12x^2(x-2)^2(x+2)$

$$= \frac{x^2+1}{12x^2(x+2)} - \frac{4x+3}{(x-2)^2} - \frac{1}{(x-2)(x+2)}$$

$$= \frac{(x^2+1)(x-2)^2-(4x+3)(12x^2)(x+2)-12x^2(x-2)}{12x^2(x-2)^2(x+2)}$$

$$= \frac{x^4-4x^3+5x^2-4x+4-48x^4-132x^3-72x^2-12x^3+24x^2}{12x^2(x-2)^2(x+2)}$$

$$= \frac{-47x^4-148x^3-43x^2-4x+4}{12x^2(x-2)^2(x+2)}$$

$$\text{or } -\frac{47x^4+148x^3+43x^2+4x-4}{12x^2(x-2)^2(x+2)}$$

42. $\dfrac{72y^3+149y^2-186y-99}{8y(y+3)^2(y-3)}$

43. LCD $= 3y(y+4)^2(y^2-4y+16)$

$$\frac{7}{3y(y^2-4y+16)} + \frac{y^2+4}{(y+4)(y^2-4y+16)} - \frac{y}{(y+4)^2}$$

$$= \frac{7(y+4)^2}{3y(y^2-4y+16)(y+4)^2} + \frac{(y^2+4)(3y)(y+4)}{3y(y^2-4y+16)(y+4)(y+4)}$$

$$- \frac{y(3y)(y^2-4y+16)}{3y(y^2-4y+16)(y+4)^2}$$

$$= \frac{7(y^2+8y+16)+3y(y^3+4y^2+4y+16)-3y^2(y^2-4y+16)}{3y(y+4)^2(y^2-4y+16)}$$

$$= \frac{24y^3-29y^2+104y+112}{3y(y+4)^2(y^2-4y+16)}$$

44. $\dfrac{12x^3+5x^2+84x+45}{2x(x+3)^2(x^2-3x+9)}$

45. LCD $= (x+1)(x-1)(x+3)(x-3)$

$$\frac{x-1}{x^2(x+1)-9(x+1)} - \frac{x+3}{x^2(x-3)-(x-3)}$$

$$= \frac{x-1}{(x+1)(x^2-9)} - \frac{x+3}{(x-3)(x^2-1)}$$

$$= \frac{x-1}{(x+1)(x+3)(x-3)} - \frac{x+3}{(x-3)(x+1)(x-1)}$$

$$= \frac{(x-1)(x-1)}{(x+1)(x+3)(x-3)(x-1)} - \frac{(x+3)(x+3)}{(x-3)(x+1)(x-1)(x+3)}$$

$$= \frac{(x^2-2x+1)-(x^2+6x+9)}{(x+1)(x+3)(x-3)(x-1)} = \frac{\cancel{x^2}-2x+1-\cancel{x^2}-6x-9}{(x+1)(x+3)(x-3)(x-1)}$$

$$= \frac{-8x-8}{(x+1)(x+3)(x-3)(x-1)} = \frac{-8\overset{1}{\cancel{(x+1)}}}{\underset{1}{\cancel{(x+1)}}(x+3)(x-3)(x-1)}$$

$$= \frac{-8}{(x+3)(x-3)(x-1)}$$

46. $\dfrac{6x+3}{(x+1)(x-1)(x+2)(x-2)}$

47. Division must be done before addition.

$$\frac{x+6}{x-5}+\frac{1}{x+4}\cdot\frac{x^2-x-20}{x+6}=\frac{x+6}{x-5}+\frac{1}{\underset{1}{\cancel{x+4}}}\cdot\frac{(x-5)\overset{1}{\cancel{(x+4)}}}{x+6}$$

$$=\frac{x+6}{x-5}+\frac{x-5}{x+6}\qquad \text{LCD}=(x-5)(x+6)$$

$$=\frac{(x+6)(x+6)}{(x-5)(x+6)}+\frac{(x-5)(x-5)}{(x+6)(x-5)}$$

$$=\frac{x^2+12x+36+x^2-10x+25}{(x-5)(x+6)}=\frac{2x^2+2x+61}{(x-5)(x+6)}$$

48. $\frac{5x^2-4x+25}{(x+4)(2x-3)}$

Exercises 6.6 (page 273)

1. $\frac{\overset{3}{\cancel{21}}m^3n}{\underset{2}{\cancel{14}}mn^2}\cdot\frac{\overset{2}{\cancel{8}}mn^3}{\underset{5}{\cancel{20}}m^2n^2}=\frac{3m}{5}$

2. $\frac{8}{3ab}$

3. $\frac{\overset{1}{\cancel{3}}\overset{1}{\cancel{(5h-2)}}}{\underset{3}{\underset{\cancel{6}}{\cancel{18}}}h}\cdot\frac{\overset{2}{\overset{\cancel{4}}{\cancel{8}}}h}{\underset{3}{\cancel{6}}h\underset{1}{\cancel{(5h-2)}}}=\frac{2}{9h}$

4. $\frac{3k}{5}$

5. $\frac{d^2}{d^2}\cdot\frac{\left(\frac{c}{d}+2\right)}{\left(\frac{c^2}{d^2}-4\right)}=\frac{d^2\left(\frac{c}{d}\right)+d^2(2)}{d^2\left(\frac{c^2}{d^2}\right)-d^2(4)}$

$$=\frac{cd+2d^2}{c^2-4d^2}=\frac{d\overset{1}{\cancel{(c+2d)}}}{(c-2d)\underset{1}{\cancel{(c+2d)}}}=\frac{d}{c-2d}$$

6. $\frac{x+y}{y}$

7. $\dfrac{\frac{(a+2)}{1}\cdot\frac{(a+2)}{1}+\frac{(a+2)}{1}\left(\frac{-9}{a+2}\right)}{\frac{(a+2)}{1}\cdot\frac{(a+1)}{1}+\frac{(a+2)}{1}\left(\frac{a-7}{a+2}\right)}$

$$=\frac{a^2+4a+4-9}{a^2+3a+2+a-7}=\frac{a^2+4a-5}{a^2+4a-5}=1$$

8. $\frac{x+3}{x+5}$

9. $\dfrac{\frac{y(x-y)}{1}\cdot\frac{(x+y)}{y}+\frac{y(x-y)}{1}\cdot\frac{y}{x-y}}{\frac{y(x-y)}{1}\cdot\frac{y}{x-y}}=\frac{x^2-y^2+y^2}{y^2}=\frac{x^2}{y^2}$

10. $-\frac{1}{a}$

11. $\dfrac{\frac{x(x+1)}{1}\cdot\frac{x}{x+1}+\frac{x(x+1)}{1}\cdot\frac{4}{x}}{\frac{x(x+1)}{1}\cdot\frac{x}{x+1}+\frac{x(x+1)}{1}\cdot\frac{(-2)}{1}}$

$$=\frac{x^2+4x+4}{x^2-2x^2-2x}=\frac{(x+2)\overset{1}{\cancel{(x+2)}}}{-x\underset{1}{\cancel{(x+2)}}}=-\frac{x+2}{x}$$

12. $\frac{2x+1}{4x}$

13. $\dfrac{\frac{x(x-1)}{1}\cdot\frac{(x+4)}{x}+\frac{x(x-1)}{1}\cdot\left(-\frac{3}{x-1}\right)}{\frac{x(x-1)}{1}\cdot\frac{(x+1)}{1}+\frac{x(x-1)}{1}\cdot\left(\frac{2x+1}{x-1}\right)}$

$$=\frac{x^2+3x-4-3x}{x^3-x+2x^2+x}=\frac{x^2-4}{x^3+2x^2}=\frac{\overset{1}{\cancel{(x+2)}}(x-2)}{x^2\underset{1}{\cancel{(x+2)}}}=\frac{x-2}{x^2}$$

14. $\frac{x-2}{(x-6)(x-4)}$

15. $\dfrac{\frac{4}{x^2}-\frac{1}{y^2}}{\frac{2}{x}+\frac{1}{y}}=\dfrac{\frac{x^2y^2}{1}\cdot\frac{4}{x^2}+\frac{x^2y^2}{1}\left(-\frac{1}{y^2}\right)}{\frac{x^2y^2}{1}\cdot\frac{2}{x}+\frac{x^2y^2}{1}\cdot\frac{1}{y}}$

$$=\frac{4y^2-x^2}{2xy^2+x^2y}=\frac{\overset{1}{\cancel{(2y+x)}}(2y-x)}{xy\underset{1}{\cancel{(2y+x)}}}=\frac{2y-x}{xy}$$

16. $\frac{3x+y}{xy}$

17. $\dfrac{\frac{(x+2)(x-2)}{1}\left(\frac{x-2}{x+2}\right)-\frac{(x+2)(x-2)}{1}\left(\frac{x+2}{x-2}\right)}{\frac{(x+2)(x-2)}{1}\left(\frac{x-2}{x+2}\right)+\frac{(x+2)(x-2)}{1}\left(\frac{x+2}{x-2}\right)}$

$$=\frac{(x-2)(x-2)-(x+2)(x+2)}{(x-2)(x-2)+(x+2)(x+2)}$$

$$=\frac{x^2-4x+4-(x^2+4x+4)}{x^2-4x+4+x^2+4x+4}=\frac{-8x}{2x^2+8}=-\frac{4x}{x^2+4}$$

18. $\frac{m^2+9}{6m}$

19. $\dfrac{\frac{xy(x-y)}{1}\cdot\frac{2x+y}{x}-\frac{xy(x-y)}{1}\cdot\frac{3x+y}{(x-y)}}{\frac{xy(x-y)}{1}\cdot\frac{x+y}{y}+\frac{xy(x-y)}{1}\cdot\frac{2(x+y)}{(x-y)}}$

$$=\frac{y(x-y)(2x+y)-xy(3x+y)}{x(x+y)(x-y)+2xy(x+y)}$$

$$=\frac{y[(x-y)(2x+y)-x(3x+y)]}{x(x+y)[(x-y)+2y]}$$

$$=\frac{y[2x^2-xy-y^2-3x^2-xy)]}{x(x+y)(x+y)}$$

$$=\frac{-y[x^2+2xy+y^2]}{x(x+y)^2}=\frac{-y(x+y)^2}{x(x+y)^2}=-\frac{y}{x}$$

20. $\frac{a}{b}$

21. $\dfrac{1}{x+\dfrac{1}{x+\frac{1}{2x}}}=\dfrac{1}{x+\dfrac{1(2x)}{\left(x+\frac{1}{2x}\right)(2x)}}=\dfrac{1}{x+\frac{2x}{2x^2+1}}$

$$=\frac{1(2x^2+1)}{\left(x+\frac{2x}{2x^2+1}\right)(2x^2+1)}=\frac{2x^2+1}{x(2x^2+1)+2x}$$

$$=\frac{2x^2+1}{2x^3+x+2x}=\frac{2x^2+1}{2x^3+3x}$$

22. $\frac{2y^2+2}{y^3+3y}$

23. $\dfrac{x+\dfrac{1(3)}{\left(2+\frac{x}{3}\right)(3)}}{x-\dfrac{3(2)}{\left(4+\frac{x}{2}\right)(2)}}=\dfrac{x+\frac{3}{6+x}}{x-\frac{6}{8+x}}=\dfrac{\left(x+\frac{3}{6+x}\right)(6+x)(8+x)}{\left(x-\frac{6}{8+x}\right)(6+x)(8+x)}$

$$=\frac{x(48+14x+x^2)+3(8+x)}{x(48+14x+x^2)-6(6+x)}=\frac{48x+14x^2+x^3+24+3x}{48x+14x^2+x^3-36-6x}$$

$$=\frac{x^3+14x^2+51x+24}{x^3+14x^2+42x-36}$$

24. $\frac{x^3+8x^2+22x+60}{x^3+8x^2-24}$

Review Exercises 6.7 (page 276)

1.
$$\begin{aligned} 20 - 45a^2 &= 0 \\ 5(4 - 9a^2) &= 0 \\ 5(2 + 3a)(2 - 3a) &= 0 \\ 2 + 3a = 0 \text{ or } 2 - 3a &= 0 \\ a = -\tfrac{2}{3} \text{ or } \quad a &= \tfrac{2}{3} \end{aligned}$$
The domain is the set of all real numbers except $-\frac{2}{3}$ and $\frac{2}{3}$.

2. All real numbers except 6 and $-5/2$

3. $\dfrac{4z(z^2 + z - 6)}{2(z^2 + 2z - 3)} = \dfrac{2z(z+3)(z-2)}{(z+3)(z-1)} = \dfrac{2z(z-2)}{z-1}$

4. $\dfrac{2k(k+1)}{k+4}$

5. $\dfrac{(a-3b)(a^2 + 3ab + 9b^2)}{(a-3b)(a+2)} = \dfrac{a^2 + 3ab + 9b^2}{a+2}$ **6.** $\dfrac{1}{4x^2 - 2x + 1}$

7. $\dfrac{-35mn^2p^2}{14m^3p^3} \cdot \dfrac{13m^4n}{52n^3p} = -\dfrac{5m^2}{8p^2}$ **8.** $-\dfrac{4c^2}{3ab^3}$

9. $\dfrac{(z+2)(z+1)}{(z+1)(z+1)} \cdot \dfrac{(z+1)(z-1)}{(z+3)(z-1)} = \dfrac{z+2}{z+3}$ **10.** 1

11. $\dfrac{(x+y)(x^2 - xy + y^2)}{3(x^2 - xy + y^2)} \cdot \dfrac{(x+2y)(x-y)}{(x+y)(x-y)} \cdot \dfrac{15x^2y}{5xy(x+2y)} = x$

12. 2

13. $\dfrac{20y - 7}{6y - 8} - \dfrac{17 + 2y}{6y - 8} = \dfrac{20y - 7 - 17 - 2y}{6y - 8}$
$$= \frac{18y - 24}{6y - 8} = \frac{6(3y-4)}{2(3y-4)} = 3$$

14. -2 **15.** $\dfrac{11}{30e^3f} \cdot \dfrac{3f}{3f} - \dfrac{7}{45e^2f^2} \cdot \dfrac{2e}{2e} = \dfrac{33f - 14e}{90e^3f^2}$

16. $\dfrac{30u^2 - 49v}{280u^4v^2}$

17. $\dfrac{a+1}{(a+1)(a-2)} - \dfrac{a-2}{(a+3)(a-2)} = \dfrac{1}{a-2} - \dfrac{1}{a+3}$
LCD $= (a-2)(a+3)$
$$\frac{(a+3)}{(a+3)} \cdot \frac{1}{a-2} + \frac{(a-2)}{(a-2)} \cdot \frac{-1}{a+3} = \frac{a + 3 - a + 2}{(a-2)(a+3)} = \frac{5}{(a-2)(a+3)}$$

18. $\dfrac{3}{(x-1)(x+2)}$

19. $\dfrac{15x}{5x(x+4)} - \dfrac{7}{3x^2} - \dfrac{3(x+4)}{(x+4)^2} = \dfrac{3}{x+4} - \dfrac{7}{3x^2} - \dfrac{3}{x+4} = -\dfrac{7}{3x^2}$

20. $-\dfrac{5}{11y^3}$

21. $\dfrac{\frac{y+1}{1}}{\frac{y+1}{1}} \cdot \dfrac{\frac{x}{y+1} + 2}{\frac{x}{y+1} - 2} = \dfrac{\left(\frac{y+1}{1}\right)\left(\frac{x}{y+1}\right) + \left(\frac{y+1}{1}\right)\left(\frac{2}{1}\right)}{\left(\frac{y+1}{1}\right)\left(\frac{x}{y+1}\right) + \left(\frac{y+1}{1}\right)\left(\frac{-2}{1}\right)}$
$$= \frac{x + (y+1)(2)}{x + (y+1)(-2)} = \frac{x + 2y + 2}{x - 2y - 2}$$

22. $\dfrac{3b - a + 6}{2b + a + 4}$

23. $\dfrac{\frac{R^3T^3}{1}}{\frac{R^3T^3}{1}} \cdot \dfrac{\frac{8}{R^3} + \frac{1}{T^3}}{\frac{4}{R^2} - \frac{1}{T^2}} = \dfrac{\frac{R^3T^3}{1} \cdot \frac{8}{R^3} + \frac{R^3T^3}{1}\left(\frac{1}{T^3}\right)}{\frac{R^3T^3}{1} \cdot \frac{4}{R^2} + \frac{R^3T^3}{1}\left(-\frac{1}{T^2}\right)}$
$$= \frac{8T^3 + R^3}{4RT^3 - R^3T} = \frac{(2T+R)(4T^2 - 2RT + R^2)}{RT(2T+R)(2T-R)}$$
$$= \frac{4T^2 - 2RT + R^2}{RT(2T-R)}$$

24. $\dfrac{mn(n + 4m)}{n^2 + 4mn + 16m^2}$

Exercises 6.8 (page 286)

Note: The checks for these exercises usually will not be shown.

1. LCD $= 4k(k-5)$
$$\frac{4k(k-5)}{1} \cdot \frac{2}{k-5} - \frac{4k(k-5)}{1} \cdot \frac{5}{k} = \frac{4k(k-5)}{1} \cdot \frac{3}{4k}$$
$$\begin{aligned} 4k(2) - 20(k-5) &= (k-5)(3) \\ 8k - 20k + 100 &= 3k - 15 \\ -15k &= -115 \\ k &= \tfrac{115}{15} = \tfrac{23}{3} \end{aligned}$$
$\left\{\frac{23}{3}\right\}$

2. $\left\{-\frac{35}{4}\right\}$

3. LCD $= x - 2$ **4.** $\{\ \}$
$$\frac{x-2}{1} \cdot \frac{x}{x-2} = \frac{x-2}{1} \cdot \frac{2}{x-2} + \frac{x-2}{1}(5)$$
$$\begin{aligned} x &= 2 + 5x - 10 \\ 8 &= 4x \\ x &= 2 \end{aligned}$$
Not in the domain; solution set: $\{\ \}$

5. LCD $= 2m - 3$
$$\frac{2m-3}{1} \cdot \frac{12m}{2m-3} = \frac{2m-3}{1} \cdot \frac{6}{1} + \frac{2m-3}{1} \cdot \frac{18}{2m-3}$$
$$\begin{aligned} 12m &= (2m-3)(6) + 18 \\ 12m &= 12m - 18 + 18 \\ 12m &= 12m \\ 0 &= 0 \quad \text{Identity} \end{aligned}$$
Solution set: all real numbers except $\frac{3}{2}$

6. Identity; solution set: all real numbers except $-\frac{6}{5}$

7.
$$\begin{aligned} 2y(3y) &= 7y + 5 \\ 6y^2 &= 7y + 5 \\ 6y^2 - 7y - 5 &= 0 \\ (2y+1)(3y-5) &= 0 \end{aligned}$$
$2y + 1 = 0 \quad | \quad 3y - 5 = 0$
$y = -\frac{1}{2} \quad | \quad y = \frac{5}{3}$
$\left\{-\frac{1}{2}, \frac{5}{3}\right\}$

8. $\left\{\frac{1}{2}, -\frac{3}{2}\right\}$

9.
$$\begin{aligned} (3e-5)(2e+3) &= 4e(e) \\ 6e^2 - e - 15 &= 4e^2 \\ 2e^2 - e - 15 &= 0 \\ (e-3)(2e+5) &= 0 \end{aligned}$$
$e - 3 = 0 \quad | \quad 2e + 5 = 0$
$e = 3 \quad | \quad e = -\frac{5}{2}$ Solution set: $\left\{3, -\frac{5}{2}\right\}$

10. $\{1, -\frac{7}{4}\}$

11. LCD $= 2x^2$

$$\overset{1}{\cancel{2}}x^2 \cdot \frac{1}{\underset{1}{\cancel{2}}} - 2x^{\cancel{2}} \cdot \frac{1}{\underset{1}{\cancel{x}}} = 2\overset{1}{\cancel{x^2}} \cdot \frac{4}{\underset{1}{\cancel{x^2}}}$$

$$x^2 - 2x = 8$$
$$x^2 - 2x - 8 = 0$$
$$(x - 4)(x + 2) = 0$$

$x - 4 = 0 \mid x + 2 = 0$

$x = 4 \mid x = -2$

$\{4, -2\}$

12. $\{3, -\frac{4}{7}\}$

13. LCD $= 15x(x + 1)$

$$\frac{15x(\overset{1}{\cancel{x+1}})}{1} \cdot \frac{4}{\underset{1}{\cancel{x+1}}} = \frac{15x(x+1)}{1} \cdot \frac{3}{x} + \frac{\overset{1}{\cancel{15}}x(x+1)}{1} \cdot \frac{1}{\underset{1}{\cancel{15}}}$$

$$60x = 45x + 45 + x^2 + x$$
$$0 = x^2 - 14x + 45$$
$$0 = (x - 5)(x - 9)$$

$x - 5 = 0 \mid x - 9 = 0$

$x = 5 \mid x = 9$

$\{5, 9\}$

14. $\{-2, -3\}$

15. LCD $= x(x + 3)(x + 4)$

$$\frac{x(x+3)(\overset{1}{\cancel{x+4}})}{1} \cdot \frac{6}{\underset{1}{\cancel{x+4}}} = \frac{x(\overset{1}{\cancel{x+3}})(x+4)}{1} \cdot \frac{5}{\underset{1}{\cancel{x+3}}} + \frac{\overset{1}{\cancel{x}}(x+3)(x+4)}{1} \cdot \frac{4}{\underset{1}{\cancel{x}}}$$

$$x(x + 3)(6) = x(x + 4)(5) + (x + 3)(x + 4)(4)$$
$$6x^2 + 18x = 5x^2 + 20x + 4x^2 + 28x + 48$$
$$0 = 3x^2 + 30x + 48$$
$$0 = 3(x^2 + 10x + 16)$$
$$0 = 3(x + 2)(x + 8)$$

$x + 2 = 0 \mid x + 8 = 0$

$x = -2 \mid x = -8$

$\{-2, -8\}$

16. $\{2, -\frac{5}{4}\}$

17. LCD $= 5(x + 3)(x - 3)$

Domain: all real numbers except -3 and 3

$$\frac{5\overset{1}{\cancel{(x+3)(x-3)}}}{1} \cdot \frac{6}{\underset{1}{\cancel{x^2-9}}} + \frac{\overset{1}{\cancel{5}}(x+3)(x-3)}{1} \cdot \frac{1}{\underset{1}{\cancel{5}}}$$

$$= \frac{5(x+3)(\overset{1}{\cancel{x-3}})}{1} \cdot \frac{1}{\underset{1}{\cancel{x-3}}}$$

$$30 + x^2 - 9 = 5x + 15$$
$$x^2 - 5x + 6 = 0$$
$$(x - 2)(x - 3) = 0$$

$x - 2 = 0 \mid x - 3 = 0$

$x = 2 \mid x = 3$ Not in the domain

Check for $x = 2$ $\quad \frac{6}{x^2 - 9} + \frac{1}{5} = \frac{1}{x - 3}$

$$\frac{6}{2^2 - 9} + \frac{1}{5} \stackrel{?}{=} \frac{1}{2 - 3}$$
$$-\frac{6}{5} + \frac{1}{5} \stackrel{?}{=} -1$$
$$-1 = -1$$

Solution set: $\{2\}$

18. $\{\frac{7}{3}\}$

19. LCD $= (x + 2)(x - 2)$

Domain: all real numbers except 2 and -2

$$(x+2)(x-2)\left(\frac{x+2}{x-2} - \frac{x-2}{x+2}\right) = \overset{1}{\cancel{(x+2)(x-2)}}\left(\frac{16}{\underset{1}{\cancel{x^2-4}}}\right)$$

$$(x+2)(\overset{1}{\cancel{x-2}})\left(\frac{x+2}{\underset{1}{\cancel{x-2}}}\right) - (\overset{1}{\cancel{x+2}})(x-2)\left(\frac{x-2}{\underset{1}{\cancel{x+2}}}\right) = 16$$

$$(x + 2)(x + 2) - (x - 2)(x - 2) = 16$$
$$x^2 + 4x + 4 - (x^2 - 4x + 4) = 16$$
$$x^2 + 4x + 4 - x^2 + 4x - 4 = 16$$
$$8x = 16$$
$$x = 2$$

But 2 is not in the domain of the variable. Therefore, the solution set is $\{\ \}$.

20. $\{\ \}$

21. LCD $= (2x - 5)(x - 3)(x + 3)$

$$(2x-5)(x-3)(x+3)\left(\frac{1}{(2x-5)(x-3)} + \frac{x-1}{(2x-5)(x+3)}\right)$$
$$= (2x-5)(x-3)(x+3)\left(\frac{-4}{(x+3)(x-3)}\right)$$

$$(x + 3) + (x - 3)(x - 1) = (2x - 5)(-4)$$
$$x + 3 + x^2 - 4x + 3 = -8x + 20$$
$$x^2 + 5x - 14 = 0$$
$$(x + 7)(x - 2) = 0$$

$x + 7 = 0 \quad x - 2 = 0$

$x = -7 \quad x = 2$

$\{2, -7\}$

22. $\{1, 10\}$

23. LCD $= (x + 4)(x^2 - 4x + 16)(x - 4)$

$$(x+4)(x^2-4x+16)(x-4)\left(\frac{8}{(x+4)(x^2-4x+16)} + \frac{3}{x^2-16}\right)$$
$$= (x+4)(x^2-4x+16)(x-4)\left(\frac{-1}{x^2-4x+16}\right)$$

$$8(x - 4) + 3(x^2 - 4x + 16) = -1(x + 4)(x - 4)$$
$$8x - 32 + 3x^2 - 12x + 48 = -(x^2 - 16)$$
$$3x^2 - 4x + 16 = -x^2 + 16$$
$$4x^2 - 4x = 0$$
$$4x(x - 1) = 0$$
$$4x = 0 \text{ or } x - 1 = 0$$
$$x = 0 \text{ or } \quad x = 1$$

$\{0, 1\}$

24. $\{2\}$

Exercises 6.9 (page 290)

1. $6x - 2y = xy - 12$

$$6x + 12 = xy + 2y$$
$$6(x + 2) = (x + 2)y$$
$$\frac{6(\overset{1}{\cancel{x+2}})}{(\underset{1}{\cancel{x+2}})} = y$$
$$y = 6$$

2. $x = -2$

3. $zs = x - m$

$$m = x - zs$$

4. $N = \dfrac{n - s^2}{1 - s^2}$

5. $(5yz)\left(\dfrac{2x}{5yz}\right) = (5yz)(z + x)$

$$2x = 5yz^2 + 5xyz$$
$$2x - 5xyz = 5yz^2$$
$$x(2 - 5yz) = 5yz^2$$
$$x = \frac{5yz^2}{2 - 5yz}$$

6. $y = \dfrac{xz}{x - z}$

7.
$$9C = 5(F - 32)$$
$$9C = 5F - 160$$
$$9C + 160 = 5F$$
$$F = \frac{9C + 160}{5}$$

8. $B = \dfrac{2A - hb}{h}$

9. $s = c + \dfrac{\overset{1}{\not c}a}{\underset{1}{\not c}}$
$$s = c + a$$
$$c = s - a$$

10. $R = \dfrac{rZ}{r - Z}$

11.
$$A = P + Prt$$
$$A - P = Prt$$
$$r = \frac{A - P}{Pt}$$

12. $t = \dfrac{2S + g}{2g}$

13. $\dfrac{ra}{1} \cdot \dfrac{v^2}{1} = \dfrac{\overset{1}{\not r}a}{1} \cdot \dfrac{2}{\underset{1}{\not r}} - \dfrac{r\overset{1}{\not a}}{1} \cdot \dfrac{1}{\underset{1}{\not a}}$
$$rav^2 = 2a - r$$
$$r = 2a - rav^2$$
$$r = a(2 - rv^2)$$
$$a = \frac{r}{2 - rv^2}$$

14. $s = \dfrac{p}{1 - p}$

15.
$$S(1 - r) = a$$
$$S - Sr = a$$
$$S - a = Sr$$
$$r = \frac{S - a}{S}$$

16. $R = \dfrac{E - Ir}{I}$

17. $Fuv\left(\dfrac{1}{F}\right) = Fuv\left(\dfrac{1}{u} + \dfrac{1}{v}\right)$
$$uv = Fuv\left(\frac{1}{u}\right) + Fuv\left(\frac{1}{v}\right)$$
$$uv = Fv + Fu$$
$$uv = F(v + u)$$
$$F = \frac{uv}{u + v}$$

18. $b = \dfrac{ac}{a - c}$

19.
$$L = a + nd - d$$
$$L - a + d = nd$$
$$n = \frac{L - a + d}{d}$$

20. $h = \dfrac{A - 2\pi r^2}{2\pi r}$

21.
$$C = \frac{\pi A \cdot a}{\pi A \cdot 1 + \dfrac{\overset{1}{\not\pi \not A}}{1} \cdot \dfrac{a}{\underset{1}{\not\pi \not A}}}$$
$$\frac{C}{1} = \frac{\pi Aa}{\pi A + a}$$
$$C(\pi A + a) = \pi Aa$$
$$C\pi A + Ca = \pi Aa$$
$$C\pi A = \pi Aa - Ca$$
$$C\pi A = a(\pi A - C)$$
$$a = \frac{C\pi A}{\pi A - C}$$

22. $v = \dfrac{c^2(R - V)}{c^2 - RV}$

Exercises 6.10 (page 294)

(The checks will not be shown.)

1. Henry's rate $= \dfrac{1 \text{ house}}{5 \text{ days}} = \dfrac{1}{5}$ house per day

Teri's rate $= \dfrac{1 \text{ house}}{4 \text{ days}} = \dfrac{1}{4}$ house per day

Let x = number of days to paint the house.

Amount Henry paints	+	Amount Teri paints	=	Amount painted together	
$\dfrac{x}{5}$	+	$\dfrac{x}{4}$	=	1	One house painted

LCD = 20; $\dfrac{20}{1} \cdot \dfrac{x}{5} + \dfrac{20}{1} \cdot \dfrac{x}{4} = \dfrac{20}{1} \cdot 1$
$$4x + 5x = 20$$
$$9x = 20$$
$$x = \tfrac{20}{9} = 2\tfrac{2}{9}$$

It will take $2\frac{2}{9}$ days to paint the house.

2. $3\frac{3}{5}$ days

3. Let x = number of hours for David to type 80 pages.

Trisha's rate $= \dfrac{100 \text{ pages}}{3 \text{ hours}} = \dfrac{100}{3}$ pages per hr

David's rate $= \dfrac{80 \text{ pages}}{x \text{ hours}} = \dfrac{80}{x}$ pages per hr

Since they both type for 10 hr to produce 500 pages:

Amount Trisha types	+	Amount David types	=	500 pages
$\dfrac{100}{3}(10)$	+	$\dfrac{80}{x}(10)$	=	500

LCD = $3x$; $\dfrac{3x}{1} \cdot \dfrac{1{,}000}{3} + \dfrac{3x}{1} \cdot \dfrac{800}{x} = \dfrac{3x}{1} \cdot \dfrac{500}{1}$
$$1{,}000x + 2{,}400 = 1{,}500x$$
$$2{,}400 = 500x$$
$$x = \frac{2{,}400}{500} = \tfrac{24}{5} = 4\tfrac{4}{5}$$

It takes David $4\frac{4}{5}$ hr to type 80 pages.

4. $49\frac{1}{2}$ min

5. Machine A's rate $= \dfrac{1 \text{ job}}{36 \text{ hr}} = \dfrac{1}{36}$ job per hr

Machine B's rate $= \dfrac{1 \text{ job}}{24 \text{ hr}} = \dfrac{1}{24}$ job per hr

Let x = number of hours for both machines to run together.
Since machine A runs for 12 hr before machine B is turned on, machine A runs for $x + 12$ hr.

Amount of job done by machine A	+	Amount of job done by machine B	=	1 job
$\dfrac{1}{36}(x + 12)$	+	$\dfrac{1}{24}(x)$	=	1

LCD = 72 $\quad \dfrac{72}{1} \cdot \dfrac{x + 12}{36} + \dfrac{72}{1} \cdot \dfrac{x}{24} = 72 \cdot 1$
$$2x + 24 + 3x = 72$$
$$5x = 48$$
$$x = \tfrac{48}{5} = 9\tfrac{3}{5}$$

It will take $9\frac{3}{5}$ hr to finish the job.

6. $6\frac{9}{11}$ hr

7. Let x = smaller number
$x + 8$ = larger number

One-fourth the larger number is one more than one-third the smaller number

$$\frac{1}{4}(x+8) = 1 + \frac{1}{3}(x)$$

LCD = 12

$$\frac{12}{1}\cdot\frac{x+8}{4} = \frac{12}{1}\cdot 1 + \frac{12}{1}\cdot\frac{x}{3}$$

$$3x + 24 = 12 + 4x$$
$$12 = x$$
$$x = 12$$
$$x + 8 = 12 + 8 = 20$$

The numbers are 12 and 20.

8. Smaller number is -20; larger number is -14.

9. Let d = distance in mi.

$$\text{time} = \frac{\text{distance}}{\text{rate}}$$

Slow plane's time $= \frac{d}{400}$ Fast plane's time $= \frac{d}{500}$

Slow plane's time $-$ Fast plane's time $= \frac{1}{2}$ *hr*

$$\frac{d}{400} - \frac{d}{500} = \frac{1}{2}$$

LCD = 2,000

$$\frac{2{,}000}{1}\cdot\frac{d}{400} - \frac{2{,}000}{1}\cdot\frac{d}{500} = \frac{2{,}000}{1}\cdot\frac{1}{2}$$

$$5d - 4d = 1{,}000$$
$$d = 1{,}000$$

The distance is 1,000 mi.

10. 6 mi

11. Let x = number of quarts of antifreeze to be drained and replaced.

$$.45(14) - .45(x) + 1.00x = .50(14)$$
$$6.3 + .55x = 7$$
$$630 + 55x = 700$$
$$55x = 70$$
$$x = \tfrac{70}{55} = \tfrac{14}{11} = 1\tfrac{3}{11}$$

$1\frac{3}{11}$ qt must be drained and replaced.

12. $3\frac{3}{7}$ qt

13. Let u = units digit
$u + 1$ = tens digit

$$\frac{(u+1)u}{u+1+u} = \frac{6}{5}$$
$$\frac{u^2+u}{2u+1} = \frac{6}{5}$$
$$5(u^2+u) = 6(2u+1)$$
$$5u^2 + 5u = 12u + 6$$
$$5u^2 - 7u - 6 = 0$$
$$(u-2)(5u+3) = 0$$

$u - 2 = 0$ | $5u + 3 = 0$
$u = 2$ | $u = -\frac{3}{5}$ Not in the domain

$u + 1 = 3$

The number is 32.

14. 34

15. Let x = number of hours for pipe 1 to fill tank
$x + 1$ = number of hours for pipe 2 to fill tank

Rate of pipe 1 $= \frac{1 \text{ tank}}{x \text{ hr}} = \frac{1}{x}$ tank per hr

Rate of pipe 2 $= \frac{1 \text{ tank}}{(x+1) \text{ hr}} = \frac{1}{x+1}$ tank per hr

Rate of pipe 3 $= \frac{1 \text{ tank}}{2 \text{ hr}} = \frac{1}{2}$ tank per hr

Amount 1 does in 3 hr + Amount 2 does in 3 hr $-$ Amount 3 does in 3 hr = 1 full tank

$$\frac{1}{x}(3) + \frac{1}{x+1}(3) - \frac{1}{2}(3) = 1$$

$$\frac{2x(x+1)}{1}\cdot\frac{3}{x} + \frac{2x(x+1)}{1}\cdot\frac{3}{x+1} - \frac{2x(x+1)}{1}\cdot\frac{3}{2} = \frac{2x(x+1)}{1}\cdot\frac{1}{1}$$

$$6(x+1) + 6x - 3x(x+1) = 2x(x+1)$$
$$6x + 6 + 6x - 3x^2 - 3x = 2x^2 + 2x$$
$$5x^2 - 7x - 6 = 0$$
$$(x-2)(5x+3) = 0$$

$x - 2 = 0$ | $5x + 3 = 0$
$x = 2$ | $x = -\frac{3}{5}$ Not in the domain

It takes 2 hr for pipe 1 to fill the tank.

16. 2 hr

17. Let x = number of hours for Sandra to proofread 60 pages.

Ruth's rate $= \frac{230 \text{ pages}}{4 \text{ hr}}$ Sandra's rate $= \frac{60 \text{ pages}}{x \text{ hr}}$

Number of pages Ruth reads in 6 hr + Number of pages Sandra reads in 6 hr = 525 pages

$$\left(\frac{230}{4}\right)(6) + \left(\frac{60}{x}\right)(6) = 525 \quad (\text{LCD} = x)$$

$$x\left(345 + \frac{360}{x}\right) = 525x$$
$$345x + 360 = 525x$$
$$360 = 180x$$
$$x = 2$$

It takes 2 hr for Sandra to proofread 60 pages.

18. 28 hr

19. Let x = number of minutes for machine B to process 4,300 ft of film.

Rate of machine A $= \frac{5{,}700 \text{ ft}}{60 \text{ min}}$ Rate of machine B $= \frac{4{,}300 \text{ ft}}{x \text{ min}}$

Amount of film done by machine A in 50 min + Amount of film done by machine B in 50 min = Amount of film done together in 50 min

$$\frac{5{,}700}{60}\cdot(50) + \frac{4{,}300}{x}\cdot(50) = 15{,}500$$
$$4{,}750 + \frac{215{,}000}{x} = 15{,}500$$
$$4.750x + 215{,}000 = 15{,}500x$$
$$215{,}000 = 10{,}750x$$
$$x = 20$$

It takes 20 min for machine B to process 4,300 ft of film.

20. 30 min

Review Exercises 6.11 (page 297)

1. $7(3z) = (2z + 5)(13)$
$21z = 26z + 65$
$-5z = 65$
$z = -13$ $\{-13\}$

2. $\{-8\}$

3. $$\begin{aligned}(5a-4)(3a+10) &= (6a)(-2)\\ 15a^2+38a-40 &= -12a\\ 15a^2+50a-40 &= 0\\ 3a^2+10a-8 &= 0\\ (a+4)(3a-2) &= 0\end{aligned}$$
$$a+4=0 \quad\Big|\quad 3a-2=0$$
$$a=-4 \quad\Big|\quad a=\frac{2}{3} \qquad \left\{-4, \frac{2}{3}\right\}$$

4. $\left\{2, \frac{7}{5}\right\}$

5. LCD $= 4x^2$; $\frac{4x^2}{1}\cdot\frac{3}{x}+\frac{4x^2}{1}\cdot\frac{-8}{x^2}=\frac{4x^2}{1}\cdot\frac{1}{4}$
$$\begin{aligned}12x-32 &= x^2\\ 0 &= x^2-12x+32\\ 0 &= (x-4)(x-8)\end{aligned}$$
$$x-4=0 \quad\Big|\quad x-8=0$$
$$x=4 \quad\Big|\quad x=8$$
$\{4, 8\}$

6. $\left\{-2, \frac{4}{5}\right\}$

7. LCD $= x(x+1)(x-1)$
$$\frac{x(x+1)(x-1)}{1}\cdot\frac{9}{(x+1)}=\frac{x(x+1)(x-1)}{1}\cdot\frac{4}{x}+\frac{x(x+1)(x-1)}{1}\cdot\frac{1}{(x-1)}$$
$$\begin{aligned}9x(x-1) &= 4(x+1)(x-1)+x(x+1)\\ 9x^2-9x &= 4x^2-4+x^2+x\\ 4x^2-10x+4 &= 0\\ 2(2x^2-5x+2) &= 0\\ 2(x-2)(2x-1) &= 0\end{aligned}$$
$$x-2=0 \qquad 2x-1=0$$
$$x=2 \qquad x=\tfrac{1}{2} \qquad \left\{2, \tfrac{1}{2}\right\}$$

8. $\left\{1, -\frac{1}{5}\right\}$

9. LCD $= x+5$; the domain is the set of all real numbers except -5.
$$\frac{x+5}{1}\cdot\frac{x}{x+5}=\frac{x+5}{1}\cdot\frac{8}{1}-\frac{x+5}{1}\cdot\frac{5}{x+5}$$
$$\begin{aligned}x &= 8(x+5)-5\\ x &= 8x+40-5\\ x &= 8x+35\\ -7x &= 35\\ x &= -5 \quad \text{Not in the domain}\end{aligned}$$
No solution $\{\ \}$

10. $\{x \mid x \text{ is any real number except 0 or 3}\}$

11. $$\begin{aligned}5x-10y &= 14+6x-3y\\ -14-x &= 7y\\ y &= \frac{-x-14}{7}=-\frac{x+14}{7}\end{aligned}$$

12. $x=\frac{y-6}{7}$

13. $$\begin{aligned}R(R_1+R_2) &= R_1R_2\\ RR_1+RR_2 &= R_1R_2\\ RR_2 &= R_1R_2-RR_1\\ RR_2 &= R_1(R_2-R)\\ R_1 &= \frac{RR_2}{R_2-R}\end{aligned}$$

14. $v=\frac{uF}{u-F}$

15. A's rate $=\frac{1\text{ job}}{20\text{ hr}}=\frac{1}{20}$ job per hr

B's rate $=\frac{1\text{ job}}{15\text{ hr}}=\frac{1}{15}$ job per hr

Let x = number of hours both tractors worked together.
Since tractor A starts working 5 hr before tractor B starts, the time tractor A works is $x+5$.

Amount A does + Amount B does = 1 job done
$$\frac{1}{20}(x+5)+\frac{1}{15}(x) = 1$$
LCD $= 60$; $\frac{60}{1}\cdot\frac{x+5}{20}+\frac{60}{1}\cdot\frac{x}{15}=60\cdot 1$
$$\begin{aligned}3x+15+4x &= 60\\ 7x &= 45\\ x &= \tfrac{45}{7}=6\tfrac{3}{7}\end{aligned}$$
It will take $6\frac{3}{7}$ hr to finish the job.

16. $5\frac{1}{7}$ min

17. Let x = numerator, $x+10$ = denominator; Original fraction $=\frac{x}{x+10}$

If 1 is added to both the numerator and denominator, the value of the new fraction is $\frac{2}{3}$
$$\frac{(x)+1}{(x+10)+1}=\frac{2}{3}$$
LCD $= 3(x+11)$ $\qquad \frac{x+1}{x+11}=\frac{2}{3}$
$$\frac{3(x+11)}{1}\cdot\frac{x+1}{(x+11)}=\frac{3(x+11)}{1}\cdot\frac{2}{3}$$
$$\begin{aligned}3(x+1) &= (x+11)(2)\\ 3x+3 &= 2x+22\\ x &= 19\\ x+10 &= 29\end{aligned}$$
Therefore, the original fraction is $\frac{19}{29}$.

18. $4\frac{8}{13}$ hr

Chapter 6 Diagnostic Test (page 301)

Following each problem number is the textbook section number (in parentheses) where that kind of problem is discussed.

1. (6.1) The values that make the denominator zero must be excluded:
$$\begin{aligned}x^2-4x &= 0\\ x(x-4) &= 0\end{aligned}$$
$$x=0 \quad\Big|\quad x-4=0$$
$$\quad\Big|\quad x=4 \qquad \text{Exclude: 0 and 4}$$
Therefore, the domain is the set of all real numbers except 0 and 4.

2. (6.1) $$\begin{aligned}3y^2-y-10 &= 0\\ (3y+5)(y-2) &= 0\end{aligned}$$
$$3y+5=0 \quad\Big|\quad y-2=0$$
$$y=-\tfrac{5}{3} \quad\Big|\quad y=2 \qquad \text{Exclude: } -\tfrac{5}{3} \text{ and } 2$$
Therefore, the domain is the set of all real numbers except $-\frac{5}{3}$ and 2.

3. (6.1) Changing the signs of the numerator and denominator:
$$\frac{4}{-h}=\frac{-4}{-(-h)}=\frac{-4}{h}$$ The missing term is h.

4. (6.1) Changing the signs of the numerator and denominator:
$$\frac{-3}{k-2}=\frac{-(-3)}{-(k-2)}=\frac{3}{2-k}$$ The missing term is 3.

5. (6.1) No; we can't get the second fraction from the first by multiplying or dividing both numerator and denominator by the same number.

6. (6.2) $$\frac{f^2+5f+6}{f^2-9}=\frac{(f+2)\overset{1}{\cancel{(f+3)}}}{(f-3)\underset{1}{\cancel{(f+3)}}}=\frac{f+2}{f-3}$$

7. (6.2) $$\frac{x^4-2x^3+5x^2-10x}{x^3-8}\quad \begin{matrix}\leftarrow \text{Factor by grouping}\\ \leftarrow \text{Difference of two cubes}\end{matrix}$$
$$=\frac{x^3(x-2)+5x(x-2)}{(x-2)(x^2+2x+4)}$$
$$=\frac{(x-2)(x^3+5x)}{(x-2)(x^2+2x+4)}=\frac{x(x^2+5)}{x^2+2x+4}$$

8. (6.3) $$\frac{z}{\underset{1}{\cancel{(2z+1)}}(z-3)}\cdot\frac{\overset{1}{\cancel{3}}\overset{1}{\cancel{(2z+1)}}(z-2)}{\underset{2}{\cancel{6}}z^2}=\frac{z-2}{2z(z-3)}$$

9. (6.3) $$\frac{\overset{1}{\cancel{(x+4)}}\overset{1}{\cancel{(x-6)}}}{\underset{1}{\cancel{(x-6)}}\underset{1}{\cancel{(x+6)}}}\cdot\frac{\overset{1}{\cancel{(x+1)}}\overset{1}{\cancel{(x+6)}}}{(x-3)\underset{1}{\cancel{(x+4)}}}\cdot\frac{1}{\underset{1}{\cancel{(x+1)}}(x^2-x+1)}$$
$$=\frac{1}{(x-3)(x^2-x+1)}$$

10. (6.3) $$\frac{3\overset{1}{\cancel{(m+n)}}}{(m-n)\underset{1}{\cancel{(m^2+mn+n^2)}}}\cdot\frac{\overset{1}{\cancel{(m^2+mn+n^2)}}}{\underset{1}{\cancel{(m+n)}}(m-n)}=\frac{3}{(m-n)^2}$$

11. (6.5) $$\frac{20a+27b}{12a-20b}+\frac{13b-44a}{12a-20b}=\frac{20a+27b+13b-44a}{12a-20b}$$
$$=\frac{-24a+40b}{12a-20b}=\frac{-\overset{2}{\cancel{8}}\overset{1}{\cancel{(3a-5b)}}}{\underset{1}{\cancel{4}}\underset{1}{\cancel{(3a-5b)}}}=-2$$

12. (6.5) $$\frac{x(x-4)}{(x+4)(x-4)}-\frac{x(x+4)}{(x-4)(x+4)}-\frac{32}{(x-4)(x+4)}$$
$$=\frac{x(x-4)-x(x+4)-32}{(x+4)(x-4)}=\frac{x^2-4x-x^2-4x-32}{(x+4)(x-4)}$$
$$=\frac{-8x-32}{(x+4)(x-4)}=\frac{-8\overset{1}{\cancel{(x+4)}}}{\underset{1}{\cancel{(x+4)}}(x-4)}=\frac{-8}{x-4}\quad\text{or}\quad -\frac{8}{x-4}$$
$$\text{or}\quad \frac{8}{4-x}$$

13. (6.5) $$\frac{3(x+2)}{(x-2)(x+3)(x+2)}-\frac{2(x+3)}{(x-2)(x+2)(x+3)}-\frac{3(x-2)}{(x+2)(x+3)(x-2)}$$
$$=\frac{3(x+2)-2(x+3)-(3)(x-2)}{(x-2)(x+3)(x+2)}$$
$$=\frac{3x+6-2x-6-3x+6}{(x-2)(x+3)(x+2)}=\frac{-2x+6}{(x-2)(x+3)(x+2)}$$

14. (6.6) $$\frac{\frac{8h^4}{5k}}{\frac{4h^2}{15k^3}}=\frac{8h^4}{5k}\div\frac{4h^2}{15k^3}=\frac{\overset{2}{\cancel{8}}h^4}{\underset{1}{\cancel{5}}k}\cdot\frac{\overset{3}{\cancel{15}}k^3}{\underset{1}{\cancel{4}}h^2}=\frac{6h^2k^2}{1}=6h^2k^2$$

15. (6.6) $$\frac{6-\frac{4}{w}}{\frac{3w}{w-2}+\frac{1}{w}}\cdot\frac{w(w-2)}{w(w-2)}=\frac{\frac{6}{1}\cdot\frac{w(w-2)}{1}-\frac{4}{\cancel{w}}\cdot\frac{\cancel{w}(w-2)}{1}}{\frac{3w}{\underset{1}{\cancel{w-2}}}\cdot\frac{w\overset{1}{\cancel{(w-2)}}}{1}+\frac{1}{\underset{1}{\cancel{w}}}\cdot\frac{\overset{1}{\cancel{w}}(w-2)}{1}}$$
$$=\frac{6w(w-2)-4(w-2)}{3w^2+w-2}=\frac{2(w-2)(3w-2)}{(w+1)(3w-2)}=\frac{2(w-2)}{w+1}$$

16. (6.8) $$\frac{\overset{1}{\cancel{(3a+5)}}(a-2)}{1}\cdot\frac{2}{\underset{1}{\cancel{3a+5}}}-\frac{(3a+5)\overset{1}{\cancel{(a-2)}}}{1}\cdot\frac{6}{\underset{1}{\cancel{a-2}}}$$
$$=\frac{(3a+5)(a-2)}{1}\cdot\frac{3}{1}$$
$$(a-2)(2)-(3a+5)(6)=(3a^2-a-10)(3)$$
$$2a-4-18a-30=9a^2-3a-30$$
$$0=9a^2+13a+4$$
$$0=(9a+4)(a+1)$$
$$9a+4=0 \quad\Big|\quad a+1=0$$
$$a=-\tfrac{4}{9} \quad\Big|\quad a=-1$$
$$\{-\tfrac{4}{9},-1\}$$

17. (6.8) The domain is the set of all real numbers except -7.
$$(x+7)\left(\frac{x}{x+7}\right)=(x+7)\left(3-\frac{7}{x+7}\right)$$
$$x=(x+7)(3)-(x+7)\left(\frac{7}{x+7}\right)$$
$$x=3x+21-7$$
$$x=3x+14$$
$$-14=2x$$
$x=-7$ But -7 is not in the domain of the variable. Therefore, there is no solution. The solution set is { }.

18. (6.8) $$\frac{\overset{3}{\cancel{6}}z^2}{1}\cdot\frac{3}{\underset{1}{\cancel{2}}z}+\frac{\overset{1}{\cancel{6z^2}}}{1}\cdot\frac{3}{\underset{1}{\cancel{z^2}}}=\frac{\overset{1}{\cancel{6}}z^2}{1}\left(-\frac{1}{\underset{1}{\cancel{6}}}\right)$$
$$9z+18=-z^2$$
$$z^2+9z+18=0$$
$$(z+3)(z+6)=0$$
$$z+3=0 \quad\Big|\quad z+6=0$$
$$z=-3 \quad\Big|\quad z=-6$$
$$\{-3,-6\}$$

19. (6.9) $$I=\frac{E}{R+r}$$
$$I(R+r)=E$$
$$IR+Ir=E$$
$$Ir=E-IR$$
$$r=\frac{E-IR}{I}$$

20. (6.10) Let x = number of hours Ruben takes to make 8 bushings.

Rachelle's rate $=\dfrac{24 \text{ bushings}}{8 \text{ hr}}=\dfrac{3}{1}$ bushings per hr

Ruben's rate $=\dfrac{8 \text{ bushings}}{x \text{ hr}}=\dfrac{8}{x}$ bushings per hr

$$\frac{3}{1}(4)+\frac{8}{x}(4)=14$$
$$\frac{x}{1}\cdot\frac{12}{1}+\frac{x}{1}\cdot\frac{32}{x}=\frac{x}{1}\cdot\frac{14}{1}$$
$$12x+32=14x$$
$$32=2x$$
$$16=x$$
It takes Ruben 16 hr to make 8 bushings.

Cumulative Review Exercises: Chapters 1–6 (page 302)

1. $(x^{4-(-2)}y^{-3})^{-2}=(x^6y^{-3})^{-2}=x^{6(-2)}y^{-3(-2)}=x^{-12}y^6=\dfrac{y^6}{x^{12}}$

2. $x=-9$ **3.** $9-2m+12\le 12-3m+5$
$$21-2m\le 17-3m$$
$$m\le -4$$
4. Yes

5. Yes **6.** $2x^3 - 13x^2 + 26x - 15$

7.

$$\begin{array}{r} 4a - 3 \quad \text{R } 10 \\ 2a + 3 \overline{) 8a^2 + 6a + 1} \\ 8a^2 + 12a \\ \hline - 6a + 1 \\ - 6a - 9 \\ \hline + 10 \end{array}$$

Answer: $4a - 3 + \dfrac{10}{2a + 3}$

8. $x^6 - 12x^5 + 60x^4 - 160x^3 + 240x^2 - 192x + 64$

9. $$\frac{\overset{1}{\cancel{x+3}}}{\underset{1}{\cancel{(2x-1)}}\underset{1}{\cancel{(x+2)}}} \cdot \frac{\overset{1}{\cancel{(2x-1)}}\overset{1}{\cancel{(x-4)}}}{(3x+1)\underset{1}{\cancel{(x+3)}}} \cdot \frac{\overset{1}{\cancel{x+2}}}{\underset{1}{\cancel{x-4}}} = \frac{1}{3x+1}$$

10. $\dfrac{3x^2 - 5x + 28}{(2x + 1)(x + 7)(x - 3)}$

11. LCD is $(a + 3)(a^2 - 3a + 9)(a - 3)$

$$\frac{a-3}{(a+3)(a^2-3a+9)} - \frac{1}{(a+3)(a-3)}$$
$$= \frac{(a-3)(a-3)}{(a+3)(a^2-3a+9)(a-3)} + \frac{-1(a^2-3a+9)}{(a+3)(a^2-3a+9)(a-3)}$$
$$= \frac{a^2-6a+9-a^2+3a-9}{(a+3)(a^2-3a+9)(a-3)} = \frac{-3a}{(a+3)(a^2-3a+9)(a-3)}$$

12. $(2x + 3y)(5x - 6y)$ **13.** $(x^2 + 3)(x^2 - 4)$
$= (x^2 + 3)(x - 2)(x + 2)$

14. $(x - 2)(x^2 + 2x + 4)$ **15.** $3x(2a + b) - y(2a + b)$
$= (2a + b)(3x - y)$

16. Length = 12 in.; width = 5 in.

17. Let x = number of 20¢ stamps
$50 - x$ = number of 18¢ stamps
$20x + 18(50 - x) = 970$
$20x + 900 - 18x = 970$
$2x = 70$
$x = 35$
$50 - x = 15$ (The check will not be shown.)
Mrs. Kishinami bought thirty-five 20¢ stamps and fifteen 18¢ stamps.

18. 485

19. Let x = the number of dollars for each of four people to invest
$x - 4{,}250$ = the number of dollars for each of five people to invest
$4x = 5(x - 4{,}250)$ The amounts invested are equal
$4x = 5x - 21{,}250$
$21{,}250 = x$ (The check will not be shown.)
The cost of the property is 4($21,250), or $85,000.

20. 5 yd by 9 yd

21. Let x = number of hours Carlos travels (before overtaking Alberto)
$x + 2$ = number of hours Alberto travels
$\left(55\frac{\text{mi}}{\cancel{\text{hr}}}\right)(x\,\cancel{\text{hr}})$ = distance Carlos travels (before overtaking Alberto)
$\left(45\frac{\text{mi}}{\cancel{\text{hr}}}\right)[(x + 2)\,\cancel{\text{hr}}]$ = distance Alberto travels (before being overtaken by Carlos)
$55x = 45(x + 2)$ The distances are equal
$55x = 45x + 90$
$10x = 90$
$x = 9$ Carlos will overtake Alberto in 9 hr. (The check will not be shown.)

Exercises 7.1A (page 306)

1. $5^{1/2}$ **2.** $7^{1/2}$ **3.** $z^{1/3}$ **4.** $x^{1/4}$ **5.** $x^{3/4}$ **6.** $y^{2/5}$
7. $(x^{2/3})^2 = x^{4/3}$ **8.** $x^{3/2}$ **9.** $x^{2n/n} = x^2$ **10.** y^5 **11.** $\sqrt{7}$
12. $\sqrt[3]{5}$ **13.** $\sqrt[5]{a^3}$ **14.** $\sqrt[3]{b^2}$ **15.** $\sqrt[n]{x^m}$ **16.** $x\sqrt[a]{x^b}$
17. $(2^3)^{1/3} = 2^1 = 2$ **18.** 3 **19.** $[(-3)^3]^{2/3} = (-3)^2 = 9$
20. 4 **21.** $4^{3/2} = (\sqrt{4})^3 = 2^3 = 8$ **22.** 27
23. $(\sqrt[4]{-16})^3$ The index is even and the radicand is negative; the number is not real.
24. Not a real number

Exercises 7.1B (page 309)

1. $x^{1/2+3/2} = x^{4/2} = x^2$ **2.** y^2 **3.** $a^{3/4+(-1/2)} = a^{3/4-2/4} = a^{1/4}$
4. $b^{1/2}$ **5.** $z^{-3/6}z^{4/6} = z^{1/6}$ **6.** $N^{5/12}$ **7.** $H^{(3/4)(2)} = H^{3/2}$
8. $s^{5/2}$ **9.** $x^{(-3/4)(1/3)} = x^{-1/4} = \dfrac{1}{x^{1/4}}$ **10.** $\dfrac{1}{y^{1/3}}$
11. $a^{3/4-1/2} = a^{3/4-2/4} = a^{1/4}$ **12.** $b^{1/2}$
13. $x^{1/2-(-1/3)} = x^{3/6+2/6} = x^{5/6}$ **14.** $z^{7/12}$
15. $x^{4/6}x^{6/6}x^{-3/6} = x^{4/6+6/6-3/6} = x^{7/6}$ **16.** $x^{17/12}$
17. $x^{-3/2}x^{4/3} = x^{-9/6+8/6} = x^{-1/6} = \dfrac{1}{x^{1/6}}$ **18.** $\dfrac{1}{x^{1/2}}$
19. $u^{1/2-(-1/4)}v^{-2/3-(-1)} = u^{2/4+1/4}v^{-2/3+1} = u^{3/4}v^{1/3}$
20. $uv^{2/5}$ **21.** $(2^4)^{3/2}x^{(-2/5)(3/2)}y^{(4/9)(3/2)} = 2^6x^{-3/5}y^{2/3} = \dfrac{64y^{2/3}}{x^{3/5}}$
22. $\dfrac{4x^{3/4}}{y^{1/2}}$ **23.** $\dfrac{a^{6(7/3)}d^{0(7/3)}}{b^{-9(7/3)}c^{3(7/3)}} = \dfrac{a^{14}d^0}{b^{-21}c^7} = \dfrac{a^{14}b^{21}}{c^7}$ **24.** $\dfrac{y^{14}w^{21}}{z^7}$
25. $\left(\dfrac{x^4}{2^5z^3}\right)^{2/5} = \dfrac{x^{8/5}}{2^2z^{6/5}} = \dfrac{x^{8/5}}{4z^{6/5}}$ **26.** $\dfrac{a^{10/3}}{4b^{10/3}}$
27. $\dfrac{x^{(-1)(-3/5)}y^{(2/3)(-3/5)}}{z^{(-5)(-3/5)}} = \dfrac{x^{3/5}y^{-2/5}}{z^3} = \dfrac{x^{3/5}}{y^{2/5}z^3}$ **28.** $\dfrac{b^{2/3}}{a^{1/2}c^{4/3}}$
29. $(3^2x^{-2/3-(-2)}y^{2/9})^{-3/2} = 3^{2(-3/2)}x^{(4/3)(-3/2)}y^{(2/9)(-3/2)}$
$= 3^{-3}x^{-2}y^{-1/3} = \dfrac{1}{27x^2y^{1/3}}$
30. $\dfrac{1}{9SR^{2/5}}$ **31.** $(2^3)^{-2/3} = 2^{3(-2/3)} = 2^{-2} = \dfrac{1}{2^2} = \frac{1}{4}$ **32.** $\frac{1}{9}$
33. $(2^2)^{1/2}(3^2)^{-3/2} = 2(3^{-3}) = \frac{2}{27}$ **34.** $\frac{1}{32}$
35. $(10^2)^{-1/2}(-2^{3(1/3)}) = 10^{-1}(-2) = -\frac{2}{10} = -\frac{1}{5}$ **36.** $-\frac{3}{100}$
37. $3x^{2/3}y^{1/6}(y - 3x)$ **38.** $4a^{3/4}b^{3/8}(3a - 2b)$
39. $3x^{1/2}y^{1/3}(6x^2 + 7y)$ **40.** $6x^{3/5}y^{1/4}(6y + 5x)$
41. $(5v^{-1})^2 - 17(v^{-1}) + 6 = (5v^{-1} - 2)(v^{-1} - 3)$
42. $(5v^{-1} - 6)(v^{-1} - 1)$ **43.** $(5k^{-2} - 1)(7k^{-2} - 1)$
44. $(3k^{-2} - 1)(6k^{-2} - 1)$ **45.** $a^{-3/4}b^{-8/3}(2a - b^3)$
46. $x^{-2/5}y^{-1/3}(3y - 5x)$ **47.** $4c^{-3/4}d^{-2/3}e^{-4/5}(2e + 3cd)$
48. $5x^{-1/2}y^{-1/3}(3x + 2y)$

Exercises 7.2A (page 314)

1. $2|x|$ **2.** $3|y|$ **3.** $2x$ **4.** $3y$
5. $\sqrt[4]{2^4x^4y^8} = 2\cdot|x|\cdot y^2 = 2|x|y^2$ **6.** $3u^2|v|$
7. $\sqrt{2^4\cdot 2} = 2^2\sqrt{2} = 4\sqrt{2}$ **8.** $3\sqrt{3}$ **9.** $\sqrt{(-2)^2} = |-2| = 2$
10. $\sqrt{(-3)^2} = |-3| = 3$ **11.** $\sqrt[3]{-3^3\cdot 3^2} = -3\sqrt[3]{3^2} = -3\sqrt[3]{9}$

12. $-4\sqrt[3]{4}$ **13.** $\sqrt[4]{2^5} = \sqrt[4]{2^4 \cdot 2} = 2\sqrt[4]{2}$ **14.** $2\sqrt[4]{3}$

15. $\sqrt[5]{-x^5x^2} = -x\sqrt[5]{x^2}$ **16.** $-z\sqrt[5]{z^3}$

17. $\sqrt{2^3a^4b^2} = \sqrt{2^2 \cdot 2a^4b^2} = 2a^2|b|\sqrt{2}$ **18.** $2m^4|u|\sqrt{5}$

19. $\sqrt{3^2 \cdot 2m^2mn^4n} = 3|m|n^2\sqrt{2mn}$ **20.** $5h^2|k|\sqrt{2hk}$

21. $5\sqrt[3]{-2^3 \cdot 3a^5b^2} = 5\sqrt[3]{-2^3 \cdot 3a^3a^2b^2} = 5(-2a)\sqrt[3]{3a^2b^2}$
$= -10a\sqrt[3]{3a^2b^2}$

22. $-18c\sqrt[3]{2cd}$ **23.** $\sqrt[5]{2^5 \cdot 2m^{10}mp^{15}u} = 2m^2p^3\sqrt[5]{2mu}$

24. $2v^2w^3\sqrt[5]{4u^4w}$ **25.** $\sqrt[3]{2^3(a+b)^3} = 2(a+b)$ **26.** $3(x-y)^2$

27. $\dfrac{3}{2abc}\sqrt[3]{2^3 \cdot 2^2a^6a^2b^9c^9c} = \dfrac{3}{2abc}(2a^2b^3c^3)\sqrt[3]{2^2a^2c} = 3ab^2c^2\sqrt[3]{4a^2c}$

28. $14abc\sqrt[3]{a^2c}$

Exercises 7.2B (page 318)

1. $\dfrac{\sqrt{16}}{\sqrt{25}} = \dfrac{4}{5}$ **2.** $\dfrac{4}{5}$ **3.** $\dfrac{\sqrt[3]{-27}}{\sqrt[3]{64}} = -\dfrac{3}{4}$ **4.** $-\dfrac{2}{5}$

5. $\dfrac{\sqrt[4]{a^4b^8}}{\sqrt[4]{16}} = \dfrac{|a|b^2}{2}$ **6.** $\dfrac{c^2|d^3|}{3}$ **7.** $\sqrt{\dfrac{4x^2}{y^2}} = \dfrac{\sqrt{4x^2}}{\sqrt{y^2}} = \left|\dfrac{2x}{y}\right|$

8. $\dfrac{x^2}{3|y|}$ **9.** $\dfrac{10}{\sqrt{5}} \cdot \dfrac{\sqrt{5}}{\sqrt{5}} = \dfrac{10\sqrt{5}}{\sqrt{5}\sqrt{5}} = \dfrac{10\sqrt{5}}{5} = 2\sqrt{5}$ **10.** $7\sqrt{2}$

11. $\dfrac{5}{\sqrt{3}} \cdot \dfrac{\sqrt{3}}{\sqrt{3}} = \dfrac{5\sqrt{3}}{\sqrt{3}\sqrt{3}} = \dfrac{5\sqrt{3}}{3}$ **12.** $\dfrac{8\sqrt{7}}{7}$

13. $\dfrac{9}{\sqrt[3]{3}} \cdot \dfrac{\sqrt[3]{3^2}}{\sqrt[3]{3^2}} = \dfrac{9\sqrt[3]{3^2}}{\sqrt[3]{3^3}} = \dfrac{9\sqrt[3]{9}}{3} = 3\sqrt[3]{9}$ **14.** $2\sqrt[3]{25}$

15. $\dfrac{8}{\sqrt[5]{4}} \cdot \dfrac{\sqrt[5]{2^3}}{\sqrt[5]{2^3}} = \dfrac{8\sqrt[5]{2^3}}{\sqrt[5]{2^5}} = \dfrac{\overset{4}{\cancel{8}}\sqrt[5]{8}}{\underset{1}{\cancel{2}}} = 4\sqrt[5]{8}$ **16.** $\dfrac{5\sqrt[4]{27}}{3}$

17. $\sqrt[3]{\dfrac{m^3m^2}{-3} \cdot \dfrac{3^2}{3^2}} = \sqrt[3]{\dfrac{m^3m^29}{-3^3}} = \dfrac{m\sqrt[3]{9m^2}}{\sqrt[3]{-3^3}} = -\dfrac{m\sqrt[3]{9m^2}}{3}$

18. $-\dfrac{k\sqrt[3]{25k}}{5}$ **19.** $\dfrac{n}{2m}\sqrt[3]{\dfrac{4m^2n}{n^3}} = \dfrac{n}{2m} \cdot \dfrac{1}{n}\sqrt[3]{4m^2n} = \dfrac{1}{2m}\sqrt[3]{4m^2n}$

20. $\dfrac{1}{3y}\sqrt[3]{9xy^2}$ **21.** $\sqrt[4]{\dfrac{3m^4}{2^2p^2} \cdot \dfrac{2^2p^2}{2^2p^2}} = \dfrac{\sqrt[4]{12m^4p^2}}{\sqrt[4]{2^4p^4}} = \dfrac{|m|\sqrt[4]{12p^2}}{|2p|}$

22. $\dfrac{|a|\sqrt[4]{40b}}{|2b|}$ **23.** $\sqrt[5]{\dfrac{5y^5}{2^3x^2} \cdot \dfrac{2^2x^3}{2^2x^3}} = \dfrac{\sqrt[5]{20x^3y^5}}{\sqrt[5]{2^5x^5}} = \dfrac{y\sqrt[5]{20x^3}}{2x}$

24. $\dfrac{p\sqrt[5]{20m^2}}{2m}$ **25.** $\dfrac{4x^2}{5y^3}\sqrt[3]{\dfrac{3y^2}{8x^3}} = \dfrac{4x^2}{5y^3} \cdot \dfrac{1}{2x}\sqrt[3]{3y^2} = \dfrac{2x}{5y^3}\sqrt[3]{3y^2}$

26. $\dfrac{1}{4y}\sqrt[3]{18xy}$

27. $\dfrac{2x^3}{y} \cdot \sqrt[3]{\dfrac{5}{2^4x^6xz^6z}} = \dfrac{\overset{1}{\cancel{2}}\overset{x}{\cancel{x^3}}}{y} \cdot \dfrac{\sqrt[3]{5}}{\underset{1}{\cancel{2}}\underset{1}{\cancel{x^2}}z^2\sqrt[3]{2xz}}$

$= \dfrac{x\sqrt[3]{5}}{yz^2\sqrt[3]{2xz}} \cdot \dfrac{\sqrt[3]{2^2x^2z^2}}{\sqrt[3]{2^2x^2z^2}} = \dfrac{\overset{1}{\cancel{x}}\sqrt[3]{20x^2z^2}}{yz^2 \cdot 2\underset{1}{\cancel{x}}z} = \dfrac{\sqrt[3]{20x^2z^2}}{2yz^3}$

28. $\dfrac{\sqrt[3]{5bc^2d^2}}{c}$

Exercises 7.2C (page 319)

1. $x^{3/6} = x^{1/2} = \sqrt{x}$ **2.** $\sqrt[3]{x}$ **3.** $a^{6/8} = a^{3/4} = \sqrt[4]{a^3}$ **4.** $\sqrt[4]{a}$

5. $(3^3b^3)^{1/6} = 3^{3/6}b^{3/6} = 3^{1/2}b^{1/2} = (3b)^{1/2} = \sqrt{3b}$ **6.** $\sqrt[3]{2b^2}$

7. $(7^2a^2)^{1/6} = 7^{2/6}a^{2/6} = (7a)^{1/3} = \sqrt[3]{7a}$ **8.** $2\sqrt{3x}$

9. $(3^4x^4z^{12})^{1/8} = 3^{1/2}x^{1/2}z^{3/2} = \sqrt{3xz^3} = z\sqrt{3xz}$ **10.** $x\sqrt{3xy}$

11. $(2^8x^8y^4z^{10})^{1/6} = 2^{8/6}x^{8/6}y^{4/6}z^{10/6} = 2^{4/3}x^{4/3}y^{2/3}z^{5/3}$
$= 2^{3/3} \cdot 2^{1/3} \cdot x^{3/3} \cdot x^{1/3} \cdot y^{2/3} \cdot z^{3/3}z^{2/3}$
$= 2xz(2xy^2z^2)^{1/3} = 2xz\sqrt[3]{2xy^2z^2}$

12. $2x^2z\sqrt{2yz}$ **13.** $\left(\dfrac{x^3}{3^3}\right)^{1/6} = \dfrac{x^{1/2}}{3^{1/2}} = \sqrt{\dfrac{x}{3}} = \sqrt{\dfrac{x(3)}{3(3)}} = \dfrac{\sqrt{3x}}{3}$

14. $\dfrac{\sqrt[3]{4x}}{2}$ **15.** $\dfrac{1}{a^{3/6}} = \dfrac{1}{a^{1/2}} = \dfrac{1}{\sqrt{a}} = \dfrac{1\sqrt{a}}{\sqrt{a}\sqrt{a}} = \dfrac{\sqrt{a}}{a}$ **16.** $\dfrac{\sqrt{x}}{x}$

Exercises 7.3 (page 322)

1. $11\sqrt{2}$ **2.** $18\sqrt{5}$ **3.** $4\sqrt{6}$ **4.** $6\sqrt{7}$ **5.** $\sqrt{15} + \sqrt{10}$

6. $\sqrt{2} + \sqrt{14}$ **7.** $2\sqrt{5}$ **8.** $3\sqrt{3}$ **9.** $3 + 2\sqrt{3}$

10. $5 + 3\sqrt{5}$ **11.** $2\sqrt{3} - 2\sqrt{2}$ **12.** $2\sqrt{7} - \sqrt{3}$

13. $3\sqrt{2} + \sqrt{2} = 4\sqrt{2}$ **14.** $2\sqrt{3}$ **15.** $7\sqrt[3]{xy}$ **16.** $10\sqrt[4]{ab}$

17. $2\sqrt{25 \cdot 2} - \sqrt{16 \cdot 2} = 2 \cdot 5\sqrt{2} - 4\sqrt{2} = 10\sqrt{2} - 4\sqrt{2} = 6\sqrt{2}$

18. $3\sqrt{6}$ **19.** $3\sqrt{16 \cdot 2x} - \sqrt{4 \cdot 2x} = 12\sqrt{2x} - 2\sqrt{2x} = 10\sqrt{2x}$

20. $6\sqrt{3y}$

21. $\sqrt{25 \cdot 5M} + \sqrt{4 \cdot 5M} - \sqrt{9 \cdot 5M} = 5\sqrt{5M} + 2\sqrt{5M} - 3\sqrt{5M}$
$= 4\sqrt{5M}$

22. $4\sqrt{3P}$ **23.** $3\sqrt[3]{x} + \dfrac{2}{2}\sqrt[3]{x} = 3\sqrt[3]{x} + \sqrt[3]{x} = 4\sqrt[3]{x}$ **24.** $6\sqrt[3]{a}$

25. $\sqrt[3]{a^3 \cdot a} + 2a\sqrt[3]{8 \cdot a} = a\sqrt[3]{a} + 4a\sqrt[3]{a} = 5a\sqrt[3]{a}$ **26.** $5H\sqrt[5]{H^2}$

27. $\sqrt[5]{x^2y^5y} + \sqrt[5]{x^5x^2y} = y\sqrt[5]{x^2y} + x\sqrt[5]{x^2y} = (y+x)\sqrt[5]{x^2y}$

28. $(b+a)\sqrt[5]{a^3b^3}$

29. $\frac{3}{1}\sqrt{\frac{1}{6} \cdot \frac{6}{6}} + \sqrt{4 \cdot 3} - \frac{5}{1}\sqrt{\frac{3}{2} \cdot \frac{2}{2}} = \frac{3}{6}\sqrt{6} + 2\sqrt{3} - \frac{5}{2}\sqrt{6}$
$= (\frac{1}{2} - \frac{5}{2})\sqrt{6} + 2\sqrt{3} = -2\sqrt{6} + 2\sqrt{3}$

30. $\sqrt{10} + 2\sqrt{5}$

31. $\dfrac{10}{2}\sqrt{5b} - \dfrac{3b}{2}\sqrt{\dfrac{4}{5b} \cdot \dfrac{5b}{5b}} = 5\sqrt{5b} - \dfrac{3b}{2} \cdot \dfrac{2}{5b}\sqrt{5b}$
$= 5\sqrt{5b} - \frac{3}{5}\sqrt{5b} = (5 - \frac{3}{5})\sqrt{5b} = (\frac{25}{5} - \frac{3}{5})\sqrt{5b} = \frac{22}{5}\sqrt{5b}$

32. $\dfrac{7x}{2}\sqrt[3]{4}$

33. $\dfrac{2k}{1}\sqrt[4]{\dfrac{3}{2^3k} \cdot \dfrac{2k^3}{2k^3}} - \dfrac{1}{k}\sqrt[4]{\dfrac{2k^3}{3^3} \cdot \dfrac{3}{3}} + \dfrac{5k^2}{1}\sqrt[4]{\dfrac{6}{k^2} \cdot \dfrac{k^2}{k^2}}$
$= \dfrac{2k}{1} \cdot \dfrac{1}{2k}\sqrt[4]{6k^3} - \dfrac{1}{k} \cdot \dfrac{1}{3}\sqrt[4]{6k^3} + \dfrac{5k^2}{1} \cdot \dfrac{1}{k}\sqrt[4]{6k^2}$
$- \sqrt[4]{6k^3} - \dfrac{1}{3k}\sqrt[4]{6k^3} + 5k\sqrt[4]{6k^2} = \left(1 - \dfrac{1}{3k}\right)\sqrt[4]{6k^3} + 5k\sqrt[4]{6k^2}$
$= \left(\dfrac{3k}{3k} - \dfrac{1}{3k}\right)\sqrt[4]{6k^3} + 5k\sqrt[4]{6k^2} = \dfrac{3k-1}{3k}\sqrt[4]{6k^3} + 5k\sqrt[4]{6k^2}$

34. $2a\sqrt[3]{4a}$ **35.** $\sqrt{(2x+1)^2} = 2x + 1$ **36.** $4a + 1$

Exercises 7.4A (page 325)

1. $\sqrt{3 \cdot 3} = 3$ **2.** 7 **3.** $\sqrt[3]{3 \cdot 9} = \sqrt[3]{3^3} = 3$ **4.** 4

5. $\sqrt[4]{9 \cdot 9} = \sqrt[4]{3^4} = 3$ **6.** 5

7. $\sqrt{100a^2b^3} = \sqrt{10^2a^2b^2b} = 10ab\sqrt{b}$ **8.** $9xy\sqrt{x}$

9. $3\cdot 2\sqrt[5]{2a^3b\cdot 2^4a^2b} = 6\sqrt[5]{2^5a^5b^2} = 6\cdot 2a\sqrt[5]{b^2} = 12a\sqrt[5]{b^2}$

10. $40b\sqrt[5]{c^3}$ **11.** $5\sqrt{7}(5\sqrt{7}) = 25\sqrt{7\cdot 7} = 25\cdot 7 = 175$

12. 96 **13.** $\sqrt{2}(\sqrt{2})+\sqrt{2}(1) = 2+\sqrt{2}$ **14.** $3+\sqrt{3}$

15. $\sqrt{x}(\sqrt{x}) - \sqrt{x}(3) = x - 3\sqrt{x}$ **16.** $4\sqrt{y}-y$

17. $\sqrt{3}(2\sqrt{3})+\sqrt{3}(1) = 2\cdot 3+\sqrt{3} = 6+\sqrt{3}$ **18.** $15+\sqrt{5}$

19. $\sqrt{3x}(\sqrt{3x}) - \sqrt{3x}(4\sqrt{12}) = \sqrt{3x\cdot 3x} - 4\sqrt{6^2x}$
$= 3x - 4\cdot 6\sqrt{x} = 3x - 24\sqrt{x}$

20. $5\sqrt{2a}+15a$ **21.** $7+3\sqrt{7}+2\sqrt{7}+6 = 13+5\sqrt{7}$

22. $11+6\sqrt{3}$ **23.** $5^2 - (\sqrt{3})^2 = 25-3 = 22$ **24.** 2

25. $(2\sqrt{3})^2 + 2(2\sqrt{3})(-5) + (-5)^2 = 12 - 20\sqrt{3}+25$
$= 37-20\sqrt{3}$

26. $59-30\sqrt{2}$

27. $2\cdot 5\cdot 2\sqrt{3\cdot 7^2x^6xy^4y} = 20\cdot 7x^3y^2\sqrt{3xy} = 140x^3y^2\sqrt{3xy}$

28. $300x^4y^4\sqrt{2xy}$ **29.** $3^2(\sqrt{2x+5})^2 = 9(2x+5) = 18x+45$

30. $|48x-32|$ **31.** $(\sqrt{2x})^2-(3)^2 = 2x-9$ **32.** $5x-49$

33. $(\sqrt{xy})^2 - 2(\sqrt{xy})(6\sqrt{y}) + (6\sqrt{y})^2 = xy - 12y\sqrt{x}+36y$

34. $ab+4a\sqrt{b}+4a$
(We use the binomial formula in 35–40.)

35. $(\sqrt[5]{x})^5 + 5(\sqrt[5]{x})^4 + 10(\sqrt[5]{x})^3 + 10(\sqrt[5]{x})^2 + 5\sqrt[5]{x}+1$
$= x + 5\sqrt[5]{x^4}+10\sqrt[5]{x^3}+10\sqrt[5]{x^2}+5\sqrt[5]{x}+1$

36. $1+3\sqrt[3]{y}+3\sqrt[3]{y^2}+y$

37. $(x^{1/4})^4 + 4(x^{1/4})^3+6(x^{1/4})^2+4(x^{1/4})^1+1$
$= x+4x^{3/4}+6x^{1/2}+4x^{1/4}+1$

38. $x+2x^{1/2}y^{1/2}+y$ or $x+2\sqrt{xy}+y$

39. $[1-4\sqrt{2}+6(\sqrt{2})^2-4(\sqrt{2})^3+(\sqrt{2})^4]-1$
$= 1-4\sqrt{2}+6(2)-4(2)\sqrt{2}+4-1$
$= -4\sqrt{2}+12-8\sqrt{2}+4 = 16-12\sqrt{2}$

40. $8+5\sqrt{2}$

Exercises 7.4B (page 328)

1. $\sqrt{\frac{32}{2}} = \sqrt{16} = 4$ **2.** 7 **3.** $\sqrt[3]{\frac{5}{2^2}\cdot\frac{2}{2}} = \sqrt[3]{\frac{10}{2^3}} = \frac{\sqrt[3]{10}}{2}$ or $\frac{1}{2}\sqrt[3]{10}$

4. $\frac{\sqrt[3]{28}}{2}$ or $\frac{1}{2}\sqrt[3]{28}$ **5.** $\frac{3}{1}\sqrt[4]{\frac{15x}{5x}} = 3\sqrt[4]{3}$ **6.** $3\sqrt[4]{6}$

7. $\sqrt[5]{\frac{128z^7}{2z}} = \sqrt[5]{64z^6} = \sqrt[5]{2^6z^6} = 2z\sqrt[5]{2z}$ **8.** $3b\sqrt[5]{3b}$

9. $\sqrt{\frac{72x^3y^2}{2xy^2}} = \sqrt{36x^2} = 6x$ **10.** $3y$

11. $\frac{\sqrt{20}}{\sqrt{5}}+\frac{5\sqrt{10}}{\sqrt{5}} = \sqrt{\frac{20}{5}}+5\sqrt{\frac{10}{5}} = \sqrt{4}+5\sqrt{2} = 2+5\sqrt{2}$

12. $2+\sqrt{3}$ **13.** $3\sqrt[4]{\frac{2^5m^2}{2m^3}} = 3\sqrt[4]{\frac{2^4}{m}\cdot\frac{m^3}{m^3}} = 3\sqrt[4]{\frac{2^4m^3}{m^4}} = \frac{6\sqrt[4]{m^3}}{m}$

14. $\frac{3\sqrt[4]{H^3}}{2H}$ **15.** $\frac{4\sqrt[3]{8x}}{2\sqrt[3]{4x}}+\frac{6\sqrt[3]{32x^4}}{2\sqrt[3]{4x}} = 2\sqrt[3]{\frac{8x}{4x}}+3\sqrt[3]{\frac{32x^4}{4x}}$
$= 2\sqrt[3]{2}+3\sqrt[3]{8x^3}$
$= 2\sqrt[3]{2}+3\cdot 2x = 2\sqrt[3]{2}+6x$

16. $6a^2+3\sqrt[3]{2}$

17. $\frac{6}{\sqrt{3}-1}\cdot\frac{\sqrt{3}+1}{\sqrt{3}+1} = \frac{6(\sqrt{3}+1)}{(\sqrt{3})^2-1^2} = \frac{6(\sqrt{3}+1)}{3-1} = \frac{6(\sqrt{3}+1)}{2}$
$= 3(\sqrt{3}+1) = 3\sqrt{3}+3$

18. $5\sqrt{3}-5$

19. $\frac{\sqrt{2}}{\sqrt{3}+\sqrt{2}}\cdot\frac{\sqrt{3}-\sqrt{2}}{\sqrt{3}-\sqrt{2}} = \frac{\sqrt{2}(\sqrt{3}-\sqrt{2})}{(\sqrt{3})^2-(\sqrt{2})^2} = \frac{\sqrt{6}-2}{3-2} = \sqrt{6}-2$

20. $\frac{7+\sqrt{14}}{5}$

21. $\frac{\sqrt{7}+\sqrt{3}}{\sqrt{7}-\sqrt{3}}\cdot\frac{\sqrt{7}+\sqrt{3}}{\sqrt{7}+\sqrt{3}} = \frac{(\sqrt{7})^2+2\sqrt{7}\sqrt{3}+(\sqrt{3})^2}{(\sqrt{7})^2-(\sqrt{3})^2}$
$= \frac{7+2\sqrt{21}+3}{7-3} = \frac{10+2\sqrt{21}}{4} = \frac{2(5+\sqrt{21})}{4} = \frac{5+\sqrt{21}}{2}$

22. $\frac{8-\sqrt{55}}{3}$

23. $\frac{4\sqrt{3}-\sqrt{2}}{4\sqrt{3}+\sqrt{2}}\cdot\frac{4\sqrt{3}-\sqrt{2}}{4\sqrt{3}-\sqrt{2}} = \frac{(4\sqrt{3}-\sqrt{2})^2}{(4\sqrt{3})^2-(\sqrt{2})^2} = \frac{48-8\sqrt{6}+2}{48-2}$
$= \frac{50-8\sqrt{6}}{46} = \frac{2(25-4\sqrt{6})}{46} = \frac{25-4\sqrt{6}}{23}$

24. $2x-2\sqrt{x^2+x}+1$

25. $\sqrt{\frac{(a+3)(a-1)}{(a+3)(a+1)}} = \sqrt{\frac{a-1}{a+1}\cdot\frac{a+1}{a+1}} = \frac{\sqrt{a^2-1}}{\sqrt{(a+1)^2}} = \frac{\sqrt{a^2-1}}{a+1}$

26. $\frac{\sqrt{m^2-1}}{m-1}$

Exercises 7.4C (page 330)

1. $a^{1/2}a^{1/4} = a^{1/2+1/4} = a^{3/4} = \sqrt[4]{a^3}$ **2.** $\sqrt[6]{b^5}$

3. $(2^3)^{1/2}(2^4)^{1/3} = 2^{3/2}2^{4/3} = 2^{3/2+4/3} = 2^{17/6}$
$= 2^{2+5/6} = 2^2\cdot 2^{5/6} = 4\sqrt[6]{32}$ **4.** $9\sqrt[6]{243}$

5. $x^{2/3}x^{3/4}x^{1/2} = x^{8/12+9/12+6/12} = x^{23/12}$
$= x^{1+11/12} = x^1x^{11/12} = x\sqrt[12]{x^{11}}$ **6.** $y\sqrt[12]{y^7}$

7. $(-2^3z^2)^{1/3}(-z)^{1/3}(2^4z^3)^{1/4} = 2^1z^{2/3}z^{1/3}2^1z^{3/4}$
$= 2^2z^{1+3/4} = 4z\sqrt[4]{z^3}$ **8.** $6w\sqrt[4]{w^3}$

9. $\frac{G^{3/4}}{G^{2/3}} = G^{3/4-2/3} = G^{9/12-8/12} = G^{1/12} = \sqrt[12]{G}$ **10.** $\sqrt[10]{H^3}$

11. $\frac{-x^{2/3}}{x^{5/6}} = -x^{2/3-5/6} = -x^{4/6-5/6} = -x^{-1/6}$
$= -\frac{1}{x^{1/6}} = -\frac{1}{x^{1/6}}\cdot\frac{x^{5/6}}{x^{5/6}} = -\frac{x^{5/6}}{x^{6/6}} = -\frac{\sqrt[6]{x^5}}{x}$ **12.** $-\frac{\sqrt[6]{y^5}}{y}$

Review Exercises 7.5 (page 332)

1. $\sqrt[4]{a^3}$ **2.** $\sqrt[4]{3^3y^3}$ or $\sqrt[4]{27y^3}$ **3.** $\sqrt[5]{2^2x^4}$ or $\sqrt[5]{4x^4}$

4. $b^{3/4}$ **5.** $(8x^4)^{1/5}$ or $8^{1/5}x^{4/5}$ **6.** $(27x^3)^{1/5}$ or $27^{1/5}x^{3/5}$

7. $(\sqrt[3]{-64})^2 = (-4)^2 = 16$ **8.** 9

9. $P^{(2/3)(2/3)}R^{(3/4)(2/3)} = P^{4/9}R^{1/2}$ **10.** $\frac{1}{27a^{1/2}b^{5/2}}$

11. $\sqrt[3]{2^3\cdot 4} = 2\sqrt[3]{4}$ **12.** $-5x$ **13.** $(2^4y^8)^{1/4} = 2y^2$

14. $m\sqrt[5]{mp}$ **15.** $\sqrt{\frac{4x^2x}{y}\cdot\frac{y}{y}} = \frac{2x\sqrt{xy}}{y}$ **16.** $\frac{5\sqrt{3y}}{y}$

17. $\sqrt[3]{8x^3x^2} = 2x\sqrt[3]{x^2}$ **18.** $75x$

19. $\dfrac{\sqrt{3}+\sqrt{7}}{\sqrt{3}-\sqrt{7}}\cdot\dfrac{\sqrt{3}+\sqrt{7}}{\sqrt{3}+\sqrt{7}}=\dfrac{3+\sqrt{21}+\sqrt{21}+7}{(\sqrt{3})^2-(\sqrt{7})^2}=\dfrac{10+2\sqrt{21}}{-4}=-\dfrac{5+\sqrt{21}}{2}$

20. $2\sqrt{5}-2$

21. $\sqrt{2}(4\sqrt{2})+11\sqrt{2}-3=8+11\sqrt{2}-3=5+11\sqrt{2}$

22. $16+2\sqrt{39}$

23. $\sqrt[3]{8\cdot 2\cdot x^3x^2}+x\sqrt[3]{27\cdot 2x^2}=2x\sqrt[3]{2x^2}+x\cdot 3\sqrt[3]{2x^2}=5x\sqrt[3]{2x^2}$

24. 9 **25.** $-z^{2/3}\cdot z^{1/2}=-z^{4/6+3/6}=-z^{7/6}=-z^{1+1/6}=-z\sqrt[6]{z}$

26. $2-4x\sqrt[4]{8}+6x^2\sqrt{2}-4x^3\sqrt[4]{2}+x^4$

27. $\dfrac{G^{2/4}}{G^{1/5}}=G^{1/2-1/5}=G^{5/10-2/10}=G^{3/10}=\sqrt[10]{G^3}$

28. $\dfrac{\sqrt[4]{a^2b}}{a}$

29. $7x^{-2/3}y^{-3/4}z^{-2/3}(2yz-x)$ **30.** $(4x^{-1}+3)(x^{-1}-1)$

Exercises 7.6 (page 340)

1. $\sqrt{3x+1}=5$
$(\sqrt{3x+1})^2=5^2$
$3x+1=25$
$3x=24$
$x=8$

Check $\sqrt{3x+1}=5$
$\sqrt{3(8)+1}\stackrel{?}{=}5$
$\sqrt{25}\stackrel{?}{=}5$
$5=5$

$\{8\}$

2. $\{4\}$

3. $\sqrt{x+1}=\sqrt{2x-7}$
$(\sqrt{x+1})^2=(\sqrt{2x-7})^2$
$x+1=2x-7$
$8=x$
$x=8$

Check $\sqrt{x+1}=\sqrt{2x-7}$
$\sqrt{8+1}\stackrel{?}{=}\sqrt{16-7}$
$\sqrt{9}=\sqrt{9}$

$\{8\}$

4. $\{3\}$ **5.** $\sqrt[4]{4x-11}=1$
$(\sqrt[4]{4x-11})^4=(1)^4$
$4x-11=1$
$4x=12$
$x=3$

Check $\sqrt[4]{4x-11}-1=0$
$\sqrt[4]{4(3)-11}-1\stackrel{?}{=}0$
$\sqrt[4]{1}-1\stackrel{?}{=}0$
$1-1\stackrel{?}{=}0$
$0=0$

$\{3\}$

6. $\{5\}$

7. $\sqrt{4x-1}=2x$
$(\sqrt{4x-1})^2=(2x)^2$
$4x-1=4x^2$
$0=4x^2-4x+1$
$0=(2x-1)(2x-1)$
$2x-1=0$
$x=\dfrac{1}{2}$

Check $\sqrt{4x-1}=2x$
$\sqrt{4\left(\dfrac{1}{2}\right)-1}\stackrel{?}{=}2\left(\dfrac{1}{2}\right)$
$\sqrt{2-1}\stackrel{?}{=}1$
$1=1$

$\left\{\dfrac{1}{2}\right\}$

8. $\left\{\dfrac{1}{3}\right\}$

9. $\sqrt{x+7}=2x-1$
$(\sqrt{x+7})^2=(2x-1)^2$
$x+7=4x^2-4x+1$
$0=4x^2-5x-6$
$0=(x-2)(4x+3)$
$x-2=0 \quad | \quad 4x+3=0$
$x=2 \quad | \quad x=-\frac{3}{4}$

Check for $x=2$
$\sqrt{x+7}=2x-1$
$\sqrt{2+7}\stackrel{?}{=}4-1$
$3=3$

Check for $x=-\frac{3}{4}$
$\sqrt{x+7}=2x-1$
$\sqrt{-\frac{3}{4}+7}\stackrel{?}{=}2(-\frac{3}{4})-1$
$\sqrt{\frac{25}{4}}\stackrel{?}{=}-\frac{3}{2}-1$
$\frac{5}{2}\neq-\frac{5}{2}$

Therefore, the solution set is $\{2\}$.

10. No solution

11. $\sqrt{3x+4}=\sqrt{2x-4}+2$
$(\sqrt{3x+4})^2=(\sqrt{2x-4}+2)^2$
$3x+4=2x-4+4\sqrt{2x-4}+4$
$x+4=4\sqrt{2x-4}$
$(x+4)^2=(4\sqrt{2x-4})^2$
$x^2+8x+16=32x-64$
$x^2-24x+80=0$
$(x-4)(x-20)=0$
$x-4=0 \quad | \quad x-20=0$
$x=4 \quad | \quad x=20$

Check for $x=4$
$\sqrt{3(4)+4}-\sqrt{2(4)-4}\stackrel{?}{=}2$
$\sqrt{16}-\sqrt{4}\stackrel{?}{=}2$
$4-2=2$

Check for $x=20$
$\sqrt{3(20)+4}-\sqrt{2(20)-4}\stackrel{?}{=}2$
$\sqrt{64}-\sqrt{36}\stackrel{?}{=}2$
$8-6\stackrel{?}{=}2$
$2=2$

$\{4, 20\}$

12. $\{5, 13\}$

13. $\sqrt[3]{2x+3}-2=0$
$(\sqrt[3]{2x+3})^3=(2)^3$
$2x+3=8$
$2x=5$
$x=\frac{5}{2}$

Check $\sqrt[3]{2\left(\dfrac{5}{2}\right)+3}-2\stackrel{?}{=}0$
$\sqrt[3]{5+3}-2\stackrel{?}{=}0$
$\sqrt[3]{8}-2\stackrel{?}{=}0$
$2-2\stackrel{?}{=}0$
$0=0$

$\{\frac{5}{2}\}$

14. $\{\frac{15}{2}\}$

15. $\sqrt{4u+1}-\sqrt{u-2}=\sqrt{u+3}$
$(\sqrt{4u+1}-\sqrt{u-2})^2=(\sqrt{u+3})^2$
$4u+1-2\sqrt{4u+1}\sqrt{u-2}+u-2=u+3$
$4u-4=2\sqrt{(4u+1)(u-2)}$
$(2u-2)^2=(\sqrt{(4u+1)(u-2)})^2$
$4u^2-8u+4=4u^2-7u-2$
$-u=-6$
$u=6$

Check $\sqrt{4(6)+1}-\sqrt{(6)-2}\stackrel{?}{=}\sqrt{(6)+3}$
$\sqrt{25}-\sqrt{4}\stackrel{?}{=}\sqrt{9}$
$5-2\stackrel{?}{=}3$
$3=3$

$\{6\}$

16. $\{1\}$ **17.** $x^{1/2}=5$
$(x^{1/2})^2=5^2$
$x=25$

Check $(25)^{1/2}\stackrel{?}{=}5$
$(5^2)^{1/2}\stackrel{?}{=}5$
$5=5$

$\{25\}$

18. $\{27\}$

19. $2x^{-5/3}=64$
$x^{-5/3}=32$
$(x^{-5/3})^{-3/5}=(32)^{-3/5}$
$x=(2^5)^{-3/5}=2^{-3}=\dfrac{1}{2^3}$ or $\dfrac{1}{8}$

$\{\frac{1}{8}\}$ (The check will not be shown.)

20. $\{\frac{1}{4}\}$

21. $R=16, I=5$
$5=\sqrt{\dfrac{P}{16}}$
$25=\dfrac{P}{16}$
$P=400$

22. 576

23. $n=50, \sigma=3\frac{1}{3}=\frac{10}{3}$
$\dfrac{10}{3}=\sqrt{50(\frac{2}{3})p}$
$\dfrac{10}{3}=\sqrt{\frac{100}{3}p}$
$\frac{100}{9}=\frac{100}{3}p$
$p=\frac{1}{3}$

24. $\dfrac{2}{5}$

Exercises 7.7 (page 345)

1. $x^2 = (\sqrt{6})^2 + (\sqrt{3})^2$
$x^2 = 6 + 3$
$x^2 = 9$
$x = \pm\sqrt{9} = \pm 3$
$x = 3$ (-3 is not in the domain.)

2. 4

3. $10^2 + 6^2 = x^2$
$100 + 36 = x^2$
$136 = x^2$
$x = \pm\sqrt{136} = \pm\sqrt{4 \cdot 34} = \pm 2\sqrt{34}$
$x = 2\sqrt{34}$ ($-2\sqrt{34}$ is not in the domain.)

4. $4\sqrt{13}$

5. $(x+1)^2 + (\sqrt{20})^2 = (x+3)^2$
$x^2 + 2x + 1 + 20 = x^2 + 6x + 9$
$12 = 4x$
$x = 3$

6. 7

7. Let x = length of diagonal.
$x^2 = 4^2 + 4^2$
$x^2 = 16 + 16$
$x^2 = 32$
$x = \pm\sqrt{32} = \pm\sqrt{16 \cdot 2} = \pm 4\sqrt{2}$
$x = 4\sqrt{2}$ ($-4\sqrt{2}$ is not in the domain.)
The diagonal is $4\sqrt{2}$ units long.

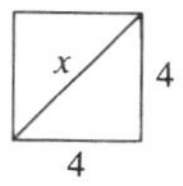

[Figure: square with sides 4, 4 and diagonal x]

8. $3\sqrt{2}$

9. Let w = width.
$(24)^2 + w^2 = (25)^2$
$576 + w^2 = 625$
$w^2 = 49$
$w = \pm\sqrt{49} = \pm 7$
$w = 7$ (-7 is not in the domain.)
The width is 7.

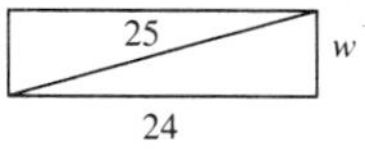

[Figure: rectangle with diagonal 25, length 24, width w]

10. 9

11. Let x = length of one leg
$2x - 4$ = length of other leg
$(10)^2 = (2x-4)^2 + x^2$
$100 = 4x^2 - 16x + 16 + x^2$
$0 = 5x^2 - 16x - 84$
$0 = (5x + 14)(x - 6)$

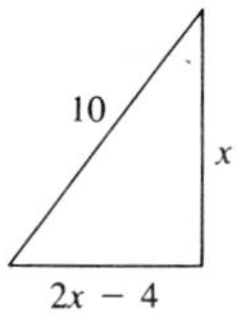

$5x + 14 = 0$	$x - 6 = 0$
$x = -\frac{14}{5}$	$x = 6$
	$2x - 4 = 12 - 4 = 8$

Not in the domain (refers to $x = -\frac{14}{5}$)
One leg is 6 units long and the other is 8 units long.

[Figure: right triangle with hypotenuse 10, legs x and $2x - 4$]

12. $\frac{9}{5}, \frac{38}{5}$

13. Let x = Nguyen's rate
$x + 7$ = Jaime's rate

Nguyen's distance is $\left(x\frac{\text{mi}}{\text{hr}}\right)(1\ \text{hr}) = x$ mi

Jaime's distance is $\left((x+7)\frac{\text{mi}}{\text{hr}}\right)(1\ \text{hr}) = (x+7)$ mi

$x^2 + (x+7)^2 = 13^2$
$x^2 + x^2 + 14x + 49 = 169$
$2x^2 + 14x - 120 = 0$
$x^2 + 7x - 60 = 0$
$(x - 5)(x + 12) = 0$

$x - 5 = 0$	$x + 12 = 0$
$x = 5$	$x = -12$ Reject

(The check will not be shown.) Nguyen's rate is 5 mph.

14. 4 mph

Exercises 7.8A (page 349)

1. $3 + \sqrt{16}\sqrt{-1} = 3 + 4i$ **2.** $4 - 5i$

3. $0 + \sqrt{64}\sqrt{-1} = 0 + 8i$ **4.** $0 + 10i$

5. $5 + \sqrt{16}\sqrt{2}\sqrt{-1} = 5 + 4i\sqrt{2}$ **6.** $6 + 3i\sqrt{2}$

7. $\sqrt{36}\sqrt{-1} + 2 = 6i + 2 = 2 + 6i$ **8.** $3 - 5i$

9. $2i - 3 = -3 + 2i$ **10.** $-4 + 3i$ **11.** $14 + 0i$

12. $-7 + 0i$ **13.** $3 - 4i = x + 2yi$ **14.** $x = 2, y = 5$

$3 = x$	$-4 = 2y$
$x = 3$	$y = -2$

15. $5x - 3i = 6 - 7yi$ **16.** $x = -\frac{3}{2}, y = -3$

$5x = 6$	$-3 = -7y$
$x = \frac{6}{5}$	$y = \frac{-3}{-7} = \frac{3}{7}$

17. $x\sqrt{3} - yi = 2 + i\sqrt{2}$ **18.** $x = \frac{3\sqrt{2}}{4}, y = -\frac{\sqrt{5}}{5}$

$x\sqrt{3} = 2$	$-y = \sqrt{2}$
$x = \frac{2}{\sqrt{3}} = \frac{2\sqrt{3}}{3}$	$y = -\sqrt{2}$

19. $\frac{3}{4}x - \frac{1}{3}yi = \frac{3}{5}x + \frac{1}{2}yi$ **20.** $x = 0, y = 0$

$\frac{3}{4}x = \frac{3}{5}x$	$-\frac{1}{3}y = \frac{1}{2}y$
$15x = 12x$	$-2y = 3y$
$3x = 0$	$-5y = 0$
$x = 0$	$y = 0$

Exercises 7.8B (page 350)

1. $4 + 3i + 5 - i = 9 + 2i$ **2.** $3 + 3i$

3. $7 - 4i - 5 - 2i = 2 - 6i$ **4.** $4 - 4i$

5. $2 + i + 3i - 2 + 4i = 8i = 0 + 8i$ **6.** $6 - 4i$

7. $2 + 3i - x - yi = (2 - x) + (3 - y)i$ **8.** $(x - 7) + (-1 - y)i$

9. $9 + \sqrt{16}\sqrt{-1} + 2 + \sqrt{25}\sqrt{-1} + 6 - \sqrt{64}\sqrt{-1}$
$= 9 + 4i + 2 + 5i + 6 - 8i$
$= (9 + 2 + 6) + (4 + 5 - 8)i$
$= 17 + i$

10. $11 + 3i$ **11.** $4 + 3i - 5 + i = 3x + 2yi + 2x - 3yi$
$-1 + 4i = 5x - yi$

Therefore, $5x = -1$	$-y = 4$
$x = -\frac{1}{5}$	$y = -4$

12. $x = \frac{3}{2}, y = 1$

13. $2 - 5i - 5 - 3i = 3x + 2yi - 5x - 3yi$ **14.** $x = -\frac{4}{3}, y = \frac{6}{5}$
$-3 - 8i = -2x - yi$

$-3 = -2x$	$-8 = -y$
$x = \frac{3}{2}$	$y = 8$

Exercises 7.8C (page 352)

1. $1^2 - i^2 = 1 - (-1) = 1 + 1 = 2 = 2 + 0i$ **2.** $13 + 0i$

3. $12 + 5i - 2i^2 = 12 + 5i + 2 = 14 + 5i$ **4.** $16 - 11i$

5. $12 - 22i + 6i^2 = 12 - 22i - 6 = 6 - 22i$ **6.** $-2 + 29i$

7. $(\sqrt{5})^2 - (2i)^2 = 5 - (-4) = 5 + 4 = 9 = 9 + 0i$ **8.** $16 + 0i$

9. $5i^2 - 10i = -5 - 10i$ **10.** $-12 - 6i$

11. $(2)^2 + 2(2)(5i) + (5i)^2 = 4 + 20i + 25i^2$
$= 4 + 20i - 25 = -21 + 20i$

12. $-7 - 24i$ **13.** $i^8 \cdot i^2 = 1 \cdot (-1) = -1 = -1 + 0i$

14. $0 - i$ **15.** $(i^4)^{21} i^3 = (1)(-i) = -i = 0 - i$ **16.** $0 + i$

17. $3^3 i^2 i = 27(-1)i = -27i = 0 - 27i$ 18. $0 - 8i$

19. $2^4 i^4 = 16(1) = 16 + 0i$ 20. $81 + 0i$

21. $(3 - 2i)(4 + 5i) = 12 + 7i - 10i^2 = 12 + 7i + 10 = 22 + 7i$

22. $58 - 14i$ 23. $(2 - i)^2 = 4 - 4i + (i)^2 = 4 - 4i - 1 = 3 - 4i$

24. $8 + 6i$ 25. $[3 + i^6]^2 = [3 - 1]^2 = 2^2 = 4 + 0i$

26. $25 + 0i$ 27. $i^{10+23} = i^{33} = (i^4)^8 \cdot i = 1 \cdot i = 0 + i$

28. $-1 + 0i$ 29. $i^{15} + i^7 = -i + (-i) = 0 - 2i$ 30. $-1 - i$

31. $[2 + (-i)^{11}]^2 = [2 + i]^2 = 4 + 4i + i^2 = 4 + 4i - 1 = 3 + 4i$

32. $8 + 6i$

33. $1 - 5i + 10i^2 - 10i^3 + 5i^4 - i^5 = 1 - 5i - 10 + 10i + 5 - i$
$= -4 + 4i$ (We used the binomial theorem.)

34. $-4 + 0i$ 35. $2^4 - 4(2^3)i + 6(2^2)i^2 - 4(2)i^3 + i^4$
$= 16 - 32i - 24 + 8i + 1 = -7 - 24i$

36. $52 - 47i$

Exercises 7.8D (page 354)

1. $3 + 2i$ 2. $5 - 4i$ 3. $-5i$ 4. $7i$ 5. 10 6. -8

7. $\dfrac{10}{1 + 3i} \cdot \dfrac{1 - 3i}{1 - 3i} = \dfrac{10(1 - 3i)}{1 - 9i^2} = \dfrac{10(1 - 3i)}{1 + 9} = \dfrac{10(1 - 3i)}{10} = 1 - 3i$

8. $1 - 2i$

9. $\dfrac{1 + i}{1 - i} \cdot \dfrac{1 + i}{1 + i} = \dfrac{1 + 2i + i^2}{1 - i^2} = \dfrac{1 + 2i - 1}{1 - (-1)} = \dfrac{2i}{2} = i = 0 + i$

10. $0 - i$ 11. $\dfrac{8 + i}{i} \cdot \dfrac{-i}{-i} = \dfrac{-8i - i^2}{-i^2} = \dfrac{-8i - (-1)}{-(-1)} = 1 - 8i$

12. $-1 - 4i$ 13. $\dfrac{3}{2i} \cdot \dfrac{i}{i} = \dfrac{3i}{2i^2} = \dfrac{3i}{-2} = 0 - \dfrac{3}{2}i$ 14. $0 - \dfrac{4}{5}i$

15. $\dfrac{15i}{1 - 2i} \cdot \dfrac{1 + 2i}{1 + 2i} = \dfrac{15i + 30i^2}{1 - 4i^2} = \dfrac{15i - 30}{1 + 4}$
$= \dfrac{15i - 30}{5} = \dfrac{5(3i - 6)}{5} = -6 + 3i$

16. $-6 + 2i$

17. $\dfrac{4 + 3i}{2 - i} \cdot \dfrac{2 + i}{2 + i} = \dfrac{8 + 10i + 3i^2}{4 - i^2} = \dfrac{8 + 10i - 3}{4 - (-1)}$
$= \dfrac{5 + 10i}{5} = \dfrac{5(1 + 2i)}{5} = 1 + 2i$

18. $\dfrac{4}{5} + \dfrac{1}{10}i$

Review Exercises 7.9 (page 356)

1. $\sqrt{x - 5} = \sqrt{3x + 8}$
Domain: $\{x \mid x - 5 \geq 0\} \cap \{x \mid 3x + 8 \geq 0\} = \{x \mid x \geq 5\}$
$(\sqrt{x - 5})^2 = (\sqrt{3x + 8})^2$
$x - 5 = 3x + 8$
$-13 = 2x$
$x = -\frac{13}{2}$ Not in the domain
Therefore, there is no solution. $\{ \}$

2. $\{7\}$

3. $\sqrt{5x - 4} = \sqrt{2x + 1} + 1$
$(\sqrt{5x - 4})^2 = (\sqrt{2x + 1} + 1)^2$
$5x - 4 = 2x + 1 + 2\sqrt{2x + 1} + 1$
$3x - 6 = 2\sqrt{2x + 1}$
$9x^2 - 36x + 36 = 4(2x + 1)$
$9x^2 - 44x + 32 = 0$
$(x - 4)(9x - 8) = 0$
$x - 4 = 0 \mid 9x - 8 = 0$
$x = 4 \mid x = \frac{8}{9}$

Check for $x = 4$
$\sqrt{5x - 4} - \sqrt{2x + 1} = 1$
$\sqrt{5(4) - 4} - \sqrt{2(4) + 1} \stackrel{?}{=} 1$
$4 - 3 \stackrel{?}{=} 1$
$1 = 1$
4 is a solution.

Check for $x = \frac{8}{9}$
$\sqrt{5x - 4} - \sqrt{2x + 1} = 1$
$\sqrt{5(\frac{8}{9}) - 4} - \sqrt{2(\frac{8}{9}) + 1} \stackrel{?}{=} 1$
$\sqrt{\dfrac{40 - 36}{9}} - \sqrt{\dfrac{16 + 9}{9}} \stackrel{?}{=} 1$
$\frac{2}{3} - \frac{5}{3} \stackrel{?}{=} 1$
$-1 \neq 1$
$\frac{8}{9}$ is not a solution.

$\{4\}$

4. $\{-\frac{36}{5}\}$ 5. $x^{5/6} = 32$
$(x^{5/6})^{6/5} = 32^{6/5}$
$x = (2^5)^{6/5}$
$= 2^6$ or 64
$\{64\}$ (The check will not be shown.)

6. $\{81\}$

7. $5 - yi = x + 6i$
$5 = x \mid -y = 6$
$x = 5 \mid y = -6$

8. $x = -1, y = -2$

9. $(3i + 2)(4 - 2i) = 12i + 8 - 6i^2 - 4i$
$= 12i + 8 + 6 - 4i = 14 + 8i$

10. $28 - 96i$

11. $\dfrac{2 + i}{1 + 3i} \cdot \dfrac{1 - 3i}{1 - 3i} = \dfrac{2 - 3i^2 - 5i}{1 - 9i^2} = \dfrac{5 - 5i}{10} = \dfrac{1}{2} - \dfrac{1}{2}i$

12. $\frac{1}{5} + \frac{7}{5}i$

13. $(4 + \sqrt{9 \cdot 3(-1)}) + (2 - \sqrt{4 \cdot 3(-1)}) - (1 - \sqrt{3(-1)})$
$= (4 + 3i\sqrt{3}) + (2 - 2i\sqrt{3}) - (1 - i\sqrt{3})$
$= 4 + 3i\sqrt{3} + 2 - 2i\sqrt{3} - 1 + i\sqrt{3}$
$= 5 + 2i\sqrt{3}$

14. Width = 3 ft; length = 5 ft

15. Let x = width
$x + 3$ = length
$x^2 + (x + 3)^2 = (\sqrt{45})^2$
$x^2 + x^2 + 6x + 9 = 45$
$2x^2 + 6x - 36 = 0$
$2(x + 6)(x - 3) = 0$
$2 \neq 0 \mid x + 6 = 0 \mid x - 3 = 0$
$x = -6 \mid x = 3$
Not in the domain $\mid x + 3 = 6$

Check $3^2 + 6^2 \stackrel{?}{=} (\sqrt{45})^2$
$9 + 36 \stackrel{?}{=} 45$
$45 = 45$
The width is 3 m and the length is 6 m.

16. $\frac{3}{5}$

17. Let x = rate of southbound car
$x - 15$ = rate of eastbound car

Distance of southbound car
$= \left(x \dfrac{\text{mi}}{\text{hr}}\right)\left(\dfrac{1}{3}\text{hr}\right) = \dfrac{x}{3}\text{ mi}$
Distance of eastbound car
$= \left([x - 15]\dfrac{\text{mi}}{\text{hr}}\right)\left(\dfrac{1}{3}\text{hr}\right)$
$= \dfrac{x - 15}{3}\text{ mi}$

$\left(\dfrac{x}{3}\right)^2 + \left(\dfrac{x - 15}{3}\right)^2 = 25^2$
$\dfrac{x^2}{9} + \dfrac{x^2 - 30x + 225}{9} = 625$
$x^2 + x^2 - 30x + 225 = 5{,}625$
$2x^2 - 30x - 5{,}400 = 0$
$x^2 - 15x - 2{,}700 = 0$
$(x - 60)(x + 45) = 0$
$x - 60 = 0 \mid x + 45 = 0$
$x = 60 \mid x = -45 \leftarrow$ Reject
(The check will not be shown.) The rate of the faster car is 60 mph.

Chapter 7 Diagnostic Test (page 361)

Following each problem number is the textbook section number (in parentheses) where that kind of problem is discussed.

1. (7.1) $x^{1/2}x^{-1/4} = x^{1/2-1/4} = x^{1/4}$

2. (7.1) $(R^{-4/3})^3 = R^{(-4/3)(3)} = R^{-4} = \dfrac{1}{R^4}$

3. (7.1) $\dfrac{a^{5/6}}{a^{1/3}} = a^{5/6-1/3} = a^{5/6-2/6} = a^{3/6} = a^{1/2}$

4. (7.1) $\left(\dfrac{x^{-2/3}y^{3/5}}{x^{1/3}y}\right)^{-5/2} = (x^{-2/3-1/3}y^{3/5-1})^{-5/2} = (x^{-1}y^{-2/5})^{-5/2}$
$= x^{(-1)(-5/2)}y^{(-2/5)(-5/2)} = x^{5/2}y$

5. (7.1) $\dfrac{b^{2/3}}{b^{-1/5}} = b^{2/3-(-1/5)} = b^{2/3+1/5} = b^{13/15}$

6. (7.2) $\sqrt[3]{54x^6y^7} = \sqrt[3]{2(27)x^6y^6y} = 3x^2y^2\sqrt[3]{2y}$

7. (7.2) $\dfrac{4xy}{\sqrt{2x}} = \dfrac{4xy}{\sqrt{2x}} \cdot \dfrac{\sqrt{2x}}{\sqrt{2x}} = \dfrac{4xy\sqrt{2x}}{2x} = 2y\sqrt{2x}$

8. (7.2) $\sqrt[6]{a^3} = a^{3/6} = a^{1/2} = \sqrt{a}$

9. (7.3) $\sqrt{40} + \sqrt{9} = 2\sqrt{10} + 3$

10. (7.4) $\sqrt{x}\sqrt[3]{x} = x^{1/2}x^{1/3} = x^{1/2+1/3} = x^{5/6} = \sqrt[6]{x^5}$

11. (7.1) $(-27)^{2/3} = [(-3)^3]^{2/3} = (-3)^{3(2/3)} = (-3)^2 = 9$

12. (7.3) $4\sqrt{8y} + 3\sqrt{32y} = 4\sqrt{4 \cdot 2y} + 3\sqrt{16 \cdot 2y}$
$= 4(2)\sqrt{2y} + 3(4)\sqrt{2y} = 20\sqrt{2y}$

13. (7.3) $3\sqrt{\dfrac{5x^2}{2}} - 5\sqrt{\dfrac{x^2}{10}} = 3\sqrt{\dfrac{5x^2}{2} \cdot \dfrac{2}{2}} - 5\sqrt{\dfrac{x^2}{10} \cdot \dfrac{10}{10}}$
$= 3\sqrt{\dfrac{10x^2}{4}} - 5\sqrt{\dfrac{10x^2}{100}}$
$= \dfrac{3x\sqrt{10}}{2} - \dfrac{5x\sqrt{10}}{10}$
$= \dfrac{3}{2}x\sqrt{10} - \dfrac{1}{2}x\sqrt{10} = x\sqrt{10}$

14. (7.4) $\sqrt{2x^4}\sqrt{8x^3} = \sqrt{16x^6x} = 4x^3\sqrt{x}$

15. (7.4) $\sqrt{2x}(\sqrt{8x} - 5\sqrt{2}) = \sqrt{2x}\sqrt{8x} + \sqrt{2x}(-5\sqrt{2})$
$= \sqrt{16x^2} - 5\sqrt{4x} = 4x - 10\sqrt{x}$

16. (7.4) $\dfrac{\sqrt{10x} + \sqrt{5x}}{\sqrt{5x}} = \dfrac{\sqrt{10x}}{\sqrt{5x}} + \dfrac{\sqrt{5x}}{\sqrt{5x}} = \sqrt{\dfrac{10x}{5x}} + 1 = \sqrt{2} + 1$

17. (7.4) $\dfrac{5}{\sqrt{7} + \sqrt{2}} = \dfrac{5}{\sqrt{7} + \sqrt{2}} \cdot \dfrac{(\sqrt{7} - \sqrt{2})}{(\sqrt{7} - \sqrt{2})}$
$= \dfrac{5(\sqrt{7} - \sqrt{2})}{7 - 2} = \sqrt{7} - \sqrt{2}$

18. (7.4) $(1 - \sqrt[3]{x})^3 = (1)^3 + 3(1)^2(-\sqrt[3]{x}) + 3(1)(-\sqrt[3]{x})^2 + (-\sqrt[3]{x})^3$
$= 1 - 3\sqrt[3]{x} + 3\sqrt[3]{x^2} - x$

19. (7.8)
$(5 - \sqrt{-8}) - (3 - \sqrt{-18}) = 5 - \sqrt{4 \cdot 2(-1)} - 3 + \sqrt{9 \cdot 2(-1)}$
$= 5 - 2i\sqrt{2} - 3 + 3i\sqrt{2} = 2 + i\sqrt{2}$

20. (7.8) $(3 + i)(2 - 5i) = 6 - 15i + 2i - 5i^2$
$= 6 - 13i - 5(-1) = 6 - 13i + 5$
$= 11 - 13i$

21. (7.8) $\dfrac{10}{1 - 3i} = \dfrac{10}{1 - 3i} \cdot \dfrac{1 + 3i}{1 + 3i} = \dfrac{10(1 + 3i)}{1 - 9i^2}$
$= \dfrac{10(1 + 3i)}{10} = 1 + 3i$

22. (7.7) $(2 - i)^3 = 2^3 - 3(2^2)i + 3(2)i^2 - i^3$
$= 8 - 12i + 6(-1) - (-i)$
$= 8 - 6 - 12i + i$
$= 2 - 11i$

23. (7.6)
$x^{3/2} = 8$
$(x^{3/2})^{2/3} = 8^{2/3} = (2^3)^{2/3}$
$x = 2^2$
$x = 4$
The solution set is $\{4\}$.

Check
$x^{3/2} = 8$
$(4)^{3/2} \stackrel{?}{=} 2^3$
$(2^2)^{3/2} \stackrel{?}{=} 2^3$
$2^3 = 2^3$

24. (7.6) $\sqrt{x - 3} + 5 = x$
$(\sqrt{x - 3})^2 = (x - 5)^2$
$x - 3 = x^2 - 10x + 25$
$0 = x^2 - 11x + 28$
$0 = (x - 4)(x - 7)$
$x - 4 = 0 \mid x - 7 = 0$
$x = 4 \mid x = 7$

Check for $x = 4$
$\sqrt{x - 3} + 5 = x$
$\sqrt{4 - 3} + 5 \stackrel{?}{=} 4$
$\sqrt{1} + 5 \stackrel{?}{=} 4$
$1 + 5 \neq 4$
Therefore, 4 is not a solution.

Check for $x = 7$
$\sqrt{x - 3} + 5 = x$
$\sqrt{7 - 3} + 5 \stackrel{?}{=} 7$
$\sqrt{4} + 5 \stackrel{?}{=} 7$
$2 + 5 \stackrel{?}{=} 7$
$7 = 7$
Therefore, 7, is a solution. The solution set is $\{7\}$.

25. (7.7) $(x + 3)^2 = (x + 1)^2 + (\sqrt{12})^2$
$x^2 + 6x + 9 = x^2 + 2x + 1 + 12$
$4x = 4$
$x = 1$

Cumulative Review Exercises: Chapter 1–7 (page 362)

1. LCD $= (a + 3)(a - 3)(a - 1)$

$\dfrac{6}{(a + 3)(a - 3)} \cdot \dfrac{a - 1}{a - 1} - \dfrac{2}{(a - 3)(a - 1)} \cdot \dfrac{a + 3}{a + 3}$
$= \dfrac{6a - 6}{(a + 3)(a - 3)(a - 1)} - \dfrac{2a + 6}{(a + 3)(a - 3)(a - 1)}$
$= \dfrac{6a - 6 - 2a - 6}{(a + 3)(a - 3)(a - 1)} = \dfrac{4a - 12}{(a + 3)(a - 3)(a - 1)}$
$= \dfrac{4(a - 3)}{(a + 3)(a - 3)(a - 1)} = \dfrac{4}{(a + 3)(a - 1)}$

2. $\dfrac{x^2 + 2x + 4}{x^2}$

3. $\dfrac{\dfrac{4}{x} \cdot \dfrac{x^2}{1} - \dfrac{8}{x^2} \cdot \dfrac{x^2}{1}}{\dfrac{1}{x} \cdot \dfrac{x^2}{1} - \dfrac{2}{x^2} \cdot \dfrac{x^2}{1}} = \dfrac{4x - 8}{x - 2} = \dfrac{4(x - 2)}{x - 2} = 4$

4. $5x + 2\sqrt{5x}$

5. $(\sqrt{26})^2 - 2\sqrt{26}\sqrt{10} + (\sqrt{10})^2 = 26 - 2\sqrt{2^2 \cdot 13 \cdot 5} + 10$
$= 36 - 4\sqrt{65}$

6. $-12 - 6\sqrt{5}$

7. $12x - 3 - 3 + 2x = 8x - 12$
$14x - 6 = 8x - 12$
$6x = -6$
$x = -1$
$\{-1\}$

8. $\{x \mid x \leq 5\}$

9.
$$\sqrt{2x+2} = 1 + \sqrt{3x-12}$$
$$(\sqrt{2x+2})^2 = (1+\sqrt{3x-12})^2$$
$$2x+2 = 1 + 2(1)(\sqrt{3x-12}) + 3x - 12$$
$$2x+2 = 3x - 11 + 2\sqrt{3x-12}$$
$$-x+13 = 2\sqrt{3x-12}$$
$$(-x+13)^2 = (2\sqrt{3x-12})^2$$
$$x^2 - 26x + 169 = 4(3x-12)$$
$$x^2 - 26x + 169 = 12x - 48$$
$$x^2 - 38x + 217 = 0$$
$$(x-7)(x-31) = 0$$
$$x-7=0 \quad | \quad x-31=0$$
$$x=7 \quad | \quad x=31$$

Check for $x = 7$
$$\sqrt{2x+2} = 1 + \sqrt{3x-12}$$
$$\sqrt{2(7)+2} \stackrel{?}{=} 1 + \sqrt{3(7)-12}$$
$$\sqrt{16} \stackrel{?}{=} 1 + \sqrt{9}$$
$$4 \stackrel{?}{=} 1 + 3$$
$$4 = 4$$

Check for $x = 31$
$$\sqrt{2x+2} = 1 + \sqrt{3x-12}$$
$$\sqrt{2(31)+2} \stackrel{?}{=} 1 + \sqrt{3(31)-12}$$
$$\sqrt{64} \stackrel{?}{=} 1 + \sqrt{81}$$
$$8 \neq 1 + 9$$

Therefore, the only solution is 7. $\{7\}$

10. $\{-3\}$

11.
$$6x^2 + 11x = 10$$
$$6x^2 + 11x - 10 = 0$$
$$(3x-2)(2x+5) = 0$$
$$3x-2=0 \quad | \quad 2x+5=0$$
$$3x = 2 \quad | \quad 2x = -5$$
$$x = \tfrac{2}{3} \quad | \quad x = -\tfrac{5}{2}$$
$$\{\tfrac{2}{3}, -\tfrac{5}{2}\}$$

12. $a = \dfrac{bc}{b-c}$

13. $|2x-5| > 3$
$$2x-5 > 3 \quad \text{or} \quad 2x-5 < -3$$
$$2x > 8 \quad \text{or} \quad 2x < 2$$
$$x > 4 \quad \text{or} \quad x < 1$$
$$\{x \mid x > 4\} \cup \{x \mid x < 1\}$$

14. $\frac{30}{11}$ hr $= 2\frac{8}{11}$ hr

15. Let $2x$ = number of centimeters in width
$3x$ = number of centimeters in length
$$(2x)(3x) = 54$$
$$6x^2 = 54$$
$$x^2 = 9$$
$$x = \pm 3$$
-3 is not in the domain; $x = 3$
$$2x = 6$$
$$3x = 9$$
Check (6 cm)(9 cm) = 54 sq cm, and the ratio of 6 cm to 9 cm is 2 to 3. The width is 6 cm and the length is 9 cm.

16. \$14,200 at 7.6%; \$12,300 at 7.3%

17. Let x = the speed of the current
$27 - x$ = speed of boat going upstream
$27 + x$ = speed of boat going downstream

$$\text{Time upstream} = \frac{120 \text{ mi}}{(27-x)\frac{\text{mi}}{\text{hr}}} \quad \left(t = \frac{d}{r}\right)$$

$$\text{Time downstream} = \frac{150 \text{ mi}}{(27+x)\frac{\text{mi}}{\text{hr}}}$$

$$\frac{120}{27-x} = \frac{150}{27+x} \quad \text{The times are equal}$$
$$120(27+x) = 150(27-x)$$
$$4(27+x) = 5(27-x) \quad \text{(Dividing both sides by 30)}$$
$$108 + 4x = 135 - 5x$$
$$9x = 27$$
$$x = 3$$
(The check will not be shown.) The speed of the current is 3 mph.

18. 9 yd by 9 yd

Exercises 8.1 (page 368)

1. a. (3, 0) **b.** (0, 5) **c.** $(-5, 2)$ **d.** $(-4, -3)$

2. a. (4, 3) **b.** $(-6, 0)$ **c.** $(3, -4)$ **d.** $(0, -3)$

3. C $(-4, 5)$; B (3, 2); A (0, 0)

4. B $(-2, 4)$; C (3, 5); A $(-2, -3)$

5. a. $d = \sqrt{[2-(-2)]^2 + [1-(-2)]^2}$
$= \sqrt{4^2 + 3^2} = \sqrt{16+9} = \sqrt{25} = 5$

b. $d = \sqrt{[3-(-3)]^2 + [-1-3]^2}$
$= \sqrt{36+16} = \sqrt{52} = \sqrt{4 \cdot 13} = 2\sqrt{13}$

c. $|-2-5| = |-7| = 7$

d. $|-2-(-5)| = |3| = 3$

e. $d = \sqrt{(4-0)^2 + (6-0)^2} = \sqrt{16+36} = \sqrt{52}$
$= \sqrt{4 \cdot 13} = 2\sqrt{13}$

6. a. 13 **b.** $\sqrt{74}$ **c.** 6 **d.** 6 **e.** $3\sqrt{13}$

7. $|AB| = \sqrt{[4-(-2)]^2 + (2-2)^2} = \sqrt{6^2} = 6$
$|BC| = \sqrt{(6-4)^2 + (8-2)^2} = \sqrt{4+36} = \sqrt{40}$
$= \sqrt{4 \cdot 10} = 2\sqrt{10}$
$|AC| = \sqrt{[6-(-2)]^2 + (8-2)^2} = \sqrt{64+36} = \sqrt{100} = 10$
Perimeter $= 6 + 2\sqrt{10} + 10 = 16 + 2\sqrt{10}$

8. $16 + 4\sqrt{5}$

9. $|AB| = \sqrt{[5-(-3)]^2 + [-1-(-2)]^2} = \sqrt{64+1} = \sqrt{65}$
$|BC| = \sqrt{(3-5)^2 + [2-(-1)]^2} = \sqrt{4+9} = \sqrt{13}$
$|AC| = \sqrt{[3-(-3)]^2 + [2-(-2)]^2} = \sqrt{36+16} = \sqrt{52}$
$(|AB|)^2 \stackrel{?}{=} (|BC|)^2 + (|AC|)^2$
$(\sqrt{65})^2 \stackrel{?}{=} (\sqrt{13})^2 + (\sqrt{52})^2$
$65 = 13 + 52$
Therefore, the triangle is a right triangle.

10. The triangle is a right triangle.

11. Domain $= \{2, 3, 0, -3\}$
Range $= \{-1, 4, 2, -2\}$

(3, 4); (0, 2); (2, −1); (−3, −2)

12. Domain $= \{-4, 0, 3, 1, -3\}$
Range $= \{0, -2, 5, -3\}$

(1, 5); (−4, 0); (0, 0); (−3, −3); (3, −2)

Exercises 8.2 (page 376)

1.

x	y
2	0
0	3
4	−3

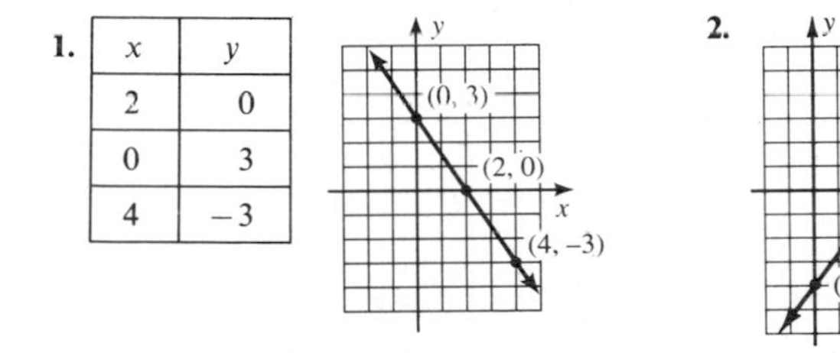

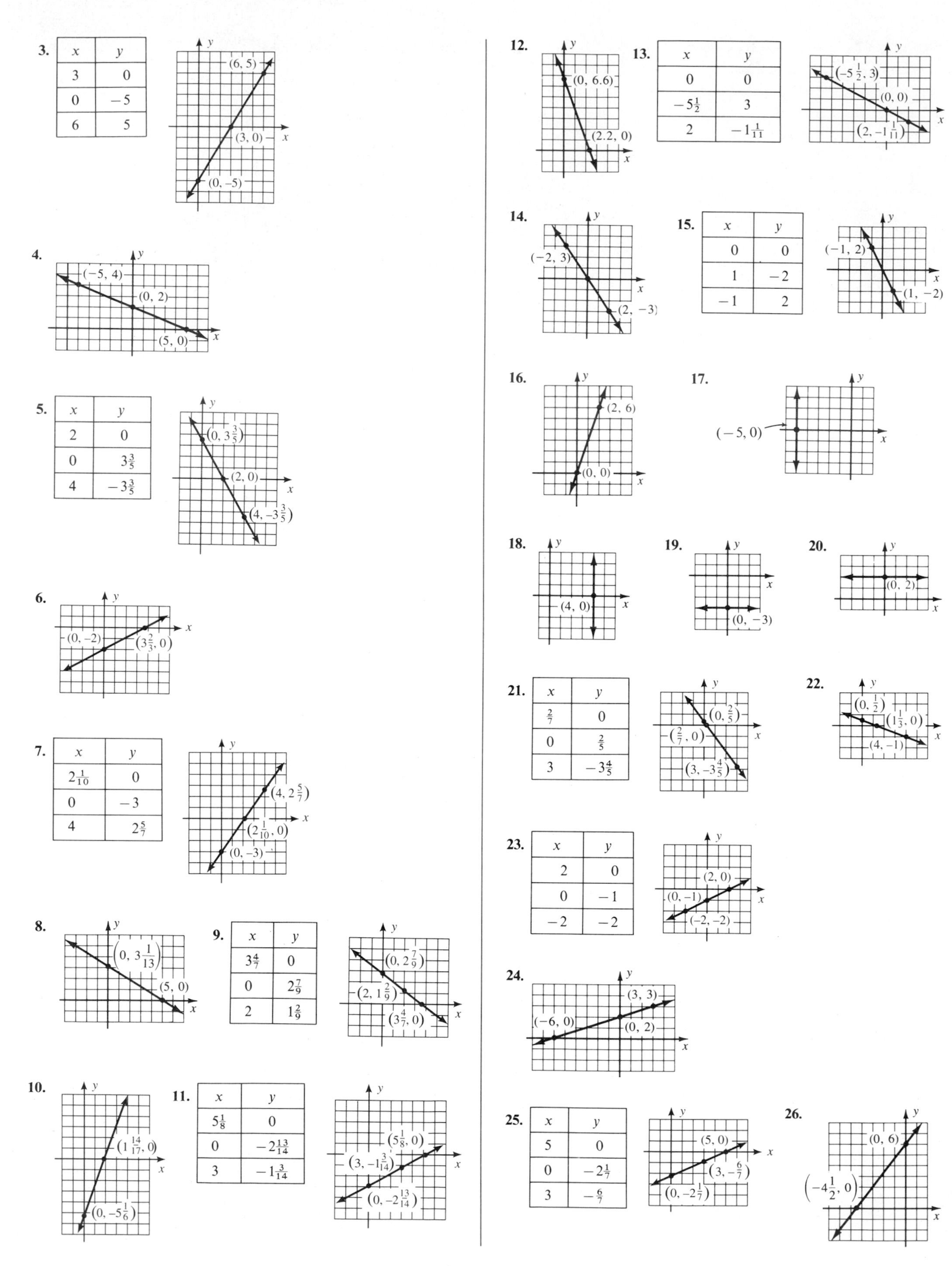

3.

x	y
3	0
0	-5
6	5

5.

x	y
2	0
0	$3\frac{3}{5}$
4	$-3\frac{3}{5}$

7.

x	y
$2\frac{1}{10}$	0
0	-3
4	$2\frac{5}{7}$

9.

x	y
$3\frac{4}{7}$	0
0	$2\frac{7}{9}$
2	$1\frac{2}{9}$

11.

x	y
$5\frac{1}{8}$	0
0	$-2\frac{13}{14}$
3	$-1\frac{3}{14}$

13.

x	y
0	0
$-5\frac{1}{2}$	3
2	$-1\frac{1}{11}$

15.

x	y
0	0
1	-2
-1	2

21.

x	y
$\frac{2}{7}$	0
0	$\frac{2}{5}$
3	$-3\frac{4}{5}$

23.

x	y
2	0
0	-1
-2	-2

25.

x	y
5	0
0	$-2\frac{1}{7}$
3	$-\frac{6}{7}$

In Exercises 27–30, it is best if the units on the two axes are not the same.

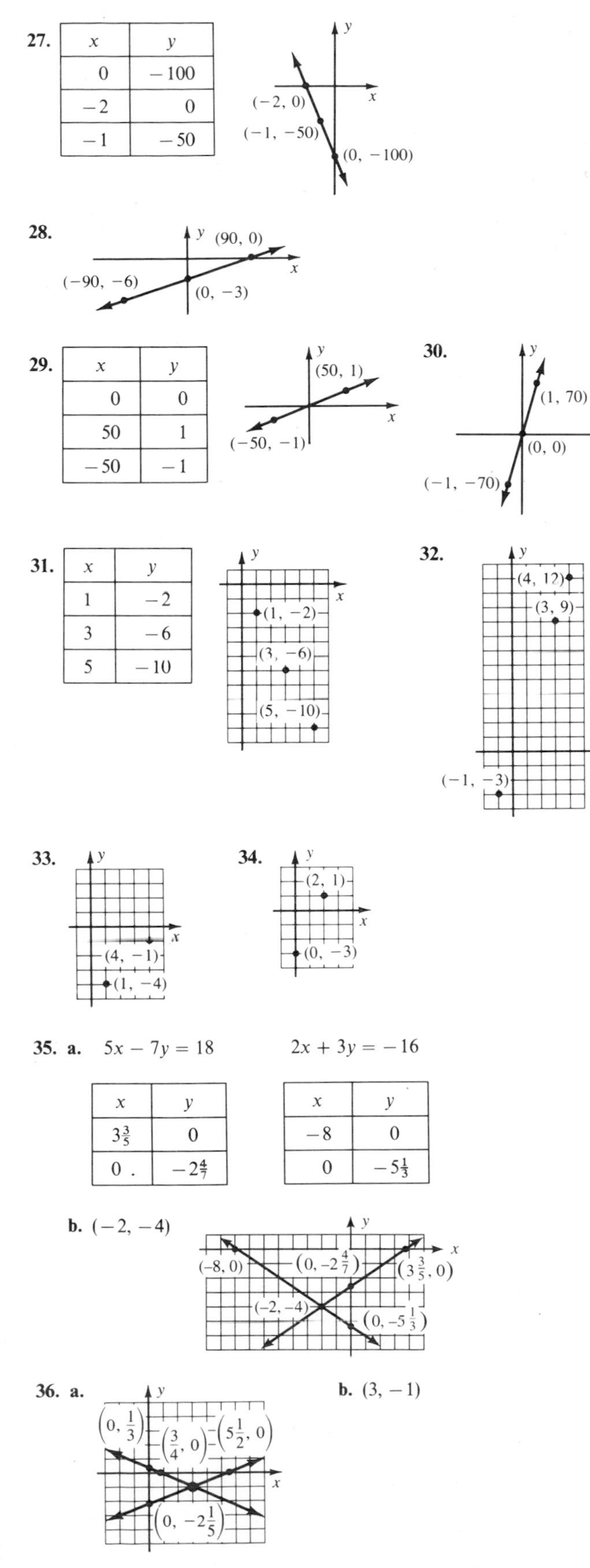

27.

x	y
0	-100
-2	0
-1	-50

28.

29.

x	y
0	0
50	1
-50	-1

30.

31.

x	y
1	-2
3	-6
5	-10

32.

33.

34.

35. a. $5x - 7y = 18$ $\qquad$ $2x + 3y = -16$

x	y
$3\frac{3}{5}$	0
0	$-2\frac{4}{7}$

x	y
-8	0
0	$-5\frac{1}{3}$

b. $(-2, -4)$

36. a.

b. $(3, -1)$

Exercises 8.3A (page 383)

1. $m = \dfrac{6-4}{10-1} = \dfrac{2}{9}$ $\qquad$ **2.** $\dfrac{5}{6}$

3. $m = \dfrac{(-7)-(-5)}{1-(-5)} = \dfrac{-7+5}{1+5} = \dfrac{-2}{6} = \dfrac{-1}{3}$ $\qquad$ **4.** $-\dfrac{1}{2}$

5. $m = \dfrac{(-5)-(-5)}{2-(-7)} = \dfrac{0}{9} = 0$ $\qquad$ **6.** 0

7. $m = \dfrac{-2-3}{-4-(-4)} = \dfrac{-5}{0}$ $\qquad$ **8.** Does not exist

Not defined; m does not exist.

9. a. $y = \frac{3}{4}x$ $\qquad$ **b.** $y = \frac{3}{4}x + 5$ $\qquad$ **c.** $y = \frac{3}{4}x - 2$

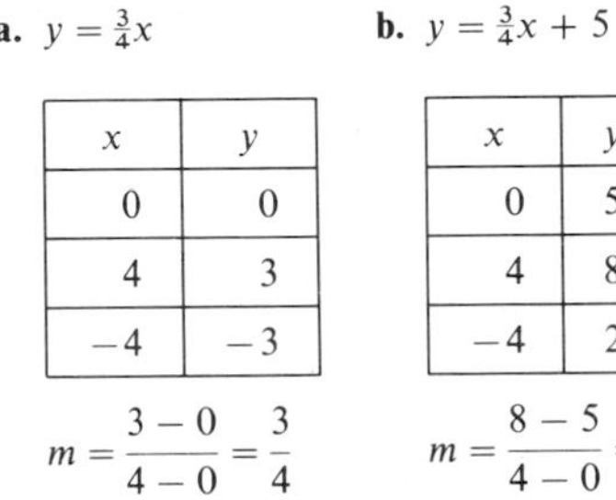

x	y
0	0
4	3
-4	-3

$m = \dfrac{3-0}{4-0} = \dfrac{3}{4}$

x	y
0	5
4	8
-4	2

$m = \dfrac{8-5}{4-0} = \dfrac{3}{4}$

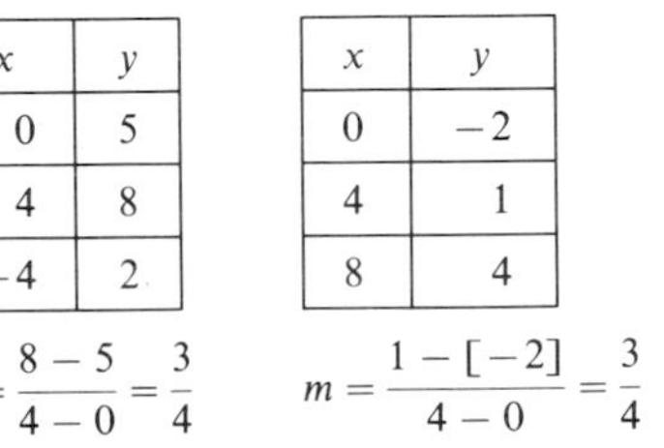

x	y
0	-2
4	1
8	4

$m = \dfrac{1-[-2]}{4-0} = \dfrac{3}{4}$

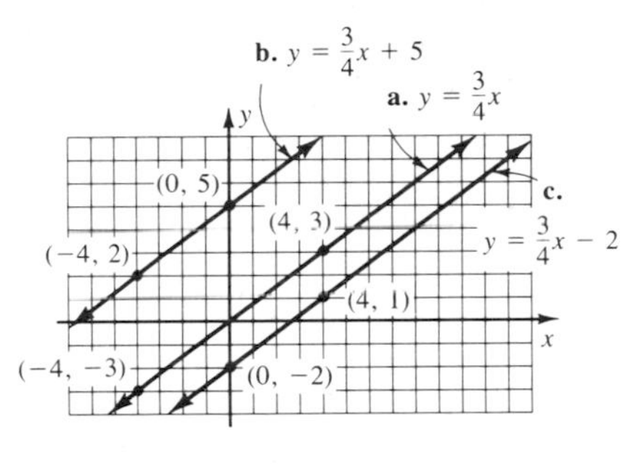

10. a. $m = 1$ $\qquad$ **b.** $m = 1$ $\qquad$ **c.** $m = 1$

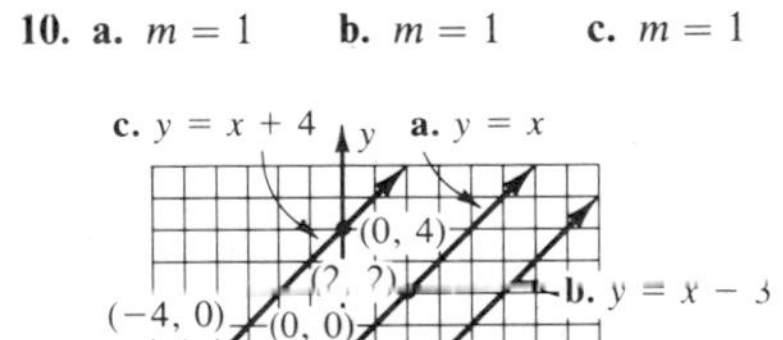

11.

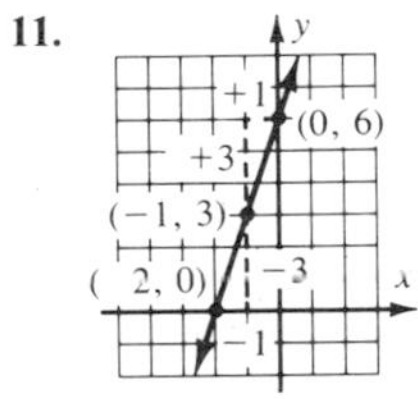

12.

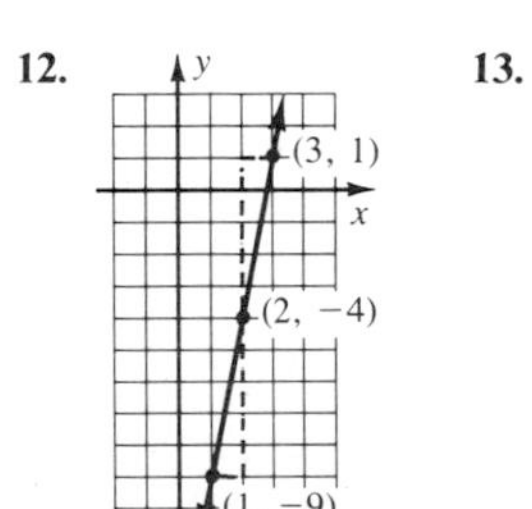

13.

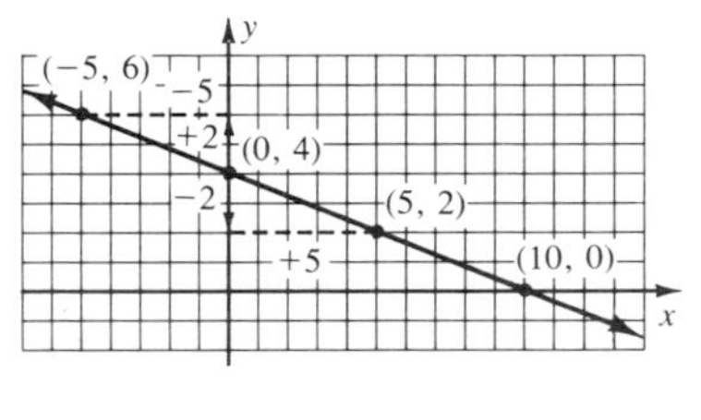

14.

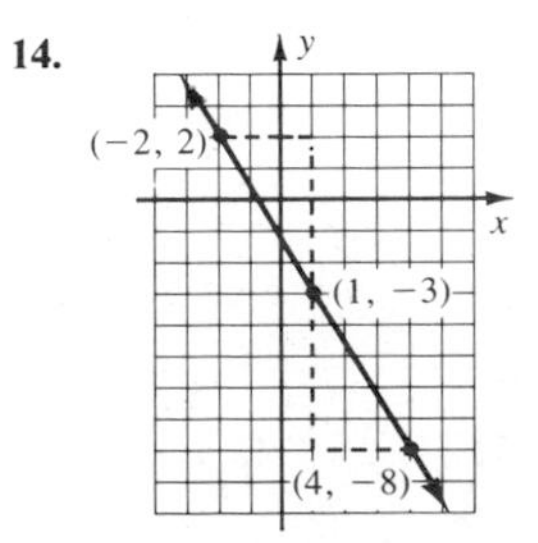

Exercises 8.3B (page 388)

1. $5x - 3y + 7 = 0$ **2.** $4x + 9y - 3 = 0$

3. $\frac{\overset{5}{\cancel{10}}}{1}\left(\frac{x}{\underset{1}{\cancel{2}}}\right) - \frac{\overset{2}{\cancel{10}}}{1}\left(\frac{y}{\underset{1}{\cancel{5}}}\right) = \left(\frac{10}{1}\right)\left(\frac{1}{1}\right)$ **4.** $6x - 7y + 42 = 0$

$5x - 2y = 10$
$5x - 2y - 10 = 0$

5. $\frac{3}{1}\left(\frac{y}{1}\right) = -\frac{\overset{1}{\cancel{3}}}{1}\left(\frac{5}{\underset{1}{\cancel{3}}}\right)x + \left(\frac{3}{1}\right)\left(\frac{4}{1}\right)$ **6.** $3x + 8y + 40 = 0$

$3y = -5x + 12$
$5x + 3y - 12 = 0$

7. $4x - 4y = 11 - 2x - 6y$ **8.** $25x - 8y - 15 = 0$
$6x + 2y - 11 = 0$

9. $y - (-3) = \frac{1}{5}(x - 4)$ **10.** $5x + 6y + 16 = 0$
$5y + 15 = x - 4$
$x - 5y - 19 = 0$

11. $y - 5 = \frac{1}{4}[x - (-6)]$ **12.** $4x + 5y + 22 = 0$
$4y - 20 = x + 6$
$x - 4y + 26 = 0$

13. $y - 3 = -4(x - [-1])$ **14.** $2x - y - 6 = 0$
$y - 3 = -4x - 4$
$4x + y + 1 = 0$

15. $y = \frac{5}{7}x - 3$ **16.** $x + 4y + 8 = 0$
$7y = 5x - 21$
$5x - 7y - 21 = 0$

17. $y = -\frac{4}{3}x + \frac{1}{2}$ **18.** $12x + 20y - 15 = 0$
$6y = -8x + 3$
$8x + 6y - 3 = 0$

19. $y = 0x + 5$ **20.** $y - 7 = 0$
$y - 5 = 0$

21. The line is horizontal; the equation is $y = 3$ or $y - 3 = 0$.

22. $y + 5 = 0$

23. The line is vertical; the equation is $x = 7$ or $x - 7 = 0$.

24. $x + 6 = 0$ **25. a.** $-5y = -4x - 20$
$y = \frac{-4x - 20}{-5}$
$y = \frac{4}{5}x + 4$
b. $m = \frac{4}{5}$
c. y-intercept $= 4$

26. a. $y = -\frac{8}{3}x + 8$ **b.** $-\frac{8}{3}$ **c.** 8

27. $2x + 9y + 15 = 0$
$9y = -2x - 15$
a. $y = -\frac{2}{9}x - \frac{5}{3}$
b. $m = -\frac{2}{9}$
c. y-intercept $= -\frac{5}{3}$

28. a. $y = \frac{18}{5}x - \frac{12}{5}$ **b.** $\frac{18}{5}$ **c.** $-\frac{12}{5}$

29. $m = \frac{4 - (-1)}{6 - 8} = \frac{5}{-2}$ **30.** $3x + 2y - 17 = 0$
$y - 4 = \frac{-5}{2}(x - 6)$
$2y - 8 = -5x + 30$
$5x + 2y - 38 = 0$

31. $m = \frac{4 - 0}{7 - 10} = \frac{4}{-3}$ **32.** $5x - 2y + 20 = 0$
$y - 0 = \frac{-4}{3}(x - 10)$
$3y = -4x + 40$
$4x + 3y - 40 = 0$

33. $m = \frac{(-1) - 3}{(-3) - (-9)} = \frac{-4}{6} = \frac{-2}{3}$ **34.** $3x + 4y + 17 = 0$
$y - (-1) = \frac{-2}{3}[x - (-3)]$
$3y + 3 = -2x - 6$
$2x + 3y + 9 = 0$

35. The line must have the same slope as $3x - 5y = 6$.
$-5y = -3x + 6$
$y = \frac{3}{5}x - \frac{6}{5}$
$m = \frac{3}{5}$
The slope of $3x - 5y = 6$ is $\frac{3}{5}$.
$y - 7 = \frac{3}{5}[x - (-4)]$
$5y - 35 = 3x + 12$
$0 = 3x - 5y + 47$

36. $7x + 4y - 36 = 0$

37. Slope of $2x + 4y = 3$: **38.** $2x + y - 9 = 0$
$4y = -2x + 3$
$y = -\frac{1}{2}x + \frac{3}{4}$ Slope is $-\frac{1}{2}$.
Slope of required line is 2.
$y - 2 = 2(x - 6)$
$y - 2 = 2x - 12$
$2x - y - 10 = 0$

39. The line must have the same slope as $3x + 5y - 12 = 0$.
$5y = -3x + 12$ x-intercept $= (4, 0)$
$y = \frac{-3}{5}x + \frac{12}{5}$
$m = \frac{-3}{5}$
$y - 0 = \frac{-3}{5}(x - 4)$
$5y = -3x + 12$
$3x + 5y - 12 = 0$

40. $9x - 14y + 27 = 0$ **41.** x-intercept $= (-6, 0)$
y-intercept $= (0, 4)$
$m = \frac{4 - 0}{0 - (-6)} = \frac{4}{6} = \frac{2}{3}$
Using $(0, 4)$
$y - 4 = \frac{2}{3}(x - 0)$
$3y - 12 = 2x$
$2x - 3y + 12 = 0$

42. $4x - 5y - 60 = 0$

Exercises 8.4 (page 394)

1. Boundary line: $4x + 5y = 20$
Dashed line because the inequality is $<$.
Half-plane includes $(0, 0)$ because:
$4(0) + 5(0) < 20$
$0 < 20$ True

x	y
5	0
0	4

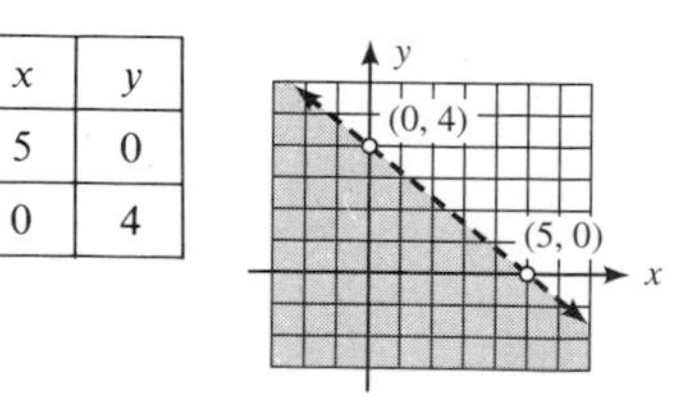

2.

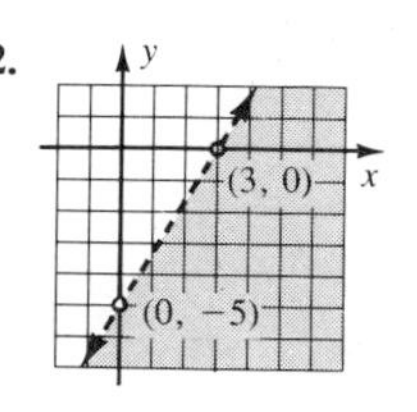

3. Boundary line: $3x - 8y = -16$
Dashed line because the inequality is $>$. Half-plane includes (0, 0) because:
$3(0) - 8(0) > -16$
$0 > -16$ True

x	y
$-5\frac{1}{3}$	0
0	2

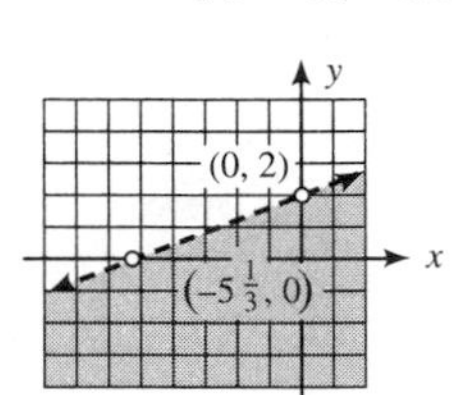

4. $(-3, 0)$; $\left(0, \frac{-18}{5}\right)$

5. Boundary line: $9x + 7y = -27$
Solid line because the inequality is $\leq$.
Half-plane does not include (0, 0) because:
$9(0) + 7(0) \leq -27$
$0 \leq -27$ False

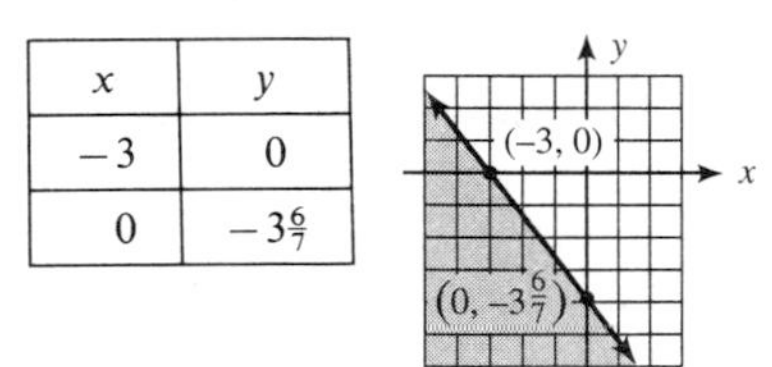

x	y
-3	0
0	$-3\frac{6}{7}$

6. $\left(\frac{-28}{5}, 0\right)$; (0, 2)

7. Boundary line: $x = -1$
Solid line because the inequality is $\geq$.
Half-plane includes (0, 0) because:
$0 \geq -1$ True

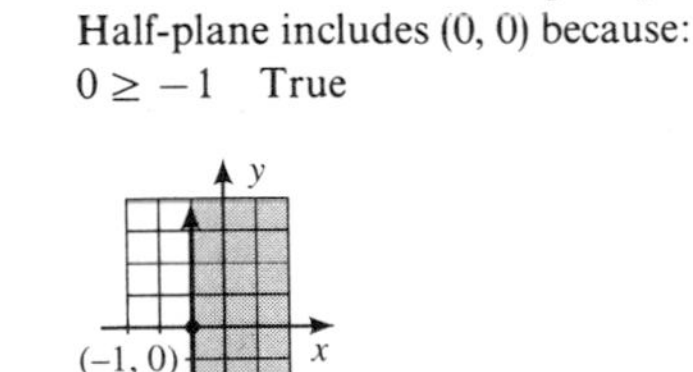

8. (0, −4)

9. Boundary line: $6x - 13y = 0$
Dashed line because the inequality is $>$.
Boundary line passes through the origin.
Half-plane does not include (1, 1) because:
$6(1) - 13(1) > 0$
$-7 > 0$ False

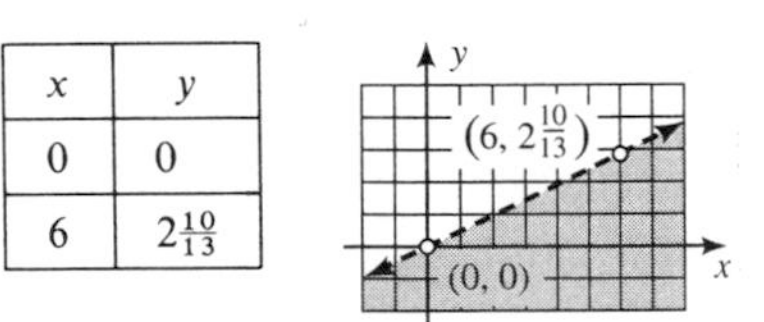

x	y
0	0
6	$2\frac{10}{13}$

10. (0, 0); $(2, -\frac{8}{9})$

11. Boundary line: $14x + 3y = 17$
Solid line because the inequality is $\leq$.
Half-plane includes (0, 0) because:
$14(0) + 3(0) \leq 17$
$0 \leq 17$ True

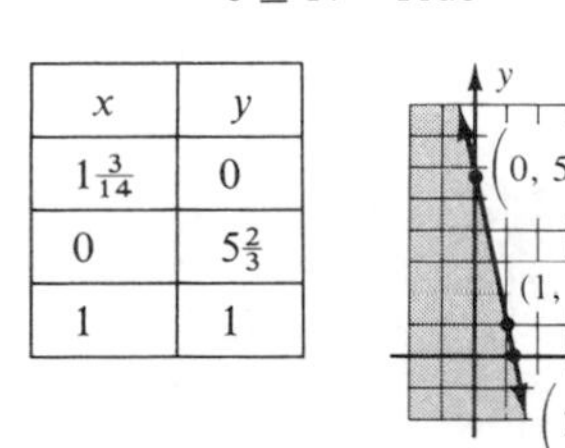

x	y
$1\frac{3}{14}$	0
0	$5\frac{2}{3}$
1	1

12.

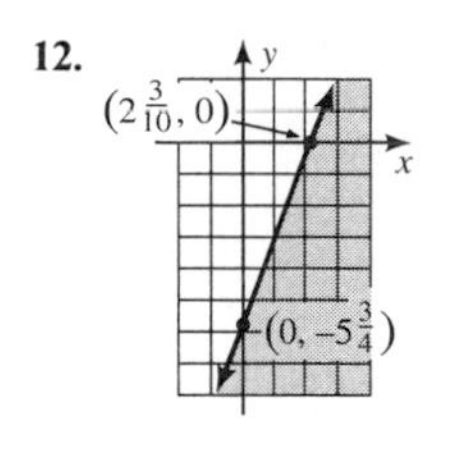

13. $\overset{1}{\cancel{4}}\left(\frac{x}{\underset{1}{\cancel{4}}}\right) - \overset{2}{\cancel{4}}\left(\frac{y}{\underset{1}{\cancel{2}}}\right) > 4(1)$
$x - 2y > 4$
Boundary line: $x - 2y = 4$
Dashed line because the inequality is $>$.
Half-plane does not include (0, 0) because:
$(0) - 2(0) > 4$
$0 > 4$ False

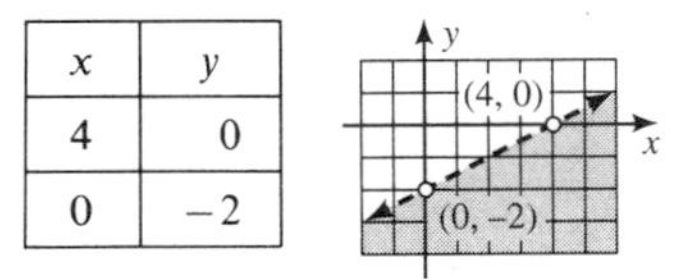

x	y
4	0
0	-2

14. (0, 5); (3, 0)

15. $4x + 8 + 7 \leq 15 - 6x$
$10x \leq 0$
$x \leq 0$
Boundary line: $x = 0$
Solid line because the inequality is $\leq$.
Half-plane does not include (1, 0) because:
$1 \leq 0$ False

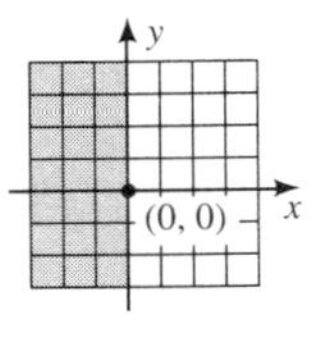

16.

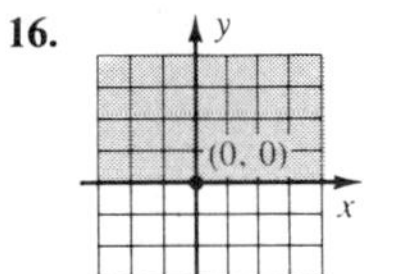

17.

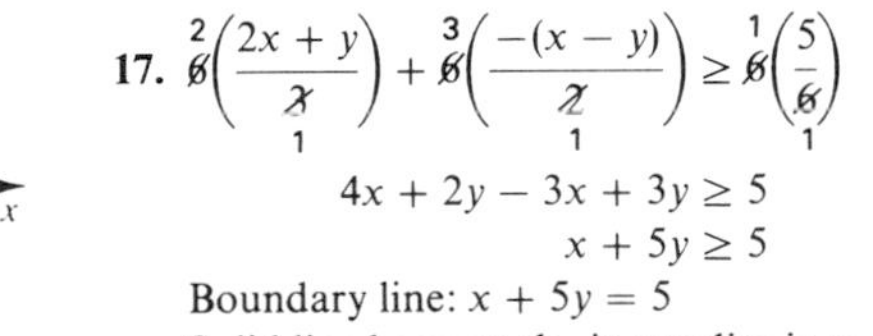

$\overset{2}{\cancel{6}}\left(\frac{2x + y}{\underset{1}{\cancel{3}}}\right) + \overset{3}{\cancel{6}}\left(\frac{-(x - y)}{\underset{1}{\cancel{2}}}\right) \geq \overset{1}{\cancel{6}}\left(\frac{5}{\underset{1}{\cancel{6}}}\right)$
$4x + 2y - 3x + 3y \geq 5$
$x + 5y \geq 5$
Boundary line: $x + 5y = 5$
Solid line because the inequality is $\geq$.
Half-plane does not include (0, 0) because:
$(0) + 5(0) \geq 5$
$0 \geq 5$ False

18.

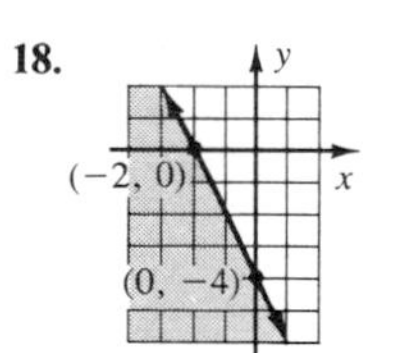

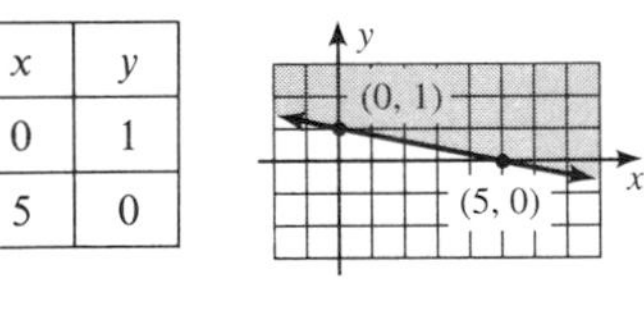

x	y
0	1
5	0

Review Exercises 8.5 (page 396)

1. a. $d = \sqrt{[2 - (-3)]^2 + [-4 - (-4)]^2} = \sqrt{5^2 + 0^2} = 5$

b. $d = \sqrt{[4 - (-2)]^2 + [1 - (-3)]^2} = \sqrt{6^2 + 4^2}$
$= \sqrt{36 + 16} = \sqrt{52} = \sqrt{4 \cdot 13} = 2\sqrt{13}$

2. Domain is $\{0, -2, 3\}$; range is $\{5, 3, -4, 0\}$.

3. Domain is $\{-2, 0, 6, 8\}$;
range is $\{-6, -3, 6, 9\}$.

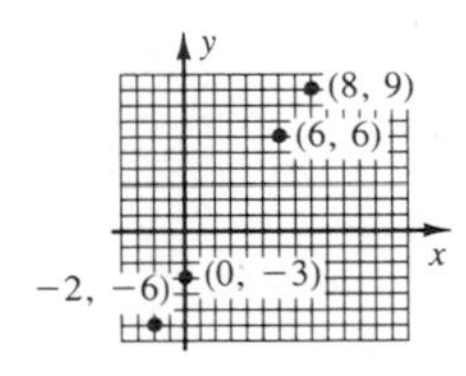

4.

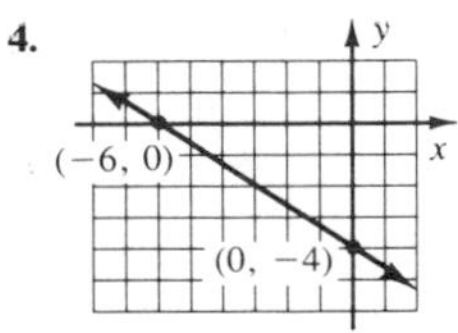

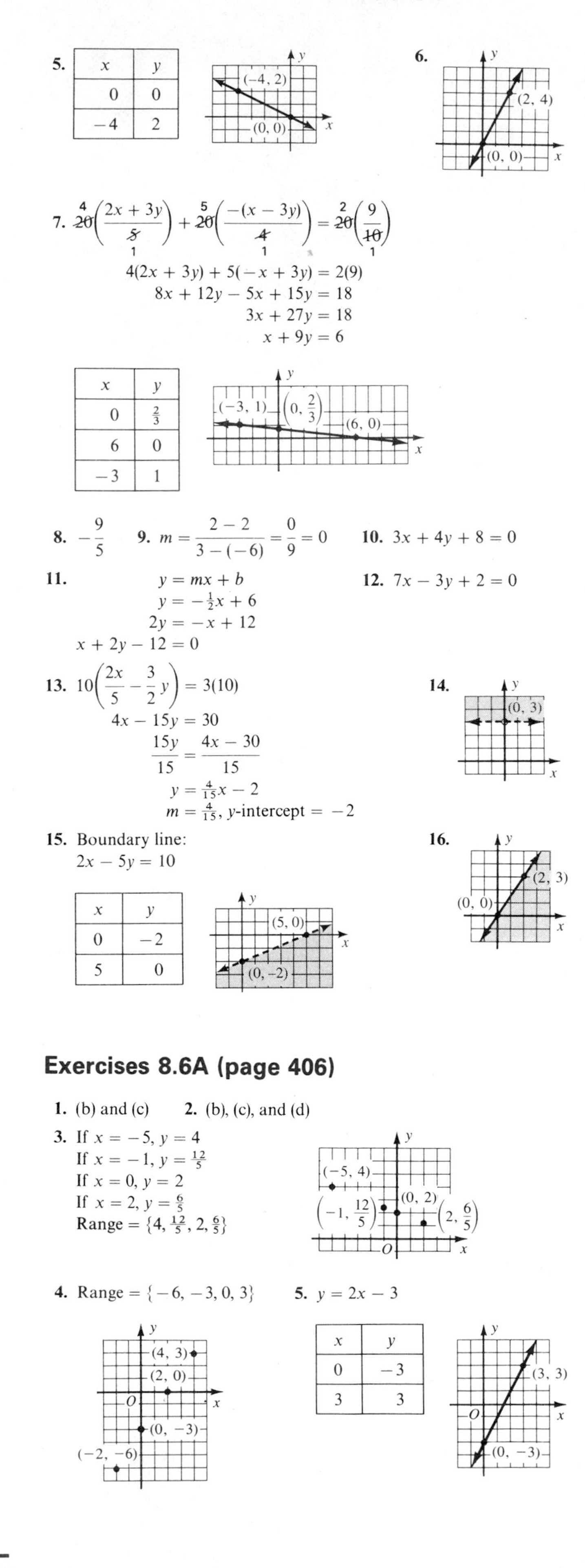

5.

x	y
0	0
−4	2

6.

7. $\overset{4}{\cancel{20}}\left(\frac{2x+3y}{\underset{1}{\cancel{5}}}\right)+\overset{5}{\cancel{20}}\left(\frac{-(x-3y)}{\underset{1}{\cancel{4}}}\right)=\overset{2}{\cancel{20}}\left(\frac{9}{\underset{1}{\cancel{10}}}\right)$

$$4(2x+3y)+5(-x+3y)=2(9)$$
$$8x+12y-5x+15y=18$$
$$3x+27y=18$$
$$x+9y=6$$

x	y
0	$\frac{2}{3}$
6	0
−3	1

8. $-\frac{9}{5}$ 9. $m=\frac{2-2}{3-(-6)}=\frac{0}{9}=0$ 10. $3x+4y+8=0$

11. $y=mx+b$
$y=-\frac{1}{2}x+6$
$2y=-x+12$
$x+2y-12=0$

12. $7x-3y+2=0$

13. $10\left(\frac{2x}{5}-\frac{3}{2}y\right)=3(10)$
$4x-15y=30$
$\frac{15y}{15}=\frac{4x-30}{15}$
$y=\frac{4}{15}x-2$
$m=\frac{4}{15}$, y-intercept $=-2$

14.

15. Boundary line:
$2x-5y=10$

x	y
0	−2
5	0

16.

Exercises 8.6A (page 406)

1. (b) and (c) 2. (b), (c), and (d)

3. If $x=-5$, $y=4$
If $x=-1$, $y=\frac{12}{5}$
If $x=0$, $y=2$
If $x=2$, $y=\frac{6}{5}$
Range $=\{4, \frac{12}{5}, 2, \frac{6}{5}\}$

4. Range $=\{-6, -3, 0, 3\}$ 5. $y=2x-3$

x	y
0	−3
3	3

6.

7. **a.** Yes **b.** No **c.** No **d.** Yes

8. **a.** No **b.** Yes **c.** Yes **d.** No

9. **a.** The domain is the set of all real numbers. (The curve extends infinitely far to the left and to the right.) The range is the set of all real numbers. (The curve extends infinitely far up and down.)
   b. The domain is the set of all real numbers, and so is the range. (The equation is the equation of a straight line that is not a horizontal or vertical line.)
   c. For the domain: $x-5\geq 0 \Rightarrow x\geq 5$. Therefore, the domain is $\{x\,|\,x\geq 5\}$. For the range, we know that the square root sign has an understood plus sign in front of it. Therefore, y cannot be negative. As x gets larger and larger, so does $x-5$, and so does $\sqrt{x-5}$. Therefore, the range is $\{y\,|\,y\geq 0\}$.
   d. The domain is $\{1, 4, 7\}$. If $x=1$, $y=0$. If $x=4$, $y=-3$. If $x=7$, $y=-6$. Therefore, the range is $\{0, -3, -6\}$.

10. **a.** Domain: $\{x\,|\,-3\leq x\leq 3\}$ Range: $\{y\,|\,-2\leq y\leq 0\}$
    b. Domain: $\{x\,|\,x\geq -3\}$ Range: $\{y\,|\,y\geq 0\}$
    c. Domain: $\{x\,|\,x\in R\}$ Range: $\{y\,|\,y\in R\}$
    d. Domain: $\{0, 1\}$ Range: $\{-4, -1\}$

Exercises 8.6B (page 411)

1. **a.** $f(2)=3(2)-1=6-1=5$
   b. $f(0)=3(0)-1=-1$
   c. $f(a-2)=3(a-2)-1=3a-6-1=3a-7$
   d. $f(x+2)=3(x+2)-1=3x+6-1=3x+5$

2. **a.** $f(3)=13$ **b.** $f(-5)=-19$ **c.** $f(0)=1$
   d. $f(x-2)=4x-7$

3. $f(5)=2(5)^2-3=50-3=47$
$f(2)=2(2)^2-3=5$
Therefore, $\frac{f(5)-f(2)}{6}=\frac{47-5}{6}=\frac{42}{6}=7.$

4. 8

5. $f(3)=3(3)^2-2(3)+4=27-6+4=25$
$f(1)=3(1)^2-2(1)+4=3-2+4=5$
$f(0)=3(0)^2-2(0)+4=4$
Therefore, $2f(3)+4f(1)-3f(0)$
$=2(25)+4(5)-3(4)=50+20-12=58$

6. 46

7. $f(-3)=(-3)^3=-27$
$g(2)=\frac{1}{2}$
$6g(2)=6\left(\frac{1}{2}\right)=3$
Therefore, $f(-3)-6g(2)=-27-3=-30.$

8. 29

9. $H(2)=3(2)^2-2(2)+4=12-4+4=12$
$K(3)=(3)-(3)^2=3-9=-6$
Therefore, $2H(2)-3K(3)=2(12)-3(-6)=24+18=42.$

10. −4

11. $D_f=\{x\,|\,x\geq 4\}$ because $\sqrt{x-4}$ is not real when $x-4<0$ or $x<4$.
$R_f=\{y\,|\,y\geq 0\}$ because $y=\sqrt{x-4}$, and the principal square root cannot be negative.

12. $D_f=\{x\,|\,x\geq \frac{16}{9}\}$; $R_f=\{y\,|\,y\leq 0\}$

13.
$$\begin{aligned} f(x) &= x^2 - x \\ f(x+h) &= (x+h)^2 - (x+h) \\ &= x^2 + 2xh + h^2 - x - h \\ \frac{f(x+h) - f(x)}{h} &= \frac{x^2 + 2xh + h^2 - x - h - x^2 + x}{h} \\ &= \frac{2xh + h^2 - h}{h} = \frac{2xh}{h} + \frac{h^2}{h} + \frac{-h}{h} = 2x + h - 1 \end{aligned}$$

14. $6x + 3h$

15. $f(x) = x^4$ ← We use the binomial theorem
$$\begin{aligned} \frac{f(x+h) - f(x)}{h} &= \frac{(x+h)^4 - x^4}{h} \\ &= \frac{\cancel{x^4} + 4x^3h + 6x^2h^2 + 4xh^3 + h^4 - \cancel{x^4}}{h} \\ &= \frac{4x^3h + 6x^2h^2 + 4xh^3 + h^4}{h} \\ &= \frac{\overset{1}{\cancel{h}}(4x^3 + 6x^2h + 4xh^2 + h^3)}{\underset{1}{\cancel{h}}} \\ &= 4x^3 + 6x^2h + 4xh^2 + h^3 \end{aligned}$$

16. $3x^2 + 3xh + h^2$

17. $A(r) = \pi r^2 \qquad C(r) = 2\pi r$

Then
$$\begin{aligned} &\frac{3A(r) - 2C(r)}{\pi r} \\ &= \frac{3\pi r^2 - 2(2\pi r)}{\pi r} \\ &= \frac{3\pi r^2}{\pi r} + \frac{-4\pi r}{\pi r} = 3r - 4 \end{aligned}$$

18. $5r + 6$

19. $h(2) = 64(2) - 16(2)^2 = 128 - 64 = 64$

20. 112

21.
$$\begin{aligned} C(100) &= 500 + 20(100) - 0.1(100^2) \\ &= 500 + 2{,}000 - 0.1(10{,}000) \\ &= 2{,}500 - 1{,}000 = 1{,}500 \end{aligned}$$

22. 400

23.
$$\begin{aligned} g(3, -4) &= 5(3)^2 - 2(-4)^2 + 7(3) - 4(-4) \\ &= 45 - 32 + 21 + 16 = 50 \end{aligned}$$

24. $z = h(-1, -2) = 10$

25.
$$\begin{aligned} f(100, 0.08, 12) &= 100(1 + 0.08)^{12} = 100(1.08)^{12} \\ &\approx 251.82 \end{aligned}$$

26. 2,979,659,632

Exercises 8.6C (page 415)

1. a.

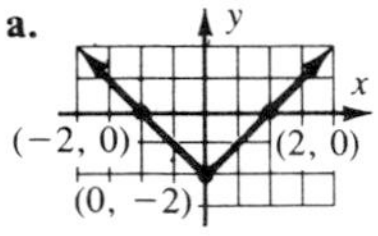

b.

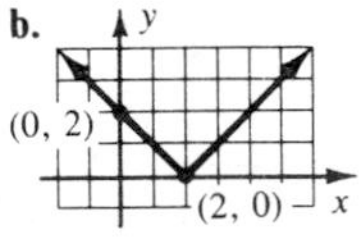

c.

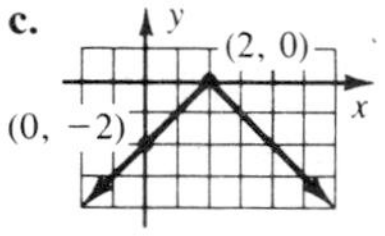

2. a.

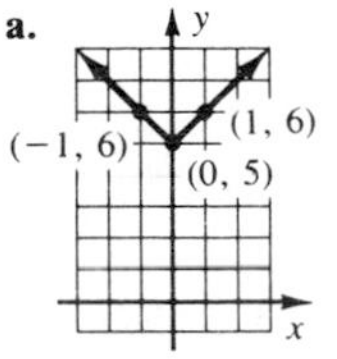

b.

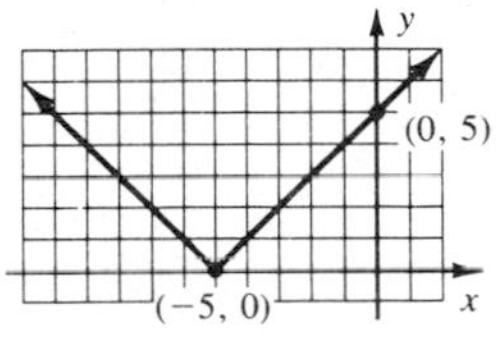

c.

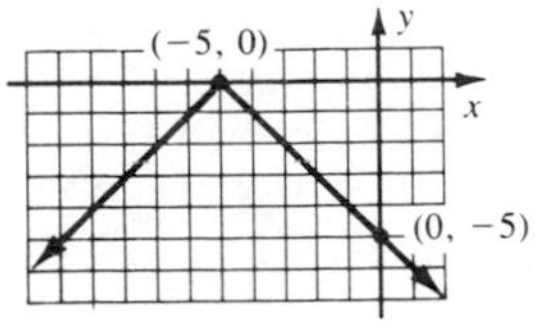

3. a.

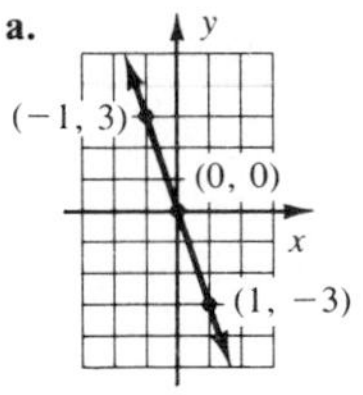

b.

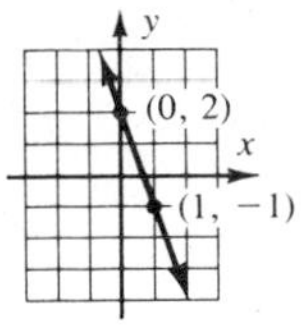

c.

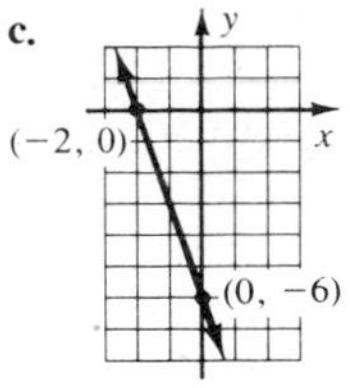

4. a.

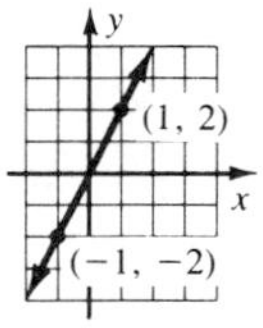

b.

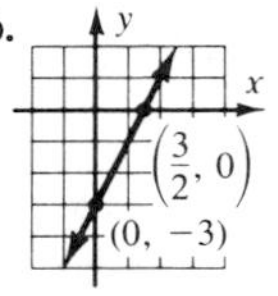

c.

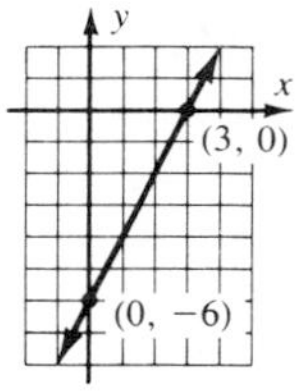

Exercises 8.7 (page 419)

1. $\mathcal{R} = \{(-10, 7), (3, -8), (-5, -4), (3, 9)\}$

$\mathcal{R}^{-1} = \{(7, -10), (-8, 3), (-4, -5), (9, 3)\}$

$D_{\mathcal{R}^{-1}} = \{7, -8, -4, 9\} = R_{\mathcal{R}}$

$R_{\mathcal{R}^{-1}} = \{-10, 3, -5, 3\} = D_{\mathcal{R}}$

$\mathcal{R}$ *is not* a function.

$\mathcal{R}^{-1}$ *is* a function.

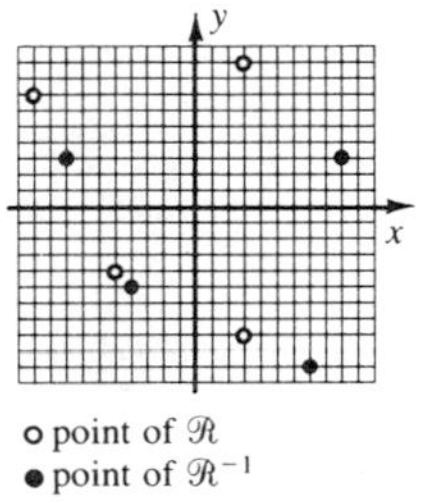

2. $\mathcal{R}^{-1} = \{(-6, 9), (11, 0), (8, 3), (-6, -2), (-4, 10)\}$

$D_{\mathcal{R}^{-1}} = \{-6, 11, 8, -4\} = R_{\mathcal{R}}$

$R_{\mathcal{R}^{-1}} = \{9, 0, 3, -2, 10\} = D_{\mathcal{R}}$

$\mathcal{R}$ is a function.

$\mathcal{R}^{-1}$ *is not* a function.

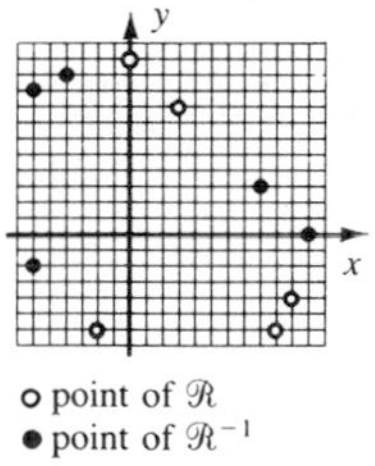

3. $x = 5 - 2y \to y = \dfrac{5 - x}{2} = f^{-1}(x)$

$f(x) = 5 - 2x$

x	y
$2\frac{1}{2}$	0
0	5

$f^{-1}(x) = \dfrac{5 - x}{2}$

x	y
5	0
0	$2\frac{1}{2}$

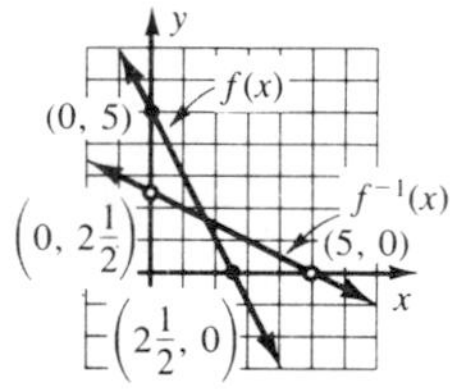

4. $f^{-1}(x) = \dfrac{x + 10}{3}$

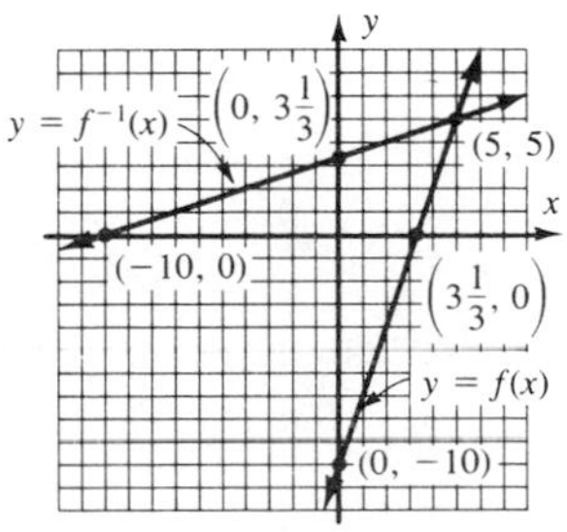

5.
$$\begin{aligned} x &= \frac{4y - 3}{5} \\ 5x &= 4y - 3 \\ 5x + 3 &= 4y \\ y &= \frac{5x + 3}{4} = f^{-1}(x) \end{aligned}$$

6. $f^{-1}(x) = \dfrac{3x + 7}{2}$

7. $x = \dfrac{5}{y+2}$

$x(y+2) = 5$

$xy + 2x = 5$

$xy = 5 - 2x$

$y = \dfrac{5 - 2x}{x} = f^{-1}(x)$

8. $f^{-1}(x) = \dfrac{x + 10}{2x}$

Exercises 8.8 (page 424)

1.

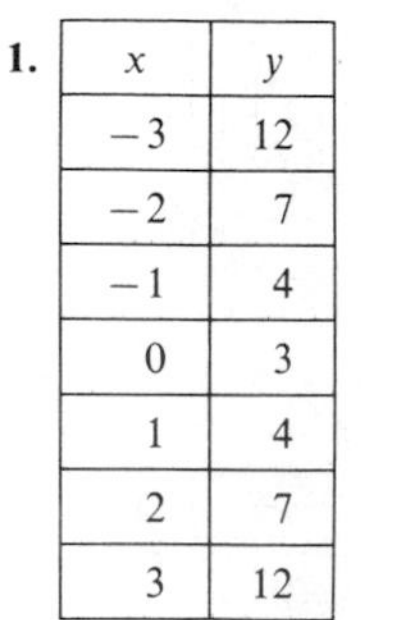

x	y
−3	12
−2	7
−1	4
0	3
1	4
2	7
3	12

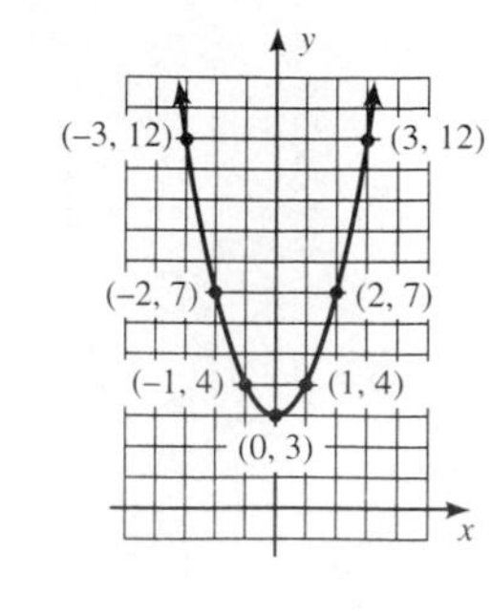

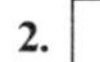

2.

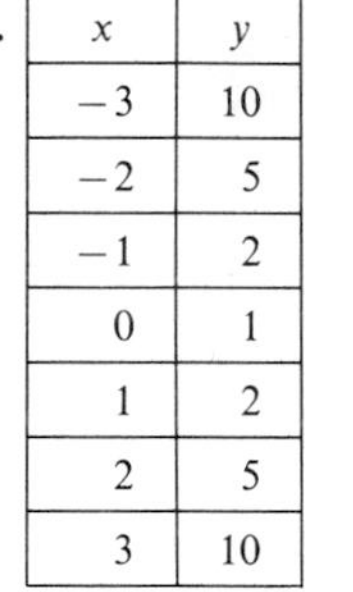

x	y
−3	10
−2	5
−1	2
0	1
1	2
2	5
3	10

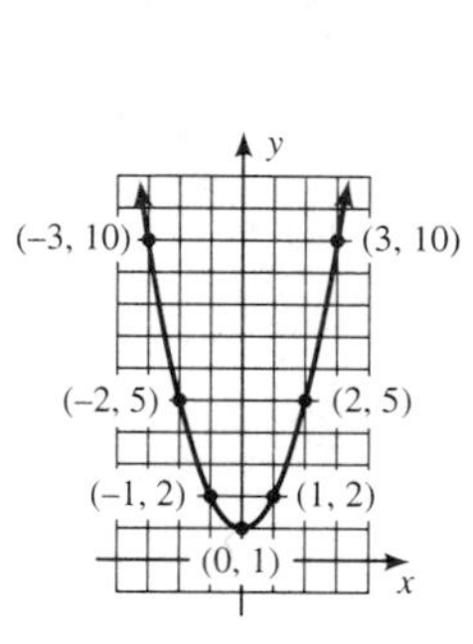

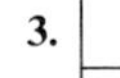

3.

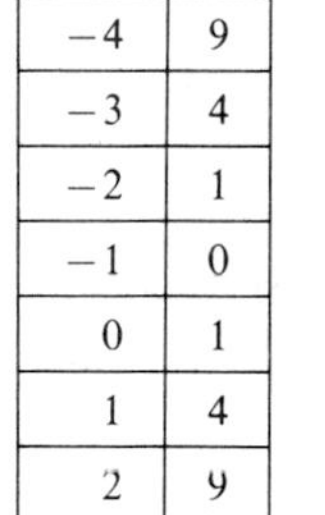

x	y
−4	9
−3	4
−2	1
−1	0
0	1
1	4
2	9

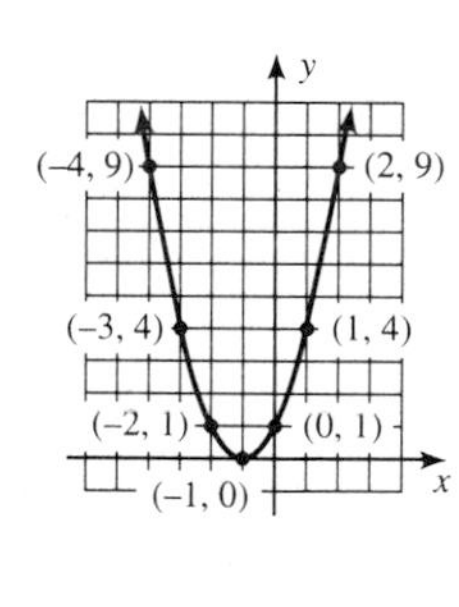

4.

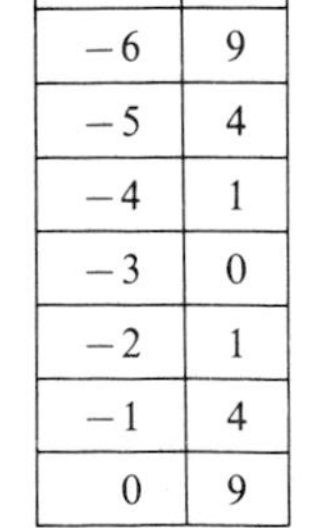

x	y
−6	9
−5	4
−4	1
−3	0
−2	1
−1	4
0	9

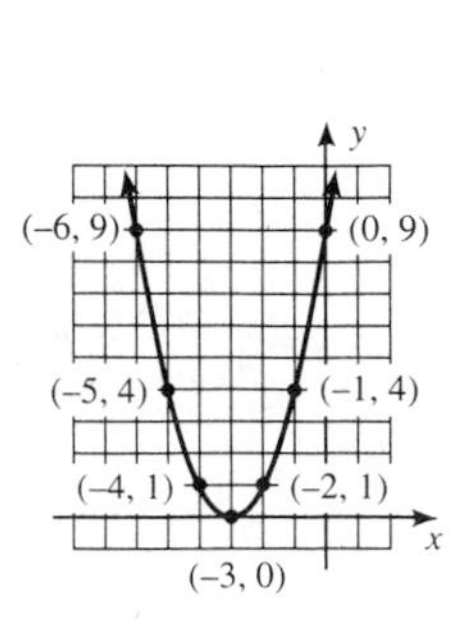

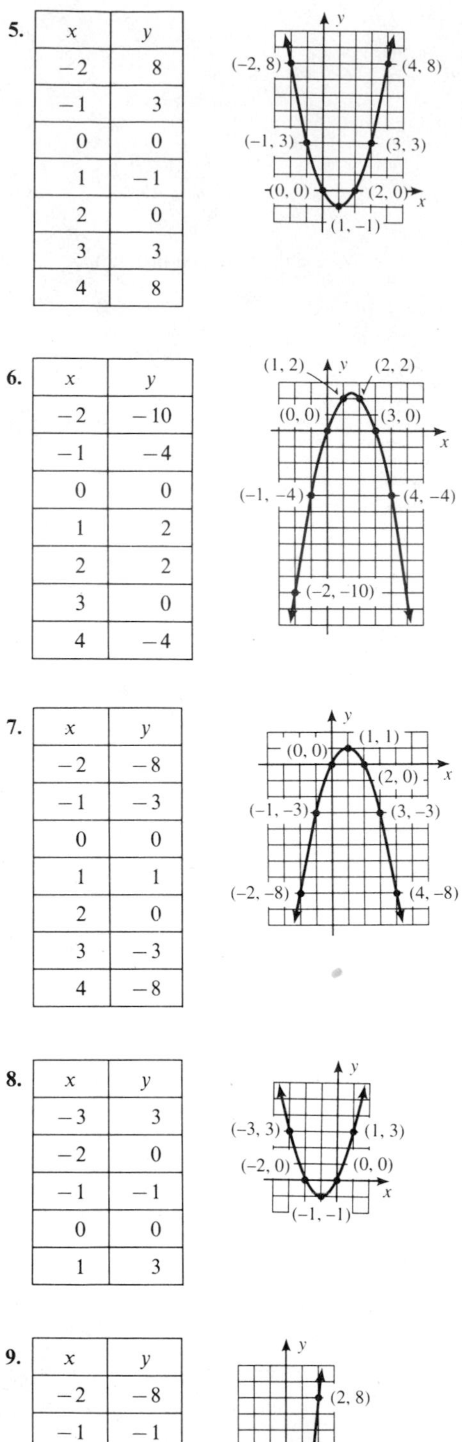

5.

x	y
−2	8
−1	3
0	0
1	−1
2	0
3	3
4	8

6.

x	y
−2	−10
−1	−4
0	0
1	2
2	2
3	0
4	−4

7.

x	y
−2	−8
−1	−3
0	0
1	1
2	0
3	−3
4	−8

8.

x	y
−3	3
−2	0
−1	−1
0	0
1	3

9.

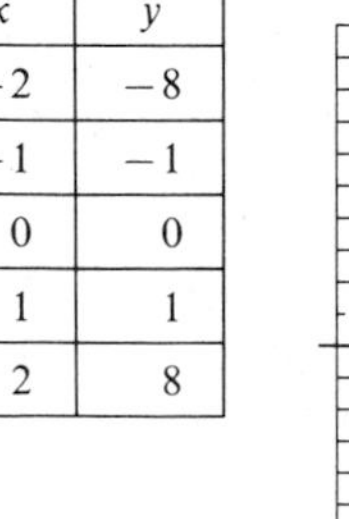

x	y
−2	−8
−1	−1
0	0
1	1
2	8

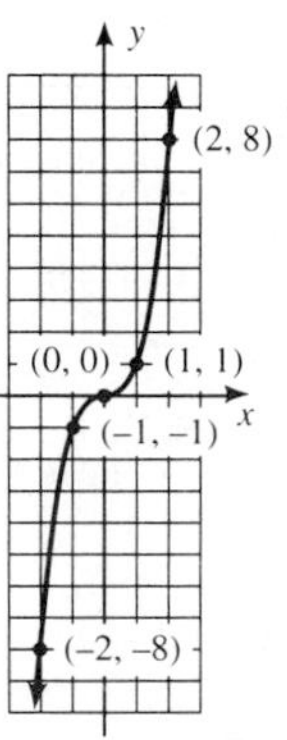

10.

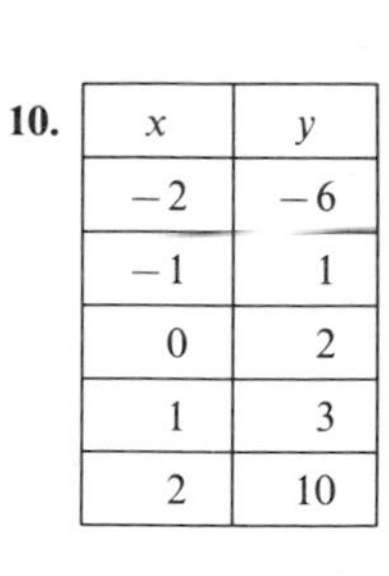

x	y
-2	-6
-1	1
0	2
1	3
2	10

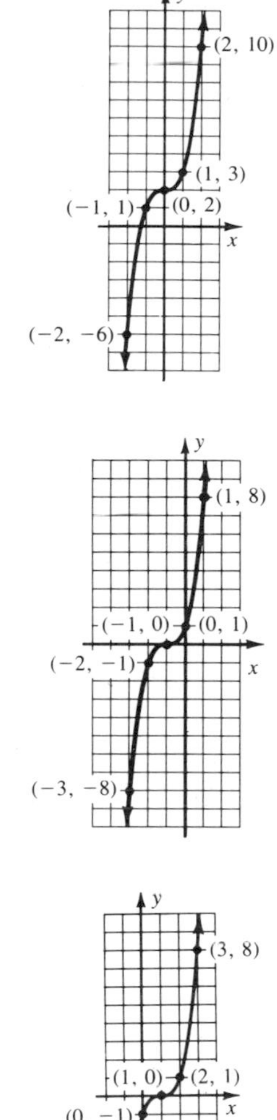

11.

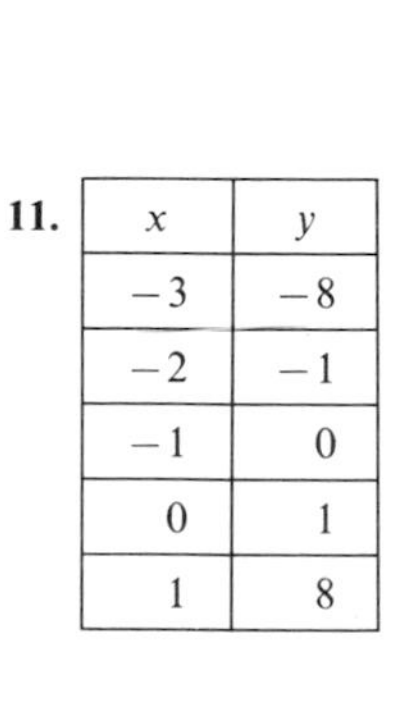

x	y
-3	-8
-2	-1
-1	0
0	1
1	8

12.

x	y
-1	-8
0	-1
1	0
2	1
3	8

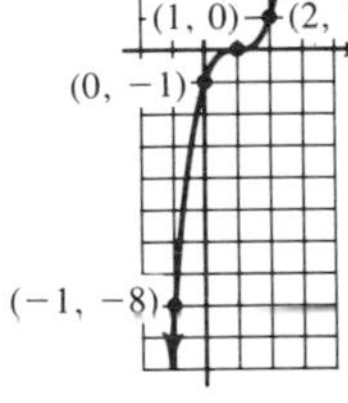

13. We find the range here by using synthetic division. Substitution can be used, instead.

x	1	-2	-13	20 $f(x)$
-4	1	-6	11	-24
-3	1	-5	2	14
-2	1	-4	-5	30
0	1	-2	-13	20
2	1	0	-13	-6
3	1	1	-10	-10
4	1	2	-5	0
5	1	3	2	30

Range $= \{-24, 14, 30, 20, -6, -10, 0\}$

14. $\{18, 32, 12, -32, -48, -52, -38, 0\}$

Exercises 8.9A (page 431)

1. $y = kx$
$12 = k(3)$
$k = 4$
$y_2 = 4(5)$
$y_2 = 20$

Condition 1 $\begin{cases} x_1 = 3 \\ y_1 = 12 \end{cases}$
Condition 2 $\begin{cases} x_2 = 5 \\ y_2 = ? \end{cases}$

Alternate method:
$$\frac{y_1}{y_2} = \frac{x_1}{x_2}$$
$$\frac{12}{y_2} = \frac{3}{5}$$
$$3y_2 = (5)(12)$$
$$y_2 = \frac{(5)(12)}{3} = 20$$

2. $k = 2.5$
$y_2 = -15$

3. Let M = number of miles on 50 gal of gasoline.
$M = kG$
$161 = k(23)$
$k = 7$
$M_2 = 7(50) = 350$

Alternate method:
$$\frac{161 \text{ mi}}{23 \text{ gal}} = \frac{M \text{ mi}}{50 \text{ gal}}$$
$$23\,M = (161)(50)$$
$$M = 350$$
Leon can expect to drive 350 mi on 50 gal of gasoline.

4. $k = 5$
$m = 375$

5. Let x = height of tree.
$$\frac{5 \text{ ft}}{3 \text{ ft}} = \frac{x \text{ ft}}{27 \text{ ft}}$$
$3x = 5(27)$
$x = 45$; the tree is 45 ft tall.

6. 33 ft

7. $C = kr$
$47.1 = k(7.5)$
$k = 6.28$
$C_2 = 6.28(4.5)$
$C_2 = 28.26$

Alternate method:
$$\frac{47.1}{C} = \frac{7.5}{4.5}$$
$7.5C = (47.1)(4.5)$
$7.5C = 211.95$
$C = 28.26$

8. $k = 0.433$
$P - 7.794$

9. $A = kr^2$
$28.26 = k(3)^2$
$$k = \frac{28.26}{9} = 3.14$$
$A_2 = 3.14(6)^2 = 113.04$

Alternate method:
$$\frac{28.26}{A} = \frac{3^2}{6^2}$$
$$A = \frac{(28.26)(36)}{9} = 113.04$$

10. $k = 12.56$; $S = 200.96$

11. Let S = amount of sediment carried by current
s = speed of current

Condition 1 $\begin{cases} S_1 = 1 \\ s_1 = 2 \end{cases}$ $\quad \dfrac{S_1}{S_2} = \dfrac{s_1^6}{s_2^6}$

Condition 2 $\begin{cases} S_2 = ? \\ s_2 = 4 \end{cases}$ $\quad \dfrac{1}{S_2} = \dfrac{2^6}{4^6}$

$S_2 = \dfrac{4^6}{2^6} = \dfrac{(2^2)^6}{2^6} = \dfrac{2^{12}}{2^6} = 2^{12-6} = 2^6 = 64$; the stream carries 64 units of sediment. This shows that when the speed of the current is doubled, its destructive power becomes 64 times as great.

12. \$213.25

Exercises 8.9B (page 434)

1. $y = \dfrac{k}{x}$
$7 = \dfrac{k}{2}$
$k = 14$
$y_2 = \dfrac{14}{-7} = -2$

2. $k = -48$
$z = -6$

3. $P = \dfrac{k}{V}$
$18 = \dfrac{k}{15}$
$k = 270$
$P_2 = \dfrac{270}{10} = 27$

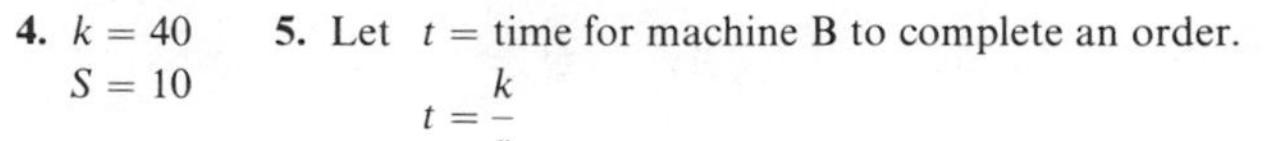

4. $k = 40$
$S = 10$

5. Let t = time for machine B to complete an order.

$$t = \frac{k}{r}$$
$$3 = \frac{k}{375}$$
$$k = 1{,}125$$
$$t_2 = \frac{1{,}125}{225} = 5$$; it would take machine B 5 hr to complete an order.

6. $k = 2{,}475$
7.5 hr

7.
$$y = \frac{k}{x^2}$$
$$9 = \frac{k}{4^2}$$
$$k = 9(16) = 144$$
$$y_2 = \frac{144}{3^2} = \frac{144}{9} = 16$$

8. $k = 72$
$C = 2$

9.
$$F = \frac{k}{d^2}$$
$$3 = \frac{k}{4^2}$$
$$k = 48$$
$$F_2 = \frac{48}{2^2} = 12$$

10. $6\frac{2}{3}$

11. a. If the length is ℓ and the width is w, $\ell = \frac{k}{w}$.

b. $15 = \frac{k}{3}$
$k = 45$
$\ell = \frac{45}{5} = 9$; the length is 9 cm.

12. a. If the altitude is h and the base is b, $h = \frac{k}{b}$.

b. $k = 60$; $h = 6$; the altitude is 6 cm.

Exercises 8.9C (page 438)

1.
$$z = kxy$$
$$-36 = k(-3)(2)$$
$$k = 6$$
$$z_2 = (6)(4)(3) = 72$$

2. $k = 4$
$A = 84$

3.
$$I = kPr$$
$$115.50 = k(880)(0.0875)$$
$$k = 1.5$$
$$I_2 = (1.5)(760)(0.0925) = 105.45$$

4. $k = 3.6084021$
$I \approx 428.50$

5.
$$F = kAV^2$$
$$1.8 = k(1)(20^2)$$
$$k = 0.0045$$
$$F_2 = 0.0045(2)(60^2) = 32.4$$
The force is 32.4 lb.

6. $P = 2{,}474.7$

7.
$$z = \frac{kx}{y}$$
$$12 = \frac{k(6)}{2}$$
$$k = 4$$
$$z_2 = \frac{(4)(-8)}{-4} = 8$$

8. $k = 4$
$R = 1.6$

9.
$$e = \frac{kPL}{A}$$
$$3 = \frac{k(2.4)(45)}{0.9}$$
$$k = 0.025$$
$$e_2 = \frac{0.025(1.5)(40)}{0.75} = 2$$

10. 300 lb

Review Exercises 8.10 (page 441)

1. (a) and (c) **2.** (b) and (c)

3. a.
$$f(2) = 3(2)^2 - 5(2) + 4 = 12 - 10 + 4 = 6$$

b.
$$f(0) = 3(0)^2 - 5(0) + 4 = 0 - 0 + 4 = 4$$

c.
$$f(x-1) = 3(x-1)^2 - 5(x-1) + 4 = 3(x^2 - 2x + 1) - 5x + 5 + 4 = 3x^2 - 11x + 12$$

4. -22

5. (graph) $(-2, 3)$, $(2, 3)$, $(-1, 2)$, $(1, 2)$, $(0, 1)$

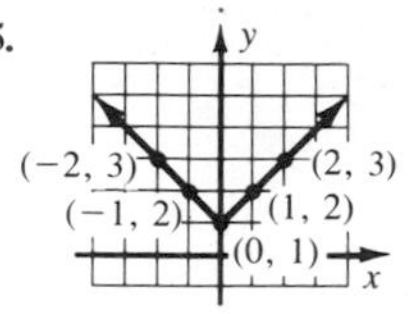

6. (graph) $(-2, 1)$, $(2, 1)$, $(-1, 0)$, $(1, 0)$, $(0, -1)$

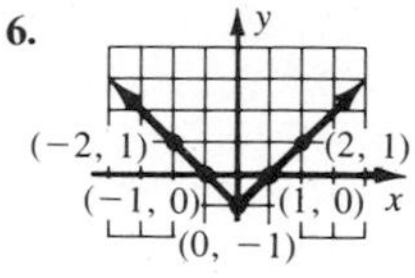

7. (graph) $(-3, 2)$, $(1, 2)$, $(-2, 1)$, $(0, 1)$, $(-1, 0)$

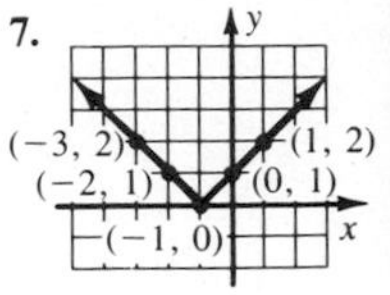

8. $\mathscr{R}^{-1} = \{(-2, -1), (4, -4), (0, 5), (-2, 4), (-3, -5)\}$
$D_{\mathscr{R}^{-1}} = \{-2, 4, 0, -3\} = R_{\mathscr{R}}$
$R_{\mathscr{R}^{-1}} = \{-1, -4, 5, 4, -5\} = D_{\mathscr{R}}$
$\mathscr{R}$ is a function.
$\mathscr{R}^{-1}$ *is not* a function.

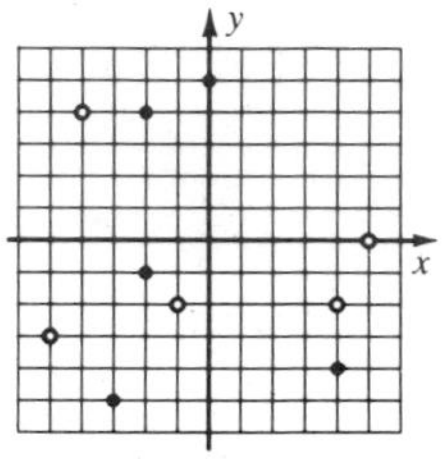

(graph) ∘ point of $\mathscr{R}$; • point of $\mathscr{R}^{-1}$

9. $3x = y + 6$
$y = 3x - 6 = f^{-1}(x)$

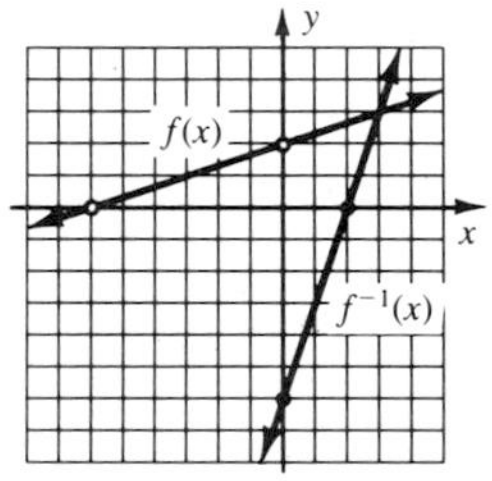

$f(x) = \frac{x+6}{3}$

x	y
-6	0
0	2

$f^{-1}(x) = 3x - 6$

x	y
2	0
0	-6

(graph of $f(x)$ and $f^{-1}(x)$)

The domain is the set of all real numbers for $y = f(x)$ and for $y = f^{-1}(x)$. The range is the set of all real numbers for $y = f(x)$ and for $y = f^{-1}(x)$.

10. $f^{-1}(x) = \frac{2 - 2x}{5}$

The domain is the set of all real numbers for $y = f(x)$ and for $y = f^{-1}(x)$. The range is the set of all real numbers for $y = f(x)$ and for $y = f^{-1}(x)$.

(graph) $f^{-1}(x)$, $f(x)$, $(0, 1)$, $(-4, 2)$, $(1, 0)$, $(2, -4)$

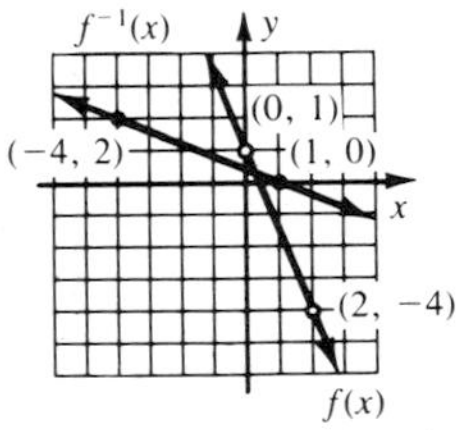

11.
$$x = \frac{9}{2 - 7y}$$
$$x(2 - 7y) = 9$$
$$2x - 7xy = 9$$
$$2x - 9 = 7xy$$
$$y = \frac{2x - 9}{7x} = f^{-1}(x)$$

12. $f^{-1}(x) = \frac{10x + 11}{8x}$

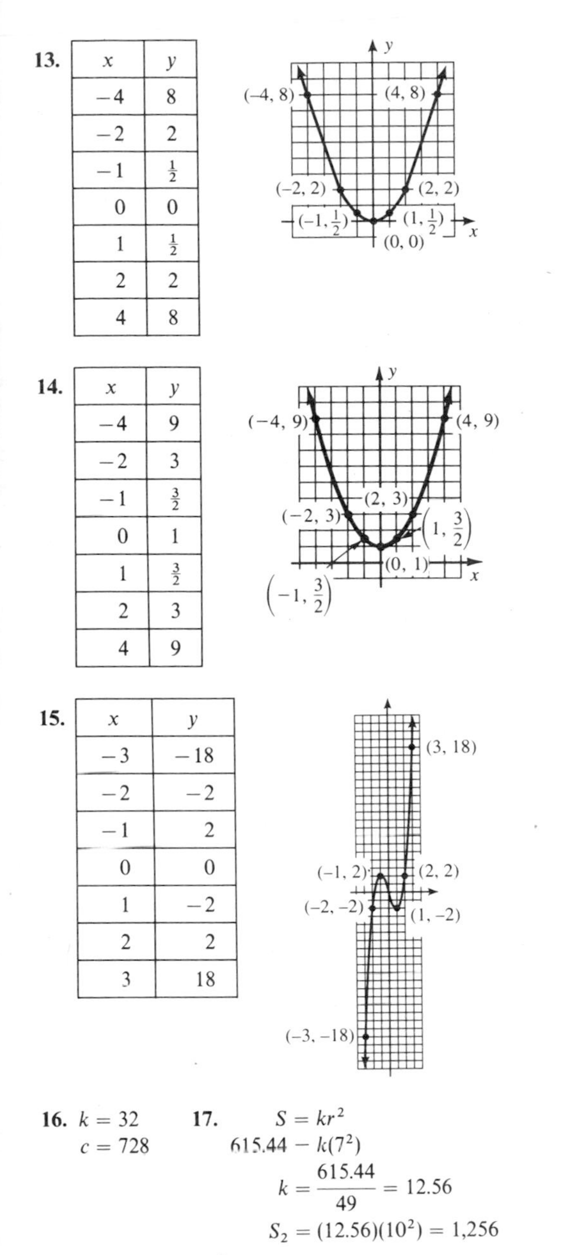

13.

x	y
-4	8
-2	2
-1	$\frac{1}{2}$
0	0
1	$\frac{1}{2}$
2	2
4	8

14.

x	y
-4	9
-2	3
-1	$\frac{3}{2}$
0	1
1	$\frac{3}{2}$
2	3
4	9

15.

x	y
-3	-18
-2	-2
-1	2
0	0
1	-2
2	2
3	18

16. $k = 32$
$c = 728$

17. $S = kr^2$
$615.44 = k(7^2)$
$k = \dfrac{615.44}{49} = 12.56$
$S_2 = (12.56)(10^2) = 1{,}256$

18. $k = 500$; $t_2 = 6\frac{2}{3}$
It would take machine B $6\frac{2}{3}$ hr to complete the order.

Chapter 8 Diagnostic Test (page 447)

Following each problem number is the textbook section number (in parentheses) where that kind of problem is discussed.

1. a. (8.1)

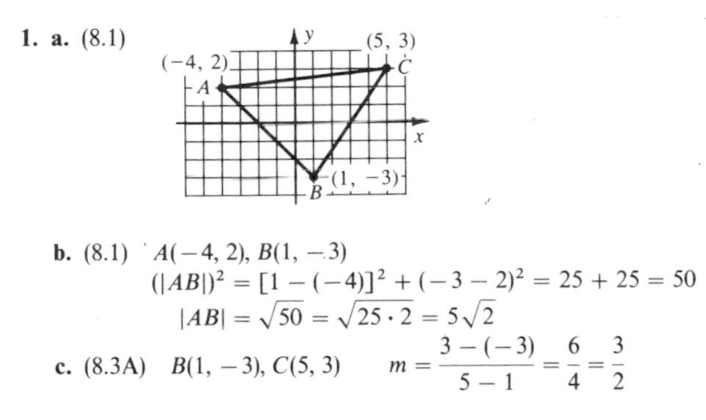

b. (8.1) $A(-4, 2)$, $B(1, -3)$
$(|AB|)^2 = [1 - (-4)]^2 + (-3 - 2)^2 = 25 + 25 = 50$
$|AB| = \sqrt{50} = \sqrt{25 \cdot 2} = 5\sqrt{2}$

c. (8.3A) $B(1, -3)$, $C(5, 3)$ $\quad m = \dfrac{3 - (-3)}{5 - 1} = \dfrac{6}{4} = \dfrac{3}{2}$

d. (8.3B) $A(-4, 2)$, $B(1, -3)$ $\quad m = \dfrac{-3 - 2}{1 - (-4)} = \dfrac{-5}{5} = -1$
$y - y_1 = m(x - x_1)$
$y - (-3) = -1(x - 1)$
$y + 3 = -x + 1$
$x + y + 2 = 0$

e. (8.3) If $y = 0$, $x = -2$
The x-intercept is $(-2, 0)$.

f. (8.3) If $x = 0$, $y = -2$
The y-intercept is $(0, -2)$.

2. (8.3B) $y = mx + b$
$y = \frac{6}{5}x - 4$
$5y = 6x - 20$
$6x - 5y - 20 = 0$

3. (8.2) $x - 2y = 6$
If $x = 0$, $y = -3$
If $y = 0$, $x = 6$

x	y
0	-3
6	0

(6, 0)
(0, −3)

4. (8.8) $y = 1 + x - x^2$
$y = 1 + (-2) - (-2)^2 = -5$
$y = 1 + (-1) - (-1)^2 = -1$
$y = 1 + (0) - (0)^2 = 1$
$y = 1 + (1) - (1)^2 = 1$
$y = 1 + (2) - (2)^2 = -1$
$y = 1 + (3) - (3)^2 = -5$

x	y
-2	-5
-1	-1
0	1
1	1
2	-1
3	-5

(0, 1) (1, 1) (−1, −1) (2, −1) (−2, −5) (3, −5)

5. (8.4) $4x - 3y \leq -12$
Boundary line: $4x - 3y = -12$
If $x = 0$, $y = 4$ $\quad$ If $y = 0$, $x = -3$
Boundary line is solid because the inequality is $\leq$.
Test point: $(0, 0)$ $\quad 4x - 3y \leq -12$
$4(0) - 3(0) \leq -12$
$0 \leq -12$ False

(0, 4)
(−3, 0)

6. (8.1, 8.6) **a.** $\{(-4, -5), (2, 4), (4, -2), (-2, 3), (2, -1)\}$
Domain: $\{-4, 2, 4, -2\}$

b. $\{(-4, -5), (2, 4), (4, -2), (-2, 3), (2, -1)\}$
Range: $\{-5, 4, -2, 3, -1\}$

c.

(2, 4) (−2, 3) (2, −1) (4, −2) (−4, −5)

d. The relation *is not* a function because (2, 4) and (2, −1) have the same first coordinate and different second coordinates.

7. (8.6) $f(x) = 3(x)^2 - 5$

a. $f(-2) = 3(-2)^2 - 5 = 12 - 5 = 7$

b. $f(4) = 3(4)^2 - 5 = 48 - 5 = 43$

c. $\dfrac{f(4) - f(-2)}{6} = \dfrac{43 - 7}{6} = \dfrac{36}{6} = 6$

8. (8.1, 8.6) $\mathscr{R} = \{(1, 4), (-5, -3), (4, -2), (-5, 2)\}$

$\mathscr{R}^{-1} = \{(4, 1), (-3, -5), (-2, 4), (2, -5)\}$

$D_{\mathscr{R}^{-1}} = \{4, -3, -2, 2\} = R_{\mathscr{R}}$

$R_{\mathscr{R}^{-1}} = \{1, -5, 4\} = D_{\mathscr{R}}$

$\mathscr{R}$ *is not* a function.

$\mathscr{R}^{-1}$ is a function.

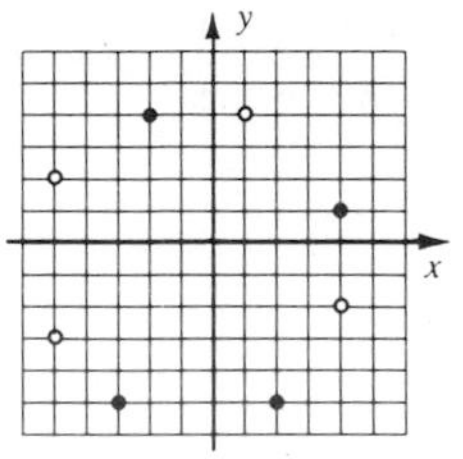

o point of $\mathscr{R}$
• point of $\mathscr{R}^{-1}$

9. a. (8.7) $y = f(x) = -\dfrac{3}{2}x + 1$

$x = -\dfrac{3}{2}y + 1$

$2x = -3y + 2$

$3y = 2 - 2x$

$y = \dfrac{2 - 2x}{3} = f^{-1}(x)$

$f(x) = -\dfrac{3}{2}x + 1$

x	y
$\frac{2}{3}$	0
0	1
4	−5

$f^{-1}(x) = \dfrac{2 - 2x}{3}$

x	y
1	0
0	$\frac{2}{3}$
−5	4

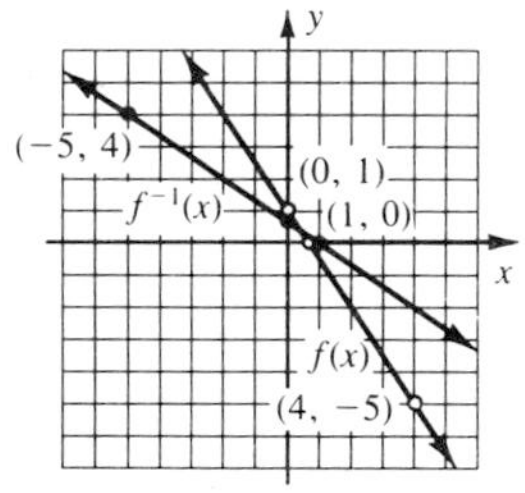

b. (8.9) $y = f(x) = \dfrac{15}{4(2 - 5x)}$

$D_{f(x)} =$ all real numbers except $\frac{2}{5} = R_{f^{-1}(x)}$

$x = \dfrac{15}{4(2 - 5y)}$

$4x(2 - 5y) = 15$

$8x - 20xy = 15$

$y = \dfrac{8x - 15}{20x} = f^{-1}(x)$

$D_{f^{-1}(x)} =$ all real numbers except $0 = R_{f(x)}$

10. (8.9) $w = \dfrac{kxy}{z^2}$

$20 = \dfrac{k(8)(6)}{(12)^2}$

$k = 60$

$w_2 = \dfrac{(60)(6)(10)}{(5)^2} = 144$

Cumulative Review Exercises: Chapters 1–8 (page 448)

1. $2x^4 - 24x^2 + 54 = 2(x^4 - 12x^2 + 27)$
$= 2(x^2 - 3)(x^2 - 9)$
$= 2(x^2 - 3)(x + 3)(x - 3)$

2. $ab(2a - 3b)(4a^2 + 6ab + 9b^2)$

3. $\sqrt[3]{-8a^5} = \sqrt[3]{-2^3a^3a^2} = -2a\sqrt[3]{a^2}$

4. $7x^2\sqrt{2} + 4x^2\sqrt{2x}$

5. $\dfrac{3 + 2i}{2 - i} \cdot \dfrac{2 + i}{2 + i} = \dfrac{6 + 7i + 2i^2}{4 - i^2} = \dfrac{6 + 7i - 2}{4 - (-1)} = \dfrac{4 + 7i}{5} = \dfrac{4}{5} + \dfrac{7}{5}i$

6. $2 + 2i$ **7.** $(3 + i)^4 = 3^4 + 4(3^3)i + 6(3^2)i^2 + 4(3)i^3 + i^4$
$= 81 + 108i + 54(-1) + 12(-i) + 1$
$= 81 - 54 + 1 + (108 - 12)i = 28 + 96i$

8. $r = \dfrac{S - a}{S}$ **9.** $x^{-1/3} = 3$
$(x^{-1/3})^{-3} = (3)^{-3}$
$x = \dfrac{1}{3^3} = \dfrac{1}{27}$
$\left\{\dfrac{1}{27}\right\}$

10. $3x^2 - 2x - 3$

11. $f(0) = 3(0)^2 - 2(0) + 7 = 7$
$f(-2) = 3(-2)^2 - 2(-2) + 7 = 3(4) + 4 + 7 = 12 + 11 = 23$

12.

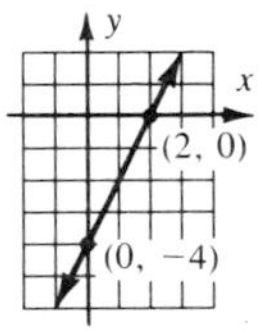

13. $m = \dfrac{4 - (-2)}{-3 - 1} = \dfrac{6}{-4} = -\dfrac{3}{2}$
$y - 4 = -\dfrac{3}{2}(x - [-3])$
$2(y - 4) = -3(x + 3)$
$2y - 8 = -3x - 9$
$3x + 2y + 1 = 0$

14. $k = -24$
$y_2 = -\dfrac{8}{9}$

15. Let $x =$ number of pounds of cashews
$60 - x =$ number of pounds of peanuts
$7.40x + 2.80(60 - x) = 4.18(60)$
$740x + 280(60 - x) = 418(60)$
$74x + 28(60 - x) = 418(6)$
$74x + 1{,}680 - 28x = 2{,}508$
$46x = 828$
$x = 18$
$60 - x = 42$

Check 18 lb @ \$7.40 = \$133.20
42 lb @ \$2.80 = \$117.60
\$250.80
60 lb @ \$4.18 = \$250.80

The grocer should use 18 lb of cashews and 42 lb of peanuts.

16. 88 mi

17. Let $x =$ length of shorter leg
$x + 7 =$ length of longer leg
$x + 9 =$ length of hypotenuse
$x^2 + (x + 7)^2 = (x + 9)^2$
$x^2 + x^2 + 14x + 49 = x^2 + 18x + 81$
$x^2 - 4x - 32 = 0$
$(x - 8)(x + 4) = 0$

$x - 8 = 0$	$x + 4 = 0$
$x = 8$	$x = -4 \leftarrow$ Reject
$x + 7 = 15$	
$x + 9 = 17$	

(The check will not be shown.) The shorter leg is 8 cm, the longer leg is 15 cm, and the hypotenuse is 17 cm.

18. $k = 6$; $S = 150$; the surface area is 150 cm^2.

Exercises 9.1A (page 452)

1. 20.0855 **2.** 1.6487 **3.** 0.0498 **4.** 0.6065

5. $y = 4^x$

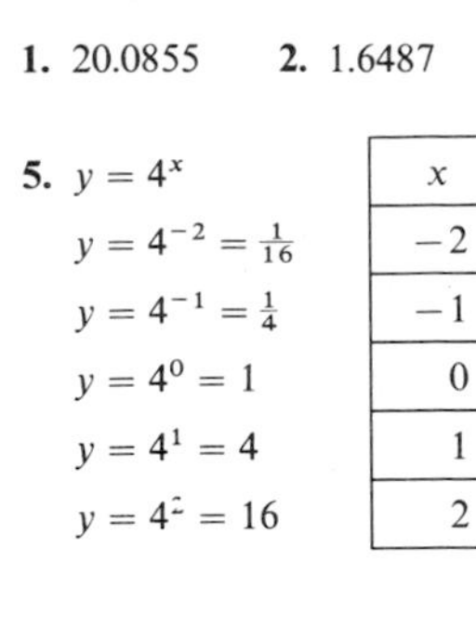

$y = 4^{-2} = \frac{1}{16}$

$y = 4^{-1} = \frac{1}{4}$

$y = 4^0 = 1$

$y = 4^1 = 4$

$y = 4^2 = 16$

x	y
-2	$\frac{1}{16}$
-1	$\frac{1}{4}$
0	1
1	4
2	16

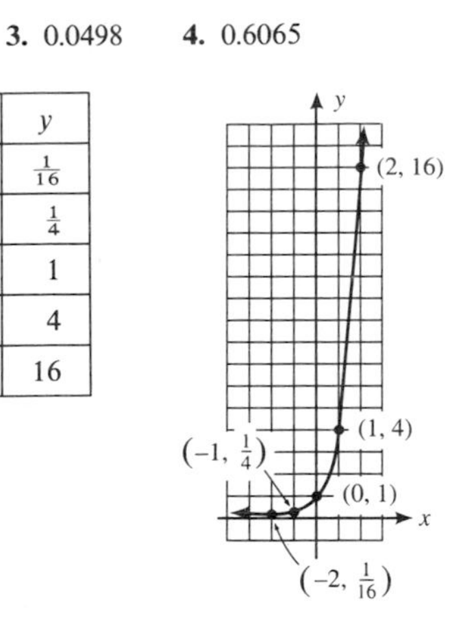

6.

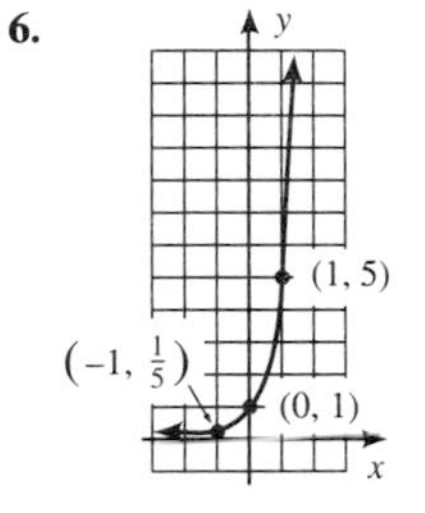

7. $y = e^x$

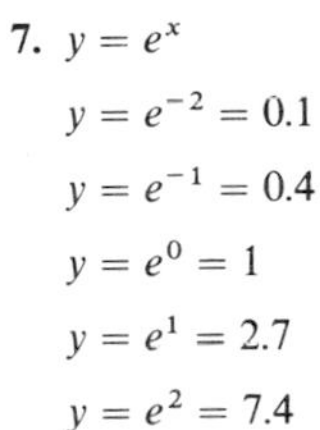

$y = e^{-2} = 0.1$

$y = e^{-1} = 0.4$

$y = e^0 = 1$

$y = e^1 = 2.7$

$y = e^2 = 7.4$

x	y
-2	0.1
-1	0.4
0	1
1	2.7
2	7.4

8.

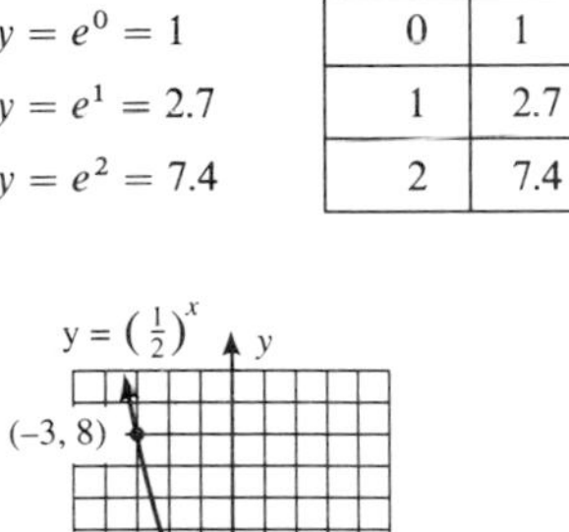

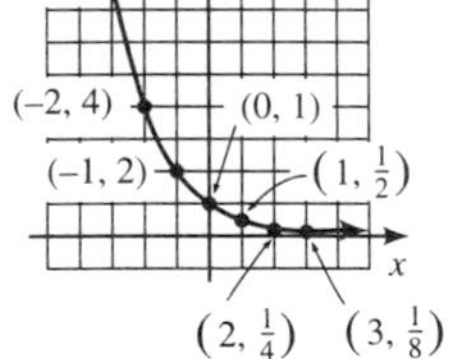

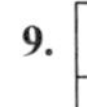

9.

x	y
-3	$\frac{8}{27}$
-2	$\frac{4}{9}$
-1	$\frac{2}{3}$
0	1
1	$\frac{3}{2}$
2	$\frac{9}{4}$
3	$\frac{27}{8}$
4	$\frac{81}{16}$
5	$\frac{243}{32}$

$y = (\frac{3}{2})^x$

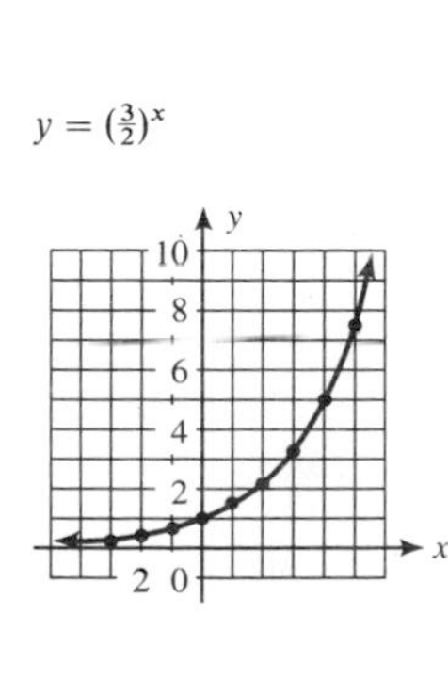

Exercises 9.1B (page 455)

1. The exponential form for $y = \log_2 x$ is $x = 2^y$.
The exponential form for $y = \log_{10} x$ is $x = 10^y$.

$x = 2^y$ $x = 10^y$

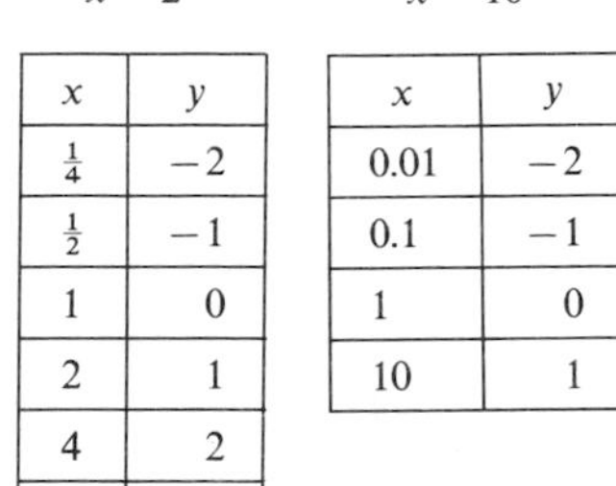

x	y
$\frac{1}{4}$	-2
$\frac{1}{2}$	-1
1	0
2	1
4	2
8	3

x	y
0.01	-2
0.1	-1
1	0
10	1

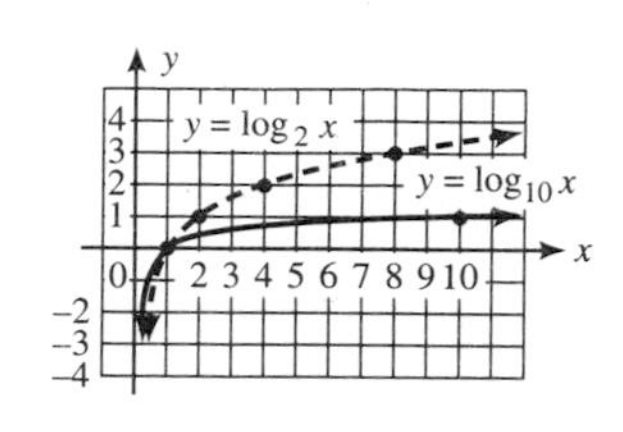

2.

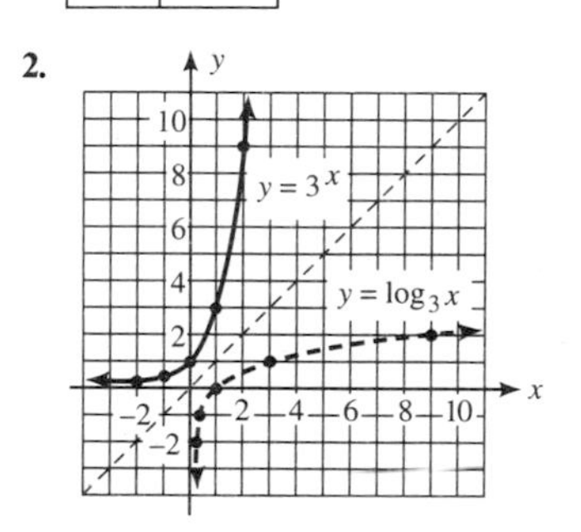

3. $y = e^x$ $y = \log_e x$

x	y
-1	0.4
0	1
1	2.7
2	7.4

x	y
0.4	-1
1	0
2.7	1
7.4	2

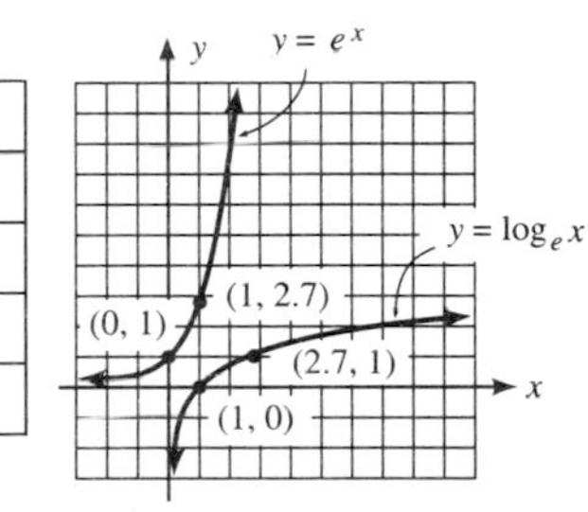

Exercises 9.2A (page 456)

1. $\log_3 9 = 2$ **2.** $\log_4 64 = 3$ **3.** $\log_{10} 1{,}000 = 3$
4. $\log_{10} 100{,}000 = 5$ **5.** $\log_2 16 = 4$ **6.** $\log_4 16 = 2$
7. $\log_3 \frac{1}{9} = -2$ **8.** $\log_2 \frac{1}{8} = -3$ **9.** $\log_{12} 1 = 0$
10. $\log_8 1 = 0$ **11.** $\log_{16} 4 = \frac{1}{2}$ **12.** $\log_8 2 = \frac{1}{3}$ **13.** $8^2 = 64$
14. $2^5 = 32$ **15.** $7^2 = 49$ **16.** $4^3 = 64$ **17.** $5^0 = 1$
18. $6^0 = 1$ **19.** $9^{1/2} = 3$ **20.** $8^{1/3} = 2$ **21.** $10^2 = 100$
22. $10^3 = 1{,}000$ **23.** $10^{-3} = \frac{1}{10^3} = \frac{1}{1{,}000} = 0.001$
24. $10^{-2} = 0.01$

Exercises 9.2B (page 458)

1. $N = 5^2 = 25$ **2.** 32 **3.** $3^x = 9 = 3^2$; $x = 2$ **4.** 4

5. $b^3 = 27 = 3^3$; $b = 3$ **6.** 3 **7.** $5^x = 125 = 5^3$; $x = 3$ **8.** 4

9. $10^x = 10^{-4}$; $x = -4$ **10.** -3 **11.** $(\frac{3}{2})^2 = N; N = \frac{9}{4}$ **12.** $\frac{125}{27}$

13.
$9^x = \frac{1}{3}$
$(3^2)^x = 3^{-1}$
$3^{2x} = 3^{-1}$
$2x = -1$
$x = -\frac{1}{2}$

14. -2

15.
$b^{1.5} = 8 = 2^3$
$b^{3/2} = 2^3$
$(b^{3/2})^{2/3} = (2^3)^{2/3}$
$b = 2^2 = 4$

16. 25 **17.** $2^{-2} = N; N = \frac{1}{2^2}; N = \frac{1}{4}$ **18.** 0.01

19. Let $x = \log_5 25 \Leftrightarrow 5^x = 25 = 5^2$ **20.** 4
$$x = 2$$
$$\log_5 25 = 2$$

21. Let $x = \log_{10} 10{,}000 \Leftrightarrow 10^x = 10{,}000 = 10^4$ **22.** 5
$$x = 4$$
$$\log_{10} 10{,}000 = 4$$

23. Let $x = \log_4 8 \Leftrightarrow 4^x = 8$ **24.** $\frac{2}{3}$
$$(2^2)^x = 2^3$$
$$2^{2x} = 2^3$$
$$2x = 3$$
$$x = \tfrac{3}{2}$$
$$\log_4 8 = \tfrac{3}{2}$$

25. Let $x = \log_3 3^4 \Leftrightarrow 3^x = 3^4$ **26.** 5
$$x = 4$$
$$\log_3 3^4 = 4$$

27. Let $x = \log_{16} 16 \Leftrightarrow 16^x = 16$ **28.** 1
$$x = 1$$
$$\log_{16} 16 = 1$$

29. Let $x = \log_8 1 \Leftrightarrow 8^x = 1 = 8^0$ **30.** 0
$$x = 0$$
$$\log_8 1 = 0$$

Exercises 9.3 (page 463)

1. $\log_{10} 31 + \log_{10} 7$ **2.** $\log_{10} 17 + \log_{10} 29$

3. $\log_{10} 41 - \log_{10} 13$ **4.** $\log_{10} 19 - \log_{10} 23$ **5.** $3 \log_{10} 19$

6. $4 \log_{10} 7$ **7.** $\log_{10} 75^{1/5} = \frac{1}{5} \log_{10} 75$ **8.** $\frac{1}{4} \log_{10} 38$

9. $\log_{10} 35\sqrt{73} - \log_{10}(1.06)^8 = \log_{10} 35 + \frac{1}{2} \log_{10} 73 - 8 \log_{10} 1.06$

10. $\log_{10} 27 + \frac{1}{2} \log_{10} 31 - 10 \log_{10} 1.03$

11. $\log_{10}(2)(7) = \log_{10} 2 + \log_{10} 7$ **12.** 1.322
$$\approx 0.301 + 0.845 = 1.146$$

13. $\log_{10} 9 - \log_{10} 7 = \log_{10} 3^2 - \log_{10} 7$ **14.** 0.243
$$= 2 \log_{10} 3 - \log_{10} 7$$
$$\approx 2(0.477) - 0.845$$
$$\approx 0.954 - 0.845 = 0.109$$

15. $\log_{10}(27)^{1/2} = \log_{10}(3^3)^{1/2}$ **16.** 0.4515
$$= \log_{10} 3^{3/2} = \tfrac{3}{2} \log_{10} 3$$
$$\approx (1.5)(0.477) = 0.7155$$

17. $\log_{10}(6^2)^2 = \log_{10} 6^4 = 4 \log_{10} 6$ **18.** 3.982
$$= 4 \log_{10}(2)(3)$$
$$= 4[\log_{10} 2 + \log_{10} 3]$$
$$\approx 4[0.301 + 0.477]$$
$$= 4(0.778) = 3.112$$

19. $\log_{10}(2)(3)(10^3) = \log_{10} 2 + \log_{10} 3 + 3 \log_{10} 10$
$$\approx 0.301 + 0.477 + 3 = 3.778$$

20. 3.146 **21.** $\log_b xy$ **22.** $\log_b x^4y^2$

23. $\log_b x^2 - \log_b y^3 = \log_b \frac{x^2}{y^3}$ **24.** $2 \log_b x$

25. $3 \log_b x - 6 \log_b y = \log_b x^3 - \log_b y^6 = \log_b \frac{x^3}{y^6}$ **26.** $\log_b y$

27. $\log_b(x^2 - y^2) - \log_b(x + y)^3 = \log_b \frac{x^2 - y^2}{(x + y)^3}$ **28.** $\log_b(x + z)$
$$= \log_b \frac{(x + y)(x - y)}{(x + y)^3} = \log_b \frac{(x - y)}{(x + y)^2}$$

29. $\log_b(2xy)^2 - \log_b 3xy^2 + \log_b 3x$
$$= \log_b 4x^2y^2 - \log_b 3xy^2 + \log_b 3x$$
$$= \log_b \frac{4x^2y^2(3x)}{3xy^2} = \log_b 4x^2 = \log_b(2x)^2 = 2 \log_b 2x$$

30. $\log_b 3y^2$

Exercises 9.4A (page 465)

1. a. 0.4771 **b.** 3.4771 **c.** 1.4771 **d.** 1.0986 **e.** 8.0064 **f.** 3.4012

2. a. 1.0000 **b.** 2.0000 **c.** 5.0000 **d.** 2.3026 **e.** 4.6052 **f.** 11.5129

3. a. −1.0000 **b.** −2.0000 **c.** −3.0000 **d.** −2.3026 **e.** −4.6052 **f.** −6.9078

4. a. −0.5229 **b.** −3.5229 **c.** −5.5229 **d.** −1.2040 **e.** −8.1117 **f.** −12.7169

Exercises 9.4B (page 468)

1. 2.35 **2.** 4.06 **3.** 0.851 **4.** 0.755 **5.** 2,560 **6.** 65.1 **7.** 0.00306 **8.** 0.000409 **9.** 15.2 **10.** 230 **11.** 0.0754 **12.** 0.00578

Exercises 9.5 (page 472)

1. $(3^3)^x = \frac{1}{3^2}$ **2.** $-\frac{4}{3}$
$$3^{3x} = 3^{-2}$$
$$3x = -2$$
$$x = -\tfrac{2}{3}$$

3. $(2^2)^x = \frac{1}{2^3}$ **4.** $-\frac{2}{3}$
$$2^{2x} = 2^{-3}$$
$$2x = -3$$
$$x = -\tfrac{3}{2}$$

5. $(5^2)^{2x+3} = 5^{x-1}$ **6.** 1
$$5^{4x+6} = 5^{x-1}$$
$$4x + 6 = x - 1$$
$$3x = -7$$
$$x = -\tfrac{7}{3}$$

7.
$$2^x = 3$$
$$\log 2^x = \log 3$$
$$x \log 2 = \log 3$$
$$x = \frac{\log 3}{\log 2} \approx \frac{0.4771}{0.3010} \approx 1.59$$
(1.58 if log 3 and log 2 are not rounded off)

8. 0.861 **9.**
$$e^x = 8$$
$$\ln e^x = \ln 8$$
$$x \ln e = \ln 8$$
$$x \approx 2.0794$$
$$x \approx 2.08$$

10. 3.00

11.
$$(7.43)^{x+1} = 9.55$$
$$\log(7.43)^{x+1} = \log 9.55$$
$$(x + 1)\log 7.43 = \log 9.55$$
$$x + 1 = \frac{\log 9.55}{\log 7.43} \approx \frac{0.9800}{0.8710}$$
$$x + 1 \approx 1.125$$
$$x \approx 0.125$$

12. 2.20

13.
$$(8.71)^{2x+1} = 8.57$$
$$\log(8.71)^{2x+1} = \log 8.57$$
$$(2x + 1)\log 8.71 = \log 8.57$$
$$2x + 1 = \frac{\log 8.57}{\log 8.71} \approx \frac{0.93298}{0.94002}$$
$$2x + 1 \approx 0.99251$$
$$x \approx -0.00374$$
(−0.00372 if tables are used)

14. 0.500

15.
$$\begin{aligned} e^{3x+4} &= 5 \\ \ln e^{3x+4} &= \ln 5 \\ (3x+4)\ln e &\approx 1.609 \\ 3x+4 &\approx 1.609 \\ 3x &\approx -2.391 \\ x &\approx -0.797 \end{aligned}$$

16. 1.93

17.
$$\begin{aligned} \log(3x-1)(4) &= \log(9x+2) \\ (3x-1)(4) &= 9x+2 \\ 12x-4 &= 9x+2 \\ 3x &= 6 \\ x &= 2 \end{aligned}$$

Check
$$\begin{aligned} \log[3(2)-1] + \log 4 &\stackrel{?}{=} \log[9(2)+2] \\ \log 5 + \log 4 &\stackrel{?}{=} \log 20 \\ \log(5)(4) &\stackrel{?}{=} \log 20 \\ \log 20 &= \log 20 \end{aligned}$$

2 is a solution.

18. 2

19.
$$\begin{aligned} \ln \frac{x+4}{3} &= \ln(x-2) \\ \frac{x+4}{3} &= x-2 \\ x+4 &= 3(x-2) \\ x+4 &= 3x-6 \\ 10 &= 2x \\ x &= 5 \end{aligned}$$

Check
$$\begin{aligned} \ln(5+4) - \ln 3 &\stackrel{?}{=} \ln(5-2) \\ \ln \tfrac{9}{3} &\stackrel{?}{=} \ln 3 \\ \ln 3 &= \ln 3 \end{aligned}$$

5 is a solution.

20. 2

21.
$$\begin{aligned} \log \frac{5x+2}{x-1} &= 0.7782 \\ \frac{5x+2}{x-1} &= 10^{0.7782} \\ \frac{5x+2}{x-1} &\approx 6 \\ 5x+2 &\approx 6x-6 \\ 8 &\approx x \end{aligned}$$

Check for $x \approx 8.00$
$$\begin{aligned} \log[5(8)+2] - \log(8-1) &\stackrel{?}{=} 0.7782 \\ \log 42 \quad \log 7 &\stackrel{?}{=} 0.7782 \\ \log \tfrac{42}{7} &\stackrel{?}{=} 0.7782 \\ \log 6 &\stackrel{?}{=} 0.7782 \\ 0.77815 &\approx 0.7782 \end{aligned}$$

8.00 is a solution.

22. 2.00

23.
$$\begin{aligned} \log x(7-x) &= \log 10 \\ x(7-x) &= 10 \\ 7x - x^2 &= 10 \\ x^2 - 7x + 10 &= 0 \\ (x-2)(x-5) &= 0 \end{aligned}$$

$x - 2 = 0 \mid x - 5 = 0$

$x = 0 \mid x = 5$

Check for $x = 2$
$$\begin{aligned} \log 2 + \log(7-2) &\stackrel{?}{=} \log 10 \\ \log(2)(5) &\stackrel{?}{=} \log 10 \\ \log 10 &= \log 10 \end{aligned}$$

Check for $x = 5$
$$\begin{aligned} \log 5 + \log(7-5) &\stackrel{?}{=} \log 10 \\ \log(5)(2) &\stackrel{?}{=} \log 10 \\ \log 10 &= \log 10 \end{aligned}$$

Both 2 and 5 are solutions.

24. 1 and 10

25.
$$\begin{aligned} \ln x(x-3) &= \ln 4 \\ x(x-3) &= 4 \\ x^2 - 3x &= 4 \\ x^2 - 3x - 4 &= 0 \\ (x-4)(x+1) &= 0 \end{aligned}$$

$x - 4 = 0 \mid x + 1 = 0$

$x = 4 \mid x = -1$

Check for $x = 4$
$$\begin{aligned} \ln 4 + \ln(4-3) &\stackrel{?}{=} \ln 4 \\ \ln 4 + \ln 1 &\stackrel{?}{=} \ln 4 \\ \ln 4 + 0 &\stackrel{?}{=} \ln 4 \\ \ln 4 &= \ln 4 \end{aligned}$$

Check for $x = -1$
$$\ln(-1) + \ln(-1-3) \stackrel{?}{=} \ln 4$$
Not real numbers

-1 *is not* a solution; 4 is the only solution.

26. 2

27.
$$\begin{aligned} \log(x+1)(x-2) &= \log 10 \\ x^2 - x - 2 &= 10 \\ x^2 - x - 12 &= 0 \\ (x-4)(x+3) &= 0 \end{aligned}$$

$x - 4 = 0 \mid x + 3 = 0$

$x = 4 \mid x = -3$

Check for $x = 4$
$$\begin{aligned} \log(4+1) + \log(4-2) &\stackrel{?}{=} 1 \\ \log 5 + \log 2 &\stackrel{?}{=} 1 \\ \log(5)(2) = \log 10 &= 1 \end{aligned}$$

Check for $x = -3$
$$\begin{aligned} \log(-3+1) + \log(-3-2) &\stackrel{?}{=} 1 \\ \log(-2) + \log(-5) &\stackrel{?}{=} 1 \end{aligned}$$
Not real numbers

-3 is *not* a solution; 4 is the only solution.

28. 4

29.
$$\begin{aligned} \log \frac{10x}{x-450} &= 2 \\ \frac{10x}{x-450} &= 10^2 \\ 10x &= 100(x-450) \\ 10x &= 100x - 45{,}000 \\ 45{,}000 &= 90x \\ x &= 500 \end{aligned}$$

Check
$$\begin{aligned} \log 10(500) - \log(500-450) &\stackrel{?}{=} 2 \\ \log(5{,}000) - \log(50) &\stackrel{?}{=} 2 \\ \log \frac{5{,}000}{50} &\stackrel{?}{=} 2 \\ \log 100 &= 2 \end{aligned}$$

500 is a solution.

30. 12

Exercises 9.6 (page 476)

Note: Your calculator displays may vary slightly from those shown below.

1. $P = \$1{,}250$, $r = 0.055$, $t = 20$

a.
$$\begin{aligned} A &= P(1+r)^t \\ A &= \$1{,}250(1+0.055)^{20} \\ &= \$1{,}250(1.055)^{20} \\ &\approx \$1{,}250(2.917757) \\ &\approx \$3{,}647.20 \end{aligned}$$

b. $A = P\left(1 + \frac{r}{k}\right)^{kt}$; $k = 12$
$$\begin{aligned} A &= \$1{,}250\left(1 + \frac{0.055}{12}\right)^{12(20)} \\ &= \$1{,}250(1+0.00458333)^{240} \\ &= \$1{,}250(1.00458333)^{240} \\ &\approx \$1{,}250(2.9966255) \\ &\approx \$3{,}745.78 \end{aligned}$$

c. $A = P\left(1 + \frac{r}{k}\right)^{kt}$; $k = 365$
$$\begin{aligned} A &= \$1{,}250\left(1 + \frac{0.055}{365}\right)^{365(20)} \\ &= \$1{,}250(1+0.000150685)^{7300} \\ &= \$1{,}250(1.000150685)^{7300} \\ &\approx \$1{,}250(3.0039164) \\ &\approx \$3{,}754.90 \end{aligned}$$

d. $A = Pe^{rt}$
$$\begin{aligned} A &= \$1{,}250e^{0.055(20)} \\ &= \$1{,}250e^{1.1} \\ &\approx \$1{,}250(3.0041660) \\ &\approx \$3{,}755.21 \end{aligned}$$

2a. \$4,372.64 **b.** \$4,436.73 **c.** \$4,442.62 **d.** \$4,442.83

3. $P = \$1{,}500,\ r = 0.0575,\ A = \$2{,}000$

$$A = Pe^{rt}$$
$$\$2{,}000 = \$1{,}500e^{0.0575t}$$
$$1.3333333 \approx e^{0.0575t}$$
$$\log_e 1.3333333 \approx 0.0575t$$
$$0.2876821 \approx 0.0575t$$
$$t \approx 5.003 \text{ years}$$

4. 8.546 years

5. The formula is $y = Ce^{0.035t}$

a. $C = 500,\ t = 3$

$$y = 500e^{0.035(3)}$$
$$= 500e^{0.105}$$
$$\approx 500(1.110711)$$
≈ 555.4; there will be about 555.4 bacteria present.

b. $y = 800;\ C = 500$

$$800 = 500e^{0.035t}$$
$$1.6 = e^{0.035t}$$
$$\ln 1.6 = \ln e^{0.035t}$$
$$0.4700036 \approx 0.035t \ln e \quad (\ln e = 1)$$
$$0.035t \approx 0.4700036$$
$t \approx 13.43$; it will take about 13.43 hr.

6. a. 946.1 bacteria **b.** 23.01 hr

7. The formula is $y = Ce^{-0.3t}$; $y = 80$; $C = 100$.

$$80 = 100e^{-0.3t}$$
$$0.8 = e^{-0.3t}$$
$$\ln 0.8 = \ln e^{-0.3t}$$
$$-0.223144 \approx -0.3t \ln e \quad (\ln e = 1)$$
$$-0.3t \approx -0.223144$$
$$t \approx 0.7438 \text{ years}$$

8. 0.7754 years

Exercises 9.7 (page 478)

1. $\log_2 156 = \dfrac{\log_{10} 156}{\log_{10} 2} \approx \dfrac{2.19312}{0.30103} \approx 7.285$ **2.** 4.954

3. $\log_{12} 7.54 = \dfrac{\log_{10} 7.54}{\log_{10} 12} \approx \dfrac{0.87737}{1.07918} \approx 0.8130$ **4.** 0.7602

5. $\log_e 3.04 \approx \dfrac{\log_{10} 3.04}{\log_{10} 2.718} \approx \dfrac{0.48287}{0.43425} \approx 1.112$ **6.** 1.406

7. $\log_{6.8} 0.507 = \dfrac{\log 0.507}{\log 6.8} \approx \dfrac{-0.29499}{0.83251} \approx -0.3543$ **8.** -1.651

Review Exercises 9.8 (page 480)

1. $\log_3 81 = 4$ **2.** $4^{-2} = 0.0625$ **3.** $10^x = 1{,}000$; $10^x = 10^3$; $x = 3$

4. -2 **5.** $9^{3/2} = N$; $N = (3^2)^{3/2} = 3^3 = 27$ **6.** 2

7.
$$10^x = 145.6$$
$$\log 10^x = \log 145.6$$
$$x \log 10 = \log 145.6 \quad (\log 10 = 1)$$
$$x \approx 2.1632$$

8. $13 \log x$

9. $\log \dfrac{3}{5} + \log \dfrac{5}{3} = \log \left(\dfrac{3}{5}\right)\left(\dfrac{5}{3}\right) = \log 1 = 0$ **10.** $-\log 2x$

11. 1.4062 **12.** -3.0966 **13.** -3.3755 **14.** 8.5755

15. 2,554 **16.** 12.00 **17.**
$$(3^4)^{x-1} = \frac{1}{3^2}$$
$$3^{4x-4} = 3^{-2}$$
$$4x - 4 = -2$$
$$4x = 2$$
$$x = \tfrac{1}{2}$$

18. $\frac{3}{2}$

19. $\log_4 75 = \dfrac{\log_{10} 75}{\log_{10} 4} \approx \dfrac{1.8751}{0.6021} \approx 3.114$

20. The inverse function of $y = 6^x$ is $y = \log_6 x$.

$y = 6^x$

x	y
-2	$\frac{1}{36}$
-1	$\frac{1}{6}$
0	1
1	6
2	36

$y = \log_6 x$

x	y
$\frac{1}{36}$	-2
$\frac{1}{6}$	-1
1	0
6	1
36	2

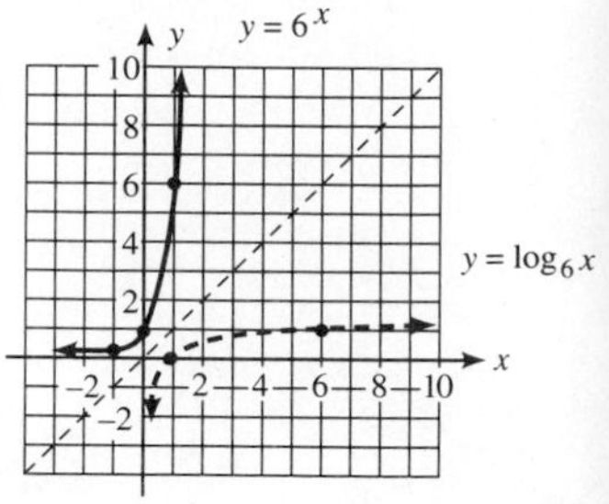

21. We use $A = Pe^{rt}$, where $A = \$2{,}500$, $P = \$2{,}000$, $r = 0.058$.

$$2{,}500 = 2{,}000e^{0.058t}$$
$$1.25 = e^{0.058t}$$
$$\ln 1.25 = \ln e^{0.058t}$$
$$0.223144 \approx 0.058t \ln e \quad (\ln e = 1)$$
$$0.058t \approx 0.223144$$
$$t \approx 3.8$$

It will be about 3.8 yr before there is \$2,500 in the account.

Chapter 9 Diagnostic Test (page 485)

Following each problem number is the textbook section number (in parentheses) where that kind of problem is discussed.

1. (9.2) **a.** *Exponential form* $2^4 = 16 \Leftrightarrow$ *Logarithmic form* $\log_2 16 = 4$

b. *Logarithmic form* $\log_{2.5} 6.25 = 2 \Leftrightarrow$ *Exponential form* $(2.5)^2 = 6.25$

2. (9.2) **a.** $\log_4 N = 3 \Leftrightarrow 4^3 = N$
Therefore, $N = 64$

b. $\log_{10} 10^{-2} = x \Leftrightarrow 10^x = 10^{-2}$
Therefore, $x = -2$

c. $\log_b 6 = 1 \Leftrightarrow b^1 = 6$
Therefore, $b = 6$

d. $\log_5 1 = x \Leftrightarrow 5^x = 1 = 5^0$
Therefore, $x = 0$

e. $\log_{0.5} N = -2 \Leftrightarrow (0.5)^{-2} = N$

$$(\tfrac{1}{2})^{-2} = N$$
$$2^2 = N$$
$$N = 4$$

3. (9.3) **a.** $\log x + \log y - \log z = \log \dfrac{xy}{z}$

b.
$$\frac{1}{2}\log x^4 + 2\log x = \log(x^4)^{1/2} + 2\log x$$
$$= \log x^2 + 2\log x$$
$$= 2\log x + 2\log x = 4\log x$$

4. (9.3)
$$\log(x^2 - 9) - \log(x - 3)$$
$$= \log(x + 3)(x - 3) - \log(x - 3)$$
$$= \log(x + 3) + \log(x - 3) - \log(x - 3) = \log(x + 3)$$

5. (9.5)
$$\log(3x + 5) - \log 7 = \log(x - 1)$$
$$\log \frac{(3x + 5)}{7} = \log(x - 1)$$
$$\frac{3x + 5}{7} = \frac{x - 1}{1}$$
$$3x + 5 = 7x - 7$$
$$12 = 4x$$
$$x = 3$$

Check

$$\log(3x + 5) - \log 7 = \log(x - 1)$$
$$\log[3(3) + 5] - \log 7 \stackrel{?}{=} \log(3 - 1)$$
$$\log 14 - \log 7 \stackrel{?}{=} \log 2$$
$$\log \tfrac{14}{7} \stackrel{?}{=} \log 2$$
$$\log 2 = \log 2$$

3 is a solution.

6. (9.5)
$$\begin{aligned} \log(x+8)+\log(x-2) &= \log 11 \\ \log(x+8)(x-2) &= \log 11 \\ x^2+6x-16 &= 11 \\ x^2+6x-27 &= 0 \\ (x-3)(x+9) &= 0 \end{aligned}$$

$x-3=0 \mid x+9=0$
$x=3 \mid x=-9$

Check for $x = 3$	*Check for* $x = -9$
$\log(3+8)+\log(3-2) \stackrel{?}{=} \log 11$	$\log(-9+8) = \log(-1)$,
$\log 11 + \log 1 \stackrel{?}{=} \log 11$	which is not real.
$\log 11 = \log 11$	

Therefore, 3 is the only solution.

7. (9.5)
$$\begin{aligned} e^{3x-4} &= 8 \\ \ln e^{3x-4} &= \ln 8 \\ (3x-4)\ln e &\approx 2.0794415 \\ 3x-4 &\approx 2.0794415 \\ 3x &\approx 6.0794415 \\ x &\approx 2.03 \end{aligned}$$

8. (9.7) Find $\log_2 718$.
$$\log_2 N = \frac{\log_{10} N}{\log_{10} 2}$$
$$\log_2 718 = \frac{\log 718}{\log 2} \approx \frac{2.8561}{0.3010} \approx 9.49$$

9. (9.1) The inverse function of $y = 7^x$ is $y = \log_7 x$.

$y = 7^x$

	x	y
$y = 7^{-2} = \frac{1}{49}$	-2	$\frac{1}{49}$
$y = 7^{-1} = \frac{1}{7}$	-1	$\frac{1}{7}$
$y = 7^0 = 1$	0	1
$y = 7^1 = 7$	1	7
$y = 7^2 = 49$	2	49

$y = \log_7 x$

$x = 7^y$

	x	y
$x = 7^{-2} = \frac{1}{49}$	$\frac{1}{49}$	-2
$x = 7^{-1} = \frac{1}{7}$	$\frac{1}{7}$	-1
$x = 7^0 = 1$	1	0
$x = 7^1 = 7$	7	1
$x = 7^2 = 49$	49	2

$y = 7^x$ $y = \log_7 x$

10. (9.6) The formula is $y = Ce^{0.05t}$. $C = 1{,}500$.

a. $t = 2$
$$\begin{aligned} y &= 1{,}500e^{0.05(2)} \\ y &= 1{,}500e^{0.1} \\ y &\approx 1{,}500(1.1051709) \\ y &\approx 1{,}658 \end{aligned}$$
There will be about 1,658 bacteria present.

b. $C = 1{,}500$; $y = 4{,}500$
$$\begin{aligned} 4{,}500 &= 1{,}500e^{0.05t} \\ 3 &= e^{0.05t} \\ \ln 3 &= \ln e^{0.05t} \quad \text{Taking ln of both sides} \\ 1.0986123 &\approx 0.05t \ln e \quad (\ln e = 1) \\ 0.05t &\approx 1.0986123 \\ t &\approx 22 \end{aligned}$$
It will take about 22 hr for the bacteria to triple.

Cumulative Review Exercises: Chapters 1–9 (page 486)

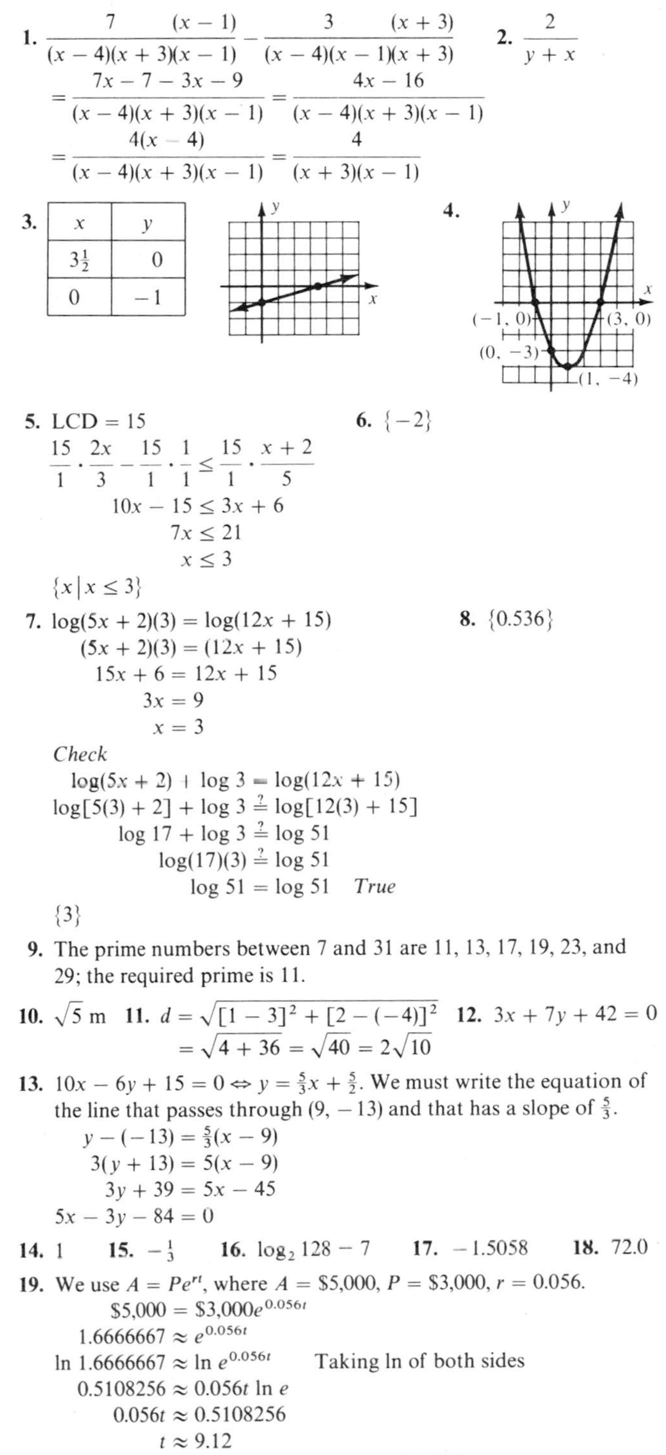

1.
$$\frac{7}{(x-4)(x+3)}\frac{(x-1)}{(x-1)} - \frac{3}{(x-4)(x-1)}\frac{(x+3)}{(x+3)}$$
$$= \frac{7x-7-3x-9}{(x-4)(x+3)(x-1)} = \frac{4x-16}{(x-4)(x+3)(x-1)}$$
$$= \frac{4(x-4)}{(x-4)(x+3)(x-1)} = \frac{4}{(x+3)(x-1)}$$

2. $\dfrac{2}{y+x}$

3.

x	y
$3\frac{1}{2}$	0
0	-1

4.

5. LCD = 15
$$\frac{15}{1}\cdot\frac{2x}{3} - \frac{15}{1}\cdot\frac{1}{1} \le \frac{15}{1}\cdot\frac{x+2}{5}$$
$$\begin{aligned} 10x - 15 &\le 3x + 6 \\ 7x &\le 21 \\ x &\le 3 \end{aligned}$$
$\{x \mid x \le 3\}$

6. $\{-2\}$

7.
$$\begin{aligned} \log(5x+2)(3) &= \log(12x+15) \\ (5x+2)(3) &= (12x+15) \\ 15x+6 &= 12x+15 \\ 3x &= 9 \\ x &= 3 \end{aligned}$$
Check
$$\begin{aligned} \log(5x+2) + \log 3 &= \log(12x+15) \\ \log[5(3)+2] + \log 3 &\stackrel{?}{=} \log[12(3)+15] \\ \log 17 + \log 3 &\stackrel{?}{=} \log 51 \\ \log(17)(3) &\stackrel{?}{=} \log 51 \\ \log 51 &= \log 51 \quad \textit{True} \end{aligned}$$
$\{3\}$

8. $\{0.536\}$

9. The prime numbers between 7 and 31 are 11, 13, 17, 19, 23, and 29; the required prime is 11.

10. $\sqrt{5}$ m **11.** $d = \sqrt{[1-3]^2 + [2-(-4)]^2} = \sqrt{4+36} = \sqrt{40} = 2\sqrt{10}$ **12.** $3x + 7y + 42 = 0$

13. $10x - 6y + 15 = 0 \Leftrightarrow y = \frac{5}{3}x + \frac{5}{2}$. We must write the equation of the line that passes through $(9, -13)$ and that has a slope of $\frac{5}{3}$.
$$\begin{aligned} y - (-13) &= \tfrac{5}{3}(x-9) \\ 3(y+13) &= 5(x-9) \\ 3y + 39 &= 5x - 45 \\ 5x - 3y - 84 &= 0 \end{aligned}$$

14. 1   **15.** $-\frac{1}{3}$ **16.** $\log_2 128 = 7$ **17.** -1.5058 **18.** 72.0

19. We use $A = Pe^{rt}$, where $A = \$5{,}000$, $P = \$3{,}000$, $r = 0.056$.
$$\begin{aligned} \$5{,}000 &= \$3{,}000e^{0.056t} \\ 1.6666667 &\approx e^{0.056t} \\ \ln 1.6666667 &\approx \ln e^{0.056t} \quad \text{Taking ln of both sides} \\ 0.5108256 &\approx 0.056t \ln e \\ 0.056t &\approx 0.5108256 \\ t &\approx 9.12 \end{aligned}$$
It will be about 9.12 yr before there is \$5,000 in the account.

20. $2\frac{1}{2}$ liters

21. Let x = number of hours for Maria to do the job

$x + 2$ = number of hours for Claudio to do the job

$\frac{1 \text{ job}}{x \text{ hr}}$ = Maria's rate Maria works for $(2 + 3\frac{1}{3})$ hr, or $\frac{16}{3}$ hr

$\frac{1 \text{ job}}{(x+2) \text{ hr}}$ = Claudio's rate Claudio works for $3\frac{1}{3}$ hr, or $\frac{10}{3}$ hr

$$\left(\frac{1 \text{ job}}{x \text{ hr}}\right)\left(5\frac{1}{3} \text{ hr}\right) + \left(\frac{1 \text{ job}}{(x+2) \text{ hr}}\right)\left(3\frac{1}{3} \text{ hr}\right) = 1 \text{ job}$$

$$\left(\frac{1}{x}\right)\left(\frac{16}{3}\right) + \left(\frac{1}{x+2}\right)\left(\frac{10}{3}\right) = 1$$

$$\frac{3x(x+2)}{1}\left(\frac{1}{x}\right)\left(\frac{16}{3}\right) + \frac{3x(x+2)}{1}\left(\frac{1}{x+2}\right)\left(\frac{10}{3}\right) = \frac{3x(x+2)}{1}(1)$$

$$(x+2)(16) + 10x = 3x(x+2)$$
$$16x + 32 + 10x = 3x^2 + 6x$$
$$0 = 3x^2 - 20x - 32$$
$$0 = (3x+4)(x-8)$$

$3x + 4 = 0$ | $x - 8 = 0$
$3x = -4$ | $x = 8$
$x = -\frac{4}{3}$ | $x + 2 = 10$

Reject

(The check will not be shown.) It would take Maria 8 hr and Claudio 10 hr to do the job.

Exercises 10.1 (page 489)

1. $3x^2 + 5x - 2 = 0$ $\begin{cases} a = 3 \\ b = 5 \\ c = -2 \end{cases}$

2. $2x^2 - 3x - 5 = 0$ $\begin{cases} a = 2 \\ b = -3 \\ c = -5 \end{cases}$

3. $3x^2 + 0x - 4 = 0$ $\begin{cases} a = 3 \\ b = 0 \\ c = -4 \end{cases}$

4. $x^2 + 0x - 16 = 0$ $\begin{cases} a = 1 \\ b = 0 \\ c = -16 \end{cases}$

5. $4x = 12 + 3x^2$
$3x^2 - 4x + 12 = 0$ $\begin{cases} a = 3 \\ b = -4 \\ c = 12 \end{cases}$

6. $2x^2 - 3x + 10 = 0$ $\begin{cases} a = 2 \\ b = -3 \\ c = 10 \end{cases}$

7. $3x^2 + 4x - 2 = 0$ $\begin{cases} a = 3 \\ b = 4 \\ c = -2 \end{cases}$

8. $5x^2 - 6x + 3 = 0$ $\begin{cases} a = 5 \\ b = -6 \\ c = 3 \end{cases}$

9. $x^2 + 2x - 4 = 0$ $\begin{cases} a = 1 \\ b = 2 \\ c = -4 \end{cases}$

10. $x^2 - 4x + 20 = 0$ $\begin{cases} a = 1 \\ b = -4 \\ c = 20 \end{cases}$

11. $3x^2 - 6x = x^2 - 4x - 5$
$2x^2 - 2x + 5 = 0$ $\begin{cases} a = 2 \\ b = -2 \\ c = 5 \end{cases}$

12. $13x^2 + 20x + 12 = 0$ $\begin{cases} a = 13 \\ b = 20 \\ c = 12 \end{cases}$

Exercises 10.2 (page 492)

1. $x^2 = 27$
$x = \pm\sqrt{27} = \pm 3\sqrt{3}$
$\{3\sqrt{3}, -3\sqrt{3}\}$

2. $\{2\sqrt{2}, -2\sqrt{2}\}$

3. $x^2 = -16$
$x = \pm\sqrt{-16}$
$x = \pm 4i$
$\{4i, -4i\}$

4. $\{9i, -9i\}$

5. $x^2 = 7$
$x = \pm\sqrt{7}$
$\{\sqrt{7}, -\sqrt{7}\}$

6. $\{\sqrt{13}, -\sqrt{13}\}$

7. $x^2 = -12$
$x = \pm\sqrt{-12}$
$x = \pm 2i\sqrt{3}$
$\{2i\sqrt{3}, -2i\sqrt{3}\}$

8. $\{5i\sqrt{3}, -5i\sqrt{3}\}$

9. $0 = 8x^3 - 12x$
$0 = 4x(2x^2 - 3)$

$4x = 0$ | $2x^2 - 3 = 0$
$x = 0$ | $2x^2 = 3$
| $x^2 = \frac{3}{2}$
| $x = \pm\sqrt{\frac{3}{2}}$
| $x = \pm\sqrt{\frac{3}{2}\cdot\frac{2}{2}} = \pm\frac{\sqrt{6}}{2}$

$\left\{0, \frac{\sqrt{6}}{2}, -\frac{\sqrt{6}}{2}\right\}$

10. $\left\{0, \frac{\sqrt{3}}{2}, -\frac{\sqrt{3}}{2}\right\}$

11. $x^2 = -\frac{4}{5}$
$x = \pm\sqrt{-\frac{4}{5}}$
$x = \pm\frac{2i}{\sqrt{5}} = \pm\frac{2i}{\sqrt{5}}\cdot\frac{\sqrt{5}}{\sqrt{5}}$
$x = \pm\frac{2\sqrt{5}}{5}i$
$\left\{\frac{2\sqrt{5}}{5}i, -\frac{2\sqrt{5}}{5}i\right\}$

12. $\left\{\frac{5\sqrt{3}}{3}i, -\frac{5\sqrt{3}}{3}i\right\}$

13. $2x^2 = 12x$
$2x^2 - 12x = 0$
$2x(x - 6) = 0$
$2x = 0$ | $x - 6 = 0$
$x = 0$ | $x = 6$
$\{0, 6\}$

14. $\left\{0, \frac{1}{10}\right\}$

15. $x^2 - 2x = 2x^2 + 3x$
$x^2 + 5x = 0$
$x(x + 5) = 0$
$x = 0$ or $x = -5$
$\{0, -5\}$

16. $\{0, 5\}$

17. $3x(x + 1) = x(x + 2)$
$3x^2 + 3x = x^2 + 2x$
$2x^2 + x = 0$
$x(2x + 1) = 0$
$x = 0$ | $2x + 1 = 0$
Not in the domain | $x = -\frac{1}{2}$
$\{-\frac{1}{2}\}$

18. $\{2\}$

19. $x^{-2/3} = 16$
$(x^{-2/3})^{-3/2} = \pm 16^{-3/2}$
$x = \pm\frac{1}{16^{3/2}}$
$= \pm\frac{1}{(\sqrt{16})^3} = \pm\frac{1}{4^3} = \pm\frac{1}{64}$
$\{\frac{1}{64}, -\frac{1}{64}\}$ (Both check.)

20. $\{32, -32\}$

21. Let x = length of one side.
$x^2 + x^2 = (\sqrt{32})^2$
$2x^2 = 32$
$x^2 = 16$
$x = 4 \mid x = -4 \leftarrow$ Not in the domain
The length of a side is 4.

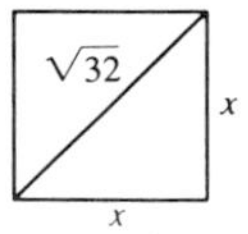

22. $9\sqrt{2}$

23. Let x = length of diagonal (in cm).

$7^2 + 10^2 = x^2$

$49 + 100 = x^2$

$\sqrt{149} = x$

($x = -\sqrt{149}$ is not the domain.)

The diagonal is $\sqrt{149}$ cm long.

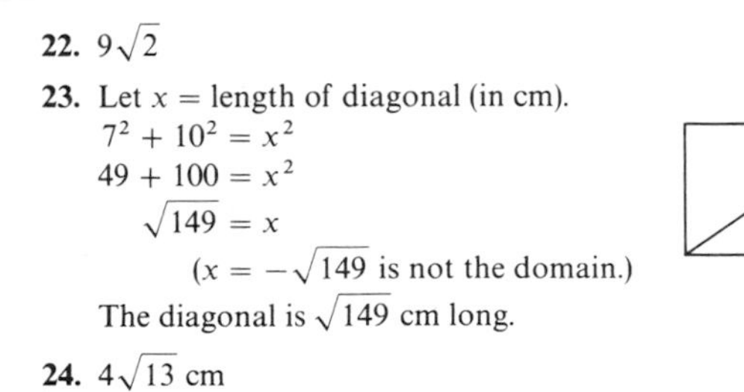

24. $4\sqrt{13}$ cm

Exercises 10.3 (page 495)

1. Let $z = x^2$: $z^2 - 37z + 36 = 0$

$(z - 36)(z - 1) = 0$

$z - 36 = 0$	$z - 1 = 0$
$z = 36$	$z = 1$
$x^2 = 36$	$x^2 = 1$
$x^2 - 36 = 0$	$x^2 - 1 = 0$
$(x - 6)(x + 6) = 0$	$(x - 1)(x + 1) = 0$
$x = 6 \quad x = -6$	$x = 1 \quad x = -1$

$\{-6, -1, 1, 6\}$

2. $\{-3, -2, 2, 3\}$

3. Let $y = z^{-2}$: $y^2 - 10y + 9 = 0$

$(y - 9)(y - 1) = 0$

$y - 9 = 0$	$y - 1 = 0$
$y = 9$	$y = 1$
$z^{-2} = 9$	$z^{-2} = 1$
$\frac{1}{z^2} = 9$	$\frac{1}{z^2} = 1$
$9z^2 = 1$	$z^2 = 1$
$z = \pm\frac{1}{3}$	$z = \pm 1$

$\{1, -1, \frac{1}{3}, -\frac{1}{3}\}$ (All check.)

4. $\{1, -1, \frac{1}{2}, -\frac{1}{2}\}$

5. Let $a = y^{1/3}$: $a^2 - 5a + 4 = 0$

$(a - 4)(a - 1) = 0$

$a - 4 = 0$	$a - 1 = 0$
$a = 4$	$a = 1$
$y^{1/3} = 4$	$y^{1/3} = 1$
$(y^{1/3})^3 = 4^3$	$(y^{1/3})^3 = 1^3$
$y = 64$	$y = 1$

$\{1, 64\}$

6. $\{1, 729\}$

7. Let $a = z^{-2}$

$a^2 - 4a = 0$

$a(a - 4) = 0$

$a = 0$	$a = 4$
$z^{-2} = 0$	$z^{-2} = 4$
$\frac{1}{z^2} = 0$	$\frac{1}{z^2} = \frac{4}{1}$
	$4z^2 = 1$
$1 = 0$	$4z^2 - 1 = 0$
False	$(2z - 1)(2z + 1) = 0$

$z = \frac{1}{2} \quad z = -\frac{1}{2}$

$\{\frac{1}{2}, -\frac{1}{2}\}$

8. $\{\frac{1}{3}, -\frac{1}{3}\}$

9. Let $a = K^{-1/3}$

$a^2 = K^{-2/3}$

$a^2 + 2a + 1 = 0$

$(a + 1)(a + 1) = 0$

Same answer for both factors

$a + 1 = 0$

$a = -1$

$K^{-1/3} = -1$

$(K^{-1/3})^{-3} = (-1)^{-3} = \frac{1}{(-1)^3} = \frac{1}{-1} = -1$

$K = -1$

$\{-1\}$

10. $\{1\}$

11. Let $a = (x^2 - 4x)$

$a^2 = (x^2 - 4x)^2$

$a^2 - a - 20 = 0$

$(a + 4)(a - 5) = 0$

$a + 4 = 0$	$a - 5 = 0$	
$a = -4$	$a = 5$	
$x^2 - 4x = -4$	$x^2 - 4x = 5$	
$x^2 - 4x + 4 = 0$	$x^2 - 4x - 5 = 0$	
$(x - 2)(x - 2) = 0$	$(x - 5)(x + 1) = 0$	
$x - 2 = 0$	$x - 5 = 0$	$x + 1 = 0$
$x = 2$	$x = 5$	$x = -1$

Solution set: $\{-1, 2, 5\}$ (All check.)

12. $\{-1, 1, 3\}$

13. Let w = width

$2w$ = length

w / $2w$

$Area = w(2w) = 2w^2$

$Perimeter = 2w + 2(2w) = 6w$

$2w^2 + 6w = 80$

$2w^2 + 6w - 80 = 0$

$w^2 + 3w - 40 = 0$

$(w + 8)(w - 5) = 0$

$w = -8$	$w = 5$
Not in the domain	$2w = 10$

The width is 5 and the length is 10.

14. Width = 4; length = 12

15. Let r = speed from Los Angeles to Mexico (in mph)

$r + 20$ = speed returning (in mph)

$\text{Time} = \frac{\text{Distance}}{\text{Rate}}$

Time to go + Time to return = 5 hr

$$\frac{120}{r} + \frac{120}{r + 20} = 5$$

$$\frac{24}{r} + \frac{24}{r + 20} = 1$$

LCD $= r(r + 20)$

$$\frac{r(r + 20)}{1} \cdot \frac{24}{r} + \frac{r(r + 20)}{1} \cdot \frac{24}{r + 20} = \frac{r(r + 20)}{1} \cdot 1$$

$24r + 480 + 24r = r^2 + 20r$

$r^2 - 28r - 480 = 0$

$(r + 12)(r - 40) = 0$

$r + 12 = 0$	$r - 40 = 0$
$r = -12$	$r = 40$
Not in the domain	Bruce's speed going to Mexico was 40 mph.

16. 30 mph

Exercises 10.4 (page 501)

1. $x^2 - 6x = 11$

$x^2 - 6x + 9 = 9 + 11$

$(x - 3)^2 = 20$

$x - 3 = \pm\sqrt{20}$

$x - 3 = \pm\sqrt{4 \cdot 5}$

$x = 3 \pm 2\sqrt{5}$

$\{3 + 2\sqrt{5}, 3 - 2\sqrt{5}\}$

2. $\{5 + 2\sqrt{3}, 5 - 2\sqrt{3}\}$

3. $x^2 - 4x = 13$

$x^2 - 4x + 4 = 13 + 4$

$(x - 2)^2 = 17$

$x - 2 = \pm\sqrt{17}$

$x = 2 \pm \sqrt{17}$

$\{2 + \sqrt{17}, 2 - \sqrt{17}\}$

4. $\{4 + 2i, 4 - 2i\}$

5.
$$x^2 - 2x = 2$$
$$x^2 - 2x + 1 = 2 + 1$$
$$(x-1)^2 = 3$$
$$x - 1 = \pm\sqrt{3}$$
$$x = 1 \pm \sqrt{3}$$
$\{1+\sqrt{3}, 1-\sqrt{3}\}$

6. $\left\{\dfrac{2+\sqrt{3}}{2}, \dfrac{2-\sqrt{3}}{2}\right\}$

7. $3x^2 - x - 2 = 0 \begin{cases} a = 3 \\ b = -1 \\ c = -2 \end{cases}$

$$x = \frac{-(-1) \pm \sqrt{(-1)^2 - 4(3)(-2)}}{2(3)}$$
$$= \frac{1 \pm \sqrt{1+24}}{6} = \frac{1 \pm \sqrt{25}}{6}$$
$$= \frac{1 \pm 5}{6} = \begin{cases} \dfrac{1+5}{6} = \dfrac{6}{6} = 1 \\ \dfrac{1-5}{6} = \dfrac{-4}{6} = -\dfrac{2}{3} \end{cases}$$
$\{1, -\frac{2}{3}\}$

8. $\{\frac{1}{2}, -2\}$

9. $x^2 - 4x + 1 = 0 \begin{cases} a = 1 \\ b = -4 \\ c = 1 \end{cases}$

$$x = \frac{-(-4) \pm \sqrt{(-4)^2 - 4(1)(1)}}{2(1)}$$
$$= \frac{4 \pm \sqrt{16-4}}{2(1)} = \frac{4 \pm \sqrt{12}}{2}$$
$$= \frac{4 \pm 2\sqrt{3}}{2} = 2 \pm \sqrt{3}$$
$\{2+\sqrt{3}, 2-\sqrt{3}\}$

10. $\{2+\sqrt{5}, 2-\sqrt{5}\}$

11. $x^2 - 4x + 2 = 0 \begin{cases} a = 1 \\ b = -4 \\ c = 2 \end{cases}$

$$x = \frac{+4 \pm \sqrt{(-4)^2 - 4(1)(2)}}{2(1)}$$
$$x = \frac{4 \pm \sqrt{16-8}}{2} = \frac{4 \pm \sqrt{8}}{2} = \frac{4 \pm 2\sqrt{2}}{2}$$
$$x = 2 \pm \sqrt{2}$$
$\{2+\sqrt{2}, 2-\sqrt{2}\}$

12. $\{1+\sqrt{3}, 1-\sqrt{3}\}$

13. $x^2 + x + 5 = 0 \begin{cases} a = 1 \\ b = 1 \\ c = 5 \end{cases}$

$$x = \frac{-(1) \pm \sqrt{(1)^2 - 4(1)(5)}}{2(1)} = \frac{-1 \pm \sqrt{1-20}}{2}$$
$$= \frac{-1 \pm \sqrt{-19}}{2} = \frac{-1 \pm i\sqrt{19}}{2}$$
$\left\{\dfrac{-1+i\sqrt{19}}{2}, \dfrac{-1-i\sqrt{19}}{2}\right\}$

14. $\left\{\dfrac{-1+3i\sqrt{3}}{2}, \dfrac{-1-3i\sqrt{3}}{2}\right\}$

15. $3x^2 + 2x + 1 = 0 \begin{cases} a = 3 \\ b = 2 \\ c = 1 \end{cases}$

$$x = \frac{-2 \pm \sqrt{(2)^2 - 4(3)(1)}}{2(3)}$$
$$= \frac{-2 \pm \sqrt{4-12}}{6} = \frac{-2 \pm \sqrt{-8}}{6}$$
$$= \frac{-2 \pm 2i\sqrt{2}}{6} = \frac{-1 \pm i\sqrt{2}}{3}$$
$\left\{\dfrac{-1+i\sqrt{2}}{3}, \dfrac{-1-i\sqrt{2}}{3}\right\}$

16. $\left\{\dfrac{-3+i\sqrt{23}}{8}, \dfrac{-3-i\sqrt{23}}{8}\right\}$

17. $2x^2 - 8x + 9 = 0 \begin{cases} a = 2 \\ b = -8 \\ c = 9 \end{cases}$

$$x = \frac{-(-8) \pm \sqrt{(-8)^2 - 4(2)(9)}}{2(2)}$$
$$= \frac{8 \pm \sqrt{64-72}}{4} = \frac{8 \pm \sqrt{-8}}{4} = \frac{8 \pm 2i\sqrt{2}}{4}$$
$$= \frac{\overset{1}{\cancel{2}}(4 \pm i\sqrt{2})}{\underset{2}{\cancel{4}}} = \frac{4 \pm i\sqrt{2}}{2}$$
$\left\{\dfrac{4+i\sqrt{2}}{2}, \dfrac{4-i\sqrt{2}}{2}\right\}$

18. $\left\{\dfrac{3+i\sqrt{3}}{3}, \dfrac{3-i\sqrt{3}}{3}\right\}$

19. $3x\left(x + \dfrac{1}{3}\right) = 3x\left(\dfrac{-1}{3x}\right)$
$$3x^2 + x = -1$$
$3x^2 + x + 1 = 0 \begin{cases} a = 3 \\ b = 1 \\ c = 1 \end{cases}$

$$x = \frac{-1 \pm \sqrt{1^2 - 4(3)(1)}}{2(3)}$$
$$= \frac{-1 \pm \sqrt{1-12}}{6} = \frac{-1 \pm \sqrt{-11}}{6} = \frac{-1 \pm i\sqrt{11}}{6}$$
Both solutions are in the domain.
$\left\{\dfrac{-1+i\sqrt{11}}{6}, \dfrac{-1-i\sqrt{11}}{6}\right\}$

20. $\left\{\dfrac{-1+i\sqrt{15}}{8}, \dfrac{-1-i\sqrt{15}}{8}\right\}$

21. $2x^2 - 5x + 7 = 0 \begin{cases} a = 2 \\ b = -5 \\ c = 7 \end{cases}$

$$x = \frac{-(-5) \pm \sqrt{(-5)^2 - 4(2)(7)}}{2(2)}$$
$$= \frac{5 \pm \sqrt{25-56}}{4} = \frac{5 \pm \sqrt{-31}}{4}$$
$$= \frac{5 \pm i\sqrt{31}}{4}$$
$\left\{\dfrac{5+i\sqrt{31}}{4}, \dfrac{5-i\sqrt{31}}{4}\right\}$

22. $\left\{\dfrac{5+i\sqrt{47}}{6}, \dfrac{5-i\sqrt{47}}{6}\right\}$

23. $x^2 - 4x + 5 = 0 \begin{cases} a = 1 \\ b = -4 \\ c = 5 \end{cases}$

$$x = \frac{-(-4) \pm \sqrt{(-4)^2 - 4(1)(5)}}{2(1)}$$

$$= \frac{4 \pm \sqrt{16 - 20}}{2} = \frac{4 \pm \sqrt{-4}}{2} = \frac{4 \pm 2i}{2} = 2 \pm i$$

Check for $x = 2 - i$:

$$\begin{aligned} x^2 - 4x + 5 &= 0 \\ (2 - i)^2 - 4(2 - i) + 5 &\stackrel{?}{=} 0 \\ 4 - 4i + i^2 - 8 + 4i + 5 &\stackrel{?}{=} 0 \\ 4 - 4i - 1 - 8 + 4i + 5 &\stackrel{?}{=} 0 \\ 9 - 9 + 4i - 4i &\stackrel{?}{=} 0 \\ 0 &= 0 \end{aligned}$$

The check for $x = 2 + i$ is left to the student.

$\{2 + i, 2 - i\}$

24. $\{3 + i, 3 - i\}$

25. Let w = width

$w + 2$ = length

Area = (length)(width) = $(w + 2)w$

$$\begin{aligned} w^2 + 2w &= 2 \\ w^2 + 2w - 2 &= 0 \end{aligned} \begin{cases} a = 1 \\ b = 2 \\ c = -2 \end{cases}$$

$$w = \frac{-(2) \pm \sqrt{(2)^2 - 4(1)(-2)}}{2(1)} = \frac{-2 \pm \sqrt{4 + 8}}{2}$$

$$= \frac{-2 \pm \sqrt{12}}{2} = \frac{-2 \pm 2\sqrt{3}}{2}$$

$$= -1 \pm \sqrt{3} = \begin{cases} -1 + \sqrt{3} \approx -1 + 1.732 \\ -1 - \sqrt{3} \approx -1 - 1.732 \end{cases}$$

$$\approx \begin{cases} 0.732 \approx 0.73 \\ -2.732 \approx -2.73 \text{ Not in the domain} \end{cases}$$

$w + 2 = -1 + \sqrt{3} + 2 = 1 + \sqrt{3} \approx 2.73$

The width of the rectangle is $-1 + \sqrt{3}$ (about 0.73), and its length is $1 + \sqrt{3}$ (about 2.73).

26. Width $= -2 + \sqrt{10} \approx 1.16$

Length $= 2 + \sqrt{10} \approx 5.16$

27. Let x = length of a side.

Perimeter $= 4x$ *Area* $= x^2$

$$\begin{aligned} 4x &= 4 + x^2 \\ x^2 - 4x + 4 &= 0 \\ (x - 2)(x - 2) &= 0 \\ x &= 2 \end{aligned}$$

The length of a side is 2, or 2.00.

28. Side $= 2 + \sqrt{6} \approx 4.45$

Exercises 10.5 (page 504)

1. $x^2 - x - 12 = 0 \begin{cases} a = 1 \\ b = -1 \\ c = -12 \end{cases}$

$$\begin{aligned} b^2 - 4ac &= (-1)^2 - 4(1)(-12) \\ &= 1 + 48 = 49 \quad \text{Perfect square} \end{aligned}$$

Therefore, the roots are real, rational, and unequal.

2. Real, rational, and unequal

3. $6x^2 - 7x - 2 = 0 \begin{cases} a = 6 \\ b = -7 \\ c = -2 \end{cases}$

$b^2 - 4ac = (-7)^2 - 4(6)(-2) = 49 + 48 = 97$

Therefore, the roots are real, unequal, irrational conjugates.

4. Real, unequal, irrational conjugates

5. $x^2 - 4x + 4 = 0 \begin{cases} a = 1 \\ b = -4 \\ c = 4 \end{cases}$

$b^2 - 4ac = (-4)^2 - 4(1)(4) = 16 - 16 = 0$

Therefore, there is one real, rational root of multiplicity two.

6. One real, rational root of multiplicity two

7. $9x^2 - 6x + 2 = 0 \begin{cases} a = 9 \\ b = -6 \\ c = 2 \end{cases}$

$$\begin{aligned} b^2 - 4ac &= (-6)^2 - 4(9)(2) \\ &= 36 - 72 = -36 \end{aligned}$$

Therefore, the roots are complex conjugates.

8. Complex conjugates

9.

$$\begin{array}{r|l} x = 4 & x = -2 \\ x - 4 = 0 & x + 2 = 0 \end{array}$$

$$\begin{aligned} (x - 4) \cdot (x + 2) &= 0 \\ x^2 - 2x - 8 &= 0 \end{aligned}$$

10. $x^2 + x - 6 = 0$

11.

$$\begin{array}{r|l} x = 0 & x = 5 \\ x = 0 & x - 5 = 0 \end{array}$$

$$\begin{aligned} x \cdot (x - 5) &= 0 \\ x^2 - 5x &= 0 \end{aligned}$$

12. $x^2 - 6x = 0$

13.

$$\begin{array}{r|l} x = 2 + \sqrt{3} & x = 2 - \sqrt{3} \\ x - 2 - \sqrt{3} = 0 & x - 2 + \sqrt{3} = 0 \end{array}$$

$$\begin{aligned} [(x - 2) - \sqrt{3}][(x - 2) + \sqrt{3}] &= 0 \\ (x - 2)^2 - (\sqrt{3})^2 &= 0 \\ x^2 - 4x + 4 - 3 &= 0 \\ x^2 - 4x + 1 &= 0 \end{aligned}$$

14. $x^2 - 6x + 4 = 0$

15.

$$\begin{array}{r|r} x = \frac{1}{2} & x = \frac{2}{3} \\ 2x = 1 & 3x = 2 \\ 2x - 1 = 0 & 3x - 2 = 0 \end{array}$$

$$\begin{aligned} (2x - 1)(3x - 2) &= 0 \\ 6x^2 - 7x + 2 &= 0 \end{aligned}$$

16. $15x^2 - 19x + 6 = 0$

17.

$$\begin{array}{r|r} x = \frac{1 + i\sqrt{3}}{2} & x = \frac{1 - i\sqrt{3}}{2} \\ 2x = 1 + i\sqrt{3} & 2x = 1 - i\sqrt{3} \\ 2x - 1 - i\sqrt{3} = 0 & 2x - 1 + i\sqrt{3} = 0 \end{array}$$

$$\begin{aligned} [(2x - 1) - i\sqrt{3}][(2x - 1) + i\sqrt{3}] &= 0 \\ (2x - 1)^2 - (i\sqrt{3})^2 &= 0 \\ 4x^2 - 4x + 1 + 3 &= 0 \\ 4x^2 - 4x + 4 &= 0 \\ x^2 - x + 1 &= 0 \end{aligned}$$

18. $3x^2 - 2x + 1 = 0$

19.

$$\begin{array}{r|r|r} x = 1 & x = 3 & x = 4 \\ x - 1 = 0 & x - 3 = 0 & x - 4 = 0 \end{array}$$

$$\begin{aligned} (x - 1)(x - 3)(x - 4) &= 0 \\ (x^2 - 4x + 3)(x - 4) &= 0 \\ x^3 - 8x^2 + 19x - 12 &= 0 \end{aligned}$$

20. $x^3 - 8x^2 + 17x - 10 = 0$

21.

$$\begin{array}{r|r|r} x = 3 & x = -2i & x = 2i \\ x - 3 = 0 & x + 2i = 0 & x - 2i = 0 \end{array}$$

$$\begin{aligned} (x - 3)(x + 2i)(x - 2i) &= 0 \\ (x - 3)(x^2 + 4) &= 0 \\ x^3 - 3x^2 + 4x - 12 &= 0 \end{aligned}$$

22. $x^3 - 2x^2 + 9x - 18 = 0$

23. If $5 - i$ is a root, $5 + i$ is a root.

$$x = 5 - i \quad \text{or} \quad x = 5 + i$$
$$x - 5 + i = 0 \quad \text{or} \quad x - 5 - i = 0$$
$$([x - 5] + i)([x - 5] - i) = 0$$
$$[x - 5]^2 - i^2 = 0$$
$$x^2 - 10x + 25 - (-1) = 0$$
$$x^2 - 10x + 26 = 0$$

24. $x^2 - 4x + 13 = 0$

25. If $1 - 2\sqrt{5}$ is a root, $1 + 2\sqrt{5}$ is a root.

$$x = 1 - 2\sqrt{5} \quad \text{or} \quad x = 1 + 2\sqrt{5}$$
$$x - 1 + 2\sqrt{5} = 0 \quad \text{or} \quad x - 1 - 2\sqrt{5} = 0$$
$$([x - 1] + 2\sqrt{5})([x - 1] - 2\sqrt{5}) = 0$$
$$[x - 1]^2 - (2\sqrt{5})^2 = 0$$
$$x^2 - 2x + 1 - 20 = 0$$
$$x^2 - 2x - 19 = 0$$

26. $x^2 - 6x - 41 = 0$

27. If $-3i$ is a root, $+3i$ is a root.

$$x = 2 \quad \Big| \quad x = -3i \quad \Big| \quad x = 3i$$
$$x - 2 = 0 \quad \Big| \quad x + 3i = 0 \quad \Big| \quad x - 3i = 0$$
$$(x - 2)(x + 3i)(x - 3i) = 0$$
$$(x - 2)(x^2 + 9) = 0$$
$$x^3 - 2x^2 + 9x - 18 = 0$$

28. $x^3 + 3x^2 + 36x + 108 = 0$

Review Exercises 10.6 (page 507)

1.

$$x^2 + x - 6 = 0$$
$$(x + 3)(x - 2) = 0$$
$$x + 3 = 0 \quad \Big| \quad x - 2 = 0$$
$$x = -3 \quad \Big| \quad x = 2$$

$\{-3, 2\}$

2. $\{5, -2\}$

3. $x^2 - 2x - 4 = 0 \begin{cases} a = 1 \\ b = -2 \\ c = -4 \end{cases}$

$$x = \frac{-(-2) \pm \sqrt{(-2)^2 - 4(1)(-4)}}{2(1)}$$
$$= \frac{2 \pm \sqrt{4 + 16}}{2} = \frac{2 \pm \sqrt{20}}{2} = \frac{2 \pm 2\sqrt{5}}{2} = 1 \pm \sqrt{5}$$

$\{1 + \sqrt{5}, 1 - \sqrt{5}\}$

4. $\{2 + \sqrt{2}, 2 - \sqrt{2}\}$

5. $x^2 - 2x + 5 = 0 \begin{cases} a = 1 \\ b = -2 \\ c = 5 \end{cases}$

$$x = \frac{-(-2) \pm \sqrt{(-2)^2 - 4(1)(5)}}{2(1)}$$
$$= \frac{2 \pm \sqrt{-16}}{2} = \frac{2 \pm 4i}{2} = 1 \pm 2i$$

$\{1 + 2i, 1 - 2i\}$

6. $\{2 + i\sqrt{3}, 2 - i\sqrt{3}\}$

7.

$$16x^2 = 9$$
$$x^2 = \frac{9}{16}$$
$$x = \pm\frac{3}{4}$$

$\left\{\frac{3}{4}, -\frac{3}{4}\right\}$

8. $\left\{\frac{5}{6}, -\frac{5}{6}\right\}$

9. LCD $= 3(x - 2)$

$$\frac{\overset{1}{\cancel{3}}(x - 2)}{1} \cdot \frac{(x + 2)}{\underset{1}{\cancel{3}}} = \frac{3\overset{1}{\cancel{(x - 2)}}}{1} \cdot \frac{1}{\underset{1}{\cancel{(x - 2)}}} + \frac{\overset{1}{\cancel{3}}(x - 2)}{1} \cdot \frac{2}{\underset{1}{\cancel{3}}}$$
$$(x - 2)(x + 2) = 3 + 2(x - 2)$$
$$x^2 - 4 = 3 + 2x - 4$$
$$x^2 - 2x - 3 = 0$$
$$(x - 3)(x + 1) = 0$$
$$x - 3 = 0 \quad \Big| \quad x + 1 = 0$$
$$x = 3 \quad \Big| \quad x = -1$$

$\{3, -1\}$

10. $\{-4, 2\}$

11.

$$x^2 + 3x - 10 = 3x - 2x^2 + 2$$
$$3x^2 = 12$$
$$x^2 = 4$$
$$x = \pm 2$$

$\{2, -2\}$

12. $\left\{\frac{3\sqrt{5}}{5}, -\frac{3\sqrt{5}}{5}\right\}$

13. Let $z = (x^2 - 4x)$; then $z^2 = (x^2 - 4x)^2$.

$$z^2 + 5z + 4 = 0$$
$$(z + 4)(z + 1) = 0$$

$$z + 4 = 0 \quad \Big| \quad z + 1 = 0$$
$$z = -4 \quad \Big| \quad z = -1$$
$$x^2 - 4x = -4 \quad \Big| \quad x^2 - 4x = -1$$
$$x^2 - 4x + 4 = 0 \quad \Big| \quad x^2 - 4x + 1 = 0$$
$$(x - 2)^2 = 0 \quad \Big| \quad x = \frac{-(-4) \pm \sqrt{(-4)^2 - 4(1)(1)}}{2(1)}$$
$$x - 2 = 0 \quad \Big| \quad = \frac{4 \pm \sqrt{16 - 4}}{2}$$
$$x = 2 \quad \Big| \quad = \frac{4 \pm \sqrt{12}}{2} = \frac{4}{2} \pm \frac{2\sqrt{3}}{2} = 2 \pm \sqrt{3}$$

$\{2, 2 + \sqrt{3}, 2 - \sqrt{3}\}$

14. $\{-3, -3 + \sqrt{5}, -3 - \sqrt{5}\}$

15.

$$(x^2 - 1)(x^2 - 64) = 0$$
$$(x - 1)(x + 1)(x + 8)(x - 8) = 0$$
$$x - 1 = 0 \quad \Big| \quad x + 1 = 0 \quad \Big| \quad x + 8 = 0 \quad \Big| \quad x - 8 = 0$$
$$x = 1 \quad \Big| \quad x = -1 \quad \Big| \quad x = -8 \quad \Big| \quad x = 8$$

$\{1, -1, 8, -8\}$

16. $\{1, -1, 9, -9\}$

17. $x^2 - 6x + 7 = 0 \begin{cases} a = 1 \\ b = -6 \\ c = 7 \end{cases}$

$b^2 - 4ac = 36 - 28 = 8 > 0$

Therefore, the roots are real, unequal, irrational conjugates.

18. Real, unequal, irrational conjugates

19. Let x = first odd integer

$x + 2$ = second odd integer

$$(x)(x + 2) - 14 = 85$$
$$x^2 + 2x - 99 = 0$$
$$(x + 11)(x - 9) = 0$$
$$x + 11 = 0 \quad \Big| \quad x - 9 = 0$$
$$x = -11 \quad \Big| \quad x = 9$$
$$x + 2 = -9 \quad \Big| \quad x + 2 = 11$$

There are two answers: The integers are -11 and -9, or the integers are 9 and 11.

20. 7 and 8, or -8 and -7

21. Let $x = \text{width}$

$x + 3 = \text{length}$

$x(x + 3) = 8$

$x^2 + 3x - 8 = 0$

$$x = \frac{-3 \pm \sqrt{9 + 32}}{2} = \frac{-3 \pm \sqrt{41}}{2} \approx \frac{-3 \pm 6.403}{2}$$

$$x = \frac{-3 - \sqrt{41}}{2} \approx \frac{-3 - 6.403}{2} \approx -4.70 \quad \text{Not in the domain}$$

$$x = \frac{-3 + \sqrt{41}}{2} \approx \frac{-3 + 6.403}{2} \approx 1.70$$

$$x + 3 = \frac{-3 + \sqrt{41}}{2} + 3 = \frac{3 + \sqrt{41}}{2} \approx \frac{3 + 6.403}{2} \approx 4.70$$

The width is $\frac{-3 + \sqrt{41}}{2}$ (about 1.70), and the length is $\frac{3 + \sqrt{41}}{2}$ (about 4.70).

22. One leg is $\frac{-2 + \sqrt{14}}{2} \approx 0.87$.

The other leg is $\frac{2 + \sqrt{14}}{2} \approx 2.87$.

23. $x = 1 - \sqrt{7}$ | $x = 1 + \sqrt{7}$

$x - 1 + \sqrt{7} = 0$ | $x - 1 - \sqrt{7} = 0$

$(x - 1 + \sqrt{7})(x - 1 - \sqrt{7}) = 0$

$([x - 1] + \sqrt{7})([x - 1] - \sqrt{7}) = 0$

$([x - 1]^2 - [\sqrt{7}]^2 = 0$

$x^2 - 2x + 1 - 7 = 0$

$x^2 - 2x - 6 = 0$

24. $x^3 - 2x^2 + 6x - 12 = 0$

Exercises 10.7 (page 519)

1. Set $y = 0$.

$x^2 - 2x - 3 = 0$

$(x - 3)(x + 1) = 0$

$x - 3 = 0$ | $x + 1 = 0$

$x = 3$ | $x = -1$

x-intercepts are -1 and 3.

y-intercept is -3.

Axis of symmetry is $x = \frac{-(-2)}{2} = 1$.

$f(1) = 1^2 - 2(1) - 3 = 1 - 2 - 3 = -4$

Vertex is at $(1, -4)$; opens upward.

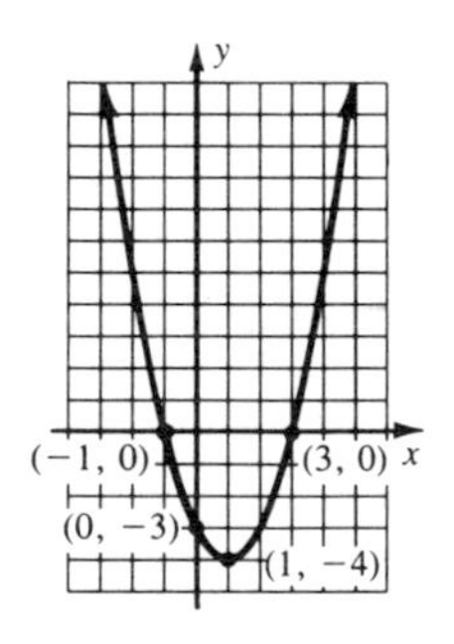

2. x-intercepts are -3 and 5.

y-intercept is -15.

Axis of symmetry is $x = 1$.

Vertex is at $(1, -16)$; opens upward.

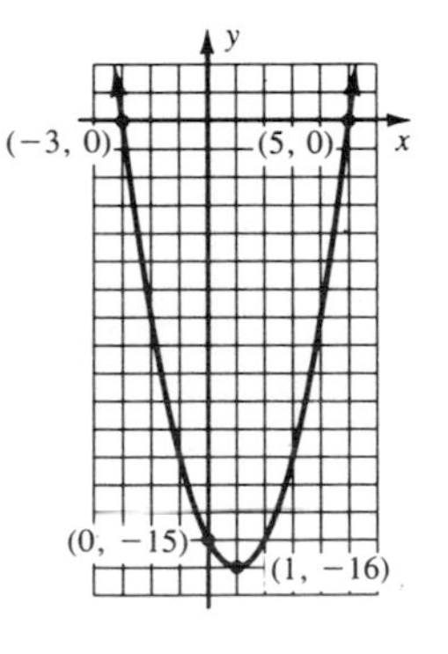

3. Set $y = 0$. Then $x^2 - 2x - 13 = 0$.

$$x = \frac{-(-2) \pm \sqrt{(-2)^2 - 4(1)(-13)}}{2(1)}$$

$$= \frac{2 \pm \sqrt{4 + 52}}{2} = \frac{2 \pm \sqrt{56}}{2}$$

$$= \frac{2 \pm \sqrt{4 \cdot 14}}{2} = \frac{2 \pm 2\sqrt{14}}{2}$$

$$= 1 \pm \sqrt{14}$$

x-intercepts are $1 + \sqrt{14}$ and $1 - \sqrt{14}$.

y-intercept is -13.

Axis of symmetry is $x = \frac{-(-2)}{2} = 1$.

$f(1) = 1^2 - 2(1) - 13 = -14$

Vertex is at $(1, -14)$; opens upward.

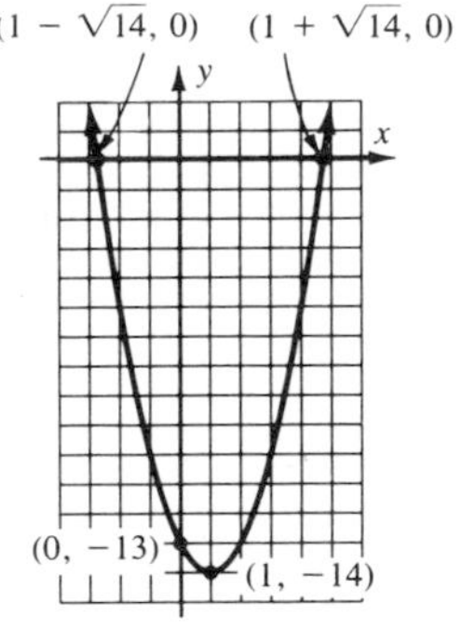

4. x-intercepts are $2 + 2\sqrt{3}$ and $2 - 2\sqrt{3}$.

y-intercept is -8.

Axis of symmetry is $x = 2$.

Vertex is at $(2, -12)$; opens upward.

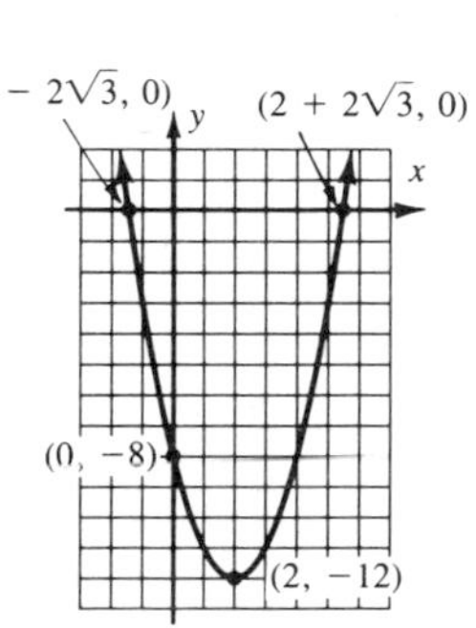

5. Set $y = 0$.

$x^2 - 4x + 3 = 0$

$(x - 3)(x - 1) = 0$

$x - 1 = 0$ | $x - 3 = 0$

$x = 1$ | $x = 3$

x-intercepts are 1 and 3.

y-intercept is 3.

Axis of symmetry is $x = \frac{-(-4)}{2} = 2$.

$f(2) = 2^2 - 4(2) + 3 = -1$

Vertex is at $(2, -1)$; opens upward.

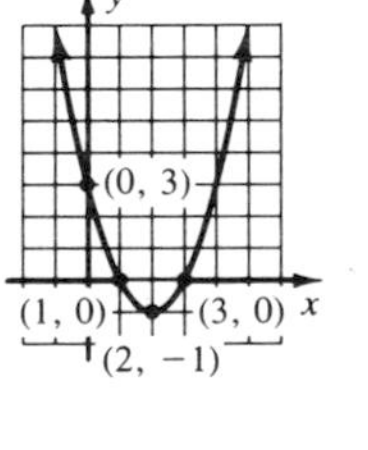

6. x-intercepts are 2 and -4.

y-intercept is -8.

Axis of symmetry is $x = -1$.

Vertex is at $(-1, -9)$; opens upward.

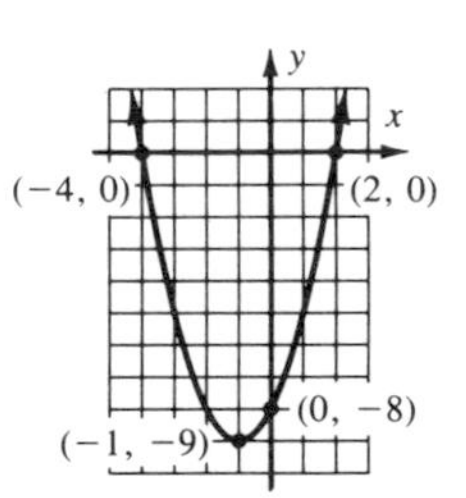

7. Set $y = 0$.

$x^2 + 3x - 10 = 0$

$(x - 2)(x + 5) = 0$

$x - 2 = 0$ | $x + 5 = 0$

$x = 2$ | $x = -5$

x-intercepts are 2 and -5.

y-intercept is -10.

Axis of symmetry is $x = -\frac{3}{2}$.

$f(-\frac{3}{2}) = (-\frac{3}{2})^2 + 3(-\frac{3}{2}) - 10 = -\frac{49}{4}$

Vertex is at $(-\frac{3}{2}, -\frac{49}{4})$; opens upward.

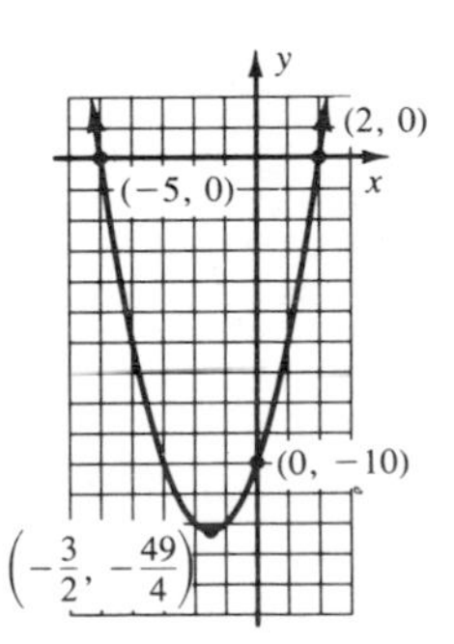

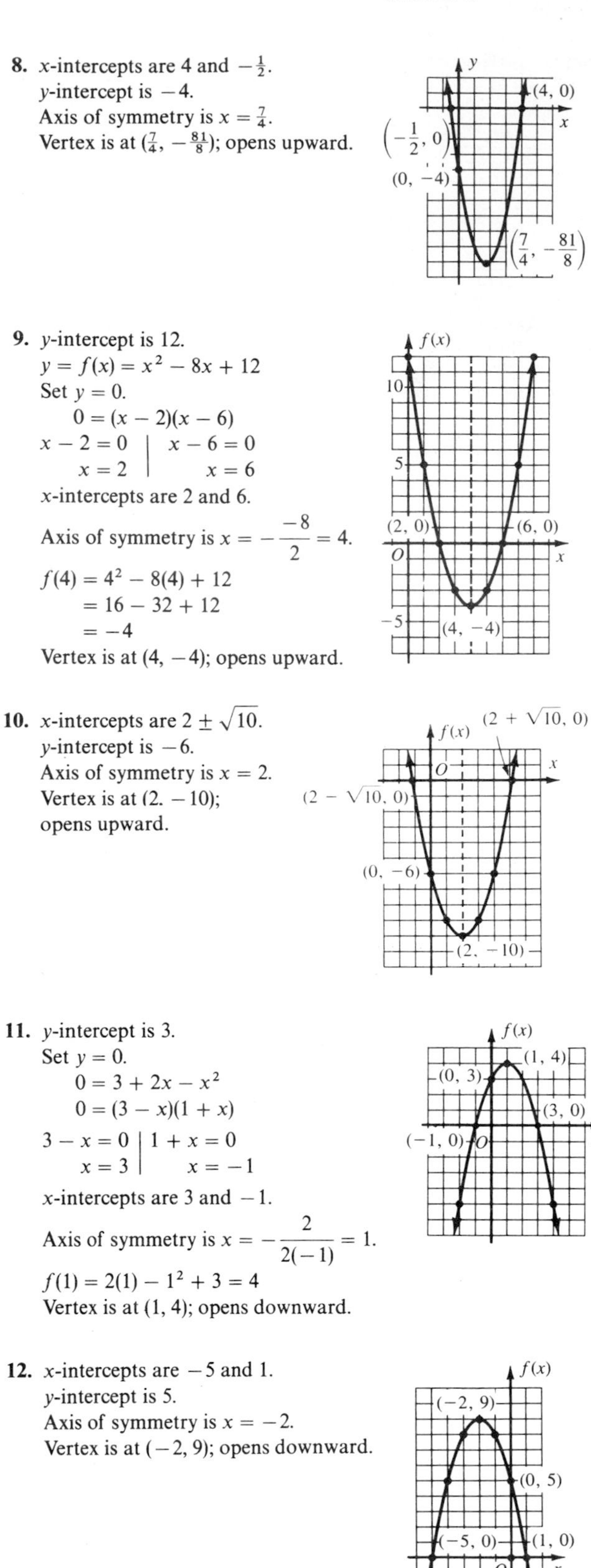

8. x-intercepts are 4 and $-\frac{1}{2}$.
y-intercept is -4.
Axis of symmetry is $x = \frac{7}{4}$.
Vertex is at $(\frac{7}{4}, -\frac{81}{8})$; opens upward.

9. y-intercept is 12.
$y = f(x) = x^2 - 8x + 12$
Set $y = 0$.
$0 = (x-2)(x-6)$
$x - 2 = 0 \quad | \quad x - 6 = 0$
$x = 2 \quad | \quad x = 6$
x-intercepts are 2 and 6.

Axis of symmetry is $x = -\dfrac{-8}{2} = 4$.

$f(4) = 4^2 - 8(4) + 12$
$= 16 - 32 + 12$
$= -4$
Vertex is at $(4, -4)$; opens upward.

10. x-intercepts are $2 \pm \sqrt{10}$.
y-intercept is -6.
Axis of symmetry is $x = 2$.
Vertex is at $(2, -10)$;
opens upward.

11. y-intercept is 3.
Set $y = 0$.
$0 = 3 + 2x - x^2$
$0 = (3 - x)(1 + x)$
$3 - x = 0 \quad | \quad 1 + x = 0$
$x = 3 \quad | \quad x = -1$
x-intercepts are 3 and -1.

Axis of symmetry is $x = -\dfrac{2}{2(-1)} = 1$.

$f(1) = 2(1) - 1^2 + 3 = 4$
Vertex is at $(1, 4)$; opens downward.

12. x-intercepts are -5 and 1.
y-intercept is 5.
Axis of symmetry is $x = -2$.
Vertex is at $(-2, 9)$; opens downward.

13. $y = f(x) = x^2 - 6x + 10$
Set $y = 0$. Then $x^2 - 6x + 10 = 0$.

$$x = \frac{-(-6) \pm \sqrt{(-6)^2 - 4(1)(10)}}{2(1)}$$
$$= \frac{6 \pm \sqrt{36 - 40}}{2} = \frac{6 \pm \sqrt{-4}}{2}$$
$$= \frac{6 \pm 2i}{2} = 3 \pm i$$

This means that the curve does not cross the x-axis.
y-intercept is 10.

Axis of symmetry is $x = -\dfrac{-6}{2} = 3$.

Vertex is at $(3, 1)$; opens upward.

14. no x-intercepts
y-intercept is 11.
Axis of symmetry is $x = 3$.
Vertex is at $(3, 2)$; opens upward.

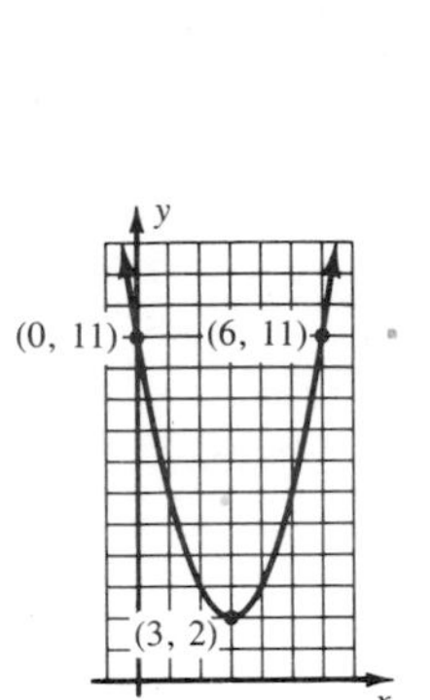

15. y-intercept is -9.
Set $y = 0$.
$0 = 4x^2 - 9$
$9 = 4x^2$
$x^2 = \dfrac{9}{4}$

$x = \pm\sqrt{\dfrac{9}{4}} = \pm\dfrac{3}{2}$

x-intercepts are $\pm\dfrac{3}{2}$.

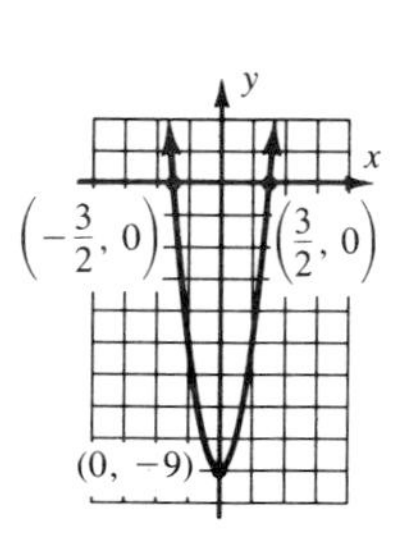

Axis of symmetry is $x = \dfrac{-0}{2(4)} = 0$.

$f(0) = -9$
Vertex is at $(0, -9)$; opens upward.

16. x-intercepts are $\pm\dfrac{2}{3}$.
y-intercept is -4.
Axis of symmetry is $x = 0$.
Vertex is at $(0, -4)$; opens upward.

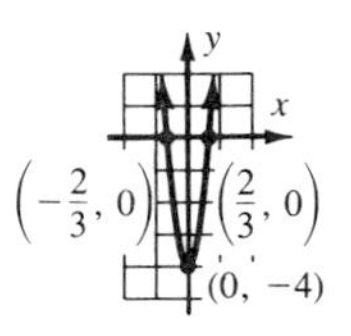

17. $y = f(x) = x^2 - 4x + 6$ If we set $y = 0$, there are no real solutions. (no x-intercepts)
y-intercept is 6.

Axis of symmetry is $x = -\dfrac{-4}{2} = 2$.

Vertex is at $(2, 2)$; opens upward.

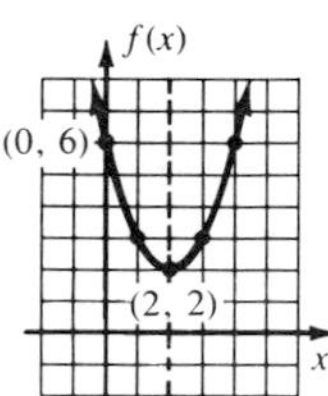

18. no x-intercepts
y-intercept is -2.
Axis of symmetry is $x = 1$.
Vertex is at $(1, -1)$; opens downward.

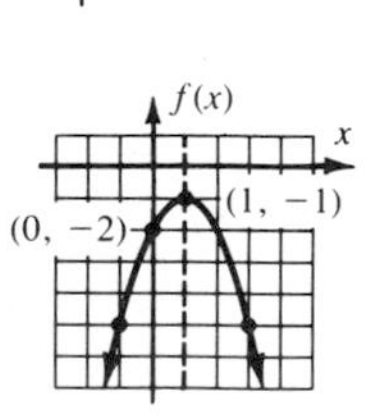

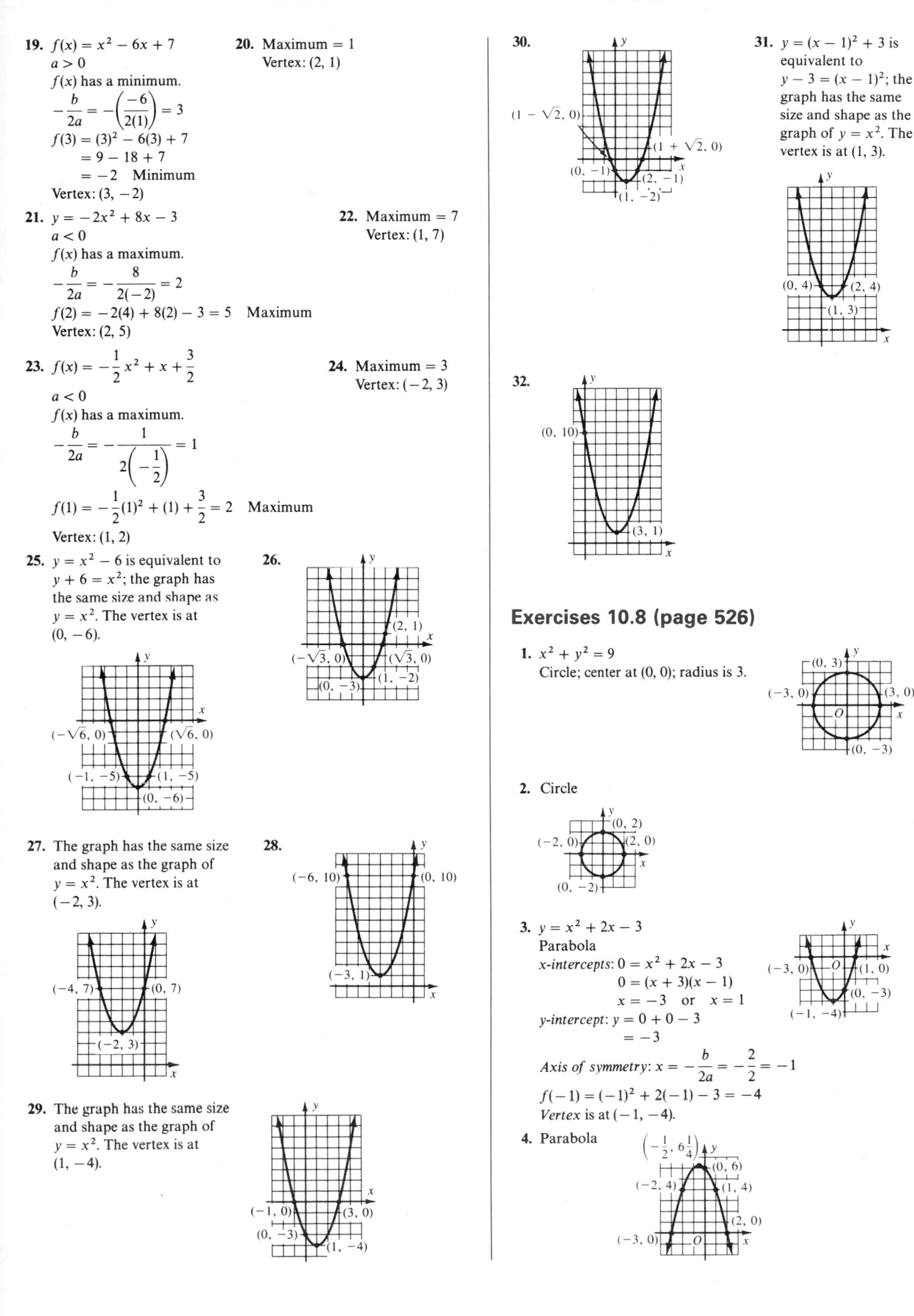

19. $f(x) = x^2 - 6x + 7$
$a > 0$
$f(x)$ has a minimum.
$-\frac{b}{2a} = -\left(\frac{-6}{2(1)}\right) = 3$
$f(3) = (3)^2 - 6(3) + 7$
$= 9 - 18 + 7$
$= -2$ Minimum
Vertex: $(3, -2)$

20. Maximum $= 1$
Vertex: $(2, 1)$

21. $y = -2x^2 + 8x - 3$
$a < 0$
$f(x)$ has a maximum.
$-\frac{b}{2a} = -\frac{8}{2(-2)} = 2$
$f(2) = -2(4) + 8(2) - 3 = 5$ Maximum
Vertex: $(2, 5)$

22. Maximum $= 7$
Vertex: $(1, 7)$

23. $f(x) = -\frac{1}{2}x^2 + x + \frac{3}{2}$
$a < 0$
$f(x)$ has a maximum.
$-\frac{b}{2a} = -\frac{1}{2\left(-\frac{1}{2}\right)} = 1$
$f(1) = -\frac{1}{2}(1)^2 + (1) + \frac{3}{2} = 2$ Maximum
Vertex: $(1, 2)$

24. Maximum $= 3$
Vertex: $(-2, 3)$

25. $y = x^2 - 6$ is equivalent to $y + 6 = x^2$; the graph has the same size and shape as $y = x^2$. The vertex is at $(0, -6)$.

26.

27. The graph has the same size and shape as the graph of $y = x^2$. The vertex is at $(-2, 3)$.

28.

29. The graph has the same size and shape as the graph of $y = x^2$. The vertex is at $(1, -4)$.

30.

31. $y = (x - 1)^2 + 3$ is equivalent to $y - 3 = (x - 1)^2$; the graph has the same size and shape as the graph of $y = x^2$. The vertex is at $(1, 3)$.

32.

Exercises 10.8 (page 526)

1. $x^2 + y^2 = 9$
Circle; center at $(0, 0)$; radius is 3.

2. Circle

3. $y = x^2 + 2x - 3$
Parabola
x-intercepts: $0 = x^2 + 2x - 3$
$0 = (x + 3)(x - 1)$
$x = -3$ or $x = 1$
y-intercept: $y = 0 + 0 - 3$
$= -3$
Axis of symmetry: $x = -\frac{b}{2a} = -\frac{2}{2} = -1$
$f(-1) = (-1)^2 + 2(-1) - 3 = -4$
Vertex is at $(-1, -4)$.

4. Parabola

5. $3x^2 + 3y^2 = 21$
$x^2 + y^2 = 7$
Circle; center at (0, 0); radius is $\sqrt{7} \approx 2.6$.

6. Ellipse

7. $4x^2 + 9y^2 = 36$
Ellipse
x-intercepts: $4x^2 + 9(0)^2 = 36$
$4x^2 = 36$
$x^2 = 9$
$x = \pm 3$
y-intercepts: $4(0)^2 + 9y^2 = 36$
$9y^2 = 36$
$y^2 = 4$
$y = \pm 2$

8. Ellipse

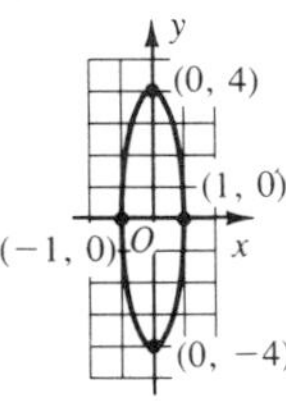

9. $9x^2 - 4y^2 = 36$
Hyperbola
x-intercepts: $9x^2 - 4(0)^2 = 36$
$9x^2 = 36$
$x^2 = \pm 4$
$x = \pm 2$
(Graph does not intersect *y*-axis.)
The rectangle of reference has vertices at (2, 3)(2, −3)(−2, 3), and (−2, −3).

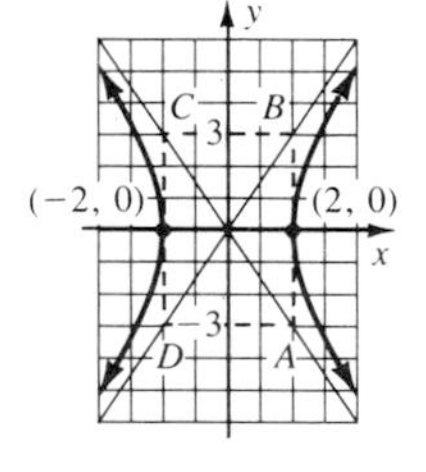

10. Hyperbola

11. $y^2 = 8x$
Parabola
When $x = 1$, $y^2 = 8$
$y = \pm\sqrt{8} \approx \pm 2.8$
When $x = 2$, $y^2 = 16$
$y = \pm 4$
When $x = 0$, $y^2 = 0$
$y = 0$
Vertex is at (0, 0)

12. Parabola

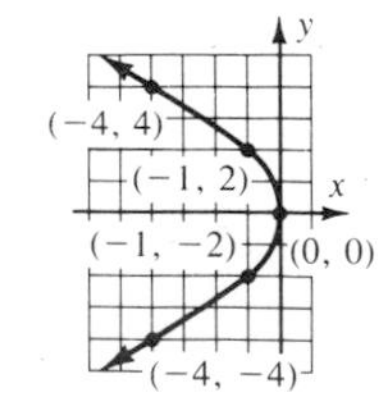

13. $(x^2 + 2x + ?) + (y^2 - 4y + ?) = -1$
$(x^2 + 2x + 1) + (y^2 - 4y + 4) = -1 + 1 + 4$
$(x + 1)^2 + (y - 2)^2 = 4$ is in the form $(x - h)^2 + (y - k)^2 = r^2$, where $h = -1$, $k = 2$, and $r = 2$. The conic is a circle with its center at (−1, 2) and with a radius of 2.

14. Circle; center at (2, −3), radius is 1.

15.
$x = -y^2 - 2y$
$x = -(y^2 + 2y + ?)$
$x - 1 = -(y^2 + 2y + 1)$
$x - 1 = -(y + 1)^2$
The conic is a parabola that opens to the left; the vertex is at (1, −1). The line of symmetry is the line $y = \dfrac{-(-2)}{2(-1)} = -1$.

16. The conic is a parabola that opens to the right; the vertex is at (−2, 1). The line of symmetry is the line $y = 1$.

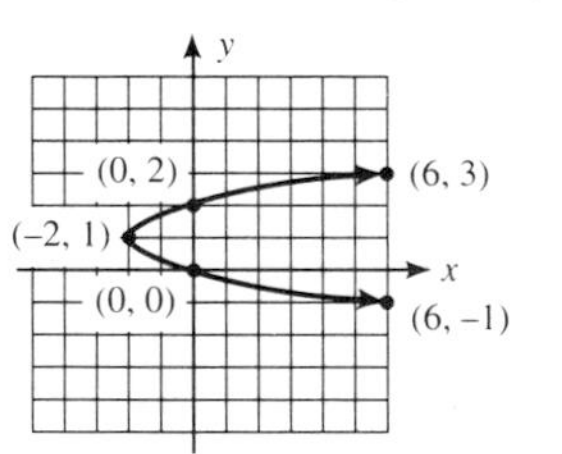

Exercises 10.9A (page 532)

(*Note*: Some exercises are shown with method 1 and some with method 2.)

1. $(x + 1)(x - 2) < 0$
Let $f(x) = (x + 1)(x - 2) = 0$.
Then $x + 1 = 0$ | $x - 2 = 0$
$x = -1$ | $x = 2$
The *x*-intercepts are −1 and 2.
$f(x) = (x + 1)(x - 2)$
$f(0) = (1)(-2) = -2$
The curve goes through (0, −2).
Solution set: $\{x \mid -1 < x < 2\}$ or $(-1, 2)$

2. $\{x \mid x < -1\} \cup \{x \mid x > 2\}$ or $(-\infty, -1) \cup (2, +\infty)$

3. $3 - 2x - x^2 \geq 0$

Let $f(x) = (3 + x)(1 - x) = 0$.

Then $3 + x = 0$ | $1 - x = 0$

$x = -3$ | $x = 1$

The critical points are -3 and 1.

		Same signs	
Sign of $3 + x$	$- - -$	$+ + + + + +$	$+ + + + + +$
Sign of $1 - x$	$+ + +$	$+ + + + + +$	$- - - - - -$

$-4 \quad -3 \quad -2 \quad -1 \quad 0 \quad 1 \quad 2 \quad 3 \quad 4$

Our inequality was $\geq$. We select the interval on which both signs are the same. The graph is:

$-4 \quad -3 \quad -2 \quad -1 \quad 0 \quad 1 \quad 2 \quad 3 \quad 4$

Solution set: $\{x \mid -3 \leq x \leq 1\}$ or $[-3, 1]$

4. $\{x \mid x \leq -2\} \cup \{x \mid x \geq 1\}$ or $(-\infty, -2] \cup [1, +\infty)$

$-3 \quad -2 \quad -1 \quad 0 \quad 1 \quad 2$

5. $x^2 - 3x - 4 < 0$

Let $f(x) = x^2 - 3x - 4 = 0$.

Then $(x - 4)(x + 1) = 0$.

x-intercepts: 4 and -1

$f(0) = 0 - 0 - 4 = -4$

The curve goes through $(0, -4)$.

Solution set: $\{x \mid -1 < x < 4\}$ or $(-1, 4)$

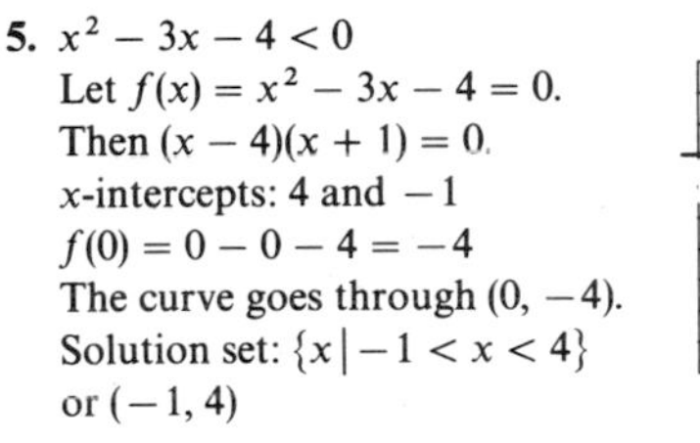

$-3 \quad -2 \quad -1 \quad 0 \quad 1 \quad 2 \quad 3 \quad 4 \quad 5$

6. $\{x \mid x < -5\} \cup \{x \mid x > 1\}$ or $(-\infty, -5) \cup (1, +\infty)$

$-6 \quad -5 \quad -4 \quad -3 \quad -2 \quad -1 \quad 0 \quad 1 \quad 2$

7. $x^2 + 7 > 6x$

$x^2 - 6x + 7 > 0$

Let $f(x) = x^2 - 6x + 7 = 0$.

$$x = \frac{-(-6) \pm \sqrt{36 - 28}}{2} = \frac{6 \pm \sqrt{8}}{2}$$

$$= \frac{6 \pm 2\sqrt{2}}{2} = 3 \pm \sqrt{2}$$

x-intercepts: $3 - \sqrt{2} \approx 1.6$ and $3 + \sqrt{2} \approx 4.4$

$f(0) = 0 - 0 + 7 = 7$

The graph goes through $(0, 7)$.

Solution set: $\{x \mid x < 3 - \sqrt{2}\} \cup \{x \mid x > 3 + \sqrt{2}\}$ or

$\approx 1.6 \qquad \approx 4.4$

$-3 \quad -2 \quad -1 \quad 0 \quad 1 \quad 2 \quad 3 \quad 4 \quad 5 \quad 6$

$(-\infty, 3 - \sqrt{2}) \cup (3 + \sqrt{2}, +\infty)$

8. $\{x \mid 4 - \sqrt{3} < x < 4 + \sqrt{3}\}$ or $(4 - \sqrt{3}, 4 + \sqrt{3})$

$\approx 2.3 \qquad \approx 5.7$

$1 \quad 2 \quad 3 \quad 4 \quad 5 \quad 6 \quad 7$

9. $x^2 \leq 5x$

Let $f(x) = x^2 - 5x = 0$.

Then $x(x - 5) = 0$.

x-intercepts: 0 and 5

$f(1) = 1 - 5 = -4$

The graph goes through $(1, -4)$.

Solution set: $\{x \mid 0 \leq x \leq 5\}$ or $[0, 5]$

$-1 \quad 0 \quad 1 \quad 2 \quad 3 \quad 4 \quad 5 \quad 6$

10. $\{x \mid x \leq 0\} \cup \{x \mid x \geq 3\}$ or $(-\infty, 0] \cup [3, +\infty)$

$-2 \quad -1 \quad 0 \quad 1 \quad 2 \quad 3 \quad 4 \quad 5$

11. $3x - x^2 > 0$

Let $f(x) = x(3 - x) = 0$.

Critical points are 0 and 3.

		Same signs	
Sign of x	$- - - -$	$+ + + + +$	$+ + + + +$
Sign of $3 - x$	$+ + + +$	$+ + + + +$	$- - - - -$

$-2 \quad -1 \quad 0 \quad 1 \quad 2 \quad 3 \quad 4 \quad 5$

We select the interval on which both signs are the same. The graph is:

$-2 \quad -1 \quad 0 \quad 1 \quad 2 \quad 3 \quad 4 \quad 5$

Solution set: $\{x \mid 0 < x < 3\}$ or $(0, 3)$

12. $\{x \mid x < 0\} \cup \{x \mid x > 7\}$ or $(-\infty, 0) \cup (7, +\infty)$

$-1 \quad 0 \quad 1 \quad 2 \quad 3 \quad 4 \quad 5 \quad 6 \quad 7 \quad 8$

13. $x^3 - 3x^2 - x + 3 > 0$

Let $f(x) = x^3 - 3x^2 - x + 3 = 0$.

Factoring by grouping, we have

$x^3 - 3x^2 - x + 3 = (x - 3)(x + 1)(x - 1)$.

Critical points are -1, 1, and 3.

		Two neg.		All positive
Sign of $x - 3$	$- - - -$	$- - -$	$- - -$	$+ + + +$
Sign of $x + 1$	$- - - -$	$+ + +$	$+ + +$	$+ + + +$
Sign of $x - 1$	$- - - -$	$- - -$	$+ + +$	$+ + + +$

$-3 \quad -2 \quad -1 \quad 0 \quad 1 \quad 2 \quad 3 \quad 4 \quad 5$

We select the intervals on which there is an *even* number of negative signs or where all the signs are positive. The graph is:

$-3 \quad -2 \quad -1 \quad 0 \quad 1 \quad 2 \quad 3 \quad 4 \quad 5$

Solution set: $\{x \mid -1 < x < 1\} \cup \{x \mid x > 3\}$ or $(-1, 1) \cup (3, +\infty)$

14. $\{x \mid -2 < x < -1\} \cup \{x \mid x > 1\}$ or $(-2, -1) \cup (1, +\infty)$

$-3 \quad -2 \quad -1 \quad 0 \quad 1 \quad 2 \quad 3 \quad 4 \quad 5$

15. Let $f(x) = (x + 1)(x - 2)^2 = 0$.
Critical points are -1 and 2.

Different signs

Sign of $(x - 2)^2$ + + + + ⋮ + + + + + + + + + + +
Sign of $x + 1$ − − − − ⋮ + + + + + + + + + + +

−3 −2 −1 0 1 2 3 4 5

We select the interval on which the signs are different.
The graph is:

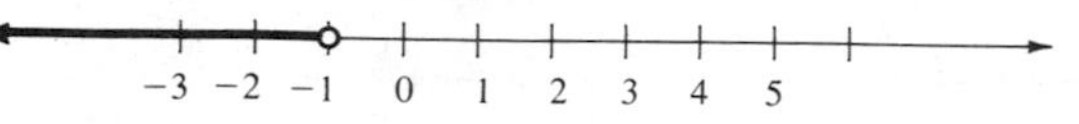

Solution set: $\{x \mid x < -1\}$ or $(-\infty, -1)$

16. $\{x \mid x < -2\} \cup \{x \mid -2 < x < 1\}$ or $(-\infty, -2) \cup (-2, 1)$

−4 −3 −2 −1 0 1 2

17. $\dfrac{x - 2}{x + 3} \geq 0$

Same signs ⋮ ⋮ Same signs

Sign of $x - 2$ − − − − ⋮ − − − − − − − − ⋮ + + + + + + + +
Sign of $x + 3$ − − − − ⋮ + + + + + + + + ⋮ + + + + + + + +

−5 −4 −3 −2 −1 0 1 2 3 4

We select the interval on which the signs are the same. (Remember that x cannot equal -3.) The graph is:

−5 −4 −3 −2 −1 0 1 2 3 4

Solution set: $\{x \mid x < -3\} \cup \{x \mid x \geq 2\}$ or $(-\infty, -3) \cup [2, +\infty)$

18. $\{x \mid -4 \leq x < 2\}$ or $[-4, 2)$

−5 −4 −3 −2 −1 0 1 2 3

19. Multiply by $(x + 1)^2$, which is positive. This clears fractions and does not change the sense of the inequality.

$$\frac{(x+1)^2}{1} \cdot \frac{x}{1} < \frac{(x+1)^2}{1} \cdot \frac{2}{x+1}$$
$$(x + 1)^2 x < 2(x + 1)$$
$$(x + 1)^2 x - 2(x + 1) < 0$$
$$(x + 1)[(x + 1)x - 2] < 0$$
$$(x + 1)(x^2 + x - 2) < 0$$
$$(x + 1)(x + 2)(x - 1) < 0$$

Let $f(x) = (x + 1)(x + 2)(x - 1) = 0$.
Critical points are -2, -1, and 1.

Three neg. ⋮ Two neg. ⋮ One neg. ⋮

Sign of $x + 1$ − − − − − ⋮ − ⋮ + + + ⋮ + + + + +
Sign of $x + 2$ − − − − − ⋮ + ⋮ + + + ⋮ + + + + +
Sign of $x - 1$ − − − − − ⋮ − ⋮ − − − ⋮ + + + + +

−4 −3 −2 −1 0 1 2 3

We select the intervals on which is an *odd* number of negative signs. The graph is:

−4 −3 −2 −1 0 1 2 3

Solution set: $\{x \mid x < -2\} \cup \{x \mid -1 < x < 1\}$ or $(-\infty, -2) \cup (-1, 1)$

20. $\{x \mid -4 < x < -3\} \cup \{x \mid x > 1\}$ or $(-4, -3) \cup (1, +\infty)$

−5 −4 −3 −2 −1 0 1 2 3

21. We must solve $x^2 - 4x - 12 \geq 0$. Let $y = f(x) = (x - 6)(x + 2) = 0$. Critical points are 6 and -2.

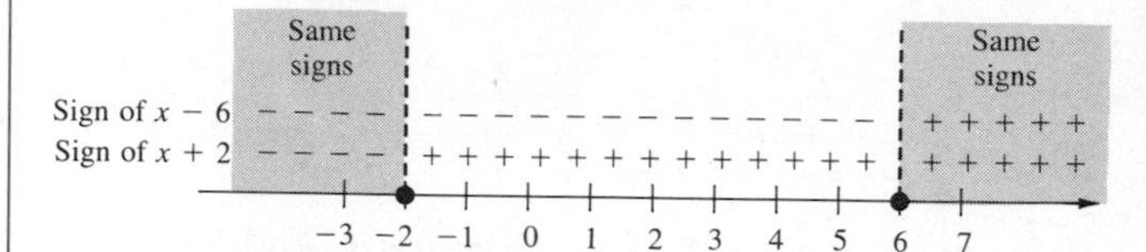

We select the interval on which the signs are the same.
The graph is:

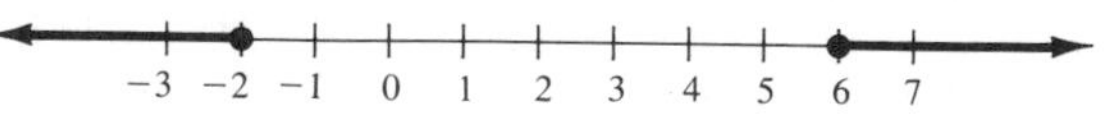

Solution set: $\{x \mid x \leq -2\} \cup \{x \mid x \geq 6\}$ or $(-\infty, -2] \cup [6, +\infty)$

22. $\{x \mid x \leq -5\} \cup \{x \mid x \geq 2\}$ or $(-\infty, -5] \cup [2, +\infty)$

23. We must solve $4 + 3x - x^2 \geq 0$. Let $y = f(x) = (4 - x)(1 + x) = 0$. Critical points are 4 and -1.

Same signs

Sign of $4 - x$ + + + + + ⋮ + + + + + + + + ⋮ − − − − − − − −
Sign of $1 + x$ − − − − − ⋮ + + + + + + + + ⋮ + + + + + + + +

−3 −2 −1 0 1 2 3 4 5 6

We select the interval on which the signs are the same.
The graph is:

−3 −2 −1 0 1 2 3 4 5 6

Solution set: $\{x \mid -1 \leq x \leq 4\}$ or $[-1, 4]$

24. $\{x \mid -5 \leq x \leq 1\}$ or $[-5, 1]$

Exercises 10.9B (page 534)

1. $y \leq x^2 + 5x + 6$. We first graph $y = x^2 + 5x + 6$; x-intercepts are -2 and -3; y-intercept is 6. We try (0, 0) in the inequality: $0 \leq 0^2 + 0 + 6$ is true. Therefore, we shade in the region that contains (0, 0).

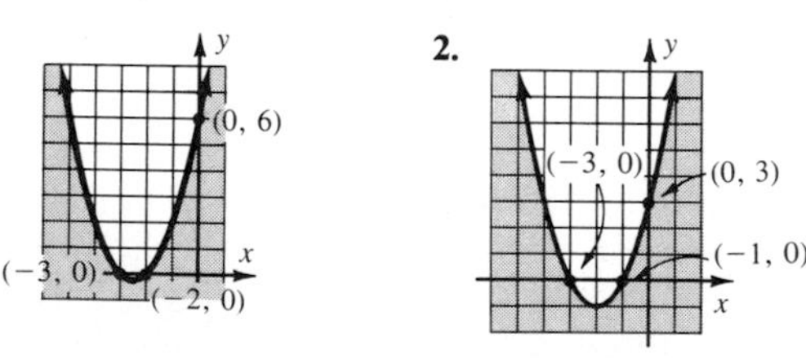

3. $x^2 + 4y^2 > 4$. We first graph $\dfrac{x^2}{4} + \dfrac{y^2}{1} = 1$, using a dotted line. It is an ellipse with x-intercepts 2 and -2 and y-intercepts 1 and -1. Then we try (0, 0) in the inequality: $0^2 + 4(0^2) > 4$ is a false statement. Therefore, we shade the region that does not contain (0, 0).

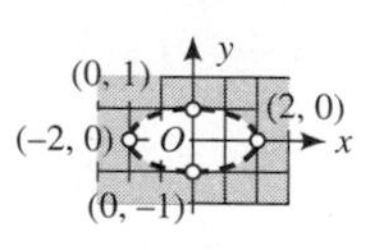

4.

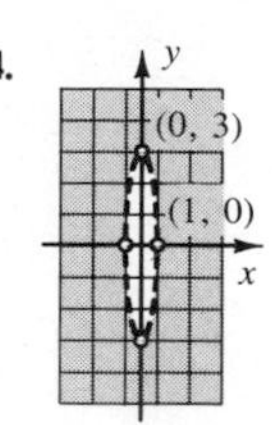

5. We first graph $(x + 2)^2 + (y - 3)^2 = 9$; this is a circle with its center at $(-2, 3)$ and with a radius of 3. Then we try $(0, 0)$ in the inequality:
$(0 + 2)^2 + (0 - 3)^2 \le 9$ is a false statement. Therefore, we shade the region that does not contain $(0, 0)$.

6.

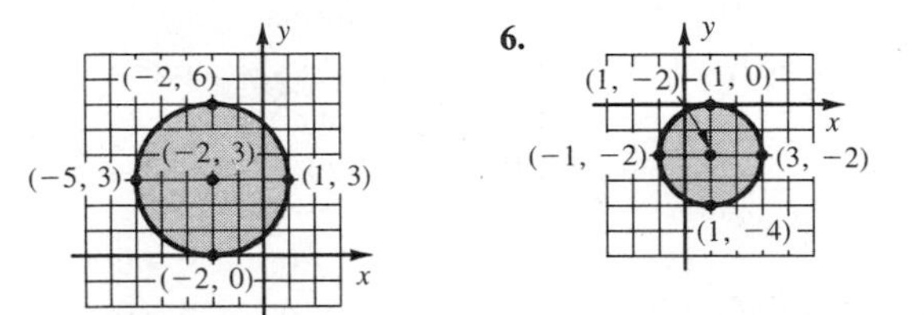

7. $4x^2 - y^2 < 4$. We first graph $4x^2 - y^2 = 4$, using a dotted line. It is a hyperbola with x-intercepts 1 and -1. We try $(0, 0)$ in the inequality:
$4(0^2) - 0^2 < 4$ is a true statement. We shade the region that contains $(0, 0)$.

8.

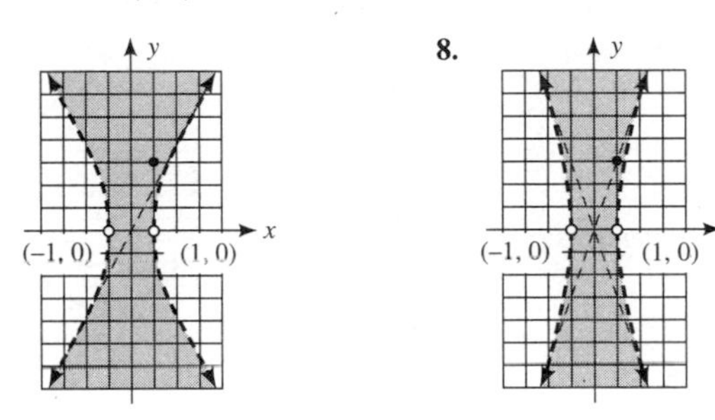

Review Exercises 10.10 (page 538)

1. a. Letting $f(x) = 0$, we have
$x^2 - 2x - 8 = 0$
$(x - 4)(x + 2) = 0$
$x - 4 = 0 \mid x + 2 = 0$
$x = 4 \mid x = -2$
x-intercepts are 4 and -2.
b. $f(0) = 0^2 - 2(0) - 8 = -8$;
y-intercept is -8.
c. Axis of symmetry: $x = -\dfrac{-2}{2} = 1$
$f(1) = 1^2 - 2(1) - 8 = -9$
Vertex is at $(1, -9)$.
d.

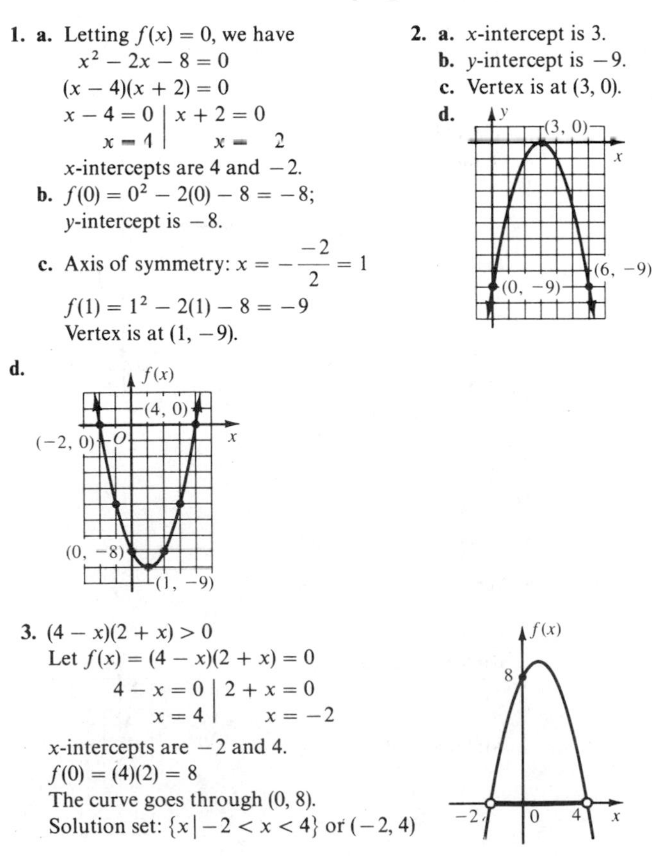

2. a. x-intercept is 3.
b. y-intercept is -9.
c. Vertex is at $(3, 0)$.
d.

3. $(4 - x)(2 + x) > 0$
Let $f(x) = (4 - x)(2 + x) = 0$
$4 - x = 0 \mid 2 + x = 0$
$x = 4 \mid x = -2$
x-intercepts are -2 and 4.
$f(0) = (4)(2) = 8$
The curve goes through $(0, 8)$.
Solution set: $\{x \mid -2 < x < 4\}$ or $(-2, 4)$

f(x)
8
−2
0
4
x

4. $\{x \mid x < 1\} \cup \{x \mid 2 < x < 3\}$ or $(-\infty, 1) \cup (2, 3)$

5.
$$x - \frac{3}{x + 2} > 0$$
$$\frac{x^2 + 2x - 3}{x + 2} > 0$$
$$\frac{(x + 3)(x - 1)}{x + 2} > 0$$
Critical points are -3, 1, and -2.

Two neg. | No negatives

Sign of $x + 3$ − − − − − | + | + + + + + | + + + +
Sign of $x - 1$ − − − − − | − | − − − − − | + + + +
Sign of $x + 2$ − − − − − | − | + + + + + | + + + +

−5 −4 −3 −2 −1 0 1 2

We select the intervals on which there is an even number of negative signs or no negative signs. The graph is:

−5 −4 −3 −2 −1 0 1 2

Solution set: $\{x \mid -3 < x < -2\} \cup \{x \mid x > 1\}$ or $(-3, -2) \cup (1, +\infty)$

6. $\{x \mid -1 \le x \le 6\}$

7. $3x^2 + 3y^2 = 27$
This is a circle with its center at the origin and a radius of 3.

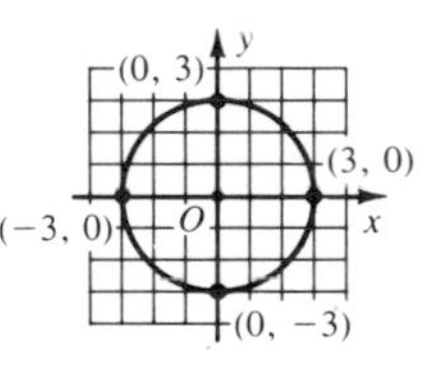

8. Hyperbola

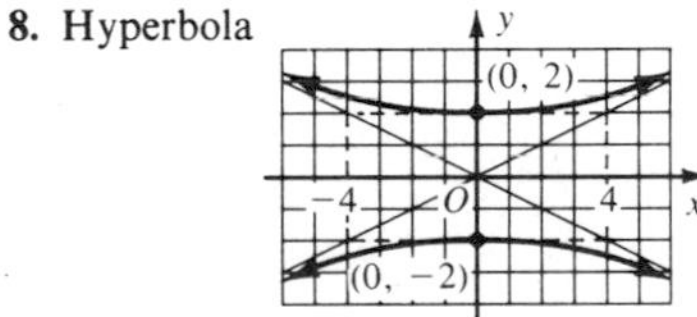

9. $9x^2 + 4y^2 = 36$ Ellipse
x-intercepts: $9x^2 + 4(0)^2 = 36$
$9x^2 = 36$
$x = \pm 2$
y-intercepts: $9(0)^2 + 4y^2 = 36$
$4y^2 = 36$
$y = \pm 3$

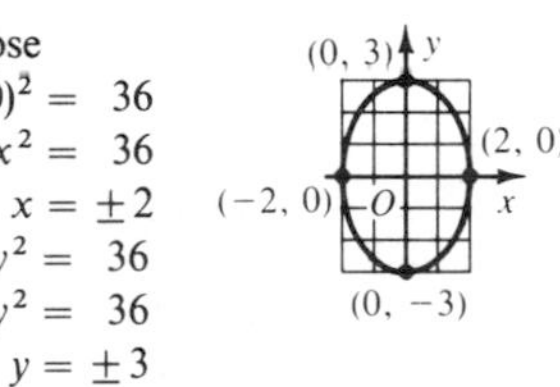

10.

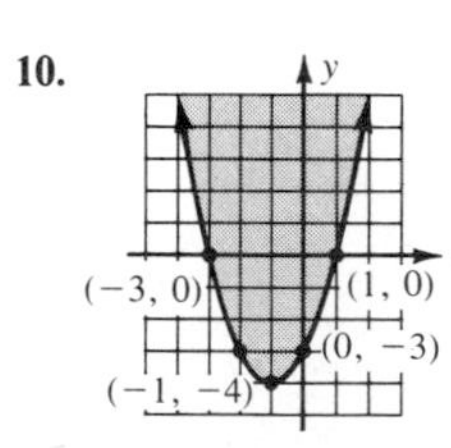

Chapter 10 Diagnostic Test (page 541)

Following each problem number is the textbook section number (in parentheses) where that kind of problem is discussed. (The checks are not always shown.)

1. (10.2) a.
$2x^2 = 6x$
$2x^2 - 6x = 0$
$2x(x - 3) = 0$
$2x = 0 \mid x - 3 = 0$
$x = 0 \mid x = 3$
$\{0, 3\}$

b.
$2x^2 = 18$
$2x^2 - 18 = 0$
$2(x^2 - 9) = 0$
$2(x + 3)(x - 3) = 0$
$x + 3 = 0 \mid x - 3 = 0$
$x = -3 \mid x = 3$
$\{-3, 3\}$

2. (10.3) $\frac{x-1}{2}+\frac{4}{x+1}=2$ LCD $= 2(x+1)$

$$\frac{2(x+1)}{1}\cdot\frac{(x-1)}{2}+\frac{2(x+1)}{1}\cdot\frac{4}{x+1}=\frac{2(x+1)}{1}\cdot\frac{2}{1}$$
$$(x+1)(x-1)+2\cdot 4=4(x+1)$$
$$x^2-1+8=4x+4$$
$$x^2-4x+3=0$$
$$(x-1)(x-3)=0$$
$$x-1=0 \mid x-3=0$$
$$x=1 \mid x=3$$

Check for $x = 1$

$$\frac{x-1}{2}+\frac{4}{x+1}=2$$
$$\frac{(1)-1}{2}+\frac{4}{(1)+1}\stackrel{?}{=}2$$
$$0+2=2$$

Check for $x = 3$

$$\frac{x-1}{2}+\frac{4}{x+1}=2$$
$$\frac{(3)-1}{2}+\frac{4}{(3)+1}\stackrel{?}{=}2$$
$$1+1=2$$

$\{1, 3\}$

3. (10.3) (1) $2x^{2/3}+3x^{1/3}=2$

Let $z = x^{1/3}$; then $z^2 = x^{2/3}$.

Equation (1) becomes $2z^2+3z-2=0$.

$$(z+2)(2z-1)=0$$

$$z+2=0 \mid 2z-1=0$$
$$z=-2 \mid z=\tfrac{1}{2}$$
$$x^{1/3}=-2 \mid x^{1/3}=\tfrac{1}{2}$$
$$(x^{1/3})^3=(-2)^3 \mid (x^{1/3})^3=(\tfrac{1}{2})^3$$
$$x=-8 \mid x=\tfrac{1}{8}$$

Check for $x = -8$

$$2(-8)^{2/3}+3(-8)^{1/3}\stackrel{?}{=}2$$
$$2(4)+3(-2)\stackrel{?}{=}2$$
$$8-6=2 \text{ True}$$

Check for $x = \frac{1}{8}$

$$2(\tfrac{1}{8})^{2/3}+3(\tfrac{1}{8})^{1/3}\stackrel{?}{=}2$$
$$2(\tfrac{1}{4})+3(\tfrac{1}{2})\stackrel{?}{=}2$$
$$\tfrac{1}{2}+\tfrac{3}{2}=2 \text{ True}$$

$\{-8, \frac{1}{8}\}$

4. (10.4) **a.** $x^2 = 6x - 7$

$x^2-6x+7=0$ will not factor; we use the quadratic formula.

$$x=\frac{-(-6)\pm\sqrt{(-6)^2-4(7)}}{2(1)}=\frac{6\pm\sqrt{36-28}}{2}$$
$$=\frac{6\pm\sqrt{8}}{2}=\frac{6\pm2\sqrt{2}}{2}=3\pm\sqrt{2}$$

$\{3+\sqrt{2}, 3-\sqrt{2}\}$

b. $3x^2+7x-1$ wil not factor; we use the quadratic formula.

$$3x^2+7x-1=0 \begin{cases} a=3 \\ b=7 \\ c=-1 \end{cases}$$

$$x=\frac{-7\pm\sqrt{7^2-4(3)(-1)}}{2(3)}=\frac{-7\pm\sqrt{49+12}}{6}$$
$$=\frac{-7\pm\sqrt{61}}{6}$$

$$\left\{\frac{-7+\sqrt{61}}{6}, \frac{-7-\sqrt{61}}{6}\right\}$$

5. (10.4) $x^2+6x+10=0$ will not factor; we use the quadratic formula.

$a = 1, b = 6, c = 10$

$$x=\frac{-6\pm\sqrt{6^2-4(1)(10)}}{2(1)}=\frac{-6\pm\sqrt{36-40}}{2}$$
$$=\frac{-6\pm\sqrt{-4}}{2}=\frac{-6\pm2i}{2}$$
$$=\frac{-6}{2}\pm\frac{2i}{2}=-3\pm i$$

$\{-3+i, -3-i\}$

6. (10.5) **a.** $25x^2-20x+7=0 \begin{cases} a=25 \\ b=-20 \\ c=7 \end{cases}$

$$b^2-4ac=(-20)^2-4(25)(7)$$
$$=400-700=-300$$

The roots are complex conjugates.

b. If $2+i\sqrt{3}$ is a root, $2-i\sqrt{3}$ is also a root.

$x=2+i\sqrt{3}$, $x=2-i\sqrt{3}$, and $x=3$

$x-2-i\sqrt{3}=0$, $x-2+i\sqrt{3}=0$, and $x-3=0$

and so

$$(x-2-i\sqrt{3})(x-2+i\sqrt{3})(x-3)=0$$
$$([x-2]-i\sqrt{3})([x-2]+i\sqrt{3})(x-3)=0$$
$$([x-2]^2-[i\sqrt{3}]^2)(x-3)=0$$
$$(x^2-4x+4-3i^2)(x-3)=0$$
$$(x^2-4x+4+3)(x-3)=0$$
$$(x^2-4x+7)(x-3)=0$$
$$x^3-7x^2+19x-21=0$$

7. (10.8) $25x^2+16y^2=400$ Ellipse

x-intercepts: Set $y = 0$:

$$25x^2+16(0)^2=400$$
$$25x^2=400$$
$$x^2=16$$
$$x=\pm4$$

y-intercepts: Set $x = 0$:

$$25(0)^2+16y^2=400$$
$$16y^2=400$$
$$y^2=25$$
$$y=\pm5$$

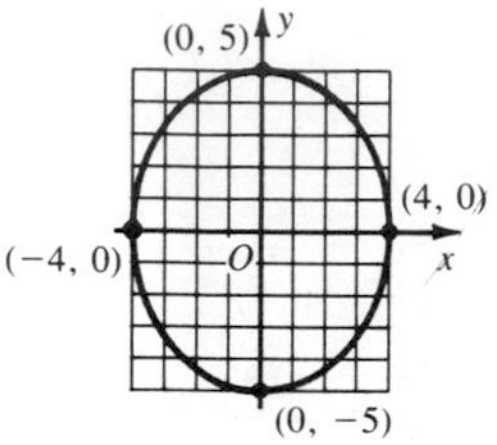

8. (10.9) **a.** $x^2+5<6x$

$$x^2-6x+5<0$$
$$(x-1)(x-5)<0$$

Let $f(x)=(x-1)(x-5)=0$.

Then $x-1=0 \mid x-5=0$

$x=1 \mid x=5$

The x-intercepts are 1 and 5.

$f(x)=(x-1)(x-5)$

$f(0)=(-1)(-5)=5$

The curve goes through (0, 5).

Solution set: $\{x \mid 1<x<5\}$.

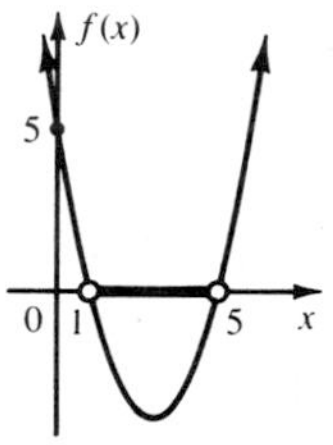

b. $x^2+2x\geq 8$

$$x^2+2x-8\geq0$$
$$(x+4)(x-2)\geq0$$

Let $f(x)=(x+4)(x-2)=0$.

Then $x+4=0 \mid x-2=0$

$x=-4 \mid x=2$

The x-intercepts are -4 and 2.

$f(x)=(x+4)(x-2)$

$f(0)=(4)(-2)=-8$

The curve goes through $(0, -8)$.

Solution set: $\{x \mid x\leq-4\}\cup\{x \mid x\geq2\}$

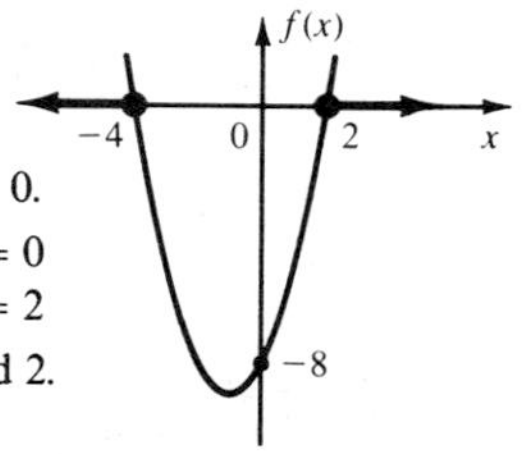

9. (10.7) $f(x)=x^2-4x$

a. Equation of axis of symmetry: $x=-\frac{b}{2a}=-\frac{-4}{2(1)}=2$

b. Vertex: $f(2)=(2)^2-4(2)$

$$=4-8=-4$$

Vertex: $(2, -4)$

c. x-intercepts: $f(x)=x(x-4)=0$

$x=0 \mid x-4=0$

$x=4$

x-intercepts: 0 and 4

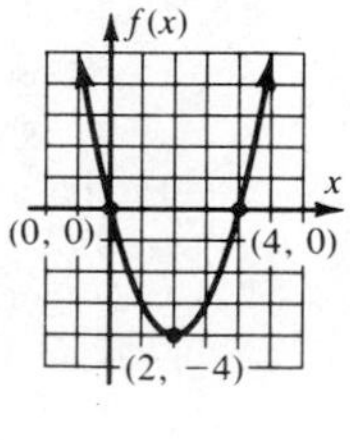

10. (10.3) Let x = number of hours for Oscar to do the job.
$x + 2$ = number of hours for Jan to do the job.
Oscar's rate is $\frac{1}{x}\frac{\text{job}}{\text{hr}}$; Jan's rate is $\frac{1}{x+2}\frac{\text{job}}{\text{hr}}$.
Jan works 4 hr and Oscar works 3 hr to do the complete job.

$$\begin{array}{ccccc}\text{Work done by Jan} & + & \text{Work done by Oscar} & = & \text{Total work done (the complete job)}\\ 4\left(\frac{1}{x+2}\right) & + & 3\left(\frac{1}{x}\right) & = & 1\end{array}$$

$$\frac{4}{x+2} + \frac{3}{x} = 1$$

LCD $= x(x + 2)$

$$\frac{x(x+2)}{1}\cdot\frac{4}{x+2} + \frac{x(x+2)}{1}\cdot\frac{3}{x} = \frac{x(x+2)}{1}\cdot\frac{1}{1}$$

$$4x + (x + 2)(3) = x(x + 2)$$

$$4x + 3x + 6 = x^2 + 2x$$

$$0 = x^2 - 5x - 6$$

$$0 = (x - 6)(x + 1)$$

$$\begin{array}{l|l} x - 6 = 0 & x + 1 = 0\\ x = 6 & x = -1\\ x + 2 = 8 & \text{Not in the domain}\end{array}$$

It takes 6 hr for Oscar to do the job and 8 hr for Jan to do the job.

Cumulative Review Exercises: Chapter 1–10 (page 542)

1. $m = \frac{-1-5}{4-2} = \frac{-6}{2} = -3$ **2.** $x + 2y - 4 = 0$

3. $|4 - 3x| \geq 10$ **4.** $\{-2\}$

$$\begin{array}{l|l} 4 - 3x \geq 10 & 4 - 3x \leq -10\\ -3x \geq 6 & -3x \leq -14\\ \frac{-3x}{-3} \leq \frac{6}{-3} & \frac{-3x}{-3} \geq \frac{-14}{-3}\\ x \leq -2 & x \geq \frac{14}{3}\end{array}$$

$\{x \mid x \leq -2\} \cup \{x \geq \frac{14}{3}\}$

5. $4x - 2y < 8$
Boundary line: $4x - 2y = 8$
Boundary line is dashed because = is not included in <.
Half-plane includes (0, 0) because
$4(0) - 2(0) < 8$
$0 < 8$ True

x	y
2	0
0	−4

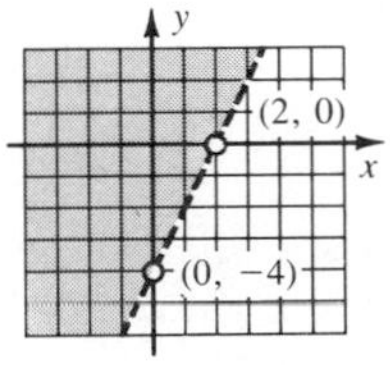

6. $f^{-1}(x) = \frac{x+3}{2}$

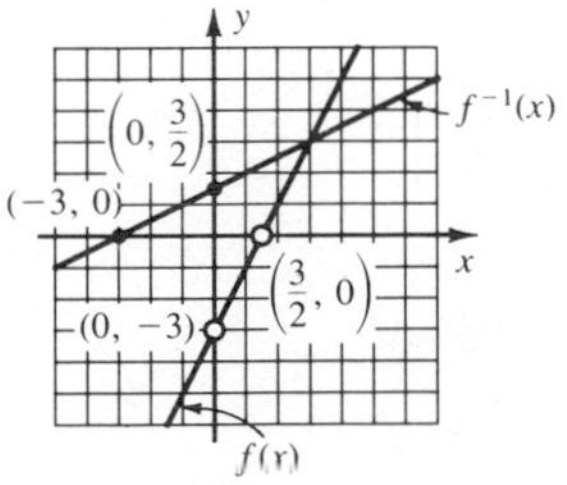

7. $\left(\frac{x^{-3}y^4}{x^2y}\right)^{-2} = (x^{-3-2}y^{4-1})^{-2} = (x^{-5}y^3)^{-2} = x^{10}y^{-6} = \frac{x^{10}}{y^6}$

8. $\frac{4}{x^{2/5}}$ **9.** $a^{1/3}\cdot a^{-1/2} = a^{1/3+(-1/2)} = a^{2/6+(-3/6)} = a^{-1/6} = \frac{1}{a^{1/6}}$

10. $\left\{\frac{1+3i\sqrt{3}}{2}, \frac{1-3i\sqrt{3}}{2}\right\}$ **11.** $3x^2 + x - 3 = 0 \begin{cases} a = 3\\ b = 1\\ c = -3\end{cases}$

$$x = \frac{-1 \pm \sqrt{1^2 - 4(3)(-3)}}{2(3)}$$

$$x = \frac{-1 \pm \sqrt{1 + 36}}{6}$$

$$x = \frac{-1 \pm \sqrt{37}}{6}$$

$$\left\{\frac{-1+\sqrt{37}}{6}, \frac{-1-\sqrt{37}}{6}\right\}$$

12. 1.956 **13.** $\log(x + 11) - \log(x + 1) = \log 6$ **14.** $b = 16$

$$\log\frac{x+11}{x+1} = \log 6$$

$$\frac{x+11}{x+1} = 6$$

$$6(x + 1) = x + 11$$

$$6x + 6 = x + 11$$

$$5x = 5$$

$$x = 1$$

15. $\log_7 \frac{1}{7} = x$
$7^x = \frac{1}{7} = 7^{-1}$
$x = -1$

16.

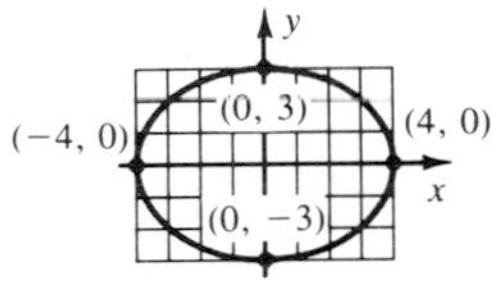

17. We must solve $4 - 3x - x^2 \geq 0$. Solving $4 - 3x - x^2 = 0$, we have $(4 + x)(1 - x) = 0$. Critical points are -4 and 1.

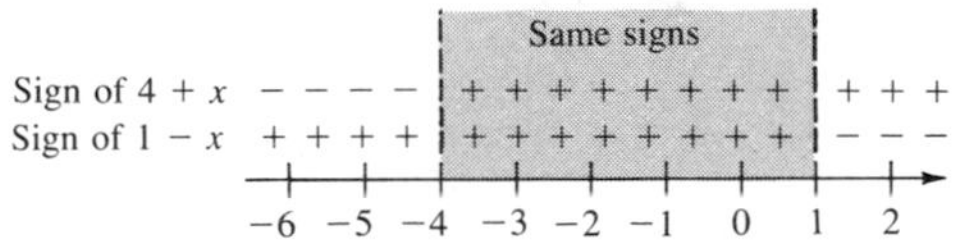

The domain of $y = f(x) = \sqrt{4 - 3x - x^2}$ is $\{x \mid -4 \leq x \leq 1\}$.

18. Length = 8; width = 6

19. Let x = number of hours for Merwin
Merwin's rate is $\frac{1}{x}\frac{\text{job}}{\text{hr}}$.
$x + 3$ = number of hours for Mina
Mina's rate is $\frac{1}{x+3}\frac{\text{job}}{\text{hr}}$.
Mina works a total of 8 hr and Merwin works 3 hr.

$$\begin{array}{ccccc}\text{Work done by Mina} & + & \text{Work done by Merwin} & = & \text{Total work done (the complete job)}\\ 8\left(\frac{1}{x+3}\right) & + & 3\left(\frac{1}{x}\right) & = & 1\end{array}$$

LCD $= x(x + 3)$

$$\frac{x(x+3)}{1}\cdot\frac{8}{x+3} + \frac{x(x+3)}{1}\cdot\frac{3}{x} = \frac{x(x+3)}{1}\cdot 1$$

$$8x + 3x + 9 = x^2 + 3x$$

$$x^2 - 8x - 9 = 0$$

$$(x - 9)(x + 1) = 0$$

$$\begin{array}{l|l} x + 1 = 0 & x - 9 = 0\\ x = -1 & x = 9\\ \text{Not in the domain} & x + 3 = 12\end{array}$$

It takes 9 hr for Merwin to do the job and 12 hr for Mina to do the job.

20. $\dfrac{-3+\sqrt{41}}{2}$ mph ≈ 1.7 mph

21.
$$A = Pe^{rt}$$
$$2{,}000 = 1{,}000e^{0.0625t}$$
$$2 = e^{0.0625t}$$
$$\ln 2 = \ln e^{0.0625t}$$
$$0.6931 \approx 0.0625t$$
$$t \approx 11$$

It will take about 11 yr for the money to double.

Exercises 11.2 (page 548)

1. (1) $2x + y = 6$
Intercepts: (3, 0), (0, 6)
(2) $x - y = 0$
Intercepts: (0, 0)
Additional point: (5, 5).

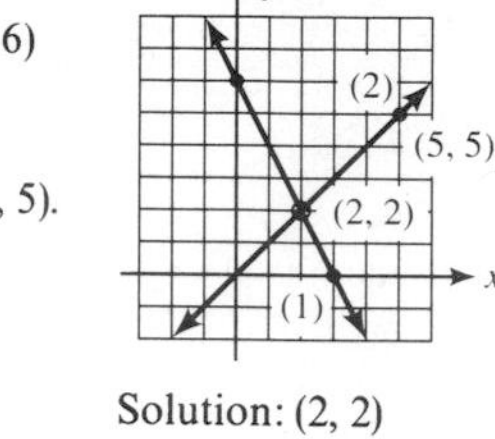

Solution: (2, 2)

2.

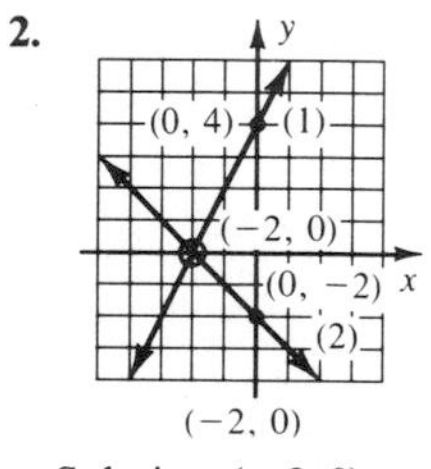

Solution: $(-2, 0)$

3. (1) $x + 2y = 3$
Intercepts:
(3, 0), $(0, \frac{3}{2})$
(2) $3x - y = -5$
Intercepts:
$(-\frac{5}{3}, 0)$, (0, 5)

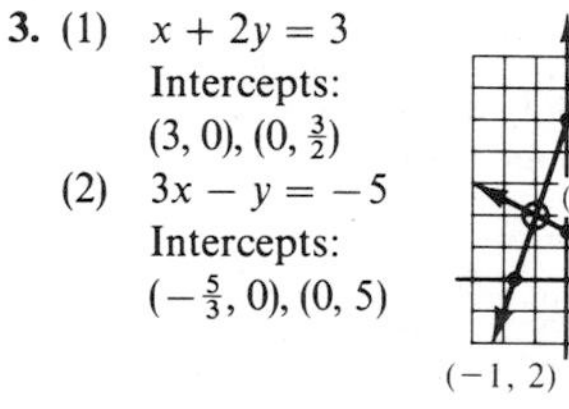

Solution: $(-1, 2)$

4.

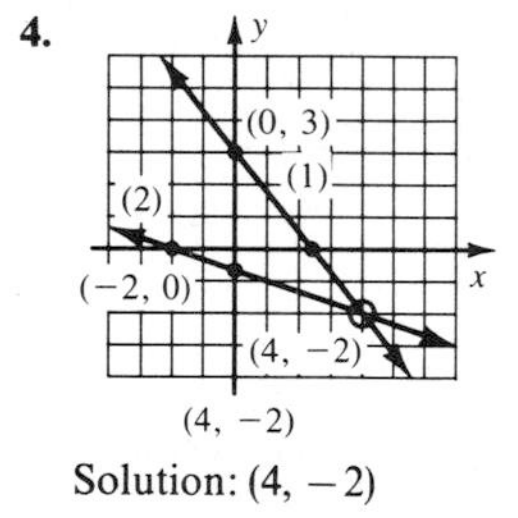

Solution: $(4, -2)$

5. (1) $x + 2y = 0$
Intercept: (0, 0)
Additional point: $(6, -3)$
(2) $2x - y = 0$
Intercept: (0, 0)
Additional point: (3, 6)
Solution: (0, 0)

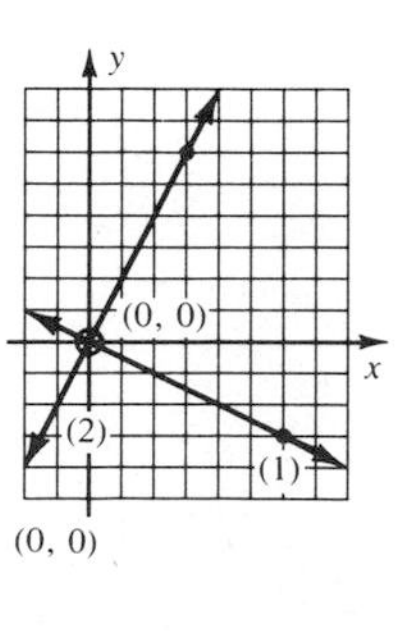

6.

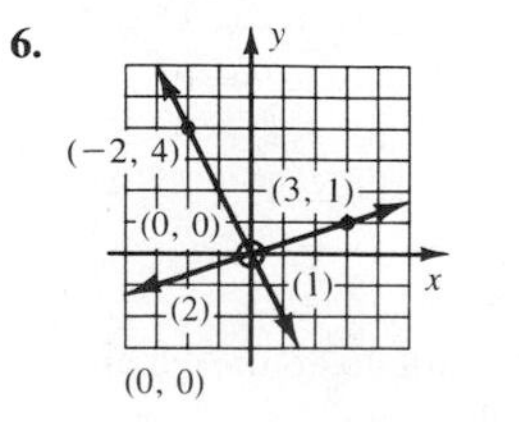

Solution: (0, 0)

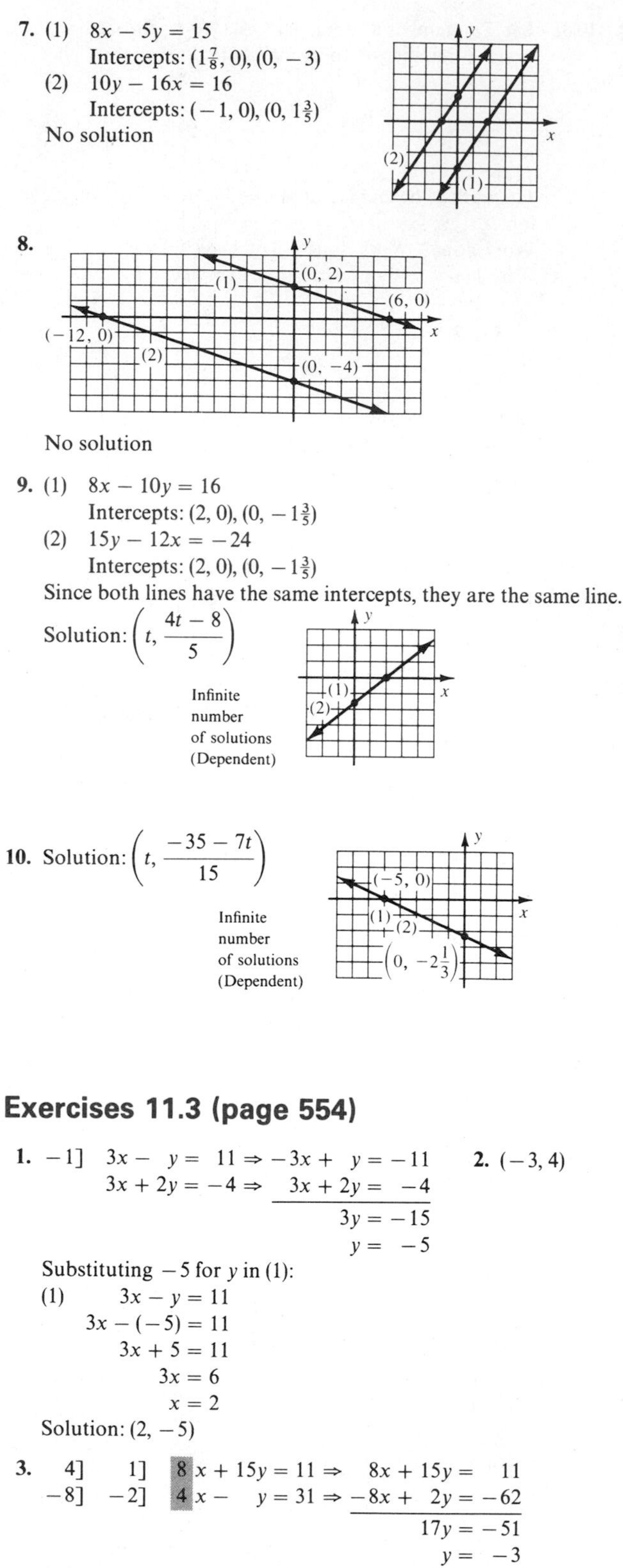

7. (1) $8x - 5y = 15$
Intercepts: $(1\frac{7}{8}, 0)$, $(0, -3)$
(2) $10y - 16x = 16$
Intercepts: $(-1, 0)$, $(0, 1\frac{3}{5})$
No solution

8.

No solution

9. (1) $8x - 10y = 16$
Intercepts: (2, 0), $(0, -1\frac{3}{5})$
(2) $15y - 12x = -24$
Intercepts: (2, 0), $(0, -1\frac{3}{5})$
Since both lines have the same intercepts, they are the same line.
Solution: $\left(t, \dfrac{4t-8}{5}\right)$

Infinite number of solutions (Dependent)

10. Solution: $\left(t, \dfrac{-35-7t}{15}\right)$

Infinite number of solutions (Dependent)

Exercises 11.3 (page 554)

1.
$$\begin{array}{rl} -1] & 3x - y = 11 \Rightarrow -3x + y = -11 \\ & 3x + 2y = -4 \Rightarrow \underline{\ \ 3x + 2y = -4} \\ & \qquad\qquad\qquad\qquad 3y = -15 \\ & \qquad\qquad\qquad\qquad\ \ y = -5 \end{array}$$

Substituting -5 for y in (1):
(1) $3x - y = 11$
$3x - (-5) = 11$
$3x + 5 = 11$
$3x = 6$
$x = 2$
Solution: $(2, -5)$

2. $(-3, 4)$

3.
$$\begin{array}{rrl} 4] & 1] & 8x + 15y = 11 \Rightarrow \ \ 8x + 15y = 11 \\ -8] & -2] & 4x - y = 31 \Rightarrow \underline{-8x + 2y = -62} \\ & & \qquad\qquad\qquad\qquad 17y = -51 \\ & & \qquad\qquad\qquad\qquad\quad y = -3 \end{array}$$

Substituting -3 for y in (2):
(2) $4x - y = 31$
$4x - (-3) = 31$
$4x + 3 = 31$
$4x = 28$
$x = 7$
Solution: $(7, -3)$

4. (6, 3)

5. $\begin{array}{rrrcl} 9] & 3] & 7x - \boxed{3}y = 3 & \Rightarrow & 21x - 9y = 9 \\ -3] & -1] & 20x - \boxed{9}y = 12 & \Rightarrow & -20x + 9y = -12 \\ & & & & \overline{x \qquad = -3} \end{array}$

Substituting -3 for x in (1):

$$\begin{aligned} (1) \quad 7x - 3y &= 3 \\ 7(-3) - 3y &= 3 \\ -21 - 3y &= 3 \\ -3y &= 24 \\ y &= -8 \end{aligned}$$

Solution: $(-3, -8)$

6. $(9, -13)$

7. $\begin{array}{rrrcl} 4] & 2] & \boxed{6}x + 5y = 0 & \Rightarrow & 12x + 10y = 0 \\ -6] & -3] & \boxed{4}x - 3y = 38 & \Rightarrow & -12x + 9y = -114 \\ & & & & \overline{19y = -114} \\ & & & & y = -6 \end{array}$

Substituting -6 for y in (1):

$$\begin{aligned} (1) \quad 6x + 5y &= 0 \\ 6x + 5(-6) &= 0 \\ 6x - 30 &= 0 \\ 6x &= 30 \\ x &= 5 \end{aligned}$$

Solution: $(5, -6)$

8. $(-4, 10)$

9. $\begin{array}{rrcl} -2] & 4x + 6y = 5 & \Rightarrow & -8x - 12y = -10 \\ & 8x + 12y = 7 & \Rightarrow & 8x + 12y = 7 \\ & & & \overline{0 = -3} \end{array}$ A *false* statement

Inconsistent; solution set: { }

10. Inconsistent: solution set: { }

11. $\begin{array}{rrcl} -2] & 7x - 3y = 5 & \Rightarrow & -14x + 6y = -10 \\ & 14x - 6y = 10 & \Rightarrow & 14x - 6y = 10 \\ & & & \overline{0 = 0} \end{array}$ A *true* statement

Dependent; $\left(t, \dfrac{7t - 5}{3}\right)$

12. Dependent; $\left(t, \dfrac{2t - 4}{3}\right)$

13. $\begin{array}{rrrcl} 6] & 3] & 9x + 4y = -4 & \Rightarrow & 27x + 12y = -12 \\ 4] & 2] & 15x - 6y = 25 & \Rightarrow & 30x - 12y = 50 \\ & & & & \overline{57x \qquad = 38} \\ & & & & x = \frac{38}{57} = \frac{2}{3} \end{array}$

Substituting $\frac{2}{3}$ for x in (1):

$$\begin{aligned} (1) \quad 9x + 4y &= -4 \\ \overset{3}{\cancel{9}}\left(\frac{2}{\underset{1}{\cancel{3}}}\right) + 4y &= -4 \\ 6 + 4y &= -4 \\ 4y &= -10 \\ y &= -\tfrac{10}{4} = -\tfrac{5}{2} \end{aligned}$$

Solution: $(\frac{2}{3}, -\frac{5}{2})$

14. $(-\frac{7}{4}, \frac{2}{5})$

15. Add $15x$ to both sides of equation (2) so that like terms are in the same column:

$\begin{array}{rrrcl} 15] & 5] & \boxed{9}x + 10y = -3 & \Rightarrow & 45x + 50y = -15 \\ -9] & -3] & \boxed{15}x + 14y = 7 & \Rightarrow & -45x - 42y = -21 \\ & & & & \overline{8y = -36} \\ & & & & y = -\frac{36}{8} = -\frac{9}{2} \end{array}$

Substituting $-\frac{9}{2}$ for y in (1):

$$\begin{aligned} (1) \quad 9x + 10y &= -3 \\ 9x + \overset{5}{\cancel{10}}\left(-\frac{9}{\underset{1}{\cancel{2}}}\right) &= -3 \\ 9x - 45 &= -3 \\ 9x &= 42 \\ x &= \tfrac{42}{9} = \tfrac{14}{3} \end{aligned}$$

Solution: $(\frac{14}{3}, -\frac{9}{2})$

16. $(\frac{3}{7}, \frac{5}{6})$

Exercises 11.4 (page 558)

1. (1) $7x + 4y = 4$

(2) $y = 6 - 3x$

Substituting $6 - 3x$ for y in (1):

$$\begin{aligned} (1) \quad 7x + 4y &= 4 \\ 7x + 4(\boxed{6 - 3x}) &= 4 \\ 7x + 24 - 12x &= 4 \\ -5x &= -20 \\ x &= 4 \end{aligned}$$

Substituting 4 for x in $y = 6 - 3x$:

$$y = 6 - 3(4) = 6 - 12 = -6$$

Solution: $(4, -6)$

2. $(-7, 3)$

3. (1) $5x - 4y = -1$

(2) $3x + y = -38 \Rightarrow y = -3x - 38$

Substituting $-3x - 38$ for y in (1):

$$\begin{aligned} (1) \quad 5x - 4y &= -1 \\ 5x - 4(\boxed{-3x - 38}) &= -1 \\ 5x + 12x + 152 &= -1 \\ 17x &= -153 \\ x &= -9 \end{aligned}$$

Substituting -9 for x in $y = -3x - 38$:

$$y = -3(-9) - 38 = 27 - 38 = -11$$

Solution: $(-9, -11)$

4. $(-5, -4)$

5. (1) $8x - 5y = 4$

(2) $x - 2y = -16 \Rightarrow x = 2y - 16$

Substituting $2y - 16$ for x in (1):

$$\begin{aligned} (1) \quad 8(\boxed{2y - 16}) - 5y &= 4 \\ 16y - 128 - 5y &= 4 \\ 11y &= 132 \\ y &= 12 \end{aligned}$$

Substituting 12 for y in $x = 2y - 16$:

$$x = 2(12) - 16 = 24 - 16 = 8$$

Solution: $(8, 12)$

6. $(10, 16)$

7. (1) $15x + 5y = 8$

(2) $6x + 2y = -10 \Rightarrow y = -3x - 5$

Substituting $-3x - 5$ for y in (1):

$$\begin{aligned} (1) \quad 15x + 5(\boxed{-3x - 5}) &= 8 \\ 15x - 15x - 25 &= 8 \\ -25 &= 8 \quad \text{A false statement} \end{aligned}$$

Inconsistent; solution set: { }

8. Inconsistent; solution set: { }

9. (1) $20x - 10y = 70$

(2) $6x - 3y = 21 \Rightarrow y = 2x - 7$

Substituting $2x - 7$ for y in (1):

$$\begin{aligned} (1) \quad 20x - 10(\boxed{2x - 7}) &= 70 \\ 20x - 20x + 70 &= 70 \\ 70 &= 70 \quad \text{A true statement} \end{aligned}$$

Dependent; $(t, 2t - 7)$

10. Dependent; $(5s + 9, s)$

11. (1) $8x + 4y = 7$

(2) $3x + 6y = 6 \Rightarrow 3x = 6 - 6y$

$$x = \frac{6 - 6y}{3} = 2 - 2y$$

Substituting $2 - 2y$ for x in (1):

$$\begin{aligned} (1) \quad 8x + 4y &= 7 \\ 8(\boxed{2 - 2y}) + 4y &= 7 \\ 16 - 16y + 4y &= 7 \\ -12y &= -9 \Rightarrow y = \tfrac{3}{4} \end{aligned}$$

Substituting $\frac{3}{4}$ for y in $x = 2 - 2y$:

$$x = 2 - 2(\tfrac{3}{4}) = \tfrac{1}{2}$$

Solution: $(\frac{1}{2}, \frac{3}{4})$

12. $(\frac{3}{5}, \frac{1}{4})$

13. (1) $4x + 4y = 3 \Rightarrow 4x = 3 - 4y \Rightarrow x = \dfrac{3 - 4y}{4}$

(2) $6x + 12y = -6$

Substituting $\dfrac{3-4y}{4}$ for x in (2):

(2) $6x + 12y = -6$

$$\overset{3}{\cancel{6}}\left(\frac{3-4y}{\underset{2}{\cancel{4}}}\right) + 12y = -6$$

LCD = 2

$$\frac{\overset{1}{\cancel{2}}}{1}\cdot 3\left(\frac{3-4y}{\underset{1}{\cancel{2}}}\right) + 2(12y) = 2(-6)$$

$$\begin{aligned} 9 - 12y + 24y &= -12 \\ 12y &= -21 \\ y &= -\tfrac{21}{12} = -\tfrac{7}{4} = -1\tfrac{3}{4} \end{aligned}$$

Substituting $-\frac{7}{4}$ for y in $x = \dfrac{3-4y}{4}$:

$$x = \frac{3 - 4(-\frac{7}{4})}{4} = \frac{3+7}{4} = \frac{10}{4} = 2\tfrac{1}{2}$$

Solution: $(2\frac{1}{2}, -1\frac{3}{4})$

14. $(2\frac{1}{2}, -2\frac{1}{3})$

Exercises 11.5 (page 561)

1.

$$\begin{array}{lrcl} (1) & 2x + y + z &=& 4 \\ (2) & x - y + 3z &=& -2 \\ (3) & x + y + 2z &=& 1 \\ \hline (1)+(2){:} & 3x \quad + 4z &=& 2 \\ (2)+(3){:} & 2x \quad + 5z && -1 \end{array}$$

Next, eliminate x:

$$\begin{array}{lrcl} 2] & 3x + 4z = 2 \Rightarrow & 6x + 8z =& 4 \\ -3] & 2x + 5z = -1 \Rightarrow & -6x - 15z =& 3 \\ \hline & & -7z =& 7 \\ & & z =& -1 \end{array}$$

Substituting -1 for z in $3x + 4z = 2$:

$$\begin{aligned} 3x - 4 &= 2 \\ 3x &= 6 \\ x &= 2 \end{aligned}$$

Substituting 2 for x and -1 for z in (3):

(3) $x + y + 2z = 1$

$$\begin{aligned} 2 + y - 2 &= 1 \\ y &= 1 \end{aligned}$$

The solution is $(2, 1, -1)$.

2. $(1, -2, 2)$

3.

$$\begin{array}{lrcll} (1) & x + 2y + 2z &=& 0 \\ (2) & 2x - y + z &=& -3 \\ (3) & 4x + 2y + 3z &=& 2 \\ \hline (1)+2(2){:} & 5x \quad + 4z &=& -6 & (4) \\ 2(2)+(3){:} & 8x \quad + 5z &=& -4 \end{array}$$

$$\begin{array}{lrcl} -5] & 5x + 4z = -6 \Rightarrow & -25x - 20z =& 30 \\ 4] & 8x + 5z = -4 \Rightarrow & 32x + 20z =& -16 \\ \hline & & 7x =& 14 \\ & & x =& 2 \end{array}$$

Substituting 2 for x in (4):

(4) $5x + 4z = -6$

$$\begin{aligned} 5(2) + 4z &= -6 \\ 4z &= -16 \\ z &= -4 \end{aligned}$$

Substituting 2 for x and -4 for z in (1):

(1) $x + 2y + 2z = 0$

$$\begin{aligned} 2 + 2y + 2(-4) &= 0 \\ 2y &= 6 \\ y &= 3 \end{aligned}$$

The solution is $(2, 3, -4)$.

4. $(3, -4, 2)$

5.

$$\begin{array}{lrcll} (1) & x \quad + 2z &=& 7 \\ (2) & 2x - y &=& 5 \\ (3) & 2y + z &=& 4 \\ \hline 2(2)+(3){:} & 4x \quad + z &=& 14 & (4) \\ (1){:} & x \quad + 2z &=& 7 \\ \hline 2(4){:} & 8x + 2z &=& 28 \\ -(1){:} & -x - 2z &=& -7 \\ \hline & 7x &=& 21 \\ & x &=& 3 \end{array}$$

Substituting 3 for x in (4):

(4) $4x + z = 14$

$$\begin{aligned} 4(3) + z &= 14 \\ z &= 2 \end{aligned}$$

Substituting 2 for z in (3):

(3) $2y + z = 4$

$$\begin{aligned} 2y + 2 &= 4 \\ 2y &= 2 \\ y &= 1 \end{aligned}$$

The solution is $(3, 1, 2)$.

6. $(2, -1, 3)$

7.

$$\begin{array}{lrcll} (1) & 2x + 3y + z &=& 7 \\ (2) & 4x \quad - 2z &=& -6 \\ (3) & 6y - z &=& 0 \\ \hline 2(3){:} & 12y - 2z &=& 0 \\ -(2){:} & -4x \quad + 2z &=& +6 \\ \hline 2(3)-(2){:} & -4x + 12y &=& 6 & (4) \\ (1)+(3){:} & 2x + 9y &=& 7 & (5) \\ \frac{1}{2}(4){:} & -2x + 6y &=& 3 \\ \hline (5)+\frac{1}{2}(4){:} & 15y &=& 10 \\ & y &=& \frac{2}{3} \end{array}$$

Substituting $\frac{2}{3}$ for y in (3):

(3) $6y - z = 0$

$$\begin{aligned} 6(\tfrac{2}{3}) - z &= 0 \\ 4 - z = 0 &\Rightarrow z = 4 \end{aligned}$$

Substituting $\frac{2}{3}$ for y and 4 for z in (1):

(1) $2x + 3y + z = 7$

$$\begin{aligned} 2x + 3(\tfrac{2}{3}) + 4 &= 7 \\ 2x + 2 + 4 &= 7 \\ 2x &= 1 \\ x &= \tfrac{1}{2} \end{aligned}$$

The solution is $(\frac{1}{2}, \frac{2}{3}, 4)$.

8. $(\frac{3}{4}, \frac{2}{5}, -1)$

9.

$$\begin{array}{lrcll} (1) & x + y + z + w &=& 5 \\ (2) & 2x - y + 2z - w &=& -2 \\ (3) & x + 2y - z - 2w &=& -1 \\ (4) & -x + 3y + 3z + w &=& 1 \\ \hline (1)+(2){:} & 3x \quad + 3z &=& 3 & (5) \\ 2(1)+(3){:} & 3x + 4y + z &=& 9 & (6) \\ (2)+(4){:} & x + 2y + 5z &=& -1 & (7) \\ \hline (6){:} & 3x + 4y + z &=& 9 \\ -2(7){:} & -2x - 4y - 10z &=& 2 \\ \hline (6)-2(7){:} & x \quad - 9z &=& 11 \\ -\frac{1}{3}(5){:} & -x \quad - z &=& -1 & (8) \\ \hline & -10z &=& 10 & \text{Adding} \\ & z &=& -1 \end{array}$$

Substituting -1 for z in $-1(8)$:

$-1(8)$ $x + z = 1$

$$\begin{aligned} x + (-1) &= 1 \\ x &= 2 \end{aligned}$$

Substituting 2 for x and -1 for z in (7):

(7) $x + 2y + 5z = -1$

$$\begin{aligned} 2 + 2y - 5 &= -1 \\ 2y &= 2 \\ y &= 1 \end{aligned}$$

Substituting 2 for x, 1 for y, and -1 for z in (1):

$$\begin{array}{ll} (1) & x + y + z + w = 5 \\ & 2 + 1 - 1 + w = 5 \\ & w = 3 \end{array}$$

The solution is $(2, 1, -1, 3)$.

10. $(1, 2, 3, -2)$

11.

$$\begin{array}{rlr} (1) & 6x + 4y + 9z + 5w = -3 & \\ (2) & 2x + 8y - 6z + 15w = 8 & \\ (3) & 4x - 4y + 3z - 10w = -3 & \\ (4) & 2x - 4y + 3z - 5w = -1 & \\ (1) + (4): & 8x + 12z = -4 & (5) \\ 2(1) + (3): & 16x + 4y + 21z = -9 & (6) \\ (2) + 3(4): & 8x - 4y + 3z = 5 & (7) \\ (5): & 8x + 12z = -4 & \\ (6) + (7): & 24x + 24z = -4 & (8) \\ 3(5) - (8): & 12z = -8 & \\ & z = -\tfrac{2}{3} & \end{array}$$

Substituting $-\frac{2}{3}$ for z in (5):

$$\begin{array}{ll} (5) & 8x + 12z = -4 \\ & 8x + 12(-\tfrac{2}{3}) = -4 \\ & 8x - 8 = -4 \\ & 8x = 4 \\ & x = \tfrac{4}{8} = \tfrac{1}{2} \end{array}$$

Substituting $\frac{1}{2}$ for x and $-\frac{2}{3}$ for z in (7):

$$\begin{array}{ll} (7) & 8x - 4y + 3z = 5 \\ & 8(\tfrac{1}{2}) - 4y + 3(-\tfrac{2}{3}) = 5 \\ & 4 - 4y - 2 = 5 \\ & -4y = 3 \\ & y = -\tfrac{3}{4} \end{array}$$

Substituting $\frac{1}{2}$ for x, $-\frac{3}{4}$ for y, and $-\frac{2}{3}$ for z in (4):

$$\begin{array}{ll} (4) & 2x - 4y + 3z - 5w = -1 \\ & 2(\tfrac{1}{2}) - 4(-\tfrac{3}{4}) + 3(-\tfrac{2}{3}) - 5w = -1 \\ & 1 + 3 - 2 - 5w = -1 \\ & -5w = -3 \\ & w = \tfrac{3}{5} \end{array}$$

The solution is $(\frac{1}{2}, -\frac{3}{4}, -\frac{2}{3}, \frac{3}{5})$.

12. $(-\frac{5}{6}, \frac{1}{3}, \frac{3}{2}, -\frac{1}{4})$

Review Exercises 11.6 (page 563)

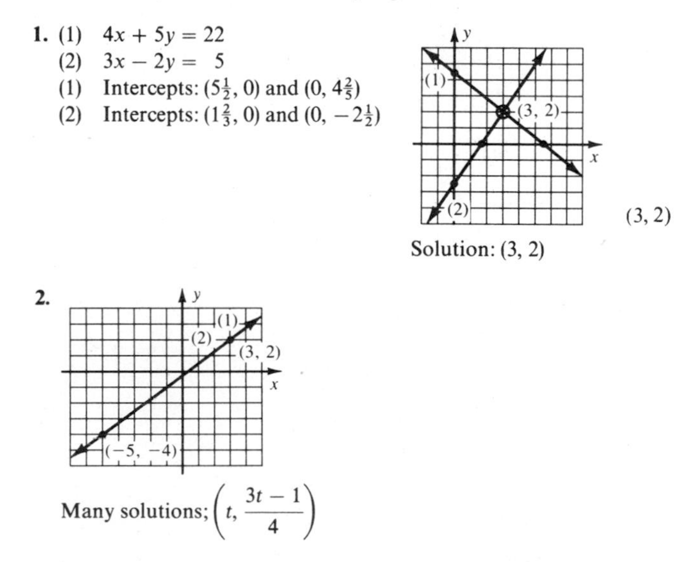

1. (1) $4x + 5y = 22$
(2) $3x - 2y = 5$
(1) Intercepts: $(5\frac{1}{2}, 0)$ and $(0, 4\frac{2}{5})$
(2) Intercepts: $(1\frac{2}{3}, 0)$ and $(0, -2\frac{1}{2})$

Solution: $(3, 2)$

$(3, 2)$

2.

Many solutions; $\left(t, \dfrac{3t - 1}{4}\right)$

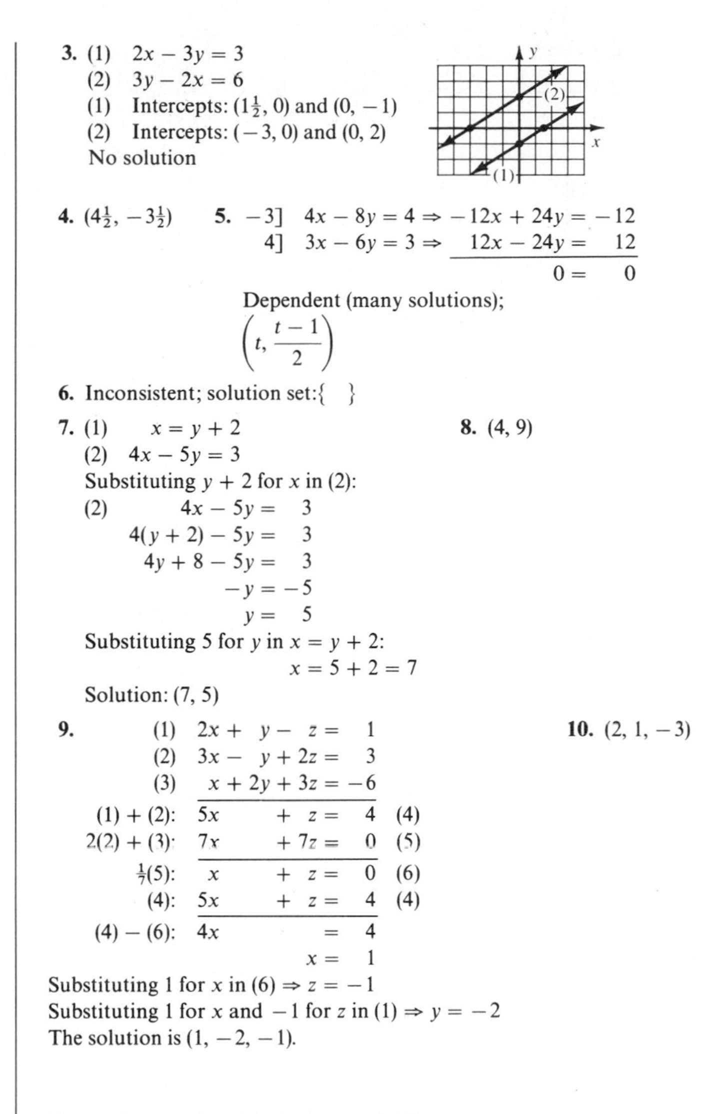

3. (1) $2x - 3y = 3$
(2) $3y - 2x = 6$
(1) Intercepts: $(1\frac{1}{2}, 0)$ and $(0, -1)$
(2) Intercepts: $(-3, 0)$ and $(0, 2)$
No solution

4. $(4\frac{1}{2}, -3\frac{1}{2})$

5.

$$\begin{array}{lll} -3] & 4x - 8y = 4 \Rightarrow & -12x + 24y = -12 \\ 4] & 3x - 6y = 3 \Rightarrow & 12x - 24y = 12 \\ & & 0 = 0 \end{array}$$

Dependent (many solutions); $\left(t, \dfrac{t - 1}{2}\right)$

6. Inconsistent; solution set: $\{\ \}$

7.

$$\begin{array}{ll} (1) & x = y + 2 \\ (2) & 4x - 5y = 3 \end{array}$$

Substituting $y + 2$ for x in (2):

$$\begin{array}{ll} (2) & 4x - 5y = 3 \\ & 4(y + 2) - 5y = 3 \\ & 4y + 8 - 5y = 3 \\ & -y = -5 \\ & y = 5 \end{array}$$

Substituting 5 for y in $x = y + 2$:

$$x = 5 + 2 = 7$$

Solution: $(7, 5)$

8. $(4, 9)$

9.

$$\begin{array}{rlr} (1) & 2x + y - z = 1 & \\ (2) & 3x - y + 2z = 3 & \\ (3) & x + 2y + 3z = -6 & \\ (1) + (2): & 5x + z = 4 & (4) \\ 2(2) + (3): & 7x + 7z = 0 & (5) \\ \tfrac{1}{7}(5): & x + z = 0 & (6) \\ (4): & 5x + z = 4 & (4) \\ (4) - (6): & 4x = 4 & \\ & x = 1 & \end{array}$$

Substituting 1 for x in (6) $\Rightarrow z = -1$
Substituting 1 for x and -1 for z in (1) $\Rightarrow y = -2$
The solution is $(1, -2, -1)$.

10. $(2, 1, -3)$

Exercises 11.7A (page 574)

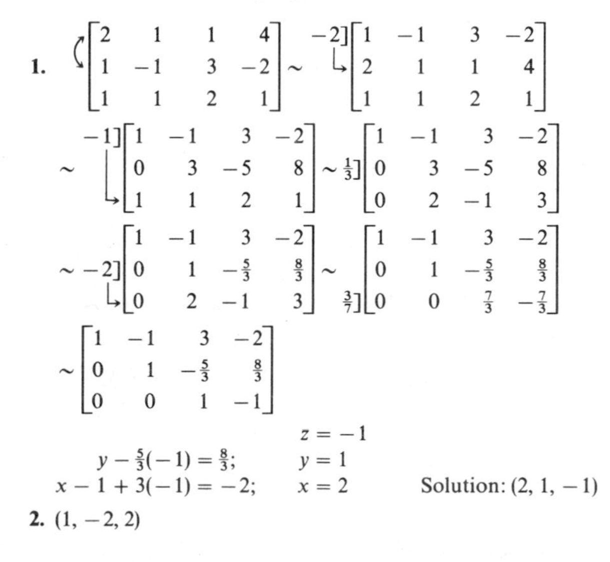

1.

$$\begin{bmatrix} 2 & 1 & 1 & 4 \\ 1 & -1 & 3 & -2 \\ 1 & 1 & 2 & 1 \end{bmatrix} \sim \begin{matrix} -2] \\ \\ \\ \end{matrix} \begin{bmatrix} 1 & -1 & 3 & -2 \\ 2 & 1 & 1 & 4 \\ 1 & 1 & 2 & 1 \end{bmatrix}$$

$$\sim \begin{matrix} -1] \\ \\ \\ \end{matrix} \begin{bmatrix} 1 & -1 & 3 & -2 \\ 0 & 3 & -5 & 8 \\ 1 & 1 & 2 & 1 \end{bmatrix} \sim \begin{matrix} \\ \tfrac{1}{3}] \\ \\ \end{matrix} \begin{bmatrix} 1 & -1 & 3 & -2 \\ 0 & 3 & -5 & 8 \\ 0 & 2 & -1 & 3 \end{bmatrix}$$

$$\sim \begin{matrix} \\ -2] \\ \\ \end{matrix} \begin{bmatrix} 1 & -1 & 3 & -2 \\ 0 & 1 & -\tfrac{5}{3} & \tfrac{8}{3} \\ 0 & 2 & -1 & 3 \end{bmatrix} \sim \begin{matrix} \\ \\ \tfrac{3}{7}] \end{matrix} \begin{bmatrix} 1 & -1 & 3 & -2 \\ 0 & 1 & -\tfrac{5}{3} & \tfrac{8}{3} \\ 0 & 0 & \tfrac{7}{3} & -\tfrac{7}{3} \end{bmatrix}$$

$$\sim \begin{bmatrix} 1 & -1 & 3 & -2 \\ 0 & 1 & -\tfrac{5}{3} & \tfrac{8}{3} \\ 0 & 0 & 1 & -1 \end{bmatrix}$$

$$\begin{array}{rl} & z = -1 \\ y - \tfrac{5}{3}(-1) = \tfrac{8}{3}; & y = 1 \\ x - 1 + 3(-1) = -2; & x = 2 \end{array}$$

Solution: $(2, 1, -1)$

2. $(1, -2, 2)$

3. $\frac{1}{2}]\begin{bmatrix}2&3&1&7\\4&0&-2&-6\\0&6&-1&0\end{bmatrix} \sim -4]\begin{bmatrix}1&\frac{3}{2}&\frac{1}{2}&\frac{7}{2}\\4&0&-2&-6\\0&6&-1&0\end{bmatrix}$

$\sim -\frac{1}{6}]\begin{bmatrix}1&\frac{3}{2}&\frac{1}{2}&\frac{7}{2}\\0&-6&-4&-20\\0&6&-1&0\end{bmatrix} \sim -6]\begin{bmatrix}1&\frac{3}{2}&\frac{1}{2}&\frac{7}{2}\\0&1&\frac{2}{3}&\frac{10}{3}\\0&6&-1&0\end{bmatrix}$

$\sim -\frac{1}{5}]\begin{bmatrix}1&\frac{3}{2}&\frac{1}{2}&\frac{7}{2}\\0&1&\frac{2}{3}&\frac{10}{3}\\0&0&-5&-20\end{bmatrix} \sim \begin{bmatrix}1&\frac{3}{2}&\frac{1}{2}&\frac{7}{2}\\0&1&\frac{2}{3}&\frac{10}{3}\\0&0&1&4\end{bmatrix}$

$z = 4$

$y + \frac{2}{3}(4) = \frac{10}{3};\quad y = \frac{2}{3}$

$x + \frac{3}{2}(\frac{2}{3}) + \frac{1}{2}(4) = \frac{7}{2};\quad x = \frac{1}{2}\quad$ Solution: $(\frac{1}{2}, \frac{2}{3}, 4)$

4. $(2, -1, 3)$

5. $-2]\begin{bmatrix}1&1&1&1&5\\2&-1&2&-1&-2\\1&2&-1&-2&-1\\-1&3&3&1&1\end{bmatrix} \sim -1]\begin{bmatrix}1&1&1&1&5\\0&-3&0&-3&-12\\1&2&-1&-2&-1\\-1&3&3&1&1\end{bmatrix}$

$\sim 1]\begin{bmatrix}1&1&1&1&5\\0&-3&0&-3&-12\\0&1&-2&-3&-6\\-1&3&3&1&1\end{bmatrix} \sim -\frac{1}{3}]\begin{bmatrix}1&1&1&1&5\\0&-3&0&-3&-12\\0&1&-2&-3&-6\\0&4&4&2&6\end{bmatrix}$

$\sim -1]\begin{bmatrix}1&1&1&1&5\\0&1&0&1&4\\0&1&-2&-3&-6\\0&4&4&2&6\end{bmatrix} \sim -4]\begin{bmatrix}1&1&1&1&5\\0&1&0&1&4\\0&0&-2&-4&-10\\0&4&4&2&6\end{bmatrix}$

$\sim -\frac{1}{2}]\begin{bmatrix}1&1&1&1&5\\0&1&0&1&4\\0&0&-2&-4&-10\\0&0&4&-2&-10\end{bmatrix} \sim -4]\begin{bmatrix}1&1&1&1&5\\0&1&0&1&4\\0&0&1&2&5\\0&0&4&-2&-10\end{bmatrix}$

$\sim -\frac{1}{10}]\begin{bmatrix}1&1&1&1&5\\0&1&0&1&4\\0&0&1&2&5\\0&0&0&-10&-30\end{bmatrix} \sim \begin{bmatrix}1&1&1&1&5\\0&1&0&1&4\\0&0&1&2&5\\0&0&0&1&3\end{bmatrix}$

$w = 3$

$z + 2(3) = 5;\quad z = -1$

$y + 0(-1) + 1(3) = 4;\quad y = 1$

$x + 1 + (-1) + 3 = 5;\quad x = 2\quad$ Solution: $(2, 1, -1, 3)$

6. $(1, 2, 3, -2)$

Exercises 11.7B (page 578)

1. $\begin{vmatrix}3&4\\2&5\end{vmatrix} = (3)(5) - (2)(4) = 15 - 8 = 7$ 2. 22

3. $\begin{vmatrix}2&-4\\5&-3\end{vmatrix} = 2(-3) - 5(-4) = -6 + 20 = 14$ 4. 40

5. $\begin{vmatrix}-7&-3\\5&8\end{vmatrix} = (-7)(8) - (5)(-3) = -56 - (-15) = -41$

6. 0 7. $\begin{vmatrix}2&-4\\3&x\end{vmatrix} = 20$ 8. 3

$2x - (3)(-4) = 20$

$2x + 12 = 20$

$2x = 8$

$x = 4$

9. a. $\begin{vmatrix}4&-1\\-3&0\end{vmatrix}$ b. $-\begin{vmatrix}4&-1\\-3&0\end{vmatrix}$ c. $\begin{vmatrix}1&3\\4&-1\end{vmatrix}$ d. $-\begin{vmatrix}1&3\\4&-1\end{vmatrix}$ e. $\begin{vmatrix}2&3\\5&-1\end{vmatrix}$ f. $\begin{vmatrix}2&3\\5&-1\end{vmatrix}$

10. a. $\begin{vmatrix}0&1\\1&5\end{vmatrix}$ b. $\begin{vmatrix}0&1\\1&5\end{vmatrix}$ c. $\begin{vmatrix}2&0\\-3&-2\end{vmatrix}$ d. $-\begin{vmatrix}2&0\\-3&-2\end{vmatrix}$ e. $\begin{vmatrix}2&1\\4&5\end{vmatrix}$ f. $-\begin{vmatrix}2&1\\4&5\end{vmatrix}$

11. $\begin{vmatrix}1&2&1\\3&1&2\\4&2&0\end{vmatrix} = (1)\begin{vmatrix}3&1\\4&2\end{vmatrix} - (2)\begin{vmatrix}1&2\\4&2\end{vmatrix} + (0)\begin{vmatrix}1&2\\3&1\end{vmatrix}$

$= 1(6 - 4) - 2(2 - 8) + 0 = 2 + 12 = 14$

12. 18

13. $\begin{vmatrix}1&3&-2\\-1&2&-3\\0&4&1\end{vmatrix} = 0\begin{vmatrix}3&-2\\2&-3\end{vmatrix} - 4\begin{vmatrix}1&-2\\-1&-3\end{vmatrix} + 1\begin{vmatrix}1&3\\-1&2\end{vmatrix}$

$= 0 - 4(-3 - 2) + 1(2 + 3)$

$= 0 + 20 + 5 = 25$

14. -10

15. $\begin{vmatrix}1&-2&3\\-3&4&0\\2&6&5\end{vmatrix} = (3)\begin{vmatrix}-3&4\\2&6\end{vmatrix} - (0) + (5)\begin{vmatrix}1&-2\\-3&4\end{vmatrix}$

$= 3(-18 - 8) - 0 + 5(4 - 6)$

$= 3(-26) + 5(-2) = -78 - 10 = -88$

16. -84 17. $\begin{vmatrix}6&7&8\\-6&7&-9\\0&0&-2\end{vmatrix} = 0 - 0 - 2\begin{vmatrix}6&7\\-6&7\end{vmatrix}$

$= 0 - 0 - 2(42 + 42) = -168$

18. -180 19. $\begin{vmatrix}x&0&1\\0&2&3\\4&-1&-2\end{vmatrix} = 6$ 20. -2

$x\begin{vmatrix}2&3\\-1&-2\end{vmatrix} + 1\begin{vmatrix}0&2\\4&-1\end{vmatrix} = 6$

$x(-1) + 1(-8) = 6$

$-x - 8 = 6$

$x = -14$

Exercises 11.7C (page 584)

1. $x = \dfrac{\begin{vmatrix}11&-1\\-4&2\end{vmatrix}}{\begin{vmatrix}3&-1\\3&2\end{vmatrix}} = \dfrac{22 - 4}{6 - (-3)} = \dfrac{18}{9} = 2;$ 2. $(-3, 4)$

$y = \dfrac{\begin{vmatrix}3&11\\3&-4\end{vmatrix}}{\begin{vmatrix}3&-1\\3&2\end{vmatrix}} = \dfrac{-12 - 33}{6 - (-3)} = \dfrac{-45}{9} = -5$

The solution is $(2, -5)$.

3. $x = \dfrac{\begin{vmatrix}11&15\\31&-1\end{vmatrix}}{\begin{vmatrix}8&15\\4&-1\end{vmatrix}} = \dfrac{-11 - 465}{-8 - 60} = \dfrac{-476}{-68} = 7;$ 4. $(6, 3)$

$y = \dfrac{\begin{vmatrix}8&11\\4&31\end{vmatrix}}{\begin{vmatrix}8&15\\4&-1\end{vmatrix}} = \dfrac{248 - 44}{-8 - 60} = \dfrac{204}{-68} = -3$

The solution is $(7, -3)$.

5. $x = \dfrac{\begin{vmatrix} 3 & -3 \\ 12 & -9 \end{vmatrix}}{\begin{vmatrix} 7 & -3 \\ 20 & -9 \end{vmatrix}} = \dfrac{-27-(-36)}{-63-(-60)} = \dfrac{9}{-3} = -3;$

$y = \dfrac{\begin{vmatrix} 7 & 3 \\ 20 & 12 \end{vmatrix}}{\begin{vmatrix} 7 & -3 \\ 20 & -9 \end{vmatrix}} = \dfrac{84-60}{-63-(-60)} = \dfrac{24}{-3} = -8$

The solution is $(-3, -8)$.

6. $(9, -13)$

7. $x = \dfrac{\begin{vmatrix} 0 & 5 \\ 38 & -3 \end{vmatrix}}{\begin{vmatrix} 6 & 5 \\ 4 & -3 \end{vmatrix}} = \dfrac{0-190}{-18-20} = \dfrac{-190}{-38} = 5;$

$y = \dfrac{\begin{vmatrix} 6 & 0 \\ 4 & 38 \end{vmatrix}}{\begin{vmatrix} 6 & 5 \\ 4 & -3 \end{vmatrix}} = \dfrac{228-0}{-18-20} = \dfrac{228}{-38} = -6$

The solution is $(5, -6)$.

8. $(-4, 10)$

9. $x = \dfrac{\begin{vmatrix} 5 & 6 \\ 7 & 12 \end{vmatrix}}{\begin{vmatrix} 4 & 6 \\ 8 & 12 \end{vmatrix}} = \dfrac{60-42}{48-48} = \dfrac{18}{0}$ Not defined

The denominator was zero and the numerator was not zero; the system is inconsistent.

10. Inconsistent

11. $x = \dfrac{\begin{vmatrix} 5 & -3 \\ 10 & -6 \end{vmatrix}}{\begin{vmatrix} 7 & -3 \\ 14 & -6 \end{vmatrix}} = \dfrac{-30-(-30)}{-42-(-42)} = \dfrac{0}{0}$ Not defined

The numerator and denominator were both zero; the system is dependent.

12. Dependent

13. $D = \begin{vmatrix} 2 & 1 & 1 \\ 1 & -1 & 3 \\ 1 & 1 & 2 \end{vmatrix} = (2)\begin{vmatrix} -1 & 3 \\ 1 & 2 \end{vmatrix} - (1)\begin{vmatrix} 1 & 3 \\ 1 & 2 \end{vmatrix} + (1)\begin{vmatrix} 1 & -1 \\ 1 & 1 \end{vmatrix}$

$= 2(-5) - 1(-1) + 1(2)$

$= -10 + 1 + 2 = -7$

$D_x = \begin{vmatrix} 4 & 1 & 1 \\ -2 & -1 & 3 \\ 1 & 1 & 2 \end{vmatrix} = (4)\begin{vmatrix} -1 & 3 \\ 1 & 2 \end{vmatrix} - (-2)\begin{vmatrix} 1 & 1 \\ 1 & 2 \end{vmatrix} + (1)\begin{vmatrix} 1 & 1 \\ -1 & 3 \end{vmatrix}$

$= 4(-5) + 2(1) + 1(4) = -20 + 2 + 4 = -14$

$D_y = \begin{vmatrix} 2 & 4 & 1 \\ 1 & -2 & 3 \\ 1 & 1 & 2 \end{vmatrix} = (2)\begin{vmatrix} -2 & 3 \\ 1 & 2 \end{vmatrix} - (4)\begin{vmatrix} 1 & 3 \\ 1 & 2 \end{vmatrix} + (1)\begin{vmatrix} 1 & -2 \\ 1 & 1 \end{vmatrix}$

$= 2(-7) - 4(-1) + 1(3) = -14 + 4 + 3 = -7$

$D_z = \begin{vmatrix} 2 & 1 & 4 \\ 1 & -1 & -2 \\ 1 & 1 & 1 \end{vmatrix}$

$= (2)\begin{vmatrix} -1 & -2 \\ 1 & 1 \end{vmatrix} - (1)\begin{vmatrix} 1 & -2 \\ 1 & 1 \end{vmatrix} + (4)\begin{vmatrix} 1 & -1 \\ 1 & 1 \end{vmatrix}$

$= 2(1) - 1(3) + 4(2) = 2 - 3 + 8 = 7$

$x = \dfrac{D_x}{D} = \dfrac{-14}{-7} = 2,\ y = \dfrac{D_y}{D} = \dfrac{-7}{-7} = 1,\ z = \dfrac{D_z}{D} = \dfrac{7}{-7} = -1$

The solution is $(2, 1, -1)$.

14. $(1, -2, 2)$

15. $D = \begin{vmatrix} 2 & 3 & 1 \\ 4 & 0 & -2 \\ 0 & 6 & -1 \end{vmatrix} = (0) - (6)\begin{vmatrix} 2 & 1 \\ 4 & -2 \end{vmatrix} + (-1)\begin{vmatrix} 2 & 3 \\ 4 & 0 \end{vmatrix}$

$= -6(-8) - 1(-12) = 48 + 12 = 60$

$D_x = \begin{vmatrix} 7 & 3 & 1 \\ -6 & 0 & -2 \\ 0 & 6 & -1 \end{vmatrix} = (0) - (6)\begin{vmatrix} 7 & 1 \\ -6 & -2 \end{vmatrix} + (-1)\begin{vmatrix} 7 & 3 \\ -6 & 0 \end{vmatrix}$

$= 0 - 6(-8) - 1(18) = 30$

$D_y = \begin{vmatrix} 2 & 7 & 1 \\ 4 & -6 & -2 \\ 0 & 0 & -1 \end{vmatrix} = +(-1)\begin{vmatrix} 2 & 7 \\ 4 & -6 \end{vmatrix} = -1(-40) = 40$

$D_z = \begin{vmatrix} 2 & 3 & 7 \\ 4 & 0 & -6 \\ 0 & 6 & 0 \end{vmatrix} = -(6)\begin{vmatrix} 2 & 7 \\ 4 & -6 \end{vmatrix} = -6(-40) = 240$

$x = \dfrac{D_x}{D} = \dfrac{30}{60} = \dfrac{1}{2},\ y = \dfrac{D_y}{D} = \dfrac{40}{60} = \dfrac{2}{3},\ z = \dfrac{D_z}{D} = \dfrac{240}{60} = 4$

The solution is $\left(\dfrac{1}{2}, \dfrac{2}{3}, 4\right)$.

16. $(2, -1, 3)$

Exercises 11.8 (page 588)

Note: The checks for these exercises will not be shown.

1. Let x = one number
y = the other number

(1) $x + y = 30$
(2) $x - y = 12$

$2x = 42$
$x = 21$

Substituting 21 for x in (1) $\Rightarrow 21 + y = 30 \Rightarrow y = 9$
The numbers are 21 and 9.

2. 125° and 55°

3. Let x = number of nickels
y = number of quarters

(1) $x + y = 15$
(2) $5x + 25y = 175$

Solving (1) for x, we have:
(1) $x + y = 15 \Rightarrow x = 15 - y$
Substituting $15 - y$ for x in (2):
(2) $5x + 25y = 175$
$5(15 - y) + 25y = 175$
$75 - 5y + 25y = 175$
$20y = 100$
$y = 5$

Substituting 5 for y in $x = 15 - y$:
$x = 15 - 5 = 10$
Beatrice has 5 quarters and 10 nickels.

4. 15 dimes, 7 half-dollars

5. Let x = numerator
y = denominator

(1) $\dfrac{x}{y} = \dfrac{2}{3}$

(2) $\dfrac{x + 10}{y - 5} = 1$

(1) $\dfrac{x}{y} = \dfrac{2}{3} \Rightarrow 3x = 2y$

(2) $\dfrac{x + 10}{y - 5} = 1 \Rightarrow x + 10 = y - 5 \Rightarrow x = y - 15$

Substituting $y - 15$ for x in $3x = 2y$:

$$\begin{aligned} 3(y - 15) &= 2y \\ 3y - 45 &= 2y \\ y &= 45 \end{aligned}$$

Substituting 45 for y in $x = y - 15$:

$$x = 45 - 15 = 30$$

The original fraction was $\frac{30}{45}$.

6. $\frac{12}{24}$

7. Let u = units digit
t = tens digit
h = hundreds digit

$$\begin{array}{lrcl}
(1) & h + t + u &=& 20 \\
(2) & t - u &=& 3 \\
(3) & h + t &=& 15 \\ \hline
(1) + (2) & h + 2t &=& 23 \quad (4) \\
(3) & h + t &=& 15 \\ \hline
(4) - (3) & t &=& 8
\end{array}$$

Substituting 8 for t in (3) $\Rightarrow h = 7$
Substituting 8 for t in (2) $\Rightarrow u = 5$
The number is 785.

8. 498

9. Let a = the number of hours for Albert to do the job
b = the number of hours for Bill to do the job
c = the number of hours for Carlos to do the job

Then Albert's rate is $\frac{1}{a}\,\frac{\text{job}}{\text{hr}}$, Bill's rate is $\frac{1}{b}\,\frac{\text{job}}{\text{hr}}$, and Carlos's rate is $\frac{1}{c}\,\frac{\text{job}}{\text{hr}}$.

$$\begin{array}{lrcl}
(1) & \frac{2}{a} + \frac{2}{b} + \frac{2}{c} &=& 1 \\
(2) & \frac{3}{b} + \frac{3}{c} &=& 1 \\
(3) & \frac{4}{a} + \frac{4}{b} &=& 1 \\ \hline
(1) & \frac{2}{a} + \frac{2}{b} + \frac{2}{c} &=& 1 \\
\frac{1}{2}(3) & \frac{2}{a} + \frac{2}{b} &=& \frac{1}{2} \\ \hline
(1) - \frac{1}{2}(3) & \frac{2}{c} &=& \frac{1}{2} \Rightarrow c = 4
\end{array}$$

Substituting 4 for c in (2) $\Rightarrow b = 12$
Substituting 12 for b in (3) $\Rightarrow a = 6$
It would take Albert 6 hr, Bill 12 hr, and Carlos 4 hr to do the job alone.

10. A: 4 hr; B: 8 hr; C 12 hr

11. Let x = average speed of plane in still air (in mph)
y = average speed of wind (in mph)
$x - y$ = average speed flying against the wind (in mph)
$x + y$ = average speed flying with the wind (in mph)
Formula to use: $rt = d$

$$\begin{array}{lll}
(1)\ 2] & (x - y)\frac{11}{2} = 2{,}750 \Rightarrow & 11(x - y) = 2(2{,}750) \\
(2) & (x + y)(5) = 2{,}750 \Rightarrow & 5(x + y) = 2{,}750 \\
(1)\ 1/11] & 11x - 11y = 5{,}500 \Rightarrow & x - y = 500 \\
\ 1/5] & 5x + 5y = 2{,}750 \Rightarrow & x + y = 550 \\ \hline
 & & 2x = 1{,}050 \\
 & & x = 525 \\
 & & 525 + y = 550 \\
 & & y = 25
\end{array}$$

The speed of the plane in still air is 525 mph and the speed of the wind is 25 mph.

12. 540 mph = speed of plane; 60 mph = speed of wind

13. Let x = number of pounds of grade A coffee
y = number of pounds of grade B coffee

$$\begin{array}{lll}
(1) & x + y = 90 \quad \Rightarrow -365] & x + y = 90 \\
(2)\ 100] & 3.85x + 3.65y = 338.90 \Rightarrow & 385x + 365y = 33{,}890 \\
 & & -365x - 365y = -32{,}850 \\
 & & 385x + 365y = 33{,}890 \\ \hline
 & & 20x = 1{,}040 \\
 & & x = 52 \\
 & & 52 + y = 90 \\
 & & y = 38
\end{array}$$

The mixture should contain 52 lb of grade A coffee and 38 lb of grade B coffee.

14. 37 lb of grade A coffee; 43 lb of grade B coffee

15. Let u = units digit
t = tens digit

$$\begin{array}{lrcl}
(1) & t + u &=& 12 \\
(2) & (10t + u) - (10u + t) &=& 54 \\
(2) & 10t + u - 10u - t &=& 54 \\
(2) & 9t - 9u &=& 54 \\
\frac{1}{9}(2) & t - u &=& 6 \\
(1) & t + u &=& 12 \\ \hline
(1) + \frac{1}{9}(2) & 2t &=& 18 \\
 & t &=& 9
\end{array}$$

Substituting 9 for t in (1) $\Rightarrow 9 + u = 12 \Rightarrow u = 3$
The number is 93.

16. 86

17. Let x = number of 18¢ stamps
y = number of 22¢ stamps
z = number of 45¢ stamps

$$\begin{array}{lrcl}
(1) & x + y + z &=& 29 \\
(2) & 18x + 22y + 45z &=& 721 \\
(3) & y &=& 2x
\end{array}$$

Substituting (3) into (1) and into (2):

$$\begin{array}{lrcl}
(4) & x + 2x + z = 29 \Rightarrow & 3x + z &= 29 \\
(5) & 18x + 44x + 45z = 721 \Rightarrow & 62x + 45z &= 721 \\
-45(4) & & -135x - 45z &= -1{,}305 \\
(5) & & 62x + 45z &= 721 \\ \hline
 & & -73x &= -584 \\
 & & x &= 8
\end{array}$$

Substituting 8 for x in (3):
$y = 2(8) = 16$
Substituting 8 for x and 16 for y in (1):
$8 + 16 + z = 29 \Rightarrow z = 5$
Tom bought eight 18¢ stamps, sixteen 22¢ stamps, and five 45¢ stamps.

18. Six 18¢ stamps; twelve 22¢ stamps; three 45¢ stamps

19. Let t = cost of tie (in cents)
p = cost of pin (in cents)
(1) $t + p = 110$
(2) $t = 100 + p$
Substituting $100 + p$ for t in (1):

$$\begin{array}{lrcl}
(1) & t + p &=& 110 \\
 & 100 + p + p &=& 110 \\
 & 2p &=& 10 \\
 & p &=& 5 \\
 & t = 100 + p &=& 100 + 5 = 105
\end{array}$$

The pin costs 5¢, and the tie costs \$1.05.

20. 7 on the upper branch; 5 on the lower branch

Exercises 11.9 (page 595)

1. (1) $x^2 = 2y$
(2) $x - y = -4$
$y = x + 4$
Substituting $x + 4$ for y in (1):
(1) $x^2 = 2y$
$x^2 = 2(x + 4)$
$x^2 = 2x + 8$
$x^2 - 2x - 8 = 0$
$(x - 4)(x + 2) = 0$
$x - 4 = 0 \mid x + 2 = 0$
$x = 4 \mid x = -2$
Substituting 4 for x in $y = x + 4$:
$y = 8$
Substituting -2 for x in $y = x + 4$:
$y = 2$
The solutions are (4, 8) and (−2, 2).

2. (2, 1) and (−4, 4)

3. (1) $x^2 = 4y$
(2) $x - y = 1$
Solve (2) for y:
$x - y = 1$
$y = x - 1$
Substituting $x - 1$ for y in (1):
(1) $x^2 = 4y$
$x^2 = 4(x - 1)$
$x^2 = 4x - 4$
$x^2 - 4x + 4 = 0$
$(x - 2)(x - 2) = 0$
Therefore, $x = 2$.
Substituting 2 for x in (2):
(2) $x - y = 1$
$2 - y = 1 \Rightarrow y = 1$
There is only one solution, (2, 1).

4. (−5, 0) and (4, 3)

5. (1) $xy = 4$
(2) $x - 2y = 2$
Solve (2) for x:
$x - 2y = 2$
$x = 2y + 2$
Substituting $2y + 2$ for x in (1):
(1) $xy = 4$
$(2y + 2)y = 4$
$2y^2 + 2y - 4 = 0$
$y^2 + y - 2 = 0$
$(y + 2)(y - 1) = 0$
$y + 2 = 0 \mid y - 1 = 0$
$y = -2 \mid y = 1$
Substituting -2 for y in $xy = 4$:
$x(-2) = 4$
$x = -2$
Substituting 1 for y in $xy = 4$:
$x(1) = 4 \Rightarrow x = 4$
The solutions are (−2, −2) and (4, 1).

6. $(-i\sqrt{3}, i\sqrt{3})$ and $(i\sqrt{3}, -i\sqrt{3})$ are the algebraic solutions. The curves do not intersect.

7. (1) $x^2 + y^2 = 61$
(2) $x^2 - y^2 = 11$
(1) + (2): $2x^2 = 72$
$x^2 = 36$
$x = \pm 6$
Substituting ± 6 for x in (1):
(1) $x^2 + y^2 = 61$
$(\pm 6)^2 + y^2 = 61$
$36 + y^2 = 61$
$y^2 = 25$
$y = \pm 5$
The solutions are (6, 5), (6, −5), (−6, 5), and (−6, −5).

8. (2, 0), (−2, 0), and (0, −1)

9. (1) $2x^2 + 3y^2 = 21$
(2) $x^2 + 2y^2 = 12$
$-1]\ 2x^2 + 3y^2 = 21 \Rightarrow -2x^2 - 3y^2 = -21$
$2]\ 1x^2 + 2y^2 = 12 \Rightarrow 2x^2 + 4y^2 = 24$
$y^2 = 3$
$y = \pm\sqrt{3}$
Substituting $\pm\sqrt{3}$ for y in (2):
(2) $x^2 + 2y^2 = 12$
$x^2 + 2(\pm\sqrt{3})^2 = 12$
$x^2 + 6 = 12$
$x^2 = 6$
$x = \pm\sqrt{6}$
The solutions are
$(\sqrt{6}, \sqrt{3}), (\sqrt{6}, -\sqrt{3}), (-\sqrt{6}, \sqrt{3}), (-\sqrt{6}, -\sqrt{3})$.

10. $(3\sqrt{2}, \sqrt{2}), (3\sqrt{2}, -\sqrt{2}), (-3\sqrt{2}, \sqrt{2})$, and $(-3\sqrt{2}, -\sqrt{2})$

Exercises 11.10A (page 597)

1. (1) $4x - 3y > -12$
(2) $y > 2$
Boundary line for (1): $4x - 3y = -12$
Intercepts: (0, 4) and (−3, 0)
Boundary line for (2): $y = 2$, a horizontal line that passes through (0, 2)

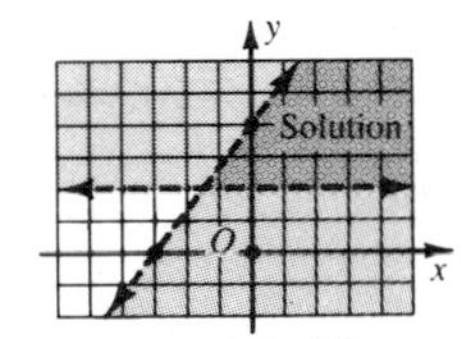

Both lines are dashed lines because equality is not included

The origin is in the correct half-plane for (1) because $4(0) - 3(0) > -12$ is true. The origin is not in the correct half-plane for (2) because $0 > 2$ is false.

2.

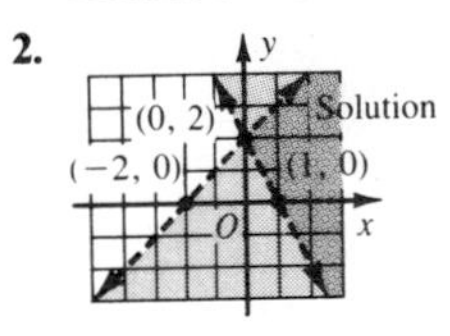

3. (1) $2x - y \le 2$
(2) $x + y \ge 5$
Boundary line for (1): $2x - y = 2$
Intercepts: (0, −2) and (1, 0)
Boundary line for (2): $x + y = 5$
Intercepts: (0, 5) and (5, 0)
The origin is in the correct half-plane for (1), but not for (2).

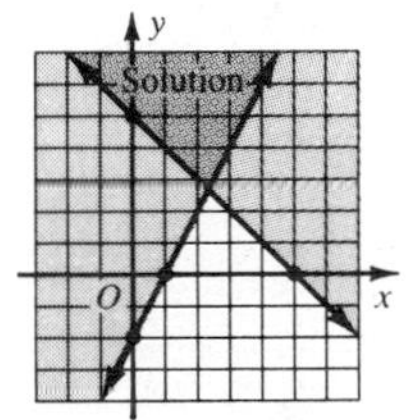

Both lines are solid because equality is included

4.

y
Solution
x
(−2, 0)
O
(4, 0)
(0, −2)
(0, −4)

5. (1) $2x + y < 0$
(2) $x - y \geq -3$
Boundary line for (1): $2x + y = 0$
Points on line: (0, 0) and $(-1, 2)$
The point $(-1, 0)$ is in the correct half-plane for (1).
Boundary line for (2): $x - y = -3$
Points on line: (0, 3) and $(-3, 0)$
The origin is in the correct half-plane for (2).

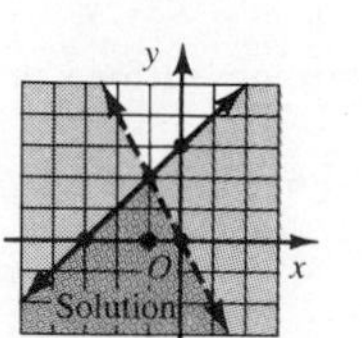

The boundary line for (1) is dashed because equality is not included; the boundary line for (2) is solid because equality is included

6.

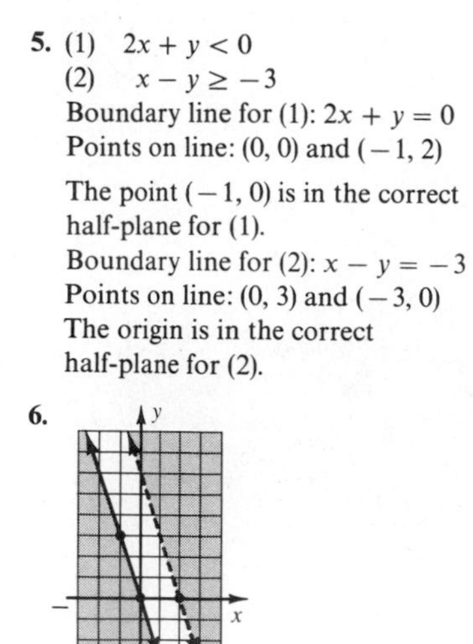

No solution

7. (1) $3x - 2y < 6$
(2) $x + 2y \leq 4$
(3) $6x + y > -6$
Boundary line for (1):
$3x - 2y = 6$
Intercepts: (2, 0) and $(0, -3)$
The half-plane for (1) includes (0, 0) because $3(0) - 2(0) < 6$.
Boundary line for (2): $x + 2y = 4$
Intercepts: (4, 0) and (0, 2)
The half-plane for (2) includes (0, 0) because $(0) + 2(0) \leq 4$.
Boundary line for (3): $6x + y = -6$
Intercepts: $(-1, 0)$ and $(0, -6)$
The half-plane for (3) includes (0, 0) because $6(0) + (0) > -6$.

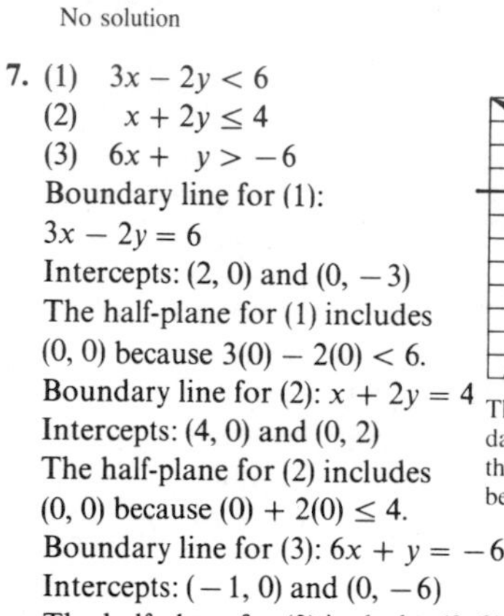

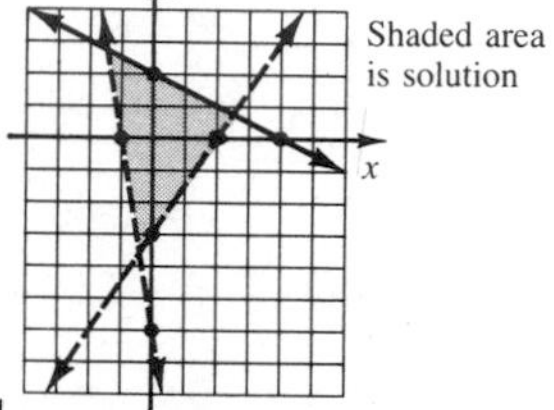

Shaded area is solution

The boundary lines for (1) and (3) are dashed because equality is not included; the boundary line for (2) is solid because equality is included

8.

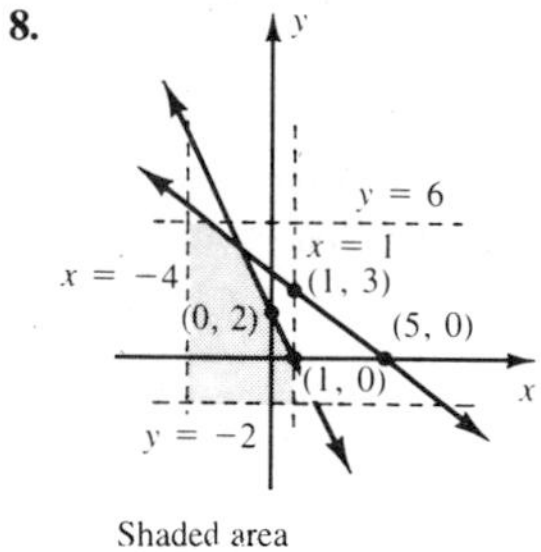

Shaded area is solution

Exercises 11.10B (page 599)

1. Boundary for $\frac{x^2}{9} + \frac{y^2}{4} < 1$ is $\frac{x^2}{9} + \frac{y^2}{4} = 1$, which is an ellipse with intercepts $(\pm 3, 0)$ and $(0, \pm 2)$. Boundary for $\frac{x^2}{4} + \frac{y^2}{9} < 1$ is $\frac{x^2}{4} + \frac{y^2}{9} = 1$, which is an ellipse with intercepts $(\pm 2, 0)$ and $(0, \pm 3)$. Both boundaries must be graphed as dashed curves. Substituting (0, 0) in both inequalities gives true statements.

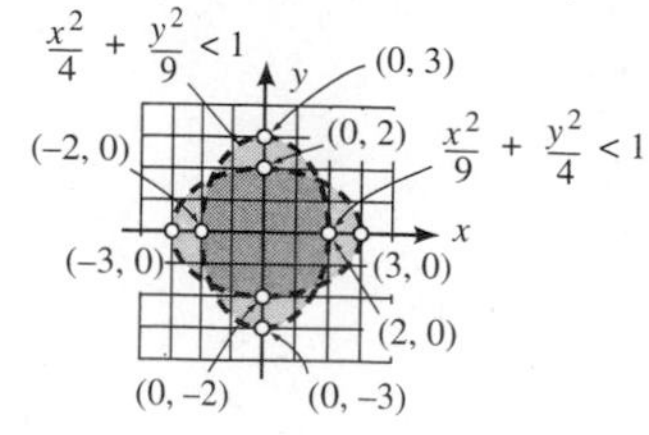

2.

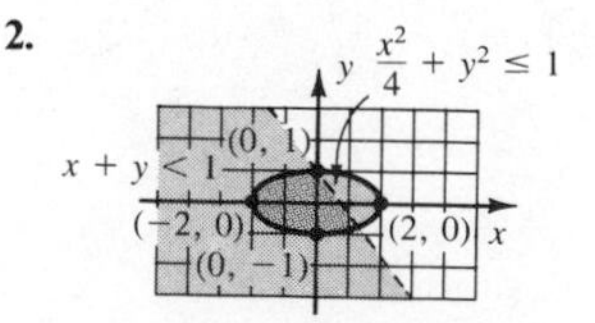

3. Boundary for $y > 1 - x^2$ is $y = 1 - x^2$, which is a parabola with intercepts $(\pm 1, 0)$ and (0, 1). Boundary for $x^2 + y^2 < 4$ is $x^2 + y^2 = 4$, which is a circle with intercepts $(\pm 2, 0)$ and $(0, \pm 2)$. Both curves must be graphed as dashed curves. Substituting (0, 0) in $y > 1 - x^2$ gives a false statement; we must shade the region that does not contain the origin. Substituting (0, 0) in $x^2 + y^2 < 4$ gives a true statement; we must shade the region that contains the origin. The final answer is heavily shaded.

4.

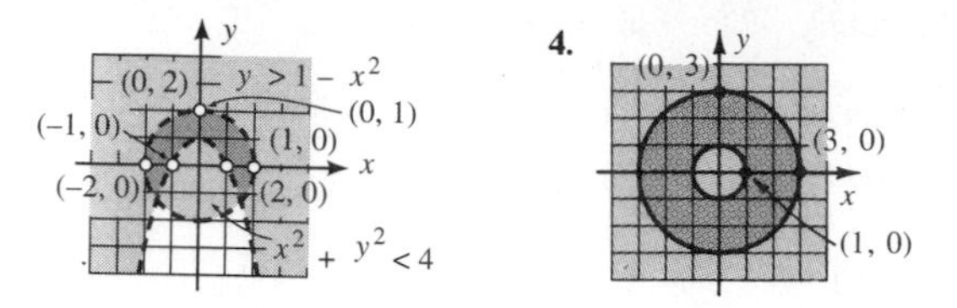

5. Boundary for $x^2 + y^2 \leq 9$ is $x^2 + y^2 = 9$, which is a circle of radius 3 with its center at the origin; it must be graphed with a solid line. Boundary for $y \geq x - 2$ is a straight line with intercepts (2, 0) and $(0, -2)$; it must be graphed with a solid line. Substituting (0, 0) in $x^2 + y^2 \leq 9$ gives a true statement; we shade the region that contains the origin (the region *inside* the circle). Substituting (0, 0) in $y \geq x - 2$ gives a true statement; we shade the half-plane that contains the origin. The final answer is heavily shaded.

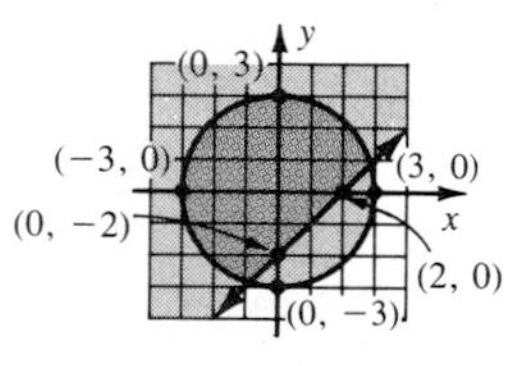

Review Exercises 11.11 (page 601)

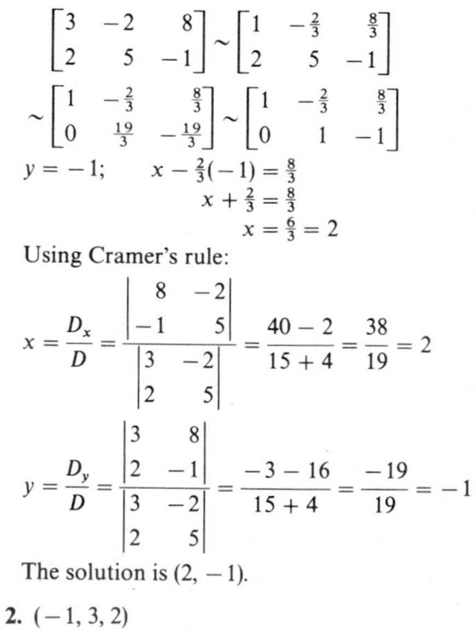

1. Using the matrix method:

$$\begin{bmatrix} 3 & -2 & 8 \\ 2 & 5 & -1 \end{bmatrix} \sim \begin{bmatrix} 1 & -\frac{2}{3} & \frac{8}{3} \\ 2 & 5 & -1 \end{bmatrix}$$

$$\sim \begin{bmatrix} 1 & -\frac{2}{3} & \frac{8}{3} \\ 0 & \frac{19}{3} & -\frac{19}{3} \end{bmatrix} \sim \begin{bmatrix} 1 & -\frac{2}{3} & \frac{8}{3} \\ 0 & 1 & -1 \end{bmatrix}$$

$y = -1$; $\quad x - \frac{2}{3}(-1) = \frac{8}{3}$

$x + \frac{2}{3} = \frac{8}{3}$

$x = \frac{6}{3} = 2$

Using Cramer's rule:

$$x = \frac{D_x}{D} = \frac{\begin{vmatrix} 8 & -2 \\ -1 & 5 \end{vmatrix}}{\begin{vmatrix} 3 & -2 \\ 2 & 5 \end{vmatrix}} = \frac{40 - 2}{15 + 4} = \frac{38}{19} = 2$$

$$y = \frac{D_y}{D} = \frac{\begin{vmatrix} 3 & 8 \\ 2 & -1 \end{vmatrix}}{\begin{vmatrix} 3 & -2 \\ 2 & 5 \end{vmatrix}} = \frac{-3 - 16}{15 + 4} = \frac{-19}{19} = -1$$

The solution is $(2, -1)$.

2. $(-1, 3, 2)$

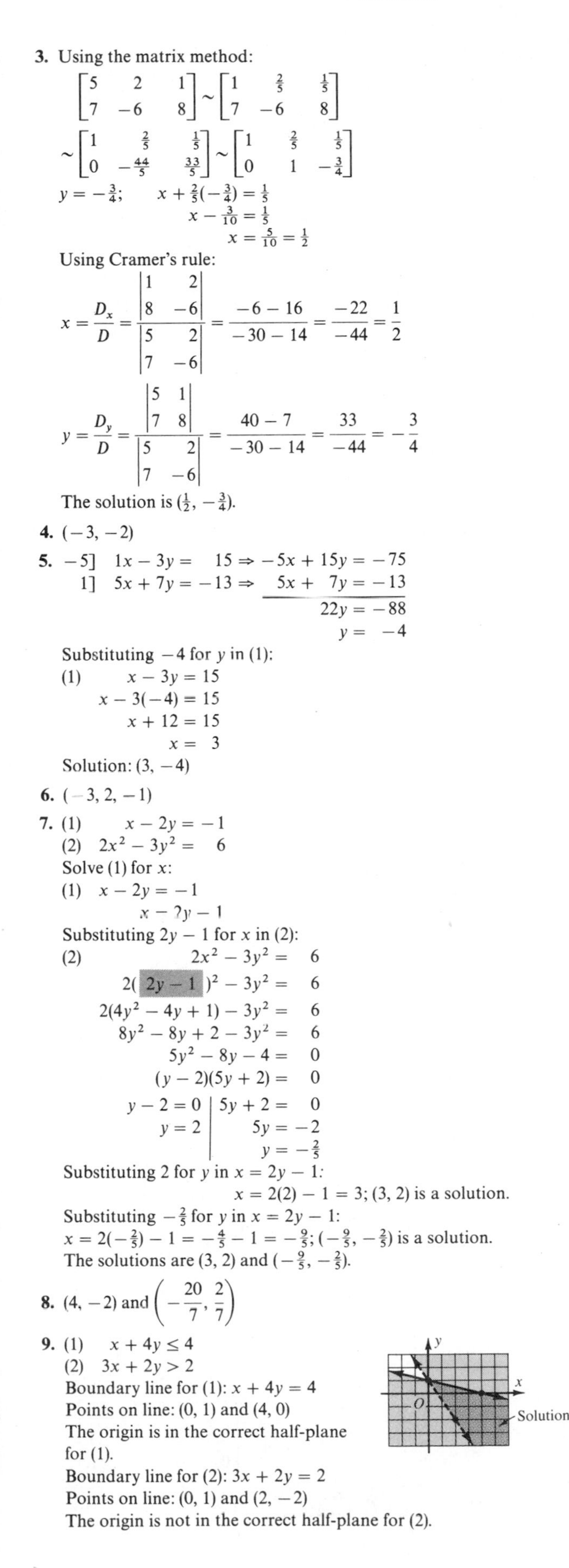

3. Using the matrix method:

$$\begin{bmatrix} 5 & 2 & 1 \\ 7 & -6 & 8 \end{bmatrix} \sim \begin{bmatrix} 1 & \frac{2}{5} & \frac{1}{5} \\ 7 & -6 & 8 \end{bmatrix}$$

$$\sim \begin{bmatrix} 1 & \frac{2}{5} & \frac{1}{5} \\ 0 & -\frac{44}{5} & \frac{33}{5} \end{bmatrix} \sim \begin{bmatrix} 1 & \frac{2}{5} & \frac{1}{5} \\ 0 & 1 & -\frac{3}{4} \end{bmatrix}$$

$y = -\frac{3}{4}$; $\quad x + \frac{2}{5}(-\frac{3}{4}) = \frac{1}{5}$

$x - \frac{3}{10} = \frac{1}{5}$

$x = \frac{5}{10} = \frac{1}{2}$

Using Cramer's rule:

$$x = \frac{D_x}{D} = \frac{\begin{vmatrix} 1 & 2 \\ 8 & -6 \end{vmatrix}}{\begin{vmatrix} 5 & 2 \\ 7 & -6 \end{vmatrix}} = \frac{-6 - 16}{-30 - 14} = \frac{-22}{-44} = \frac{1}{2}$$

$$y = \frac{D_y}{D} = \frac{\begin{vmatrix} 5 & 1 \\ 7 & 8 \end{vmatrix}}{\begin{vmatrix} 5 & 2 \\ 7 & -6 \end{vmatrix}} = \frac{40 - 7}{-30 - 14} = \frac{33}{-44} = -\frac{3}{4}$$

The solution is $(\frac{1}{2}, -\frac{3}{4})$.

4. $(-3, -2)$

5.
$$\begin{aligned} -5]\quad 1x - 3y &= 15 \Rightarrow -5x + 15y = -75 \\ 1]\quad 5x + 7y &= -13 \Rightarrow \;\; 5x + 7y = -13 \\ && 22y = -88 \\ && y = -4 \end{aligned}$$

Substituting -4 for y in (1):

(1) $\quad x - 3y = 15$

$x - 3(-4) = 15$

$x + 12 = 15$

$x = 3$

Solution: $(3, -4)$

6. $(-3, 2, -1)$

7. (1) $\quad x - 2y = -1$

(2) $\quad 2x^2 - 3y^2 = 6$

Solve (1) for x:

(1) $\quad x - 2y = -1$

$x = 2y - 1$

Substituting $2y - 1$ for x in (2):

(2) $\quad 2x^2 - 3y^2 = 6$

$2(2y - 1)^2 - 3y^2 = 6$

$2(4y^2 - 4y + 1) - 3y^2 = 6$

$8y^2 - 8y + 2 - 3y^2 = 6$

$5y^2 - 8y - 4 = 0$

$(y - 2)(5y + 2) = 0$

$y - 2 = 0 \quad | \quad 5y + 2 = 0$

$y = 2 \quad | \quad 5y = -2$

$\quad | \quad y = -\frac{2}{5}$

Substituting 2 for y in $x = 2y - 1$:

$x = 2(2) - 1 = 3$; $(3, 2)$ is a solution.

Substituting $-\frac{2}{5}$ for y in $x = 2y - 1$:

$x = 2(-\frac{2}{5}) - 1 = -\frac{4}{5} - 1 = -\frac{9}{5}$; $(-\frac{9}{5}, -\frac{2}{5})$ is a solution.

The solutions are $(3, 2)$ and $(-\frac{9}{5}, -\frac{2}{5})$.

8. $(4, -2)$ and $\left(-\frac{20}{7}, \frac{2}{7}\right)$

9. (1) $\quad x + 4y \le 4$

(2) $\quad 3x + 2y > 2$

Boundary line for (1): $x + 4y = 4$

Points on line: $(0, 1)$ and $(4, 0)$

The origin is in the correct half-plane for (1).

Boundary line for (2): $3x + 2y = 2$

Points on line: $(0, 1)$ and $(2, -2)$

The origin is not in the correct half-plane for (2).

Solution

10.

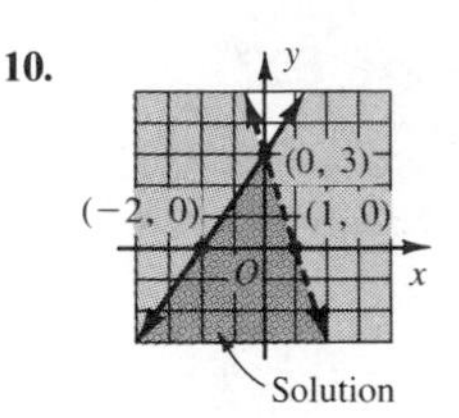

11. Boundary for $y \le 12 + 4x - x^2$ is $y = 12 + 4x - x^2$, which is a parabola. Its intercepts are $(-2, 0)$, $(6, 0)$, and $(0, 12)$, and its vertex is at $(2, 16)$; it is graphed with a solid line. Boundary for $x^2 + y^2 \le 16$ is $x^2 + y^2 = 16$, which is a circle of radius 4 with its center at the origin; it must be graphed with a solid line. Substituting $(0, 0)$ in $y \le 12 + 4x - x^2$ gives a true statement; we shade the region that contains the origin (the region "inside" the parabola). Substituting $(0, 0)$ in $x^2 + y^2 \le 16$ gives a true statement; we shade the region that contains the origin (the region *inside* the circle). The final answer is the heavily shaded region.

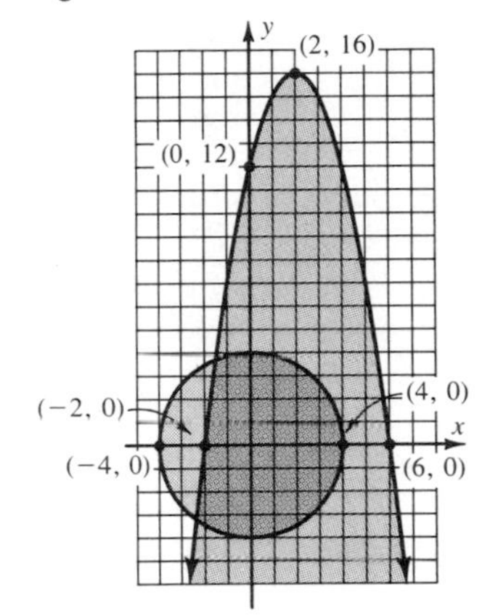

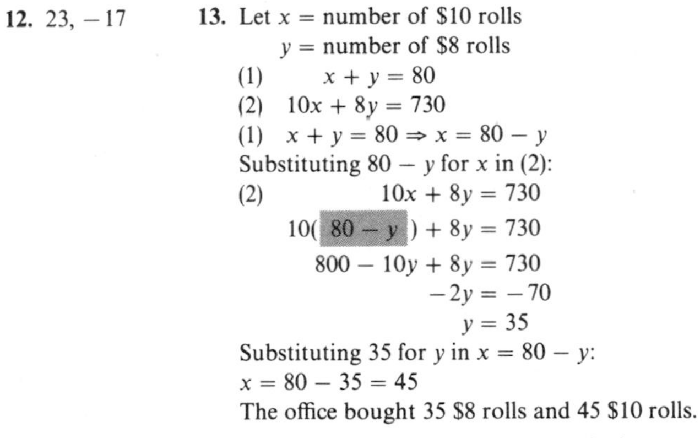

12. 23, −17

13. Let x = number of \$10 rolls

y = number of \$8 rolls

(1) $\quad x + y = 80$

(2) $\quad 10x + 8y = 730$

(1) $\quad x + y = 80 \Rightarrow x = 80 - y$

Substituting $80 - y$ for x in (2):

(2) $\quad 10x + 8y = 730$

$10(80 - y) + 8y = 730$

$800 - 10y + 8y = 730$

$-2y = -70$

$y = 35$

Substituting 35 for y in $x = 80 - y$:

$x = 80 - 35 = 45$

The office bought 35 \$8 rolls and 45 \$10 rolls.

14. 24 mph is the average speed of Jennifer's boat; 8 mph is the average speed of the river.

15. Let u = units digit

t = tens digit

h = hundreds digit

(1) $\quad u + t + h = 12$

(2) $\quad h = 3u$

(3) $\quad t = 2 + u$

Substituting (2) and (3) into (1):

$u + (2 + u) + 3u = 12$

$5u + 2 = 12$

$5u = 10$

$u = 2$

Substituting 2 for u in (2): $h = 3(2) = 6$

and (3): $t = 2 + 2 = 4$

The number is 642.

16. One pair is −4, 3, and another is −3, 4.

Chapter 11 Diagnostic Test (page 607)

Following each problem number is the textbook section number (in parentheses) where that kind of problem is discussed.

1. (11.2) (1) $3x + 2y = 4$

If $x = 0$, $3(0) + 2y = 4 \Rightarrow y = 2$

If $y = 0$, $3x + 2(0) = 4 \Rightarrow x = \frac{4}{3}$

(2) $x - y = 3$

If $x = 0$, $0 - y = 3 \Rightarrow y = -3$

If $y = 0$, $x - 0 = 3 \Rightarrow x = 3$

Solution: $(2, -1)$

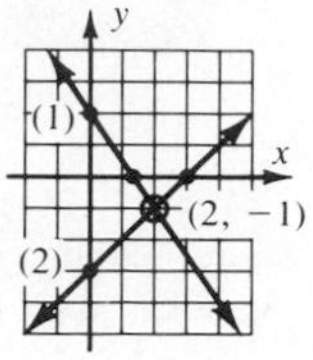

2. (11.3)

$$\begin{array}{rrl} -2] & 4x - 3y = 13 \Rightarrow & -8x + 6y = -26 \\ 3] & 5x - 2y = 4 \Rightarrow & 15x - 6y = 12 \\ \hline & & 7x = -14 \\ & & x = -2 \end{array}$$

Substituting -2 for x in (2):

$$\begin{aligned} (2) \qquad 5x - 2y &= 4 \\ 5(-2) - 2y &= 4 \\ -10 - 2y &= 4 \\ -2y &= 14 \\ y &= -7 \end{aligned}$$

Solution: $(-2, -7)$

3. (11.3 or 11.4) (1) $5x + 4y = 23$

(2) $3x + 2y = 9 \Rightarrow 2y = 9 - 3x$

$$y = \frac{9 - 3x}{2}$$

Substituting $\frac{9 - 3x}{2}$ for y in (1):

$$\begin{aligned} (1) \qquad 5x + 4y &= 23 \\ 5x + \frac{\overset{2}{\cancel{4}}}{1}\left(\frac{9 - 3x}{\underset{1}{\cancel{2}}}\right) &= 23 \\ 5x + 2(9 - 3x) &= 23 \\ 5x + 18 - 6x &= 23 \\ -x &= 5 \\ x &= -5 \end{aligned}$$

Substituting -5 for x in $y = \frac{9 - 3x}{2}$ gives $\frac{9 - 3(-5)}{2}$

$$= \frac{9 + 15}{2} = \frac{24}{2} = 12$$

Solution: $(-5, 12)$

4. (11.3)

$$\begin{array}{rrrl} 16] & 2] & 15x + 8y = -18 \Rightarrow & 30x + 16y = -36 \\ -8] & -1] & 9x + 16y = -8 \Rightarrow & -9x - 16y = +8 \\ \hline & & & 21x = -28 \\ & & & x = \frac{-28}{21} \\ & & & = \frac{-4}{3} \end{array}$$

Substituting $\frac{-4}{3}$ for x in (1):

$$\begin{aligned} (1) \qquad 15x + 8y &= -18 \\ \frac{\overset{5}{\cancel{15}}}{1}\left(\frac{-4}{\underset{1}{\cancel{3}}}\right) + 8y &= -18 \\ -20 + 8y &= -18 \\ 8y &= 2 \\ y &= \tfrac{2}{8} = \tfrac{1}{4} \end{aligned}$$

Solution: $(-\frac{4}{3}, \frac{1}{4})$

5. (11.3)

$$\begin{array}{rrrl} 4] & 2] & -10x + 35y = -18 \Rightarrow & -20x + 70y = -36 \\ 10] & 5] & 4x - 14y = 8 \Rightarrow & 20x - 70y = 40 \\ \hline & & & 0 = 4 \\ & & & \text{False} \end{array}$$

No solution; inconsistent

6. (11.5)

$$\begin{array}{rrr} (1) & x + y + z = & 0 \\ (2) & 2x \qquad - 3z = & 5 \\ (3) & 3y + 4z = & 3 \\ \hline 2(1) - (2) & 2y + 5z = & -5 \quad (4) \\ (3) & 3y + 4z = & 3 \\ \hline 3(4) - 2(3) & 7z = & -21 \\ & z = & -3 \end{array}$$

Substituting -3 for z in (2):

$2x - 3(-3) = 5 \Rightarrow x = -2$

Substituting -3 for z in (3):

$3y + 4(-3) = 3 \Rightarrow y = 5$

Solution: $(-2, 5, -3)$

7. (11.9) (1) $y^2 = 8x$

(2) $3x + y = 2 \Rightarrow y = 2 - 3x$

Substituting $2 - 3x$ for y in (1):

$$\begin{aligned} (1) \qquad y^2 &= 8x \\ (2 - 3x)^2 &= 8x \\ 4 - 12x + 9x^2 &= 8x \\ 9x^2 - 20x + 4 &= 0 \\ (x - 2)(9x - 2) &= 0 \end{aligned}$$

$$\begin{array}{c|c} x - 2 = 0 & 9x - 2 = 0 \\ x = 2 & x = \frac{2}{9} \end{array}$$

If $x = 2$, $y = 2 - 3x = 2 - 3(2) = -4$.

One solution is $(2, -4)$.

If $x = \frac{2}{9}$, $y = 2 - 3(\frac{2}{9}) = \frac{4}{3}$.

A second solution is $(\frac{2}{9}, \frac{4}{3})$.

8. (11.10A) (1) $2x + 3y \le 6$

Boundary line $2x + 3y = 6$ is solid because equality is included.

If $y = 0$, $2x + 3(0) = 6 \Rightarrow x = 3$

If $x = 0$, $2(0) + 3y = 6 \Rightarrow y = 2$

Substituting (0, 0) in (1):

$2(0) + 3(0) \le 6$

$0 \le 6$ *True*

The half-plane containing (0, 0) is the solution of (1).

(2) $y - 2x < 2$

Boundary line $y - 2x = 2$ is dashed because equality is *not* included.

If $y = 0$, $0 - 2x = 2 \Rightarrow x = -1$

If $x = 0$, $y - 2(0) = 2 \Rightarrow y = 2$

Substituting (0, 0) in (2):

$0 - 2(0) < 2$

$0 < 2$ *True*

The half-plane containing (0, 0) is a solution of (2).

The final answer is the heavily shaded region.

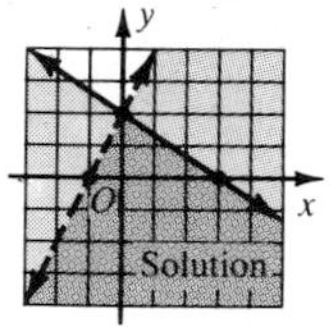

9. (11.7) Using the matrix method:

$$\begin{matrix}-1] \\ \\ \\ \end{matrix}\begin{bmatrix} 1 & -1 & -1 & 0 \\ 1 & 3 & 1 & 4 \\ 7 & -2 & -5 & 2 \end{bmatrix} \sim \begin{matrix}-7] \\ \\ \\ \end{matrix}\begin{bmatrix} 1 & -1 & -1 & 0 \\ 0 & 4 & 2 & 4 \\ 7 & -2 & -5 & 2 \end{bmatrix}$$

$$\sim \begin{matrix} \\ \frac{1}{4}] \\ \\ \end{matrix}\begin{bmatrix} 1 & -1 & -1 & 0 \\ 0 & 4 & 2 & 4 \\ 0 & 5 & 2 & 2 \end{bmatrix} \sim \begin{matrix} \\ -5] \\ \\ \end{matrix}\begin{bmatrix} 1 & -1 & -1 & 0 \\ 0 & 1 & \frac{1}{2} & 1 \\ 0 & 5 & 2 & 2 \end{bmatrix}$$

$$\sim \begin{matrix} \\ \\ -2] \end{matrix}\begin{bmatrix} 1 & -1 & -1 & 0 \\ 0 & 1 & \frac{1}{2} & 1 \\ 0 & 0 & -\frac{1}{2} & -3 \end{bmatrix} \sim \begin{bmatrix} 1 & -1 & -1 & 0 \\ 0 & 1 & \frac{1}{2} & 1 \\ 0 & 0 & 1 & 6 \end{bmatrix}$$

$z = 6$

$y + \frac{1}{2}(6) = 1$; $\quad y = -2$

$x - 1(-2) - 1(6) = 0$; $\quad x = 4$

Using Cramer's rule:

$$D = \begin{vmatrix} 1 & -1 & -1 \\ 1 & 3 & 1 \\ 7 & -2 & -5 \end{vmatrix} = -2; \quad D_x = \begin{vmatrix} 0 & -1 & -1 \\ 4 & 3 & 1 \\ 2 & -2 & -5 \end{vmatrix} = -8;$$

$$D_y = \begin{vmatrix} 1 & 0 & -1 \\ 1 & 4 & 1 \\ 7 & 2 & -5 \end{vmatrix} = 4; \quad D_z = \begin{vmatrix} 1 & -1 & 0 \\ 1 & 3 & 4 \\ 7 & -2 & 2 \end{vmatrix} = -12$$

$$x = \frac{D_x}{D} = \frac{-8}{-2} = 4; \quad y = \frac{D_y}{D} = \frac{4}{-2} = -2;$$

$$z = \frac{D_z}{D} = \frac{-12}{-2} = 6$$

Solution: $(4, -2, 6)$

10. (11.8) Let b = speed of boat (in mph)
c = speed of stream (in mph)
Then $b - c$ = speed upstream (in mph)
$b + c$ = speed downstream (in mph)

$(b - c)\frac{\text{mi}}{\text{hr}}(2\ \text{hr})$ = distance upstream (in mi)

$(b + c)\frac{\text{mi}}{\text{hr}}(1\ \text{hr})$ = distance downstream (in mi)

(1) $2(b - c) = 30 \Rightarrow b - c = 15$
(2) $1(b + c) = 30 \Rightarrow b + c = 30$

$2b = 45$
$b = \frac{45}{2}$, or $22\frac{1}{2}$

Substituting $\frac{45}{2}$ for b in (2): $\frac{45}{2} + c = 30$
$c = \frac{60}{2} - \frac{45}{2} = \frac{15}{2}$, or $7\frac{1}{2}$

Check: Speed downstream is $(22\frac{1}{2} + 7\frac{1}{2})$ mph, or 30 mph; speed upstream is $(22\frac{1}{2} - 7\frac{1}{2})$ mph, or 15 mph. Distance upstream is $15\frac{\text{mi}}{\text{hr}}(2\ \text{hr}) = 30$ mi. Distance downstream is $30\frac{\text{mi}}{\text{hr}}(1\ \text{hr}) = 30$ mi. Distances are equal. Therefore, the speed of the boat in still water is $22\frac{1}{2}$ mph, and the speed of the river is $7\frac{1}{2}$ mph.

Cumulative Review Exercises: Chapters 1-11 (page 608)

1. $$\frac{x}{y(x-y)} + \frac{y}{x(y-x)} = \frac{x}{y(x-y)}\cdot\frac{x}{x} + \frac{-y}{x(x-y)}\cdot\frac{y}{y} = \frac{x^2 - y^2}{xy(x-y)} = \frac{(x+y)(x-y)}{xy(x-y)} = \frac{x+y}{xy}$$

2. $x^2 - 2x + 3 + \dfrac{29}{x-3}$

3. $(2^2x^2)^{1/6} = 2^{2/6}x^{2/6} = 2^{1/3}x^{1/3} = \sqrt[3]{2x}$ **4.** $\dfrac{|b|}{4|a|}\sqrt{6b}$

5. $m = \dfrac{4-2}{2-(-1)} = \dfrac{2}{3}$ **6.** $\{x \mid 2 < x < 6\}$

$y - y_1 = m(x - x_1)$
$y - 2 = \frac{2}{3}[x - (-1)]$
$3(y - 2) = 2(x + 1)$
$3y - 6 = 2x + 2$
$0 = 2x - 3y + 8$

7. $4x^2 = 25$ **8.** $\{5, 7\}$
$x^2 = \frac{25}{4}$
$x = \pm\sqrt{\frac{25}{4}}$
$x = \pm\frac{5}{2}$
$\{\frac{5}{2}, -\frac{5}{2}\}$

9. $x^2 - 4x + 5 = 0 \begin{cases} a = 1 \\ b = -4 \\ c = 5 \end{cases}$

$$x = \frac{-(-4) \pm \sqrt{(-4)^2 - 4(1)(5)}}{2(1)} = \frac{4 \pm \sqrt{-4}}{2} = \frac{4 \pm 2i}{2} = 2 \pm i$$

$\{2 + i, 2 - i\}$

10. No **11.** $\frac{3}{5}$ **12.** $-\frac{5}{3}$

13. $-2]$ $4x + 3y = 8 \Rightarrow -8x - 6y = -16$ **14.** Inconsistent
$8x + 7y = 12 \Rightarrow 8x + 7y = 12$
$y = -4$

$4x + 3(-4) = 8$
$4x - 12 = 8$
$4x = 20$
$x = 5$
$(5, -4)$

15.
(1) $2x + 3y + z = 4$
(2) $x + 4y - z = 0$
(3) $3x + y - z = -5$
(1) + (2) $3x + 7y = 4$ (4)
(1) + (3) $5x + 4y = -1$ (5)
5] $3x + 7y = 4 \Rightarrow 15x + 35y = 20$ (6)
$-3]$ $5x + 4y = -1 \Rightarrow -15x - 12y = 3$ (7)
(6) + (7) $23y = 23$
$y = 1$

Substituting 1 for y in (4), we have $3x + 7(1) = 4$
$3x = -3$
$x = -1$

Substituting 1 for y and -1 for x in (1), we have
$2(-1) + 3(1) + z = 4$
$-2 + 3 + z = 4$
$z = 3$

(The check is left to the student.) The solution is $(-1, 1, 3)$.

16.

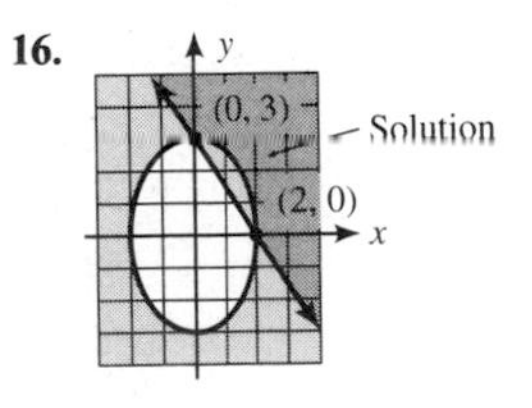

17. We must solve the inequality $12 - 4x - x^2 \geq 0$. Let $y = f(x) = (6 + x)(2 - x) = 0$. Critical points are -6 and 2.

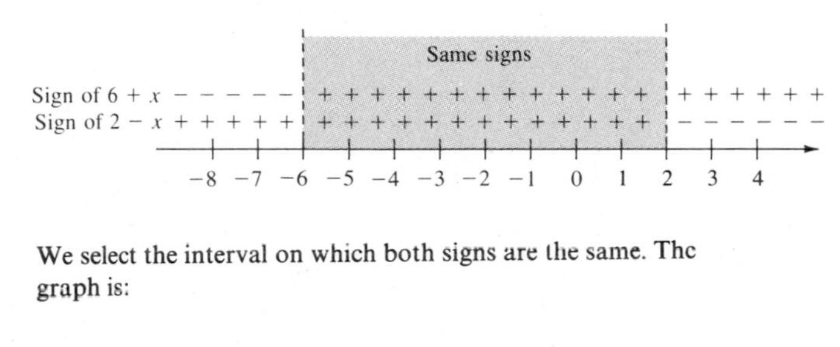

We select the interval on which both signs are the same. The graph is:

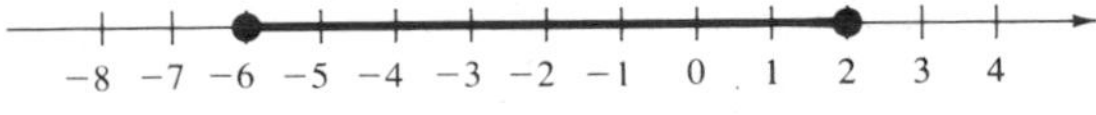

Solution set: $\{x \mid -6 \leq x \leq 2\}$

18. Length: 12; width: 5

19. Let x = number of hours for Jeannie to do the job alone
$x + 2$ = number of hours for Darryl to do the job alone

Then $\frac{1}{x}$ = Jeannie's rate (in jobs per hour)

$\frac{1}{x+2}$ = Darryl's rate (in jobs per hour)

$$\left(\frac{1}{x}\right)(3) + \left(\frac{1}{x+2}\right)(4) = 1$$
$$\frac{3}{x} + \frac{4}{x+2} = 1$$
$$x(x+2)\left(\frac{3}{x} + \frac{4}{x+2}\right) = 1(x)(x+2)$$
$$3(x+2) + 4x = x^2 + 2x$$
$$3x + 6 + 4x = x^2 + 2x$$
$$7x + 6 = x^2 + 2x$$
$$0 = x^2 - 5x - 6$$
$$0 = (x-6)(x+1)$$

$x - 6 = 0$	$x + 1 = 0$
$x = 6$	$x = -1$ (Not in the domain)
$x + 2 = 8$	

It would take 8 hr for Darryl and 6 hr for Jeannie to do the job alone.

20. Barbara's average speed: 9 mph; Pat's average speed: 12 mph

21. Let x = number of liters of pure alcohol.
There will be $(x + 3)\ell$ of the mixture.
$$(1.00)x + 0.50(3) = 0.70(x+3)$$
$$10x + 5(3) = 7(x+3)$$
$$10x + 15 = 7x + 21$$
$$3x = 6$$
$$x = 2$$
(The check will not be shown.) 2ℓ of pure alcohol should be added.

Exercises 12.1 (page 613)

1. 25, 30, 35 (Add 5 to preceding term.) **2.** 17, 20, 23

3. 9, 7, 5 (Add -2 to preceding term.) **4.** $2, \frac{7}{3}, \frac{8}{3}$

5. $a_n = n + 4$
$a_1 = 1 + 4 = 5$
$a_2 = 2 + 4 = 6$
$a_3 = 3 + 4 = 7$
The first three terms are 5, 6, 7.

6. 3, 5, 7, 9

7. $a_n = \frac{1-n}{n}$
$a_1 = \frac{1-(1)}{(1)} = 0$
$a_2 = \frac{1-(2)}{(2)} = -\frac{1}{2}$
$a_3 = \frac{1-(3)}{(3)} = -\frac{2}{3}$
$a_4 = \frac{1-(4)}{(4)} = -\frac{3}{4}$
The first four terms are $0, -\frac{1}{2}, -\frac{2}{3}, -\frac{3}{4}$.

8. 0, 1, 3

9. $a_n = n^2 - 1$
$a_1 = 1^2 - 1 = 0$
$a_2 = 2^2 - 1 = 3$
$a_3 = 3^2 - 1 = 8$
The first three terms are 0, 3, 8.

10. 2, 9, 28, 65

11. $a_n = 2n - 3$
$a_1 = 2(1) - 3 = -1$
$a_2 = 2(2) - 3 = 1$
$a_3 = 2(3) - 3 = 3$
$a_4 = 2(4) - 3 = 5$
$S_4 = a_1 + a_2 + a_3 + a_4$
$= -1 + 1 + 3 + 5 = 8$

12. $10\frac{3}{4}$

13. $a_n = \frac{n-1}{n+1}$
$a_1 = \frac{1-1}{1+1} = 0$
$a_2 = \frac{2-1}{2+1} = \frac{1}{3}$
$a_3 = \frac{3-1}{3+1} = \frac{1}{2}$
$S_3 = 0 + \frac{1}{3} + \frac{1}{2} = \frac{5}{6}$

14. 45 **15.** $\frac{1}{7} + \frac{2}{7} + \frac{3}{7} + \frac{4}{7} + \frac{5}{7}$ **16.** $\frac{1}{11} + \frac{2}{11} + \frac{3}{11} + \frac{4}{11} + \frac{5}{11}$

17. $(2 \cdot 1 + 1)^2 + (2 \cdot 2 + 1)^2 + (2 \cdot 3 + 1)^2 + \cdots = 3^2 + 5^2 + 7^2 + \cdots$

18. $3^3 + 4^3 + 5^3 + \cdots$

Exercises 12.2 (page 618)

1. 3, 8, 13, 18
$18 - 13 = 5$
$13 - 8 = 5$
$8 - 3 = 5$
AP, because all the differences are 5

2. AP; $d = 4$

3. 7, 4, 1, -2
$-2 - 1 = -3$
$1 - 4 = -3$
$4 - 7 = -3$
AP, because all the differences are -3

4. AP; $d = -5$

5. 4, $5\frac{1}{2}$, 7, 9
$9 - 7 = 2$
$7 - 5\frac{1}{2} = 1\frac{1}{2}$
Not an AP, because the differences are not all the same

6. Not an AP

7. $2x - 1, x, 1, -x + 2$
$-x + 2 - 1 = -x + 1$
$1 - x = -x + 1$
$x - (2x - 1) = -x + 1$
AP, because all the differences are $-x + 1$

8. AP; $d = x - 1$

9. $a_1, a_1 + d, a_1 + 2d, a_1 + 3d$
$5, 5 + (-7), 5 + 2(-7), 5 + 3(-7)$
$5, -2, -9, -16$

10. $4, -1, -6, -11, -16$

11. $a_5 = a_1 + (5 - 1)d$
$31 = 7 + 4d \Rightarrow 24 = 4d \Rightarrow d = 6$
AP: 7, 7 + 6, 7 + 2(6), 7 + 3(6), 7 + 4(6)
7, 13, 19, 25, 31

12. 6, 15, 24, 33, 42, 51

13. $-8, -2, 4, \ldots$
$4 - (-2) = 6 = d$
$-2 - (-8) = 6$
$a_{31} = a_1 + (31 - 1)d$
$a_{31} = -8 + 30(6) = 172$

14. 155

15. $x, 2x + 1, 3x + 2, \ldots$
$d = (2x + 1) - x = x + 1$
$a_{11} = a_1 + (11 - 1)d = x + 10(x + 1)$
$a_{11} = 11x + 10$

16. $10z - 7$

17. $2 + 4 + \cdots + 100$
$S_{50} = \frac{50}{2}(2 + 100) = 2{,}550$

18. 2,500

19. $a_1 = 2$, $a_{100} = 200$, and $n = 100$;
$S_{100} = \frac{\overset{50}{\cancel{100}}(2 + 200)}{\underset{1}{\cancel{2}}} = 10{,}100$

20. 2,500

21. $a_1 = 5(3) = 15$, $a_8 = 5(8) = 40$, and $n = 8 - 3 + 1 = 6$;

$$S_6 = \frac{\overset{3}{\cancel{6}}(15 + 40)}{\underset{1}{\cancel{2}}} = 165$$

22. 270 **23.**

$$\begin{aligned} a_6 &= a_1 + (6-1)d \Rightarrow a_1 + 5d = 15 \quad (1)\\ a_{12} &= a_1 + (12-1)d \Rightarrow a_1 + 11d = 39 \quad (2)\\ & 6d = 24\\ & d = 4 \end{aligned}$$

Substituting 4 for d in (1):

$a_1 + 5(4) = 15$
$a_1 + 20 = 15$
$a_1 = -5$

24. $d = 5$
$a_1 = -8$

25. $16 = -5 + (n-1)(3)$
$21 = 3n - 3 \Rightarrow 3n = 24 \Rightarrow n = 8$
$S_8 = \frac{8}{2}(-5 + 16) = 4(11) = 44$

26. $n = 9$
$S_9 = 81$

27. $S_n = \frac{n}{2}(a_1 + a_n)$
$44 = \frac{n}{2}(5 + 17)$
$44 = \frac{n}{2}(22) \Rightarrow n = 4$
$a_4 = 5 + 3d = 17$
$3d = 12 \Rightarrow d = 4$

28. $n = 8$
$d = \frac{39}{7}$

29. $a_n = a_1 + 8(\frac{3}{2}) \Rightarrow a_n = a_1 + 12$
$S_n = \frac{n}{2}(a_1 + a_n)$
$-\frac{9}{4} = \frac{9}{2}(a_1 + a_n)$
Substituting $a_1 + 12$ for a_n:
$-\frac{9}{4} = \frac{9}{2}(a_1 + a_1 + 12)$
$-\frac{9}{4} = \frac{9}{2}(2a_1 + 12)$
$-\frac{9}{4} = 9a_1 + 54$
$-9 = 36a_1 + 216 \Rightarrow a_1 = -\frac{25}{4}$
Substituting $-\frac{25}{4}$ for a_1 in $a_n = a_1 + 12$:
$a_n = -\frac{25}{4} + \frac{48}{4} = \frac{23}{4}$

30. $a_1 = -\frac{3}{2}$
$a_n = a_7 = 3$

31. $\left.\begin{aligned} a_1 &= 16\\ a_2 &= 48 \end{aligned}\right\} d = 48 - 16 = 32$
$a_3 = 80$
$\vdots$
$a_{10} = a_1 + 9d = 16 + 9(32) = 16 + 288 = 304$
$a_{12} = 304 + 64 = 368$
$S_{12} = \frac{12}{2}(16 + 368) = 6(384) = 2,304$
The rock falls 304 ft during the tenth second and 2,304 ft during the first 12 sec.

32. \$12.40

Exercises 12.3 (page 624)

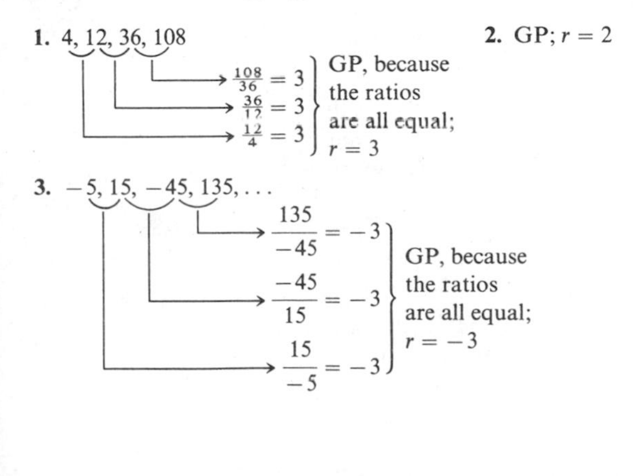

2. GP; $r = 2$

4. GP; $r = -4$

5. $2, \frac{1}{2}, \frac{1}{8}, \frac{1}{16}$

$\frac{1}{16} \div \frac{1}{8} = \frac{1}{16} \cdot \frac{8}{1} = \frac{1}{2}$
$\frac{1}{8} \div \frac{1}{2} = \frac{1}{8} \cdot \frac{2}{1} = \frac{1}{4}$
Not a GP, because all ratios are not equal

6. Not a GP

7. $5x, 10xy, 20xy^2, 40xy^3, \ldots$

$\frac{40xy^3}{20xy^2} = 2y$
$\frac{20xy^2}{10xy} = 2y$
$\frac{10xy}{5x} = 2y$
GP, because the ratios are all equal; $r = 2y$

8. GP; $r = 3z$

9. $a_1 = 12$
$a_2 = 12(\frac{1}{3}) = 4$
$a_3 = 4(\frac{1}{3}) = \frac{4}{3}$
$a_4 = \frac{4}{3}(\frac{1}{3}) = \frac{4}{9}$
$a_5 = \frac{4}{9}(\frac{1}{3}) = \frac{4}{27}$
$12, 4, \frac{4}{3}, \frac{4}{9}, \frac{4}{27}$

10. 8, 12, 18, 27

11. $a_5 = a_1r^4$
$54 = \frac{2}{3}r^4$
$81 = r^4$
$r = \pm 3$
There are two answers.
One answer: $r = 3$
$\frac{2}{3}, 2, 6, 18, 54$
Second answer: $r = -3$
$\frac{2}{3}, -2, 6, -18, 54$

12. *One answer*: $r = 5$
$\frac{3}{25}, \frac{3}{5}, 3, 15, 75$
Second answer: $r = -5$
$\frac{3}{25}, -\frac{3}{5}, 3, -15, 75$

13. $-9, -6, -4, \ldots$
$r = \frac{-4}{-6} = \frac{2}{3}$
$a_7 = a_1r^6 = -9(\frac{2}{3})^6 = -\frac{64}{81}$

14. $-\frac{6,561}{32}$

15. $16x, 8xy, 4xy^2, \ldots$
$\frac{8xy}{16x} = \frac{y}{2} = r$
$a_8 = a_1r^7 = 16x\left(\frac{y}{2}\right)^7 = \frac{xy^7}{8}$

16. $\frac{x^{17}y}{27}$

17. $a_1 = 3$, $r = \frac{1}{3}$, and $n = 5$;

$$S_5 = \frac{3[1 - (\frac{1}{3})^5]}{1 - \frac{1}{3}} = \frac{3(1 - \frac{1}{243})}{\frac{2}{3}} = \frac{3(\frac{242}{243})}{\frac{2}{3}} = \frac{242}{81} \cdot \frac{3}{2} = \frac{121}{27}$$

18. $\frac{635}{64}$

19. $a_5 = a_1r^4$
$80 = a_1(\frac{2}{3})^4$
$80 = \frac{16}{81}a_1$
$a_1 = 405$
$S_5 = \frac{a_1(1 - r^5)}{1 - r}$

$$= \frac{405[1 - (\frac{2}{3})^5]}{1 - \frac{2}{3}} = \frac{405[1 - \frac{32}{243}]}{\frac{1}{3}} = \frac{405}{1}\left(\frac{211}{243}\right)\frac{3}{1} = 1,055$$

20. $a_1 = 5$
$S_5 = 155$

21. (2) $a_5 = a_1 r^4 = \frac{112}{9}$
(1) $a_3 = a_1 r^2 = 28$

Dividing (2) by (1): $r^2 = \dfrac{112}{9} \div 28 = \dfrac{112}{9} \cdot \dfrac{1}{28} = \dfrac{4}{9}$

$r = \pm\frac{2}{3}$

One solution: $r_1 = \frac{2}{3}$:

(1) $a_1(\frac{2}{3})^2 = 28$

$\frac{4}{9}a_1 = 28$

$a_1 = 63$

$$S_5 = \frac{63[1-(\frac{2}{3})^5]}{1-\frac{2}{3}} = \frac{63-\frac{224}{27}}{\frac{1}{3}} = \frac{1{,}477}{27}\cdot\frac{3}{1} = \frac{1{,}477}{9}$$

The other solution: $r_2 = -\frac{2}{3}$:

(1) $a_1(-\frac{2}{3})^2 = 28$

$a_1 = 63$

$$S_5 = \frac{63[1-(-\frac{2}{3})^5]}{1-(-\frac{2}{3})} = \frac{63-(-\frac{224}{27})}{\frac{5}{3}} = \frac{1{,}925}{27}\cdot\frac{3}{5} = \frac{385}{9}$$

22. One solution: $r_1 = \frac{1}{4}$, $a_1 = 1{,}536$, $S_4 = 2{,}040$
The other solution: $r_2 = -\frac{1}{4}$, $a_1 = -1{,}536$, $S_4 = -1{,}224$

23. $a_n = a_1 r^{n-1}$

$3 = a_1(\frac{1}{2})^{n-1}$

(1) $\frac{3}{2} = a_1(\frac{1}{2})^n$

$$S_n = \frac{a_1(1-r^n)}{1-r}$$

$$189 = \frac{a_1[1-(\frac{1}{2})^n]}{1-\frac{1}{2}} = \frac{a_1 - a_1(\frac{1}{2})^n}{\frac{1}{2}}$$

$$189 = \frac{a_1 - \frac{3}{2}}{\frac{1}{2}} = 2a_1 - 3 \quad (\leftarrow \text{From (1)})$$

$192 = 2a_1$

$a_1 = 96$

Substituting 96 for a_1 in (1):

$\frac{3}{2} = 96(\frac{1}{2})^n$

$(\frac{1}{2})^6 = (\frac{1}{2})^n$ $\quad \frac{3}{2} \div 96 = \frac{1}{64} = (\frac{1}{2})^6$

$6 = n$

24. $a_1 = 108$
$n = 4$

25. $$S_3 = \frac{a_1(1-[\frac{6}{5}]^3)}{1-\frac{6}{5}}$$

$= a_1(-\frac{91}{125})(-\frac{5}{1}) = \frac{91}{25}a_1$

$\frac{91}{25}a_1 = 22{,}750$

$a_1 = \$6{,}250$

$a_5 = a_1 r^4$
$= 6{,}250(\frac{6}{5})^4 = 6{,}250(\frac{1{,}296}{625})$
$= \$12{,}960$

26. \$6,250

27. $a_n = a_1 r^{n-1}$

a. $a_{10} = 1(2)^9 = 512¢ = \$5.12$

b. $a_{31} = 1(2)^{30} = 1{,}073{,}741{,}824¢ = \$10{,}737{,}418.24$
or $\log a_{31} = 30 \log 2 \approx 30(0.3010) = 9.0300$
$a_{31} = \text{antilog } 9.0300 \approx 1{,}072{,}000{,}000¢ = \$10{,}720{,}000$

c. $$S_{31} = \frac{a_1(1-r^n)}{1-r} = \frac{1(1-2^{31})}{1-2} = 2^{31}-1$$
$= 2{,}147{,}483{,}646¢ = \$21{,}474{,}836.46$
If logs are used: \$21,430,000

Exercises 12.4 (page 628)

1. $3 + 1 + \frac{1}{3} + \cdots$ $\quad \frac{1}{3} = r$

$$S_\infty = \frac{a_1}{1-r} = \frac{3}{\frac{2}{3}} = \frac{9}{2}$$

2. $\frac{81}{10}$

3. $\frac{4}{3} + 1 + \frac{3}{4} + \cdots$ $\quad \frac{\frac{3}{4}}{1} = \frac{3}{4} = r$

$$S_\infty = \frac{a_1}{1-r} = \frac{\frac{4}{3}}{1-\frac{3}{4}} = \frac{\frac{4}{3}}{\frac{1}{4}} = \frac{16}{3}$$

4. $\frac{1}{9}$

5. $-6 - 4 - \frac{8}{3} - \cdots$ $\quad \frac{-4}{-6} = \frac{2}{3} = r$

$$S_\infty = \frac{a_1}{1-r} = \frac{-6}{1-\frac{2}{3}} = \frac{-6}{\frac{1}{3}} = -18$$

6. $-\frac{343}{2}$

7. $0.\bar{2} = 0.2222\ldots$
$= 0.2 + 0.02 + 0.002 + 0.0002 + \cdots$ $\quad \frac{0.02}{0.2} = 0.1$

$$S_\infty = \frac{a_1}{1-r} = \frac{0.2}{1-0.1} = \frac{0.2}{0.9} = \frac{2}{9}$$

8. $\frac{7}{33}$

9. $0.0\overline{54} = 0.05454\ldots$
$= 0.054 + 0.00054 + \cdots$ $\quad \frac{0.00054}{0.054} = 0.01 = r$

$$S_\infty = \frac{a_1}{1-r} = \frac{0.054}{1-0.01} = \frac{0.054}{0.99} = \frac{54}{990} = \frac{3}{55}$$

10. $\frac{13}{330}$

11. $8.6\bar{4} = 8.6444\ldots$
$= 8.6 + 0.04 + 0.004 + 0.0004 + \cdots$

$$r = 0.1,\ S_\infty = \frac{0.04}{1-0.1} = \frac{0.04}{0.9} = \frac{4}{90} = \frac{2}{45}$$

$8.6444\ldots = \frac{8}{1} + \frac{6}{10} + \frac{2}{45}$
$= \frac{720}{90} + \frac{54}{90} + \frac{4}{90} = \frac{778}{90} = \frac{389}{45}$

12. $\frac{79}{15}$

13. $a_1 = 5$, $r = -\frac{3}{4}$, and $|-\frac{3}{4}| < 1$; $\quad S_\infty = \dfrac{5}{1-(-\frac{3}{4})} = \dfrac{5}{\frac{7}{4}} = \dfrac{20}{7}$

14. $\frac{5}{3}$

15. 9 ft 6 ...

The geometric series of the *heavy* lines is

$6 + 4 + \frac{8}{3} + \cdots$

$$S_\infty = \frac{a_1}{1-r} = \frac{6}{1-\frac{2}{3}} = \frac{6}{\frac{1}{3}} = 18$$

This distance is doubled to include all but the first drop: $2(18) = 36$
The total distance traveled = 9 ft + 36 ft = 45 ft

16. 40 ft

17. 12 in.

From the sketch, you see that we have a situation that is different from the bouncing ball described in Exercise 15. This is a single geometric series.

$$S_\infty = \frac{12}{1-\frac{9}{10}} = \frac{12}{\frac{1}{10}} = 120$$

The total distance is 120 in.

Review Exercises 12.5 (page 630)

1. $7, 5, 3, \ldots$
$3 - 5 = -2$
$5 - 7 = -2$
AP, because -2 is the common difference

2. GP

3. $\frac{1}{2}, \frac{1}{4}, \frac{1}{6}, \ldots$
$\frac{1}{6} - \frac{1}{4} = -\frac{1}{12}$
$\frac{1}{4} - \frac{1}{2} = -\frac{1}{4}$
Therefore, *not* an AP

$\frac{1}{2}, \frac{1}{4}, \frac{1}{6}, \ldots$
$\frac{\frac{1}{6}}{\frac{1}{4}} = \frac{2}{3}$
$\frac{\frac{1}{4}}{\frac{1}{2}} = \frac{1}{2}$
Therefore, *not* a GP

Neither

4. GP

5. $3x - 2, 2x - 1, x$
$(2x-1) - (3x-2) = -x + 1$
$x - (2x-1) = -x + 1$
AP

6. $-4, -2, 0, 2$
$d = 2$
$a_{30} = 54$
$S_{30} = 750$

7. $a_n = (\frac{1}{2})^n$
$a_1 = (\frac{1}{2})^1 = \frac{1}{2}$
$a_2 = (\frac{1}{2})^2 = \frac{1}{4}$
$a_3 = (\frac{1}{2})^3 = \frac{1}{8}$

$$S_n = \frac{a_1(1 - r^n)}{1 - r}$$

$$S_5 = \frac{\frac{1}{2}(1 - \frac{1}{32})}{\frac{1}{2}} = \frac{31}{32}$$

$$S_\infty = \frac{a_1}{1 - r} = \frac{\frac{1}{2}}{1 - \frac{1}{2}} = 1$$

$\frac{1}{2}, \frac{1}{4}, \frac{1}{8}$

8. $n = 16$
$d = \frac{4}{5}$

9. $a_n = a_1 + 6(\frac{3}{2}) = a_1 + 9$
$S_n = \frac{n(a_1 + a_n)}{2}$
$\frac{7}{2} = \frac{7(2a_1 + 9)}{2}$
$1 = 2a_1 + 9$
$-8 = 2a_1$
$-4 = a_1$
$a_n = a_7 = -4 + 9 = 5$

10. $a_1 = \frac{81}{16}$
$S_6 = \frac{133}{48}$

11. $a_5 = \frac{32}{9} = a_1r^4$ (1)
$a_3 = 8 = a_1r^2$ (2)
$\frac{4}{9} = r^2$ Dividing (1) by (2)
$\pm\frac{2}{3} = r$
If $r = \frac{2}{3}$: $a_3 = a_1r^2$
$8 = a_1 \cdot \frac{4}{9}$
$18 = a_1$

$$S_5 = \frac{a_1(1 - r^5)}{1 - r} = \frac{18[1 - (\frac{2}{3})^5]}{1 - \frac{2}{3}} = \frac{18(1 - \frac{32}{243})}{\frac{1}{3}} = \frac{18}{1}\left(\frac{211}{243}\right) \cdot \frac{3}{1}$$
$= \frac{422}{9}$

If $r = -\frac{2}{3}$: $a_3 = a_1r^2$
$8 = a_1 \cdot \frac{4}{9}$
$18 = a_1$

$$S_5 = \frac{18[1 - (-\frac{2}{3})^5]}{1 - (-\frac{2}{3})} = \frac{18(1 + \frac{32}{243})}{\frac{5}{3}} = \frac{18}{1}\left(\frac{275}{243}\right) \cdot \frac{3}{5} = \frac{110}{9}$$

12. $\frac{32,813}{9,990}$

13. 8 ft, 6, ... Heavy lines: $6 + \frac{9}{2} + \cdots$

$$S_\infty = \frac{6}{1 - \frac{3}{4}} = \frac{6}{\frac{1}{4}} = 24$$

Total distance = 8 ft + 2(24 ft) = 56 ft

14. $\frac{9}{11} + \frac{12}{11} + \frac{15}{11} + \frac{18}{11} + \frac{21}{11}$

15. The series is geometric; $a_1 = 2$, $r = \frac{1}{4}$, and $n = 6$;

$$S_6 = \frac{2[1 - (\frac{1}{4})^6]}{1 - \frac{1}{4}} = \frac{2(1 - \frac{1}{4,096})}{\frac{3}{4}} = \frac{2(\frac{4,095}{4,096})}{\frac{3}{4}} = \frac{4,095}{2,048} \cdot \frac{4}{3} = \frac{1,365}{512}$$

Chapter 12 Diagnostic Test (page 635)

Following each problem number is the textbook section number (in parentheses) where that kind of problem is discussed.

1. (12.1) $a_n = \frac{2n - 1}{n}$ (The sequence is not arithmetic and it is not geometric.)

$$S_4 = \frac{2(1) - 1}{1} + \frac{2(2) - 1}{2} + \frac{2(3) - 1}{3} + \frac{2(4) - 1}{4}$$
$$= 1 + \frac{3}{2} + \frac{5}{3} + \frac{7}{4} = \frac{71}{12}$$

2. (12.2, 12.3) **a.** $8, -20, 50, \cdots$

$\frac{50}{-20} = -\frac{5}{2}$, $\frac{-20}{8} = -\frac{5}{2}$ } GP, $r = -\frac{5}{2}$

b. $\frac{1}{2}, \frac{3}{4}, 1, \frac{5}{4}, \frac{3}{2}, \ldots$

$1 - \frac{3}{4} = \frac{1}{4}$, $\frac{3}{4} - \frac{1}{2} = \frac{1}{4}$ } AP, $d = \frac{1}{4}$ (Remaining differences are also $\frac{1}{4}$.)

c. $2x - 1, 3x, 4x + 2, \ldots$

$4x + 2 - 3x = x + 2$, $3x - (2x - 1) = x + 1$ } Therefore, *not* an AP

$\frac{4x + 2}{3x} \neq \frac{3x}{2x - 1}$ } Therefore, *not* a GP

Neither

d. $\frac{c^4}{16}, -\frac{c^3}{8}, \frac{c^2}{4}, \cdots$

$\frac{c^2}{4} \div \left(-\frac{c^3}{8}\right) = \frac{c^2}{4} \cdot \left(-\frac{8}{c^3}\right) = -\frac{2}{c}$

$-\frac{c^3}{8} \div \frac{c^4}{16} = -\frac{c^3}{8} \cdot \frac{16}{c^4} = -\frac{2}{c}$ } GP, $r = -\frac{2}{c}$

3. (12.2)
a. $x + 1, (x + 1) + (x - 1), (x + 1) + 2(x - 1),$
$(x + 1) + 3(x - 1), (x + 1) + 4(x - 1)$
$= x + 1, 2x, 3x - 1, 4x - 2, 5x - 3$

b. $a_5 = a_1 + 4d$
$-2 = 2 + 4d$
$-4 = 4d$
$-1 = d$

$2, 2 + (-1), 2 + 2(-1), 2 + 3(-1), 2 + 4(-1)$
$= 2, 1, 0, -1, -2$

c. $1 - 6h, 2 - 4h, 3 - 2h, \ldots$

$(2 - 4h) - (1 - 6h) = 1 + 2h = d$

$a_{15} = a_1 + 14d = (1 - 6h) + 14(1 + 2h) = 15 + 22h$

4. (12.2) **a.** $a_7 = a_1 + 6d \Rightarrow 2 = a_1 + 6d$ (1)
$a_3 = a_1 + 2d \Rightarrow 1 = a_1 + 2d$ (2)
$1 = 4d$ Subtracting (2) from (1)
$\frac{1}{4} = d$
Substituting $\frac{1}{4}$ for d in (2):
$1 = a_1 + 2\left(\frac{1}{4}\right)$
$1 = a_1 + \frac{1}{2}$
$\frac{1}{2} = a_1$

b. $S_n = \frac{n}{2}(a_1 + a_n)$
$7 = \frac{n}{2}(10 - 8)$
$7 = n$

$a_n = a_1 + (n - 1)d$
$-8 = 10 + 6d$
$-18 = 6d$
$-3 = d$

5. (12.3) $\frac{c^4}{16}, \frac{c^4}{16}\left(-\frac{2}{c}\right), \frac{c^4}{16}\left(-\frac{2}{c}\right)^2, \frac{c^4}{16}\left(-\frac{2}{c}\right)^3, \frac{c^4}{16}\left(-\frac{2}{c}\right)^4$
$= \frac{c^4}{16}, -\frac{c^3}{8}, \frac{c^2}{4}, -\frac{c}{2}, 1$

6. (12.3) $a_4 = a_1r^3 \Rightarrow -8 = a_1r^3$ (1)
$a_2 = a_1r \Rightarrow -18 = a_1r$ (2)
$\frac{8}{18} = r^2$ Dividing (1) by (2)
$\pm\frac{2}{3} = r$

If $r = \frac{2}{3}$:
Substituting $\frac{2}{3}$ for r in (2):
$-18 = a_1(\frac{2}{3}) \Rightarrow a_1 = -27$
$S_4 = \frac{a_1(1 - r^4)}{1 - r} = \frac{-27(1 - \frac{16}{81})}{1 - (\frac{2}{3})}$
$S_4 = \frac{-27(\frac{65}{81})}{\frac{1}{3}} = -65$

If $r = -\frac{2}{3}$:
Substituting $-\frac{2}{3}$ for r in (2):
$-18 = a_1(-\frac{2}{3}) \Rightarrow a_1 = 27$
$S_4 = \frac{a_1(1 - r^4)}{1 - r} = \frac{27(1 - \frac{16}{81})}{1 - (-\frac{2}{3})}$
$S_4 = \frac{27(\frac{65}{81})}{\frac{5}{3}} = \frac{65}{5} = 13$

7. (12.3)

$$a_n = a_1 r^{n-1}$$
$$a_n r = a_1 r^n$$
$$-16(-\tfrac{2}{3}) = a_1(-\tfrac{2}{3})^n$$
(1) $$\tfrac{32}{3} = a_1(-\tfrac{2}{3})^n$$
$$S_n = \frac{a_1(1-r^n)}{1-r}$$
$$26 = \frac{a_1[1-(-\tfrac{2}{3})^n]}{1-(-\tfrac{2}{3})}$$

From (1)

$$26 = \frac{a_1 - a_1(-\tfrac{2}{3})^n}{\tfrac{5}{3}} = \frac{a_1 - \tfrac{32}{3}}{\tfrac{5}{3}}$$
$$26 = \frac{3a_1 - 32}{5}$$
$$130 = 3a_1 - 32$$
$$54 = a_1$$

Substituting 54 for a_1 in (1):

$$\tfrac{32}{3} = 54(-\tfrac{2}{3})^n$$
$$\tfrac{16}{81} = (-\tfrac{2}{3})^n \Rightarrow (\tfrac{2}{3})^4 = (-\tfrac{2}{3})^n$$
$$4 = n$$

8. (12.4) $8, -4, 2, -1, \ldots$

$$\frac{-1}{2} = -\frac{1}{2}, \quad \frac{2}{-4} = -\frac{1}{2}, \quad \frac{-4}{8} = -\frac{1}{2} \quad \text{GP} \quad r = -\frac{1}{2}$$

$$S_\infty = \frac{a_1}{1-r} = \frac{8}{1-(-\tfrac{1}{2})} = \frac{8}{\tfrac{3}{2}} = \frac{16}{3}$$

9. (12.4) $3.0\overline{3} = 3.0333\ldots = 3 + 0.0333\ldots$

$$0.0333\ldots = 0.03 + 0.003 + 0.0003 + \cdots$$
$$\frac{0.0003}{0.003} = \frac{1}{10} = r$$
$$0.0333\ldots = \frac{0.03}{1-\tfrac{1}{10}} = \frac{0.03}{0.9} = \frac{3}{90} = \frac{1}{30}$$
$$3.0333\ldots = 3 + \tfrac{1}{30} = \tfrac{91}{30}$$

10. (12.4) 6 ft, 4 ft

The series of heavy lines:

$$S_\infty = \frac{4}{1-\tfrac{2}{3}} = \frac{4}{\tfrac{1}{3}} = 12$$

Total distance = 6 ft + 2(12 ft) = 6 ft + 24 ft = 30 ft

Cumulative Review Exercises: Chapters 1–12 (page 636)

1. $\left(\dfrac{x^5z^2}{y^4}\right)^{-2} = \left(\dfrac{y^4}{x^5z^2}\right)^2 = \dfrac{y^8}{x^{10}z^4}$ **2.** $\dfrac{1}{64}a^{1/5}$

3. $\sqrt{27x^3y^4} + 3xy^2\sqrt{12x} - x^2y^2\sqrt{\dfrac{3\cdot x}{x\cdot x}}$

$$= 3xy^2\sqrt{3x} + 3xy^2(2)\sqrt{3x} - \frac{x^2y^2}{x}\sqrt{3x}$$
$$= 3xy^2\sqrt{3x} + 6xy^2\sqrt{3x} - xy^2\sqrt{3x} = 8xy^2\sqrt{3x}$$

4. $3(\sqrt{7} + \sqrt{2})$, or $3\sqrt{7} + 3\sqrt{2}$

5. $\dfrac{(6-5i)(6-5i)}{(6+5i)(6-5i)} = \dfrac{36 - 60i + 25i^2}{36 + 25}$

$$= \frac{(36-25) - 60i}{61} = \frac{11}{61} - \frac{60}{61}i$$

6. $\dfrac{(x+1)(x-1)}{x^2 - 2x + 4}$

7. $\dfrac{x(2)(x+3) - 9(2)(x-3) - 1}{2(x-1)(x-3)}$

$$= \frac{2x^2 + 6x - 18x + 54 - 1}{2(x-3)(x+3)}$$
$$= \frac{2x^2 - 12x + 53}{2x^2 - 18}$$

8. $x^2 + 4x - 3 + \dfrac{-9x + 4}{x^2 - x + 1}$

9. $\dfrac{x+1}{(3x-1)(x+5)} \cdot \dfrac{(3x-1)(2x-3)}{(2x-3)(x+1)} = \dfrac{1}{x+5}$

10. $x^5 + 10x^4 + 40x^3 + 80x^2 + 80x + 32$

11. $(x^2 - 4)\left(\dfrac{6-x}{x^2-4} - \dfrac{x}{x+2}\right) = 2(x^2 - 4)$

$$6 - x - x(x-2) = 2x^2 - 8$$
$$6 - x - x^2 + 2x = 2x^2 - 8$$
$$0 = 3x^2 - x - 14$$
$$0 = (3x - 7)(x + 2)$$

$3x - 7 = 0$, $3x = 7$, $x = \tfrac{7}{3}$ | $x + 2 = 0$, $x = -2$ Not in the domain

(The check for $\tfrac{7}{3}$ will not be shown.) The solution set is $\{\tfrac{7}{3}\}$.

12. $\{3\}$

13.

$$x - 3 = \sqrt{x+9}$$
$$(x-3)^2 = (\sqrt{x+9})^2$$
$$x^2 - 6x + 9 = x + 9$$
$$x^2 - 7x = 0$$
$$x(x-7) = 0$$

$x = 0$ | $x - 7 = 0$, $x = 7$

Check for $x = 0$:

$$0 \stackrel{?}{=} \sqrt{0+9} + 3$$
$$0 \stackrel{?}{=} \sqrt{9} + 3$$
$$0 \neq 3 + 3$$

Check for $x = 7$:

$$7 \stackrel{?}{=} \sqrt{7+9} + 3$$
$$7 \stackrel{?}{=} \sqrt{16} + 3$$
$$7 = 4 + 3$$

The solution set is $\{7\}$.

14. $\{x \mid x \le -5 \text{ or } x \ge 3\}$ or, in interval notation, $(-\infty, -5] \cup [3, +\infty)$

15. a. $\log_5 \tfrac{1}{25} = x$

$$5^x = \tfrac{1}{25} = 5^{-2}$$
$$x = -2$$
$$\{-2\}$$

b. $\log_b 4 = \tfrac{1}{2}$

$$b^{1/2} = 4$$
$$(b^{1/2})^2 = 4^2$$
$$b = 16$$
$$\{16\}$$

(The checks will not be shown.)

16. $\{2 + 2\sqrt{3}, 2 - 2\sqrt{3}\}$

17. 3] $2x + 4y = 0 \Rightarrow 6x + 12y = 0$

4] $5x - 3y = 13 \Rightarrow 20x - 12y = 52$

$$26x = 52$$
$$x = 2$$

$$2x + 4y = 0$$
$$2(2) + 4y = 0$$
$$4 + 4y = 0$$
$$4y = -4$$
$$y = -1$$

$(2, -1)$

18. $(1, -2, 2)$

19. $\left|\dfrac{x}{2} + 3\right| < 4$

$$-4 < \frac{x}{2} + 3 < 4$$
$$\underline{-3} \qquad \underline{-3} < \underline{-3}$$
$$-7 < \frac{x}{2} < 1$$
$$-14 < x < 2$$

$\{x \mid -14 < x < 2\}$ or $(-14, 2)$

20. y, x, (−1, 0), (3, 0), (0, −3), (1, −4)

21. (1) $2x - 5y \geq 10$
(2) $5x + y < 5$
Boundary line for (1): $2x - 5y = 10$
Intercepts: $(0, -2)$ and $(5, 0)$
Boundary line for (2): $5x + y = 5$
Intercepts: $(0, 5)$ and $(1, 0)$

Origin is the correct half-plane for (2), but not for (1).

22. a. All real numbers except $\frac{5}{3}$

b. $f(x+h) = \dfrac{2}{3(x+h)-5}$ or $\dfrac{2}{3x+3h-5}$

23. $y = f(x) = \dfrac{2}{3x-5}$

$x = \dfrac{2}{3y-5}$

$x(3y-5) = 2$

$3xy - 5x = 2$

$3xy = 5x + 2$

$y = f^{-1}(x) = \dfrac{5x+2}{3x}$

24. a. $2\sqrt{26}$ **b.** $m = -5$

25. $y - (-4) = -5(x - 3)$

$y + 4 = -5x + 15$

$5x + y - 11 = 0$

26. $d = \frac{9}{2}; n = 5$

27. $a_4 = a_1 r^3$

$-1 = a_1(-\frac{2}{7})^3$

$a_1 = \frac{343}{8}$

$S_5 = \dfrac{a_1(1 - r^5)}{1 - r} = \dfrac{\frac{343}{8}[1 - (-\frac{2}{7})^5]}{1 - (-\frac{2}{7})} = \dfrac{\frac{343}{8}[1 + \frac{32}{16,807}]}{\frac{9}{7}}$

$= \dfrac{7}{9}\left(\dfrac{343}{8}\right)\left(\dfrac{16,839}{16,807}\right) = \dfrac{1,871}{56}$

28. Two answers: 6, 7, and 8 or -3, -2, and -1

29. Let x = number of pounds of cheaper coffee
y = number of pounds of more expensive coffee

$$\begin{cases} x + y = 30 & (1) \\ 3.50x + 4.25y = 4.00(30) & (2) \end{cases}$$

$-350x - 350y = -10,500$ Multiplying (1) by -350
$350x + 425y = 12,000$ Multiplying (2) by 100
$75y = 1,500$ Adding
$y = 20$
$x = 30 - y = 30 - 20 = 10$

(The check will not be shown.) 10 lb of the cheaper coffee and 20 lb of the more expensive coffee should be used.

30. John's rate is 9 mph, and Roberta's rate is 12 mph.

Exercises B.1 (page 644)

1. $3_{\wedge}86.$ Characteristic is 2. **2.** 1
3. $5_{\wedge}67$ Characteristic is 0. **4.** 1
5. $0.5_{\wedge}16$ Characteristic is -1, written as $9 - 10$.
6. -2, written as $8 - 10$ **7.** $9_{\wedge}3000000.$ Characteristic is 7.
8. 5 **9.** $0.00008_{\wedge}06$ Characteristic is -5, written as $5 - 10$.
10. -4, written as $6 - 10$ **11.** $7_{\wedge}8000.$ Characteristic is 4.
12. 3 **13.** 5 (same as exponent of 10) **14.** 4
15. -3 (same as exponent of 10), written as $7 - 10$
16. -5, written as $5 - 10$

Exercises B.2 (page 646)

1. $\log 7_{\wedge}54. \approx 2.8774$ **2.** 2.2695 **3.** $\log 1_{\wedge}7. \approx 1.2304$
4. 1.4624 **5.** $\log 3_{\wedge}350. \approx 3.5250$ **6.** 3.6637
7. $\log 7_{\wedge}000. \approx 3.8451$ **8.** 2.3010
9. $\log 0.06_{\wedge}04 \approx 8.7810 - 10$ **10.** $8.2695 - 10$
11. $\log(5.64 \times 10^3) \approx 3.7513$ **12.** $6.3304 - 10$

Exercises B.3 (page 647)

1. 1.2809 **2.** 1.6487 **3.** 2.0919 **4.** 2.0149
5. $\ln 8.3 + \ln 10 \approx 2.1163 + 2.3026 = 4.4189$ **6.** 4.1271
7. $\ln 0.2 - \ln 100 \approx (8.3906 - 10) - 4.6052 = -6.2146$
8. -5.1160

Exercises B.4 (page 649)

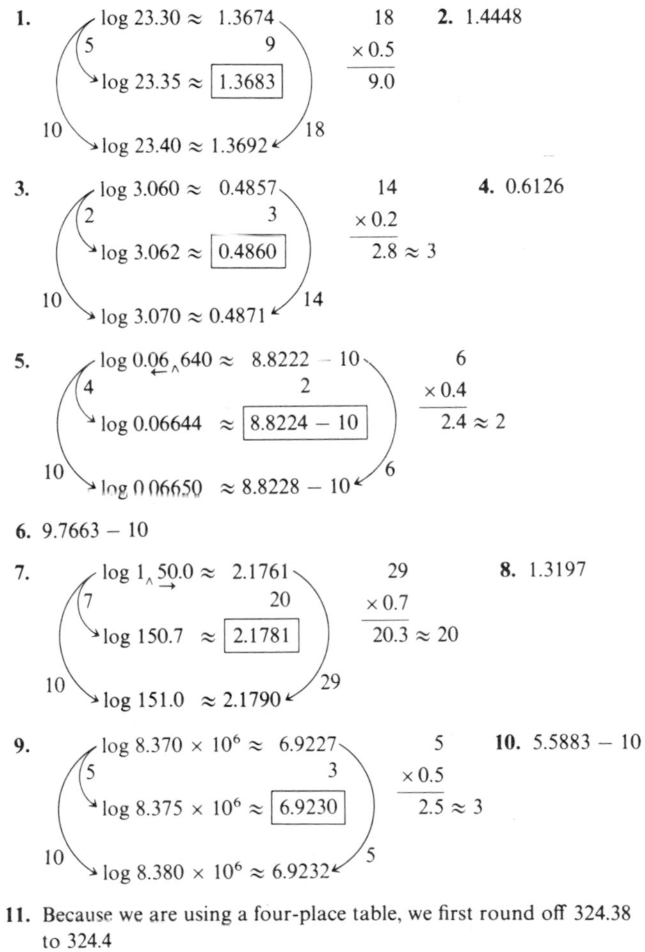

11. Because we are using a four-place table, we first round off 324.38 to 324.4

$\log 3_{\wedge}24.0 \approx 2.5105$
4 | 6
$\log 324.4 \approx$ **2.5111**
10 | 14
$\log 325.0 \approx 2.5119$

14
$\times 0.4$
$5.6 \approx 6$

12. 1.8754

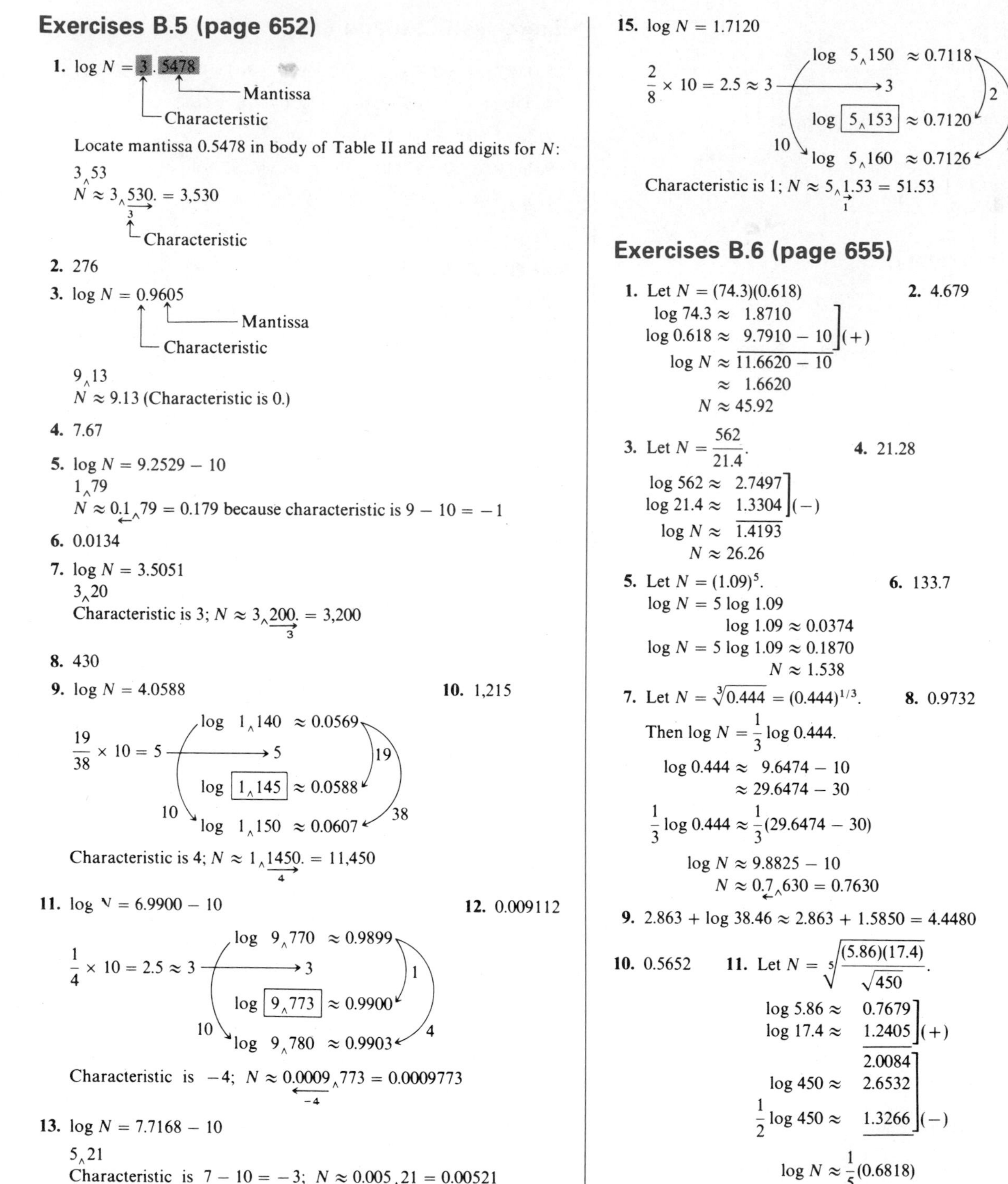

Exercises B.5 (page 652)

1. $\log N = 3.5478$

3 — Characteristic; .5478 — Mantissa

Locate mantissa 0.5478 in body of Table II and read digits for N:

$3_\wedge 53$

$N \approx 3_\wedge 530. = 3{,}530$ (decimal moved 3 places — Characteristic)

2. 276

3. $\log N = 0.9605$

0 — Characteristic; .9605 — Mantissa

$9_\wedge 13$

$N \approx 9.13$ (Characteristic is 0.)

4. 7.67

5. $\log N = 9.2529 - 10$

$1_\wedge 79$

$N \approx 0.1_\wedge 79 = 0.179$ because characteristic is $9 - 10 = -1$

6. 0.0134

7. $\log N = 3.5051$

$3_\wedge 20$

Characteristic is 3; $N \approx 3_\wedge 200. = 3{,}200$ (3)

8. 430

9. $\log N = 4.0588$

$\log 1_\wedge 140 \approx 0.0569$

$\frac{19}{38} \times 10 = 5 \rightarrow 5$ (19, 38)

$\log \boxed{1_\wedge 145} \approx 0.0588$

10 $\quad \log 1_\wedge 150 \approx 0.0607$

Characteristic is 4; $N \approx 1_\wedge 1450. = 11{,}450$ (4)

10. 1,215

11. $\log N = 6.9900 - 10$

$\log 9_\wedge 770 \approx 0.9899$

$\frac{1}{4} \times 10 = 2.5 \approx 3 \rightarrow 3$ (1, 4)

$\log \boxed{9_\wedge 773} \approx 0.9900$

10 $\quad \log 9_\wedge 780 \approx 0.9903$

Characteristic is -4; $N \approx 0.0009_\wedge 773 = 0.0009773$ (-4)

12. 0.009112

13. $\log N = 7.7168 - 10$

$5_\wedge 21$

Characteristic is $7 - 10 = -3$; $N \approx 0.005_\wedge 21 = 0.00521$ (-3)

14. 0.00000873

15. $\log N = 1.7120$

$\log 5_\wedge 150 \approx 0.7118$

$\frac{2}{8} \times 10 = 2.5 \approx 3 \rightarrow 3$ (2, 8)

$\log \boxed{5_\wedge 153} \approx 0.7120$

10 $\quad \log 5_\wedge 160 \approx 0.7126$

Characteristic is 1; $N \approx 5_\wedge 1.53 = 51.53$ (1)

16. 97,120

Exercises B.6 (page 655)

1. Let $N = (74.3)(0.618)$

$$\begin{aligned} \log 74.3 &\approx 1.8710 \\ \log 0.618 &\approx 9.7910 - 10 \quad (+) \\ \log N &\approx 11.6620 - 10 \\ &\approx 1.6620 \\ N &\approx 45.92 \end{aligned}$$

2. 4.679

3. Let $N = \frac{562}{21.4}$.

$$\begin{aligned} \log 562 &\approx 2.7497 \\ \log 21.4 &\approx 1.3304 \quad (-) \\ \log N &\approx 1.4193 \\ N &\approx 26.26 \end{aligned}$$

4. 21.28

5. Let $N = (1.09)^5$.

$$\begin{aligned} \log N &= 5 \log 1.09 \\ \log 1.09 &\approx 0.0374 \\ \log N &= 5 \log 1.09 \approx 0.1870 \\ N &\approx 1.538 \end{aligned}$$

6. 133.7

7. Let $N = \sqrt[3]{0.444} = (0.444)^{1/3}$.

Then $\log N = \frac{1}{3} \log 0.444$.

$$\begin{aligned} \log 0.444 &\approx 9.6474 - 10 \\ &\approx 29.6474 - 30 \\ \frac{1}{3} \log 0.444 &\approx \frac{1}{3}(29.6474 - 30) \\ \log N &\approx 9.8825 - 10 \\ N &\approx 0.7_\wedge 630 = 0.7630 \end{aligned}$$

8. 0.9732

9. $2.863 + \log 38.46 \approx 2.863 + 1.5850 = 4.4480$

10. 0.5652

11. Let $N = \sqrt[5]{\frac{(5.86)(17.4)}{\sqrt{450}}}$.

$$\begin{aligned} \log 5.86 &\approx 0.7679 \\ \log 17.4 &\approx 1.2405 \quad (+) \\ & 2.0084 \\ \log 450 &\approx 2.6532 \\ \frac{1}{2} \log 450 &\approx 1.3266 \quad (-) \\ \log N &\approx \frac{1}{5}(0.6818) \\ &\approx 0.1364 \\ N &\approx 1.369 \end{aligned}$$

12. 0.4027

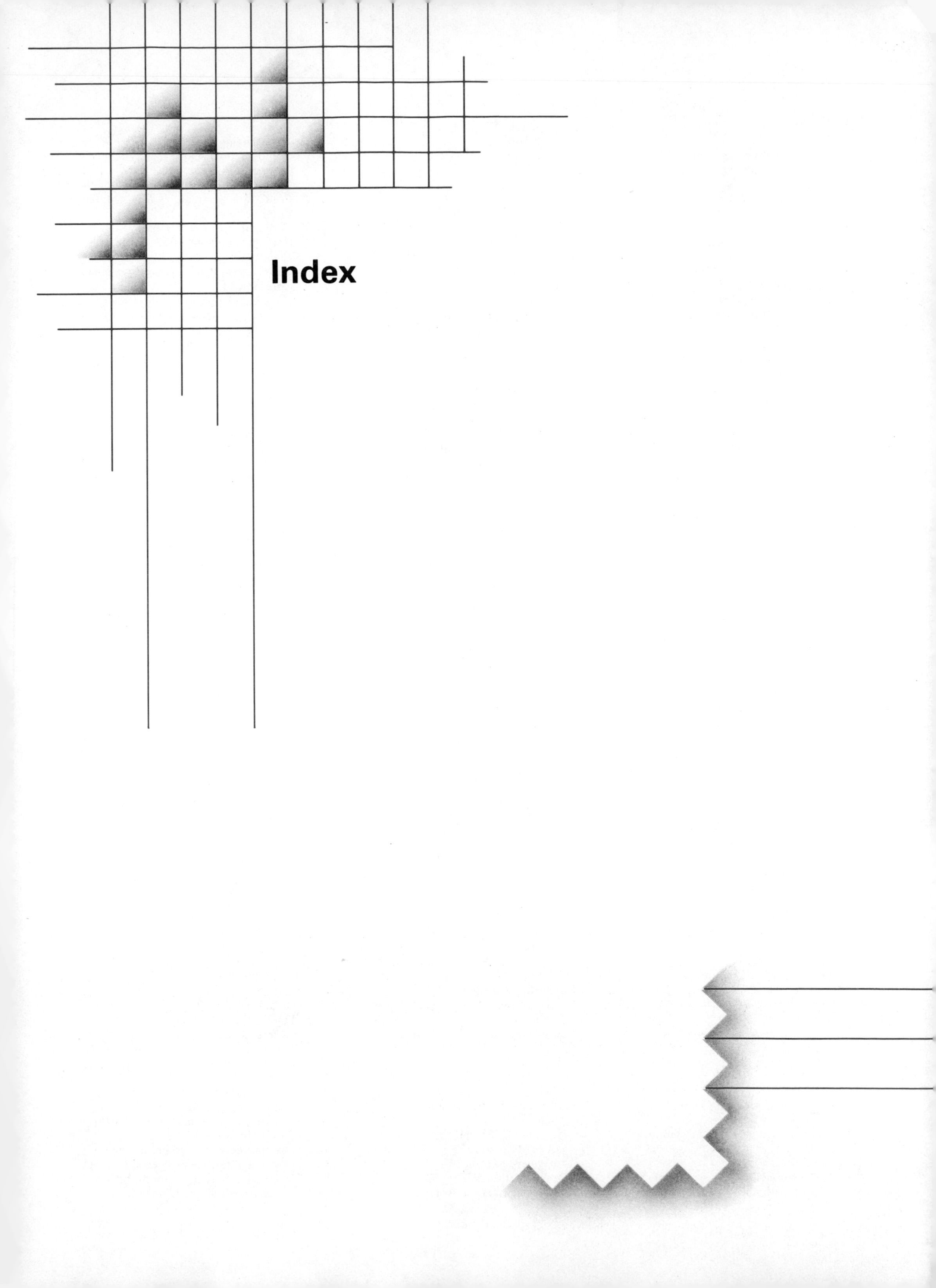

Index